CRC Handbook
of
Mathematical
Sciences

6th Edition

Editor

William H. Beyer, Ph.D.
Chairman
Department of Mathematical Sciences
University of Akron
Akron, Ohio

CRC Press, Inc.
Boca Raton, Florida

161612··

Library of Congress Cataloging-in-Publication Data

Main entry under title:

CRC Handbook of mathematical sciences.
    5th ed. published under title: Handbook of tables for mathematics.

    Bibliography: p.
    Includes index.
    I. Mathematics—Tables, etc. I. Beyer, William H. II. Title: Handbook of tables for mathematics.
QA47.H324   1978   510′.21′2   78-10602
ISBN 0-8493-0656-6

Direct all inquiries to CRC Press, Inc., 2000 Corporate Blvd., N.W., Boca Raton, Florida, 33431.

© 1987 by CRC Press, Inc.
Second Printing, 1987
Third Printing, 1988
Fourth Printing 1988
Fifth Printing, 1991

International Standard Book Number 0-8493-0656-6

Library of Congress Card Number 78-10602
Printed in the United States

PREFACE

Numerical tables of mathematical and statistical functions are in continual demand by professional scientists, by those in the teaching profession, and by students of mathematics and related sciences. The *CRC Handbook of Mathematical Sciences,* published by CRC Press, Inc., contains the most up-to-date, authoritative, logically arranged, and ready-usable collection of reference material available. Prior to the preparation of this 6th Edition of the *Handbook*, the contents of predecessor editions were carefully examined to determine if certain tables and/or reference materials should be altered, expanded, or deleted. The net result is that this 6th Edition has been prepared to provide an adequately broad spectrum of traditional and modern mathematical sciences data necessary for today's scientific needs, even in light of today's computer technology.

The same successful format which characterized the 5th Edition has been retained. Material is presented in a multisectional format, with each section containing a valuable collection of fundamental reference material — both expository and tabular in form. The format is such that existing sections can be expanded or reduced as necessary, and new sections can be developed as warranted. The 6th Edition has been vastly improved by the addition of material on numerical solutions of nonlinear equations, statistical tests of hypotheses, statistical confidence intervals, and analysis of variance tables. Omissions include tables involving squares and square roots, cube and cube roots; logarithms of the binomial coefficients; reciprocals, circumferences and areas of circles; natural trigonometric functions for angles in radius; and financial tables. It is hoped that these changes will prove to be beneficial to the users of the *Handbook*.

The editor gratefully acknowledges the services of Paul Gottehrer, Editor, for the handling of the detail work which is so essential in the final production of this edition.

All errors called to our attention have, to the best of the Editor's knowledge, been corrected. As in the past CRC Press, Inc., and the Editor invite and welcome regular input from the many users of this handbook. Since the inception, some 20 years ago, of *The CRC Handbook of Mathematical Sciences* (formerly called *The CRC Handbook of Mathematics*), users of the *Handbook* have forwarded data, advice, guidance, and constructive criticisms. This information has provided a most effective means for keeping the editions of the *Handbook* updated, accurate, and abreast of the times.

William H. Beyer, Editor

MATHEMATICS ADVISORY BOARD

ACKNOWLEDGMENTS

Acknowledgment is made to the following authors, editors, and publishers whose material has been used in the Handbook of Mathematical Sciences and for which permission has been received.

AMERICAN SOCIETY FOR TESTING MATERIALS STP-15C;
    ASTM Manual on Quality Control of Materials (1951)
        Factors for Computing Control Limits

ROLF E. BARGMANN, PH.D.
    Matrices and Determinants

BIOMETRIKA TRUSTEES, E. S. PEARSON
    Cambridge University Press
        Biometrika Tables for Statisticians and Biometrika, Vol. 32,
            Percentage Points, Chi-Square Distribution
            Percentage Points, F-Distribution

E. RICHARD COHEN, A.B., M.S., PH.D.
    Fundamental Physical Constants

W. E. DESKINS, PH.D.
    Basic Concepts in Algebra

HOWARD EVES, PH.D.
    Analytic Geometry
    Curves and Surfaces
    Mensuration Formula
    Trigonometry

R. E. GASKELL, PH.D.
    The Laplace Transform

BRIAN GIRLING, M.Sc., F.I.M.A.
    Differentiation Formulas
    Finite Differences
    Sine, Cosine, and Exponential Integrals
    The Z Transforms

MADHU-SUDAN GUPTA, M.Sc. TECH., M.S., A.M.
    Inverse Hyperbolic Function Tables

HARPER & ROWE, NEW YORK, N.Y.
    Concepts of Calculus by A. H. Lighston
        Bessel Function J_0, J_1

SAMUEL HERRICK, B.A., Sc.D., PH.D.
    Astrodynamics: Basic Orbital Equations
    Astrodynamical Terminology. Notation, and Usage
        Condensed from Astrodynamics: N.Y. an Nostrand Reinhold, 1970

INTERNATIONAL BUSINESS MACHINES CORPORATION
    IBM Brochure "360 Principles of Operation," Form A 22-6821-3
        Hexadecimal-Direct Conversion Table

McGRAW-HILL, NEW YORK
    Lazenga Diagram-Interpolation Coefficients for Orthogonal Polynomials
        and for x^n in Terms of Orthogonal Polynomials
            Table of Real and Imaginary Parts, Zeros, and Singularities
            Table of Transformations of Regions

OLIVER AND BOYD, LTD., EDINBURGH, SCOTLAND
    Statistical Tables for Biological, Agricultural and Medical Research of Fisher & Yates
        Percentage Points, Student's t-Distribution

RICHARD PRATT, A.M.
    Explanations to Tables
    Use of Logarithms (Law of Exponents)
    Integral Tables

THE ROYAL SOCIETY, LONDON, ENGLAND
    Vol. 3 (1954) 2 Royal Society Mathematical Tables
        Number of Combinations

E. N. SICKAFUS and N. A. MACKIE
    The Interstitial Sphere

SPRINGER-VERLAG NEW YORK, INC.
    Funfstellige Funktionentafeln (1930), Hayashi, K.
        Number of Permutations

TABLE OF CONTENTS

GREEK ALPHABET

Greek letter	Greek name	English equivalent	Greek letter	Greek name	English equivalent
A α	Alpha	a	N ν	Nu	n
B β	Beta	b	Ξ ξ	Xi	x
Γ γ	Gamma	g	O o	Omicron	ŏ
Δ δ	Delta	d	Π π	Pi	p
E ε	Epsilon	ĕ	P ρ	Rho	r
Z ζ	Zeta	z	Σ σ ς	Sigma	s
H η	Eta	ē	T τ	Tau	t
Θ θ ϑ	Theta	th	Υ υ	Upsilon	u
I ι	Iota	i	Φ φ φ	Phi	ph
K κ	Kappa	k	X χ	Chi	ch
Λ λ	Lambda	l	Ψ ψ	Psi	ps
M μ	Mu	m	Ω ω	Omega	ō

THE NUMBER OF EACH DAY OF THE YEAR

Day of Mo.	Jan.	Feb.	Mar.	Apr.	May	Jun.	Jul.	Aug.	Sep.	Oct.	Nov.	Dec.	Day of Mo.
1	1	32	60	91	121	152	182	213	244	274	305	335	1
2	2	33	61	92	122	153	183	214	245	275	306	336	2
3	3	34	62	93	123	154	184	215	246	276	307	337	3
4	4	35	63	94	124	155	185	216	247	277	308	338	4
5	5	36	64	95	125	156	186	217	248	278	309	339	5
6	6	37	65	96	126	157	187	218	249	279	310	340	6
7	7	38	66	97	127	158	188	219	250	280	311	341	7
8	8	39	67	98	128	159	189	220	251	281	312	342	8
9	9	40	68	99	129	160	190	221	252	282	313	343	9
10	10	41	69	100	130	161	191	222	253	283	314	344	10
11	11	42	70	101	131	162	192	223	254	284	315	345	11
12	12	43	71	102	132	163	193	224	255	285	316	346	12
13	13	44	72	103	133	164	194	225	256	286	317	347	13
14	14	45	73	104	134	165	195	226	257	287	318	348	14
15	15	46	74	105	135	166	196	227	258	288	319	349	15
16	16	47	75	106	136	167	197	228	259	289	320	350	16
17	17	48	76	107	137	168	198	229	260	290	321	351	17
18	18	49	77	108	138	169	199	230	261	291	322	352	18
19	19	50	78	109	139	170	200	231	262	292	323	353	19
20	20	51	79	110	140	171	201	232	263	293	324	354	20
21	21	52	80	111	141	172	202	233	264	294	325	355	21
22	22	53	81	112	142	173	203	234	265	295	326	356	22
23	23	54	82	113	143	174	204	235	266	296	327	357	23
24	24	55	83	114	144	175	205	236	267	297	328	358	24
25	25	56	84	115	145	176	206	237	268	298	329	359	25
26	26	57	85	116	146	177	207	238	269	299	330	360	26
27	27	58	86	117	147	178	208	239	270	300	331	361	27
28	28	59	87	118	148	179	209	240	271	301	332	362	28
29	29	*	88	119	149	180	210	241	272	302	333	363	29
30	30		89	120	150	181	211	242	273	303	334	364	30
31	31		90		151		212	243		304		365	31

* In leap years, after February 28, add 1 to the tabulated number.

I. CONSTANTS AND CONVERSION FACTORS

SI SYSTEM OF MEASUREMENT

SI, which is the abbreviation of the French words "Systeme Internationale d'Unites," is the accepted abbreviation for the International Metric System, which has seven base units, as shown below.

UNITS FOR A SYSTEM OF MEASURES AS USED INTERNATIONALLY

Quantity measured	Unit	Abbreviation
Length	meter	m
Mass	kilogram	kg
Time	second	s
Electric current	ampere	A
Temperature	degree Kelvin	K
Luminous intensity	candela	cd
Amount of substance	mole	mol

Supplementary and Derived Units
From Base Units
as used Internationally

Supplementary Units

Plane angle	radian	rad
Solid angle	steradian	sr

Derived Units

Area	square meter	m^2	
Volume	cubic meter	m^3	
Frequency	hertz	Hz	(s^{-1})
Density	kilogram per cubic meter	kg/m^3	
Velocity	meter per second	m/s	
Angular velocity	radian per second	rad/s	
Acceleration	meter per second squared	m/s^2	
Angular acceleration	radian per second squared	rad/s^2	
Force	newton	N	$(kg \cdot m/s^2)$
Pressure	newton per sq meter	N/m^2	
Kinematic viscosity	sq meter per second	m^2/s	
Dynamic viscosity	newton-second per sq meter	$N \cdot s/m^2$	
Work, energy, quantity of heat	joule	J	$(N \cdot m)$
Power	watt	W	(J/s)
Electric charge	coulomb	C	$(A \cdot s)$
Voltage, potential difference, electromotive force	volt	V	(W/A)
Electric field strength	volt per meter	V/m	
Electric resistance	ohm	Ω	(V/A)
Electric capacitance	farad	F	$(A \cdot s/V)$
Magnetic flux	weber	Wb	$(V \cdot s)$
Inductance	henry	H	$(V \cdot s/A)$
Magnetic flux density	tesla	T	(Wb/m^2)
Magnetic field strength	ampere per meter	A/m	
Magnetomotive force	ampere	A	
Luminous flux	lumen	lm	$(cd \cdot sr)$
Luminance	candela per sq meter	cd/m^2	
Illumination	lux	lx	(lm/m^2)

RECOMMENDED UNIT PREFIXES

Multiples and submultiples	Prefixes	Symbols
10^{18}	exa	E
10^{15}	peca	P
10^{12}	tera	T
10^9	giga	G
10^6	mega	M
10^3	kilo	k
10^2	hecto	h
10	deka	da
10^{-1}	deci	d
10^{-2}	centi	c
10^{-3}	milli	m
10^{-6}	micro	μ (greek mu)
10^{-9}	nano	n
10^{-12}	pico	p
10^{-15}	femto	f
10^{-18}	atto	a

DEFINED VALUES AND EQUIVALENTS

Meter .... (m)	1 650 763.73 wave lengths in vacuo of the unperturbed transition $2p_{10} - 5d_5$ in ^{86}Kr
Kilogram .... (kg)	mass of the international kilogram at Sèvres, France
Second .... (s)	1/31 556 925 974 7 of the tropical year at 12^h ET, 0 January 1900
Degree Kelvin .... ($^\circ$K)	defined in the thermodynamic scale by assigning 273.16°K to the triple point of water (freezing point, 273.15°K = 0°C)
Unified atomic mass unit .... (u)	1/12 the mass of an atom of the ^{12}C nuclide
Mole .... (mol)	amount of substance containing the same number of atoms as 12 g of pure ^{12}C
Standard acceleration of free fall .... (g_n)	9.806 65 M s^{-2}, 980.665 cm s^{-2}
Normal atmospheric pressure .... (atm)	101 325 N m^{-2}, 1 013 250 dyn cm^{-2}
Thermochemical calorie .... (cal$_{th}$)	4.1840 J, 4.1840 $\times$ 10^7 erg
International Steam Table calorie .... (cal$_{IT}$)	4.1868 J, 4.1868 $\times$ 10^7 erg
Liter .... (l)	0.001 000 028 m^3, 1 000.028 cm^3 (recommended by CIPM, 1950)
	0.001 m^3, 1000 cm^3 (recommended by GCWM 1964)
Inch .... (in)	0.0254 m, 2.54 cm
Pound (avdp) .... (lb)	0.453 592 37 kg, 453.592 37 g

CONVERSION FACTORS

Conversion Factors — Metric to English

To obtain	Multiply	By
Inches	Centimeters	0.3937007874
Feet	Meters	3.280839895
Yards	Meters	1.093613298
Miles	Kilometers	0.6213711922
Ounces	Grams	$3.527396195 \times 10^{-2}$
Pounds	Kilograms	2.204622622
Gallons	Liters	0.2641720524
Fluid ounces	Milliliters (cc)	$3.381402270 \times 10^{-2}$
Square inches	Square centimeters	0.1550003100
Square feet	Square meters	10.76391042
Square yards	Square meters	1.195990046
Cubic inches	Milliliters (cc)	$6.102374409 \times 10^{-2}$
Cubic feet	Cubic meters	35.31466672
Cubic yards	Cubic meters	1.307950619

Conversion Factors — English to Metric*

To obtain	Multiply	By
Microns	Mils	**25.4**
Centimeters	Inches	**2.54**
Meters	Feet	**0.3048**
Meters	Yards	**0.9144**
Kilometers	Miles	**1.609344**
Grams	Ounces	28.34952313
Kilograms	Pounds	**0.45359237**
Liters	Gallons	**3.785411784**
Milliliters (cc)	Fluid ounces	29.57352956
Square centimeters	Square inches	**6.4516**
Square meters	Square feet	**0.09290304**
Square meters	Square yards	**0.83612736**
Milliliters (cc)	Cubic inches	**16.387064**
Cubic meters	Cubic feet	$2.831684659 \times 10^{-2}$
Cubic meters	Cubic yards	0.764554858

Conversion Factors — General*

To obtain	Multiply	By
Atmospheres	Feet of water @ 4°C	2.950×10^{-5}
Atmospheres	Inches of mercury @ 0°C	3.342×10^{-2}
Atmospheres	Pounds per square inch	6.804×10^{-2}
BTU	Foot-pounds	1.285×10^{-3}
BTU	Joules	9.480×10^{-4}
Cords	Cubic feet	**128**

*Boldface numbers are exact; others are given to ten significant figures where so indicated by the multiplier factor.

Conversion Factors – General (Continued)

To obtain	Multiply	By
Degree (angle)	Radians	57.2958
Ergs	Foot-pounds	1.356×10^7
Feet	Miles	**5280**
Feet of water @ 4°C	Atmospheres	33.90
Foot-pounds	Horsepower-hours	1.98×10^6
Foot-pounds	Kilowatt-hours	2.655×10^6
Food-pounds per min	Horsepower	3.3×10^4
Horsepower	Foot-pounds per sec	1.818×10^{-3}
Inches of mercury @ 0°C	Pounds per square inch	2.036
Joules	BTU	1054.8
Joules	Foot-pounds	1.35582
Kilowatts	BTU per min	1.758×10^{-2}
Kilowatts	Foot-pounds per min	2.26×10^{-5}
Kilowatts	Horsepower	0.745712
Knots	Miles per hour	0.86897624
Miles	Feet	1.894×10^{-4}
Nautical miles	Miles	0.86897624
Radians	Degrees	1.745×10^{-2}
Square feet	Acres	**43560**
Watts	BTU per min	17.5796

Temperature Factors

$$°F = 9/5 \ (°C) + 32$$

Fahrenheit temperature = 1.8 (temperature in kelvins) −459.67

$$°C = 5/9 \ [(°F) - 32]$$

Celsius temperature = temperature in kelvins −273.15
Fahrenheit temperature = 1.8 (Celsius temperature) +32

CONVERSION FACTORS
U. S. AND METRIC UNITS

Each unit in bold face type is followed by its equivalent in other units of the same quantity.

Acre—0.0015625 square mile (statute); 4.3560×10^4 square feet; 0.40468564 hectare.

Bushel—(U.S.)—1.244456 cubic feet; 2150.42 cubic inches; 0.035239 cubic meter; 35.23808 liters.

Centimeter—0.0328084 foot; 0.393701 inch.

Circular Mil—7.853982×10^{-7} square inches; 5.067075×10^{-6} square centimeters.

Cubic Centimeter—0.061024 cubic inch; 0.270512 dram (U. S. fluid); 16.230664 minims (U. S.); 0.999972 milliliter.

Cubic Foot—0.803564 bushel (U. S.); 7.480520 gallons (U. S. liquid); 0.028317 cubic meter; 28.31605 liters.

Cubic Inch—16.387064 cubic centimeters.

Cubic Meter—35.314667 cubic feet; 264.17205 gallons (U. S. liquid).

Foot—0.3048 meter.

Gallon (U. S. liquid)—0.1336816 cubic foot; 0.832675 gallon (British); 231 cubic inches; 0.0037854 cubic meter; 3.785306 liters.

Grain—0.06479891 gram.

Gram—0.00220462 pound (avoirdupois); 0.035274 ounce (avoirdupois); 15.432358 grains.

Hectare—2.471054 acres; 1.07639×10^5 square feet.

Inch—2.54 centimeters.

Kilogram—2.204623 pounds (avoirdupois).

Kilometer—0.621371 mile (statute).

Liter—0.264179 gallon (U. S. liquid); 0.0353157 cubic foot; 1.056718 quarts (U. S. liquid).

Meter—1.093613 yards; 3.280840 feet; 39.37008 inches.

Mile (statute)—1.609344 kilometers.

Ounce (U. S. fluid)—1.804688 cubic inches; 29.573730 cubic centimeters.

Ounce (avoirdupois)—28.349523 grams.

Ounce (apothecary or troy)—31.103486 grams.

Pint (U. S. liquid)—0.473163 liter; 473.17647 cubic centimeters.

Pound (avoirdupois)—0.453592 kilogram; 453.59237 grams.

Pound (apothecary or troy)—0.3732417 kilogram, 373.24172 grams.

Quart (U. S. dry)—1.10119 liters.

Quart (liquid)—0.946326 liter.

Radian—57.295779 degrees.

Rod—5.0292 meters.

Square Centimeter—0.155000 square inch.

Square Foot—0.09290304 square meter.

Square Inch—645.16 square millimeters.

Square Meter—10.763910 square feet.

Square Yard—0.836127 square meter.

Ton (short)—907.18474 kilograms.

Yard—0.9144 meter.

See Index for extensive Conversion Factors.

METRIC CONVERSION TABLE

Inches	Centimeters	Centimeters	Inches
1	2.54	1	0.393701
2	5.0	2	0.787402
3	7.62	3	1.181103
4	10.16	4	1.574804
5	12.70	5	1.968505
6	15.24	6	2.362206
7	17.78	7	2.755907
8	20.32	8	3.149608
9	22.86	9	3.543309

Feet	Meters	Meters	Feet
1	0.3048	1	3.280840
2	0.6096	2	6.561680
3	0.9144	3	9.842520
4	1.2192	4	13.123360
5	1.5240	5	16.404200
6	1.8288	6	19685040
7	2.1336	7	22.965880
8	2.4384	8	26.246720
9	2.7432	9	29.527560

Yards	Meters	Meters	Yards
1	0.9144	1	1.0936133
2	1.8288	2	2.1872266
3	2.7432	3	3.2808399
4	3.6576	4	4.3744532
5	4.5720	5	5.4680665
6	5.4864	6	6.5616798
7	6.4008	7	7.6552931
8	7.3152	8	8.7489064
9	8.2296	9	9.8425197

Miles (statute)	Kilometers	Kilometers	Miles
1	1.609344	1	0.6213712
2	3.218688	2	1.2427424
3	4.828032	3	1.8641136
4	6.437376	4	2.4854848
5	8.046720	5	3.1068560
6	9.656064	6	3.7282272
7	11.265408	7	4.3495984
8	12.874752	8	4.9709696
9	14.484096	9	5.5923408

Square inches	Square centimeters	Square centimeters	Square inches
1	6.45	1	0.155
2	12.90	2	0.310
3	19.36	3	0.465
4	25.81	4	0.620
5	32.26	5	0.775
6	38.71	6	0.930
7	45.16	7	1.085
8	51.61	8	1.240
9	58.06	9	1.395

METRIC CONVERSION TABLE (continued)

Square feet	Square meters	Square meters	Square feet
1	0.0929	1	10.76
2	0.1858	2	21.53
3	0.2787	3	32.29
4	0.3716	4	43.06
5	0.4645	5	53.82
6	0.5574	6	64.58
7	0.6503	7	75.35
8	0.7432	8	86.11
9	0.8361	9	96.88

Cubic inches	Cubic centimeters	Cubic centimeters	Cubic Inches
1	16.39	1	0.0610
2	32.77	2	0.1221
3	49.16	3	0.1831
4	65.55	4	0.2441
5	81.94	5	0.3051
6	98.32	6	0.3661
7	114.71	7	0.4272
8	131.10	8	0.4882
9	147.48	9	0.5492

Cubic feet	Cubic meters	Cubic meters	Cubic feet
1	0.0283	1	35.3
2	0.0566	2	70.6
3	0.0850	3	105.9
4	0.1133	4	141.3
5	0.1416	5	176.6
6	0.1699	6	211.9
7	0.1982	7	247.2
8	0.2265	8	282.5
9	0.2549	9	317.8

Cubic feet	Liters	Liters	Cubic feet
1	28.32	1	0.0353
2	56.63	2	0.0706
3	84.95	3	0.1060
4	113.26	4	0.1413
5	141.58	5	0.1766
6	169.90	6	0.2119
7	198.21	7	0.2472
8	226.53	8	0.2825
9	254.84	9	0.3178

U.S. gallons	Liters	Liters	U.S. gallons
1	3.785306	1	0.264179
2	7.570612	2	0.528358
3	11.355918	3	0.792537
4	15.141224	4	1.056716
5	18.926530	5	1.320895
6	22.711836	6	1.585074
7	26.497142	7	1.849253
8	30.282448	8	2.113432
9	34.067754	9	2.377611

METRIC CONVERSION TABLE (continued)

British or Imperial gallons	Liters	Liters	British or Imperial gallons
1	4.546	1	0.220
2	9.092	2	0.440
3	13.638	3	0.660
4	18.184	4	0.880
5	22.730	5	1.100
6	27.276	6	1.320
7	31.822	7	1.540
8	36.368	8	1.760
9	40.914	9	1.980

Pounds (av)	Kilograms	Kilograms	Pounds (av)
1	0.45359237	1	2.2046226
2	0.90718474	2	4.4092452
3	1.36077711	3	6.6138678
4	1.81436948	4	8.8184904
5	2.26796185	5	11.0231130
6	2.72155422	6	13.2277356
7	3.17514659	7	15.4323582
8	3.62873896	8	17.6369808
9	4.08233133	9	19.8416034

Ounces Avoirdupois	Grams	Grams	Ounces Avoirdupois
1	28.350	1	0.035274
2	56.699	2	0.070548
3	85.049	3	0.10582
4	113.40	4	0.14110
5	141.75	5	0.17637
6	170.10	6	0.21164
7	198.45	7	0.24692
8	226.80	8	0.28219
9	255.15	9	0.31747

Pounds per foot	Kilograms per meter	Kilograms per meter	Pounds per foot
1	1.4882	1	0.6720
2	2.9763	2	1.3439
3	4.4645	3	2.0159
4	5.9527	4	2.6879
5	7.4408	5	3.3598
6	8.9290	6	4.0318
7	10.4171	7	4.7038
8	11.9053	8	5.3758
9	13.3935	9	6.0477

Pounds per square inch	Kilograms per square centimeter	Kilograms per square centimeter	Pounds per square inch
1	0.0703	1	14.22
2	0.1406	2	28.45
3	0.2109	3	42.67
4	0.2812	4	56.89
5	0.3515	5	71.12
6	0.4218	6	85.34
7	0.4922	7	99.96
8	0.5625	8	113.79
9	0.6328	9	128.01

METRIC CONVERSION TABLE (continued)

Pounds per square inch	Kilonewtons per square meter	Kilonewtons per square meter	Pounds per square inch
1	6.895	1	0.145
2	13.790	2	0.290
3	20.684	3	0.435
4	27.579	4	0.580
5	34.474	5	0.725
6	41.369	6	0.870
7	48.263	7	1.015
8	55.158	8	1.160
9	62.053	9	1.305

Pounds per square foot	Kilograms per square meter	Kilograms per square meter	Pounds per square foot
1	4.88	1	0.2048
2	9.77	2	0.4096
3	14.65	3	0.6144
4	19.53	4	0.8193
5	24.41	5	1.0241
6	29.30	6	1.2289
7	34.18	7	1.4337
8	39.06	8	1.6385
9	43.94	9	1.8433

Pound feet	Kilogram meters	Kilogram meters	Pound feet
1	0.138	1	7.23
2	0.277	2	14.47
3	0.415	3	21.70
4	0.553	4	28.93
5	0.691	5	36.17
6	0.830	6	43.40
7	0.968	7	50.63
8	1.106	8	57.87
9	1.244	9	65.10

Foot pounds force	Joules	Joules	Foot pounds force
1	1.356	1	0.7376
2	2.712	2	1.4751
3	4.068	3	2.2127
4	5.423	4	2.9502
5	6.779	5	3.6878
6	8.135	6	4.4254
7	9.491	7	5.1629
8	10.847	8	5.9005
9	12.202	9	6.6381

British thermal units	Kilojoules	Kilojoules	British thermal units
1	105.51	1	94.78
2	211.01	2	189.56
3	316.52	3	284.35
4	422.02	4	379.13
5	527.53	5	473.91
6	633.03	6	568.69
7	738.54	7	663.47
8	844.04	8	758.25
9	949.55	9	853.04

METRIC CONVERSION TABLE (continued)

Horsepower	Kilowatts	Kilowatts	Horsepower
1	0.746	1	1.341
2	1.491	2	2.682
3	2.237	3	4.023
4	2.983	4	5.364
5	3.729	5	6.705
6	4.474	6	8.046
7	5.220	7	9.387
8	5.966	8	10.728
9	6.711	9	12.069

Pounds force	Newtons	Newtons	Pounds force
1	4.448	1	0.22481
2	8.896	2	0.44962
3	13.345	3	0.67443
4	17.793	4	0.89924
5	22.241	5	1.12404
6	26.689	6	1.34885
7	31.138	7	1.57366
8	35.586	8	1.79847
9	40.034	9	2.02328

CONVERSION FACTORS

L. P. Buseth

To convert from	To	Multiply by
Abampere	Ampere	*10*
Abcoulomb	Coulomb	*10*
Abfarad	Farad	1×10^9
Abhenry	Henry	1×10^{-9}
Abmho	Siemens (mho)	1×10^9
Abohm	Ohm	1×10^{-9}
Abvolt	Volt	1×10^{-8}
Acre	Hectare	0.40468564
	Square foot	*43560*
	Square kilometer	4.046856×10^{-3}
	Square meter	4046.85642
	Square mile	1.5625×10^{-3} *(1/640)*
	Square yard	*4840*
Acre (U.S. Survey)	Square meter	4046.872610
Acre-foot	Cubic meter	1233.482
	Cubic yard	1613.333
Acre-inch	Cubic foot	*3630*
	Cubic meter	102.7902
	Gallon (Brit.)	22610.67
	Gallon (U.S.)	27154.29
Ampere (int., mean)	Ampere	0.99985
Ampere (int., U.S.)	Ampere	0.999835
Ampere/square centimeter	Ampere/square inch	*6.4516*
Ampere/square inch	Ampere/square centimeter	0.1550003
Ampere-hour	Coulomb	*3600*
Ampere(-turn)	Gilbert	1.256637
Ångström	Nanometer	*0.1*
Apostilb	Candela/square meter	0.3183099 *(1/π)*
Are	Square foot	1076.391
	Square meter	*100*
Astronomical unit	Kilometer	1.4959787×10^8
Atmosphere	Atmosphere (tech.)	1.033227
	Bar	*1.01325*
	Foot of H$_2$O (conv.)	33.89854
	Inch of Hg (conv.)	29.92126
	Kilogram-force/square centimeter	1.033227
	Kilopascal	*101.325*
	Meter of H$_2$O (conv.)	10.33227
	Millibar	*1013.25*
	Millimeter of Hg (conv.)	760
	Newton/square centimeter	*10.1325*
	Pascal (N/square meter)	1.01325×10^5
	Pound-force/square foot	2116.22
	Pound-force/square inch	14.69595
	Ton-force (long)/square foot	0.944740
	Ton-force (short)/square foot	1.058108
	Ton-force (long)/square inch	6.56069×10^{-3}
	Ton-force (short)/square inch	7.34797×10^{-3}
	Torr	*760*
Atmosphere (tech.)	Atmosphere	0.967841
	Bar	*0.980665*
	Foot of H$_2$O (conv.)	32.8084
	Inch of Hg (conv.)	28.9590
	Kilogram-force/square centimeter	*1*
	Kilopascal	*98.0665*
	Meter of H$_2$O (conv.)	*10*
	Millibar	*980.665*
	Millimeter of Hg (conv.)	735.559
	Newton/square centimeter	9.80665
	Pascal (N/m^2)	98066.5
	Pound-force/square inch	14.22334
Bag (Brit.)	Gallon (Brit.)	*24*
Bar	Atmosphere	0.9869233

To convert from	To	Multiply by
	Atmosphere (tech.)	1.019716
	Dyne/square centimeter	1×10^6
	Foot of H$_2$O (conv.)	33.4553
	Inch of Hg (conv.)	29.5300
	Kilogram-force/square centimeter	1.019716
	Kilopascal	*100*
	Meter of H$_2$O (conv.)	10.19716
	Millibar	*1000*
	Millimeter of Hg (conv.)	750.062
	Newton/square centimeter	*10*
	Pascal (N/m^2)	1×10^5
	Pound-force/square foot	2088.54
	Pound-force/square inch	14.50377
	Ton-force (long)/square foot	0.932385
	Ton-force (short)/square foot	1.04427
	Ton-force (long)/square inch	6.47490×10^{-3}
	Ton-force (short)/square inch	7.25189×10^{-3}
	Torr	750.062
Barleycorn (Brit.)	Inch	0.333333 *(1/3)*
Barn	Square meter	1×10^{-28}
Barrel (Brit., beer)	Gallon (Brit.)	*36*
	Liter	163.6592
Barrel (Brit., wine)	Gallon (Brit.)	*31.5*
	Liter	143.2018
Barrel (petroleum)	Cubic foot	5.614583
	Cubic meter	0.1589873
	Gallon (Brit.)	34.97232
	Gallon (U.S.)	*42*
	Liter	158.9873
Barrel (U.S., dry)	Bushel (U.S.)	3.281219
	Cubic foot	4.083333
	Cubic inch	*7056*
	Cubic meter	0.1156271
	Liter	115.6271
	Pint (U.S., dry)	209.998
	Quart (U.S., dry)	104.9990
Barrel (U.S., cranb.)	Cubic inch	*5826*
	Liter	95.4710
Barrel (U.S., liquid)	Cubic foot	4.2109375
	Cubic inch	7276.5
	Cubic meter	0.1192405
	Gallon (Brit.)	26.22925
	Gallon (U.S.)	*31.5*
	Liter	119.2405
Barye	Bar	1×10^{-6}
	Dyne/square centimeter	*1*
Becquerel	Curie	2.702703×10^{11}
Biot	Ampere	*10*
Board foot	Cubic foot	0.083333 *(1/12)*
Bolt (cloth)	Foot	*120*
Btu	Calorie	251.996
	Cubic foot-atmosphere	*0.367717*
	Foot-poundal	25036.9
	Foot-pound-force	778.169
	Horsepower-hour	3.93015×10^{-4}
	Horsepower-hour (metric)	3.98464×10^{-4}
	Joule	1055.056
	Kilocalorie	0.251996
	Kilogram-force-meter	107.586
	Kilowatt-hour	2.93071×10^{-4}
	Liter-atmosphere	10.4126
	Watt-hour	0.293071
Btu (39 °F, 4 °C)	Joule	1059.67
Btu (60 °F, 15.6 °C)	Joule	1054.68
Btu (mean)	Joule	1055.87
Btu (thermochemical)	Joule	1054.350
Btu/cubic foot	Joule/cubic meter	37258.9

CONVERSION FACTORS (continued)

To convert from	To	Multiply by	To convert from	To	Multiply by
	kilocalorie/cubic meter	8.89915		Mile (statute)	0.1363636
Btu/°F	Calorie/°C	453.592	Caliber	Inch	0.01
	Joule/°C	1899.10		Millimeter	0.254
Btu/hour	Btu/minute	0.0166667 (1/60)	Calorie	Btu	3.96832×10^{-3}
	Btu/second	2.77778×10^{-4}		Cubic foot-atmosphere	1.45922×10^{-3}
	Calorie/second	0.0699988		Foot-poundal	99.3543
	Foot-pound-force/ second	0.216158		Foot-pound-force	3.08803
	Horsepower	3.93015×10^{-4}		Horsepower-hour	1.55961×10^{-6}
	Watt	0.293071		Horsepower-hour (metric)	1.58124×10^{-6}
Btu/(hour × square foot)	Watt/square meter	3.15459		Joule	4.1868
Btu/ (hour × square foot × °F)	Calorie/ second × square meter × °C)	1.35623		Kilocalorie	0.001
				Kilogram-force-meter	0.426935
	Watt/(square meter × °C)	5.67826		Kilowatt-hour	1.163×10^{-6}
				Liter-atmosphere	0.0413205
Btu/ (hour × square foot × °F/foot)	Watt/(meter × °C)	1.73073		Watt-hour	1.163×10^{-3}
			Calorie (15°C)	Joule	4.1855
			Calorie (20°C)	Joule	4.18190
Btu/ (hour × square foot × °F/inch)	Watt/(meter × °C)	0.144228	Calorie (mean)	Joule	4.19002
			Calorie (thermochem.)	Joule	4.184
			Calorie/°C	Btu/°F	2.20462×10^{-3}
Btu/minute	Calorie/second	4.19993		Joule/°F	2.326
	Horsepower	0.0235809	Calorie/gram	Btu/pound	1.8
	Watt	17.5843		Joule/kilogram	4186.8
Btu/(minute × square foot)	Watt/square meter	189.273	Calorie/(gram × °C)	Btu/(pound × °F)	1
Btu/pound	Calorie/gram	0.555556		Joule/(kilogram × °C)	4186.8
	Joule/kilogram	2326	Calorie/minute	Watt	0.06978
	Kilocalorie/kilogram	0.555556	Calorie/(minute × square centimeter)	Watt/square meter	697.8
	Watt-hour/kilogram	0.646111			
Btu/(pound × °F)	Calorie/(gram × °C)	1	Calorie/second	Watt	4.1868
	Joule/(kilogram × °C)	4186.8	Calorie/ (second × square centimeter)	Kilowatt/square meter	41.868
Btu/second	Horsepower	1.41485			
	Kilowatt	1.055056			
Btu/(second × square foot)	Kilowatt/square meter	11.3565	Calorie/ (second × square centimeter × °C)	Kilowatt/ (square meter × °C)	41.868
Btu/(second × square foot × °F)	Kilowatt/ (square meter × °C)	20.4417	Calorie/ (second × square centimeter × °C/ centimeter)	Watt/(meter × °C)	418.68
Btu/ (second × square foot × °F/foot)	Kilowatt/(meter × °C)	6.23064			
			Calorie/ square centimter	Kilojoule/square meter	41.868
Btu/ (second × square foot × °F /inch)	Watt/(meter × °C)	519.220	Candela	Hefner unit	1.11
				Lumen/steradian	1
			Candela/ square centimeter	Candela/square foot	929.0304
				Candela/square inch	6.4516
Btu/square foot	Joule/square meter	11356.5		Lambert	3.141593 (π)
	watt-hour/square meter	3.15459	Candela/square foot	Candela/square inch	6.944444×10^{-3} (1/144)
Bucket (Brit.)	Gallon (Brit.)	4			
Bushel (Brit.)	Bushel (U.S.)	1.032057		Candela/square meter	10.76391
	Gallon (Brit.)	8		Foot-lambert	3.141593 (π)
	Liter	36.36872		Lambert	3.381582×10^{-3}
Bushel (U.S.)	Barrel (U.S., dry)	0.3047647	Candela/square inch	Candela/ square centimeter	0.1550003
	Bushel (Brit.)	0.9689390			
	Cubic foot	1.244456		Candela/square foot	144
	Cubic inch	2150.42		Foot-lambert	452.3893
	Gallon (Brit.)	7.751512		Lambert	0.4869478
	Gallon (U.S., liquid)	9.309177	Candela/square meter	Candela/square foot	0.09290304
	Liter	35.23907		Lambert	3.141593×10^{-4}
	Peck (U.S.)	4	Carat (metric)	Gram	0.2
	Pint (U.S., dry)	64	Cental	Kilogram	45.359237
	Quart (U.S., dry)	32		Pound	100
Butt (Brit.)	Gallon (Brit.)	108 or 126	°C heat unit (chu)	Btu	1.8
Cable length (int.)	Foot	607.6115		Calorie	453.592
	Meter	185.2		Joule	1899.10
	Mile (nautical)	0.1	Centiliter	Cubic centimeter	10
Cable length (U.S.)	Foot	720		Cubic inch	0.6102374
	Meter	219.456		Drachm (Brit., fliud)	2.815606
	Mile (nautical)	0.1184968		Dram (U.S., fluid)	2.705122

CONVERSION FACTORS (continued)

To convert from	To	Multiply by
Centimeter	Ounce (Brit., fluid)	0.3519508
	Ounce (U.S., fluid)	0.3381402
	Foot	0.03280840
	Inch	0.3937008
	Micrometer	10000
	Mil	393.7008
	Millimeter	10
	Yard	0.01093613
Centimeter of Hg (conv.)	Atmosphere	0.0131579
	Millibar	13.3322
	Millimeter of H_2O (conv.)	135.951
	Pascal	1333.22
	Pound-force/square inch	0.193368
Centimeter of H_2O (conv.)	Atmosphere	9.67841×10^{-4}
	Millibar	0.980665
	Millimeter of Hg (conv.)	0.735559
	Kilogram-force/square centimeer	0.001
	Pascal	98.0665
	Pound-force/square inch	0.0142233
Centimeter/second	Foot/minute	1.968504
	Foot/second	0.03280840
	Kilometer/hour	0.036
	Meter/minute	0.6
	Mile/hour	0.02236936
Centimeter/square second	Foot/square second	0.03280840
	Kilometer/(hour × second)	0.036
	Meter/square second	0.01
	Mile/(hour × second)	0.02236936
Centipoise	Pascal-second	0.001
Centistokes	Square meter/second	1×10^{-6}
Chain (Gunter's)	Foot	66
Chain (Ramsden's)	Foot	100
Circular inch	Circular mil	1×10^{6}
	Square centimeter	5.067075
	Square inch	0.7853982
Circular mil	Square inch	7.853982×10^{-7}
	Square micrometer	506.7075
	Square mil	0.7853982
Circular millimeter	Square millimeter	0 7853982
Circumference	Degree	360
	Gon (grade)	400
	Radian	6.283185 (2π)
Clo	(°C × square meter)/watt	0.2003712
Cord	Cord-foot	8
	Cubic foot	128
Cord-foot	Cord	0.125 (1/8)
	Cubic foot	16
Coulomb	Ampere-second	1
Cubic centimeter	Cubic foot	3.531467×10^{-5}
	Cubic inch	0.06102374
	Cubic meter	1×10^{-6}
	Cubic millimeter	1000
	Cubic yard	1.307951×10^{-6}
	Drachm (Brit., fluid)	0.2815606
	Dram (U.S., fluid)	0.2705122
	Gallon (Brit.)	2.199692×10^{-4}
	Gallon (U.S.)	2.641721×10^{-4}
	Gill (Brit.)	7.039016×10^{-3}
	Gill (U.S.)	8.453506×10^{-3}
	Liter	0.001
	Milliliter	1
	Minim (Brit.)	16.89364
	Minim (U.S.)	16.23073
	Ounce (Brit., fluid)	0.03519508
	Ounce (U.S., fluid)	0.03381402
	Pint (Brit.)	1.759754×10^{-3}
	Pint (U.S., dry)	1.816166×10^{-3}

To convert from	To	Multiply by
	Pint (U.S., liquid)	2.113376×10^{-3}
	Quart (Brit.)	8.798770×10^{-4}
	Quart (U.S., dry)	9.080830×10^{-4}
	Quart (U.S., liquid)	1.056688×10^{-3}
Cubic centimeter/gram	Cubic foot/pound	0.0160185
Cubic centimeter/second	Cubic foot/minute	2.118880×10^{-3}
	Liter/hour	3.6
Cubic centimeter-atmosphere	Joule	0.101325
	Watt-hour	2.814583×10^{-5}
Cubic decimeter	Cubic centimeter	1000
	Cubic foot	0.03531467
	Cubic inch	61.02374
	Cubic meter	0.001
	Liter	1
Cubic foot	Acre-foot	2.295684×10^{-5}
	Board foot	12
	Bushel (Brit.)	0.7786044
	Bushel (U.S.)	0.8035640
	Cord	7.8125×10^{-3} (1/128)
	Cord-foot	0.0625 (1/16)
	Cubic centimeter	28316.847
	Cubic inch	1728
	Cubic meter	0 028316847
	Cubic yard	0.03703704 (1/27)
	Gallon (Brit.)	6.228835
	Gallon (U.S.)	7.480519
	Liter	28.316847
	Pint (Brit.)	49.83068
	Pint (U.S., dry)	51.42809
	Pint (U.S., liquid)	59.84416
	Quart (Brit.)	24.91534
	Quart (U.S., dry)	25.71405
	Quart (U.S., liquid)	29.92208
Cubic foot/hour	Cubic centimeter/second	7.865791
	Liter/minute	0.4719474
Cubic foot/minute	Cubic centimeter/second	471.9474
	Gallon (Brit.)/second	0.1038139
	Gallon (U.S.)/second	0.1246753
Cubic foot/pound	Cubic meter/kilogram	0.06242796
Cubic foot/second	Cubic meter/hour	101.9406
	Cubic yard/minute	2.222222
	Gallon (Brit.)/minute	373.7301
	Gallon (U.S.)/minute	448.8312
	Liter/minute	1699.011
Cubic foot-atmosphere	Btu	2.71948
	Calorie	685.298
	Foot-pound-force	2116.22
	Joule	2869.205
	Kilogram-force-meter	292.577
	Liter-atmosphere	28.31685
	Watt-hour	0.7970012
Cubic foot (pound-force/ square inch)	Btu	0.7970012
	Calorie	
	Joule	0.185050
	Watt-hour	46.6317
		195.238
		0.0542327
Cubic inch	Board foot	6.944444×10^{-3} (1/144)
	Bushel (Brit.)	4.505813×10^{-4}
	Bushel (U.S.)	4.650254×10^{-4}
	Cubic centimeter	16.387064
	Cubic foot	5.787037×10^{-4} (1/1728)
	Cubic meter	1.6387064×10^{-5}
	Cubic yard	2.143347×10^{-5}
	Drachm (Brit., fluid)	4.613952
	Dram (U.S., fluid)	4.432900
	Gallon (Brit.)	3.604650×10^{-3}
	Gallon (U.S.)	4.329004×10^{-3} (1/231)
	Liter	0.016387064
	Milliliter	16.387064
	Ounce (Brit., fluid)	0.5767440

CONVERSION FACTORS (continued)

To convert from	To	Multiply by
	Ounce (U.S., fluid)	0.5541126
	Pint (Brit.)	0.02883720
	Pint (U.S., dry)	0.02976163
	Pint (U.S., liquid)	0.03463203
	Quart (Brit.)	0.01441860
	Quart (U.S., dry)	0.01488081
	Quart (U.S., liquid)	0.01731602
Cubic inch/minute	Cubic centimeter/second	0.2731177
Cubic kilometer	Cubic mile	0.2399128
Cubic meter	Barrel (petroleum)	6.289811
	Barrel (U.S., dry)	8.648490
	Barrel (U.S., liquid)	8.386414
	Bushel (U.S.)	28.37759
	Cubic centimeter	1×10^6
	Cubic decimeter	1000
	Cubic foot	35.31467
	Cubic inch	61023.74
	Cubic yard	1.307951
	Gallon (Brit.)	219.9692
	Gallon (U.S.)	264.1721
	Liter	1000
	Pint (Brit.)	1759.754
	Pint (U.S., dry)	1816.166
	Pint (U.S., liquid)	2113.376
	Quart (Brit.)	879.8770
	Quart (U.S., dry)	908.0830
	Quart (U.S., liquid)	1056.688
	Register ton	0.3531467
Cubic meter/kilogram	Cubic foot/pound	16.01846
Cubic mile	Cubic kilometer	4.168182
Cubic millimeter	Cubic centimeter	0.001
	Cubic inch	6.102374×10^{-5}
	Minim (Brit.)	0.01689364
	Minim (U.S.)	0.01623073
Cubic yard	Bushel (Brit.)	21.02232
	Bushel (U.S.)	21.69623
	Cubic foot	27
	Cubic inch	46656
	Cubic meter	0.76455486
	Gallon (Brit.)	168.1786
	Gallon (U.S.)	201.9740
	Liter	764.5549
Cubic yard/minute	Cubic foot/second	0.45
	Gallon (Brit.)/second	2.802976
	Gallon (U.S.)/second	3.366234
	Liter/second	12.74258
Cubit	Inch	18
Cup (metric)	Milliliter	200
Cup (U.S.)	Milliliter	236.588
	Ounce (U.S. fluid)	8
Curie	Becquerel	3.7×10^{10}
Darcy	Square meter	9.869233×10^{-13}
Day (mean solar)	Hour	24
	Minute	1440
	Second	86400
Day (sidereal)	Second	86164.09
Decibel	Neper	0.115129255
Degree (angular)	Circumference	2.777778×10^{-3} (1/360)
	Gon (grade)	1.111111
	Minute (angular)	60
	Quadrant	0.01111111 (1/90)
	Radian	0.01745329
	Second (angular)	3600
Degree/foot	Radian/meter	0.05726146
Degree/inch	Radian/meter	0.6871375
Degree/second	revolution/minute	0.1666667 (1/6)
°C (temp. interval)	°Fahrenheit	1.8
	°Rankine	1.8
	Kelvin	1
(°C × hour)/kilocalorie	°C/watt	0.859845
(°C × hour × square meter)/kilocalorie	(°C × square meter)/watt	0.859845
°F (temp. interval)	°Celsius	0.5555556 (5/9)
	°Rankine	1

To convert from	To	Multiply by
	Kelvin	0.5555556 (5/9)
(°F × hour)/Btu	°C/watt	1.89563
(°F × hour × square foot)/Btu	(°C × square meter)/watt	0.176110
(°F/inch × hour × square foot)/Btu	(°C × meter)/watt	6.93347
Denier	Tex	0.111111 (1/9)
Drachm (Brit. fluid)	Dram (U.S., fluid)	0.9607599
	Milliliter	3.551633
	Minim (Brit.)	60
	Ounce (Brit. fluid)	0.125 (1/8)
Dram (apoth. or troy)	Dram (avoirdupois)	2.1942857
	Grain	60
	Gram	3.8879346
	Ounce (apoth. or troy)	0.125 (1/8)
	Pennyweight	2.5
	Scruple	3
Dram (avoirdupois)	Grain	27.34375
	Gram	1.7718452
	Ounce (avoirdupois)	0.0625 (1/16)
Dram (U.S., fluid)	Cubic centimeter	3.696691
	Cubic inch	0.2255859
	Drachm (Brit. fluid)	1.040843
	Gallon (U.S.)	9.765625×10^{-4} (1/1024)
	Gill (U.S.)	0.03125 (1/32)
	Milliliter	3.696691
	Minim (U.S.)	60
	Ounce (U.S., fluid)	0.125 (1/8)
	Pint (U.S., liquid)	7.8125×10^{-3} (1/128)
	Quart (U.S., liquid)	3.90625×10^{-3} (1/256)
Dyne	Kilogram-force	1.019716×10^{-6}
	Newton	1×10^{-5}
	Poundal	7.233014×10^{-5}
	Pound-force	2.248089×10^{-6}
Dyne/centimeter	Newton/meter	0.001
Dyne/square centimeter	Bar	1×10^{-6}
	Kilogram-force/square centimeter	1.019716×10^{-6}
	Millimeter of Hg (conv.)	7.50062×10^{-4}
	Millimeter of H_2O (conv.)	0.01019716
	Pascal (N/square meter)	0.1
	Pound-force/square inch	1.450377×10^{-5}
Dyne-centimeter	Erg	1
	Foot-poundal	2.37304×10^{-6}
	Foot-pound-force	7.37562×10^{-8}
	Joule	1×10^{-7}
	Kilogram-force-meter	1.019716×10^{-8}
	Newton-meter	1×10^{-7}
Dyne-second/square centimeter	Poise	1
	Pascal-second	0.1
Electronvolt	Erg	1.60219×10^{-12}
	Joule	1.60219×10^{-19}
Ell	Inch	45
Erg	Dyne-centimeter	1
	Joule	1×10^{-7}
	Watt-hour	2.777778×10^{-11}
Erg/(square centimeter × second)	Watt/square meter	0.001
Farad (int. mean)	Farad	0.999510
Farad (int. U.S.)	Farad	0.999505
Fathom	Foot	6
Fermi	Meter	1×10^{-15}
Firkin (Brit.)	Gallon (Brit.)	9
Firkin (U.S.)	Gallon (U.S.)	9
Foot	Centimeter	30.48
	Foot (U.S. Survey)	0.999998
	Inch	12
	Meter	0.3048
	Millimeter	304.8
	Mile (nautical)	1.645788×10^{-4}
	Mile (statute)	1.893939×10^{-4}

CONVERSION FACTORS (continued)

To convert from	To	Multiply by	To convert from	To	Multiply by
Foot (U.S. Survey)	Yard	0.333333 (1/3)		Cubic centimeter	4546.09
Foot (U.S. Survey)	Foot	1.000002		Cubic foot	0.1605437
	Meter	0.30480060960		Cubic inch	277.4194
Foot of H_2O (conv.)	Atmosphere	0.0294998		Cubic yard	5.946061×10^{-3}
	Bar	0.0298907		Drachm (Brit., fluid)	1280
	Inch of Hg (conv.)	0.882671		Gallon (U.S.)	1.200950
	Kilogram-force/square centimeter	0.03048		Gill (Brit.)	32
	Millimeter of Hg (conv.)	22.4198		Liter	4.54609
	Pascal (N/square meter)	2989.07		Minim (Brit.)	76800
	Pound-force/square inch	0.433527		Ounce (Brit., fluid)	160
Foot/°F	Meter/°C	0.54864		Peck (Brit.)	0.5
Foot/hour	Meter/second	8.466667×10^{-5}		Pint (Brit.)	8
Foot/minute	Kilometer/hour	0.018288		Quart (Brit.)	4
	Knot	9.87413×10^{-3}	Gallon (U.S., dry)	Bushel (U.S.)	0.125 (1/8)
	Meter/second	5.08×10^{-3}		Cubic inch	268.8025
	Mile/hour	0.01136364 (1/88)		Liter	4.404884
Foot/second	Kilometer/hour	1.09728	Gallon (U.S., liquid)	Barrel (petroleum)	0.02380952 (1/42)
	Knot	0.5924838		Cubic centimeter	3785.412
	Meter/minute	18.288		Cubic foot	0.13368056
	Meter/second	0.3048		Cubic inch	231
	Mile/hour	0.6818182		Cubic yard	4.951132×10^{-3}
Foot/square second	Kilometer/(hour × second)	1.09728		Dram (U.S., fluid)	1024
	Meter/square second	0.3048		Gallon (Brit.)	0.8326742
	Mile/(hour × second)	0.6818182		Gill (U.S.)	32
Foot to the fourth power	Meter to the fourth power	8.630975×10^{-3}		Liter	3.785412
				Minim (U.S.)	61440
Foot-candle	Lumen/square foot	1		Ounce (U.S., fluid)	128
	Lumen/square meter	10.76391		Pint (U.S., liquid)	8
	Lux	10.76391		Quart (U.S., liquid)	4
Foot-lambert	Candela/square centimeter	3.426259×10^{-4}	Gallon (Brit.)/minute	Cubic foot/hour	9.632619
	Candela/square foot	0.3183099 (1/π)		Cubic foot/second	2.675728×10^{-3}
	Candela/square meter	3.426259		Cubic meter/hour	0.2727654
	Lambert	1.076391×10^{-3}		Liter/second	0.07576817
	Meter-lambert	10.76391	Gallon (U.S.)/minute	Cubic foot/hour	8.020834
Foot-poundal	Btu	3.99411×10^{-5}		Cubic foot/second	2.228009×10^{-3}
	Calorie	0.0100650		Cubic meter/hour	0.2271247
	Foot-pound-force	0.0310810		Liter/second	0.06309020
	Joule	0.0421401	Gamma	Tesla	1×10^{-9}
	Kilogram-force-meter	4.29710×10^{-3}	Gauss	Tesla	1×10^{-4}
	Liter-atmosphere	4.15891×10^{-4}		Weber/square meter	1×10^{-4}
	Watt-hour	1.17056×10^{-5}	Geepound	Slug	1
Foot-pound-force	Btu	1.28507×10^{-3}	Gigawatt-hour	Kilowatt-hour	1×10^{6}
	Calorie	0.323832	Gilbert	Ampere	0.7957747
	Cubic foot-atmosphere	4.72541×10^{-4}	Gill (Brit.)	Cubic centimeter	142.0653
	Foot-poundal	32.1740		Cubic inch	8.669357
	Horsepower-hour	5.05051×10^{-7}		Gallon (Brit.)	0.03125 (1/32)
	Horsepower-hour (metric)	5.12055×10^{-7}		Gill (U.S.)	1.200950
	Joule	1.35582		Milliliter	142.0653
	Kilogram-force-meter	0.138255		Ounce (Brit. fluid)	5
	Liter-atmosphere	0.0133809		Pint (Brit.)	0.25 (1/4)
	Newton-meter	1.35582		Quart (Brit.)	0.125 (1/8)
	Watt-hour	3.76616×10^{-4}	Gill (U.S.)	Cubic centimeter	118.2941
Foot-pound-force/hour	Watt	3.76616×10^{-4}		Cubic inch	7.21875
Foot-pound-force/minute	Horsepower	3.03030×10^{-5}		Gallon (U.S.)	0.03125 (1/32)
	Horsepower (metric)	3.07233×10^{-5}		Gill (Brit.)	0.8326742
	Watt	0.0225970		Milliliter	118.2941
Foot-pound-force/second	Horsepower	1.81818×10^{-3} (1/550)		Ounce (U.S., fluid)	4
	Horsepower (metric)	1.84340×10^{-3}		Pint (U.S. liquid)	0.25 (1/4)
	Watt	1.355818		Quart (U.S., liquid)	0.125 (1/8)
			Gon (grade)	Circumference	0.002 5 (1/400)
				Degree (angular)	0.9
Franklin	Coulomb	3.335641×10^{-10}		Minute (angular)	54
Furlong	Foot	660		Radian	0.01570796
	Meter	201.168		Second (angular)	3240
	Mile (statute)	0.125 (1/8)	Grain	Carat (metric)	0.32399455
	Yard	220		Dram	0.03657143
Gal	Centimeter/square second	1		Milligram	64.79891
	Meter/square second	0.01		Ounce (avoirdupois)	2.285714×10^{-3}
				Ounce (troy)	2.083333×10^{-3} (1/480)
				Pennyweight	0.04166667 (1/24)
				Pound	1.428571×10^{-4} (1/7000)
				Scruple	0.05 (1/20)
Gallon (Brit.)	Bushel (Brit.)	0.125 (1/8)	Grain/cubic foot	Milligram/liter	2.288352

CONVERSION FACTORS (continued)

To convert from	To	Multiply by	To convert from	To	Multiply by
Grain/gallon (Brit.)	Milligram/liter	14.25377		Horsepower (metric)	1.01387
Grain/gallon (U.S.)	Milligram/liter	17.11806		Joule/second	745.700
	Pound/million gallons	142.8571		Kilocalorie/hour	641.186
Gram	Carat (metric)	5		Kilocalorie/minute	10.6864
	Dram	0.56438339		Kilocalorie/second	0.178107
	Grain	15.432358		Kilogram-force-meter/ second	76.0402
	Kilogram	0.001		Kilowatt	0.745700
	Milligram	1000	Horsepower (boiler)	Kilowatt	9.80950
	Ounce (avoirdupois)	0.035273962	Horsepower (electric)	Kilowatt	0.746
	Ounce (troy)	0.032150747	Horsepower (metric)	Foot-pound-force/ second	542.476
	Pennyweight	0.64301493		Horsepower	0.986320
	Pound	2.2046226×10^{-3}		Kilocalorie/hour	632.415
	Scruple	0.77161792		Kilocalorie/minute	10.54025
	Ton (metric)	1×10^{-6}		Kilocalorie/second	0.175671
Gram/(centimeter × second)	Poise	1		Kilogram-force-meter/ second	75
Gram/cubic centimeter	Kilogram/cubic decimeter	1		Kilowatt	0.735499
	Kilogram/cubic meter	1000	Horsepower (water)	Kilowatt	0.746043
	Kilogram/liter	1	Horsepower-hour	Btu	2544.43
	Pound/cubic foot	62.42796		Foot-pound-force	1.98×10^{6}
	Pound/cubic inch	0.03612729		Horsepower-hour (metric)	1.01387
	Pound/gallon (Brit.)	10.02241		Joule	2.68452×10^{6}
	Pound/gallon (U.S.)	8.345404		Kilocalorie	641.186
Gram/cubic meter	Grain/cubic foot	0.4369957		Kilogram-force-meter	2.73745×10^{5}
Gram/liter	Grain/gallon (Brit.)	70.15689		Kilowatt-hour	0.745700
	Grain/gallon (U.S.)	58.41783		Megajoule	2.68452
	Gram/cubic centimeter	0.001	Horsepower-hour (metric)	Horsepower-hour	0.986320
	Kilogram/cubic meter	1		Joule	2.64780×10^{6}
	Pound/cubic foot	0.0624280		Kilocalorie	632.415
	Pound/gallon (Brit.)	0.0100224		Kilogram-force-meter	2.7×10^{5}
	Pound/gallon (U.S.)	8.34540×10^{-3}		Kilowatt-hour	0.735499
Gram/meter	Ounce/yard	0.03225451		Megajoule	2.64780
Gram/milliliter	Gram/cubic centimeter	1	Hour (mean solar)	Day	0.04166667 (1/24)
Gram/square meter	Ounce/square foot	0.3277058		Minute	60
	Ounce/square yard	0.02949352		Second	3600
Gram/ton (long)	Gram/ton (metric)	0.9842065		Week	5.952381×10^{-3} (1/168)
	Gram/ton (short)	0.8928571	Hundredweight (long)	Hundredweight (short)	1.12
	Milligram/kilogram	0.9842065		Kilogram	50.80234544
Gram/ton (metric)	Gram/ton (long)	1.016047		Pound	112
	Gram/ton (short)	0.9071847		Ton (long)	0.05
	Milligram/kilogram	1		Ton (metric)	0.050802345
Gram/ton (short)	Gram/ton (long)	1.12		Ton (short)	0.056
	Gram/ton (metric)	1.102311	Hundredweight (short)	Hundredweight (long)	0.89285714
	Milligram/kilogram	1.102311		Kilogram	45.359237
Gram-force	Dyne	980.665		Pound	100
	Newton	9.80665×10^{-3}		Ton (long)	0.044642857
Gram-force/ square centimeter	Pascal	98.0665		Ton (metric)	0.045359237
Gram-force-centimeter	Erg	980.665		Ton (short)	0.05
	Joule	9.80665×10^{-5}	Inch	Centimeter	2.54
Gray	Joule/kilogram	1		Foot	0.08333333 (1/12)
Hand	Inch	4		Mil	1000
Hectare	Acre	2.471054		Millimeter	25.4
	Are	100		Yard	0.02777778 (1/36)
	Square foot	1.076391×10^{5}	Inch of Hg (conv.)	Atmosphere	0.0334211
	Square kilometer	0.01		Foot of H_2O (conv.)	1.132925
	Square meter	10000		Inch of H_2O (conv.)	13.5951
	Square mile	3.861022×10^{-3}		Kilogram-force/ square centimeter	0.0345316
	Square yard	11959.90		Millibar	33.8639
Hectogram	Kilogram	0.1		Millimeter of H_2O (conv.)	345.316
Hectoliter	Cubic meter	0.1		Pascal	3386.39
Hefner unit	Candela	0.903		Pound-force/square inch	0.491154
Henry (int. mean)	Henry	1.00049	Inch of H_2O (conv.)	Inch of Hg (conv.)	0.0735559
Henry (int. U.S.)	Henry	1.000495		Kilogram-force/ square centimeter	2.54×10^{-3}
Hogshead (U.S.)	Gallon (U.S.)	63		Millibar	2.49089
Horsepower	Btu/hour	2544.43		Millimeter of Hg (conv.)	1.86832
	Btu/minute	42.4072		Pascal	249.089
	Btu/second	0.706787		Pound-force/square inch	0.0361273
	Foot-pound-force/hour	1.98×10^{6}			
	Foot-pound-force/ minute	33000			
	Foot-pound-force/ second	550			

CONVERSION FACTORS (continued)

To convert from	To	Multiply by	To convert from	To	Multiply by
Inch/°F	Millimeter/°C	45.72		Ton (metric)	0.001
Inch/hour	Millimeter/minute	0.4233333		Ton (short)	1.1023113×10^{-3}
	Millimeter/second	7.055556×10^{-3}	Kilogram/cubic meter	Gram/cubic centimeter	0.001
	Foot/minute	1.388889×10^{-3}		Gram/liter	1
Inch/minute	Foot/hour	5		Pound/cubic foot	0.06242796
	Meter/hour	1.524		Pound/cubic inch	3.612729×10^{-5}
	Millimeter/second	0.4233333	Kilogram/meter	Gram/centimeter	10
Inch/second	Foot/hour	300		Pound/foot	0.6719690
	Meter/minute	1.524		Pound/inch	0.05599741
Inch to the fourth power	Meter to the fourth power	4.162314×10^{-7}	Kilogram-force	Dyne	9.80665×10^5
Joule	Btu	9.47817×10^{-4}		Newton	9.80665
	Calorie	0.238846		Pound-force	2.20462
	Centigrade heat unit	5.26565		Poundal	70.9316
	Cubic foot-atmosphere	3.48529×10^{-4}	Kilogram-force/square centimeter	Atmosphere	0.967841
	Cubic foot-pound-force/ square inch	3.12196×10^{-3}		Atmosphere (technical)	1
	Erg	1×10^7		Bar	0.980665
	Foot-poundal	23.7304		Foot of H$_2$O (conv.)	32.8084
	Foot-pound-force	0.737562		Inch of Hg (conv.)	28.9590
	Horsepower-hour	3.72506×10^{-7}		Kilogram-force/ square millimeter	0.01
	Horsepower-hour (metric)	3.77673×10^{-7}		Meter of H$_2$O (conv.)	10
	Kilogram-force-meter	0.101972		Millimeter of Hg (conv.)	735.559
	Liter-atmosphere	9.86923×10^{-3}		Newton/square millimeter	0.0980665
	Newton-meter	1		Pascal (N/square meter)	98066.5
	Watt-hour	2.777778×10^{-4} (1/3600)		Pound-force/square foot	2048.16
	Watt-second	1		Pound-force/square inch	14.22334
Joule/°C	Btu/°F	5.26565×10^{-4}		Ton-force (long)/ square foot	0.914358
Joule/gram	Btu/pound	0.429923		Ton-force (short)/ square foot	1.02408
	Kilocalorie/kilogram	0.238846		Ton-force (long)/ square inch	6.34971×10^{-3}
Joule/(gram × °C)	Btu/(pound × °F)	0.238846		Ton-force (short)/ square inch	7.11167×10^{-3}
	Kilocalorie/ (kilogram × °C)	0.238846	Kilogram-force/ square meter	Pascal	9.80665
Joule/hour	Watt	2.777778×10^{-4} (1/3600)	Kilogram-force/ square millimeter	Newton/square millimeter	9.80665
Joule/minute	Watt	0.01666667 (1/60)		Megapascal	9.80665
Joule/second	Watt	1		Pound-force/square inch	1422.334
Kelvin (temp. interval)	°Celsius	1	Kilogram-force-meter	Btu	9.29491×10^{-3}
	°Fahrenheit	1.8		Calorie	2.34228
	°Rankine	1.8		Cubic foot-atmosphere	3.41790×10^{-3}
Kilderkin (Brit.)	Gallon (Brit.)	18		Erg	9.80665×10^7
Kilocalorie	Btu	3.96832		Foot-poundal	232.715
	Calorie	1000		Foot-pound-force	7.23301
	Joule	4186.8		Horsepower-hour	3.65304×10^{-6}
Kilocalorie/cubic meter	Btu/cubic foot	0.112370		Horsepower-hour (metric)	3.70370×10^{-6}
	Kilojoule/cubic meter	4.1868		Joule	9.80665
Kilocalorie/hour	Watt	1.163		Liter-atmosphere	0.0967841
Kilocalorie/(hour × square meter)	Watt/square meter	1.163		Newton-meter	9.80665
Kilocalorie/(hour × square meter × °C)	Watt/(square meter × °C)	1.163		Watt-hour	2.72407×10^{-3}
Kilocalorie/(hour × square meter × °C/ centimeter)	Watt/(meter × °C)	0.01163	Kilometer	Astronomical unit	6.68459×10^{-9}
Kilocalorie/kilogram	Btu/pound	1.8		Foot	3280.840
	Joule/gram	4.1868		Light year	1.05702×10^{-13}
Kilocalorie/(kilogram × °C)	Btu/(pound × °F)	1		Mile (nautical)	0.5399568
	Kilojoule/(kg × °C)	4.1868		Mile (statute)	0.6213712
Kilocalorie/minute	Foot-pound-force/ second	51.4671		Yard	1093.613
	Horsepower	0.0935765	Kilometer/hour	Foot/minute	54.68066
	Horsepower (metric)	0.0948744		Foot/second	0.9113444
	Watt	69.78		Inch/second	10.93613
Kilocalorie/second	Kilowatt	4.1868		Knot	0.5399568
Kilogram	Grain	15432.358		Meter/minute	16.66667
	Gram	1000		Meter/second	0.2777778
	Hundredweight (long)	0.019684131		Mile/hour	0.6213712
	Hundredweight (short)	0.022046226	Kilometer/(hour × second)	Centimeter/square second	27.77778
	Ounce (avoirdupois)	35.273962			
	Ounce (troy)	32.150747		Foot/square second	0.9113444
	Pound	2.2046226		Meter/square second	0.2777778
	Ton (long)	9.8420653×10^{-4}		Mile/(hour × second)	0.6213712

CONVERSION FACTORS (continued)

To convert from	To	Multiply by
Kilopascal	Pound-force/square foot	20.8854
	Pound-force/square inch	0.1450377
Kilopond	Kilogram-force	1
	Newton	9.80665
Kilowatt	Btu/hour	3412.14
	Btu/minute	56.8690
	Btu/second	0.947817
	Foot-pound-force/hour	2.65522×10^6
	Foot-pound-force/minute	44253.7
	Foot-pound-force/second	737.562
	Horsepower	1.34102
	Horsepower (metric)	1.35962
	Joule/hour	3.6×10^6
	Joule/minute	60000
	Joule/second	1000
	Kilocalorie/hour	859.845
	Kilocalorie/minute	14.3308
	Kilocalorie/second	0.238846
	Kilogram-force-meter/hour	3.67098×10^5
	Kilogram-force-meter/minute	6118.30
	Kilogram-force-meter/second	101.972
Kilowatt-hour	Btu	3412.14
	Foot-pound-force	2.65522×10^6
	Horsepower-hour	1.34102
	Horsepower-hour (metric)	1.35962
	Joule	3.6×10^6
	Kilocalorie	859.845
	Kilogram-force-meter	3.67098×10^5
	Megajoule	3.6
Kilowatt-hour/pound	Btu/pound	3412.14
	Joule/gram	7936.641
	Kilocalorie/kilogram	1895.63
Kilowatt-hour/kilogram	Btu/pound	1547.72
Kip	Pound-force	1000
Kip/square inch	Newton/square millimeter	6.89476
	Megapascal	6.89476
Knot	Foot/minute	101.2686
	Foot/second	1.687810
	Kilometer/hour	1.852
	Meter/minute	30.86667
	Meter/second	0.5144444
	Mile (nautical)/hour	1
	Mile (statute)/hour	1.150779
Lambert	Candela/square centimeter	0.3183099 $(1/\pi)$
	Candela/square foot	295.7196
	Candela/square inch	2.053608
	Candela/square meter	3183.099
	Foot-lambert	929.0304
Langley	Joule/square meter	41840
Last (Brit.)	Gallon (Brit.)	640
League (nautical)	Mile (nautical)	3
League (statute)	Mile (statute)	3
Light year	Astronomical unit	63239.7
	Kilometer	9.46053×10^{12}
	Mile	5.87850×10^{12}
	Parsec	0.306595
Line	Inch	0.1 or 0.083333 $(1/12)$
	Millimeter	2.54 or 2.116667
Line	Weber	1×10^{-8}
Link	Chain	0.01
Liter	Bushel (Brit.)	0.027496156
	Bushel (U.S.)	0.02837759
	Cubic centimeter	1000
	Cubic decimeter	1
	Cubic foot	0.03531467
	Cubic inch	61.02374
	Cubic meter	0.001

To convert from	To	Multiply by
	Cubic yard	1.307951×10^{-3}
	Drachm (Brit., fluid)	281.5606
	Dram (U.S., fluid)	270.5122
	Gallon (Brit.)	0.21996925
	Gallon (U.S.)	0.26417205
	Gill (Brit.)	7.039016
	Gill (U.S.)	8.453506
	Milliliter	1000
	Minim (Brit.)	16893.64
	Minim (U.S.)	16230.73
	Ounce (Brit., fluid)	35.19508
	Ounce (U.S., fluid)	33.81402
	Pint (Brit.)	1.759754
	Pint (U.S., dry)	1.816166
	Pint (U.S., liquid)	2.113376
	Quart (Brit.)	0.8798770
	Quart (U.S., dry)	0.9080830
	Quart (U.S., liquid)	1.056688
Liter (1901—1964)	Cubic decimeter	1.000028
Liter/minute	Cubic foot/hour	2.118880
	Cubic foot/second	5.885778×10^{-4}
	Gallon (Brit.)/hour	13.19815
	Gallon (Brit.)/second	3.666154×10^{-3}
	Gallon (U.S.)/hour	15.85032
	Gallon (U.S.)/second	4.402868×10^{-3}
Liter/second	Cubic foot/hour	127.1328
	Cubic foot/minute	2.118880
	Gallon (Brit.)/hour	791.8893
	Gallon (Brit.)/minute	13.19815
	Gallon (U.S.)/hour	951.0194
	Gallon (U.S.)/minute	15.85032
Liter-atmosphere	Btu	0.0960376
	Calorie	24.2011
	Cubic foot-atmosphere	0.0353147
	Cubic foot-pound-force/square inch	0.518983
	Foot-poundal	2404.48
	Foot-pound-force	74.7335
	Horsepower-hour	3.77442×10^{-5}
	Horsepower-hour (metric)	3.82677×10^{-5}
	Joule	101.325
	Kilogram-force-meter	10.3323
	Watt-hour	0.0281458
Liter-bar	Joule	100
Lumen/square centimeter	Lux	10000
	Phot	1
Lumen/square foot	Lux	10.76391
Lumen/square meter	Lumen/square foot	0.09290304
	Lux	1
Lux	Lumen/square meter	1
	Phot	1×10^{-4}
Maxwell	Weber	1×10^{-8}
Megajoule	Kilowatt-hour	0.2777778
Megapascal	Bar	10
	Newton/square millimeter	1
Megohm	Ohm	1×10^6
Meter	Ångström	1×10^{10}
	Fathom	0.5468066
	Foot	3.2808399
	Foot (U.S. Survey)	3.2808333
	Inch	39.37007874
	Micrometer	1×10^6
	Mile (nautical)	5.399568×10^{-4}
	Mile (statute)	6.213712×10^{-4}
	Nanometer	1×10^9
	Yard	1.093613298
Meter/hour	Foot/minute	0.05468066
	Foot/second	9.113444×10^{-4}
	Millimeter/minute	16.66667
	Millimeter/second	0.2777778
Meter/minute	Foot/second	0.05468066
	Kilometer/hour	0.06

CONVERSION FACTORS (continued)

To convert from	To	Multiply by	To convert from	To	Multiply by
	Knot	0.03239741		Pound/ton (short)	0.002
	Mile (statute)/hour	0.03728227	Milligram/liter	Grain/gallon (Brit.)	0.07015689
	Millimeter/second	16.66667		Grain/gallon (U.S.)	0.05841783
Meter/second	Foot/minute	196.8504		Gram/cubic meter	1
	Kilometer/hour	3.6		Pound/cubic foot	6.242796×10^{-5}
	Kilometer/minute	0.06	Milligram/cubic meter	Grain/cubic foot	4.369957×10^{-4}
	Knot	1.943844	Milligram-force	Dyne	0.980665
	Mile (statute)/hour	2.236936		Newton	9.80665×10^{-6}
	Mile (statute)/minute	0.03728227	Milligram-force/ centimeter	Dyne/centimeter	0.980665
Meter/square second	Foot/square second	3.280840			
	Kilometer/(hour × second)	3.6		Newton/meter	9.80665×10^{-4}
	Mile/(hour × second)	2.236936	Milligram-force/inch	Dyne/centimeter	0.386089
Meter-candle	Lux	1		Newton/meter	3.86089×10^{-4}
Mho (ohm^{-1})	Siemens	1	Milliliter	Cubic centimeter	1
Microfarad	Farad	1×10^{-6}	Millimeter	Ångström	1×10^{7}
Microgram	Grain	1.5432358×10^{-5}		Inch	0.03937008
	Gram	1×10^{-6}		Micrometer	1000
Micrometer	Ångström	10000	Millimeter of Hg (conv.)	Atmosphere	1.315789×10^{-3}
	Mil	0.03937008			
	Millimeter	0.001		Dyne/square centimeter	1333.224
	Nanometer	1000		foot of H_2O (conv.)	0.0446033
Micron	Micrometer	1		Gram-force/square centimeter	1.35951
Mil	Inch	0.001			
	Micrometer	25.4		Millibar	1.333224
	Millimeter	0.0254		Millimeter of H_2O (conv.)	13.5951
Mile (nautical)	Foot	6076.1155			
	Kilometer	1.852		Pascal	133.3224
	Mile (statute)	1.150779		Pound-force/square foot	2.78450
	Yard	2025.372		Pound-force/square inch	0.0193368
Mile (statute)	Chain (Gunter's)	80		Torr	1
	Chain (Ramsden's)	52.8	Millimeter of H_2O (conv.)	Atmosphere	9.67841×10^{-3}
	Foot	5280			
	Furlong	8		Gram-force/square centimeter	0.1
	Inch	63360			
	Kilometer	1.609344		Millibar	0.0980665
	Light year	1.70111×10^{-13}		Millimeter of Hg (conv.)	0.0735559
	Meter	1609.344			
	Mile (nautical)	0.86897624		Pascal	9.80665
	Parsec	5.21552×10^{-14}		Pound force/square inch	1.42233×10^{-3}
	Rod	320	Millimicron	Nanometer	1
	Yard	1760	Minim (Brit.)	Drachm (Brit., fluid)	0.01666667 (1/60)
Mile (U.S. Survey)	Meter	1609.3472187		Milliliter	0.05919388
Mile/gallon (Brit.)	Kilometer/liter	0.354006		Minim (U.S.)	0.9607599
Mile/gallon (U.S.)	Kilometer/liter	0.425144		Ounce (Brit., fluid)	2.083333×10^{-3}
Mile/hour	Foot/minute	88			(1/480)
	Foot/second	1.466667	Minim (U.S.)	Dram (U.S., fluid)	0.01666667 (1/60)
	Kilometer/hour	1.609344		Milliliter	0.06161152
	Knot	0.8689762		Minim (Brit.)	1.040843
	Meter/minute	26.8224		Ounce (U.S., fluid)	2.083333×10^{-3}
	Meter/second	0.44704			(1/480)
Mile/(hour × minute)	Centimeter/square second	0.7450667	Minute	Day	6.944444×10^{-4}
					(1/1440)
Mile/(hour × second)	Centimeter/square second	44.704		Hour	0.01666667 (1/60)
				Second	60
Mile/minute	Foot/second	88		Week	9.920635×10^{-5}
	Kilometer/hour	96.56064	Minute (angular)	Circumference	4.629630×10^{-5}
	Knot	52.13857		Degree (angular)	0.01666667 (1/60)
	Meter/second	26.8224		Gon (grade)	0.01851852 (1/54)
Millibar	Pascal	100		Quadrant	1.851852×10^{-4}
Milligram	Carat (metric)	0.005		Radian	2.908882×10^{-4}
	Dram	5.6438339×10^{-4}		Second (angular)	60
	Grain	0.015432358	Month (mean of 4-year period)	Day	30.4375
	Ounce (avoirdupois)	3.5273962×10^{-5}			
	Ounce (troy)	3.2150747×10^{-5}		Hour	730.5
	Pennyweight	6.4301493×10^{-4}		Minute	43830
	Pound	2.2046226×10^{-6}		Second	2.6298×10^{6}
	Scruple	7.7161792×10^{-4}		Week	4.348214
Milligram/assay ton (Brit.)	Milligram/kilogram	30.612245	Nail (Brit.)	Inch	2.25
			Nanometer	Ångström	10
	Ounce (troy)/ton (long)	1		Micrometer	0.001
Milligram/assay ton (U.S.)	Milligram/kilogram	34.285714		Mil	3.937008×10^{-5}
			Neper	Decibel	8.685890
	Ounce(troy)/ton (short)	1	Newton	Dyne	1×10^{5}
Milligram/kilogram	Gram/ton (metric)	1		Kilogram-force	0.1019716
				Poundal	7.23301

CONVERSION FACTORS (continued)

To convert from	To	Multiply by	To convert from	To	Multiply by
	Pound-force	0.224809		Milligram/kilogram	27.90179
Newton/square centimeter	Newton/square millimeter	*0.01*	Ounce (avoirdupois)/ ton(short)	Gram/ton (metric)	*31.25*
	Pascal	*10000*		Milligram/kilogram	*31.25*
Newton/square meter	Pascal	*1*	Ounce (avoirdupois)/ yard	Gram/meter	31.00342
Newton/square millimeter	Kilogram-force/ square millimeter	0.1019716	Ounce-force (avoirdupois)	Newton	0.2780139
	Megapascal	*1*	Ounce-force (avoirdu- pois)/square inch	Pascal	430.922
	Ton-force (metric)/ square meter	101.9716	Ounce-force (avoirdu- pois)-inch	Newton-meter	7.06155×10^{-3}
Newton-meter	Foot-pound-force	0.737562	Pace	Foot	*2.5*
	Joule	*1*	Palm	Inch	*3*
	Kilogram-force-meter	0.1019716	Parsec	Astronomical unit	2.06265×10^{5}
	Watt-hour	2.777778×10^{-4}		Kilometer	3.0857×10^{13}
	Watt-second	*1*		Light year	3.26164
Nit	Candela/square meter	*1*		Mile (statute)	1.91735×10^{13}
Noggin (Brit.)	Gill (Brit.)	*1*	Part per million	Gram/ton (metric)	*1*
Nox	Lux	*0.001*		Milligram/kilogram	*1*
Oersted	Ampere/meter	79.57747		Milliliter/cubic meter	*1*
Ohm (int. mean)	Ohm.	1.00049		Ounce(avoirdupois)/ton (long)	*0.03584*
Ohm (int. U.S.)	Ohm	1.000495			
Ohm/foot	Ohm/meter	3.280840		Ounce (avoirdupois)/ ton(short)	*0.032*
Ohm-centimeter	Ohm-meter	*0.01*			
Ohm-circular mil/foot	Ohm-meter	1.662426×10^{-9}		Ounce(troy)/ton(long)	0.03266667
Ohm-meter	Ohm-square millimeter/ meter	1×10^{6}		Ounce(troy)/ton(short)	0.02916667
			Pascal	Atmosphere	9.869233×10^{-6}
Ohm-square millimeter/ meter	Ohm-meter	1×10^{-6}		Bar	1×10^{-5}
Ounce (avoirdupois)	Dram	*16*		Dyne/square centimeter	*10*
	Grain	*437.5*		Foot of H_2O (conv.)	3.34552×10^{-4}
	Gram	28.349523		Inch of Hg (conv.)	2.95300×10^{-4}
	Ounce (apot. or troy)	0.91145833		Inch of H_2O (conv.)	4.01463×10^{-3}
	Pennyweight	18.229167		Kilogram-force/square centimeter	1.01972×10^{-5}
	Pound	*0.0625 (1/16)*			
	Scruple	21.875		Millibar	*0.01*
Ounce (troy or ap.)	Grain	*480*		Millimeter of Hg (conv.)	7.50062×10^{-3}
	Gram	*31.1034768*			
	Ounce (avoirdupois)	1.0971429		Millimeter of H_2O (conv.)	0.101972
	Penneyweight	*20*			
	Pound (avoirdupois)	0.068571429		Newton/square meter	*1*
	Scruple	*24*		Newton/square millimeter	1×10^{-6}
Ounce (Brit. fluid)	Cubic centimeter	28.41306			
	Cubic inch	1.733871		Poundal/square foot	0.671969
	Drachm (Brit., fluid)	*8*		Pound-force/square foot	0.0208854
	Dram (U.S., fluid)	7.686079		Pound-force/square inch	1.45038×10^{-4}
	Gallon (Brit.)	6.25×10^{-3} *(1/160)*		Torr	7.50062×10^{-3}
	Gill (Brit.)	*0.2*	Pascal-second	Poise	*10*
	Milliliter	28.41306	Peck (Brit.)	Gallon (Brit.)	*2*
	Minim (Brit.)	*480*	Peck (U.S.)	Bushel (U.S.)	*0.25*
	Ounce (U.S., fluid)	0.9607599		Quart (U.S., dry)	*8*
	Pint (Brit.)	*0.05*	Pennyweight	Dram	0.87771429
	Quart (Brit.)	*0.025 (1/40)*		Grain	*24*
Ounce (U.S., fluid)	Cubic centimeter	29.57353		Gram	*1.55517384*
	Cubic inch	1.8046875		Ounce (avoirdupois)	0.054857143
	Dram (U.S., fluid)	*8*		Ounce (apoth. or troy)	*0.05*
	Gallon (U.S.)	7.8125×10^{-3} *(1/128)*		Pound	3.4285714×10^{-3}
	Gill (U.S.)	*0.25*	Perch	Foot	*16.5*
	Milliliter	29.57353	Phot	Lux	*10000*
	Minim (U.S.)	*480*	Pica (printer's)	Point (printer's)	*12*
	Ounce (Brit., fluid)	1.040843	Picofarad	Farad	1×10^{-12}
	Pint (U.S., liquid)	*0.0625 (1/16)*	Pint (Brit.)	Cubic centimeter	*568.26125*
	Quart (U.S., liquid)	*0.03125 (1/32)*		Cubic inch	34.67743
Ounce (avoirdupois)/cu- bic foot	Kilogram/cubic meter	1.001154		Gallon (Brit.)	*0.125 (1/8)*
				Gill (Brit.)	*4*
Ounce (avoirdupois)/cu- bic inch	Kilogram/cubic meter	1729.994		Liter	*0.56826125*
				Milliliter	*568.26125*
Ounce (avoirdupois)/ gallon (Brit.)	Kilogram/cubic meter	6.236023		Ounce (Brit., fluid)	*20*
				Pint (U.S., dry)	1.032057
Ounce (avoirdupois)/ gallon (U.S.)	Kilogram/cubic meter	7.489152		Pint (U.S. liquid)	1.200950
				Quart (Brit.)	*0.5*
Ounce (avoirdupois)/ square foot	Gram/square meter	305.1517	Pint (U.S., dry)	Bushel (U.S.)	*0.015625 (1/64)*
				Cubic centimeter	550.6105
Ounce (avoirdupois)/ square yard	Gram/square meter	33.90575		Cubic inch	*33.6003125*
				Liter	0.5506105
Ounce (avoirdupois)/ ton(long)	Gram/ton (metric)	27.90179		Milliliter	550.6105

CONVERSION FACTORS (continued)

To convert from	To	Multiply by
	Peck (U.S.)	0.0625 (1/16)
	Pint (Brit.)	0.9689390
	Quart (U.S. dry)	0.5
Pint (U.S., liquid)	Cubic centimeter	473.1765
	Cubic inch	28.875
	Gallon (U.S.)	0.125 (1/8)
	Gill (U.S.)	4
	Liter	0.4731765
	Milliliter	473.1765
	Ounce (U.S., fluid)	16
	Pint (Brit.)	0.8326742
	Quart (U.S., liquid)	0.5
Point (printer's, Didot)	Millimeter	0.3760650
Point (printer's, U.S.)	Inch	0.013837
	Millimeter	0.3514598
Poise	Dyne-second/square centimeter	1
	Gram/(centimeter × second)	1
	Pascal-second	0.1
Pole (Brit.)	Foot	16.5
Pond	Gram-force	1
Pottle (Brit.)	Gallon (Brit.)	0.5
Pound (avoirdupois)	Dram	256
	Grain	7000
	Gram	453.59237
	Hundredweight (long)	8.9285714×10^{-3}
	Hundredweight (short)	0.01
	Kilogram	0.45359237
	Ounce (avoirdupois)	16
	Ounce (troy)	14.583333
	Pennyweight	291.66667
	Pound (troy)	1.2152778
	Scruple	350
	Stone (Brit.)	0.07142857 (1/14)
	Ton (long)	4.4642857×10^{-4}
	Ton (metric)	4.5359237×10^{-4}
	Ton (short)	5×10^{-4} (1/2000)
Pound (troy)	Dram (troy)	96
	Grain	5760
	Gram	373.2417216
	Ounce (troy)	12
	Pennyweight	240
	Pound (avoirdupois)	0.82285714
	Scruple	288
Pound/acre	Kilogram/hectare	1.120851
Pound/cubic foot	Gram/liter	16.01846
	Kilogram/cubic meter	16.01846
	Pound/cubic inch	5.787037×10^{-4}
Pound/cubic inch	Gram/cubic centimeter	27.679905
	Pound/cubic foot	1728
Pound/cubic yard	Kilogram/cubic meter	0.5932764
Pound/foot	Kilogram/meter	1.488164
Pound/(foot × hour)	Pascal-second	4.133789×10^{-4}
Pound/(foot × second)	Pascal-second	1.488164
Pound/gallon (Brit.)	Gram/cubic centimeter	0.09977637
	Gram/liter	99.77637
	Kilogram/cubic meter	99.77637
	Pound/cubic foot	6.228835
	Ton(long)/cubic yard	0.07507968
Pound/gallon (U.S.)	Gram/cubic centimeter	0.1198264
	Gram/liter	119.8264
	Kilogram/cubic meter	119.8264
	Pound/cubic foot	7.480519
	Ton(short)/cubic yard	0.1009870
Pound/hour	Gram/minute	7.559873
	Gram/second	0.1259979
	Kilogram/day	10.88622
Pound/horsepower-hour	Kilogram/megajoule	0.1689659
	Kilogram/kilowatt-hour	0.6082774
Pound/inch	Kilogram/meter	17.85797
Pound/minute	Gram/second	7.559873
	Kilogram/hour	27.21554
Pound/second	Kilogram/hour	1632.932
	Kilogram/minute	27.21554

To convert from	To	Multiply by
Pound/square foot	Kilogram/square meter	4.882428
Poundal	Gram-force	14.0981
	Newton	0.1382550
	Pound-force	0.0310810
	Pascal	1.488164
Poundal-foot	Newton-meter	0.0421401
Poundal-second/square foot	Pascal-second	1.488164
Pound-force	Kilogram-force	0.453592
	Newton	4.44822
	Poundal	32.1740
Pound-force/foot	Newton/meter	14.5939
Pound-force/inch	Newton/meter	175.127
Pound-force/square foot	Atmosphere	4.72541×10^{-4}
	Bar	4.78803×10^{-4}
	Foot of H_2O (conv.)	0.0160185
	Gram-force/square centimeter	0.488243
	Inch of Hg (conv.)	0.0141390
	Millimeter of Hg (conv.)	0.359131
	Millimeter of H_2O (conv.)	4.88243
	Pascal	47.8803
	Pound-force/square inch	6.944444×10^{-3} (1/144)
Pound-force/square inch	Atmosphere	0.0680460
	Bar	0.0689476
	Foot of H_2O (conv.)	2.30666
	Inch of Hg (conv.)	2.03602
	Kilogram-force/square centimeter	0.0703070
	Meter of H_2O (conv.)	0.703070
	Millibar	68.9476
	Millimeter of Hg (conv.)	51.7149
	Pascal	6894.76
	Pound-force/square foot	144
Pound-force-foot	Newton-meter	1.35582
Pound-force-foot/inch	Newton-meter/meter	53.3787
Pound-force-inch	Newton-meter	0.112985
Pound-force-inch/inch	Newton-meter/meter	4.44822
Pound-force-second/square foot	Pascal-second	47.8803
Pound-force-second/square inch	Pascal-second	6894.76
Psi	Pound-force/square inch	1
Puncheon (Brit.)	Gallon (Brit.)	70
Quadrant	Degree (angular)	90
	Gon (grade)	100
	Minute (angular)	5400
	Radian	1.570796 (π/2)
Quart (Brit.)	Cubic centimeter	1136.5225
	Cubic foot	0.04013591
	Cubic inch	69.35486
	Gallon (Brit.)	0.25 (1/4)
	Gill (Brit.)	8
	Liter	1.1365225
	Ounce (Brit., fluid)	40
	Pint (Brit.)	2
	Quart (U.S., dry)	1.032057
	Quart (U.S., liquid)	1.200950
Quart (U.S., dry)	Bushel (U.S.)	0.03125 (1/32)
	Cubic centimeter	1101.221
	Cubic foot	0.03888925
	Cubic inch	67.200625
	Liter	1.101221
	Peck (U.S.)	0.125 (1/8)
	Pint (U.S., dry)	2
	Quart (U.S., liquid)	1.163647
Quart (U.S., liquid)	Cubic centimeter	946.35295
	Cubic foot	0.03342014
	Cubic inch	57.75
	Dram (U.S., fluid)	256
	Gallon (U.S.)	0.25 (1/4)

CONVERSION FACTORS (continued)

To convert from	To	Multiply by	To convert from	To	Multiply by
	Gill (U.S.)	8		Square foot	4356
	Liter	0.94635295		Square meter	404.6856
	Ounce (U.S., fluid)	32	Square chain (Ramsden's)	Square foot	10000
	Pint (U.S., liquid)	2	Square chain(U.S. Survey)	Square meter	404.687261
	Quart (Brit.)	0.8326742	Square degree	Steradian	3.046174×10^{-4}
	Quart (U.S., dry)	0.8593670	Square foot	Acre	2.295684×10^{-5}
Quarter (Brit., cap.)	Gallon (Brit.)	64		Square centimeter	929.0304
Quarter (Brit., mass)	Pound	28		Square chain (Gunter's)	2.295684×10^{-4}
Quarter (U.S., long)	Pound	560		Square chain (Ramsden's)	1×10^{-4}
Quarter (U.S., short)	Pound	500		Square inch	144
Quintal	Kilogram	100		Square link (Gunter's)	2.295684
Rad	Gray	0.01		Square meter	0.09290304
	Joule/kilogram	0.01		Square mile	3.587006×10^{-8}
Radian	Circumference	0.1591549 (1/2 π)		Square rod	3.673095×10^{-3}
	Degree (angular)	57.295780		Square yard	0.1111111 (1/9)
	Gon (grade)	63.66198	Square foot (U.S. Survey)	Square meter	0.092903412
	Minute (angular)	3437.747	Square foot/hour	Square meter/second	2.58064×10^{-5}
	Quadrant	0.6366198 (2/π)	Square inch	Circular mil	1.273240×10^{6}
	Revolution	0.1591549		Circular millimeter	821.4432
	Second (angular)	2.062648×10^{5}		Square centimeter	6.4516
Radian/centimeter	Degree/millimeter	5.729578		Square foot	6.944444×10^{-3} (1/144)
	Degree/foot	1746.375		Square millimeter	645.16
	Degree/inch	145.5313	Square inch/second	Square foot/minute	0.4166667
Radian/second	Revolution/minute	9.549297		Square meter/hour	2.322576
Radian/square second	Revolution/square minute	572.9578	Square kilometer	Acre	247.1054
Register ton	Cubic foot	100		Hectare	100
	Cubic meter	2.831685		Square foot	1.076391×10^{7}
Rem	Sievert	0.01		Square meter	1×10^{6}
Revolution	Degree (angular)	360		Square mile	0.38610216
	Gon (Grade)	400		Square yard	1.195990×10^{6}
	Radian	6.283185 (2 π)	Square link(Gunter's)	Square foot	0.4356
Revolution/minute	Degree/second	6	Square link(Ramsden's)	Square foot	1
Reyn	Pascal-second	6894.76	Square meter	Acre	2.471054×10^{-4}
Rhe	1/pascal-second	10		Are	0.01
Right angle	Degree	90		Hectare	1×10^{-4}
	Gon (grade)	100		Square centimeter	10000
Rod	Foot	16.5		Square chain (Gunter's)	2.471054×10^{-3}
Roentgen	Coulomb/kilogram	2.58×10^{-4}		Square foot	10.76391
Rood (Brit.)	Acre	0.25 (1/4)		Square inch	1550.003
	Square meter	1011.7141		Square kilometer	1×10^{-6}
Rope (Brit.)	Foot	20		Square link (Gunter's)	24.71054
Scruple	Dram (apoth. or troy)	0.3333333 (1/3)		Square mile	3.861022×10^{-7}
	Grain	20		Square yard	1.195990
	Gram	1.2959782	Square mil	Circular mil	1.273240
	Ounce (avoirdupois)	0.045714286		Square inch	1×10^{-6}
	Ounce (apoth. or troy)	0.04166667 (1/24)		Square micrometer	645.16
	Pennyweight	0.83333333 (10/12)		Square millimeter	6.4516×10^{-4}
	Pound	2.857143×10^{-3} (1/350)	Square mile	Acre	640
Scruple (Brit. fluid)	Minim (Brit.)	20		Square chain (Gunter's)	6400
Seam (Brit.)	Gallon (Brit.)	64		Square foot	2.78784×10^{7}
Second (angular)	Degree	2.777778×10^{-4} (1/3600)		Square kilometer	2.589988110
	Gon (grade)	3.086420×10^{-4} (1/3240)		Square meter	2.589988×10^{6}
	Minute (angular)	0.01666667 (1/60)		Square rod	1.024×10^{5}
	Radian	4.848137×10^{-6}		Square yard	3.0976×10^{6}
Shake	Second	1×10^{-8}		Township	0.02777778 (1/36)
Siemens	Mho (ohm^{-1})	1	Square mile (U.S. Survey)	Square kilometer	2.589998470
Slug	Geepound	1	Square millimeter	Circular mil	1973.525
	Kilogram	14.5939		Circular millimeter	1.273240
	Pound	32.1740		Square centimeter	0.01
Slug/cubic foot	Kilogram/cubic meter	515.379		Square inch	1.550003×10^{-3}
Slug/(foot × second)	Pascal-second	47.8803		Square mil	1550.003
Span	Inch	9	Square rod	Acre	0.00625 (1/160)
Sphere	Steradian	12.56637 (4 π)		Square foot	272.25
Square centimeter	Circular mil	1.973525×10^{5}		Square meter	25.29285
	Circular millimeter	127.3240	Square yard	Acre	2.066116×10^{-4}
	Square foot	1.076391×10^{-3}		Square foot	9
	Square inch	0.1550003		Square inch	1296
	Square meter	1×10^{-4}		Square meter	0.83612736
	Square millimeter	100		Square mile	3.228306×10^{-7}
	Square yard	1.195990×10^{-4}			
Square chain(Gunter's)	Acre	0.1			

CONVERSION FACTORS (continued)

To convert from	To	Multiply by
Standard (Petrograd)	Cubic foot	*165*
Statampere	Ampere	3.335641×10^{-10}
Statcoulomb	Coulomb	3.335641×10^{-10}
Statfarad	Farad	1.112650×10^{-12}
Stathenry	Henry	8.987552×10^{11}
Statmho	Siemens	1.112650×10^{-12}
Statohm	Ohm	8.987552×10^{11}
Statvolt	Volt	299.7925
Steradian	Sphere	0.07957747 (1/4 π)
	Spherical right angle	0.6366198 (2/π)
	Square degree	3282.806
Stere	Cubic meter	*1*
Stilb	Candela/square centimeter	*1*
Stokes	Square meter/second	1×10^{-4}
Stone	Pound	*14*
Tablespoon (metric)	Milliliter	*15*
Tablespoon (U.S.)	Milliliter	14.79
Teaspoon (metric)	Milliliter	5
Teaspoon (U.S.)	Milliliter	4.93
Terawatt-hour	Kilowatt-hour	1×10^{9}
Tesla	Weber/square meter	*1*
Tex	Denier	9
	Gram/kilometer	*1*
Therm	Btu	1×10^{5}
Thou	Mil	*1*
Ton (assay, (Brit.)	Gram	32.66667
Ton (assay, U.S.)	Gram	29.16667
Ton (long)	Hundredweight (long)	*20*
	Hundredweight (short)	*22.4*
	Kilogram	*1016.0469088*
	Pound	*2240*
	Ton (metric)	1.016047
	Ton (short)	*1.12*
Ton (metric)	Hundredweight (long)	19.684131
	Hundredweight (short)	22.046226
	Kilogram	*1000*
	Pound	2204.6226
	Ton (long)	0.98420653
	Ton (short)	1.1023113
Ton (short)	Hundredweight (long)	17.857143
	Hundredweight (short)	*20*
	Kilogram	*907.18474*
	Pound	*2000*
	Ton (long)	0.89285714
	Ton (metric)	0.90718474
Ton(long)/cubic yard	Kilogram/cubic meter	1328.939
Ton (metric)/cubic meter	Gram/cubic centimeter	*1*
	Kilogram/cubic decimeter	*1*
Ton(short)/cubic yard	Kilogram/cubic meter	1186.553
Ton-force (long)	Newton	9964.02
Ton-force (metric)	Newton	9806.65
Ton-force (short)	Newton	8896.44
Ton-force(long)/square foot	Atmosphere	1.05849
	Bar	1.07252
	Kilogram-force/square centimeter	1.09366
	Newton/square millimeter	0.107252
	Pascal	1.07252×10^{5}
	Pound-force/square inch	15.5556
Ton-force(long)/square inch	Atmosphere	152.423
	Bar	154.443
	Kilogram-force/ square centimeter	157.488
	Newton/square millimeter	15.4443
	Pascal	1.54443×10^{7}
	Pound-force/square inch	*2240*
Ton-force(metric)/ square meter	Atmosphere	0.0967841

To convert from	To	Multiply by
	Bar	*0.0980665*
	Kilogram-force/square centimeter	*0.1*
	Newton/square millimeter	9.80665×10^{-3}
	Pascal	9806.65
	Pound-force/square inch	1.42233
Ton-force (short)/square foot	Atmosphere	0.945083
	Bar	0.957605
	Kilogram-force/square centimeter	0.976486
	Newton/square millimeter	0.0957605
	Pascal	9.57605×10^{4}
	Pound-force/square inch	13.8889
Ton-force (short)/square inch	Atmosphere	136.092
	Bar	137.895
	Kilogram-force/square centimeter	140.614
	Newton/square millimeter	13.7895
	Pascal	1.37895×10^{7}
	Pound-force/square inch	*2000*
Tonne	Kilogram	*1000*
Torr	Millibar	1.333224
	Millimeter of Hg (conv.)	1
	Pascal	133.3224
Township (U.S.)	Square kilometer	93.23957
	Square mile	*36*
Unit pole	Weber	1.256637×10^{-7}
Volt (int. mean)	Volt	1.00034
Volt (int. U.S.)	Volt	1.000330
Volt/inch	Volt/meter	39.37008
Volt-second	Weber	*1*
Watt	Btu/hour	3.41214
	Btu/minute	0.0568690
	Calorie/minute	14.3308
	Calorie/second	0.238846
	Erg/second	1×10^{7}
	Foot-pound-force/ minute	44.2537
	Foot-pound-force/ second	0.737562
	Horsepower	1.34102×10^{-3}
	Horsepower (metric)	1.35962×10^{-3}
	Joule/second	*1*
	Kilocalorie/hour	0.859845
	Kilogram-force-meter/ second	0.101972
Watt (int. mean)	Watt	1.00019
Watt (int. U.S.)	Watt	1.000165
Watt/square inch	Btu/(hour × square foot)	491.348
	Kilocalorie/(hour × square meter)	1332.76
	Watt/square meter	1550.003
Watt/square meter	Kilocalorie/(hour × square meter)	0.859845
Watt-hour	Btu	3.41214
	Calorie	859.845
	Foot-pound-force	2655.22
	Horsepower-hour	1.34102×10^{-3}
	Horsepower-hour (metric)	1.35962×10^{-3}
	Joule	*3600*
	Kilogram-force-meter	367.098
	Liter-atmosphere	35.5292
Watt-second	Erg	1×10^{7}
	Joule	*1*
	Newton-meter	*1*
Weber	Maxwell	1×10^{8}
Weber/square meter	Gauss	*10000*

CONVERSION FACTORS (continued)

To convert from	To	Multiply by	To convert from	To	Multiply by
Week	Day	7		Hour	8766
	Hour	168		Minute	5.2596×10^5
	Minute	10080		Second	3.15576×10^7
	Month	0.2299795		Week	52.17857
	Second	6.048×10^5	Year (leap)	Day	366
X-unit	Meter	1.00202×10^{-13}	Year (normal calendar)	Day	365
Yard	Centimeter	91.44		Hour	8760
	Fathom	0.5		Minute	5.256×10^5
	Foot	3		Second	3.1536×10^7
	Inch	36		Week	52.14286
	Meter	0.9144	Year (sidereal)	Day	365.25636
	Mile	5.681818×10^{-4}		Second	3.155815×10^7
Year (calendar, mean of 4-year period)	Day	365.25		Year (tropical)	1.0000388
			Year (tropical)	Day	365.24220
				Second	3.1556926×10^7
				Year (sidereal)	0.9999612

DECIMAL EQUIVALENTS OF COMMON FRACTIONS

		1/64	= 0.015 625		11/32	22/64	= 0.343 75			43/64	= 0.671 875
	1/32	2/64	= .031 25			23/64	= .359 375	11/16	22/32	44/64	= .687 5
		3/64	= .046 875	3/8	12/32	24/64	= .375			45/64	= .703 125
1/16	2/32	4/64	= .062 5			25/64	= .390 625		23/32	46/64	= .718 75
		5/64	= .078 125		13/32	26/64	= .406 25			47/64	= .734 375
	3/32	6/64	= .093 75			27/64	= .421 875	3/4	24/32	48/64	= .75
		7/64	= .109 375	7/16	14/32	28/64	= .437 5			49/64	= .765 625
1/8	4/32	8/64	= .125			29/64	= .453 125		25/32	50/64	= .781 25
		9/64	= .140 625		15/32	30/64	= .468 75			51/64	= .796 875
	5/32	10/64	= .156 25			31/64	= .484 375	13/16	26/32	52/64	= .812 5
		11/64	= .171 875	1/2	16/32	32/64	= .50			53/64	= .828 125
3/16	6/32	12/64	= .187 5			33/64	= .515 625		27/32	54/64	= .843 75
		13/64	= .203 125		17/32	34/64	= .531 25			55/64	= .859 375
	7/32	14/64	= .218 75			35/64	= .546 875	7/8	28/32	56/64	= .875
		15/64	= .234 375	9/16	18/32	36/64	= .562 5			57/64	= .890 625
1/4	8/32	16/64	= .25			37/64	= .578 125		29/32	58/64	= .906 25
		17/64	= .265 625		19/32	38/64	= .593 75			59/64	= .921 875
	9/32	18/64	= .281 25			39/64	= .609 375	15/16	30/32	60/64	= .937 5
		19/64	= .296 875	5/8	20/32	40/64	= .625			61/64	= .953 125
5/16	10/32	20/64	= .312 5			41/64	= .640 625		31/32	62/64	= .968 75
		21/64	= .328 125		21/32	42/64	= .656 25			63/64	= .984 375

FUNDAMENTAL PHYSICAL CONSTANTS

Dr. E. Richard Cohen

The following table contains data which are a tentative revision of the 1963 values of the fundamental physical constants.

It has become increasingly clear in the last several years that the 1963 analysis of the fundamental physical constants by Cohen and DuMond must be revised and that the values recommended at the time are in error by as much as 100 ppm. The strongest evidence for this revision came from the measurement in 1967 of macroscopic phase coherence in superconductors by Langenberg, Parker and Taylor at the University of Pennsylvania. Their measured value of the quantum of magnetic flux ($h/2e$), measured with an accuracy of 4 ppm, was inconsistent with the 1963 recommendation by 10 times that amount. This verified the growing evidence from spectroscopic and microwave data that the value of the fine structure constant needed a revision of 20 ppm.

It is therefore clear that a complete revision of the 1963 recommendation is necessary. Such a revision will of course include experimental data in addition to that on the fine structure constant, including careful attention to the electrical standards maintained by each national standards laboratory as recalibrated with respect to BIPM in 1968, effective January 1, 1969.

The following table of numerical values of the physical constants is intended as a general indication of the extent of the revision required in the 1963 values. Because of the tentative nature, and since the full effect of experimental correlations between data have not been calculated, no errors are quoted for these values. The numerical values, although tentative, and not representing a full reassessment of the available data are believed to be more reliable than the 1963 values. A more recent discussion of the status of the physical constants as of approximately January, 1969, is given by B. N. Taylor, W. H. Parker, and D. N. Langenberg in Reviews of Modern Physics, July 1969.

FUNDAMENTAL PHYSICAL CONSTANTS

Constant	Symbol	Old value	New value		*Correction ppm
Speed of light in vacuum	c	2.997925_1	2.997925	$\times 10^8$ ms^{-1}	0
Gravitational constant	G	6.670_5	6.670	10^{-11} Nm^{2}kg^{-2}	0
Elementary charge	e	1.60210_2	1.6022	10^{-19}C	+60
		4.80298_7	4.8032	10^{-10} esu	+60
Avogadro constant	N_A	6.02252_9	6.0222	10^{26} kmole^{-1}	−60
Mass unit	u	1.66043_2	1.66053	10^{-27} kg	+60
Electron rest mass	m_e	9.10908_{13}	9.1096	10^{-31} kg	+60
		5.48597_3	5.48593	10^{-4} u	0
Proton rest mass	m_p	1.67252_3	1.67262	10^{-27} kg	+60
		1.00727663_8	1.00727661	u	0
Neutron rest mass	m_n	1.67482_3	1.67492	10^{-27} kg	+60
		1.0086654_4	1.0086652	u	0
Faraday constant	F	9.64870_5	9.6487	10^4C mole^{-1}	0
		2.89261_{25}	2.8926	10^{14} esu	0
Planck constant	h	6.62559_{16}	6.6262	10^{-34} Js	+100
	$h/2\pi$	1.054494_{25}	1.05459	10^{-34} Js	+100
Fine-structure constant	α	7.29720_3	7.29735	10^{-3}	+20
$2\pi e^2/hc$	$1/\alpha$	137.0388_6	137.0360		−20
Charge-to-mass ratio	e/m_e	1.758796_6	1.75880	10^{11}C kg^{-1}	0
for electron		5.27274_2	5.27276	10^{17} esu	0
Quantum of magnetic flux	hc/e	4.13556_4	4.13571	10^{-11} Wb	+40
		1.379474_{13}	1.37952	10^{-17} esu	+40
Rydberg constant	R_∞	1.0973731_1	1.0973731	10^7 m^{-1}	0
Bohr radius	a_0	5.29167_2	5.29177	10^{-11} m	+20
Compton wavelength of	$\lambda_c = h/m_e c$	2.42621_2	2.42631	10^{-12} m	+40
electron	$\lambda_c/2\pi$	3.86144_3	3.86159	10^{-13} m	+40
Gyromagnetic ratio of	γ	2.675192_7	2.67519	10^8 rad s^{-1}T^{-1}	0
proton	$\gamma/2\pi$	4.25770_1	4.2577	10^7 Hz T^{-1}	0
(Uncorrected for	γ'	2.675123_7	†2.67512	10^8 s^{-1}T^{-1}	0
diamagnetism H$_2$O)	$\gamma'/2\pi$	4.25759_1	†4.257586	10^7 Hz T^{-1}	0
Bohr magneton	μ_B	9.2732_2	9.2741	10^{-24} J T^{-1}	+100
Nuclear magneton	μ_N	5.05050_{13}	5.0510	10^{-27} J T^{-1}	+100
Proton Moment	μ_p	1.41049_4	1.4106	10^{-26} J T^{-1}	+80
	μ_p/μ_N	2.79276_2	2.79278		0
(Uncorrected for diamagnetism in H$_2$O sample)		2.79268_2	2.79271		0
Gas constant	R_0	8.31434_{35}	8.3143	J deg^{-1} mole^{-1}	0
Boltzmann constant	k	1.38054_6	1.3806	10^{-23} J deg^{-1}	+60
First radiation constant ($2\pi hc^2$)	c_1	3.74150_9	3.7418	10^{-16} W m^2	+80
Second radiation constant (hc/k)	c_2	1.43879_6	1.4388	10^{-2} m deg	0
Stephan-Boltzmann constant	σ	5.6697_{10}	5.6696	10^{-8} W m^{-2} deg^{-4}	−20

* This column gives the correction resulting only from the increase of 20 ppm in the value of the fine structure constant, not the total change from 1963 to the tentative new value.

† The value for the gyromagnetic ratio of the proton has been recommended by the Comité International des Poids et Mesures in their meeting of 14–17 October 1968 for international metrological usage. This value is based on the 1969 BIPM scales of resistance and electromotive force which are in agreement, as exactly as is possible, with the (absolute) definitions of electrical units adopted by the Conférence Generale des Poids et Mesures.

MISCELLANEOUS CONSTANTS

PHYSICAL CONSTANTS

Equatorial radius of the earth = 6378.388 km = 3963.34 miles (statute).
Polar radius of the earth, 6356.912 km = 3949.99 miles (statute).
1 degree of latitude at 40° = 69 miles.
1 international nautical mile = 1.15078 miles (statute) = 1852 m = 6076.115 ft.
Mean density of the earth = 5.522 g/cm^3 = 344.7 lb/ft^3.
Constant of gravitation, $(6.673 \pm 0.003) \times 10^3$ cm^3 gm^{-1} s^{-2}.
Acceleration due to gravity at sea level, latitude 45° = 980.6194cm/s^2 = 32,1726 ft/sec^2
Length of seconds pendulum at sea level, latitude 45° = 99.3575 cm = 39.1171 in.
1 knot (international) = 101.269 ft/min = 1.6878 ft/sec = 1.1508 miles (statute)/hr.
1 micron = 10^{-4} cm.
1 angstrom = 10^{-8} cm.
Mass of hydrogen atom = $(1.67339 \pm 0.0031) \times 10^{-24}$ g.
Density of mercury at 0°C = 13.5955 g/ml.
Density of water at 3.98°C = 1.000000 g/ml.
Density, maximum, of water, at 3.98°C = 0.999973 g/cm^3.
Density of dry air at 0°C, 760 mm = 1.2929 g/liter.
Velocity of sound in dry air at 0°C = 331.36 m/s - 1087.1 ft/sec.
Velocity of light in vacuum = $(2.997925 \pm 0.000002) \times 10^{10}$ cm/s.
Heat of fusion of water 0°C = 79.71 cal/g.
Heat of vaporization of water 100°C = 539.55 cal/g.
Electrochemical equivalent of silver 0.001118 g/sec international amp.
Absolute wave length of red cadmium light in air at 15°C, 760 mm pressure = 6438.4696 A.
Wave length of orange-red line of krypton 86 = 6057.802 A.

π CONSTANTS

π = 3.14159 26535 89793 23846 26433 83279 50288 41971 69399 37511
$1/\pi$ = 0.31830 98861 83790 67153 77675 26745 02872 40689 19291 48091
π^2 = 9.86960 44010 89358 61883 44909 99876 15113 53136 99407 24079
$\log_e \pi$ = 1.14472 98858 49400 17414 34273 51353 05871 16472 94812 91531
$\log_{10} \pi$ = 0.49714 98726 94133 85435 12682 88290 89887 36516 78324 38044
$\log_{10} \sqrt{2\pi}$ = 0.39908 99341 79057 52478 25035 91507 69595 02099 34102 92128

CONSTANTS INVOLVING e

e = 2.71828 18284 59045 23536 02874 71352 66249 77572 47093 69996
$1/e$ = 0.36787 94411 71442 32159 55237 70161 46086 74458 11131 03177
e^2 = 7.38905 60989 30650 22723 04274 60575 00781 31803 15570 55185
M = $\log_{10} e$ = 0.43429 44819 03251 82765 11289 18916 60508 22943 97005 80367
$1/M$ = $\log_e 10$ = 2.30258 50929 94045 68401 79914 54684 36420 76011 01488 62877
$\log_{10} M$ = 9.63778 43113 00536 78912 29674 98645 − 10

π^e AND e^π CONSTANTS

π^e = 22.45915 77183 61045 47342 71522
e^π = 23.14069 26327 79269 00572 90864
$e^{-\pi}$ = 0.04321 39182 63772 24977 44177
$e^{\frac{1}{2}\pi}$ = 4.81047 73809 65351 65547 30357
i^i = $e^{-\frac{1}{2}\pi}$ = 0.20787 95763 50761 90854 69556

NUMERICAL CONSTANTS

$\sqrt{2}$ = 1.41421 35623 73095 04880 16887 24209 69807 85696 71875 37695
$\sqrt[3]{2}$ = 1.25992 10498 94873 16476 72106 07278 22835 05702 51464 70151
$\log_e 2$ = 0.69314 71805 59945 30941 72321 21458 17656 80755 00134 36026
$\log_{10} 2$ = 0.30102 99956 63981 19521 37388 94724 49302 67681 89881 46211
$\sqrt{3}$ = 1.73205 08075 68877 29352 74463 41505 87236 69428 05253 81039
$\sqrt[3]{3}$ = 1.44224 95703 07408 38232 16383 10780 10958 83918 69253 49935
$\log_e 3$ = 1.09861 22886 68109 69139 52452 36922 52570 46474 90557 82275
$\log_{10} 3$ = 0.47712 12547 19662 43729 50279 03255 11530 92001 28864 19070

OTHER CONSTANTS

Euler's Constant γ = 0.57721 56649 01532 86061
$\log_e \gamma$ = −0.54953 93129 81644 82234
Golden Ratio ϕ = 1.61803 39887 49894 84820 45868 34365 63811 77203 09180

NUMBERS CONTAINING π

	Number	Logarithm		Number	Logarithm
π	3.1415 927	0.4971 499	$2\pi^2$	19.7392 088	1.2953 297
2π	6.2831 853	0.7981 799	$\pi/180$	0.0174 533	8.2418 774 $-$ 10
3π	9.4247 780	0.9742 711	$180/\pi$	57.2957 795	1.7581 226
4π	12.5663 706	1.0992 099	$4\pi^2$	39.4784 176	1.5963 597
8π	25.1327 412	1.4002 399	$1/\pi^2$	0.1013 212	9.0057 003 $-$ 10
$\pi/2$	1.5707 963	0.1961 199	$1/(2\pi^2)$	0.0506 606	8.7046 703 $-$ 10
$\pi/3$	1.0471 976	0.0200 286	$1/(4\pi^2)$	0.0253 303	8.4036 403 $-$ 10
$\pi/4$	0.7853 982	9.8950 899 $-$ 10	$\sqrt{\pi}$	1.7724 539	0.2485 749
$\pi/6$	0.5235 988	9.7189 986 $-$ 10	$\sqrt{\frac{\pi}{2}}$	0.8862 269	9.9475 449 $-$ 10
$\pi/8$	0.3926 991	9.5940 599 $-$ 10	$\sqrt{\frac{\pi}{4}}$	0.4431 135	9.6465 149 $-$ 10
$2\pi/3$	2.0943 951	0.3210 586	$\sqrt{\frac{\pi}{2}}$	1.2533 141	0.0980 599
$4\pi/3$	4.1887 902	0.6220 886	$\sqrt{\frac{2}{\pi}}$	0.7978 846	9.9019 401 $-$ 10
$1/\pi$	0.3183 099	9.5028 501 $-$ 10		31.0062 767	1.4914 496
$2/\pi$	0.6366 198	9.8038 801 $-$ 10	π^3	1.4645 919	0.1657 166
$4/\pi$	1.2732 395	0.1049 101	$\sqrt[3]{\pi}$	0.6827 841	9.8342 834 $-$ 10
$1/(2\pi)$	0.1591 549	9.2018 201 $-$ 10	$1/\sqrt[3]{\pi}$	2.1450 294	0.3314 332
$1/(4\pi)$	0.0795 775	8.9007 901 $-$ 10	$\sqrt[3]{\pi^2}$	0.5641 896	9.7514 251 $-$ 10
$1/(6\pi)$	0.0530 516	8.7246 989 $-$ 10	$1/\sqrt{\pi}$	0.3989 423	9.6009 101 $-$ 10
$1/(8\pi)$	0.0397 887	8.5997 601 $-$ 10	$1/\sqrt{2\pi}$		
π^2	9.8696 044	0.9942 997	$2/\sqrt{\pi}$	1.1283 792	0.0524 551

MULTIPLES OF $\frac{\pi}{2}$

n	$n\frac{\pi}{2}$	n	$n\frac{\pi}{2}$	n	$n\frac{\pi}{2}$	n	$n\frac{\pi}{2}$
1	1.57079 63268	26	40.84070 44967	51	80.11061 26665	76	119.38502 08364
2	3.14159 26536	27	42.41150 08235	52	81.68140 89933	77	120.95131 71632
3	4.71238 89804	28	43.98229 71503	53	83.25220 53201	78	122.52211 34900
4	6.28318 53072	29	45.55309 34771	54	84.82300 16469	79	124.09290 98168
5	7.85398 16340	30	47.12388 98038	55	86.39379 79737	80	125.66370 61436
6	9.42477 79608	31	48.69468 61306	56	87.96459 43005	81	127.23450 24704
7	10.99557 42876	32	50.26548 24574	57	89.53539 06273	82	128.80529 87972
8	12.56637 06144	33	51.83627 87842	58	91.10618 69541	83	130.37609 51240
9	14.13716 69412	34	53.40707 51110	59	92.67698 32809	84	131.94689 14508
10	15.70796 32679	35	54.97787 14378	60	94.24777 96077	85	133.51768 77776
11	17.27875 95947	36	56.54866 77646	61	95.81857 59345	86	135.08848 41044
12	18.84955 59215	37	58.11946 40914	62	97.38937 22613	87	136.65928 04312
13	20.42035 22483	38	59.69026 04182	63	98.96016 85881	88	138.23007 67580
14	21.99114 85751	39	61.26105 67450	64	100.53096 49149	89	139.80087 30847
15	23.56194 49019	40	62.83185 30718	65	102.10176 12417	90	141.37166 94115
16	25.13274 12287	41	64.40264 93986	66	103.67255 75685	91	142.94246 57383
17	26.70533 75555	42	65.97344 57254	67	105.24335 38953	92	144.51326 20651
18	28.27433 38823	43	67.54424 20522	68	106.81415 02221	93	146.08405 83919
19	29.84513 02091	44	69.11503 83790	69	108.38494 65488	94	147.65485 47187
20	31.41592 65359	45	70.68583 47058	70	109.95574 28765	95	149.22565 10455
21	32.98672 28627	46	72.25663 10326	71	111.52653 92024	96	150.79644 73723
22	34.55751 91895	47	73.82742 73594	72	113.09733 55292	97	152.36724 36991
23	36.12831 55163	48	75.39822 36862	73	114.66813 18560	98	153.93804 00259
24	37.69911 18431	49	76.96902 00129	74	116.23892 81828	99	155.50883 63527
25	39.26990 81699	50	78.53981 63397	75	117.80972 45096	100	157.07963 26795

II. ALGEBRA

FACTORS AND EXPANSIONS

$(a \pm b)^2 = a^2 \pm 2ab + b^2.$

$(a \pm b)^3 = a^3 \pm 3a^2b + 3ab^2 \pm b^3.$

$(a \pm b)^4 = a^4 \pm 4a^3b + 6a^2b^2 \pm 4ab^3 + b^4.$

$a^2 - b^2 = (a - b)(a + b).$

$a^2 + b^2 = (a + b\sqrt{-1})(a - b\sqrt{-1}).$

$a^3 - b^3 = (a - b)(a^2 + ab + b^2).$

$a^3 + b^3 = (a + b)(a^2 - ab + b^2).$

$a^4 + b^4 = (a^2 + ab\sqrt{2} + b^2)(a^2 - ab\sqrt{2} + b^2).$

$a^n - b^n = (a - b)(a^{n-1} + a^{n-2}b + \ldots + b^{n-1}).$

$a^n - b^n = (a + b)(a^{n-1} - a^{n-2}b + \ldots - b^{n-1}),$

for even values of **n**.

$a^n + b^n = (a + b)(a^{n-1} - a^{n-2}b + \ldots + b^{n-1}),$

for odd values of **n**.

$a^4 + a^2b^2 + b^4 = (a^2 + ab + b^2)(a^2 - ab + b^2).$

$(a + b + c)^2 = a^2 + b^2 + c^2 + 2ab + 2ac + 2bc.$

$(a + b + c)^3 = a^3 + b^3 + c^3 + 3a^2(b + c) + 3b^2(a + c) + 3c^2(a + b) + 6abc.$

$(a + b + c + d + \ldots)^2 = a^2 + b^2 + c^2 + d^2 + \ldots + 2a(b + c + d + \ldots) + 2b(c + d + \ldots) + 2c(d + \ldots) + \ldots$

See also under Series.

POWERS AND ROOTS

$a^x \times a^y = a^{(x+y)}.$

$a^0 = 1 \,[\text{if } a \neq 0]$

$(ab)^x = a^x b^x.$

$\dfrac{a^x}{a^y} = a^{(x-y)}.$

$a^{-x} = \dfrac{1}{a^x}.$

$\left(\dfrac{a}{b}\right)^x = \dfrac{a^x}{b^x}.$

$(a^x)^y = a^{xy}.$

$a^{\frac{1}{x}} = \sqrt[x]{a}.$

$\sqrt[x]{ab} = \sqrt[x]{a}\,\sqrt[x]{b}.$

$\sqrt[x]{\sqrt[y]{a}} = \sqrt[xy]{a}.$

$a^{\frac{x}{y}} = \sqrt[y]{a^x}.$

$\sqrt[x]{\dfrac{a}{b}} = \dfrac{\sqrt[x]{a}}{\sqrt[x]{b}}.$

PROPORTION

If $\dfrac{a}{b} = \dfrac{c}{d},$ then $\dfrac{a + b}{b} = \dfrac{c + d}{d},$

$\dfrac{a - b}{b} = \dfrac{c - d}{d},$ $\dfrac{a - b}{a + b} = \dfrac{c - d}{c + d}.$

*ARITHMETIC PROGRESSION

An arithmetic progression is a sequence of numbers such that each number differs from the previous number by a constant amount, called the *common difference*.

If a_1 is the first term; a_n the nth term; d the common difference; n the number of terms; and s_n the sum of n terms—

$$a_n = a_1 + (n - 1)d, \quad s_n = \frac{n}{2} [a_1 + a_n].$$

$$s_n = \frac{n}{2} [2a_1 + (n - 1)d].$$

The arithmetic mean between a and b is given by $\dfrac{a + b}{2}$.

*GEOMETRIC PROGRESSION

A geometric progression is a sequence of numbers such that each number bears a constant ratio, called the *common ratio*, to the previous number.

If a_1 is the first term; a_n the nth term; r the common ratio; n the number of terms; and s_n the sum of n terms

$$a_n = a_1 r^{n-1}; \ s_n = a_1 \frac{1 - r^n}{1 - r}$$

$$= a_1 \frac{r^n - 1}{r - 1}, \quad r \neq 1.$$

$$= \frac{a_1 - r a_n}{1 - r}$$

$$= \frac{r a_n - a_1}{r - 1}$$

If $|r| < 1$, then the sum of an infinite geometrical progression converges to the limiting value

$$\frac{a_1}{1 - r}, \quad \left[s_\infty = \lim_{n \to \infty} \frac{a_1(1 - r^n)}{1 - r} = \frac{a_1}{1 - r} \right]$$

The geometric mean between a and b is given by $\sqrt{ab}$.

*It is customary to represent a_n by l in a finite progression and refer to it as the last term.

HARMONIC PROGRESSION

A sequence of numbers whose reciprocals form an arithmetic progression is called an harmonic progression. Thus

$$\frac{1}{a_1}, \ \frac{1}{a_1 + d}, \ \frac{1}{a_1 + 2d}, \cdots, \frac{1}{a_1 + (n - 1)d}, \cdots,$$

where

$$\frac{1}{a_n} = \frac{1}{a_1 + (n - 1)d}$$

forms an harmonic progression. The harmonic mean between a and b is given by $\dfrac{2ab}{a + b}$.

If A, G, H respectively represent the arithmetic mean, geometric mean, and harmonic mean between a and b, then $G^2 = AH$.

QUADRATIC EQUATIONS

Any quadratic equation may be reduced to the form,—

$$ax^2 + bx + c = 0.$$

Then

$$x = \frac{-b \pm \sqrt{b^2 - 4ac}}{2a}.$$

If a, b, and c are real then:

If $b^2 - 4ac$ is positive, the roots are real and unequal;

If $b^2 - 4ac$ is zero, the roots are real and equal;

If $b^2 - 4ac$ is negative, the roots are imaginary and unequal.

CUBIC EQUATIONS

A cubic equation, $y^3 + py^2 + qy + r = 0$ may be reduced to the form,—

$$x^3 + ax + b = 0$$

by substituting for y the value, $x - \dfrac{p}{3}$. Here

$$a = \tfrac{1}{3}(3q - p^2) \text{ and } b = \tfrac{1}{27}(2p^3 - 9pq + 27r).$$

For solution let,—

$$A = \sqrt[3]{-\frac{b}{2} + \sqrt{\frac{b^2}{4} + \frac{a^3}{27}}}, \qquad B = \sqrt[3]{+\frac{b}{2} + \sqrt{\frac{b^2}{4} + \frac{a^3}{27}}},$$

then the values of x will be given by,

$$x = A + B, \quad -\frac{A + B}{2} + \frac{A - B}{2}\sqrt{-3}, \quad -\frac{A + B}{2} - \frac{A - B}{2}\sqrt{-3}.$$

If p, q, r are real, then:

If $\dfrac{b^2}{4} + \dfrac{a^3}{27} > 0$, there will be one real root and two conjugate complex roots;

If $\dfrac{b^2}{4} + \dfrac{a^3}{27} = 0$, there will be three real roots of which at least two are equal;

If $\dfrac{b^2}{4} + \dfrac{a^3}{27} < 0$, there will be three real and unequal roots.

Trigonometric Solution of the Cubic Equation

The form $x^3 + ax + b = 0$ with $ab \neq 0$ can always be solved by transforming it to the trigonometric identity

$$4 \cos^3 \theta - 3 \cos \theta - \cos (3\theta) \equiv 0.$$

Let $x = m \cos \theta$, then

$$x^3 + ax + b \equiv m^3 \cos^3 \theta + am \cos \theta + b \equiv 4 \cos^3 \theta - 3 \cos \theta - \cos (3\theta) \equiv 0.$$

Hence

$$\frac{4}{m^3} = -\frac{3}{am} = \frac{-\cos(3\theta)}{b},$$

from which follows that

$$m = 2 \sqrt{-\frac{a}{3}}, \quad \cos (3\theta) = \frac{3b}{am}.$$

Any solution θ_1 which satisfies $\cos (3\theta) = \dfrac{3b}{am}$, will also have the solutions

$$\theta_1 + \frac{2\pi}{3} \quad \text{and} \quad \theta_1 + \frac{4\pi}{3}.$$

The roots of the cubic $x^3 + ax + b = 0$ are

$$2 \sqrt{-\frac{a}{3}} \cos \theta_1, \quad 2 \sqrt{-\frac{a}{3}} \cos \left(\theta_1 + \frac{2\pi}{3}\right), \quad 2 \sqrt{-\frac{a}{3}} \cos \left(\theta_1 + \frac{4\pi}{3}\right).$$

Example where hyperbolic functions are necessary for solution with latter procedure

The roots of the equation $x^3 - x + 2 = 0$ may be found as follows:

Here

$$a = -1, \quad b = 2, \quad m = 2 \sqrt{\tfrac{1}{3}} = 1.155$$

$$\cos (3\theta) = \frac{6}{-1.155} = -5.196$$

$$\cos (3\theta) = -\cos (3\theta - \pi) = -\cosh [i(3\theta - \pi)] = -5.196.$$

Using hyperbolic function tables for cosh $[i(3\theta - \pi)] = 5.196$, it is found that

$$i(3\theta - \pi) = 2.332.$$

Thus

$$3\theta - \pi = -i(2.332).$$
$$3\theta = \pi - i(2.332)$$
$$\theta_1 = \frac{\pi}{3} - i(0.777)$$
$$\theta_1 + \frac{2\pi}{3} = \pi - i(0.777)$$
$$\theta_1 + \frac{4\pi}{3} = \frac{5\pi}{3} - i(0.777)$$

$$\cos \theta_1 = \cos\left[\frac{\pi}{3} - i(0.777)\right]$$

$$= \left(\cos \frac{\pi}{3}\right)[\cos i(0.777)] + \left(\sin \frac{\pi}{3}\right)[\sin i(0.777)]$$

$$= \left(\cos \frac{\pi}{3}\right)(\cosh 0.777) + i\left(\sin \frac{\pi}{3}\right)(\sinh 0.777)$$

$$= (0.5)(1.317) + i(0.866)(0.858) = 0.659 + i(0.743).$$

Note that

$$\cos \mu = \cosh (i\mu) \quad \text{and} \quad \sin \mu = -i \sinh (i\mu).$$

Similarly

$$\cos\left(\theta_1 + \frac{2\pi}{3}\right) = \cos [\pi - i(0.777)]$$

$$= (\cos \pi)(\cosh 0.777) + i(\sin \pi)(\sinh 0.777)$$

$$= -1.317,$$

and

$$\cos\left(\theta_1 + \frac{4\pi}{3}\right) = \cos\left[\frac{5\pi}{3} - i(0.777)\right]$$

$$= \left(\cos \frac{5\pi}{3}\right)(\cosh 0.777) + i\left(\sin \frac{5\pi}{3}\right)(\sinh 0.777)$$

$$= (0.5)(1.317) - i(0.866)(0.858) = 0.659 - i(0.743).$$

The required roots are

$$1.155[0.659 + i(0.743)] = 0.760 + i(0.858)$$
$$(1.155)(-1.317) = -1.520$$
$$(1.155)[0.659 - i(0.743)] = 0.760 - i(0.858).$$

QUARTIC OR BIQUADRATIC EQUATION

A quartic equation,

$$x^4 + ax^3 + bx^2 + cx + d = 0,$$

has the *resolvent cubic equation*

$$y^3 - by^2 + (ac - 4d)y - a^2d + 4bd - c^2 = 0.$$

Let y be any root of this equation, and

$$R = \sqrt{\frac{a^2}{4} - b + y}.$$

If $R \neq 0$, then let

$$D = \sqrt{\frac{3a^2}{4} - R^2 - 2b + \frac{4ab - 8c - a^3}{4R}}$$

and

$$E = \sqrt{\frac{3a^2}{4} - R^2 - 2b - \frac{4ab - 8c - a^3}{4R}}$$

If $R = 0$, then let

$$D = \sqrt{\frac{3a^2}{4} - 2b + 2\sqrt{y^2 - 4d}}$$

and

$$E = \sqrt{\frac{3a^2}{4} - 2b - 2\sqrt{y^2 - 4d}}.$$

Then the four roots of the original equation are given by

$$x = -\frac{a}{4} + \frac{R}{2} \pm \frac{D}{2}$$

and

$$x = -\frac{a}{4} - \frac{R}{2} \pm \frac{E}{2}.$$

EQUATION $x^n = c$

Using DeMoivre's theorem:

$$(\cos\theta + i\sin\theta)^n = \cos n\theta + i\sin n\theta; \quad i = \sqrt{-1},$$

the equation $x^n = c$ has n roots given by

$$x = \sqrt[n]{c}\left(\cos\frac{2m\pi}{n} + i\sin\frac{2m\pi}{n}\right) \text{if } c > 0,$$

or

$$x = \sqrt[n]{-c}\left(\cos\frac{(2m+1)\pi}{n} + i\sin\frac{(2m+1)\pi}{n}\right) \text{ if } c < 0,$$

where m takes the n values $0, 1, 2, \ldots (n-1)$ giving n roots.

PARTIAL FRACTIONS

This section applies only to rational algebraic fractions with numerator of lower degree than the denominator. Improper fractions can be reduced to proper fractions by long division.

Every fraction may be expressed as the sum of component fractions whose denominators are factors of the denominator of the original fraction.

Let $N(x)$ = numerator, a polynomial of the form

$$n_0 + n_1 x + n_2 x^2 + \cdots + n_i x^i$$

I. *Non-repeated Linear Factors*

$$\frac{N(x)}{(x - a)G(x)} = \frac{A}{x - a} + \frac{F(x)}{G(x)}$$

$$A = \left[\frac{N(x)}{G(x)}\right]_{x=a}$$

$F(x)$ determined by methods discussed in the following sections.

Example:

$$\frac{x^2 + 3}{x(x - 2)(x^2 + 2x + 4)} = \frac{A}{x} + \frac{B}{x - 2} + \frac{F(x)}{x^2 + 2x + 4}$$

$$A = \left[\frac{x^2 + 3}{(x - 2)(x^2 + 2x + 4)}\right]_{x=0} = -\frac{3}{8}$$

$$B = \left[\frac{x^2 + 3}{x(x^2 + 2x + 4)}\right]_{x=2} = \frac{4 + 3}{2(4 + 4 + 4)} = \frac{7}{24}$$

II. *Repeated Linear Factors*

$$\frac{N(x)}{x^m G(x)} = \frac{A_0}{x^m} + \frac{A_1}{x^{m-1}} + \cdots + \frac{A_{m-1}}{x} + \frac{F(x)}{G(x)}$$

$$F(x) = f_0 + f_1 x + f_2 x^2 + \cdots, \quad G(x) = g_0 + g_1 x + g_2 x^2 + \cdots$$

$$A_0 = \frac{n_0}{g_0}, \quad A_1 = \frac{n_1 - A_0 g_1}{g_0}, \quad A_2 = \frac{n_2 - A_0 g_2 - A_1 g_1}{g_0}$$

General term:

$$A_k = \frac{1}{g_0}\left[n_k - \sum_{i=0}^{k-1} A_i g_{k-i}\right]$$

$$*m = 1 \begin{cases} f_0 = n_1 - A_0 g_1 \\ f_1 = n_2 - A_0 g_2 \\ f_j = n_{j+1} - A_0 g_{j+1} \end{cases}$$

$$m = 2 \begin{cases} f_0 = n_2 - A_0 g_2 - A_1 g_1 \\ f_1 = n_3 - A_0 g_3 - A_1 g_2 \\ f_j = n_{j+2} - [A_0 g_{j+2} + A_1 g_{j+1}] \end{cases}$$

$$m = 3 \begin{cases} f_0 = n_3 - A_0 g_3 - A_1 g_2 - A_2 g_1 \\ f_1 = n_4 - A_0 g_4 - A_1 g_3 - A_2 g_2 \\ f_j = n_{j+3} - [A_0 g_{j+3} + A_1 g_{j+2} + A_2 g_{j+1}] \end{cases}$$

$$\text{any } m: \ f_j = n_{m+j} - \sum_{i=0}^{m-1} A_i g_{m+j-i}$$

Example:

$$\frac{x^2 + 1}{x^3(x^2 - 3x + 6)} = \frac{A_0}{x^3} + \frac{A_1}{x^2} + \frac{A_2}{x} + \frac{f_1 x + f_0}{x^2 - 3x + 6}$$

$$A_0 = \frac{1}{6}, \quad A_1 = \frac{0 - (\frac{1}{6})(-3)}{6} = \frac{1}{12},$$

$$A_2 = \frac{1 - (\frac{1}{6})(1) - (\frac{1}{12})(-3)}{6} = \frac{13}{72},$$

$$m = 3 \begin{cases} f_0 = 0 - \frac{1}{6}(0) + \frac{1}{12}(1) - \frac{13}{72}(-3) = \frac{11}{24} \\ f_1 = 0 - \frac{1}{6}(0) - \frac{1}{12}(0) - \frac{13}{72}(1) = -\frac{13}{72} \end{cases}$$

*Note: If $G(x)$ contains linear factors, $F(x)$ may be determined by previous section I.

III. *Repeated Linear Factors*

$$\frac{N(x)}{(x - a)^m G(x)} = \frac{A_0}{(x - a)^m} + \frac{A_1}{(x - a)^{m-1}} + \cdots + \frac{A_{m-1}}{(x - a)} + \frac{F(x)}{G(x)}$$

Change to form $\dfrac{N'(y)}{y^m G'(y)}$ by substitution of $x = y + a$.   Resolve into partial fractions in terms of y as described in Section II.   Then express in terms of x by substitution $y = x - a$.

Example:

$$\frac{x - 3}{(x - 2)^2(x^2 + x + 1)}$$

Let $x - 2 = y$, $x = y + 2$

$$\frac{(y + 2) - 3}{y^2[(y + 2)^2 + (y + 2) + 1]} = \frac{y - 1}{y^2(y^2 + 5y + 7)} = \frac{A_0}{y^2} + \frac{A_1}{y} + \frac{f_1 y + f_0}{y^2 + 5y + 7}$$

$$A_0 = -\frac{1}{7}, \quad A_1 = \frac{1 - (-\frac{1}{7})(5)}{7} = \frac{12}{49},$$

$$m = 2 \begin{cases} f_0 = 0 - (-\frac{1}{7})(1) - (\frac{12}{49})(5) = -\frac{53}{49} \\ f_1 = 0 - (-\frac{1}{7})(0) - (\frac{12}{49})(1) = -\frac{12}{49} \end{cases}$$

$$\therefore \ \frac{y - 1}{y^2(y^2 + 5y + 7)} = \frac{-\frac{1}{7}}{y^2} + \frac{\frac{12}{49}}{y} + \frac{-\frac{12}{49}y - \frac{53}{49}}{y^2 + 5y + 7}$$

Let $y = x - 2$, then

$$\frac{x - 3}{(x - 2)^2(x^2 + x + 1)} = \frac{-\frac{1}{7}}{(x - 2)^2} + \frac{\frac{12}{35}}{(x - 2)} + \frac{-\frac{12}{49}(x - 2) - \frac{53}{49}}{x^2 + x + 1}$$

$$= -\frac{1}{7(x - 2)^2} + \frac{12}{35(x - 2)} + \frac{-12x - 29}{49(x^2 + x + 1)}$$

IV. *Repeated Linear Factors*

Alternative method of determining coefficients:

$$\frac{N(x)}{(x-a)^m G(x)} = \frac{A_0}{(x-a)^m} + \cdots + \frac{A_k}{(x-a)^{m-k}} + \cdots + \frac{A_{m-1}}{x-a} + \frac{F(x)}{G(x)}$$

$$A_k = \frac{1}{k!}\left\{ D_x^k \left[\frac{N(x)}{G(x)} \right] \right\}_{x=a}$$

where D_x^k is the differentiating operator, and the derivative of zero order is defined as:

$$D_x^0 u = u.$$

V. *Factors of Higher Degree*

Factors of higher degree have the corresponding numerators indicated.

$$\frac{N(x)}{(x^2 + h_1 x + h_0) G(x)} = \frac{a_1 x + a_0}{x^2 + h_1 x + h_0} + \frac{F(x)}{G(x)}$$

$$\frac{N(x)}{(x^2 + h_1 x + h_0)^2 G(x)} = \frac{a_1 x + a_0}{(x^2 + h_1 x + h_0)^2} + \frac{b_1 x + b_0}{(x^2 + h_1 x + h_0)} + \frac{F(x)}{G(x)}$$

$$\frac{N(x)}{(x^3 + h_2 x^2 + h_1 x + h_0) G(x)} = \frac{a_2 x^2 + a_1 x + a_0}{x^3 + h_2 x^2 + h_1 x + h_0} + \frac{F(x)}{G(x)}$$

etc.

Problems of this type are determined first by solving for the coefficients due to linear factors as shown above, and then determining the remaining coefficients by the general methods given below.

VI. *General Methods for Evaluating Coefficients*

1.

$$\frac{N(x)}{D(x)} = \frac{N(x)}{G(x) H(x) L(x)} = \frac{A(x)}{G(x)} + \frac{B(x)}{H(x)} + \frac{C(x)}{L(x)} + \cdots$$

Multiply both sides of equation by $D(x)$ to clear fractions. Then collect terms, equate like powers of x, and solve the resulting simultaneous equations for the unknown coefficients.

2. Clear fractions as above. Then let x assume certain convenient values ($x = 1, 0, -1, \ldots$). Solve the resulting equations for the unknown coefficients.

3.

$$\frac{N(x)}{G(x) H(x)} = \frac{A(x)}{G(x)} + \frac{B(x)}{H(x)}$$

Then

$$\frac{N(x)}{G(x) H(x)} - \frac{A(x)}{G(x)} = \frac{B(x)}{H(x)}$$

If $A(x)$ can be determined, such as by Method I, then $B(x)$ can be found as above.

BASIC CONCEPTS IN ALGEBRA

Dr. W. E. Deskins

I. ALGEBRA OF SETS

1. Intuitively a set is a collection of objects called the elements of the set. Set and set membership are generally accepted as basic, undefined terms used to define and construct mathematical systems.

   The notation $a \in A$ indicates that a is an element of the set A. The notation $a \notin A$ means that a is not a member of A.

   A set is sometimes specified by listing its elements within a set of braces: $\{a\}$ is the set containing only the element a.

2. Set A is a subset of set B provided $a \in A$ implies $a \in B$. This is denoted by $A \subseteq B$. Every set has as a subset the empty or null set, denoted by ϕ, which has no elements.

3. Set A equals set B, written $A = B$, if and only if $A \subseteq B$ and $B \subseteq A$. A is a proper subset of B, sometimes indicated by $A \subset B$, if and only if $A \subseteq B$ and $A \neq B$; then B has at least one element which does not belong to A.

4. The Cartesian product of sets A and B, denoted by $A \times B$, is the set of all ordered pairs (a, b) where $a \in A$ and $b \in B$. A subset R of $A \times A$ is a binary relation on A, and this is an equivalence relation on A provided (i) $(a, a) \in R$ for every $a \in A$, (ii) $(a, b) \in R$ implies $(b, a) \in R$, and (iii) $(a, b) \in R$ and $(b, c) \in R$ imply $(a, c) \in R$. Ordinary equality of numbers, equality of sets, and congruence of plane figures are examples of equivalence relations.

5. A subset F of $A \times B$ is a *function* from A to B provided each element of A appears exactly once as the first element of a pair in F. A function F from A to B is *onto* provided each element of B appears at least once as the second element of a pair in F. It is *one-to-one* provided each element of B appears at most once as the second element of a pair in F. A function from $A \times A$ to A is a *binary operation* on A. Addition and multiplication of ordinary numbers are examples of binary operations.

6. If consideration is restricted to elements and subsets of a particular set I, then I is the universal set.

7. Common binary operations on subsets of I are: $A \cup B$, the union or join of sets A and B, is the set of all elements of I which belong to either A or B or both A and B.

   $A \cap B$, the intersection or meet of sets A and B, is the set of all elements of I which belong to both A and B.

   $A \setminus B$, the difference of sets A and B, is the set of elements of I which belong to A but not B.

   The difference $I \setminus A$ is denoted by A' and called the complement of A (relative to I). Except in dealing with the concept of complementation the use of a universal set is not essential to the above ideas.

8. Some theorems basic to the Algebra of Sets:

   Let A, B, and C be arbitrary subsets of a universal set I.

   (a) (Commutativity) $A \cup B = B \cup A$ and $A \cap B = B \cap A$.

   (b) (Associativity) $(A \cup B) \cup C = A \cup (B \cup C)$ and
   $(A \cap B) \cap C = A \cap (B \cap C)$.

(c) (Distributivity) $\quad A \cap (B \cup C) = (A \cap B) \cup (A \cap C)$ and

$$A \cup (B \cap C) = (A \cup B) \cap (A \cup C).$$

(d) (Idempotency) $\quad A \cup A = A \cap A = A.$

(e) Properties of I and ϕ: $\quad A \cap I = A \cup \phi = A,$

$$A \cup I = I, \text{ and}$$

$$A \cap \phi = \phi.$$

(f) $(A \cap B) \cup (A \backslash B) = A.$

(g) $(A \backslash B) \cup B = A \cup B.$

(h) $A \subseteq A \cup B.$

(i) $A \cap B \subseteq A.$

(j) $A \cup B = A$ if and only if $B \subseteq A.$

(k) $A \cap B = A$ if and only if $A \subseteq B.$

(m) $A \backslash B = A \backslash (A \cap B).$

(n) (DeMorgan's Theorem) $(A \backslash B) \cap (A \backslash C) = A \backslash (B \cup C)$ and

$$(A \backslash B) \cup (A \backslash C) = A \backslash (B \cap C).$$

(o) $(A \cup B)' = A' \cap B'$ and $(A \cap B)' = A' \cup B'.$

(p) $A \cup A' = I$ and $A \cap A' = \phi.$

9. A mathematical system S is a set $S = \{E, O, A\}$ where E is a nonempty set of elements, O is a set of relations and operations on E, and A is a set of axioms, postulates, or assumptions concerning the elements of E and O.

10. The Algebra of Sets provides an example of a mathematical system called a Boolean Algebra (or Boolean Ring) which is defined as:

Set E of elements $a, b, c, \ldots$;

Set O of 2 binary operations $\oplus$ and $\otimes$; (Here $a \oplus b$ denotes the image of (a, b) under the binary operation.)

Set A of axioms for all a, b, c of E:

A_1. The binary operations are commutative; i.e.,

$$a \oplus b = b \oplus a \quad \text{and} \quad a \otimes b = b \otimes a.$$

A_2. Each binary operation is distributive over the other; i.e.,

$$a \oplus (b \otimes c) = (a \oplus b) \otimes (a \oplus c) \quad \text{and} \quad a \otimes (b \oplus c) = (a \otimes b) + (a \otimes c).$$

A_3. There exist elements e and z in E such that for each $a \in E$, $a \oplus z = a$ and $a \otimes e = a.$

A_4. For each $a \in E$ there exists an element $a' \in E$ such that $a \otimes a' = e$ and $a \oplus a' = z.$

In the algebra of subsets of a (universal) set I, ϕ plays the role of z, I that of e, $\cup$ that of $\oplus$, and $\cap$ that of $\otimes$.

11. A Boolean Algebra has the Principle of Duality: If the interchanges of $\begin{Bmatrix} \oplus \text{ and } \otimes \\ e \text{ and } z \end{Bmatrix}$ are made in a correct statement, then the result is also a correct statement.

12. In addition to the Algrebra of Sets which is a Boolean Algebra, other representations of Boolean Algebra that are interesting of themselves and valuable for their applications are:

(a) The Algebra of Symbolic Logic

(b) The Algebra of Switching Currents

Algebra of Sets	*Binary Operator*	*Symbolic Logic*	*Binary Operator*	*Switching Circuits*	*Binary Operator*
Union of 2 sets	$\cup$	Disjunction of 2 propositions	$\vee$	2 switches in $\parallel$	$+$
Intersection of 2 sets	$\cap$	Conjunction of 2 propositions	$\wedge$	2 switches in series	$\times$
Complement of a set A		Negation of a proposition T, F	$\sim$	on-off, or 1, 0	1 or $^-$

Both in symbolic logic and in switching circuits, we can consider "0" and "1" as the elements, in the former representing "False" and "True"; in the latter, "Off" and "On", satisfying the following "rules":

$$\left.\begin{array}{l} 0 + 0 = 0 \\ 1 + 1 = 1 \end{array}\right\} \text{ i.e. } a + a = a$$

$$\left.\begin{array}{l} 0 \times 0 = 0 \\ 1 \times 1 = 1 \end{array}\right\} \text{ i.e. } a \times a = a$$

$$0' = 1$$

$$1' = 0$$

$$\left.\begin{array}{l} 0 + 0' = 0 + 1 = 1 \\ 1 + 1' = 1 + 0 = 1 \end{array}\right\} \text{ i.e. } a + a' = 1$$

$$\left.\begin{array}{l} 0 \times 0' = 0 \times 1 = 0 \\ 1 \times 1' = 1 \times 0 = 0 \end{array}\right\} a \times a' = 0$$

In switching circuits:

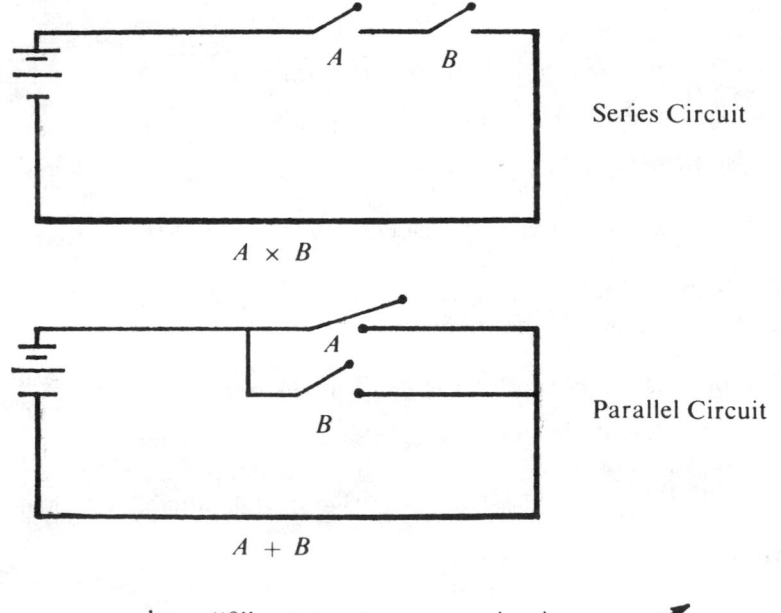

$A \times B$

Series Circuit

$A + B$

Parallel Circuit

here, "0" represents an open circuit: $\bullet\!\!-\!\!-\!\!\nearrow\!\!-\!\!\bullet$ and "1"

$A = 0$

a closed circuit: $\bullet\!\!-\!\!\nearrow\!\!\bullet\!\!-\!\!\bullet$

$A = 1$

In the Algebra of Symbolic Logic, we use *Truth Tables* to define the operations $\wedge$, $\vee$, $\sim$ as follows:

					Other Operators Used	
p	q	$p \wedge q$	$p \vee q$	$\sim p$	$p \to q$	$p \leftrightarrow q$
T	T	T	T	F	T	T
T	F	F	T	F	F	F
F	T	F	T	T	T	F
F	F	F	F	T	T	T

13. In order to re-emphasize the use of switching circuits and their relation to truth tables the following is included. Conventionally a "1" represents "True" and a "0" represents "False." The switching circuit symbols are $-$, $\cdot$, $+$, $\to$, $\equiv$ representing "Not," "And," "Or," "Implies," "Equivalent" respectively and their Truth Table Definitions are

p	q	$p \cdot q$	$p + q$	$-p$	$p \to q$	$p \equiv q$
0	0	0	0	1	1	1
0	1	0	1	1	1	0
1	0	0	1	0	0	0
1	1	1	1	0	1	1

The comparison with the Algebra of Symbolic Logic being obvious. The "rules" for these circuits are as follows:

$$0 + 0 = 0$$
$$1 + 1 = 1$$
$$1 + 0 = 0 + 1 = 1$$
$$0 \cdot 0 = 0$$
$$1 \cdot 1 = 1$$
$$0 \cdot 1 = 1 \cdot 0 = 0$$
$$\bar{0} = 1$$
$$\bar{1} = 0$$

Mechanical switches or relays are represented by

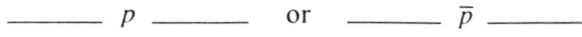

the former indicating that the circuit is closed, i.e. the switch is made, when $p = 1$ and the latter indicating the converse namely that the circuit is closed when $\bar{p} = 1$ or, what amounts to the same thing, when $p = 0$.

Electronic switches or gates are represented by more complex symbols—four in all, three of which are independent and can stand alone

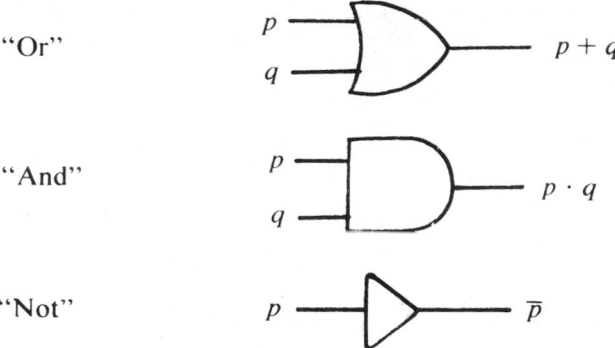

and one which represents the negation of an input or an output and is used with one of the above

"Not"

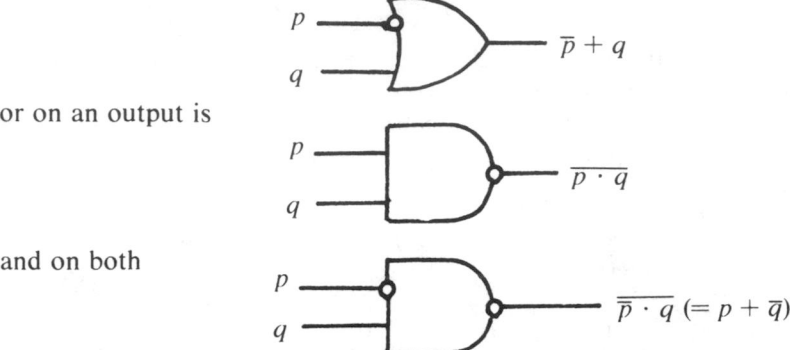

An example of its use on an input line is

$$\bar{p} + q$$

or on an output is

$$\overline{p \cdot q}$$

and on both

$$\overline{\bar{p} \cdot q} \,(= p + \bar{q})$$

The basic functions obtained from the two types of switching circuits are

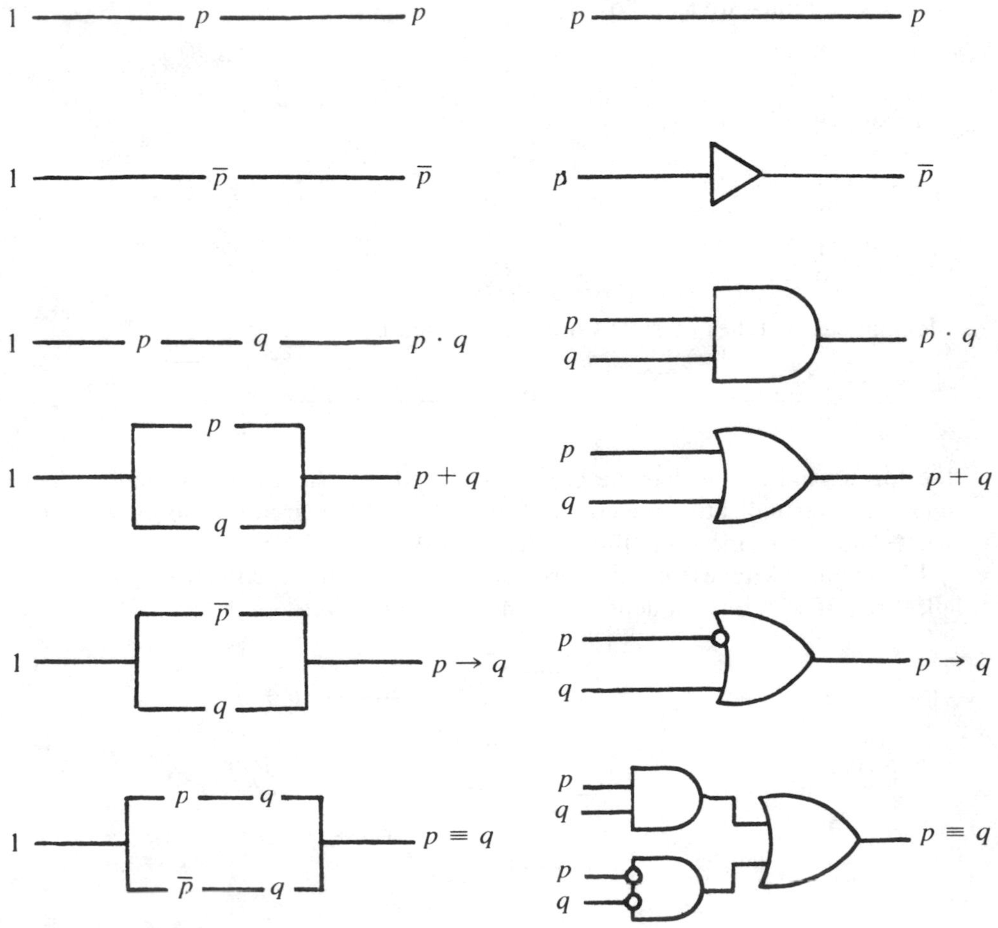

All the above electronic circuits can be negated by simply adding a negating circle to the output as for example in

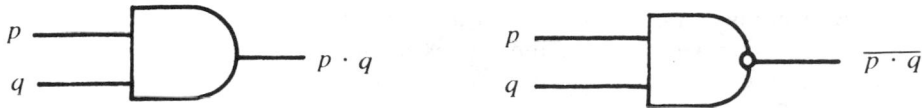

Alternative circuits however, which are direct analogues of their relay switching counterparts, are

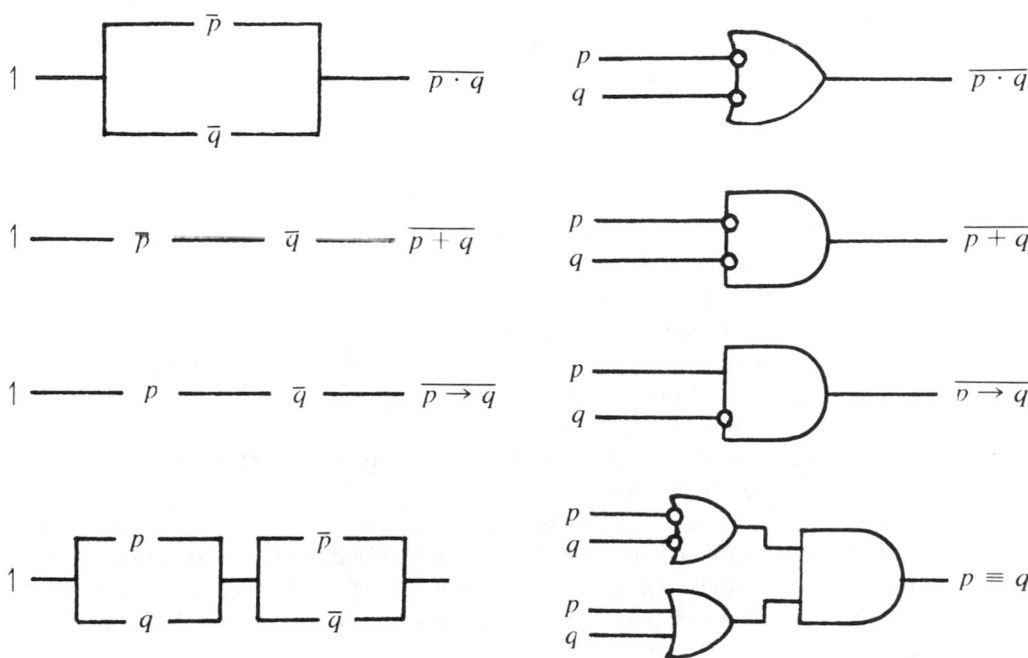

The operation $+$ is sometimes referred to as the "Inclusive Or" and $\neq$ as the "Exclusive Or", the former having the value "True" when both the inputs are "True"—see the truth table. Note that $p \neq q$ is a shorthand for $\overline{p \equiv q}$.

14. Two sets A and B are equivalent (have same cardinal) if and only if there exists a one-to-one correspondence between the elements of the two sets. This is an equivalence relation on the collection of subsets of set I.

    A set is infinite if and only if it is equivalent with a proper subset of itself.

    A set is called countably (denumerably) infinite if it is equivalent with the set of all positive integers. The set of all rational numbers is countably infinite but the set of all real numbers is noncountably infinite. The cardinal of the set of all rational numbers is denoted by (aleph null); the cardinal of the set of reals is denoted by (aleph).

II. ABSTRACT ALGEBRAIC SYSTEMS

1. *Semigroup.* A semigroup is a system $\{S, \theta, A\}$; S is a nonempty set $\{a, b, c, \ldots\}$, θ consists of one binary operation on S, denoted by $*$, and A consists of the axiom

    A_1.   Associativity: $a * (b * c) = (a * b) * c$ for all $a, b \circ c \in S$.

        Basic Theorem. *(Generalized Associativity).* If $a_1, a_2, \ldots, a_n$ are elements of S then all associations of the n elements yield the same "product". (For example,

$$a * ((b * c) * d) = a * (b * (c * d)) = (a * b) * (c * d), \text{ etc.)}$$

2. *Group.*   A group is a system $\{G, \theta, A\}$; G is a nonempty set $\{a, b, c, \ldots\}$, θ consists of one binary operation denoted by $\circ$, and A consists of the axioms:

A_1.   Associativity: $a \circ (b \circ c) = (a \circ b) \circ c$ for all $a, b, c \in G$.

A_2.   Identity Element: G contains an element e having the property, $a \circ e = e \circ a = a$ for every $a \in G$.

A_3.   Inverse Element: For each $a \in G$ there is an element $a' \in G$ with the property, $a \circ a' = a' \circ a = e$.

   If the following additional axiom belongs to A,

A_4.   Commutativity: $a \circ b = b \circ a$ for all $a, b \in G$. Then the group is called Abelian (after Niels Henrik Abel). Some basic theorems:

   (a) The element e (Axiom A_2) is unique. Then e is *the* identity element of G.

   (b) The element a' (Axiom A_3) is unique for each $a \in G$.   Then a' is *the* inverse of a in G.

   (c) The equation $a \circ x = b$ has a unique solution in G, viz., $x = a' \circ b$.

   (d) $(a')' = a$ and $(a \circ b)' = b' \circ a'$.

   (e) $a \circ b = a \circ c$ if and only if $b = c$.

      If a nonempty subset H of G satisfies the two conditions:

         H_1. $a \circ b \in H$ whenever $a, b \in H$.   (Closure)

         H_2. $a \in H$ if and only if $a' \in H$.

      then H is a *subgroup* of G.

      (Lagrange).   If G is a finite set then the number of elements in H divides the number of elements in G.

      Example of group.   Let G be the set of all one-to-one functions from a nonempty S onto itself.   For any $f, g \in G$, define the function $f \circ g$ as the function which maps s onto $f(g(s))$, for each $s \in S$.   Relative to this binary operation G is a group, the *symmetric group* of all permutations on S.

      Each group is essentially a subgroup of the symmetric group of some set S.

3. *Ring.*   A ring is a system $\{R, \theta, A\}$; R is a nonempty set $\{a, b, c, \ldots\}$, θ consists of two binary operations denoted by $+$ and $\times$, and A consists of the axioms:

A_0.   Relative to addition (i.e., $+$) R is an Abelian group in which the identity element is denoted by z and the inverse of a is denoted by $-a$.

M_0.   Relative to multiplication (i.e., $\times$) R is a semigroup.

D_1.   Left distributive: $a \times (b + c) = (a \times b) + (a \times c)$, all $a, b, c \in R$.

D_2.   Right distributive: $(b + c)a = (b \times a) + (c \times a)$, all $a, b, c \in R$.

EXAMPLE 1.   The set of all integers (whole numbers) and ordinary addition and multiplication.

EXAMPLE 2.   The set of all real functions continuous on the interval $0 \leq y \leq 1$, with addition and multiplication defined by $(f + g)(y) = f(y) + g(y)$, sum of real numbers, and $(f \times g)(y) = f(y) \times g(y)$, product of real numbers.

Special types of rings have been studied extensively.

3.1   *Integral Domain.*   An integral domain is a ring R in which multiplication ($\times$) satisfies the additional assumptions:

M_1.   Commutativity: $a \times b = b \times a$ for all a and b in R.

M_2.   Multiplicative identity: R contains an element $e \neq z$ with the property $a \times e = e \times a = a$ for all a in R.

M_3.   Cancellation: $a \times b = a \times c$ if and only if $b = c$.

   An element u of integral domain R is a *unit* provided R contains v such that $u \times v = e$.

An element p of integral domain R is a *prime* (irreducible element) provided $p = a \times b$ implies that exactly one of the elements a or b is a unit.

The elements of integral domain R which differ from z and are neither units nor primes are *composites*.

In some integral domains (such as the ring of integers) each composite can be factored uniquely (up to unit factors) as the product of a finite set of primes. However in the integral domain of all entire functions this is not true.

3.2 *Field.* A field is an integral domain in which every element except z is a unit. In other words, the non-z elements form an Abelian group relative to multiplication ($\times$).

EXAMPLE 1. The rational field consisting of ordinary fractions, addition, and multiplication.

EXAMPLE 2. The set of all real numbers $a + b\sqrt{2}$, a and b rational. Then

$$(a + b\sqrt{2}) + (c + d\sqrt{2}) = (a + c) + (b + d)\sqrt{2} \text{ and}$$
$$(a + b\sqrt{2}) \times (c + d\sqrt{2}) = (ac + 2bd) + (ad + bc)\sqrt{2}.$$

Besides these well-known examples there exist finite fields (sometimes called Galois fields).

EXAMPLE 3. Let p be a prime integer. Denote by $GF(p)$ the p integers $0, 1, \ldots, p - 1$. Define addition($\oplus$) of two of these elements a and b as the remainder of $a + b$ (ordinary addition) after division by p. (Thus $1 \oplus (p - 1) = 0$.)

Define $a \otimes b$, the product, to be the remainder of ab (ordinary multiplication) after division by p. (Thus, when $p = 3$, $2 \otimes 2 = 1$.) The resulting system $\{GF(p), \oplus, \otimes\}$ is the (modular) field of integers modulo p.

3.3 *Skew Field or Division Ring.* A skew field is a ring in which the non-z elements form a group relative to multiplication ($\times$).

The classical example of a skew field is the ring of real *quaternions*, first described by W. R. Hamilton. A quaternion is expressible in the form $ae + bi + cj + dk$ where $a, b, c,$ and d are real numbers and $e, i, j,$ and k are elements which commute with all real numbers and multiply as follows:

$$e \times e = e, \quad e \times i = i \times e = i, \quad e \times j = j \times e = j, \quad e \times k = k \times e = k;$$
$$i \times i = -e, \quad i \times j = k, \quad j \times i = -k, \quad i \times k = -j, \quad k \times i = j,$$
$$j \times j = -e, \quad j \times k = i, \quad k \times j = -i, \quad k \times k = -e.$$

These elements distribute over addition. e is generally identified with and written as the real number 1.

3.4 *Matric Ring.* The matric ring $M_n(R)$ over the ring R, where n is a positive integer, consists of all doubly-ordered sets of n^2 elements of R, written as an array

$$\begin{pmatrix} a_{1,1} & a_{1,2} \cdots a_{1,n} \\ a_{2,1} & a_{2,2} \cdots a_{2,n} \\ \vdots \\ a_{n,1} & \cdots \quad a_{n,n} \end{pmatrix} = (a_{i,j})$$

with addition and multiplication defined as follows:

$$(a_{i,j}) + (b_{i,j}) = (a_{i,j} + b_{i,j})$$

$$(a_{i,j}) \times (b_{i,j}) = (c_{i,j})$$

where

$$c_{i,j} = \sum_{k=1}^{n} a_{i,k} b_{k,j}, \quad i = 1,\ldots,n \quad \text{and} \quad j = 1,\ldots,n.$$

If $n > 1$, then multiplication is noncommutative in general; i.e., $(a_{i,j}) \times (b_{i,j})$ can differ from $(b_{i,j}) \times (a_{i,j})$. Moreover, the product of two nonzero matrices can be the zero matrix (which consists of only the element z in all n^2 positions).

A similar useful method for forming a new ring from a known ring utilizes sequences.

3.5   *Power Series and Polynomial Ring.*   Let R be a ring in which multiplication ($\times$) is commutative. The set $PS(R)$ of all sequences $(a_0,a_1,\ldots)$ with $a_i \in R$ is the power series ring of R, with addition and multiplication defined as

$$(a_0,a_1,\ldots) \oplus (b_0,b_1,\ldots) = (a_0 + b_0, \ a_1 + b_1,\ldots) \text{ and}$$

$$(a_0,a_1,\ldots) \oplus (b_0,b_1,\ldots) = (c_0,c_1,\ldots)$$

where

$$c_0 = a_0 \times b_0, \quad c_1 = a_0 \times b_1 + a_1 \times b_0,\ldots, \text{ and,}$$

generally,

$$c_n = a_0 \times b_n + a_1 \times b_{n-1} + \cdots + a_n \times b_0.$$

The subset $P(R)$ of $PS(R)$ consisting of those sequences $(a_0,a_1,\ldots)$ in which at most only finitely many of the a_i differ from z, form a ring relative to the addition and multiplication just defined. This ring $\{P(R),\oplus,\otimes\}$, is the polynomial ring of R.

Some theorems for rings, fields, etc.
(a) In a ring R, if $a = b$ and $c = d$, then $a + c = b + d$ and $a \times c = b \times d$.
(b) In a ring R, $-(-a) = a$; $(-a) \times b = a \times (-b) = -(a \times b)$; and $(-a) \times (-b) = a \times b$, for all $a,b \in R$.
(c) In a ring R, $a \times z = z \times a = z$, for all $a \in R$.
(d) In a ring R the equation $a + x = b$ has a unique solution, viz., $x = -a + b$.
(f) In a field, skew field, or integral domain, $a \times b = z$ if and only if a and/or b equals z.
(g) A finite integral domain is a field.
(h) The polynomial ring of an integral domain is also an integral domain.
(i) The power series ring of an integral domain is also an integral domain.
(j) A ring is a field provided it is both an integral domain and a skew field.
(k) If R is a (skew) field, then the equation $a \times y = b$, $a \neq z$, has a unique solution $y = a' \times b$.
(l) The polynomial ring and the power series ring of a field are unique factorization domains.

4. *Vector Space.*   A vector space $V(F)$ over a field F consists of a nonempty set V (the vectors), a binary operation ($\oplus$) on V, a function (called *scalar multiplication*) from the product set $F \times V$ onto V with the image of (a,v) denoted by $a \circ v$, and the following axioms:
A_0. Relative to addition ($\oplus$) V is an Abelian group in which the identity element (vector) is denoted by z and the inverse of v is denoted by $-v$.
M_1. $a \circ (b \circ v) = (ab) \circ v$ for all $a,b \in F$ and $v \in F$. (Here ab denotes the product of a and b in F.)
M_2. $1 \circ v = v$ for all $v \in V$. (Here 1 denotes the multiplicative indentity element of F.)

D_1. $a \circ (\mu \oplus \nu) = (a \circ \mu) \oplus (a \circ \nu)$ for all $a \in F, \mu, \nu \in V$.

D_2. $(a + b) \circ \nu = (a \circ \nu) \oplus (b \circ \nu)$ for all $a, b \in F, \nu \in V$. (Here + denotes addition in the field F.)

The elements of F are referred to as *scalars*.

EXAMPLE 1. The polynomial ring $P(F)$ of a field F is a vector space over F. In this example scalar multiplication is a special case of the multiplication defined for $P(F)$.

EXAMPLE 2. Denote by $C_n(F)$ the set of all n-tuples, $(a_1, a_2, \ldots, a_n), n$ a positive integer, with all $a_i \in F$. Define

$$(a_1, \ldots, a_n) \oplus (b_1, \ldots, b_n) = (a_1 + b_1, \ldots, a_n + b_n) \quad \text{and}$$
$$c \circ (a_1, \ldots, a_n) = (c \times a_1, \ldots, c \times a_n),$$

where + and $\times$ denote the addition and multiplication, respectively, of the field F. Relative to $\oplus$ and $\circ$, $C_n(F)$ is a vector space, called the *n-dimensional coordinate space* over F.

A vector space $V(F)$ is *n-dimensional* over F provided V contains n elements $\nu_1, \nu_2, \ldots, \nu_n$ such that each element $\nu \in V$ is uniquely expressible in the form

$$\nu = a_1 \circ \nu_1 \oplus a_2 \circ \nu_2 \oplus \cdots \oplus a_n \circ \nu_n$$

for some $a_1, a_2, \ldots, a_n \in F$.

Two vector spaces $V(F)$ and $W(F)$ over the field of scalars F are *isomorphic* provided there is a one-to-one correspondence between the elements of V and the elements of W which is preserved under the arithmetic of the two spaces.

Basic Theorem. An n-dimensional vector space $V(F)$ is isomorphic with the coordinate space $C_n(F)$ (of Example 2, above).

MATRICES AND DETERMINANTS

Dr. R. E. Bargmann

1. GENERAL DEFINITIONS

1.1. A matrix is an array of numbers, consisting of m rows and n columns. It is usually denoted by a bold-face capital letter, e.g.,

$$\mathbf{A} \qquad \Sigma \qquad \mathbf{M}$$

1.2. The (i, j) element of a matrix is the element occurring in row i and column j. It is usually denoted by a lower-case letter with subscripts, e.g.,

$$a_{ij} \qquad \sigma_{ij} \qquad m_{ij}$$

Exceptions to this convention will be stated where required.

1.3. A matrix is called rectangular if m (number of rows) $\neq n$ (number of columns).

1.4. A matrix is called square if $m = n$.

1.5a. In the transpose of a matrix $\mathbf{A}$, denoted by $\mathbf{A}'$, the element in the j'th row and i'th column of $\mathbf{A}$ is equal to the element in the i'th row and j'th column of $\mathbf{A}'$. Formally $(\mathbf{A}')_{ij} = (\mathbf{A})_{ji}$ where the symbol $(\mathbf{A}')_{ij}$ denotes the (i, j) element of $\mathbf{A}'$.

1.5b. The Hermitian conjugate of a matrix $\mathbf{A}$, denoted by $\mathbf{A}^H$ or $\mathbf{A}^\dagger$ is obtained by transposing $\mathbf{A}$ and replacing each element by its conjugate complex. Hence if

$$a_{kl} = u_{kl} + iv_{kl}$$

then

$$(\mathbf{A}^H)_{kl} = u_{lk} - iv_{lk}$$

where typical elements have been denoted by (k, l) to avoid confusion with $i = \sqrt{-1}$.

1.6a. A square matrix is called symmetric if $\mathbf{A} = \mathbf{A}'$.

1.6b. A square matrix is called Hermitian if $\mathbf{A} = \mathbf{A}^H$.

1.7. A matrix with m rows and 1 column is called a column vector and is usually denoted by bold faced, lower-case letters, e.g.,

$$\beta \qquad \mathbf{x} \qquad \mathbf{a}$$

1.8. A matrix with one row and n columns is called a row vector and is usually denoted by a primed, bold faced, lower-case letter, e.g.,

$$\mathbf{a}' \qquad \mathbf{c}' \qquad \mu'$$

1.9. A matrix with one row and one column is called a scalar, and is usually denoted by a lower-case letter, occasionally italicized.

1.10. The diagonal extending from upper left (NW) to lower right (SE) is called the principal diagonal of a square matrix.

1.11a. A matrix with all elements above the principal diagonal equal to zero is called a lower triangular matrix.

Example

$$\mathbf{T} = \begin{bmatrix} t_{11} & 0 & 0 \\ t_{21} & t_{22} & 0 \\ t_{31} & t_{32} & t_{33} \end{bmatrix} \quad \text{is lower triangular}$$

1.11b. The transpose of a lower triangular matrix is called an upper triangular matrix.

1.12. A square matrix with all off-diagonal elements equal to zero is called a diagonal matrix, denoted by the letter **D** with subscript indicating the typical element in the principal diagonal.

Example

$$\mathbf{D}_a = \begin{bmatrix} a_1 & 0 & 0 \\ 0 & a_2 & 0 \\ 0 & 0 & a_3 \end{bmatrix} \text{ is diagonal}$$

2. ADDITION, SUBTRACTION, AND MULTIPLICATION

2.1. Two matrices **A** and **B** can be added (subtracted) if the number of rows (columns) in **A** equals the number of rows (columns) in **B**.

$$\mathbf{A} \pm \mathbf{B} = \mathbf{C}$$

implies

$$a_{ij} \pm b_{ij} = c_{ij} \qquad i = 1, 2, \ldots m$$
$$j = 1, 2, \ldots n$$

2.2. Multiplication of a matrix or vector by a scalar implies multiplication of each element by the scalar. If

$$\mathbf{B} = \gamma \mathbf{A}$$

then

$$b_{ij} = \gamma a_{ij}$$

for all elements.

2.3a. Two matrices, **A** and **B**, can be multiplied if the number of columns in **A** equals the number of rows in **B**.

2.3b. Let **A** be of order $(m \times n)$ (have m rows and n columns) and **B** of order $(n \times p)$. Then the product of two matrices **C** = **AB**, is a matrix of order $(m \times p)$ with elements

$$c_{ij} = \sum_{k=1}^{n} a_{ik} b_{kj}$$

This states that c_{ij} is the scalar product of the i'th row vector of **A** and the j'th column vector of **B**.

Example

$$\begin{bmatrix} 3 & 4 & 2 \\ 2 & 3 & -1 \end{bmatrix} \begin{bmatrix} 1 & -2 & -4 \\ 0 & -1 & 2 \\ 6 & -3 & 9 \end{bmatrix} = \begin{bmatrix} 15 & -16 & 14 \\ -4 & -4 & -11 \end{bmatrix}$$

e.g.,

$$c_{23} = \begin{bmatrix} 2 & 3 & -1 \end{bmatrix} \begin{bmatrix} -4 \\ 2 \\ 9 \end{bmatrix}$$

$$= 2 \times (-4) + 3 \times 2 + (-1) \times 9 = -11$$

2.3c. In general, matrix multiplication is not commutative

$$AB \neq BA$$

2.3d. Matrix multiplication is associative

$$A(BC) = (AB)C$$

2.3e. The distributive law for multiplication and addition holds as in the case of scalars,

$$(A + B)C = AC + BC$$
$$C(A + B) = CA + CB$$

2.4. In some applications, the term-by-term product of two matrices **A** and **B** of identical order is defined as

$$C = A * B$$

where

$$c_{ij} = a_{ij} b_{ij}$$

2.5. $(ABC)' = C'B'A'$

2.6. $(ABC)^H = C^H B^H A^H$

2.7. If both **A** and **B** are symmetric, then $(AB)' = BA$. Note that the product of two symmetric matrices is generally not symmetric.

3. RECOGNITION RULES AND SPECIAL FORMS

3.1. A column (row) vector with all elements equal to zero is called a null vector, and usually denoted by the symbol **0**.

3.2. A null matrix has all elements equal to zero.

3.3a. A diagonal matrix with all elements equal to one in the principal diagonal is called the identity matrix **I**.

3.3b. γI, i.e., a diagonal matrix with all diagonal elements equal to a constant γ, is called a scalar matrix.

3.4. A matrix which has only one element equal to one and all others equal to zero is called an elementary matrix $(EL)_{ij}$.

Example

$$(EL)_{23} = \begin{bmatrix} 0 & 0 & 0 & 0 & 0 \\ 0 & 0 & 1 & 0 & 0 \\ 0 & 0 & 0 & 0 & 0 \\ 0 & 0 & 0 & 0 & 0 \end{bmatrix}$$

The order of the matrix is usually implicit.

3.5a. The symbol **j** is reserved for a column vector with all elements equal to 1.

3.5b. The symbol **j'** is reserved for a row vector with all elements equal to 1.

3.6. An expression ending with a column vector is a column vector.

Example

$$ABx = y$$

(It is assumed that rule 2.3a is satisfied, else matrix multiplication would not be defined.)

3.7. An expression beginning with a row vector is a row vector.

Example

$$y'(A + BC) = d'$$

3.8. An expression beginning with a row vector and ending with a column vector, is a scalar.

Example

$$\mathbf{a'Bc} = \gamma$$

3.9a. If $\mathbf{Q}$ is a square matrix, the scalar $\mathbf{x'Qx}$ is called a quadratic form. If $\mathbf{Q}$ is non-symmetric, one can always find a symmetric matrix $\mathbf{Q^*}$ such that

$$\mathbf{x'Qx} = \mathbf{x'Q^*x}$$

where

$$(\mathbf{Q^*})_{ij} = \tfrac{1}{2}(q_{ij} + q_{ji})$$

3.9b. If $\mathbf{Q}$ is a square matrix the scalar $\mathbf{x^H Q x}$ is called a Hermitian form.

3.10. A scalar $\mathbf{x'Qy}$ is called a bilinear form.

3.11. The scalar $\mathbf{x'x} = \Sigma x_i^2$, i.e., the sum of squares of all elements of $\mathbf{x}$.

3.12. The scalar $\mathbf{x'y} = \Sigma x_i y_i$, i.e., the sum of products of elements in $\mathbf{x}$ by those in $\mathbf{y}$. $\mathbf{x}$ and $\mathbf{y}$ have the same number of elements.

3.13. The scalar $\mathbf{x'D_w x} = \Sigma w_i x_i^2$ is called a weighted sum of squares.

3.14. The scalar $\mathbf{x'D_w y} = \Sigma w_i x_i y_i$ is called a weighted sum of products.

3.15a. The vector $\mathbf{Aj}$ is a column vector whose elements are the row sums of $\mathbf{A}$.

3.15b. The vector $\mathbf{j'A}$ is a row vector whose elements are the column sums of $\mathbf{A}$.

3.15c. The scalar $\mathbf{j'Aj}$ is the sum of all elements in $\mathbf{A}$. Schematically

$\mathbf{A}$	$\mathbf{Aj}$
$\mathbf{j'A}$	$\mathbf{j'Aj}$

3.16a. If $\mathbf{B} = \mathbf{D_w A}$; then $b_{ij} = w_i a_{ij}$.

3.16b. If $\mathbf{B} = \mathbf{A D_w}$; then $b_{ij} = a_{ij} w_j$,

3.17. Interchanging summation and matrix notation:

If

$$\mathbf{ABCD} = \mathbf{E}$$

then

$$e_{ij} = \sum_k \sum_l \sum_m a_{ik} b_{kl} c_{lm} d_{mj}$$

The second subscript of an element must coincide with the first of the next one. Reordering and transposing may be required.

Example

If

$$e_{ij} = \sum_k \sum_l \sum_m a_{kl} b_{ki} c_{jm} d_{ml}$$

$$= \sum_k \sum_l \sum_m b_{ki} a_{kl} d_{ml} c_{jm}$$

Then

$$\mathbf{E} = \mathbf{B'AD'C'}$$

3.18a. $\mathbf{A'A}$ is a symmetric matrix whose (i,j) element is the scalar product of the i'th column vector and the j'th column vector of $\mathbf{A}$.

3.18b. $\mathbf{AA'}$ is a symmetric matrix whose (i,j) element is the scalar product of the i'th row vector and the j'th row vector of $\mathbf{A}$.

4. DETERMINANTS

4.1a. A determinant $|\mathbf{A}|$ or $\det(\mathbf{A})$ is a scalar function of a square matrix defined in such a way that

$$|\mathbf{A}|\,|\mathbf{B}| = |\mathbf{AB}|$$

and

$$\begin{vmatrix} a_{11} & a_{12} \\ a_{21} & a_{22} \end{vmatrix} = a_{11}a_{22} - a_{12}a_{21}$$

4.1b. $|\mathbf{A}| = |\mathbf{A}'|$

4.2.

$$\begin{vmatrix} a_{11} & a_{12} & a_{13} \\ a_{21} & a_{22} & a_{23} \\ a_{31} & a_{32} & a_{33} \end{vmatrix} = \begin{aligned} & a_{11}a_{22}a_{33} + a_{12}a_{23}a_{31} + a_{13}a_{21}a_{32} \\ & - a_{13}a_{22}a_{31} - a_{11}a_{23}a_{32} - a_{12}a_{21}a_{33} \end{aligned}$$

4.3.

$$\begin{vmatrix} a_{11} & a_{12} & \cdots & a_{1n} \\ a_{21} & a_{22} & \cdots & a_{2n} \\ a_{n1} & a_{n2} & \cdots & a_{nm} \end{vmatrix} = \sum (-1)^{\delta} a_{1i_1} a_{2i_2} \cdots a_{ni_n}$$

where the sum is over all permutations

$$i_1 \neq i_2 \neq \cdots i_n$$

and δ denotes the number of exchanges necessary to bring the sequence $(i_1, i_2, \ldots i_n)$ back into the natural order $(1, 2, \ldots n)$.

4.4. If two rows (columns) in a matrix are exchanged, the determinant will change its sign.

4.5. A determinant does not change its value if a linear combination of other rows (columns) is added to any given row (column).

Example

$$\begin{vmatrix} a_{11} & a_{12} & a_{13} & a_{14} \\ b_{21} & b_{22} & b_{23} & b_{24} \\ a_{31} & a_{32} & a_{33} & a_{34} \\ a_{41} & a_{42} & a_{43} & a_{44} \end{vmatrix} = \begin{vmatrix} a_{11} & a_{12} & a_{13} & a_{14} \\ a_{21} & a_{22} & a_{23} & a_{24} \\ a_{31} & a_{32} & a_{33} & a_{34} \\ a_{41} & a_{42} & a_{43} & a_{44} \end{vmatrix}$$

where

$$b_{2i} = a_{2i} + \gamma_1 a_{1i} + \gamma_3 a_{3i} + \gamma_4 a_{4i}$$
$$i = 1, 2, 3, 4$$

$\gamma_1, \gamma_3, \gamma_4$ arbitrary.

4.6. If the i'th row (column) equals (a constant times) the j'th row (column) of a matrix, its determinant is equal to zero, $(i \neq j)$.

4.7. If, in a matrix $\mathbf{A}$, each element of a row (column) is multiplied by a constant γ, the determinant is multiplied by γ.

4.8. $|\gamma\mathbf{A}| = \gamma^n |\mathbf{A}|$ assuming that $\mathbf{A}$ is of order $(n \times n)$.

4.9. The cofactor of a square matrix $\mathbf{A}$, $\mathrm{cof}_{ij}(\mathbf{A})$ **is** the determinant of a matrix obtained by striking the i'th row and j'th column of $\mathbf{A}$ and choosing positive (negative) sign if $i + j$ is even (odd).

Example

$$\text{cof}_{23} \begin{bmatrix} 2 & 4 & 3 \\ 6 & 1 & 5 \\ -2 & 1 & 3 \end{bmatrix} = - \begin{vmatrix} 2 & 4 \\ -2 & 1 \end{vmatrix}$$

$$= -(2 + 8) = -10$$

4.10. (Laplace Development)

$$|\mathbf{A}| = a_{i1}\text{cof}_{i1}(\mathbf{A}) + a_{i2}\text{cof}_{i2}(\mathbf{A}) + \cdots + a_{in}\text{cof}_{in}(\mathbf{A})$$

$$= a_{1j}\text{cof}_{1j}(\mathbf{A}) + a_{2j}\text{cof}_{2j}(\mathbf{A}) + \cdots + a_{nj}\text{cof}_{nj}(\mathbf{A})$$

for any row i or any column j.

4.11. Numerical Evaluation of the determinant of a symmetric matrix.

Note: If **A** is non-symmetric, form **A'A** or **AA'** by rule 3.18, obtain its determinant, and take the square root.

("Forward Doolittle Scheme", "left side")
Let

$$p_{11} = a_{11}, \quad p_{12} = a_{12} = a_{21}, \ldots p_{1n} = a_{1n}$$

$$
\begin{array}{ccccc}
p_{11} & p_{12} & p_{13} & \cdots & p_{1n} \\
\underline{1} & u_{12} & u_{13} & \cdots & u_{1n} \\
& a_{22} & a_{23} & \cdots & a_{2n} \\
& p_{22} & p_{23} & \cdots & p_{2n} \\
& \underline{1} & u_{23} & \cdots & u_{2n} \\
& & a_{33} & \cdots & a_{3n} \\
& & p_{33} & \cdots & p_{3n} \\
& & 1 & \cdots & u_{3n} \\
& & \cdot & \cdots & \cdot \\
& & & & a_{nn} \\
& & & & p_{nn} \\
& & & & 1
\end{array}
$$

$$u_{1i} = p_{1i}/p_{11} \qquad\qquad\qquad i = 1, 2, \ldots n$$

$$p_{2i} = a_{2i} - u_{12}p_{1i} \qquad\qquad i = 2, 3, \ldots n$$

$$u_{2i} = p_{2i}/p_{22}$$

$$p_{3i} = a_{3i} - u_{13}p_{1i} - u_{23}p_{2i} \qquad i = 3, 4, \ldots n$$

$$u_{3i} = p_{3i}/p_{33}$$

$$p_{ki} = a_{ki} - u_{1k}p_{1i} - u_{2k}p_{2i} - \cdots - u_{k-1,k}p_{k-1,i} \qquad \begin{array}{l} i = k, k+1, \ldots n \\ k = 2, 3, \ldots n \end{array}$$

$$u_{ki} = p_{ki}/p_{kk}$$

If, at some stage, $p_{kk} = 0$, reordering of rows and columns may be required. If the matrix is positive-definite (see 8.16) (always true for **AA'** or **A'A**, see rule 10.24), none of the

p_{kk} will be zero. The p_{ii} are called pivots. Then

$$|\mathbf{A}| = \prod_{i=1}^{n} p_{ii}$$

Further, if $\mathbf{A}$ is partitioned

$$\mathbf{A} = \begin{bmatrix} \mathbf{A}_{11} & \mathbf{A}_{12} \\ \mathbf{A}'_{12} & \mathbf{A}_{22} \end{bmatrix}$$

where $\mathbf{A}_{11}$ is of order $(k \times k)$, then

$$|\mathbf{A}_{11}| = \prod_{i=1}^{k} p_{ii}$$

(Numerical Examples: see 6.14.)

5. SINGULARITY AND RANK

5.1. A matrix $\mathbf{A}$ is called singular if there exists a vector $\mathbf{x} \neq \mathbf{0}$ such that $\mathbf{Ax} = \mathbf{0}$ or $\mathbf{A'x} = \mathbf{0}$. Note $\mathbf{x} \neq \mathbf{0}$ if a single element of $\mathbf{x}$ is unequal 0. If a matrix is not singular, it is called non-singular.

5.2. If a matrix $\mathbf{A}_1$ can be formed by selection of r rows and columns of $\mathbf{A}$ such that $\mathbf{A}_1\mathbf{x} \neq \mathbf{0}$ or $\mathbf{A}'_1\mathbf{x} \neq \mathbf{0}$ for every $\mathbf{x} \neq \mathbf{0}$, and if addition of an $(r + 1)st$ row and column would produce a singular matrix, r is called the rank of $\mathbf{A}$.

Example

$$\mathbf{A} = \begin{bmatrix} 2 & 4 & 6 \\ 1 & 3 & 7 \\ 3 & 7 & 13 \\ 1 & 1 & -1 \end{bmatrix}$$

Note that

$$[1, \quad 1, \quad -1] \begin{bmatrix} 2 & 4 & 6 \\ 1 & 3 & 7 \\ 3 & 7 & 13 \end{bmatrix} = [0 \quad 0 \quad 0]$$

and

$$[1, \quad -1, \quad -1] \begin{bmatrix} 2 & 4 & 6 \\ 1 & 3 & 7 \\ 1 & 1 & -1 \end{bmatrix} = [0 \quad 0 \quad 0]$$

but

$$\begin{bmatrix} 2 & 4 \\ 1 & 3 \end{bmatrix} \begin{bmatrix} x_1 \\ x_2 \end{bmatrix} \neq \begin{bmatrix} 0 \\ 0 \end{bmatrix}$$

or

$$[x_1 \quad x_2] \begin{bmatrix} 2 & 4 \\ 1 & 3 \end{bmatrix} \neq [0, \quad 0]$$

for any arbitrary

$$[x_1, \quad x_2] \neq [0, \quad 0].$$

Hence the matrix has rank 2.

5.3. If **A** has rank r and if $\mathbf{A_1}$ is a non-singular submatrix consisting of r rows and columns of **A**, then $\mathbf{A_1}$ is called a basis of **A**.

5.4a. The determinant of a square singular matrix is 0.

5.4b. The determinant of a non-singular matrix is $\neq 0$.

5.5. rank $(\mathbf{AB}) \leq$ min [rank $(\mathbf{A})$, rank $(\mathbf{B})$].

5.6. rank $(\mathbf{AA'})$ = rank $(\mathbf{A'A})$ = rank $(\mathbf{A})$.

5.7. $|\mathbf{A'A}| = |\mathbf{AA'}| = |\mathbf{A}|^2$ if **A** is square.

5.8. $|\mathbf{A'A}| = |\mathbf{AA'}| \geq 0$ for every **A** with real elements.

6.   INVERSION

(Regular Case, non-singular matrices)

6.1. If **A** is square and non-singular ($|\mathbf{A}| \neq 0$) there exists a unique matrix $\mathbf{A}^{-1}$ such that $\mathbf{AA}^{-1} = \mathbf{A}^{-1}\mathbf{A} = \mathbf{I}$.

6.2. $(\mathbf{ABC})^{-1} = \mathbf{C}^{-1}\mathbf{B}^{-1}\mathbf{A}^{-1}$ (provided that all inverses exist).

6.3. $(\mathbf{A}^{-1})' = (\mathbf{A}')^{-1}$

6.4. $\mathbf{Ax} = \mathbf{b}$ is a system of linear equations. If **A** is square and non-singular, there exists a unique solution

$$\mathbf{x} = \mathbf{A}^{-1}\mathbf{b}$$

6.5. $(\gamma\mathbf{A})^{-1} = (1/\gamma)\mathbf{A}^{-1}$

6.6. $|\mathbf{A}^{-1}| = 1/|\mathbf{A}|$

6.7. $\mathbf{D}_w^{-1} = \mathbf{D}_{1/w}$ where **D** is a diagonal matrix.

6.8. If

$$\mathbf{A} = \mathbf{B} + \mathbf{uv'}$$

then

$$\mathbf{A}^{-1} = \mathbf{B}^{-1} - \lambda\mathbf{yz'}$$

where

$$\mathbf{y} = \mathbf{B}^{-1}\mathbf{u}, \quad \mathbf{z'} = \mathbf{v'B}^{-1},$$

and

$$\lambda = 1/(1 + \mathbf{z'u})$$

Example 6.8.1

$$\mathbf{A} = \begin{bmatrix} 4 & 2 & 4 & 5 \\ 3 & 9 & 12 & 15 \\ 2 & 4 & 11 & 10 \\ 1 & 2 & 4 & 10 \end{bmatrix}$$

This matrix can be written as

$$\begin{bmatrix} 3 & 0 & 0 & 0 \\ 0 & 3 & 0 & 0 \\ 0 & 0 & 3 & 0 \\ 0 & 0 & 0 & 5 \end{bmatrix} + \begin{bmatrix} 1 \\ 3 \\ 2 \\ 1 \end{bmatrix} [1 \quad 2 \quad 4 \quad 5] = \mathbf{B} + \mathbf{uv'}$$

$$\mathbf{B}^{-1} = \begin{bmatrix} 1/3 & 0 & 0 & 0 \\ 0 & 1/3 & 0 & 0 \\ 0 & 0 & 1/3 & 0 \\ 0 & 0 & 0 & 1/5 \end{bmatrix}$$

$$\mathbf{y} = \mathbf{B}^{-1}\mathbf{u} = \begin{bmatrix} 1/3 \\ 1 \\ 2/3 \\ 1/5 \end{bmatrix}$$

$$\mathbf{z}' = \mathbf{v}'\mathbf{B}^{-1} = [1/3 \quad 2/3 \quad 4/3 \quad 1]$$

$$\mathbf{z}'\mathbf{u} = 1/3 \times 1 + 2/3 \times 3 + 4/3 \times 2 + 1 \times 1 = 6$$

$$\lambda = 1/7$$

$$\mathbf{A}^{-1} = \begin{bmatrix} 1/3 & 0 & 0 & 0 \\ 0 & 1/3 & 0 & 0 \\ 0 & 0 & 1/3 & 0 \\ 0 & 0 & 0 & 1/5 \end{bmatrix} - (1/7)\begin{bmatrix} 1/3 \\ 1 \\ 2/3 \\ 1/5 \end{bmatrix}[1/3 \quad 2/3 \quad 4/3 \quad 1]$$

$$= (1/315)\begin{bmatrix} 100 & -10 & -20 & -15 \\ -15 & 75 & -60 & -45 \\ -10 & -20 & 65 & -30 \\ -3 & -6 & -12 & 54 \end{bmatrix}$$

(This rule is especially useful if all off-diagonal elements are equal, then $\mathbf{u} = k\mathbf{j}$ and $\mathbf{v}' = \mathbf{j}'$ and $\mathbf{B}$ is diagonal.)

6.9. Let $\mathbf{B}$ (elements b_{ij}) have a known inverse, $\mathbf{B}^{-1}$ (elements b^{ij}). Let $\mathbf{A} = \mathbf{B}$ except for one element $a_{rs} = b_{rs} + k$. Then the elements of $\mathbf{A}^{-1}$ are

$$a^{ij} = b^{ij} - \frac{kb^{ir}b^{sj}}{1 + kb^{sr}}.$$

6.10. (Partitioning)

Let

$$\mathbf{A} = \begin{matrix} (p) \\ (q) \end{matrix} \begin{matrix} (p) & (q) \\ \begin{bmatrix} \mathbf{B} & \mathbf{C} \\ \mathbf{D} & \mathbf{E} \end{bmatrix} \end{matrix}$$

(letters in parentheses denote order of the submatrices)

Let $\mathbf{B}^{-1}$ and $\mathbf{E}^{-1}$ exist. Then

$$\mathbf{A}^{-1} = \begin{bmatrix} \mathbf{X} & \mathbf{Y} \\ \mathbf{Z} & \mathbf{U} \end{bmatrix}$$

where

$$\mathbf{X} = (\mathbf{B} - \mathbf{C}\mathbf{E}^{-1}\mathbf{D})^{-1}$$
$$\mathbf{U} = (\mathbf{E} - \mathbf{D}\mathbf{B}^{-1}\mathbf{C})^{-1}$$
$$\mathbf{Y} = -\mathbf{B}^{-1}\mathbf{C}\mathbf{U}$$
$$\mathbf{Z} = -\mathbf{E}^{-1}\mathbf{D}\mathbf{X}$$

6.11. (Partitioning of Determinants)

Let

$$|\mathbf{A}| = \begin{vmatrix} \mathbf{B} & \mathbf{C} \\ \mathbf{D} & \mathbf{E} \end{vmatrix} \qquad \text{(same structure as in 6.10)}$$

Then

$$|\mathbf{A}| = |\mathbf{E}| \, |(\mathbf{B} - \mathbf{CE}^{-1}\mathbf{D})| = |\mathbf{B}| \, |(\mathbf{E} - \mathbf{DB}^{-1}\mathbf{C})|$$

6.12. Let

$$\mathbf{A} = \mathbf{B} + \mathbf{UV}$$

where $\mathbf{B}(n \times n)$ has an inverse

$\mathbf{U}$ is of order $(n \times k)$, with k usually very small

$\mathbf{V}$ is of order $(k \times n)$

(the special case for $k = 1$ is treated in 6.8).

Then

$$\mathbf{A}^{-1} = \mathbf{B}^{-1} - \mathbf{Y}\mathbf{\Lambda}\mathbf{Z}$$

where

$$\mathbf{Y} = \mathbf{B}^{-1}\mathbf{U}(n \times k)$$
$$\mathbf{Z} = \mathbf{VB}^{-1}(k \times n)$$

and

$$\mathbf{\Lambda}(k \times k) = [\mathbf{I} + \mathbf{ZU}]^{-1}$$

6.13. Let a_{ij} denote the elements of $\mathbf{A}$ and a^{ij} those of $\mathbf{A}^{-1}$. Then

$$a^{ij} = \text{cof}_{ji}(\mathbf{A})/|\mathbf{A}|$$

where cof is the determinant defined in 4.9.

6.14. "Doolittle" Method of inverting symmetric matrices (see also 4.11). Let

$$p_{11} = a_{11}, \quad p_{12} = a_{12} = a_{21}, \dots p_{1n} = a_{1n} = a_{n1}$$

Forward Solution

p_{11}	p_{12}	p_{13}	$\cdots$	p_{1n}	1				
1	u_{12}	u_{13}	$\cdots$	u_{1n}	u_{11}				
	a_{22}	a_{23}	$\cdots$	a_{2n}	0	1			
	p_{22}	p_{23}	$\cdots$	p_{2n}	p_{21}	$p_{2\text{II}}$			
	1	u_{23}	$\cdots$	u_{2n}	u_{21}	$u_{2\text{II}}$			
		a_{33}	$\cdots$	a_{3n}	0	0	1		
		p_{33}	$\cdots$	p_{3n}	p_{31}	$p_{3\text{II}}$	$p_{3\text{III}}$		
		1	$\cdots$	u_{3n}	u_{31}	$u_{3\text{II}}$	$u_{3\text{III}}$		
			$\cdots$	$\cdot$	$\cdot$	$\cdots$	$\cdot$		
				a_{nn}	0	0	0	$\cdots$	1
				p_{nn}	p_{n1}	$p_{n\text{II}}$	$p_{n\text{III}}$	$\cdots$	p_{nN}
				1	u_{n1}	$u_{n\text{II}}$	$u_{n\text{III}}$	$\cdots$	u_{nN}

$$u_{1i} = p_{1i}/p_{11} \qquad\qquad i = 1, 2, \ldots n, \mathrm{I}$$

$$p_{2i} = a_{2i} - u_{12}p_{1i} \qquad\qquad i = 2, 3, \ldots n, \mathrm{I}, \mathrm{II}$$

$$u_{2i} = p_{2i}/p_{22}$$

$$p_{3i} = a_{3i} - u_{13}p_{1i} - u_{23}p_{2i} \qquad\qquad i = 3, 4, \ldots n, \mathrm{I}, \mathrm{II}, \mathrm{III}$$

$$u_{3i} = p_{3i}/p_{33}$$

$$p_{ki} = a_{ki} - u_{1k}p_{1i} - u_{2k}p_{2i} - \cdots - u_{k-1,k}p_{k-1,i} \quad i = k, k+1, \ldots n, \mathrm{I}, \mathrm{II}, \ldots K$$
$$k = 2, 3, \ldots n$$

$$u_{ki} = p_{ki}/p_{kk}$$

Backward Solution

(j refers to Arabic, J refers to Roman numerals)

The elements of $\mathbf{A}^{-1}$ are a^{ij}

$$a^{nj} = u_{nJ} \qquad\qquad \begin{aligned} j &= 1, 2, \ldots n; \\ J &= \mathrm{I}, \mathrm{II}, \ldots N \end{aligned}$$

$$a^{n-1j} = u_{n-1,J} - u_{n-1,n}a^{nj} \qquad\qquad \begin{aligned} j &= 1, 2, \ldots (n-1); \\ J &= \mathrm{I}, \mathrm{II}, \ldots (N-1) \end{aligned}$$

$$a^{n-2,j} = u_{n-2,J} - u_{n-2,n}a^{nj} - u_{n-2,n-1}a^{n-1,j} \qquad\qquad \begin{aligned} j &= 1, 2, \ldots (n-2); \\ J &= \mathrm{I}, \mathrm{II}, \ldots (N-2) \end{aligned}$$

$$a^{n-k,j} = u_{n-k,J} - u_{n-k,n}a^{nj} - u_{n-k,n-1}a^{n-1,j} - \cdots - u_{n-k,n-k+1}a^{n-k+1,j} \qquad \begin{aligned} j &= 1, 2, \ldots (n-k); \\ J &= \mathrm{I}, \mathrm{II}, \ldots (N-k); \\ k &= 1, 2, \ldots (n-1), \end{aligned}$$

and $a^{ji} = a^{ij}$.

Numerical Example 6.14.1.

Invert the Matrix

$$\begin{bmatrix} 25 & 30 & -10 \\ 30 & 40 & -6 \\ -10 & -6 & 17 \end{bmatrix}$$

a_1	25	30	-10	1		
u_1	1	1.2	-0.4	0.04		
	a_2	40	-6	0	1	
	p_2	4	6	-1.2	1	
	u_2	1	1.5	-0.3	0.25	
		a_3	17	0	0	1
		p_3	4	2.2	-1.5	1
		u_3	1	0.55	-0.375	0.25

1.61	-1.125	0.55
-1.125	0.8125	-0.375
0.55	-0.375	0.25

Enter row a_1.
Elements in u_1 = Elements in a_1 divided by $a_{11}(=25)$.
Enter row a_2.

$$p_{22} = 40 - 1.2 \times 30 = 4$$
$$p_{23} = -6 - 1.2 \times (-10) = 6$$
$$p_{21} = 0 - 1.2 \times 1 = -1.2$$
$$p_{211} = 1$$

Elements in u_2 = Elements in p_2 divided by $p_{22}(=4)$.
Enter row a_3.

$$p_{33} = 17 - (-0.4) \times (-10) - 1.5 \times 6 = 4$$
$$p_{31} = 0 - (-0.4) \times 1 - 1.5 \times (-1.2) = 2.2$$
$$p_{311} = 0 - 1.5 \times 1 = -1.5$$
$$p_{3111} = 1$$

Elements in u_3 = Elements in p_3 divided by $p_{33}(=4)$.
  Copy the right-hand side of the last (third) u – row as the last column below the double line.

$$a^{21} = -0.3 - 1.5 \times 0.55 = -1.125$$
$$a^{22} = 0.25 - 1.5 \times (-0.375) = 0.8125$$
$$a^{23} = 0 - 1.5 \times 0.25 = -0.375 \quad \text{(check against } a^{32}).$$

These are entered in the next to last (second) column below.

$$a^{11} = 0.04 - (-0.4) \times 0.55 - 1.2 \times (-1.125) = 1.61$$
$$a^{12} = 0 - (-0.4) \times (-0.375) - 1.2 \times 0.8125 = -1.125 \quad \text{(check against } a^{21})$$
$$a^{13} = 0 - (-0.4) \times (0.25) - 1.2 \times (-0.375) = 0.55 \quad \text{(check against } a^{31}).$$

6.15. A matrix is called orthogonal if $\mathbf{A}' = \mathbf{A}^{-1}$ (or $\mathbf{AA}' = \mathbf{I}$).

7. TRACES

7.1. If $\mathbf{A}$ is a square matrix then the trace of $\mathbf{A}$ is $tr\,\mathbf{A} = \sum_i a_{ii}$, i.e., the sum of the diagonal elements.
  7.2. If $\mathbf{A}$ is of order $(m \times k)$ and $\mathbf{B}$ of order $(k \times m)$ then $tr(\mathbf{AB}) = tr(\mathbf{BA})$.
  7.3. If $\mathbf{A}$ is of order $(m \times k)$, $\mathbf{B}$ of order $(k \times r)$ and $\mathbf{C}$ of order $(r \times m)$, then

$$tr(\mathbf{ABC}) = tr(\mathbf{BCA}) = tr(\mathbf{CAB}).$$

7.3a. If $\mathbf{b}$ is a column vector and $\mathbf{c}'$ a row vector, then

$$tr(\mathbf{Abc}') = tr(\mathbf{bc}'\mathbf{A}) = \mathbf{c}'\mathbf{Ab}$$

since the trace of a scalar is the scalar.
  7.4. $tr(\mathbf{A} + \gamma\mathbf{B}) = tr\mathbf{A} + \gamma tr\mathbf{B}$; where γ is a scalar.
  7.5. $tr(\mathbf{EL})_{ij}\mathbf{A} = tr\mathbf{A}(\mathbf{EL})_{ij} = a_{ji}$; where $(\mathbf{EL})_{ij}$ is an elementary matrix as defined in 3.4.
  7.6. $tr(\mathbf{EL})_{ij}\mathbf{A}(\mathbf{EL})_{rs}\mathbf{B} = a_{jr}b_{si}$
  (These rules are useful in matrix differentiation)

  7.7. The trace of the second order of a square matrix $\mathbf{A}$ is the sum of the determinants of all $\binom{n}{2}$ matrices of order (2×2) which can be formed by intersecting rows i and j with columns i and j.

$$tr_2 \mathbf{A} = \begin{vmatrix} a_{11} & a_{12} \\ a_{21} & a_{22} \end{vmatrix} + \begin{vmatrix} a_{11} & a_{13} \\ a_{31} & a_{33} \end{vmatrix}$$

$$+ \cdots + \begin{vmatrix} a_{11} & a_{1n} \\ a_{n1} & a_{nn} \end{vmatrix} + \begin{vmatrix} a_{22} & a_{23} \\ a_{32} & a_{33} \end{vmatrix}$$

$$+ \cdots + \begin{vmatrix} a_{22} & a_{2n} \\ a_{n2} & a_{nn} \end{vmatrix} + \cdots + \begin{vmatrix} a_{n-1,n-1} & a_{n-1,n} \\ a_{n,n-1} & a_{nn} \end{vmatrix}$$

7.8. The trace of the k'th order of a square matrix is the sum of the determinants of all $\binom{n}{k}$ matrices of order $(k \times k)$ which can be formed by intersecting any k rows of $\mathbf{A}$ with the same k columns.

$$tr_k \mathbf{A} = \sum \begin{vmatrix} a_{i_1 i_1} & a_{i_1 i_2} & \cdots & a_{i_1 i_k} \\ a_{i_2 i_1} & a_{i_2 i_2} & \cdots & a_{i_2 i_k} \\ \cdot & \cdot & \cdots & \cdot \\ a_{i_k i_1} & a_{i_k i_2} & \cdots & a_{i_k i_k} \end{vmatrix}$$

where the sum extends over all combinations of n elements taken k at a time in order

$$i_1 < i_2 < \cdots < i_k.$$

7.9. Rules 7.2 and 7.3 (cyclic exchange) are valid for trace of k'th order.

7.10. $tr_n \mathbf{A} = |\mathbf{A}|$ if $\mathbf{A}$ is of order $(n \times n)$.

8. CHARACTERISTIC ROOTS AND VECTORS

8.1. If $\mathbf{A}$ is a square matrix of order $(n \times n)$, then $|\mathbf{A} - \lambda \mathbf{I}| = 0$ is called the characteristic equation of the matrix $\mathbf{A}$. It is a polynomial of the n'th degree in λ.

8.2. The n roots of the characteristic equation (not necessarily distinct) are called the characteristic roots of $\mathbf{A}$

$$ch(\mathbf{A}) = \lambda_1, \lambda_2, \ldots \lambda_n$$

8.3. The characteristic equation of $\mathbf{A}$ can be obtained by the relation

$$\lambda^n - (tr\mathbf{A})\lambda^{n-1} + (tr_2\mathbf{A})\lambda^{n-2} - (tr_3\mathbf{A})\lambda^{n-3} \cdots - (-1)^n (tr_{n-1}\mathbf{A})\lambda + (-1)^n |\mathbf{A}| = 0$$

where tr_k is defined in 7.8.

Example 8.3.1

$$\mathbf{A} = \begin{bmatrix} 25 & 30 & -10 \\ 30 & 40 & -6 \\ -10 & -6 & 17 \end{bmatrix}$$

$tr\mathbf{A} = 25 + 40 + 17 = 82$

$tr_2\mathbf{A} = (25 \times 40 - 30 \times 30) + (25 \times 17 - 10 \times 10) + (40 \times 17 - 6 \times 6) = 1069$

$tr_3\mathbf{A} = |\mathbf{A}| = 25 \times 4 \times 4 = 400$

(cf. 6.14 and procedure stated in 4.11)

Hence

$$\lambda^3 - 82\lambda^2 + 1069\lambda - 400 = 0$$

The solutions (by Newton iteration) are

$$\lambda_1 = 65.86108$$
$$\lambda_2 = 15.75339$$
$$\lambda_3 = 0.38553$$

These are the characteristic roots of **A**.

8.4. $ch(\mathbf{A} + \gamma\mathbf{I}) = \gamma + ch(\mathbf{A})$

8.5. $ch(\mathbf{AB}) = ch(\mathbf{BA})$

(except that **AB** or **BA** may have additional roots equal to zero).

8.6. $ch(\mathbf{A}^{-1}) = 1/ch(\mathbf{A})$

8.7. If $\lambda_1, \lambda_2, \ldots \lambda_n$ are the roots of **A** then

$$\sum_i \lambda_i = tr\mathbf{A}$$

$$\sum_{i<j} \lambda_i\lambda_j = tr_2\mathbf{A}$$

$$\sum_{i<j<k} \lambda_i\lambda_j\lambda_k = tr_3\mathbf{A}$$

$$\prod_i \lambda_i = |\mathbf{A}|$$

8.8. If $\mathbf{x}'$ denotes the radius vector (running coordinates $[x, y, z]$) and if a matrix **Q** is positive-definite, then

$$(\mathbf{x}' - \mathbf{x}_0')\mathbf{Q}^{-1}(\mathbf{x} - \mathbf{x}_0) = 1$$

is the equation of an ellipsoid with center at $[x_0, y_0, z_0] = \mathbf{x}_0'$ and semi-axes equal to the square roots of the characteristic roots of **Q**.

8.9. The characteristic roots of a triangular (or diagonal) matrix are the diagonal elements of the matrix.

8.10. If **A** is a real matrix with positive roots, then

$$ch_{min}(\mathbf{AA}') \leq [ch_{min}(\mathbf{A})]^2 \leq [ch_{max}(\mathbf{A})]^2 \leq ch_{max}(\mathbf{AA}')$$

where ch_{min} denotes the smallest and ch_{max} the largest root.

8.11. The ratio of two quadratic forms (**B** non-singular)

$$u = \frac{\mathbf{x}'\mathbf{Ax}}{\mathbf{x}'\mathbf{Bx}}$$

attains stationary values at the roots of $\mathbf{B}^{-1}\mathbf{A}$. In particular

$$u_{max} = ch_{max}(\mathbf{B}^{-1}\mathbf{A}) \qquad \text{and} \qquad u_{min} = ch_{min}(\mathbf{B}^{-1}\mathbf{A})$$

8.12. The equation system

$$\mathbf{Ax} = \lambda\mathbf{x}$$

permits non-zero solutions only if λ is one of the characteristic roots of **A**. Such a solution **x** is called a characteristic vector.

8.13. If **x** is a solution to 8.12, so is $\gamma\mathbf{x}$ for an arbitrary scalar γ.

8.14. A solution **x** which has unit length ($\mathbf{x}'\mathbf{x} = 1$) is called the eigenvector associated with the characteristic root λ of **A**. The vector is frequently denoted by **e**.

8.15. A real symmetric matrix has real roots.

8.16. A matrix **A** is called positive-definite (abbreviated p.d.) if the quadratic form $\mathbf{x}'\mathbf{Ax} > 0$ for every $\mathbf{x} \neq \mathbf{0}$.

8.17. A matrix **A** is called positive-semidefinite (abbreviated p.s.d.) if the quadratic form $\mathbf{x'Ax} > 0$ and/or $\mathbf{x'Ax} = 0$ for some $\mathbf{x} \neq \mathbf{0}$.

8.18. A positive-definite real symmetric matrix has only positive characteristic roots.

8.19. If a real symmetric matrix is positive-semidefinite, it has no negative roots. The number of non-zero roots equals the rank of the matrix.

8.20. If all roots of a real symmetric matrix are distinct, the associated eigenvectors are distinct.

8.21. The matrix of eigenvectors

$$\mathbf{E} = [\mathbf{e}_1, \mathbf{e}_2, \dots \mathbf{e}_n]$$

of a real symmetric matrix is (or can be chosen to be) orthogonal.

8.22. $\mathbf{AE} = \mathbf{ED}_\lambda$.

8.23. For a real symmetric matrix, $\mathbf{A} = \mathbf{ED}_\lambda \mathbf{E'}$ (decomposition into matrices of unit rank)

$$\mathbf{E'AE} = \mathbf{D}_\lambda$$

where $\mathbf{D}_\lambda$ denotes the diagonal matrix of characteristic roots ordered in the same way as the eigenvector columns in $\mathbf{E}$.

8.24. If $f(\lambda)$ is a polynomial in λ, then

$$f(\mathbf{A}) = \mathbf{ED}_{f(\lambda)}\mathbf{E}^{-1}$$

where λ are the characteristic roots of **A** and **E** is the matrix of associated eigenvectors. If **A** is symmetric, $\mathbf{E}^{-1} = \mathbf{E'}$.

Example 8.24.1

Consider the matrix in 8.3 (and 6.14).

$$\mathbf{A} = \begin{bmatrix} 25 & 30 & -10 \\ 30 & 40 & -6 \\ -10 & -6 & 17 \end{bmatrix}$$

The characteristic roots were found in Example 8.3.1,

$$\lambda_1 = 65.86108 \qquad \lambda_2 = 15.75339 \qquad \lambda_3 = 0.38553.$$

To find some **x** such that $\mathbf{Ax} = \lambda_1 \mathbf{x}$, we arbitrarily set the first element of **x** equal to 1. Using only the first two rows of **A** we solve the equation system

$$25 + 30x_2 - 10x_3 = 65.86108$$
$$30 + 40x_2 - 6x_3 = 65.86108x_2$$

which yields $x_2 = 1.24294$ and $x_3 = -0.35729$. Substitution of these values into the third equation

$$-10 - 6x_2 + 17x_3 = 65.86108x_3$$

yields zero to five decimal places, indicating the accuracy of the first characteristic root. To reduce to unit length the characteristic vector

$$[1 \qquad 1.24294 \qquad -0.35729]$$

we divide each element by

$$\sqrt{1 + 1.24294^2 + 0.35729^2}$$

and thus obtain the first eigenvector

$$[0.61170 \qquad 0.76030 \qquad -0.21855]$$

This, written as a column vector, is e_1. Repeating the same process for the second and third eigenvectors we obtain

$$e_2 = \begin{bmatrix} -0.08659 \\ 0.33896 \\ 0.93681 \end{bmatrix} \qquad e_3 = \begin{bmatrix} 0.78634 \\ -0.55412 \\ 0.27318 \end{bmatrix}$$

The three vectors can be placed into the eigenvector matrix **E**, which is easily seen to be orthogonal.

9. CONDITIONAL INVERSES

9.1. Any matrix **A** (singular or non-singular, rectangular or square) has some conditional or generalized inverse $\mathbf{A}^{(-1)}$ defined by the relation

$$\mathbf{A}\mathbf{A}^{(-1)}\mathbf{A} = \mathbf{A}.$$

9.2. If (and only if) **A** is square and non-singular, $\mathbf{A}^{(-1)}$ is unique and equals $\mathbf{A}^{-1}$. Otherwise there will be infinitely many matrices $\mathbf{A}^{(-1)}$ which satisfy the defining relation 9.1.

9.3a. If **A** is rectangular ($n \times m$) of rank m, with $m < n$, then $\mathbf{A}^{(-1)}$ is of order ($m \times n$) and $\mathbf{A}^{(-1)}\mathbf{A} = \mathbf{I}(m \times m)$. Then $\mathbf{A}^{(-1)}$ is called an inverse from the left. $\mathbf{A}\mathbf{A}^{(-1)} \neq \mathbf{I}$ in this case.

9.3b. If **A** is rectangular ($n \times m$) of rank n, with $m > n$, then $\mathbf{A}^{(-1)}$ is of order ($m \times n$) and $\mathbf{A}\mathbf{A}^{(-1)} = \mathbf{I}(n \times n)$. Then $\mathbf{A}^{(-1)}$ is called an inverse to the right. In this case,

$$\mathbf{A}^{(-1)}\mathbf{A} \neq \mathbf{I}.$$

9.3c. For a square, singular matrix, $\mathbf{A}\mathbf{A}^{(-1)} \neq \mathbf{I}$ and $\mathbf{A}^{(-1)}\mathbf{A} \neq \mathbf{I}$.

Example 9.3.1

$$\mathbf{A} = \begin{bmatrix} 3 \\ 2 \\ 1 \end{bmatrix}$$

The row vector [1/3　0　0] is an inverse from the left. The row vector

$$[x \quad y \quad (1 - 3x - 2y)]$$

is a conditional inverse of the above matrix **A** for any values of x and y. It is called the generalized inverse of **A**.

Example 9.3.2

$$\mathbf{A} = \begin{bmatrix} 1 & 2 & 3 \\ 2 & 5 & 6 \\ 3 & 7 & 9 \end{bmatrix}$$

A conditional inverse is

$$\mathbf{A}^{(-1)} = \begin{bmatrix} 5 & -2 & 0 \\ -2 & 1 & 0 \\ 0 & 0 & 0 \end{bmatrix}$$

Here it was obtained by inversion of the basis (the 2×2 matrix in the upper left-hand corner) and replacement of the other elements by zeros.

9.4. A square matrix **A** is called idempotent if $\mathbf{AA} = \mathbf{A}^2 = \mathbf{A}$.

9.5. $\mathbf{AA}^{(-1)}$ and $\mathbf{A}^{(-1)}\mathbf{A}$ are idempotent.

9.6. All characteristic roots of idempotent matrices are either zero or one.

9.7. A system of linear equations (*m* equations in *n* unknowns)

$$\mathbf{Ax} = \mathbf{b}$$

is called consistent if there exists some solution **x** which satisfies the equation system.

Example 9.7.1

The system

$$x + y = 2$$
$$2x + 2y = 4 \qquad \text{is consistent.}$$

Example 9.7.2

The system

$$x + y = 2$$
$$2x + 2y = 5 \qquad \text{is inconsistent,}$$

for no pair of values (x, y) will satisfy this system.

9.8. If, in a system of equations (rectangular or square)

$$\mathbf{Ax} = \mathbf{b}$$

$\mathbf{AA}^{(-1)}\mathbf{b} = \mathbf{b}$ for some conditional inverse $\mathbf{A}^{(-1)}$, then $\mathbf{AA}^{(-1)}\mathbf{b} = \mathbf{b}$ for every conditional inverse of **A**, and $\mathbf{Ax} = \mathbf{b}$ is consistent. Conversely, if $\mathbf{AA}^{(-1)}\mathbf{b} \neq \mathbf{b}$ for some conditional inverse $\mathbf{A}^{(-1)}$ then $\mathbf{AA}^{(-1)}\mathbf{b} \neq \mathbf{b}$ for every conditional inverse of **A**, and $\mathbf{Ax} = \mathbf{b}$ is inconsistent.

9.9. If $\mathbf{Ax} = \mathbf{b}$ is consistent, then $\mathbf{x} = \mathbf{A}^{(-1)}\mathbf{b}$ is a solution (generally a different one for each $\mathbf{A}^{(-1)}$).

9.10. Let **y** $(p \times 1)$ be a set of linear functions of the solutions **x** $(n \times 1)$ of a consistent system of equations $\mathbf{Ax} = \mathbf{b}$, given by the relation $\mathbf{y} = \mathbf{Cx}$. Then $\mathbf{y} = \mathbf{Cx}$ is called unique if the same values of **y** will result regardless which solution **x** is used.

Example 9.10.1

$$3x + 4y + 5z = 22$$
$$x + y + z = 6$$

is a consistent system. One solution would be

$$x = 3 \quad y = 2 \quad z = 1$$

Another solution is

$$x = 2 \quad y = 4 \quad z = 0$$

The linear function

$$[7 \quad 9 \quad 11] \begin{bmatrix} x \\ y \\ z \end{bmatrix} = u$$

$(7x + 9y + 11z = u)$ will have the same value (50) regardless which of the two (or any other) solutions is substituted. Thus *u* is unique.

9.11. Let $\mathbf{Ax} = \mathbf{b}$ be a consistent system of equations. For $\mathbf{Cx} = \mathbf{y}$ to be a unique linear combination of the solution $\mathbf{x}$, it is necessary and sufficient that $\mathbf{CA}^{(-1)}\mathbf{A} = \mathbf{C}$. If this relation holds for some $\mathbf{A}^{(-1)}$ it will hold for every conditional inverse of $\mathbf{A}$. If it is violated for some $\mathbf{A}^{(-1)}$ it will be violated for every $\mathbf{A}^{(-1)}$, and $\mathbf{y}$ will be non-unique.

9.12. Let $\mathbf{A}$ be of rank r and select r rows and r columns which form a basis of $\mathbf{A}$. Then a conditional inverse of $\mathbf{A}$ can be obtained as follows: Invert the $(r \times r)$ matrix, place the inverse (without transposing) into the r rows corresponding to the column numbers and the r columns corresponding to the row numbers of the basis, and place zero into all remaining elements. Thus, if $\mathbf{A}$ is of order (5×4) and rank 3, and if rows $1, 2, 4$ and columns $2, 3, 4$ are selected as a basis, $\mathbf{A}^{(-1)}$, of order (4×5) will contain the inverse elements of the basis in rows $2, 3, 4$ and columns $1, 2, 4$, and zeros elsewhere. (See example 9.3.2)

9.13. If $\mathbf{A}$ is a square, singular matrix of order $(n \times n)$ and rank r, let $\mathbf{M}$ be a matrix of order $[n \times (n - r)]$ and $\mathbf{K}$ another matrix of order $[(n - r) \times n]$ chosen in such a way that $\mathbf{A} + \mathbf{MK}$ is non-singular. Then $(\mathbf{A} + \mathbf{MK})^{-1}$ is a conditional inverse of $\mathbf{A}$.

Example 9.13.1

$$\mathbf{A} = \begin{bmatrix} 3 & -1 & -1 & -1 \\ -1 & 3 & -1 & -1 \\ -1 & -1 & 3 & -1 \\ -1 & -1 & -1 & 3 \end{bmatrix}$$

is of order (4×4) and rank 3. Take $\mathbf{M} = \mathbf{j}$ (column vector of ones) and $\mathbf{K} = \mathbf{j}'$ (row vector of ones). Then $\mathbf{A} + \mathbf{MK} = \mathbf{A} + \mathbf{jj}' = 4\mathbf{I}$. Hence $(1/4)\,\mathbf{I}$ is a conditional inverse of $\mathbf{A}$.

9.14. The "Doolittle" method (see 6.14) can be employed to obtain a conditional inverse of a symmetric matrix. If, at any stage, the leading element of the p-row is zero, that cycle is disregarded.

Example 9.14.1
Invert, conditionally, the matrix

$$\mathbf{A} = \begin{bmatrix} 4 & 2 & -2 & 4 \\ 2 & 17 & 11 & 6 \\ -2 & 11 & 10 & 1 \\ 4 & 6 & 1 & 30 \end{bmatrix}$$

4	2	−2	4	1			
1	.5	−.5	1	.25			
	17	11	6	0	1		
	16	12	4	−.5	1		
	1	.75	.25	−.03125	.0625		
		10	1	0	0	1	
		0					
			30	0	0	0	1
			25	−.875	−.25	0	1
			1	−.035	−.01	0	.04

$$\begin{bmatrix} .29625 & -.0225 & 0 & -.035 \\ -.0225 & .065 & 0 & -.01 \\ 0 & 0 & 0 & 0 \\ -.035 & -.01 & 0 & .04 \end{bmatrix} = \mathbf{A}^{(-1)}$$

10.   MATRIX DIFFERENTIATION

10.1a. If the elements of a matrix $\mathbf{Y}$ $(m \times n)$ are functions of a scalar, x, the expression

$$\partial \mathbf{Y}/\partial x$$

denotes a matrix of order $(m \times n)$ with elements $\partial y_{ij}/\partial x$.

10.1b. If the elements of a column (row) vector $\mathbf{y}$ $(\mathbf{y}')$ are functions of a scalar, x, the expression

$$\partial \mathbf{y}/\partial x \quad (\partial \mathbf{y}'/\partial x)$$

denotes a column (row) vector with elements $\partial y_i/\partial x$.

10.2a. If y is a scalar function of $m \times n$ variables, x_{ij}, arranged into a matrix $\mathbf{X}$, the expression

$$\partial y/\partial \mathbf{X}$$

denotes a matrix with elements $\partial y/\partial x_{ij}$.

(Note: Partial differentiation is performed with respect to the element in row i and column j of $\mathbf{X}$. If the same x-variable occurs in another place as, e.g., in a symmetric matrix, differentiation with respect to the distinct (repeated) variable is performed in two stages.)

Example 10.2.1

If $y = \mathbf{j}'\mathbf{Xj}$ (sum of all elements of a square matrix), $\partial y/\partial \mathbf{X}$ is a matrix of ones. If $\mathbf{X}$ is symmetric, one can introduce a new notation $x_{ij} = x_{ji} = z_{ij}$. Then

$$\begin{aligned} \partial y/\partial z_{ij} &= (\partial y/\partial x_{ij})(\partial x_{ij}/\partial z_{ij}) \\ &\quad + (\partial y/\partial x_{ji})(\partial x_{ji}/\partial z_{ij}) \\ &= 1 + 1 = 2 \quad (\text{if } i \neq j) \\ &= 1 \quad\quad\quad\ (\text{if } i = j). \end{aligned}$$

10.2b. If y is a scalar function of n variables, x_i, arranged into a column (row) vector $\mathbf{x}$ $(\mathbf{x}')$, the expression

$$\partial y/\partial \mathbf{x} \quad (\partial y/\partial \mathbf{x}')$$

denotes a column (row) vector with elements $\partial y/\partial x_i$.

10.3. If $\mathbf{y}$ is a column vector with m elements, each a function of n variables, x_i, arranged into a row vector $\mathbf{x}'$, the expression $\partial \mathbf{y}/\partial \mathbf{x}'$ denotes a matrix with m rows and n columns, with elements $\partial y_i/\partial x_j$.

10.4. $\partial \mathbf{Y}/\partial y_{ij} = (\mathbf{EL})_{ij}$ (see definition of $(\mathbf{EL})$ in 3.4).

10.5. $\partial \mathbf{UV}/\partial x = (\partial \mathbf{U}/\partial x)\mathbf{V} + \mathbf{U}(\partial \mathbf{V}/\partial x)$.

10.6. $\partial \mathbf{AY}/\partial x = \mathbf{A}(\partial \mathbf{Y}/\partial x)$ (if elements of $\mathbf{A}$ are not functions of x).

10.7. $\partial \mathbf{Y}'/\partial y_{ij} = (\mathbf{EL})_{ji}$

10.8. $\partial \mathbf{A}'\mathbf{YA}/\partial x = \mathbf{A}'(\partial \mathbf{Y}/\partial x)\mathbf{A}$

10.9. $\partial \mathbf{Y}'\mathbf{AY}/\partial x = (\partial \mathbf{Y}'/\partial x)\mathbf{AY} + \mathbf{Y}'\mathbf{A}(\partial \mathbf{Y}/\partial x)$

10.10. $\partial \mathbf{a}'\mathbf{x}/\partial \mathbf{x} = \mathbf{a}$

10.11. $\partial \mathbf{x}'\mathbf{x}/\partial \mathbf{x} = 2\mathbf{x}$

10.12. $\partial \mathbf{x}'\mathbf{A}\mathbf{x}/\partial \mathbf{x} = \mathbf{A}\mathbf{x} + \mathbf{A}'\mathbf{x}$

10.13. (Chain Rule No. 1) $\quad \partial \mathbf{y}/\partial \mathbf{x}' = (\partial \mathbf{y}/\partial \mathbf{z}')(\partial \mathbf{z}/\partial \mathbf{x}')$

10.14. $\partial \mathbf{A}\mathbf{x}/\partial \mathbf{x}' = \mathbf{A}$

10.15. $\partial tr\mathbf{X}/\partial \mathbf{X} = \mathbf{I}$

10.16. $\partial tr\mathbf{A}\mathbf{X}/\partial \mathbf{X} = \partial tr\mathbf{X}\mathbf{A}/\partial \mathbf{X} = \mathbf{A}'$

10.17. $\partial tr\mathbf{A}\mathbf{X}\mathbf{B}/\partial \mathbf{X} = \mathbf{A}'\mathbf{B}'$

10.18. $\partial tr\mathbf{X}'\mathbf{A}\mathbf{X}/\partial \mathbf{X} = \mathbf{A}\mathbf{X} + \mathbf{A}'\mathbf{X}$

10.19. $\partial \log |\mathbf{X}| /\partial \mathbf{X} = (\mathbf{X}')^{-1}$ $\quad$ (log to base e).

10.20. $\partial \mathbf{Y}^{-1}/\partial x = -\mathbf{Y}^{-1}(\partial \mathbf{Y}/\partial x)\mathbf{Y}^{-1}$

10.21. (Chain Rule No. 2)

$$\partial y/\partial x = tr(\partial y/\partial \mathbf{Z})(\partial \mathbf{Z}'/\partial x)$$

where y and x are scalars. The scalar y is a function of $m \times n$ variables z_{ij}, and each of the z_{ij} is a function of x.

Example 10.21.1

Obtain $\log |\mathbf{R} - \mathbf{F}\mathbf{F}'| /\partial \mathbf{F}$, where $\mathbf{R}$ is symmetric.

By Chain Rule No. 2:

$\partial \log |\mathbf{R} - \mathbf{F}\mathbf{F}'| /\partial f_{ij}$

$\quad = tr[\partial \log |\mathbf{R} - \mathbf{F}\mathbf{F}'| /\partial(\mathbf{R} - \mathbf{F}\mathbf{F}')][\partial(\mathbf{R} - \mathbf{F}\mathbf{F}')/\partial f_{ij}]$ $\quad$ (since $\mathbf{R}$ and $\mathbf{F}\mathbf{F}'$ are symmetric)

$\quad = tr(\mathbf{R} - \mathbf{F}\mathbf{F}')^{-1}[\partial(\mathbf{R} - \mathbf{F}\mathbf{F}')/\partial f_{ij}]$ $\quad$ (by 10.19)

$\quad = tr(\mathbf{R} - \mathbf{F}\mathbf{F}')^{-1}[-(\partial \mathbf{F}/\partial f_{ij})\mathbf{F}' - \mathbf{F}(\partial \mathbf{F}'/\partial f_{ij})]$ $\quad$ (by 10.5)

$\quad = tr(\mathbf{R} - \mathbf{F}\mathbf{F}')^{-1}[-(\mathbf{E}\mathbf{L})_{ij}\mathbf{F}' - \mathbf{F}(\mathbf{E}\mathbf{L})_{ji}]$ $\quad$ (by 10.4 and 10.7)

$\quad = -tr(\mathbf{R} - \mathbf{F}\mathbf{F}')^{-1}(\mathbf{E}\mathbf{L})_{ij}\mathbf{F}' - tr(\mathbf{R} - \mathbf{F}\mathbf{F}')^{-1}\mathbf{F}(\mathbf{E}\mathbf{L})_{ji}$

$\quad = -tr(\mathbf{E}\mathbf{L})_{ij}\mathbf{F}'(\mathbf{R} - \mathbf{F}\mathbf{F}')^{-1} - tr(\mathbf{E}\mathbf{L})_{ji}(\mathbf{R} - \mathbf{F}\mathbf{F}')^{-1}\mathbf{F}$ $\quad$ (by 7.3)

$\quad = -[\mathbf{F}'(\mathbf{R} - \mathbf{F}\mathbf{F}')^{-1}]_{ji} - [(\mathbf{R} - \mathbf{F}\mathbf{F}')^{-1}\mathbf{F}]_{ij},$

where $[\ ]_{ij}$ denotes the (i,j) element of the matrix in brackets (by 7.5),

$\quad = -[(\mathbf{R} - \mathbf{F}\mathbf{F}')^{-1}\mathbf{F}]_{ij} - [(\mathbf{R} - \mathbf{F}\mathbf{F}')^{-1}\mathbf{F}]_{ij}$ $\quad$ (since $\mathbf{R} - \mathbf{F}\mathbf{F}'$ is symmetric)

$\quad = -2[(\mathbf{R} - \mathbf{F}\mathbf{F}')^{-1}\mathbf{F}]_{ij}.$

Hence, by definition 10.2a

$$\partial \log |\mathbf{R} - \mathbf{F}\mathbf{F}'| /\partial \mathbf{F} = -2(\mathbf{R} - \mathbf{F}\mathbf{F}')^{-1}\mathbf{F}.$$

10.22. $|\partial \mathbf{y}/\partial \mathbf{x}'| = J(\mathbf{y};\mathbf{x})$ is called the Jacobian or Functional Determinant used in variable transformation of multiple integrals. Formally, if $\mathbf{y}$ is a column vector with m elements, each a function of m variables x_i arranged into a row vector $\mathbf{x}'$,

$$dx_1 dx_2 \ldots dx_m = |\partial \mathbf{y}/\partial \mathbf{x}'|^{-1} dy_1 dy_2 \ldots dy_m.$$

10.23. For a scalar y (a function of m variables x_i) to attain a stationary value, it is necessary that

$$\partial y/\partial \mathbf{x} = \mathbf{0}.$$

10.24. For a stationary value to be a minimum (maximum) it is necessary that

$$\partial(\partial y/\partial \mathbf{x})/\partial \mathbf{x}' \quad (-\partial(\partial y/\partial \mathbf{x})/\partial \mathbf{x}')$$

be a positive-definite matrix for the value of $\mathbf{x}$ satisfying 10.23.

Example 10.24.1

Find the values of $\boldsymbol{\beta}$ which minimize $u = \mathbf{x}'\mathbf{x}$ (the sum of squares of x_i) where $\mathbf{x} = \mathbf{y} - \mathbf{A}\boldsymbol{\beta}$ (with $\mathbf{y}$ and $\mathbf{A}$ known and fixed).

$$\partial u / \partial \beta' = (\partial u / \partial x')(\partial x / \partial \beta') \quad \text{(by Chain Rule No. 1)}$$
$$= -2x'A \quad \text{(by 10.11 and 10.14)}$$

Hence

$$\partial u / \partial \beta = -2A'x$$
$$= -2A'(y - A\beta).$$

Hence, for a stationary value, by 10.23, it is necessary that

$$A'A\hat{\beta} = A'y$$

where $\hat{\beta}$ denotes the values which make u stationary. Now,

$$\partial(\partial u / \partial \beta) / \partial \beta' = 2\partial(A'A\beta) / \partial \beta' = 2A'A.$$

If A has real elements, and if $A'A$ is non-singular, then it is positive-definite (since, given an arbitrary real $x \neq 0$, $x'A'Ax = z'z$, with $z = Ax$; thus this is a sum of squares). Hence $\hat{\beta}$ minimizes u.

10.25. (Generalized Newton Iteration)

Let x_0' be an initial estimate (m elements) of the roots of the m equations

$$f(x') = 0$$

where the m elements of the column vector f are each functions of $x_1, x_2, \ldots x_m$. Then an improved root is

$$x_1 = x_0 - Q_0^{-1}f(x_0'),$$

where Q_0 is the matrix of derivatives $\partial f / \partial x'$ evaluated at $x = x_0$. The usual procedure consists of evaluating $f(x_0')$, then solving $Q_0 u = f(x_0')$ for u. Then $x_1 = x_0 - u$.

Example 10.25.1

Solve

$$f_1(x, y) = x^3 - x^2 y + y^2 - 3.526 = 0$$
$$f_2(x, y) = x^3 + y^3 - 14.911 = 0$$

$$Q = \begin{bmatrix} 3x^2 - 2xy & 2y - x^2 \\ 3x^2 & 3y^2 \end{bmatrix}$$

Take $x_0 = 1, \quad y_0 = 2$

$$f_1(x_0, y_0) = -0.526$$
$$f_2(x_0, y_0) = -5.911$$

$$Q_0 = \begin{bmatrix} -1 & 3 \\ 3 & 12 \end{bmatrix}$$

$$-u + 3v = -0.526$$
$$3u + 12v = -5.911$$

$$\text{yields } u = -0.55, \quad v = -0.36.$$

Then,

$$x_1 = x_0 - u = 1.55$$
$$y_1 = y_0 - v = 2.36$$
$$f_1(x_1, y_1) = 0.0976$$
$$f_2(x_1, y_1) = 1.9572$$

$$\mathbf{Q}_1 = \begin{bmatrix} -0.1085 & 2.3175 \\ 7.2075 & 16.7088 \end{bmatrix}$$

$$-0.1085u + 2.3175v = 0.0976$$

$$7.2075u + 16.7088v = 1.9572$$

yields $u = 0.157, \quad v = 0.049.$

Then

$$x_2 = x_1 - u = 1.393$$

$$y_2 = y_1 - v = 2.311$$

$$f_1(x_2, y_2) = 0.03337$$

$$f_2(x_2, y_2) = 0.13443$$

$$\mathbf{Q}_2 = \begin{bmatrix} -0.61710 & 2.68155 \\ 5.82135 & 16.02216 \end{bmatrix}$$

$$-0.61710u + 2.68155v = 0.03337$$

$$5.82135u + 16.02216v = 0.13443$$

yields $u = -0.0068, \quad v = 0.0109.$

Then,

$$x_3 = x_2 - u = 1.3998$$

$$y_3 = y_2 - v = 2.3001$$

(The exact roots are $x = 1.4$ and $y = 2.3$).

11. STATISTICAL MATRIX FORMS

11.1. Let E denote the expectation operator, and let **y** be a set of p random variables. Then

$$E(\mathbf{y}) = \boldsymbol{\mu}$$

states that $E(y_i) = \mu_i \quad (i = 1, 2, \ldots p).$

11.2. Let var denote variance. Then

$$\mathrm{var}(\mathbf{y}) = \boldsymbol{\Sigma}$$

denotes a $p \times p$ symmetric matrix whose elements are $\mathrm{cov}(y_i, y_j)$, and whose diagonal elements are $\mathrm{var}(y_i)$, where cov denotes covariance.

11.3. $E(\mathbf{AY} + \mathbf{b}) = \mathbf{A}E(\mathbf{y}) + \mathbf{b} = \mathbf{A}\boldsymbol{\mu} + \mathbf{b}$

11.4. $\mathrm{var}(\mathbf{Ay} + \mathbf{b}) = \mathbf{A}\,\mathrm{var}(\mathbf{y})\,\mathbf{A}' = \mathbf{A}\boldsymbol{\Sigma}\mathbf{A}'$

11.5. $\mathrm{cov}(\mathbf{y}, \mathbf{z}')$ denotes a matrix with elements $\mathrm{cov}(y_i, z_j)$.
   $\mathrm{cov}(\mathbf{z}, \mathbf{y}') = [\mathrm{cov}(\mathbf{y}, \mathbf{z}')]'.$

11.6. $\mathrm{cov}(\mathbf{Ay} + \mathbf{b}, \mathbf{z}'\mathbf{C} + \mathbf{d}') = \mathbf{A}\,\mathrm{cov}(\mathbf{y}, \mathbf{z}')\,\mathbf{C}$

11.7. $\mathrm{var}(\mathbf{y}) = E(\mathbf{yy}') - E(\mathbf{y})E(\mathbf{y}')$

11.8. $\mathrm{cov}(\mathbf{y}, \mathbf{z}') = E(\mathbf{yz}') - E(\mathbf{y})E(\mathbf{z}')$

11.9. (Expected "sum of squares")

$$E(\mathbf{y}'\mathbf{Q}\mathbf{y}) = tr[\mathbf{Q}\,\mathrm{var}\,(\mathbf{y})] + E(\mathbf{y}')\mathbf{Q}E(\mathbf{y}).$$

11.10. If a matrix $\mathbf{Q}$ is symmetric and positive-definite, one can find a lower triangular matrix $\mathbf{T}$ (with positive diagonal terms, for uniqueness) such that $\mathbf{TT}' = \mathbf{Q}$. The matrices $\mathbf{T}$ and $\mathbf{T}^{-1}$ can be obtained from the Doolittle pattern (6.14) (Gauss elimination or square-root method) as follows: In each cycle, divide the p-row (left and right hand side) by $\sqrt{p_{ii}}$ (instead of p_{ii} for the u-row). Thus obtain rows designated as t-rows. The left-hand side (Arabic subscripts) is $\mathbf{T}'$, and the right-hand side (Roman subscripts) is $\mathbf{T}^{-1}$.

11.11. If a coordinate system $\mathbf{x}$ is oblique, and if the cosines between reference vectors (scalar products of basis vectors of unit length) are stated in a symmetric matrix $\mathbf{Q}$, then $\mathbf{T}^{-1}\mathbf{x} = \mathbf{y}$ is an orthogonal system, where $\mathbf{T}$ is obtained from $\mathbf{Q}$ by 11.10.

11.12. The likelihood function of a sample of size n from a multivariate normal distribution (p responses), with common variance-covariance matrix $\Sigma(p \times p)$, and with means or main effects replaced by maximum-likelihood or least-squares estimates, can be written as

$$\log L = -\frac{np}{2} \log 2\pi - \frac{n}{2} \log |\Sigma| - \frac{n}{2} tr\, \Sigma^{-1} \mathbf{S}$$

where $\Sigma(p \times p)$ is the common variance-covariance matrix, and $\mathbf{S}$ is its maximum-likelihood estimate (matrix of sums of squares and products due to error, divided by sample size n).

11.13. If Σ has a structure under a model or null hypothesis, and if elements of Σ are to be estimated, by maximum-likelihood, two cases can be distinguished: (11.14) Σ^{-1} has the same structure (intraclass correlation, mixed model, compound symmetry, factor analysis). (11.15) Σ^{-1} has a different structure (autocorrelation, Simplex structure).

11.14. If the structure of Σ and Σ^{-1} is identical, and if u and v are elements (or functions of elements) of Σ^{-1} then estimates of Σ can be obtained from the relations (usually requiring Newton iteration, see 10.25):

$$\partial \log L / \partial u = \frac{n}{2} tr\, \mathbf{A}(\Sigma - \mathbf{S})$$

where $\mathbf{A} = \partial \Sigma^{-1}/\partial u$, is frequently an elementary matrix (see 3.4 and, especially, the rules 7.5 and 7.6).

$$\partial^2 \log L / \partial u \partial v = \frac{n}{2} tr(\partial \mathbf{A}/\partial v)(\Sigma - \mathbf{S}) + \frac{n}{2} tr\, \mathbf{A}\Sigma^{-1}\mathbf{B}\Sigma^{-1}$$

where $\mathbf{B} = \partial \Sigma^{-1}/\partial v$. These rules are useful to obtain Newton iterations and asymptotic variance-covariance matrices of the estimates.

11.15. If the structures of Σ and Σ^{-1} are different, then an estimate of Σ can be obtained from the relations

$$\partial \log L / \partial x = -\frac{n}{2} tr\, \mathbf{A}(\Sigma^{-1} - \mathbf{Q}),$$

where

$$\mathbf{Q} = \Sigma^{-1}\mathbf{S}\Sigma^{-1}$$

and

$$\mathbf{A} = \partial \Sigma / \partial x \qquad \text{(see comments in 11.14)}.$$

$$\partial^2 \log L / \partial x \partial y = -\frac{n}{2} tr(\partial \mathbf{A}/\partial y)(\Sigma^{-1} - \mathbf{Q}) + \frac{n}{2} tr\, \mathbf{A}\Sigma^{-1}\mathbf{B}(\Sigma^{-1} - \mathbf{Q}) - \frac{n}{2} tr\, \mathbf{A}\mathbf{Q}\mathbf{B}\Sigma^{-1},$$

where

$$\mathbf{B} = \partial \Sigma / \partial y$$

x and y are elements (or functions of elements) of Σ. The comments of 11.14 apply, but the iterative procedure is considerably more complex.

Suggestions for further reading:

A. S. Householder, *The Theory of Matrices in Numerical Analysis*, Blaisdell, 1964.

S. R. Searle, *Matrix Algebra for the Biological Sciences*, Wiley, 1966.

P. H. Schonemann, Matrix differentiation of traces and determinants, *Psychometrika*, 1966.

III. COMBINATORIAL ANALYSIS

POWERS OF NUMBERS

The larger numbers are expressed exponentially to at least seven significant figures. The approximate value written as a whole number may be obtained by shifting the decimal point to the right by the number of places indicated in the exponent of 10 shown at the head of each group of values. For example: the approximate value of 33^8 is found in the table as 14.064086×10^{11}. Written as a whole number it is 1 406 408 600 000.

n	n^4	n^5	n^6	n^7	n^8	n^9
1	1	1	1	1	1	1
2	16	32	64	128	256	512
3	81	243	729	2187	6561	19683
4	256	1024	4096	16384	65536	262144
5	625	3125	15625	78125	390625	1953125
6	1296	7776	46656	279936	1679616	10077696
7	2401	16807	117649	823543	5764801	40353607
8	4096	32768	262144	2097152	16777216	134217728
9	6561	59049	531441	4782969	43046721	387420489
					$\times 10^8$	$\times 10^9$
10	10000	100000	1000000	10000000	1.000000	1.000000
11	14641	161051	1771561	19487171	2.143589	2.357948
12	20736	248832	2985984	35831808	4.299817	5.159780
13	28561	371293	4826809	62748517	8.157307	10.604499
14	38416	537824	7529536	105413504	14.757891	20.661047
15	50625	759375	11390625	170859375	25.628906	38.443359
16	65536	1048576	16777216	268435456	42.949673	68.719477
17	83521	1419857	24137569	410338673	69.757574	118.587876
18	104976	1889568	34012224	612220032	110.199606	198.359290
19	130321	2476099	47045881	893871739	169.835630	322.687698
				$\times 10^9$	$\times 10^{10}$	$\times 10^{11}$
20	160000	3200000	64000000	1.280000	2.560000	5.120000
21	194481	4084101	85766121	1.801080	3.782286	7.942800
22	234256	5153632	113379904	2.494358	5.487587	12.072692
23	279841	6436343	148035889	3.404825	7.831099	18.011527
24	331776	7962624	191102976	4.586471	11.007531	26.418075
25	390625	9765625	244140625	6.103516	15.258789	38.146973
26	456976	11881376	308915776	8.031810	20.882706	54.295037
27	531441	14348907	387420489	10.460353	28.242954	76.255975
28	614656	17210368	481890304	13.492929	37.780200	105.784560
29	707281	20511149	594823321	17.249876	50.024641	145.071460
			$\times 10^8$	$\times 10^{10}$	$\times 10^{11}$	$\times 10^{13}$
30	810000	24300000	7.290000	2.187000	6.561000	1.968300
31	923521	28629151	8.875037	2.751261	8.528910	2.643962
32	1048576	33554432	10.737418	3.435974	10.995116	3.518437
33	1185921	39135393	12.914680	4.261844	14.064086	4.641148
34	1336336	45435424	15.448044	5.252335	17.857939	6.071699
35	1500625	52521875	18.382656	6.433930	22.518754	7.881564
36	1679616	60466176	21.767823	7.836416	28.211099	10.155996
37	1874161	69343957	25.657264	9.493188	35.124795	12.996174
38	2085136	79235168	30.109364	11.441558	43.477921	16.521610
39	2313441	90224199	35.187438	13.723101	53.520093	20.872836
			$\times 10^9$	$\times 10^{10}$	$\times 10^{12}$	$\times 10^{14}$
40	2560000	102400000	4.096000	16.384000	6.553600	2.621440
41	2825761	115856201	4.750104	19.475427	7.984925	3.273819
42	3111696	130691232	5.489032	23.053933	9.682652	4.066714
43	3418801	147008443	6.321363	27.181861	11.688200	5.025926
44	3748096	164916224	7.256314	31.927781	14.048224	6.181218
45	4100625	184528125	8.303766	37.366945	16.815125	7.566806
46	4477456	205962976	9.474297	43.581766	20.047612	9.221902
47	4879681	229345007	10.779215	50.662312	23.811287	11.191305
48	5308416	254803968	12.230590	58.706834	28.179280	13.526055
49	5764801	282475249	13.841287	67.822307	33.232931	16.284136
50	6250000	312500000	15.625000	78.125000	39.062500	19.531250

POWERS OF NUMBERS (Continued)

n	n^4	n^5	n^6	n^7	n^8	n^9
			$\times 10^9$	$\times 10^{11}$	$\times 10^{13}$	$\times 10^{14}$
50	6250000	312500000	15.625000	7.812500	3.906250	19.531250
51	6765201	345025251	17.596288	8.974107	4.576794	23.341652
52	7311616	380204032	19.770610	10.280717	5.345973	27.799059
53	7890481	418195493	22.164361	11.747111	6.225969	32.997636
54	8503056	459165024	24.794911	13.389252	7.230196	39.043059
55	9150625	503284375	27.680641	15.224352	8.373394	46.053666
56	9834496	550731776	30.840979	17.270948	9.671731	54.161694
57	10556001	601692057	34.296447	19.548975	11.142916	63.514620
58	11316496	656356768	38.068693	22.079842	12.806308	74.276587
59	12117361	714924299	42.180534	24.886515	14.683044	86.629958
		$\times 10^8$	$\times 10^{10}$	$\times 10^{11}$	$\times 10^{13}$	$\times 10^{16}$
60	12960000	7.776000	4.665600	27.993600	16.796160	1.007770
61	13845841	8.445963	5.152037	31.427428	19.170731	1.169415
62	14776336	9.161328	5.680024	35.216146	21.834011	1.353709
63	15752961	9.924365	6.252350	39.389806	24.815578	1.563381
64	16777216	10.737418	6.871948	43.980465	28.147498	1.801440
65	17850625	11.602906	7.541889	49.022279	31.864481	2.071191
66	18974736	12.523326	8.265395	54.551607	36.004061	2.376268
67	20151121	13.501251	9.045838	60.607116	40.606768	2.720653
68	21381376	14.539336	9.886748	67.229888	45.716324	3.108710
69	22667121	15.640313	10.791816	74.463533	51.379837	3.545209
		$\times 10^8$	$\times 10^{10}$	$\times 10^{12}$	$\times 10^{14}$	$\times 10^{16}$
70	24010000	16.807000	11.764900	8.235430	5.764801	4.035361
71	25411681	18.042294	12.810028	9.095120	6.457535	4.584850
72	26873856	19.349176	13.931407	10.030613	7.222041	5.199870
73	28398241	20.730716	15.133423	11.047399	8.064601	5.887159
74	29986576	22.190066	16.420649	12.151280	8.991947	6.654041
75	31640625	23.730469	17.797852	13.348389	10.011292	7.508469
76	33362176	25.355254	19.269993	14.645195	11.130348	8.459064
77	35153041	27.067842	20.842238	16.048523	12.357363	9.515169
78	37015056	28.871744	22.519960	17.565569	13.701144	10.686892
79	38950081	30.770564	24.308746	19.203909	15.171088	11.985160
		$\times 10^8$	$\times 10^{10}$	$\times 10^{12}$	$\times 10^{14}$	$\times 10^{16}$
80	40960000	32.768000	26.214400	20.971520	16.777216	13.421773
81	43046721	34.867844	28.242954	22.876792	18.530202	15.009464
82	45212176	37.073984	30.400667	24.928547	20.441409	16.761955
83	47458321	39.390406	32.694037	27.136051	22.522922	18.694026
84	49787136	41.821194	35.129803	29.509035	24.787589	20.821575
85	52200625	44.370531	37.714952	32.057709	27.249053	23.161695
86	54700816	47.042702	40.456724	34.792782	29.921793	25.732742
87	57289761	49.842092	43.362620	37.725479	32.821167	28.554415
88	59969536	52.773192	46.440409	40.867560	35.963452	31.647838
89	62742241	55.840594	49.698129	44.231335	39.365888	35.035640
		$\times 10^9$	$\times 10^{11}$	$\times 10^{13}$	$\times 10^{15}$	$\times 10^{17}$
90	65610000	5.904900	5.314410	4.782969	4.304672	3.874205
91	68574961	6.240321	5.678693	5.167610	4.702525	4.279298
92	71639296	6.590815	6.063550	5.578466	5.132189	4.721614
93	74805201	6.956884	6.469902	6.017009	5.595818	5.204111
94	78074896	7.339040	6.898698	6.484776	6.095689	5.729948
95	81450625	7.737809	7.350919	6.983373	6.634204	6.302494
96	84934656	8.153727	7.827578	7.514475	7.213896	6.925340
97	88529281	8.587340	8.329720	8.079828	7.837434	7.602311
98	92236816	9.039208	8.858424	8.681255	8.507630	8.337478
99	96059601	9.509900	9.414801	9.320653	9.227447	9.135172
100	100000000	10.000000	10.000000	10.000000	10.000000	10.000000

POSITIVE POWERS OF TWO

n	2ⁿ			
1	2			
2	4			
3	8			
4	16			
5	32			
6	64			
7	128			
8	256			
9	512			
10	1024			
11	2048			
12	4096			
13	8192			
14	16384			
15	32768			
16	65536			
17	13107	2		
18	26214	4		
19	52428	8		
20	10485	76		
21	20971	52		
22	41943	04		
23	83886	08		
24	16777	216		
25	33554	432		
26	67108	864		
27	13421	7728		
28	26843	5456		
29	53687	0912		
30	10737	41824		
31	21474	83648		
32	42949	67296		
33	85899	34592		
34	17179	86918	4	
35	34359	73836	8	
36	68719	47673	6	
37	13743	89534	72	
38	27487	79069	44	
39	54975	58138	88	
40	10995	11627	776	
41	21990	23255	552	
42	43980	46511	104	
43	87960	93022	208	
44	17592	18604	4416	
45	35184	37208	8832	
46	70368	74417	7664	
47	14073	74883	55328	
48	28147	49767	10656	
49	56294	99534	21312	
50	11258	99906	84262	4

n	2ⁿ						
51	22517	99813	68524	8			
52	45035	99627	37049	6			
53	90071	99254	74099	2			
54	18014	39850	94819	84			
55	36028	79701	89639	68			
56	72057	59403	79279	36			
57	14411	51880	75855	872			
58	28823	03761	51711	744			
59	57646	07523	03423	488			
60	11529	21504	60684	6976			
61	23058	43009	21369	3952			
62	46116	86018	42738	7904			
63	92233	72036	85477	5808			
64	18446	74407	37095	51616			
65	36893	48814	74191	03232			
66	73786	97629	48382	06464			
67	14757	39525	89676	41292	8		
68	29514	79051	79352	82585	6		
69	59029	58103	58705	65171	2		
70	11805	91620	71741	13034	24		
71	23611	83241	43482	26068	48		
72	47223	66482	86964	52136	96		
73	94447	32965	73929	04273	92		
74	18889	46593	14785	80854	784		
75	37778	93186	29571	61709	568		
76	75557	86372	59143	23419	136		
77	15111	57274	51828	64683	8272		
78	30223	14549	03657	29367	6544		
79	60446	29098	07314	58735	3088		
80	12089	25819	61462	91747	06176		
81	24178	51639	22925	83494	12352		
82	48357	03278	45851	66988	24704		
83	96714	06556	91703	33976	49408		
84	19342	81311	38340	66795	29881	6	
85	38685	62622	76681	33590	59763	2	
86	77371	25245	53362	67181	19526	4	
87	15474	25049	10672	53436	23905	28	
88	30948	50098	21345	06872	47810	56	
89	61897	00196	42690	13744	95621	12	
90	12379	40039	28538	02748	99124	224	
91	24758	80078	57076	05497	98248	448	
92	49517	60157	14152	10995	96496	896	
93	99035	20314	28304	21991	92993	792	
94	19807	04062	85660	84398	38598	7584	
95	39614	08125	71321	68796	77197	5168	
96	79228	16251	42643	37593	54395	0336	
97	15845	63250	28528	67518	70879	00672	
98	31691	26500	57057	35037	41758	01344	
99	63382	53001	14114	70074	83516	02688	
100	12676	50600	22822	94014	96703	20537	6
101	25353	01200	45645	88029	93406	41075	2

NEGATIVE POWERS OF TWO

n	2^{-n}								
0	1.0								
1	0.5								
2	0.25								
3	0.125								
4	0.0625								
5	0.03125								
6	0.01562	5							
7	0.00781	25							
8	0.00390	625							
9	0.00195	3125							
10	0.00097	65625							
11	0.00048	82812	5						
12	0.00024	41406	25						
13	0.00012	20703	125						
14	0.00006	10351	5625						
15	0.00003	05175	78125						
16	0.00001	52587	89062	5					
17	0.00000	76293	94531	25					
18	0.00000	38146	97265	625					
19	0.00000	19073	48632	8125					
20	0.00000	09536	74316	40625					
21	0.00000	04768	37158	20312	5				
22	0.00000	02384	18579	10156	25				
23	0.00000	01192	09289	55078	125				
24	0.00000	00596	04644	77539	0625				
25	0.00000	00298	02322	38769	53125				
26	0.00000	00149	01161	19384	76562	5			
27	0.00000	00074	50580	59692	38281	25			
28	0.00000	00037	25290	29846	19140	625			
29	0.00000	00018	62645	14923	09570	3125			
30	0.00000	00009	31322	57461	54785	15625			
31	0.00000	00004	65661	28730	77392	57812	5		
32	0.00000	00002	32830	64365	38696	28906	25		
33	0.00000	00001	16415	32182	69348	14453	125		
34	0.00000	00000	58207	66091	34674	07226	5625		
35	0.00000	00000	29103	83045	67337	03613	28125		
36	0.00000	00000	14551	91522	83668	51806	64062	5	
37	0.00000	00000	07275	95761	41834	25903	32031	25	
38	0.00000	00000	03637	97880	70917	12951	66015	625	
39	0.00000	00000	01818	98940	35458	56475	83007	8125	
40	0.00000	00000	00909	49470	17729	28237	91503	90625	
41	0.00000	00000	00454	74735	08864	64118	95751	95312	5
42	0.00000	00000	00227	37367	54432	32059	47875	97656	25
43	0.00000	00000	00113	68683	77216	16029	73937	98828	125
44	0.00000	00000	00056	84341	88608	08014	86968	99414	0625
45	0.00000	00000	00028	42170	94304	04007	43484	49707	03125
46	0.00000	00000	00014	21085	47152	02003	71742	24853	51562 5
47	0.00000	00000	00007	10542	73576	01001	85871	12426	75781 25
48	0.00000	00000	00003	55271	36788	00500	92935	56213	37890 625
49	0.00000	00000	00001	77635	68394	00250	46467	78106	68945 3125
50	0.00000	00000	00000	88817	84197	00125	23233	89053	34472 65625

SUMS OF POWERS OF INTEGERS, $\sum_{k=1}^{n} k^m$

$(m = 1, 2, 3, 4); 1 \leq n \leq 40$

n	Σk	Σk^2	Σk^3	Σk^4
1	1	1	1	1
2	3	5	9	17
3	6	14	36	98
4	10	30	100	354
5	15	55	225	979
6	21	91	441	2275
7	28	140	784	4676
8	36	204	1296	8772
9	45	285	2025	15333
10	55	385	3025	25333
11	66	506	4356	39974
12	78	650	6084	60710
13	91	819	8281	89271
14	105	1015	11025	127687
15	120	1240	14400	178312
16	136	1496	18496	243848
17	153	1785	23409	327369
18	171	2109	29241	432345
19	190	2470	36100	562666
20	210	2870	44100	722666
21	231	3311	53361	917147
22	253	3795	64009	1151403
23	276	4324	76176	1431244
24	300	4900	90000	1763020
25	325	5525	105625	2153645
26	351	6201	123201	2610621
27	378	6930	142884	3142062
28	406	7714	164836	3756718
29	435	8555	189225	4463999
30	465	9455	216225	5273999
31	496	10416	246016	6197520
32	528	11440	278784	7246096
33	561	12529	314721	8432017
34	595	13685	354025	9768353
35	630	14910	396900	11268978
36	666	16206	443556	12948594
37	703	17575	494209	14822755
38	741	19019	549081	16907891
39	780	20540	608400	19221332
40	820	22140	672400	21781332

SUMS OF POWERS OF THE FIRST n INTEGERS*

$$\sum_{k=1}^{n} k = 1 + 2 + 3 + \cdots + n = \frac{n(n+1)}{2}$$

$$\sum_{k=1}^{n} k^2 = 1^2 + 2^2 + 3^2 + \cdots + n^2 = \frac{n(n+1)(2n+1)}{6}$$

$$\sum_{k=1}^{n} k^3 = \frac{n^2(n+1)^2}{4}$$

$$\sum_{k=1}^{n} k^4 = \frac{n}{30}(n+1)(2n+1)(3n^2+3n-1).$$

$$\sum_{k=1}^{n} k^5 = \frac{n^2}{12}(n+1)^2(2n^2+2n-1).$$

$$\sum_{k=1}^{n} k^6 = \frac{n}{42}(n+1)(2n+1)(3n^4+6n^3-3n+1).$$

$$\sum_{k=1}^{n} k^7 = \frac{n^2}{24}(n+1)^2(3n^4+6n^3-n^2-4n+2).$$

$$\sum_{k=1}^{n} k^8 = \frac{n}{90}(n+1)(2n+1)(5n^6+15n^5+5n^4-15n^3-n^2+9n-3).$$

$$\sum_{k=1}^{n} k^9 = \frac{n^2}{20}(n+1)^2(2n^6+6n^5+n^4-8n^3+n^2+6n-3).$$

$$\sum_{k=1}^{n} k^{10} = \frac{n}{66}(n+1)(2n+1)(3n^8+12n^7+8n^6-18n^5$$
$$-10n^4+24n^3+2n^2-15n+5).$$

Note that

$\sum_{k=1}^{n} k^p = 1^p + 2^p + 3^p + \cdots + n^p$ is a function of n which can be conveniently generated

by use of the *following proposition*

If
$$\sum_{k=1}^{n} k^p = a_1 n^{p+1} + a_2 n^p + a_3 n^{p-1} + \cdots + a_{p+1} n$$

then

$$\sum_{k=1}^{n} k^{p+1} = \frac{p+1}{p+2}a_1 n^{p+2} + \frac{p+1}{p+1}a_2 n^{p+1} + \frac{p+1}{p}a_3 n^p$$
$$+ \cdots + \frac{p+1}{2}a_{p+1}n^2 + \left[1 - (p+1)\sum_{k=1}^{p+1}\frac{a_k}{(p+3-k)}\right]n$$

Example *Since* $\sum\limits_{k=1}^{n} k = \frac{1}{2}n^2 + \frac{1}{2}n$, then

$$\sum_{k=1}^{n} k^2 = \frac{1}{3}n^3 + \frac{1}{2}n^2 + \frac{1}{6}n \text{ and from this result}$$

$$\sum_{k=1}^{n} k^3 = \frac{1}{4}n^4 + \frac{1}{2}n^3 + \frac{1}{4}n^2 \quad \text{etc.}$$

This proposition is extracted from a paper written by Michael A. Budin and Arnold J. Cantor entitled "Simplified Computation of Sums of Powers of Integers."

*SUMS OF RECIPROCAL POWERS

n	$\zeta(n) = \displaystyle\sum_{k=1}^{\infty} k^{-n}$				$\displaystyle\sum_{k=1}^{\infty} (-1)^{k-1} k^{-n}$			
1	∞				0.69314	71805	59945	30942
2	1.64493	40668	48226	43637	0.82246	70334	24113	21824
3	1.20205	69031	59594	28540	0.90154	26773	69695	71405
4	1.08232	32337	11138	19152	0.94703	28294	97245	91758
5	1.03692	77551	43369	92633	0.97211	97704	46909	30594
6	1.01734	30619	84449	13971	0.98555	10912	97435	10410
7	1.00834	92773	81922	82684	0.99259	38199	22830	28267
8	1.00407	73561	97944	33938	0.99623	30018	52647	89923
9	1.00200	83928	26082	21442	0.99809	42975	41605	33077
10	1.00099	45751	27818	08534	0.99903	95075	98271	56564
11	1.00049	41886	04119	46456	0.99951	71434	98060	75414
12	1.00024	60865	53308	04830	0.99975	76851	43858	19085
13	1.00012	27133	47578	48915	0.99987	85427	63265	11549
14	1.00006	12481	35058	70483	0.99993	91703	45979	71817
15	1.00003	05882	36307	02049	0.99996	95512	13099	23808
16	1.00001	52822	59408	65187	0.99998	47642	14906	10644
17	1.00000	76371	97637	89976	0.99999	23782	92041	01198
18	1.00000	38172	93264	99984	0.99999	61878	69610	11348
19	1.00000	19082	12716	55394	0.99999	80935	08171	67511
20	1.00000	09539	62033	87280	0.99999	90466	11581	52212
21	1.00000	04769	32986	78781	0.99999	95232	58215	54282
22	1.00000	02384	50502	72773	0.99999	97616	13230	82255
23	1.00000	01192	19925	96531	0.99999	98808	01318	43950
24	1.00000	00596	08189	05126	0.99999	99403	98892	39463
25	1.00000	00298	03503	51465	0.99999	99701	98856	96283
26	1.00000	00149	01554	82837	0.99999	99850	99231	99657
27	1.00000	00074	50711	78984	0.99999	99925	49550	48496
28	1.00000	00037	25334	02479	0.99999	99962	74753	40011
29	1.00000	00018	62659	72351	0.99999	99981	37369	41811
30	1.00000	00009	31327	43242	0.99999	99990	68682	28145
31	1.00000	00004	65662	90650	0.99999	99995	34340	33145
32	1.00000	00002	32831	18337	0.99999	99997	67169	89595
33	1.00000	00001	16415	50173	0.99999	99998	83584	85805
34	1.00000	00000	58207	72088	0.99999	99999	41792	39905
35	1.00000	00000	29103	85044	0.99999	99999	70896	18953
36	1.00000	00000	14551	92189	0.99999	99999	85448	09143
37	1.00000	00000	07275	95984	0.99999	99999	92724	04461
38	1.00000	00000	03637	97955	0.99999	99999	96362	02193
39	1.00000	00000	01818	98965	0.99999	99999	98181	01084
40	1.00000	00000	00909	49478	0.99999	99999	99090	50538
41	1.00000	00000	00454	74738	0.99999	99999	99545	25268
42	1.00000	00000	00227	37368	0.99999	99999	99772	62633

For $n > 42$, $\displaystyle\sum_{k=1}^{\infty} k^{-(n+1)} = \frac{1}{2}\left[1 + \sum_{k=1}^{\infty} k^{-n}\right]$, $\displaystyle\sum_{k=1}^{\infty} (-1)^{k-1} k^{-(n+1)} = \frac{1}{2}\left[1 + \sum_{k=1}^{\infty} (-1)^{k-1} k^{-n}\right]$

*Note: By definition, Riemann's Zeta Function is

$$\zeta(p) = \text{Zeta}(p) = 1 + \frac{1}{2^p} + \frac{1}{3^p} + \frac{1}{4^p} + \cdots$$

*SUMS OF RECIPROCAL POWERS

n	$\sum_{k=0}^{\infty} (2k + 1)^{-n}$				$\sum_{k=0}^{\infty} (-1)^k (2k + 1)^{-n}$			
1		∞			0.78539	81633	97448	310
2	1.23370	05501	36169	82735	0.91596	55941	77219	015
3	1.05179	97902	64644	99972	0.96894	61462	59369	380
4	1.01467	80316	04192	05455	0.98894	45517	41105	336
5	1.00452	37627	95139	61613	0.99615	78280	77088	064
6	1.00144	70766	40942	12191	0.99868	52222	18438	135
7	1.00047	15486	52376	55476	0.99955	45078	90539	909
8	1.00015	51790	25296	11930	0.99984	99902	46829	657
9	1.00005	13451	83843	77259	0.99994	96841	87220	090
10	1.00001	70413	63044	82549	0.99998	31640	26196	877
11	1.00000	56660	51090	10935	0.99999	43749	73823	699
12	1.00000	18858	48583	11958	0.99999	81223	50587	882
13	1.00000	06280	55421	80232	0.99999	93735	83771	841
14	1.00000	02092	40519	21150	0.99999	97910	87248	735
15	1.00000	00697	24703	12929	0.99999	99303	40842	624
16	1.00000	00232	37157	37916	0.99999	99767	75950	903
17	1.00000	00077	44839	45587	0.99999	99922	57782	104
18	1.00000	00025	81437	55666	0.99999	99974	19086	745
19	1.00000	00008	60444	11452	0.99999	99991	39660	745
20	1.00000	00002	86807	69746	0.99999	99997	13213	274
21	1.00000	00000	95601	16531	0.99999	99999	04403	029
22	1.00000	00000	31866	77514	0.99999	99999	68134	064
23	1.00000	00000	10622	20241	0.99999	99999	89377	965
24	1.00000	00000	03540	72294	0.99999	99999	96459	311
25	1.00000	00000	01180	23874	0.99999	99999	98819	768
26	1.00000	00000	00393	41247	0.99999	99999	99606	589
27	1.00000	00000	00131	13740	0.99999	99999	99868	863
28	1.00000	00000	00043	71245	0.99999	99999	99956	288
29	1.00000	00000	00014	57081	0.99999	99999	99985	429
30	1.00000	00000	00004	85694	0.99999	99999	99995	143
31	1.00000	00000	00001	61898	0.99999	99999	99998	381
32	1.00000	00000	00000	53966	0.99999	99999	99999	460
33	1.00000	00000	00000	17989	0.99999	99999	99999	820
34	1.00000	00000	00000	05996	0.99999	99999	99999	940
35	1.00000	00000	00000	01999	0.99999	99999	99999	980
36	1.00000	00000	00000	00666	0.99999	99999	99999	993
37	1.00000	00000	00000	00222	0.99999	99999	99999	998
38	1.00000	00000	00000	00074	0.99999	99999	99999	999
39	1.00000	00000	00000	00025				
40	1.00000	00000	00000	00008				
41	1.00000	00000	00000	00003				
42	1.00000	00000	00000	00001				

*This table is related to Riemann's Zeta Function, where

$$1 + \frac{1}{2^p} + \frac{1}{3^p} + \frac{1}{4^p} + \cdots = \zeta(p) = \text{Zeta } (p)$$

Factorials, Exact Values

n	$n!$
0	1 (by definition)
1	1
2	2
3	6
4	24
5	120
6	720
7	5040
8	40,320
9	362,880
10	3,628,800
11	39,916,800
12	479,001,600
13	6,227,020,800
14	87,178,291,200
15	1,307,674,368,000
16	20,922,789,888,000
17	355,687,428,096,000
18	6,402,373,705,728,000
19	121,645,100,408,832,000
20	2,432,902,008,176,640,000
21	51,090,942,171,709,440,000
22	1,124,000,727,777,607,680,000
23	25,852,016,738,884,976,640,000
24	620,448,401,733,239,439,360,000
25	15,511,210,043,330,985,984,000,000
26	403,291,461,126,605,635,584,000,000
27	10,888,869,450,418,352,160,768,000,000
28	304,888,344,611,713,860,501,504,000,000
29	8,841,761,993,739,701,954,543,616,000,000
30	265,252,859,812,191,058,636,308,480,000,000
31	8.22284×10^{33}
32	2.63131×10^{35}
33	8.68332×10^{36}
34	2.95233×10^{38}
35	1.03331×10^{40}
36	3.71993×10^{41}
37	1.37638×10^{43}
38	5.23023×10^{44}
39	2.03979×10^{46}

$\underline{n} = n! = e^{-n} n^n \sqrt{2\pi n}$. approximately, known as Stirling's formula

$\log_e n! = n \log_e n - n$, approximately.

FACTORIALS AND THEIR COMMON LOGARITHMS

This table presents values of $n! = n(n-1)(n-2) \ldots 2 \cdot 1$ and its logarithm for numbers from 1 to 100. The values of $n!$ are expressed exponentially to 5 significant figures.

n	$n!$	$\log n!$	n	$n!$	$\log n!$
			50	3.0414×10^{64}	64.48307
1	1,0000	0.00000	51	1.5511×10^{66}	66.19065
2	2,0000	0.30103	52	8.0658×10^{67}	67.90665
3	6,0000	0.77815	53	4.2749×10^{69}	69.63092
4	2.4000×10	1.38021	54	2.3084×10^{71}	71.36332
5	1.2000×10^{2}	2.07918	55	1.2696×10^{73}	73.10368
6	7.2000×10^{2}	2.85733	56	7.1100×10^{74}	74.85187
7	5.0400×10^{3}	3.70243	57	4.0527×10^{76}	76.60774
8	4.0320×10^{4}	4.60552	58	2.3506×10^{78}	78.37117
9	3.6288×10^{5}	5.55976	59	1.3868×10^{80}	80.14202
10	3.6288×10^{6}	6.55976	60	8.3210×10^{81}	81.92017
11	3.9917×10^{7}	7.60116	61	5.0758×10^{83}	83.70550
12	4.7900×10^{8}	8.68034	62	3.1470×10^{85}	85.49790
13	6.2270×10^{9}	9.79428	63	1.9826×10^{87}	87.29724
14	8.7178×10^{10}	10.94041	64	1.2689×10^{89}	89.10342
15	1.3077×10^{12}	12.11650	65	8.2477×10^{90}	90.91633
16	2.0923×10^{13}	13.32062	66	5.4434×10^{92}	92.73587
17	3.5569×10^{14}	14.55107	67	3.6471×10^{94}	94.56195
18	6.4024×10^{15}	15.80634	68	2.4800×10^{96}	96.39446
19	1.2165×10^{17}	17.08509	69	1.7112×10^{98}	98.23331
20	2.4329×10^{18}	18.38612	70	1.1979×10^{100}	100.07841
21	5.1091×10^{19}	19.70834	71	8.5048×10^{101}	101.92966
22	1.1240×10^{21}	21.05077	72	6.1234×10^{103}	103.78700
23	2.5852×10^{22}	22.41249	73	4.4701×10^{105}	105.65032
24	6.2045×10^{23}	23.79271	74	3.3079×10^{107}	107.51955
25	1.5511×10^{25}	25.19065	75	2.4809×10^{109}	109.39461
26	4.0329×10^{26}	26.60562	76	1.8855×10^{111}	111.27543
27	1.0889×10^{28}	28.03698	77	1.4518×10^{113}	113.16192
28	3.0489×10^{29}	29.48414	78	1.1324×10^{115}	115.05401
29	8.8418×10^{30}	30.94654	79	8.9462×10^{116}	116.95164
30	2.6525×10^{32}	32.42366	80	7.1569×10^{118}	118.85473
31	8.2228×10^{33}	33.91502	81	5.7971×10^{120}	120.76321
32	2.6313×10^{35}	35.42017	82	4.7536×10^{122}	122.67703
33	8.6833×10^{36}	36.93869	83	3.9455×10^{124}	124.59610
34	2.9523×10^{38}	38.47016	84	3.3142×10^{126}	126.52038
35	1.0333×10^{40}	40.01423	85	2.8171×10^{128}	128.44980
36	3.7199×10^{41}	41.57054	86	2.4227×10^{130}	130.38430
37	1.3764×10^{43}	43.13874	87	2.1078×10^{132}	132.32382
38	5.2302×10^{44}	44.71852	88	1.8548×10^{134}	134.26830
39	2.0398×10^{46}	46.30959	89	1.6508×10^{136}	136.21769
40	8.1592×10^{47}	47.91165	90	1.4857×10^{138}	138.17194
41	3.3453×10^{49}	49.52443	91	1.3520×10^{140}	140.13098
42	1.4050×10^{51}	51.14768	92	1.2438×10^{142}	142.09476
43	6.0415×10^{52}	52.78115	93	1.1568×10^{144}	144.06325
44	2.6583×10^{54}	54.42460	94	1.0874×10^{146}	146.03638
45	1.1962×10^{56}	56.07781	95	1.0330×10^{148}	148.01410
46	5.5026×10^{57}	57.74057	96	9.9168×10^{149}	149.99637
47	2.5862×10^{59}	59.41267	97	9.6193×10^{151}	151.98314
48	1.2414×10^{61}	61.09391	98	9.4269×10^{153}	153.97437
49	6.0828×10^{62}	62.78410	99	9.3326×10^{155}	155.97000
50	3.0414×10^{64}	64.48307	100	9.3326×10^{157}	157.97000

$$n! = \left(\frac{n}{e}\right)^{n} \sqrt{2n\pi} + h; \; n = 1, 2, 3, \ldots \quad \left[0 < \frac{h}{n!} < \frac{1}{12n}\right], \quad \lim_{n \to \infty} \frac{n! e^{n}}{n^{n+\frac{1}{2}}} = \sqrt{2\pi}, \quad \lim_{n \to \infty} \frac{(n!)^{\frac{1}{n}}}{n} = \frac{1}{e}$$

RECIPROCALS OF FACTORIALS AND THEIR COMMON LOGARITHMS

This table presents the reciprocals of the factorials and their logarithms for numbers from 1 to 100.

n	$1/n!$	$\log(1/n!)$	n	$1/n!$	$\log(1/n!)$
1	1.	.00000	51	$.64470 \times 10^{-66}$	$\overline{67}.80935$
2	0.5	$\overline{1}.69897$	52	$.12398 \times 10^{-67}$	$\overline{68}.09335$
3	.16667	$\overline{1}.22185$	53	$.23392 \times 10^{-69}$	$\overline{70}.36908$
•4	$.41667 \times 10^{-1}$	$\overline{2}.61979$	54	$.43319 \times 10^{-71}$	$\overline{72}.63668$
5	$.83333 \times 10^{-2}$	$\overline{3}.92082$	55	$.78762 \times 10^{-73}$	$\overline{74}.89632$
6	$.13889 \times 10^{-2}$	$\overline{3}.14267$	56	$.14065 \times 10^{-74}$	$\overline{75}.14813$
7	$.19841 \times 10^{-3}$	$\overline{4}.29757$	57	$.24675 \times 10^{-76}$	$\overline{77}.39226$
8	$.24802 \times 10^{-4}$	$\overline{5}.39448$	58	$.42543 \times 10^{-78}$	$\overline{79}.62883$
9	$.27557 \times 10^{-5}$	$\overline{6}.44024$	59	$.72107 \times 10^{-80}$	$\overline{81}.85798$
10	$.27557 \times 10^{-6}$	$\overline{7}.44024$	60	$.12018 \times 10^{-81}$	$\overline{82}.07983$
11	$.25052 \times 10^{-7}$	$\overline{8}.39884$	61	$.19701 \times 10^{-83}$	$\overline{84}.29450$
12	$.20877 \times 10^{-8}$	$\overline{9}.31966$	62	$.31776 \times 10^{-85}$	$\overline{86}.50210$
13	$.16059 \times 10^{-9}$	$\overline{10}.20572$	63	$.50439 \times 10^{-87}$	$\overline{88}.70276$
14	$.11471 \times 10^{-10}$	$\overline{11}.05959$	64	$.78810 \times 10^{-89}$	$\overline{90}.89658$
15	$.76472 \times 10^{-12}$	$\overline{13}.88350$	65	$.12125 \times 10^{-90}$	$\overline{91}.08367$
16	$.47795 \times 10^{-13}$	$\overline{14}.67938$	66	$.18371 \times 10^{-92}$	$\overline{93}.26413$
17	$.28115 \times 10^{-14}$	$\overline{15}.44893$	67	$.27419 \times 10^{-94}$	$\overline{95}.43805$
18	$.15619 \times 10^{-15}$	$\overline{16}.19366$	68	$.40322 \times 10^{-96}$	$\overline{97}.60554$
19	$.82206 \times 10^{-17}$	$\overline{18}.91491$	69	$.58438 \times 10^{-98}$	$\overline{99}.76669$
20	$.41103 \times 10^{-18}$	$\overline{19}.61388$	70	$.83482 \times 10^{-100}$	$\overline{101}.92159$
21	$.19573 \times 10^{-19}$	$\overline{20}.29166$	71	$.11758 \times 10^{-101}$	$\overline{102}.07034$
22	$.88968 \times 10^{-21}$	$\overline{22}.94923$	72	$.16331 \times 10^{-103}$	$\overline{104}.21300$
23	$.38682 \times 10^{-22}$	$\overline{23}.58751$	73	$.22371 \times 10^{-105}$	$\overline{106}.34968$
24	$.16117 \times 10^{-23}$	$\overline{24}.20729$	74	$.30231 \times 10^{-107}$	$\overline{108}.48045$
25	$.64470 \times 10^{-25}$	$\overline{26}.80935$	75	$.40308 \times 10^{-109}$	$\overline{110}.60539$
26	$.24796 \times 10^{-26}$	$\overline{27}.39438$	76	$.53036 \times 10^{-111}$	$\overline{112}.72457$
27	$.91837 \times 10^{-28}$	$\overline{29}.96302$	77	$.68879 \times 10^{-113}$	$\overline{114}.83808$
28	$.32799 \times 10^{-29}$	$\overline{30}.51586$	78	$.88306 \times 10^{-115}$	$\overline{116}.94599$
29	$.11310 \times 10^{-30}$	$\overline{31}.05346$	79	$.11178 \times 10^{-116}$	$\overline{117}.04836$
30	$.37700 \times 10^{-32}$	$\overline{33}.57634$	80	$.13972 \times 10^{-118}$	$\overline{119}.14527$
31	$.12161 \times 10^{-33}$	$\overline{34}.08498$	81	$.17250 \times 10^{-120}$	$\overline{121}.23679$
32	$.38004 \times 10^{-35}$	$\overline{36}.57983$	82	$.21036 \times 10^{-122}$	$\overline{123}.32297$
33	$.11516 \times 10^{-36}$	$\overline{37}.06131$	83	$.25345 \times 10^{-124}$	$\overline{125}.40390$
34	$.33872 \times 10^{-38}$	$\overline{39}.52984$	84	$.30173 \times 10^{-126}$	$\overline{127}.47962$
35	$.96776 \times 10^{-40}$	$\overline{41}.98577$	85	$.35497 \times 10^{-128}$	$\overline{129}.55020$
36	$.26882 \times 10^{-41}$	$\overline{42}.42946$	86	$.41276 \times 10^{-130}$	$\overline{131}.61570$
37	$.72655 \times 10^{-43}$	$\overline{44}.86126$	87	$.47444 \times 10^{-132}$	$\overline{133}.67618$
38	$.19120 \times 10^{-44}$	$\overline{45}.28148$	88	$.53913 \times 10^{-134}$	$\overline{135}.73170$
39	$.49025 \times 10^{-46}$	$\overline{47}.69041$	89	$.60577 \times 10^{-136}$	$\overline{137}.78231$
40	$.12256 \times 10^{-47}$	$\overline{48}.08835$	90	$.67308 \times 10^{-138}$	$\overline{139}.82806$
41	$.29893 \times 10^{-49}$	$\overline{50}.47557$	91	$.73964 \times 10^{-140}$	$\overline{141}.86902$
42	$.71174 \times 10^{-51}$	$\overline{52}.85232$	92	$.80396 \times 10^{-142}$	$\overline{143}.90524$
43	$.16552 \times 10^{-52}$	$\overline{53}.21885$	93	$.86447 \times 10^{-144}$	$\overline{145}.93675$
44	$.37618 \times 10^{-54}$	$\overline{55}.57540$	94	$.91965 \times 10^{-146}$	$\overline{147}.96362$
45	$.83597 \times 10^{-56}$	$\overline{57}.92219$	95	$.96806 \times 10^{-148}$	$\overline{149}.98590$
46	$.18173 \times 10^{-57}$	$\overline{58}.25943$	96	$.10084 \times 10^{-149}$	$\overline{150}.00363$
47	$.38666 \times 10^{-59}$	$\overline{60}.58733$	97	$.10396 \times 10^{-151}$	$\overline{152}.01686$
48	$.80555 \times 10^{-61}$	$\overline{62}.90609$	98	$.10608 \times 10^{-153}$	$\overline{154}.02563$
49	$.16440 \times 10^{-62}$	$\overline{63}.21590$	99	$.10715 \times 10^{-155}$	$\overline{156}.03000$
50	$.32879 \times 10^{-64}$	$\overline{65}.51693$	100	$.10715 \times 10^{-157}$	$\overline{158}.03000$

• For example $\log \frac{1}{4!} = \overline{2}.61979 = .61979 - 2 = 8.61979 - 10.$

NUMBER OF PERMUTATIONS $P(n,m)$

This table contains the number of permutations of n distinct things taken m at a time, given by

$$P(n,m) = \frac{n!}{(n-m)!} = n(n-1)\cdots(n-m+1)$$

n \ m	0	1	2	3	4	5	6	7	8	9	10
0	1										
1	1	1									
2	1	2	2								
3	1	3	6	6							
4	1	4	12	24	24						
5	1	5	20	60	120	120					
6	1	6	30	120	360	720	720				
7	1	7	42	210	840	2520	5040	5040			
8	1	8	56	336	1680	6720	20160	40320	40320		
9	1	9	72	504	3024	15120	60480	1 81440	3 62880	3 62880	
10	1	10	90	720	5040	30240	1 51200	6 04800	18 14400	36 28800	36 28800
11	1	11	110	990	7920	55440	3 32640	16 63200	66 52800	199 58400	399 16800
12	1	12	132	1320	11880	95040	6 65280	39 91680	199 58400	798 33600	2395 00800
13	1	13	156	1716	17160	1 54440	12 35520	86 48640	518 91840	2594 59200	10378 36800
14	1	14	182	2184	24024	2 40240	21 62160	172 97280	1210 80960	7264 85760	36324 28800
15	1	15	210	2730	32760	3 60360	36 03600	324 32400	2594 59200	18162 14400	1 08972 86400

n \ m	11	12	13	14	15
8					
9					
10					
11	399 16800				
12	4790 01600	4790 01600			
13	31135 10400	62270 20800	62270 20800		
14	1 45297 15200	4 35891 45600	8 71782 91200	8 71782 91200	
15	5 44864 32000	21 79457 28000	65 38371 84000	130 76743 68000	130 76743 68000

NUMBER OF COMBINATIONS

$$\binom{n}{m} = C(n, m)$$

Properties of Binomial Coefficients

$$(1 + x)^n = \binom{n}{0} x^0 + \binom{n}{1} x^n + \binom{n}{2} x^2 + \ldots + \binom{n}{n} x^n$$

$$\binom{n}{m} + \binom{n}{m+1} = \binom{n+1}{m+1}, \binom{n}{m} = \binom{n}{n-m}$$

This leads to Pascal's triangle

$$\binom{n}{0} + \binom{n}{1} + \binom{n}{2} + \ldots + \binom{n}{n} = 2^n$$

$$\binom{n}{0} - \binom{n}{1} + \binom{n}{2} - \ldots (-1)^n \binom{n}{n} = 0$$

$$\binom{n}{n} + \binom{n+1}{n} + \binom{n+2}{n} + \ldots + \binom{n+m}{n} = \binom{n+m+1}{n+1}$$

$$\binom{n}{0} + \binom{n}{2} + \binom{n}{4} + \ldots = 2^{n-1}$$

$$\binom{n}{1} + \binom{n}{3} + \binom{n}{5} + \ldots = 2^{n-1}$$

$$\binom{n}{0}^2 + \binom{n}{1}^2 + \binom{n}{2}^2 + \ldots + \binom{n}{n}^2 = \binom{2n}{n}$$

$$\binom{m}{0}\binom{n}{p} + \binom{m}{1}\binom{n}{p-1} + \ldots + \binom{m}{p}\binom{n}{0} = \binom{m+n}{p}$$

$$(1)\binom{n}{1} + (2)\binom{n}{2} + (3)\binom{n}{3} + \ldots + (n)\binom{n}{n} = n2^{n-1}$$

$$(1)\binom{n}{1} - (2)\binom{n}{2} + (3)\binom{n}{3} - \ldots (-1)^{n+1} (n)\binom{n}{n} = 0$$

NUMBER OF COMBINATIONS $C(n,m)$

This table contains the number of combinations of n distinct things taken m at a time, given by

$$\binom{n}{m} = C(n,m) = \frac{n!}{m!(n-m)!} = \frac{P(n,r)}{m!} \ .$$

For values missing from the above table, use the relation $\binom{n}{m} = \binom{n}{n-m}$, e.g. $\binom{20}{12} = \binom{20}{8} = 125970$. $\binom{n}{m}$ is also referred to as a binomial coefficient. A recursion relation for the binomial coefficients is

$$\binom{n+1}{m+1} = \binom{n}{m} + \binom{n}{m+1}$$

NUMBER OF COMBINATIONS

$$\binom{n}{m} = C(n,m)$$

n \\ m	0	1	2	3	4	5	6	7	8
1	1	1							
2	1	2	1						
3	1	3	3	1					
4	1	4	6	4	1				
5	1	5	10	10	5	1			
6	1	6	15	20	15	6	1		
7	1	7	21	35	35	21	7	1	
8	1	8	28	56	70	56	28	8	1
9	1	9	36	84	126	126	84	36	9
10	1	10	45	120	210	252	210	120	45
11	1	11	55	165	330	462	462	330	165
12	1	12	66	220	495	792	924	792	495
13	1	13	78	286	715	1287	1716	1716	1287
14	1	14	91	364	1001	2002	3003	3432	3003
15	1	15	105	455	1365	3003	5005	6435	6435
16	1	16	120	560	1820	4368	8008	11440	12870
17	1	17	136	680	2380	6188	12376	19448	24310
18	1	18	153	816	3060	8568	18564	31824	43758
19	1	19	171	969	3876	11628	27132	50388	75582
20	1	20	190	1140	4845	15504	38760	77520	1 25970
21	1	21	210	1330	5985	20349	54264	1 16280	2 03490
22	1	22	231	1540	7315	26334	74613	1 70544	3 19770
23	1	23	253	1771	8855	33649	1 00947	2 45157	4 90314
24	1	24	276	2024	10626	42504	1 34596	3 46104	7 35471
25	1	25	300	2300	12650	53130	1 77100	4 80700	10 81575
26	1	26	325	2600	14950	65780	2 30230	6 57800	15 62275
27	1	27	351	2925	17550	80730	2 96010	8 88030	22 20075
28	1	28	378	3276	20475	98280	3 76740	11 84040	31 08105
29	1	29	406	3654	23751	1 18755	4 75020	15 60780	42 92145
30	1	30	435	4060	27405	1 42506	5 93775	20 35800	58 52925
31	1	31	465	4495	31465	1 69911	7 36281	26 29575	78 88725
32	1	32	496	4960	35960	2 01376	9 06192	33 65856	105 18300
33	1	33	528	5456	40920	2 37336	11 07568	42 72048	138 84156
34	1	34	561	5984	46376	2 78256	13 44904	53 79616	181 56204
35	1	35	595	6545	52360	3 24632	16 23160	67 24520	235 35820
36	1	36	630	7140	58905	3 76992	19 47792	83 47680	302 60340
37	1	37	666	7770	66045	4 35897	23 24784	102 95472	386 08020
38	1	38	703	8436	73815	5 01942	27 60681	126 20256	489 03492
39	1	39	741	9139	82251	5 75757	32 62623	153 80937	615 23748
40	1	40	780	9880	91390	6 58008	38 38380	186 43560	769 04685
41	1	41	820	10660	101270	7 49398	44 96388	224 81940	955 48245
42	1	42	861	11480	111930	8 50668	52 45786	269 78328	1180 30185
43	1	43	903	12341	123410	9 62598	60 96454	322 24114	1450 08513
44	1	44	946	13244	135751	10 86008	70 59052	383 20568	1772 32627
45	1	45	990	14190	148995	12 21759	81 45060	453 79620	2155 53195
46	1	46	1035	15180	163185	13 70754	93 66819	535 24680	2609 32815
47	1	47	1081	16215	178365	15 33939	107 37573	628 91499	3144 57495
48	1	48	1128	17296	194580	17 12304	122 71512	736 29072	3773 48994
49	1	49	1176	18424	211876	19 06884	139 83816	859 00584	4509 78066
50	1	50	1225	19600	230300	21 18760	158 90700	998 84400	5368 78650

NUMBER OF COMBINATIONS

$$\binom{n}{m} = C(n, m)$$

m / n	9	10	11	12	13
9	1				
10	10	1			
11	55	11	1		
12	220	66	12	1	
13	715	286	78	13	1
14	2002	1001	364	91	14
15	5005	3003	1365	455	105
16	11440	8008	4368	1820	560
17	24310	19448	12376	6188	2380
18	48620	43758	31824	18564	8568
19	92378	92378	75582	50388	27132
20	1 67960	1 84756	1 67960	1 25970	77520
21	2 93930	3 52716	3 52716	2 93930	2 03490
22	4 97420	6 46646	7 05432	6 46646	4 97420
23	8 17190	11 44066	13 52078	13 52078	11 44066
24	13 07504	19 61256	24 96144	27 04156	24 96144
25	20 42975	32 68760	44 57400	52 00300	52 00300
26	31 24550	53 11735	77 26160	96 57700	104 00600
27	46 86825	84 36285	130 37895	173 83860	200 58300
28	69 06900	131 23110	214 74180	304 21755	374 42160
29	100 15005	200 30010	345 97290	518 95935	678 63915
30	143 07150	300 45015	546 27300	864 93225	1197 59850
31	201 60075	443 52165	846 72315	1411 20525	2062 53075
32	280 48800	645 12240	1290 24480	2257 92840	3474 73600
33	385 67100	925 61040	1935 36720	3548 17320	5731 66440
34	524 51256	1311 28140	2860 97760	5483 54040	9279 83760
35	706 07460	1835 79396	4172 25900	8344 51800	14763 37800
36	941 43280	2541 86856	6008 05296	12516 77700	23107 89600
37	1244 03620	3483 30136	8549 92152	18524 82996	35624 67300
38	1630 11640	4727 33756	12033 22288	27074 75148	54149 50296
39	2119 15132	6357 45396	16760 56044	39107 97436	81224 25444
40	2734 38880	8476 60528	23118 01440	55868 53480	1 20332 22880
41	3503 43565	11210 99408	31594 61968	78986 54920	1 76200 76360
42	4458 91810	14714 42973	42805 61376	1 10581 16888	2 55187 31280
43	5639 21995	19173 34783	57520 04349	1 53386 78264	3 65768 48168
44	7098 30508	24812 56778	76693 39132	2 10906 82613	5 19155 26432
45	8861 63135	31901 87286	1 01505 95910	2 87600 21745	7 30062 09045
46	11017 16330	40763 50421	1 33407 83196	3 89106 17655	10 17662 30790
47	13626 49145	51780 66751	1 74171 33617	5 22514 00851	14 06768 48445
48	16771 06640	65407 15896	2 25952 00368	6 96685 34468	19 29282 49296
49	20544 55634	82178 22536	2 91359 16264	9 22637 34836	26 25967 83764
50	25054 33700	1 02722 78170	3 73537 38800	12 13996 51100	35 48605 18600

NUMBER OF COMBINATIONS

$$\binom{n}{m} = C(n,m)$$

n \ m	14	15	16	17	18	19
14	1					
15	15	1				
16	120	16	1			
17	680	136	17	1		
18	3060	816	153	18	1	
19	11628	3876	969	171	19	1
20	38760	15504	4845	1140	190	20
21	1 16280	54264	20349	5985	1330	210
22	3 19770	1 70544	74613	26334	7315	1540
23	8 17190	4 90314	2 45157	1 00947	33649	8855
24	19 61256	13 07504	7 35471	3 46104	1 34596	42504
25	44 57400	32 68760	20 42975	10 81575	4 80700	1 77100
26	96 57700	77 26180	53 11735	31 24550	15 62275	6 57800
27	200 58300	173 83860	130 37895	84 36285	46 86825	22 20075
28	401 16600	374 42160	304 21755	214 74180	131 23110	69 06900
29	775 58760	775 58760	678 63915	518 95935	345 97290	200 30010
30	1454 22675	1551 17520	1454 22675	1197 59850	864 93225	546 27300
31	2651 82525	3005 40195	3005 40195	2651 82525	2062 53075	1411 20525
32	4714 35600	5657 22720	6010 80390	5657 22720	4714 35600	3473 73600
33	8188 09200	10371 58320	11668 03110	11668 03110	10371 58320	8188 09200
34	13919 75640	18559 67520	22039 61430	23336 06220	22039 61430	18559 67520
35	23199 59400	32479 43160	40599 28950	45375 67650	45375 67650	40599 28950
36	37962 97200	55679 02560	73078 72110	85974 96600	90751 35300	85974 96600
37	61070 86800	93641 99760	1 28757 74670	1 59053 68710	1 76726 31900	1 76726 31900
38	96695 54100	1 54712 86560	2 22399 74430	2 87811 43380	3 35780 00610	3 53452 63800
39	1 50845 04396	2 51408 40660	3 77112 60990	5 10211 17810	6 23591 43990	6 89232 64410
40	2 32069 29840	4 02253 45056	6 28521 01650	8 87323 78800	11 33802 61800	13 12824 08400
41	3 52401 52720	6 34322 74896	10 30774 46706	15 15844 80450	20 21126 40600	24 46626 70200
42	5 28602 29080	9 86724 27616	16 65097 21602	25 46619 27156	35 36971 21050	44 67753 10800
43	7 83789 60360	15 15326 56696	26 51821 49218	42 11716 48758	60 83590 48206	80 04724 31850
44	11 49558 08528	22 99116 17056	41 67148 05914	68 63537 97976	102 95306 96964	140 88314 80056
45	16 68713 34960	34 48674 25584	64 66264 22970	110 30686 03890	171 58844 94940	243 83621 77020
46	23 98775 44005	51 17387 60544	99 14938 48554	174 96950 26860	281 89530 98830	415 42466 71960
47	34 16437 74795	75 16163 04549	150 32326 09098	274 11888 75414	456 86481 25690	697 31997 70790
48	48 23206 23240	109 32600 79344	225 48489 13647	424 44214 84512	730 98370 01104	1154 18478 96480
49	67 52488 72536	157 55807 02584	334 81089 92991	649 92703 98159	1155 42584 85616	1885 16848 97584
50	93 78456 56300	225 08295 75120	492 36896 95575	984 73793 91150	1805 35288 83775	3040 59433 83200

NUMBER OF COMBINATIONS

$$\binom{n}{m} = C(n,m)$$

$n \backslash m$	20	21	22	23	24	25
20	1					
21	21	1				
22	231	22	1			
23	1771	253	23	1		
24	10626	2024	276	24	1	
25	53130	12650	2300	300	25	1
26	2 30230	65780	14950	2600	325	26
27	8 88030	2 96010	80730	17550	2925	351
28	31 08105	11 84040	3 76740	98280	20475	3276
29	100 15005	42 92145	15 60780	4 75620	1 18755	23751
30	300 45015	143 07150	58 52925	20 35800	5 93775	1 42506
31	846 72315	443 52165	201 60075	78 88725	26 29575	7 36281
32	2257 92840	1290 24480	645 12240	280 48800	105 18300	33 65856
33	5731 66440	3548 17320	1935 36720	925 61040	385 67100	138 84156
34	13919 75640	9279 83760	5483 54040	2860 97760	1311 28140	524 51256
35	32479 43160	23199 59400	14763 37800	8344 51800	4172 25900	1835 79396
36	73078 72110	55679 02560	37962 97200	23107 89600	12516 77700	6008 05296
37	1 59053 68710	1 28757 74670	93641 99760	61070 86800	35624 67300	18524 82996
38	3 35786 00610	2 87811 43380	2 22399 74430	1 54712 86560	96695 54100	54149 50296
39	6 89232 64410	6 23591 43990	5 10211 17810	3 77112 60990	2 51408 40660	1 50845 04396
40	13 78465 28820	13 12824 08400	11 33802 61800	8 87323 78800	6 28521 01650	4 02253 45056
41	26 91289 37220	26 91289 37220	24 46626 70200	20 21126 40600	15 15844 80450	10 30774 46706
42	51 37916 07420	53 82578 74440	51 37916 07420	44 67753 10800	35 36971 21050	25 46619 27156
43	96 05669 18220	105 20494 81860	105 20494 81860	96 05669 18220	80 04724 31850	60 83590 48206
44	176 10393 50070	201 26164 00080	210 40989 63720	201 26164 00080	176 10393 50070	140 88314 80056
45	316 98708 30126	377 36557 50150	411 67153 63800	411 67153 63800	377 36557 50150	316 98708 30126
46	560 82330 07146	694 35265 80276	789 03711 13950	823 34307 27600	789 03711 13950	694 35265 80276
47	976 24796 79106	1255 17595 87422	1483 38976 94226	1612 38018 41550	1612 38018 41550	1483 38976 94226
48	1673 56794 49896	2231 42392 66528	2738 56572 81648	3095 76995 35776	3224 76036 83100	3095 76995 35776
49	2827 75273 46376	3904 99187 16424	4969 98965 48176	5834 33568 17424	6320 53032 18876	6320 53032 18876
50	4712 92122 43960	6732 74460 62800	8874 98152 64600	10804 32533 65600	12154 86600 36300	12641 06064 37752

POSITIONAL NOTATION

In our ordinary system of writing numbers, the value of any digit depends on its position in the number. The value of a digit in any position is ten times the value of the same digit one position to the right, or one-tenth the value of the same digit one position to the left. Thus, for example,

$$173.246 = 1 \times 10^2 + 7 \times 10^1 + 3 + 2 \times \frac{1}{10} + 4 \times \frac{1}{10^2} + 6 \times \frac{1}{10^3}.$$

There is no reason that a number other than 10 cannot be used as the *base*, or *radix*, of the number system. In fact, bases of 2, 8, and 16 are commonly used in working with digital computers. When the base used is not clear from the context, it is usually indicated as a parenthesized subscript or merely as a subscript. Thus

$$743_{(8)} = 7 \times 8^2 + 4 \times 8 + 3 = 7 \times 64 + 4 \times 8 + 3 = 448 + 32 + 3 = 483_{(10)}$$

$$1011.101_{(2)} = 1 \times 2^3 + 0 \times 2^2 + 1 \times 2 + 1 + 1 \times \tfrac{1}{2} + 0 \times \tfrac{1}{4} + 1 \times \tfrac{1}{8} = 11.625_{(10)}$$

CHANGE OF BASE

In this section, it is assumed that all calculations will be performed in base 10, since this is the only base in which most people can easily compute. However, there is no logical reason that some other base could not be used for the computations.

To convert a number from another base into base 10:

Simply write down the digits of the number, with each one multiplied by its appropriate positional value. Then perform the indicated computations in base 10, and write down the answer.

For examples, see the two examples in the previous section.

To convert a number from base 10 into another base:

The part of the number to the left of the point and the part to the right must be operated on separately. For the integer part (the part to the left of the point):

a. Divide the number by the new base, getting an integer quotient and remainder.
b. Write down the remainder as the last digit of the number in the new base.
c. Using the quotient from the last division in place of the original number, repeat the above two steps until the quotient becomes zero.

For the fractional part (the part to the right of the point):

a. Multiply the number by the new base.
b. Write down the integral part of the product as the first digit of the fractional part in the new base.
c. Using the fractional part of the last product in place of the original number, repeat the above two steps until the product becomes an integer, or until the desired number of places have been computed.

Examples:

These examples show a convenient method of arranging the computations.

1. Convert $103.118_{(10)}$ to base 8.

$$
\begin{array}{ll}
8 & \underline{|103|} \quad 7 \\
8 & \quad \underline{|12|} \quad 4 \\
& \qquad 1
\end{array}
\qquad 147.074324\ldots
$$

	.118
	8
	.944
	8
	7.552
	8
	4.416
	8
	3.328
	8
	2.624
	8
	4.992

The calculation of the fractional part could be carried out as far as desired. It is a non-terminating fraction which will eventually repeat itself.

$$103.118_{(10)} = 147.074324\ldots_{(8)}$$

The calculations may be further shortened by not writing down the multiplier and divisor at each step of the algorithm, as shown in the next example.

2. Convert $275.824_{(10)}$ to base 5.

$$
\begin{array}{ll}
5 & \underline{|275|} \quad 0 \\
& \quad \underline{|55|} \quad 0 \\
& \quad \underline{|11|} \quad 1 \\
& \qquad 2
\end{array}
\qquad
\begin{array}{l}
.824 \\
4.120 \\
0.600 \\
3.000
\end{array}
$$

$$275.824_{(10)} = 2100.403_{(5)}$$

To convert from one base to another (neither of which is 10):

The easiest procedure is usually to convert first to base 10, and then to the desired base. However, there are two exceptions to this:

1. If computational facility is possessed in either of the bases, it may be used instead of base 10, and the appropriate one of the above methods applied.
2. If the two bases are different powers of the same number, the conversion may be done digit-by-digit to the base which is the common root of both bases, and then digit-by-digit back to the other base.

Example: Convert $127.653_{(8)}$ to base 16. (For base 16, the letters A–F are used for the digits $10_{(10)}-15_{(10)}$.)

The first step is to convert the number to base 2, simply by converting each digit to its binary equivalent:

$$127.653_{(8)} = 001\ 010\ 111 \cdot 110\ 101\ 011_{(2)}$$

Now by simply regrouping the binary number into groups of four binary digits, starting at the point, we convert to base 16:

$$127.653_{(8)} = 101\ 0111 \cdot 1101\ 0101\ 1_{(2)} = 57.D58_{(16)}$$

10$^{\pm n}$ IN OCTAL SCALE

10^n	n	10^{-n}	10^n	n	10^{-n}
1	0	1.000 000 000 000 000	112 402 762 000	10	0.000 000 000 006 676
12	1	0.063 146 314 631 463	1 351 035 564 000	11	0.000 000 000 000 538
144	2	0.005 075 341 217 270	16 432 451 210 000	12	0.000 000 000 000 043
1 750	3	0.000 406 111 564 571	221 441 634 520 000	13	0.000 000 000 000 003
23 420	4	0.000 032 155 613 531	2 657 142 036 440 000	14	0.000 000 000 000 000
303 240	5	0.000 002 476 132 611	34 327 724 461 500 000	15	0.000 000 000 000 000
3 641 100	6	0.000 000 206 157 364	434 157 115 760 200 000	16	0.000 000 000 000 000
46 113 200	7	0.000 000 015 327 745	5 432 127 413 542 400 000	17	0.000 000 000 000 000
575 360 400	8	0.000 000 001 257 144	67 405 553 164 731 000 000	18	0.000 000 000 000 000
7 346 545 000	9	0.000 000 000 104 560			

2^n IN DECIMAL SCALE

n	2^n	n	2^n	n	2^n
0.001	1.00069 33874 62581	0.01	1.00695 55500 56719	0.1	1.07177 34625 36293
0.002	1.00138 72557 11335	0.02	1.01395 94797 90029	0.2	1.14869 83549 97035
0.003	1.00208 16050 79633	0.03	1.02101 21257 07193	0.3	1.23114 44133 44916
0.004	1.00277 64359 01078	0.04	1.02811 38266 56067	0.4	1.31950 79107 72894
0.005	1.00347 17485 09503	0.05	1.03526 49238 41377	0.5	1.41421 35623 73095
0.006	1.00416 75432 38973	0.06	1.04246 57608 41121	0.6	1.51571 65665 10398
0.007	1.00486 38204 23785	0.07	1.04971 66836 23067	0.7	1.62450 47927 12471
0.008	1.00556 05803 98468	0.08	1.05701 80405 61380	0.8	1.74110 11265 92248
0.009	1.00625 78234 97782	0.09	1.06437 01824 53360	0.9	1.86606 59830 73615

$n \log_{10} 2$, $n \log_2 10$ IN DECIMAL SCALE

n	$n \log_{10} 2$	$n \log_2 10$	n	$n \log_{10} 2$	$n \log_2 10$
1	0.30102 99957	3.32192 80949	6	1.80617 99740	19.93156 85693
2	0.60205 99913	6.64385 61898	7	2.10720 99696	23.25349 66642
3	0.90308 99870	9.96578 42847	8	2.40823 99653	26.57542 47591
4	1.20411 99827	13.28771 23795	9	2.70926 99610	29.89735 28540
5	1.50514 99783	16.60964 04744	10	3.01029 99566	33.21928 09489

ADDITION AND MULTIPLICATION TABLES

Addition   Multiplication

Binary Scale

$$0 + 0 = 0 \qquad\qquad 0 \times 0 = 0$$
$$0 + 1 = 1 + 0 = 1 \qquad 0 \times 1 = 1 \times 0 = 0$$
$$1 + 1 = 10 \qquad\qquad 1 \times 1 = 1$$

Octal Scale

Addition

0	01	02	03	04	05	06	07
1	02	03	04	05	06	07	10
2	03	04	05	06	07	10	11
3	04	05	06	07	10	11	12
4	05	06	07	10	11	12	13
5	06	07	10	11	12	13	14
6	07	10	11	12	13	14	15
7	10	11	12	13	14	15	16

Multiplication

1	02	03	04	05	06	07
2	04	06	10	12	14	16
3	06	11	14	17	22	25
4	10	14	20	24	30	34
5	12	17	24	31	36	43
6	14	22	30	36	44	52
7	16	25	34	43	52	61

MATHEMATICAL CONSTANTS IN OCTAL SCALE

$\pi = (3.11037\ 552421)_{(8)}$ $e = (2.55760\ 521305)_{(8)}$ $\gamma = (0.44742\ 147707)_{(8)}$

$\pi^{-1} = (0.24276\ 301556)_{(8)}$ $e^{-1} = (0.27426\ 530661)_{(8)}$ $\log_e \gamma = -(0.43127\ 233602)_{(8)}$

$\sqrt{\pi} = (1.61337\ 611067)_{(8)}$ $\sqrt{e} = (1.51411\ 230704)_{(8)}$ $\log_2 \gamma = -(0.62573\ 030645)_{(8)}$

$\log_e \pi = (1.11206\ 404435)_{(8)}$ $\log_{10} e = (0.33626\ 754251)_{(8)}$ $\sqrt{2} = (1.32404\ 746320)_{(8)}$

$\log_2 \pi = (1.51544\ 163223)_{(8)}$ $\log_2 e = (1.34252\ 166245)_{(8)}$ $\log_e 2 = (0.54271\ 027760)_{(8)}$

$\sqrt{10} = (3.12305\ 407267)_{(8)}$ $\log_2 10 = (3.24464\ 741136)_{(8)}$ $\log_e 10 = (2.23273\ 067355)_{(8)}$

OCTAL-DECIMAL INTEGER CONVERSION TABLE

	0	1	2	3	4	5	6	7
0000	0000	0001	0002	0003	0004	0005	0006	0007
0010	0008	0009	0010	0011	0012	0013	0014	0015
0020	0016	0017	0018	0019	0020	0021	0022	0023
0030	0024	0025	0026	0027	0028	0029	0030	0031
0040	0032	0033	0034	0035	0036	0037	0038	0039
0050	0040	0041	0042	0043	0044	0045	0046	0047
0060	0048	0049	0050	0051	0052	0053	0054	0055
0070	0056	0057	0058	0059	0060	0061	0062	0063
0100	0064	0065	0066	0067	0068	0069	0070	0071
0110	0072	0073	0074	0075	0076	0077	0078	0079
0120	0080	0081	0082	0083	0084	0085	0086	0087
0130	0088	0089	0090	0091	0092	0093	0094	0095
0140	0096	0097	0098	0099	0100	0101	0102	0103
0150	0104	0105	0106	0107	0108	0109	0110	0111
0160	0112	0113	0114	0115	0116	0117	0118	0119
0170	0120	0121	0122	0123	0124	0125	0126	0127
0200	0128	0129	0130	0131	0132	0133	0134	0135
0210	0136	0137	0138	0139	0140	0141	0142	0143
0220	0144	0145	0146	0147	0148	0149	0150	0151
0230	0152	0153	0154	0155	0156	0157	0158	0159
0240	0160	0161	0162	0163	0164	0165	0166	0167
0250	0168	0169	0170	0171	0172	0173	0174	0175
0260	0176	0177	0178	0179	0180	0181	0182	0183
0270	0184	0185	0186	0187	0188	0189	0190	0191
0300	0192	0193	0194	0195	0196	0197	0198	0199
0310	0200	0201	0202	0203	0204	0205	0206	0207
0320	0208	0209	0210	0211	0212	0213	0214	0215
0330	0216	0217	0218	0219	0220	0221	0222	0223
0340	0224	0225	0226	0227	0228	0229	0230	0231
0350	0232	0233	0234	0235	0236	0237	0238	0239
0360	0240	0241	0242	0243	0244	0245	0246	0247
0370	0248	0249	0250	0251	0252	0253	0254	0255

	0	1	2	3	4	5	6	7
0400	0256	0257	0258	0259	0260	0261	0262	0263
0410	0264	0265	0266	0267	0268	0269	0270	0271
0420	0272	0273	0274	0275	0276	0277	0278	0279
0430	0280	0281	0282	0283	0284	0285	0286	0287
0440	0288	0289	0290	0291	0292	0293	0294	0295
0450	0296	0297	0298	0299	0300	0301	0302	0303
0460	0304	0305	0306	0307	0308	0309	0310	0311
0470	0312	0313	0314	0315	0316	0317	0318	0319
0500	0320	0321	0322	0323	0324	0325	0326	0327
0510	0328	0329	0330	0331	0332	0333	0334	0335
0520	0336	0337	0338	0339	0340	0341	0342	0343
0530	0344	0345	0346	0347	0348	0349	0350	0351
0540	0352	0353	0354	0355	0356	0357	0358	0359
0550	0360	0361	0362	0363	0364	0365	0366	0367
0560	0368	0369	0370	0371	0372	0373	0374	0375
0570	0376	0377	0378	0379	0380	0381	0382	0383
0600	0384	0385	0386	0387	0388	0389	0390	0391
0610	0392	0393	0394	0395	0396	0397	0398	0399
0620	0400	0401	0402	0403	0404	0405	0406	0407
0630	0408	0409	0410	0411	0412	0413	0414	0415
0640	0416	0417	0418	0419	0420	0421	0422	0423
0650	0424	0425	0426	0427	0428	0429	0430	0431
0660	0432	0433	0434	0435	0436	0437	0438	0439
0670	0440	0441	0442	0443	0444	0445	0446	0447
0700	0448	0449	0450	0451	0452	0453	0454	0455
0710	0456	0457	0458	0459	0460	0461	0462	0463
0720	0464	0465	0466	0467	0468	0469	0470	0471
0730	0472	0473	0474	0475	0476	0477	0478	0479
0740	0480	0481	0482	0483	0484	0485	0486	0487
0750	0488	0489	0490	0491	0492	0493	0494	0495
0760	0496	0497	0498	0499	0500	0501	0502	0503
0770	0504	0505	0506	0507	0508	0509	0510	0511

0000	0000
to	to
0777	0511
(Octal)	(Decimal)

Octal	Decimal
10000-	4096
20000-	8192
30000-	12288
40000-	16384
50000-	20480
60000-	24576
70000-	28672

	0	1	2	3	4	5	6	7
1000	0512	0513	0514	0515	0516	0517	0518	0519
1010	0520	0521	0522	0523	0524	0525	0526	0527
1020	0528	0529	0530	0531	0532	0533	0534	0535
1030	0536	0537	0538	0539	0540	0541	0542	0543
1040	0544	0545	0546	0547	0548	0549	0550	0551
1050	0552	0553	0554	0555	0556	0557	0558	0559
1060	0560	0561	0562	0563	0564	0565	0566	0567
1070	0568	0569	0570	0571	0572	0573	0574	0575
1100	0576	0577	0578	0579	0580	0581	0582	0583
1110	0584	0585	0586	0587	0588	0589	0590	0591
1120	0592	0593	0594	0595	0596	0597	0598	0599
1130	0600	0601	0602	0603	0604	0605	0606	0607
1140	0608	0609	0610	0611	0612	0613	0614	0615
1150	0616	0617	0618	0619	0620	0621	0622	0623
1160	0624	0625	0626	0627	0628	0629	0630	0631
1170	0632	0633	0634	0635	0636	0637	0638	0639
1200	0640	0641	0642	0643	0644	0645	0646	0647
1210	0648	0649	0650	0651	0652	0653	0654	0655
1220	0656	0657	0658	0659	0660	0661	0662	0663
1230	0664	0665	0666	0667	0668	0669	0670	0671
1240	0672	0673	0674	0675	0676	0677	0678	0679
1250	0680	0681	0682	0683	0684	0685	0686	0687
1260	0688	0689	0690	0691	0692	0693	0694	0695
1270	0696	0697	0698	0699	0700	0701	0702	0703
1300	0704	0705	0706	0707	0708	0709	0710	0711
1310	0712	0713	0714	0715	0716	0717	0718	0719
1320	0720	0721	0722	0723	0724	0725	0726	0727
1330	0728	0729	0730	0731	0732	0733	0734	0735
1340	0736	0737	0738	0739	0740	0741	0742	0743
1350	0744	0745	0746	0747	0748	0749	0750	0751
1360	0752	0753	0754	0755	0756	0757	0758	0759
1370	0760	0761	0762	0763	0764	0765	0766	0767

	0	1	2	3	4	5	6	7
1400	0768	0769	0770	0771	0772	0773	0774	0775
1410	0776	0777	0778	0779	0780	0781	0782	0783
1420	0784	0785	0786	0787	0788	0789	0790	0791
1430	0792	0793	0794	0795	0796	0797	0798	0799
1440	0800	0801	0802	0803	0804	0805	0806	0807
1450	0808	0809	0810	0811	0812	0813	0814	0815
1460	0816	0817	0818	0819	0820	0821	0822	0823
1470	0824	0825	0826	0827	0828	0829	0830	0831
1500	0832	0833	0834	0835	0836	0837	0838	0839
1510	0840	0841	0842	0843	0844	0845	0846	0847
1520	0848	0849	0850	0851	0852	0853	0854	0855
1530	0856	0857	0858	0859	0860	0861	0862	0863
1540	0864	0865	0866	0867	0868	0869	0870	0871
1550	0872	0873	0874	0875	0876	0877	0878	0879
1560	0880	0881	0882	0883	0884	0885	0886	0887
1570	0888	0889	0890	0891	0892	0893	0894	0895
1600	0896	0897	0898	0899	0900	0901	0902	0903
1610	0904	0905	0906	0907	0908	0909	0910	0911
1620	0912	0913	0914	0915	0916	0917	0918	0919
1630	0920	0921	0922	0923	0924	0925	0926	0927
1640	0928	0929	0930	0931	0932	0933	0934	0935
1650	0936	0937	0938	0939	0940	0941	0942	0943
1660	0944	0945	0946	0947	0948	0949	0950	0951
1670	0952	0953	0954	0955	0956	0957	0958	0959
1700	0960	0961	0962	0963	0964	0965	0966	0967
1710	0968	0969	0970	0971	0972	0973	0974	0975
1720	0976	0977	0978	0979	0980	0981	0982	0983
1730	0984	0985	0986	0987	0988	0989	0990	0991
1740	0992	0993	0994	0995	0996	0997	0998	0999
1750	1000	1001	1002	1003	1004	1005	1006	1007
1760	1008	1009	1010	1011	1012	1013	1014	1015
1770	1016	1017	1018	1019	1020	1021	1022	1023

1000	0512
to	to
1777	1023
(Octal)	(Decimal)

OCTAL-DECIMAL INTEGER CONVERSION TABLE (Continued)

	0	1	2	3	4	5	6	7
2000	1024	1025	1026	1027	1028	1029	1030	1031
2010	1032	1033	1034	1035	1036	1037	1038	1039
2020	1040	1041	1042	1043	1044	1045	1046	1047
2030	1048	1049	1050	1051	1052	1053	1054	1055
2040	1056	1057	1058	1059	1060	1061	1062	1063
2050	1064	1065	1066	1067	1068	1069	1070	1071
2060	1072	1073	1074	1075	1076	1077	1078	1079
2070	1080	1081	1082	1083	1084	1085	1086	1087
2100	1088	1089	1090	1091	1092	1093	1094	1095
2110	1096	1097	1098	1099	1100	1101	1102	1103
2120	1104	1105	1106	1107	1108	1109	1110	1111
2130	1112	1113	1114	1115	1116	1117	1118	1119
2140	1120	1121	1122	1123	1124	1125	1126	1127
2150	1128	1129	1130	1131	1132	1133	1134	1135
2160	1136	1137	1138	1139	1140	1141	1142	1143
2170	1144	1145	1146	1147	1148	1149	1150	1151
2200	1152	1153	1154	1155	1156	1157	1158	1159
2210	1160	1161	1162	1163	1164	1165	1166	1167
2220	1168	1169	1170	1171	1172	1173	1174	1175
2230	1176	1177	1178	1179	1180	1181	1182	1183
2240	1184	1185	1186	1187	1188	1189	1190	1191
2250	1192	1193	1194	1195	1196	1197	1198	1199
2260	1200	1201	1202	1203	1204	1205	1206	1207
2270	1208	1209	1210	1211	1212	1213	1214	1215
2300	1216	1217	1218	1219	1220	1221	1222	1223
2310	1224	1225	1226	1227	1228	1229	1230	1231
2320	1232	1233	1234	1235	1236	1237	1238	1239
2330	1240	1241	1242	1243	1244	1245	1246	1247
2340	1248	1249	1250	1251	1252	1253	1254	1255
2350	1256	1257	1258	1259	1260	1261	1262	1263
2360	1264	1265	1266	1267	1268	1269	1270	1271
2370	1272	1273	1274	1275	1276	1277	1278	1279

	0	1	2	3	4	5	6	7
2400	1280	1281	1282	1283	1284	1285	1286	1287
2410	1288	1289	1290	1291	1292	1293	1294	1295
2420	1296	1297	1298	1299	1300	1301	1302	1303
2430	1304	1305	1306	1307	1308	1309	1310	1311
2440	1312	1313	1314	1315	1316	1317	1318	1319
2450	1320	1321	1322	1323	1324	1325	1326	1327
2460	1328	1329	1330	1331	1332	1333	1334	1335
2470	1336	1337	1338	1339	1340	1341	1342	1343
2500	1344	1345	1346	1347	1348	1349	1350	1351
2510	1352	1353	1354	1355	1356	1357	1358	1359
2520	1360	1361	1362	1363	1364	1365	1366	1367
2530	1368	1369	1370	1371	1372	1373	1374	1375
2540	1376	1377	1378	1379	1380	1381	1382	1383
2550	1384	1385	1386	1387	1388	1389	1390	1391
2560	1392	1393	1394	1395	1396	1397	1398	1399
2570	1400	1401	1402	1403	1404	1405	1406	1407
2600	1408	1409	1410	1411	1412	1413	1414	1415
2610	1416	1417	1418	1419	1420	1421	1422	1423
2620	1424	1425	1426	1427	1428	1429	1430	1431
2630	1432	1433	1434	1435	1436	1437	1438	1439
2640	1440	1441	1442	1443	1444	1445	1446	1447
2650	1448	1449	1450	1451	1452	1453	1454	1455
2660	1456	1457	1458	1459	1460	1461	1462	1463
2670	1464	1465	1466	1467	1468	1469	1470	1471
2700	1472	1473	1474	1475	1476	1477	1478	1479
2710	1480	1481	1482	1483	1484	1485	1486	1487
2720	1488	1489	1490	1491	1492	1493	1494	1495
2730	1496	1497	1498	1499	1500	1501	1502	1503
2740	1504	1505	1506	1507	1508	1509	1510	1511
2750	1512	1513	1514	1515	1516	1517	1518	1519
2760	1520	1521	1522	1523	1524	1525	1526	1527
2770	1528	1529	1530	1531	1532	1533	1534	1535

Side reference (left margin):

2000 | 1024
to | to
2777 | 1535
(Octal) | (Decimal)

Octal	Decimal
10000-	4096
20000-	8192
30000-	12288
40000-	16384
50000-	20480
60000-	24576
70000-	28672

	0	1	2	3	4	5	6	7
3000	1536	1537	1538	1539	1540	1541	1542	1543
3010	1544	1545	1546	1547	1548	1549	1550	1551
3020	1552	1553	1554	1555	1556	1557	1558	1559
3030	1560	1561	1562	1563	1564	1565	1566	1567
3040	1568	1569	1570	1571	1572	1573	1574	1575
3050	1576	1577	1578	1579	1580	1581	1582	1583
3060	1584	1585	1586	1587	1588	1589	1590	1591
3070	1592	1593	1594	1595	1596	1597	1598	1599
3100	1600	1601	1602	1603	1604	1605	1606	1607
3110	1608	1609	1610	1611	1612	1613	1614	1615
3120	1616	1617	1618	1619	1620	1621	1622	1623
3130	1624	1625	1626	1627	1628	1629	1630	1631
3140	1632	1633	1634	1635	1636	1637	1638	1639
3150	1640	1641	1642	1643	1644	1645	1646	1647
3160	1648	1649	1650	1651	1652	1653	1654	1655
3170	1656	1657	1658	1659	1660	1661	1662	1663
3200	1664	1665	1666	1667	1668	1669	1670	1671
3210	1672	1673	1674	1675	1676	1677	1678	1679
3220	1680	1681	1682	1683	1684	1685	1686	1687
3230	1688	1689	1690	1691	1692	1693	1694	1695
3240	1696	1697	1698	1699	1700	1701	1702	1703
3250	1704	1705	1706	1707	1708	1709	1710	1711
3260	1712	1713	1714	1715	1716	1717	1718	1719
3270	1720	1721	1722	1723	1724	1725	1726	1727
3300	1728	1729	1730	1731	1732	1733	1734	1735
3310	1736	1737	1738	1739	1740	1741	1742	1743
3320	1744	1745	1746	1747	1748	1749	1750	1751
3330	1752	1753	1754	1755	1756	1757	1758	1759
3340	1760	1761	1762	1763	1764	1765	1766	1767
3350	1768	1769	1770	1771	1772	1773	1774	1775
3360	1776	1777	1778	1779	1780	1781	1782	1783
3370	1784	1785	1786	1787	1788	1789	1790	1791

Side reference (left margin):

3000 | 1536
to | to
3777 | 2047
(Octal) | (Decimal)

	0	1	2	3	4	5	6	7
3400	1792	1793	1794	1795	1796	1797	1798	1799
3410	1800	1801	1802	1803	1804	1805	1806	1807
3420	1808	1809	1810	1811	1812	1813	1814	1815
3430	1816	1817	1818	1819	1820	1821	1822	1823
3440	1824	1825	1826	1827	1828	1829	1830	1831
3450	1832	1833	1834	1835	1836	1837	1838	1839
3460	1840	1841	1842	1843	1844	1845	1846	1847
3470	1848	1849	1850	1851	1852	1853	1854	1855
3500	1856	1857	1858	1859	1860	1861	1862	1863
3510	1864	1865	1866	1867	1868	1869	1870	1871
3520	1872	1873	1874	1875	1876	1877	1878	1879
3530	1880	1881	1882	1883	1884	1885	1886	1887
3540	1888	1889	1890	1891	1892	1893	1894	1895
3550	1896	1897	1898	1899	1900	1901	1902	1903
3560	1904	1905	1906	1907	1908	1909	1910	1911
3570	1912	1913	1914	1915	1916	1917	1918	1919
3600	1920	1921	1922	1923	1924	1925	1926	1927
3610	1928	1929	1930	1931	1932	1933	1934	1935
3620	1936	1937	1938	1939	1940	1941	1942	1943
3630	1944	1945	1946	1947	1948	1949	1950	1951
3640	1952	1953	1954	1955	1956	1957	1958	1959
3650	1960	1961	1962	1963	1964	1965	1966	1967
3660	1968	1969	1970	1971	1972	1973	1974	1975
3670	1976	1977	1978	1979	1980	1981	1982	1983
3700	1984	1985	1986	1987	1988	1989	1990	1991
3710	1992	1993	1994	1995	1996	1997	1998	1999
3720	2000	2001	2002	2003	2004	2005	2006	2007
3730	2008	2009	2010	2011	2012	2013	2014	2015
3740	2016	2017	2018	2019	2020	2021	2022	2023
3750	2024	2025	2026	2027	2028	2029	2030	2031
3760	2032	2033	2034	2035	2036	2037	2038	2039
3770	2040	2041	2042	2043	2044	2045	2046	2047

OCTAL-DECIMAL INTEGER CONVERSION TABLE (Continued)

	0	1	2	3	4	5	6	7
4000	2048	2049	2050	2051	2052	2053	2054	2055
4010	2056	2057	2058	2059	2060	2061	2062	2063
4020	2064	2065	2066	2067	2068	2069	2070	2071
4030	2072	2073	2074	2075	2076	2077	2078	2079
4040	2080	2081	2082	2083	2084	2085	2086	2087
4050	2088	2089	2090	2091	2092	2093	2094	2095
4060	2096	2097	2098	2099	2100	2101	2102	2103
4070	2104	2105	2106	2107	2108	2109	2110	2111
4100	2112	2113	2114	2115	2116	2117	2118	2119
4110	2120	2121	2122	2123	2124	2125	2126	2127
4120	2128	2129	2130	2131	2132	2133	2134	2135
4130	2136	2137	2138	2139	2140	2141	2142	2143
4140	2144	2145	2146	2147	2148	2149	2150	2151
4150	2152	2153	2154	2155	2156	2157	2158	2159
4160	2160	2161	2162	2163	2164	2165	2166	2167
4170	2168	2169	2170	2171	2172	2173	2174	2175
4200	2176	2177	2178	2179	2180	2181	2182	2183
4210	2184	2185	2186	2187	2188	2189	2190	2191
4220	2192	2193	2194	2195	2196	2197	2198	2199
4230	2200	2201	2202	2203	2204	2205	2206	2207
4240	2208	2209	2210	2211	2212	2213	2214	2215
4250	2216	2217	2218	2219	2220	2221	2222	2223
4260	2224	2225	2226	2227	2228	2229	2230	2231
4270	2232	2233	2234	2235	2236	2237	2238	2239
4300	2240	2241	2242	2243	2244	2245	2246	2247
4310	2248	2249	2250	2251	2252	2253	2254	2255
4320	2256	2257	2258	2259	2260	2261	2262	2263
4330	2264	2265	2266	2267	2268	2269	2270	2271
4340	2272	2273	2274	2275	2276	2277	2278	2279
4350	2280	2281	2282	2283	2284	2285	2286	2287
4360	2288	2289	2290	2291	2292	2293	2294	2295
4370	2296	2297	2298	2299	2300	2301	2302	2303

	0	1	2	3	4	5	6	7
4400	2304	2305	2306	2307	2308	2309	2310	2311
4410	2312	2313	2314	2315	2316	2317	2318	2319
4420	2320	2321	2322	2323	2324	2325	2326	2327
4430	2328	2329	2330	2331	2332	2333	2334	2335
4440	2336	2337	2338	2339	2340	2341	2342	2343
4450	2344	2345	2346	2347	2348	2349	2350	2351
4460	2352	2353	2354	2355	2356	2357	2358	2359
4470	2360	2361	2362	2363	2364	2365	2366	2367
4500	2368	2369	2370	2371	2372	2373	2374	2375
4510	2376	2377	2378	2379	2380	2381	2382	2383
4520	2384	2385	2386	2387	2388	2389	2390	2391
4530	2392	2393	2394	2395	2396	2397	2398	2399
4540	2400	2401	2402	2403	2404	2405	2406	2407
4550	2408	2409	2410	2411	2412	2413	2114	2415
4560	2416	2417	2418	2419	2420	2421	2422	2423
4570	2424	2425	2426	2427	2428	2429	2430	2431
4600	2432	2433	2434	2435	2436	2437	2438	2439
4610	2440	2441	2442	2443	2444	2445	2446	2447
4620	2448	2449	2450	2451	2452	2453	2454	2455
4630	2456	2457	2458	2459	2460	2461	2462	2463
4640	2464	2465	2466	2467	2468	2469	2470	2471
4650	2472	2473	2474	2475	2476	2477	2478	2479
4660	2480	2481	2482	2483	2484	2485	2486	2487
4670	2488	2489	2490	2491	2492	2493	2494	2495
4700	2496	2497	2498	2499	2500	2501	2502	2503
4710	2504	2505	2506	2507	2508	2509	2510	2511
4720	2512	2513	2514	2515	2516	2517	2518	2519
4730	2520	2521	2522	2523	2524	2525	2526	2527
4740	2528	2529	2530	2531	2532	2533	2534	2535
4750	2536	2537	2538	2539	2540	2541	2542	2543
4760	2544	2545	2546	2547	2548	2549	2550	2551
4770	2552	2553	2554	2555	2556	2557	2558	2559

4000	2048
to	to
4777	2559
(Octal)	(Decimal)

Octal	Decimal
10000-	4096
20000-	8192
30000-	12288
40000-	16384
50000-	20480
60000-	24576
70000-	28672

	0	1	2	3	4	5	6	7
5000	2560	2561	2562	2563	2564	2565	2566	2567
5010	2568	2569	2570	2571	2572	2573	2574	2575
5020	2576	2577	2578	2579	2580	2581	2582	2583
5030	2584	2585	2586	2587	2588	2589	2590	2591
5040	2592	2593	2594	2595	2596	2597	2598	2599
5050	2600	2601	2602	2603	2604	2605	2606	2607
5060	2608	2609	2610	2611	2612	2613	2614	2615
5070	2616	2617	2618	2619	2620	2621	2622	2623
5100	2624	2625	2626	2627	2628	2629	2630	2631
5110	2632	2633	2634	2635	2636	2637	2638	2639
5120	2640	2641	2642	2643	2644	2645	2646	2647
5130	2648	2649	2650	2651	2652	2653	2654	2655
5140	2656	2657	2658	2659	2660	2661	2662	2663
5150	2664	2665	2666	2667	2668	2669	2670	2671
5160	2672	2673	2674	2675	2676	2677	2678	2679
5170	2680	2681	2682	2683	2684	2685	2686	2687
5200	2688	2689	2690	2691	2692	2693	2694	2695
5210	2696	2697	2698	2699	2700	2701	2702	2703
5220	2704	2705	2706	2707	2708	2709	2710	2711
5230	2712	2713	2714	2715	2716	2717	2718	2719
5240	2720	2721	2722	2723	2724	2725	2726	2727
5250	2728	2729	2730	2731	2732	2733	2734	2735
5260	2736	2737	2738	2739	2740	2741	2742	2743
5270	2744	2745	2746	2747	2748	2749	2750	2751
5300	2752	2753	2754	2755	2756	2757	2758	2759
5310	2760	2761	2762	2763	2764	2765	2766	2767
5320	2768	2769	2770	2771	2772	2773	2774	2775
5330	2776	2777	2778	2779	2780	2781	2782	2783
5340	2784	2785	2786	2787	2788	2789	2790	2791
5350	2792	2793	2794	2795	2796	2797	2798	2799
5360	2800	2801	2802	2803	2804	2805	2806	2807
5370	2808	2809	2810	2811	2812	2813	2814	2815

	0	1	2	3	4	5	6	7
5400	2816	2817	2818	2819	2820	2821	2822	2823
5410	2824	2825	2826	2827	2828	2829	2830	2831
5420	2832	2833	2834	2835	2836	2837	2838	2839
5430	2840	2841	2842	2843	2844	2845	2846	2847
5440	2848	2849	2850	2851	2852	2853	2854	2855
5450	2856	2857	2858	2859	2860	2861	2862	2863
5460	2864	2865	2866	2867	2868	2869	2870	2871
5470	2872	2873	2874	2875	2876	2877	2878	2879
5500	2880	2881	2882	2883	2884	2885	2886	2887
5510	2888	2889	2890	2891	2892	2893	2894	2895
5520	2896	2897	2898	2899	2900	2901	2902	2903
5530	2904	2905	2906	2907	2908	2909	2910	2911
5540	2912	2913	2914	2915	2916	2917	2918	2919
5550	2920	2921	2922	2923	2924	2925	2926	2927
5560	2928	2929	2930	2931	2932	2933	2934	2935
5570	2936	2937	2938	2939	2940	2941	2942	2943
5600	2944	2945	2946	2947	2948	2949	2950	2951
5610	2952	2953	2954	2955	2956	2957	2958	2959
5620	2960	2961	2962	2963	2964	2965	2966	2967
5630	2968	2969	2970	2971	2972	2973	2974	2975
5640	2976	2977	2978	2979	2980	2981	2982	2983
5650	2984	2985	2986	2987	2988	2989	2990	2991
5660	2992	2993	2994	2995	2996	2997	2998	2999
5670	3000	3001	3002	3003	3004	3005	3006	3007
5700	3008	3009	3010	3011	3012	3013	3014	3015
5710	3016	3017	3018	3019	3020	3021	3022	3023
5720	3024	3025	3026	3027	3028	3029	3030	3031
5730	3032	3033	3034	3035	3036	3037	3038	3039
5740	3040	3041	3042	3043	3044	3045	3046	3047
5750	3048	3049	3050	3051	3052	3053	3054	3055
5760	3056	3057	3058	3059	3060	3061	3062	3063
5770	3064	3065	3066	3067	3068	3069	3070	3071

5000	2560
to	to
5777	3071
(Octal)	(Decimal)

OCTAL-DECIMAL INTEGER CONVERSION TABLE (Continued)

	0	1	2	3	4	5	6	7
6000	3072	3073	3074	3075	3076	3077	3078	3079
6010	3080	3081	3082	3083	3084	3085	3086	3087
6020	3088	3089	3090	3091	3092	3093	3094	3095
6030	3096	3097	3098	3099	3100	3101	3102	3103
6040	3104	3105	3106	3107	3108	3109	3110	3111
6050	3112	3113	3114	3115	3116	3117	3118	3119
6060	3120	3121	3122	3123	3124	3125	3126	3127
6070	3128	3129	3130	3131	3132	3133	3134	3135
6100	3136	3137	3138	3139	3140	3141	3142	3143
6110	3144	3145	3146	3147	3148	3149	3150	3151
6120	3152	3153	3154	3155	3156	3157	3158	3159
6130	3160	3161	3162	3163	3164	3165	3166	3167
6140	3168	3169	3170	3171	3172	3173	3174	3175
6150	3176	3177	3178	3179	3180	3181	3182	3183
6160	3184	3185	3186	3187	3188	3189	3190	3191
6170	3192	3193	3194	3195	3196	3197	3198	3199
6200	3200	3201	3202	3203	3204	3205	3206	3207
6210	3208	3209	3210	3211	3212	3213	3214	3215
6220	3216	3217	3218	3219	3220	3221	3222	3223
6230	3224	3225	3226	3227	3228	3229	3230	3231
6240	3232	3233	3234	3235	3236	3237	3238	3239
6250	3240	3241	3442	3243	3244	3245	3246	3247
6260	3248	3249	3250	3251	3252	3253	3254	3255
6270	3256	3257	3258	3259	3260	3261	3262	3263
6300	3264	3265	3266	3267	3268	3269	3270	3871
6310	3972	3273	3274	3275	3276	3277	3278	3279
6320	3280	3281	3282	3283	3284	3285	3286	3287
6330	3288	3289	3290	3291	3292	3293	3294	3295
6340	3296	3297	3298	3299	3300	3301	3302	3003
6350	3304	3305	3306	3307	3308	3309	3310	3311
6360	3312	3313	3314	3315	3316	3317	3318	3319
6370	3320	3321	3322	3323	3324	3325	3326	3327

	0	1	2	3	4	5	6	7
6400	3328	3329	3330	3331	3332	3333	3334	3335
6410	3336	3337	3338	3339	3340	3341	3342	3343
6420	3344	3345	3346	3347	3348	3349	3350	3351
6430	3352	3353	3354	3355	3356	3357	3358	3359
6440	3360	3361	3362	3363	3364	3365	3366	3367
6450	3368	3369	3370	3371	3372	3373	3374	3375
6460	3376	3377	3378	3379	3380	3381	3382	3383
6470	3384	3385	3386	3387	3388	3389	3390	3391
6500	3392	3393	3394	3395	3396	3397	3398	3399
6510	3400	3401	3402	3403	3404	3405	3406	3407
6520	3408	3409	3410	3411	3412	3413	3414	3415
6530	3416	3417	3418	3419	3420	3421	3422	3423
6540	3424	3425	3426	3427	3428	3429	3430	3431
6550	3432	3433	3434	3435	3436	3437	3438	3439
6560	3440	3441	3442	3443	3444	3445	3446	3447
6570	3448	3449	3450	3451	3452	3453	3454	3455
6600	3456	3457	3458	3459	3460	3461	3462	3463
6610	3464	3465	3466	3467	3468	3469	3470	3471
6620	3472	3473	3474	3475	3476	3477	3478	3479
6630	3480	3481	3482	3483	3484	3485	3486	3487
6640	3488	3489	3490	3491	3492	3493	3494	3495
6650	3496	3497	3498	3499	3500	3501	3502	3503
6660	3504	3505	3506	3507	3508	3509	3510	3511
6670	3512	3513	3514	3515	3516	3517	3518	3519
6700	3520	3521	3522	3523	3524	3525	3526	3527
6710	3528	3529	3530	3531	3532	3533	3534	3535
6720	3536	3537	3538	3539	3540	3541	3542	3543
6730	3544	3545	3546	3547	3548	3549	3550	3551
6740	3552	3553	3554	3555	3556	3557	3558	3559
6750	3560	3561	3562	3563	3564	3655	3566	3567
6760	3568	3569	3570	3571	3572	3573	3574	3575
6770	3576	3577	3578	3579	3580	3581	3582	3583

6000 to 6777 (Octal) | 3072 to 3583 (Decimal)

Octal	Decimal
10000	4096
20000	8192
30000	12288
40000	16384
50000	20480
60000	24576
70000	28672

	0	1	2	3	4	5	6	7
7000	3584	3585	3586	3587	3588	3589	3590	3591
7010	3592	3593	3594	3595	3596	3597	3598	3599
7020	3600	3601	3602	3603	3604	3605	3606	3607
7030	3608	3609	3610	3611	3612	3613	3614	3615
7040	3616	3617	3618	3619	3620	3621	3622	3623
7050	3624	3625	3626	3627	3628	3629	3630	3631
7060	3632	3633	3634	3635	3636	3637	3638	3639
7070	3640	3641	3642	3643	3644	3645	3646	3647
7100	3648	3649	3650	3651	3652	3653	3654	3655
7110	3656	3657	3658	3659	3660	3661	3662	3663
7120	3664	3665	3666	3667	3668	3669	3670	3671
7130	3672	3673	3674	3675	3676	3677	3678	3679
7140	3680	3681	3682	3683	3684	3685	3686	3687
7150	3688	3689	3690	3691	3692	3693	3694	3695
7160	3696	3697	3698	3699	3700	3701	3702	3703
7170	3704	3705	3706	3707	3708	3709	3710	3711
7200	3712	3713	3714	3715	3716	3717	3718	3719
7210	3720	3721	3722	3723	3724	3725	3726	3727
7220	3728	3729	3730	3731	3732	3733	3734	3735
7230	3736	3737	3738	3739	3740	3741	3742	3743
7240	3744	3745	3746	3747	3748	3749	3750	3751
7250	3752	3753	3754	3755	3756	3757	3758	3759
7260	3760	3761	3762	3763	3764	3765	3766	3767
7270	3768	3769	3770	3771	3772	3773	3774	3775
7300	3776	3777	3778	3779	3780	3781	3782	3783
7310	3784	3785	3786	3787	3788	3789	3790	3791
7320	3792	3893	3794	3795	3796	3797	3798	3799
7330	3800	3801	3802	3803	3804	3805	3806	3807
7340	3808	3809	3810	3811	3812	3813	3814	3815
7350	3816	3817	3818	3819	3820	3821	3822	3823
7360	3824	3825	3826	3827	3828	3829	3830	3831
7370	3832	3833	3834	3835	3836	3837	3838	3839

7000 to 7777 (Octal) | 3584 to 4095 (Decimal)

	0	1	2	3	4	5	6	7
7400	3840	3841	3482	3843	3844	3845	3846	3847
7410	3848	3849	3850	3851	3852	3853	3854	3855
7420	3856	3857	3858	3859	3860	3861	3862	3863
7430	3864	3865	3866	3867	3868	3869	3870	3871
7440	3872	3873	3874	3875	3876	3877	3878	3879
7450	3880	3881	3882	3883	3884	3885	3886	3887
7460	3888	3889	3890	3891	3892	3893	3894	3895
7470	3896	3897	3898	3899	3900	3901	3902	3903
7500	3904	3905	3906	3907	3908	3909	3910	3911
7510	3912	3913	3914	3915	3916	3917	3918	3919
7520	3920	3921	3922	3923	3924	3925	3926	3927
7530	3928	3929	3930	3931	3932	3933	3934	3935
7540	3936	3937	3938	3939	3940	3941	3942	3943
7550	3944	3945	3946	3947	3948	3949	3950	3951
7560	3952	3953	3954	3955	3956	3957	3958	3959
7570	3960	3961	3962	3963	3964	3965	3966	3967
7600	3968	3969	3970	3971	3972	3973	3974	3975
7610	3976	3977	3978	3979	3980	3981	3982	3983
7620	3984	3985	3986	3987	3988	3989	3990	3991
7630	3992	3993	3994	3995	3996	3997	3998	3999
7640	4000	4001	4002	4003	4004	4005	4006	4007
7650	4008	4009	4010	4011	4012	4013	4014	4015
7660	4016	4017	4018	4019	4020	4021	4022	4023
7670	4024	4025	4026	4027	4028	4029	4030	4031
7700	4032	4033	4034	4035	4036	4037	4038	4039
7710	4040	4041	4042	4043	4044	4045	4046	4047
7720	4048	4049	4050	4051	4052	4053	4054	4055
7730	4056	4057	4058	4059	4060	4061	4062	4063
7740	4064	4065	4066	4067	4068	4069	4070	4071
7750	4072	4073	4074	4075	4076	4077	4078	4079
7760	4080	4081	4082	4083	4084	4085	4086	4087
7770	4088	4089	4090	4091	4092	4093	4094	4095

OCTAL-DECIMAL FRACTION CONVERSION TABLE

This table covers the entries from $(.000)_8$ to $(.377)_8$. For entries from $(.400)_8$ to $(.777)_8$, cognizance should be made of the fact that $(.400)_8$ is $(.500)_{10}$. Hence if $(.637)_8$ is desired, find $(.237)_8$ in table, namely, $(.310456)_{10}$ and add $(.50000)_{10}$ for $(.400)_8$. Thus

$(.637)_8 = (.237)_8 + (.400)_8$

$\quad\quad = (.310456)_{10} + (.50000)_{10}$

$\quad\quad = (.810456)_{10}.$

OCTAL	DEC.	OCTAL	DEC.	OCTAL	DEC.	OCTAL	DEC.
.000	.000000	.100	.125000	.200	.250000	.300	.375000
.001	.001953	.101	.126953	.201	.251953	.301	.376953
.002	.003906	.102	.128906	.202	.253906	.302	.378906
.003	.005859	.103	.130859	.203	.255859	.303	.380859
.004	.007812	.104	.132812	.204	.257812	.304	.382812
.005	.009765	.105	.134765	.205	.259765	.305	.384765
.006	.011718	.106	.136718	.206	.261718	.306	.386718
.007	.013671	.107	.138671	.207	.263671	.307	.388671
.010	.015625	.110	.140625	.210	.265625	.310	.390625
.011	.017578	.111	.142578	.211	.267578	.311	.392578
.012	.019531	.112	.144531	.212	.269531	.312	.394531
.013	.021484	.113	.146484	.213	.271484	.313	.396484
.014	.023437	.114	.148437	.214	.273437	.314	.398437
.015	.025390	.115	.150390	.215	.275390	.315	.400490
.016	.027343	.116	.152343	.216	.277343	.316	.402343
.017	.029296	.117	.154296	.217	.279296	.317	.404296
.020	.031250	.120	.156250	.220	.281250	.320	.406250
.021	.033203	.121	.158203	.221	.283203	.321	.408203
.022	.035156	.122	.160156	.222	.285156	.322	.410156
.023	.037109	.123	.162109	.223	.287109	.323	.412109
.024	.039062	.124	.164062	.224	.289062	.324	.414062
.025	.041015	.125	.166015	.225	.291015	.325	.416015
.026	.042968	.126	.167968	.226	.292968	.326	.417968
.027	.044921	.127	.169921	.227	.294921	.327	.419921
.030	.046875	.130	.171875	.230	.294875	.330	.421875
.031	.048828	.131	.173828	.231	.298828	.331	.423828
.032	.050781	.132	.175781	.232	.300781	.332	.425781
.033	.052734	.133	.177734	.233	.302734	.333	.427734
.034	.054687	.134	.179687	.234	.304687	.334	.429687
.035	.056640	.135	.181640	.235	.306640	.335	.431640
.036	.058593	.136	.183593	.236	.308593	.336	.433593
.037	.060546	.137	.185546	.237	.310546	.337	.435546
.040	.062500	.140	.187500	.240	.312500	.340	.437500
.041	.064453	.141	.189453	.241	.314453	.341	.439453
.042	.066406	.142	.191406	.242	.316406	.342	.441406
.043	.068359	.143	.193359	.243	.318359	.343	.443359
.044	.070312	.144	.195312	.244	.320312	.344	.445312
.045	.072265	.145	.197265	.245	.322265	.345	.447265
.046	.074218	.146	.199218	.246	.324218	.346	.449218
.047	.076171	.147	.201171	.247	.326171	.347	.451171
.050	.078125	.150	.203125	.250	.328125	.350	.453125
.051	.080078	.151	.205078	.251	.330078	.351	.455078
.052	.082031	.152	.207031	.252	.332031	.352	.457031
.053	.083984	.153	.208984	.253	.333984	.353	.458984
.054	.085937	.154	.210937	.254	.335937	.354	.460937
.055	.087890	.155	.212890	.255	.337890	.355	.462890
.056	.089843	.156	.214843	.256	.339843	.356	.464843
.057	.091796	.157	.216796	.257	.341796	.357	.466796
.060	.093750	.160	.218750	.260	.343750	.360	.468750
.061	.095703	.161	.220703	.261	.345703	.361	.470703
.062	.097656	.162	.222656	.262	.347656	.362	.472656
.063	.099609	.163	.224609	.263	.349609	.363	.474609
.064	.101562	.164	.226562	.264	.351562	.364	.476562
.065	.103515	.165	.228515	.265	.353515	.365	.478515
.066	.105468	.166	.230468	.266	.355468	.366	.480468
.067	.107421	.167	.232421	.267	.357421	.367	.482421
.070	.109375	.170	.234375	.270	.359375	.370	.484375
.071	.111328	.171	.236328	.271	.361328	.371	.486328
.072	.113281	.172	.238281	.272	.363281	.372	.488281
.073	.115234	.173	.240234	.273	.365234	.373	.490234
.074	.117187	.174	.242187	.274	.367187	.374	.492187
.075	.119140	.175	.244140	.275	.369140	.375	.494140
.076	.121093	.176	.246093	.276	.371093	.376	.496093
.077	.123046	.177	.248046	.277	.373046	.377	.498046

OCTAL-DECIMAL FRACTION CONVERSION TABLE (Continued)

OCTAL	DEC.	OCTAL	DEC.	OCTAL	DEC.	OCTAL	DEC.
.000000	.000000	.000100	.000244	.000200	.000488	.000300	.000732
.000001	.000004	.000101	.000247	.000201	.000492	.000301	.000736
.000002	.000007	.000102	.000251	.000202	.000495	.000302	.000740
.000003	.000011	.000103	.000255	.000203	.000499	.000303	.000743
.000004	.000015	.000104	.000259	.000204	.000503	.000304	.000747
.000005	.000019	.000105	.000263	.000205	.000507	.000305	.000751
.000006	.000022	.000106	.000267	.000206	.000511	.000306	.000755
.000007	.000026	.000107	.000270	.000207	.000514	.000307	.000759
.000010	.000030	.000110	.000274	.000210	.000518	.000310	.000762
.000011	.000034	.000111	.000278	.000211	.000522	.000311	.000766
.000012	.000038	.000112	.000282	.000212	.000526	.000312	.000770
.000013	.000041	.000113	.000286	.000213	.000530	.000313	.000774
.000014	.000045	.000114	.000289	.000214	.000534	.000314	.000778
.000015	.000049	.000115	.000293	.000215	.000537	.000315	.000782
.000016	.000053	.000116	.000297	.000216	.000541	.000316	.000785
.000017	.000057	.000117	.000301	.000217	.000545	.000317	.000789
.000020	.000061	.000120	.000305	.000220	.000549	.000320	.000793
.000021	.000064	.000121	.000308	.000221	.000553	.000321	.000797
.000022	.000068	.000122	.000312	.000222	.000556	.000322	.000801
.000023	.000072	.000123	.000316	.000223	.000560	.000323	.000805
.000024	.000076	.000124	.000320	.000224	.000564	.000324	.000808
.000025	.000080	.000125	.000324	.000225	.000568	.000325	.000812
.000026	.000083	.000126	.000328	.000226	.000572	.000326	.000816
.000027	.000087	.000127	.000331	.000227	.000576	.000327	.000820
.000030	.000091	.000130	.000335	.000230	.000579	.000330	.000823
.000031	.000095	.000131	.000339	.000231	.000583	.000331	.000827
.000032	.000099	.000132	.000343	.000232	.000587	.000332	.000831
.000033	.000102	.000133	.000347	.000233	.000591	.000333	.000835
.000034	.000106	.000134	.000350	.000234	.000595	.000334	.000839
.000035	.000110	.000135	.000354	C00235	.000598	.000335	.000843
.000036	.000114	.000136	.000358	.000236	.000602	.000336	.000846
.000037	.000118	.000137	.000362	.000237	.000606	.000337	.000850
.000040	.000122	.000140	.000366	.000240	.000610	.000340	.000854
.000041	.000125	.000141	.000370	.000241	.000614	.000341	.000858
.000042	.000129	.000142	.000373	.000242	.000617	.000342	.000862
.000043	.000133	.000143	.000377	.000243	.000621	.000343	.000865
.000044	.000137	.000144	.000381	.000244	.000625	.000344	.000869
.000045	.000141	.000145	.000385	.000245	.000629	.000345	.000873
.000046	.000144	.000146	.000389	.000246	.000633	.000346	.000877
.000047	.000148	.000147	.000392	.000247	.000637	.000347	.000881
.000050	.000152	.000150	.000396	.000250	.000640	.000350	.000885
.000051	.000156	.000151	.000400	.000251	.000644	.000351	.000888
.000052	.000160	.000152	.000404	.000252	.000648	.000352	.000892
.000053	.000164	.000153	.000408	.000253	.000652	.000353	.000896
.000054	.000167	.000154	.000411	.000254	.000656	.000354	.000900
.000055	.000171	.000155	.000415	.000255	.000659	.000355	.000904
.000056	.000175	.000156	.000419	.000256	.000663	.000356	.000907
.000057	.000179	.000157	.000423	.000257	.000667	.000357	.000911
.000060	.000183	.000160	.000427	.000260	.000671	.000360	.000915
.000061	.000186	.000161	.000431	.000261	.000675	.000361	.000919
.000062	.000190	.000162	.000434	.000262	.000679	.000362	.000923
.000063	.000194	.000163	.000438	.000263	.000682	.000363	.000926
.000064	.000198	.000164	.000442	.000264	.000686	.000364	.000930
.000065	.000202	.000165	.000446	.000265	.000690	.000365	.000934
.000066	.000205	.000166	.000450	.000266	.000694	.000366	.000938
.000067	.000209	.000167	.000453	.000267	.000698	.000367	.000942
.000070	.000213	.000170	.000457	.000270	.000701	.000370	.000946
.000071	.000217	.000171	.000461	.000271	.000705	.000371	.000949
.000072	.000221	.000172	.000465	.000272	.000709	.000372	.000953
.000073	.000225	.000173	.000469	.000273	.000713	.000373	.000957
.000074	.000228	.000174	.000473	.000274	.000717	.000374	.000961
.000075	.000232	.000175	.000476	.000275	.000720	.000375	.000965
.000076	.000326	.000176	.000480	.000276	.000724	.000376	.000968
.000077	.000240	.000177	.000484	.000277	.000728	.000377	.000972

OCTAL-DECIMAL FRACTION CONVERSION TABLE (Continued)

OCTAL	DEC.	OCTAL	DEC.	OCTAL	DEC.	OCTAL	DEC.
.000400	.000976	.000500	.001220	.000600	.001464	.000700	.001708
.000401	.000980	.000501	.001224	.000601	.001468	.000701	.001712
.000402	.000984	.000502	.001228	.000602	.001472	.000702	.001716
.000403	.000988	.000503	.001232	.000603	.001476	.000703	.001720
.000404	.000991	.000504	.001235	.000604	.001480	.000704	.001724
.000405	.000995	.000505	.001239	.000605	.001483	.000705	.001728
.000406	.000999	.000506	.001243	.000606	.001487	.000706	.001731
.000407	.001003	.000507	.001247	.000607	.001491	.000707	.001735
.000410	.001007	.000510	.001251	.000610	.001495	.000710	.001739
.000411	.001010	.000511	.001255	.000611	.001499	.000711	.001743
.000412	.001014	.000512	.001258	.000612	.001502	.000712	.001747
.000413	.001018	.000513	.001262	.000613	.001506	.000713	.001750
.000414	.001022	.000514	.001266	.000614	.001510	.000714	.001754
.000415	.001026	.000515	.001270	.000615	.001514	.000715	.001758
.000416	.001029	.000516	.001274	.000616	.001518	.000716	.001762
.000417	.001033	.000517	.001277	.000617	.001522	.000717	.001766
.000420	.001037	.000520	.001281	.000620	.001525	.000720	.001770
.000421	.001041	.000521	.001285	.000621	.001529	.000721	.001733
.000422	.001045	.000522	.001289	.000622	.001533	.000722	.001777
.000423	.001049	.000523	.001293	.000623	.001537	.000723	.001781
.000424	.001052	.000524	.001296	.000624	.001541	.000724	.001785
.000425	.001056	.000525	.001300	.000625	.001544	.000725	.001789
.000426	.001060	.000526	.001304	.000626	.001548	.000726	.001792
.000427	.001064	.000527	.001308	.000627	.001552	.000727	.001796
.000430	.001068	.000530	.001312	.000630	.001556	.000730	.001800
.000431	.001071	.000531	.001316	.000631	.001560	.000731	.001804
.000432	.001075	.000532	.001319	.000632	.001564	.000732	.001808
.000433	.001079	.000533	.001323	.000633	.001567	.000733	.001811
.000434	.001083	.000534	.001327	.000634	.001571	.000734	.001815
.000435	.001087	.000535	.001331	.000635	.001575	.000735	.001819
.000436	.001091	.000536	.001335	.000636	.001579	.000736	.001823
.000437	.001094	.000537	.001338	.000637	.001583	.000737	.001827
.000440	.001098	.000540	.001342	.000640	.001586	.000740	.001831
.000441	.001102	.000541	.001346	.000641	.001590	.000741	.001834
.000442	.001106	.000542	.001350	.000642	.001594	.000742	.001838
.000443	.001110	.000543	.001354	.000643	.001598	.000743	.001842
.000444	.001113	.000544	.001358	.000644	.001602	.000744	.001846
.000445	.001117	.000545	.001361	.000645	.001605	.000745	.001850
.000446	.001121	.000546	.001365	.000646	.001609	.000746	.001853
.000447	.001125	.000547	.001369	.000647	.001613	.000747	.001857
.000450	.001129	.000550	.001373	.000650	.001617	.000750	.001861
.000451	.001132	.000551	.001377	.000651	.001621	.000751	.001865
.000452	.001136	.000552	.001380	.000652	.001625	.000752	.001869
.000453	.001140	.000553	.001384	.000653	.001628	.000753	.001873
.000454	.001144	.000554	.001388	.000654	.001632	.000754	.001876
.000455	.001148	.000555	.001392	.000655	.001636	.000755	.001880
.000456	.001152	.000556	.001396	.000656	.001640	.000756	.001884
.000157	.001155	.000557	.001399	.000657	.001644	.000757	.001888
.000460	.001159	.000560	.001403	.000660	.001647	.000760	.001892
.000461	.001163	.000561	.001407	.000661	.001651	.000761	.001895
.000462	.001167	.000562	.001411	.000662	.001655	.000762	.001899
.000463	.001171	.000563	.001415	.000663	.001659	.000763	.001903
.000464	.001174	.000564	.001419	.000664	.001663	.000764	.001907
.000465	.001178	.000565	.001422	.000665	.001667	.000765	.001911
.000466	.001182	.000566	.001426	.000666	.001670	.000766	.001914
.000467	.001186	.000567	.001430	.000667	.001674	.000767	.001918
.000470	.001190	.000570	.001434	.000670	.001678	.000770	.001922
.000471	.001194	.000571	.001438	.000671	.001682	.000771	.001926
.000472	.001197	.000572	.001441	.000672	.001686	.000772	.001930
.000473	.001201	.000573	.001445	.000673	.001689	.000773	.001934
.000474	.001205	.000574	.001449	.000674	.001693	.000774	.001937
.000475	.001209	.000575	.001453	.000675	.001697	.000775	.001941
.000476	.001213	.000576	.001457	.000676	.001701	.000776	.001945
.000477	.001216	.000577	.001461	.000677	.001705	.000777	.001949

I. HEXADECIMAL AND DECIMAL DIRECT CONVERSION TABLE

The following tables aid in converting hexadecimal (base 16) numbers to decimal, and the reverse. Note that the base 16 digits for the decimal values 10–15 are represented by the letters A–F, respectively.

This table provides direct conversion of decimal and hexadecimal numbers in these ranges:

HEXADECIMAL	DECIMAL
000 to FFF	0000 to 4095

For numbers outside the range of the table, add the following values to the table figures:

HEXADECIMAL	DECIMAL
1000	4096
2000	8192
3000	12288
4000	16384
5000	20480
6000	24576
7000	28672
8000	32768
9000	36864
A000	40960
B000	45056
C000	49152
D000	53248
E000	57344
F000	61440

	0	1	2	3	4	5	6	7	8	9	A	B	C	D	E	F
00__	0000	0001	0002	0003	0004	0005	0006	0007	0008	0009	0010	0011	0012	0013	0014	0015
01__	0016	0017	0018	0019	0020	0021	0022	0023	0024	0025	0026	0027	0028	0029	0030	0031
02__	0032	0033	0034	0035	0036	0037	0038	0039	0040	0041	0042	0043	0044	0045	0046	0047
03__	0048	0049	0050	0051	0052	0053	0054	0055	0056	0057	0058	0059	0060	0061	0062	0063
04__	0064	0065	0066	0067	0068	0069	0070	0071	0072	0073	0074	0075	0076	0077	0078	0079
05__	0080	0081	0082	0083	0084	0085	0086	0087	0088	0089	0090	0091	0092	0093	0094	0095
06__	0096	0097	0098	0099	0100	0101	0102	0103	0104	0105	0106	0107	0108	0109	0110	0111
07__	0112	0113	0114	0115	0116	0117	0118	0119	0120	0121	0122	0123	0124	0125	0126	0127
08__	0128	0129	0130	0131	0132	0133	0134	0135	0136	0137	0138	0139	0140	0141	0142	0143
09__	0144	0145	0146	0147	0148	0149	0150	0151	0152	0153	0154	0155	0156	0157	0158	0159
0A__	0160	0161	0162	0163	0164	0165	0166	0167	0168	0169	0170	0171	0172	0173	0174	0175
0B__	0176	0177	0178	0179	0180	0181	0182	0183	0184	0185	0186	0187	0188	0189	0190	0191
0C__	0192	0193	0194	0195	0196	0197	0198	0199	0200	0201	0202	0203	0204	0205	0206	0207
0D__	0208	0209	0210	0211	0212	0213	0214	0215	0216	0217	0218	0219	0220	0221	0222	0223
0E__	0224	0225	0226	0227	0228	0229	0230	0231	0232	0233	0234	0235	0236	0237	0238	0239
0F__	0240	0241	0242	0243	0244	0245	0246	0247	0248	0249	0250	0251	0252	0253	0254	0255
10__	0256	0257	0258	0259	0260	0261	0262	0263	0264	0265	0266	0267	0268	0269	0270	0271
11__	0272	0273	0274	0275	0276	0277	0278	0279	0280	0281	0282	0283	0284	0285	0286	0287
12__	0288	0289	0290	0291	0292	0293	0294	0295	0296	0297	0298	0299	0300	0301	0302	0303
13__	0304	0305	0306	0307	0308	0309	0310	0311	0312	0313	0314	0315	0316	0317	0318	0319
14__	0320	0321	0322	0323	0324	0325	0326	0327	0328	0329	0330	0331	0332	0333	0334	0335
15__	0336	0337	0338	0339	0340	0341	0342	0343	0344	0345	0346	0347	0348	0349	0350	0351
16__	0352	0353	0354	0355	0356	0357	0358	0359	0360	0361	0362	0363	0364	0365	0366	0367
17__	0368	0369	0370	0371	0372	0373	0374	0375	0376	0377	0378	0379	0380	0381	0382	0383
18__	0384	0385	0386	0387	0388	0389	0390	0391	0392	0393	0394	0395	0396	0397	0398	0399
19__	0400	0401	0402	0403	0404	0405	0406	0407	0408	0409	0410	0411	0412	0413	0414	0415
1A__	0416	0417	0418	0419	0420	0421	0422	0423	0424	0425	0426	0427	0428	0429	0430	0431
1B__	0432	0433	0434	0435	0436	0437	0438	0439	0440	0441	0442	0443	0444	0445	0446	0447
1C__	0448	0449	0450	0451	0452	0453	0454	0455	0456	0457	0458	0459	0460	0461	0462	0463
1D__	0464	0465	0466	0467	0468	0469	0470	0471	0472	0473	0474	0475	0476	0477	0478	0479
1E__	0480	0481	0482	0483	0484	0485	0486	0487	0488	0489	0490	0491	0492	0493	0494	0495
1F__	0496	0497	0498	0499	0500	0501	0502	0503	0504	0505	0506	0507	0508	0509	0510	0511

DIRECT CONVERSION TABLE (Continued)

	0	1	2	3	4	5	6	7	8	9	A	B	C	D	E	F
20__	0512	0513	0514	0515	0516	0517	0518	0519	0520	0521	0522	0523	0524	0525	0526	0527
21__	0528	0529	0530	0531	0532	0533	0534	0535	0536	0537	0538	0539	0540	0541	0542	0543
22__	0544	0545	0546	0547	0548	0549	0550	0551	0552	0553	0554	0555	0556	0557	0558	0559
23__	0560	0561	0562	0563	0564	0565	0566	0567	0568	0569	0570	0571	0572	0573	0574	0575
24__	0576	0577	0578	0579	0580	0581	0582	0583	0584	0585	0586	0587	0588	0589	0590	0591
25__	0592	0593	0594	0595	0596	0597	0598	0599	0600	0601	0602	0603	0604	0605	0606	0607
26__	0608	0609	0610	0611	0612	0613	0614	0615	0616	0617	0618	0619	0620	0621	0622	0623
27__	0624	0625	0626	0627	0628	0629	0630	0631	0632	0633	0634	0635	0636	0637	0638	0639
28__	0640	0641	0642	0643	0644	0645	0646	0647	0648	0649	0650	0651	0652	0653	0654	0655
29__	0656	0657	0658	0659	0660	0661	0662	0663	0664	0665	0666	0667	0668	0669	0670	0671
2A__	0672	0673	0674	0675	0676	0677	0678	0679	0680	0681	0682	0683	0684	0685	0686	0687
2B__	0688	0689	0690	0691	0692	0693	0694	0695	0696	0697	0698	0699	0700	0701	0702	0703
2C__	0704	0705	0706	0707	0708	0709	0710	0711	0712	0713	0714	0715	0716	0717	0718	0719
2D__	0720	0721	0722	0723	0724	0725	0726	0727	0728	0729	0730	0731	0732	0733	0734	0735
2E__	0736	0737	0738	0739	0740	0741	0742	0743	0744	0745	0746	0747	0748	0749	0750	0751
2F__	0752	0753	0754	0755	0756	0757	0758	0759	0760	0761	0762	0763	0764	0765	0766	0767
30__	0768	0769	0770	0771	0772	0773	0774	0775	0776	0777	0778	0779	0780	0781	0782	0783
31__	0784	0785	0786	0787	0788	0789	0790	0791	0792	0793	0794	0795	0796	0797	0798	0799
32__	0800	0801	0802	0803	0804	0805	0806	0807	0808	0809	0810	0811	0812	0813	0814	0815
33__	0816	0817	0818	0819	0820	0821	0822	0823	0824	0825	0826	0827	0828	0829	0830	0831
34__	0832	0833	0834	0835	0836	0837	0838	0839	0840	0841	0842	0843	0844	0845	0846	0847
35__	0848	0849	0850	0851	0852	0853	0854	0855	0856	0857	0858	0859	0860	0861	0862	0863
36__	0864	0865	0866	0867	0868	0869	0870	0871	0872	0873	0874	0875	0876	0877	0878	0879
37__	0880	0881	0882	0883	0884	0885	0886	0887	0888	0889	0890	0891	0892	0893	0894	0895
38__	0896	0897	0898	0899	0900	0901	0902	0903	0904	0905	0906	0907	0908	0909	0910	0911
39__	0912	0913	0914	0915	0916	0917	0918	0919	0920	0921	0922	0923	0924	0925	0926	0927
3A__	0928	0929	0930	0931	0932	0933	0934	0935	0936	0937	0938	0939	0940	0941	0942	0943
3B__	0944	0945	0946	0947	0948	0949	0950	0951	0952	0953	0954	0955	0956	0957	0958	0959
3C__	0960	0961	0962	0963	0964	0965	0966	0967	0968	0969	0970	0971	0972	0973	0974	0975
3D__	0976	0977	0978	0979	0980	0981	0982	0983	0984	0985	0986	0987	0988	0989	0990	0991
3E__	0992	0993	0994	0995	0996	0997	0998	0999	1000	1001	1002	1003	1004	1005	1006	1007
3F__	1008	1009	1010	1011	1012	1013	1014	1015	1016	1017	1018	1019	1020	1021	1022	1023
40__	1024	1025	1026	1027	1028	1029	1030	1031	1032	1033	1034	1035	1036	1037	1038	1039
41__	1040	1041	1042	1043	1044	1045	1046	1047	1048	1049	1050	1051	1052	1053	1054	1055
42__	1056	1057	1058	1059	1060	1061	1062	1063	1064	1065	1066	1067	1068	1069	1070	1071
43__	1072	1073	1074	1075	1076	1077	1078	1079	1080	1081	1082	1083	1084	1085	1086	1087
44__	1088	1089	1090	1091	1092	1093	1094	1095	1096	1097	1098	1099	1100	1101	1102	1103
45__	1104	1105	1106	1107	1108	1109	1110	1111	1112	1113	1114	1115	1116	1117	1118	1119
46__	1120	1121	1122	1123	1124	1125	1126	1127	1128	1129	1130	1131	1132	1133	1134	1135
47__	1136	1137	1138	1139	1140	1141	1142	1143	1144	1145	1146	1147	1148	1149	1150	1151
48__	1152	1153	1154	1155	1156	1157	1158	1159	1160	1161	1162	1163	1164	1165	1166	1167
49__	1168	1169	1170	1171	1172	1173	1174	1175	1176	1177	1178	1179	1180	1181	1182	1183
4A__	1184	1185	1186	1187	1188	1189	1190	1191	1192	1193	1194	1195	1196	1197	1198	1199
4B__	1200	120f	1202	1203	1204	1205	1206	1207	1208	1209	1210	1211	1212	1213	1214	1215
4C__	1216	1217	1218	1219	1220	1221	1222	1223	1224	1225	1226	1227	1228	1229	1230	1231
4D__	1232	1233	1234	1235	1236	1237	1238	1239	1240	1241	1242	1243	1244	1245	1246	1247
4E__	1248	1249	1250	1251	1252	1253	1254	1255	1256	1257	1258	1259	1260	1261	1262	1263
4F__	1264	1265	1266	1267	1268	1269	1270	1271	1272	1273	1274	1275	1276	1277	1278	1279
50__	1280	1281	1282	1283	1284	1285	1286	1287	1288	1289	1290	1291	1292	1293	1294	1295
51__	1296	1297	1298	1299	1300	1301	1302	1303	1304	1305	1306	1307	1308	1309	1310	1311
52__	1312	1313	1314	1315	1316	1317	1318	1319	1320	1321	1322	1323	1324	1325	1326	1327
53__	1328	1329	1330	1331	1332	1333	1334	1335	1336	1337	1338	1339	1340	1341	1342	1343
54__	1344	1345	1346	1347	1348	1349	1350	1351	1352	1353	1354	1355	1356	1357	1358	1359
55__	1360	1361	1362	1363	1364	1365	1366	1367	1368	1369	1370	1371	1372	1373	1374	1375
56__	1376	1377	1378	1379	1380	1381	1382	1383	1384	1385	1386	1387	1388	1389	1390	1391
57__	1392	1393	1394	1395	1396	1397	1398	1399	1400	1401	1402	1403	1404	1405	1406	1407
58__	1408	1409	1410	1411	1412	1413	1414	1415	1416	1417	1418	1419	1420	1421	1422	1423
59__	1424	1425	1426	1427	1428	1429	1430	1431	1432	1433	1434	1435	1436	1437	1438	1439
5A__	1440	1441	1442	1443	1444	1445	1446	1447	1448	1449	1450	1451	1452	1453	1454	1455
5B__	1456	1457	1458	1459	1460	1461	1462	1463	1464	1465	1466	1467	1468	1469	1470	1471
5C__	1472	1473	1474	1475	1476	1477	1478	1479	1480	1481	1482	1483	1484	1485	1486	1487
5D__	1488	1489	1490	1491	1492	1493	1494	1495	1496	1497	1498	1499	1500	1501	1502	1503
5E__	1504	1505	1506	1507	1508	1509	1510	1511	1512	1513	1514	1515	1516	1517	1518	1519
5F__	1520	1521	1522	1523	1524	1525	1526	1527	1528	1529	1530	1531	1532	1533	1534	1535

DIRECT CONVERSION TABLE (Continued)

	0	1	2	3	4	5	6	7	8	9	A	B	C	D	E	F
60__	1536	1537	1538	1539	1540	1541	1542	1543	1544	1545	1546	1547	1548	1549	1550	1551
61__	1552	1553	1554	1555	1556	1557	1558	1559	1560	1561	1562	1563	1564	1565	1566	1567
62__	1568	1569	1570	1571	1572	1573	1574	1575	1576	1577	1578	1579	1580	1581	1582	1583
63__	1584	1585	1586	1587	1588	1589	1590	1591	1592	1593	1594	1595	1596	1597	1598	1599
64__	1600	1601	1602	1603	1604	1605	1606	1607	1608	1609	1610	1611	1612	1613	1614	1615
65__	1616	1617	1618	1619	1620	1621	1622	1623	1624	1625	1626	1627	1628	1629	1630	1631
66__	1632	1633	1634	1635	1636	1637	1638	1639	1640	1641	1642	1643	1644	1645	1646	1647
67__	1648	1649	1650	1651	1652	1653	1654	1655	1656	1657	1658	1659	1660	1661	1662	1663
68__	1664	1665	1666	1667	1668	1669	1670	1671	1672	1673	1674	1675	1676	1677	1678	1679
69__	1680	1681	1682	1683	1684	1685	1686	1687	1688	1689	1690	1691	1692	1693	1694	1695
6A__	1696	1697	1698	1699	1700	1701	1702	1703	1704	1705	1706	1707	1708	1709	1710	1711
6B__	1712	1713	1714	1715	1716	1717	1718	1719	1720	1721	1722	1723	1724	1725	1726	1727
6C__	1728	1729	1730	1731	1732	1733	1734	1735	1736	1737	1738	1739	1740	1741	1742	1743
6D__	1744	1745	1746	1747	1748	1749	1750	1751	1752	1753	1754	1755	1756	1757	1758	1759
6E__	1760	1761	1762	1763	1764	1765	1766	1767	1768	1769	1770	1771	1772	1773	1774	1775
6F__	1776	1777	1778	1779	1780	1781	1782	1783	1784	1785	1786	1787	1788	1789	1790	1791
70__	1792	1793	1794	1795	1796	1797	1798	1799	1800	1801	1802	1803	1804	1805	1806	1807
71__	1808	1809	1810	1811	1812	1813	1814	1815	1816	1817	1818	1819	1820	1821	1822	1823
72__	1824	1825	1826	1827	1828	1829	1830	1831	1832	1833	1834	1835	1836	1837	1838	1839
73__	1840	1841	1842	1843	1844	1845	1846	1847	1848	1849	1850	1851	1852	1853	1854	1855
74__	1856	1857	1858	1859	1860	1861	1862	1863	1864	1865	1866	1867	1868	1869	1870	1871
75__	1872	1873	1874	1875	1876	1877	1878	1879	1880	1881	1882	1883	1884	1885	1886	1887
76__	1888	1889	1890	1891	1892	1893	1894	1895	1896	1897	1898	1899	1900	1901	1902	1903
77__	1904	1905	1906	1907	1908	1909	1910	1911	1912	1913	1914	1915	1916	1917	1918	1919
78__	1920	1921	1922	1923	1924	1925	1926	1927	1928	1929	1930	1931	1932	1933	1934	1935
79__	1936	1937	1938	1939	1940	1941	1942	1943	1944	1945	1946	1947	1948	1949	1950	1951
7A__	1952	1953	1954	1955	1956	1957	1958	1959	1960	1961	1962	1963	1964	1965	1966	1967
7B__	1968	1969	1970	1971	1972	1973	1974	1975	1976	1977	1978	1979	1980	1981	1982	1983
7C__	1984	1985	1986	1987	1988	1989	1990	1991	1992	1993	1994	1995	1996	1997	1998	1999
7D__	2000	2001	2002	2003	2004	2005	2006	2007	2008	2009	2010	2011	2012	2013	2014	2015
7E__	2016	2017	2018	2019	2020	2021	2022	2023	2024	2025	2026	2027	2028	2029	2030	2031
7F__	2032	2033	2034	2035	2036	2037	2038	2039	2040	2041	2042	2043	2044	2045	2046	2047
80__	2048	2049	2050	2051	2052	2053	2054	2055	2056	2057	2058	2059	2060	2061	2062	2063
81__	2064	2065	2066	2067	2068	2069	2070	2071	2072	2073	2074	2075	2076	2077	2078	2079
82__	2080	2081	2082	2083	2084	2085	2086	2087	2088	2089	2090	2091	2092	2093	2094	2095
83__	2096	2097	2098	2099	2100	2101	2102	2103	2104	2105	2106	2107	2108	2109	2110	2111
84__	2112	2113	2114	2115	2116	2117	2118	2119	2120	2121	2122	2123	2124	2125	2126	2127
85__	2128	2129	2130	2131	2132	2133	2134	2135	2136	2137	2138	2139	2140	2141	2142	2143
86__	2144	2145	2146	2147	2148	2149	2150	2151	2152	2153	2154	2155	2156	2157	2158	2159
87__	2160	2161	2162	2163	2164	2165	2166	2167	2168	2169	2170	2171	2172	2173	2174	2175
88__	2176	2177	2178	2179	2180	2181	2182	2183	2184	2185	2186	2187	2188	2189	2190	2191
89__	2192	2193	2194	2195	2196	2197	2198	2199	2200	2201	2202	2203	2204	2205	2206	2207
8A__	2208	2209	2210	2211	2212	2213	2214	2215	2216	2217	2218	2219	2220	2221	2222	2223
8B__	2224	2225	2226	2227	2228	2229	2230	2231	2232	2233	2234	2235	2236	2237	2238	2239
8C__	2240	2241	2242	2243	2244	2245	2246	2247	2248	2249	2250	2251	2252	2253	2254	2255
8D__	2256	2257	2258	2259	2260	2261	2262	2263	2264	2265	2266	2267	2268	2269	2270	2271
8E__	2272	2273	2274	2275	2276	2277	2278	2279	2280	2281	2282	2283	2284	2285	2286	2287
8F__	2288	2289	2290	2291	2292	2293	2294	2295	2296	2297	2298	2299	2300	2301	2302	2303
90__	2304	2305	2306	2307	2308	2309	2310	2311	2312	2313	2314	2315	2316	2317	2318	2319
91__	2320	2321	2322	2323	2324	2325	2326	2327	2328	2329	2330	2331	2332	2333	2334	2335
92__	2336	2337	2338	2339	2340	2341	2342	2343	2344	2345	2346	2347	2348	2349	2350	2351
93__	2352	2353	2354	2355	2356	2357	2358	2359	2360	2361	2362	2363	2364	2365	2366	2367
94__	2368	2369	2370	2371	2372	2373	2374	2375	2376	2377	2378	2379	2380	2381	2382	2383
95__	2384	2385	2386	2387	2388	2389	2390	2391	2392	2393	2394	2395	2396	2397	2398	2399
96__	2400	2401	2402	2403	2404	2405	2406	2407	2408	2409	2410	2411	2412	2413	2414	2415
97__	2416	2417	2418	2419	2420	2421	2422	2423	2424	2425	2426	2427	2428	2429	2430	2431
98__	2432	2433	2434	2435	2436	2437	2438	2439	2440	2441	2442	2443	2444	2445	2446	2447
99__	2448	2449	2450	2451	2452	2453	2454	2455	2456	2457	2458	2459	2460	2461	2462	2463
9A__	2464	2465	2466	2467	2468	2469	2470	2471	2472	2473	2474	2475	2476	2477	2478	2479
9B__	2480	2481	2482	2483	2484	2485	2486	2487	2488	2489	2490	2491	2492	2493	2494	2495
9C__	2496	2497	2498	2499	2500	2501	2502	2503	2504	2505	2506	2507	2508	2509	2510	2511
9D__	2512	2513	2514	2515	2516	2517	2518	2519	2520	2521	2522	2523	2524	2525	2526	2527
9E__	2528	2529	2530	2531	2532	2533	2534	2535	2536	2537	2538	2539	2540	2541	2542	2543
9F__	2544	2545	2546	2547	2548	2549	2550	2551	2552	2553	2554	2555	2556	2557	2558	2559

DIRECT CONVERSION TABLE (Continued)

	0	1	2	3	4	5	6	7	8	9	A	B	C	D	E	F
A0__	2560	2561	2562	2563	2564	2565	2566	2567	2568	2569	2570	2571	2572	2573	2574	2575
A1__	2576	2577	2578	2579	2580	2581	2582	2583	2584	2585	2586	2587	2588	2589	2590	2591
A2__	2592	2593	2594	2595	2596	2597	2598	2599	2600	2601	2602	2603	2604	2605	2606	2607
A3__	2608	2609	2610	2611	2612	2613	2614	2615	2616	2617	2618	2619	2620	2621	2622	2623
A4__	2624	2625	2626	2627	2628	2629	2630	2631	2632	2633	2634	2635	2636	2637	2638	2639
A5__	2640	2641	2642	2643	2644	2645	2646	2647	2648	2649	2650	2651	2652	2653	2654	2655
A6__	2656	2657	2658	2659	2660	2661	2662	2663	2664	2665	2666	2667	2668	2669	2670	2671
A7__	2672	2673	2674	2675	2676	2677	2678	2679	2680	2681	2682	2683	2684	2685	2686	2687
A8__	2688	2689	2690	2691	2692	2693	2694	2695	2696	2697	2698	2699	2700	2701	2702	2703
A9__	2704	2705	2706	2707	2708	2709	2710	2711	2712	2713	2714	2715	2716	2717	2718	2719
AA__	2720	2721	2722	2723	2724	2725	2726	2727	2728	2729	2730	2731	2732	3733	2734	2735
AB__	2736	2737	2738	2739	2740	2741	2742	2743	2744	2745	2746	2747	2748	2749	2750	2751
AC__	2752	2753	2754	2755	2756	2757	2758	2759	2760	2761	2762	2763	2764	2765	2766	2767
AD__	2768	2769	2770	2771	2772	2773	2774	2775	2776	2777	2778	2779	2780	2781	2782	2783
AE__	2784	2785	2786	2787	2788	2789	2790	2791	2792	2793	2794	2795	2796	2797	2798	2799
AF__	2800	2801	2802	2803	2804	2805	2806	2807	2808	2809	2810	2811	2812	2813	2814	2815
B0__	2816	2817	2818	2819	2820	2821	2822	2823	2824	2825	2826	2827	2828	2829	2830	2831
B1__	2832	2833	2834	2835	2836	2837	2838	2839	2840	2841	2842	2843	2844	2845	2846	2847
B2__	2848	2849	2850	2851	2852	2853	2854	2855	2856	2857	2858	2859	2860	2861	2862	2863
B3__	2864	2865	2866	2867	2868	2869	2870	2871	2872	2873	2874	2875	2876	2877	2878	2879
B4__	2880	2881	2882	2883	2884	2885	2886	2887	2888	2889	2890	2891	2892	2893	2894	2895
B5__	2896	2897	2898	2899	2900	2901	2902	2903	2904	2905	2906	2907	2908	2909	2910	2911
B6__	2912	2913	2914	2915	2916	2917	2918	2919	2920	2921	2922	2923	2924	2925	2926	2927
B7__	2928	2929	2930	2931	2932	2933	2934	2935	2936	2937	2938	2939	2940	2941	2942	2943
B8__	2944	2945	2946	2947	2948	2949	2950	2951	2952	2953	2954	2955	2956	2957	2958	2959
B9__	2960	2961	2962	2963	2964	2965	2966	2967	2968	2969	2970	2971	2972	2973	2974	2975
BA__	2976	2977	2978	2979	2980	2981	2982	2983	2984	2985	2986	2987	2988	2989	2990	2991
BB__	2992	2993	2994	2995	2996	2997	2998	2999	3000	3001	3002	3003	3004	3005	3006	3007
BC__	3008	3009	3010	3011	3012	3013	3014	3015	3016	3017	3018	3019	3020	3021	3022	3023
BD__	3024	3025	3026	3027	3028	3029	3030	3031	3032	3033	3034	3035	3036	3037	3038	3039
BE__	3040	3041	3042	3043	3044	3045	3046	3047	3048	3049	3050	3051	3052	3053	3054	3055
BF__	3056	3057	3058	3059	3060	3061	3062	3063	3064	3065	3066	3067	3068	3069	3070	3071
C0__	3072	3073	3074	3075	3076	3077	3078	3079	3080	3081	3082	3083	3084	3085	3086	3087
C1__	3088	3089	3090	3091	3092	3093	3094	3095	3096	3097	3098	3099	3100	3101	3102	3103
C2__	3104	3105	3106	3107	3108	3109	3110	3111	3112	3113	3114	3115	3116	3117	3118	3119
C3__	3120	3121	3122	3123	3124	3125	3126	3127	3128	3129	3130	3131	3132	3133	3134	3135
C4__	3136	3137	3138	3139	3140	3141	3142	3143	3144	3145	3146	3147	3148	3149	3150	3151
C5__	3152	3153	3154	3155	3156	3157	3158	3159	3160	3161	3162	3613	3164	3165	3166	3167
C6__	3168	3169	3170	3171	3172	3173	3174	3175	3176	3177	3178	3179	3180	3181	3182	3183
C7__	3184	3185	3186	3187	3188	3189	3190	3191	3192	3193	3194	3195	3196	3197	3198	3199
C8__	3200	3201	3202	3203	3204	3205	3206	3207	3208	3209	3210	3211	3212	3213	3214	3215
C9__	3216	3217	3218	3219	3220	3221	3222	3223	3224	3225	3226	3227	3228	3229	3230	3231
CA__	3232	3233	3234	3235	3236	3237	3238	3239	3240	3241	3242	3243	3244	3245	3246	3247
CB__	3248	3249	3250	3251	3252	3253	3254	3255	3256	3257	3258	3259	3260	3261	3262	3263
CC__	3264	3265	3266	3267	3268	3269	3270	3271	3272	3273	3274	3275	3276	3277	3278	3279
CD__	3280	3281	3282	3283	3284	3285	3286	3287	3288	3289	3290	3291	3292	3293	3294	3295
CE__	3296	3297	3298	3299	3300	3301	3302	3303	3304	3305	3306	3307	3308	3309	3310	3311
CF__	3312	3313	3314	3315	3316	3317	3318	3319	3320	3321	3322	3323	3324	3325	3326	3327
D0__	3328	3329	3330	3331	3332	3333	3334	3335	3336	3337	3338	3339	3340	3341	3342	3343
D1__	3344	3345	3346	3347	3348	3349	3350	3351	3352	3353	3354	3355	3356	3357	3358	3359
D2__	3360	3361	3362	3363	3364	3365	3366	3367	3368	3369	3370	3371	3372	3373	3374	3375
D3__	3376	3377	3378	3379	3380	3381	3382	3383	3384	3385	3386	3387	3388	3389	3390	3391
D4__	3392	3393	3394	3395	3396	3397	3398	3399	3400	3401	3402	3403	3404	3405	3406	3407
D5__	3408	3409	3410	3411	3412	3413	3414	3415	3416	3417	3418	3419	3420	3421	3422	3423
D6__	3424	3425	3426	3427	3428	3429	3430	3431	3432	3433	3434	3435	3436	3437	3438	3439
D7__	3440	3441	3442	3443	3444	3445	3446	3447	3448	3449	3450	3451	3452	3453	3454	3455
D8__	3456	3457	3458	3459	3460	3461	3462	3463	3464	3465	3466	3467	3468	3469	3470	3471
D9__	3472	3473	3474	3475	3476	3477	3478	3479	3480	3481	3482	3483	3484	3485	3486	3487
DA__	3488	3489	3490	3491	3492	3493	3494	3495	3496	3497	3498	3499	3500	3501	3502	3503
DB__	3504	3505	3506	3507	3508	3509	3510	3511	3512	3513	3514	3515	3516	3517	3518	3519
DC__	3520	3521	3522	3523	3524	3525	3526	3527	3528	3529	3530	3531	3532	3533	3534	3535
DD__	3536	3537	3538	3539	3540	3541	3542	3543	3544	3545	3546	3547	3548	3549	3550	3551
DE__	3552	3553	3554	3555	3556	3557	3558	3559	3560	3561	3562	3563	3564	3565	3566	3567
DF__	3568	3569	3570	3571	3572	3573	3574	3575	3576	3577	3578	3579	3580	3581	3582	3583

DIRECT CONVERSION TABLE (Continued)

	0	1	2	3	4	5	6	7	8	9	A	B	C	D	E	F
E0__	3584	3585	3586	3587	3588	3589	3590	3591	3592	3593	3594	3595	3596	3597	3598	3599
E1__	3600	3601	3602	3603	3604	3605	3606	3607	3608	3609	3610	3611	3612	3613	3614	3615
E2__	3616	3617	3618	3619	3620	3621	3622	3623	3624	3625	3626	3627	3628	3629	3630	3631
E3__	3632	3633	3634	3635	3636	3637	3638	3639	3640	3641	3642	3643	3644	3645	3646	3647
E4__	3648	3649	3650	3651	3652	3653	3654	3655	3656	3657	3658	3659	3660	3661	3662	3663
E5__	3664	3665	3666	3667	3668	3669	3670	3671	3672	3673	3674	3675	3676	3677	3678	3679
E6__	3680	3681	3682	3683	3684	3685	3686	3687	3688	3689	3690	3691	3692	3693	3694	3695
E7__	3696	3697	3698	3699	3700	3701	3702	3703	3704	3705	3706	3707	3708	3709	3710	3711
E8__	3712	3713	3714	3715	3716	3717	3718	3719	3720	3721	3722	3723	3724	3725	3726	3727
E9__	3728	3729	3730	3731	3732	3733	3734	3735	3736	3737	3738	3739	3740	3741	3742	3743
EA__	3744	3745	3746	3747	3748	3749	3750	3751	3752	3753	3754	3755	3756	3757	3758	3759
EB__	3760	3761	3762	3763	3764	3765	3766	3767	3768	3769	3770	3771	3772	3773	3774	3775
EC__	3776	3777	3778	3779	3780	3781	3782	3783	3784	3785	3786	3787	3788	3789	3790	3791
ED__	3792	3793	3794	3795	3796	3797	3798	3799	3800	3801	3802	3803	3804	3805	3806	3807
EE__	3808	3809	3810	3811	3812	3813	3814	3815	3816	3817	3818	3819	3820	3821	3822	3823
EF__	3824	3825	3826	3827	3828	3829	3830	3831	3832	3833	3834	3835	3836	3837	3838	3839
F0__	3840	3841	3842	3843	3844	3845	3846	3847	3848	3849	3850	3851	3852	3853	3854	3855
F1__	3856	3857	3858	3859	3860	3861	3862	3863	3864	3865	3866	3867	3868	3869	3870	3871
F2__	3872	3873	3874	3875	3876	3877	3878	3879	3880	3881	3882	3883	3884	3885	3886	3887
F3__	3888	3889	3890	3891	3892	3893	3894	3895	3896	3897	3898	3899	3900	3901	3902	3903
F4__	3904	3905	3906	3907	3908	3909	3910	3911	3912	3913	3914	3915	3916	3917	3918	3919
F5__	3920	3921	3922	3923	3924	3925	3926	3927	3928	3929	3930	3931	3932	3933	3934	3935
F6__	3936	3937	3938	3939	3940	3941	3942	3943	3944	3945	3946	3947	3948	3949	3950	3951
F7__	3952	3953	3954	3955	3956	3957	3958	3959	3960	3961	3962	3963	3964	3965	3966	3967
F8__	3968	3969	3970	3971	3972	3973	3974	3975	3976	3977	3978	3979	3980	3981	3982	3983
F9__	3984	3985	3986	3987	3988	3989	3990	3991	3992	3993	3994	3995	3996	3997	3998	3999
FA__	4000	4001	4002	4003	4004	4005	4006	4007	4008	4009	4010	4011	4012	4013	4014	4015
FB__	4016	4017	4018	4019	4020	4021	4022	4023	4024	4025	4026	4027	4028	4029	4030	4031
FC__	4032	4033	4034	4035	4036	4037	4038	4039	4040	4041	4042	4043	4044	4045	4046	4047
FD__	4048	4049	4050	4051	4052	4053	4054	4055	4056	4057	4058	4059	4060	4061	4062	4063
FE__	4064	4065	4066	4067	4068	4069	4070	4071	4072	4073	4074	4075	4076	4077	4078	4079
FF__	4080	4081	4082	4083	4084	4085	4086	4087	4088	4089	4090	4091	4092	4093	4094	4095

II. HEXADECIMAL AND DECIMAL INTEGER CONVERSION TABLE

	8		7		6		5		4		3		2		1
Hex	Decimal	Hex	Decimal	Hex	Decimal	Hex	Decimal	Hex	Decimal	Hex	Decimal	Hex	Decimal	Hex	Decimal
0	0	0	0	0	0	0	0	0	0	0	0	0	0	0	0
1	268,435,456	1	16,777,216	1	1,048,576	1	65,536	1	4,096	1	256	1	16	1	1
2	536,870,912	2	33,554,432	2	2,097,152	2	131,072	2	8,192	2	512	2	32	2	2
3	805,306,368	3	50,331,648	3	3,145,728	3	196,608	3	12,288	3	768	3	48	3	3
4	1,073,741,824	4	67,108,864	4	4,194,304	4	262,144	4	16,384	4	1,024	4	64	4	4
5	1,342,177,280	5	83,886,080	5	5,242,880	5	327,680	5	20,480	5	1,280	5	80	5	5
6	1,610,612,736	6	100,663,296	6	6,291,456	6	393,216	6	24,576	6	1,536	6	96	6	6
7	1,879,048,192	7	117,440,512	7	7,340,032	7	458,752	7	28,672	7	1,792	7	112	7	7
8	2,147,483,648	8	134,217,728	8	8,388,608	8	524,288	8	32,768	8	2,048	8	128	8	8
9	2,415,919,104	9	150,994,944	9	9,437,184	9	589,824	9	36,864	9	2,304	9	144	9	9
A	2,684,354,560	A	167,772,160	A	10,485,760	A	655,360	A	40,960	A	2,560	A	160	A	10
B	2,952,790,016	B	184,549,376	B	11,534,336	B	720,896	B	45,056	B	2,816	B	176	B	11
C	3,221,225,472	C	201,326,592	C	12,582,912	C	786,432	C	49,152	C	3,072	C	192	C	12
D	3,489,660,928	D	218,103,808	D	13,631,488	D	851,968	D	53,248	D	3,328	D	208	D	13
E	3,758,096,384	E	234,881,024	E	14,680,064	E	917,504	E	57,344	E	3,584	E	224	E	14
F	4,026,531,840	F	251,658,240	F	15,728,640	F	983,040	F	61,440	F	3,840	F	240	F	15
	8		7		6		5		4		3		2		1

INTEGER CONVERSION TABLE (Continued)

TO CONVERT HEXADECIMAL TO DECIMAL

1. Locate the column of decimal numbers corresponding to the left-most digit or letter of the hexadecimal; select from this column and record the number that corresponds to the position of the hexadecimal digit or letter.

2. Repeat step 1 for the next (second from the left) position.

3. Repeat step 1 for the units (third from the left) position.

4. Add the numbers selected from the table to form the decimal number.

To convert integer numbers greater than the capacity of table, use the techniques below:

HEXADECIMAL TO DECIMAL

Successive cumulative multiplication from left to right, adding units position.

Example: $D34_{16} = 3380_{10}$

$$
\begin{array}{rr}
D = & 13 \\
& \times 16 \\
\hline
& 208 \\
3 = & +3 \\
\hline
& 211 \\
& \times 16 \\
\hline
& 3376 \\
4 = & +4 \\
\hline
& 3380
\end{array}
$$

EXAMPLE	
Conversion of Hexadecimal Value	D34
1. D	3328
2. 3	48
3. 4	4
4. Decimal	3380

TO CONVERT DECIMAL TO HEXADECIMAL

1. (a) Select from the table the highest decimal number that is equal to or less than the number to be converted.
   (b) Record the hexadecimal of the column containing the selected number.
   (c) Subtract the selected decimal from the number to be converted.

2. Using the remainder from step 1(c) repeat all of step 1 to develop the second position of the hexadecimal (and a remainder).

3. Using the remainder from step 2 repeat all of step 1 to develop the units position of the hexadecimal.

4. Combine terms to form the hexadecimal number.

EXAMPLE	
Conversion of Decimal Value	3380
1. D	-3328
	52
2. 3	-48
	4
3. 4	-4
4. Hexa-decimal	D34

DECIMAL TO HEXADECIMAL

Divide and collect the remainder in reverse order.

Example: $3380_{10} = X_{16}$

```
16 | 3380        remainder
16 | 211      4
16 | 13       3
           D
```

$3380_{10} = D34_{16}$

INTEGER CONVERSION TABLE (Continued)

<u>POWERS OF 16 TABLE</u>

Example: $268,435,456_{10} = (2.68435456 \times 10^8)_{10} = 1000\ 0000_{16} = (10^7)_{16}$

16^n	n
1	0
16	1
256	2
4 096	3
65 536	4
1 048 576	5
16 777 216	6
268 435 456	7
4 294 967 296	8
68 719 476 736	9
1 099 511 627 776	10 = A
17 592 186 044 416	11 = B
281 474 976 710 656	12 = C
4 503 599 627 370 496	13 = D
72 057 594 037 927 936	14 = E
1 152 921 504 606 846 976	15 = F

Decimal Values

III. HEXADECIMAL AND DECIMAL FRACTION CONVERSION TABLE

1		2			3				4					
Hex	Decimal	Hex	Decimal		Hex	Decimal			Hex	Decimal Equivalent				
.0	.0000	.00	.0000	0000	.000	.0000	0000	0000	.0000	.0000	0000	0000	0000
.1	.0625	.01	.0039	0625	.001	.0002	4414	0625	.0001	.0000	1525	8789	0625
.2	.1250	.02	.0078	1250	.002	.0004	8828	1250	.0002	.0000	3051	7578	1250
.3	.1875	.03	.0117	1875	.003	.0007	3242	1875	.0003	.0000	4577	6367	1875
.4	.2500	.04	.0156	2500	.004	.0009	7656	2500	.0004	.0000	6103	5156	2500
.5	.3125	.05	.0195	3125	.005	.0012	2070	3125	.0005	.0000	7629	3945	3125
.6	.3750	.06	.0234	3750	.006	.0014	6484	3750	.0006	.0000	9155	2734	3750
.7	.4375	.07	.0273	4375	.007	.0017	0898	4375	.0007	.0001	0681	1523	4375
.8	.5000	.08	.0312	5000	.008	.0019	5312	5000	.0008	.0001	2207	0312	5000
.9	.5625	.09	.0351	5625	.009	.0021	9726	5625	.0009	.0001	3732	9101	5625
.A	.6250	.0A	.0390	6250	.00A	.0024	4140	6250	.000A	.0001	5258	7890	6250
.B	.6875	.0B	.0429	6875	.00B	.0026	8554	6875	.000B	.0001	6784	6679	6875
.C	.7500	.0C	.0468	7500	.00C	.0029	2968	7500	.000C	.0001	8310	5468	7500
.D	.8125	.0D	.0507	8125	.00D	.0031	7382	8125	.000D	.0001	9836	4257	8125
.E	.8750	.0E	.0546	8750	.00E	.0034	1796	8750	.000E	.0002	1362	3046	8750
.F	.9375	.0F	.0585	9375	.00F	.0036	6210	9375	.000F	.0002	2888	1835	9375
	1		2			3				4				

<u>TO CONVERT .ABC HEXADECIMAL TO DECIMAL</u>

Find .A   in position 1   .6250
Find .0B   in position 2   .0429 6875
Find .00C in position 3   <u>.0029 2968 7500</u>
  .ABC Hex is equal to   .6708 9843 7500

FRACTION CONVERSION TABLE (Continued)

TO CONVERT .13 DECIMAL TO HEXADECIMAL

1. Find .1250 next lowest to   .1300
            subtract $-.1250$ = .2 Hex
2. Find .0039 0625 next lowest to   .0050 0000
                                     $-.0039\ 0625$ = .01
3. Find .0009 7656 2500   .0010 9375 0000
                                     $-.0009\ 7656\ 2500$ = .004
4. Find .0001 0681 1523 4375   .0001 1718 7500 0000
                                     $-.0001\ 0681\ 1523\ 4375$ = .0007
                                        .0000 1037 5976 5625 = .2147 Hex

5. 13 Decimal is approximately equal to ——————————————↑

To convert fractions beyond the capacity of table, use techniques below:

HEXADECIMAL FRACTION TO DECIMAL

Convert the hexadecimal fraction to its decimal equivalent using the same technique as for integer numbers. Divide the results by 16^n (n is the number of fraction positions).

Example: $.8A7_{16} = .540771_{10}$

$$8A7_{16} = 2215_{10}$$
$$16^3 = 4096$$

$$4096 \overline{\smash{\big)}\ 2215.000000} \quad .540771$$

DECIMAL FRACTION TO HEXADECIMAL

Collect integer parts of product in the order of calculation.

Example: $.5408_{10} = .8A7_{16}$

$$
\begin{array}{r}
.5408 \\
\times 16 \\
\hline
8 \leftarrow \boxed{8}\ .6528 \\
\times 16 \\
\hline
A \leftarrow \boxed{10}\ .4448 \\
\times 16 \\
\hline
7 \leftarrow \boxed{7}\ .1168
\end{array}
$$

HEXADECIMAL ADDITION AND SUBTRACTION TABLE

Example: 6 + 2 = 8, 8 − 2 = 6, and 8 − 6 = 2

	1	2	3	4	5	6	7	8	9	A	B	C	D	E	F
1	02	03	04	05	06	07	08	09	0A	0B	0C	0D	0E	0F	10
2	03	04	05	06	07	08	09	0A	0B	0C	0D	0E	0F	10	11
3	04	05	06	07	08	09	0A	0B	0C	0D	0E	0F	10	11	12
4	05	06	07	08	09	0A	0B	0C	0D	0E	0F	10	11	12	13
5	06	07	08	09	0A	0B	0C	0D	0E	0F	10	11	12	13	14
6	07	08	09	0A	0B	0C	0D	0E	0F	10	11	12	13	14	15
7	08	09	0A	0B	0C	0D	0E	0F	10	11	12	13	14	15	16
8	09	0A	0B	0C	0D	0E	0F	10	11	12	13	14	15	16	17
9	0A	0B	0C	0D	0E	0F	10	11	12	13	14	15	16	17	18
A	0B	0C	0D	0E	0F	10	11	12	13	14	15	16	17	18	19
B	0C	0D	0E	0F	10	11	12	13	14	15	16	17	18	19	1A
C	0D	0E	0F	10	11	12	13	14	15	16	17	18	19	1A	1B
D	0E	0F	10	11	12	13	14	15	16	17	18	19	1A	1B	1C
E	0F	10	11	12	13	14	15	16	17	18	19	1A	1B	1C	1D
F	10	11	12	13	14	15	16	17	18	19	1A	1B	1C	1D	1E

HEXADECIMAL MULTIPLICATION TABLE

Example: 2 × 4 = 08, F × 2 = 1E

	1	2	3	4	5	6	7	8	9	A	B	C	D	E	F
1	01	02	03	04	05	06	07	08	09	0A	0B	0C	0D	0E	0F
2	02	04	06	08	0A	0C	0E	10	12	14	16	18	1A	1C	1E
3	03	06	09	0C	0F	12	15	18	1B	1E	21	24	27	2A	2D
4	04	08	0C	10	14	18	1C	20	24	28	2C	30	34	38	3C
5	05	0A	0F	14	19	1E	23	28	2D	32	37	3C	41	46	4B
6	06	0C	12	18	1E	24	2A	30	36	3C	42	48	4E	54	5A
7	07	0E	15	1C	23	2A	31	38	3F	46	4D	54	5B	62	69
08	08	10	18	20	28	30	38	40	48	50	58	60	68	70	78
9	09	12	1B	24	2D	36	3F	48	51	5A	63	6C	75	7E	87
A	0A	14	1E	28	32	3C	46	50	5A	64	6E	78	82	8C	96
B	0B	16	21	2C	37	42	4D	58	63	6E	79	84	8F	9A	A5
C	0C	18	24	30	3C	48	54	60	6C	78	84	90	9C	A8	B4
D	0D	1A	27	34	41	4E	5B	68	75	82	8F	9C	A9	B6	C3
E	0E	1C	2A	38	46	54	62	70	7E	8C	9A	A8	B6	C4	D2
F	0F	1E	2D	3C	4B	5A	69	78	87	96	A5	B4	C3	D2	E1

TOTIENT FUNCTION $\phi(n)$

Introductory facts

$\phi(n)$ is the number of integers not exceeding and relatively prime to n.

σ_k is the sum of the k'th powers of divisors of n.

σ_0 is usually denoted by $d(n)$ and specifies the number of divisors of n.

For example if $n = 10$, then the divisors are 1, 2, 5, 10 and the sum of the first powers is 18, i.e. $\sigma_0 = d(10) = 4$; $\sigma_1 = 18$; $\phi(10) = 4$.

n	$\phi(n)$	σ_0	σ_1	n	$\phi(n)$	σ_0	σ_1	n	$\phi(n)$	σ_0	σ_1	n	$\phi(n)$	σ_0	σ_1
1	1	1	1	41	40	2	42	81	54	5	121	121	110	3	133
2	1	2	3	42	12	8	96	82	40	4	126	122	60	4	186
3	2	2	4	43	42	2	44	83	82	2	84	123	80	4	168
4	2	3	7	44	20	6	84	84	24	12	224	124	60	6	224
5	4	2	6	45	24	6	78	85	64	4	108	125	100	4	156
6	2	4	12	46	22	4	72	86	42	4	132	126	36	12	312
7	6	2	8	47	46	2	48	87	56	4	120	127	126	2	128
8	4	4	15	48	16	10	124	88	40	8	180	128	64	8	255
9	6	3	13	49	42	3	57	89	88	2	90	129	84	4	176
10	4	4	18	50	20	6	93	90	24	12	234	130	48	8	252
11	10	2	12	51	32	4	72	91	72	4	112	131	130	2	132
12	4	6	28	52	24	6	98	92	44	6	168	132	40	12	336
13	12	2	14	53	52	2	54	93	60	4	128	133	108	4	160
14	6	4	24	54	18	8	120	94	46	4	144	134	66	4	204
15	8	4	24	55	40	4	72	95	72	4	120	135	72	8	240
16	8	5	31	56	24	8	120	96	32	12	252	136	64	8	270
17	16	2	18	57	36	4	80	97	96	2	98	137	136	2	138
18	6	6	39	58	28	4	90	98	42	6	171	138	44	8	288
19	18	2	20	59	58	2	60	99	60	6	156	139	138	2	140
20	8	6	42	60	16	12	168	100	40	9	217	140	48	12	336
21	12	4	32	61	60	2	62	101	100	2	102	141	92	4	192
22	10	4	36	62	30	4	96	102	32	8	216	142	70	4	216
23	22	2	24	63	36	6	104	103	102	2	104	143	120	4	168
24	8	8	60	64	32	7	127	104	48	8	210	144	48	15	403
25	20	3	31	65	48	4	84	105	48	8	192	145	112	4	180
26	12	4	42	66	20	8	144	106	52	4	162	146	72	4	222
27	18	4	40	67	66	2	68	107	106	2	108	147	84	6	228
28	12	6	56	68	32	6	126	108	36	12	280	148	72	6	266
29	28	2	30	69	44	4	96	109	108	2	110	149	148	2	150
30	8	8	72	70	24	8	144	110	40	8	216	150	40	12	372
31	30	2	32	71	70	2	72	111	72	4	152	151	150	2	152
32	16	6	63	72	24	12	195	112	48	10	248	152	72	8	300
33	20	4	48	73	72	2	74	113	112	2	114	153	96	6	234
34	16	4	54	74	36	4	114	114	36	8	240	154	60	8	288
35	24	4	48	75	40	6	124	115	88	4	144	155	120	4	192
36	12	9	91	76	36	6	140	116	56	6	210	156	48	12	392
37	36	2	38	77	60	4	96	117	72	6	182	157	156	2	158
38	18	4	60	78	24	8	168	118	58	4	180	158	78	4	240
39	24	4	56	79	78	2	80	119	96	4	144	159	104	4	216
40	16	8	90	80	32	10	186	120	32	16	360	160	64	12	378

TOTIENT FUNCTION $\phi(n)$ (Continued)

n	$\phi(n)$	σ_0	σ_1	n	$\phi(n)$	σ_0	σ_1	n	$\phi(n)$	σ_0	σ_1	n	$\phi(n)$	σ_0	σ_1
161	132	4	192	211	210	2	212	261	168	6	390	311	310	2	312
162	54	10	363	212	104	6	378	262	130	4	396	312	96	16	840
163	162	2	164	213	140	4	288	263	262	2	264	313	312	2	314
164	80	6	294	214	106	4	324	264	80	16	720	314	156	4	474
165	80	8	288	215	168	4	264	265	208	4	324	315	144	12	624
166	82	4	252	216	72	16	600	266	108	8	480	316	156	6	560
167	166	2	168	217	180	4	256	267	176	4	360	317	316	2	318
168	48	16	480	218	108	4	330	268	132	6	476	318	104	8	648
169	156	3	183	219	144	4	296	269	268	2	270	319	280	4	360
170	64	8	324	220	80	12	504	270	72	16	720	320	128	14	762
171	108	6	260	221	192	4	252	271	270	2	272	321	212	4	432
172	84	6	308	222	72	8	456	272	128	10	558	322	132	8	576
173	172	2	174	223	222	2	224	273	144	8	448	323	288	4	360
174	56	8	360	224	96	12	504	274	136	4	414	324	108	15	847
175	120	6	248	225	120	9	403	275	200	6	372	325	240	6	434
176	80	10	372	226	112	4	342	276	88	12	672	326	162	4	492
177	116	4	240	227	226	2	228	277	276	2	278	327	216	4	440
178	88	4	270	228	72	12	560	278	138	4	420	328	160	8	630
179	178	2	180	229	228	2	230	279	180	6	416	329	276	4	384
180	48	18	546	230	88	8	432	280	96	16	720	330	80	16	864
181	180	2	182	231	120	8	384	281	280	2	282	331	330	2	332
182	72	8	336	232	112	8	450	282	92	8	576	332	164	6	588
183	120	4	248	233	232	2	234	283	282	2	284	333	216	6	494
184	88	8	360	234	72	12	546	284	140	6	504	334	166	4	504
185	144	4	228	235	184	4	288	285	144	8	480	335	264	4	408
186	60	8	384	236	116	6	420	286	120	8	504	336	96	20	992
187	160	4	216	237	156	4	320	287	240	4	336	337	336	2	338
188	92	6	336	238	96	8	432	288	96	18	819	338	156	6	549
189	108	8	320	239	238	2	240	289	272	3	307	339	224	4	456
190	72	8	360	240	64	20	744	290	112	8	540	340	128	12	756
191	190	2	192	241	240	2	242	291	192	4	392	341	300	4	384
192	64	14	508	242	110	6	399	292	144	6	518	342	108	12	780
193	192	2	194	243	162	6	364	293	292	2	294	343	294	4	400
194	96	4	294	244	120	6	434	294	84	12	684	344	168	8	660
195	96	8	336	245	168	6	342	295	232	4	360	345	176	8	576
196	84	9	399	246	80	8	504	296	144	8	570	346	172	4	522
197	196	2	198	247	216	4	280	297	180	8	480	347	346	2	348
198	60	12	468	248	120	8	480	298	148	4	450	348	112	12	840
199	198	2	200	249	164	4	336	299	264	4	336	349	348	2	350
200	80	12	465	250	100	8	468	300	80	18	868	350	120	12	744
201	132	4	272	251	250	2	252	301	252	4	352	351	216	8	560
202	100	4	306	252	72	18	728	302	150	4	456	352	160	12	756
203	168	4	240	253	220	4	288	303	200	4	408	353	352	2	354
204	64	12	504	254	126	4	384	304	144	10	620	354	116	8	720
205	160	4	252	255	128	8	432	305	240	4	372	355	280	4	432
206	102	4	312	256	128	9	511	306	96	12	702	356	176	6	630
207	132	6	312	257	256	2	258	307	306	2	308	357	192	8	576
208	96	10	434	258	84	8	528	308	120	12	672	358	178	4	540
209	180	4	240	259	216	4	304	309	204	4	416	359	358	2	360
210	48	16	576	260	96	12	588	310	120	8	576	360	96	24	1170

TOTIENT FUNCTION $\phi(n)$ (Continued)

n	$\phi(n)$	σ_0	σ_1	n	$\phi(n)$	σ_0	σ_1	n	$\phi(n)$	σ_0	σ_1	n	$\phi(n)$	σ_0	σ_1
361	342	3	381	411	272	4	552	461	460	2	462	511	432	4	592
362	180	4	546	412	204	6	728	462	120	16	1152	512	256	10	1023
363	220	6	532	413	348	4	480	463	462	2	464	513	324	8	800
364	144	12	784	414	132	12	936	464	224	10	930	514	256	4	774
365	288	4	444	415	328	4	504	465	240	8	768	515	408	4	624
366	120	8	744	416	192	12	882	466	232	4	702	516	168	12	1232
367	366	2	368	417	276	4	560	467	466	2	468	517	460	4	576
368	176	10	744	418	180	8	720	468	144	18	1274	518	216	8	912
369	240	6	546	419	418	2	420	469	396	4	544	519	344	4	696
370	144	8	684	420	96	24	1344	470	184	8	864	520	192	16	1260
371	312	4	432	421	420	2	422	471	312	4	632	521	520	2	522
372	120	12	896	422	210	4	636	472	232	8	900	522	168	12	1170
373	372	2	374	423	276	6	624	473	420	4	528	523	522	2	524
374	160	8	648	424	208	8	810	474	156	8	960	524	260	6	924
375	200	8	624	425	320	6	558	475	360	6	620	525	240	12	992
376	184	8	720	426	140	8	864	476	192	12	1008	526	262	4	792
377	336	4	420	427	360	4	496	477	312	6	702	527	480	4	576
378	108	16	960	428	212	6	756	478	238	4	720	528	160	20	1488
379	378	2	380	429	240	8	672	479	478	2	480	529	506	3	553
380	144	12	840	430	168	8	792	480	128	24	1512	530	208	8	972
381	252	4	512	431	430	2	432	481	432	4	532	531	348	6	780
382	190	4	576	432	144	20	1240	482	240	4	726	532	216	12	1120
383	382	2	384	433	432	2	434	483	264	8	768	533	480	4	588
384	128	16	1020	434	180	8	768	484	220	9	931	534	176	8	1080
385	240	8	576	435	224	8	720	485	384	4	588	535	424	4	648
386	192	4	582	436	216	6	770	486	162	12	1092	536	264	8	1020
387	252	6	572	437	396	4	480	487	486	2	488	537	356	4	720
388	192	6	686	438	144	8	888	488	240	8	930	538	268	4	810
389	388	2	390	439	438	2	440	489	324	4	656	539	420	6	684
390	96	16	1008	440	160	16	1080	490	168	12	1026	540	144	24	1680
391	352	4	432	441	252	9	741	491	490	2	492	541	540	2	542
392	168	12	855	442	192	8	756	492	160	12	1176	542	270	4	816
393	260	4	528	443	442	2	444	493	448	4	540	543	360	4	728
394	196	4	594	444	144	12	1064	494	216	8	840	544	256	12	1134
395	312	4	480	445	352	4	540	495	240	12	936	545	432	4	660
396	120	18	1092	446	222	4	672	496	240	10	992	546	144	16	1344
397	396	2	398	447	296	4	600	497	420	4	576	547	546	2	548
398	198	4	600	448	192	14	1016	498	164	8	1008	548	272	6	966
399	216	8	640	449	448	2	450	499	498	2	500	549	360	6	806
400	160	15	961	450	120	18	1209	500	200	12	1092	550	200	12	1116
401	400	2	402	451	400	4	504	501	332	4	672	551	504	4	600
402	132	8	816	452	224	6	798	502	250	4	756	552	176	16	1440
403	360	4	448	453	300	4	608	503	502	2	504	553	468	4	640
404	200	6	714	454	226	4	684	504	144	24	1560	554	276	4	834
405	216	10	726	455	288	8	672	505	400	4	612	555	288	8	912
406	168	8	720	456	144	16	1200	506	220	8	864	556	276	6	980
407	360	4	456	457	456	2	458	507	312	6	732	557	556	2	558
408	128	16	1080	458	228	4	690	508	252	6	896	558	180	12	1248
409	408	2	410	459	288	8	720	509	508	2	510	559	504	4	616
410	160	8	756	460	176	12	1008	510	128	16	1296	560	192	20	1488

TOTIENT FUNCTION $\phi(n)$ (Continued)

n	$\phi(n)$	σ_0	σ_1	n	$\phi(n)$	σ_0	σ_1	n	$\phi(n)$	σ_0	σ_1	n	$\phi(n)$	σ_0	σ_1
561	320	8	864	611	552	4	672	661	660	2	662	711	468	6	1040
562	280	4	846	612	192	18	1638	662	330	4	996	712	352	8	1350
563	562	2	564	613	612	2	614	663	384	8	1008	713	660	4	768
564	184	12	1344	614	306	4	924	664	328	8	1260	714	192	16	1728
565	448	4	684	615	320	8	1008	665	432	8	960	715	480	8	1008
566	282	4	852	616	240	16	1440	666	216	12	1482	716	356	6	1260
567	324	10	968	617	616	2	618	667	616	4	720	717	476	4	960
568	280	8	1080	618	204	8	1248	668	332	6	1176	718	358	4	1080
569	568	2	570	619	618	2	620	669	444	4	896	719	718	2	720
570	144	16	1440	620	240	12	1344	670	264	8	1224	720	192	30	2418
571	570	2	572	621	396	8	960	671	600	4	744	721	612	4	832
572	240	12	1176	622	310	4	936	672	192	24	2016	722	342	6	1143
573	380	4	768	623	528	4	720	673	672	2	674	723	480	4	968
574	240	8	1008	624	192	20	1736	674	336	4	1014	724	360	6	1274
575	440	6	744	625	500	5	781	675	360	12	1240	725	560	6	930
576	192	21	1651	626	312	4	942	676	312	9	1281	726	220	12	1596
577	576	2	578	627	360	8	960	677	676	2	678	727	726	2	728
578	272	6	921	628	312	6	1106	678	224	8	1368	728	288	16	1680
579	384	4	776	629	576	4	684	679	576	4	784	729	486	7	1093
580	224	12	1260	630	144	24	1872	680	256	16	1620	730	288	8	1332
581	492	4	672	631	630	2	632	681	452	4	912	731	672	4	792
582	192	8	1176	632	312	8	1200	682	300	8	1152	732	240	12	1736
583	520	4	648	633	420	4	848	683	682	2	684	733	732	2	734
584	288	8	1110	634	316	4	954	684	216	18	1820	734	366	4	1104
585	288	12	1092	635	504	4	768	685	544	4	828	735	336	12	1368
586	292	4	882	636	208	12	1512	686	294	8	1200	736	352	12	1512
587	586	2	588	637	504	6	798	687	456	4	920	737	660	4	816
588	168	18	1596	638	280	8	1080	688	336	10	1364	738	240	12	1638
589	540	4	640	639	420	6	936	689	624	4	756	739	738	2	740
590	232	8	1080	640	256	16	1530	690	176	16	1728	740	288	12	1596
591	392	4	792	641	640	2	642	691	690	2	692	741	432	8	1120
592	288	10	1178	642	212	8	1296	692	344	6	1218	742	312	8	1296
593	592	2	594	643	642	2	644	693	360	12	1248	743	742	2	744
594	180	16	1440	644	264	12	1344	694	346	4	1044	744	240	16	1920
595	384	8	864	645	336	8	1056	695	552	4	840	745	592	4	900
596	296	6	1050	646	288	8	1080	696	224	16	1800	746	372	4	1122
597	396	4	800	647	646	2	648	697	640	4	756	747	492	6	1092
598	264	8	1008	648	216	20	1815	698	348	4	1050	748	320	12	1512
599	598	2	600	649	580	4	720	699	464	4	936	749	636	4	864
600	160	24	1860	650	240	12	1302	700	240	18	1736	750	200	16	1872
601	600	2	602	651	360	8	1024	701	700	2	702	751	750	2	752
602	252	8	1056	652	324	6	1148	702	216	16	1680	752	368	10	1488
603	396	6	884	653	652	2	654	703	648	4	760	753	500	4	1008
604	300	6	1064	654	216	8	1320	704	320	14	1524	754	336	8	1260
605	440	6	798	655	520	4	792	705	368	8	1152	755	600	4	912
606	200	8	1224	656	320	10	1302	706	352	4	1062	756	216	24	2240
607	606	2	608	657	432	6	962	707	600	4	816	757	756	2	758
608	288	12	1260	658	276	8	1152	708	232	12	1680	758	378	4	1140
609	336	8	960	659	658	2	660	709	708	2	710	759	440	8	1152
610	240	8	1116	660	160	24	2016	710	280	8	1296	760	288	16	1800

TOTIENT FUNCTION $\phi(n)$ (Continued)

n	$\phi(n)$	σ_0	σ_1	n	$\phi(n)$	σ_0	σ_1	n	$\phi(n)$	σ_0	σ_1	n	$\phi(n)$	σ_0	σ_1
761	760	2	762	811	810	2	812	861	480	8	1344	911	910	2	912
762	252	8	1536	812	336	12	1680	862	430	4	1296	912	288	20	2480
763	648	4	880	813	540	4	1088	863	862	2	864	913	820	4	1008
764	380	6	1344	814	360	8	1368	864	288	24	2520	914	456	4	1374
765	384	12	1404	815	648	4	984	865	688	4	1044	915	480	8	1488
766	382	4	1152	816	256	20	2232	866	432	4	1302	916	456	6	1610
767	696	4	840	817	756	4	880	867	544	6	1228	917	780	4	1056
768	256	18	2044	818	408	4	1230	868	360	12	1792	918	288	16	2160
769	768	2	770	819	432	12	1456	869	780	4	960	919	918	2	920
770	240	16	1728	820	320	12	1764	870	224	16	2160	920	352	16	2160
771	512	4	1032	821	820	2	822	871	792	4	952	921	612	4	1232
772	384	6	1358	822	272	8	1656	872	432	8	1650	922	460	4	1386
773	772	2	774	823	822	2	824	873	576	6	1274	923	840	4	1008
774	252	12	1716	824	408	8	1560	874	396	8	1440	924	240	24	2688
775	600	6	992	825	400	12	1488	875	600	8	1248	925	720	6	1178
776	384	8	1470	826	348	8	1440	876	288	12	2072	926	462	4	1392
777	432	8	1216	827	826	2	828	877	876	2	878	927	612	6	1352
778	388	4	1170	828	264	18	2184	878	438	4	1320	928	448	12	1890
779	720	4	840	829	828	2	830	879	584	4	1176	929	928	2	930
780	192	24	2352	830	328	8	1512	880	320	20	2232	930	240	16	2304
781	700	4	864	831	552	4	1112	881	880	2	882	931	756	6	1140
782	352	8	1296	832	384	14	1778	882	252	18	2223	932	464	6	1638
783	504	8	1200	833	672	6	1026	883	882	2	884	933	620	4	1248
784	336	15	1767	834	276	8	1680	884	384	12	1764	924	466	4	1404
785	624	4	948	835	664	4	1008	885	464	8	1440	935	640	8	1296
786	260	8	1584	836	360	12	1680	886	442	4	1332	936	288	24	2730
787	786	2	788	837	540	8	1280	887	886	2	888	937	936	2	938
788	392	6	1386	838	418	4	1260	888	288	16	2280	938	396	8	1632
789	524	4	1056	839	838	2	840	889	756	4	1024	939	624	4	1256
790	312	8	1440	840	192	32	2880	890	352	8	1620	940	368	12	2016
791	672	4	912	841	812	3	871	891	540	10	1452	941	940	2	942
792	240	24	2340	842	420	4	1266	892	444	6	1568	942	312	8	1896
793	720	4	868	843	560	4	1128	893	828	4	960	943	880	4	1008
794	396	4	1194	844	420	6	1484	894	296	8	1800	944	464	10	1860
795	416	8	1296	845	624	6	1098	895	712	4	1080	945	432	16	1920
796	396	6	1400	846	276	12	1872	896	384	16	2040	946	420	8	1584
797	796	2	798	847	660	6	1064	897	528	8	1344	947	946	2	948
798	216	16	1920	848	416	10	1674	898	448	4	1350	948	312	12	2240
799	736	4	864	849	564	4	1136	899	840	4	960	949	864	4	1036
800	320	18	1953	850	320	12	1674	900	240	27	2821	950	360	12	1860
801	528	6	1170	851	792	4	912	901	832	4	972	951	632	4	1272
802	400	4	1206	852	280	12	2016	902	400	8	1512	952	384	16	2160
803	720	4	888	853	852	2	854	903	504	8	1408	953	952	2	954
804	264	12	1904	854	360	8	1488	904	448	8	1710	954	312	12	2106
805	528	8	1152	855	432	12	1560	905	720	4	1092	955	760	4	1152
806	360	8	1344	856	424	8	1620	906	300	8	1824	956	476	6	1680
807	536	4	1080	857	856	2	858	907	906	2	908	957	560	8	1440
808	400	8	1530	858	240	16	2016	908	452	6	1596	958	478	4	1440
809	808	2	810	859	858	2	860	909	600	6	1326	959	816	4	1104
810	216	20	2178	860	336	12	1848	910	288	16	2016	960	256	28	3048

TOTIENT FUNCTION $\phi(n)$ (Continued)

n	$\phi(n)$	σ_0	σ_1	n	$\phi(n)$	σ_0	σ_1	n	$\phi(n)$	σ_0	σ_1	n	$\phi(n)$	σ_0	σ_1
961	930	3	993	971	970	2	972	981	648	6	1430	991	990	2	992
962	432	8	1596	972	324	18	2548	982	490	4	1476	992	480	12	2016
963	636	6	1404	973	828	4	1120	983	982	2	984	993	660	4	1328
964	480	6	1694	974	486	4	1464	984	320	16	2520	994	420	8	1728
965	768	4	1164	975	480	12	1736	985	784	4	1188	995	792	4	1200
966	264	16	2304	976	480	10	1922	986	448	8	1620	996	328	12	2352
967	966	2	968	977	976	2	978	987	552	8	1536	997	996	2	998
968	440	12	1995	978	324	8	1968	988	432	12	1960	998	498	4	1500
969	576	8	1440	979	880	4	1080	989	924	4	1056	999	648	8	1520
970	384	8	1764	980	336	18	2394	990	240	24	2808	1000	400	16	2340

From British Association for the Advancement of Science, Mathematical Tables, vol. VIII, Number-divisor tables, by permission of Cambridge University Press.

TABLES OF INDICES AND POWER RESIDUES

H. C. Williams, Ph.D.

1. Introduction

Let a and b be any two positive integers. If the largest integer that divides both a and b is unity, a and b are said to be *relatively prime*. If m is any positive integer, the *totient of m*, denoted by $\phi(m)$, is defined to be the number of positive integers less than m and relatively prime to m. If one writes

$$m = p_1^{\alpha_1} p_2^{\alpha_2} \dots p_n^{\alpha_n},$$

where $p_1, p_2, \dots, p_n$ are distinct primes, it can be shown that

$$\phi(m) = p_1^{\alpha_1 - 1} p_2^{\alpha_2 - 1} \dots p_n^{\alpha_n - 1}(p_1 - 1)(p_2 - 1) \dots (p_n - 1).$$

Two integers a and b are said to be *congruent modulo* an integer m if m is an integer divisor of $a - b$. This relation between a and b is denoted by

$$a \equiv b \pmod{m}.$$

If a is congruent to b modulo m and $0 \le b < m$, b is called the residue of a modulo m. a and b are said to be distinct modulo m if m is not an integer divisor of $a - b$. A well-known result is the

Theorem. If a and m (≥ 0) are relatively prime integers, then

$$a^{\phi(m)} \equiv 1 \pmod{m}.$$

Corollary. If p is a prime which is not a divisor of an integer a, then

$$a^{p-1} \equiv 1 \pmod{p}.$$

If a and m (> 0) are relatively prime integers, and μ is the least positive integer for which

$$a^{\mu} \equiv 1 \pmod{m},$$

μ is called the *exponent* to which a belongs modulo m. If the exponent to which a belongs modulo m is $\phi(m)$, a is defined to be a *primitive root* modulo m (or a primitive root of m).
Theorem. The only integers which possess primitive roots are 2, 4, p^n, $2p^n$, where p is an odd prime.

If m does have a primitive root, it has $\phi(\phi(m))$ distinct primitive roots modulo m.

Let p be any prime, and let g be any primitive root of p. To each integer a relatively prime to p there corresponds a unique integer i such that

$$a \equiv g^i \pmod{p} \qquad (0 < i < p-1).$$

i is called the *index* to base g of a modulo p. This is written

$$i = \text{ind}_g a.$$

For example, consider the prime 11 which has a primitive root $g = 2$. Construct the table below, where

$$r_i \equiv g^i \pmod{p} \qquad (0 < r_i < p).$$

i	0	1	2	3	4	5	6	7	8	9	10
r_i	1	2	4	8	5	10	9	7	3	6	1

Note that the bottom row runs through all possible residues modulo 11, except zero. The index, then, of any number in this row is the corresponding entry in the top row. That is, $\text{ind}_2 1 = 0$, $\text{ind}_2 2 = 1$, $\text{ind}_2 3 = 8$, $\text{ind}_2 4 = 2$, $\text{ind}_2 5 = 4$, $\text{ind}_2 6 = 9$, etc. Indices possess the following important properties (analogous to those of logarithms).

TABLES OF INDICES AND POWER RESIDUES (Continued)

(1) $\text{ind}_g 1 = 0$

(2) $\text{ind}_g(-1) = (p-1)/2$

(3) $\text{ind}_g(ab) \equiv \text{ind}_g a + \text{ind}_g b \pmod{p-1}$

(4) $\text{ind}_g a^n \equiv n\,\text{ind}_g a \pmod{p-1}$

(5) $\text{ind}_g a \equiv \text{ind}_g g' \cdot \text{ind}_{g'} a \pmod{p-1}$,

where g' is any other primitive root of p.

The *power residue* of an integer b to base g is simply the integer r such that

$$r \equiv g^b \pmod{p} \qquad (0 < r < p).$$

Since the power residue of $\text{ind}_g a$ is congruent to a modulo p, it is clear that the concept of indices and power residues is analogous to that of logarithms and antilogarithms.

2. Use of the tables

In the tables that follow, the indices and power residues are given for all primes less than 100. (A table of indices and power residues for all primes and prime powers less than 2000 is given in [1].) In the tables which follow, the base g for the indices and power residues is chosen as the smallest primitive root of p.

To determine the index or power residue of an integer a modulo p, look at the appropriate table for prime p and select the entry in the column under j and in the row under i, where

$$a \equiv 10i + j \pmod{p} \qquad (0 \le i,j \le 9).$$

For example, the index modulo 97 of 134 ($\equiv 37 \pmod{97}$) is the entry under the number 7 and across from the number 3. This is found to be 91.

Indices and power residues can be used to simplify number-theoretic calculations. The following two examples illustrate the use of indices and power residues.

Example. Find an integer congruent to

$$(134)^{92}\,(54)^{67} \quad \text{modulo (97)}.$$

Let

$$a \equiv (134)^{92}\,(54)^{67} \pmod{97};$$

then

$$\text{ind}_g a \equiv 92\,\text{ind}_g 37 + 67\,\text{ind}_g 54 \pmod{96}$$

$$\equiv 92\,(91) + 67\,(52) \pmod{96}$$

$$\equiv 48 \pmod{96}.$$

Hence

$$a \equiv g^{48} \equiv 96 \pmod{97}.$$

This is obtained by looking in the body of the table of indices under $p = 97$ and locating the index 48, and then reading the appropriate quadratic residue from the first column and top row.

Example. Find an integer x such that

$$44x^{21} \equiv 53 \pmod{73}.$$

If we take indices of both sides, we get

$$\text{ind}_g 44 + 21\,\text{ind}_g x \equiv \text{ind}_g 53 \pmod{72}$$

$$71 + 21\,\text{ind}_g x \equiv 53 \pmod{72}$$

$$21\,\text{ind}_g x \equiv 54 \pmod{72}$$

$$7\,\text{ind}_g x \equiv 18 \pmod{24}$$

TABLES OF INDICES AND POWER RESIDUES (Continued)

Multiplying by 7, we get

$$\text{ind}_g x \equiv 7(18) \equiv 6 \quad (\text{mod } 24).$$

Hence

$$x \equiv g^6, g^{30}, g^{54} \quad (\text{mod } 73)$$

and

$$x \equiv 3, 24, 46 \quad (\text{mod } 73).$$

REFERENCE

1. *A Table of Indices and Power Residues,* University of Oklahoma Mathematical Tables Project, W. W. Norton and Co., New York, N.Y., 1962.

INDICES FOR PRIMES 3 to 97

Prime 3
$g = 2$

Indices

	0	1	2	3	4	5	6	7	8	9
0		0	1							

Prime 5
$g = 2$

Indices

	0	1	2	3	4	5	6	7	8	9
0		0	1	3	2					

Prime 7
$g = 3$

Indices

	0	1	2	3	4	5	6	7	8	9
0		0	2	1	4	5	3			

Prime 11
$g = 2$

Indices

	0	1	2	3	4	5	6	7	8	9
0		0	1	8	2	4	9	7	3	6
1	5									

Prime 13
$g = 2$

Indices

	0	1	2	3	4	5	6	7	8	9
0		0	1	4	2	9	5	11	3	8
1	10	7	6							

Prime 17
$g = 3$

Indices

	0	1	2	3	4	5	6	7	8	9
0		0	14	1	12	5	15	11	10	2
1	3	7	13	4	9	6	8			

Prime 19
$g = 2$

Indices

	0	1	2	3	4	5	6	7	8	9
0		0	1	13	2	16	14	6	3	8
1	17	12	15	5	7	11	4	10	9	

Prime 23
$g = 5$

Indices

	0	1	2	3	4	5	6	7	8	9
0		0	2	16	4	1	18	19	6	10
1	3	9	20	14	21	17	8	7	12	15
2	5	13	11							

Prime 29
$g = 2$

Indices

	0	1	2	3	4	5	6	7	8	9
0		0	1	5	2	22	6	12	3	10
1	23	25	7	18	13	27	4	21	11	9
2	24	17	26	20	8	16	19	15	14	

Prime 31
$g = 3$

Indices

	0	1	2	3	4	5	6	7	8	9
0		0	24	1	18	20	25	28	12	2
1	14	23	19	11	22	21	6	7	26	4
2	8	29	17	27	13	10	5	3	16	9
3	14									

Prime 37
$g = 2$

Indices

	0	1	2	3	4	5	6	7	8	9
0		0	1	26	2	23	27	32	3	16
1	24	30	28	11	33	13	4	7	17	35
2	25	22	31	15	29	10	12	6	34	21
3	14	9	5	20	8	19	18			

Prime 41
$g = 6$

Indices

	0	1	2	3	4	5	6	7	8	9
0		0	26	15	12	22	1	39	38	30
1	8	3	27	31	25	37	24	33	16	9
2	34	14	29	36	13	4	17	5	11	7
3	23	28	10	18	19	21	2	32	35	6
4	20									

Prime 43
$g = 3$

Indices

	0	1	2	3	4	5	6	7	8	9
0		0	27	1	12	25	28	35	39	2
1	10	30	13	32	20	26	24	38	29	19
2	37	36	15	16	40	8	17	3	5	41
3	11	34	9	31	23	18	14	7	4	33
4	22	6	21							

Prime 47
$g = 5$

Indices

	0	1	2	3	4	5	6	7	8	9
0		0	18	20	36	1	38	32	8	40
1	19	7	10	11	4	21	26	16	12	45
2	37	6	25	5	28	2	29	14	22	35
3	39	3	44	27	34	33	30	42	17	31
4	9	15	24	13	43	41	23			

Prime 53
$g = 2$

Indices

	0	1	2	3	4	5	6	7	8	9
0		0	1	17	2	47	18	14	3	34
1	48	6	19	24	15	12	4	10	35	37
2	49	31	7	39	20	42	25	51	16	46
3	13	33	5	23	11	9	36	30	38	41
4	50	45	32	22	8	29	40	44	21	28
5	43	27	26							

Prime 59
$g = 2$

Indices

	0	1	2	3	4	5	6	7	8	9
0		0	1	50	2	6	51	18	3	42
1	7	25	52	45	19	56	4	40	43	38
2	8	10	26	15	53	12	46	34	20	28
3	57	49	5	17	41	24	44	55	39	37
4	9	14	11	33	27	48	16	23	54	36
5	13	32	47	22	35	31	21	30	29	

Prime 61
$g = 2$

Indices

	0	1	2	3	4	5	6	7	8	9
0		0	1	6	2	22	7	49	3	12
1	23	15	8	40	50	28	4	47	13	26
2	24	55	16	57	9	44	41	18	51	35
3	29	59	5	21	48	11	14	39	27	46
4	25	54	56	43	17	34	58	20	10	38
5	45	53	42	33	19	37	52	32	36	31
6	30									

INDICES FOR PRIMES 3 to 97 (Continued)

Prime 67
$g = 2$

Indices

	0	1	2	3	4	5	6	7	8	9
0		0	1	39	2	15	40	23	3	12
1	16	59	41	19	24	54	4	64	13	10
2	17	62	60	28	42	30	20	51	25	44
3	55	47	5	32	65	38	14	22	11	58
4	18	53	63	9	61	27	29	50	43	46
5	31	37	21	57	52	8	26	49	45	36
6	56	7	48	35	6	34	33			

Prime 71
$g = 7$

Indices

	0	1	2	3	4	5	6	7	8	9
0		0	6	26	12	28	32	1	18	52
1	34	31	38	39	7	54	24	49	58	16
2	40	27	37	15	44	56	45	8	13	68
3	60	11	30	57	55	29	64	20	22	65
4	46	25	33	48	43	10	21	9	50	2
5	62	5	51	23	14	59	19	42	4	3
6	66	69	17	53	36	67	63	47	61	41
7	35									

Prime 73
$g = 5$

Indices

	0	1	2	3	4	5	6	7	8	9
0		0	8	6	16	1	14	33	24	12
1	9	55	22	59	41	7	32	21	20	62
2	17	39	63	46	30	2	67	18	49	35
3	15	11	40	61	29	34	28	64	70	65
4	25	4	47	51	71	13	54	31	38	66
5	10	27	3	53	26	56	57	68	43	5
6	23	58	19	45	48	60	69	50	37	52
7	42	44	36							

Prime 79
$g = 3$

Indices

	0	1	2	3	4	5	6	7	8	9
0		0	4	1	8	62	5	43	12	2
1	66	68	9	34	57	63	16	21	6	32
2	70	54	72	26	13	46	38	3	61	11
3	67	56	20	69	25	37	10	19	36	35
4	74	75	58	49	76	64	30	59	17	28
5	50	22	42	77	7	52	65	33	15	31
6	71	45	60	55	24	18	73	48	29	27
7	41	51	14	44	23	47	40	43	39	

Prime 83
$g = 2$

Indices

	0	1	2	3	4	5	6	7	8	9
0		0	1	72	2	27	73	8	3	62
1	28	24	74	77	9	17	4	56	63	47
2	29	80	25	60	75	54	78	52	10	12
3	18	38	5	14	57	35	64	20	48	67
4	30	40	81	71	26	7	61	23	76	16
5	55	46	79	59	53	51	11	37	3	34
6	19	66	39	70	6	22	15	45	58	50
7	36	33	65	69	21	44	49	32	68	43
8	31	42	41							

Prime 89
$g = 3$

Indices

	0	1	2	3	4	5	6	7	8	9
0		0	16	1	32	70	17	81	48	2
1	86	84	33	23	9	71	64	6	18	35
2	14	82	12	57	49	52	39	3	25	59
3	87	31	80	85	22	63	34	11	51	24
4	30	21	10	29	28	72	73	54	65	74
5	68	7	55	78	19	66	41	36	75	43
6	15	69	47	83	8	5	13	56	38	58
7	79	62	50	20	27	53	67	77	40	42
8	46	4	37	61	26	76	45	60	44	

Prime 97
$g = 5$

Indices

	0	1	2	3	4	5	6	7	8	9
0		0	34	70	68	1	8	31	6	44
1	35	86	42	25	65	71	40	89	78	81
2	69	5	24	77	76	2	59	18	3	13
3	9	46	74	60	27	32	16	91	19	95
4	7	85	39	4	58	45	15	84	14	62
5	36	63	93	10	52	87	37	55	47	67
6	43	64	80	75	12	26	94	57	61	51
7	66	11	50	28	29	72	53	21	33	30
8	41	88	23	17	73	90	38	83	92	54
9	76	56	47	20	22	82	48			

PRIMITIVE ROOTS FOR PRIMES 3 to 5003

In this table
  g denotes the least primitive root of p
  G denotes the least negative primitive root of p
  ϵ denotes whether 10, -10 both or neither are primitive roots of p

Introductory Facts

As noted in the preceding Totient and Indices Tables, the number of integers not exceeding and relatively prime to a fixed integer n is represented by $\phi(n)$. These integers form a group; the group is cyclic if and only if $n = 2, 4$, or n is of the form p^k or $2p^k$ where p is an odd prime. We refer to g as a primitive root of n if it generates that group i.e. if $g, g^2, \ldots, g^{\phi(n)}$ are distinct modulo n. There are $\phi(\phi(n))$ primitive roots of n. If g is a primitive root of p and $g^{p-1} \not\equiv 1 \pmod{p^2}$, then g is a primitive root of p^k for all k. If $g^{p-1} \equiv 1 \pmod{p^2}$ then $g + p$ is a primitive root of p^k for all k.

If g is a primitive root of p^k then either g or $g + p^k$, whichever is odd, is a primitive root of $2p^k$.

If g is a primitive root of n, then g^k is a primitive root of n if and only if k and $\phi(n)$ are relatively prime, and each primitive root of n is of this form, i.e. $(k, \phi(n)) = 1$.

p	$p-1$	g	$-G$	ϵ	p	$p-1$	g	$-G$	ϵ
3	2	2	1	-10	167	$2 \cdot 83$	5	2	10
5	2^2	2	2	—	173	$2^2 \cdot 43$	2	2	—
7	$2 \cdot 3$	3	2	10	179	$2 \cdot 89$	2	3	10
11	$2 \cdot 5$	2	3	—	181	$2^2 \cdot 3^2 \cdot 5$	2	2	± 10
13	$2^2 \cdot 3$	2	2	—	191	$2 \cdot 5 \cdot 19$	19	2	-10
17	2^4	3	3	± 10	193	$2^6 \cdot 3$	5	5	± 10
19	$2 \cdot 3^2$	2	4	10	197	$2^2 \cdot 7^2$	2	2	—
23	$2 \cdot 11$	5	2	10	199	$2 \cdot 3^2 \cdot 11$	3	2	-10
29	$2^2 \cdot 7$	2	2	± 10	211	$2 \cdot 3 \cdot 5 \cdot 7$	2	4	—
31	$2 \cdot 3 \cdot 5$	3	7	-10	223	$2 \cdot 3 \cdot 37$	3	9	10
37	$2^2 \cdot 3^2$	2	2	—	227	$2 \cdot 113$	2	3	-10
41	$2^3 \cdot 5$	6	6	—	229	$2^2 \cdot 3 \cdot 19$	6	6	± 10
43	$2 \cdot 3 \cdot 7$	3	9	-10	233	$2^3 \cdot 29$	3	3	± 10
47	$2 \cdot 23$	5	2	10	239	$2 \cdot 7 \cdot 17$	7	2	—
53	$2^2 \cdot 13$	2	2	—	241	$2^4 \cdot 3 \cdot 5$	7	7	—
59	$2 \cdot 29$	2	3	10	251	$2 \cdot 5^3$	6	3	—
61	$2^2 \cdot 3 \cdot 5$	2	2	± 10	257	2^8	3	3	± 10
67	$2 \cdot 3 \cdot 11$	2	4	-10	263	$2 \cdot 131$	5	2	10
71	$2 \cdot 5 \cdot 7$	7	2	-10	269	$2^2 \cdot 67$	2	2	± 10
73	$2^3 \cdot 3^2$	5	5	—	271	$2 \cdot 3^3 \cdot 5$	6	2	—
79	$2 \cdot 3 \cdot 13$	3	2	—	277	$2^2 \cdot 3 \cdot 23$	5	5	—
83	$2 \cdot 41$	2	3	-10	281	$2^3 \cdot 5 \cdot 7$	3	3	—
89	$2^3 \cdot 11$	3	3	—	283	$2 \cdot 3 \cdot 47$	3	6	-10
97	$2^5 \cdot 3$	5	5	± 10	293	$2^2 \cdot 73$	2	2	—
101	$2^2 \cdot 5^2$	2	2	—	307	$2 \cdot 3^2 \cdot 17$	5	7	-10
103	$2 \cdot 3 \cdot 17$	5	2	—	311	$2 \cdot 5 \cdot 31$	17	2	-10
107	$2 \cdot 53$	2	3	-10	313	$2^3 \cdot 3 \cdot 13$	10	10	± 10
109	$2^2 \cdot 3^3$	6	6	± 10	317	$2^2 \cdot 79$	2	2	—
113	$2^4 \cdot 7$	3	3	± 10	331	$2 \cdot 3 \cdot 5 \cdot 11$	3	5	—
127	$2 \cdot 3^2 \cdot 7$	3	9	—	337	$2^4 \cdot 3 \cdot 7$	10	10	± 10
131	$2 \cdot 5 \cdot 13$	2	3	10	347	$2 \cdot 173$	2	3	-10
137	$2^3 \cdot 17$	3	3	—	349	$2^2 \cdot 3 \cdot 29$	2	2	—
139	$2 \cdot 3 \cdot 23$	2	4	—	353	$2^5 \cdot 11$	3	3	—
149	$2^2 \cdot 37$	2	2	± 10	359	$2 \cdot 179$	7	2	-10
151	$2 \cdot 3 \cdot 5^2$	6	5	-10	367	$2 \cdot 3 \cdot 61$	6	2	10
157	$2^2 \cdot 3 \cdot 13$	5	5	—	373	$2^2 \cdot 3 \cdot 31$	2	2	—
163	$2 \cdot 3^4$	2	4	-10	379	$2 \cdot 3^3 \cdot 7$	2	4	10

PRIMITIVE ROOTS FOR PRIMES 3 to 5003 (Continued)

p	$p-1$	g	$-G$	ϵ	p	$p-1$	g	$-G$	ϵ
383	$2 \cdot 191$	5	2	10	769	$2^8 \cdot 3$	11	11	—
389	$2^2 \cdot 97$	2	2	± 10	773	$2^2 \cdot 193$	2	2	—
397	$2^2 \cdot 3^2 \cdot 11$	5	5	—	787	$2 \cdot 3 \cdot 131$	2	4	-10
401	$2^4 \cdot 5^2$	3	3	—	797	$2^2 \cdot 199$	2	2	—
409	$2^3 \cdot 3 \cdot 17$	21	21	—	809	$2^3 \cdot 101$	3	3	—
419	$2 \cdot 11 \cdot 19$	2	3	10	811	$2 \cdot 3^4 \cdot 5$	3	5	10
421	$2^2 \cdot 3 \cdot 5 \cdot 7$	2	2	—	821	$2^2 \cdot 5 \cdot 41$	2	2	± 10
431	$2 \cdot 5 \cdot 43$	7	5	-10	823	$2 \cdot 3 \cdot 137$	3	2	10
433	$2^4 \cdot 3^3$	5	5	± 10	827	$2 \cdot 7 \cdot 59$	2	3	-10
439	$2 \cdot 3 \cdot 73$	15	5	-10	829	$2^2 \cdot 3^2 \cdot 23$	2	2	—
443	$2 \cdot 13 \cdot 17$	2	3	-10	839	$2 \cdot 419$	11	2	-10
449	$2^6 \cdot 7$	3	3	—	853	$2^2 \cdot 3 \cdot 71$	2	2	—
457	$2^3 \cdot 3 \cdot 19$	13	13	—	857	$2^3 \cdot 107$	3	3	± 10
461	$2^2 \cdot 5 \cdot 23$	2	2	$\perp 10$	859	$2 \cdot 3 \cdot 11 \cdot 13$	2	4	—
463	$2 \cdot 3 \cdot 7 \cdot 11$	3	2	—	863	$2 \cdot 431$	5	2	10
467	$2 \cdot 233$	2	3	-10	877	$2^2 \cdot 3 \cdot 73$	2	2	—
479	$2 \cdot 239$	13	2	-10	881	$2^4 \cdot 5 \cdot 11$	3	3	—
487	$2 \cdot 3^5$	3	2	10	883	$2 \cdot 3^2 \cdot 7^2$	2	4	-10
491	$2 \cdot 5 \cdot 7^2$	2	4	10	887	$2 \cdot 443$	5	2	10
499	$2 \cdot 3 \cdot 83$	7	5	10	907	$2 \cdot 3 \cdot 151$	2	4	—
503	$2 \cdot 251$	5	2	10	911	$2 \cdot 5 \cdot 7 \cdot 13$	17	3	-10
509	$2^2 \cdot 127$	2	2	± 10	919	$2 \cdot 3^3 \cdot 17$	7	5	-10
521	$2^3 \cdot 5 \cdot 13$	3	3	—	929	$2^5 \cdot 29$	3	3	—
523	$2 \cdot 3^2 \cdot 29$	2	4	-10	937	$2^3 \cdot 3^2 \cdot 13$	5	5	± 10
541	$2^2 \cdot 3^3 \cdot 5$	2	2	± 10	941	$2^2 \cdot 5 \cdot 47$	2	2	± 10
547	$2 \cdot 3 \cdot 7 \cdot 13$	2	4	—	947	$2 \cdot 11 \cdot 43$	2	3	-10
557	$2^2 \cdot 139$	2	2	—	953	$2^3 \cdot 7 \cdot 17$	3	3	± 10
563	$2 \cdot 281$	2	3	-10	967	$2 \cdot 3 \cdot 7 \cdot 23$	5	2	—
569	$2^3 \cdot 71$	3	3	—	971	$2 \cdot 5 \cdot 97$	6	3	10
571	$2 \cdot 3 \cdot 5 \cdot 19$	3	5	10	977	$2^4 \cdot 61$	3	3	± 10
577	$2^6 \cdot 3^2$	5	5	± 10	983	$2 \cdot 491$	5	2	10
587	$2 \cdot 293$	2	3	-10	991	$2 \cdot 3^2 \cdot 5 \cdot 11$	6	2	-10
593	$2^4 \cdot 37$	3	3	± 10	997	$2^2 \cdot 3 \cdot 83$	7	7	—
599	$2 \cdot 13 \cdot 23$	7	2	-10	1009	$2^4 \cdot 3^2 \cdot 7$	11	11	—
601	$2^3 \cdot 3 \cdot 5^2$	7	7	—	1013	$2^2 \cdot 11 \cdot 23$	3	3	—
607	$2 \cdot 3 \cdot 101$	3	2	—	1019	$2 \cdot 509$	2	3	10
613	$2^2 \cdot 3^2 \cdot 17$	2	2	—	1021	$2^2 \cdot 3 \cdot 5 \cdot 17$	10	10	± 10
617	$2^3 \cdot 7 \cdot 11$	3	3	—	1031	$2 \cdot 5 \cdot 103$	14	2	—
619	$2 \cdot 3 \cdot 103$	2	4	10	1033	$2^3 \cdot 3 \cdot 43$	5	5	± 10
631	$2 \cdot 3^2 \cdot 5 \cdot 7$	3	9	-10	1039	$2 \cdot 3 \cdot 173$	3	2	-10
641	$2^7 \cdot 5$	3	3	—	1049	$2^3 \cdot 131$	3	3	—
643	$2 \cdot 3 \cdot 107$	11	7	—	1051	$2 \cdot 3 \cdot 5^2 \cdot 7$	7	5	10
647	$2 \cdot 17 \cdot 19$	5	2	10	1061	$2^2 \cdot 5 \cdot 53$	2	2	—
653	$2^2 \cdot 163$	2	2	—	1063	$2 \cdot 3^2 \cdot 59$	3	2	10
659	$2 \cdot 7 \cdot 47$	2	3	10	1069	$2^2 \cdot 3 \cdot 89$	6	6	± 10
661	$2^2 \cdot 3 \cdot 5 \cdot 11$	2	2	—	1087	$2 \cdot 3 \cdot 181$	3	2	10
673	$2^5 \cdot 3 \cdot 7$	5	5	—	1091	$2 \cdot 5 \cdot 109$	2	4	10
677	$2^2 \cdot 13^2$	2	2	—	1093	$2^2 \cdot 3 \cdot 7 \cdot 13$	5	5	—
683	$2 \cdot 11 \cdot 31$	5	10	-10	1097	$2^3 \cdot 137$	3	3	± 10
691	$2 \cdot 3 \cdot 5 \cdot 23$	3	6	—	1103	$2 \cdot 19 \cdot 29$	5	3	10
701	$2^2 \cdot 5^2 \cdot 7$	2	2	± 10	1109	$2^2 \cdot 277$	2	2	± 10
709	$2^2 \cdot 3 \cdot 59$	2	2	± 10	1117	$2^2 \cdot 3^2 \cdot 31$	2	2	—
719	$2 \cdot 359$	11	2	-10	1123	$2 \cdot 3 \cdot 11 \cdot 17$	2	4	-10
727	$2 \cdot 3 \cdot 11^2$	5	7	10	1129	$2^3 \cdot 3 \cdot 47$	11	11	—
733	$2^2 \cdot 3 \cdot 61$	6	6	—	1151	$2 \cdot 5^2 \cdot 23$	17	2	-10
739	$2 \cdot 3^2 \cdot 41$	3	6	—	1153	$2^7 \cdot 3^2$	5	5	± 10
743	$2 \cdot 7 \cdot 53$	5	2	10	1163	$2 \cdot 7 \cdot 83$	5	3	-10
751	$2 \cdot 3 \cdot 5^3$	3	2	—	1171	$2 \cdot 3^2 \cdot 5 \cdot 13$	2	4	10
757	$2^2 \cdot 3^3 \cdot 7$	2	2	—	1181	$2^2 \cdot 5 \cdot 59$	7	7	± 10
761	$2^3 \cdot 5 \cdot 19$	6	6	—	1187	$2 \cdot 593$	2	3	-10

PRIMITIVE ROOTS FOR PRIMES 3 to 5003 (Continued)

p	$p-1$	g	$-G$	ϵ	p	$p-1$	g	$-G$	ϵ
1193	$2^3 \cdot 149$	3	3	± 10	1619	$2 \cdot 809$	2	3	10
1201	$2^4 \cdot 3 \cdot 5^2$	11	11	—	1621	$2^2 \cdot 3^4 \cdot 5$	2	2	± 10
1213	$2^2 \cdot 3 \cdot 101$	2	2	—	1627	$2 \cdot 3 \cdot 271$	3	6	—
1217	$2^6 \cdot 19$	3	3	± 10	1637	$2^2 \cdot 409$	2	2	—
1223	$2 \cdot 13 \cdot 47$	5	2	10	1657	$2^3 \cdot 3^2 \cdot 23$	11	11	—
1229	$2^2 \cdot 307$	2	2	± 10	1663	$2 \cdot 3 \cdot 277$	3	2	10
1231	$2 \cdot 3 \cdot 5 \cdot 41$	3	2	—	1667	$2 \cdot 7^2 \cdot 17$	2	3	-10
1237	$2^2 \cdot 3 \cdot 103$	2	2	—	1669	$2^2 \cdot 3 \cdot 139$	2	2	—
1249	$2^5 \cdot 3 \cdot 13$	7	7	—	1693	$2^2 \cdot 3^2 \cdot 47$	2	2	—
1259	$2 \cdot 17 \cdot 37$	2	3	10	1697	$2^5 \cdot 53$	3	3	± 10
1277	$2^2 \cdot 11 \cdot 29$	2	2	—	1699	$2 \cdot 3 \cdot 283$	3	6	—
1279	$2 \cdot 3^2 \cdot 71$	3	2	-10	1709	$2^2 \cdot 7 \cdot 61$	3	3	± 10
1283	$2 \cdot 641$	2	3	-10	1721	$2^3 \cdot 5 \cdot 43$	3	3	—
1289	$2^3 \cdot 7 \cdot 23$	6	6	—	1723	$2 \cdot 3 \cdot 7 \cdot 41$	3	6	—
1291	$2 \cdot 3 \cdot 5 \cdot 43$	2	4	10	1733	$2^2 \cdot 433$	2	2	—
1297	$2^4 \cdot 3^4$	10	10	± 10	1741	$2^2 \cdot 3 \cdot 5 \cdot 29$	2	2	± 10
1301	$2^2 \cdot 5^2 \cdot 13$	2	2	± 10	1747	$2 \cdot 3^2 \cdot 97$	2	4	—
1303	$2 \cdot 3 \cdot 7 \cdot 31$	6	2	10	1753	$2^3 \cdot 3 \cdot 73$	7	7	—
1307	$2 \cdot 653$	2	3	-10	1759	$2 \cdot 3 \cdot 293$	6	2	-10
1319	$2 \cdot 659$	13	2	-10	1777	$2^4 \cdot 3 \cdot 37$	5	5	± 10
1321	$2^3 \cdot 3 \cdot 5 \cdot 11$	13	13	—	1783	$2 \cdot 3^4 \cdot 11$	10	2	10
1327	$2 \cdot 3 \cdot 13 \cdot 17$	3	9	10	1787	$2 \cdot 19 \cdot 47$	2	3	-10
1361	$2^4 \cdot 5 \cdot 17$	3	3	—	1789	$2^2 \cdot 3 \cdot 149$	6	6	± 10
1367	$2 \cdot 683$	5	2	10	1801	$2^3 \cdot 3^2 \cdot 5^2$	11	11	—
1373	$2^2 \cdot 7^3$	2	2	—	1811	$2 \cdot 5 \cdot 181$	6	3	10
1381	$2^2 \cdot 3 \cdot 5 \cdot 23$	2	2	± 10	1823	$2 \cdot 911$	5	2	10
1399	$2 \cdot 3 \cdot 233$	13	5	-10	1831	$2 \cdot 3 \cdot 5 \cdot 61$	3	9	—
1409	$2^7 \cdot 11$	3	3	—	1847	$2 \cdot 13 \cdot 71$	5	2	10
1423	$2 \cdot 3^2 \cdot 79$	3	9	—	1861	$2^2 \cdot 3 \cdot 5 \cdot 31$	2	2	± 10
1427	$2 \cdot 23 \cdot 31$	2	3	-10	1867	$2 \cdot 3 \cdot 311$	2	4	-10
1429	$2^2 \cdot 3 \cdot 7 \cdot 17$	6	6	± 10	1871	$2 \cdot 5 \cdot 11 \cdot 17$	14	2	-10
1433	$2^3 \cdot 179$	3	3	± 10	1873	$2^4 \cdot 3^2 \cdot 13$	10	10	± 10
1439	$2 \cdot 719$	7	2	-10	1877	$2^2 \cdot 7 \cdot 67$	2	2	—
1447	$2 \cdot 3 \cdot 241$	3	2	10	1879	$2 \cdot 3 \cdot 313$	6	2	—
1451	$2 \cdot 5^2 \cdot 29$	2	3	—	1889	$2^5 \cdot 59$	3	3	—
1453	$2^2 \cdot 3 \cdot 11^2$	2	2	—	1901	$2^2 \cdot 5^2 \cdot 19$	2	2	—
1459	$2 \cdot 3^6$	3	6	—	1907	$2 \cdot 953$	2	3	-10
1471	$2 \cdot 3 \cdot 5 \cdot 7^2$	6	5	-10	1913	$2^3 \cdot 239$	3	3	± 10
1481	$2^3 \cdot 5 \cdot 37$	3	3	—	1931	$2 \cdot 5 \cdot 193$	2	3	—
1483	$2 \cdot 3 \cdot 13 \cdot 19$	2	4	—	1933	$2^2 \cdot 3 \cdot 7 \cdot 23$	5	5	—
1487	$2 \cdot 743$	5	2	10	1949	$2^2 \cdot 487$	2	2	± 10
1489	$2^4 \cdot 3 \cdot 31$	14	14	—	1951	$2 \cdot 3 \cdot 5^2 \cdot 13$	3	2	—
1493	$2^2 \cdot 373$	2	2	—	1973	$2^2 \cdot 17 \cdot 29$	2	2	—
1499	$2 \cdot 7 \cdot 107$	2	3	—	1979	$2 \cdot 23 \cdot 43$	2	3	10
1511	$2 \cdot 5 \cdot 151$	11	2	-10	1987	$2 \cdot 3 \cdot 331$	2	4	—
1523	$2 \cdot 761$	2	3	-10	1993	$2^3 \cdot 3 \cdot 83$	5	5	—
1531	$2 \cdot 3^2 \cdot 5 \cdot 17$	2	4	10	1997	$2^2 \cdot 499$	2	2	—
1543	$2 \cdot 3 \cdot 257$	5	2	10	1999	$2 \cdot 3^3 \cdot 37$	3	5	-10
1549	$2^2 \cdot 3^2 \cdot 43$	2	2	± 10	2003	$2 \cdot 7 \cdot 11 \cdot 13$	5	3	-10
1553	$2^4 \cdot 97$	3	3	± 10	2011	$2 \cdot 3 \cdot 5 \cdot 67$	3	5	—
1559	$2 \cdot 19 \cdot 41$	19	2	-10	2017	$2^5 \cdot 3^2 \cdot 7$	5	5	± 10
1567	$2 \cdot 3^3 \cdot 29$	3	2	10	2027	$2 \cdot 1013$	2	3	-10
1571	$2 \cdot 5 \cdot 157$	2	3	10	2029	$2^2 \cdot 3 \cdot 13^2$	2	2	± 10
1579	$2 \cdot 3 \cdot 263$	3	5	10	2039	$2 \cdot 1019$	7	2	-10
1583	$2 \cdot 7 \cdot 113$	5	2	10	2053	$2^2 \cdot 3^3 \cdot 19$	2	2	—
1597	$2^2 \cdot 3 \cdot 7 \cdot 19$	11	11	—	2063	$2 \cdot 1031$	5	2	10
1601	$2^6 \cdot 5^2$	3	3	—	2069	$2^3 \cdot 11 \cdot 47$	2	2	± 10
1607	$2 \cdot 11 \cdot 73$	5	2	10	2081	$2^5 \cdot 5 \cdot 13$	3	3	—
1609	$2^3 \cdot 3 \cdot 67$	7	7	—	2083	$2 \cdot 3 \cdot 347$	2	4	-10
1613	$2^2 \cdot 13 \cdot 31$	3	3	—	2087	$2 \cdot 7 \cdot 149$	5	2	—

PRIMITIVE ROOTS FOR PRIMES 3 to 5003 (Continued)

p	$p-1$	g	$-G$	ϵ	p	$p-1$	g	$-G$	ϵ
2089	$2^3 \cdot 3^2 \cdot 29$	7	7	—	2579	$2 \cdot 1289$	2	3	10
2099	$2 \cdot 1049$	2	3	10	2591	$2 \cdot 5 \cdot 7 \cdot 37$	7	2	—
2111	$2 \cdot 5 \cdot 211$	7	2	-10	2593	$2^5 \cdot 3^4$	7	7	± 10
2113	$2^6 \cdot 3 \cdot 11$	5	5	± 10	2609	$2^4 \cdot 163$	3	3	—
2129	$2^4 \cdot 7 \cdot 19$	3	3	—	2617	$2^3 \cdot 3 \cdot 109$	5	5	± 10
2131	$2 \cdot 3 \cdot 5 \cdot 71$	2	4	—	2621	$2^2 \cdot 5 \cdot 131$	2	2	± 10
2137	$2^3 \cdot 3 \cdot 89$	10	10	± 10	2633	$2^3 \cdot 7 \cdot 47$	3	3	± 10
2141	$2^2 \cdot 5 \cdot 107$	2	2	± 10	2647	$2 \cdot 3^3 \cdot 7^2$	3	2	—
2143	$2 \cdot 3^2 \cdot 7 \cdot 17$	3	9	10	2657	$2^5 \cdot 83$	3	3	± 10
2153	$2^3 \cdot 269$	3	3	± 10	2659	$2 \cdot 3 \cdot 443$	2	4	—
2161	$2^4 \cdot 3^3 \cdot 5$	23	23	—	2663	$2 \cdot 11^3$	5	2	10
2179	$2 \cdot 3^2 \cdot 11^2$	7	5	10	2671	$2 \cdot 3 \cdot 5 \cdot 89$	7	5	-10
2203	$2 \cdot 3 \cdot 367$	5	7	-10	2677	$2^2 \cdot 3 \cdot 223$	2	2	—
2207	$2 \cdot 1103$	5	2	10	2683	$2 \cdot 3^2 \cdot 149$	2	4	—
2213	$2^2 \cdot 7 \cdot 79$	2	2	—	2687	$2 \cdot 17 \cdot 79$	5	3	10
2221	$2^2 \cdot 3 \cdot 5 \cdot 37$	2	2	± 10	2689	$2^7 \cdot 3 \cdot 7$	19	19	—
2237	$2^2 \cdot 13 \cdot 43$	2	2	—	2693	$2^2 \cdot 673$	2	2	—
2239	$2 \cdot 3 \cdot 373$	3	2	-10	2699	$2 \cdot 19 \cdot 71$	2	3	10
2243	$2 \cdot 19 \cdot 59$	2	3	-10	2707	$2 \cdot 3 \cdot 11 \cdot 41$	2	4	-10
2251	$2 \cdot 3^2 \cdot 5^3$	7	5	10	2711	$2 \cdot 5 \cdot 271$	7	2	-10
2267	$2 \cdot 11 \cdot 103$	2	3	-10	2713	$2^3 \cdot 3 \cdot 113$	5	5	± 10
2269	$2^2 \cdot 3^4 \cdot 7$	2	2	± 10	2719	$2 \cdot 3^2 \cdot 151$	3	2	-10
2273	$2^5 \cdot 71$	3	3	± 10	2729	$2^3 \cdot 11 \cdot 31$	3	3	—
2281	$2^3 \cdot 3 \cdot 5 \cdot 19$	7	7	—	2731	$2 \cdot 3 \cdot 5 \cdot 7 \cdot 13$	3	5	10
2287	$2 \cdot 3^2 \cdot 127$	19	7	—	2741	$2^2 \cdot 5 \cdot 137$	2	2	± 10
2293	$2^2 \cdot 3 \cdot 191$	2	2	—	2749	$2^2 \cdot 3 \cdot 229$	6	6	—
2297	$2^3 \cdot 7 \cdot 41$	5	5	± 10	2753	$2^6 \cdot 43$	3	3	± 10
2309	$2^2 \cdot 577$	2	2	± 10	2767	$2 \cdot 3 \cdot 461$	3	9	10
2311	$2 \cdot 3 \cdot 5 \cdot 7 \cdot 11$	3	2	—	2777	$2^3 \cdot 347$	3	3	± 10
2333	$2^2 \cdot 11 \cdot 53$	2	2	—	2789	$2^2 \cdot 17 \cdot 41$	2	2	± 10
2339	$2 \cdot 7 \cdot 167$	2	3	10	2791	$2 \cdot 3^2 \cdot 5 \cdot 31$	6	7	—
2341	$2^2 \cdot 3^2 \cdot 5 \cdot 13$	7	7	± 10	2797	$2^2 \cdot 3 \cdot 233$	2	2	—
2347	$2 \cdot 3 \cdot 17 \cdot 23$	3	6	-10	2801	$2^4 \cdot 5^2 \cdot 7$	3	3	—
2351	$2 \cdot 5^2 \cdot 47$	13	3	-10	2803	$2 \cdot 3 \cdot 467$	2	4	-10
2357	$2^2 \cdot 19 \cdot 31$	2	2	—	2819	$2 \cdot 1409$	2	3	10
2371	$2 \cdot 3 \cdot 5 \cdot 79$	2	4	10	2833	$2^4 \cdot 3 \cdot 59$	5	5	± 10
2377	$2^3 \cdot 3^3 \cdot 11$	5	5	—	2837	$2^2 \cdot 709$	2	2	—
2381	$2^2 \cdot 5 \cdot 7 \cdot 17$	3	3	—	2843	$2 \cdot 7^2 \cdot 29$	2	4	-10
2383	$2 \cdot 3 \cdot 397$	5	13	10	2851	$2 \cdot 3 \cdot 5^2 \cdot 19$	2	4	10
2389	$2^2 \cdot 3 \cdot 199$	2	2	± 10	2857	$2^3 \cdot 3 \cdot 7 \cdot 17$	11	11	—
2393	$2^3 \cdot 13 \cdot 23$	3	3	—	2861	$2^2 \cdot 5 \cdot 11 \cdot 13$	2	2	± 10
2399	$2 \cdot 11 \cdot 109$	11	2	-10	2879	$2 \cdot 1439$	7	2	-10
2411	$2 \cdot 5 \cdot 241$	6	3	10	2887	$2 \cdot 3 \cdot 13 \cdot 37$	5	2	10
2417	$2^4 \cdot 151$	3	3	± 10	2897	$2^4 \cdot 181$	3	3	± 10
2423	$2 \cdot 7 \cdot 173$	5	2	10	2903	$2 \cdot 1451$	5	2	10
2437	$2^2 \cdot 3 \cdot 7 \cdot 29$	2	2	—	2909	$2^2 \cdot 727$	2	2	± 10
2441	$2^3 \cdot 5 \cdot 61$	6	6	—	2917	$2^2 \cdot 3^6$	5	5	—
2447	$2 \cdot 1223$	5	2	10	2927	$2 \cdot 7 \cdot 11 \cdot 19$	5	2	10
2459	$2 \cdot 1229$	2	3	10	2939	$2 \cdot 13 \cdot 113$	2	3	10
2467	$2 \cdot 3^2 \cdot 137$	2	4	—	2953	$2^3 \cdot 3^2 \cdot 41$	13	13	—
2473	$2^3 \cdot 3 \cdot 103$	5	5	± 10	2957	$2^2 \cdot 739$	2	2	—
2477	$2^2 \cdot 619$	2	2	—	2963	$2 \cdot 1481$	2	3	-10
2503	$2 \cdot 3^2 \cdot 139$	3	2	—	2969	$2^3 \cdot 7 \cdot 53$	3	3	—
2521	$2^3 \cdot 3^2 \cdot 5 \cdot 7$	17	17	—	2971	$2 \cdot 3^3 \cdot 5 \cdot 11$	10	5	10
2531	$2 \cdot 5 \cdot 11 \cdot 23$	2	3	—	2999	$2 \cdot 1499$	17	2	-10
2539	$2 \cdot 3^3 \cdot 47$	2	4	10	3001	$2^3 \cdot 3 \cdot 5^3$	14	14	—
2543	$2 \cdot 31 \cdot 41$	5	2	10	3011	$2 \cdot 5 \cdot 7 \cdot 43$	2	3	10
2549	$2^2 \cdot 7^2 \cdot 13$	2	2	± 10	3019	$2 \cdot 3 \cdot 503$	2	4	10
2551	$2 \cdot 3 \cdot 5^2 \cdot 17$	6	2	—	3023	$2 \cdot 1511$	5	2	10
2557	$2^2 \cdot 3^2 \cdot 71$	2	2	—	3037	$2^2 \cdot 3 \cdot 11 \cdot 23$	2	2	—

PRIMITIVE ROOTS FOR PRIMES 3 to 5003 (Continued)

p	$p-1$	g	$-G$	ϵ	p	$p-1$	g	$-G$	ϵ
3041	$2^5 \cdot 5 \cdot 19$	3	3	—	3541	$2^2 \cdot 3 \cdot 5 \cdot 59$	7	7	—
3049	$2^3 \cdot 3 \cdot 127$	11	11	—	3547	$2 \cdot 3^2 \cdot 197$	2	4	-10
3061	$2^2 \cdot 3^2 \cdot 5 \cdot 17$	6	6	—	3557	$2^2 \cdot 7 \cdot 127$	2	2	—
3067	$2 \cdot 3 \cdot 7 \cdot 73$	2	4	-10	3559	$2 \cdot 3 \cdot 593$	3	2	-10
3079	$2 \cdot 3^4 \cdot 19$	6	2	-10	3571	$2 \cdot 3 \cdot 5 \cdot 7 \cdot 17$	2	4	10
3083	$2 \cdot 23 \cdot 67$	2	3	-10	3581	$2^2 \cdot 5 \cdot 179$	2	2	± 10
3089	$2^4 \cdot 193$	3	3	—	3583	$2 \cdot 3^2 \cdot 199$	3	2	—
3109	$2^2 \cdot 3 \cdot 7 \cdot 37$	6	6	—	3593	$2^3 \cdot 449$	3	3	± 10
3119	$2 \cdot 1559$	7	2	-10	3607	$2 \cdot 3 \cdot 601$	5	11	10
3121	$2^4 \cdot 3 \cdot 5 \cdot 13$	7	7	—	3613	$2^2 \cdot 3 \cdot 7 \cdot 43$	2	2	—
3137	$2^6 \cdot 7^2$	3	3	± 10	3617	$2^5 \cdot 113$	3	3	± 10
3163	$2 \cdot 3 \cdot 17 \cdot 31$	3	6	-10	3623	$2 \cdot 1811$	5	2	10
3167	$2 \cdot 1583$	5	2	10	3631	$2 \cdot 3 \cdot 5 \cdot 11^2$	15	10	-10
3169	$2^5 \cdot 3^2 \cdot 11$	7	7	—	3637	$2^2 \cdot 3^2 \cdot 101$	2	2	—
3181	$2^2 \cdot 3 \cdot 5 \cdot 53$	7	7	—	3643	$2 \cdot 3 \cdot 607$	2	4	-10
3187	$2 \cdot 3^3 \cdot 59$	2	4	—	3659	$2 \cdot 31 \cdot 59$	2	3	10
3191	$2 \cdot 5 \cdot 11 \cdot 29$	11	5	—	3671	$2 \cdot 5 \cdot 367$	13	2	—
3203	$2 \cdot 1601$	2	3	-10	3673	$2^3 \cdot 3^3 \cdot 17$	5	5	± 10
3209	$2^3 \cdot 401$	3	3	—	3677	$2^2 \cdot 919$	2	2	—
3217	$2^4 \cdot 3 \cdot 67$	5	5	—	3691	$2 \cdot 3^2 \cdot 5 \cdot 41$	2	4	—
3221	$2^2 \cdot 5 \cdot 7 \cdot 23$	10	10	± 10	3697	$2 \cdot 43^2$	5	5	—
3229	$2^2 \cdot 3 \cdot 269$	6	6	—	3701	$2^2 \cdot 5^2 \cdot 37$	2	2	± 10
3251	$2 \cdot 5^3 \cdot 13$	6	3	10	3709	$2^2 \cdot 3^2 \cdot 103$	2	2	± 10
3253	$2^2 \cdot 3 \cdot 271$	2	2	—	3719	$2 \cdot 11 \cdot 13^2$	7	2	-10
3257	$2^3 \cdot 11 \cdot 37$	3	3	± 10	3727	$2 \cdot 3^4 \cdot 23$	3	2	10
3259	$2 \cdot 3^2 \cdot 181$	3	5	10	3733	$2^2 \cdot 3 \cdot 311$	2	2	—
3271	$2 \cdot 3 \cdot 5 \cdot 109$	3	5	-10	3739	$2 \cdot 3 \cdot 7 \cdot 89$	7	5	—
3299	$2 \cdot 17 \cdot 97$	2	3	10	3761	$2^4 \cdot 5 \cdot 47$	3	3	—
3301	$2^2 \cdot 3 \cdot 5^2 \cdot 11$	6	6	± 10	3767	$2 \cdot 7 \cdot 269$	5	2	10
3307	$2 \cdot 3 \cdot 19 \cdot 29$	2	4	-10	3769	$2^3 \cdot 3 \cdot 157$	7	7	—
3313	$2^4 \cdot 3^2 \cdot 23$	10	10	± 10	3779	$2 \cdot 1889$	2	3	10
3319	$2 \cdot 3 \cdot 7 \cdot 79$	6	2	—	3793	$2^4 \cdot 3 \cdot 79$	5	5	—
3323	$2 \cdot 11 \cdot 151$	2	3	-10	3797	$2^2 \cdot 13 \cdot 73$	2	2	—
3329	$2^8 \cdot 13$	3	3	—	3803	$2 \cdot 1901$	2	3	-10
3331	$2 \cdot 3^2 \cdot 5 \cdot 37$	3	5	10	3821	$2^2 \cdot 5 \cdot 191$	3	3	± 10
3343	$2 \cdot 3 \cdot 557$	5	11	10	3823	$2 \cdot 3 \cdot 7^2 \cdot 13$	3	9	—
3347	$2 \cdot 7 \cdot 239$	2	3	-10	3833	$2^3 \cdot 479$	3	3	± 10
3359	$2 \cdot 23 \cdot 73$	11	2	-10	3847	$2 \cdot 3 \cdot 641$	5	2	10
3361	$2^5 \cdot 3 \cdot 5 \cdot 7$	22	22	—	3851	$2 \cdot 5^2 \cdot 7 \cdot 11$	2	4	—
3371	$2 \cdot 5 \cdot 337$	2	3	10	3853	$2^2 \cdot 3^2 \cdot 107$	2	2	—
3373	$2^2 \cdot 3 \cdot 281$	5	5	—	3863	$2 \cdot 1931$	5	2	10
3389	$2^2 \cdot 7 \cdot 11^2$	3	3	± 10	3877	$2^2 \cdot 3 \cdot 17 \cdot 19$	2	2	—
3391	$2 \cdot 3 \cdot 5 \cdot 113$	3	5	-10	3881	$2^3 \cdot 5 \cdot 97$	13	13	—
3407	$2 \cdot 13 \cdot 131$	5	2	10	3889	$2^4 \cdot 3^5$	11	11	—
3413	$2^2 \cdot 853$	2	2	—	3907	$2 \cdot 3^2 \cdot 7 \cdot 31$	2	4	-10
3433	$2^3 \cdot 3 \cdot 11 \cdot 13$	5	5	± 10	3911	$2 \cdot 5 \cdot 17 \cdot 23$	13	2	-10
3449	$2^3 \cdot 431$	3	3	—	3917	$2^2 \cdot 11 \cdot 89$	2	2	—
3457	$2^7 \cdot 3^3$	7	7	—	3919	$2 \cdot 3 \cdot 653$	3	2	—
3461	$2^2 \cdot 5 \cdot 173$	2	2	± 10	3923	$2 \cdot 37 \cdot 53$	2	3	-10
3463	$2 \cdot 3 \cdot 577$	3	9	10	3929	$2^3 \cdot 491$	3	3	—
3467	$2 \cdot 1733$	2	3	-10	3931	$2 \cdot 3 \cdot 5 \cdot 131$	2	4	—
3469	$2^2 \cdot 3 \cdot 17^2$	2	2	± 10	3943	$2 \cdot 3^3 \cdot 73$	3	9	10
3491	$2 \cdot 5 \cdot 349$	2	3	—	3947	$2 \cdot 1973$	2	3	-10
3499	$2 \cdot 3 \cdot 11 \cdot 53$	2	4	—	3967	$2 \cdot 3 \cdot 661$	6	2	10
3511	$2^3 \cdot 3^2 \cdot 5 \cdot 13$	7	2	-10	3989	$2^2 \cdot 997$	2	2	± 10
3517	$2^2 \cdot 3 \cdot 293$	2	2	—	4001	$2^5 \cdot 5^3$	3	3	—
3527	$2 \cdot 41 \cdot 43$	5	2	10	4003	$2 \cdot 3 \cdot 23 \cdot 29$	2	4	—
3529	$2^3 \cdot 3^2 \cdot 7^2$	17	17	—	4007	$2 \cdot 2003$	5	2	10
3533	$2^2 \cdot 883$	2	2	—	4013	$2^2 \cdot 17 \cdot 59$	2	2	—
3539	$2 \cdot 29 \cdot 61$	2	3	10	4019	$2 \cdot 7^2 \cdot 41$	2	4	10

PRIMITIVE ROOTS FOR PRIMES 3 to 5003 (Continued)

p	$p-1$	g	$-G$	ϵ	p	$p-1$	g	$-G$	ϵ
4021	$2^2 \cdot 3 \cdot 5 \cdot 67$	2	2		4519	$2 \cdot 3^2 \cdot 251$	3	9	—
4027	$2 \cdot 3 \cdot 11 \cdot 61$	3	6	-10	4523	$2 \cdot 7 \cdot 17 \cdot 19$	5	3	-10
4049	$2^4 \cdot 11 \cdot 23$	3	3	—	4547	$2 \cdot 2273$	2	3	-10
4051	$2 \cdot 3^4 \cdot 5^2$	10	5	10	4549	$2^2 \cdot 3 \cdot 379$	6	6	—
4057	$2^3 \cdot 3 \cdot 13^2$	5	5	± 10	4561	$2^4 \cdot 3 \cdot 5 \cdot 19$	11	11	—
4073	$2^3 \cdot 509$	3	3	± 10	4567	$2 \cdot 3 \cdot 761$	3	7	10
4079	$2 \cdot 2039$	11	2	-10	4583	$2 \cdot 29 \cdot 79$	5	2	10
4091	$2 \cdot 5 \cdot 409$	2	3	10	4591	$2 \cdot 3^3 \cdot 5 \cdot 17$	11	2	-10
4093	$2^2 \cdot 3 \cdot 11 \cdot 31$	2	2	—	4597	$2^2 \cdot 3 \cdot 383$	5	5	—
4099	$2 \cdot 3 \cdot 683$	2	4	10	4603	$2 \cdot 3 \cdot 13 \cdot 59$	2	4	-10
4111	$2 \cdot 3 \cdot 5 \cdot 137$	12	2	-10	4621	$2^2 \cdot 3 \cdot 5 \cdot 7 \cdot 11$	2	2	—
4127	$2 \cdot 2063$	5	2	10	4637	$2^2 \cdot 19 \cdot 61$	2	2	—
4129	$2^5 \cdot 3 \cdot 43$	13	13	—	4639	$2 \cdot 3 \cdot 773$	3	2	-10
4133	$2^2 \cdot 1033$	2	2	—	4643	$2 \cdot 11 \cdot 211$	5	3	-10
4139	$2 \cdot 2069$	2	3	10	4649	$2^3 \cdot 7 \cdot 83$	3	3	—
4153	$2^3 \cdot 3 \cdot 173$	5	5	± 10	4651	$2 \cdot 3 \cdot 5^2 \cdot 31$	3	5	10
4157	$2^2 \cdot 1039$	2	2	—	4657	$2^4 \cdot 3 \cdot 97$	15	15	—
4159	$2 \cdot 3^3 \cdot 7 \cdot 11$	3	2	—	4663	$2 \cdot 3^2 \cdot 7 \cdot 37$	3	9	—
4177	$2^4 \cdot 3^2 \cdot 29$	5	5	± 10	4673	$2^6 \cdot 73$	3	3	± 10
4201	$2^3 \cdot 3 \cdot 5^2 \cdot 7$	11	11	—	4679	$2 \cdot 2339$	11	2	-10
4211	$2 \cdot 5 \cdot 421$	6	3	10	4691	$2 \cdot 5 \cdot 7 \cdot 67$	2	3	10
4217	$2^3 \cdot 17 \cdot 31$	3	3	± 10	4703	$2 \cdot 2351$	5	2	10
4219	$2 \cdot 3 \cdot 19 \cdot 37$	2	4	10	4721	$2^4 \cdot 5 \cdot 59$	6	6	—
4229	$2^2 \cdot 7 \cdot 151$	2	2	± 10	4723	$2 \cdot 3 \cdot 787$	2	4	-10
4231	$2 \cdot 3^2 \cdot 5 \cdot 47$	3	2	-10	4729	$2^3 \cdot 3 \cdot 197$	17	17	—
4241	$2^4 \cdot 5 \cdot 53$	3	3	—	4733	$2^2 \cdot 7 \cdot 13^2$	5	5	—
4243	$2 \cdot 3 \cdot 7 \cdot 101$	2	4	-10	4751	$2 \cdot 5^3 \cdot 19$	19	3	-10
4253	$2^2 \cdot 1063$	2	2	—	4759	$2 \cdot 3 \cdot 13 \cdot 61$	3	5	-10
4259	$2 \cdot 2129$	2	3	10	4783	$2 \cdot 3 \cdot 797$	6	2	10
4261	$2^2 \cdot 3 \cdot 5 \cdot 71$	2	2	± 10	4787	$2 \cdot 2393$	2	3	-10
4271	$2 \cdot 5 \cdot 7 \cdot 61$	7	3	-10	4789	$2^2 \cdot 3^2 \cdot 7 \cdot 19$	2	2	—
4273	$2^4 \cdot 3 \cdot 89$	5	5	—	4793	$2^3 \cdot 599$	3	3	± 10
4283	$2 \cdot 2141$	2	3	-10	4799	$2 \cdot 2399$	7	2	-10
4289	$2^6 \cdot 67$	3	3	—	4801	$2^6 \cdot 3 \cdot 5^2$	7	7	—
4297	$2^3 \cdot 3 \cdot 179$	5	5	—	4813	$2^2 \cdot 3 \cdot 401$	2	2	—
4327	$2 \cdot 3 \cdot 7 \cdot 103$	3	2	10	4817	$2^4 \cdot 7 \cdot 43$	3	3	± 10
4337	$2^4 \cdot 271$	3	3	± 10	4831	$2 \cdot 3 \cdot 5 \cdot 7 \cdot 23$	3	2	—
4339	$2 \cdot 3^2 \cdot 241$	10	5	10	4861	$2^2 \cdot 3^5 \cdot 5$	11	11	—
4349	$2^2 \cdot 1087$	2	2	± 10	4871	$2 \cdot 5 \cdot 487$	11	3	-10
4357	$2^2 \cdot 3^2 \cdot 11^2$	2	2	—	4877	$2^2 \cdot 23 \cdot 53$	2	2	—
4363	$2 \cdot 3 \cdot 727$	2	4	-10	4889	$2^3 \cdot 13 \cdot 47$	3	3	—
4373	$2^2 \cdot 1093$	2	2	—	4903	$2 \cdot 3 \cdot 19 \cdot 43$	3	2	—
4391	$2 \cdot 5 \cdot 439$	14	2	-10	4909	$2^2 \cdot 3 \cdot 409$	6	6	—
4397	$2^2 \cdot 7 \cdot 157$	2	2	—	4919	$2 \cdot 2459$	13	2	-10
4409	$2^3 \cdot 19 \cdot 29$	3	3	—	4931	$2 \cdot 5 \cdot 17 \cdot 29$	6	3	10
4421	$2^2 \cdot 5 \cdot 13 \cdot 17$	3	3	± 10	4933	$2^2 \cdot 3^2 \cdot 137$	2	2	—
4423	$2 \cdot 3 \cdot 11 \cdot 67$	3	7	10	4937	$2^3 \cdot 617$	3	3	± 10
4441	$2^3 \cdot 3 \cdot 5 \cdot 37$	21	21	—	4943	$2 \cdot 7 \cdot 353$	7	2	10
4447	$2 \cdot 3^2 \cdot 13 \cdot 19$	3	2	10	4951	$2 \cdot 3^2 \cdot 5^2 \cdot 11$	6	2	-10
4451	$2 \cdot 5^2 \cdot 89$	2	3	10	4957	$2^2 \cdot 3 \cdot 7 \cdot 59$	2	2	—
4457	$2^3 \cdot 557$	3	3	± 10	4967	$2 \cdot 13 \cdot 191$	5	2	10
4463	$2 \cdot 23 \cdot 97$	5	2	10	4969	$2^3 \cdot 3^3 \cdot 23$	11	11	—
4481	$2^7 \cdot 5 \cdot 7$	3	3	—	4973	$2^2 \cdot 11 \cdot 113$	2	2	—
4483	$2 \cdot 3^3 \cdot 83$	2	4	—	4987	$2 \cdot 3^2 \cdot 277$	2	4	-10
4493	$2^2 \cdot 1123$	2	2	—	4993	$2^7 \cdot 3 \cdot 13$	5	5	—
4507	$2 \cdot 3 \cdot 751$	2	4	—	4999	$2 \cdot 3 \cdot 7^2 \cdot 17$	3	9	—
4513	$2^5 \cdot 3 \cdot 47$	7	7	—	5003	$2 \cdot 41 \cdot 61$	2	3	-10
4517	$2^2 \cdot 1129$	2	2	—					

From Applied Mathematics Series—55, Primitive Roots, Factorization of $p-1$, pages 864, 866, 867. By permission of United States Department of Commerce, National Bureau of Standards, Washington, D.C.

PRIMES

The following table contains all primes from 1 to 100,000. Taken from the Handbook of Mathematical Functions, Applied Mathematics Series 55, Permission received from National Bureau of Standards, Washington, D. C.

PRIMES

	24	23	22	21	20	19	18	17	16	15	14	13	12	11	10	9	8	7	6	5	4	3	2	1	0
1	21391	20359	19427	18329	17393	16411	15413	14533	13513	12569	11677	10663	9739	8837	7927	7001	6143	5281	4421	3581	2749	1993	1229	547	2
2	21397	20369	19429	18341	17401	16417	15427	14537	13523	12577	11681	10667	9743	8839	7933	7013	6151	5297	4423	3583	2753	1997	1231	557	3
3	21401	20389	19433	18353	17417	16421	15439	14543	13537	12583	11689	10687	9749	8849	7937	7019	6163	5303	4441	3593	2767	1999	1237	563	5
4	21407	20393	19441	18367	17419	16427	15443	14549	13553	12589	11699	10691	9767	8861	7949	7027	6173	5309	4447	3607	2777	2003	1249	569	7
5	21419	20399	19447	18371	17431	16433	15451	14551	13567	12601	11701	10709	9769	8863	7951	7039	6197	5323	4451	3613	2789	2011	1259	571	11
6	21433	20407	19457	18379	17443	16447	15461	14557	13577	12611	11717	10711	9781	8867	7963	7043	6199	5333	4457	3617	2791	2017	1277	577	13
7	21467	20411	19463	18397	17449	16451	15467	14561	13591	12613	11719	10723	9787	8887	7993	7057	6203	5347	4463	3623	2797	2027	1279	587	17
8	21481	20431	19469	18401	17467	16453	15473	14563	13597	12619	11731	10729	9791	8893	8009	7069	6211	5351	4481	3631	2801	2029	1283	593	19
9	21487	20441	19471	18413	17471	16477	15493	14591	13613	12637	11743	10733	9803	8923	8011	7079	6217	5381	4483	3637	2803	2039	1289	599	23
10	21491	20443	19477	18427	17477	16481	15497	14593	13619	12641	11777	10739	9811	8929	8017	7103	6221	5387	4493	3643	2819	2053	1291	601	29
11	21493	20477	19483	18433	17483	16487	15511	14621	13627	12647	11779	10753	9817	8933	8039	7109	6229	5393	4507	3659	2833	2063	1297	607	31
12	21499	20479	19489	18439	17489	16493	15527	14627	13633	12653	11783	10771	9829	8941	8053	7121	6247	5399	4513	3671	2837	2069	1301	613	37
13	21503	20483	19501	18443	17491	16519	15541	14629	13649	12659	11789	10781	9833	8951	8059	7127	6257	5407	4517	3673	2843	2081	1303	617	41
14	21517	20507	19507	18451	17497	16529	15551	14633	13669	12671	11801	10789	9839	8963	8069	7129	6263	5413	4519	3677	2851	2083	1307	619	43
15	21521	20509	19531	18457	17509	16547	15559	14639	13679	12689	11807	10799	9851	8969	8081	7151	6269	5417	4523	3691	2857	2087	1319	631	47
16	21523	20521	19541	18461	17519	16553	15569	14653	13681	12697	11813	10831	9857	8971	8087	7159	6271	5419	4547	3697	2861	2089	1321	641	53
17	21529	20533	19543	18481	17539	16561	15581	14657	13687	12703	11821	10837	9859	8999	8089	7177	6277	5431	4549	3701	2879	2099	1327	643	59
18	21557	20543	19553	18493	17551	16567	15583	14669	13691	12713	11827	10847	9871	9001	8093	7187	6287	5437	4561	3709	2887	2111	1361	647	61
19	21559	20549	19559	18503	17569	16573	15601	14683	13693	12721	11831	10853	9883	9007	8101	7193	6299	5441	4567	3719	2897	2113	1367	653	67
20	21563	20551	19571	18517	17573	16603	15607	14699	13697	12739	11833	10859	9887	9011	8111	7207	6301	5443	4583	3727	2903	2129	1373	659	71
21	21569	20563	19577	18521	17579	16607	15619	14713	13709	12743	11839	10861	9901	9013	8117	7211	6311	5449	4591	3733	2909	2131	1381	661	73
22	21577	20593	19583	18523	17581	16619	15629	14717	13711	12757	11863	10867	9907	9029	8123	7213	6317	5471	4597	3739	2917	2137	1399	673	79
23	21587	20599	19597	18539	17597	16631	15641	14723	13721	12763	11867	10883	9923	9041	8147	7219	6323	5477	4603	3761	2927	2141	1409	677	83
24	21589	20611	19603	18541	17599	16633	15643	14731	13723	12781	11887	10889	9929	9043	8161	7229	6329	5479	4621	3767	2939	2143	1423	683	89
25	21599	20627	19609	18553	17609	16649	15647	14737	13729	12791	11897	10891	9931	9049	8167	7237	6337	5483	4637	3769	2953	2153	1427	691	97
26	21601	20639	19661	18583	17623	16651	15649	14741	13751	12799	11903	10903	9941	9059	8171	7243	6343	5501	4639	3779	2957	2161	1429	701	101
27	21611	20641	19681	18587	17627	16657	15661	14747	13757	12809	11909	10909	9949	9067	8179	7247	6353	5503	4643	3793	2963	2179	1433	709	103
28	21613	20663	19687	18593	17657	16661	15667	14753	13759	12821	11923	10937	9967	9091	8191	7253	6359	5507	4649	3797	2969	2203	1439	719	107
29	21617	20681	19697	18617	17659	16673	15671	14759	13763	12823	11927	10939	9973	9103	8209	7283	6361	5519	4651	3803	2971	2207	1447	727	109
30	21647	20693	19699	18637	17669	16691	15679	14767	13781	12829	11933	10949	10007	9109	8219	7297	6367	5521	4657	3821	2999	2213	1451	733	113
31	21649	20707	19709	18661	17681	16693	15683	14771	13789	12841	11939	10957	10009	9127	8221	7307	6373	5527	4663	3823	3001	2221	1453	739	127
32	21661	20717	19717	18671	17683	16699	15727	14779	13799	12853	11941	10973	10037	9133	8231	7309	6379	5531	4673	3833	3011	2237	1459	743	131
33	21673	20719	19727	18679	17707	16703	15731	14783	13807	12889	11953	10979	10039	9137	8233	7321	6389	5557	4679	3847	3019	2239	1471	751	137
34	21683	20731	19739	18691	17713	16729	15733	14797	13829	12893	11959	10987	10061	9151	8237	7331	6397	5563	4691	3851	3023	2243	1481	757	139
35	21701	20743	19751	18701	17729	16741	15737	14813	13831	12899	11969	10993	10067	9157	8243	7333	6421	5569	4703	3853	3037	2251	1483	761	149
36	21713	20747	19753	18713	17737	16747	15739	14821	13841	12907	11971	11003	10069	9161	8263	7349	6427	5573	4721	3863	3041	2267	1487	769	151
37	21727	20749	19759	18719	17747	16759	15749	14827	13859	12911	11981	11027	10079	9173	8269	7351	6449	5581	4723	3877	3049	2269	1489	773	157
38	21737	20753	19763	18731	17749	16763	15761	14831	13873	12917	11987	11047	10091	9181	8273	7369	6451	5591	4729	3881	3061	2273	1493	787	163
39	21739	20759	19777	18743	17761	16787	15767	14843	13877	12919	12007	11057	10093	9187	8287	7393	6469	5623	4733	3889	3067	2281	1499	797	167
40	21751	20771	19793	18749	17783	16811	15773	14851	13879	12923	12011	11059	10099	9199	8291	7411	6473	5639	4751	3907	3079	2287	1511	809	173
41	21757	20773	19801	18757	17789	16823	15787	14867	13883	12941	12037	11069	10103	9203	8293	7417	6481	5641	4759	3911	3083	2293	1523	811	179
42	21767	20789	19813	18773	17791	16829	15791	14869	13901	12953	12041	11071	10111	9209	8297	7433	6491	5647	4783	3917	3089	2297	1531	821	181
43	21773	20807	19819	18787	17807	16831	15797	14879	13903	12959	12043	11083	10133	9221	8311	7451	6521	5651	4787	3919	3109	2309	1543	823	191
44	21787	20809	19841	18793	17827	16843	15803	14887	13907	12967	12049	11087	10139	9227	8317	7457	6529	5653	4789	3923	3119	2311	1549	827	193
45	21799	20849	19843	18797	17837	16871	15809	14891	13913	12973	12071	11093	10141	9239	8329	7459	6547	5657	4793	3929	3121	2333	1553	829	197
46	21803	20857	19853	18803	17839	16879	15817	14897	13921	12979	12073	11113	10151	9241	8353	7477	6551	5659	4799	3931	3137	2339	1559	839	199
47	21817	20873	19861	18839	17851	16883	15823	14923	13931	12983	12097	11117	10159	9257	8363	7481	6553	5669	4801	3943	3163	2341	1567	853	211
48	21821	20879	19867	18859	17863	16889	15859	14929	13933	13001	12101	11119	10163	9277	8369	7487	6563	5683	4813	3947	3167	2347	1571	857	223
49	21839	20887	19889	18869	17881	16901	15877	14939	13963	13003	12107	11131	10169	9281	8377	7489	6569	5689	4817	3967	3169	2351	1579	859	227
50	21841	20897	19891	18899	17891	16903	15881	14947	13967	13007	12109	11149	10177	9283	8387	7499	6571	5693	4831	3989	3181	2357	1583	863	229

PRIMES (continued)

PRIMES (continued)

	0	1	2	3	4	5	6	7	8	9	10	11	12	13	14	15	16	17	18	19	20	21	22	23	24
51	233	877	1597	2371	3187	4001	4861	5701	6577	7507	8389	9293	10181	11159	12113	13009	13997	14951	15887	16921	17903	18911	19913	20899	21851
52	239	881	1601	2377	3191	4003	4871	5711	6581	7517	8419	9311	10193	11161	12119	13033	13999	14957	15889	16927	17909	18913	19919	20903	21859
53	241	883	1607	2381	3203	4007	4877	5717	6599	7523	8423	9319	10211	11171	12143	13037	14009	14969	15901	16931	17911	18917	19927	20921	21863
54	251	887	1609	2383	3209	4013	4889	5737	6607	7529	8429	9323	10223	11173	12149	13043	14011	14983	15907	16937	17921	18919	19937	20929	21871
55	257	907	1613	2389	3217	4019	4903	5741	6619	7537	8431	9337	10243	11177	12157	13049	14029	15013	15913	16943	17923	18947	19949	20939	21881
56	263	911	1619	2393	3221	4021	4909	5743	6637	7541	8443	9341	10247	11197	12161	13063	14033	15017	15919	16963	17929	18959	19961	20947	21893
57	269	919	1621	2399	3229	4027	4919	5749	6653	7547	8447	9343	10253	11213	12163	13093	14051	15031	15923	16979	17939	18973	19963	20959	21911
58	271	929	1627	2411	3251	4049	4931	5779	6659	7549	8461	9349	10259	11239	12197	13099	14057	15053	15937	16981	17957	18979	19973	20963	21929
59	277	937	1637	2417	3253	4051	4933	5783	6661	7559	8467	9371	10267	11243	12203	13103	14071	15061	15959	16987	17959	19001	19979	20981	21937
60	281	941	1657	2423	3257	4057	4937	5791	6673	7561	8501	9377	10271	11251	12211	13109	14081	15073	15971	16993	17971	19009	19991	20983	21943
61	283	947	1663	2437	3259	4073	4943	5801	6679	7573	8513	9391	10273	11257	12227	13121	14083	15077	15973	17011	17977	19013	19993	21001	21961
62	293	953	1667	2441	3271	4079	4951	5807	6689	7577	8521	9397	10289	11261	12239	13127	14087	15083	15991	17021	17981	19031	19997	21011	21977
63	307	967	1669	2447	3299	4091	4957	5813	6691	7583	8527	9403	10301	11273	12241	13147	14107	15091	16001	17027	17987	19037	20011	21013	21991
64	311	971	1693	2459	3301	4093	4967	5821	6701	7589	8537	9413	10303	11279	12251	13151	14143	15101	16007	17029	17989	19051	20021	21017	21997
65	313	977	1697	2467	3307	4099	4969	5827	6703	7591	8539	9419	10313	11287	12253	13159	14149	15107	16033	17033	18013	19069	20023	21019	22003
66	317	983	1699	2473	3313	4111	4973	5839	6709	7603	8543	9421	10321	11299	12263	13163	14153	15121	16057	17041	18041	19073	20029	21023	22013
67	331	991	1709	2477	3319	4127	4987	5843	6719	7607	8563	9431	10331	11311	12269	13171	14159	15131	16061	17047	18043	19079	20047	21031	22027
68	337	997	1721	2503	3323	4129	4993	5849	6733	7621	8573	9433	10333	11317	12277	13177	14173	15137	16063	17053	18047	19081	20051	21059	22031
69	347	1009	1723	2521	3329	4133	4999	5851	6737	7639	8581	9437	10337	11321	12281	13183	14177	15139	16067	17077	18049	19087	20063	21061	22037
70	349	1013	1733	2531	3331	4139	5003	5857	6761	7643	8597	9439	10343	11329	12289	13187	14197	15149	16069	17093	18059	19121	20071	21067	22039
71	353	1019	1741	2539	3343	4153	5009	5861	6763	7649	8599	9461	10357	11351	12301	13217	14207	15161	16073	17099	18061	19139	20089	21089	22051
72	359	1021	1747	2543	3347	4157	5011	5867	6779	7669	8609	9463	10369	11353	12323	13219	14221	15173	16087	17107	18077	19141	20101	21101	22063
73	367	1031	1753	2549	3359	4159	5021	5869	6781	7673	8623	9467	10391	11369	12329	13229	14243	15187	16091	17117	18089	19157	20107	21107	22067
74	373	1033	1759	2551	3361	4177	5023	5879	6791	7681	8627	9473	10399	11383	12343	13241	14249	15193	16097	17123	18097	19163	20113	21121	22073
75	379	1039	1777	2557	3371	4201	5039	5881	6793	7687	8629	9479	10427	11393	12347	13249	14251	15199	16103	17137	18119	19181	20117	21139	22079
76	383	1049	1783	2579	3373	4211	5051	5897	6803	7691	8641	9491	10429	11399	12373	13259	14281	15217	16111	17159	18121	19183	20123	21143	22091
77	389	1051	1787	2591	3389	4217	5059	5903	6823	7699	8647	9497	10433	11411	12377	13267	14293	15227	16127	17167	18127	19207	20129	21149	22093
78	397	1061	1789	2593	3391	4219	5077	5923	6827	7703	8663	9511	10453	11423	12379	13291	14303	15233	16139	17183	18131	19211	20143	21157	22109
79	401	1063	1801	2609	3407	4229	5081	5927	6829	7717	8669	9521	10457	11437	12391	13297	14321	15241	16141	17189	18133	19213	20147	21163	22111
80	409	1069	1811	2617	3413	4231	5087	5939	6833	7723	8677	9533	10459	11443	12401	13309	14323	15259	16183	17191	18143	19219	20149	21169	22123
81	419	1087	1823	2621	3433	4241	5099	5953	6841	7727	8681	9539	10463	11447	12409	13313	14327	15263	16187	17203	18149	19231	20161	21179	22129
82	421	1091	1831	2633	3449	4243	5101	5981	6857	7741	8689	9547	10477	11467	12413	13327	14341	15269	16189	17207	18169	19237	20173	21187	22133
83	431	1093	1847	2647	3457	4253	5107	5987	6863	7753	8693	9551	10487	11471	12421	13331	14347	15271	16193	17209	18181	19249	20177	21191	22147
84	433	1097	1861	2657	3461	4259	5113	6007	6869	7757	8699	9587	10499	11483	12433	13337	14369	15277	16217	17231	18191	19259	20183	21193	22153
85	439	1103	1867	2659	3463	4261	5119	6011	6871	7759	8707	9601	10501	11489	12437	13339	14387	15287	16223	17239	18199	19267	20201	21211	22157
86	443	1109	1871	2663	3467	4271	5147	6029	6883	7789	8713	9613	10513	11491	12451	13367	14389	15289	16229	17257	18211	19273	20219	21221	22159
87	449	1117	1873	2671	3469	4273	5153	6037	6899	7793	8719	9619	10529	11497	12457	13381	14401	15299	16231	17291	18217	19289	20231	21227	22171
88	457	1123	1877	2677	3491	4283	5167	6043	6907	7817	8731	9623	10531	11503	12473	13397	14407	15307	16249	17293	18223	19301	20233	21247	22189
89	461	1129	1879	2683	3499	4289	5171	6047	6911	7823	8737	9629	10559	11519	12479	13399	14411	15313	16253	17299	18229	19309	20249	21269	22193
90	463	1151	1889	2687	3511	4297	5179	6053	6917	7829	8741	9631	10567	11527	12487	13411	14419	15319	16267	17317	18233	19319	20261	21277	22229
91	467	1153	1901	2689	3517	4327	5189	6067	6947	7841	8747	9643	10589	11549	12491	13417	14423	15329	16273	17321	18251	19333	20269	21283	22247
92	479	1163	1907	2693	3527	4337	5197	6073	6949	7853	8753	9649	10597	11551	12497	13421	14431	15331	16301	17327	18253	19373	20287	21313	22259
93	487	1171	1913	2699	3529	4339	5209	6079	6959	7867	8761	9661	10601	11579	12503	13441	14437	15349	16319	17333	18257	19379	20297	21317	22271
94	491	1181	1931	2707	3533	4349	5227	6089	6961	7873	8779	9677	10607	11587	12511	13451	14447	15359	16333	17341	18269	19381	20323	21319	22273
95	499	1187	1933	2711	3539	4357	5231	6091	6967	7877	8783	9679	10613	11593	12517	13457	14449	15361	16339	17351	18287	19387	20327	21323	22277
96	503	1193	1949	2713	3541	4363	5233	6101	6971	7879	8803	9689	10627	11597	12527	13463	14461	15373	16349	17359	18289	19391	20333	21341	22279
97	509	1201	1951	2719	3547	4373	5237	6113	6977	7883	8807	9697	10631	11617	12539	13469	14479	15377	16361	17377	18301	19403	20341	21347	22283
98	521	1213	1973	2729	3557	4391	5261	6121	6983	7901	8819	9719	10639	11621	12541	13477	14489	15383	16363	17383	18307	19417	20347	21377	22291
99	523	1217	1979	2731	3559	4397	5273	6131	6991	7907	8821	9721	10651	11633	12547	13487	14503	15391	16369	17387	18311	19421	20353	21379	22303
100	541	1223	1987	2741	3571	4409	5279	6133	6997	7919	8831	9733	10657	11657	12553	13499	14519	15401	16381	17389	18313	19423	20357	21383	22307

PRIMES (continued)

n	25	26	27	28	29	30	31	32	33	34	35	36	37	38	39	40	41	42	43	44	45	46	47	48	49
1	22343	23327	24317	25409	26407	27457	28513	29453	30577	31607	32611	33617	34651	35771	36787	37831	38923	39979	41113	42083	43063	44203	45317	46451	47533
2	22349	23333	24329	25411	26417	27479	28517	29473	30593	31627	32621	33619	34667	35797	36791	37847	38933	39983	41117	42089	43067	44207	45319	46457	47543
3	22367	23339	24337	25423	26423	27481	28537	29483	30631	31643	32633	33623	34673	35801	36793	37853	38953	39989	41131	42101	43093	44221	45329	46471	47563
4	22369	23357	24359	25439	26431	27487	28541	29501	30637	31649	32647	33629	34679	35803	36809	37861	38959	40009	41141	42131	43103	44249	45337	46477	47569
5	22381	23369	24371	25447	26437	27509	28547	29527	30643	31657	32653	33637	34687	35809	36821	37871	38971	40013	41143	42139	43117	44257	45341	46489	47581
6	22391	23371	24373	25453	26449	27527	28549	29531	30649	31663	32687	33641	34693	35831	36833	37879	38977	40031	41149	42157	43133	44263	45343	46499	47591
7	22397	23399	24379	25457	26459	27529	28559	29537	30661	31667	32693	33647	34703	35837	36847	37889	38993	40037	41161	42169	43151	44267	45361	46507	47599
8	22409	23417	24391	25463	26479	27539	28571	29567	30671	31687	32707	33679	34721	35839	36857	37897	39019	40039	41177	42179	43159	44269	45377	46511	47609
9	22433	23431	24407	25469	26489	27541	28573	29569	30677	31699	32713	33703	34729	35851	36871	37907	39023	40063	41179	42181	43177	44273	45389	46523	47623
10	22441	23447	24413	25471	26497	27551	28579	29573	30689	31721	32717	33713	34739	35863	36877	37951	39041	40087	41183	42187	43189	44279	45403	46549	47629
11	22447	23459	24419	25523	26501	27581	28591	29581	30697	31723	32719	33721	34747	35869	36887	37957	39043	40093	41189	42193	43201	44281	45413	46559	47639
12	22453	23473	24439	25537	26513	27583	28597	29587	30703	31727	32749	33739	34757	35879	36899	37963	39047	40099	41201	42197	43207	44293	45427	46567	47653
13	22469	23497	24443	25541	26539	27611	28603	29599	30707	31729	32771	33749	34759	35897	36901	37967	39079	40111	41203	42209	43223	44351	45433	46573	47657
14	22481	23509	24469	25561	26557	27617	28607	29611	30713	31741	32779	33751	34763	35899	36913	37987	39089	40123	41213	42221	43237	44357	45439	46589	47659
15	22483	23531	24473	25577	26561	27631	28619	29629	30727	31751	32783	33757	34781	35911	36919	37991	39097	40127	41221	42223	43261	44371	45481	46591	47681
16	22501	23537	24481	25579	26573	27647	28621	29633	30757	31769	32789	33767	34807	35923	36923	37993	39103	40129	41227	42227	43271	44381	45491	46601	47699
17	22511	23539	24499	25583	26591	27653	28627	29641	30763	31771	32797	33769	34819	35933	36929	37997	39107	40151	41231	42239	43283	44383	45497	46619	47701
18	22531	23549	24509	25589	26597	27673	28631	29663	30773	31793	32801	33773	34841	35951	36931	38011	39113	40153	41233	42257	43291	44389	45503	46633	47711
19	22541	23557	24517	25601	26627	27689	28643	29669	30781	31799	32803	33791	34843	35963	36943	38039	39119	40163	41243	42281	43313	44417	45523	46639	47713
20	22543	23561	24527	25603	26641	27691	28649	29671	30803	31817	32831	33797	34847	35969	36947	38047	39133	40169	41257	42283	43319	44449	45533	46643	47717
21	22549	23563	24533	25609	26647	27697	28657	29683	30809	31847	32833	33809	34849	35977	36973	38053	39139	40177	41263	42293	43321	44453	45541	46649	47737
22	22567	23567	24547	25621	26669	27701	28661	29717	30817	31849	32839	33811	34871	35983	36979	38069	39157	40189	41269	42299	43331	44483	45553	46663	47741
23	22571	23581	24551	25633	26683	27733	28663	29723	30829	31859	32843	33827	34877	35993	36997	38083	39161	40193	41281	42307	43391	44491	45557	46679	47743
24	22573	23593	24571	25639	26687	27737	28669	29741	30839	31873	32869	33829	34883	35999	37003	38113	39163	40213	41299	42323	43397	44497	45569	46681	47777
25	22613	23599	24593	25643	26693	27739	28687	29753	30841	31883	32887	33851	34897	36007	37013	38119	39181	40231	41333	42331	43399	44501	45587	46687	47779
26	22619	23603	24611	25657	26699	27743	28697	29759	30851	31891	32909	33857	34913	36011	37019	38149	39191	40237	41341	42337	43403	44507	45589	46691	47791
27	22621	23609	24623	25667	26701	27749	28703	29761	30853	31907	32911	33863	34919	36013	37021	38153	39199	40241	41351	42349	43411	44519	45599	46703	47797
28	22637	23623	24631	25673	26711	27751	28711	29789	30859	31957	32917	33871	34939	36017	37039	38177	39209	40253	41357	42359	43427	44531	45613	46723	47807
29	22639	23627	24659	25679	26713	27763	28723	29803	30869	31963	32933	33889	34949	36037	37049	38189	39217	40277	41381	42373	43441	44533	45631	46727	47809
30	22643	23629	24671	25693	26717	27767	28729	29819	30871	31973	32939	33893	34961	36061	37057	38197	39227	40283	41387	42379	43451	44537	45641	46747	47819
31	22651	23633	24677	25703	26723	27773	28751	29833	30881	31981	32941	33911	34963	36067	37061	38201	39229	40289	41389	42391	43457	44543	45659	46751	47837
32	22669	23663	24683	25717	26729	27779	28753	29837	30893	31991	32957	33923	34981	36073	37087	38219	39233	40343	41399	42397	43481	44549	45667	46757	47843
33	22679	23669	24691	25733	26731	27791	28759	29851	30911	32003	32969	33931	35023	36083	37097	38231	39239	40351	41411	42403	43487	44563	45673	46769	47857
34	22691	23671	24697	25741	26737	27793	28771	29863	30931	32009	32971	33937	35027	36097	37117	38237	39241	40357	41413	42407	43499	44579	45677	46771	47869
35	22697	23677	24709	25747	26759	27799	28789	29867	30937	32027	32983	33941	35051	36107	37123	38239	39251	40361	41443	42409	43517	44587	45691	46807	47881
36	22699	23687	24733	25759	26777	27803	28793	29873	30941	32029	32987	33961	35053	36109	37139	38261	39293	40387	41453	42433	43541	44617	45697	46811	47903
37	22709	23689	24749	25763	26783	27809	28807	29879	30949	32051	32993	33967	35059	36131	37159	38273	39301	40423	41467	42437	43543	44621	45707	46817	47911
38	22717	23719	24763	25771	26801	27817	28813	29881	30971	32057	32999	33997	35069	36137	37171	38281	39313	40427	41479	42443	43573	44623	45737	46819	47917
39	22721	23741	24767	25793	26813	27823	28817	29917	30977	32059	33013	34019	35081	36151	37181	38287	39317	40429	41491	42451	43577	44633	45751	46829	47933
40	22727	23743	24781	25799	26821	27827	28837	29921	30983	32063	33023	34031	35083	36161	37189	38299	39323	40433	41507	42457	43579	44641	45757	46831	47939
41	22739	23747	24793	25801	26833	27847	28843	29927	31013	32069	33029	34033	35089	36187	37199	38303	39341	40459	41513	42461	43591	44647	45763	46853	47947
42	22741	23753	24799	25819	26839	27851	28859	29947	31019	32077	33037	34039	35099	36191	37201	38317	39343	40471	41519	42463	43597	44651	45767	46861	47951
43	22751	23767	24809	25841	26849	27883	28867	29959	31033	32083	33049	34057	35107	36209	37217	38321	39359	40483	41521	42467	43607	44657	45779	46867	47963
44	22769	23773	24821	25847	26861	27893	28871	29983	31039	32089	33053	34061	35111	36217	37223	38327	39367	40487	41539	42473	43609	44683	45817	46877	47969
45	22777	23789	24841	25849	26863	27901	28879	29989	31051	32099	33071	34123	35117	36229	37243	38329	39371	40493	41543	42487	43613	44687	45821	46889	47977
46	22783	23801	24847	25867	26879	27917	28901	30011	31063	32117	33073	34127	35129	36241	37253	38333	39373	40499	41549	42491	43627	44699	45823	46901	47981
47	22787	23813	24851	25873	26881	27919	28909	30013	31069	32119	33083	34129	35141	36251	37273	38351	39383	40507	41579	42499	43633	44701	45827	46919	48017
48	22807	23819	24859	25889	26891	27941	28921	30029	31079	32141	33091	34141	35149	36263	37277	38371	39397	40519	41593	42509	43649	44711	45833	46933	48023
49	22811	23827	24877	25903	26893	27943	28927	30047	31081	32143	33107	34147	35153	36269	37307	38377	39409	40529	41597	42533	43651	44729	45841	46957	48029
50	22817	23831	24889	25913	26903	27947	28933	30059	31091	32159	33113	34157	35159	36277	37309	38393	39419	40531	41603	42557	43661	44741	45853	46993	48049

PRIMES (continued)

	25	26	27	28	29	30	31	32	33	34	35	36	37	38	39	40	41	42	43	44	45	46	47	48	49
51	22853	23831	24889	25919	26893	27953	28949	30071	31121	32173	33119	34159	35171	36293	37313	38377	39439	40543	41609	42569	43669	44753	45863	46997	48073
52	22859	23833	24907	25931	26903	27961	28961	30089	31123	32183	33149	34171	35201	36299	37321	38393	39443	40559	41611	42571	43691	44771	45869	47017	48079
53	22861	23857	24917	25933	26921	27967	28979	30091	31139	32189	33151	34183	35221	36307	37337	38431	39451	40577	41617	42577	43711	44773	45887	47041	48091
54	22871	23869	24919	25939	26927	27983	29009	30097	31147	32191	33161	34211	35227	36313	37339	38447	39461	40583	41621	42589	43717	44777	45893	47051	48109
55	22877	23873	24923	25943	26947	27997	29017	30103	31151	32203	33179	34213	35251	36319	37357	38449	39499	40591	41627	42611	43721	44789	45943	47057	48119
56	22901	23879	24943	25951	26951	28001	29021	30109	31153	32213	33181	34217	35257	36341	37361	38453	39503	40597	41641	42641	43753	44797	45949	47059	48121
57	22907	23887	24953	25969	26953	28019	29023	30113	31159	32233	33191	34231	35267	36343	37363	38459	39509	40609	41647	42643	43759	44809	45953	47087	48131
58	22921	23893	24967	25981	26959	28027	29027	30119	31177	32237	33199	34253	35279	36353	37369	38461	39511	40627	41651	42649	43777	44819	45959	47093	48157
59	22937	23899	24971	25997	26981	28031	29033	30133	31181	32251	33203	34259	35281	36373	37379	38501	39521	40637	41659	42667	43781	44839	45971	47111	48163
60	22943	23909	24977	25999	26987	28051	29059	30137	31183	32257	33211	34261	35291	36383	37397	38543	39541	40639	41669	42677	43783	44843	45979	47119	48179
61	22961	23911	24979	26003	26993	28057	29063	30139	31189	32261	33223	34267	35311	36389	37409	38557	39551	40693	41681	42683	43787	44851	45989	47123	48187
62	22963	23917	24989	26017	27011	28069	29077	30161	31193	32297	33247	34273	35317	36433	37423	38561	39563	40697	41687	42689	43789	44867	46021	47129	48193
63	22973	23929	25013	26021	27017	28081	29101	30169	31219	32299	33287	34283	35323	36451	37441	38567	39569	40699	41719	42697	43793	44879	46027	47137	48197
64	22993	23957	25031	26029	27031	28087	29123	30181	31223	32303	33289	34297	35327	36457	37447	38569	39581	40709	41729	42701	43801	44887	46049	47143	48221
65	23003	23971	25033	26041	27043	28097	29129	30187	31231	32309	33301	34301	35339	36467	37463	38593	39607	40739	41737	42703	43853	44893	46051	47147	48239
66	23011	23977	25037	26053	27059	28099	29131	30197	31237	32321	33311	34303	35353	36469	37483	38603	39619	40751	41759	42709	43867	44909	46061	47149	48247
67	23017	23981	25057	26083	27061	28109	29137	30203	31247	32323	33317	34313	35363	36473	37489	38609	39623	40759	41761	42719	43889	44917	46073	47161	48259
68	23021	23993	25073	26099	27067	28111	29147	30211	31249	32327	33329	34319	35381	36479	37493	38611	39631	40763	41771	42727	43891	44927	46091	47189	48271
69	23027	24001	25087	26107	27073	28123	29153	30223	31253	32341	33331	34327	35393	36493	37501	38629	39659	40771	41777	42737	43913	44939	46093	47207	48281
70	23029	24007	25097	26111	27077	28151	29167	30241	31259	32353	33343	34337	35401	36497	37507	38639	39667	40787	41801	42743	43933	44953	46099	47221	48299
71	23039	24019	25111	26113	27091	28163	29173	30253	31267	32359	33347	34351	35407	36523	37511	38651	39671	40801	41809	42751	43943	44959	46103	47237	48311
72	23041	24023	25117	26119	27103	28181	29179	30259	31271	32363	33349	34361	35419	36527	37517	38653	39679	40813	41813	42767	43951	44963	46133	47251	48313
73	23053	24029	25121	26141	27107	28183	29191	30269	31277	32369	33353	34367	35423	36529	37529	38669	39703	40819	41843	42773	43961	44971	46141	47269	48337
74	23057	24043	25127	26153	27109	28201	29201	30271	31307	32371	33359	34369	35437	36541	37537	38671	39709	40823	41849	42787	43963	44983	46147	47279	48341
75	23059	24049	25147	26161	27127	28211	29207	30293	31319	32377	33377	34381	35447	36551	37547	38677	39719	40829	41851	42793	43969	44987	46153	47287	48353
76	23063	24061	25153	26171	27143	28219	29209	30307	31321	32381	33391	34403	35449	36559	37549	38693	39727	40841	41863	42797	43973	45007	46171	47293	48371
77	23071	24071	25163	26177	27179	28229	29221	30313	31327	32401	33403	34421	35461	36563	37561	38699	39733	40847	41879	42821	43987	45013	46181	47297	48383
78	23081	24077	25169	26183	27191	28277	29231	30319	31333	32411	33409	34429	35491	36571	37567	38707	39749	40849	41887	42829	43991	45053	46183	47303	48397
79	23087	24083	25171	26189	27197	28279	29243	30323	31337	32413	33413	34439	35507	36583	37571	38711	39761	40853	41893	42839	43997	45061	46187	47309	48407
80	23099	24091	25183	26203	27211	28283	29251	30341	31357	32423	33427	34457	35509	36587	37573	38713	39769	40867	41897	42841	44017	45077	46199	47317	48409
81	23117	24097	25189	26209	27239	28289	29269	30347	31379	32429	33457	34469	35521	36599	37579	38723	39779	40879	41903	42853	44021	45083	46219	47339	48413
82	23131	24103	25219	26227	27241	28297	29287	30367	31387	32441	33461	34471	35527	36607	37589	38729	39791	40883	41911	42859	44027	45119	46229	47351	48437
83	23143	24107	25229	26237	27253	28307	29297	30389	31391	32443	33469	34483	35531	36629	37591	38737	39799	40897	41927	42863	44029	45121	46237	47353	48449
84	23159	24109	25237	26249	27259	28309	29303	30391	31393	32467	33479	34487	35533	36637	37607	38747	39821	40903	41941	42899	44041	45127	46261	47363	48463
85	23167	24113	25243	26251	27271	28319	29311	30403	31397	32479	33487	34499	35537	36643	37619	38749	39827	40927	41947	42901	44053	45131	46271	47381	48473
86	23173	24121	25247	26261	27277	28349	29327	30427	31469	32491	33493	34501	35543	36653	37633	38767	39829	40933	41953	42923	44059	45137	46273	47387	48479
87	23189	24133	25253	26263	27281	28351	29333	30431	31477	32497	33503	34511	35569	36671	37643	38783	39839	40939	41957	42929	44071	45139	46279	47389	48481
88	23197	24137	25261	26267	27283	28387	29339	30449	31481	32503	33521	34513	35573	36677	37649	38791	39841	40949	41959	42937	44087	45161	46301	47407	48487
89	23201	24151	25301	26293	27299	28393	29347	30467	31489	32507	33529	34519	35591	36683	37657	38803	39847	40961	41969	42943	44089	45179	46307	47417	48491
90	23203	24169	25303	26297	27329	28403	29363	30469	31511	32531	33533	34537	35593	36691	37663	38821	39857	40973	41981	42953	44101	45181	46309	47419	48497
91	23209	24179	25307	26309	27337	28409	29383	30491	31513	32533	33547	34543	35597	36697	37691	38833	39863	40993	41983	42961	44111	45191	46327	47431	48523
92	23227	24181	25309	26317	27361	28411	29387	30493	31517	32537	33563	34549	35603	36709	37693	38839	39869	41011	41999	42967	44119	45197	46337	47441	48527
93	23251	24197	25321	26321	27367	28429	29389	30497	31531	32561	33569	34583	35617	36713	37699	38851	39883	41017	42013	42979	44123	45233	46349	47459	48533
94	23269	24203	25339	26339	27397	28433	29399	30509	31541	32563	33577	34591	35671	36721	37717	38861	39887	41023	42017	42989	44129	45247	46351	47491	48539
95	23279	24223	25343	26347	27407	28439	29401	30517	31543	32569	33581	34603	35677	36739	37747	38867	39901	41039	42019	43003	44131	45259	46381	47497	48541
96	23291	24229	25349	26357	27409	28447	29411	30529	31547	32573	33587	34607	35729	36749	37781	38873	39929	41047	42023	43013	44159	45263	46399	47501	48563
97	23293	24239	25357	26371	27427	28463	29423	30539	31567	32579	33589	34613	35731	36761	37783	38891	39937	41051	42043	43019	44171	45281	46411	47507	48571
98	23297	24247	25367	26387	27431	28477	29429	30553	31573	32587	33599	34631	35747	36767	37799	38903	39953	41057	42061	43037	44179	45289	46439	47513	48589
99	23311	24251	25373	26393	27437	28493	29437	30557	31583	32603	33601	34649	35753	36779	37811	38917	39971	41077	42071	43049	44189	45293	46441	47521	48593
100	23321	24281	25391	26399	27449	28499	29443	30559	31601	32609	33613	34651	35759	36781	37813	38921	39979	41081	42073	43051	44201	45307	46447	47527	48611

PRIMES (continued)

PRIMES (continued)

PRIMES (continued)

	50	51	52	53	54	55	56	57	58	59	60	61	62	63	64	65	66	67	68	69	70	71	72	73	74
1	48619	49667	50767	51817	52937	54001	55109	56197	57193	58243	59369	60509	61637	62791	63823	65071	66107	67247	68389	69497	70663	71719	72859	73999	75083
2	48623	49669	50773	51827	52951	54011	55117	56207	57203	58271	59377	60521	61643	62801	63839	65089	66109	67261	68399	69499	70667	71741	72869	74017	75109
3	48647	49681	50777	51829	52957	54013	55127	56209	57221	58309	59387	60527	61651	62819	63841	65099	66137	67271	68437	69539	70687	71761	72871	74021	75133
4	48649	49697	50789	51839	52963	54037	55147	56237	57223	58313	59393	60539	61657	62827	63853	65101	66161	67273	68443	69557	70709	71777	72883	74027	75149
5	48661	49711	50821	51853	52967	54049	55163	56239	57241	58321	59399	60589	61667	62851	63857	65111	66169	67289	68447	69593	70717	71789	72889	74047	75161
6	48673	49727	50833	51859	52973	54059	55171	56249	57251	58337	59407	60601	61673	62861	63863	65119	66173	67307	68449	69623	70729	71807	72893	74051	75167
7	48677	49739	50839	51869	52981	54083	55201	56263	57259	58363	59417	60607	61681	62869	63901	65123	66179	67339	68473	69653	70753	71809	72901	74071	75169
8	48679	49741	50849	51871	52999	54091	55207	56267	57269	58367	59419	60611	61687	62873	63907	65129	66191	67343	68477	69661	70769	71821	72907	74077	75181
9	48731	49747	50857	51893	53003	54101	55213	56269	57271	58369	59441	60617	61703	62897	63913	65141	66221	67349	68483	69677	70783	71837	72911	74093	75193
10	48733	49757	50867	51899	53017	54121	55217	56299	57283	58379	59443	60623	61717	62903	63929	65147	66239	67369	68489	69691	70793	71843	72923	74099	75209
11	48751	49783	50873	51907	53047	54133	55219	56311	57287	58391	59447	60631	61723	62921	63949	65167	66271	67391	68491	69697	70823	71849	72931	74101	75211
12	48757	49787	50891	51913	53051	54139	55229	56333	57301	58393	59453	60637	61729	62927	63977	65171	66293	67399	68501	69709	70841	71861	72937	74131	75217
13	48761	49789	50893	51929	53069	54151	55243	56359	57329	58403	59467	60647	61751	62939	63997	65173	66301	67409	68507	69737	70843	71867	72949	74143	75223
14	48767	49801	50909	51941	53077	54163	55249	56369	57331	58411	59471	60649	61757	62969	64007	65179	66337	67411	68521	69739	70849	71879	72953	74149	75227
15	48779	49807	50923	51949	53087	54167	55259	56377	57347	58417	59473	60659	61781	62971	64013	65183	66343	67421	68531	69761	70853	71881	72959	74159	75239
16	48781	49811	50929	51971	53089	54181	55291	56383	57349	58427	59497	60661	61813	62981	64019	65203	66347	67427	68539	69763	70867	71887	72973	74161	75253
17	48787	49823	50951	51973	53093	54193	55313	56393	57367	58439	59509	60679	61819	62983	64033	65213	66359	67429	68543	69767	70877	71899	72977	74167	75269
18	48799	49831	50957	51977	53101	54217	55331	56401	57373	58441	59513	60689	61837	62987	64037	65239	66361	67433	68567	69779	70879	71909	72997	74177	75277
19	48809	49843	50969	51991	53113	54251	55333	56417	57383	58451	59539	60703	61843	62989	64063	65257	66373	67447	68581	69809	70891	71917	73009	74189	75289
20	48817	49853	50971	52009	53117	54269	55337	56431	57389	58453	59557	60719	61861	63029	64067	65267	66377	67453	68597	69821	70901	71933	73013	74197	75307
21	48821	49871	50989	52021	53129	54277	55339	56437	57397	58477	59561	60727	61871	63031	64081	65269	66383	67477	68611	69827	70913	71941	73019	74201	75323
22	48823	49877	50993	52027	53147	54287	55343	56443	57413	58481	59567	60733	61879	63059	64091	65287	66403	67481	68633	69829	70919	71947	73037	74203	75329
23	48847	49891	51001	52051	53149	54293	55351	56453	57427	58511	59581	60737	61909	63067	64109	65293	66413	67489	68639	69833	70921	71963	73039	74209	75337
24	48857	49919	51031	52057	53161	54311	55373	56467	57457	58537	59611	60757	61927	63073	64123	65309	66431	67493	68659	69847	70937	71971	73043	74219	75347
25	48859	49921	51043	52067	53171	54319	55381	56473	57467	58543	59617	60761	61933	63079	64151	65323	66449	67499	68669	69857	70949	71983	73061	74231	75353
26	48869	49927	51047	52069	53173	54323	55399	56477	57487	58549	59621	60763	61949	63097	64153	65327	66457	67511	68683	69859	70951	71987	73063	74257	75367
27	48871	49937	51059	52081	53189	54331	55411	56479	57493	58567	59627	60773	61961	63103	64157	65353	66463	67523	68687	69877	70957	71993	73079	74279	75377
28	48883	49939	51061	52103	53197	54347	55439	56489	57503	58573	59629	60779	61967	63113	64171	65357	66467	67531	68699	69899	70969	71999	73091	74287	75389
29	48889	49943	51071	52121	53201	54361	55441	56501	57527	58579	59651	60793	61979	63127	64187	65371	66491	67537	68711	69911	70979	72019	73121	74293	75391
30	48907	49957	51109	52127	53231	54367	55457	56503	57529	58601	59659	60811	61981	63131	64189	65381	66499	67547	68713	69929	70981	72031	73127	74297	75401
31	48947	49991	51131	52147	53233	54371	55469	56509	57557	58603	59663	60821	61987	63149	64217	65393	66509	67559	68729	69931	70991	72043	73133	74311	75403
32	48953	49993	51133	52153	53239	54377	55487	56519	57559	58613	59669	60859	61991	63179	64223	65407	66523	67567	68737	69941	70997	72047	73141	74317	75407
33	48973	49999	51137	52163	53267	54401	55501	56527	57571	58631	59671	60869	62003	63197	64231	65413	66529	67577	68743	69959	70999	72053	73181	74323	75431
34	48989	50021	51151	52177	53269	54403	55511	56531	57587	58657	59693	60887	62011	63199	64237	65419	66533	67579	68749	69991	71011	72073	73189	74353	75437
35	48991	50023	51157	52181	53279	54409	55529	56533	57593	58661	59699	60889	62017	63211	64271	65423	66541	67589	68767	69997	71023	72077	73237	74357	75479
36	49003	50033	51169	52183	53281	54413	55541	56543	57601	58679	59707	60899	62039	63241	64279	65437	66553	67601	68771	70001	71039	72089	73243	74363	75503
37	49009	50047	51193	52189	53299	54419	55547	56569	57637	58687	59723	60901	62047	63247	64283	65447	66569	67607	68777	70003	71059	72091	73259	74377	75511
38	49019	50051	51197	52201	53309	54421	55579	56591	57641	58693	59729	60913	62053	63277	64301	65449	66571	67619	68791	70009	71069	72101	73277	74381	75521
39	49031	50053	51199	52223	53323	54437	55589	56597	57649	58699	59743	60917	62057	63281	64303	65479	66587	67631	68813	70019	71081	72103	73291	74383	75527
40	49033	50069	51203	52237	53327	54443	55603	56599	57653	58711	59747	60919	62071	63299	64319	65497	66593	67651	68819	70039	71089	72109	73303	74411	75533
41	49037	50077	51217	52249	53353	54449	55609	56611	57667	58727	59753	60923	62081	63311	64327	65519	66601	67679	68821	70051	71119	72139	73309	74413	75539
42	49043	50087	51229	52253	53359	54469	55619	56629	57679	58733	59771	60937	62099	63313	64333	65521	66617	67699	68863	70061	71129	72161	73327	74419	75541
43	49057	50093	51239	52259	53377	54493	55621	56633	57689	58741	59779	60943	62119	63317	64373	65537	66629	67709	68881	70067	71143	72167	73331	74441	75553
44	49069	50101	51241	52267	53381	54497	55631	56659	57697	58757	59791	60953	62129	63331	64381	65539	66643	67723	68891	70079	71147	72169	73351	74449	75557
45	49081	50111	51257	52289	53401	54499	55633	56663	57709	58763	59797	60961	62131	63337	64399	65543	66653	67733	68897	70099	71153	72173	73361	74453	75571
46	49103	50119	51263	52291	53407	54503	55639	56671	57713	58771	59809	61001	62137	63347	64403	65551	66683	67741	68903	70111	71161	72211	73363	74471	75577
47	49109	50123	51283	52301	53411	54517	55661	56681	57719	58787	59833	61007	62141	63353	64433	65557	66697	67751	68909	70117	71167	72221	73369	74489	75583
48	49117	50129	51287	52313	53419	54521	55663	56687	57727	58789	59863	61027	62143	63361	64439	65563	66701	67757	68917	70121	71171	72223	73379	74507	75611
49	49121	50131	51307	52321	53437	54539	55667	56701	57731	58831	59879	61031	62171	63367	64451	65579	66713	67759	68927	70123	71191	72227	73387	74509	75617
50	49123	50147	51329	52361	53441	54541	55673	56711	57737	58889	59887	61043	62189	63377	64453	65581	66721	67763	68947	70139	71209	72229	73417	74521	75619

PRIMES (continued)

	50	51	52	53	54	55	56	57	58	59	60	61	62	63	64	65	66	67	68	69	70	71	72	73	74
51	49139	50153	51341	52363	53453	54547	55681	56713	57751	58897	59921	61051	62191	63377	64483	65587	66733	67777	68927	70141	71233	72251	73421	74527	75629
52	49157	50159	51343	52369	53479	54559	55691	56731	57773	58901	59929	61057	62201	63389	64489	65599	66739	67783	68947	70157	71237	72253	73433	74531	75641
53	49169	50177	51347	52379	53503	54563	55697	56737	57781	58907	59951	61091	62207	63391	64499	65609	66749	67789	68963	70163	71249	72269	73453	74551	75653
54	49171	50207	51349	52387	53507	54577	55711	56747	57787	58909	59957	61099	62213	63397	64513	65617	66751	67801	68993	70177	71257	72271	73459	74561	75659
55	49177	50221	51361	52391	53527	54581	55717	56767	57791	58913	59971	61121	62219	63409	64553	65629	66763	67807	69001	70181	71261	72277	73471	74567	75679
56	49193	50227	51383	52433	53549	54583	55721	56773	57793	58921	59981	61129	62233	63419	64567	65633	66791	67819	69011	70183	71263	72287	73477	74573	75683
57	49199	50231	51407	52453	53551	54601	55733	56779	57803	58937	59999	61141	62273	63421	64577	65647	66797	67829	69019	70199	71287	72307	73483	74587	75689
58	49201	50261	51413	52457	53569	54617	55763	56783	57809	58943	60013	61151	62297	63439	64591	65651	66809	67843	69029	70201	71293	72313	73517	74597	75703
59	49207	50263	51419	52489	53591	54623	55787	56807	57829	58963	60017	61153	62299	63443	64601	65657	66821	67853	69031	70207	71317	72337	73523	74609	75707
60	49211	50273	51421	52501	53593	54629	55793	56809	57839	58967	60029	61169	62303	63463	64609	65677	66841	67867	69061	70223	71327	72341	73529	74611	75709
61	49223	50287	51427	52511	53597	54631	55799	56813	57847	58979	60037	61211	62311	63467	64613	65687	66851	67883	69067	70229	71329	72353	73547	74623	75721
62	49253	50291	51437	52517	53609	54647	55807	56821	57853	58991	60041	61223	62323	63473	64621	65699	66853	67891	69073	70237	71333	72367	73553	74653	75731
63	49261	50311	51439	52529	53611	54667	55813	56827	57859	58997	60077	61231	62327	63487	64627	65701	66863	67901	69109	70241	71339	72379	73561	74687	75743
64	49277	50321	51449	52541	53617	54673	55817	56843	57881	59009	60083	61253	62347	63493	64633	65707	66877	67927	69119	70249	71341	72383	73571	74699	75767
65	49279	50329	51461	52543	53623	54679	55819	56857	57899	59011	60089	61261	62351	63499	64661	65713	66883	67931	69127	70271	71347	72421	73583	74707	75773
66	49297	50333	51473	52553	53629	54709	55823	56873	57901	59021	60091	61283	62383	63521	64663	65717	66889	67933	69143	70289	71353	72431	73589	74713	75781
67	49307	50341	51479	52561	53633	54713	55829	56891	57917	59023	60101	61291	62401	63527	64667	65719	66919	67939	69149	70297	71359	72461	73597	74717	75787
68	49331	50359	51481	52567	53639	54721	55837	56893	57923	59029	60103	61297	62417	63533	64679	65729	66923	67943	69151	70309	71363	72467	73607	74719	75793
69	49333	50363	51487	52571	53653	54727	55843	56897	57943	59051	60107	61331	62423	63541	64693	65731	66931	67957	69163	70313	71387	72469	73609	74729	75797
70	49339	50377	51503	52579	53657	54751	55849	56909	57947	59053	60127	61333	62459	63559	64709	65761	66943	67961	69191	70321	71389	72481	73613	74731	75821
71	49363	50383	51511	52583	53681	54767	55871	56911	57973	59063	60133	61339	62467	63577	64717	65777	66947	67967	69193	70327	71399	72493	73637	74747	75833
72	49367	50387	51517	52609	53693	54773	55889	56921	57977	59069	60139	61343	62473	63587	64747	65789	66949	67979	69197	70351	71411	72497	73643	74759	75853
73	49369	50411	51521	52627	53699	54779	55897	56923	57991	59077	60149	61357	62477	63589	64763	65809	66959	67987	69203	70373	71413	72503	73651	74761	75869
74	49391	50417	51539	52631	53717	54787	55901	56929	58013	59083	60161	61363	62483	63599	64781	65827	66973	67993	69221	70379	71419	72533	73673	74771	75883
75	49393	50423	51551	52639	53719	54799	55903	56941	58027	59093	60167	61379	62497	63601	64783	65831	66977	68023	69233	70381	71429	72547	73679	74779	75913
76	49409	50441	51563	52667	53731	54829	55921	56951	58031	59107	60169	61381	62501	63607	64793	65837	67003	68041	69239	70393	71437	72551	73681	74797	75931
77	49411	50459	51577	52673	53759	54833	55927	56957	58043	59113	60209	61403	62507	63611	64811	65839	67021	68053	69247	70423	71443	72559	73693	74821	75937
78	49417	50461	51581	52691	53773	54851	55931	56963	58049	59119	60217	61409	62533	63617	64817	65843	67033	68059	69257	70429	71453	72577	73699	74827	75941
79	49429	50497	51593	52697	53777	54869	55933	56983	58057	59123	60223	61417	62539	63629	64849	65851	67043	68071	69259	70439	71471	72613	73709	74831	75967
80	49433	50503	51599	52709	53783	54877	55949	56989	58061	59141	60251	61441	62549	63647	64853	65867	67049	68087	69263	70451	71473	72617	73721	74843	75979
81	49451	50513	51607	52711	53791	54881	55967	56993	58067	59149	60257	61463	62563	63649	64871	65881	67057	68099	69313	70457	71479	72623	73727	74857	75983
82	49459	50527	51613	52721	53813	54907	55987	56999	58073	59159	60259	61469	62581	63659	64877	65899	67061	68111	69317	70459	71483	72643	73751	74861	75989
83	49463	50539	51631	52727	53819	54917	55997	57037	58099	59167	60271	61471	62591	63667	64879	65921	67073	68113	69337	70481	71503	72647	73757	74869	75991
84	49477	50543	51637	52733	53831	54919	56003	57041	58109	59183	60289	61483	62597	63671	64891	65927	67079	68141	69341	70487	71527	72649	73771	74873	75997
85	49481	50549	51647	52747	53849	54941	56009	57047	58111	59197	60293	61487	62603	63689	64901	65929	67103	68147	69371	70489	71537	72661	73783	74887	76001
86	49499	50551	51659	52757	53857	54949	56039	57059	58129	59207	60317	61493	62617	63691	64919	65951	67121	68161	69379	70501	71549	72671	73819	74891	76003
87	49523	50581	51673	52769	53861	54959	56041	57073	58147	59209	60331	61507	62627	63697	64921	65957	67139	68171	69383	70507	71551	72673	73823	74897	76031
88	49529	50587	51679	52783	53881	54973	56053	57077	58151	59219	60337	61511	62633	63703	64927	65963	67141	68207	69389	70529	71563	72679	73847	74903	76039
89	49531	50591	51683	52807	53887	54979	56081	57089	58153	59221	60343	61519	62639	63709	64937	65981	67153	68209	69401	70537	71569	72689	73849	74923	76079
90	49537	50593	51691	52813	53891	54983	56087	57097	58169	59233	60353	61543	62653	63719	64951	65983	67157	68213	69403	70549	71593	72701	73859	74929	76081
91	49547	50599	51713	52817	53897	55001	56093	57107	58171	59239	60373	61547	62659	63727	64969	65993	67169	68219	69427	70571	71597	72707	73867	74933	76091
92	49549	50627	51719	52837	53899	55009	56099	57119	58189	59243	60383	61553	62683	63737	64997	65997	67181	68227	69431	70573	71633	72719	73877	74941	76099
93	49559	50647	51721	52859	53917	55021	56101	57131	58193	59263	60397	61559	62687	63743	65003	66037	67187	68239	69439	70583	71647	72727	73883	74959	76103
94	49597	50651	51749	52861	53923	55049	56113	57139	58199	59273	60413	61561	62701	63761	65011	66041	67189	68261	69457	70589	71663	72733	73897	75011	76123
95	49603	50671	51767	52879	53927	55051	56123	57143	58207	59281	60427	61583	62723	63773	65027	66047	67211	68279	69463	70607	71671	72739	73907	75013	76129
96	49613	50683	51769	52883	53939	55057	56131	57149	58211	59333	60443	61603	62731	63781	65029	66067	67213	68281	69467	70619	71693	72763	73939	75017	76147
97	49627	50707	51787	52889	53951	55061	56149	57163	58217	59341	60449	61609	62743	63793	65033	66071	67217	68311	69473	70621	71699	72767	73943	75029	76157
98	49633	50723	51797	52901	53959	55073	56167	57173	58229	59351	60457	61613	62753	63799	65053	66083	67219	68329	69481	70627	71707	72797	73951	75037	76159
99	49639	50741	51803	52903	53987	55079	56171	57179	58231	59357	60493	61627	62761	63803	65063	66089	67231	68351	69491	70639	71711	72811	73961	75041	76163
100	49663	50753	51817	52919	53993	55103	56179	57191	58237	59359	60497	61631	62773	63809	65071	66103	67247	68371	69493	70657	71713	72823	73973	75079	76207

PRIMES (continued)

PRIMES (continued)

	75	76	77	78	79	80	81	82	83	84	85	86	87	88	89	90	91	92	93	94	95
1	76213	77359	78487	79627	80737	81817	82903	84131	85243	86381	87557	88807	89867	90989	92177	93187	94351	95443	96587	97829	98953
2	76231	77369	78497	79631	80747	81839	82913	84137	85247	86389	87559	88811	89891	90997	92179	93199	94379	95461	96589	97841	98963
3	76243	77377	78509	79633	80749	81847	82939	84143	85259	86399	87583	88813	89897	91009	92189	93229	94397	95467	96601	97843	98981
4	76249	77383	78511	79657	80761	81853	82963	84163	85297	86413	87587	88817	89899	91019	92203	93239	94399	95471	96643	97847	98993
5	76253	77417	78517	79669	80777	81869	82981	84179	85303	86423	87589	88819	89909	91033	92219	93241	94421	95479	96661	97849	98999
6	76259	77419	78539	79687	80779	81883	82997	84181	85313	86441	87613	88843	89917	91079	92221	93251	94427	95483	96667	97859	99013
7	76261	77431	78541	79691	80783	81899	83003	84191	85331	86453	87623	88853	89923	91081	92227	93253	94433	95507	96671	97861	99017
8	76283	77447	78553	79693	80789	81901	83009	84199	85333	86461	87629	88861	89939	91097	92233	93257	94439	95527	96697	97871	99023
9	76289	77471	78569	79697	80803	81919	83023	84211	85361	86467	87631	88867	89959	91099	92237	93281	94441	95531	96703	97879	99041
10	76303	77477	78571	79699	80809	81929	83047	84221	85363	86477	87641	88873	89963	91121	92243	93283	94447	95539	96731	97883	99053
11	76333	77479	78577	79757	80819	81931	83059	84223	85369	86491	87643	88883	89977	91127	92251	93287	94463	95549	96737	97919	99079
12	76343	77489	78583	79769	80831	81937	83063	84229	85381	86501	87649	88897	89983	91129	92269	93307	94477	95561	96739	97927	99083
13	76367	77491	78593	79777	80833	81943	83071	84239	85411	86509	87671	88903	89989	91139	92297	93319	94483	95569	96749	97931	99089
14	76369	77509	78607	79801	80849	81953	83077	84247	85427	86531	87679	88919	90001	91141	92311	93323	94513	95581	96757	97943	99103
15	76379	77513	78623	79811	80863	81967	83089	84263	85429	86533	87683	88937	90007	91151	92317	93337	94529	95597	96763	97961	99109
16	76387	77521	78643	79813	80897	81971	83093	84299	85439	86539	87691	88951	90011	91153	92333	93371	94531	95603	96769	97967	99119
17	76403	77527	78649	79817	80909	81973	83101	84307	85447	86561	87697	88969	90017	91159	92347	93377	94541	95617	96779	97973	99131
18	76421	77543	78653	79823	80911	82007	83117	84313	85451	86573	87701	88993	90019	91163	92353	93383	94543	95621	96787	97987	99133
19	76423	77549	78691	79829	80917	82009	83137	84317	85453	86579	87719	88997	90023	91183	92357	93407	94547	95629	96797	98009	99137
20	76441	77551	78697	79841	80923	82013	83177	84319	85469	86587	87721	89003	90031	91193	92363	93419	94559	95633	96799	98011	99139
21	76463	77557	78707	79843	80929	82021	83203	84347	85487	86599	87739	89009	90053	91199	92369	93427	94561	95651	96821	98017	99149
22	76471	77563	78713	79847	80933	82031	83207	84349	85513	86627	87743	89017	90059	91229	92377	93463	94573	95701	96823	98041	99173
23	76481	77569	78721	79861	80953	82037	83219	84377	85517	86629	87751	89021	90067	91237	92381	93479	94583	95707	96827	98047	99181
24	76487	77573	78737	79867	80963	82039	83221	84389	85523	86677	87767	89041	90071	91243	92383	93481	94597	95713	96847	98057	99191
25	76493	77587	78779	79873	80989	82051	83227	84391	85531	86689	87793	89051	90073	91249	92387	93487	94603	95717	96851	98081	99223
26	76507	77591	78781	79889	81001	82067	83231	84401	85549	86693	87797	89057	90089	91253	92399	93491	94613	95723	96857	98101	99233
27	76511	77611	78787	79901	81013	82073	83233	84407	85571	86711	87803	89069	90107	91283	92401	93493	94621	95731	96893	98123	99241
28	76519	77617	78791	79903	81017	82129	83243	84421	85577	86719	87811	89071	90121	91291	92413	93497	94649	95737	96907	98129	99251
29	76537	77621	78797	79907	81019	82139	83257	84431	85597	86729	87833	89083	90127	91297	92419	93503	94651	95747	96911	98143	99257
30	76541	77641	78803	79939	81023	82141	83267	84437	85601	86743	87853	89087	90149	91303	92431	93523	94687	95773	96931	98179	99259
31	76543	77647	78809	79943	81031	82153	83269	84443	85607	86753	87869	89101	90163	91309	92459	93529	94693	95783	96953	98207	99277
32	76561	77659	78823	79967	81041	82163	83273	84449	85619	86767	87877	89107	90173	91331	92461	93553	94709	95789	96959	98213	99289
33	76579	77681	78839	79973	81043	82171	83299	84457	85621	86771	87881	89113	90187	91367	92467	93557	94723	95791	96973	98221	99317
34	76597	77687	78853	79979	81047	82183	83311	84463	85627	86777	87887	89119	90191	91369	92479	93559	94727	95801	96979	98227	99347
35	76603	77689	78857	79987	81049	82189	83339	84467	85639	86783	87911	89123	90197	91373	92489	93563	94747	95803	96989	98251	99349
36	76607	77699	78877	79997	81071	82193	83341	84481	85643	86813	87917	89137	90199	91381	92503	93581	94771	95813	96997	98257	99367
37	76631	77711	78887	79999	81077	82207	83357	84499	85661	86837	87931	89153	90203	91387	92507	93601	94777	95819	97001	98269	99371
38	76649	77713	78889	80021	81083	82217	83383	84503	85667	86843	87943	89189	90217	91393	92551	93607	94781	95857	97003	98297	99377
39	76651	77719	78893	80039	81097	82219	83389	84509	85669	86851	87959	89203	90227	91397	92557	93629	94789	95869	97007	98299	99391
40	76667	77723	78901	80051	81101	82223	83399	84521	85691	86857	87961	89209	90239	91411	92567	93637	94793	95873	97021	98317	99397
41	76673	77731	78919	80077	81119	82231	83401	84523	85703	86861	87973	89213	90263	91423	92569	93683	94811	95881	97039	98321	99401
42	76679	77743	78929	80107	81131	82237	83407	84533	85711	86869	87977	89227	90271	91453	92581	93701	94819	95891	97073	98323	99409
43	76697	77747	78941	80111	81157	82241	83417	84551	85717	86923	87991	89231	90281	91457	92593	93703	94823	95911	97081	98327	99431
44	76717	77761	78977	80141	81163	82261	83423	84559	85733	86927	88001	89237	90289	91459	92623	93719	94837	95917	97103	98347	99439
45	76733	77773	78979	80147	81173	82267	83431	84589	85751	86929	88003	89261	90313	91463	92627	93739	94841	95923	97117	98369	99469
46	76753	77783	78989	80149	81181	82279	83437	84629	85781	86939	88007	89269	90353	91493	92639	93761	94847	95929	97127	98377	99487
47	76757	77797	79031	80153	81197	82301	83443	84631	85793	86951	88019	89273	90359	91499	92641	93763	94849	95947	97151	98387	99497
48	76771	77801	79039	80167	81199	82307	83449	84649	85817	86959	88037	89293	90371	91513	92647	93787	94873	95957	97157	98389	99523
49	76777	77813	79043	80173	81203	82339	83459	84653	85819	86969	88069	89303	90373	91529	92657	93811	94889	95959	97159	98407	99527
50	76781	77839	79063	80177	81223	82349	83471	84659	85829	86993	88079	89317	90379	91541	92669	93827	94903	95971	97169	98411	99529

PRIMES (continued)

PRIMES (continued)

	75	76	77	78	79	80	81	82	83	84	85	86	87	88	89	90	91	92	93	94	95
51	76801	77849	79087	80177	81233	82349	83477	84673	85831	87011	88093	89329	90379	91541	92671	93809	94907	95987	97171	98419	99551
52	76819	77863	79103	80191	81239	82351	83497	84691	85837	87013	88117	89353	90397	91571	92681	93811	94933	95989	97177	98429	99559
53	76829	77867	79111	80207	81281	82361	83537	84697	85843	87037	88129	89371	90401	91573	92683	93827	94949	96001	97187	98443	99563
54	76831	77893	79133	80209	81283	82373	83557	84701	85847	87041	88169	89381	90403	91577	92693	93851	94951	96013	97213	98453	99571
55	76837	77899	79139	80221	81293	82387	83561	84713	85853	87049	88177	89387	90407	91583	92699	93871	94961	96017	97231	98459	99577
56	76847	77929	79147	80231	81299	82393	83563	84719	85889	87071	88211	89393	90437	91591	92707	93887	94993	96043	97241	98467	99581
57	76871	77933	79151	80233	81307	82421	83579	84731	85903	87083	88223	89399	90439	91621	92717	93889	94999	96053	97259	98473	99607
58	76873	77951	79153	80239	81331	82457	83591	84737	85909	87103	88237	89413	90469	91631	92723	93893	95003	96059	97283	98479	99611
59	76883	77969	79159	80251	81343	82463	83597	84751	85931	87107	88241	89417	90473	91639	92737	93901	95009	96079	97301	98491	99623
60	76907	77977	79181	80263	81349	82469	83609	84761	85933	87119	88259	89431	90481	91673	92753	93911	95021	96097	97303	98507	99643
61	76913	77983	79187	80273	81353	82471	83617	84787	85991	87121	88261	89443	90499	91691	92761	93913	95027	96137	97327	98519	99661
62	76919	77999	79193	80279	81359	82483	83621	84793	85999	87133	88289	89449	90511	91703	92767	93923	95063	96149	97367	98533	99667
63	76943	78007	79201	80287	81371	82487	83639	84809	86011	87149	88301	89459	90523	91711	92779	93937	95071	96157	97369	98543	99679
64	76949	78017	79229	80309	81373	82493	83641	84811	86017	87151	88321	89477	90527	91733	92789	93941	95083	96167	97373	98561	99689
65	76961	78031	79231	80317	81401	82499	83653	84827	86027	87179	88327	89491	90529	91753	92791	93949	95087	96179	97379	98563	99707
66	76963	78041	79241	80329	81409	82507	83663	84857	86029	87181	88337	89501	90533	91757	92801	93967	95089	96181	97381	98573	99709
67	76991	78049	79259	80341	81421	82529	83689	84859	86069	87187	88379	89513	90547	91771	92809	93971	95093	96199	97387	98597	99713
68	77003	78059	79273	80347	81439	82531	83701	84869	86077	87211	88397	89519	90583	91781	92821	93979	95101	96211	97397	98621	99719
69	77017	78079	79279	80363	81457	82549	83717	84871	86083	87221	88411	89521	90599	91801	92831	93983	95107	96221	97423	98627	99721
70	77023	78101	79283	80369	81463	82559	83719	84913	86111	87223	88423	89527	90617	91807	92849	93997	95111	96223	97429	98639	99733
71	77029	78121	79301	80387	81509	82561	83737	84919	86113	87251	88427	89533	90619	91811	92857	94007	95131	96233	97441	98641	99761
72	77041	78137	79309	80407	81517	82567	83761	84947	86117	87253	88463	89561	90631	91813	92861	94009	95143	96259	97453	98663	99767
73	77047	78139	79319	80429	81527	82571	83773	84961	86131	87257	88469	89563	90641	91823	92863	94033	95153	96263	97459	98669	99787
74	77069	78157	79333	80447	81533	82591	83777	84967	86137	87277	88471	89567	90647	91837	92867	94049	95177	96269	97463	98689	99793
75	77081	78163	79337	80449	81547	82601	83791	84977	86143	87281	88493	89591	90659	91841	92893	94057	95189	96281	97499	98711	99809
76	77093	78167	79349	80471	81551	82609	83813	84979	86161	87293	88499	89597	90677	91867	92899	94063	95191	96289	97501	98713	99817
77	77101	78173	79357	80473	81553	82613	83833	84991	86171	87299	88513	89599	90679	91873	92921	94079	95203	96293	97511	98717	99823
78	77137	78179	79367	80489	81559	82619	83843	85009	86179	87313	88523	89603	90697	91909	92927	94099	95213	96323	97523	98729	99829
79	77141	78191	79379	80491	81563	82633	83857	85021	86183	87317	88547	89611	90703	91921	92941	94109	95219	96329	97547	98731	99833
80	77153	78193	79393	80513	81569	82651	83869	85027	86197	87323	88589	89627	90709	91939	92951	94111	95231	96331	97549	98737	99839
81	77167	78203	79397	80527	81611	82657	83873	85037	86201	87337	88591	89633	90731	91943	92957	94117	95233	96337	97553	98773	99859
82	77171	78229	79399	80537	81619	82699	83891	85049	86209	87359	88607	89653	90749	91951	92959	94121	95239	96353	97561	98779	99871
83	77191	78233	79411	80557	81629	82721	83903	85061	86239	87383	88609	89657	90787	91957	92987	94153	95257	96377	97571	98801	99877
84	77201	78241	79423	80567	81637	82723	83911	85081	86243	87403	88643	89659	90793	91961	92993	94169	95261	96401	97577	98807	99881
85	77213	78259	79427	80599	81647	82727	83921	85087	86249	87407	88651	89669	90803	91967	93001	94201	95267	96419	97579	98809	99901
86	77237	78277	79433	80603	81649	82729	83933	85091	86257	87421	88657	89671	90821	91969	93047	94207	95273	96431	97583	98837	99907
87	77239	78283	79451	80611	81667	82757	83939	85093	86263	87427	88661	89681	90823	91997	93053	94219	95279	96443	97607	98849	99923
88	77243	78301	79481	80621	81677	82759	83969	85103	86269	87433	88663	89689	90833	92003	93059	94229	95287	96451	97609	98867	99929
89	77249	78307	79493	80627	81689	82763	83983	85109	86287	87443	88667	89753	90841	92009	93077	94253	95311	96457	97613	98869	99961
90	77261	78311	79531	80629	81701	82781	83987	85121	86291	87473	88681	89759	90847	92033	93083	94261	95317	96461	97649	98873	99971
91	77263	78317	79537	80651	81703	82787	84011	85133	86293	87481	88721	89767	90863	92041	93089	94273	95327	96469	97651	98887	99989
92	77267	78341	79549	80657	81707	82793	84017	85147	86297	87491	88729	89779	90887	92051	93097	94291	95339	96479	97673	98893	99991
93	77269	78347	79559	80669	81727	82799	84047	85159	86311	87509	88741	89783	90901	92077	93103	94307	95369	96487	97687	98897	
94	77279	78367	79561	80671	81737	82811	84053	85193	86323	87511	88747	89797	90907	92083	93113	94309	95383	96493	97711	98899	
95	77291	78401	79579	80677	81749	82813	84059	85199	86341	87517	88771	89809	90911	92107	93131	94321	95393	96497	97729	98909	
96	77317	78427	79589	80681	81761	82837	84061	85201	86351	87523	88789	89819	90917	92111	93133	94327	95401	96517	97771	98911	
97	77323	78437	79601	80683	81769	82847	84067	85213	86353	87539	88793	89821	90931	92119	93139	94331	95413	96527	97777	98927	
98	77339	78439	79609	80687	81773	82883	84089	85223	86357	87541	88799	89833	90947	92143	93151	94343	95419	96553	97787	98929	
99	77347	78467	79613	80701	81799	82889	84121	85229	86369	87547	88801	89839	90971	92153	93169	94349	95429	96557	97789	98939	
100	77351	78479	79621	80713	81817	82891	84127	85237	86371	87553	88807	89849	90977	92173	93179	94351	95441	96581	97813	98947	

FACTORS AND PRIMES

This table presents the prime factors of all factorable numbers and the mantissas of the common logarithms of all prime numbers from 1 to 2,000. The table runs across two facing pages. Thus, the factors of 258 are found on a line with 25 and under vertical column 8 to be $2 \cdot 3 \cdot 43$. If n is prime, the mantissa of its common logarithm is given. If n is not prime its prime factors are given.

n	0	1	2	3	4
0	....	0000000	3010300	4771213	2^3
1	$2 \cdot 5$	0413927	$2^2 \cdot 3$	1139434	$2 \cdot 7$
2	$2^2 \cdot 5$	$3 \cdot 7$	$2 \cdot 11$	3617278	$2^3 \cdot 3$
3	$2 \cdot 3 \cdot 5$	4913617	2^5	$3 \cdot 11$	$2 \cdot 17$
4	$2^2 \cdot 5$	6127839	$2 \cdot 3 \cdot 7$	6334685	$2^2 \cdot 11$
5	$2 \cdot 5^2$	$3 \cdot 17$	$2^2 \cdot 13$	7242759	$2 \cdot 3^3$
6	$2^2 \cdot 3 \cdot 5$	7853298	$2 \cdot 31$	$3^2 \cdot 7$	2^6
7	$2 \cdot 5 \cdot 7$	8512583	$2^3 \cdot 3^2$	8633229	$2 \cdot 37$
8	$2^4 \cdot 5$	3^4	$2 \cdot 41$	9190781	$2^2 \cdot 3 \cdot 7$
9	$2 \cdot 3^2 \cdot 5$	$7 \cdot 13$	$2^2 \cdot 23$	$3 \cdot 31$	$2 \cdot 47$
10	$2^3 \cdot 5^2$	0043214	$2 \cdot 3 \cdot 17$	0128372	$2^3 \cdot 13$
11	$2 \cdot 5 \cdot 11$	$3 \cdot 37$	$2^4 \cdot 7$	0530784	$2 \cdot 3 \cdot 19$
12	$2^2 \cdot 3 \cdot 5$	11^2	$2 \cdot 61$	$3 \cdot 41$	$2^2 \cdot 31$
13	$2 \cdot 5 \cdot 13$	1172713	$2^2 \cdot 3 \cdot 11$	$7 \cdot 19$	$2 \cdot 67$
14	$2^2 \cdot 5 \cdot 7$	$3 \cdot 47$	$2 \cdot 71$	$11 \cdot 13$	$2^4 \cdot 3^2$
15	$2 \cdot 3 \cdot 5^2$	1789769	$2^3 \cdot 19$	$3^2 \cdot 17$	$2 \cdot 7 \cdot 11$
16	$2^6 \cdot 5$	$7 \cdot 23$	$2 \cdot 3^4$	2121876	$2^2 \cdot 41$
17	$2 \cdot 5 \cdot 17$	$3^2 \cdot 19$	$2^2 \cdot 43$	2380461	$2 \cdot 3 \cdot 29$
18	$2^2 \cdot 3^2 \cdot 5$	2576786	$2 \cdot 7 \cdot 13$	$3 \cdot 61$	$2^3 \cdot 23$
19	$2 \cdot 5 \cdot 19$	2810334	$2^6 \cdot 3$	2855573	$2 \cdot 97$
20	$2^3 \cdot 5^2$	$3 \cdot 67$	$2 \cdot 101$	$7 \cdot 29$	$2^2 \cdot 3 \cdot 17$
21	$2 \cdot 3 \cdot 5 \cdot 7$	3242825	$2^2 \cdot 53$	$3 \cdot 71$	$2 \cdot 107$
22	$2^2 \cdot 5 \cdot 11$	$13 \cdot 17$	$2 \cdot 3 \cdot 37$	3483049	$2^5 \cdot 7$
23	$2 \cdot 5 \cdot 23$	$3 \cdot 7 \cdot 11$	$2^3 \cdot 29$	3673559	$2 \cdot 3^2 \cdot 13$
24	$2^4 \cdot 3 \cdot 5$	3820170	$2 \cdot 11^2$	3^5	$2^2 \cdot 61$
25	$2 \cdot 5^3$	3996737	$2^2 \cdot 3^2 \cdot 7$	$11 \cdot 23$	$2 \cdot 127$
26	$2^2 \cdot 5 \cdot 13$	$3^2 \cdot 29$	$2 \cdot 131$	4199557	$2^2 \cdot 3 \cdot 11$
27	$2 \cdot 3^3 \cdot 5$	4329693	$2^4 \cdot 17$	$3 \cdot 7 \cdot 13$	$2 \cdot 137$
28	$2^2 \cdot 5 \cdot 7$	4487063	$2 \cdot 3 \cdot 47$	4517864	$2^3 \cdot 71$
29	$2 \cdot 5 \cdot 29$	$3 \cdot 97$	$2^2 \cdot 73$	4668676	$2 \cdot 3 \cdot 7^2$
30	$2^2 \cdot 3 \cdot 5^2$	$7 \cdot 43$	$2 \cdot 151$	$3 \cdot 101$	$2^4 \cdot 19$
31	$2 \cdot 5 \cdot 31$	4927604	$2^2 \cdot 3 \cdot 13$	4955443	$2 \cdot 157$
32	$2^6 \cdot 5$	$3 \cdot 107$	$2 \cdot 7 \cdot 23$	$17 \cdot 19$	$2^2 \cdot 3^4$
33	$2 \cdot 3 \cdot 5 \cdot 11$	5198280	$2^3 \cdot 83$	$3^2 \cdot 37$	$2 \cdot 167$
34	$2^2 \cdot 5 \cdot 17$	$11 \cdot 31$	$2 \cdot 3^2 \cdot 19$	7^3	$2^3 \cdot 43$
35	$2 \cdot 5^2 \cdot 7$	$3^2 \cdot 13$	$2^5 \cdot 11$	5477747	$2 \cdot 3 \cdot 59$
36	$2^3 \cdot 3^2 \cdot 5$	19^2	$2 \cdot 181$	$3 \cdot 11^2$	$2^2 \cdot 7 \cdot 13$
37	$2 \cdot 5 \cdot 37$	$7 \cdot 53$	$2^2 \cdot 3 \cdot 31$	5717088	$2 \cdot 11 \cdot 17$
38	$2^2 \cdot 5 \cdot 19$	$3 \cdot 127$	$2 \cdot 191$	5831988	$2^7 \cdot 3$
39	$2 \cdot 3 \cdot 5 \cdot 13$	$17 \cdot 23$	$2^3 \cdot 7^2$	$3 \cdot 131$	$2 \cdot 197$
40	$2^4 \cdot 5^2$	6031444	$2 \cdot 3 \cdot 67$	$13 \cdot 31$	$2^3 \cdot 101$
41	$2 \cdot 5 \cdot 41$	$3 \cdot 137$	$2^2 \cdot 103$	$7 \cdot 59$	$2 \cdot 3^2 \cdot 23$
42	$2^2 \cdot 3 \cdot 5 \cdot 7$	6242821	$2 \cdot 211$	$3^2 \cdot 47$	$2^2 \cdot 53$
43	$2 \cdot 5 \cdot 43$	6344773	$2^4 \cdot 3^3$	6364879	$2 \cdot 7 \cdot 31$
44	$2^2 \cdot 5 \cdot 11$	$3^3 \cdot 7^2$	$2 \cdot 13 \cdot 17$	6464037	$2^2 \cdot 3 \cdot 37$
45	$2 \cdot 3^2 \cdot 5^2$	$11 \cdot 41$	$2^2 \cdot 113$	$3 \cdot 151$	$2 \cdot 227$
46	$2^2 \cdot 5 \cdot 23$	6637009	$2 \cdot 3 \cdot 7 \cdot 11$	6655810	$2^4 \cdot 29$
47	$2 \cdot 5 \cdot 47$	$3 \cdot 157$	$2^3 \cdot 59$	$11 \cdot 43$	$2 \cdot 3 \cdot 79$
48	$2^5 \cdot 3 \cdot 5$	$13 \cdot 37$	$2 \cdot 241$	$3 \cdot 7 \cdot 23$	$2^2 \cdot 11^2$
49	$2 \cdot 5 \cdot 7^2$	6910815	$2^2 \cdot 3 \cdot 41$	$17 \cdot 29$	$2 \cdot 13 \cdot 19$

FACTORS AND PRIMES (Continued)

n	5	6	7	8	9
0	**6989700**	$2 \cdot 3$	**8450980**	2^3	3^2
1	$3 \cdot 5$	2^4	**2304489**	$2 \cdot 3^2$	**2787536**
2	5^2	$2 \cdot 13$	3^3	$2^2 \cdot 7$	**4623980**
3	$5 \cdot 7$	$2^2 \cdot 3^2$	**5682017**	$2 \cdot 19$	$3 \cdot 13$
4	$3^2 \cdot 5$	$2 \cdot 23$	**6720979**	$2^4 \cdot 3$	7^2
5	$5 \cdot 11$	$2^3 \cdot 7$	$3 \cdot 19$	$2 \cdot 29$	**7708520**
6	$5 \cdot 13$	$2 \cdot 3 \cdot 11$	**8260748**	$2^2 \cdot 17$	$3 \cdot 23$
7	$3 \cdot 5^2$	$2^2 \cdot 19$	$7 \cdot 11$	$2 \cdot 3 \cdot 13$	**8976271**
8	$5 \cdot 17$	$2 \cdot 43$	$3 \cdot 29$	$2^3 \cdot 11$	**9493900**
9	$5 \cdot 19$	$2^5 \cdot 3$	**9867717**	$2 \cdot 7^2$	$3^2 \cdot 11$
10	$3 \cdot 5 \cdot 7$	$2 \cdot 53$	**0293838**	$2^2 \cdot 3^3$	**0374265**
11	$5 \cdot 23$	$2^2 \cdot 29$	$3^2 \cdot 13$	$2 \cdot 59$	$7 \cdot 17$
12	5^3	$2 \cdot 3^2 \cdot 7$	**1038037**	2^7	$3 \cdot 43$
13	$3^3 \cdot 5$	$2^3 \cdot 17$	**1367206**	$2 \cdot 3 \cdot 23$	**1430148**
14	$5 \cdot 29$	$2 \cdot 73$	$3 \cdot 7^2$	$2^2 \cdot 37$	**1731863**
15	$5 \cdot 31$	$2^2 \cdot 3 \cdot 13$	**1958997**	$2 \cdot 79$	$3 \cdot 53$
16	$3 \cdot 5 \cdot 11$	$2 \cdot 83$	**2227165**	$2^3 \cdot 3 \cdot 7$	13^2
17	$5^2 \cdot 7$	$2^4 \cdot 11$	$3 \cdot 59$	$2 \cdot 89$	**2528530**
18	$5 \cdot 37$	$2 \cdot 3 \cdot 31$	$11 \cdot 17$	$2^2 \cdot 47$	$3^3 \cdot 7$
19	$3 \cdot 5 \cdot 13$	$2^2 \cdot 7^2$	**2944662**	$2 \cdot 3^2 \cdot 11$	**2988531**
20	$5 \cdot 41$	$2 \cdot 103$	$3^2 \cdot 23$	$2^4 \cdot 13$	$11 \cdot 19$
21	$5 \cdot 43$	$2^3 \cdot 3^3$	$7 \cdot 31$	$2 \cdot 109$	$3 \cdot 73$
22	$3^2 \cdot 5^2$	$2 \cdot 113$	**3560259**	$2^2 \cdot 3 \cdot 19$	**3598355**
23	$5 \cdot 47$	$2^2 \cdot 59$	$3 \cdot 79$	$2 \cdot 7 \cdot 17$	**3783979**
24	$5 \cdot 7^2$	$2 \cdot 3 \cdot 41$	$13 \cdot 19$	$2^3 \cdot 31$	$3 \cdot 83$
25	$3 \cdot 5 \cdot 17$	2^5	**4099331**	$2 \cdot 3 \cdot 43$	$7 \cdot 37$
26	$5 \cdot 53$	$2 \cdot 7 \cdot 19$	$3 \cdot 89$	$2^2 \cdot 67$	**4297523**
27	$5^2 \cdot 11$	$2^3 \cdot 3 \cdot 23$	**4424798**	$2 \cdot 139$	$3^2 \cdot 31$
28	$3 \cdot 5 \cdot 19$	$2 \cdot 11 \cdot 13$	$7 \cdot 41$	$2^5 \cdot 3^2$	17^2
29	$5 \cdot 59$	$2^3 \cdot 37$	$3^3 \cdot 11$	$2 \cdot 149$	$13 \cdot 23$
30	$5 \cdot 61$	$2 \cdot 3^2 \cdot 17$	**4871384**	$2^2 \cdot 7 \cdot 11$	$3 \cdot 103$
31	$3^2 \cdot 5 \cdot 7$	$2^3 \cdot 79$	**5010593**	$2 \cdot 3 \cdot 53$	$11 \cdot 29$
32	$5^2 \cdot 13$	$2 \cdot 163$	$3 \cdot 109$	$2^3 \cdot 41$	$7 \cdot 47$
33	$5 \cdot 67$	$2^4 \cdot 3 \cdot 7$	**5276299**	$2 \cdot 13^2$	$3 \cdot 113$
34	$3 \cdot 5 \cdot 23$	$2 \cdot 173$	**5403295**	$2^2 \cdot 3 \cdot 29$	**5428254**
35	$5 \cdot 71$	$2^3 \cdot 89$	$3 \cdot 7 \cdot 17$	$2 \cdot 179$	**5550944**
36	$5 \cdot 73$	$2 \cdot 3 \cdot 61$	**5646661**	$2^4 \cdot 23$	$3^2 \cdot 41$
37	$3 \cdot 5^2$	$2^3 \cdot 47$	$13 \cdot 29$	$2 \cdot 3^3 \cdot 7$	**5786392**
38	$5 \cdot 7 \cdot 11$	$2 \cdot 193$	$3^2 \cdot 43$	$2^2 \cdot 97$	**5899496**
39	$5 \cdot 79$	$2^2 \cdot 3^2 \cdot 11$	**5987905**	$2 \cdot 199$	$3 \cdot 7 \cdot 19$
40	$3^4 \cdot 5$	$2 \cdot 7 \cdot 29$	$11 \cdot 37$	$2^3 \cdot 3 \cdot 17$	**6117233**
41	$5 \cdot 83$	$2^5 \cdot 13$	$3 \cdot 139$	$2 \cdot 11 \cdot 19$	**6222140**
42	$5^2 \cdot 17$	$2 \cdot 3 \cdot 71$	$7 \cdot 61$	$2^2 \cdot 107$	$3 \cdot 11 \cdot 13$
43	$3 \cdot 5 \cdot 29$	$2^2 \cdot 109$	$19 \cdot 23$	$2 \cdot 3 \cdot 73$	**6424645**
44	$5 \cdot 89$	$2 \cdot 223$	$3 \cdot 149$	$2^6 \cdot 7$	**6522463**
45	$5 \cdot 7 \cdot 13$	$2^3 \cdot 3 \cdot 19$	**6599162**	$2 \cdot 229$	$3^3 \cdot 17$
46	$3 \cdot 5 \cdot 31$	$2 \cdot 233$	**6693169**	$2^2 \cdot 3^2 \cdot 13$	$7 \cdot 67$
47	$5^2 \cdot 19$	$2^2 \cdot 7 \cdot 17$	$3^2 \cdot 53$	$2 \cdot 239$	**6803355**
48	$5 \cdot 97$	$2 \cdot 3^5$	**6875290**	$2^3 \cdot 61$	$3 \cdot 163$
49	$3^2 \cdot 5 \cdot 11$	$2^4 \cdot 31$	$7 \cdot 71$	$2 \cdot 3 \cdot 83$	**6981005**

FACTORS AND PRIMES (Continued)

n	0	1	2	3	4
50	$2^2 \cdot 5^3$	$3 \cdot 167$	$2 \cdot 251$	**7015680**	$2^3 \cdot 3^2 \cdot 7$
51	$2 \cdot 3 \cdot 5 \cdot 17$	$7 \cdot 73$	2^9	$3^3 \cdot 19$	$2 \cdot 257$
52	$2^3 \cdot 5 \cdot 13$	**7168377**	$2 \cdot 3^2 \cdot 29$	**7185017**	$2^2 \cdot 131$
53	$2 \cdot 5 \cdot 53$	$3^2 \cdot 59$	$2^2 \cdot 7 \cdot 19$	$13 \cdot 41$	$2 \cdot 3 \cdot 89$
54	$2^2 \cdot 3^3 \cdot 5$	**7331973**	$2 \cdot 271$	$3 \cdot 181$	$2^5 \cdot 17$
55	$2 \cdot 5^2 \cdot 11$	$19 \cdot 29$	$2^3 \cdot 3 \cdot 23$	$7 \cdot 79$	$2 \cdot 277$
56	$2^4 \cdot 5 \cdot 7$	$3 \cdot 11 \cdot 17$	$2 \cdot 281$	**7505084**	$2^2 \cdot 3 \cdot 47$
57	$2 \cdot 3 \cdot 5 \cdot 19$	**7566361**	$2^2 \cdot 11 \cdot 13$	$3 \cdot 191$	$2 \cdot 7 \cdot 41$
58	$2^2 \cdot 5 \cdot 29$	$7 \cdot 83$	$2 \cdot 3 \cdot 97$	$11 \cdot 53$	$2^3 \cdot 73$
59	$2 \cdot 5 \cdot 59$	$3 \cdot 197$	$2^4 \cdot 37$	**7730547**	$2 \cdot 3^3 \cdot 11$
60	$2^3 \cdot 3 \cdot 5^2$	**7788745**	$2 \cdot 7 \cdot 43$	$3^2 \cdot 67$	$2^2 \cdot 151$
61	$2 \cdot 5 \cdot 61$	$13 \cdot 47$	$2^2 \cdot 3^2 \cdot 17$	**7874605**	$2 \cdot 307$
62	$2^2 \cdot 5 \cdot 31$	$3^3 \cdot 23$	$2 \cdot 311$	$7 \cdot 89$	$2^4 \cdot 3 \cdot 13$
63	$2 \cdot 3^2 \cdot 5 \cdot 7$	**8000294**	$2^3 \cdot 79$	$3 \cdot 211$	$2 \cdot 317$
64	$2^7 \cdot 5$	**8068580**	$2 \cdot 3 \cdot 107$	**8082110**	$2^2 \cdot 7 \cdot 23$
65	$2 \cdot 5^2 \cdot 13$	$3 \cdot 7 \cdot 31$	$2^2 \cdot 163$	**8149132**	$2 \cdot 3 \cdot 109$
66	$2^3 \cdot 3 \cdot 5 \cdot 11$	**8202015**	$2 \cdot 331$	$3 \cdot 13 \cdot 17$	$2^3 \cdot 83$
67	$2 \cdot 5 \cdot 67$	$11 \cdot 61$	$2^5 \cdot 3 \cdot 7$	**8280151**	$2 \cdot 337$
68	$2^3 \cdot 5 \cdot 17$	$3 \cdot 227$	$2 \cdot 11 \cdot 31$	**8344207**	$2^2 \cdot 3^2 \cdot 19$
69	$2 \cdot 3 \cdot 5 \cdot 23$	**8394780**	$2^2 \cdot 173$	$3^2 \cdot 7 \cdot 11$	$2 \cdot 347$
70	$2^2 \cdot 5^2 \cdot 7$	**8457180**	$2 \cdot 3^3 \cdot 13$	$19 \cdot 37$	$2^6 \cdot 11$
71	$2 \cdot 5 \cdot 71$	$3^2 \cdot 79$	$2^3 \cdot 89$	$23 \cdot 31$	$2 \cdot 3 \cdot 7 \cdot 17$
72	$2^4 \cdot 3^2 \cdot 5$	$7 \cdot 103$	$2 \cdot 19^2$	$3 \cdot 241$	$2^2 \cdot 181$
73	$2 \cdot 5 \cdot 73$	$17 \cdot 43$	$2^2 \cdot 3 \cdot 61$	**8651040**	$2 \cdot 367$
74	$2^2 \cdot 5 \cdot 37$	$3 \cdot 13 \cdot 19$	$2 \cdot 7 \cdot 53$	**8709888**	$2^3 \cdot 3 \cdot 31$
75	$2 \cdot 3 \cdot 5^3$	**8756399**	$2^4 \cdot 47$	$3 \cdot 251$	$2 \cdot 13 \cdot 29$
76	$2^3 \cdot 5 \cdot 19$	**8813847**	$2 \cdot 3 \cdot 127$	$7 \cdot 109$	$2^2 \cdot 191$
77	$2 \cdot 5 \cdot 7 \cdot 11$	$3 \cdot 257$	$2^2 \cdot 193$	**8881795**	$2 \cdot 3^2 \cdot 43$
78	$2^2 \cdot 3 \cdot 5 \cdot 13$	$11 \cdot 71$	$2 \cdot 17 \cdot 23$	$3^3 \cdot 29$	$2^4 \cdot 7^2$
79	$2 \cdot 5 \cdot 79$	$7 \cdot 113$	$2^3 \cdot 3^2 \cdot 11$	$13 \cdot 61$	$2 \cdot 397$
80	$2^5 \cdot 5^2$	$3^2 \cdot 89$	$2 \cdot 401$	$11 \cdot 73$	$2^2 \cdot 3 \cdot 67$
81	$2 \cdot 3^4 \cdot 5$	**9090209**	$2^2 \cdot 7 \cdot 29$	$3 \cdot 271$	$2 \cdot 11 \cdot 37$
82	$2^2 \cdot 5 \cdot 41$	**9143432**	$2 \cdot 3 \cdot 137$	**9153998**	$2^3 \cdot 103$
83	$2 \cdot 5 \cdot 83$	$3 \cdot 277$	$2^6 \cdot 13$	$7^2 \cdot 17$	$2 \cdot 3 \cdot 139$
84	$2^3 \cdot 3 \cdot 5 \cdot 7$	29^2	$2 \cdot 421$	$3 \cdot 281$	$2^2 \cdot 211$
85	$2 \cdot 5^2 \cdot 17$	$23 \cdot 37$	$2^3 \cdot 3 \cdot 71$	**9309490**	$2 \cdot 7 \cdot 61$
86	$2^2 \cdot 5 \cdot 43$	$3 \cdot 7 \cdot 41$	$2 \cdot 431$	**9360108**	$2^5 \cdot 3^3$
87	$2 \cdot 3 \cdot 5 \cdot 29$	$13 \cdot 67$	$2^3 \cdot 109$	$3^2 \cdot 97$	$2 \cdot 19 \cdot 23$
88	$2^4 \cdot 5 \cdot 11$	**9449759**	$2 \cdot 3^2 \cdot 7^2$	**9459607**	$2^2 \cdot 13 \cdot 17$
89	$2 \cdot 5 \cdot 89$	$3^4 \cdot 11$	$2^2 \cdot 223$	$19 \cdot 47$	$2 \cdot 3 \cdot 149$
90	$2^2 \cdot 3^2 \cdot 5^2$	$17 \cdot 53$	$2 \cdot 11 \cdot 41$	$3 \cdot 7 \cdot 43$	$2^3 \cdot 113$
91	$2 \cdot 5 \cdot 7 \cdot 13$	**9595184**	$2^4 \cdot 3 \cdot 19$	$11 \cdot 83$	$2 \cdot 457$
92	$2^3 \cdot 5 \cdot 23$	$3 \cdot 307$	$2 \cdot 461$	$13 \cdot 71$	$2^2 \cdot 3 \cdot 7 \cdot 11$
93	$2 \cdot 3 \cdot 5 \cdot 31$	$7^2 \cdot 19$	$2^2 \cdot 233$	$3 \cdot 311$	$2 \cdot 467$
94	$2^2 \cdot 5 \cdot 47$	**9735896**	$2 \cdot 3 \cdot 157$	$23 \cdot 41$	$2^4 \cdot 59$
95	$2 \cdot 5^2 \cdot 19$	$3 \cdot 317$	$2^3 \cdot 7 \cdot 17$	**9790929**	$2 \cdot 3^2 \cdot 53$
96	$2^6 \cdot 3 \cdot 5$	31^2	$2 \cdot 13 \cdot 37$	$3^2 \cdot 107$	$2^3 \cdot 241$
97	$2 \cdot 5 \cdot 97$	**9872192**	$2^2 \cdot 3^5$	$7 \cdot 139$	$2 \cdot 487$
98	$2^2 \cdot 5 \cdot 7^2$	$3^2 \cdot 109$	$2 \cdot 491$	**9925535**	$2^3 \cdot 3 \cdot 41$
99	$2 \cdot 3^2 \cdot 5 \cdot 11$	**9960737**	$2^5 \cdot 31$	$3 \cdot 331$	$2 \cdot 7 \cdot 71$

FACTORS AND PRIMES (Continued)

n	5	6	7	8	9
50	$5 \cdot 101$	$2 \cdot 11 \cdot 23$	$3 \cdot 13^2$	$2^2 \cdot 127$	**7067178**
51	$5 \cdot 103$	$2^2 \cdot 3 \cdot 43$	$11 \cdot 47$	$2 \cdot 7 \cdot 37$	$3 \cdot 173$
52	$3 \cdot 5^2 \cdot 7$	$2 \cdot 263$	$17 \cdot 31$	$2^4 \cdot 3 \cdot 11$	23^2
53	$5 \cdot 107$	$2^3 \cdot 67$	$3 \cdot 179$	$2 \cdot 269$	$7^2 \cdot 11$
54	$5 \cdot 109$	$2 \cdot 3 \cdot 7 \cdot 13$	**7379873**	$2^2 \cdot 137$	$3^2 \cdot 61$
55	$3 \cdot 5 \cdot 37$	$2^2 \cdot 139$	**7458552**	$2 \cdot 3^2 \cdot 31$	$13 \cdot 43$
56	$5 \cdot 113$	$2 \cdot 283$	$3^4 \cdot 7$	$2^3 \cdot 71$	**7551123**
57	$5^2 \cdot 23$	$2^6 \cdot 3^2$	**7611758**	$2 \cdot 17^2$	$3 \cdot 193$
58	$3^2 \cdot 5 \cdot 13$	$2 \cdot 293$	**7686381**	$2^2 \cdot 3 \cdot 7^2$	$19 \cdot 31$
59	$5 \cdot 7 \cdot 17$	$2^2 \cdot 149$	$3 \cdot 199$	$2 \cdot 13 \cdot 23$	**7774268**
60	$5 \cdot 11^2$	$2 \cdot 3 \cdot 101$	**7831887**	$2^5 \cdot 19$	$3 \cdot 7 \cdot 29$
61	$3 \cdot 5 \cdot 41$	$2^2 \cdot 7 \cdot 11$	**7902852**	$2 \cdot 3 \cdot 103$	**7916906**
62	5^4	$2 \cdot 313$	$3 \cdot 11 \cdot 19$	$2^2 \cdot 157$	$17 \cdot 37$
63	$5 \cdot 127$	$2^2 \cdot 3 \cdot 53$	$7^2 \cdot 13$	$2 \cdot 11 \cdot 29$	$3^2 \cdot 71$
64	$3 \cdot 5 \cdot 43$	$2 \cdot 17 \cdot 19$	**8109043**	$2^3 \cdot 3^4$	$11 \cdot 59$
65	$5 \cdot 131$	$2^4 \cdot 41$	$3^2 \cdot 73$	$2 \cdot 7 \cdot 47$	**8188854**
66	$5 \cdot 7 \cdot 19$	$2 \cdot 3^2 \cdot 37$	$23 \cdot 29$	$2^2 \cdot 167$	$3 \cdot 223$
67	$3^3 \cdot 5^2$	$2^2 \cdot 13^2$	**8305887**	$2 \cdot 3 \cdot 113$	$7 \cdot 97$
68	$5 \cdot 137$	$2 \cdot 7^3$	$3 \cdot 229$	$2^4 \cdot 43$	$13 \cdot 53$
69	$5 \cdot 139$	$2^2 \cdot 3 \cdot 29$	$17 \cdot 41$	$2 \cdot 349$	$3 \cdot 233$
70	$3 \cdot 5 \cdot 47$	$2 \cdot 353$	$7 \cdot 101$	$2^2 \cdot 3 \cdot 59$	**8506462**
71	$5 \cdot 11 \cdot 13$	$2^2 \cdot 179$	$3 \cdot 239$	$2 \cdot 359$	**8567289**
72	$5^2 \cdot 29$	$2 \cdot 3 \cdot 11^2$	**8615344**	$2^3 \cdot 7 \cdot 13$	3^6
73	$3 \cdot 5 \cdot 7^2$	$2^5 \cdot 23$	$11 \cdot 67$	$2 \cdot 3^2 \cdot 41$	**8686444**
74	$5 \cdot 149$	$2 \cdot 373$	$3^2 \cdot 83$	$2^2 \cdot 11 \cdot 17$	$7 \cdot 107$
75	$5 \cdot 151$	$2^2 \cdot 3^2 \cdot 7$	**8790959**	$2 \cdot 379$	$3 \cdot 11 \cdot 23$
76	$3^2 \cdot 5 \cdot 17$	$2 \cdot 383$	$13 \cdot 59$	$2^6 \cdot 3$	**8859263**
77	$5^2 \cdot 31$	$2^5 \cdot 97$	$3 \cdot 7 \cdot 37$	$2 \cdot 389$	$19 \cdot 41$
78	$5 \cdot 157$	$2 \cdot 3 \cdot 131$	**8959747**	$2^2 \cdot 197$	$3 \cdot 263$
79	$3 \cdot 5 \cdot 53$	$2^2 \cdot 199$	**9014583**	$2 \cdot 3 \cdot 7 \cdot 19$	$17 \cdot 47$
80	$5 \cdot 7 \cdot 23$	$2 \cdot 13 \cdot 31$	$3 \cdot 269$	$2^3 \cdot 101$	**9079485**
81	$5 \cdot 163$	$2^4 \cdot 3 \cdot 17$	$19 \cdot 43$	$2 \cdot 409$	$3^2 \cdot 7 \cdot 13$
82	$3 \cdot 5^2 \cdot 11$	$2 \cdot 7 \cdot 59$	**9175055**	$2^2 \cdot 3^2 \cdot 23$	**9185545**
83	$5 \cdot 167$	$2^2 \cdot 11 \cdot 19$	$3^3 \cdot 31$	$2 \cdot 419$	**9237620**
84	$5 \cdot 13^2$	$2 \cdot 3^2 \cdot 47$	$7 \cdot 11^2$	$2^4 \cdot 53$	$3 \cdot 283$
85	$3^2 \cdot 5 \cdot 19$	$2^3 \cdot 107$	**9329808**	$2 \cdot 3 \cdot 11 \cdot 13$	**9339932**
86	$5 \cdot 173$	$2 \cdot 433$	$3 \cdot 17^2$	$2^2 \cdot 7 \cdot 31$	$11 \cdot 79$
87	$5^3 \cdot 7$	$2^2 \cdot 3 \cdot 73$	**9429996**	$2 \cdot 439$	$3 \cdot 293$
88	$3 \cdot 5 \cdot 59$	$2 \cdot 443$	**9479236**	$2^3 \cdot 3 \cdot 37$	$7 \cdot 127$
89	$5 \cdot 179$	$2^7 \cdot 7$	$3 \cdot 13 \cdot 23$	$2 \cdot 449$	$29 \cdot 31$
90	$5 \cdot 181$	$2 \cdot 3 \cdot 151$	**9576073**	$2^2 \cdot 227$	$3^2 \cdot 101$
91	$3 \cdot 5 \cdot 61$	$2^2 \cdot 229$	$7 \cdot 131$	$2 \cdot 3^3 \cdot 17$	**9633155**
92	$5^2 \cdot 37$	$2 \cdot 463$	$3^2 \cdot 103$	$2^5 \cdot 29$	**9680157**
93	$5 \cdot 11 \cdot 17$	$2^2 \cdot 3^2 \cdot 13$	**9717396**	$2 \cdot 7 \cdot 67$	$3 \cdot 313$
94	$3^2 \cdot 5 \cdot 7$	$2 \cdot 11 \cdot 43$	**9763500**	$2^2 \cdot 3 \cdot 79$	$13 \cdot 73$
95	$5 \cdot 191$	$2^2 \cdot 239$	$3 \cdot 11 \cdot 29$	$2 \cdot 479$	$7 \cdot 137$
96	$5 \cdot 193$	$2 \cdot 3 \cdot 7 \cdot 23$	**9854265**	$2^3 \cdot 11^2$	$3 \cdot 17 \cdot 19$
97	$3 \cdot 5^2 \cdot 13$	$2^4 \cdot 61$	**9898946**	$2 \cdot 3 \cdot 163$	$11 \cdot 89$
98	$5 \cdot 197$	$2 \cdot 17 \cdot 29$	$3 \cdot 7 \cdot 47$	$2^2 \cdot 13 \cdot 19$	$23 \cdot 43$
99	$5 \cdot 199$	$2^2 \cdot 3 \cdot 83$	**9986952**	$2 \cdot 499$	$3^2 \cdot 37$

FACTORS AND PRIMES (Continued)

n	0	1	2	3	4
100	$2^3 \cdot 5^3$	$7 \cdot 11 \cdot 13$	$2 \cdot 3 \cdot 167$	$17 \cdot 59$	$2^3 \cdot 251$
101	$2 \cdot 5 \cdot 101$	$3 \cdot 337$	$2^2 \cdot 11 \cdot 23$	**0056094**	$2 \cdot 3 \cdot 13^2$
102	$2^2 \cdot 3 \cdot 5 \cdot 17$	**0090257**	$2 \cdot 7 \cdot 73$	$3 \cdot 11 \cdot 31$	2^{10}
103	$2 \cdot 5 \cdot 103$	**0132587**	$2^3 \cdot 3 \cdot 43$	**0141003**	$2 \cdot 11 \cdot 47$
104	$2^4 \cdot 5 \cdot 13$	$3 \cdot 347$	$2 \cdot 521$	$7 \cdot 149$	$2^2 \cdot 3^2 \cdot 29$
105	$2 \cdot 3 \cdot 5^2 \cdot 7$	**0216027**	$2^3 \cdot 263$	$3^4 \cdot 13$	$2 \cdot 17 \cdot 31$
106	$2^2 \cdot 5 \cdot 53$	**0257154**	$2 \cdot 3^2 \cdot 59$	**0265333**	$2^3 \cdot 7 \cdot 19$
107	$2 \cdot 5 \cdot 107$	$3^2 \cdot 7 \cdot 17$	$2^4 \cdot 67$	$29 \cdot 37$	$2 \cdot 3 \cdot 179$
108	$2^2 \cdot 3^3 \cdot 5$	$23 \cdot 47$	$2 \cdot 541$	$3 \cdot 19^2$	$2^3 \cdot 271$
109	$2 \cdot 5 \cdot 109$	**0378248**	$2^2 \cdot 3 \cdot 7 \cdot 13$	**0386202**	$2 \cdot 547$
110	$2^2 \cdot 5^2 \cdot 11$	$3 \cdot 367$	$2 \cdot 19 \cdot 29$	**0425755**	$2^4 \cdot 3 \cdot 23$
111	$2 \cdot 3 \cdot 5 \cdot 37$	$11 \cdot 101$	$2^3 \cdot 139$	$3 \cdot 7 \cdot 53$	$2 \cdot 557$
112	$2^5 \cdot 5 \cdot 7$	$19 \cdot 59$	$2 \cdot 3 \cdot 11 \cdot 17$	**0503798**	$2^2 \cdot 281$
113	$2 \cdot 5 \cdot 113$	$3 \cdot 13 \cdot 29$	$2^2 \cdot 283$	$11 \cdot 103$	$2 \cdot 3^4 \cdot 7$
114	$2^2 \cdot 3 \cdot 5 \cdot 19$	$7 \cdot 163$	$2 \cdot 571$	$3^2 \cdot 127$	$2^3 \cdot 11 \cdot 13$
115	$2 \cdot 5^2 \cdot 23$	**0610753**	$2^7 \cdot 3^2$	**0618293**	$2 \cdot 577$
116	$2^2 \cdot 5 \cdot 29$	$3^3 \cdot 43$	$2 \cdot 7 \cdot 83$	**0655797**	$2^2 \cdot 3 \cdot 97$
117	$2 \cdot 3^2 \cdot 5 \cdot 13$	**0685569**	$2^2 \cdot 293$	$3 \cdot 17 \cdot 23$	$2 \cdot 587$
118	$2^2 \cdot 5 \cdot 59$	**0722499**	$2 \cdot 3 \cdot 197$	$7 \cdot 13^2$	$2^5 \cdot 37$
119	$2 \cdot 5 \cdot 7 \cdot 17$	$3 \cdot 397$	$2^3 \cdot 149$	**0766404**	$2 \cdot 3 \cdot 199$
120	$2^4 \cdot 3 \cdot 5^2$	**0795430**	$2 \cdot 601$	$3 \cdot 401$	$2^2 \cdot 7 \cdot 43$
121	$2 \cdot 5 \cdot 11^2$	$7 \cdot 173$	$2^2 \cdot 3 \cdot 101$	**0838608**	$2 \cdot 607$
122	$2^2 \cdot 5 \cdot 61$	$3 \cdot 11 \cdot 37$	$2 \cdot 13 \cdot 47$	**0874265**	$2^3 \cdot 3^2 \cdot 17$
123	$2 \cdot 3 \cdot 5 \cdot 41$	**0902581**	$2^4 \cdot 7 \cdot 11$	$3^2 \cdot 137$	$2 \cdot 617$
124	$2^2 \cdot 5 \cdot 31$	$17 \cdot 73$	$2 \cdot 3^3 \cdot 23$	$11 \cdot 113$	$2^2 \cdot 311$
125	$2 \cdot 5^4$	$3^2 \cdot 139$	$2^2 \cdot 313$	$7 \cdot 179$	$2 \cdot 3 \cdot 11 \cdot 19$
126	$2^2 \cdot 3^2 \cdot 5 \cdot 7$	$13 \cdot 97$	$2 \cdot 631$	$3 \cdot 421$	$2^4 \cdot 79$
127	$2 \cdot 5 \cdot 127$	$31 \cdot 41$	$2^3 \cdot 3 \cdot 53$	$19 \cdot 67$	$2 \cdot 7^2 \cdot 13$
128	$2^8 \cdot 5$	$3 \cdot 7 \cdot 61$	$2 \cdot 641$	**1082267**	$2^2 \cdot 3 \cdot 107$
129	$2 \cdot 3 \cdot 5 \cdot 43$	**1109262**	$2^2 \cdot 17 \cdot 19$	$3 \cdot 431$	$2 \cdot 647$
130	$2^2 \cdot 5^2 \cdot 13$	**1142773**	$2 \cdot 3 \cdot 7 \cdot 31$	**1149444**	$2^3 \cdot 163$
131	$2 \cdot 5 \cdot 131$	$3 \cdot 19 \cdot 23$	$2^5 \cdot 41$	$13 \cdot 101$	$2 \cdot 3^2 \cdot 73$
132	$2^2 \cdot 3 \cdot 5 \cdot 11$	**1209028**	$2 \cdot 661$	$3^3 \cdot 7^2$	$2^2 \cdot 331$
133	$2 \cdot 5 \cdot 7 \cdot 19$	11^3	$2^2 \cdot 3^2 \cdot 37$	$31 \cdot 43$	$2 \cdot 23 \cdot 29$
134	$2^2 \cdot 5 \cdot 67$	$3^2 \cdot 149$	$2 \cdot 11 \cdot 61$	$17 \cdot 79$	$2^4 \cdot 3 \cdot 7$
135	$2 \cdot 3^3 \cdot 5^2$	$7 \cdot 193$	$2^3 \cdot 13^2$	$3 \cdot 11 \cdot 41$	$2 \cdot 677$
136	$2^4 \cdot 5 \cdot 17$	**1338581**	$2 \cdot 3 \cdot 227$	$29 \cdot 47$	$2^2 \cdot 11 \cdot 31$
137	$2 \cdot 5 \cdot 137$	$3 \cdot 457$	$2^2 \cdot 7^3$	**1376705**	$2 \cdot 3 \cdot 229$
138	$2^2 \cdot 3 \cdot 5 \cdot 23$	**1401937**	$2 \cdot 691$	$3 \cdot 461$	$2^3 \cdot 173$
139	$2 \cdot 5 \cdot 139$	$13 \cdot 107$	$2^4 \cdot 3 \cdot 29$	$7 \cdot 199$	$2 \cdot 17 \cdot 41$
140	$2^3 \cdot 5^2 \cdot 7$	$3 \cdot 467$	$2 \cdot 701$	$23 \cdot 61$	$2^2 \cdot 3^2 \cdot 13$
141	$2 \cdot 3 \cdot 5 \cdot 47$	$17 \cdot 83$	$2^2 \cdot 353$	$3^2 \cdot 157$	$2 \cdot 7 \cdot 101$
142	$2^2 \cdot 5 \cdot 71$	$7^2 \cdot 29$	$2 \cdot 3^2 \cdot 79$	**1532049**	$2^4 \cdot 89$
143	$2 \cdot 5 \cdot 11 \cdot 13$	$3^3 \cdot 53$	$2^2 \cdot 179$	**1562462**	$2 \cdot 3 \cdot 239$
144	$2^5 \cdot 3^2 \cdot 5$	$11 \cdot 131$	$2 \cdot 7 \cdot 103$	$3 \cdot 13 \cdot 37$	$2^2 \cdot 19^2$
145	$2 \cdot 5^2 \cdot 29$	**1616674**	$2^2 \cdot 3 \cdot 11^2$	**1622656**	$2 \cdot 727$
146	$2^2 \cdot 5 \cdot 73$	$3 \cdot 487$	$2 \cdot 17 \cdot 43$	$7 \cdot 11 \cdot 19$	$2^3 \cdot 3 \cdot 61$
147	$2 \cdot 3 \cdot 5 \cdot 7^2$	**1676127**	$2^6 \cdot 23$	$3 \cdot 491$	$2 \cdot 11 \cdot 67$
148	$2^3 \cdot 5 \cdot 37$	**1705551**	$2 \cdot 3 \cdot 13 \cdot 19$	**1711412**	$2^2 \cdot 7 \cdot 53$
149	$2 \cdot 5 \cdot 149$	$3 \cdot 7 \cdot 71$	$2^2 \cdot 373$	**1740598**	$2 \cdot 3^2 \cdot 83$

FACTORS AND PRIMES (Continued)

n	5	6	7	8	9
100	3·5·67	2·503	19·53	2⁴·3²·7	**0038912**
101	5·7·29	2³·127	3²·113	2·509	**0081742**
102	5²·41	2·3²·19	13·79	2²·257	3·7³
103	3²·5·23	2²·7·37	17·61	2·3·173	**0166155**
104	5·11·19	2·523	3·349	2³·131	**0207755**
105	5·211	2⁵·3·11	7·151	2·23²	3·353
106	3·5·71	2·13·41	11·97	2²·3·89	**0289777**
107	5²·43	2²·269	3·359	2·7²·11	13·83
108	5·7·31	2·3·181	**0362295**	2⁶·17	3²·11²
109	3·5·73	2³·137	**0402066**	2·3²·61	7·157
110	5·13·17	2·7·79	3²·41	2²·277	**0449315**
111	5·223	2³·3²·31	**0480532**	2·13·43	3·373
112	3²·5³	2·563	7²·23	2³·3·47	**0526939**
113	5·227	2⁴·71	3·379	2·569	17·67
114	5·229	2·3·191	31·37	2²·7·41	3·383
115	3·5·7·11	2²·17²	13·89	2·3·193	19·61
116	5·233	2·11·53	3·389	2⁴·73	7·167
117	5²·47	2³·3·7²	11·107	2·19·31	3²·131
118	3·5·79	2·593	**0744507**	2²·3³·11	29·41
119	5·239	2²·13·23	3²·7·19	2·599	11·109
120	5·241	2·3²·67	17·71	2³·151	3·13·31
121	3⁵·5	2⁶·19	**0852906**	2·3·7·29	23·53
122	5²·7²	2·613	3·409	2²·307	**0895519**
123	5·13·19	2²·3·103	**0923697**	2·619	3·7·59
124	3·5·83	2·7·89	29·43	2⁵·3·13	**0965624**
125	5·251	2³·157	3·419	2·17·37	**1000257**
126	5·11·23	2·3·211	7·181	2²·317	3³·47
127	3·5²·17	2²·11·29	**1061909**	2·3²·71	**1068705**
128	5·257	2·643	3²·11·13	2³·7·23	**1102529**
129	5·7·37	2⁴·3⁴	**1129400**	2·11·59	3·433
130	3²·5·29	2·653	**1162756**	2²·3·109	7·11·17
131	5·263	2²·7·47	3·439	2·659	**1202448**
132	5²·53	2·3·13·17	**1228709**	2⁴·83	3·443
133	3·5·89	2³·167	7·191	2·3·223	13·103
134	5·269	2·673	3·449	2²·337	19·71
135	5·271	2²·3·113	23·59	2·7·97	3²·151
136	3·5·7·13	2·683	**1357685**	2³·3²·19	37²
137	5³·11	2⁵·43	3⁴·17	2·13·53	7·197
138	5·277	2·3²·7·11	19·73	2²·347	3·463
139	3²·5·31	2²·349	11·127	2·3·233	**1458177**
140	5·281	2·19·37	3·7·67	2⁷·11	**1489110**
141	5·283	2³·3·59	13·109	2·709	3·11·43
142	3·5²·19	2·23·31	**1544240**	2²·3·7·17	**1550322**
143	5·7·41	2²·359	3·479	2·719	**1580608**
144	5·17²	2·3·241	**1604685**	2³·181	3²·7·23
145	3·5·97	2⁴·7·13	31·47	2·3⁶	**1640553**
146	5·293	2·733	3²·163	2²·367	13·113
147	5²·59	2²·3²·41	7·211	2·739	3·17·29
148	3³·5·11	2·743	**1723110**	2⁴·3·31	**1728947**
149	5·13·23	2³·11·17	3·499	2·7·107	**1758016**

FACTORS AND PRIMES (Continued)

n	0	1	2	3	4
150	$2^3 \cdot 3 \cdot 5^3$	$19 \cdot 79$	$2 \cdot 751$	$3^2 \cdot 167$	$2^5 \cdot 47$
151	$2 \cdot 5 \cdot 151$	**1792645**	$2^3 \cdot 3^3 \cdot 7$	$17 \cdot 89$	$2 \cdot 757$
152	$2^4 \cdot 5 \cdot 19$	$3^2 \cdot 13^2$	$2 \cdot 761$	**1826999**	$2^3 \cdot 3 \cdot 127$
153	$2 \cdot 3^2 \cdot 5 \cdot 17$	**1849752**	$2^3 \cdot 383$	$3 \cdot 7 \cdot 73$	$2 \cdot 13 \cdot 59$
154	$2^3 \cdot 5 \cdot 7 \cdot 11$	$23 \cdot 67$	$2 \cdot 3 \cdot 257$	**1883659**	$2^3 \cdot 193$
155	$2 \cdot 5^3 \cdot 31$	$3 \cdot 11 \cdot 47$	$2^4 \cdot 97$	**1911715**	$2 \cdot 3 \cdot 7 \cdot 37$
156	$2^3 \cdot 3 \cdot 5 \cdot 13$	$7 \cdot 223$	$2 \cdot 11 \cdot 71$	$3 \cdot 521$	$2^3 \cdot 17 \cdot 23$
157	$2 \cdot 5 \cdot 157$	**1961762**	$2^3 \cdot 3 \cdot 131$	$11^3 \cdot 13$	$2 \cdot 787$
158	$2^3 \cdot 5 \cdot 79$	$3 \cdot 17 \cdot 31$	$2 \cdot 7 \cdot 113$	**1994809**	$2^4 \cdot 3^2 \cdot 11$
159	$2 \cdot 3 \cdot 5 \cdot 53$	$37 \cdot 43$	$2^3 \cdot 199$	$3^3 \cdot 59$	$2 \cdot 797$
160	$2^6 \cdot 5^2$	**2043913**	$2 \cdot 3^2 \cdot 89$	$7 \cdot 229$	$2^2 \cdot 401$
161	$2 \cdot 5 \cdot 7 \cdot 23$	$3^2 \cdot 179$	$2^3 \cdot 13 \cdot 31$	**2076344**	$2 \cdot 3 \cdot 269$
162	$2^3 \cdot 3^4 \cdot 5$	**2097830**	$2 \cdot 811$	$3 \cdot 541$	$2^3 \cdot 7 \cdot 29$
163	$2 \cdot 5 \cdot 163$	$7 \cdot 233$	$2^5 \cdot 3 \cdot 17$	$23 \cdot 71$	$2 \cdot 19 \cdot 43$
164	$2^3 \cdot 5 \cdot 41$	$3 \cdot 547$	$2 \cdot 821$	$31 \cdot 53$	$2^3 \cdot 3 \cdot 137$
165	$2 \cdot 3 \cdot 5^2 \cdot 11$	$13 \cdot 127$	$2^3 \cdot 7 \cdot 59$	$3 \cdot 19 \cdot 29$	$2 \cdot 827$
166	$2^2 \cdot 5 \cdot 83$	$11 \cdot 151$	$2 \cdot 3 \cdot 277$	**2208922**	$2^7 \cdot 13$
167	$2 \cdot 5 \cdot 167$	$3 \cdot 557$	$2^3 \cdot 11 \cdot 19$	$7 \cdot 239$	$2 \cdot 3^2 \cdot 31$
168	$2^4 \cdot 3 \cdot 5 \cdot 7$	41^2	$2 \cdot 29^2$	$3^2 \cdot 11 \cdot 17$	$2^2 \cdot 421$
169	$2 \cdot 5 \cdot 13^2$	$19 \cdot 89$	$2^3 \cdot 3^2 \cdot 47$	**2286570**	$2 \cdot 7 \cdot 11^3$
170	$2^2 \cdot 5^2 \cdot 17$	$3^5 \cdot 7$	$2 \cdot 23 \cdot 37$	$13 \cdot 131$	$2^3 \cdot 3 \cdot 71$
171	$2 \cdot 3^2 \cdot 5 \cdot 19$	$29 \cdot 59$	$2^4 \cdot 107$	$3 \cdot 571$	$2 \cdot 857$
172	$2^3 \cdot 5 \cdot 43$	**2357809**	$2 \cdot 3 \cdot 7 \cdot 41$	**2362853**	$2^3 \cdot 431$
173	$2 \cdot 5 \cdot 173$	$3 \cdot 577$	$2^3 \cdot 433$	**2387986**	$2 \cdot 3 \cdot 17^2$
174	$2^2 \cdot 3 \cdot 5 \cdot 29$	**2407988**	$2 \cdot 13 \cdot 67$	$3 \cdot 7 \cdot 83$	$2^4 \cdot 109$
175	$2 \cdot 5^3 \cdot 7$	$17 \cdot 103$	$2^3 \cdot 3 \cdot 73$	**2437819**	$2 \cdot 877$
176	$2^5 \cdot 5 \cdot 11$	$3 \cdot 587$	$2 \cdot 881$	$41 \cdot 43$	$2^2 \cdot 3^2 \cdot 7^2$
177	$2 \cdot 3 \cdot 5 \cdot 59$	$7 \cdot 11 \cdot 23$	$2^3 \cdot 443$	$3^2 \cdot 197$	$2 \cdot 887$
178	$2^2 \cdot 5 \cdot 89$	$13 \cdot 137$	$2 \cdot 3^4 \cdot 11$	**2511513**	$2^3 \cdot 223$
179	$2 \cdot 5 \cdot 179$	$3^2 \cdot 199$	$2^8 \cdot 7$	$11 \cdot 163$	$2 \cdot 3 \cdot 13 \cdot 23$
180	$2^3 \cdot 3^2 \cdot 5^2$	**2555137**	$2 \cdot 17 \cdot 53$	$3 \cdot 601$	$2^3 \cdot 11 \cdot 41$
181	$2 \cdot 5 \cdot 181$	**2579185**	$2^3 \cdot 3 \cdot 151$	$7^2 \cdot 37$	$2 \cdot 907$
182	$2^3 \cdot 5 \cdot 7 \cdot 13$	$3 \cdot 607$	$2 \cdot 911$	**2607867**	$2^5 \cdot 3 \cdot 19$
183	$2 \cdot 3 \cdot 5 \cdot 61$	**2626883**	$2^3 \cdot 229$	$3 \cdot 13 \cdot 47$	$2 \cdot 7 \cdot 131$
184	$2^4 \cdot 5 \cdot 23$	$7 \cdot 263$	$2 \cdot 3 \cdot 307$	$19 \cdot 97$	$2^2 \cdot 461$
185	$2 \cdot 5^2 \cdot 37$	$3 \cdot 617$	$2^3 \cdot 463$	$17 \cdot 109$	$2 \cdot 3^2 \cdot 103$
186	$2^2 \cdot 3 \cdot 5 \cdot 31$	**2697464**	$2 \cdot 7^2 \cdot 19$	$3^4 \cdot 23$	$2^3 \cdot 233$
187	$2 \cdot 5 \cdot 11 \cdot 17$	**2720738**	$2^4 \cdot 3^2 \cdot 13$	**2725378**	$2 \cdot 937$
188	$2^3 \cdot 5 \cdot 47$	$3^2 \cdot 11 \cdot 19$	$2 \cdot 941$	$7 \cdot 269$	$2^3 \cdot 3 \cdot 157$
189	$2 \cdot 3^2 \cdot 5 \cdot 7$	$31 \cdot 61$	$2^3 \cdot 11 \cdot 43$	$3 \cdot 631$	$2 \cdot 947$
190	$2^2 \cdot 5^2 \cdot 19$	**2789821**	$2 \cdot 3 \cdot 317$	$11 \cdot 173$	$2^4 \cdot 7 \cdot 17$
191	$2 \cdot 5 \cdot 191$	$3 \cdot 7^2 \cdot 13$	$2^3 \cdot 239$	**2817150**	$2 \cdot 3 \cdot 11 \cdot 29$
192	$2^7 \cdot 3 \cdot 5$	$17 \cdot 113$	$2 \cdot 31^3$	$3 \cdot 641$	$2^3 \cdot 13 \cdot 37$
193	$2 \cdot 5 \cdot 193$	**2857823**	$2^3 \cdot 3 \cdot 7 \cdot 23$	**2862319**	$2 \cdot 967$
194	$2^2 \cdot 5 \cdot 97$	$3 \cdot 647$	$2 \cdot 971$	$29 \cdot 67$	$2^3 \cdot 3^5$
195	$2 \cdot 3 \cdot 5^2 \cdot 13$	**2902573**	$2^5 \cdot 61$	$3^2 \cdot 7 \cdot 31$	$2 \cdot 977$
196	$2^3 \cdot 5 \cdot 7^2$	$37 \cdot 53$	$2 \cdot 3^2 \cdot 109$	$13 \cdot 151$	$2^2 \cdot 491$
197	$2 \cdot 5 \cdot 197$	$3^3 \cdot 73$	$2^3 \cdot 17 \cdot 29$	**2951271**	$2 \cdot 3 \cdot 7 \cdot 47$
198	$2^2 \cdot 3^2 \cdot 5 \cdot 11$	$7 \cdot 283$	$2 \cdot 991$	$3 \cdot 661$	$2^6 \cdot 31$
199	$2 \cdot 5 \cdot 199$	$11 \cdot 181$	$2^3 \cdot 3 \cdot 83$	**2995073**	$2 \cdot 997$

FACTORS AND PRIMES (Continued)

n	5	6	7	8	9
150	$5 \cdot 7 \cdot 43$	$2 \cdot 3 \cdot 251$	$11 \cdot 137$	$2^2 \cdot 13 \cdot 29$	$3 \cdot 503$
151	$3 \cdot 5 \cdot 101$	$2^2 \cdot 379$	$37 \cdot 41$	$2 \cdot 3 \cdot 11 \cdot 23$	$7^2 \cdot 31$
152	$5^3 \cdot 61$	$2 \cdot 7 \cdot 109$	$3 \cdot 509$	$2^3 \cdot 191$	$11 \cdot 139$
153	$5 \cdot 307$	$2^9 \cdot 3$	$29 \cdot 53$	$2 \cdot 769$	$3^4 \cdot 19$
154	$3 \cdot 5 \cdot 103$	$2 \cdot 773$	$7 \cdot 13 \cdot 17$	$2^2 \cdot 3^2 \cdot 43$	1900514
155	$5 \cdot 311$	$2^3 \cdot 389$	$3^2 \cdot 173$	$2 \cdot 19 \cdot 41$	1928461
156	$5 \cdot 313$	$2 \cdot 3^3 \cdot 29$	1950690	$2^5 \cdot 7^2$	$3 \cdot 523$
157	$3^3 \cdot 5^3 \cdot 7$	$2^3 \cdot 197$	$19 \cdot 83$	$2 \cdot 3 \cdot 263$	1983821
158	$5 \cdot 317$	$2 \cdot 13 \cdot 61$	$3 \cdot 23^2$	$2^3 \cdot 397$	$7 \cdot 227$
159	$5 \cdot 11 \cdot 29$	$2^2 \cdot 3 \cdot 7 \cdot 19$	2033049	$2 \cdot 17 \cdot 47$	$3 \cdot 13 \cdot 41$
160	$3 \cdot 5 \cdot 107$	$2 \cdot 11 \cdot 73$	2060159	$2^3 \cdot 3 \cdot 67$	2065560
161	$5 \cdot 17 \cdot 19$	$2^4 \cdot 101$	$3 \cdot 7^2 \cdot 11$	$2 \cdot 809$	2092468
162	$5^3 \cdot 13$	$2 \cdot 3 \cdot 271$	2113876	$2^3 \cdot 11 \cdot 37$	$3^2 \cdot 181$
163	$3 \cdot 5 \cdot 109$	$2^3 \cdot 409$	2140487	$2 \cdot 3^2 \cdot 7 \cdot 13$	$11 \cdot 149$
164	$5 \cdot 7 \cdot 47$	$2 \cdot 823$	$3^3 \cdot 61$	$2^4 \cdot 103$	$17 \cdot 97$
165	$5 \cdot 331$	$2^3 \cdot 3^2 \cdot 23$	2193225	$2 \cdot 829$	$3 \cdot 7 \cdot 79$
166	$3^2 \cdot 5 \cdot 37$	$2 \cdot 7^2 \cdot 17$	2219356	$2^2 \cdot 3 \cdot 139$	2224563
167	$5^2 \cdot 67$	$2^3 \cdot 419$	$3 \cdot 13 \cdot 43$	$2 \cdot 839$	$23 \cdot 73$
168	$5 \cdot 337$	$2 \cdot 3 \cdot 281$	$7 \cdot 241$	$2^3 \cdot 211$	$3 \cdot 563$
169	$3 \cdot 5 \cdot 113$	$2^5 \cdot 53$	2296818	$2 \cdot 3 \cdot 283$	2301934
170	$5 \cdot 11 \cdot 31$	$2 \cdot 853$	$3 \cdot 569$	$2^3 \cdot 7 \cdot 61$	2327421
171	$5 \cdot 7^3$	$2^2 \cdot 3 \cdot 11 \cdot 13$	$17 \cdot 101$	$2 \cdot 859$	$3^2 \cdot 191$
172	$3 \cdot 5^2 \cdot 23$	$2 \cdot 863$	$11 \cdot 157$	$2^6 \cdot 3^3$	$7 \cdot 13 \cdot 19$
173	$5 \cdot 347$	$2^3 \cdot 7 \cdot 31$	$3^2 \cdot 193$	$2 \cdot 11 \cdot 79$	$37 \cdot 47$
174	$5 \cdot 349$	$2 \cdot 3^2 \cdot 97$	2422929	$2^3 \cdot 19 \cdot 23$	$3 \cdot 11 \cdot 53$
175	$3^3 \cdot 5 \cdot 13$	$2^3 \cdot 439$	$7 \cdot 251$	$2 \cdot 3 \cdot 293$	2452658
176	$5 \cdot 353$	$2 \cdot 883$	$3 \cdot 19 \cdot 31$	$2^3 \cdot 13 \cdot 17$	$29 \cdot 01$
177	$5^2 \cdot 71$	$2^4 \cdot 3 \cdot 37$	2496874	$2 \cdot 7 \cdot 127$	$3 \cdot 593$
178	$3 \cdot 5 \cdot 7 \cdot 17$	$2 \cdot 19 \cdot 47$	2521246	$2^3 \cdot 3 \cdot 149$	2526103
179	$5 \cdot 359$	$2^3 \cdot 449$	$3 \cdot 599$	$2 \cdot 29 \cdot 31$	$7 \cdot 257$
180	$5 \cdot 19^2$	$2 \cdot 3 \cdot 7 \cdot 43$	$13 \cdot 139$	$2^4 \cdot 113$	$3^3 \cdot 67$
181	$3 \cdot 5 \cdot 11^2$	$2^3 \cdot 227$	$23 \cdot 79$	$2 \cdot 3^2 \cdot 101$	$17 \cdot 107$
182	$5^2 \cdot 73$	$2 \cdot 11 \cdot 83$	$3^2 \cdot 7 \cdot 29$	$2^2 \cdot 457$	$31 \cdot 59$
183	$5 \cdot 367$	$2^3 \cdot 3^2 \cdot 17$	$11 \cdot 167$	$2 \cdot 919$	$3 \cdot 613$
184	$3^3 \cdot 5 \cdot 41$	$2 \cdot 13 \cdot 71$	2664669	$2^2 \cdot 3 \cdot 7 \cdot 11$	43^3
185	$5 \cdot 7 \cdot 53$	$2^6 \cdot 29$	$3 \cdot 619$	$2 \cdot 929$	$11 \cdot 13^3$
186	$5 \cdot 373$	$2 \cdot 3 \cdot 311$	2711443	$2^3 \cdot 467$	$3 \cdot 7 \cdot 89$
187	$3 \cdot 5^4$	$2^3 \cdot 7 \cdot 67$	2734643	$2 \cdot 3 \cdot 313$	2739268
188	$5 \cdot 13 \cdot 29$	$2 \cdot 23 \cdot 41$	$3 \cdot 17 \cdot 37$	$2^5 \cdot 59$	2762320
189	$5 \cdot 379$	$2^3 \cdot 3 \cdot 79$	$7 \cdot 271$	$2 \cdot 13 \cdot 73$	$3^3 \cdot 211$
190	$3 \cdot 5 \cdot 127$	$2 \cdot 953$	2803507	$2^2 \cdot 3^2 \cdot 53$	$23 \cdot 83$
191	$5 \cdot 383$	$2^3 \cdot 479$	$3^3 \cdot 71$	$2 \cdot 7 \cdot 137$	$19 \cdot 101$
192	$5^2 \cdot 7 \cdot 11$	$2 \cdot 3^3 \cdot 107$	$41 \cdot 47$	$2^3 \cdot 241$	$3 \cdot 643$
193	$3^2 \cdot 5 \cdot 43$	$2^4 \cdot 11^3$	$13 \cdot 149$	$2 \cdot 3 \cdot 17 \cdot 19$	$7 \cdot 277$
194	$5 \cdot 389$	$2 \cdot 7 \cdot 139$	$3 \cdot 11 \cdot 59$	$2^3 \cdot 487$	2898118
195	$5 \cdot 17 \cdot 23$	$2^2 \cdot 3 \cdot 163$	$19 \cdot 103$	$2 \cdot 11 \cdot 89$	$3 \cdot 653$
196	$3 \cdot 5 \cdot 131$	$2 \cdot 983$	$7 \cdot 281$	$2^4 \cdot 3 \cdot 41$	$11 \cdot 179$
197	$5^2 \cdot 79$	$2^3 \cdot 13 \cdot 19$	$3 \cdot 659$	$2 \cdot 23 \cdot 43$	2964458
198	$5 \cdot 397$	$2 \cdot 3 \cdot 331$	2981979	$2^3 \cdot 7 \cdot 71$	$3^2 \cdot 13 \cdot 17$
199	$3 \cdot 5 \cdot 7 \cdot 19$	$2^3 \cdot 499$	3003781	$2 \cdot 3^3 \cdot 37$	3008128

EXTENDED TABLES OF FACTORS AND PRIMES

The following procedure makes possible the determination of a number between 2009 and 19,949 as being either prime, or if not, what its factors will be. It is herewith included with the kind permission of its author, Professor Leonard Caners.

The following symbols will be used:

N is the number whose factors, if any, are to be determined.

A is N with the last digit dropped. Thus if $N = 17{,}873$, $A = 1787$.

R is the range and is the integral part of the square root of N with its last two digits dropped. Thus R of $17{,}873 = 13$.

K is the key number and is found in the table below.

P_1 is a possible factor corresponding to K.

The twin series are given in the table for the sake of completeness but are not written down in actual practice.

There are two steps as outlined below.

Step I

N	K	P_1	Series	Procedure
Any number ending in 1, 3, 7, or 9, within limits stated above	$A - R$	$10R + $ last digit	$K + n; P_1 - 10n$ $n = 0, 1, \ldots, 2R + 1$	Beginning with K read table of Factors and Primes from left to right and look for corresponding possible factors

Step II

N ending in	K	P_1	Series	Procedure
1	$(A - 2) - 3R$	$10R + 7$	$K + 3n; P_1 - 10n$ $n = 0, 1, \ldots, 2R + 1$	Same as in Step I except that only every *third* entry in table of Factors and Primes is examined for corresponding possible factors
3	$A - 3R$	$10R + 1$	$n = 0, 1, \ldots, 2R + 1$	
7	$(A - 2) - 3R$	$10R + 9$	$n = 0, 1, \ldots, 2R + 1$	
9	$A - 3R$	$10R + 3$	$n = 0, 1, \ldots, 2R + 1$	

As an example consider 7519. Using tables on preceding pages proceed as follows:

Step I

From Table $N = 7519$; $A = 751$; $R = 8$, $K = 743$; $P_1 = 89$ and the series obtained:

743	89
744	79
745	69
746	59
—	—
751	9
752	1
753	11
—	—
760	81

EXTENDED TABLES OF FACTORS AND PRIMES (Continued)

Examining the Table of Factors and Primes beginning with 743 note whether 743 has 89 as a factor; whether 744 has 79 as a factor, etc. Then continue to read from left to right keeping in mind the possible factors 69, 59, . . . 81.

Since none are found one concludes that 7519 has no factors within the given range ending in 9 or 1.

Step II

From Table $K = 727$ and $P_1 = 83$; the series obtained is:

727	83
730	73
733	63
—	—
751	3
754	7
—	—
778	87

Proceed exactly as in Step I. Begin with 727 and note whether it has 83 as a factor. Thereafter examine every *third* entry for the remaining corresponding possible factors 73, 63 . . . 3, 7, . . . 87. Since 730 yields the factor 73 one concludes that 73 is the factor of 7519. By division 7519 = 73 × 103. Had no factor been found in either Step I or Step II the conclusion would be that the number under consideration was prime.

Proof for Step I

Let N end in 1. Then, if N has a factor ending in 1, write

$$(10a + 1)(10b + 1) = N = 10A + 1$$
$$10ab + a + b = A$$

or

$$b(10a + 1) = (A - a)$$

This proves that any factor $10a + 1$ of N is also a factor of $A - a$. Let $a = (R - n)$, then $(10a + 1)$ becomes $(10R + 1) - 10n$ and $(A - a)$ becomes $(A - R) + n$, i.e., $P_1 - 10n$ and $K + n$ respectively as given in the table. Since $P_1 - 10n$ becomes a number ending in 9 when $P_1 < 10n$, all possible factors ending in 9 are also provided for.

The proof for Step II is similar. For possible factors ending in 3 or 7 write

$$(10a + 7)(10b + 3) = N = 10A + 1.$$

By identical reasoning this results in $P_1 - 10n$ and $K + 3n$ respectively given in the table. This completes the proof for numbers ending in 1.

Identical reasoning establishes the key numbers and series for ending in 3, 7, or 9.

DIOPHANTINE EQUATIONS

H. C. Williams

Let $f(x_1, x_2, \ldots, x_n)$ be a given polynomial in the n variables $x_1, x_2, \ldots, x_n$ with integral coefficients. The equation

$$(*) \quad f(x_1, x_2, \ldots, x_n) = 0$$

is said to be a *Diophantine equation* when it is to be solved with integral (or rational) values for the unknowns $x_1, x_2, \ldots, x_n$.

The general or complete integral (rational) solution of (*) is that solution which contains all integral (rational) solutions of (*). For the equations that follow, we shall only be concerned with integral solutions. When it is possible, the general solution of these equations will be presented; however, many of these equations have not yet been completely solved. In the course of discussing such equations, some partial results will be given.

Much more information about Diophantine equations can be found in the following works: (1) Z. I. Borevich and I. R. Shafarevich, *Number Theory*, Academic Press, New York and London (1966); (2) R. D. Carmichael, *The Theory of Numbers and Diophantine Analysis*, Dover, New York, N.Y., 1959; (3) B. N. Delone and D. K. Faddeev, *The Theory of Irrationalities of the Third Degree*, Vol. 10, Translations of Mathematical Monographs, American Mathematical Society, Providence, R.I., 1964; (4) L. E. Dickson, *History of the Theory of Numbers*, Vol. II, Chelsea, New York, N.Y., 1952; (5) L. E. Dickson, *Introduction to the Theory of Numbers*, Dover, New York, N.Y., 1957; (6) L. J. Mordell, *Diophantine Equations*, Academic Press, London and New York, 1969. The book by Mordell is especially valuable for modern developments in the theory of Diophantine Equations.

Diophantine Equations of Degree Two

1. All solutions with even y of the Pythagorean Diophantine equation

$$(1) \quad x^2 + y^2 = z^2$$

are given by

$$x = r(m^2 - n^2), \, y = 2rmn, \, z = r(m^2 + n^2),$$

where r,m,n, are integer parameters.

DIOPHANTINE EQUATIONS

2. All solutions of

(2) $x^2 + y^2 = z^2 + w^2$

are given by

$2x = mn + pq, \ 2z = mp + nq,$
$2y = mp - nq, \ 2w = mn - pq,$

where m,q, are both even; n,p, are both even; or all four are odd.

3. The general integer solution of the equation

(3) $x_1^2 + x_2^2 + \ldots + x_n^2 = x^2$, $(x_1, x_2, \ldots, x_n) = 1$, x > 0, is given by
$dx_i = 2q_i q_n \ (i = 1, 2, \ldots, n-1),$
$dx_n = q_n^2 - q_1^2 - \ldots - q_{n-1}^2,$
$dx = q_n^2 + q_1^2 + \ldots + q_{n-1}^2,$

where the q's are arbitrary integers with $(q_1, q_2, \ldots, q_n) = 1$ and d > 0 is chosen so that $(x_1, x_2, \ldots, x_n) = 1$.

4. If $e \neq 0$ and $d = b^2 - 4ac$ is not the square of an integer, and if (u,v,w) is a given solution (u, v, w not all zero) of

(4) $ax^2 + bxy + cy^2 = ez^2,$

then all of its integer solutions are

$x = kr, \ y = ks, \ z = kt,$
$r = -(au+bv)m^2 - 2cvmn + cun^2,$
$s = avm^2 - 2aumn - (bu+cv)n^2, \ t = wT, \ T = am^2 + bmn + cn^2$

where m and n are relatively prime and k is an irreducible fraction whose denominator is gD, where g divides w, and D divides both T and the expression 2aum + 2cvn + b(un+vm) and hence also $(b^2 - 4ac)eg^2$.

5. If a,b,c, are square free, (a,b) = (b,c) = (a,c) = 1, and a,b,c, do not have the same sign, then the equation

(5) $ax^2 + by^2 + cz^2 = 0$

has non-trivial solutions if and only if −bc, −ca, −ab, are quadratic residues of a,b,c, respectively, and if

$$ax^2 + by^2 + cz^2 \equiv 0 \ (\text{mod } 8)$$

is solvable. In fact, if (5) is put in its canonical form, i.e., a > 0, b > 0, c < 0, there exists a non-trivial solution with

$$|x| < \sqrt{b|c|}, \ |y| \leqslant \sqrt{|c|a}, \ |z| \leqslant \sqrt{ab},$$

DIOPHANTINE EQUATIONS

and $(x,y) = (z,y) = (z,x) = 1$.

If a,b,c, are coprime in pairs and $abc \neq 0$, and if u,v,w, is a solution of (5) such that $(u,v) = (u,w) = 1$, we may assume that au is even and determine integers r,s,t, such that $aur + bvs + cwt = 1$ (r even). Express abc as a product $k\ell$ of two integers in all ways. Select integers d,m,n, subject to the following conditions:

$$(n,m) = (n,\ell) = (m,k) = 1,$$

and

d is even; or d,m,n, all odd, $k \equiv \ell \pmod 2$.

Put

$$t_1 = d\ell m^2, \ t_2 = dkn^2, \ t_3 = dmn, \ h = ar^2 + bs^2 + ct^2, \ U = 2r-hu,$$
$$V = 2s-hv, \ W = 2t-hw, \ 2u_1 = vW-wV, \ 2v_1 = wU-uW, \ 2w_1 = uV-vU.$$

Then

$$x = (ut_1 + Ut_2 - 2bcu_1 t_3)/2,$$
$$y = (vt_1 + Vt_2 - 2cav_1 t_3)/2,$$
$$z = (wt_1 + Wt_2 - 2baw_1 t_3)/2,$$

satisfy (5), and all integer solutions of (5) are so obtained.

Equations of Degree Three

Apart from the trivial solutions $x=y=0$, $u=-w$, $x=z$, $y=w$, the general solution of

$$(6) \quad x^3 + y^3 = z^3 + w^3$$

is given by

$$x = r(HQ-M^2), \ y = r(GQ+M^2),$$
$$z = r(Q^2-MH), \ w = r(Q^2+MG),$$

where

$$M = m^2 + 3n^2,$$
$$Q = q^2 + 3p^2,$$
$$H = 3mp + 3np - mq + 3nq,$$
$$G = 3mp - 3np + mq + 3nq, \text{ and } r,m,n,p,q, \text{ are integer parameters.}$$

7. The complete solution of

$$(7) \quad z^2 = x^3 + y^3$$

such that y is odd and prime to x is given by

DIOPHANTINE EQUATIONS

$$x = -4p^3q + 4q^3, y = p^4 + 8pq^3,$$
$$x = -p^4 + 6p^2q^2 + 3q^4, y = p^4 + 6p^2q^2 - 3q^4,$$
$$x = p^4 + 6p^2q^2 - 3q^4, y = -p^4 + 6p^2q^2 + 3q^4,$$
$$x = 2p^4 - 4p^3q - 4pq^3 + 2q^4, y = p^4 + 4p^3q - 6p^2q^2 + 4pq^3 + q^4,$$
$$x = 4p^3q + 24p^2q^2 + 48pq^3 + 36q^4, y = p^4 + 8p^3q + 24p^2q^2 + 24pq^3,$$

where p and q are selected so that y is odd and prime to x.

8. The equation

$$(8) \qquad x^3 + y^3 = Az^3$$

has no non-trivial integer solutions if

$$A = p, 2p, 9p, p^2, 9p^2, 4p^2, pq, p_1 p_2^2,$$

or if

$$A = q, 4q, 9q, 2q^2, q^2, 9q^2, q_1 q_2^2, p^2 q^2,$$

where p and q are primes with $p \equiv 5$ (mod 18) and $q \equiv 11$ (mod 18). When (8) does have non-trivial solutions, it can be shown that all of these solutions can be derived by applying certain operations to a finite number of basic solutions. A table of these basic solutions for all $A \leqslant 500$ is given in Selmer [5,6].
9. The equation

$$(9) \qquad ax^3 + by^3 = c$$

with a>b>1, c=1 or 3, (ab,c) = 1, b=1 if c=3, has at most one integer solution (x,y), and for this solution $c^{-1} (x a^{1/3} + y b^{1/3})^3$ is either the fundamental unit or its square in the cubic field $Q(d^{1/3})$ defined by $Q(a^{1/3} b^{2/3})$. The only exception is the equation $2x^3 + y^3 = 3$, which has the two solutions (1,1) and (4,–5).
10. The equation

$$(10) \qquad x^3 + dy^3 = 1 \ (d>1 , d \neq 19, 20, 28)$$

has at most one solution with $xy \neq 0$. This is given by the fundamental unit in the cubic field $Q(d^{1/3})$ when it is a binomial unit, i.e., when the unit has the form $x + y \, d^{1/3}$. If d = 19, the only solution of (10) is (-8, 3); if d = 20, the only solution of (10) is (-19, 7); and if d = 28, the only solution of (10) is (-3, 1). A table of fundamental units of pure cubic fields $K(d^{1/3})$, $d \leqslant 1000$, is given in [2].
11. If the equation

$$(11) \qquad y^2 = ax^3 + by^2 + cx + d, a \neq 0,$$

has a solution (x,y), then max $(|x|,|y|) < \exp[(AH)^A]$ where $A = 10^6$ and $H = \max(|a|,|b|,|c|,|d|)$ (see Baker [1]).

DIOPHANTINE EQUATIONS

12. The special case of (11)

$$(12) \qquad y^2 - k = x^3$$

is discussed extensively in Mordell's book. A table of all solutions of (12) for $0 < k \leqslant 100$, as well as all (see Coghlan & Stevens [3]) solutions for $0 < -k < 100$, can be found in Hemer.[4]

Equations of Degree Four

13. The equation

$$(13) \qquad x^4 + y^4 = z^4 + w^4$$

has an infinitude of solutions given by

$$x = m^7 + m^5 n^2 - 2m^3 n^4 + 3m^2 n^5 + mn^6,$$
$$y = m^6 n - 3m^5 n^2 - 2m^4 n^3 + m^2 n^5 + n^7,$$
$$z = m^7 + m^5 n^2 - 2m^3 n^4 - 3m^2 n^5 + mn^6,$$
$$w = m^6 n + 3m^5 n^2 - 2m^4 n^3 + m^2 n^5 + n^7.$$

14. The equation

$$(14) \qquad x^4 + y^4 + z^4 = w^4$$

has no solution for $w < 220{,}000$.

15. The equation

$$(15) \qquad x^4 + y^4 + z^4 = 2w^4$$

has an infinitude of solutions given by

$$x = m^2 - n^2, \ y = 2mn + n^2,$$
$$z = m^2 + 2mn, \ w = m^2 + mn + n^2.$$

16. If the equation

$$(16) \qquad x^4 + ay^4 + bz^4 = w^2$$

has a solution x,t,u,v, then a second solution is given by

$$x = s^4 - at^4 - bu^4, \ y = 2stv,$$
$$z = 2suv, \ w = v^4 + 4as^4 t^4 + 4bs^4 u^4.$$

17. The special case

$$(17) \qquad x^4 + y^4 + z^4 = w^2$$

DIOPHANTINE EQUATIONS

of (16) has an infinitude of solutions given by

$$x = 2mn(m^2-n^2), \quad y = 2mn(m^2+n^2),$$
$$z = m^4-n^4, \quad w = m^8 + 14m^4 n^4 + n^8.$$

18. If the equation

$$(18) \qquad\qquad ax^4 + bx^2 y^2 + cy^4 = ez^2$$

has a solution (r,s,t), it has a second solution given by

$$x = r(4ces^4 t^2 - q^2), \quad y = s(4aer^4 t^2 - q^2),$$
$$z = t[4dr^4 s^4 q^2 - (e^2 t^4 - dr^4 s^4)^2],$$

where

$$q = ar^4 - cs^4, \quad d = b^2 - 4ac.$$

19. The equations

$$(19) \qquad x^4 \pm y^4 = z^2$$

have no solutions such that $xyz \neq 0$.
20. The equation

$$(20) \qquad ax^4 - by^4 = c$$

has at most one solution in positive integers when $c = 1,2,4,8$.
21. The equations

$$(21) \qquad x^4 - dy^4 = \pm 1$$

have at most one solution in positive integers. This will be given by $x=a$, $y=b$, if the fundamental unit of the ring of integers in $Q[(-4d)^{1/4}]$ takes the form $a^2 + ab(-4d)^{1/4} + b^2(-d)^{1/2}$.

Let $\epsilon = a + bd^{1/2}$ be the fundamental unit in the ring $Z[d^{1/2}]$ formed by adjoining $d^{1/2}$ to the rational integers. If $a^2 - db^2 = -1$, the equation $x^4 - dy^4 = 1$ has no solution unless $d=5$. If $d=5$ the equation has the solution $(3,2)$. If $a^2 - db^2 = 1$ and (r,s) is a solution of $x^4 - dy^4 = 1$ ($d \neq 7140$), than $a=r^2$ and $b=s^2$. If $d = 7140$, the equation has the solution $(239, 26)$.

Equations of Degree Greater than Four
22. The equations

$$(22) \qquad y^2 = x^r \pm 1 \quad (r>3)$$

have no solutions for which $xy \neq 0$.

DIOPHANTINE EQUATIONS

23. The equations

(23) $y^3 = x^r \pm 1$ $(r>2)$

have no non-trivial solutions.

24. The equation

(24) $1 + x^2 = 2y^n$ $(n>2, n\neq4)$

has no solution (r,s) for which $|rs| \neq 1$. When $n=4$, the only solutions of (24) in positive integers are $(1,1)$ and $(239, 13)$.

25. The equations

(25) $ax^n - by^n = \pm 1$,

where $a>0, b>0$, and $n\geqslant5$, have at most two solutions in positive integers.

26. The equations

(26) $x^n - dy^n = \pm 1$,

where $d>0$ and $n\geqslant5$, has at most one solution in positive integers x,y, except possibly when $d=2$ or when $n=5$ or 6 and $d = 2^n\pm1$. If $d > 1250\cdot20^{1/6}$, the equation $x^5 + dy^5 = 1$ has at most one solution in non-zero integers.

27. If

$$f(x) = a_0 x^n + a_1 x^{n-1} + \ldots + a_n,$$

has at least three simple zeros, then all integer solutions of

(27) $y^2 = f(x)$

satisfy the inequality

$$\max (|x|,|y|) < \exp \exp \exp (n^{10n^3} A^{n^2}),$$

where

$$A = \max|a_i| \; (i = 0,1, \ldots ,n).$$

28. The equation

(28) $x^n + y^n = z^n$ $(n>2)$

has no solution x,y,z, where $xyz \neq 0$, if n is a prime $<25,000$ and n divides z; or if n is a prime $<253,747,888$ and $(n,z) = 1$.

DIOPHANTINE EQUATIONS

REFERENCES

1. A. Baker, *The Diophantine Equation* $y^2 = ax^3 + bx^2 + cx + d$, *J. Lond. Math. Soc.*, Vol. 43 (1968), pp. 1-9.
2. B. D. Beach, H. C. Williams, C. R. Zarnke, *Some Computer Results on Units in Quadratic and Cubic Fields*, Proc. 25th Summer Meeting Can. Math. Congress, 1971, pp. 609-648.
3. F. B. Coghlan and N. M. Stevens, *The Diophantine Equation* $x^3 - y^2 = k$, Computers in Number Theory, Academic Press, London and New York, 1971, pp. 199-205.
4. O. Hemer, *Notes on the Diophantine Equation* $y^2 - k = x^3$, *Arkiv för Math.*, 3, 67, 1954.
5. E. S. Selmer, *The Diophantine Equation* $ax^3 + by^3 + cz^3 = 0$, *Acta Math.*, 85, 203, 1951.
6. *The Diophantine Equation* $ax^3 + by^3 + cz^3 = 0$, Completion of the tables, *Acta Math.*, 91, 191, 1954.

IV. GEOMETRY

MENSURATION FORMULAS

DR. HOWARD EVES

TRIANGLES

In the following: K = area, r = radius of the inscribed circle, R = radius of the circumscribed circle.

Right Triangle

$A + B = C = 90°$

$c^2 = a^2 + b^2$ (*Pythagorean relation*)

$a = \sqrt{(c + b)(c - b)}$

$K = \frac{1}{2}ab$

$r = \dfrac{ab}{a + b + c}$, $R = \frac{1}{2}c$

$h = \dfrac{ab}{c}$, $m = \dfrac{b^2}{c}$, $n = \dfrac{a^2}{c}$

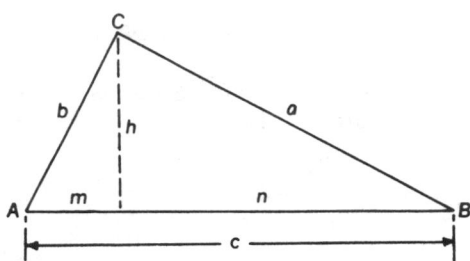

Equilateral Triangle

$A = B = C = 60°$

$K = \frac{1}{4}a^2\sqrt{3}$

$r = \frac{1}{6}a\sqrt{3}$, $R = \frac{1}{3}a\sqrt{3}$

$h = \frac{1}{2}a\sqrt{3}$

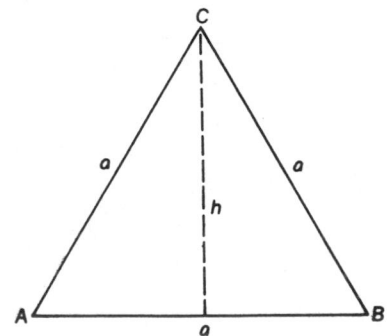

General Triangle

Let $s = \frac{1}{2}(a + b + c)$, h_c = length of altitude on side c, t_c = length of bisector of angle C, m_c = length of median to side c.

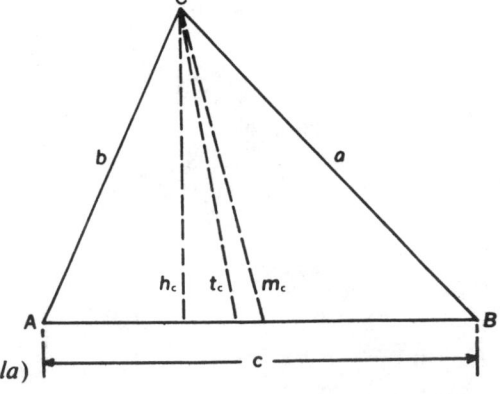

$A + B + C = 180°$

$c^2 = a^2 + b^2 - 2ab \cos C$

$\qquad$ (*law of cosines*)

$K = \frac{1}{2}h_c c = \frac{1}{2}ab \sin C$

$ = \dfrac{c^2 \sin A \sin B}{2 \sin C}$

$ = rs = \dfrac{abc}{4R}$

$ = \sqrt{s(s - a)(s - b)(s - c)}$ (*Heron's formula*)

Mensuration Formulas

$$r = c \sin \frac{A}{2} \sin \frac{B}{2} \sec \frac{C}{2} = \frac{ab \sin C}{2s} = (s - c) \tan \frac{C}{2}$$

$$= \sqrt{\frac{(s - a)(s - b)(s - c)}{s}} = \frac{K}{s} = 4R \sin \frac{A}{2} \sin \frac{B}{2} \sin \frac{C}{2}$$

$$R = \frac{c}{2 \sin C} = \frac{abc}{4 \sqrt{s(s - a)(s - b)(s - c)}} = \frac{abc}{4K}$$

$$h_c = a \sin B = b \sin A = \frac{2K}{c}$$

$$t_c = \frac{2ab}{a + b} \cos \frac{C}{2} = \sqrt{ab \left\{ 1 - \frac{c^2}{(a + b)^2} \right\}}$$

$$m_c = \sqrt{\frac{a^2}{2} + \frac{b^2}{2} - \frac{c^2}{4}}$$

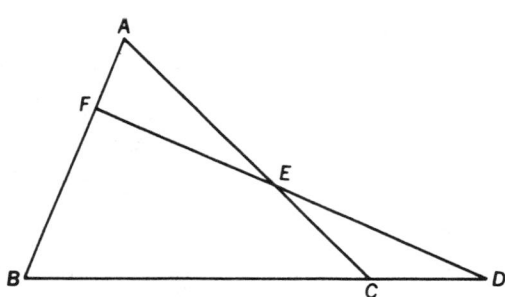 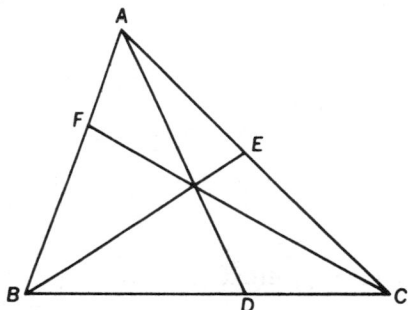

Menelaus' Theorem. A necessary and sufficient condition for points D, E, F on the respective side lines BC, CA, AB of a triangle ABC to be collinear is that

$$BD \cdot CE \cdot AF = - DC \cdot EA \cdot FB,$$

where all segments in the formula are directed segments.

Ceva's Theorem. A necessary and sufficient condition for AD, BE, CF, where D, E, F are points on the respective side lines BC, CA, AB of a triangle ABC, to be concurrent is that

$$BD \cdot CE \cdot AF = + DC \cdot EA \cdot FB,$$

where all segments in the formula are directed segments.

QUADRILATERALS

In the following: K = area, p and q are diagonals.

Rectangle

$A = B = C = D = 90°$
$K = ab, \quad p = \sqrt{a^2 + b^2}$

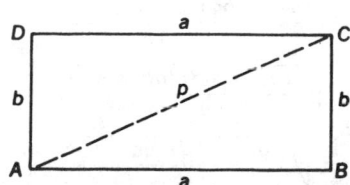

Mensuration Formulas

Parallelogram

$A = C, \quad B = D, \quad A + B = 180°$

$K = bh = ab \sin A = ab \sin B$

$h = a \sin A = a \sin B$

$p = \sqrt{a^2 + b^2 - 2ab \cos A}$

$q = \sqrt{a^2 + b^2 - 2ab \cos B} = \sqrt{a^2 + b^2 + 2ab \cos A}$

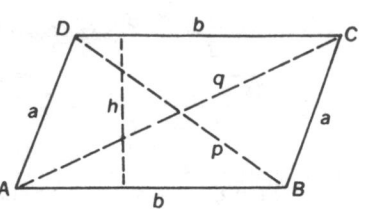

Rhombus

$p^2 + q^2 = 4a^2$

$K = \tfrac{1}{2}pq$

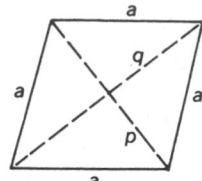

Trapezoid

$m = \tfrac{1}{2}(a + b)$

$K = \tfrac{1}{2}(a + b)h = mh$

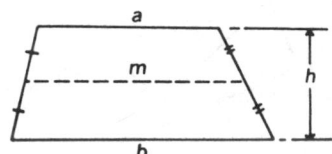

General quadrilateral

Let $s = \tfrac{1}{2}(a + b + c + d)$.

$$K = \tfrac{1}{2}pq \sin \theta$$

$$= \tfrac{1}{4}(b^2 + d^2 - a^2 - c^2) \tan \theta$$

$$= \tfrac{1}{4}\sqrt{4p^2q^2 - (b^2 + d^2 - a^2 - c^2)^2}$$

(Bretschneider's formula)

$$= \sqrt{(s - a)(s - b)(s - c)(s - d) - abcd \cos^2 \left(\frac{A + B}{2}\right)}$$

Theorem. The diagonals of a quadrilateral with consecutive sides a, b, c, d are perpendicular if and only if $a^2 + c^2 = b^2 + d^2$.

Cyclic Quadrilateral

Let R = radius of the circumscribed circle.

$A + C = B + D = 180°$

$K = \sqrt{(s - a)(s - b)(s - c)(s - d)}$

(Brahmagupta's formula)

$$= \frac{\sqrt{(ac + bd)(ad + bc)(ab + cd)}}{4R}$$

$$p = \sqrt{\frac{(ac + bd)(ab + cd)}{ad + bc}}, \quad q = \sqrt{\frac{(ac + bd)(ad + bc)}{ab + cd}}$$

$$R = \frac{1}{4}\sqrt{\frac{(ac + bd)(ad + bc)(ab + cd)}{(s - a)(s - b)(s - c)(s - d)}}, \quad \sin \theta = \frac{2K}{ac + bd}$$

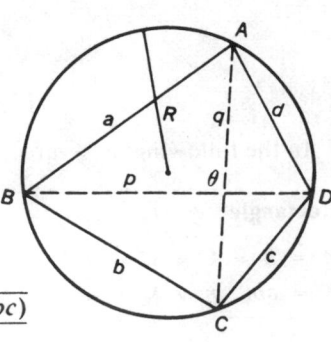

Mensuration Formulas

Ptolemy's Theorem.   A convex quadrilateral with consecutive sides a, b, c, d and diagonals p and q is cyclic if and only if $ac + bd = pq$.

Cyclic-inscriptable Quadrilateral

Let r = radius of the inscribed circle, R = radius of the circumscribed circle, m = distance between the centers of the inscribed and the circumscribed circles.

$$A + C = B + D = 180°$$
$$a + c = b + d$$
$$K = \sqrt{abcd}$$
$$\frac{1}{(R - m)^2} + \frac{1}{(R + m)^2} = \frac{1}{r^2}$$
$$r = \frac{\sqrt{abcd}}{s}$$
$$R = \frac{1}{4}\sqrt{\frac{(ac + bd)(ad + bc)(ab + cd)}{abcd}}$$

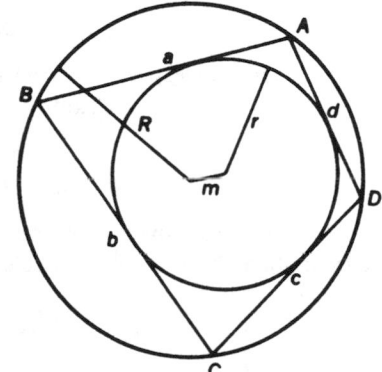

REGULAR POLYGONS

In the following: n = number of sides, s = length of each side, p = perimeter, θ = one of the vertex angles, r = radius of the inscribed circle, R = radius of the circumscribed circle, K = area.

$$\theta = \left(\frac{n - 2}{n}\right)180°$$
$$s = 2r\tan\frac{180°}{n} = 2R\sin\frac{180°}{n}$$
$$p = ns$$
$$K = \tfrac{1}{4}ns^2\cot\frac{180°}{n}$$
$$= nr^2\tan\frac{180°}{n}$$
$$= \tfrac{1}{2}nR^2\sin\frac{360°}{n}$$
$$r = \tfrac{1}{2}s\cot\frac{180°}{n}, \quad R = \tfrac{1}{2}s\csc\frac{180°}{n}$$

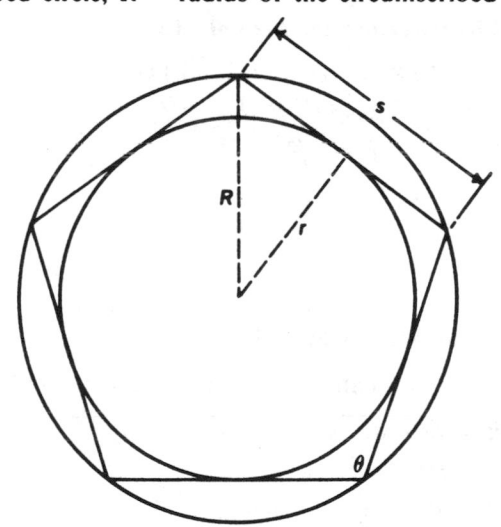

Polygon	n	K	r	R
Triangle (equilateral)	3	$0.43301s^2$	$0.28868s$	$0.57735s$
Square	4	$1.00000s^2$	$0.50000s$	$0.70711s$
Pentagon	5	$1.72048s^2$	$0.68819s$	$0.85065s$
Hexagon	6	$2.59808s^2$	$0.86603s$	$1.00000s$
Heptagon	7	$3.63391s^2$	$1.0383s$	$1.1524s$
Octagon	8	$4.82843s^2$	$1.2071s$	$1.3066s$
Nonagon	9	$6.18182s^2$	$1.3737s$	$1.4619s$
Decagon	10	$7.69421s^2$	$1.5388s$	$1.6180s$
Undecagon	11	$9.36564s^2$	$1.7028s$	$1.7747s$
Dodecagon	12	$11.19615s^2$	$1.8660s$	$1.9319s$

Mensuration Formulas

If s_k denotes the side of a regular polygon of k sides inscribed in a circle of radius R, then

$$s_{2n} = \sqrt{2R^2 - R\sqrt{4R^2 - s_n^2}}$$

If S_k denotes the side of a regular polygon of k sides circumscribed about a circle of radius r, then

$$S_{2n} = \frac{2rS_n}{2r + \sqrt{4r^2 + S_n^2}}$$

If p_k and P_k denote, respectively, the perimeters of regular polygons of k sides inscribed in and circumscribed about the same circle, then

$$P_{2n} = \frac{2p_n P_n}{p_n + P_n} \quad \text{and} \quad p_{2n} = \sqrt{p_n P_{2n}}.$$

If a_k and A_k denote, respectively, the areas of regular polygons of k sides inscribed in and circumscribed about the same circle, then

$$a_{2n} = \sqrt{a_n A_n} \quad \text{and} \quad A_{2n} = \frac{2a_{2n} A_n}{a_{2n} + A_n}.$$

CIRCLES

In the following: R = radius, D = diameter, C = circumference, K = area.

Circumference and Area of a Circle

$$C = 2\pi R = \pi D \qquad (\pi = 3.14159 \cdots)$$
$$K = \pi R^2 = \tfrac{1}{4}\pi D^2 = 0.7854 D^2$$
$$C = 2\sqrt{\pi K} = \frac{2K}{R}$$

$$K = \frac{C^2}{4\pi} = \tfrac{1}{2}CR$$

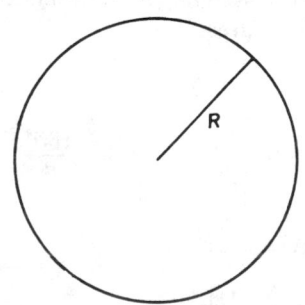

Sector and Segment of a Circle

Let the central angle θ be measured in radians ($\theta < \pi$).

$$h = R - d, \quad d = R - h$$
$$s = R\theta$$
$$d = R \cos \frac{\theta}{2} = \tfrac{1}{2}c \cot \frac{\theta}{2}$$
$$\quad = \tfrac{1}{2}\sqrt{4R^2 - c^2}$$
$$c = 2R \sin \frac{\theta}{2} = 2d \tan \frac{\theta}{2}$$
$$\quad = 2\sqrt{R^2 - d^2} = \sqrt{4h(2R - h)}$$
$$\theta = \frac{s}{R} = 2\,\mathrm{Cos}^{-1}\frac{d}{R} = 2\,\mathrm{Tan}^{-1}\frac{c}{2d} = 2\,\mathrm{Sin}^{-1}\frac{c}{2R}$$
$$K\text{ (sector)} = \tfrac{1}{2}Rs = \tfrac{1}{2}R^2\theta$$
$$K\text{ (segment)} = \tfrac{1}{2}R^2(\theta - \sin\theta) = \tfrac{1}{2}(Rs - cd) = R^2\,\mathrm{Cos}^{-1}\frac{d}{R} - d\sqrt{R^2 - d^2}$$

$$\quad = R^2\,\mathrm{Cos}^{-1}\frac{R - h}{R} - (R - h)\sqrt{2Rh - h^2}$$

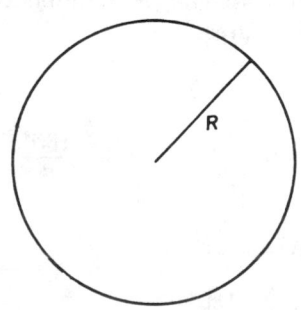

Mensuration Formulas

Sector of an Annulus

$h = R_1 - R_2$
$K = \frac{1}{2}\theta(R_1 + R_2)(R_1 - R_2)$
$\quad = \frac{1}{2}\theta h(R_1 + R_2)$
$\quad = \frac{1}{2}h(s_1 + s_2)$

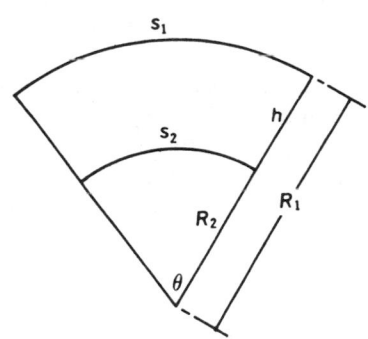

CONIC SECTIONS

Ellipse

Let p = circumference, K = area

$p \approx 2\pi\sqrt{\dfrac{a^2 + b^2}{2}}$ (approximately)

$\quad = 4aE$ (exactly)   See table of elliptic integral for E, using $k = \sqrt{a^2 - b^2}/a$.

$K = \pi ab$

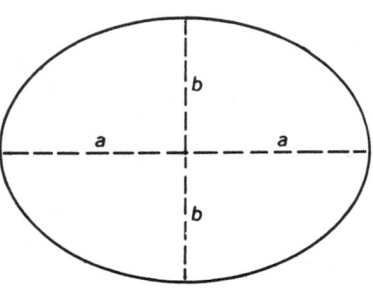

Parabolic Segment

$s = \sqrt{4x^2 + y^2} + \dfrac{y^2}{2x}\log_e\left[\dfrac{2x + \sqrt{4x^2 + y^2}}{y}\right]$

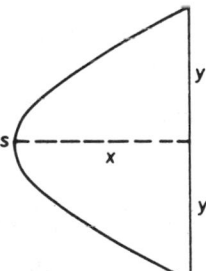

K (right segment) $= \frac{4}{3}xy$
K (oblique segment) $= \frac{4}{3}T$, where T is the area of the triangle with base along the chord of the segment and with opposite vertex at the point on the parabola at which the tangent to the parabola is parallel to the chord of the segment.

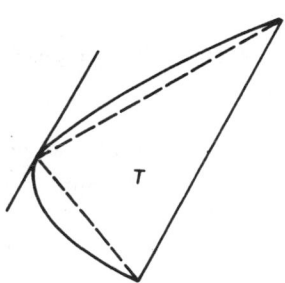

CAVALIERI'S THEOREM FOR THE PLANE

If two planar areas are included between a pair of parallel lines, and if the two segments cut off by the areas on any line parallel to the including lines are equal in length, then the two planar areas are equal.

Mensuration Formulas

PLANAR AREAS BY APPROXIMATION

Divide the planar area K into n strips by equidistant parallel chords of lengths $y_0, y_1, y_2, \ldots, y_n$ (where y_0 and/or y_n may be zero), and let h denote the common distance between the chords. Then, approximately:

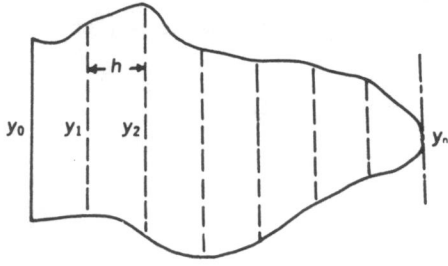

Trapezoidal Rule

$$K = h(\tfrac{1}{2}y_0 + y_1 + y_2 + \cdots + y_{n-1} + \tfrac{1}{2}y_n)$$

Durand's Rule

$$K = h(\tfrac{4}{10}y_0 + \tfrac{11}{10}y_1 + y_2 + y_3 + \cdots + y_{n-2} + \tfrac{11}{10}y_{n-1} + \tfrac{4}{10}y_n)$$

Simpson's rule (n even)

$$K = \tfrac{1}{3}h(y_0 + 4y_1 + 2y_2 + 4y_3 + 2y_4 + \cdots + 2y_{n-2} + 4y_{n-1} + y_n)$$

Weddle's Rule ($n = 6$)

$$K = \tfrac{3}{10}h(y_0 + 5y_1 + y_2 + 6y_3 + y_4 + 5y_5 + y_6)$$

SOLIDS BOUNDED BY PLANES

In the following: S = lateral surface, T = total surface, V = volume.

Cube

Let a = length of each edge.

$$T = 6a^2, \quad \text{diagonal of face} = a\sqrt{2}$$
$$V = a^3, \quad \text{diagonal of cube} = a\sqrt{3}$$

Rectangular Parallelepiped (or box)

Let a, b, c be the lengths of its edges.

$$T = 2(ab + bc + ca), \quad V = abc$$
$$\text{diagonal} = \sqrt{a^2 + b^2 + c^2}$$

Prism

$$S = (\text{perimeter of right section}) \times (\text{lateral edge})$$
$$V = (\text{area of right section}) \times (\text{lateral edge})$$
$$= (\text{area of base}) \times (\text{altitude})$$

Truncated Triangular Prism

$$V = (\text{area of right section}) \times \tfrac{1}{3}(\text{sum of the three lateral edges})$$

Mensuration Formulas

Pyramid

S of regular pyramid $= \frac{1}{2}$(perimeter of base) $\times$ (slant height)

$V = \frac{1}{3}$(area of base) $\times$ (altitude)

Frustum of Pyramid

Let B_1 = area of lower base, B_2 = area of upper base, h = altitude.

S of regular figure $= \frac{1}{2}$(sum of perimeters of bases) $\times$ (slant height)

$V = \frac{1}{3}h(B_1 + B_2 + \sqrt{B_1 B_2})$

Prismatoid

A *prismatoid* is a polyhedron having for bases two polygons in parallel planes, and for lateral faces triangles or trapezoids with one side lying in one base, and the opposite vertex or side lying in the other base, of the polyhedron. Let B_1 = area of lower base, M = area of midsection, B_2 = area of upper base, h = altitude.

$$V = \frac{1}{6}h(B_1 + 4M + B_2) \quad \text{(the } \textit{prismoidal formula}\text{)}$$

Note: Since cubes, rectangular parallelepipeds, prisms, pyramids, and frustums of pyramids are all examples of prismatoids, the formula for the volume of a prismatoid subsumes most of the above volume formulae.

Regular Polyhedra

Let v = number of vertices, e = number of edges, f = number of faces, α = each dihedral angle, a = length of each edge, r = radius of the inscribed sphere, R = radius of the circumscribed sphere, A = area of each face, T = total area, V = volume.

$v - e + f = 2$ (the *Euler-Descartes formula*—actually holds for *any* convex polyhedron)

$$T = fA$$
$$V = \frac{1}{3}rfA = \frac{1}{3}rT$$

Name	Nature of Surface	T	V
Tetrahedron	4 equilateral triangles	$1.73205a^2$	$0.11785a^3$
Hexahedron (cube)	6 squares	$6.00000a^2$	$1.00000a^3$
Octahedron	8 equilateral triangles	$3.46410a^2$	$0.47140a^3$
Dodecahedron	12 regular pentagons	$20.64573a^2$	$7.66312a^3$
Icosahedron	20 equilateral triangles	$8.66025a^2$	$2.18169a^3$

Name	v	e	f	α	a	r
Tetrahedron	4	6	4	70° 32′	$1.633R$	$0.333R$
Hexahedron	8	12	6	90°	$1.155R$	$0.577R$
Octahedron	6	12	8	109° 28′	$1.414R$	$0.577R$
Dodecahedron	20	30	12	116° 34′	$0.714R$	$0.795R$
Icosahedron	12	30	20	138° 11′	$1.051R$	$0.795R$

Mensuration Formulas

Name	A	r	R	V
Tetrahedron	$\frac{1}{4}a^2\sqrt{3}$	$\frac{1}{12}a\sqrt{6}$	$\frac{1}{4}a\sqrt{6}$	$\frac{1}{12}a^3\sqrt{2}$
Hexahedron	a^2	$\frac{1}{2}a$	$\frac{1}{2}a\sqrt{3}$	a^3
Octahedron	$\frac{1}{4}a^2\sqrt{3}$	$\frac{1}{6}a\sqrt{6}$	$\frac{1}{2}a\sqrt{2}$	$\frac{1}{3}a^3\sqrt{2}$
Dodecahedron	$\frac{1}{4}a^2\sqrt{25+10\sqrt{5}}$	$\frac{1}{20}a\sqrt{250+110\sqrt{5}}$	$\frac{1}{4}a(\sqrt{15}+\sqrt{3})$	$\frac{1}{4}a^3(15+7\sqrt{5})$
Icosahedron	$\frac{1}{4}a^2\sqrt{3}$	$\frac{1}{12}a\sqrt{42+18\sqrt{5}}$	$\frac{1}{4}a\sqrt{10+2\sqrt{5}}$	$\frac{5}{12}a^3(3+\sqrt{5})$

CYLINDERS AND CONES

In the following: B_1 = area of lower base, B_2 = area of upper base, h = altitude, S = lateral surface, T = total surface, V = volume.

Cylinder

S = (perimeter of right section) × (lateral edge)
V = (area of right section) × (lateral edge)

Right Circular Cylinder

Let R = radius of base.

$$S = 2\pi Rh, \quad T = 2\pi R(R + h), \quad V = \pi R^2 h$$

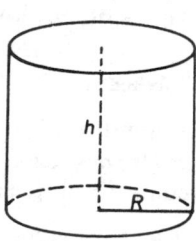

Cone

$$V = \tfrac{1}{3}B_1 h$$

Right Circular Cone

Let R = radius of base, s = slant height.

$$s = \sqrt{R^2 + h^2}$$
$$S = \pi Rs = \pi R\sqrt{R^2 + h^2}$$
$$T = \pi R(R + s) = \pi R(R + \sqrt{R^2 + h^2})$$
$$V = \tfrac{1}{3}\pi R^2 h$$

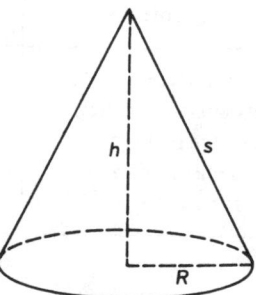

Frustum of Cone

$V = \tfrac{1}{3}h(B_1 + B_2 + \sqrt{B_1 B_2})$, where B_1 and B_2 are the areas of the bases.

Frustum of Right Circular Cone

Let R_1 = radius of lower base, R_2 = radius of upper base, s = slant height.

Mensuration Formulas

$$s = \sqrt{(R_1 - R_2)^2 + h^2}$$
$$S = \pi(R_1 + R_2)\,s$$
$$\quad = \pi(R_1 + R_2)\sqrt{(R_1 - R_2)^2 + h^2}$$
$$T = \pi[R_1^2 + R_2^2 + (R_1 + R_2)\,s]$$
$$\quad = \pi[R_1^2 + R_2^2 + (R_1 + R_2)\sqrt{(R_1 - R_2)^2 + h^2}]$$
$$V = \tfrac{1}{3}\pi h(R_1^2 + R_2^2 + R_1 R_2)$$

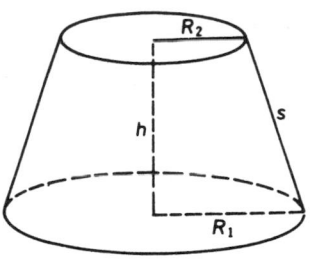

SPHERICAL FIGURES

In the following: R = radius of sphere, D = diameter of sphere, S = surface area, V = volume.

Sphere

$$D = 2R$$
$$S = 4\pi R^2 = \pi D^2 = 12.57R^2$$
$$V = \tfrac{4}{3}\pi R^3 = \tfrac{1}{6}\pi D^3 = 4.189R^3$$

Zone and Segment of One Base

$$S = 2\pi Rh = \pi Dh = \pi p^2$$
$$V = \tfrac{1}{3}\pi h^2(3R - h) = \tfrac{1}{6}\pi h(3a^2 + h^2)$$

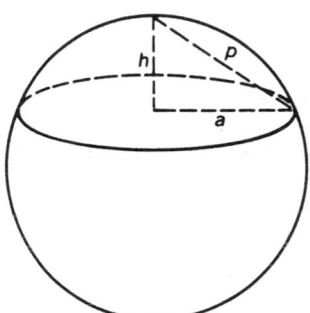

Zone and Segment of Two Bases

$$S = 2\pi Rh = \pi Dh$$
$$V = \tfrac{1}{6}\pi h(3a^2 + 3b^2 + h^2)$$

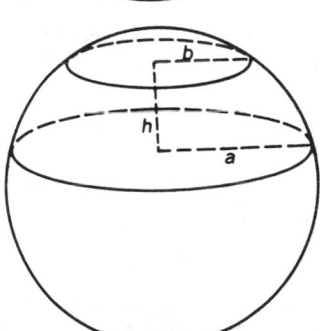

Lune

$$S = 2R^2\theta, \quad \theta \text{ in radians}$$

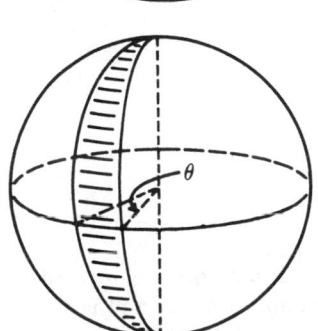

Mensuration Formulas

Spherical Sector

$$V = \tfrac{2}{3}\pi R^2 h = \tfrac{1}{6}\pi D^2 h$$

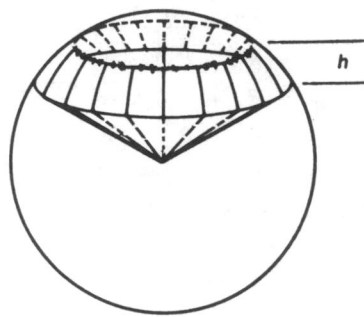

Spherical Triangle and Polygon

Let A, B, C be the angles, in radians, of the triangle; let θ = sum of angles, in radians, of a spherical polygon of n sides.

$$S = (A + B + C - \pi)R^2$$
$$S = [\theta - (n - 2)\pi]R^2$$

SPHEROIDS

Ellipsoid

Let a, b, c be the lengths of the semiaxes.

$$V = \tfrac{4}{3}\pi abc$$

Oblate Spheroid

An *oblate spheroid* is formed by the rotation of an ellipse about its minor axis. Let a and b be the major and minor semiaxes, respectively, and ϵ the eccentricity, of the revolving ellipse.

$$S = 2\pi a^2 + \pi \frac{b^2}{\epsilon} \log_\epsilon \frac{1 + \epsilon}{1 - \epsilon}$$
$$V = \tfrac{4}{3}\pi a^2 b$$

Prolate Spheroid

A *prolate spheroid* is formed by the rotation of an ellipse about its major axis. Let a and b be the major and minor semiaxes, respectively, and ϵ the eccentricity, of the revolving ellipse.

$$S = 2\pi b^2 + 2\pi \frac{ab}{\epsilon} \sin^{-1} \epsilon$$
$$V = \tfrac{4}{3}\pi ab^2$$

CIRCULAR TORUS

A *circular torus* is formed by the rotation of a circle about an axis in the plane of the circle and not cutting the circle. Let r be the radius of the revolving circle and let R be the distance of its center from the axis of rotation.

Mensuration Formulas

$$S = 4\pi^2 Rr$$
$$V = 2\pi^2 Rr^2$$

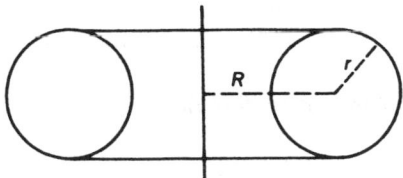

PAPPUS-GULDINUS THEOREMS

1. If a planar arc be revolved about an axis in its plane, but not cutting the arc, the area of the surface of revolution so formed is equal to the product of the length of the arc and the length of the path traced by the centroid of the arc.

2. If a planar area be revolved about an axis in its plane, but not intersecting the area, the volume of the solid of revolution so formed is equal to the product of the area and the length of the path traced by the centroid of the area.

CAVALIERI'S THEOREM FOR SPACE

If two solids are included between a pair of parallel planes, and if the two sections cut by them on any plane parallel to the including planes are equal in area, then the volumes of the solids are equal.

GENERAL PRISMATOID

A *general prismatoid* is a solid such that the area A_y of any section parallel to and distant y from a fixed plane can be expressed as a polynomial in y of degreee not higher than the third. That is,

$$A_y = ay^3 + by^2 + cy + d,$$

where a, b, c, d are constants which may be positive, zero, or negative. Let B_1 = area of lower base, M = area of midsection, B_2 = area of upper base, h = altitude.

$$V = \tfrac{1}{6}h(B_1 + 4M + B_2)$$

Note. All prismatoids, cylinders, cones, spheres, spheroids, and many other solids are general prismatoids.

CENTROIDS

If a geometrical figure possesses a center of symmetry, that point is the centroid of the figure.

If a geometrical figure possesses an axis of symmetry, the centroid of the figure lies on that axis.

Mensuration Formulae

Geometrical Figure	Location of Centroid
Perimeter of triangle	Center of the inscribed circle of the triangle whose vertices are the midpoints of the sides of the given triangle.
Arc of semicircle of radius R	Distance from diameter $= \dfrac{2R}{\pi}$
Arc of 2α radians of a circle of radius R	Distance from center of circle $= \dfrac{R \sin \alpha}{\alpha}$
Area of triangle	Intersection of the medians
Area of quadrilateral	Intersection of the diagonals of the parallelogram whose sides pass through adjacent trisection points of pairs of consecutive sides of the quadrilateral
Area of semicircle of radius R	Distance from diameter $= \dfrac{4R}{3\pi}$
Area of circular sector of radius R and central angle 2α radians	Distance from center of circle $= \dfrac{2R \sin \alpha}{3\alpha}$
Area of semiellipse of altitude h	Distance from base $= \dfrac{4h}{3\pi}$
Area of a quadrant of an ellipse of major and minor semiaxes a and b	Distance from minor axis $= \dfrac{4a}{3\pi}$, distance from major axis $= \dfrac{4b}{3\pi}$
Area of right parabolic segment of altitude h	Distance from base $= \frac{2}{5}h$
Lateral area of regular pyramid or right circular cone	Distance from base $= \frac{1}{3}h$
Area of hemisphere of radius R	Distance from base $= \frac{1}{2}R$
Volume of pyramid or cone	One fourth the way from the centroid of the base to the vertex of the pyramid or cone
Volume of frustum of pyramid or cone with d as the distance between the centroids of its bases and k as the ratio of similarity of the upper base to the lower base	On the line joining the centroids of the two bases at a distance from the centroid of the lower base $= \frac{1}{4}d\left(\dfrac{1 + 2k + 3k^2}{1 + k + k^2}\right)$
Volume of hemisphere of radius R	Distance from base $= \frac{3}{8}R$
Volume of revolution of altitude h obtained by revolving a semiellipse about its axis of symmetry	Distance from base $= \frac{3}{8}h$
Volume of paraboloid of revolution of altitude h	Distance from base $= \frac{1}{3}h$

V. TRIGONOMETRY

TRIGONOMETRY

Dr. Howard Eves

PLANE TRIGONOMETRY

Angle

That part of a straight line lying entirely to one side of a point O on the line is called a *ray* (or a *half-line*); the point O is called the *origin* of the ray. A ray of origin O is identified by the notation OA, where A is any point of the ray.

If a ray OA is rotated, in a plane, about its origin O onto ray OB, an *angle AOB* is said to be generated. Ray OA is called the *initial side*, ray OB the *terminal side*, and point O the *vertex* of the angle. The angle is said to be *positive* or *negative* according as the generating rotation is counterclockwise or clockwise.

An angle is said to be in *standard position* if its vertex is at the origin O and its initial side is on the positive x-axis of a rectangular Cartesian coordinate system (see p.344). If the terminal side of the angle falls on a coordinate axis, the angle is called a *quadrantal angle*; otherwise the angle is called a *first, second, third,* or *fourth quadrant angle* according as the terminal side falls in the first, second, third, or fourth quadrant of the coordinate system.

An angle of one *degree* is an angle in which the rotation is 1/360 of one complete rotation.

A *straight angle* is an angle of 180° (180 degrees).

A *right angle* is an angle of 90°.

An *acute angle* is an angle between 0° and 90°.

An *obtuse angle* is an angle between 90° and 180°.

A *radian* is an angle subtended at the center of a circle by an arc whose length is equal to that of the radius.

$$180° = \pi \text{ radians}; \quad 1° = \frac{\pi}{180} \text{ radians}; \quad 1 \text{ radian} = \frac{180}{\pi} \text{ degrees.}$$

The Trigonometric Functions of an Acute Angle

In the right triangle ABC,

sine A = sin A = a/c

cosine A = cos A = b/c

tangent A = tan A = a/b

cosecant A = csc A = c/a

secant A = sec A = c/b

cotangent A = cot A = ctn A = b/a

exsecant A = exsec A = sec A − 1

versine A = vers A = 1 − cos A

coversine A = covers A = 1 − sin A

haversine A = hav A = $\frac{1}{2}$ vers A

Formulas for Use in Trigonometry

The Trigonometric Functions of an Arbitrary Angle

Let α be any angle in standard position and let $P(x,y)$ be any point on the terminal side of the angle. Denote the positive distance OP by r. Then

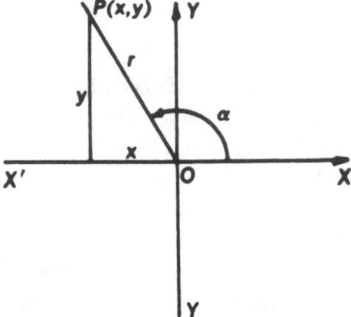

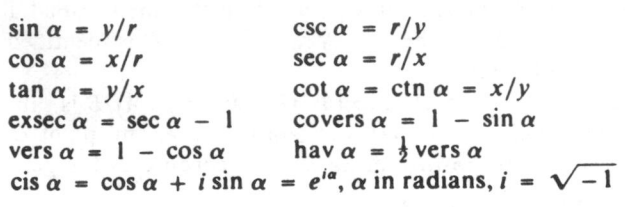

$\sin \alpha = y/r$ $\csc \alpha = r/y$

$\cos \alpha = x/r$ $\sec \alpha = r/x$

$\tan \alpha = y/x$ $\cot \alpha = \text{ctn } \alpha = x/y$

$\text{exsec } \alpha = \sec \alpha - 1$ $\text{covers } \alpha = 1 - \sin \alpha$

$\text{vers } \alpha = 1 - \cos \alpha$ $\text{hav } \alpha = \frac{1}{2} \text{vers } \alpha$

$\text{cis } \alpha = \cos \alpha + i \sin \alpha = e^{i\alpha}$, α in radians, $i = \sqrt{-1}$

RELATIONS BETWEEN CIRCULAR (OR INVERSE CIRCULAR) FUNCTIONS

$$\left(0 \le x \le \frac{\pi}{2}\right)$$

	$\sin x = a$	$\cos x = a$	$\tan x = a$
$\sin x$	a	$(1 - a^2)^{1/2}$	$a(1 + a^2)^{-1/2}$
$\cos x$	$(1 - a^2)^{1/2}$	a	$(1 + a^2)^{-1/2}$
$\tan x$	$a(1 - a^2)^{-1/2}$	$a^{-1}(1 - a^2)^{1/2}$	a
$\csc x$	a^{-1}	$(1 - a^2)^{-1/2}$	$a^{-1}(1 + a^2)^{1/2}$
$\sec x$	$(1 - a^2)^{-1/2}$	a^{-1}	$(1 + a^2)^{1/2}$
$\cot x$	$a^{-1}(1 - a^2)^{1/2}$	$a(1 - a^2)^{-1/2}$	a^{-1}

	$\csc x = a$	$\sec x = a$	$\cot x = a$
$\sin x$	a^{-1}	$a^{-1}(a^2 - 1)^{1/2}$	$(1 + a^2)^{-1/2}$
$\cos x$	$a^{-1}(a^2 - 1)^{1/2}$	a^{-1}	$a(1 + a^2)^{-1/2}$
$\tan x$	$(a^2 - 1)^{-1/2}$	$(a^2 - 1)^{1/2}$	a^{-1}
$\csc x$	a	$a(a^2 - 1)^{-1/2}$	$(1 + a^2)^{1/2}$
$\sec x$	$a(a^2 - 1)^{-1/2}$	a	$a^{-1}(1 + a^2)^{1/2}$
$\cot x$	$(a^2 - 1)^{1/2}$	$(a^2 - 1)^{-1/2}$	a

Examples:

If $\sec x = a$, then $\tan x = (a^2 - 1)^{1/2}$

$\arctan a = \arccos (1 + a^2)^{-1/2}$

Formulas for Use in Trigonometry

SIGNS OF THE TRIGONOMETRIC FUNCTIONS

Quadrant	sin	cos	tan	cot	sec	csc
I	+	+	+	+	+	+
II	+	−	−	−	−	+
III	−	−	+	+	−	−
IV	−	+	−	−	+	−

VARIATIONS OF THE TRIGONOMETRIC FUNCTIONS

Quadrant	sin	cos	tan	cot	sec	csc
I	$0 \to +1$	$+1 \to 0$	$0 \to +\infty$	$+\infty \to 0$	$+1 \to +\infty$	$+\infty \to +1$
II	$+1 \to 0$	$0 \to -1$	$-\infty \to 0$	$0 \to -\infty$	$-\infty \to -1$	$+1 \to +\infty$
III	$0 \to -1$	$-1 \to 0$	$0 \to +\infty$	$+\infty \to 0$	$-1 \to -\infty$	$-\infty \to -1$
IV	$-1 \to 0$	$0 \to +1$	$-\infty \to 0$	$0 \to -\infty$	$+\infty \to +1$	$-1 \to -\infty$

TRIGONOMETRIC FUNCTIONS OF SOME SPECIAL ANGLES

Angle	sin	cos	tan	cot	sec	csc
$0° = 0$	0	1	0	$\ldots$	1	$\ldots$
$15° = \dfrac{\pi}{12}$	$\dfrac{\sqrt{2}}{4}(\sqrt{3}-1)$	$\dfrac{\sqrt{2}}{4}(\sqrt{3}+1)$	$2-\sqrt{3}$	$2+\sqrt{3}$	$\sqrt{2}(\sqrt{3}-1)$	$\sqrt{2}(\sqrt{3}+1)$
$30° = \dfrac{\pi}{6}$	$1/2$	$\sqrt{3}/2$	$\sqrt{3}/3$	$\sqrt{3}$	$2\sqrt{3}/3$	2
$45° = \dfrac{\pi}{4}$	$\sqrt{2}/2$	$\sqrt{2}/2$	1	1	$\sqrt{2}$	$\sqrt{2}$
$60° = \dfrac{\pi}{3}$	$\sqrt{3}/2$	$1/2$	$\sqrt{3}$	$\sqrt{3}/3$	2	$2\sqrt{3}/3$
$75° = \dfrac{5\pi}{12}$	$\dfrac{\sqrt{2}}{4}(\sqrt{3}+1)$	$\dfrac{\sqrt{2}}{4}(\sqrt{3}-1)$	$2+\sqrt{3}$	$2-\sqrt{3}$	$\sqrt{2}(\sqrt{3}+1)$	$\sqrt{2}(\sqrt{3}-1)$
$90° = \dfrac{\pi}{2}$	1	0	$\ldots$	0	$\ldots$	1
$105° = \dfrac{7\pi}{12}$	$\dfrac{\sqrt{2}}{4}(\sqrt{3}+1)$	$-\dfrac{\sqrt{2}}{4}(\sqrt{3}-1)$	$-(2+\sqrt{3})$	$-(2-\sqrt{3})$	$-2(\sqrt{3}+1)$	$\sqrt{2}(\sqrt{3}-1)$
$120° = \dfrac{2\pi}{3}$	$\sqrt{3}/2$	$-1/2$	$-\sqrt{3}$	$-\sqrt{3}/3$	-2	$2\sqrt{3}/3$
$135° = \dfrac{3\pi}{4}$	$\sqrt{2}/2$	$-\sqrt{2}/2$	-1	-1	$-\sqrt{2}$	$\sqrt{2}$
$150° = \dfrac{5\pi}{6}$	$1/2$	$-\sqrt{3}/2$	$-\sqrt{3}/3$	$-\sqrt{3}$	$-2\sqrt{3}/3$	2
$165° = \dfrac{11\pi}{12}$	$\dfrac{\sqrt{2}}{4}(\sqrt{3}-1)$	$-\dfrac{\sqrt{2}}{4}(\sqrt{3}+1)$	$-(2-\sqrt{3})$	$-(2+\sqrt{3})$	$-\sqrt{2}(\sqrt{3}-1)$	$\sqrt{2}(\sqrt{3}+1)$
$180° = \pi$	0	-1	0	$\ldots$	-1	$\ldots$
$270° = \dfrac{3\pi}{2}$	-1	0	$\ldots$	0	$\ldots$	-1

Formulas for Use in Trigonometry

Relations among the Functions

$$\sin x = \frac{1}{\csc x}$$

$$\csc x = \frac{1}{\sin x}$$

$$\cos x = \frac{1}{\sec x}$$

$$\sec x = \frac{1}{\cos x}$$

$$\tan x = \frac{1}{\cot x} = \frac{\sin x}{\cos x}$$

$$\sin^2 x + \cos^2 x = 1$$

$$1 + \tan^2 x = \sec^2 x$$

$$\cot x = \frac{1}{\tan x} = \frac{\cos x}{\sin x}$$

$$1 + \cot^2 x = \csc^2 x$$

* $\sin x = \pm \sqrt{1 - \cos^2 x}$

* $\tan x = \pm \sqrt{\sec^2 x - 1}$

* $\cot x = \pm \sqrt{\csc^2 x - 1}$

* $\cos x = \pm \sqrt{1 - \sin^2 x}$

* $\sec x = \pm \sqrt{\tan^2 x + 1}$

* $\csc x = \pm \sqrt{\cot^2 x + 1}$

$$\sin x = \cos (90° - x) = \sin (180° - x)$$

$$\cos x = \sin (90° - x) = -\cos (180° - x)$$

$$\tan x = \cot (90° - x) = -\tan (180° - x)$$

$$\cot x = \tan (90° - x) = -\cot (180° - x)$$

$$\csc x = \cot \frac{x}{2} - \cot x$$

*The sign in front of radical depends on quadrant in which x falls.

Reduction Formulas

$$\sin \alpha = + \cos(\alpha - 90°) = - \sin(\alpha - 180°) = - \cos(\alpha - 270°)$$
$$\cos \alpha = - \sin(\alpha - 90°) = - \cos(\alpha - 180°) = + \sin(\alpha - 270°)$$
$$\tan \alpha = - \cot(\alpha - 90°) = + \tan(\alpha - 180°) = - \cot(\alpha - 270°)$$
$$\cot \alpha = - \tan(\alpha - 90°) = + \cot(\alpha - 180°) = - \tan(\alpha - 270°)$$
$$\sec \alpha = - \csc(\alpha - 90°) = - \sec(\alpha - 180°) = + \csc(\alpha - 270°)$$
$$\csc \alpha = + \sec(\alpha - 90°) = - \csc(\alpha - 180°) = - \sec(\alpha - 270°)$$

FURTHER REDUCTION FORMULAS

	sin	cos	tan	cot	sec	csc
$- \alpha$	$- \sin \alpha$	$+ \cos \alpha$	$- \tan \alpha$	$- \cot \alpha$	$+ \sec \alpha$	$- \csc \alpha$
$90° + \alpha$	$+ \cos \alpha$	$- \sin \alpha$	$- \cot \alpha$	$- \tan \alpha$	$- \csc \alpha$	$+ \sec \alpha$
$90° - \alpha$	$+ \cos \alpha$	$+ \sin \alpha$	$+ \cot \alpha$	$+ \tan \alpha$	$+ \csc \alpha$	$+ \sec \alpha$
$180° + \alpha$	$- \sin \alpha$	$- \cos \alpha$	$+ \tan \alpha$	$+ \cot \alpha$	$- \sec \alpha$	$- \csc \alpha$
$180° - \alpha$	$+ \sin \alpha$	$- \cos \alpha$	$- \tan \alpha$	$- \cot \alpha$	$- \sec \alpha$	$+ \csc \alpha$
$270° + \alpha$	$- \cos \alpha$	$+ \sin \alpha$	$- \cot \alpha$	$- \tan \alpha$	$+ \csc \alpha$	$- \sec \alpha$
$270° - \alpha$	$- \cos \alpha$	$- \sin \alpha$	$+ \cot \alpha$	$+ \tan \alpha$	$- \csc \alpha$	$- \sec \alpha$
$360° + \alpha$	$+ \sin \alpha$	$+ \cos \alpha$	$+ \tan \alpha$	$+ \cot \alpha$	$+ \sec \alpha$	$+ \csc \alpha$
$360° - \alpha$	$- \sin \alpha$	$+ \cos \alpha$	$- \tan \alpha$	$- \cot \alpha$	$+ \sec \alpha$	$- \csc \alpha$

Formulas for Use in Trigonometry

The previous table may be summarized and extended by the following easily remembered rule:

$$f(\pm\alpha + n90°) = \pm g(\alpha)$$

where n may be any integer, positive, negative, or zero

f is any one of the six trigonometric functions: sin, cos, tan, cot, sec, or csc

α may be any real angle measure

If n is even, then g is the same function as f. If n is odd, then g is the *cofunction* of f.

(Sine and cosine, tangent and cotangent, secant and cosecant, are cofunctions of each other.)

The second $\pm$ sign is not necessarily the same as the first one, but is determined as follows: For a given function f, a given value of n, and a given choice of the first $\pm$ sign, the second $\pm$ sign will be the same for all values of α. Thus it is only necessary to check the sign for any one value of α, and the formula will be complete.

EXAMPLES:

$\tan(\alpha + 270°) = \pm\cot\alpha$. Since $n(=3)$ is odd, we use the cofunction. To determine the sign, assume a value of α in the first quadrant. Then $\alpha + 270°$ is in the fourth quadrant, where the tangent is negative, so a minus sign is required. Thus the formula becomes

$$\tan(\alpha + 270°) = -\cot\alpha, \quad \text{valid for all values of } \alpha.$$

$\cos(\alpha - 450°) = \pm\sin\alpha$. Again assuming a value of α in the first quadrant, we find that $\alpha - 450°$ is in the fourth quadrant. Thus $\cos(\alpha - 450°)$ would be positive, and no minus sign is needed. Hence the formula becomes

$$\cos(\alpha - 450°) = \sin\alpha.$$

$\sec(180° - \alpha) = \pm\sec\alpha$. Here $n(=2)$ is even, so we use the same function. To determine the sign, again assume a value of α in the first quadrant. Then $180° - \alpha$ is in the second quadrant, where the secant is negative. Thus the formula, valid for all values of α, is

$$\sec(180° - \alpha) = -\sec\alpha.$$

Fundamental Identities

Where a double sign appears in the following, the choice of sign depends upon the quadrant in which the angle terminates.

Reciprocal relations

$$\sin\alpha = \frac{1}{\csc\alpha}, \qquad \cos\alpha = \frac{1}{\sec\alpha}, \qquad \tan\alpha = \frac{1}{\cot\alpha}$$

$$\csc\alpha = \frac{1}{\sin\alpha}, \qquad \sec\alpha = \frac{1}{\cos\alpha}, \qquad \cot\alpha = \frac{1}{\tan\alpha}$$

Product relations

$$\sin\alpha = \tan\alpha\cos\alpha, \qquad \cos\alpha = \cot\alpha\sin\alpha$$
$$\tan\alpha = \sin\alpha\sec\alpha, \qquad \cot\alpha = \cos\alpha\csc\alpha$$
$$\sec\alpha = \csc\alpha\tan\alpha, \qquad \csc\alpha = \sec\alpha\cot\alpha$$

Quotient relations

$$\sin\alpha = \frac{\tan\alpha}{\sec\alpha}, \qquad \cos\alpha = \frac{\cot\alpha}{\csc\alpha}, \qquad \tan\alpha = \frac{\sin\alpha}{\cos\alpha}$$

$$\csc\alpha = \frac{\sec\alpha}{\tan\alpha}, \qquad \sec\alpha = \frac{\csc\alpha}{\cot\alpha}, \qquad \cot\alpha = \frac{\cos\alpha}{\sin\alpha}$$

Formulas for Use in Trigonometry

Pythagorean relations

$$\sin^2\alpha + \cos^2\alpha = 1, \qquad 1 + \tan^2\alpha = \sec^2\alpha, \qquad 1 + \cot^2\alpha = \csc^2\alpha$$

Angle-sum and angle-difference relations

$$\sin(\alpha + \beta) = \sin\alpha\cos\beta + \cos\alpha\sin\beta$$
$$\sin(\alpha - \beta) = \sin\alpha\cos\beta - \cos\alpha\sin\beta$$
$$\cos(\alpha + \beta) = \cos\alpha\cos\beta - \sin\alpha\sin\beta$$
$$\cos(\alpha - \beta) = \cos\alpha\cos\beta + \sin\alpha\sin\beta$$
$$\tan(\alpha + \beta) = \frac{\tan\alpha + \tan\beta}{1 - \tan\alpha\tan\beta}$$
$$\tan(\alpha - \beta) = \frac{\tan\alpha - \tan\beta}{1 + \tan\alpha\tan\beta}$$
$$\cot(\alpha + \beta) = \frac{\cot\beta\cot\alpha - 1}{\cot\beta + \cot\alpha}$$
$$\cot(\alpha - \beta) = \frac{\cot\beta\cot\alpha + 1}{\cot\beta - \cot\alpha}$$
$$\sin(\alpha + \beta)\sin(\alpha - \beta) = \sin^2\alpha - \sin^2\beta = \cos^2\beta - \cos^2\alpha$$
$$\cos(\alpha + \beta)\cos(\alpha - \beta) = \cos^2\alpha - \sin^2\beta = \cos^2\beta - \sin^2\alpha$$

Double-angle relations

$$\sin 2\alpha = 2\sin\alpha\cos\alpha = \frac{2\tan\alpha}{1 + \tan^2\alpha}$$

$$\cos 2\alpha = \cos^2\alpha - \sin^2\alpha = 2\cos^2\alpha - 1 = 1 - 2\sin^2\alpha = \frac{1 - \tan^2\alpha}{1 + \tan^2\alpha}$$

$$\tan 2\alpha = \frac{2\tan\alpha}{1 - \tan^2\alpha}, \qquad \cot 2\alpha = \frac{\cot^2\alpha - 1}{2\cot\alpha}$$

Multiple-angle relations

$$\sin 3\alpha = 3\sin\alpha - 4\sin^3\alpha$$
$$\cos 3\alpha = 4\cos^3\alpha - 3\cos\alpha$$
$$\sin 4\alpha = 4\sin\alpha\cos\alpha - 8\sin^3\alpha\cos\alpha$$
$$\cos 4\alpha = 8\cos^4\alpha - 8\cos^2\alpha + 1$$
$$\sin 5\alpha = 5\sin\alpha - 20\sin^3\alpha + 16\sin^5\alpha$$
$$\cos 5\alpha = 16\cos^5\alpha - 20\cos^3\alpha + 5\cos\alpha$$
$$\sin 6\alpha = 32\cos^5\alpha\sin\alpha - 32\cos^3\alpha\sin\alpha + 6\cos\alpha\sin\alpha$$
$$\cos 6\alpha = 32\cos^6\alpha - 48\cos^4\alpha + 18\cos^2\alpha - 1$$
$$\sin n\alpha = 2\sin(n - 1)\alpha\cos\alpha - \sin(n - 2)\alpha$$
$$\cos n\alpha = 2\cos(n - 1)\alpha\cos\alpha - \cos(n - 2)\alpha$$
$$\tan 3\alpha = \frac{3\tan\alpha - \tan^3\alpha}{1 - 3\tan^2\alpha}$$
$$\tan 4\alpha = \frac{4\tan\alpha - 4\tan^3\alpha}{1 - 6\tan^2\alpha + \tan^4\alpha}$$
$$\tan n\alpha = \frac{\tan(n - 1)\alpha + \tan\alpha}{1 - \tan(n - 1)\alpha\tan\alpha}$$

Formulas for Use in Trigonometry

Function-product relations

$$\sin \alpha \sin \beta = \tfrac{1}{2} \cos(\alpha - \beta) - \tfrac{1}{2} \cos(\alpha + \beta)$$
$$\cos \alpha \cos \beta = \tfrac{1}{2} \cos(\alpha - \beta) + \tfrac{1}{2} \cos(\alpha + \beta)$$
$$\sin \alpha \cos \beta = \tfrac{1}{2} \sin(\alpha + \beta) + \tfrac{1}{2} \sin(\alpha - \beta)$$
$$\cos \alpha \sin \beta = \tfrac{1}{2} \sin(\alpha + \beta) - \tfrac{1}{2} \sin(\alpha - \beta)$$

Function-sum and function-difference relations

$$\sin \alpha + \sin \beta = 2 \sin \tfrac{1}{2}(\alpha + \beta) \cos \tfrac{1}{2}(\alpha - \beta)$$
$$\sin \alpha - \sin \beta = 2 \cos \tfrac{1}{2}(\alpha + \beta) \sin \tfrac{1}{2}(\alpha - \beta)$$
$$\cos \alpha + \cos \beta = 2 \cos \tfrac{1}{2}(\alpha + \beta) \cos \tfrac{1}{2}(\alpha - \beta)$$
$$\cos \alpha - \cos \beta = -2 \sin \tfrac{1}{2}(\alpha + \beta) \sin \tfrac{1}{2}(\alpha - \beta)$$

$$\tan \alpha + \tan \beta = \frac{\sin(\alpha + \beta)}{\cos \alpha \cos \beta}, \qquad \tan \alpha - \tan \beta = \frac{\sin(\alpha - \beta)}{\cos \alpha \cos \beta}$$

$$\cot \alpha + \cot \beta = \frac{\sin(\alpha + \beta)}{\sin \alpha \sin \beta}, \qquad \cot \alpha - \cot \beta = \frac{\sin(\beta - \alpha)}{\sin \alpha \sin \beta}$$

$$\frac{\sin \alpha + \sin \beta}{\sin \alpha - \sin \beta} = \frac{\tan \tfrac{1}{2}(\alpha + \beta)}{\tan \tfrac{1}{2}(\alpha - \beta)}, \qquad \frac{\sin \alpha + \sin \beta}{\cos \alpha - \cos \beta} = \cot \tfrac{1}{2}(\beta - \alpha)$$

$$\frac{\sin \alpha + \sin \beta}{\cos \alpha + \cos \beta} = \tan \tfrac{1}{2}(\alpha + \beta), \qquad \frac{\sin \alpha - \sin \beta}{\cos \alpha + \cos \beta} = \tan \tfrac{1}{2}(\alpha - \beta)$$

Half-angle relations

$$\sin \frac{\alpha}{2} = \pm \sqrt{\frac{1 - \cos \alpha}{2}}, \qquad \cos \frac{\alpha}{2} = \pm \sqrt{\frac{1 + \cos \alpha}{2}}$$

$$\tan \frac{\alpha}{2} = \pm \sqrt{\frac{1 - \cos \alpha}{1 + \cos \alpha}} = \frac{1 - \cos \alpha}{\sin \alpha} = \frac{\sin \alpha}{1 + \cos \alpha}$$

$$\cot \frac{\alpha}{2} = \pm \sqrt{\frac{1 + \cos \alpha}{1 - \cos \alpha}} = \frac{1 + \cos \alpha}{\sin \alpha} = \frac{\sin \alpha}{1 - \cos \alpha}$$

Power relations

$$\sin^2 \alpha = \tfrac{1}{2}(1 - \cos 2\alpha), \qquad \sin^3 \alpha = \tfrac{1}{4}(3 \sin \alpha - \sin 3\alpha)$$
$$\sin^4 \alpha = \tfrac{1}{8}(3 - 4 \cos 2\alpha + \cos 4\alpha)$$
$$\cos^2 \alpha = \tfrac{1}{2}(1 + \cos 2\alpha), \qquad \cos^3 \alpha = \tfrac{1}{4}(3 \cos \alpha + \cos 3\alpha)$$
$$\cos^4 \alpha = \tfrac{1}{8}(3 + 4 \cos 2\alpha + \cos 4\alpha)$$

$$\tan^2 \alpha = \frac{1 - \cos 2\alpha}{1 + \cos 2\alpha}, \qquad \cot^2 \alpha = \frac{1 + \cos 2\alpha}{1 - \cos 2\alpha}$$

Exponential relations (α in radians), Euler's equation

$$e^{i\alpha} = \cos \alpha + i \sin \alpha, \qquad i = \sqrt{-1}$$

$$\sin \alpha = \frac{e^{i\alpha} - e^{-i\alpha}}{2i}, \qquad \cos \alpha = \frac{e^{i\alpha} + e^{-i\alpha}}{2}$$

$$\tan \alpha = -i \left(\frac{e^{i\alpha} - e^{-i\alpha}}{e^{i\alpha} + e^{-i\alpha}} \right) = -i \left(\frac{e^{2i\alpha} - 1}{e^{2i\alpha} + 1} \right)$$

Formulas for Use in Trigonometry

THE TRIGONOMETRIC FUNCTIONS IN TERMS OF ONE ANOTHER

Function	$\sin \alpha$	$\cos \alpha$	$\tan \alpha$	$\cot \alpha$	$\sec \alpha$	$\csc \alpha$
$\sin \alpha$	$\sin \alpha$	$\pm \sqrt{1 - \cos^2\alpha}$	$\dfrac{\tan \alpha}{\pm \sqrt{1 + \tan^2\alpha}}$	$\dfrac{1}{\pm \sqrt{1 + \cot^2\alpha}}$	$\dfrac{\pm \sqrt{\sec^2\alpha - 1}}{\sec \alpha}$	$\dfrac{1}{\csc \alpha}$
$\cos \alpha$	$\pm \sqrt{1 - \sin^2\alpha}$	$\cos \alpha$	$\dfrac{1}{\pm \sqrt{1 + \tan^2\alpha}}$	$\dfrac{\cot \alpha}{\pm \sqrt{1 + \cot^2\alpha}}$	$\dfrac{1}{\sec \alpha}$	$\dfrac{\pm \sqrt{\csc^2\alpha - 1}}{\csc \alpha}$
$\tan \alpha$	$\dfrac{\sin \alpha}{\pm \sqrt{1 - \sin^2\alpha}}$	$\dfrac{\pm \sqrt{1 - \cos^2\alpha}}{\cos \alpha}$	$\tan \alpha$	$\dfrac{1}{\cot \alpha}$	$\pm \sqrt{\sec^2\alpha - 1}$	$\dfrac{1}{\pm \sqrt{\csc^2\alpha - 1}}$
$\cot \alpha$	$\dfrac{\pm \sqrt{1 - \sin^2\alpha}}{\sin \alpha}$	$\dfrac{\cos \alpha}{\pm \sqrt{1 - \cos^2\alpha}}$	$\dfrac{1}{\tan \alpha}$	$\cot \alpha$	$\dfrac{1}{\pm \sqrt{\sec^2\alpha - 1}}$	$\pm \sqrt{\csc^2\alpha - 1}$
$\sec \alpha$	$\dfrac{1}{\pm \sqrt{1 - \sin^2\alpha}}$	$\dfrac{1}{\cos \alpha}$	$\pm \sqrt{1 + \tan^2\alpha}$	$\dfrac{\pm \sqrt{1 + \cot^2\alpha}}{\cot \alpha}$	$\sec \alpha$	$\dfrac{\csc \alpha}{\pm \sqrt{\csc^2\alpha - 1}}$
$\csc \alpha$	$\dfrac{1}{\sin \alpha}$	$\dfrac{1}{\pm \sqrt{1 - \cos^2\alpha}}$	$\dfrac{\pm \sqrt{1 + \tan^2\alpha}}{\tan \alpha}$	$\pm \sqrt{1 + \cot^2\alpha}$	$\dfrac{\sec \alpha}{\pm \sqrt{\sec^2\alpha - 1}}$	$\csc \alpha$

Note. The choice of sign depends upon the quadrant in which the angle terminates.

Principal Values of the Inverse Trigonometric Functions

The notation arcsin x (or $\sin^{-1}x$) is used to denote any angle whose sine is x; Arcsin x (or $\mathrm{Sin}^{-1}x$) is usually used to denote the *principal value*. Similar notation is used for the other inverse trigonometric functions. The principal values of the inverse trigonometric functions are defined as follows:

$$-\pi/2 \leq \text{Arcsin } x \leq \pi/2, \qquad -1 \leq x \leq 1$$

$$0 \leq \text{Arccos } x \leq \pi, \qquad -1 \leq x \leq 1$$

$$-\pi/2 < \text{Arctan } x < \pi/2, \qquad -\infty < x < \infty$$

$$0 < \text{Arccsc } x \leq \pi/2, \qquad x \geq 1$$
$$-\pi < \text{Arccsc } x \leq -\pi/2, \qquad x \leq -1$$

$$0 \leq \text{Arcsec } x < \pi/2, \qquad x \geq 1$$
$$-\pi \leq \text{Arcsec } x < -\pi/2, \qquad x \leq -1$$

$$0 < \text{Arccot } x < \pi, \qquad -\infty < x < \infty$$

Note. There is no uniform agreement on the definitions of Arccsc x, Arcsec x, Arccot x for negative values of x.

Fundamental Identities Involving Principal Values

$$\text{Arcsin } x + \text{Arccos } x = \pi/2$$
$$\text{Arctan } x + \text{Arccot } x = \pi/2$$

If $\alpha = \text{Arcsin } x$, then

$$\sin \alpha = x, \qquad \cos \alpha = \sqrt{1 - x^2}, \qquad \tan \alpha = \frac{x}{\sqrt{1 - x^2}}$$

$$\csc \alpha = \frac{1}{x}, \qquad \sec \alpha = \frac{1}{\sqrt{1 - x^2}}, \qquad \cot \alpha = \frac{\sqrt{1 - x^2}}{x}$$

Formulas for Use in Trigonometry

If $\alpha = \text{Arccos } x$, then

$$\sin \alpha = \sqrt{1 - x^2}, \qquad \cos \alpha = x, \qquad\qquad \tan \alpha = \frac{\sqrt{1 - x^2}}{x}$$

$$\csc \alpha = \frac{1}{\sqrt{1 - x^2}}, \qquad \sec \alpha = \frac{1}{x}, \qquad\qquad \cot \alpha = \frac{x}{\sqrt{1 - x^2}}$$

If $\alpha = \text{Arctan } x$, then

$$\sin \alpha = \frac{x}{\sqrt{1 + x^2}}, \qquad \cos \alpha = \frac{1}{\sqrt{1 + x^2}}, \qquad \tan \alpha = x$$

$$\csc \alpha = \frac{\sqrt{1 + x^2}}{x}, \qquad \sec \alpha = \sqrt{1 + x^2}, \qquad \cot \alpha - \frac{1}{x}$$

Relations Between Principal Values of Inverse Trigonometric Functions

The following additional relations between principal values of the inverse trigonometric functions are useful:

$$\text{Arcsin } x = \text{Arccos}(-x) - \frac{\pi}{2} = \frac{\pi}{2} - \text{Arccos } x$$

$$= -\text{Arcsin}(-x) = \text{Arctan } \frac{x}{\sqrt{1 - x^2}}$$

$$= \frac{\pi}{2} - \text{Arccot } \frac{x}{\sqrt{1 - x^2}}$$

$$\text{Arctan } x = \text{Arccot}(-x) - \frac{\pi}{2} = \frac{\pi}{2} - \text{Arccot } x$$

$$= -\text{Arctan}(-x) = \text{Arcsin } \frac{x}{\sqrt{x^2 + 1}}$$

$$= \frac{\pi}{2} - \text{Arccos } \frac{x}{\sqrt{x^2 + 1}}$$

$$\text{Arccos } x = \frac{\pi}{2} + \text{Arcsin}(-x) = \frac{\pi}{2} - \text{Arcsin } x$$

$$= \pi - \text{Arccos}(-x) = \text{Arccot } \frac{x}{\sqrt{1 - x^2}}$$

$$= \frac{\pi}{2} - \text{Arctan } \frac{x}{\sqrt{1 - x^2}}$$

$$\text{Arccot } x = \frac{\pi}{2} + \text{Arctan}(-x) = \frac{\pi}{2} - \text{Arctan } x$$

$$= \pi - \text{Arccot}(-x) = \text{Arccos } \frac{x}{\sqrt{x^2 + 1}}$$

$$= \frac{\pi}{2} - \text{Arcsin } \frac{x}{\sqrt{x^2 + 1}}$$

Formulas for Use in Trigonometry

$$\text{Arccsc } x = \text{Arcsec } \frac{x}{\sqrt{x^2 - 1}}$$

$$\text{Arcsec } x = \text{Arccsc } \frac{x}{\sqrt{x^2 - 1}}$$

$$\cos(\text{Arcsin } x) = \sin(\text{Arccos } x) = \sqrt{1 - x^2}$$

$$\sec(\text{Arctan } x) = \csc(\text{Arccot } x) = \sqrt{x^2 + 1}$$

$$\tan(\text{Arcsec } x) = \cot(\text{Arccsc } x) = \sqrt{x^2 - 1}$$

The above are valid for x positive or negative. The following are valid only for $x \geq 0$.

$$\text{Arcsin } x = \text{Arccos } \sqrt{1 - x^2} = \text{Arccot } \frac{\sqrt{1 - x^2}}{x}$$

$$= \text{Arcsec } \frac{1}{\sqrt{1 - x^2}} = \text{Arccsc } \frac{1}{x}$$

$$= \frac{\pi}{2} - \text{Arcsin } \sqrt{1 - x^2}$$

$$\text{Arccos } x = \text{Arcsin } \sqrt{1 - x^2} = \text{Arctan } \frac{\sqrt{1 - x^2}}{x}$$

$$= \text{Arcsec } \frac{1}{x} = \text{Arccsc } \frac{1}{\sqrt{1 - x^2}}$$

$$= \frac{\pi}{2} - \text{Arccos } \sqrt{1 - x^2}$$

$$\text{Arctan } x = \text{Arccot } \frac{1}{x} = \text{Arccos } \frac{1}{\sqrt{x^2 + 1}}$$

$$= \text{Arcsec } \sqrt{x^2 + 1} = \text{Arccsc } \frac{\sqrt{x^2 + 1}}{x}$$

$$= \frac{\pi}{2} - \text{Arctan } \frac{1}{x}$$

$$\text{Arccot } x = \text{Arctan } \frac{1}{x} = \text{Arcsin } \frac{1}{\sqrt{x^2 + 1}}$$

$$= \text{Arcsec } \frac{\sqrt{x^2 + 1}}{x} = \text{Arccsc } \sqrt{x^2 + 1}$$

$$= \frac{\pi}{2} - \text{Arccot } \frac{1}{x}$$

$$\text{Arcsec } x = \text{Arctan } \sqrt{x^2 - 1} = \text{Arccot } \frac{1}{\sqrt{x^2 - 1}}$$

$$= \text{Arcsin } \frac{\sqrt{x^2 - 1}}{x} = \text{Arccos } \frac{1}{x} = \pi + \text{Arcsec}(-x)$$

$$= \frac{\pi}{2} - \text{Arccsc } x = -\frac{\pi}{2} - \text{Arccsc}(-x)$$

Formulas for Use in Trigonometry

$$\text{Arccsc } x = \text{Arctan } \frac{1}{\sqrt{x^2 - 1}} = \text{Arccot } \sqrt{x^2 - 1}$$

$$= \text{Arcsin } \frac{1}{x} = \text{Arccos } \frac{\sqrt{x^2 - 1}}{x} = \pi + \text{Arccsc}(-x)$$

$$= \frac{\pi}{2} - \text{Arcsec } x = -\frac{\pi}{2} - \text{Arcsec}(-x)$$

The following are valid if $x < 0$.

$$\text{Arcsin } x = -\text{Arccos } \sqrt{1 - x^2} = \text{Arccot } \frac{\sqrt{1 - x^2}}{x} - \pi$$

$$= -\pi - \text{Arccsc } \frac{1}{x} = -\text{Arcsec } \frac{1}{\sqrt{1 - x^2}}$$

$$= \text{Arcsin } \sqrt{1 - x^2} - \frac{\pi}{2}$$

$$\text{Arccos } x = \pi - \text{Arcsin } \sqrt{1 - x^2} = \pi + \text{Arctan } \frac{\sqrt{1 - x^2}}{x}$$

$$= \pi - \text{Arccsc } \frac{1}{\sqrt{1 - x^2}} = -\text{Arcsec } \frac{1}{x}$$

$$= \text{Arccos } \sqrt{1 - x^2} + \frac{\pi}{2}$$

$$\text{Arctan } x = \text{Arccot } \frac{1}{x} - \pi = -\text{Arccos } \frac{1}{\sqrt{x^2 + 1}}$$

$$= -\text{Arcsec } \sqrt{x^2 + 1} = -\pi - \text{Arccsc } \frac{\sqrt{x^2 + 1}}{x}$$

$$= -\frac{\pi}{2} - \text{Arctan } \frac{1}{x}$$

$$\text{Arccot } x = \pi + \text{Arctan } \frac{1}{x} = \pi - \text{Arcsin } \frac{1}{\sqrt{x^2 + 1}}$$

$$= -\text{Arcsec } \frac{\sqrt{x^2 + 1}}{x} = \pi - \text{Arccsc } \sqrt{x^2 + 1}$$

$$= \frac{3\pi}{2} - \text{Arccot } \frac{1}{x}$$

$$\text{Arcsec } x = \text{Arctan } \sqrt{x^2 - 1} - \pi = \text{Arccot } \frac{1}{\sqrt{x^2 - 1}} - \pi$$

$$= -\pi - \text{Arcsin } \frac{\sqrt{x^2 - 1}}{x} = -\text{Arccos } \frac{1}{x} = \text{Arcsec}(-x) - \pi$$

$$= -\frac{3\pi}{2} - \text{Arccsc } x = -\frac{\pi}{2} - \text{Arccsc}(-x)$$

Formulas for Use in Trigonometry

$$\text{Arccsc } x = \text{Arctan } \frac{1}{\sqrt{x^2 - 1}} - \pi = \text{Arccot } \sqrt{x^2 - 1} - \pi$$

$$= -\pi - \text{Arcsin } \frac{1}{x} = -\text{Arccos } \frac{\sqrt{x^2 - 1}}{x} = \text{Arccsc}(-x) - \pi$$

$$= -\frac{3\pi}{2} - \text{Arcsec } x = -\frac{\pi}{2} - \text{Arcsec}(-x)$$

Plane Triangle Formulas

In the following, A, B, C denote the angles of any plane triangle, a, b, c the corresponding opposite sides, and $s = \frac{1}{2}(a + b + c)$.

Radius of inscribed circle:

$$r = \sqrt{\frac{(s - a)(s - b)(s - c)}{s}}$$

Radius of circumscribed circle:

$$R = \frac{a}{2 \sin A} = \frac{b}{2 \sin B} = \frac{c}{2 \sin C}$$

Law of sines:

$$\frac{a}{\sin A} = \frac{b}{\sin B} = \frac{c}{\sin C}$$

Law of cosines:

$$a^2 = b^2 + c^2 - 2bc \cos A, \qquad \cos A = \frac{b^2 + c^2 - a^2}{2bc}$$

$$b^2 = c^2 + a^2 - 2ca \cos B, \qquad \cos B = \frac{c^2 + a^2 - b^2}{2ca}$$

$$c^2 = a^2 + b^2 - 2ab \cos C, \qquad \cos C = \frac{a^2 + b^2 - c^2}{2ab}$$

Law of tangents:

$$\frac{b - c}{b + c} = \frac{\tan \frac{1}{2}(B - C)}{\tan \frac{1}{2}(B + C)}, \qquad \frac{c - a}{c + a} = \frac{\tan \frac{1}{2}(C - A)}{\tan \frac{1}{2}(C + A)}$$

$$\frac{a - b}{a + b} = \frac{\tan \frac{1}{2}(A - B)}{\tan \frac{1}{2}(A + B)}$$

Half-angle formulas:

$$\tan \tfrac{1}{2}A = \frac{r}{s - a}, \qquad \tan \tfrac{1}{2}B = \frac{r}{s - b}, \qquad \tan \tfrac{1}{2}C = \frac{r}{s - c}$$

$$\sin \tfrac{1}{2}A = \sqrt{\frac{(s - b)(s - c)}{bc}}, \qquad \cos \tfrac{1}{2}A = \sqrt{\frac{s(s - a)}{bc}}$$

$$\sin \tfrac{1}{2}B = \sqrt{\frac{(s - c)(s - a)}{ca}}, \qquad \cos \tfrac{1}{2}B = \sqrt{\frac{s(s - b)}{ca}}$$

$$\sin \tfrac{1}{2}C = \sqrt{\frac{(s - a)(s - b)}{ab}}, \qquad \cos \tfrac{1}{2}C = \sqrt{\frac{s(s - c)}{ab}}$$

Formulas for Use in Trigonometry

Area:

$$K = \tfrac{1}{2}bc \sin A = \tfrac{1}{2}ca \sin B = \tfrac{1}{2}ab \sin C$$

$$K = \frac{a^2 \sin B \sin C}{2 \sin A} = \frac{b^2 \sin C \sin A}{2 \sin B} = \frac{c^2 \sin A \sin B}{2 \sin C}$$

$$K = \sqrt{s(s - a)(s - b)(s - c)} = rs = \frac{abc}{4R}$$

Mollweide's formulas:

$$\frac{b - c}{a} = \frac{\sin \tfrac{1}{2}(B - C)}{\cos \tfrac{1}{2}A}, \quad \frac{c - a}{b} = \frac{\sin \tfrac{1}{2}(C - A)}{\cos \tfrac{1}{2}B}$$

$$\frac{a - b}{c} = \frac{\sin \tfrac{1}{2}(A - B)}{\cos \tfrac{1}{2}C}$$

Newton's formulas:

$$\frac{b + c}{a} = \frac{\cos \tfrac{1}{2}(B - C)}{\sin \tfrac{1}{2}A}, \quad \frac{c + a}{b} = \frac{\cos \tfrac{1}{2}(C - A)}{\sin \tfrac{1}{2}B}$$

$$\frac{a + b}{c} = \frac{\cos \tfrac{1}{2}(A - B)}{\sin \tfrac{1}{2}C}$$

Solution of Right Triangles

(**a**) Given acute angle A and opposite leg a.

$$B = 90° - A, \quad b = a/\tan A = a \cot A, \quad c = a/\sin A = a \csc A$$

(**b**) Given acute angle A and adjacent leg b.

$$B = 90° - A, \quad a = b \tan A, \quad c = b/\cos A = b \sec A$$

(**c**) Given acute angle A and hypotenuse c.

$$B = 90° - A, \quad a = c \sin A, \quad b = c \cos A$$

(**d**) Given legs a and b.

$$c = \sqrt{a^2 + b^2}, \quad \tan A = a/b, \quad B = 90° - A$$

(**e**) Given hypotenuse c and leg a.

$$b = \sqrt{(c + a)(c - a)}, \quad \sin A = a/c, \quad B = 90° - A$$

Solution of Oblique Triangles

(**a**) Given sides b and c and included angle A.

Nonlogarithmic solution

$$a^2 = b^2 + c^2 - 2bc \cos A, \quad \cos B = (c^2 + a^2 - b^2)/2ca,$$
$$\cos C = (a^2 + b^2 - c^2)/2ab$$

Logarithmic solution

$$\tfrac{1}{2}(B + C) = 90° - \tfrac{1}{2}A, \quad \tan \tfrac{1}{2}(B - C) = \frac{b - c}{b + c} \tan \tfrac{1}{2}(B + C),$$
$$B = \tfrac{1}{2}(B + C) + \tfrac{1}{2}(B - C), \quad C = \tfrac{1}{2}(B + C) - \tfrac{1}{2}(B - C),$$
$$a = (b \sin A)/\sin B, \quad K = \tfrac{1}{2}bc \sin A$$

Formulas for Use in Spherical Trigonometry

Check. $A + B + C = 180°$, or use Newton's formula or law of sines.

(b) Given angles B and C and included side a.

$$A = 180° - (B + C), \quad b = (a \sin B)/\sin A,$$
$$c = (a \sin C)/\sin A, \quad K = \frac{a^2 \sin B \sin C}{2 \sin A}$$

Check. $a = b \cos C + c \cos B$, or use Newton's formula or law of tangents.

(c) Given sides a and c and opposite angle A.

$$\sin C = (c \sin A)/a, \quad B = 180° - (A + C),$$
$$b = (a \sin B)/\sin A, \quad K = \tfrac{1}{2}ac \sin B$$

Check. $a = b \cos C + c \cos B$, or use Newton's formula or law of tangents.
Note. In this case there may be two solutions, for C may have two values:

$C_1 < 90°$ and $C_2 = 180° - C_1 > 90°$. If $A + C_2 > 180°$, use only C_1.

(d) Given the three sides a, b, c.

Nonlogarithmic solution

$$\cos A = (b^2 + c^2 - a^2)/2bc, \quad \cos B = (c^2 + a^2 - b^2)/2ca,$$
$$\cos C = (a^2 + b^2 - c^2)/2ab$$

Logarithmic solution

$$s = \tfrac{1}{2}(a + b + c), \quad r = \sqrt{\frac{(s - a)(s - b)(s - c)}{s}},$$
$$\tan \tfrac{1}{2}A = \frac{r}{s - a}, \quad \tan \tfrac{1}{2}B = \frac{r}{s - b}, \quad \tan \tfrac{1}{2}C = \frac{r}{s - c},$$
$$K = \sqrt{s(s - a)(s - b)(s - c)}$$

Check. $A + B + C = 180°$.

Relations Between Accuracy of Computed Lengths and Angles

When solving a triangle for any of its parts, the following should be observed:

Significant figures for sides	Angles to the nearest
2	degree
3	ten minutes
4	minute
5	tenth of a minute

SPHERICAL TRIGONOMETRY

Right Spherical Triangles

Let a, b, c be the sides of a right spherical triangle with opposite angles $A, B, C = 90°$, respectively, where each side is measured by the angle subtended at the center of the sphere.

$$\sin a = \tan b \cot B, \qquad \sin a = \sin A \sin c$$
$$\sin b = \tan a \cot A, \qquad \sin b = \sin B \sin c$$
$$\cos A = \tan b \cot c, \qquad \cos A = \cos a \sin B$$
$$\cos B = \tan a \cot c, \qquad \cos B = \cos b \sin A$$
$$\cos c = \cot A \cot B, \qquad \cos c = \cos a \cos b$$

Formulas for Use in Spherical Trigonometry

Napier's Rules of Circular Parts

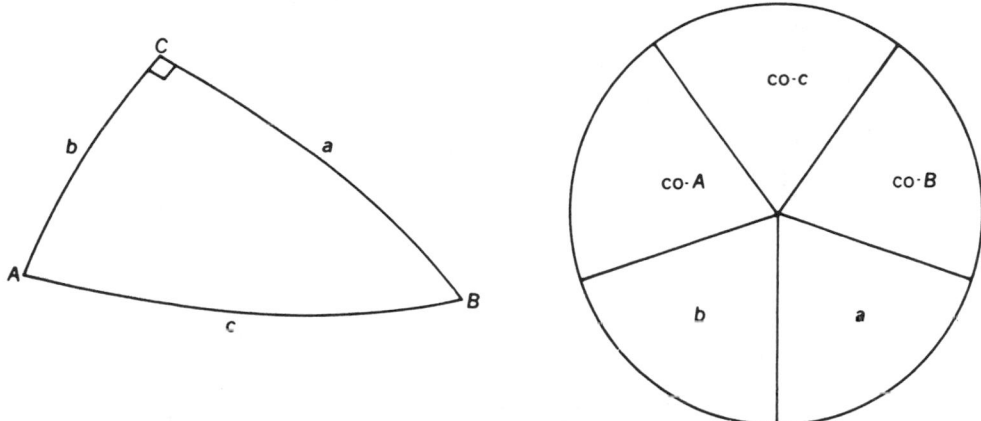

Arrange the five quantities a, b, co-A (complement of A), co-c, co-B of a right spherical triangle right-angled at C in cyclic order as pictured. If any one of these quantities is designated a *middle* part, then two of the other parts are *adjacent* to it, and the remaining two parts are *opposite* to it. The above formulas for a right spherical triangle may be recalled by the following two rules:

(**a**) The sine of any middle part is equal to the product of the *tangents* of the two *adjacent* parts.

(**b**) The sine of any middle part is equal to the product of the *cosines* of the two *opposite* parts.

Rules for Determining the Quadrant of a Calculated Part of a Right Spherical Triangle

(**a**) A leg and the angle opposite it are always of the same quadrant.

(**b**) If the hypotenuse if less than 90° the legs are of the same quadrant.

(**c**) If the hypotenuse is greater than 90°, the legs are of unlike quadrants.

Oblique Spherical Triangles

In the following, a, b, c represent the sides of any spherical triangle, A, B, C the corresponding opposite angles, $s = \frac{1}{2}(a + b + c)$, $S = \frac{1}{2}(A + B + C)$, Δ = area of triangle, E = spherical excess of triangle, R = radius of the sphere upon which the triangle lies, and a', b', c', A', B', C' are the corresponding parts of the polar triangle.

$$0° < a + b + c < 360°, \qquad 180° < A + B + C < 540°$$
$$E = A + B + C - 180°, \qquad \Delta = \pi R^2 E/180$$
$$\tan \tfrac{1}{4}E = \sqrt{\tan \tfrac{s}{2} \tan \tfrac{1}{2}(s - a) \tan \tfrac{1}{2}(s - b) \tan \tfrac{1}{2}(s - c)}$$
$$A = 180° - a', \qquad B = 180° - b', \qquad C = 180° - c'$$
$$a = 180° - A', \qquad b = 180° - B', \qquad c = 180° - C'$$

Law of sines:

$$\frac{\sin a}{\sin A} = \frac{\sin b}{\sin B} = \frac{\sin c}{\sin C}$$

Law of cosines for sides:

$$\cos a = \cos b \cos c + \sin b \sin c \cos A$$
$$\cos b = \cos c \cos a + \sin c \sin a \cos B$$
$$\cos c = \cos a \cos b + \sin a \sin b \cos C$$

Formulas for Use in Spherical Trigonometry

Law of cosines for angles:

$$\cos A = -\cos B \cos C + \sin B \sin C \cos a$$
$$\cos B = -\cos C \cos A + \sin C \sin A \cos b$$
$$\cos C = -\cos A \cos B + \sin A \sin B \cos c$$

Law of tangents:

$$\frac{\tan \frac{1}{2}(B - C)}{\tan \frac{1}{2}(B + C)} = \frac{\tan \frac{1}{2}(b - c)}{\tan \frac{1}{2}(b + c)}, \quad \frac{\tan \frac{1}{2}(C - A)}{\tan \frac{1}{2}(C + A)} = \frac{\tan \frac{1}{2}(c - a)}{\tan \frac{1}{2}(c + a)}$$

$$\frac{\tan \frac{1}{2}(A - B)}{\tan \frac{1}{2}(A + B)} = \frac{\tan \frac{1}{2}(a - b)}{\tan \frac{1}{2}(a + b)}$$

Half-angle formulas:

$$\tan \tfrac{1}{2}A = \frac{k}{\sin(s - a)}, \quad \tan \tfrac{1}{2}B = \frac{k}{\sin(s - b)}, \quad \tan \tfrac{1}{2}C = \frac{k}{\sin(s - c)},$$

where

$$k^2 = \frac{\sin(s - a)\sin(s - b)\sin(s - c)}{\sin s} = (\tan r)^2$$

Half-side formulas:

$$\tan \tfrac{1}{2}a = K \cos(S - A), \quad \tan \tfrac{1}{2}b = K \cos(S - B),$$
$$\tan \tfrac{1}{2}c = K \cos(S - C),$$

where

$$K^2 = \frac{-\cos S}{\cos(S - A)\cos(S - B)\cos(S - C)} = (\tan R)^2$$

Gauss's formulas:

$$\frac{\sin \frac{1}{2}(a - b)}{\sin \frac{1}{2}c} = \frac{\sin \frac{1}{2}(A - B)}{\cos \frac{1}{2}C}, \quad \frac{\cos \frac{1}{2}(a - b)}{\cos \frac{1}{2}c} = \frac{\sin \frac{1}{2}(A + B)}{\cos \frac{1}{2}C}$$

$$\frac{\sin \frac{1}{2}(a + b)}{\sin \frac{1}{2}c} = \frac{\cos \frac{1}{2}(A - B)}{\sin \frac{1}{2}C}, \quad \frac{\cos \frac{1}{2}(a + b)}{\cos \frac{1}{2}c} = \frac{\cos \frac{1}{2}(A + B)}{\sin \frac{1}{2}C}$$

Napier's analogies:

$$\frac{\sin \frac{1}{2}(A - B)}{\sin \frac{1}{2}(A + B)} = \frac{\tan \frac{1}{2}(a - b)}{\tan \frac{1}{2}c}, \quad \frac{\sin \frac{1}{2}(a - b)}{\sin \frac{1}{2}(a + b)} = \frac{\tan \frac{1}{2}(A - B)}{\cot \frac{1}{2}C}$$

$$\frac{\cos \frac{1}{2}(A - B)}{\cos \frac{1}{2}(A + B)} = \frac{\tan \frac{1}{2}(a + b)}{\tan \frac{1}{2}c}, \quad \frac{\cos \frac{1}{2}(a - b)}{\cos \frac{1}{2}(a + b)} = \frac{\tan \frac{1}{2}(A + B)}{\cot \frac{1}{2}C}$$

Haversine formulas:

$$\text{hav } a = \text{hav}(b - c) + \sin b \sin c \text{ hav } A$$

$$\text{hav } A = \frac{\sin(s - b)\sin(s - c)}{\sin b \sin c}$$

$$= \frac{\text{hav } a - \text{hav}(b - c)}{\sin b \sin c}$$

$$= \text{hav}[180° - (B + C)] + \sin B \sin C \text{ hav } a$$

Formulas for Use in Spherical Trigonometry

Rules for Determining the Quadrant of a Calculated Part of an Oblique Spherical Triangle

(a) If $A > B > C$, then $a > b > c$.

(b) A side (angle) which differs more from 90° than does another side (angle) is in the same quadrant as its opposite angle (side).

(c) Half the sum of any two sides and half the sum of the opposite angles are in the same quadrant.

SUMMARY OF SOLUTION OF OBLIQUE SPHERICAL TRIANGLES

Given	Solution	Check
Three sides	Half-angle formulas.	Law of sines
Three angles	Half-side formulas.	Law of sines
Two sides and included angle	Napier's analogies (to find sum and difference of unknown angles), then law of sines (to find remaining side).	Gauss's formula
Two angles and included side	Napier's analogies (to find sum and difference of unknown sides), then law of sines (to find remaining angle).	Gauss's formula
Two sides and an opposite angle	Law of sines (to find an angle), then Napier's analogies (to find remaining angle and side). Note number of solutions.	Gauss's formula
Two angles and an opposite side	Law of sines (to find a side), then Napier's analogies (to find remaining side and angle). Note number of solutions.	Gauss's formula

DEGREES, MINUTES, AND SECONDS TO RADIANS

Units in degrees, minutes or seconds	Degrees to Radians	Minutes to Radians	Seconds to Radians
10	0.174 5329	0.002 9089	0.000 0485
20	0.349 0659	0.005 8178	0.000 0970
30	0.523 5988	0.008 7266	0.000 1454
40	0.698 1317	0.011 6355	0.000 1939
50	0.872 6646	0.014 5444	0.000 2424
60	1.047 1976	0.017 4533	0.000 2909
70	1.221 7305	(0.020 3622)	(0.000 3394)
80	1.396 2634	(0.023 2711)	(0.000 3879)
90	1.570 7963	(0.026 1799)	(0.000 4363)
100	1.745 3293	....	....
200	3.490 6585	....	....
300	5.235 9878	....	....

where n = 1, 2, 3, 4, etc. n (100°) = n (1.745 3293)

RADIANS TO DEGREES, MINUTES, AND SECONDS

Radians	1.0	0.1	0.01	0.001	0.0001
1	57° 17′ 44.8″	5° 43′ 46.5″	0° 34′ 22.6″	0° 03′ 26.3″	0° 00′ 20.6″
2	114° 35′ 29.6″	11° 27′ 33.0″	1° 08′ 45.3″	0° 06′ 52.5″	0° 00′ 41.3″
3	171° 53′ 14.4″	17° 11′ 19.4″	1° 43′ 07.9′	0° 10′ 18.8″	0° 01′ 01.9″
4	229° 10′ 59.2″	22° 55′ 05.9″	2° 17′ 30.6″	0° 13′ 45.1′	0° 01′ 22.5″
5	286° 28′ 44.0″	28° 38′ 52.4″	2° 51′ 53.2″	0° 17′ 11.3″	0° 01′ 43.1″
6	343° 46′ 28.8″	34° 22′ 38.9″	3° 26′ 15.9″	0° 20′ 37.6″	0° 02′ 03.8″
7	401° 04′ 13.6″	40° 06′ 25.4″	4° 00′ 38.5″	0° 24′ 03.9″	0° 02′ 24.4″
8	458° 21′ 58.4″	45° 50′ 11.8″	4° 35′ 01.2″	0° 27′ 30.1″	0° 02′ 45.0″
9	515° 39′ 43.3″	51° 33′ 58.3″	5° 09′ 23.8″	0° 30′ 56.4″	0° 03′ 05.6″

Example: If 3.214 is desired in degrees, minutes and seconds it is obtained as follows:

$$3 = 171° \ 53′ \ 14.4″$$
$$.2 = \ 11° \ 27′ \ 33.0″$$
$$.01 = \ \ 0° \ 34′ \ 22.6″$$
$$.004 = \ \ 0° \ 13′ \ 45.1″$$
$$\overline{3.214 = 184° \ \ 8′ \ 55.1″}$$

MILS—RADIANS—DEGREES

1 mil = 0.00098175 radians = 0.05625° = 3.375′ = 202.5″
1000 mils = 0.98175 radians = 56.25°
6400 mils = 360° = 2π radians
1 radian = 1018.6 mils
1° = 17.777778 mils
1′ = 0.296296 mils
1″ = 0.0049383 mils

DEGREES—RADIANS
1 radian = 57° 17′ 44″ .80625

		log
1 radian = 57.29577 95131 degrees		1.75812 26324
1 radian = 3437.74677 07849 minutes		3.53627 38828
1 radian = 206264.80625 seconds		5.31442 51332
1 degree = 0.01745 32925 19943 radians		8.24187 73676—10
1 minute = 0.00029 08882 08666 radians		6.46372 61172—10
1 second = 0.00000 48481 36811 radians		4.68557 48668—10

DEGREES AND DECIMAL FRACTIONS TO RADIANS

The table below facilitates conversion of an angle expressed in degrees and decimal fractions into radians. To convert 25.78 into radians, find the equivalents, successively, of 20°, 5°, 0°.7, 0°.08 and add.

Deg.	Radians	Deg.	Radians	Deg.	Radians	Deg.	Radians	Deg.	Radians
10	0.174533	1	0.017453	0.1	0.001745	0.01	0.000175	0.001	0.000017
20	0.349066	2	.034907	.2	.003491	.02	.000349	.002	.000035
30	0.523599	3	.052360	.3	.005236	.03	.000524	.003	.000052
40	0.698132	4	.069813	.4	.006981	.04	.000698	.004	.000070
50	0.872665	5	.087266	.5	.008727	.05	.000873	.005	.000087
60	1.047198	6	.104720	.6	.010472	.06	.001047	.006	.000105
70	1.221730	7	.122173	.7	.012217	.07	.001222	.007	.000122
80	1.396263	8	.139626	.8	.013963	.08	.001396	.008	.000140
90	1.570796	9	.157080	.9	.015708	.09	.001571	.009	.000157

RADIANS TO DEGREES AND DECIMALS

Radians	Degrees	Radians	Degrees	Radians	Degrees	Radians	Degrees
1	57.2958	0.1	5.7296	0.01	0.5730	0.001	0.0573
2	114.5916	.2	11.4592	.02	1.1459	.002	.1146
3	171.8873	.3	17.1887	.03	1.7189	.003	.1719
4	229.1831	.4	22.9183	.04	2.2918	.004	.2292
5	286.4789	.5	28.6479	.05	2.8648	.005	.2865
6	343.7747	.6	34.3775	.06	3.4377	.006	.3438
7	401.0705	.7	40.1070	.07	4.0107	.007	.4011
8	458.3662	.8	45.8366	.08	4.5837	.008	.4584
9	515.6620	.9	51.5662	.09	5.1566	.009	.5157
10	572.9578	1.0	57.2958	.10	5.7296	.010	.5730

RADIANS—DEGREES

Multiples and Fractions of π Radians in Degrees

Radians	Radians	Deg.	Radians	Radians	Deg.	Radians	Radians	Deg.
π	3.1416	180	$\pi/2$	1.5708	90	$2\pi/3$	2.0944	120
2π	6.2832	360	$\pi/3$	1.0472	60	$3\pi/4$	2.3562	135
3π	9.4248	540	$\pi/4$	0.7854	45	$5\pi/6$	2.6180	150
4π	12.5664	720	$\pi/5$	0.6283	36	$7\pi/6$	3.6652	210
5π	15.7080	900	$\pi/6$	0.5236	30	$5\pi/4$	3.9270	225
6π	18.8496	1080	$\pi/7$	0.4488	25.714	$4\pi/3$	4.1888	240
7π	21.9911	1260	$\pi/8$	0.3927	22.5	$3\pi/2$	4.7124	270
8π	25.1327	1440	$\pi/9$	0.3491	20	$5\pi/3$	5.2360	300
9π	28.2743	1620	$\pi/10$	0.3142	18	$7\pi/4$	5.4978	315
10π	31.4159	1800	$\pi/12$	0.2618	15	$11\pi/6$	5.7596	330

CONVERSION OF ANGLES FROM ARC TO TIME

Arc	Time	Arc	Time	Arc	Time	Arc	Time
°	h m	°	h m				
′	m s	′	m s	″	s	″	s
0	0 00	20	1 20	0	0.00	8	0.53
1	0 04	30	2 00	1	0.07	9	0.60
2	0 08	40	2 40	2	0.13	10	0.67
3	0 12	50	3 20	3	0.20	20	1.33
4	0 16	60	4 00	4	0.27	30	2.00
5	0 20	70	4 40	5	0.33	40	2.67
6	0 24	80	5 20	6	0.40	50	3.33
7	0 28	90	6 00	7	0.47	60	4.00
8	0 32	100	6 40				
9	0 36	200	13 20				
10	0 40	300	20 00				

NATURAL TRIGONOMETRIC FUNCTIONS TO FIVE PLACES

For degrees indicated at the top of the page use the top column headings. For degrees indicated at the bottom use the column indications at the bottom. With degrees at the left of each block (top or bottom), use the minute column at the left; with degrees at the right of each block, use the minute column at the right. Appropriate signs for the functions must be supplied in accordance with the quadrant in which the angle measure belongs. Linear interpolation may be used except in regions where the functions are rapidly changing. See "Use of Logarithm Tables" for further discussion in interpolation practice.

0° (180°) (359)° 179°

′	Sin	Tan	Cot	Cos	Sec	Csc	′
0	.00000	.00000	———	1.0000	1.0000	———	60
1	.00029	.00029	3437.7	1.0000	1.0000	3437.7	59
2	.00058	.00058	1718.9	1.0000	1.0000	1718.9	58
3	.00087	.00087	1145.9	1.0000	1.0000	1145.9	57
4	.00116	.00116	859.44	1.0000	1.0000	859.44	56
5	.00145	.00145	687.55	1.0000	1.0000	687.55	55
6	.00175	.00175	572.96	1.0000	1.0000	572.96	54
7	.00204	.00204	491.11	1.0000	1.0000	491.11	53
8	.00233	.00233	429.72	1.0000	1.0000	429.72	52
9	.00262	.00262	381.97	1.0000	1.0000	381.97	51
10	.00291	.00291	343.77	1.0000	1.0000	343.78	50
11	.00320	.00320	312.52	.99999	1.0000	312.52	49
12	.00349	.00349	286.48	.99999	1.0000	286.48	48
13	.00378	.00378	264.44	.99999	1.0000	264.44	47
14	.00407	.00407	245.55	.99999	1.0000	245.55	46
15	.00436	.00436	229.18	.99999	1.0000	229.18	45
16	.00465	.00465	214.86	.99999	1.0000	214.86	44
17	.00495	.00495	202.22	.99999	1.0000	202.22	43
18	.00524	.00524	190.98	.99999	1.0000	190.99	42
19	.00553	.00553	180.93	.99998	1.0000	180.93	41
20	.00582	.00582	171.89	.99998	1.0000	171.89	40
21	.00611	.00611	163.70	.99998	1.0000	163.70	39
22	.00640	.00640	156.26	.99998	1.0000	156.26	38
23	.00669	.00669	149.47	.99998	1.0000	149.47	37
24	.00698	.00698	143.24	.99998	1.0000	143.24	36
25	.00727	.00727	137.51	.99997	1.0000	137.51	35
26	.00756	.00756	132.22	.99997	1.0000	132.22	34
27	.00785	.00785	127.32	.99997	1.0000	127.33	33
28	.00814	.00815	122.77	.99997	1.0000	122.78	32
29	.00844	.00844	118.54	.99996	1.0000	118.54	31
30	.00873	.00873	114.59	.99996	1.0000	114.59	30
31	.00902	.00902	110.89	.99996	1.0000	110.90	29
32	.00931	.00931	107.43	.99996	1.0000	107.43	28
33	.00960	.00960	104.17	.99995	1.0000	104.18	27
34	.00989	.00989	101.11	.99995	1.0000	101.11	26
35	.01018	.01018	98.218	.99995	1.0001	98.223	25
36	.01047	.01047	95.489	.99995	1.0001	95.495	24
37	.01076	.01076	92.908	.99994	1.0001	92.914	23
38	.01105	.01105	90.463	.99994	1.0001	90.469	22
39	.01134	.01135	88.144	.99994	1.0001	88.149	21
40	.01164	.01164	85.940	.99993	1.0001	85.946	20
41	.01193	.01193	83.844	.99993	1.0001	83.849	19
42	.01222	.01222	81.847	.99993	1.0001	81.853	18
43	.01251	.01251	79.943	.99992	1.0001	79.950	17
44	.01280	.01280	78.126	.99992	1.0001	78.133	16
45	.01309	.01309	76.390	.99991	1.0001	76.397	15
46	.01338	.01338	74.729	.99991	1.0001	74.736	14
47	.01367	.01367	73.139	.99991	1.0001	73.146	13
48	.01396	.01396	71.615	.99990	1.0001	71.622	12
49	.01425	.01425	70.153	.99990	1.0001	70.160	11
50	.01454	.01455	68.750	.99989	1.0001	68.757	10
51	.01483	.01484	67.402	.99989	1.0001	67.409	9
52	.01513	.01513	66.105	.99989	1.0001	66.113	8
53	.01542	.01542	64.858	.99988	1.0001	64.866	7
54	.01571	.01571	63.657	.99988	1.0001	63.665	6
55	.01600	.01600	62.499	.99987	1.0001	62.507	5
56	.01629	.01629	61.383	.99987	1.0001	61.391	4
57	.01658	.01658	60.306	.99986	1.0001	60.314	3
58	.01687	.01687	59.266	.99986	1.0001	59.274	2
59	.01716	.01716	58.261	.99985	1.0001	58.270	1
60	.01745	.01746	57.290	.99985	1.0002	57.299	0
′	Cos	Cot	Tan	Sin	Csc	Sec	′

90° (270°) (269)° 89°

1° (181°) (358°) 178°

′	Sin	Tan	Cot	Cos	Sec	Csc	′
0	.01745	.01746	57.290	.99985	1.0002	57.299	60
1	.01774	.01775	56.351	.99984	1.0002	56.359	59
2	.01803	.01804	55.442	.99984	1.0002	55.451	58
3	.01832	.01833	54.561	.99983	1.0002	54.570	57
4	.01862	.01862	53.709	.99983	1.0002	53.718	56
5	.01891	.01891	52.882	.99982	1.0002	52.892	55
6	.01920	.01920	52.081	.99982	1.0002	52.090	54
7	.01949	.01949	51.303	.99981	1.0002	51.313	53
8	.01978	.01978	50.549	.99980	1.0002	50.558	52
9	.02007	.02007	49.816	.99980	1.0002	49.826	51
10	.02036	.02036	49.104	.99979	1.0002	49.114	50
11	.02065	.02066	48.412	.99979	1.0002	48.422	49
12	.02094	.02095	47.740	.99978	1.0002	47.750	48
13	.02123	.02124	47.085	.99977	1.0002	47.096	47
14	.02152	.02153	46.449	.99977	1.0002	46.460	46
15	.02181	.02182	45.829	.99976	1.0002	45.840	45
16	.02211	.02211	45.226	.99976	1.0002	45.237	44
17	.02240	.02240	44.639	.99975	1.0003	44.650	43
18	.02269	.02269	44.066	.99974	1.0003	44.077	42
19	.02298	.02298	43.508	.99974	1.0003	43.520	41
20	.02327	.02328	42.964	.99973	1.0003	42.976	40
21	.02356	.02357	42.433	.99972	1.0003	42.445	39
22	.02385	.02386	41.916	.99972	1.0003	41.928	38
23	.02414	.02415	41.411	.99971	1.0003	41.423	37
24	.02443	.02444	40.917	.99970	1.0003	40.930	36
25	.02472	.02473	40.436	.99969	1.0003	40.448	35
26	.02501	.02502	39.965	.99969	1.0003	39.978	34
27	.02530	.02531	39.506	.99968	1.0003	39.519	33
28	.02560	.02560	39.057	.99967	1.0003	39.070	32
29	.02589	.02589	38.618	.99966	1.0003	38.631	31
30	.02618	.02619	38.188	.99966	1.0003	38.202	30
31	.02647	.02648	37.769	.99965	1.0004	37.782	29
32	.02676	.02677	37.358	.99964	1.0004	37.371	28
33	.02705	.02706	36.956	.99963	1.0004	36.970	27
34	.02734	.02735	36.563	.99963	1.0004	36.576	26
35	.02763	.02764	36.178	.99962	1.0004	36.191	25
36	.02792	.02793	35.801	.99961	1.0004	35.815	24
37	.02821	.02822	35.431	.99960	1.0004	35.445	23
38	.02850	.02851	35.070	.99959	1.0004	35.084	22
39	.02879	.02881	34.715	.99959	1.0004	34.730	21
40	.02908	.02910	34.368	.99958	1.0004	34.382	20
41	.02938	.02939	34.027	.99957	1.0004	34.042	19
42	.02967	.02968	33.694	.99956	1.0004	33.708	18
43	.02996	.02997	33.366	.99955	1.0005	33.381	17
44	.03025	.03026	33.045	.99954	1.0005	33.060	16
45	.03054	.03055	32.730	.99953	1.0005	32.746	15
46	.03083	.03084	32.421	.99952	1.0005	32.437	14
47	.03112	.03114	32.118	.99952	1.0005	32.134	13
48	.03141	.03143	31.821	.99951	1.0005	31.836	12
49	.03170	.03172	31.528	.99950	1.0005	31.544	11
50	.03199	.03201	31.242	.99949	1.0005	31.258	10
51	.03228	.03230	30.960	.99948	1.0005	30.976	9
52	.03257	.03259	30.683	.99947	1.0005	30.700	8
53	.03286	.03288	30.412	.99946	1.0005	30.428	7
54	.03316	.03317	30.145	.99945	1.0006	30.161	6
55	.03345	.03346	29.882	.99944	1.0006	29.899	5
56	.03374	.03376	29.624	.99943	1.0006	29.641	4
57	.03403	.03405	29.371	.99942	1.0006	29.388	3
58	.03432	.03434	29.122	.99941	1.0006	29.139	2
59	.03461	.03463	28.877	.99940	1.0006	28.894	1
60	.03490	.03492	28.636	.99939	1.0006	28.654	0
′	Cos	Cot	Tan	Sin	Csc	Sec	′

91° (271°) (268°) 88°

NATURAL TRIGONOMETRIC FUNCTIONS TO FIVE PLACES (continued)

2° (182°) (357°) 177°

′	Sin	Tan	Cot	Cos	Sec	Csc	′
0	.03490	.03492	28.636	.99939	1.0006	28.654	60
1	.03519	.03521	28.399	.99938	1.0006	28.417	59
2	.03548	.03550	28.166	.99937	1.0006	28.184	58
3	.03577	.03579	27.937	.99936	1.0006	27.955	57
4	.03606	.03609	27.712	.99935	1.0007	27.730	56
5	.03635	.03638	27.490	.99934	1.0007	27.508	55
6	.03664	.03667	27.271	.99933	1.0007	27.290	54
7	.03693	.03696	27.057	.99932	1.0007	27.075	53
8	.03723	.03725	26.845	.99931	1.0007	26.864	52
9	.03752	.03754	26.637	.99930	1.0007	26.655	51
10	.03781	.03783	26.432	.99929	1.0007	26.451	50
11	.03810	.03812	26.230	.99927	1.0007	26.249	49
12	.03839	.03842	26.031	.99926	1.0007	26.050	48
13	.03868	.03871	25.835	.99925	1.0007	25.854	47
14	.03897	.03900	25.642	.99924	1.0008	25.661	46
15	.03926	.03929	25.452	.99923	1.0008	25.471	45
16	.03955	.03958	25.264	.99922	1.0008	25.284	44
17	.03984	.03987	25.080	.99921	1.0008	25.100	43
18	.04013	.04016	24.898	.99919	1.0008	24.918	42
19	.04042	.04046	24.719	.99918	1.0008	24.739	41
20	.04071	.04075	24.542	.99917	1.0008	24.562	40
21	.04100	.04104	24.368	.99916	1.0008	24.388	39
22	.04129	.04133	24.196	.99915	1.0009	24.216	38
23	.04159	.04162	24.026	.99913	1.0009	24.047	37
24	.04188	.04191	23.859	.99912	1.0009	23.880	36
25	.04217	.04220	23.695	.99911	1.0009	23.716	35
26	.04246	.04250	23.532	.99910	1.0009	23.553	34
27	.04275	.04279	23.372	.99909	1.0009	23.393	33
28	.04304	.04308	23.214	.99907	1.0009	23.235	32
29	.04333	.04337	23.058	.99906	1.0009	23.079	31
30	.04362	.04366	22.904	.99905	1.0010	22.926	30
31	.04391	.04395	22.752	.99904	1.0010	22.774	29
32	.04420	.04424	22.602	.99902	1.0010	22.624	28
33	.04449	.04454	22.454	.99901	1.0010	22.476	27
34	.04478	.04483	22.308	.99900	1.0010	22.330	26
35	.04507	.04512	22.164	.99898	1.0010	22.187	25
36	.04536	.04541	22.022	.99897	1.0010	22.044	24
37	.04565	.04570	21.881	.99896	1.0010	21.904	23
38	.04594	.04599	21.743	.99894	1.0011	21.766	22
39	.04623	.04628	21.606	.99893	1.0011	21.629	21
40	.04653	.04658	21.470	.99892	1.0011	21.494	20
41	.04682	.04687	21.337	.99890	1.0011	21.360	19
42	.04711	.04716	21.205	.99889	1.0011	21.229	18
43	.04740	.04745	21.075	.99888	1.0011	21.098	17
44	.04769	.04774	20.946	.99886	1.0011	20.970	16
45	.04798	.04803	20.819	.99885	1.0012	20.843	15
46	.04827	.04833	20.693	.99883	1.0012	20.717	14
47	.04856	.04862	20.569	.99882	1.0012	20.593	13
48	.04885	.04891	20.446	.99881	1.0012	20.471	12
49	.04914	.04920	20.325	.99879	1.0012	20.350	11
50	.04943	.04949	20.206	.99878	1.0012	20.230	10
51	.04972	.04978	20.087	.99876	1.0012	20.112	9
52	.05001	.05007	19.970	.99875	1.0013	19.995	8
53	.05030	.05037	19.855	.99873	1.0013	19.880	7
54	.05059	.05066	19.740	.99872	1.0013	19.766	6
55	.05088	.05095	19.627	.99870	1.0013	19.653	5
56	.05117	.05124	19.516	.99869	1.0013	19.541	4
57	.05146	.05153	19.405	.99867	1.0013	19.431	3
58	.05175	.05182	19.296	.99866	1.0013	19.322	2
59	.05205	.05212	19.188	.99864	1.0014	19.214	1
60	.05234	.05241	19.081	.99863	1.0014	19.107	0
′	Cos	Cot	Tan	Sin	Csc	Sec	′

3° (183°) (356°) 176°

′	Sin	Tan	Cot	Cos	Sec	Csc	′
0	.05234	.05241	19.081	.99863	1.0014	19.107	60
1	.05263	.05270	18.976	.99861	1.0014	19.002	59
2	.05292	.05299	18.871	.99860	1.0014	18.898	58
3	.05321	.05328	18.768	.99858	1.0014	18.794	57
4	.05350	.05357	18.666	.99857	1.0014	18.692	56
5	.05379	.05387	18.564	.99855	1.0014	18.591	55
6	.05408	.05416	18.464	.99854	1.0015	18.492	54
7	.05437	.05445	18.366	.99852	1.0015	18.393	53
8	.05466	.05474	18.268	.99851	1.0015	18.295	52
9	.05495	.05503	18.171	.99849	1.0015	18.198	51
10	.05524	.05533	18.075	.99847	1.0015	18.103	50
11	.05553	.05562	17.980	.99846	1.0015	18.008	49
12	.05582	.05591	17.886	.99844	1.0016	17.914	48
13	.05611	.05620	17.793	.99842	1.0016	17.822	47
14	.05640	.05649	17.702	.99841	1.0016	17.730	46
15	.05669	.05678	17.611	.99839	1.0016	17.639	45
16	.05698	.05708	17.521	.99838	1.0016	17.549	44
17	.05727	.05737	17.431	.99836	1.0016	17.460	43
18	.05756	.05766	17.343	.99834	1.0017	17.372	42
19	.05785	.05795	17.256	.99833	1.0017	17.285	41
20	.05814	.05824	17.169	.99831	1.0017	17.198	40
21	.05844	.05854	17.084	.99829	1.0017	17.113	39
22	.05873	.05883	16.999	.99827	1.0017	17.028	38
23	.05902	.05912	16.915	.99826	1.0017	16.945	37
24	.05931	.05941	16.832	.99824	1.0018	16.862	36
25	.05960	.05970	16.750	.99822	1.0018	16.779	35
26	.05989	.05999	16.668	.99821	1.0018	16.698	34
27	.06018	.06029	16.587	.99819	1.0018	16.618	33
28	.06047	.06058	16.507	.99817	1.0018	16.538	32
29	.06076	.06087	16.428	.99815	1.0019	16.459	31
30	.06105	.06116	16.350	.99813	1.0019	16.380	30
31	.06134	.06145	16.272	.99812	1.0019	16.303	29
32	.06163	.06175	16.195	.99810	1.0019	16.226	28
33	.06192	.06204	16.119	.99808	1.0019	16.150	27
34	.06221	.06233	16.043	.99806	1.0019	16.075	26
35	.06250	.06262	15.969	.99804	1.0020	16.000	25
36	.06279	.06291	15.895	.99803	1.0020	15.926	24
37	.06308	.06321	15.821	.99801	1.0020	15.853	23
38	.06337	.06350	15.748	.99799	1.0020	15.780	22
39	.06366	.06379	15.676	.99797	1.0020	15.708	21
40	.06395	.06408	15.605	.99795	1.0021	15.637	20
41	.06424	.06438	15.534	.99793	1.0021	15.566	19
42	.06453	.06467	15.464	.99792	1.0021	15.496	18
43	.06482	.06496	15.394	.99790	1.0021	15.427	17
44	.06511	.06525	15.325	.99788	1.0021	15.358	16
45	.06540	.06554	15.257	.99786	1.0021	15.290	15
46	.06569	.06584	15.189	.99784	1.0022	15.222	14
47	.06598	.06613	15.122	.99782	1.0022	15.155	13
48	.06627	.06642	15.056	.99780	1.0022	15.089	12
49	.06656	.06671	14.990	.99778	1.0022	15.023	11
50	.06685	.06700	14.924	.99776	1.0022	14.958	10
51	.06714	.06730	14.860	.99774	1.0023	14.893	9
52	.06743	.06759	14.795	.99772	1.0023	14.829	8
53	.06773	.06788	14.732	.99770	1.0023	14.766	7
54	.06802	.06817	14.669	.99768	1.0023	14.703	6
55	.06831	.06847	14.606	.99766	1.0023	14.640	5
56	.06860	.06876	14.544	.99764	1.0024	14.578	4
57	.06889	.06905	14.482	.99762	1.0024	14.517	3
58	.06918	.06934	14.421	.99760	1.0024	14.456	2
59	.06947	.06963	14.361	.99758	1.0024	14.395	1
60	.06976	.06993	14.301	.99756	1.0024	14.336	0
′	Cos	Cot	Tan	Sin	Csc	Sec	′

NATURAL TRIGONOMETRIC FUNCTIONS TO FIVE PLACES (continued)

4° (184°)   (355°) 175°

′	Sin	Tan	Cot	Cos	Sec	Csc	′
0	.06976	.06993	14.301	.99756	1.0024	14.336	60
1	.07005	.07022	14.241	.99754	1.0025	14.276	59
2	.07034	.07051	14.182	.99752	1.0025	14.217	58
3	.07063	.07080	14.124	.99750	1.0025	14.159	57
4	.07092	.07110	14.065	.99748	1.0025	14.101	56
5	.07121	.07139	14.008	.99746	1.0025	14.044	55
6	.07150	.07168	13.951	.99744	1.0026	13.987	54
7	.07179	.07197	13.894	.99742	1.0026	13.930	53
8	.07208	.07227	13.838	.99740	1.0026	13.874	52
9	.07237	.07256	13.782	.99738	1.0026	13.818	51
10	.07266	.07285	13.727	.99736	1.0027	13.763	50
11	.07295	.07314	13.672	.99734	1.0027	13.708	49
12	.07324	.07344	13.617	.99731	1.0027	13.654	48
13	.07353	.07373	13.563	.99729	1.0027	13.600	47
14	.07382	.07402	13.510	.99727	1.0027	13.547	46
15	.07411	.07431	13.457	.99725	1.0028	13.494	45
16	.07440	.07461	13.404	.99723	1.0028	13.441	44
17	.07469	.07490	13.352	.99721	1.0028	13.389	43
18	.07498	.07519	13.300	.99719	1.0028	13.337	42
19	.07527	.07548	13.248	.99716	1.0028	13.286	41
20	.07556	.07578	13.197	.99714	1.0029	13.235	40
21	.07585	.07607	13.146	.99712	1.0029	13.184	39
22	.07614	.07636	13.096	.99710	1.0029	13.134	38
23	.07643	.07665	13.046	.99708	1.0029	13.084	37
24	.07672	.07695	12.996	.99705	1.0030	13.035	36
25	.07701	.07724	12.947	.99703	1.0030	12.985	35
26	.07730	.07753	12.898	.99701	1.0030	12.937	34
27	.07759	.07782	12.850	.99699	1.0030	12.888	33
28	.07788	.07812	12.801	.99696	1.0030	12.840	32
29	.07817	.07841	12.754	.99694	1.0031	12.793	31
30	.07846	.07870	12.706	.99692	1.0031	12.745	30
31	.07875	.07899	12.659	.99689	1.0031	12.699	29
32	.07904	.07929	12.612	.99687	1.0031	12.652	28
33	.07933	.07958	12.566	.99685	1.0032	12.606	27
34	.07962	.07987	12.520	.99683	1.0032	12.560	26
35	.07991	.08017	12.474	.99680	1.0032	12.514	25
36	.08020	.08046	12.429	.99678	1.0032	12.469	24
37	.08049	.08075	12.384	.99676	1.0033	12.424	23
38	.08078	.08104	12.339	.99673	1.0033	12.379	22
39	.08107	.08134	12.295	.99671	1.0033	12.335	21
40	.08136	.08163	12.251	.99668	1.0033	12.291	20
41	.08165	.08192	12.207	.99666	1.0034	12.248	19
42	.08194	.08221	12.163	.99664	1.0034	12.204	18
43	.08223	.08251	12.120	.99661	1.0034	12.161	17
44	.08252	.08280	12.077	.99659	1.0034	12.119	16
45	.08281	.08309	12.035	.99657	1.0034	12.076	15
46	.08310	.08339	11.992	.99654	1.0035	12.034	14
47	.08339	.08368	11.950	.99652	1.0035	11.992	13
48	.08368	.08397	11.909	.99649	1.0035	11.951	12
49	.08397	.08427	11.867	.99647	1.0035	11.909	11
50	.08426	.08456	11.826	.99644	1.0036	11.868	10
51	.08455	.08485	11.785	.99642	1.0036	11.828	9
52	.08484	.08514	11.745	.99639	1.0036	11.787	8
53	.08513	.08544	11.705	.99637	1.0036	11.747	7
54	.08542	.08573	11.664	.99635	1.0037	11.707	6
55	.08571	.08602	11.625	.99632	1.0037	11.668	5
56	.08600	.08632	11.585	.99630	1.0037	11.628	4
57	.08629	.08661	11.546	.99627	1.0037	11.589	3
58	.08658	.08690	11.507	.99625	1.0038	11.551	2
59	.08687	.08720	11.468	.99622	1.0038	11.512	1
60	.08716	.08749	11.430	.99619	1.0038	11.474	0
′	Cos	Cot	Tan	Sin	Csc	Sec	′

5° (185°)   (354°) 174°

′	Sin	Tan	Cot	Cos	Sec	Csc	′
0	.08716	.08749	11.430	.99619	1.0038	11.474	60
1	.08745	.08778	11.392	.99617	1.0038	11.436	59
2	.08774	.08807	11.354	.99614	1.0039	11.398	58
3	.08803	.08837	11.316	.99612	1.0039	11.360	57
4	.08831	.08866	11.279	.99609	1.0039	11.323	56
5	.08860	.08895	11.242	.99607	1.0039	11.286	55
6	.08889	.08925	11.205	.99604	1.0040	11.249	54
7	.08918	.08954	11.168	.99602	1.0040	11.213	53
8	.08947	.08983	11.132	.99599	1.0040	11.176	52
9	.08976	.09013	11.095	.99596	1.0041	11.140	51
10	.09005	.09042	11.059	.99594	1.0041	11.105	50
11	.09034	.09071	11.024	.99591	1.0041	11.069	49
12	.09063	.09101	10.988	.99588	1.0041	11.034	48
13	.09092	.09130	10.953	.99586	1.0042	10.998	47
14	.09121	.09159	10.918	.99583	1.0042	10.963	46
15	.09150	.09189	10.883	.99580	1.0042	10.929	45
16	.09179	.09218	10.848	.99578	1.0042	10.894	44
17	.09208	.09247	10.814	.99575	1.0043	10.860	43
18	.09237	.09277	10.780	.99572	1.0043	10.826	42
19	.09266	.09306	10.746	.99570	1.0043	10.792	41
20	.09295	.09335	10.712	.99567	1.0043	10.758	40
21	.09324	.09365	10.678	.99564	1.0044	10.725	39
22	.09353	.09394	10.645	.99562	1.0044	10.692	38
23	.09382	.09423	10.612	.99559	1.0044	10.659	37
24	.09411	.09453	10.579	.99556	1 0045	10.626	36
25	.09440	.09482	10.546	.99553	1.0045	10.593	35
26	.09469	.09511	10.514	.99551	1.0045	10.561	34
27	.09498	.09541	10.481	.99548	1.0045	10.529	33
28	.09527	.09570	10.449	.99545	1.0046	10.497	32
29	.09556	.09600	10.417	.99542	1.0046	10.465	31
30	.09585	.09629	10.385	.99540	1.0046	10.433	30
31	.09614	.09658	10.354	.99537	1.0047	10.402	29
32	.09642	.09688	10.322	.99534	1.0047	10.371	28
33	.09671	.09717	10.291	.99531	1.0047	10.340	27
34	.09700	.09746	10.260	.99528	1.0047	10.309	26
35	.09729	.09776	10.229	.99526	1.0048	10.278	25
36	.09758	.09805	10.199	.99523	1.0048	10.248	24
37	.09787	.09834	10.168	.99520	1.0048	10.217	23
38	.09816	.09864	10.138	.99517	1.0049	10.187	22
39	.09845	.09893	10.108	.99514	1.0049	10.157	21
40	.09874	.09923	10.078	.99511	1.0049	10.128	20
41	.09903	.09952	10.048	.99508	1.0049	10.098	19
42	.09932	.09981	10.019	.99506	1.0050	10.068	18
43	.09961	.10011	9.9893	.99503	1.0050	10.039	17
44	.09990	.10040	9.9601	.99500	1.0050	10.010	16
45	.10019	.10069	9.9310	.99497	1.0051	9.9812	15
46	.10048	.10099	9.9021	.99494	1.0051	9.9525	14
47	.10077	.10128	9.8734	.99491	1.0051	9.9239	13
48	.10106	.10158	9.8448	.99488	1.0051	9.8955	12
49	.10135	.10187	9.8164	.99485	1.0052	9.8672	11
50	.10164	.10216	9.7882	.99482	1.0052	9.8391	10
51	.10192	.10246	9.7601	.99479	1.0052	9.8112	9
52	.10221	.10275	9.7322	.99476	1.0053	9.7834	8
53	.10250	.10305	9.7044	.99473	1.0053	9.7558	7
54	.10279	.10334	9.6768	.99470	1.0053	9.7283	6
55	.10308	.10363	9.6493	.99467	1.0054	9.7010	5
56	.10337	.10393	9.6220	.99464	1.0054	9.6739	4
57	.10366	.10422	9.5949	.99461	1.0054	9.6469	3
58	.10395	.10452	9.5679	.99458	1.0054	9.6200	2
59	.10424	.10481	9.5411	.99455	1.0055	9.5933	1
60	.10453	.10510	9.5144	.99452	1.0055	9.5668	0
′	Cos	Cot	Tan	Sin	Csc	Sec	′

NATURAL TRIGONOMETRIC FUNCTIONS TO FIVE PLACES (continued)

6° (186°)   (353°) **173°**   **7° (187°)**   (352°) **172°**

′	Sin	Tan	Cot	Cos	Sec	Csc	′		′	Sin	Tan	Cot	Cos	Sec	Csc	′
0	.10453	.10510	9.5144	.99452	1.0055	9.5668	60		0	.12187	.12278	8.1443	.99255	1.0075	8.2055	60
1	.10482	.10540	9.4878	.99449	1.0055	9.5404	59		1	.12216	.12308	8.1248	.99251	1.0075	8.1861	59
2	.10511	.10569	9.4614	.99446	1.0056	9.5141	58		2	.12245	.12338	8.1054	.99248	1.0076	8.1668	58
3	.10540	.10599	9.4352	.99443	1.0056	9.4880	57		3	.12274	.12367	8.0860	.99244	1.0076	8.1476	57
4	.10569	.10628	9.4090	.99440	1.0056	9.4620	56		4	.12302	.12397	8.0667	.99240	1.0077	8.1285	56
5	.10597	.10657	9.3831	.99437	1.0057	9.4362	55		5	.12331	.12426	8.0476	.99237	1.0077	8.1095	55
6	.10626	.10687	9.3572	.99434	1.0057	9.4105	54		6	.12360	.12456	8.0285	.99233	1.0077	8.0905	54
7	.10655	.10716	9.3315	.99431	1.0057	9.3850	53		7	.12389	.12485	8.0095	.99230	1.0078	8.0717	53
8	.10684	.10746	9.3060	.99428	1.0058	9.3596	52		8	.12418	.12515	7.9906	.99226	1.0078	8.0529	52
9	.10713	.10775	9.2806	.99424	1.0058	9.3343	51		9	.12447	.12544	7.9718	.99222	1.0078	8.0342	51
10	.10742	.10805	9.2553	.99421	1.0058	9.3092	50		10	.12476	.12574	7.9530	.99219	1.0079	8.0156	50
11	.10771	.10834	9.2302	.99418	1.0059	9.2842	49		11	.12504	.12603	7.9344	.99215	1.0079	7.9971	49
12	.10800	.10863	9.2052	.99415	1.0059	9.2593	48		12	.12533	.12633	7.9158	.99211	1.0079	7.9787	48
13	.10829	.10893	9.1803	.99412	1.0059	9.2346	47		13	.12562	.12662	7.8973	.99208	1.0080	7.9604	47
14	.10858	.10922	9.1555	.99409	1.0059	9.2100	46		14	.12591	.12692	7.8789	.99204	1.0080	7.9422	46
15	.10887	.10952	9.1309	.99406	1.0060	9.1855	45		15	.12620	.12722	7.8606	.99200	1.0081	7.9240	45
16	.10916	.10981	9.1065	.99402	1.0060	9.1612	44		16	.12649	.12751	7.8424	.99197	1.0081	7.9059	44
17	.10945	.11011	9.0821	.99399	1.0060	9.1370	43		17	.12678	.12781	7.8243	.99193	1.0081	7.8879	43
18	.10973	.11040	9.0579	.99396	1.0061	9.1129	42		18	.12706	.12810	7.8062	.99189	1.0082	7.8700	42
19	.11002	.11070	9.0338	.99393	1.0061	9.0890	41		19	.12735	.12840	7.7882	.99186	1.0082	7.8522	41
20	.11031	.11099	9.0098	.99390	1.0061	9.0652	40		20	.12764	.12869	7.7704	.99182	1.0082	7.8344	40
21	.11060	.11128	8.9860	.99386	1.0062	9.0415	39		21	.12793	.12899	7.7525	.99178	1.0083	7.8168	39
22	.11089	.11158	8.9623	.99383	1.0062	9.0179	38		22	.12822	.12929	7.7348	.99175	1.0083	7.7992	38
23	.11118	.11187	8.9387	.99380	1.0062	8.9944	37		23	.12851	.12958	7.7171	.99171	1.0084	7.7817	37
24	.11147	.11217	8.9152	.99377	1.0063	8.9711	36		24	.12880	.12988	7.6996	.99167	1.0084	7.7642	36
25	.11176	.11246	8.8919	.99374	1.0063	8.9479	35		25	.12908	.13017	7.6821	.99163	1.0084	7.7469	35
26	.11205	.11276	8.8686	.99370	1.0063	8.9248	34		26	.12937	.13047	7.6647	.99160	1.0085	7.7296	34
27	.11234	.11305	8.8455	.99367	1.0064	8.9019	33		27	.12966	.13076	7.6473	.99156	1.0085	7.7124	33
28	.11263	.11335	8.8225	.99364	1.0064	8.8790	32		28	.12995	.13106	7.6301	.99152	1.0086	7.6953	32
29	.11291	.11364	8.7996	.99360	1.0064	8.8563	31		29	.13024	.13136	7.6129	.99148	1.0086	7.6783	31
30	.11320	.11394	8.7769	.99357	1.0065	8.8337	30		30	.13053	.13165	7.5958	.99144	1.0086	7.6613	30
31	.11349	.11423	8.7542	.99354	1.0065	8.8112	29		31	.13081	.13195	7.5787	.99141	1.0087	7.6444	29
32	.11378	.11452	8.7317	.99351	1.0065	8.7888	28		32	.13110	.13224	7.5618	.99137	1.0087	7.6276	28
33	.11407	.11482	8.7093	.99347	1.0066	8.7665	27		33	.13139	.13254	7.5449	.99133	1.0087	7.6109	27
34	.11436	.11511	8.6870	.99344	1.0066	8.7444	26		34	.13168	.13284	7.5281	.99129	1.0088	7.5942	26
35	.11465	.11541	8.6648	.99341	1.0066	8.7223	25		35	.13197	.13313	7.5113	.99125	1.0088	7.5776	25
36	.11494	.11570	8.6427	.99337	1.0067	8.7004	24		36	.13226	.13343	7.4947	.99122	1.0089	7.5611	24
37	.11523	.11600	8.6208	.99334	1.0067	8.6786	23		37	.13254	.13372	7.4781	.99118	1.0089	7.5446	23
38	.11552	.11629	8.5989	.99331	1.0067	8.6569	22		38	.13283	.13402	7.4615	.99114	1.0089	7.5282	22
39	.11580	.11659	8.5772	.99327	1.0068	8.6353	21		39	.13312	.13432	7.4451	.99110	1.0090	7.5119	21
40	.11609	.11688	8.5555	.99324	1.0068	8.6138	20		40	.13341	.13461	7.4287	.99106	1.0090	7.4957	20
41	.11638	.11718	8.5340	.99320	1.0068	8.5924	19		41	.13370	.13491	7.4124	.99102	1.0091	7.4795	19
42	.11667	.11747	8.5126	.99317	1.0069	8.5711	18		42	.13399	.13521	7.3962	.99098	1.0091	7.4635	18
43	.11696	.11777	8.4913	.99314	1.0069	8.5500	17		43	.13427	.13550	7.3800	.99094	1.0091	7.4474	17
44	.11725	.11806	8.4701	.99310	1.0069	8.5289	16		44	.13456	.13580	7.3639	.99091	1.0092	7.4315	16
45	.11754	.11836	8.4490	.99307	1.0070	8.5079	15		45	.13485	.13609	7.3479	.99087	1.0092	7.4156	15
46	.11783	.11865	8.4280	.99303	1.0070	8.4871	14		46	.13514	.13639	7.3319	.99083	1.0093	7.3998	14
47	.11812	.11895	8.4071	.99300	1.0070	8.4663	13		47	.13543	.13669	7.3160	.99079	1.0093	7.3840	13
48	.11840	.11924	8.3863	.99297	1.0071	8.4457	12		48	.13572	.13698	7.3002	.99075	1.0093	7.3684	12
49	.11869	.11954	8.3656	.99293	1.0071	8.4251	11		49	.13600	.13728	7.2844	.99071	1.0094	7.3527	11
50	.11898	.11983	8.3450	.99290	1.0072	8.4047	10		50	.13629	.13758	7.2687	.99067	1.0094	7.3372	10
51	.11927	.12013	8.3245	.99286	1.0072	8.3843	9		51	.13658	.13787	7.2531	.99063	1.0095	7.3217	9
52	.11956	.12042	8.3041	.99283	1.0072	8.3641	8		52	.13687	.13817	7.2375	.99059	1.0095	7.3063	8
53	.11985	.12072	8.2838	.99279	1.0073	8.3439	7		53	.13716	.13846	7.2220	.99055	1.0095	7.2909	7
54	.12014	.12101	8.2636	.99276	1.0073	8.3238	6		54	.13744	.13876	7.2066	.99051	1.0096	7.2757	6
55	.12043	.12131	8.2434	.99272	1.0073	8.3039	5		55	.13773	.13906	7.1912	.99047	1.0096	7.2604	5
56	.12071	.12160	8.2234	.99269	1.0074	8.2840	4		56	.13802	.13935	7.1759	.99043	1.0097	7.2453	4
57	.12100	.12190	8.2035	.99265	1.0074	8.2642	3		57	.13831	.13965	7.1607	.99039	1.0097	7.2302	3
58	.12129	.12219	8.1837	.99262	1.0074	8.2446	2		58	.13860	.13995	7.1455	.99035	1.0097	7.2152	2
59	.12158	.12249	8.1640	.99258	1.0075	8.2250	1		59	.13889	.14024	7.1304	.99031	1.0098	7.2002	1
60	.12187	.12278	8.1443	.99255	1.0075	8.2055	0		60	.13917	.14054	7.1154	.99027	1.0098	7.1853	0
′	Cos	Cot	Tan	Sin	Csc	Sec	′		′	Cos	Cot	Tan	Sin	Csc	Sec	′

96° (276°)   (263°) **83°**   **97° (277°)**   (262°) **82°**

NATURAL TRIGONOMETRIC FUNCTIONS TO FIVE PLACES (continued)

8° (188°)   (351°) 171°

′	Sin	Tan	Cot	Cos	Sec	Csc	′
0	.13917	.14054	7.1154	.99027	1.0098	7.1853	60
1	.13946	.14084	7.1004	.99023	1.0099	7.1705	59
2	.13975	.14113	7.0855	.99019	1.0099	7.1557	58
3	.14004	.14143	7.0706	.99015	1.0100	7.1410	57
4	.14033	.14173	7.0558	.99011	1.0100	7.1263	56
5	.14061	.14202	7.0410	.99006	1.0100	7.1117	55
6	.14090	.14232	7.0264	.99002	1.0101	7.0972	54
7	.14119	.14262	7.0117	.98998	1.0101	7.0827	53
8	.14148	.14291	6.9972	.98994	1.0102	7.0683	52
9	.14177	.14321	6.9827	.98990	1.0102	7.0539	51
10	.14205	.14351	6.9682	.98986	1.0102	7.0396	50
11	.14234	.14381	6.9538	.98982	1.0103	7.0254	49
12	.14263	.14410	6.9395	.98978	1.0103	7.0112	48
13	.14292	.14440	6.9252	.98973	1.0104	6.9971	47
14	.14320	.14470	6.9110	.98969	1.0104	6.9830	46
15	.14349	.14499	6.8969	.98965	1.0105	6.9690	45
16	.14378	.14529	6.8828	.98961	1.0105	6.9550	44
17	.14407	.14559	6.8687	.98957	1.0105	6.9411	43
18	.14436	.14588	6.8548	.98953	1.0106	6.9273	42
19	.14464	.14618	6.8408	.98948	1.0106	6.9135	41
20	.14493	.14648	6.8269	.98944	1.0107	6.8998	40
21	.14522	.14678	6.8131	.98940	1.0107	6.8861	39
22	.14551	.14707	6.7994	.98936	1.0108	6.8725	38
23	.14580	.14737	6.7856	.98931	1.0108	6.8589	37
24	.14608	.14767	6.7720	.98927	1.0108	6.8454	36
25	.14637	.14796	6.7584	.98923	1.0109	6.8320	35
26	.14666	.14826	6.7448	.98919	1.0109	6.8186	34
27	.14695	.14856	6.7313	.98914	1.0110	6.8052	33
28	.14723	.14886	6.7179	.98910	1.0110	6.7919	32
29	.14752	.14915	6.7045	.98906	1.0111	6.7787	31
30	.14781	.14945	6.6912	.98902	1.0111	6.7655	30
31	.14810	.14975	6.6779	.98897	1.0112	6.7523	29
32	.14838	.15005	6.6646	.98893	1.0112	6.7392	28
33	.14867	.15034	6.6514	.98889	1.0112	6.7262	27
34	.14896	.15064	6.6383	.98884	1.0113	6.7132	26
35	.14925	.15094	6.6252	.98880	1.0113	6.7003	25
36	.14954	.15124	6.6122	.98876	1.0114	6.6874	24
37	.14982	.15153	6.5992	.98871	1.0114	6.6745	23
38	.15011	.15183	6.5863	.98867	1.0115	6.6618	22
39	.15040	.15213	6.5734	.98863	1.0115	6.6490	21
40	.15069	.15243	6.5606	.98858	1.0116	6.6363	20
41	.15097	.15272	6.5478	.98854	1.0116	6.6237	19
42	.15126	.15302	6.5350	.98849	1.0116	6.6111	18
43	.15155	.15332	6.5223	.98845	1.0117	6.5986	17
44	.15184	.15362	6.5097	.98841	1.0117	6.5861	16
45	.15212	.15391	6.4971	.98836	1.0118	6.5736	15
46	.15241	.15421	6.4846	.98832	1.0118	6.5612	14
47	.15270	.15451	6.4721	.98827	1.0119	6.5489	13
48	.15299	.15481	6.4596	.98823	1.0119	6.5366	12
49	.15327	.15511	6.4472	.98818	1.0120	6.5243	11
50	.15356	.15540	6.4348	.98814	1.0120	6.5121	10
51	.15385	.15570	6.4225	.98809	1.0120	6.4999	9
52	.15414	.15600	6.4103	.98805	1.0121	6.4878	8
53	.15442	.15630	6.3980	.98800	1.0121	6.4757	7
54	.15471	.15660	6.3859	.98796	1.0122	6.4637	6
55	.15500	.15689	6.3737	.98791	1.0122	6.4517	5
56	.15529	.15719	6.3617	.98787	1.0123	6.4398	4
57	.15557	.15749	6.3496	.98782	1.0123	6.4279	3
58	.15586	.15779	6.3376	.98778	1.0124	6.4160	2
59	.15615	.15809	6.3257	.98773	1.0124	6.4042	1
60	.15643	.15838	6.3138	.98769	1.0125	6.3925	0
′	Cos	Cot	Tan	Sin	Csc	Sec	′

98° (278°)   (261°) 81°

9° (189°)   (350°) 170°

′	Sin	Tan	Cot	Cos	Sec	Csc	′
0	.15643	.15838	6.3138	.98769	1.0125	6.3925	60
1	.15672	.15868	6.3019	.98764	1.0125	6.3807	59
2	.15701	.15898	6.2901	.98760	1.0126	6.3691	58
3	.15730	.15928	6.2783	.98755	1.0126	6.3574	57
4	.15758	.15958	6.2666	.98751	1.0127	6.3458	56
5	.15787	.15988	6.2549	.98746	1.0127	6.3343	55
6	.15816	.16017	6.2432	.98741	1.0127	6.3228	54
7	.15845	.16047	6.2316	.98737	1.0128	6.3113	53
8	.15873	.16077	6.2200	.98732	1.0128	6.2999	52
9	.15902	.16107	6.2085	.98728	1.0129	6.2885	51
10	.15931	.16137	6.1970	.98723	1.0129	6.2772	50
11	.15959	.16167	6.1856	.98718	1.0130	6.2659	49
12	.15988	.16196	6.1742	.98714	1.0130	6.2546	48
13	.16017	.16226	6.1628	.98709	1.0131	6.2434	47
14	.16046	.16256	6.1515	.98704	1.0131	6.2323	46
15	.16074	.16286	6.1402	.98700	1.0132	6.2211	45
16	.16103	.16316	6.1290	.98695	1.0132	6.2100	44
17	.16132	.16346	6.1178	.98690	1.0133	6.1990	43
18	.16160	.16376	6.1066	.98686	1.0133	6.1880	42
19	.16189	.16405	6.0955	.98681	1.0134	6.1770	41
20	.16218	.16435	6.0844	.98676	1.0134	6.1661	40
21	.16246	.16465	6.0734	.98671	1.0135	6.1552	39
22	.16275	.16495	6.0624	.98667	1.0135	6.1443	38
23	.16304	.16525	6.0514	.98662	1.0136	6.1335	37
24	.16333	.16555	6.0405	.98657	1.0136	6.1227	36
25	.16361	.16585	6.0296	.98652	1.0137	6.1120	35
26	.16390	.16615	6.0188	.98648	1.0137	6.1013	34
27	.16419	.16645	6.0080	.98643	1.0138	6.0906	33
28	.16447	.16674	5.9972	.98638	1.0138	6.0800	32
29	.16476	.16704	5.9865	.98633	1.0139	6.0694	31
30	.16505	.16734	5.9758	.98629	1.0139	6.0589	30
31	.16533	.16764	5.9651	.98624	1.0140	6.0483	29
32	.16562	.16794	5.9545	.98619	1.0140	6.0379	28
33	.16591	.16824	5.9439	.98614	1.0141	6.0274	27
34	.16620	.16854	5.9333	.98609	1.0141	6.0170	26
35	.16648	.16884	5.9228	.98604	1.0142	6.0067	25
36	.16677	.16914	5.9124	.98600	1.0142	5.9963	24
37	.16706	.16944	5.9019	.98595	1.0143	5.9860	23
38	.16734	.16974	5.8915	.98590	1.0143	5.9758	22
39	.16763	.17004	5.8811	.98585	1.0144	5.9656	21
40	.16792	.17033	5.8708	.98580	1.0144	5.9554	20
41	.16820	.17063	5.8605	.98575	1.0145	5.9452	19
42	.16849	.17093	5.8502	.98570	1.0145	5.9351	18
43	.16878	.17123	5.8400	.98565	1.0146	5.9250	17
44	.16906	.17153	5.8298	.98561	1.0146	5.9150	16
45	.16935	.17183	5.8197	.98556	1.0147	5.9049	15
46	.16964	.17213	5.8095	.98551	1.0147	5.8950	14
47	.16992	.17243	5.7994	.98546	1.0148	5.8850	13
48	.17021	.17273	5.7894	.98541	1.0148	5.8751	12
49	.17050	.17303	5.7794	.98536	1.0149	5.8652	11
50	.17078	.17333	5.7694	.98531	1.0149	5.8554	10
51	.17107	.17363	5.7594	.98526	1.0150	5.8456	9
52	.17136	.17393	5.7495	.98521	1.0150	5.8358	8
53	.17164	.17423	5.7396	.98516	1.0151	5.8261	7
54	.17193	.17453	5.7297	.98511	1.0151	5.8164	6
55	.17222	.17483	5.7199	.98506	1.0152	5.8067	5
56	.17250	.17513	5.7101	.98501	1.0152	5.7970	4
57	.17279	.17543	5.7004	.98496	1.0153	5.7874	3
58	.17308	.17573	5.6906	.98491	1.0153	5.7778	2
59	.17336	.17603	5.6809	.98486	1.0154	5.7683	1
60	.17365	.17633	5.6713	.98481	1.0154	5.7588	0
′	Cos	Cot	Tan	Sin	Csc	Sec	′

99° (279°)   (260°) 80°

NATURAL TRIGONOMETRIC FUNCTIONS TO FIVE PLACES (continued)

10° (190°) (349°) 169°

′	Sin	Tan	Cot	Cos	Sec	Csc	′
0	.17365	.17633	5.6713	.98481	1.0154	5.7588	60
1	.17393	.17663	5.6617	.98476	1.0155	5.7493	59
2	.17422	.17693	5.6521	.98471	1.0155	5.7398	58
3	.17451	.17723	5.6425	.98466	1.0156	5.7304	57
4	.17479	.17753	5.6329	.98461	1.0156	5.7210	56
5	.17508	.17783	5.6234	.98455	1.0157	5.7117	55
6	.17537	.17813	5.6140	.98450	1.0157	5.7023	54
7	.17565	.17843	5.6045	.98445	1.0158	5.6930	53
8	.17594	.17873	5.5951	.98440	1.0158	5.6838	52
9	.17623	.17903	5.5857	.98435	1.0159	5.6745	51
10	.17651	.17933	5.5764	.98430	1.0160	5.6653	50
11	.17680	.17963	5.5671	.98425	1.0160	5.6562	49
12	.17708	.17993	5.5578	.98420	1.0161	5.6470	48
13	.17737	.18023	5.5485	.98414	1.0161	5.6379	47
14	.17766	.18053	5.5393	.98409	1.0162	5.6288	46
15	.17794	.18083	5.5301	.98404	1.0162	5.6198	45
16	.17823	.18113	5.5209	.98399	1.0163	5.6107	44
17	.17852	.18143	5.5118	.98394	1.0163	5.6017	43
18	.17880	.18173	5.5026	.98389	1.0164	5.5928	42
19	.17909	.18203	5.4936	.98383	1.0164	5.5838	41
20	.17937	.18233	5.4845	.98378	1.0165	5.5749	40
21	.17966	.18263	5.4755	.98373	1.0165	5.5660	39
22	.17995	.18293	5.4665	.98368	1.0166	5.5572	38
23	.18023	.18323	5.4575	.98362	1.0166	5.5484	37
24	.18052	.18353	5.4486	.98357	1.0167	5.5396	36
25	.18081	.18384	5.4397	.98352	1.0168	5.5308	35
26	.18109	.18414	5.4308	.98347	1.0168	5.5221	34
27	.18138	.18444	5.4219	.98341	1.0169	5.5134	33
28	.18166	.18474	5.4131	.98336	1.0169	5.5047	32
29	.18195	.18504	5.4043	.98331	1.0170	5.4960	31
30	.18224	.18534	5.3955	.98325	1.0170	5.4874	30
31	.18252	.18564	5.3868	.98320	1.0171	5.4788	29
32	.18281	.18594	5.3781	.98315	1.0171	5.4702	28
33	.18309	.18624	5.3694	.98310	1.0172	5.4617	27
34	.18338	.18654	5.3607	.98304	1.0173	5.4532	26
35	.18367	.18684	5.3521	.98299	1.0173	5.4447	25
36	.18395	.18714	5.3435	.98294	1.0174	5.4362	24
37	.18424	.18745	5.3349	.98288	1.0174	5.4278	23
38	.18452	.18775	5.3263	.98283	1.0175	5.4194	22
39	.18481	.18805	5.3178	.98277	1.0175	5.4110	21
40	.18509	.18835	5.3093	.98272	1.0176	5.4026	20
41	.18538	.18865	5.3008	.98267	1.0176	5.3943	19
42	.18567	.18895	5.2924	.98261	1.0177	5.3860	18
43	.18595	.18925	5.2839	.98256	1.0178	5.3777	17
44	.18624	.18955	5.2755	.98250	1.0178	5.3695	16
45	.18652	.18986	5.2672	.98245	1.0179	5.3612	15
46	.18681	.19016	5.2588	.98240	1.0179	5.3530	14
47	.18710	.19046	5.2505	.98234	1.0180	5.3449	13
48	.18738	.19076	5.2422	.98229	1.0180	5.3367	12
49	.18767	.19106	5.2339	.98223	1.0181	5.3286	11
50	.18795	.19136	5.2257	.98218	1.0181	5.3205	10
51	.18824	.19166	5.2174	.98212	1.0182	5.3124	9
52	.18852	.19197	5.2092	.98207	1.0183	5.3044	8
53	.18881	.19227	5.2011	.98201	1.0183	5.2963	7
54	.18910	.19257	5.1929	.98196	1.0184	5.2883	6
55	.18938	.19287	5.1848	.98190	1.0184	5.2804	5
56	.18967	.19317	5.1767	.98185	1.0185	5.2724	4
57	.18995	.19347	5.1686	.98179	1.0185	5.2645	3
58	.19024	.19378	5.1606	.98174	1.0186	5.2566	2
59	.19052	.19408	5.1526	.98168	1.0187	5.2487	1
60	.19081	.19438	5.1446	.98163	1.0187	5.2408	0
′	Cos	Cot	Tan	Sin	Csc	Sec	′

100° (280°) (259°) 79°

11° (191°) (348°) 168°

′	Sin	Tan	Cot	Cos	Sec	Csc	′
0	.19081	.19438	5.1446	.98163	1.0187	5.2408	60
1	.19109	.19468	5.1366	.98157	1.0188	5.2330	59
2	.19138	.19498	5.1286	.98152	1.0188	5.2252	58
3	.19167	.19529	5.1207	.98146	1.0189	5.2174	57
4	.19195	.19559	5.1128	.98140	1.0189	5.2097	56
5	.19224	.19589	5.1049	.98135	1.0190	5.2019	55
6	.19252	.19619	5.0970	.98129	1.0191	5.1942	54
7	.19281	.19649	5.0892	.98124	1.0191	5.1865	53
8	.19309	.19680	5.0814	.98118	1.0192	5.1789	52
9	.19338	.19710	5.0736	.98112	1.0192	5.1712	51
10	.19366	.19740	5.0658	.98107	1.0193	5.1636	50
11	.19395	.19770	5.0581	.98101	1.0194	5.1560	49
12	.19423	.19801	5.0504	.98096	1.0194	5.1484	48
13	.19452	.19831	5.0427	.98090	1.0195	5.1409	47
14	.19481	.19861	5.0350	.98084	1.0195	5.1333	46
15	.19509	.19891	5.0273	.98079	1.0196	5.1258	45
16	.19538	.19921	5.0197	.98073	1.0197	5.1183	44
17	.19566	.19952	5.0121	.98067	1.0197	5.1109	43
18	.19595	.19982	5.0045	.98061	1.0198	5.1034	42
19	.19623	.20012	4.9969	.98056	1.0198	5.0960	41
20	.19652	.20042	4.9894	.98050	1.0199	5.0886	40
21	.19680	.20073	4.9819	.98044	1.0199	5.0813	39
22	.19709	.20103	4.9744	.98039	1.0200	5.0739	38
23	.19737	.20133	4.9669	.98033	1.0201	5.0666	37
24	.19766	.20164	4.9594	.98027	1.0201	5.0593	36
25	.19794	.20194	4.9520	.98021	1.0202	5.0520	35
26	.19823	.20224	4.9446	.98016	1.0202	5.0447	34
27	.19851	.20254	4.9372	.98010	1.0203	5.0375	33
28	.19880	.20285	4.9298	.98004	1.0204	5.0302	32
29	.19908	.20315	4.9225	.97998	1.0204	5.0230	31
30	.19937	.20345	4.9152	.97992	1.0205	5.0159	30
31	.19965	.20376	4.9078	.97987	1.0205	5.0087	29
32	.19994	.20406	4.9006	.97981	1.0206	5.0016	28
33	.20022	.20436	4.8933	.97975	1.0207	4.9944	27
34	.20051	.20466	4.8860	.97969	1.0207	4.9873	26
35	.20079	.20497	4.8788	.97963	1.0208	4.9803	25
36	.20108	.20527	4.8716	.97958	1.0209	4.9732	24
37	.20136	.20557	4.8644	.97952	1.0209	4.9662	23
38	.20165	.20588	4.8573	.97946	1.0210	4.9591	22
39	.20193	.20618	4.8501	.97940	1.0210	4.9521	21
40	.20222	.20648	4.8430	.97934	1.0211	4.9452	20
41	.20250	.20679	4.8359	.97928	1.0212	4.9382	19
42	.20279	.20709	4.8288	.97922	1.0212	4.9313	18
43	.20307	.20739	4.8218	.97916	1.0213	4.9244	17
44	.20336	.20770	4.8147	.97910	1.0213	4.9175	16
45	.20364	.20800	4.8077	.97905	1.0214	4.9106	15
46	.20393	.20830	4.8007	.97899	1.0215	4.9037	14
47	.20421	.20861	4.7937	.97893	1.0215	4.8969	13
48	.20450	.20891	4.7867	.97887	1.0216	4.8901	12
49	.20478	.20921	4.7798	.97881	1.0217	4.8833	11
50	.20507	.20952	4.7729	.97875	1.0217	4.8765	10
51	.20535	.20982	4.7659	.97869	1.0218	4.8697	9
52	.20563	.21013	4.7591	.97863	1.0218	4.8630	8
53	.20592	.21043	4.7522	.97857	1.0219	4.8563	7
54	.20620	.21073	4.7453	.97851	1.0220	4.8496	6
55	.20649	.21104	4.7385	.97845	1.0220	4.8429	5
56	.20677	.21134	4.7317	.97839	1.0221	4.8362	4
57	.20706	.21164	4.7249	.97833	1.0222	4.8296	3
58	.20734	.21195	4.7181	.97827	1.0222	4.8229	2
59	.20763	.21225	4.7114	.97821	1.0223	4.8163	1
60	.20791	.21256	4.7046	.97815	1.0223	4.8097	0
′	Cos	Cot	Tan	Sin	Csc	Sec	′

101° (281°) (258°) 78°

NATURAL TRIGONOMETRIC FUNCTIONS TO FIVE PLACES (continued)

12° (192°)   (347°) 167°

′	Sin	Tan	Cot	Cos	Sec	Csc	′
0	.20791	.21256	4.7046	.97815	1.0223	4.8097	60
1	.20820	.21286	4.6979	.97809	1.0224	4.8032	59
2	.20848	.21316	4.6912	.97803	1.0225	4.7966	58
3	.20877	.21347	4.6845	.97797	1.0225	4.7901	57
4	.20905	.21377	4.6779	.97791	1.0226	4.7836	56
5	.20933	.21408	4.6712	.97784	1.0227	4.7771	55
6	.20962	.21438	4.6646	.97778	1.0227	4.7706	54
7	.20990	.21469	4.6580	.97772	1.0228	4.7641	53
8	.21019	.21499	4.6514	.97766	1.0228	4.7577	52
9	.21047	.21529	4.6448	.97760	1.0229	4.7512	51
10	.21076	.21560	4.6382	.97754	1.0230	4.7448	50
11	.21104	.21590	4.6317	.97748	1.0230	4.7384	49
12	.21132	.21621	4.6252	.97742	1.0231	4.7321	48
13	.21161	.21651	4.6187	.97735	1.0232	4.7257	47
14	.21189	.21682	4.6122	.97729	1.0232	4.7194	46
15	.21218	.21712	4.6057	.97723	1.0233	4.7130	45
16	.21246	.21743	4.5993	.97717	1.0234	4.7067	44
17	.21275	.21773	4.5928	.97711	1.0234	4.7004	43
18	.21303	.21804	4.5864	.97705	1.0235	4.6942	42
19	.21331	.21834	4.5800	.97698	1.0236	4.6879	41
20	.21360	.21864	4.5736	.97692	1.0236	4.6817	40
21	.21388	.21895	4.5673	.97686	1.0237	4.6755	39
22	.21417	.21925	4.5609	.97680	1.0238	4.6693	38
23	.21445	.21956	4.5546	.97673	1.0238	4.6631	37
24	.21474	.21986	4.5483	.97667	1.0239	4.6569	36
25	.21502	.22017	4.5420	.97661	1.0240	4.6507	35
26	.21530	.22047	4.5357	.97655	1.0240	4.6446	34
27	.21559	.22078	4.5294	.97648	1.0241	4.6385	33
28	.21587	.22108	4.5232	.97642	1.0241	4.6324	32
29	.21616	.22139	4.5169	.97636	1.0242	4.6263	31
30	.21644	.22169	4.5107	.97630	1.0243	4.6202	30
31	.21672	.22200	4.5045	.97623	1.0243	4.6142	29
32	.21701	.22231	4.4983	.97617	1.0244	4.6081	28
33	.21729	.22261	4.4922	.97611	1.0245	4.6021	27
34	.21758	.22292	4.4860	.97604	1.0245	4.5961	26
35	.21786	.22322	4.4799	.97598	1.0246	4.5901	25
36	.21814	.22353	4.4737	.97592	1.0247	4.5841	24
37	.21843	.22383	4.4676	.97585	1.0247	4.5782	23
38	.21871	.22414	4.4615	.97579	1.0248	4.5722	22
39	.21899	.22444	4.4555	.97573	1.0249	4.5663	21
40	.21928	.22475	4.4494	.97566	1.0249	4.5604	20
41	.21956	.22505	4.4434	.97560	1.0250	4.5545	19
42	.21985	.22536	4.4373	.97553	1.0251	4.5486	18
43	.22013	.22567	4.4313	.97547	1.0251	4.5428	17
44	.22041	.22597	4.4253	.97541	1.0252	4.5369	16
45	.22070	.22628	4.4194	.97534	1.0253	4.5311	15
46	.22098	.22658	4.4134	.97528	1.0253	4.5253	14
47	.22126	.22689	4.4075	.97521	1.0254	4.5195	13
48	.22155	.22719	4.4015	.97515	1.0255	4.5137	12
49	.22183	.22750	4.3956	.97508	1.0256	4.5079	11
50	.22212	.22781	4.3897	.97502	1.0256	4.5022	10
51	.22240	.22811	4.3838	.97496	1.0257	4.4964	9
52	.22268	.22842	4.3779	.97489	1.0258	4.4907	8
53	.22297	.22872	4.3721	.97483	1.0258	4.4850	7
54	.22325	.22903	4.3662	.97476	1.0259	4.4793	6
55	.22353	.22934	4.3604	.97470	1.0260	4.4736	5
56	.22382	.22964	4.3546	.97463	1.0260	4.4679	4
57	.22410	.22995	4.3488	.97457	1.0261	4.4623	3
58	.22438	.23026	4.3430	.97450	1.0262	4.4566	2
59	.22467	.23056	4.3372	.97444	1.0262	4.4510	1
60	.22495	.23087	4.3315	.97437	1.0263	4.4454	0
′	Cos	Cot	Tan	Sin	Csc	Sec	′

13° (193°)   (346°) 166°

′	Sin	Tan	Cot	Cos	Sec	Csc	′
0	.22495	.23087	4.3315	.97437	1.0263	4.4454	60
1	.22523	.23117	4.3257	.97430	1.0264	4.4398	59
2	.22552	.23148	4.3200	.97424	1.0264	4.4342	58
3	.22580	.23179	4.3143	.97417	1.0265	4.4287	57
4	.22608	.23209	4.3086	.97411	1.0266	4.4231	56
5	.22637	.23240	4.3029	.97404	1.0266	4.4176	55
6	.22665	.23271	4.2972	.97398	1.0267	4.4121	54
7	.22693	.23301	4.2916	.97391	1.0268	4.4066	53
8	.22722	.23332	4.2859	.97384	1.0269	4.4011	52
9	.22750	.23363	4.2803	.97378	1.0269	4.3956	51
10	.22778	.23393	4.2747	.97371	1.0270	4.3901	50
11	.22807	.23424	4.2691	.97365	1.0271	4.3847	49
12	.22835	.23455	4.2635	.97358	1.0271	4.3792	48
13	.22863	.23485	4.2580	.97351	1.0272	4.3738	47
14	.22892	.23516	4.2524	.97345	1.0273	4.3684	46
15	.22920	.23547	4.2468	.97338	1.0273	4.3630	45
16	.22948	.23578	4.2413	.97331	1.0274	4.3576	44
17	.22977	.23608	4.2358	.97325	1.0275	4.3522	43
18	.23005	.23639	4.2303	.97318	1.0276	4.3469	42
19	.23033	.23670	4.2248	.97311	1.0276	4.3415	41
20	.23062	.23700	4.2193	.97304	1.0277	4.3362	40
21	.23090	.23731	4.2139	.97298	1.0278	4.3309	39
22	.23118	.23762	4.2084	.97291	1.0278	4.3256	38
23	.23146	.23793	4.2030	.97284	1.0279	4.3203	37
24	.23175	.23823	4.1976	.97278	1.0280	4.3150	36
25	.23203	.23854	4.1922	.97271	1.0281	4.3098	35
26	.23231	.23885	4.1868	.97264	1.0281	4.3045	34
27	.23260	.23916	4.1814	.97257	1.0282	4.2993	33
28	.23288	.23946	4.1760	.97251	1.0283	4.2941	32
29	.23316	.23977	4.1706	.97244	1.0283	4.2889	31
30	.23345	.24008	4.1653	.97237	1.0284	4.2837	30
31	.23373	.24039	4.1600	.97230	1.0285	4.2785	29
32	.23401	.24069	4.1547	.97223	1.0286	4.2733	28
33	.23429	.24100	4.1493	.97217	1.0286	4.2681	27
34	.23458	.24131	4.1441	.97210	1.0287	4.2630	26
35	.23486	.24162	4.1388	.97203	1.0288	4.2579	25
36	.23514	.24193	4.1335	.97196	1.0288	4.2527	24
37	.23542	.24223	4.1282	.97189	1.0289	4.2476	23
38	.23571	.24254	4.1230	.97182	1.0290	4.2425	22
39	.23599	.24285	4.1178	.97176	1.0291	4.2375	21
40	.23627	.24316	4.1126	.97169	1.0291	4.2324	20
41	.23656	.24347	4.1074	.97162	1.0292	4.2273	19
42	.23684	.24377	4.1022	.97155	1.0293	4.2223	18
43	.23712	.24408	4.0970	.97148	1.0294	4.2173	17
44	.23740	.24439	4.0918	.97141	1.0294	4.2122	16
45	.23769	.24470	4.0867	.97134	1.0295	4.2072	15
46	.23797	.24501	4.0815	.97127	1.0296	4.2022	14
47	.23825	.24532	4.0764	.97120	1.0297	4.1973	13
48	.23853	.24562	4.0713	.97113	1.0297	4.1923	12
49	.23882	.24593	4.0662	.97106	1.0298	4.1873	11
50	.23910	.24624	4.0611	.97100	1.0299	4.1824	10
51	.23938	.24655	4.0560	.97093	1.0299	4.1774	9
52	.23966	.24686	4.0509	.97086	1.0300	4.1725	8
53	.23995	.24717	4.0459	.97079	1.0301	4.1676	7
54	.24023	.24747	4.0408	.97072	1.0302	4.1627	6
55	.24051	.24778	4.0358	.97065	1.0302	4.1578	5
56	.24079	.24809	4.0308	.97058	1.0303	4.1529	4
57	.24108	.24840	4.0257	.97051	1.0304	4.1481	3
58	.24136	.24871	4.0207	.97044	1.0305	4.1432	2
59	.24164	.24902	4.0158	.97037	1.0305	4.1384	1
60	.24192	.24933	4.0108	.97030	1.0306	4.1336	0
′	Cos	Cot	Tan	Sin	Csc	Sec	′

NATURAL TRIGONOMETRIC FUNCTIONS TO FIVE PLACES (continued)

14° (194°)   (345°) 165°   **15° (195°)**   (344°) 164°

′	Sin	Tan	Cot	Cos	Sec	Csc	′		′	Sin	Tan	Cot	Cos	Sec	Csc	′
0	.24192	.24933	4.0108	.97030	1.0306	4.1336	60		0	.25882	.26795	3.7321	.96593	1.0353	3.8637	60
1	.24220	.24964	4.0058	.97023	1.0307	4.1287	59		1	.25910	.26826	3.7277	.96585	1.0354	3.8595	59
2	.24249	.24995	4.0009	.97015	1.0308	4.1239	58		2	.25938	.26857	3.7234	.96578	1.0354	3.8553	58
3	.24277	.25026	3.9959	.97008	1.0308	4.1191	57		3	.25966	.26888	3.7191	.96570	1.0355	3.8512	57
4	.24305	.25056	3.9910	.97001	1.0309	4.1144	56		4	.25994	.26920	3.7148	.96562	1.0356	3.8470	56
5	.24333	.25087	3.9861	.96994	1.0310	4.1096	55		5	.26022	.26951	3.7105	.96555	1.0357	3.8428	55
6	.24362	.25118	3.9812	.96987	1.0311	4.1048	54		6	.26050	.26982	3.7062	.96547	1.0358	3.8387	54
7	.24390	.25149	3.9763	.96980	1.0311	4.1001	53		7	.26079	.27013	3.7019	.96540	1.0358	3.8346	53
8	.24418	.25180	3.9714	.96973	1.0312	4.0954	52		8	.26107	.27044	3.6976	.96532	1.0359	3.8304	52
9	.24446	.25211	3.9665	.96966	1.0313	4.0906	51		9	.26135	.27076	3.6933	.96524	1.0360	3.8263	51
10	.24474	.25242	3.9617	.96959	1.0314	4.0859	50		10	.26163	.27107	3.6891	.96517	1.0361	3.8222	50
11	.24503	.25273	3.9568	.96952	1.0314	4.0812	49		11	.26191	.27138	3.6848	.96509	1.0362	3.8181	49
12	.24531	.25304	3.9520	.96945	1.0315	4.0765	48		12	.26219	.27169	3.6806	.96502	1.0363	3.8140	48
13	.24559	.25335	3.9471	.96937	1.0316	4.0718	47		13	.26247	.27201	3.6764	.96494	1.0363	3.8100	47
14	.24587	.25366	3.9423	.96930	1.0317	4.0672	46		14	.26275	.27232	3.6722	.96486	1.0364	3.8059	46
15	.24615	.25397	3.9375	.96923	1.0317	4.0625	45		15	.26303	.27263	3.6680	.96479	1.0365	3.8018	45
16	.24644	.25428	3.9327	.96916	1.0318	4.0579	44		16	.26331	.27294	3.6638	.96471	1.0366	3.7978	44
17	.24672	.25459	3.9279	.96909	1.0319	4.0532	43		17	.26359	.27326	3.6596	.96463	1.0367	3.7937	43
18	.24700	.25490	3.9232	.96902	1.0320	4.0486	42		18	.26387	.27357	3.6554	.96456	1.0367	3.7897	42
19	.24728	.25521	3.9184	.96894	1.0321	4.0440	41		19	.26415	.27388	3.6512	.96448	1.0368	3.7857	41
20	.24756	.25552	3.9136	.96887	1.0321	4.0394	40		20	.26443	.27419	3.6470	.96440	1.0369	3.7817	40
21	.24784	.25583	3.9089	.96880	1.0322	4.0348	39		21	.26471	.27451	3.6429	.96433	1.0370	3.7777	39
22	.24813	.25614	3.9042	.96873	1.0323	4.0302	38		22	.26500	.27482	3.6387	.96425	1.0371	3.7737	38
23	.24841	.25645	3.8995	.96866	1.0324	4.0256	37		23	.26528	.27513	3.6346	.96417	1.0372	3.7697	37
24	.24869	.25676	3.8947	.96858	1.0324	4.0211	36		24	.26556	.27545	3.6305	.96410	1.0372	3.7657	36
25	.24897	.25707	3.8900	.96851	1.0325	4.0165	35		25	.26584	.27576	3.6264	.96402	1.0373	3.7617	35
26	.24925	.25738	3.8854	.96844	1.0326	4.0120	34		26	.26612	.27607	3.6222	.96394	1.0374	3.7577	34
27	.24954	.25769	3.8807	.96837	1.0327	4.0075	33		27	.26640	.27638	3.6181	.96386	1.0375	3.7538	33
28	.24982	.25800	3.8760	.96829	1.0327	4.0029	32		28	.26668	.27670	3.6140	.96379	1.0376	3.7498	32
29	.25010	.25831	3.8714	.96822	1.0328	3.9984	31		29	.26696	.27701	3.6100	.96371	1.0377	3.7459	31
30	.25038	.25862	3.8667	.96815	1.0329	3.9939	30		30	.26724	.27732	3.6059	.96363	1.0377	3.7420	30
31	.25066	.25893	3.8621	.96807	1.0330	3.9894	29		31	.26752	.27764	3.6018	.96355	1.0378	3.7381	29
32	.25094	.25924	3.8575	.96800	1.0331	3.9850	28		32	.26780	.27795	3.5978	.96347	1.0379	3.7341	28
33	.25122	.25955	3.8528	.96793	1.0331	3.9805	27		33	.26808	.27826	3.5937	.96340	1.0380	3.7302	27
34	.25151	.25986	3.8482	.96786	1.0332	3.9760	26		34	.26836	.27858	3.5897	.96332	1.0381	3.7263	26
35	.25179	.26017	3.8436	.96778	1.0333	3.9716	25		35	.26864	.27889	3.5856	.96324	1.0382	3.7225	25
36	.25207	.26048	3.8391	.96771	1.0334	3.9672	24		36	.26892	.27921	3.5816	.96316	1.0382	3.7186	24
37	.25235	.26079	3.8345	.96764	1.0334	3.9627	23		37	.26920	.27952	3.5776	.96308	1.0383	3.7147	23
38	.25263	.26110	3.8299	.96756	1.0335	3.9583	22		38	.26948	.27983	3.5736	.96301	1.0384	3.7108	22
39	.25291	.26141	3.8254	.96749	1.0336	3.9539	21		39	.26976	.28015	3.5696	.96293	1.0385	3.7070	21
40	.25320	.26172	3.8208	.96742	1.0337	3.9495	20		40	.27004	.28046	3.5656	.96285	1.0386	3.7032	20
41	.25348	.26203	3.8163	.96734	1.0338	3.9451	19		41	.27032	.28077	3.5616	.96277	1.0387	3.6993	19
42	.25376	.26235	3.8118	.96727	1.0338	3.9408	18		42	.27060	.28109	3.5576	.96269	1.0388	3.6955	18
43	.25404	.26266	3.8073	.96719	1.0339	3.9364	17		43	.27088	.28140	3.5536	.96261	1.0388	3.6917	17
44	.25432	.26297	3.8028	.96712	1.0340	3.9320	16		44	.27116	.28172	3.5497	.96253	1.0389	3.6879	16
45	.25460	.26328	3.7983	.96705	1.0341	3.9277	15		45	.27144	.28203	3.5457	.96246	1.0390	3.6840	15
46	.25488	.26359	3.7938	.96697	1.0342	3.9234	14		46	.27172	.28234	3.5418	.96238	1.0391	3.6803	14
47	.25516	.26390	3.7893	.96690	1.0342	3.9190	13		47	.27200	.28266	3.5379	.96230	1.0392	3.6765	13
48	.25545	.26421	3.7848	.96682	1.0343	3.9147	12		48	.27228	.28297	3.5339	.96222	1.0393	3.6727	12
49	.25573	.26452	3.7804	.96675	1.0344	3.9104	11		49	.27256	.28329	3.5300	.96214	1.0394	3.6689	11
50	.25601	.26483	3.7760	.96667	1.0345	3.9061	10		50	.27284	.28360	3.5261	.96206	1.0394	3.6652	10
51	.25629	.26515	3.7715	.96660	1.0346	3.9018	9		51	.27312	.28391	3.5222	.96198	1.0395	3.6614	9
52	.25657	.26546	3.7671	.96653	1.0346	3.8976	8		52	.27340	.28423	3.5183	.96190	1.0396	3.6575	8
53	.25685	.26577	3.7627	.96645	1.0347	3.8933	7		53	.27368	.28454	3.5144	.96182	1.0397	3.6539	7
54	.25713	.26608	3.7583	.96638	1.0348	3.8890	6		54	.27396	.28486	3.5105	.96174	1.0398	3.6502	6
55	.25741	.26639	3.7539	.96630	1.0349	3.8848	5		55	.27424	.28517	3.5067	.96166	1.0399	3.6465	5
56	.25769	.26670	3.7495	.96623	1.0350	3.8806	4		56	.27452	.28549	3.5028	.96158	1.0400	3.6427	4
57	.25798	.26701	3.7451	.96615	1.0350	3.8763	3		57	.27480	.28580	3.4989	.96150	1.0400	3.6390	3
58	.25826	.26733	3.7408	.96608	1.0351	3.8721	2		58	.27508	.28612	3.4951	.96142	1.0401	3.6353	2
59	.25854	.26764	3.7364	.96600	1.0352	3.8679	1		59	.27536	.28643	3.4912	.96134	1.0402	3.6316	1
60	.25882	.26795	3.7321	.96593	1.0353	3.8637	0		60	.27564	.28675	3.4874	.96126	1.0403	3.6280	0
′	Cos	Cot	Tan	Sin	Csc	Sec	′		′	Cos	Cot	Tan	Sin	Csc	Sec	′

104° (284°)   (255°) 75°   **105° (285°)**   (254°) 74°

NATURAL TRIGONOMETRIC FUNCTIONS TO FIVE PLACES (continued)

16° (196°)   (343°) 163°

′	Sin	Tan	Cot	Cos	Sec	Csc	′
0	.27564	.28675	3.4874	.96126	1.0403	3.6280	60
1	.27592	.28706	3.4836	.96118	1.0404	3.6243	59
2	.27620	.28738	3.4798	.96110	1.0405	3.6206	58
3	.27648	.28769	3.4760	.96102	1.0406	3.6169	57
4	.27676	.28801	3.4722	.96094	1.0406	3.6133	56
5	.27704	.28832	3.4684	.96086	1.0407	3.6097	55
6	.27731	.28864	3.4646	.96078	1.0408	3.6060	54
7	.27759	.28895	3.4608	.96070	1.0409	3.6024	53
8	.27787	.28927	3.4570	.96062	1.0410	3.5988	52
9	.27815	.28958	3.4533	.96054	1.0411	3.5951	51
10	.27843	.28990	3.4495	.96046	1.0412	3.5915	50
11	.27871	.29021	3.4458	.96037	1.0413	3.5879	49
12	.27899	.29053	3.4420	.96029	1.0413	3.5843	48
13	.27927	.29084	3.4383	.96021	1.0414	3.5808	47
14	.27955	.29116	3.4346	.96013	1.0415	3.5772	46
15	.27983	.29147	3.4308	.96005	1.0416	3.5736	45
16	.28011	.29179	3.4271	.95997	1.0417	3.5700	44
17	.28039	.29210	3.4234	.95989	1.0418	3.5665	43
18	.28067	.29242	3.4197	.95981	1.0419	3.5629	42
19	.28095	.29274	3.4160	.95972	1.0420	3.5594	41
20	.28123	.29305	3.4124	.95964	1.0421	3.5559	40
21	.28150	.29337	3.4087	.95956	1.0421	3.5523	39
22	.28178	.29368	3.4050	.95948	1.0422	3.5488	38
23	.28206	.29400	3.4014	.95940	1.0423	3.5453	37
24	.28234	.29432	3.3977	.95931	1.0424	3.5418	36
25	.28262	.29463	3.3941	.95923	1.0425	3.5383	35
26	.28290	.29495	3.3904	.95915	1.0426	3.5348	34
27	.28318	.29526	3.3868	.95907	1.0427	3.5313	33
28	.28346	.29558	3.3832	.95898	1.0428	3.5279	32
29	.28374	.29590	3.3796	.95890	1.0429	3.5244	31
30	.28402	.29621	3.3759	.95882	1.0429	3.5209	30
31	.28429	.29653	3.3723	.95874	1.0430	3.5175	29
32	.28457	.29685	3.3687	.95865	1.0431	3.5140	28
33	.28485	.29716	3.3652	.95857	1.0432	3.5106	27
34	.28513	.29748	3.3616	.95849	1.0433	3.5072	26
35	.28541	.29780	3.3580	.95841	1.0434	3.5037	25
36	.28569	.29811	3.3544	.95832	1.0435	3.5003	24
37	.28597	.29843	3.3509	.95824	1.0436	3.4969	23
38	.28625	.29875	3.3473	.95816	1.0437	3.4935	22
39	.28652	.29906	3.3438	.95807	1.0438	3.4901	21
40	.28680	.29938	3.3402	.95799	1.0439	3.4867	20
41	.28708	.29970	3.3367	.95791	1.0439	3.4833	19
42	.28736	.30001	3.3332	.95782	1.0440	3.4799	18
43	.28764	.30033	3.3297	.95774	1.0441	3.4766	17
44	.28792	.30065	3.3261	.95766	1.0442	3.4732	16
45	.28820	.30097	3.3226	.95757	1.0443	3.4699	15
46	.28847	.30128	3.3191	.95749	1.0444	3.4665	14
47	.28875	.30160	3.3156	.95740	1.0445	3.4632	13
48	.28903	.30192	3.3122	.95732	1.0446	3.4598	12
49	.28931	.30224	3.3087	.95724	1.0447	3.4565	11
50	.28959	.30255	3.3052	.95715	1.0448	3.4532	10
51	.28987	.30287	3.3017	.95707	1.0449	3.4499	9
52	.29015	.30319	3.2983	.95698	1.0450	3.4465	8
53	.29042	.30351	3.2948	.95690	1.0450	3.4432	7
54	.29070	.30382	3.2914	.95681	1.0451	3.4399	6
55	.29098	.30414	3.2879	.95673	1.0452	3.4367	5
56	.29126	.30446	3.2845	.95664	1.0453	3.4334	4
57	.29154	.30478	3.2811	.95656	1.0454	3.4301	3
58	.29182	.30509	3.2777	.95647	1.0455	3.4268	2
59	.29209	.30541	3.2743	.95639	1.0456	3.4236	1
60	.29237	.30573	3.2709	.95630	1.0457	3.4203	0
′	Cos	Cot	Tan	Sin	Csc	Sec	′

106° (286°)   (253°) 73°

17° (197°)   (342°) 162°

′	Sin	Tan	Cot	Cos	Sec	Csc	′
0	.29237	.30573	3.2709	.95630	1.0457	3.4203	60
1	.29265	.30605	3.2675	.95622	1.0458	3.4171	59
2	.29293	.30637	3.2641	.95613	1.0459	3.4138	58
3	.29321	.30669	3.2607	.95605	1.0460	3.4106	57
4	.29348	.30700	3.2573	.95596	1.0461	3.4073	56
5	.29376	.30732	3.2539	.95588	1.0462	3.4041	55
6	.29404	.30764	3.2506	.95579	1.0463	3.4009	54
7	.29432	.30796	3.2472	.95571	1.0463	3.3977	53
8	.29460	.30828	3.2438	.95562	1.0464	3.3945	52
9	.29487	.30860	3.2405	.95554	1.0465	3.3913	51
10	.29515	.30891	3.2371	.95545	1.0466	3.3881	50
11	.29543	.30923	3.2338	.95536	1.0467	3.3849	49
12	.29571	.30955	3.2305	.95528	1.0468	3.3817	48
13	.29599	.30987	3.2272	.95519	1.0469	3.3785	47
14	.29626	.31019	3.2238	.95511	1.0470	3.3754	46
15	.29654	.31051	3.2205	.95502	1.0471	3.3722	45
16	.29682	.31083	3.2172	.95493	1.0472	3.3691	44
17	.29710	.31115	3.2139	.95485	1.0473	3.3659	43
18	.29737	.31147	3.2106	.95476	1.0474	3.3628	42
19	.29765	.31178	3.2073	.95467	1.0475	3.3596	41
20	.29793	.31210	3.2041	.95459	1.0476	3.3565	40
21	.29821	.31242	3.2008	.95450	1.0477	3.3534	39
22	.29849	.31274	3.1975	.95441	1.0478	3.3502	38
23	.29876	.31306	3.1943	.95433	1.0479	3.3471	37
24	.29904	.31338	3.1910	.95424	1.0480	3.3440	36
25	.29932	.31370	3.1878	.95415	1.0480	3.3409	35
26	.29960	.31402	3.1845	.95407	1.0481	3.3378	34
27	.29987	.31434	3.1813	.95398	1.0482	3.3347	33
28	.30015	.31466	3.1780	.95389	1.0483	3.3317	32
29	.30043	.31498	3.1748	.95380	1.0484	3.3286	31
30	.30071	.31530	3.1716	.95372	1.0485	3.3255	30
31	.30098	.31562	3.1684	.95363	1.0486	3.3224	29
32	.30126	.31594	3.1652	.95354	1.0487	3.3194	28
33	.30154	.31626	3.1620	.95345	1.0488	3.3163	27
34	.30182	.31658	3.1588	.95337	1.0489	3.3133	26
35	.30209	.31690	3.1556	.95328	1.0490	3.3102	25
36	.30237	.31722	3.1524	.95319	1.0491	3.3072	24
37	.30265	.31754	3.1492	.95310	1.0492	3.3042	23
38	.30292	.31786	3.1460	.95301	1.0493	3.3012	22
39	.30320	.31818	3.1429	.95293	1.0494	3.2981	21
40	.30348	.31850	3.1397	.95284	1.0495	3.2951	20
41	.30376	.31882	3.1366	.95275	1.0496	3.2921	19
42	.30403	.31914	3.1334	.95266	1.0497	3.2891	18
43	.30431	.31946	3.1303	.95257	1.0498	3.2861	17
44	.30459	.31978	3.1271	.95248	1.0499	3.2831	16
45	.30486	.32010	3.1240	.95240	1.0500	3.2801	15
46	.30514	.32042	3.1209	.95231	1.0501	3.2772	14
47	.30542	.32074	3.1178	.95222	1.0502	3.2742	13
48	.30570	.32106	3.1146	.95213	1.0503	3.2712	12
49	.30597	.32139	3.1115	.95204	1.0504	3.2683	11
50	.30625	.32171	3.1084	.95195	1.0505	3.2653	10
51	.30653	.32203	3.1053	.95186	1.0506	3.2624	9
52	.30680	.32235	3.1022	.95177	1.0507	3.2594	8
53	.30708	.32267	3.0991	.95168	1.0508	3.2565	7
54	.30736	.32299	3.0961	.95159	1.0509	3.2535	6
55	.30763	.32331	3.0930	.95150	1.0510	3.2506	5
56	.30791	.32363	3.0899	.95142	1.0511	3.2477	4
57	.30819	.32396	3.0868	.95133	1.0512	3.2448	3
58	.30846	.32428	3.0838	.95124	1.0513	3.2419	2
59	.30874	.32460	3.0807	.95115	1.0514	3.2390	1
60	.30902	.32492	3.0777	.95106	1.0515	3.2361	0
′	Cos	Cot	Tan	Sin	Csc	Sec	′

107° (287°)   (252°) 72°

NATURAL TRIGONOMETRIC FUNCTIONS TO FIVE PLACES (continued)

18° (198°) (341°) **161°**

′	Sin	Tan	Cot	Cos	Sec	Csc	′
0	.30902	.32492	3.0777	.95106	1.0515	3.2361	60
1	.30929	.32524	3.0746	.95097	1.0516	3.2332	59
2	.30957	.32556	3.0716	.95088	1.0517	3.2303	58
3	.30985	.32588	3.0686	.95079	1.0518	3.2274	57
4	.31012	.32621	3.0655	.95070	1.0519	3.2245	56
5	.31040	.32653	3.0625	.95061	1.0520	3.2217	55
6	.31068	.32685	3.0595	.95052	1.0521	3.2188	54
7	.31095	.32717	3.0565	.95043	1.0522	3.2159	53
8	.31123	.32749	3.0535	.95033	1.0523	3.2131	52
9	.31151	.32782	3.0505	.95024	1.0524	3.2102	51
10	.31178	.32814	3.0475	.95015	1.0525	3.2074	50
11	.31206	.32846	3.0445	.95006	1.0526	3.2045	49
12	.31233	.32878	3.0415	.94997	1.0527	3.2017	48
13	.31261	.32911	3.0385	.94988	1.0528	3.1989	47
14	.31289	.32943	3.0356	.94979	1.0529	3.1960	46
15	.31316	.32975	3.0326	.94970	1.0530	3.1932	45
16	.31344	.33007	3.0296	.94961	1.0531	3.1904	44
17	.31372	.33040	3.0267	.94952	1.0532	3.1876	43
18	.31399	.33072	3.0237	.94943	1.0533	3.1848	42
19	.31427	.33104	3.0208	.94933	1.0534	3.1820	41
20	.31454	.33136	3.0178	.94924	1.0535	3.1792	40
21	.31482	.33169	3.0149	.94915	1.0536	3.1764	39
22	.31510	.33201	3.0120	.94906	1.0537	3.1736	38
23	.31537	.33233	3.0090	.94897	1.0538	3.1708	37
24	.31565	.33266	3.0061	.94888	1.0539	3.1681	36
25	.31593	.33298	3.0032	.94878	1.0540	3.1653	35
26	.31620	.33330	3.0003	.94869	1.0541	3.1625	34
27	.31648	.33363	2.9974	.94860	1.0542	3.1598	33
28	.31675	.33395	2.9945	.94851	1.0543	3.1570	32
29	.31703	.33427	2.9916	.94842	1.0544	3.1543	31
30	.31730	.33460	2.9887	.94832	1.0545	3.1515	30
31	.31758	.33492	2.9858	.94823	1.0546	3.1488	29
32	.31786	.33524	2.9829	.94814	1.0547	3.1461	28
33	.31813	.33557	2.9800	.94805	1.0548	3.1433	27
34	.31841	.33589	2.9772	.94795	1.0549	3.1406	26
35	.31868	.33621	2.9743	.94786	1.0550	3.1379	25
36	.31896	.33654	2.9714	.94777	1.0551	3.1352	24
37	.31923	.33686	2.9686	.94768	1.0552	3.1325	23
38	.31951	.33718	2.9657	.94758	1.0553	3.1298	22
39	.31979	.33751	2.9629	.94749	1.0554	3.1271	21
40	.32006	.33783	2.9600	.94740	1.0555	3.1244	20
41	.32034	.33816	2.9572	.94730	1.0556	3.1217	19
42	.32061	.33848	2.9544	.94721	1.0557	3.1190	18
43	.32089	.33881	2.9515	.94712	1.0558	3.1163	17
44	.32116	.33913	2.9487	.94702	1.0559	3.1137	16
45	.32144	.33945	2.9459	.94693	1.0560	3.1110	15
46	.32171	.33978	2.9431	.94684	1.0561	3.1083	14
47	.32199	.34010	2.9403	.94674	1.0563	3.1057	13
48	.32227	.34043	2.9375	.94665	1.0564	3.1030	12
49	.32254	.34075	2.9347	.94656	1.0565	3.1004	11
50	.32282	.34108	2.9319	.94646	1.0566	3.0977	10
51	.32309	.34140	2.9291	.94637	1.0567	3.0951	9
52	.32337	.34173	2.9263	.94627	1.0568	3.0925	8
53	.32364	.34205	2.9235	.94618	1.0569	3.0898	7
54	.32392	.34238	2.9208	.94609	1.0570	3.0872	6
55	.32419	.34270	2.9180	.94599	1.0571	3.0846	5
56	.32447	.34303	2.9152	.94590	1.0572	3.0820	4
57	.32474	.34335	2.9125	.94580	1.0573	3.0794	3
58	.32502	.34368	2.9097	.94571	1.0574	3.0768	2
59	.32529	.34400	2.9070	.94561	1.0575	3.0742	1
60	.32557	.34433	2.9042	.94552	1.0576	3.0716	0
′	Cos	Cot	Tan	Sin	Csc	Sec	′

19° (190°) (340°) **160°**

′	Sin	Tan	Cot	Cos	Sec	Csc	′
0	.32557	.34433	2.9042	.94552	1.0576	3.0716	60
1	.32584	.34465	2.9015	.94542	1.0577	3.0690	59
2	.32612	.34498	2.8987	.94533	1.0578	3.0664	58
3	.32639	.34530	2.8960	.94523	1.0579	3.0638	57
4	.32667	.34563	2.8933	.94514	1.0580	3.0612	56
5	.32694	.34596	2.8905	.94504	1.0582	3.0586	55
6	.32722	.34628	2.8878	.94495	1.0583	3.0561	54
7	.32749	.34661	2.8851	.94485	1.0584	3.0535	53
8	.32777	.34693	2.8824	.94476	1.0585	3.0509	52
9	.32804	.34726	2.8797	.94466	1.0586	3.0484	51
10	.32832	.34758	2.8770	.94457	1.0587	3.0458	50
11	.32859	.34791	2.8743	.94447	1.0588	3.0433	49
12	.32887	.34824	2.8716	.94438	1.0589	3.0407	48
13	.32914	.34856	2.8689	.94428	1.0590	3.0382	47
14	.32942	.34889	2.8662	.94418	1.0591	3.0357	46
15	.32969	.34922	2.8636	.94409	1.0592	3.0331	45
16	.32997	.34954	2.8609	.94399	1.0593	3.0306	44
17	.33024	.34987	2.8582	.94390	1.0594	3.0281	43
18	.33051	.35020	2.8556	.94380	1.0595	3.0256	42
19	.33079	.35052	2.8529	.94370	1.0597	3.0231	41
20	.33106	.35085	2.8502	.94361	1.0598	3.0206	40
21	.33134	.35118	2.8476	.94351	1.0599	3.0181	39
22	.33161	.35150	2.8449	.94342	1.0600	3.0156	38
23	.33189	.35183	2.8423	.94332	1.0601	3.0131	37
24	.33216	.35216	2.8397	.94322	1.0602	3.0106	36
25	.33244	.35248	2.8370	.94313	1.0603	3.0081	35
26	.33271	.35281	2.8344	.94303	1.0604	3.0056	34
27	.33298	.35314	2.8318	.94293	1.0605	3.0031	33
28	.33326	.35346	2.8291	.94284	1.0606	3.0007	32
29	.33353	.35379	2.8265	.94274	1.0607	2.9982	31
30	.33381	.35412	2.8239	.94264	1.0608	2.9957	30
31	.33408	.35445	2.8213	.94254	1.0610	2.9933	29
32	.33436	.35477	2.8187	.94245	1.0611	2.9908	28
33	.33463	.35510	2.8161	.94235	1.0612	2.9884	27
34	.33490	.35543	2.8135	.94225	1.0613	2.9859	26
35	.33518	.35576	2.8109	.94215	1.0614	2.9835	25
36	.33545	.35608	2.8083	.94206	1.0615	2.9811	24
37	.33573	.35641	2.8057	.94196	1.0616	2.9786	23
38	.33600	.35674	2.8032	.94186	1.0617	2.9762	22
39	.33627	.35707	2.8006	.94176	1.0618	2.9738	21
40	.33655	.35740	2.7980	.94167	1.0619	2.9713	20
41	.33682	.35772	2.7955	.94157	1.0621	2.9689	19
42	.33710	.35805	2.7929	.94147	1.0622	2.9665	18
43	.33737	.35838	2.7903	.94137	1.0623	2.9641	17
44	.33764	.35871	2.7878	.94127	1.0624	2.9617	16
45	.33792	.35904	2.7852	.94118	1.0625	2.9593	15
46	.33819	.35937	2.7827	.94108	1.0626	2.9569	14
47	.33846	.35969	2.7801	.94098	1.0627	2.9545	13
48	.33874	.36002	2.7776	.94088	1.0628	2.9521	12
49	.33901	.36035	2.7751	.94078	1.0629	2.9498	11
50	.33929	.36068	2.7725	.94068	1.0631	2.9474	10
51	.33956	.36101	2.7700	.94058	1.0632	2.9450	9
52	.33983	.36134	2.7675	.94049	1.0633	2.9426	8
53	.34011	.36167	2.7650	.94039	1.0634	2.9403	7
54	.34038	.36199	2.7625	.94029	1.0635	2.9379	6
55	.34065	.36232	2.7600	.94019	1.0636	2.9355	5
56	.34093	.36265	2.7575	.94009	1.0637	2.9332	4
57	.34120	.36298	2.7550	.93999	1.0638	2.9308	3
58	.34147	.36331	2.7525	.93989	1.0640	2.9285	2
59	.34175	.36364	2.7500	.93979	1.0641	2.9261	1
60	.34202	.36397	2.7475	.93969	1.0642	2.9238	0
′	Cos	Cot	Tan	Sin	Csc	Sec	′

NATURAL TRIGONOMETRIC FUNCTIONS TO FIVE PLACES (continued)

20° (200°)   **(339°) 159°**

′	Sin	Tan	Cot	Cos	Sec	Csc	′
0	.34202	.36397	2.7475	.93969	1.0642	2.9238	60
1	.34229	.36430	2.7450	.93959	1.0643	2.9215	59
2	.34257	.36463	2.7425	.93949	1.0644	2.9191	58
3	.34284	.36496	2.7400	.93939	1.0645	2.9168	57
4	.34311	.36529	2.7376	.93929	1.0646	2.9145	56
5	.34339	.36562	2.7351	.93919	1.0647	2.9122	55
6	.34366	.36595	2.7326	.93909	1.0649	2.9099	54
7	.34393	.36628	2.7302	.93899	1.0650	2.9075	53
8	.34421	.36661	2.7277	.93889	1.0651	2.9052	52
9	.34448	.36694	2.7253	.93879	1.0652	2.9029	51
10	.34475	.36727	2.7228	.93869	1.0653	2.9006	50
11	.34503	.36760	2.7204	.93859	1.0654	2.8983	49
12	.34530	.36793	2.7179	.93849	1.0655	2.8960	48
13	.34557	.36826	2.7155	.93839	1.0657	2.8938	47
14	.34584	.36859	2.7130	.93829	1.0658	2.8915	46
15	.34612	.36892	2.7106	.93819	1.0659	2.8892	45
16	.34639	.36925	2.7082	.93809	1.0660	2.8869	44
17	.34666	.36958	2.7058	.93799	1.0661	2.8846	43
18	.34694	.36991	2.7034	.93789	1.0662	2.8824	42
19	.34721	.37024	2.7009	.93779	1.0663	2.8801	41
20	.34748	.37057	2.6985	.93769	1.0665	2.8779	40
21	.34775	.37090	2.6961	.93759	1.0666	2.8756	39
22	.34803	.37123	2.6937	.93748	1.0667	2.8733	38
23	.34830	.37157	2.6913	.93738	1.0668	2.8711	37
24	.34857	.37190	2.6889	.93728	1.0669	2.8688	36
25	.34884	.37223	2.6865	.93718	1.0670	2.8666	35
26	.34912	.37256	2.6841	.93708	1.0671	2.8644	34
27	.34939	.37289	2.6818	.93698	1.0673	2.8621	33
28	.34966	.37322	2.6794	.93688	1.0674	2.8599	32
29	.34993	.37355	2.6770	.93677	1.0675	2.8577	31
30	.35021	.37388	2.6746	.93667	1.0676	2.8555	30
31	.35048	.37422	2.6723	.93657	1.0677	2.8532	29
32	.35075	.37455	2.6699	.93647	1.0678	2.8510	28
33	.35102	.37488	2.6675	.93637	1.0680	2.8488	27
34	.35130	.37521	2.6652	.93626	1.0681	2.8466	26
35	.35157	.37554	2.6628	.93616	1.0682	2.8444	25
36	.35184	.37588	2.6605	.93606	1.0683	2.8422	24
37	.35211	.37621	2.6581	.93596	1.0684	2.8400	23
38	.35239	.37654	2.6558	.93585	1.0685	2.8378	22
39	.35266	.37687	2.6534	.93575	1.0687	2.8356	21
40	.35293	.37720	2.6511	.93565	1.0688	2.8334	20
41	.35320	.37754	2.6488	.93555	1.0689	2.8312	19
42	.35347	.37787	2.6464	.93544	1.0690	2.8291	18
43	.35375	.37820	2.6441	.93534	1.0691	2.8269	17
44	.35402	.37853	2.6418	.93524	1.0692	2.8247	16
45	.35429	.37887	2.6395	.93514	1.0694	2.8225	15
46	.35456	.37920	2.6371	.93503	1.0695	2.8204	14
47	.35484	.37953	2.6348	.93493	1.0696	2.8182	13
48	.35511	.37986	2.6325	.93483	1.0697	2.8161	12
49	.35538	.38020	2.6302	.93472	1.0698	2.8139	11
50	.35565	.38053	2.6279	.93462	1.0700	2.8117	10
51	.35592	.38086	2.6256	.93452	1.0701	2.8096	9
52	.35619	.38120	2.6233	.93441	1.0702	2.8075	8
53	.35647	.38153	2.6210	.93431	1.0703	2.8053	7
54	.35674	.38186	2.6187	.93420	1.0704	2.8032	6
55	.35701	.38220	2.6165	.93410	1.0705	2.8010	5
56	.35728	.38253	2.6142	.93400	1.0707	2.7989	4
57	.35755	.38286	2.6119	.93389	1.0708	2.7968	3
58	.35782	.38320	2.6096	.93379	1.0709	2.7947	2
59	.35810	.38353	2.6074	.93368	1.0710	2.7925	1
60	.35837	.38386	2.6051	.93358	1.0711	2.7904	0
′	Cos	Cot	Tan	Sin	Csc	Sec	′

110° (290°)   **(249°) 69°**

21° (201°)   **(338°) 158°**

′	Sin	Tan	Cot	Cos	Sec	Csc	′
0	.35837	.38386	2.6051	.93358	1.0711	2.7904	60
1	.35864	.38420	2.6028	.93348	1.0713	2.7883	59
2	.35891	.38453	2.6006	.93337	1.0714	2.7862	58
3	.35918	.38487	2.5983	.93327	1.0715	2.7841	57
4	.35945	.38520	2.5961	.93316	1.0716	2.7820	56
5	.35973	.38553	2.5938	.93306	1.0717	2.7799	55
6	.36000	.38587	2.5916	.93295	1.0719	2.7778	54
7	.36027	.38620	2.5893	.93285	1.0720	2.7757	53
8	.36054	.38654	2.5871	.93274	1.0721	2.7736	52
9	.36081	.38687	2.5848	.93264	1.0722	2.7715	51
10	.36108	.38721	2.5826	.93253	1.0723	2.7695	50
11	.36135	.38754	2.5804	.93243	1.0725	2.7674	49
12	.36162	.38787	2.5782	.93232	1.0726	2.7653	48
13	.36190	.38821	2.5759	.93222	1.0727	2.7632	47
14	.36217	.38854	2.5737	.93211	1.0728	2.7612	46
15	.36244	.38888	2.5715	.93201	1.0730	2.7591	45
16	.36271	.38921	2.5693	.93190	1.0731	2.7570	44
17	.36298	.38955	2.5671	.93180	1.0732	2.7550	43
18	.36325	.38988	2.5649	.93169	1.0733	2.7529	42
19	.36352	.39022	2.5627	.93159	1.0734	2.7509	41
20	.36379	.39055	2.5605	.93148	1.0736	2.7488	40
21	.36406	.39089	2.5583	.93137	1.0737	2.7468	39
22	.36434	.39122	2.5561	.93127	1.0738	2.7447	38
23	.36461	.39156	2.5539	.93116	1.0739	2.7427	37
24	.36488	.39190	2.5517	.93106	1.0740	2.7407	36
25	.36515	.39223	2.5495	.93095	1.0742	2.7386	35
26	.36542	.39257	2.5473	.93084	1.0743	2.7366	34
27	.36569	.39290	2.5452	.93074	1.0744	2.7346	33
28	.36596	.39324	2.5430	.93063	1.0745	2.7325	32
29	.36623	.39357	2.5408	.93052	1.0747	2.7305	31
30	.36650	.39391	2.5386	.93042	1.0748	2.7285	30
31	.36677	.39425	2.5365	.93031	1.0749	2.7265	29
32	.36704	.39458	2.5343	.93020	1.0750	2.7245	28
33	.36731	.39492	2.5322	.93010	1.0752	2.7225	27
34	.36758	.39526	2.5300	.92999	1.0753	2.7205	26
35	.36785	.39559	2.5279	.92988	1.0754	2.7185	25
36	.36812	.39593	2.5257	.92978	1.0755	2.7165	24
37	.36839	.39626	2.5236	.92967	1.0757	2.7145	23
38	.36867	.39660	2.5214	.92956	1.0758	2.7125	22
39	.36894	.39694	2.5193	.92945	1.0759	2.7105	21
40	.36921	.39727	2.5172	.92935	1.0760	2.7085	20
41	.36948	.39761	2.5150	.92924	1.0761	2.7065	19
42	.36975	.39795	2.5129	.92913	1.0763	2.7046	18
43	.37002	.39829	2.5108	.92902	1.0764	2.7026	17
44	.37029	.39862	2.5086	.92892	1.0765	2.7006	16
45	.37056	.39896	2.5065	.92881	1.0766	2.6986	15
46	.37083	.39930	2.5044	.92870	1.0768	2.6967	14
47	.37110	.39963	2.5023	.92859	1.0769	2.6947	13
48	.37137	.39997	2.5002	.92849	1.0770	2.6927	12
49	.37164	.40031	2.4981	.92838	1.0771	2.6908	11
50	.37191	.40065	2.4960	.92827	1.0773	2.6888	10
51	.37218	.40098	2.4939	.92816	1.0774	2.6869	9
52	.37245	.40132	2.4918	.92805	1.0775	2.6849	8
53	.37272	.40166	2.4897	.92794	1.0777	2.6830	7
54	.37299	.40200	2.4876	.92784	1.0778	2.6811	6
55	.37326	.40234	2.4855	.92773	1.0779	2.6791	5
56	.37353	.40267	2.4834	.92762	1.0780	2.6772	4
57	.37380	.40301	2.4813	.92751	1.0782	2.6752	3
58	.37407	.40335	2.4792	.92740	1.0783	2.6733	2
59	.37434	.40369	2.4772	.92729	1.0784	2.6714	1
60	.37461	.40403	2.4751	.92718	1.0785	2.6695	0
′	Cos	Cot	Tan	Sin	Csc	Sec	′

111° (291°)   **(248°) 68°**

NATURAL TRIGONOMETRIC FUNCTIONS TO FIVE PLACES (continued)

22° (202°)   (337°) 157°

′	Sin	Tan	Cot	Cos	Sec	Csc	′
0	.37461	.40403	2.4751	.92718	1.0785	2.6695	60
1	.37488	.40436	2.4730	.92707	1.0787	2.6675	59
2	.37515	.40470	2.4709	.92697	1.0788	2.6656	58
3	.37542	.40504	2.4689	.92686	1.0789	2.6637	57
4	.37569	.40538	2.4668	.92675	1.0790	2.6618	56
5	.37595	.40572	2.4648	.92664	1.0792	2.6599	55
6	.37622	.40606	2.4627	.92653	1.0793	2.6580	54
7	.37649	.40640	2.4606	.92642	1.0794	2.6561	53
8	.37676	.40674	2.4586	.92631	1.0796	2.6542	52
9	.37703	.40707	2.4566	.92620	1.0797	2.6523	51
10	.37730	.40741	2.4545	.92609	1.0798	2.6504	50
11	.37757	.40775	2.4525	.92598	1.0799	2.6485	49
12	.37784	.40809	2.4504	.92587	1.0801	2.6466	48
13	.37811	.40843	2.4484	.92576	1.0802	2.6447	47
14	.37838	.40877	2.4464	.92565	1.0803	2.6429	46
15	.37865	.40911	2.4443	.92554	1.0804	2.6410	45
16	.37892	.40945	2.4423	.92543	1.0806	2.6391	44
17	.37919	.40979	2.4403	.92532	1.0807	2.6372	43
18	.37946	.41013	2.4383	.92521	1.0808	2.6354	42
19	.37973	.41047	2.4362	.92510	1.0810	2.6335	41
20	.37999	.41081	2.4342	.92499	1.0811	2.6316	40
21	.38026	.41115	2.4322	.92488	1.0812	2.6298	39
22	.38053	.41149	2.4302	.92477	1.0814	2.6279	38
23	.38080	.41183	2.4282	.92466	1.0815	2.6260	37
24	.38107	.41217	2.4262	.92455	1.0816	2.6242	36
25	.38134	.41251	2.4242	.92444	1.0817	2.6223	35
26	.38161	.41285	2.4222	.92432	1.0819	2.6205	34
27	.38188	.41319	2.4202	.92421	1.0820	2.6186	33
28	.38215	.41353	2.4182	.92410	1.0821	2.6168	32
29	.38241	.41387	2.4162	.92399	1.0823	2.6150	31
30	.38268	.41421	2.4142	.92388	1.0824	2.6131	30
31	.38295	.41455	2.4122	.92377	1.0825	2.6113	29
32	.38322	.41490	2.4102	.92366	1.0827	2.6095	28
33	.38310	.41524	2.4083	.92355	1.0828	2.6076	27
34	.38376	.41558	2.4063	.92343	1.0829	2.6058	26
35	.38403	.41592	2.4043	.92332	1.0830	2.6040	25
36	.38430	.41626	2.4023	.92321	1.0832	2.6022	24
37	.38456	.41660	2.4004	.92310	1.0833	2.6003	23
38	.38483	.41694	2.3984	.92299	1.0834	2.5985	22
39	.38510	.41728	2.3964	.92287	1.0836	2.5967	21
40	.38537	.41763	2.3945	.92276	1.0837	2.5949	20
41	.38564	.41797	2.3925	.92265	1.0838	2.5931	19
42	.38591	.41831	2.3906	.92254	1.0840	2.5913	18
43	.38617	.41865	2.3886	.92243	1.0841	2.5895	17
44	.38644	.41899	2.3867	.92231	1.0842	2.5877	16
45	.38671	.41933	2.3847	.92220	1.0844	2.5859	15
46	.38698	.41968	2.3828	.92209	1.0845	2.5841	14
47	.38725	.42002	2.3808	.92198	1.0846	2.5823	13
48	.38752	.42036	2.3789	.92186	1.0848	2.5805	12
49	.38778	.42070	2.3770	.92175	1.0849	2.5788	11
50	.38805	.42105	2.3750	.92164	1.0850	2.5770	10
51	.38832	.42139	2.3731	.92152	1.0852	2.5752	9
52	.38859	.42173	2.3712	.92141	1.0853	2.5734	8
53	.38886	.42207	2.3693	.92130	1.0854	2.5716	7
54	.38912	.42242	2.3673	.92119	1.0856	2.5699	6
55	.38939	.42276	2.3654	.92107	1.0857	2.5681	5
56	.38966	.42310	2.3635	.92096	1.0858	2.5663	4
57	.38993	.42345	2.3616	.92085	1.0860	2.5646	3
58	.39020	.42379	2.3597	.92073	1.0861	2.5628	2
59	.39046	.42413	2.3578	.92062	1.0862	2.5611	1
60	.39073	.42447	2.3559	.92050	1.0864	2.5593	0
′	Cos	Cot	Tan	Sin	Csc	Sec	′

112° (292°)   (247°) 67°

23° (203°)   (336°) 156°

′	Sin	Tan	Cot	Cos	Sec	Csc	′
0	.39073	.42447	2.3559	.92050	1.0864	2.5593	60
1	.39100	.42482	2.3539	.92039	1.0865	2.5576	59
2	.39127	.42516	2.3520	.92028	1.0866	2.5558	58
3	.39153	.42551	2.3501	.92016	1.0868	2.5541	57
4	.39180	.42585	2.3483	.92005	1.0869	2.5523	56
5	.39207	.42619	2.3464	.91994	1.0870	2.5506	55
6	.39234	.42654	2.3445	.91982	1.0872	2.5488	54
7	.39260	.42688	2.3426	.91971	1.0873	2.5471	53
8	.39287	.42722	2.3407	.91959	1.0874	2.5454	52
9	.39314	.42757	2.3388	.91948	1.0876	2.5436	51
10	.39341	.42791	2.3369	.91936	1.0877	2.5419	50
11	.39367	.42826	2.3351	.91925	1.0878	2.5402	49
12	.39394	.42860	2.3332	.91914	1.0880	2.5384	48
13	.39421	.42894	2.3313	.91902	1.0881	2.5367	47
14	.39448	.42929	2.3294	.91891	1.0883	2.5350	46
15	.39474	.42963	2.3276	.91879	1.0884	2.5333	45
16	.39501	.42998	2.3257	.91868	1.0885	2.5316	44
17	.39528	.43032	2.3238	.91856	1.0887	2.5299	43
18	.39555	.43067	2.3220	.91845	1.0888	2.5282	42
19	.39581	.43101	2.3201	.91833	1.0889	2.5264	41
20	.39608	.43136	2.3183	.91822	1.0891	2.5247	40
21	.39635	.43170	2.3164	.91810	1.0892	2.5230	39
22	.39661	.43205	2.3146	.91799	1.0893	2.5213	38
23	.39688	.43239	2.3127	.91787	1.0895	2.5196	37
24	.39715	.43274	2.3109	.91775	1.0896	2.5180	36
25	.39741	.43308	2.3090	.91764	1.0898	2.5163	35
26	.39768	.43343	2.3072	.91752	1.0899	2.5146	34
27	.39795	.43378	2.3053	.91741	1.0900	2.5129	33
28	.39822	.43412	2.3035	.91729	1.0902	2.5112	32
29	.39848	.43447	2.3017	.91718	1.0903	2.5095	31
30	.39875	.43481	2.2998	.91706	1.0904	2.5078	30
31	.39902	.43516	2.2980	.91694	1.0906	2.5062	29
32	.39928	.43550	2.2962	.91683	1.0907	2.5045	28
33	.39955	.43585	2.2944	.91671	1.0909	2.5028	27
34	.39982	.43620	2.2925	.91660	1.0910	2.5012	26
35	.40008	.43654	2.2907	.91648	1.0911	2.4995	25
36	.40035	.43689	2.2889	.91636	1.0913	2.4978	24
37	.40062	.43724	2.2871	.91625	1.0914	2.4962	23
38	.40088	.43758	2.2853	.91613	1.0915	2.4945	22
39	.40115	.43793	2.2835	.91601	1.0917	2.4928	21
40	.40141	.43828	2.2817	.91590	1.0918	2.4912	20
41	.40168	.43862	2.2799	.91578	1.0920	2.4895	19
42	.40195	.43897	2.2781	.91566	1.0921	2.4879	18
43	.40221	.43932	2.2763	.91555	1.0922	2.4862	17
44	.40248	.43966	2.2745	.91543	1.0924	2.4846	16
45	.40275	.44001	2.2727	.91531	1.0925	2.4830	15
46	.40301	.44036	2.2709	.91519	1.0927	2.4813	14
47	.40328	.44071	2.2691	.91508	1.0928	2.4797	13
48	.40355	.44105	2.2673	.91496	1.0929	2.4780	12
49	.40381	.44140	2.2655	.91484	1.0931	2.4764	11
50	.40408	.44175	2.2637	.91472	1.0932	2.4748	10
51	.40434	.44210	2.2620	.91461	1.0934	2.4731	9
52	.40461	.44244	2.2602	.91449	1.0935	2.4715	8
53	.40488	.44279	2.2584	.91437	1.0936	2.4699	7
54	.40514	.44314	2.2566	.91425	1.0938	2.4683	6
55	.40541	.44349	2.2549	.91414	1.0939	2.4667	5
56	.40567	.44384	2.2531	.91402	1.0941	2.4650	4
57	.40594	.44418	2.2513	.91390	1.0942	2.4634	3
58	.40621	.44453	2.2496	.91378	1.0944	2.4618	2
59	.40647	.44488	2.2478	.91366	1.0945	2.4602	1
60	.40674	.44523	2.2460	.91355	1.0946	2.4586	0
′	Cos	Cot	Tan	Sin	Csc	Sec	′

113° (293°)   (246°) 66°

NATURAL TRIGONOMETRIC FUNCTIONS TO FIVE PLACES (continued)

24° (204°)   (335°) 155°   **25° (205°)**   (334°) 154°

′	Sin	Tan	Cot	Cos	Sec	Csc	′		′	Sin	Tan	Cot	Cos	Sec	Csc	′
0	.40674	.44523	2.2460	.91355	1.0946	2.4586	60		0	.42262	.46631	2.1445	.90631	1.1034	2.3662	60
1	.40700	.44558	2.2443	.91343	1.0948	2.4570	59		1	.42288	.46666	2.1429	.90618	1.1035	2.3647	59
2	.40727	.44593	2.2425	.91331	1.0949	2.4554	58		2	.42315	.46702	2.1413	.90606	1.1037	2.3633	58
3	.40753	.44627	2.2408	.91319	1.0951	2.4538	57		3	.42341	.46737	2.1396	.90594	1.1038	2.3618	57
4	.40780	.44662	2.2390	.91307	1.0952	2.4522	56		4	.42367	.46772	2.1380	.90582	1.1040	2.3603	56
5	.40806	.44697	2.2373	.91295	1.0953	2.4506	55		5	.42394	.46808	2.1364	.90569	1.1041	2.3588	55
6	.40833	.44732	2.2355	.91283	1.0955	2.4490	54		6	.42420	.46843	2.1348	.90557	1.1043	2.3574	54
7	.40860	.44767	2.2338	.91272	1.0956	2.4474	53		7	.42446	.46879	2.1332	.90545	1.1044	2.3559	53
8	.40886	.44802	2.2320	.91260	1.0958	2.4458	52		8	.42473	.46914	2.1315	.90532	1.1046	2.3545	52
9	.40913	.44837	2.2303	.91248	1.0959	2.4442	51		9	.42499	.46950	2.1299	.90520	1.1047	2.3530	51
10	.40939	.44872	2.2286	.91236	1.0961	2.4426	50		10	.42525	.46985	2.1283	.90507	1.1049	2.3515	50
11	.40966	.44907	2.2268	.91224	1.0962	2.4411	49		11	.42552	.47021	2.1267	.90495	1.1050	2.3501	49
12	.40992	.44942	2.2251	.91212	1.0963	2.4395	48		12	.42578	.47056	2.1251	.90483	1.1052	2.3486	48
13	.41019	.44977	2.2234	.91200	1.0965	2.4379	47		13	.42604	.47092	2.1235	.90470	1.1053	2.3472	47
14	.41045	.45012	2.2216	.91188	1.0966	2.4363	46		14	.42631	.47128	2.1219	.90458	1.1055	2.3457	46
15	.41072	.45047	2.2199	.91176	1.0968	2.4348	45		15	.42657	.47163	2.1203	.90446	1.1056	2.3443	45
16	.41098	.45082	2.2182	.91164	1.0969	2.4332	44		16	.42683	.47199	2.1187	.90433	1.1058	2.3428	44
17	.41125	.45117	2.2165	.91152	1.0971	2.4316	43		17	.42709	.47234	2.1171	.90421	1.1059	2.3414	43
18	.41151	.45152	2.2148	.91140	1.0972	2.4300	42		18	.42736	.47270	2.1155	.90408	1.1061	2.3400	42
19	.41178	.45187	2.2130	.91128	1.0974	2.4285	41		19	.42762	.47305	2.1139	.90396	1.1062	2.3385	41
20	.41204	.45222	2.2113	.91116	1.0975	2.4269	40		20	.42788	.47341	2.1123	.90383	1.1064	2.3371	40
21	.41231	.45257	2.2096	.91104	1.0976	2.4254	39		21	.42815	.47377	2.1107	.90371	1.1066	2.3356	39
22	.41257	.45292	2.2079	.91092	1.0978	2.4238	38		22	.42841	.47412	2.1092	.90358	1.1067	2.3342	38
23	.41284	.45327	2.2062	.91080	1.0979	2.4222	37		23	.42867	.47448	2.1076	.90346	1.1069	2.3328	37
24	.41310	.45362	2.2045	.91068	1.0981	2.4207	36		24	.42894	.47483	2.1060	.90334	1.1070	2.3314	36
25	.41337	.45397	2.2028	.91056	1.0982	2.4191	35		25	.42920	.47519	2.1044	.90321	1.1072	2.3299	35
26	.41363	.45432	2.2011	.91044	1.0984	2.4176	34		26	.42946	.47555	2.1028	.90309	1.1073	2.3285	34
27	.41390	.45467	2.1994	.91032	1.0985	2.4160	33		27	.42972	.47590	2.1013	.90296	1.1075	2.3271	33
28	.41416	.45502	2.1977	.91020	1.0987	2.4145	32		28	.42999	.47626	2.0997	.90284	1.1076	2.3257	32
29	.41443	.45538	2.1960	.91008	1.0988	2.4130	31		29	.43025	.47662	2.0981	.90271	1.1078	2.3242	31
30	.41469	.45573	2.1943	.90996	1.0989	2.4114	30		30	.43051	.47698	2.0965	.90259	1.1079	2.3228	30
31	.41496	.45608	2.1926	.90984	1.0991	2.4099	29		31	.43077	.47733	2.0950	.90246	1.1081	2.3214	29
32	.41522	.45643	2.1909	.90972	1.0992	2.4083	28		32	.43104	.47769	2.0934	.90233	1.1082	2.3200	28
33	.41549	.45678	2.1892	.90960	1.0994	2.4068	27		33	.43130	.47805	2.0918	.90221	1.1084	2.3186	27
34	.41575	.45713	2.1876	.90948	1.0995	2.4053	26		34	.43156	.47840	2.0903	.90208	1.1085	2.3172	26
35	.41602	.45748	2.1859	.90936	1.0997	2.4038	25		35	.43182	.47876	2.0887	.90196	1.1087	2.3158	25
36	.41628	.45784	2.1842	.90924	1.0998	2.4022	24		36	.43209	.47912	2.0872	.90183	1.1089	2.3144	24
37	.41655	.45819	2.1825	.90911	1.1000	2.4007	23		37	.43235	.47948	2.0856	.90171	1.1090	2.3130	23
38	.41681	.45854	2.1808	.90899	1.1001	2.3992	22		38	.43261	.47984	2.0840	.90158	1.1092	2.3115	22
39	.41707	.45889	2.1792	.90887	1.1003	2.3977	21		39	.43287	.48019	2.0825	.90146	1.1093	2.3101	21
40	.41734	.45924	2.1775	.90875	1.1004	2.3961	20		40	.43313	.48055	2.0809	.90133	1.1095	2.3088	20
41	.41760	.45960	2.1758	.90863	1.1006	2.3946	19		41	.43340	.48091	2.0794	.90120	1.1096	2.3074	19
42	.41787	.45995	2.1742	.90851	1.1007	2.3931	18		42	.43366	.48127	2.0778	.90108	1.1098	2.3060	18
43	.41813	.46030	2.1725	.90839	1.1009	2.3916	17		43	.43392	.48163	2.0763	.90095	1.1099	2.3046	17
44	.41840	.46065	2.1708	.90826	1.1010	2.3901	16		44	.43418	.48198	2.0748	.90082	1.1101	2.3032	16
45	.41866	.46101	2.1692	.90814	1.1011	2.3886	15		45	.43445	.48234	2.0732	.90070	1.1102	2.3018	15
46	.41892	.46136	2.1675	.90802	1.1013	2.3871	14		46	.43471	.48270	2.0717	.90057	1.1104	2.3004	14
47	.41919	.46171	2.1659	.90790	1.1014	2.3856	13		47	.43497	.48306	2.0701	.90045	1.1106	2.2990	13
48	.41945	.46206	2.1642	.90778	1.1016	2.3841	12		48	.43523	.48342	2.0686	.90032	1.1107	2.2976	12
49	.41972	.46242	2.1625	.90766	1.1017	2.3826	11		49	.43549	.48378	2.0671	.90019	1.1109	2.2962	11
50	.41998	.46277	2.1609	.90753	1.1019	2.3811	10		50	.43575	.48414	2.0655	.90007	1.1110	2.2949	10
51	.42024	.46312	2.1592	.90741	1.1020	2.3796	9		51	.43602	.48450	2.0640	.89994	1.1112	2.2935	9
52	.42051	.46348	2.1576	.90729	1.1022	2.3781	8		52	.43628	.48486	2.0625	.89981	1.1113	2.2921	8
53	.42077	.46383	2.1560	.90717	1.1023	2.3766	7		53	.43654	.48521	2.0609	.89968	1.1115	2.2907	7
54	.42104	.46418	2.1543	.90704	1.1025	2.3751	6		54	.43680	.48557	2.0594	.89956	1.1117	2.2894	6
55	.42130	.46454	2.1527	.90692	1.1026	2.3736	5		55	.43706	.48593	2.0579	.89943	1.1118	2.2880	5
56	.42156	.46489	2.1510	.90680	1.1028	2.3721	4		56	.43733	.48629	2.0564	.89930	1.1120	2.2866	4
57	.42183	.46525	2.1494	.90668	1.1029	2.3706	3		57	.43759	.48665	2.0549	.89918	1.1121	2.2853	3
58	.42209	.46560	2.1478	.90655	1.1031	2.3692	2		58	.43785	.48701	2.0533	.89905	1.1123	2.2839	2
59	.42235	.46595	2.1461	.90643	1.1032	2.3677	1		59	.43811	.48737	2.0518	.89892	1.1124	2.2825	1
60	.42262	.46631	2.1445	.90631	1.1034	2.3662	0		60	.43837	.48773	2.0503	.89879	1.1126	2.2812	0
′	Cos	Cot	Tan	Sin	Csc	Sec	′		′	Cos	Cot	Tan	Sin	Csc	Sce	′

114° (294°)   (245°) 65°   **115° (295°)**   (244°) 64°

NATURAL TRIGONOMETRIC FUNCTIONS TO FIVE PLACES (continued)

26° (206°)   (333°) **153°**

′	Sin	Tan	Cot	Cos	Sec	Csc	′
0	.43837	.48773	2.0503	.89879	1.1126	2.2812	60
1	.43863	.48809	2.0488	.89867	1.1128	2.2798	59
2	.43889	.48845	2.0473	.89854	1.1129	2.2785	58
3	.43916	.48881	2.0458	.89841	1.1131	2.2771	57
4	.43942	.48917	2.0443	.89828	1.1132	2.2757	56
5	.43968	.48953	2.0428	.89816	1.1134	2.2744	55
6	.43994	.48989	2.0413	.89803	1.1136	2.2730	54
7	.44020	.49026	2.0398	.89790	1.1137	2.2717	53
8	.44046	.49062	2.0383	.89777	1.1139	2.2703	52
9	.44072	.49098	2.0368	.89764	1.1140	2.2690	51
10	.44098	.49134	2.0353	.89752	1.1142	2.2677	50
11	.44124	.49170	2.0338	.89739	1.1143	2.2663	49
12	.44151	.49206	2.0323	.89726	1.1145	2.2650	48
13	.44177	.49242	2.0308	.89713	1.1147	2.2636	47
14	.44203	.49278	2.0293	.89700	1.1148	2.2623	46
15	.44229	.49315	2.0278	.89687	1.1150	2.2610	45
16	.44255	.49351	2.0263	.89674	1.1151	2.2596	44
17	.44281	.49387	2.0248	.89662	1.1153	2.2583	43
18	.44307	.49423	2.0233	.89649	1.1155	2.2570	42
19	.44333	.49459	2.0219	.89636	1.1156	2.2556	41
20	.44359	.49495	2.0204	.89623	1.1158	2.2543	40
21	.44385	.49532	2.0189	.89610	1.1159	2.2530	39
22	.44411	.49568	2.0174	.89597	1.1161	2.2517	38
23	.44437	.49604	2.0160	.89584	1.1163	2.2504	37
24	.44464	.49640	2.0145	.89571	1.1164	2.2490	36
25	.44490	.49677	2.0130	.89558	1.1166	2.2477	35
26	.44516	.49713	2.0115	.89545	1.1168	2.2464	34
27	.44542	.49749	2.0101	.89532	1.1169	2.2451	33
28	.44568	.49786	2.0086	.89519	1.1171	2.2438	32
29	.44594	.49822	2.0072	.89506	1.1172	2.2425	31
30	.44620	.49858	2.0057	.89493	1.1174	2.2412	30
31	.44646	.49894	2.0042	.89480	1.1176	2.2399	29
32	.44672	.49931	2.0028	.89467	1.1177	2.2385	28
33	.44698	.49967	2.0013	.89454	1.1179	2.2372	27
34	.44724	.50004	1.9999	.89441	1.1180	2.2359	26
35	.44750	.50040	1.9984	.89428	1.1182	2.2346	25
36	.44776	.50076	1.9970	.89415	1.1184	2.2333	24
37	.44802	.50113	1.9955	.89402	1.1185	2.2320	23
38	.44828	.50149	1.9941	.89389	1.1187	2.2308	22
39	.44854	.50185	1.9926	.89376	1.1189	2.2295	21
40	.44880	.50222	1.9912	.89363	1.1190	2.2282	20
41	.44906	.50258	1.9897	.89350	1.1192	2.2269	19
42	.44932	.50295	1.9883	.89337	1.1194	2.2256	18
43	.44958	.50331	1.9868	.89324	1.1195	2.2243	17
44	.44984	.50368	1.9854	.89311	1.1197	2.2230	16
45	.45010	.50404	1.9840	.89298	1.1198	2.2217	15
46	.45036	.50441	1.9825	.89285	1.1200	2.2205	14
47	.45062	.50477	1.9811	.89272	1.1202	2.2192	13
48	.45088	.50514	1.9797	.89259	1.1203	2.2179	12
49	.45114	.50550	1.9782	.89245	1.1205	2.2166	11
50	.45140	.50587	1.9768	.89232	1.1207	2.2153	10
51	.45166	.50623	1.9754	.89219	1.1208	2.2141	9
52	.45192	.50660	1.9740	.89206	1.1210	2.2128	8
53	.45218	.50696	1.9725	.89193	1.1212	2.2115	7
54	.45243	.50733	1.9711	.89180	1.1213	2.2103	6
55	.45269	.50769	1.9697	.89167	1.1215	2.2090	5
56	.45295	.50806	1.9683	.89153	1.1217	2.2077	4
57	.45321	.50843	1.9669	.89140	1.1218	2.2065	3
58	.45347	.50879	1.9654	.89127	1.1220	2.2052	2
59	.45373	.50916	1.9640	.89114	1.1222	2.2039	1
60	.45399	.50953	1.9626	.89101	1.1223	2.2027	0
′	Cos	Cot	Tan	Sin	Csc	Sec	′

27° (207°)   (332°) **152°**

′	Sin	Tan	Cot	Cos	Sec	Csc	′
0	.45399	.50953	1.9626	.89101	1.1223	2.2027	60
1	.45425	.50989	1.9612	.89087	1.1225	2.2014	59
2	.45451	.51026	1.9598	.89074	1.1227	2.2002	58
3	.45477	.51063	1.9584	.89061	1.1228	2.1989	57
4	.45503	.51099	1.9570	.89048	1.1230	2.1977	56
5	.45529	.51136	1.9556	.89035	1.1232	2.1964	55
6	.45554	.51173	1.9542	.89021	1.1233	2.1952	54
7	.45580	.51209	1.9528	.89008	1.1235	2.1939	53
8	.45606	.51246	1.9514	.88995	1.1237	2.1927	52
9	.45632	.51283	1.9500	.88981	1.1238	2.1914	51
10	.45658	.51319	1.9486	.88968	1.1240	2.1902	50
11	.45684	.51356	1.9472	.88955	1.1242	2.1890	49
12	.45710	.51393	1.9458	.88942	1.1243	2.1877	48
13	.45736	.51430	1.9444	.88928	1.1245	2.1865	47
14	.45762	.51467	1.9430	.88915	1.1247	2.1852	46
15	.45787	.51503	1.9416	.88902	1.1248	2.1840	45
16	.45813	.51540	1.9402	.88888	1.1250	2.1828	44
17	.45839	.51577	1.9388	.88875	1.1252	2.1815	43
18	.45865	.51614	1.9375	.88862	1.1253	2.1803	42
19	.45891	.51651	1.9361	.88848	1.1255	2.1791	41
20	.45917	.51688	1.9347	.88835	1.1257	2.1779	40
21	.45942	.51724	1.9333	.88822	1.1259	2.1766	39
22	.45968	.51761	1.9319	.88808	1.1260	2.1754	38
23	.45994	.51798	1.9306	.88795	1.1262	2.1742	37
24	.46020	.51835	1.9292	.88782	1.1264	2.1730	36
25	.46046	.51872	1.9278	.88768	1.1265	2.1718	35
26	.46072	.51909	1.9265	.88755	1.1267	2.1705	34
27	.46097	.51946	1.9251	.88741	1.1269	2.1693	33
28	.46123	.51983	1.9237	.88728	1.1270	2.1681	32
29	.46149	.52020	1.9223	.88715	1.1272	2.1669	31
30	.46175	.52057	1.9210	.88701	1.1274	2.1657	30
31	.46201	.52094	1.9196	.88688	1.1276	2.1645	29
32	.46226	.52131	1.9183	.88674	1.1277	2.1633	28
33	.46252	.52168	1.9169	.88661	1.1279	2.1621	27
34	.46278	.52205	1.9155	.88647	1.1281	2.1609	26
35	.46304	.52242	1.9142	.88634	1.1282	2.1596	25
36	.46330	.52279	1.9128	.88620	1.1284	2.1584	24
37	.46355	.52316	1.9115	.88607	1.1286	2.1572	23
38	.46381	.52353	1.9101	.88593	1.1288	2.1560	22
39	.46407	.52390	1.9088	.88580	1.1289	2.1549	21
40	.46433	.52427	1.9074	.88566	1.1291	2.1537	20
41	.46458	.52464	1.9061	.88553	1.1293	2.1525	19
42	.46484	.52501	1.9047	.88539	1.1294	2.1513	18
43	.46510	.52538	1.9034	.88526	1.1296	2.1501	17
44	.46536	.52575	1.9020	.88512	1.1298	2.1489	16
45	.46561	.52613	1.9007	.88499	1.1300	2.1477	15
46	.46587	.52650	1.8993	.88485	1.1301	2.1465	14
47	.46613	.52687	1.8980	.88472	1.1303	2.1453	13
48	.46639	.52724	1.8967	.88458	1.1305	2.1441	12
49	.46664	.52761	1.8953	.88445	1.1307	2.1430	11
50	.46690	.52798	1.8940	.88431	1.1308	2.1418	10
51	.46716	.52836	1.8927	.88417	1.1310	2.1406	9
52	.46742	.52873	1.8913	.88404	1.1312	2.1394	8
53	.46767	.52910	1.8900	.88390	1.1313	2.1382	7
54	.46793	.52947	1.8887	.88377	1.1315	2.1371	6
55	.46819	.52985	1.8873	.88363	1.1317	2.1359	5
56	.46844	.53022	1.8860	.88349	1.1319	2.1347	4
57	.46870	.53059	1.8847	.88336	1.1320	2.1336	3
58	.46896	.53096	1.8834	.88322	1.1322	2.1324	2
59	.46921	.53134	1.8820	.88308	1.1324	2.1312	1
60	.46947	.53171	1.8807	.88295	1.1326	2.1301	0
′	Cos	Cot	Tan	Sin	Csc	Sec	′

NATURAL TRIGONOMETRIC FUNCTIONS TO FIVE PLACES (continued)

28° (208°)   (331°) 151°

′	Sin	Tan	Cot	Cos	Sec	Csc	′
0	.46947	.53171	1.8807	.88295	1.1326	2.1301	60
1	.46973	.53208	1.8794	.88281	1.1327	2.1289	59
2	.46999	.53246	1.8781	.88267	1.1329	2.1277	58
3	.47024	.53283	1.8768	.88254	1.1331	2.1266	57
4	.47050	.53320	1.8755	.88240	1.1333	2.1254	56
5	.47076	.53358	1.8741	.88226	1.1334	2.1242	55
6	.47101	.53395	1.8728	.88213	1.1336	2.1231	54
7	.47127	.53432	1.8715	.88199	1.1338	2.1219	53
8	.47153	.53470	1.8702	.88185	1.1340	2.1208	52
9	.47178	.53507	1.8689	.88172	1.1342	2.1196	51
10	.47204	.53545	1.8676	.88158	1.1343	2.1185	50
11	.47229	.53582	1.8663	.88144	1.1345	2.1173	49
12	.47255	.53620	1.8650	.88130	1.1347	2.1162	48
13	.47281	.53657	1.8637	.88117	1.1349	2.1150	47
14	.47306	.53694	1.8624	.88103	1.1350	2.1139	46
15	.47332	.53732	1.8611	.88089	1.1352	2.1127	45
16	.47358	.53769	1.8598	.88075	1.1354	2.1116	44
17	.47383	.53807	1.8585	.88062	1.1356	2.1105	43
18	.47409	.53844	1.8572	.88048	1.1357	2.1093	42
19	.47434	.53882	1.8559	.88034	1.1359	2.1082	41
20	.47460	.53920	1.8546	.88020	1.1361	2.1070	40
21	.47486	.53957	1.8533	.88006	1.1363	2.1059	39
22	.47511	.53995	1.8520	.87993	1.1365	2.1048	38
23	.47537	.54032	1.8507	.87979	1.1366	2.1036	37
24	.47562	.54070	1.8495	.87965	1.1368	2.1025	36
25	.47588	.54107	1.8482	.87951	1.1370	2.1014	35
26	.47614	.54145	1.8469	.87937	1.1372	2.1002	34
27	.47639	.54183	1.8456	.87923	1.1374	2.0991	33
28	.47665	.54220	1.8443	.87909	1.1375	2.0980	32
29	.47690	.54258	1.8430	.87896	1.1377	2.0969	31
30	.47716	.54296	1.8418	.87882	1.1379	2.0957	30
31	.47741	.54333	1.8405	.87868	1.1381	2.0946	29
32	.47767	.54371	1.8392	.87854	1.1383	2.0935	28
33	.47793	.54409	1.8379	.87840	1.1384	2.0924	27
34	.47818	.54446	1.8367	.87826	1.1386	2.0913	26
35	.47844	.54484	1.8354	.87812	1.1388	2.0901	25
36	.47869	.54522	1.8341	.87798	1.1390	2.0890	24
37	.47895	.54560	1.8329	.87784	1.1392	2.0879	23
38	.47920	.54597	1.8316	.87770	1.1393	2.0868	22
39	.47946	.54635	1.8303	.87756	1.1395	2.0857	21
40	.47971	.54673	1.8291	.87743	1.1397	2.0846	20
41	.47997	.54711	1.8278	.87729	1.1399	2.0835	19
42	.48022	.54748	1.8265	.87715	1.1401	2.0824	18
43	.48048	.54786	1.8253	.87701	1.1402	2.0813	17
44	.48073	.54824	1.8240	.87687	1.1404	2.0802	16
45	.48099	.54862	1.8228	.87673	1.1406	2.0791	15
46	.48124	.54900	1.8215	.87659	1.1408	2.0779	14
47	.48150	.54938	1.8202	.87645	1.1410	2.0768	13
48	.48175	.54975	1.8190	.87631	1.1412	2.0757	12
49	.48201	.55013	1.8177	.87617	1.1413	2.0747	11
50	.48226	.55051	1.8165	.87603	1.1415	2.0736	10
51	.48252	.55089	1.8152	.87589	1.1417	2.0725	9
52	.48277	.55127	1.8140	.87575	1.1419	2.0714	8
53	.48303	.55165	1.8127	.87561	1.1421	2.0703	7
54	.48328	.55203	1.8115	.87546	1.1423	2.0692	6
55	.48354	.55241	1.8103	.87532	1.1424	2.0681	5
56	.48379	.55279	1.8090	.87518	1.1426	2.0670	4
57	.48405	.55317	1.8078	.87504	1.1428	2.0659	3
58	.48430	.55355	1.8065	.87490	1.1430	2.0648	2
59	.48456	.55393	1.8053	.87476	1.1432	2.0637	1
60	.48481	.55431	1.8040	.87462	1.1434	2.0627	0
′	Cos	Cot	Tan	Sin	Csc	Sec	′

118° (298°)   (241°) 61°

29° (209°)   (330)° 150°

′	Sin	Tan	Cot	Cos	Sec	Csc	′
0	.48481	.55431	1.8040	.87462	1.1434	2.0627	60
1	.48506	.55469	1.8028	.87448	1.1435	2.0616	59
2	.48532	.55507	1.8016	.87434	1.1437	2.0605	58
3	.48557	.55545	1.8003	.87420	1.1439	2.0594	57
4	.48583	.55583	1.7991	.87406	1.1441	2.0583	56
5	.48608	.55621	1.7979	.87391	1.1443	2.0573	55
6	.48634	.55659	1.7966	.87377	1.1445	2.0562	54
7	.48659	.55697	1.7954	.87363	1.1446	2.0551	53
8	.48684	.55736	1.7942	.87349	1.1448	2.0540	52
9	.48710	.55774	1.7930	.87335	1.1450	2.0530	51
10	.48735	.55812	1.7917	.87321	1.1452	2.0519	50
11	.48761	.55850	1.7905	.87306	1.1454	2.0508	49
12	.48786	.55888	1.7893	.87292	1.1456	2.0498	48
13	.48811	.55926	1.7881	.87278	1.1458	2.0487	47
14	.48837	.55964	1.7868	.87264	1.1460	2.0476	46
15	.48862	.56003	1.7856	.87250	1.1461	2.0466	45
16	.48888	.56041	1.7844	.87235	1.1463	2.0455	44
17	.48913	.56079	1.7832	.87221	1.1465	2.0445	43
18	.48938	.56117	1.7820	.87207	1.1467	2.0434	42
19	.48964	.56156	1.7808	.87193	1.1469	2.0423	41
20	.48989	.56194	1.7796	.87178	1.1471	2.0413	40
21	.49014	.56232	1.7783	.87164	1.1473	2.0402	39
22	.49040	.56270	1.7771	.87150	1.1474	2.0392	38
23	.49065	.56309	1.7759	.87136	1.1476	2.0381	37
24	.49090	.56347	1.7747	.87121	1.1478	2.0371	36
25	.49116	.56385	1.7735	.87107	1.1480	2.0360	35
26	.49141	.56424	1.7723	.87093	1.1482	2.0350	34
27	.49166	.56462	1.7711	.87079	1.1484	2.0339	33
28	.49192	.56501	1.7699	.87064	1.1486	2.0329	32
29	.49217	.56539	1.7687	.87050	1.1488	2.0318	31
30	.49242	.56577	1.7675	.87036	1.1490	2.0308	30
31	.49268	.56616	1.7663	.87021	1.1491	2.0297	29
32	.49293	.56654	1.7651	.87007	1.1493	2.0287	28
33	.49318	.56693	1.7639	.86993	1.1495	2.0276	27
34	.49344	.56731	1.7627	.86978	1.1497	2.0266	26
35	.49369	.56769	1.7615	.86964	1.1499	2.0256	25
36	.49394	.56808	1.7603	.86949	1.1501	2.0245	24
37	.49419	.56846	1.7591	.86935	1.1503	2.0235	23
38	.49445	.56885	1.7579	.86921	1.1505	2.0225	22
39	.49470	.56923	1.7567	.86906	1.1507	2.0214	21
40	.49495	.56962	1.7556	.86892	1.1509	2.0204	20
41	.49521	.57000	1.7544	.86878	1.1510	2.0194	19
42	.49546	.57039	1.7532	.86863	1.1512	2.0183	18
43	.49571	.57078	1.7520	.86849	1.1514	2.0173	17
44	.49596	.57116	1.7508	.86834	1.1516	2.0163	16
45	.49622	.57155	1.7496	.86820	1.1518	2.0152	15
46	.49647	.57193	1.7485	.86805	1.1520	2.0142	14
47	.49672	.57232	1.7473	.86791	1.1522	2.0132	13
48	.49697	.57271	1.7461	.86777	1.1524	2.0122	12
49	.49723	.57309	1.7449	.86762	1.1526	2.0112	11
50	.49748	.57348	1.7437	.86748	1.1528	2.0101	10
51	.49773	.57386	1.7426	.86733	1.1530	2.0091	9
52	.49798	.57425	1.7414	.86719	1.1532	2.0081	8
53	.49824	.57464	1.7402	.86704	1.1533	2.0071	7
54	.49849	.57503	1.7391	.86690	1.1535	2.0061	6
55	.49874	.57541	1.7379	.86675	1.1537	2.0051	5
56	.49899	.57580	1.7367	.86661	1.1539	2.0040	4
57	.49924	.57619	1.7355	.86646	1.1541	2.0030	3
58	.49950	.57657	1.7344	.86632	1.1543	2.0020	2
59	.49975	.57696	1.7332	.86617	1.1545	2.0010	1
60	.50000	.57735	1.7321	.86603	1.1547	2.0000	0
′	Cos	Cot	Tan	Sin	Csc	Sec	′

119° (299°)   (240°) 60°

NATURAL TRIGONOMETRIC FUNCTIONS TO FIVE PLACES (continued)

30° (210°)   (329°) 149°

′	Sin	Tan	Cot	Cos	Sec	Csc	′
0	.50000	.57735	1.7321	.86603	1.1547	2.0000	60
1	.50025	.57774	1.7309	.86588	1.1549	1.9990	59
2	.50050	.57813	1.7297	.86573	1.1551	1.9980	58
3	.50076	.57851	1.7286	.86559	1.1553	1.9970	57
4	.50101	.57890	1.7274	.86544	1.1555	1.9960	56
5	.50126	.57929	1.7262	.86530	1.1557	1.9950	55
6	.50151	.57968	1.7251	.86515	1.1559	1.9940	54
7	.50176	.58007	1.7239	.86501	1.1561	1.9930	53
8	.50201	.58046	1.7228	.86486	1.1563	1.9920	52
9	.50227	.58085	1.7216	.86471	1.1565	1.9910	51
10	.50252	.58124	1.7205	.86457	1.1566	1.9900	50
11	.50277	.58162	1.7193	.86442	1.1568	1.9890	49
12	.50302	.58201	1.7182	.86427	1.1570	1.9880	48
13	.50327	.58240	1.7170	.86413	1.1572	1.9870	47
14	.50352	.58279	1.7159	.86398	1.1574	1.9860	46
15	.50377	.58318	1.7147	.86384	1.1576	1.9850	45
16	.50403	.58357	1.7136	.86369	1.1578	1.9840	44
17	.50428	.58396	1.7124	.86354	1.1580	1.9830	43
18	.50453	.58435	1.7113	.86340	1.1582	1.9821	42
19	.50478	.58474	1.7102	.86325	1.1584	1.9811	41
20	.50503	.58513	1.7090	.86310	1.1586	1.9801	40
21	.50528	.58552	1.7079	.86295	1.1588	1.9791	39
22	.50553	.58591	1.7067	.86281	1.1590	1.9781	38
23	.50578	.58631	1.7056	.86266	1.1592	1.9771	37
24	.50603	.58670	1.7045	.86251	1.1594	1.9762	36
25	.50628	.58709	1.7033	.86237	1.1596	1.9752	35
26	.50654	.58748	1.7022	.86222	1.1598	1.9742	34
27	.50679	.58787	1.7011	.86207	1.1600	1.9732	33
28	.50704	.58826	1.6999	.86192	1.1602	1.9722	32
29	.50729	.58865	1.6988	.86178	1.1604	1.9713	31
30	.50754	.58905	1.6977	.86163	1.1606	1.9703	30
31	.50779	.58944	1.6965	.86148	1.1608	1.9693	29
32	.50804	.58983	1.6954	.86133	1.1610	1.9684	28
33	.50829	.59022	1.6943	.86119	1.1612	1.9674	27
34	.50854	.59061	1.6932	.86104	1.1614	1.9664	26
35	.50879	.59101	1.6920	.86089	1.1616	1.9654	25
36	.50904	.59140	1.6909	.86074	1.1618	1.9645	24
37	.50929	.59179	1.6898	.86059	1.1620	1.9635	23
38	.50954	.59218	1.6887	.86045	1.1622	1.9625	22
39	.50979	.59258	1.6875	.86030	1.1624	1.9616	21
40	.51004	.59297	1.6864	.86015	1.1626	1.9606	20
41	.51029	.59336	1.6853	.86000	1.1628	1.9597	19
42	.51054	.59376	1.6842	.85985	1.1630	1.9587	18
43	.51079	.59415	1.6831	.85970	1.1632	1.9577	17
44	.51104	.59454	1.6820	.85956	1.1634	1.9568	16
45	.51129	.59494	1.6808	.85941	1.1636	1.9558	15
46	.51154	.59533	1.6797	.85926	1.1638	1.9549	14
47	.51179	.59573	1.6786	.85911	1.1640	1.9539	13
48	.51204	.59612	1.6775	.85896	1.1642	1.9530	12
49	.51229	.59651	1.6764	.85881	1.1644	1.9520	11
50	.51254	.59691	1.6753	.85866	1.1646	1.9511	10
51	.51279	.59730	1.6742	.85851	1.1648	1.9501	9
52	.51304	.59770	1.6731	.85836	1.1650	1.9492	8
53	.51329	.59809	1.6720	.85821	1.1652	1.9482	7
54	.51354	.59849	1.6709	.85806	1.1654	1.9473	6
55	.51379	.59888	1.6698	.85792	1.1656	1.9463	5
56	.51404	.59928	1.6687	.85777	1.1658	1.9454	4
57	.51429	.59967	1.6676	.85762	1.1660	1.9444	3
58	.51454	.60007	1.6665	.85747	1.1662	1.9435	2
59	.51479	.60046	1.6654	.85732	1.1664	1.9425	1
60	.51504	.60086	1.6643	.85717	1.1666	1.9416	0
′	Cos	Cot	Tan	Sin	Csc	Sec	′

120° (300°)   (239°) 59°

31° (211°)   (328°) 148°

′	Sin	Tan	Cot	Cos	Sec	Csc	′
0	.51504	.60086	1.6643	.85717	1.1666	1.9416	60
1	.51529	.60126	1.6632	.85702	1.1668	1.9407	59
2	.51554	.60165	1.6621	.85687	1.1670	1.9397	58
3	.51579	.60205	1.6610	.85672	1.1672	1.9388	57
4	.51604	.60245	1.6599	.85657	1.1675	1.9379	56
5	.51628	.60284	1.6588	.85642	1.1677	1.9369	55
6	.51653	.60324	1.6577	.85627	1.1679	1.9360	54
7	.51678	.60364	1.6566	.85612	1.1681	1.9351	53
8	.51703	.60403	1.6555	.85597	1.1683	1.9341	52
9	.51728	.60443	1.6545	.85582	1.1685	1.9332	51
10	.51753	.60483	1.6534	.85567	1.1687	1.9323	50
11	.51778	.60522	1.6523	.85551	1.1689	1.9313	49
12	.51803	.60562	1.6512	.85536	1.1691	1.9304	48
13	.51828	.60602	1.6501	.85521	1.1693	1.9295	47
14	.51852	.60642	1.6490	.85506	1.1695	1.9285	46
15	.51877	.60681	1.6479	.85491	1.1697	1.9276	45
16	.51902	.60721	1.6469	.85476	1.1699	1.9267	44
17	.51927	.60761	1.6458	.85461	1.1701	1.9258	43
18	.51952	.60801	1.6447	.85446	1.1703	1.9249	42
19	.51977	.60841	1.6436	.85431	1.1705	1.9239	41
20	.52002	.60881	1.6426	.85416	1.1707	1.9230	40
21	.52026	.60921	1.6415	.85401	1.1710	1.9221	39
22	.52051	.60960	1.6404	.85385	1.1712	1.9212	38
23	.52076	.61000	1.6393	.85370	1.1714	1.9203	37
24	.52101	.61040	1.6383	.85355	1.1716	1.9194	36
25	.52126	.61080	1.6372	.85340	1.1718	1.9184	35
26	.52151	.61120	1.6361	.85325	1.1720	1.9175	34
27	.52175	.61160	1.6351	.85310	1.1722	1.9166	33
28	.52200	.61200	1.6340	.85294	1.1724	1.9157	32
29	.52225	.61240	1.6329	.85279	1.1726	1.9148	31
30	.52250	.61280	1.6319	.85264	1.1728	1.9139	30
31	.52275	.61320	1.6308	.85249	1.1730	1.9130	29
32	.52299	.61360	1.6297	.85234	1.1732	1.9121	28
33	.52324	.61400	1.6287	.85218	1.1735	1.9112	27
34	.52349	.61440	1.6276	.85203	1.1737	1.9103	26
35	.52374	.61480	1.6265	.85188	1.1739	1.9094	25
36	.52399	.61520	1.6255	.85173	1.1741	1.9084	24
37	.52423	.61561	1.6244	.85157	1.1743	1.9075	23
38	.52448	.61601	1.6234	.85142	1.1745	1.9066	22
39	.52473	.61641	1.6223	.85127	1.1747	1.9057	21
40	.52498	.61681	1.6212	.85112	1.1749	1.9048	20
41	.52522	.61721	1.6202	.85096	1.1751	1.9039	19
42	.52547	.61761	1.6191	.85081	1.1753	1.9031	18
43	.52572	.61801	1.6181	.85066	1.1756	1.9022	17
44	.52597	.61842	1.6170	.85051	1.1758	1.9013	16
45	.52621	.61882	1.6160	.85035	1.1760	1.9004	15
46	.52646	.61922	1.6149	.85020	1.1762	1.8995	14
47	.52671	.61962	1.6139	.85005	1.1764	1.8986	13
48	.52696	.62003	1.6128	.84989	1.1766	1.8977	12
49	.52720	.62043	1.6118	.84974	1.1768	1.8968	11
50	.52745	.62083	1.6107	.84959	1.1770	1.8959	10
51	.52770	.62124	1.6097	.84943	1.1773	1.8950	9
52	.52794	.62164	1.6087	.84928	1.1775	1.8941	8
53	.52819	.62204	1.6076	.84913	1.1777	1.8933	7
54	.52844	.62245	1.6066	.84897	1.1779	1.8924	6
55	.52869	.62285	1.6055	.84882	1.1781	1.8915	5
56	.52893	.62325	1.6045	.84866	1.1783	1.8906	4
57	.52918	.62366	1.6034	.84851	1.1785	1.8897	3
58	.52943	.62406	1.6024	.84836	1.1788	1.8888	2
59	.52967	.62446	1.6014	.84820	1.1790	1.8880	1
60	.52992	.62487	1.6003	.84805	1.1792	1.8871	0
′	Cos	Cot	Tan	Sin	Csc	Sec	′

121° (301°)   (238°) 58°

NATURAL TRIGONOMETRIC FUNCTIONS TO FIVE PLACES (continued)

32° (212°)   **(327°) 147°**

′	Sin	Tan	Cot	Cos	Sec	Csc	′
0	.52992	.62487	1.6003	.84805	1.1792	1.8871	60
1	.53017	.62527	1.5993	.84789	1.1794	1.8862	59
2	.53041	.62568	1.5983	.84774	1.1796	1.8853	58
3	.53066	.62608	1.5972	.84759	1.1798	1.8844	57
4	.53091	.62649	1.5962	.84743	1.1800	1.8836	56
5	.53115	.62689	1.5952	.84728	1.1803	1.8827	55
6	.53140	.62730	1.5941	.84712	1.1805	1.8818	54
7	.53164	.62770	1.5931	.84697	1.1807	1.8810	53
8	.53189	.62811	1.5921	.84681	1.1809	1.8801	52
9	.53214	.62852	1.5911	.84666	1.1811	1.8792	51
10	.53238	.62892	1.5900	.84650	1.1813	1.8783	50
11	.53263	.62933	1.5890	.84635	1.1815	1.8775	49
12	.53288	.62973	1.5880	.84619	1.1818	1.8766	48
13	.53312	.63014	1.5869	.84604	1.1820	1.8757	47
14	.53337	.63055	1.5859	.84588	1.1822	1.8749	46
15	.53361	.63095	1.5849	.84573	1.1824	1.8740	45
16	.53386	.63136	1.5839	.84557	1.1826	1.8731	44
17	.53411	.63177	1.5829	.84542	1.1828	1.8723	43
18	.53435	.63217	1.5818	.84526	1.1831	1.8714	42
19	.53460	.63258	1.5808	.84511	1.1833	1.8706	41
20	.53484	.63299	1.5798	.84495	1.1835	1.8697	40
21	.53509	.63340	1.5788	.84480	1.1837	1.8688	39
22	.53534	.63380	1.5778	.84464	1.1830	1.8680	38
23	.53558	.63421	1.5768	.84448	1.1842	1.8671	37
24	.53583	.63462	1.5757	.84433	1.1844	1.8663	36
25	.53607	.63503	1.5747	.84417	1.1846	1.8654	35
26	.53632	.63544	1.5737	.84402	1.1848	1.8646	34
27	.53656	.63584	1.5727	.84386	1.1850	1.8637	33
28	.53681	.63625	1.5717	.84370	1.1852	1.8629	32
29	.53705	.63666	1.5707	.84355	1.1855	1.8620	31
30	.53730	.63707	1.5697	.84339	1.1857	1.8612	30
31	.53754	.63748	1.5687	.84324	1.1859	1.8603	29
32	.53779	.63789	1.5677	.84308	1.1861	1.8595	28
33	.53804	.63830	1.5667	.84292	1.1863	1.8586	27
34	.53828	.63871	1.5657	.84277	1.1866	1.8578	26
35	.53853	.63912	1.5647	.84261	1.1868	1.8569	25
36	.53877	.63953	1.5637	.84245	1.1870	1.8561	24
37	.53902	.63994	1.5627	.84230	1.1872	1.8552	23
38	.53926	.64035	1.5617	.84214	1.1875	1.8544	22
39	.53951	.64076	1.5607	.84198	1.1877	1.8535	21
40	.53975	.64117	1.5597	.84182	1.1879	1.8527	20
41	.54000	.64158	1.5587	.84167	1.1881	1.8519	19
42	.54024	.64199	1.5577	.84151	1.1883	1.8510	18
43	.54049	.64240	1.5567	.84135	1.1886	1.8502	17
44	.54073	.64281	1.5557	.84120	1.1888	1.8494	16
45	.54097	.64322	1.5547	.84104	1.1890	1.8485	15
46	.54122	.64363	1.5537	.84088	1.1892	1.8477	14
47	.54146	.64404	1.5527	.84072	1.1895	1.8468	13
48	.54171	.64446	1.5517	.84057	1.1897	1.8460	12
49	.54195	.64487	1.5507	.84041	1.1899	1.8452	11
50	.54220	.64528	1.5497	.84025	1.1901	1.8443	10
51	.54244	.64569	1.5487	.84009	1.1903	1.8435	9
52	.54269	.64610	1.5477	.83994	1.1906	1.8427	8
53	.54293	.64652	1.5468	.83978	1.1908	1.8419	7
54	.54317	.64693	1.5458	.83962	1.1910	1.8410	6
55	.54342	.64734	1.5448	.83946	1.1912	1.8402	5
56	.54366	.64775	1.5438	.83930	1.1915	1.8394	4
57	.54391	.64817	1.5428	.83915	1.1917	1.8385	3
58	.54415	.64858	1.5418	.83899	1.1919	1.8377	2
59	.54440	.64899	1.5408	.83883	1.1921	1.8369	1
60	.54464	.64941	1.5399	.83867	1.1924	1.8361	0
′	Cos	Cot	Tan	Sin	Csc	Sec	′

122° (302°)   **(237°) 57°**

33° (213°)   **(326°) 146°**

′	Sin	Tan	Cot	Cos	Sec	Csc	′
0	.54464	.64941	1.5399	.83867	1.1924	1.8361	60
1	.54488	.64982	1.5389	.83851	1.1926	1.8353	59
2	.54513	.65024	1.5379	.83835	1.1928	1.8344	58
3	.54537	.65065	1.5369	.83819	1.1930	1.8336	57
4	.54561	.65106	1.5359	.83804	1.1933	1.8328	56
5	.54586	.65148	1.5350	.83788	1.1935	1.8320	55
6	.54610	.65189	1.5340	.83772	1.9137	1.8312	54
7	.54635	.65231	1.5330	.83756	1.1939	1.8303	53
8	.54659	.65272	1.5320	.83740	1.1942	1.8295	52
9	.54683	.65314	1.5311	.83724	1.1944	1.8287	51
10	.54708	.65355	1.5301	.83708	1.1946	1.8279	50
11	.54732	.65397	1.5291	.83692	1.1949	1.8271	49
12	.54756	.65438	1.5282	.83676	1.1951	1.8263	48
13	.54781	.65480	1.5272	.83660	1.1953	1.8255	47
14	.54805	.65521	1.5262	.83645	1.1955	1.8247	46
15	.54829	.65563	1.5253	.83629	1.1958	1.8238	45
16	.54854	.65604	1.5243	.83613	1.1960	1.8230	44
17	.54878	.65646	1.5233	.83597	1.1962	1.8222	43
18	.54902	.65688	1.5224	.83581	1.1964	1.8214	42
19	.54927	.65729	1.5214	.83565	1.1967	1.8206	41
20	.54951	.65771	1.5204	.83549	1.1969	1.8198	40
21	.54975	.65813	1.5195	.83533	1.1971	1.8190	39
22	.54999	.65854	1.5185	.83517	1.1974	1.8182	38
23	.55024	.65896	1.5175	.83501	1.1976	1.8174	37
24	.55048	.65938	1.5166	.83485	1.1978	1.8166	36
25	.55072	.65980	1.5156	.83469	1.1981	1.8158	35
26	.55097	.66021	1.5147	.83453	1.1983	1.8150	34
27	.55121	.66063	1.5137	.83437	1.1985	1.8142	33
28	.55145	.66105	1.5127	.83421	1.1987	1.8134	32
29	.55169	.66147	1.5118	.83405	1.1990	1.8126	31
30	.55194	.66189	1.5108	.83389	1.1992	1.8118	30
31	.55218	.66230	1.5099	.83373	1.1994	1.8110	29
32	.55242	.66272	1.5089	.83356	1.1997	1.8102	28
33	.55266	.66314	1.5080	.83340	1.1999	1.8094	27
34	.55291	.66356	1.5070	.83324	1.2001	1.8086	26
35	.55315	.66398	1.5061	.83308	1.2004	1.8078	25
36	.55339	.66440	1.5051	.83292	1.2006	1.8070	24
37	.55363	.66482	1.5042	.83276	1.2008	1.8062	23
38	.55388	.66524	1.5032	.83260	1.2011	1.8055	22
39	.55412	.66566	1.5023	.83244	1.2013	1.8047	21
40	.55436	.66608	1.5013	.83228	1.2015	1.8039	20
41	.55460	.66650	1.5004	.83212	1.2018	1.8031	19
42	.55484	.66692	1.4994	.83195	1.2020	1.8023	18
43	.55509	.66734	1.4985	.83179	1.2022	1.8015	17
44	.55533	.66776	1.4975	.83163	1.2025	1.8007	16
45	.55557	.66818	1.4966	.83147	1.2027	1.8000	15
46	.55581	.66860	1.4957	.83131	1.2029	1.7992	14
47	.55605	.66902	1.4947	.83115	1.2032	1.7984	13
48	.55630	.66944	1.4938	.83098	1.2034	1.7976	12
49	.55654	.66986	1.4928	.83082	1.2036	1.7968	11
50	.55678	.67028	1.4919	.83066	1.2039	1.7960	10
51	.55702	.67071	1.4910	.83050	1.2041	1.7953	9
52	.55726	.67113	1.4900	.83034	1.2043	1.7945	8
53	.55750	.67155	1.4891	.83017	1.2046	1.7937	7
54	.55775	.67197	1.4882	.83001	1.2048	1.7929	6
55	.55799	.67239	1.4872	.82985	1.2050	1.7922	5
56	.55823	.67282	1.4863	.82969	1.2053	1.7914	4
57	.55847	.67324	1.4854	.82953	1.2055	1.7906	3
58	.55871	.67366	1.4844	.82936	1.2057	1.7898	2
59	.55895	.67409	1.4835	.82920	1.2060	1.7891	1
60	.55919	.67451	1.4826	.82904	1.2062	1.7883	0
′	Cos	Cot	Tan	Sin	Csc	Sec	′

123° (303°)   **(236°) 56°**

NATURAL TRIGONOMETRIC FUNCTIONS TO FIVE PLACES (continued)

34° (214°) (325°) 145°

′	Sin	Tan	Cot	Cos	Sec	Csc	′
0	.55919	.67451	1.4826	.82904	1.2062	1.7883	60
1	.55943	.67493	1.4816	.82887	1.2065	1.7875	59
2	.55968	.67536	1.4807	.82871	1.2067	1.7868	58
3	.55992	.67578	1.4798	.82855	1.2069	1.7860	57
4	.56016	.67620	1.4788	.82839	1.2072	1.7852	56
5	.56040	.67663	1.4779	.82822	1.2074	1.7844	55
6	.56064	.67705	1.4770	.82806	1.2076	1.7837	54
7	.56088	.67748	1.4761	.82790	1.2079	1.7829	53
8	.56112	.67790	1.4751	.82773	1.2081	1.7821	52
9	.56136	.67832	1.4742	.82757	1.2084	1.7814	51
10	.56160	.67875	1.4733	.82741	1.2086	1.7806	50
11	.56184	.67917	1.4724	.82724	1.2088	1.7799	49
12	.56208	.67960	1.4715	.82708	1.2091	1.7791	48
13	.56232	.68002	1.4705	.82692	1.2093	1.7783	47
14	.56256	.68045	1.4696	.82675	1.2096	1.7776	46
15	.56280	.68088	1.4687	.82659	1.2098	1.7768	45
16	.56305	.68130	1.4678	.82643	1.2100	1.7761	44
17	.56329	.68173	1.4669	.82626	1.2103	1.7753	43
18	.56353	.68215	1.4659	.82610	1.2105	1.7745	42
19	.56377	.68258	1.4650	.82593	1.2108	1.7738	41
20	.56401	.68301	1.4641	.82577	1.2110	1.7730	40
21	.56425	.68343	1.4632	.82561	1.2112	1.7723	39
22	.56449	.68386	1.4623	.82544	1.2115	1.7715	38
23	.56473	.68429	1.4614	.82528	1.2117	1.7708	37
24	.56497	.68471	1.4605	.82511	1.2120	1.7700	36
25	.56521	.68514	1.4596	.82495	1.2122	1.7693	35
26	.56545	.68557	1.4586	.82478	1.2124	1.7685	34
27	.56569	.68600	1.4577	.82462	1.2127	1.7678	33
28	.56593	.68642	1.4568	.82446	1.2129	1.7670	32
29	.56617	.68685	1.4559	.82429	1.2132	1.7663	31
30	.56641	.68728	1.4550	.82413	1.2134	1.7655	30
31	.56665	.68771	1.4541	.82396	1.2136	1.7648	29
32	.56689	.68814	1.4532	.82380	1.2139	1.7640	28
33	.56713	.68857	1.4523	.82363	1.2141	1.7633	27
34	.56736	.68900	1.4514	.82347	1.2144	1.7625	26
35	.56760	.68942	1.4505	.82330	1.2146	1.7618	25
36	.56784	.68985	1.4496	.82314	1.2149	1.7610	24
37	.56808	.69028	1.4487	.82297	1.2151	1.7603	23
38	.56832	.69071	1.4478	.82281	1.2154	1.7596	22
39	.56856	.69114	1.4469	.82264	1.2156	1.7588	21
40	.56880	.69157	1.4460	.82248	1.2158	1.7581	20
41	.56904	.69200	1.4451	.82231	1.2161	1.7573	19
42	.56928	.69243	1.4442	.82214	1.2163	1.7566	18
43	.56952	.69286	1.4433	.82198	1.2166	1.7559	17
44	.56976	.69329	1.4424	.82181	1.2168	1.7551	16
45	.57000	.69372	1.4415	.82165	1.2171	1.7544	15
46	.57024	.69416	1.4406	.82148	1.2173	1.7537	14
47	.57047	.69459	1.4397	.82132	1.2176	1.7529	13
48	.57071	.69502	1.4388	.82115	1.2178	1.7522	12
49	.57095	.69545	1.4379	.82098	1.2181	1.7515	11
50	.57119	.69588	1.4370	.82082	1.2183	1.7507	10
51	.57143	.69631	1.4361	.82065	1.2185	1.7500	9
52	.57167	.69675	1.4352	.82048	1.2188	1.7493	8
53	.57191	.69718	1.4344	.82032	1.2190	1.7485	7
54	.57215	.69761	1.4335	.82015	1.2193	1.7478	6
55	.57238	.69804	1.4326	.81999	1.2195	1.7471	5
56	.57262	.69847	1.4317	.81982	1.2198	1.7463	4
57	.57286	.69891	1.4308	.81965	1.2200	1.7456	3
58	.57310	.69934	1.4299	.81949	1.2203	1.7449	2
59	.57334	.69977	1.4290	.81932	1.2205	1.7442	1
60	.57358	.70021	1.4281	.81915	1.2208	1.7434	0
′	Cos	Cot	Tan	Sin	Csc	Sec	′

35° (215°) (324°) 144°

′	Sin	Tan	Cot	Cos	Sec	Csc	′
0	.57358	.70021	1.4281	.81915	1.2208	1.7434	60
1	.57381	.70064	1.4273	.81899	1.2210	1.7427	59
2	.57405	.70107	1.4264	.81882	1.2213	1.7420	58
3	.57429	.70151	1.4255	.81865	1.2215	1.7413	57
4	.57453	.70194	1.4246	.81848	1.2218	1.7406	56
5	.57477	.70238	1.4237	.81832	1.2220	1.7398	55
6	.57501	.70281	1.4229	.81815	1.2223	1.7391	54
7	.57524	.70325	1.4220	.81798	1.2225	1.7384	53
8	.57548	.70368	1.4211	.81782	1.2228	1.7377	52
9	.57572	.70412	1.4202	.81765	1.2230	1.7370	51
10	.57596	.70455	1.4193	.81748	1.2233	1.7362	50
11	.57619	.70499	1.4185	.81731	1.2235	1.7355	49
12	.57643	.70542	1.4176	.81714	1.2238	1.7348	48
13	.57667	.70586	1.4167	.81698	1.2240	1.7341	47
14	.57691	.70629	1.4158	.81681	1.2243	1.7334	46
15	.57715	.70673	1.4150	.81664	1.2245	1.7327	45
16	.57738	.70717	1.4141	.81647	1.2248	1.7320	44
17	.57762	.70760	1.4132	.81631	1.2250	1.7312	43
18	.57786	.70804	1.4124	.81614	1.2253	1.7305	42
19	.57810	.70848	1.4115	.81597	1.2255	1.7298	41
20	.57833	.70891	1.4106	.81580	1.2258	1.7291	40
21	.57857	.70935	1.4097	.81563	1.2260	1.7284	39
22	.57881	.70979	1.4089	.81546	1.2263	1.7277	38
23	.57904	.71023	1.4080	.81530	1.2265	1.7270	37
24	.57928	.71066	1.4071	.81513	1.2268	1.7263	36
25	.57952	.71110	1.4063	.81496	1.2271	1.7256	35
26	.57976	.71154	1.4054	.81479	1.2273	1.7249	34
27	.57999	.71198	1.4045	.81462	1.2276	1.7242	33
28	.58023	.71242	1.4037	.81445	1.2278	1.7235	32
29	.58047	.71285	1.4028	.81428	1.2281	1.7228	31
30	.58070	.71329	1.4019	.81412	1.2283	1.7221	30
31	.58094	.71373	1.4011	.81395	1.2286	1.7213	29
32	.58118	.71417	1.4002	.81378	1.2288	1.7206	28
33	.58141	.71461	1.3994	.81361	1.2291	1.7199	27
34	.58165	.71505	1.3985	.81344	1.2293	1.7192	26
35	.58189	.71549	1.3976	.81327	1.2296	1.7185	25
36	.58212	.71593	1.3968	.81310	1.2299	1.7179	24
37	.58236	.71637	1.3959	.81293	1.2301	1.7172	23
38	.58260	.71681	1.3951	.81276	1.2304	1.7165	22
39	.58283	.71725	1.3942	.81259	1.2306	1.7158	21
40	.58307	.71769	1.3934	.81242	1.2309	1.7151	20
41	.58330	.71813	1.3925	.81225	1.2311	1.7144	19
42	.58354	.71857	1.3916	.81208	1.2314	1.7137	18
43	.58378	.71901	1.3908	.81191	1.2317	1.7130	17
44	.58401	.71946	1.3899	.81174	1.2319	1.7123	16
45	.58425	.71990	1.3891	.81157	1.2322	1.7116	15
46	.58449	.72034	1.3882	.81140	1.2324	1.7109	14
47	.58472	.72078	1.3874	.81123	1.2327	1.7102	13
48	.58496	.72122	1.3865	.81106	1.2329	1.7095	12
49	.58519	.72167	1.3857	.81089	1.2332	1.7088	11
50	.58543	.72211	1.3848	.81072	1.2335	1.7081	10
51	.58567	.72255	1.3840	.81055	1.2337	1.7075	9
52	.58590	.72299	1.3831	.81038	1.2340	1.7068	8
53	.58614	.72344	1.3823	.81021	1.2342	1.7061	7
54	.58637	.72388	1.3814	.81004	1.2345	1.7054	6
55	.58661	.72432	1.3806	.80987	1.2348	1.7047	5
56	.58684	.72477	1.3798	.80970	1.2350	1.7040	4
57	.58708	.72521	1.3789	.80953	1.2353	1.7033	3
58	.58731	.72565	1.3781	.80936	1.2355	1.7027	2
59	.58755	.72610	1.3772	.80919	1.2358	1.7020	1
60	.58779	.72654	1.3764	.80902	1.2361	1.7013	0
′	Cos	Cot	Tan	Sin	Csc	Sec	′

NATURAL TRIGONOMETRIC FUNCTIONS TO FIVE PLACES (continued)

36° (216°)   **(323°) 143°**

′	Sin	Tan	Cot	Cos	Sec	Csc	′
0	.58779	.72654	1.3764	.80902	1.2361	1.7013	60
1	.58802	.72699	1.3755	.80885	1.2363	1.7006	59
2	.58826	.72743	1.3747	.80867	1.2366	1.6999	58
3	.58849	.72788	1.3739	.80850	1.2369	1.6993	57
4	.58873	.72832	1.3730	.80833	1.2371	1.6986	56
5	.58896	.72877	1.3722	.80816	1.2374	1.6979	55
6	.58920	.72921	1.3713	.80799	1.2376	1.6972	54
7	.58943	.72966	1.3705	.80782	1.2379	1.6966	53
8	.58967	.73010	1.3697	.80765	1.2382	1.6959	52
9	.58990	.73055	1.3688	.80748	1.2384	1.6952	51
10	.59014	.73100	1.3680	.80730	1.2387	1.6945	50
11	.59037	.73144	1.3672	.80713	1.2390	1.6939	49
12	.59061	.73189	1.3663	.80696	1.2392	1.6932	48
13	.59084	.73234	1.3655	.80679	1.2395	1.6925	47
14	.59108	.73278	1.3647	.80662	1.2397	1.6918	46
15	.59131	.73323	1.3638	.80644	1.2400	1.6912	45
16	.59154	.73368	1.3630	.80627	1.2403	1.6905	44
17	.59178	.73413	1.3622	.80610	1.2405	1.6898	43
18	.59201	.73457	1.3613	.80593	1.2408	1.6892	42
19	.59225	.73502	1.3605	.80576	1.2411	1.6885	41
20	.59248	.73547	1.3597	.80558	1.2413	1.6878	40
21	.59272	.73592	1.3588	.80541	1.2416	1.6871	39
22	.59295	.73637	1.3580	.80524	1.2419	1.6865	38
23	.59318	.73681	1.3572	.80507	1.2421	1.6858	37
24	.59342	.73726	1.3564	.80489	1.2424	1.6852	36
25	.59365	.73771	1.3555	.80472	1.2427	1.6845	35
26	.59389	.73816	1.3547	.80455	1.2429	1.6838	34
27	.59412	.73861	1.3539	.80438	1.2432	1.6832	33
28	.59436	.73906	1.3531	.80420	1.2435	1.6825	32
29	.59459	.73951	1.3522	.80403	1.2437	1.6818	31
30	.59482	.73996	1.3514	.80386	1.2440	1.6812	30
31	.59506	.74041	1.3506	.80368	1.2443	1.6805	29
32	.59529	.74086	1.3498	.80351	1.2445	1.6799	28
33	.59552	.74131	1.3490	.80334	1.2448	1.6792	27
34	.59576	.74176	1.3481	.80316	1.2451	1.6785	26
35	.59599	.74221	1.3473	.80299	1.2453	1.6779	25
36	.59622	.74267	1.3465	.80282	1.2456	1.6772	24
37	.59646	.74312	1.3457	.80264	1.2459	1.6766	23
38	.59669	.74357	1.3449	.80247	1.2462	1.6759	22
39	.59693	.74402	1.3440	.80230	1.2464	1.6753	21
40	.59716	.74447	1.3432	.80212	1.2467	1.6746	20
41	.59739	.74492	1.3424	.80195	1.2470	1.6739	19
42	.59763	.74538	1.3416	.80178	1.2472	1.6733	18
43	.59786	.74583	1.3408	.80160	1.2475	1.6726	17
44	.59809	.74628	1.3400	.80143	1.2478	1.6720	16
45	.59832	.74674	1.3392	.80125	1.2480	1.6713	15
46	.59856	.74719	1.3384	.80108	1.2483	1.6707	14
47	.59879	.74764	1.3375	.80091	1.2486	1.6700	13
48	.59902	.74810	1.3367	.80073	1.2489	1.6694	12
49	.59926	.74855	1.3359	.80056	1.2491	1.6687	11
50	.59949	.74900	1.3351	.80038	1.2494	1.6681	10
51	.59972	.74946	1.3343	.80021	1.2497	1.6674	9
52	.59995	.74991	1.3335	.80003	1.2499	1.6668	8
53	.60019	.75037	1.3327	.79986	1.2502	1.6661	7
54	.60042	.75082	1.3319	.79968	1.2505	1.6655	6
55	.60065	.75128	1.3311	.79951	1.2508	1.6649	5
56	.60089	.75173	1.3303	.79934	1.2510	1.6642	4
57	.60112	.75219	1.3295	.79916	1.2513	1.6636	3
58	.60135	.75264	1.3287	.79899	1.2516	1.6629	2
59	.60158	.75310	1.3278	.79881	1.2519	1.6623	1
60	.60182	.75355	1.3270	.79864	1.2521	1.6616	0
′	Cos	Cot	Tan	Sin	Csc	Sec	′

126° (306°)   **(233°) 53°**

37° (217°)   **(322°) 142°**

′	Sin	Tan	Cot	Cos	Sec	Csc	′
0	.60182	.75355	1.3270	.79864	1.2521	1.6616	60
1	.60205	.75401	1.3262	.79846	1.2524	1.6610	59
2	.60228	.75447	1.3254	.79829	1.2527	1.6604	58
3	.60251	.75492	1.3246	.79811	1.2530	1.6597	57
4	.60274	.75538	1.3238	.79793	1.2532	1.6591	56
5	.60298	.75584	1.3230	.79776	1.2535	1.6584	55
6	.60321	.75629	1.3222	.79758	1.2538	1.6578	54
7	.60344	.75675	1.3214	.79741	1.2541	1.6572	53
8	.60367	.75721	1.3206	.79723	1.2543	1.6565	52
9	.60390	.75767	1.3198	.79706	1.2546	1.6559	51
10	.60414	.75812	1.3190	.79688	1.2549	1.6553	50
11	.60437	.75858	1.3182	.79671	1.2552	1.6546	49
12	.60460	.75904	1.3175	.79653	1.2554	1.6540	48
13	.60483	.75950	1.3167	.79635	1.2557	1.6534	47
14	.60506	.75996	1.3159	.79618	1.2560	1.6527	46
15	.60529	.76042	1.3151	.79600	1.2563	1.6521	45
16	.60553	.76088	1.3143	.79583	1.2566	1.6515	44
17	.60576	.76134	1.3135	.79565	1.2568	1.6508	43
18	.60599	.76180	1.3127	.79547	1.2571	1.6502	42
19	.60622	.76226	1.3119	.79530	1.2574	1.6496	41
20	.60645	.76272	1.3111	.79512	1.2577	1.6489	40
21	.60668	.76318	1.3103	.79494	1.2579	1.6483	39
22	.60691	.76364	1.3095	.79477	1.2582	1.6477	38
23	.60714	.76410	1.3087	.79459	1.2585	1.6471	37
24	.60738	.76456	1.3079	.79441	1.2588	1.6464	36
25	.60761	.76502	1.3072	.79424	1.2591	1.6458	35
26	.60784	.76548	1.3064	.79406	1.2593	1.6452	34
27	.60807	.76594	1.3056	.79388	1.2596	1.6446	33
28	.60830	.76640	1.3048	.79371	1.2599	1.6439	32
29	.60853	.76686	1.3040	.79353	1.2602	1.6433	31
30	.60876	.76733	1.3032	.79335	1.2605	1.6427	30
31	.60899	.76779	1.3024	.79318	1.2608	1.6421	29
32	.60922	.76825	1.3017	.79300	1.2610	1.6414	28
33	.60945	.76871	1.3009	.79282	1.2613	1.6408	27
34	.60968	.76918	1.3001	.79264	1.2616	1.6402	26
35	.60991	.76964	1.2993	.79247	1.2619	1.6396	25
36	.61015	.77010	1.2985	.79229	1.2622	1.6390	24
37	.61038	.77057	1.2977	.79211	1.2624	1.6383	23
38	.61061	.77103	1.2970	.79193	1.2627	1.6377	22
39	.61084	.77149	1.2962	.79176	1.2630	1.6371	21
40	.61107	.77196	1.2954	.79158	1.2633	1.6365	20
41	.61130	.77242	1.2946	.79140	1.2636	1.6359	19
42	.61153	.77289	1.2938	.79122	1.2639	1.6353	18
43	.61176	.77335	1.2931	.79105	1.2641	1.6346	17
44	.61199	.77382	1.2923	.79087	1.2644	1.6340	16
45	.61222	.77428	1.2915	.79069	1.2647	1.6334	15
46	.61245	.77475	1.2907	.79051	1.2650	1.6328	14
47	.61268	.77521	1.2900	.79033	1.2653	1.6322	13
48	.61291	.77568	1.2892	.79016	1.2656	1.6316	12
49	.61314	.77615	1.2884	.78998	1.2659	1.6310	11
50	.61337	.77661	1.2876	.78980	1.2661	1.6303	10
51	.61360	.77708	1.2869	.78962	1.2664	1.6297	9
52	.61383	.77754	1.2861	.78944	1.2667	1.6291	8
53	.61406	.77801	1.2853	.78926	1.2670	1.6285	7
54	.61429	.77848	1.2846	.78908	1.2673	1.6279	6
55	.61451	.77895	1.2838	.78891	1.2676	1.6273	5
56	.61474	.77941	1.2830	.78873	1.2679	1.6267	4
57	.61497	.77988	1.2822	.78855	1.2682	1.6261	3
58	.61520	.78035	1.2815	.78837	1.2684	1.6255	2
59	.61543	.78082	1.2807	.78819	1.2687	1.6249	1
60	.61566	.78129	1.2799	.78801	1.2690	1.6243	0
′	Cos	Cot	Tan	Sin	Csc	Sec	′

127° (307°)   **(232°) 52°**

NATURAL TRIGONOMETRIC FUNCTIONS TO FIVE PLACES (continued)

38° (218°)   **(321°) 141°**

'	Sin.	Tan	Cot	Cos	Sec	Csc	'
0	.61566	.78129	1.2799	.78801	1.2690	1.6243	60
1	.61589	.78175	1.2792	.78783	1.2693	1.6237	59
2	.61612	.78222	1.2784	.78765	1.2696	1.6231	58
3	.61635	.78269	1.2776	.78747	1.2699	1.6225	57
4	.61658	.78316	1.2769	.78729	1.2702	1.6219	56
5	.61681	.78363	1.2761	.78711	1.2705	1.6213	55
6	.61704	.78410	1.2753	.78694	1.2708	1.6207	54
7	.61726	.78457	1.2746	.78676	1.2710	1.6201	53
8	.61749	.78504	1.2738	.78658	1.2713	1.6195	52
9	.61772	.78551	1.2731	.78640	1.2716	1.6189	51
10	.61795	.78598	1.2723	.78622	1.1719	1.6183	50
11	.61818	.78645	1.2715	.78604	1.2722	1.6177	49
12	.61841	.78692	1.2708	.78586	1.2725	1.6171	48
13	.61804	.78739	1.2700	.78568	1.2728	1.6165	47
14	.61887	.78786	1.2693	.78550	1.2731	1.6159	46
15	.61909	.78834	1.2685	.78532	1.2734	1.6153	45
16	.61932	.78881	1.2677	.78514	1.2737	1.6147	44
17	.61955	.78928	1.2670	.78496	1.2740	1.6141	43
18	.61978	.78975	1.2662	.78478	1.2742	1.6135	42
19	.62001	.79022	1.2655	.78460	1.2745	1.6129	41
20	.62024	.79070	1.2647	.78442	1.2748	1.6123	40
21	.62046	.79117	1.2640	.78424	1.2751	1.6117	39
22	.62069	.79164	1.2632	.78405	1.2754	1.6111	38
23	.62092	.79212	1.2624	.78387	1.2757	1.6105	37
24	.62115	.79259	1.2617	.78369	1.2760	1.6099	36
25	.62138	.79306	1.2609	.78351	1.2763	1.6093	35
26	.62160	.79354	1.2602	.78333	1.2766	1.6087	34
27	.62183	.79401	1.2594	.78315	1.2769	1.6082	33
28	.62206	.79449	1.2587	.78297	1.2772	1.6076	32
29	.62229	.79496	1.2579	.78279	1.2775	1.6070	31
30	.62251	.79544	1.2572	.78261	1.2778	1.6064	30
31	.62274	.79591	1.2564	.78243	1.2781	1.6058	29
32	.62297	.79639	1.2557	.78225	1.2784	1.6052	28
33	.62320	.79686	1.2549	.78206	1.2787	1.6046	27
34	.62342	.79734	1.2542	.78188	1.2790	1.6040	26
35	.62365	.79781	1.2534	.78170	1.2793	1.6035	25
36	.62388	.79829	1.2527	.78152	1.2796	1.6029	24
37	.62411	.79877	1.2519	.78134	1.2799	1.6023	23
38	.62433	.79924	1.2512	.78116	1.2802	1.6017	22
39	.62456	.79972	1.2504	.78098	1.2804	1.6011	21
40	.62479	.80020	1.2497	.78079	1.2807	1.6005	20
41	.62502	.80067	1.2489	.78061	1.2810	1.6000	19
42	.62524	.80115	1.2482	.78043	1.2813	1.5994	18
43	.62547	.80163	1.2475	.78025	1.2816	1.5988	17
44	.62570	.80211	1.2467	.78007	1.2819	1.5982	16
45	.62592	.80258	1.2460	.77988	1.2822	1.5976	15
46	.62615	.80306	1.2452	.77970	1.2825	1.5971	14
47	.62638	.80354	1.2445	.77952	1.2828	1.5965	13
48	.62660	.80402	1.2437	.77934	1.2831	1.5959	12
49	.62683	.80450	1.2430	.77916	1.2834	1.5953	11
50	.62706	.80498	1.2423	.77897	1.2837	1.5948	10
51	.62728	.80546	1.2415	.77879	1.2840	1.5942	9
52	.62751	.80594	1.2408	.77861	1.2843	1.5936	8
53	.62774	.80642	1.2401	.77843	1.2846	1.5930	7
54	.62796	.80690	1.2393	.77824	1.2849	1.5925	6
55	.62819	.80738	1.2386	.77806	1.2852	1.5919	5
56	.62842	.80786	1.2378	.77788	1.2855	1.5913	4
57	.62864	.80834	1.2371	.77769	1.2859	1.5907	3
58	.62887	.80882	1.2364	.77751	1.2862	1.5902	2
59	.62909	.80930	1.2356	.77733	1.2865	1.5896	1
60	.62932	.80978	1.2349	.77715	1.2868	1.5890	0
'	Cos	Cot	Tan	Sin	Csc	Sec	'

128° (308°)   **(231°) 51°**

39° (219°)   **(320°) 140°**

'	Sin	Tan	Cot	Cos	Sec	Csc	'
0	.62932	.80978	1.2349	.77715	1.2868	1.5890	60
1	.62955	.81027	1.2342	.77696	1.2871	1.5884	59
2	.62977	.81075	1.2334	.77678	1.2874	1.5879	58
3	.63000	.81123	1.2327	.77660	1.2877	1.5873	57
4	.63022	.81171	1.2320	.77641	1.2880	1.5867	56
5	.63045	.81220	1.2312	.77623	1.2883	1.5862	55
6	.63068	.81268	1.2305	.77605	1.2886	1.5856	54
7	.63090	.81316	1.2298	.77586	1.2889	1.5850	53
8	.63113	.81364	1.2290	.77568	1.2892	1.5845	52
9	.63135	.81413	1.2283	.77550	1.2895	1.5839	51
10	.63158	.81461	1.2276	.77531	1.2898	1.5833	50
11	.63180	.81510	1.2268	.77513	1.2901	1.5828	49
12	.63203	.81558	1.2261	.77494	1.2904	1.5822	48
13	.63225	.81606	1.2254	.77476	1.2907	1.5816	47
14	.63248	.81655	1.2247	.77458	1.2910	1.5811	46
15	.63271	.81703	1.2239	.77439	1.2913	1.5805	45
16	.63293	.81752	1.2232	.77421	1.2916	1.5800	44
17	.63316	.81800	1.2225	.77402	1.2919	1.5794	43
18	.63338	.81849	1.2218	.77384	1.2923	1.5788	42
19	.63361	.81898	1.2210	.77366	1.2926	1.5783	41
20	.63383	.81946	1.2203	.77347	1.2929	1.5777	40
21	.63406	.81995	1.2196	.77329	1.2932	1.5771	39
22	.63428	.82044	1.2189	.77310	1.2935	1.5766	38
23	.63451	.82092	1.2181	.77292	1.2938	1.5760	37
24	.63473	.82141	1.2174	.77273	1.2941	1.5755	36
25	.63496	.82190	1.2167	.77255	1.2944	1.5749	35
26	.63518	.82238	1.2160	.77236	1.2947	1.5744	34
27	.63540	.82287	1.2153	.77218	1.2950	1.5738	33
28	.63563	.82336	1.2145	.77199	1.2953	1.5732	32
29	.63585	.82385	1.2138	.77181	1.2957	1.5727	31
30	.63608	.82434	1.2131	.77162	1.2960	1.5721	30
31	.63630	.82483	1.2124	.77144	1.2963	1.5716	29
32	.63653	.82531	1.2117	.77125	1.2966	1.5710	28
33	.63675	.82580	1.2109	.77107	1.2969	1.5705	27
34	.63698	.82629	1.2102	.77088	1.2972	1.5699	26
35	.63720	.82678	1.2095	.77070	1.2975	1.5694	25
36	.63742	.82727	1.2088	.77051	1.2978	1.5688	24
37	.63765	.82776	1.2081	.77033	1.2981	1.5683	23
38	.63787	.82825	1.2074	.77014	1.2985	1.5677	22
39	.63810	.82874	1.2066	.76996	1.2988	1.5672	21
40	.63832	.82923	1.2059	.76977	1.2991	1.5666	20
41	.63854	.82972	1.2052	.76959	1.2994	1.5661	19
42	.63877	.83022	1.2045	.76940	1.2997	1.5655	18
43	.63899	.83071	1.2038	.76921	1.3000	1.5650	17
44	.63922	.83120	1.2031	.76903	1.3003	1.5644	16
45	.63944	.83169	1.2024	.76884	1.3007	1.5639	15
46	.63966	.83218	1.2017	.76866	1.3010	1.5633	14
47	.63989	.83268	1.2009	.76847	1.3013	1.5628	13
48	.64011	.83317	1.2002	.76828	1.3016	1.5622	12
49	.64033	.83366	1.1995	.76810	1.3019	1.5617	11
50	.64056	.83415	1.1988	.76791	1.3022	1.5611	10
51	.64078	.83465	1.1981	.76772	1.3026	1.5606	9
52	.64100	.83514	1.1974	.76754	1.3029	1.5601	8
53	.64123	.83564	1.1967	.76735	1.3032	1.5595	7
54	.64145	.83613	1.1960	.76717	1.3035	1.5590	6
55	.64167	.83662	1.1953	.76698	1.3038	1.5584	5
56	.64190	.83712	1.1946	.76679	1.3041	1.5579	4
57	.64212	.83761	1.1939	.76661	1.3045	1.5573	3
58	.64234	.83811	1.1932	.76642	1.3048	1.5568	2
59	.64256	.83860	1.1925	.76623	1.3051	1.5563	1
60	.64279	.83910	1.1918	.76604	1.3054	1.5557	0
'	Cos	Cot	Tan	Sin	Csc	Sec	'

129° (309°)   **(230°) 50°**

NATURAL TRIGONOMETRIC FUNCTIONS TO FIVE PLACES (continued)

40° (220°)   (319°) 139°   **41° (221°)**   (318°) **138°**

′	Sin	Tan	Cot	Cos	Sec	Csc	′	′	Sin	Tan	Cot	Cos	Sec	Csc	′
0	.64279	.83910	1.1918	.76604	1.3054	1.5557	60	0	.65606	.86929	1.1504	.75471	1.3250	1.5243	60
1	.64301	.83960	1.1910	.76586	1.3057	1.5552	59	1	.65628	.86980	1.1497	.75452	1.3253	1.5237	59
2	.64323	.84009	1.1903	.76567	1.3060	1.5546	58	2	.65650	.87031	1.1490	.75433	1.3257	1.5232	58
3	.64346	.84059	1.1896	.76548	1.3064	1.5541	57	3	.65672	.87082	1.1483	.75414	1.3260	1.5227	57
4	.64368	.84108	1.1889	.76530	1.3067	1.5536	56	4	.65694	.87133	1.1477	.75395	1.3264	1.5222	56
5	.64390	.84158	1.1882	.76511	1.3070	1.5530	55	5	.65716	.87184	1.1470	.75375	1.3267	1.5217	55
6	.64412	.84208	1.1875	.76492	1.3073	1.5525	54	6	.65738	.87236	1.1463	.75356	1.3270	1.5212	54
7	.64435	.84258	1.1868	.76473	1.3076	1.5520	53	7	.65759	.87287	1.1456	.75337	1.3274	1.5207	53
8	.64457	.84307	1.1861	.76455	1.3080	1.5514	52	8	.65781	.87338	1.1450	.75318	1.3277	1.5202	52
9	.64479	.84357	1.1854	.76436	1.3083	1.5509	51	9	.65803	.87389	1.1443	.75299	1.3280	1.5197	51
10	.64501	.84407	1.1847	.76417	1.3086	1.5504	50	10	.65825	.87441	1.1436	.75280	1.3284	1.5192	50
11	.64524	.84457	1.1840	.76398	1.3089	1.5498	49	11	.65847	.87492	1.1430	.75261	1.3287	1.5187	49
12	.64546	.84507	1.1833	.76380	1.3093	1.5493	48	12	.65869	.87543	1.1423	.75241	1.3291	1.5182	48
13	.64568	.84556	1.1826	.76361	1.3096	1.5488	47	13	.65891	.87595	1.1416	.75222	1.3294	1.5177	47
14	.64590	.84606	1.1819	.76342	1.3099	1.5482	46	14	.65913	.87646	1.1410	.75203	1.3297	1.5172	46
15	.64612	.84656	1.1812	.76323	1.3102	1.5477	45	15	.65935	.87698	1.1403	.75184	1.3301	1.5167	45
16	.64635	.84706	1.1806	.76304	1.3105	1.5472	44	16	.65956	.87749	1.1396	.75165	1.3304	1.5162	44
17	.64657	.84756	1.1799	.76286	1.3109	1.5466	43	17	.65978	.87801	1.1389	.75146	1.3307	1.5156	43
18	.64679	.84806	1.1792	.76267	1.3112	1.5461	42	18	.66000	.87852	1.1383	.75126	1.3311	1.5151	42
19	.64701	.84856	1.1785	.76248	1.3115	1.5456	41	19	.66022	.87904	1.1376	.75107	1.3314	1.5146	41
20	.64723	.84906	1.1778	.76229	1.3118	1.5450	40	20	.66044	.87955	1.1369	.75088	1.3318	1.5141	40
21	.64746	.84956	1.1771	.76210	1.3122	1.5445	39	21	.66066	.88007	1.1363	.75069	1.3321	1.5136	39
22	.64768	.85006	1.1764	.76192	1.3125	1.5440	38	22	.66088	.88059	1.1356	.75050	1.3325	1.5131	38
23	.64790	.85057	1.1757	.76173	1.3128	1.5435	37	23	.66109	.88110	1.1349	.75030	1.3328	1.5126	37
24	.64812	.85107	1.1750	.76154	1.3131	1.5429	36	24	.66131	.88162	1.1343	.75011	1.3331	1.5121	36
25	.64834	.85157	1.1743	.76135	1.3135	1.5424	35	25	.66153	.88214	1.1336	.74992	1.3335	1.5116	35
26	.64856	.85207	1.1736	.76116	1.3138	1.5419	34	26	.66175	.88265	1.1329	.74973	1.3338	1.5111	34
27	.64878	.85257	1.1729	.76097	1.3141	1.5413	33	27	.66197	.88317	1.1323	.74953	1.3342	1.5107	33
28	.64901	.85308	1.1722	.76078	1.3144	1.5408	32	28	.66218	.88369	1.1316	.74934	1.3345	1.5102	32
29	.64923	.85358	1.1715	.76059	1.3148	1.5403	31	29	.66240	.88421	1.1310	.74915	1.3348	1.5097	31
30	.64945	.85408	1.1708	.76041	1.3151	1.5398	30	30	.66262	.88473	1.1303	.74896	1.3352	1.5092	30
31	.64967	.85458	1.1702	.76022	1.3154	1.5392	29	31	.66284	.88524	1.1296	.74876	1.3355	1.5087	29
32	.64989	.85509	1.1695	.76003	1.3157	1.5387	28	32	.66306	.88576	1.1290	.74857	1.3359	1.5082	28
33	.65011	.85559	1.1688	.75984	1.3161	1.5382	27	33	.66327	.88628	1.1283	.74838	1.3362	1.5077	27
34	.65033	.85609	1.1681	.75965	1.3164	1.5377	26	34	.66349	.88680	1.1276	.74818	1.3366	1.5072	26
35	.65055	.85660	1.1674	.75946	1.3167	1.5372	25	35	.66371	.88732	1.1270	.74799	1.3369	1.5067	25
36	.65077	.85710	1.1667	.75927	1.3171	1.5366	24	36	.66393	.88784	1.1263	.74780	1.3373	1.5062	24
37	.65100	.85761	1.1660	.75908	1.3174	1.5361	23	37	.66414	.88836	1.1257	.74760	1.3376	1.5057	23
38	.65122	.85811	1.1653	.75889	1.3177	1.5356	22	38	.66436	.88888	1.1250	.74741	1.3380	1.5052	22
39	.65144	.85862	1.1647	.75870	1.3180	1.5351	21	39	.66458	.88940	1.1243	.74722	1.3383	1.5047	21
40	.65166	.85912	1.1640	.75851	1.3184	1.5345	20	40	.66480	.88992	1.1237	.74703	1.3386	1.5042	20
41	.65188	.85963	1.1633	.75832	1.3187	1.5340	19	41	.66501	.89045	1.1230	.74683	1.3390	1.5037	19
42	.65210	.86014	1.1626	.75813	1.3190	1.5335	18	42	.66523	.89097	1.1224	.74664	1.3393	1.5032	18
43	.65232	.86064	1.1619	.75794	1.3194	1.5330	17	43	.66545	.89149	1.1217	.74644	1.3397	1.5027	17
44	.65254	.86115	1.1612	.75775	1.3197	1.5325	16	44	.66566	.89201	1.1211	.74625	1.3400	1.5023	16
45	.65276	.86166	1.1606	.75756	1.3200	1.5320	15	45	.66588	.89253	1.1204	.74606	1.3404	1.5018	15
46	.65298	.86216	1.1599	.75738	1.3203	1.5314	14	46	.66610	.89306	1.1197	.74586	1.3407	1.5013	14
47	.65320	.86267	1.1592	.75719	1.3207	1.5309	13	47	.66632	.89358	1.1191	.74567	1.3411	1.5008	13
48	.65342	.86318	1.1585	.75700	1.3210	1.5304	12	48	.66653	.89410	1.1184	.74548	1.3414	1.5003	12
49	.65364	.86368	1.1578	.75680	1.3213	1.5299	11	49	.66675	.89463	1.1178	.74528	1.3418	1.4998	11
50	.65386	.86419	1.1571	.75661	1.3217	1.5294	10	50	.66697	.89515	1.1171	.74509	1.3421	1.4993	10
51	.65408	.86470	1.1565	.75642	1.3220	1.5289	9	51	.66718	.89567	1.1165	.74489	1.3425	1.4988	9
52	.65430	.86521	1.1558	.75623	1.3223	1.5283	8	52	.66740	.89620	1.1158	.74470	1.3428	1.4984	8
53	.65452	.86572	1.1551	.75604	1.3227	1.5278	7	53	.66762	.89672	1.1152	.74451	1.3432	1.4979	7
54	.65474	.86623	1.1544	.75585	1.3230	1.5273	6	54	.66783	.89725	1.1145	.74431	1.3435	1.4974	6
55	.65496	.86674	1.1538	.75566	1.3233	1.5268	5	55	.66805	.89777	1.1139	.74412	1.3439	1.4969	5
56	.65518	.86725	1.1531	.75547	1.3237	1.5263	4	56	.66827	.89830	1.1132	.74392	1.3442	1.4964	4
57	.65540	.86776	1.1524	.75528	1.3240	1.5258	3	57	.66848	.89883	1.1126	.74373	1.3446	1.4959	3
58	.65562	.86827	1.1517	.75509	1.3243	1.5253	2	58	.66870	.89935	1.1119	.74353	1.3449	1.4954	2
59	.65584	.86878	1.1510	.75490	1.3247	1.5248	1	59	.66891	.89988	1.1113	.74334	1.3453	1.4950	1
60	.65606	.86929	1.1504	.75471	1.3250	1.5243	0	60	.66913	.90040	1.1106	.74314	1.3456	1.4945	0
′	Cos	Cot	Tan	Sin	Csc	Sec	′	′	Cos	Cot	Tan	Sin	Csc	Sec	′

130° (310°)   (229°) **49°**   **131° (311°)**   (228°) **48°**

NATURAL TRIGONOMETRIC FUNCTIONS TO FIVE PLACES (continued)

42° (222°)   **(317°) 137°**

′	Sin	Tan	Cot	Cos	Sec	Csc	′
0	.66913	.90040	1.1106	.74314	1.3456	1.4945	60
1	.66935	.90093	1.1100	.74295	1.3460	1.4940	59
2	.66956	.90146	1.1093	.74276	1.3463	1.4935	58
3	.66978	.90199	1.1087	.74256	1.3467	1.4930	57
4	.66999	.90251	1.1080	.74237	1.3470	1.4925	56
5	.67021	.90304	1.1074	.74217	1.3474	1.4921	55
6	.67043	.90357	1.1067	.74198	1.3478	1.4916	54
7	.67064	.90410	1.1061	.74178	1.3481	1.4911	53
8	.67086	.90463	1.1054	.74159	1.3485	1.4906	52
9	.67107	.90516	1.1048	.74139	1.3488	1.4901	51
10	.67129	.90569	1.1041	.74120	1.3492	1.4897	50
11	.67151	.90621	1.1035	.74100	1.3495	1.4892	49
12	.67172	.90674	1.1028	.74080	1.3499	1.4887	48
13	.67194	.90727	1.1022	.74061	1.3502	1.4882	47
14	.67215	.90781	1.1016	.74041	1.3506	1.4878	46
15	.67237	.90834	1.1009	.74022	1.3510	1.4873	45
16	.67258	.90887	1.1003	.74002	1.3513	1.4868	44
17	.67280	.90940	1.0996	.73983	1.3517	1.4863	43
18	.67301	.90993	1.0990	.73963	1.3520	1.4859	42
19	.67323	.91046	1.0983	.73944	1.3524	1.4854	41
20	.67344	.91099	1.0977	.73924	1.3527	1.4849	40
21	.67366	.91153	1.0971	.73904	1.3531	1.4844	39
22	.67387	.91206	1.0964	.73885	1.3535	1.4840	38
23	.67409	.91259	1.0958	.73865	1.3538	1.4835	37
24	.67430	.91313	1.0951	.73846	1.3542	1.4830	36
25	.67452	.91366	1.0945	.73826	1.3545	1.4825	35
26	.67473	.91419	1.0939	.73806	1.3549	1.4821	34
27	.67495	.91473	1.0932	.73787	1.3553	1.4816	33
28	.67516	.91526	1.0926	.73767	1.3556	1.4811	32
29	.67538	.91580	1.0919	.73747	1.3560	1.4807	31
30	.67559	.91633	1.0913	.73728	1.3563	1.4802	30
31	.67580	.91687	1.0907	.73708	1.3567	1.4797	29
32	.67602	.91740	1.0900	.73688	1.3571	1.4792	28
33	.67623	.91794	1.0894	.73669	1.3574	1.4788	27
34	.67645	.91847	1.0888	.73649	1.3578	1.4783	26
35	.67666	.91901	1.0881	.73629	1.3582	1.4778	25
36	.67688	.91955	1.0875	.73610	1.3585	1.4774	24
37	.67709	.92008	1.0869	.73590	1.3589	1.4769	23
38	.67730	.92062	1.0862	.73570	1.3592	1.4764	22
39	.67752	.92116	1.0856	.73551	1.3596	1.4760	21
40	.67773	.92170	1.0850	.73531	1.3600	1.4755	20
41	.67795	.92224	1.0843	.73511	1.3603	1.4750	19
42	.67816	.92277	1.0837	.73491	1.3607	1.4746	18
43	.67837	.92331	1.0831	.73472	1.3611	1.4741	17
44	.67859	.92385	1.0824	.73452	1.3614	1.4737	16
45	.67880	.92439	1.0818	.73432	1.3618	1.4732	15
46	.67901	.92493	1.0812	.73413	1.3622	1.4727	14
47	.67923	.92547	1.0805	.73393	1.3625	1.4723	13
48	.67944	.92601	1.0799	.73373	1.3629	1.4718	12
49	.67965	.92655	1.0793	.73353	1.3633	1.4713	11
50	.67987	.92709	1.0786	.73333	1.3636	1.4709	10
51	.68008	.92763	1.0780	.73314	1.3640	1.4704	9
52	.68029	.92817	1.0774	.73294	1.3644	1.4700	8
53	.68051	.92872	1.0768	.73274	1.3647	1.4695	7
54	.68072	.92926	1.0761	.73254	1.3651	1.4690	6
55	.68093	.92980	1.0755	.73234	1.3655	1.4686	5
56	.68115	.93034	1.0749	.73215	1.3658	1.4681	4
57	.68136	.93088	1.0742	.73195	1.3662	1.4677	3
58	.68157	.93143	1.0736	.73175	1.3666	1.4672	2
59	.68179	.93197	1.0730	.73155	1.3670	1.4667	1
60	.68200	.93252	1.0724	.73135	1.3673	1.4663	0
′	Cos	Cot	Tan	Sin	Csc	Sec	′

132° (312°)   **(227°) 47°**

43° (223°)   **(316°) 136°**

′	Sin	Tan	Cot	Cos	Sec	Csc	′
0	.68200	.93252	1.0724	.73135	1.3673	1.4663	60
1	.68221	.93306	1.0717	.73116	1.3677	1.4658	59
2	.68242	.93360	1.0711	.73096	1.3681	1.4654	58
3	.68264	.93415	1.0705	.73076	1.3684	1.4649	57
4	.68285	.93469	1.0699	.73056	1.3688	1.4645	56
5	.68306	.93524	1.0692	.73036	1.3692	1.4640	55
6	.68327	.93578	1.0686	.73016	1.3696	1.4635	54
7	.68349	.93633	1.0680	.72996	1.3699	1.4631	53
8	.68370	.93688	1.0674	.72976	1.3703	1.4626	52
9	.68391	.93742	1.0668	.72957	1.3707	1.4622	51
10	.68412	.93797	1.0661	.72937	1.3711	1.4617	50
11	.68434	.93852	1.0655	.72917	1.3714	1.4613	49
12	.68455	.93906	1.0649	.72897	1.3718	1.4608	48
13	.68476	.93961	1.0643	.72877	1.3722	1.4604	47
14	.68497	.94016	1.0637	.72857	1.3726	1.4599	46
15	.68518	.94071	1.0630	.72837	1.3729	1.4595	45
16	.68539	.94125	1.0624	.72817	1.3733	1.4590	44
17	.68561	.94180	1.0618	.72797	1.3737	1.4586	43
18	.68582	.94235	1.0612	.72777	1.3741	1.4581	42
19	.68603	.94290	1.0606	.72757	1.3744	1.4577	41
20	.68624	.94345	1.0599	.72737	1.3748	1.4572	40
21	.68645	.94400	1.0593	.72717	1.3752	1.4568	39
22	.68666	.94455	1.0587	.72697	1.3756	1.4563	38
23	.68688	.94510	1.0581	.72677	1.3759	1.4559	37
24	.68709	.94565	1.0575	.72657	1.3763	1.4554	36
25	.68730	.94620	1.0569	.72637	1.3767	1.4550	35
26	.68751	.94676	1.0562	.72617	1.3771	1.4545	34
27	.68772	.94731	1.0556	.72597	1.3775	1.4541	33
28	.68793	.94786	1.0550	.72577	1.3778	1.4536	32
29	.68814	.94841	1.0544	.72557	1.3782	1.4532	31
30	.68835	.94896	1.0538	.72537	1.3786	1.4527	30
31	.68857	.94952	1.0532	.72517	1.3790	1.4523	29
32	.68878	.95007	1.0526	.72497	1.3794	1.4518	28
33	.68899	.95062	1.0519	.72477	1.3797	1.4514	27
34	.68920	.95118	1.0513	.72457	1.3801	1.4510	26
35	.68941	.95173	1.0507	.72437	1.3805	1.4505	25
36	.68962	.95229	1.0501	.72417	1.3809	1.4501	24
37	.68983	.95284	1.0495	.72397	1.3813	1.4496	23
38	.69004	.95340	1.0489	.72377	1.3817	1.4492	22
39	.69025	.95395	1.0483	.72357	1.3820	1.4487	21
40	.69046	.95451	1.0477	.72337	1.3824	1.4483	20
41	.69067	.95506	1.0470	.72317	1.3828	1.4479	19
42	.69088	.95562	1.0464	.72297	1.3832	1.4474	18
43	.69109	.95618	1.0458	.72277	1.3836	1.4470	17
44	.69130	.95673	1.0452	.72257	1.3840	1.4465	16
45	.69151	.95729	1.0446	.72236	1.3843	1.4461	15
46	.69172	.95785	1.0440	.72216	1.3847	1.4457	14
47	.69193	.95841	1.0434	.72196	1.3851	1.4452	13
48	.69214	.95897	1.0428	.72176	1.3855	1.4448	12
49	.69235	.95952	1.0422	.72156	1.3859	1.4443	11
50	.69256	.96008	1.0416	.72136	1.3863	1.4439	10
51	.69277	.96064	1.0410	.72116	1.3867	1.4435	9
52	.69298	.96120	1.0404	.72095	1.3871	1.4430	8
53	.69319	.96176	1.0398	.72075	1.3874	1.4426	7
54	.69340	.96232	1.0392	.72055	1.3878	1.4422	6
55	.69361	.96288	1.0385	.72035	1.3882	1.4417	5
56	.69382	.96344	1.0379	.72015	1.3886	1.4413	4
57	.69403	.96400	1.0373	.71995	1.3890	1.4409	3
58	.69424	.96457	1.0367	.71974	1.3894	1.4404	2
59	.69445	.96513	1.0361	.71954	1.3898	1.4400	1
60	.69466	.96569	1.0355	.71934	1.3902	1.4396	0
′	Cos	Cot	Tan	Sin	Csc	Sec	′

133° (313°)   **(226°) 46°**

NATURAL TRIGONOMETRIC FUNCTIONS TO FIVE PLACES (continued)

44° (224°)   (315°) 135°

′	Sin	Tan	Cot	Cos	Sec	Csc	′
0	.69466	.96569	1.0355	.71934	1.3902	1.4396	60
1	.69487	.96625	1.0349	.71914	1.3906	1.4391	59
2	.69508	.96681	1.0343	.71894	1.3909	1.4387	58
3	.69529	.96738	1.0337	.71873	1.3913	1.4383	57
4	.69549	.96794	1.0331	.71853	1.3917	1.4378	56
5	.69570	.96850	1.0325	.71833	1.3921	1.4374	55
6	.69591	.96907	1.0319	.71813	1.3925	1.4370	54
7	.69612	.96963	1.0313	.71792	1.3929	1.4365	53
8	.69633	.97020	1.0307	.71772	1.3933	1.4361	52
9	.69654	.97076	1.0301	.71752	1.3937	1.4357	51
10	.69675	.97133	1.0295	.71732	1.3941	1.4352	50
11	.69696	.97189	1.0289	.71711	1.3945	1.4348	49
12	.69717	.97246	1.0283	.71691	1.3949	1.4344	48
13	.69737	.97302	1.0277	.71671	1.3953	1.4340	47
14	.69758	.97359	1.0271	.71650	1.3957	1.4335	46
15	.69779	.97416	1.0265	.71630	1.3961	1.4331	45
16	.69800	.97472	1.0259	.71610	1.3965	1.4327	44
17	.69821	.97529	1.0253	.71590	1.3969	1.4322	43
18	.69842	.97586	1.0247	.71569	1.3972	1.4318	42
19	.69862	.97643	1.0241	.71549	1.3976	1.4314	41
20	.69883	.97700	1.0235	.71529	1.3980	1.4310	40
21	.69904	.97756	1.0230	.71508	1.3984	1.4305	39
22	.69925	.97813	1.0224	.71488	1.3988	1.4301	38
23	.69946	.97870	1.0218	.71468	1.3992	1.4297	37
24	.69966	.97927	1.0212	.71447	1.3996	1.4293	36
25	.69987	.97984	1.0206	.71427	1.4000	1.4288	35
26	.70008	.98041	1.0200	.71407	1.4004	1.4284	34
27	.70029	.98098	1.0194	.71386	1.4008	1.4280	33
28	.70049	.98155	1.0188	.71366	1.4012	1.4276	32
29	.70070	.98213	1.0182	.71345	1.4016	1.4271	31
30	.70091	.98270	1.0176	.71325	1.4020	1.4267	30
31	.70112	.98327	1.0170	.71305	1.4024	1.4263	29
32	.70132	.98384	1.0164	.71284	1.4028	1.4259	28
33	.70153	.98441	1.0158	.71264	1.4032	1.4255	27
34	.70174	.98499	1.0152	.71243	1.4036	1.4250	26
35	.70195	.98556	1.0147	.71223	1.4040	1.4246	25
36	.70215	.98613	1.0141	.71203	1.4044	1.4242	24
37	.70236	.98671	1.0135	.71182	1.4048	1.4238	23
38	.70257	.98728	1.0129	.71162	1.4052	1.4234	22
39	.70277	.98786	1.0123	.71141	1.4057	1.4229	21
40	.70298	.98843	1.0117	.71121	1.4061	1.4225	20
41	.70319	.98901	1.0111	.71100	1.4065	1.4221	19
42	.70339	.98958	1.0105	.71080	1.4069	1.4217	18
43	.70360	.99016	1.0099	.71059	1.4073	1.4213	17
44	.70381	.99073	1.0094	.71039	1.4077	1.4208	16
45	.70401	.99131	1.0088	.71019	1.4081	1.4204	15
46	.70422	.99189	1.0082	.70998	1.4085	1.4200	14
47	.70443	.99247	1.0076	.70978	1.4089	1.4196	13
48	.70463	.99304	1.0070	.70957	1.4093	1.4192	12
49	.70484	.99362	1.0064	.70937	1.4097	1.4188	11
50	.70505	.99420	1.0058	.70916	1.4101	1.4183	10
51	.70525	.99478	1.0052	.70896	1.4105	1.4179	9
52	.70546	.99536	1.0047	.70875	1.4109	1.4175	8
53	.70567	.99594	1.0041	.70855	1.4113	1.4171	7
54	.70587	.99652	1.0035	.70834	1.4118	1.4167	6
55	.70608	.99710	1.0029	.70813	1.4122	1.4163	5
56	.70628	.99768	1.0023	.70793	1.4126	1.4159	4
57	.70649	.99826	1.0017	.70772	1.4130	1.4154	3
58	.70670	.99884	1.0012	.70752	1.4134	1.4150	2
59	.70690	.99942	1.0006	.70731	1.4138	1.4146	1
60	.70711	1.0000	1.0000	.70711	1.4142	1.4142	0
′	Cos	Cot	Tan	Sin	Csc	Sec	′

134° (314°)   (225°) 45°

NATURAL TRIGONOMETRIC FUNCTIONS

FOR ANGLES IN RADIANS

x	Sin	Tan	Cot	Cos	x	Sin	Tan	Cot	Cos
.00	.00000	.00000	∞	1.00000	**.50**	.47943	.54630	1.8305	.87758
.01	.01000	.01000	99.997	0.99995	.51	.48818	.55936	1.7878	.87274
.02	.02000	.02000	49.993	.99980	.52	.49688	.57256	1.7465	.86782
.03	.03000	.03001	33.323	.99955	.53	.50553	.58592	1.7067	.86281
.04	.03999	.04002	24.987	.99920	.54	.51414	.59943	1.6683	.85771
.05	.04998	.05004	19.983	.99875	.55	.52269	.61311	1.6310	.85252
.06	.05996	.06007	16.647	.99820	.56	.53119	.62695	1.5950	.84726
.07	.06994	.07011	14.262	.99755	.57	.53963	.64097	1.5601	.84190
.08	.07991	.08017	12.473	.99680	.58	.54802	.65517	1.5263	.83646
.09	.08988	.09024	11.081	.99595	.59	.55636	.66956	1.4935	.83094
.10	.09983	.10033	9.9666	.99500	**.60**	.56464	.68414	1.4617	.82534
.11	.10978	.11045	9.0542	.99396	.61	.57287	.69892	1.4308	.81965
.12	.11971	.12058	8.2933	.99281	.62	.58104	.71391	1.4007	.81388
.13	.12963	.13074	7.6489	.99156	.63	.58914	.72911	1.3715	.80803
.14	.13954	.14092	7.0961	.99022	.64	.59720	.74454	1.3431	.80210
.15	.14944	.15114	6.6166	.98877	.65	.60519	.76020	1.3154	.79608
.16	.15932	.16138	6.1966	.98723	.66	.61312	.77610	1.2885	.78999
.17	.16918	.17166	5.8256	.98558	.67	.62099	.79225	1.2622	.78382
.18	.17903	.18197	5.4954	.98384	.68	.62879	.80866	1.2366	.77757
.19	.18886	.19232	5.1997	.98200	.69	.63654	.82534	1.2116	.77125
.20	.19867	.20271	4.9332	.98007	**.70**	.64422	.84229	1.1872	.76484
.21	.20846	.21314	4.6917	.97803	.71	.65183	.85953	1.1634	.75836
.22	.21823	.22362	4.4719	.97590	.72	.65938	.87707	1.1402	.75181
.23	.22798	.23414	4.2709	.97367	.73	.66687	.89492	1.1174	.74517
.24	.23770	.24472	4.0864	.97134	.74	.67429	.91309	1.0952	.73847
.25	.24740	.25534	3.9163	.96891	.75	.68164	.93160	1.0734	.73169
.26	.25708	.26602	3.7591	.96639	.76	.68892	.95045	1.0521	.72484
.27	.26673	.27676	3.6133	.96377	.77	.69614	.96967	1.0313	.71791
.28	.27636	.28755	3.4776	.96106	.78	.70328	.98926	1.0109	.71091
.29	.28595	.29841	3.3511	.95824	.79	.71035	1.0092	.99084	.70385
.30	.29552	.30934	3.2327	.95534	**.80**	.71736	1.0296	.97121	.69671
.31	.30506	.32033	3.1218	.95233	.81	.72429	1.0505	.95197	.68950
.32	.31457	.33139	3.0176	.94924	.82	.73115	1.0717	.93309	.68222
.33	.32404	.34252	2.9195	.94604	.83	.73793	1.0934	.91455	.67488
.34	.33349	.35374	2.8270	.94275	.84	.74464	1.1156	.89635	.66746
.35	.34290	.36503	2.7395	.93937	.85	.75128	1.1383	.87848	.65998
.36	.35227	.37640	2.6567	.93590	.86	.75784	1.1616	.86091	.65244
.37	.36162	.38786	2.5782	.93233	.87	.76433	1.1853	.84365	.64483
.38	.37092	.39941	2.5037	.92866	.88	.77074	1.2097	.82668	.63715
.39	.38019	.41105	2.4328	.92491	.89	.77707	1.2346	.80998	.62941
.40	.38942	.42279	2.3652	.92106	**.90**	.78333	1.2602	.79355	.62161
.41	.39861	.43463	2.3008	.91712	.91	.78950	1.2864	.77738	.61375
.42	.40776	.44657	2.2393	.91309	.92	.79560	1.3133	.76146	.60582
.43	.41687	.45862	2.1804	.90897	.93	.80162	1.3409	.74578	.59783
.44	.42594	.47078	2.1241	.90475	.94	.80756	1.3692	.73034	.58979
.45	.43497	.48306	2.0702	.90045	.95	.81342	1.3984	.71511	.58168
.46	.44395	.49545	2.0184	.89605	.96	.81919	1.4284	.70010	.57352
.47	.45289	.50797	1.9686	.89157	.97	.82489	1.4592	.68531	.56530
.48	.46178	.52061	1.9208	.88699	.98	.83050	1.4910	.67071	.55702
.49	.47063	.53339	1.8748	.88233	.99	.83603	1.5237	.65631	.54869
.50	.47943	.54630	1.8305	.87758	**1.00**	.84147	1.5574	.64209	.54030
x	Sin	Tan	Cot	Cos	x	Sin	Tan	Cot	Cos

NATURAL TRIGONOMETRIC FUNCTIONS
FOR ANGLES IN RADIANS (Continued)

x	Sin	Tan	Cot	Cos	x	Sin	Tan	Cot	Cos
1.00	.84147	1.5574	.64209	.54030	1.50	.99749	14.101	.07091	.07074
1.01	.84683	1.5922	.62806	.53186	1.51	.99815	16.428	.06087	.06076
1.02	.85211	1.6281	.61420	.52337	1.52	.99871	19.670	.05084	.05077
1.03	.85730	1.6652	.60051	.51482	1.53	.99917	24.498	.04082	.04079
1.04	.86240	1.7036	.58699	.50622	1.54	.99953	32.461	.03081	.03079
1.05	.86742	1.7433	.57362	.49757	1.55	.99978	48.078	.02080	.02079
1.06	.87236	1.7844	.56040	.48887	1.56	.99994	92.620	.01080	.01080
1.07	.87720	1.8270	.54734	.48012	1.57	1.00000	1255.8	.00080	.00080
1.08	.88196	1.8712	.53441	.47133	1.58	.99996	−108.65	−.00920	−.00920
1.09	.88663	1.9171	.52162	.46249	1.59	.99982	−52.067	−.01921	−.01920
1.10	.89121	1.9648	.50897	.45360	1.60	.99957	−34.233	−.02921	−.02920
1.11	.89570	2.0143	.49644	.44466	1.61	.99923	−25.495	−.03922	−.03919
1.12	.90010	2.0660	.48404	.43568	1.62	.99879	−20.307	−.04924	−.04918
1.13	.90441	2.1198	.47175	.42666	1.63	.99825	−16.871	−.05927	−.05917
1.14	.90863	2.1759	.45959	.41759	1.64	.99761	−14.427	−.06931	−.06915
1.15	.91276	2.2345	.44753	.40849	1.65	.99687	−12.599	−.07397	−.07912
1.16	.91680	2.2958	.43558	.39934	1.66	.99602	−11.181	−.08944	−.08909
1.17	.92075	2.3600	.42373	.39015	1.67	.99508	−10.047	−.09953	−.09904
1.18	.92461	2.4273	.41199	.38092	1.68	.99404	− 9.1208	−.10964	−.10899
1.19	.92837	2.4979	.40034	.37166	1.69	.99290	− 8.3492	−.11977	−.11892
1.20	.93204	2.5722	.38878	.36236	1.70	.99166	− 7.6966	−.12993	−.12884
1.21	.93562	2.6503	.37731	.35302	1.71	.99033	− 7.1373	−.14011	−.13875
1.22	.93910	2.7328	.36593	.34365	1.72	.98889	− 6.6524	−.15032	−.14865
1.23	.94249	2.8198	.35463	.33424	1.73	.98735	− 6.2281	−.16056	−.15853
1.24	.94578	2.9119	.34341	.32480	1.74	.98572	− 5.8535	−.17084	−.16840
1.25	.94898	3.0096	.33227	.31532	1.75	.98399	− 5.5204	−.18115	−.17825
1.26	.95209	3.1133	.32121	.30582	1.76	.98215	− 5.2221	−.19149	−.18808
1.27	.95510	3.2236	.31021	.29628	1.77	.98022	− 4.9534	−.20188	−.19789
1.28	.95802	3.3413	.29928	.28672	1.78	.97820	− 4.7101	−.21231	−.20768
1.29	.96084	3.4672	.28842	.27712	1.79	.97607	− 4.4887	−.22278	−.21745
1.30	.96356	3.6021	.27762	.26750	1.80	.97385	− 4.2863	−.23330	−.22720
1.31	.96618	3.7471	.26687	.25785	1.81	.97153	− 4.1005	−.24387	−.23693
1.32	.96872	3.9033	.25619	.24818	1.82	.96911	− 3.9294	−.25449	−.24663
1.33	.97115	4.0723	.24556	.23848	1.83	.96659	− 3.7712	−.26517	−.25631
1.34	.97348	4.2556	.23498	.22875	1.84	.96398	− 3.6245	−.27590	−.26596
1.35	.97572	4.4552	.22446	.21901	1.85	.96128	− 3.4881	−.28669	−.27559
1.36	.97786	4.6734	.21398	.20924	1.86	.95847	− 3.3608	−.29755	−.28519
1.37	.97991	4.9131	.20354	.19945	1.87	.95557	− 2.2419	−.30846	−.29476
1.38	.98185	5.1774	.19315	.18964	1.88	.95258	− 3.1304	−.31945	−.30430
1.39	.98370	5.4707	.18279	.17981	1.89	.94949	− 3.0257	−.33051	−.31381
1.40	.98545	5.7979	.17248	.16997	1.90	.94630	− 2.9271	−.34164	−.32329
1.41	.98710	6.1654	.16220	.16010	1.91	.94302	− 2.8341	−.35284	−.33274
1.42	.98865	6.5811	.15195	.15023	1.92	.93965	− 2.7463	−.36413	−.34215
1.43	.99010	7.0555	.14173	.14033	1.93	.93618	− 2.6632	−.37549	−.35153
1.44	.99146	7.6018	.13155	.13042	1.94	.93262	− 2.5843	−.38695	−.36087
1.45	.99271	8.2381	.12139	.12050	1.95	.92896	− 2.5095	−.39849	−.37018
1.46	.99387	8.9886	.11125	.11057	1.96	.92521	− 2.4383	−.41012	−.37945
1.47	.99492	9.8874	.10114	.10063	1.97	.92137	− 2.3705	−.42185	−.38868
1.48	.99588	10.983	.09105	.09067	1.98	.91744	− 2.3058	−.43368	−.39788
1.49	.99674	12.350	.08097	.08071	1.99	.91341	− 2.2441	−.44562	−.40703
1.50	.99749	14.101	.07091	.07074	2.00	.90930	− 2.1850	−.45766	−.41615
x	Sin	Tan	Cot	Cos	x	Sin	Tan	Cot	Cos

RADIX TABLE FOR CIRCULAR SINES AND COSINES IN RADIANS

x	n	$\sin [x(10^{-n})]$	$\cos [x(10^{-n})]$
1	10	0.00000 00001 00000 00000 00000	0.99999 99999 99999 99999 50000
2	10	0.00000 00002 00000 00000 00000	0.99999 99999 99999 99998 00000
3	10	0.00000 00003 00000 00000 00000	0.99999 99999 99999 99995 50000
4	10	0.00000 00004 00000 00000 00000	0.99999 99999 99999 99992 00000
5	10	0.00000 00005 00000 00000 00000	0.99999 99999 99999 99987 50000
6	10	0.00000 00006 00000 00000 00000	0.99999 99999 99999 99982 00000
7	10	0.00000 00007 00000 00000 00000	0.99999 99999 99999 99975 50000
8	10	0.00000 00008 00000 00000 00000	0.99999 99999 99999 99968 00000
9	10	0.00000 00009 00000 00000 00000	0.99999 99999 99999 99959 50000
1	9	0.00000 00010 00000 00000 00000	0.99999 99999 99999 99950 00000
2	9	0.00000 00020 00000 00000 00000	0.99999 99999 99999 99800 00000
3	9	0.00000 00030 00000 00000 00000	0.99999 99999 99999 99550 00000
4	9	0.00000 00040 00000 00000 00000	0.99999 99999 99999 99200 00000
5	9	0.00000 00050 00000 00000 00000	0.99999 99999 99999 98750 00000
6	9	0.00000 00060 00000 00000 00000	0.99999 99999 99999 98200 00000
7	9	0.00000 00069 99999 99999 99999	0.99999 99999 99999 97550 00000
8	9	0.00000 00079 99999 99999 99999	0.99999 99999 99999 96800 00000
9	9	0.00000 00089 99999 99999 99999	0.99999 99999 99999 95950 00000
1	8	0.00000 00099 99999 99999 99998	0.99999 99999 99999 95000 00000
2	8	0.00000 00199 99999 99999 99987	0.99999 99999 99999 80000 00000
3	8	0.00000 00299 99999 99999 99955	0.99999 99999 99999 55000 00000
4	8	0.00000 00399 99999 99999 99893	0.99999 99999 99999 20000 00000
5	8	0.00000 00499 99999 99999 99792	0.99999 99999 99998 75000 00000
6	8	0.00000 00599 99999 99999 99640	0.99999 99999 99998 20000 00000
7	8	0.00000 00699 99999 99999 99428	0.99999 99999 99997 55000 00000
8	8	0.00000 00799 99999 99999 99147	0.99999 99999 99996 80000 00000
9	8	0.00000 00899 99999 99999 98785	0.99999 99999 99995 95000 00000
1	7	0.00000 00999 99999 99999 98333	0.99999 99999 99995 00000 00000
2	7	0.00000 01999 99999 99999 86667	0.99999 99999 99980 00000 00000
3	7	0.00000 02999 99999 99999 55000	0.99999 99999 99955 00000 00000
4	7	0.00000 03999 99999 99998 93333	0.99999 99999 99920 00000 00000
5	7	0.00000 04999 99999 99997 91667	0.99999 99999 99875 00000 00000
6	7	0.00000 05999 99999 99996 40000	0.99999 99999 99820 00000 00000
7	7	0.00000 06999 99999 99994 28333	0.99999 99999 99755 00000 00000
8	7	0.00000 07999 99999 99991 46667	0.99999 99999 99680 00000 00000
9	7	0.00000 08999 99999 99987 85000	0.99999 99999 99595 00000 00000
1	6	0.00000 09999 99999 99983 33333	0.99999 99999 99500 00000 00000
2	6	0.00000 19999 99999 99866 66667	0.99999 99999 98000 00000 00007
3	6	0.00000 29999 99999 99550 00000	0.99999 99999 95500 00000 00034
4	6	0.00000 39999 99999 98933 33333	0.99999 99999 92000 00000 00107
5	6	0.00000 49999 99999 97916 66667	0.99999 99999 87500 00000 00260
6	6	0.00000 59999 99999 96400 00000	0.99999 99999 82000 00000 00540
7	6	0.00000 69999 99999 94283 33333	0.99999 99999 75500 00000 01000
8	6	0.00000 79999 99999 91466 66667	0.99999 99999 68000 00000 01707
9	6	0.00000 89999 99999 87850 00000	0.99999 99999 59500 00000 02734

For $n > 10$, $\sin [x(10^{-n})] = x(10^{-n})$; $\cos [x(10^{-n})] = 1 - \tfrac{1}{2} x^2 (10^{-2n})$; to 25 decimals.

RADIX TABLE FOR CIRCULAR SINES AND COSINES IN RADIANS

x	n	$\sin [x(10^{-n})]$	$\cos [x(10^{-n})]$
1	5	0.00000 99999 99999 83333 33333	0.99999 99999 50000 00000 04167
2	5	0.00001 99999 99998 66666 66667	0.99999 99998 00000 00000 66667
3	5	0.00002 99999 99995 50000 00002	0.99999 99995 50000 00003 37500
4	5	0.00003 99999 99989 33333 33342	0.99999 99992 00000 00010 66667
5	5	0.00004 99999 99979 16666 66693	0.99999 99987 50000 00026 04167
6	5	0.00005 99999 99964 00000 00065	0.99999 99982 00000 00054 00000
7	5	0.00006 99999 99942 83333 33473	0.99999 99975 50000 00100 04167
8	5	0.00007 99999 99914 66666 66940	0.99999 99968 00000 00170 66667
9	5	0.00008 99999 99878 50000 00492	0.99999 99959 50000 00273 37500
1	4	0.00009 99999 99833 33333 34167	0.99999 99950 00000 00416 66667
2	4	0.00019 99999 98666 66666 93333	0.99999 99800 00000 06666 66666
3	4	0.00029 99999 95500 00002 02500	0.99999 99550 00000 33749 99990
4	4	0.00039 99999 89333 33341 86667	0.99999 99200 00001 06666 66610
5	4	0.00049 99999 79166 66692 70833	0.99999 98750 00002 60416 66450
6	4	0.00059 99999 64000 00064 80000	0.99999 98200 00005 39999 99352
7	4	0.00069 99999 42833 33473 39167	0.99999 97550 00010 00416 65033
8	4	0.00079 99999 14666 66939 73333	0.99999 96800 00017 06666 63026
9	4	0.00089 99998 78500 00492 07499	0.99999 95950 00027 33749 92619
1	3	0.00099 99998 33333 34166 66665	0.99999 95000 00041 66666 52778

For $n > 10$, $\sin [x(10^{-n})] = x(10^{-n})$; $\cos [x(10^{-n})] = 1 - \frac{1}{2}x^2(10^{-2n})$; to 25 decimals.

HAVERSINES

The table that follows gives —values of $\frac{1}{2}(1 - \cos \theta)$ for angles between 0 and 180° are given to five significant figures. The five-place mantissas of the logarithms of the haversines are also given. The correct characteristic must be provided in each case.

The listed values of the haversines were derived from values which were computed to seven significant figures. The logarithms were independently derived from the more exact value of the haversines and are, therefore, in many cases not the exact value of the logarithm of the haversine as listed. This is notably true at the beginning of the table where the logarithm can be given with more exactness than the function.

HAVERSINES

hav $\theta = \frac{1}{2}$ vers $\theta = \frac{1}{2}(1 - \cos \theta) = \sin^2 \frac{1}{2}\theta$
hav $(-\theta) =$ hav θ
hav $(180^\circ - \theta) =$ hav $(180^\circ + \theta) = 1 -$ hav θ
Characteristics of the logarithms are omitted.

θ°	0' Value	0' Log	10' Value	10' Log	20' Value	20' Log	30' Value	30' Log	40' Value	40' Log	50' Value	50' Log
0	.00000	—	.00000	$\bar{6}$.32539	.00001	$\bar{6}$.92745	.00002	.27963	.00003	.52951	.00005	.72332
1	.00008	.88168	.00010	.01557	.00014	.13155	.00017	.23385	.00021	.32536	.00026	.40814
2	.00030	.48371	.00036	.55323	.00041	.61759	.00048	.67751	.00054	.73355	.00061	.78620
3	.00069	.83584	.00076	.88279	.00085	.92733	.00093	.96970	.00102	.01009	.00112	.04869
4	.00122	.08564	.00132	.12108	.00143	.15513	.00154	.18790	.00166	.21947	.00178	.24993
5	.00190	.27936	.00203	.30782	.00216	.33538	.00230	.36209	.00244	.38800	.00259	.41315
6	.00274	.43760	.00289	.46138	.00305	.48452	.00321	.50706	.00338	.52902	.00355	.55045
7	.00373	.57135	.00391	.59176	.00409	.61170	.00428	.63120	.00447	.65026	.00467	.66891
8	.00487	.68717	.00507	.70505	.00528	.72257	.00549	.73974	.00571	.75657	.00593	.77308
9	.00616	.78929	.00639	.80519	.00662	.82081	.00686	.83615	.00710	.85122	.00735	.86603
10	.00760	.88059	.00785	.89491	.00811	.90900	.00837	.92286	.00864	.93650	.00891	.94992
11	.90919	.96315	.00947	.97617	.00975	.98899	.01004	.00163	.01033	.01409	.01063	.02636
12	.01093	.03847	.01123	.05041	.01154	.06218	.01185	.07379	.01217	.08525	.01249	.09656
13	.01281	.10772	.01314	.11873	.01348	.12961	.01382	.14035	.01416	.15096	.01450	.16144
14	.01485	.17179	.01521	.18202	.01556	.19212	.01593	.20211	.01629	.21198	.01666	.22175
15	.01704	.23140	.01742	.24094	.01780	.25037	.01818	.25971	.01858	.26894	.01897	.27807
16	.01937	.28711	.01977	.29605	.02018	.30490	.02059	.31366	.02101	.32233	.02142	.33091
17	.02185	.33940	.02227	.34782	.02271	.35614	.02314	.36439	.02358	.37256	.02402	.38065
18	.02447	.38866	.02492	.39660	.02538	.40447	.02584	.41226	.02630	.41998	.02677	.42764
19	.02724	.43522	.02772	.44273	.02820	.45018	.02868	.45757	.02917	.46489	.02966	.47215
20	.03015	.47934	.03065	.48647	.03116	.49355	.03166	.50056	.03218	.50752	.03269	.51442
21	.03321	.52127	.03373	.52805	.03426	.53479	.03479	.54147	.03533	.54810	.03587	.55467
22	.03641	.56120	.03695	.56767	.03751	.57410	.03806	.58047	.03862	.58680	.03918	.59308
23	.03975	.59931	.04032	.60550	.04089	.61164	.04147	.61773	.04205	.62379	.04264	.62979
24	.04323	.63576	.04382	.64168	.04442	.64756	.04502	.65340	.04562	.65920	.04623	.66496
25	.04685	.67067	.04746	.67635	.04808	.68199	.04871	.68759	.04934	.69316	.04997	.69869
26	.05060	.70418	.05124	.70963	.05189	.71505	.05253	.72043	.05318	.72578	.05384	.73109
27	.05450	.73637	.05516	.74162	.05582	.74683	.05649	.75201	.05717	.75715	.05785	.76227
28	.05853	.76735	.05921	.77240	.05990	.77742	.06059	.78241	.06129	.78737	.06199	.79230
29	.06269	.79720	.06340	.80207	.06411	.80691	.06482	.81172	.06554	.81651	.06626	.82126
30	.06699	.82599	.06772	.83069	.06845	.83537	.06919	.84001	.06993	.84464	.07067	.84923
31	.07142	.85380	.07217	.85834	.07292	.86286	.07368	.86735	.07444	.87182	.07521	.87626
32	.07598	.88068	.07675	.88507	.07752	.88944	.07830	.89379	.07909	.89811	.07987	.90241
33	.08066	.90668	.08146	.91094	.08226	.91517	.08306	.91938	.08386	.92356	.08467	.92773
34	.08548	.93187	.08630	.93599	.08711	.94009	.08794	.94417	.08876	.94823	.08959	.95227
35	.09042	.95628	.09126	.96028	.09210	.96426	.09294	.96821	.09379	.97215	.09464	.97607
36	.09549	.97996	.09635	.98384	.09721	.98770	.09807	.99154	.09894	.99536	.09981	.99917
37	.10068	.00295	.10156	.00672	.10244	.01047	.10332	.01420	.10421	.01791	.10510	.02161
38	.10599	.02528	.10689	.02894	.10779	.03259	.10870	.03621	.10960	.03982	.11051	.04341
39	.11143	.04699	.11234	.05055	.11326	.05409	.11419	.05762	.11511	.06113	.11604	.06462
40	.11698	.06810	.11791	.07157	.11885	.07501	.11980	.07845	.12074	.08186	.12169	.08526
41	.12265	.08865	.12360	.09202	.12456	.09538	.12552	.09872	.12649	.10205	.12746	.10536
42	.12843	.10866	.12940	.11194	.13038	.11521	.13136	.11847	.13235	.12171	.13333	.12494
43	.13432	.12815	.13532	.13135	.13631	.13454	.13731	.13771	.13832	.14087	.13932	.14402
44	.14033	.14715	.14134	.15027	.14236	.15338	.14337	.15647	.14440	.15955	.14542	.16262
45	.14645	.16568	.14748	.16872	.14851	.17175	.14955	.17477	.15058	.17778	.15163	.18077
46	.15267	.18376	.15372	.18673	.15477	.18968	.15582	.19263	.15688	.19557	.15794	.19849
47	.15900	.20140	.16007	.20430	.16113	.20719	.16220	.21006	.16328	.21293	.16436	.21578
48	.16543	.21863	.16652	.22146	.16760	.22428	.16869	.22709	.16978	.22989	.17087	.23268
49	.17197	.23545	.17307	.23822	.17417	.24098	.17528	.24372	.17638	.24646	.17749	.24918
50	.17861	.25190	.17972	.25460	.18084	.25729	.18196	.25998	.18308	.26265	.18421	.26532
51	.18534	.26797	.18647	.27061	.18761	.27325	.18874	.27587	.18988	.27848	.19102	.28109
52	.19217	.28368	.19332	.28627	.19447	.28885	.19562	.29141	.19677	.29397	.19793	.29652
53	.19909	.29905	.20026	.30158	.20142	.30410	.20259	.30662	.20376	.30912	.20493	.31161
54	.20611	.31409	.20729	.31657	.20847	.31903	.20965	.32149	.21083	.32394	.21202	.32638
55	.21321	.32881	.21440	.33123	.21560	.33365	.21680	.33605	.21800	.33845	.21920	.34084
56	.22040	.34322	.22161	.34559	.22282	.34795	.22403	.35031	.22525	.35266	.22646	.35500
57	.22768	.35733	.22890	.35965	.23012	.36196	.23135	.36427	.23258	.36657	.23381	.36886
58	.23504	.37114	.23627	.37342	.23751	.37569	.23875	.37794	.23999	.38020	.24124	.38244
59	.24248	.38468	.24373	.38691	.24498	.38913	.24623	.39134	.24749	.39355	.24874	.39575
60	.25000	.39794	.25126	.40012	.25252	.40230	.25379	.40447	.25506	.40663	.25632	.40879

HAVERSINES (Continued)

Characteristics of the logarithms are omitted.

$\theta°$	0' Value	Log	10' Value	Log	20' Value	Log	30' Value	Log	40' Value	Log	50' Value	Log
60	.25000	.39794	.25126	.40012	.25252	.40230	.25379	.40447	.25506	.40663	.25632	.40879
61	.25760	.41094	.25887	.41308	.26014	.41521	.26142	.41734	.26270	.41946	.26398	.42157
62	.26526	.42368	.26655	.42578	.26784	.42787	.26913	.42996	.27042	.43203	.27171	.43411
63	.27300	.43617	.27430	.43823	.27560	.44028	.27690	.44232	.27820	.44436	.27951	.44639
64	.28081	.44842	.28212	.45044	.28343	.45245	.28474	.45446	.28606	.45645	.28737	.45845
65	.28869	.46043	.29001	.46241	.29133	.46439	.29265	.46635	.29398	.46831	.29530	.47027
66	.29663	.47222	.29796	.47416	.29929	.47610	.30063	.47803	.30196	.47995	.30330	.48187
67	.30463	.48378	.30597	.48568	.30732	.48758	.30866	.48948	.31000	.49137	.31135	.49325
68	.31270	.49512	.31405	.49699	.31540	.49886	.31675	.50072	.31810	.50257	.31946	.50442
69	.32082	.50626	.32217	.50809	.32353	.50992	.32490	.51174	.32626	.51356	.32762	.51538
70	.32899	.51718	.33036	.51898	.33173	.52078	.33310	.52257	.33447	.52435	.33584	.52613
71	.33722	.52791	.33859	.52968	.33997	.53144	.34135	.53320	.34273	.53495	.34411	.53670
72	.34549	.53844	.34688	.54017	.34826	.54190	.34965	.54363	.35103	.54535	.35242	.54707
73	.35381	.54878	.35521	.55048	.35660	.55218	.35799	.55387	.35939	.55556	.36078	.55725
74	.36218	.55893	.36358	.56060	.36498	.56227	.36638	.56393	.36778	.56559	.36919	.56725
75	.37059	.56889	.37200	.57054	.37340	.57218	.37481	.57381	.37622	.57544	.37763	.57706
76	.37904	.57868	.38045	.58030	.38186	.58191	.38328	.58351	.38469	.58511	.38611	.58671
77	.38752	.58830	.38894	.58988	.39036	.59147	.39178	.59304	.39320	.59461	.39462	.59618
78	.39604	.59774	.39747	.59930	.39889	.60085	.40032	.60240	.40174	.60395	.40317	.60549
79	.40460	.60702	.40602	.60855	.40745	.61008	.40888	.61160	.41031	.61311	.41174	.61463
80	.41318	.61613	.41461	.61764	.41604	.61914	.41748	.62063	.41891	.62212	.42035	.62361
81	.42178	.62509	.42322	.62657	.42466	.62804	.42610	.62951	.42753	.63097	.42897	.63243
82	.43041	.63389	.43185	.63534	.43330	.63678	.43474	.63823	.43618	.63966	.43762	.64110
83	.43907	.64253	.44051	.64395	.44195	.64538	.44340	.64679	.44484	.64821	.44629	.64962
84	.44774	.65102	.44918	.65242	.45063	.65382	.45208	.65521	.45353	.65660	.45497	.65799
85	.45642	.65937	.45787	.66074	.45932	.66212	.46077	.66348	.46222	.66485	.46367	.66621
86	.46512	.66757	.46657	.66892	.46802	.67027	.46948	.67161	.47093	.67295	.47238	.67429
87	.47383	.67562	.47528	.67695	.47674	.67828	.47819	.67960	.47964	.68092	.48110	.68223
88	.48255	.68354	.48400	.68485	.48546	.68615	.48691	.68745	.48837	.68874	.48982	.69004
89	.49127	.69132	.49273	.69261	.49418	.69389	.49564	.69516	.49709	.69644	.49855	.69770
90	.50000	.69897	.50145	.70023	.50291	.70149	.50436	.70274	.50582	.70399	.50727	.70524
91	.50873	.70648	.51018	.70772	.51163	.70896	.51309	.71019	.51454	.71142	.51600	.71265
92	.51745	.71387	.51890	.71509	.52036	.71630	.52181	.71751	.52326	.71872	.52472	.71992
93	.52617	.72112	.52762	.72232	.52907	.72352	.53052	.72471	.53198	.72589	.53343	.72708
94	.53488	.72825	.53633	.72943	.53778	.73060	.53923	.73177	.54068	.73294	.54213	.73410
95	.54358	.73526	.54503	.73642	.54647	.73757	.54792	.73872	.54937	.73987	.55082	.74101
96	.55226	.74215	.55371	.74328	.55516	.74442	.55660	.74554	.55805	.74667	.55949	.74779
97	.56093	.74891	.56238	.75003	.56382	.75114	.56526	.75225	.56670	.75336	.56815	.75446
98	.56959	.75556	.57103	.75666	.57247	.75775	.57390	.75884	.57534	.75993	.57678	.76101
99	.57822	.76209	.57965	.76317	.58109	.76424	.58252	.76531	.58396	.76638	.58539	.76745
100	.58682	.76851	.58826	.76957	.58969	.77062	.59112	.77167	.59255	.77272	.59398	.77377
101	.59540	.77481	.59683	.77585	.59826	.77689	.59968	.77792	.60111	.77895	.60253	.77998
102	.60396	.78101	.60538	.78203	.60680	.78305	.60822	.78406	.60964	.78507	.61106	.78608
103	.61248	.78709	.61389	.78809	.61531	.78909	.61672	.79009	.61814	.79108	.61955	.79208
104	.62096	.79306	.62237	.79405	.62378	.79503	.62519	.79601	.62660	.79699	.62800	.79796
105	.62941	.79893	.63081	.79990	.63222	.80087	.63362	.80183	.63502	.80279	.63642	.80374
106	.63782	.80470	.63922	.80565	.64061	.80660	.64201	.80754	.64340	.80848	.64479	.80942
107	.64619	.81036	.64758	.81129	.64897	.81222	.65035	.81315	.65174	.81407	.65312	.81500
108	.65451	.81592	.65589	.81683	.65727	.81775	.65865	.81866	.66003	.81956	.66141	.82047
109	.66278	.82137	.66416	.82227	.66553	.82317	.66690	.82406	.66827	.82495	.66964	.82584
110	.67101	.82673	.67238	.82761	.67374	.82849	.67510	.82937	.67647	.83025	.67783	.83112
111	.67918	.83199	.68054	.83285	.68190	.83372	.68325	.83458	.68460	.83544	.68595	.83629
112	.68730	.83715	.68865	.83800	.69000	.83885	.69134	.83969	.69268	.84054	.69403	.84138
113	.69537	.84221	.69670	.84305	.69804	.84388	.69937	.84471	.70071	.84554	.70204	.84636
114	.70337	.84718	.70470	.84800	.70602	.84882	.70735	.84963	.70867	.85044	.70999	.85125
115	.71131	.85206	.71263	.85286	.71394	.85366	.71526	.85446	.71657	.85526	.71788	.85605
116	.71919	.85684	.72049	.85763	.72180	.85841	.72310	.85920	.72440	.85998	.72570	.86076
117	.72700	.86153	.72829	.86230	.72958	.86307	.73087	.86384	.73216	.86461	.73345	.86537
118	.73474	.86613	.73602	.86689	.73730	.86764	.73858	.86840	.73986	.86915	.74113	.86990
119	.74240	.87064	.74368	.87138	.74494	.87212	.74621	.87286	.74748	.87360	.74874	.87433
120	.75000	.87506	.75126	.87579	.75251	.87652	.75377	.87724	.75502	.87796	.75627	.87868

HAVERSINES (Continued)

Characteristics of the logarithms are omitted.

θ°	0′ Value	Log	10′ Value	Log	20′ Value	Log	30′ Value	Log	40′ Value	Log	50′ Value	Log
120	.75000	.87506	.75126	.87579	.75251	.87652	.75377	.87724	.75502	.87796	.75627	.87868
121	.75752	.87939	.75876	.88011	.76001	.88082	.76125	.88153	.76249	.88223	.76373	.88294
122	.76496	.88364	.76619	.88434	.76742	.88503	.76865	.88573	.76988	.88642	.77110	.88711
123	.77232	.88780	.77354	.88848	.77475	.88916	.77597	.88984	.77718	.89052	.77839	.89120
124	.77960	.89187	.78080	.89254	.78200	.89321	.78320	.89387	.78440	.89454	.78560	.89520
125	.78679	.89586	.78798	.89651	.78917	.89717	.79035	.89782	.79153	.89847	.79271	.89912
126	.79389	.89976	.79507	.90040	.79624	.90104	.79741	.90168	.79858	.90232	.79974	.90295
127	.80091	.90358	.80207	.90421	.80323	.90484	.80438	.90546	.80553	.90608	.80668	.90670
128	.80783	.90732	.80898	.90794	.81012	.90855	.81126	.90916	.81239	.90977	.81353	.91037
129	.81466	.91098	.81579	.91158	.81692	.91218	.81804	.91277	.81916	.91337	.82028	.91396
130	.82139	.91455	.82251	.91514	.82362	.91573	.82472	.91631	.82583	.91689	.82693	.91747
131	.82803	.91805	.82913	.91862	.83022	.91919	.83131	.91976	.83240	.92033	.83348	.92090
132	.83457	.92146	.83564	.92202	.83672	.92258	.83780	.92314	.83887	.92369	.83993	.92425
133	.84100	.92480	.84206	.92534	.84312	.92589	.84418	.92643	.84523	.92698	.84628	.92751
134	.84733	.92805	.84837	.92859	.84942	.92912	.85045	.92965	.85149	.93018	.85252	.93071
135	.85355	.93123	.85458	.93175	.85560	.93227	.85663	.93279	.85764	.93331	.85866	.93382
136	.85967	.93433	.86068	.93484	.86168	.93535	.86269	.93585	.86369	.93636	.86468	.93686
137	.86568	.93736	.86667	.93785	.86765	.93835	.86864	.93884	.86962	.93933	.87060	.93982
138	.87157	.94030	.87254	.94079	.87351	.94127	.87448	.94175	.87544	.94223	.87640	.94270
139	.87735	.94318	.87831	.94365	.87926	.94412	.88020	.94458	.88115	.94505	.88209	.94551
140	.88302	.94597	.88396	.94643	.88489	.94689	.88581	.94734	.88674	.94779	.88766	.94824
141	.88857	.94869	.88949	.94914	.89040	.94958	.89130	.95003	.89221	.95047	.89311	.95090
142	.89401	.95134	.89490	.95177	.89579	.95221	.89668	.95264	.89756	.95306	.89844	.95349
143	.89932	.95391	.90019	.95433	.90106	.95475	.90193	.95517	.90279	.95559	.90365	.95600
144	.90451	.95641	.90536	.95682	.90621	.95723	.90706	.95763	.90790	.95804	.90874	.95844
145	.90958	.95884	.91041	.95924	.91124	.95963	.91206	.96002	.91289	.96042	.91370	.96081
146	.91452	.96119	.91533	.96158	.91614	.96196	.91694	.96234	.91774	.96272	.91854	.96310
147	.91934	.96347	.92013	.96385	.92091	.96422	.92170	.96459	.92248	.96495	.92325	.96532
148	.92402	.96568	.92479	.96604	.92556	.96640	.92632	.96676	.92708	.96712	.92783	.96747
149	.92858	.96782	.92933	.96817	.93007	.96852	.93081	.96886	.93155	.96921	.93228	.96955
150	.93301	.96989	.93374	.97023	.93446	.97089	.93518	.97089	.93589	.97123	.93660	.97156
151	.93731	.97188	.93801	.97221	.93871	.97253	.93941	.97285	.94010	.97317	.94079	.97349
152	.94147	.97381	.94215	.97412	.94283	.97443	.94351	.97474	.94418	.97505	.94484	.97536
153	.94550	.97566	.94616	.97597	.94682	.97627	.94747	.97656	.94811	.97686	.94876	.97716
154	.94940	.97745	.95003	.97774	.95066	.97803	.95129	.97831	.95192	.97860	.95254	.97888
155	.95315	.97916	.95377	.97944	.95438	.97972	.95498	.97999	.95558	.98027	.95618	.98054
156	.95677	.98081	.95736	.98108	.95795	.98134	.95853	.98161	.95911	.98187	.95968	.98213
157	.96025	.98239	.96082	.98264	.96138	.98290	.96194	.98315	.96249	.98340	.96305	.98365
158	.96359	.98389	.96413	.98414	.96467	.98438	.96521	.98462	.96574	.98486	.96627	.98510
159	.96679	.98533	.96731	.98557	.96782	.98580	.96834	.98603	.96884	.98625	.96935	.98648
160	.96985	.98670	.97034	.98692	.97083	.98714	.97132	.98736	.97180	.98758	.97228	.98779
161	.97276	.98801	.97323	.98822	.97370	.98842	.97416	.98863	.97462	.98884	.97508	.98904
162	.97553	.98924	.97598	.98944	.97642	.98964	.97686	.98983	.97729	.99003	.97773	.99022
163	.97815	.99041	.97858	.99059	.97899	.99078	.97941	.99096	.97982	.99115	.98023	.99133
164	.98063	.99151	.98103	.99168	.98142	.99186	.98182	.99203	.98220	.99220	.98258	.99237
165	.98296	.99254	.98334	.99270	.98371	.99287	.98407	.99303	.98444	.99319	.98479	.99335
166	.98515	.99350	.98550	.99366	.98584	.99381	.98618	.99396	.98652	.99411	.98686	.99425
167	.98719	.99440	.98751	.99454	.98783	.99468	.98815	.99482	.98846	.99496	.98877	.99509
168	.98907	.99523	.98937	.99536	.98967	.99549	.98996	.99562	.99025	.99574	.99053	.99587
169	.99081	.99599	.99109	.99611	.99136	.99623	.99163	.99635	.99189	.99646	.99215	.99658
170	.99240	.99669	.99265	.99680	.99290	.99691	.99314	.99701	.99338	.99712	.99361	.99722
171	.99384	.99732	.99407	.99742	.99429	.99751	.99451	.99761	.99472	.99770	.99493	.99779
172	.99513	.99788	.99533	.99797	.99553	.99805	.99572	.99814	.99591	.99822	.99609	.99830
173	.99627	.99838	.99645	.99845	.99662	.99853	.99679	.99860	.99695	.99867	.99711	.99874
174	.99726	.99881	.99741	.99887	.99756	.99894	.99770	.99900	.99784	.99906	.99797	.99912
175	.99810	.99917	.99822	.99923	.99834	.99928	.99846	.99933	.99857	.99938	.99868	.99943
176	.99878	.99947	.99888	.99951	.99898	.99956	.99907	.99959	.99915	.99963	.99924	.99967
177	.99931	.99970	.99939	.99973	.99946	.99976	.99952	.99979	.99959	.99982	.99964	.99984
178	.99970	.99987	.99974	.99989	.99979	.99991	.99983	.99993	.99986	.99994	.99990	.99995
179	.99992	.99997	.99995	.99998	.99997	.99999	.99998	.99999	.99999	.00000	1.00000	.00000
180	1.00000	.00000										

VI. LOGARITHMIC, EXPONENTIAL, AND HYPERBOLIC FUNCTIONS
LAWS OF EXPONENTS

For a any real number and m a positive integer, the exponential a^m is defined as

$$\underbrace{a \cdot a \cdot a \cdots \cdot a}_{m \text{ terms}}$$

Using this definition, it is easy to show that the following three *laws of exponents* hold:

I. $a^m \cdot a^n = a^{m+n}$

II. $\dfrac{a^m}{a^n} = \begin{cases} a^{m-n} & \text{if } m > n \\ 1 & \text{if } m = n \\ \dfrac{1}{a^{n-m}} & \text{if } m < n \end{cases}$

III. $(a^m)^n = a^{mn}$

The n-th root function is defined as the inverse of the n-th power function; that is, if

$$b^n = a, \quad \text{then} \quad b = \sqrt[n]{a}.$$

If n is odd, there will be a unique real number satisfying the above definition for $\sqrt[n]{a}$, for any real value of a. If n is even, for positive values of a there will be two real values for $\sqrt[n]{a}$, one positive and one negative. By convention, the symbol $\sqrt[n]{a}$ is understood to mean the positive value in this case. If n is even and a is negative, there are no real values for $\sqrt[n]{a}$.

If we now attempt to extend the definition of the exponential a^t to all rational values of the exponent t, in such a way that the three laws of exponents continue to hold, it is easily shown that the required definitions are:

$$a^0 = 1$$
$$a^{p/q} = \sqrt[q]{a^p}$$
$$a^{-t} = \frac{1}{a^t}$$

In order to avoid difficulties with imaginary numbers and division by zero, a must now be restricted to be positive.

With this extended definition, it is possible to restate the second law of exponents in a simpler form:

II$'$. $\dfrac{a^m}{a^n} = a^{m-n}$

It is shown in advanced calculus that this definition may be further extended so that the exponent may be any real number, and the laws of exponents continue to hold. When the quantity a^x thus defined is viewed as a function of the exponent x, with the base a held constant, it is a continuous function. Also, if $a > 1$, the exponential function is monotone increasing, and if $0 < a < 1$, it is monotone decreasing.

IV. $a^{x^{y^z}} = a^u$, where $u = x_y^v$ and $v = y^z$

LOGARITHMS

Any monotone function has a single-valued inverse function, which is also monotone. Furthermore, if the original function is continuous, so is the inverse. Therefore, the inverse function to the exponential function a^x exists for all positive values of a, except $a = 1$. This function is given the name *logarithm to the base* a, abbreviated $\log_a$. That is, if

$$x = a^y, \quad \text{then} \quad y = \log_a x.$$

This function is defined and continuous for all positive values of x. It is monotone increasing if $a > 1$, and monotone decreasing if $a < 1$.

If the laws of exponents are rewritten in terms of logarithms, they become the *laws of logarithms*:

I. $\log_a(xy) = \log_a x + \log_a y$

II. $\log_a\left(\dfrac{x}{y}\right) = \log_a x - \log_a y$

III. $\log_a(x^n) = n \log_a x$

Logarithms derive their main usefulness in computation from the above laws, since they allow multiplication, division, and exponentiation to be replaced by the simpler operations of addition, subtraction, and multiplication, respectively. See the examples which follow.

Further recourse to the definition of logarithm leads to the following formula for change of base

$$\log_a x = \log_b x / \log_b a = (\log_b x) \cdot (\log_a b)$$

Two numbers are commonly used as bases for logarithms. Logarithms to the base 10 are most convenient for use in computation. These logarithm are called common or Briggsian logarithms.

The other usual base for logarithms is an irrational number denoted by e, whose value is approximately 2.71828 .... These logarithms are called natural, Naperian, or hyperbolic logarithms, and occur in many formulas of higher mathematics. The abbreviation ln is frequently used for the natural logarithm function.

Other bases for logarithms, such as 2 and 3, occur in certain applications. These applications are rather specialized and separate tables for these bases are not given. Instead, the formulas for change of base are applied to common or natural logarithms.

If the formulas for change of base are applied to the two usual bases, the following formulas result:

$$\log_{10} x = \log_e x / \log_e 10 = (\log_{10} e)(\log_e x) = M \log_e x$$
$$= 0.43429\ 44819 \log_e x$$
$$\log_e x = \log_{10} x / \log_{10} e = (\log_e 10)(\log_{10} x) = \frac{1}{M} \log_{10} x$$
$$= 2.30258\ 50930 \log_{10} x$$

The following remarks apply to common logarithms.

Since most numbers are irrational powers of ten, a common logarithm, in general, consists of an integer, which is called the characteristic, and an endless decimal, the mantissa.

It is to be observed that the common logarithms of all numbers expressed by the same figures in the same order with the decimal point in different positions have different characteristics but the same mantissa. To illustrate:—if the decimal point stand after the first figure of a number, counting from the left, the characteristic is 0; if after two figures, it is 1; if after three figures, it is 2; and so forth. If the decimal point stand before the first significant figure the characteristic is -1, usually written $\bar{1}$; if there is one zero between the decimal point and the first significant figure it is $\bar{2}$, and so on. For example: log 256 = 2.408240, log 2.56 = 0.408240, log 0.256 = $\bar{1}$.408240, log 0.00256 = $\bar{3}$.408240. The two latter are often written log 0.256 = 9.408240 $-$ 10, log 0.00256 = 7.408240 $-$ 10.

Notice that, although the common logarithm of a number less than one is a negative number, it is customarily written as a negative characteristic and a *positive* mantissa, since the mantissas are usually given in tables as positive numbers. This is the reason that the negative sign is written above the characteristic, since it does not apply to the mantissa. Thus log 0.00256 = $\bar{3}$.408240 = 7.408240 $-$ 10 = -2.591760.

A method of determining characteristics of logarithms is to write the number with one figure to the left of the decimal point multiplied by the appropriate power of 10. The characteristic is then the exponent used. For example:

$$256\ 000\ 000 = 2.56 \times 10^8 \qquad \log = 8.408240$$
$$0.000\ 000\ 256 = 2.56 \times 10^{-7} \qquad \log = \bar{7}.408240 \quad \text{or} \quad 3.408240 - 10$$

Inasmuch as the characteristic may be determined by inspection, the mantissas only are given in tables of common logarithms.

USE OF LOGARITHM TABLES

To find the common logarithm of a number:

(*Note:* This description and examples refer specifically to the table entitled "Six-Place Mantissas for Common Logarithms." For the other tables, there will be minor differences from this description. Most of these differences are obvious. Notes with the individual tables explain any differences which are not immediately obvious.)

For a number of four figures, take out the tabular mantissa on a line with the first three figures of the number and under its fourth figure. The characteristic is determined as previously explained.

For a number of less than four figures, supply zeros to make a four figure number and take the value of the mantissa from the tables as before. For example: log 2 = log 2.000 = 0.301030.

(Notice that in some of the tables not all of the digits of the logarithm are given for every value. For example, in the table of six-place common logarithms, the first two digits of each mantissa are given only once for each line. The remaining four digits of each mantissa are given in the correct place in the table. When the leading two digits are not given on a line, they should be taken from the last line above it on which they do appear. When a mantissa is marked with an asterisk, it indicates that the value for the leading digits is to be taken from the *next* line instead of the present line. Similar remarks apply to the other tables in which this method of presenting the values are used.)

For a number of more than four figures, interpolation must be used. There are several precautions that must be observed when interpolating:

1. Linear interpolation, as described below, may only be used to add one extra digit to the argument (i.e., in the present case, for a five-digit argument).
2. Even though the mantissas given in the table are accurate to six decimal places, interpolated values are accurate only to the same number of places as in the argument, i.e. five places.
3. Because of the rapidly changing values in this region of the table, interpolation is not accurate if the first two digits of the argument are 11 or 12. For this reason, the table has been extended at the end so that such values may be read directly from

the table with five-digit arguments, without interpolation. (The four-place table is not so extended. If interpolation is required in this section of the four-place table, the value should be read instead from the six-place table without interpolation.)

If the above precautions cannot be observed. then higher order interpolation should be used.

Where applicable, linear interpolation is carried out as follows:

Take the tabular value of the mantissa for the first four figures; find the difference between the mantissa and the next greater tabular mantissa and multiply the difference so found by the remaining figures of the number as a decimal and add the product to the mantissa of the first four figures. For example, to find log 46.762:

$$\log 46.76 = 1.669875$$

Tabular difference between this mantissa and that for 4677 is .000092

$$\therefore \log 46.762 = 1.669875 + .2 \times .000092$$
$$= 1.669875 + .000018$$
$$= 1.669893$$

The last digit is not accurate, so must be rounded out. Thus

$$\log 46.762 = 1.66989$$

The accuracy will not ordinarily be affected by more than 1 in the last place if the mantissas are rounded to five decimal places before interpolation, and this makes the computation somewhat easier.

In the four-place logarithm table, a column of proportional parts is given at the end of each line. The number in the column under the fourth digit of the argument is the amount that must be added to any mantissa in that line to interpolate for the fourth digit. This number is to be added to the last place of the mantissa. These numbers are averages for the entire line, so may be off by 1 in the last place.

For example, to find log 33.74

$$\log 33.7 = 1.5276$$
$$\text{proportional part for } 4 = \underline{\qquad 5}$$
$$\therefore \log 33.74 = 1.5281$$

To find the number corresponding to a given logarithm:

(*Note*: This number is called the antilogarithm, and is denoted by $\log^{-1}$. Since the logarithm function is the inverse of the exponential function, $\log_a^{-1} x = a^x$. Therefore, any procedure or table which calculates antilogarithms may also be used to calculate exponentials, and vice-versa. In particular, tables of e^x may be used to compute antilogarithms to the base e.)

The procedure given below refers to the six-place logarithm table. As before, any significant deviation for other tables will be noted.

If the mantissa is found exactly in the table, join the figure at the top which is directly above the given mantissa to the three figures on the line at the left and place the decimal point according to the characteristic of the logarithm. For example,

$$\log^{-1} 3.399674 = \text{antilog } 3.399674 = 2510.$$

If the mantissa is not found exactly in the table it is necessary to interpolate. For example, to find antilog 3.400280, we find in the table

$$\text{antilog } 3.400192 = 2513.$$
$$\text{antilog } \underline{3.400365} = 2514.$$
$$\text{tabular difference} \qquad 173$$

The required difference is 88, so we must add $\frac{88}{173}$ = .5.

$$\therefore \text{ antilog } 3.400280 = 2513.5$$

The same precautions must be observed for interpolation in finding antilogarithms as in finding logarithms.

A four-place antilogarithm table is also provided. When using this table, the mantissa of the logarithm is looked up on the margins of the table, and the significant digits of the antilogarithm are read from the body of the table. Lookup and interpolation are done in the same manner as when looking up logarithms in a table of logarithms.

Tables of natural logarithms are used in the same way as tables of common logarithms, except that they contain both the characteristics and the mantissas of the logarithms.

Examples of the use of logarithms in computation follow. Almost all computation with logarithms is done with common logarithms, since the computation of the characteristic is simpler, and since only the significant digits of the argument need be given in the table, without regard for the decimal point location. These examples all use the table of six-place common logarithms.

1. $52600 \times 0.00381 \times 2.74 = 549.11$

log 52600	=	4.720986
log 0.00381	=	3.580925
log 2.74	=	0.437751
adding	=	2.739662
antilog	=	549.11

The sum is the logarithm of the product, the mantissa of which is 739662. On looking up this mantissa in the logarithm tables we see that it corresponds to the digits 54911. The characteristic is 2, hence there are three figures before the decimal point. The number corresponding to the logarithm, called the antilogarithm, is 549.11.

2. $0.00123 \div 52.7 = 0.000\ 023\ 34$ An alternative method:

log 0.00123	= $\bar{3}.089905$		log 0.00123	=	7.089905 − 10
log 52.7	= 1.721811		log 52.7	=	1.721811
subtracting	= $\bar{5}.368094$				5.368094 − 10
antilog	= 0.000 023 34				

The characteristic 5($\bar{5}$. − 10) shows four zeros after the decimal point before the first significant figure.

3. $\dfrac{273 \times 780}{292 \times 760} \times 15 \times 0.09 = 1.2954$

log 273	= 2.436163		log 292	= 2.465383
log 780	= 2.892095		log 760	= 2.880814
log 15	= 1.176091		log denominator	= 5.346197
log 0.09	= $\bar{2}.954243$			
log numerator	= 5.458592			

log numerator	=	5.458592
log denominator	=	5.346197
subtracting	=	0.112395
antilog	=	1.2954

As division may be accomplished by multiplying by the reciprocal of a number, the above may be considerably simplified. The logarithm of the reciprocal of a number, called the cologarithm, is readily obtained from the table by subtracting the logarithm of the number from zero. This may be readily read off from the table of mantissas. Change the sign of the characteristic algebraically adding to it -1, then mentally subtract each figure of the mantissa from 9 proceeding from left to right, the last figure being subtracted from 10. The example then is:

$$
\begin{array}{lll}
\log 273 & = & 2.436163 \\
\log 780 & = & 2.892095 \\
\log\ \ 15 & = & 1.176091 \\
\log\ \ \ 0.09 & = & \bar{2}.954243 \\
\text{colog } 292 & = & \bar{3}.534617 \\
\text{colog } 760 & = & \underline{\bar{3}.119186} \\
& & 0.112395
\end{array}
$$

4. $(0.00098)^4 = 9.2237 \times 10^{-13}$ An alternative method:

$$
\begin{array}{lll}
\log 0.00098 & = & \bar{4}.991226 \\
& & \underline{\qquad\quad 4} \\
& & 3.964904 \quad \text{(a)} \\
\bar{4} \times 4 & = & \underline{\bar{16}.\qquad} \quad \text{(b)} \\
\log (0.00098)^4 & = & \overline{13}.964904 \quad \text{(c)} \\
\text{antilog} & = & 9.2237 \times 10^{-13}
\end{array}
$$

$$
\begin{array}{lll}
\log 0.00098 & = & 6.991226 - 10 \\
& & \underline{\qquad\qquad 4} \\
& & 27.964904 - 40 \\
\text{or} & & 7.964904 - 20 \\
\text{or} & & \overline{13}.964904 \\
\text{antilog} & = & 9.2237 \times 10^{-13}
\end{array}
$$

In the above it will be noted that the mantissa is always positive, hence the multiplication of the mantissa shown at (a), while (b) shows the multiplication of the characteristic. (c) is the algebraic sum.

5. $\sqrt[5]{492} = 3.4546$
   $\log 492 = 2.691965$

Dividing the logarithm by 5 gives as the logarithm of the root 0.538393 the antilogarithm of which is 3.4546 both characteristic and mantissa being positive. When the characteristic is negative and not evenly divisible by the root to be taken, a modification of the logarithm is necessary, as the following example shows:

6. $\sqrt[3]{0.000372} = 0.07192$
   $\log 3.72 \times 10^{-4} = \bar{4}.570543$ (a)
                    $= 26.570543 - 30$ (b)

Dividing (b) by 3 gives $8.856848 - 10$ which may be written $\bar{2}.856848$ and is the logarithm of the root sought, the antilogarithm of which is 0.07192.

7. $(0.000\ 372)^{1.2} = 0.000\ 076\ 675$

$$
\begin{array}{lll}
\log 0.000\ 372 & = & \bar{4}.570543 \\
& & \text{or } 6.570543 - 10 \\
& & \underline{\qquad\qquad\quad 1.2} \\
& & 7.884652 - 12 \\
\text{antilog} & = & 0.000\ 076\ 675
\end{array}
$$

8. $(0.000372)^{-1.32} = 33642$

$$
\begin{array}{lll}
\text{colog } 0.000372 & = & 3.429457 \\
& & \underline{\qquad\quad 1.32} \\
& & 4.526883 \\
\text{antilog} & = & 33642
\end{array}
$$

SIX-PLACE MANTISSAS FOR COMMON LOGARITHMS

N	0	1	2	3	4	5	6	7	8	9
100	00 0000	0434	0868	1301	1734	2166	2598	3029	3461	3891
01	4321	4751	5181	5609	6038	6466	6894	7321	7748	8174
02	00 8600	9026	9451	9876	*0300	*0724	*1147	*1570	*1993	*2415
03	01 2837	3259	3680	4100	4521	4940	5360	5779	6197	6616
04	01 7033	7451	7868	8284	8700	9116	9532	9947	*0361	*0775
05	02 1189	1603	2016	2428	2841	3252	3664	4075	4486	4896
06	5306	5715	6125	6533	6942	7350	7757	8164	8571	8978
07	02 9384	9789	*0195	*0600	*1004	*1408	*1812	*2216	*2619	*3021
08	03 3424	3826	4227	4628	5029	5430	5830	6230	6629	7028
09	03 7426	7825	8223	8620	9017	9414	9811	*0207	*0602	*0998
110	04 1393	1787	2182	2576	2969	3362	3755	4148	4540	4932
11	5323	5714	6105	6495	6885	7275	7664	8053	8442	8830
12	04 9218	9606	9993	*0380	*0766	*1153	*1538	*1924	*2309	*2694
13	05 3078	3463	3846	4230	4613	4996	5378	5760	6142	6524
14	05 6905	7286	7666	8046	8426	8805	9185	9563	9942	*0320
15	06 0698	1075	1452	1829	2206	2582	2958	3333	3709	4083
16	4458	4832	5206	5580	5953	6326	6699	7071	7443	7815
17	06 8186	8557	8928	9298	9668	*0038	*0407	*0776	*1145	*1514
18	07 1882	2250	2617	2985	3352	3718	4085	4451	4816	5182
19	5547	5912	6276	6640	7004	7368	7731	8094	8457	8819
120	07 9181	9543	9904	*0266	*0626	*0987	*1347	*1707	*2067	*2426
21	08 2785	3144	3503	3861	4219	4576	4934	5291	5647	6004
22	6360	6716	7071	7426	7781	8136	8490	8845	9198	9552
23	08 9905	*0258	*0611	*0963	*1315	*1667	*2018	*2370	*2721	*3071
24	09 3422	3772	4122	4471	4820	5169	5518	5866	6215	6562
25	09 6910	7257	7604	7951	8298	8644	8990	9335	9681	*0026
26	10 0371	0715	1059	1403	1747	2091	2434	2777	3119	3462
27	3804	4146	4487	4828	5169	5510	5851	6191	0531	6871
28	10 7210	7549	7888	8227	8565	8903	9241	9579	9916	*0253
29	11 0590	0926	1263	1599	1934	2270	2605	2940	3275	3609
130	3943	4277	4611	4944	5278	5611	5943	6276	6608	6940
31	11 7271	7603	7934	8265	8595	8926	9256	9586	9915	*0245
32	12 0574	0903	1231	1560	1888	2216	2544	2871	3198	3525
33	3852	4178	4504	4830	5156	5481	5806	6131	6456	6781
34	12 7105	7429	7753	8076	8399	8722	9045	9368	9690	*0012
35	13 0334	0655	0977	1298	1619	1939	2260	2580	2900	3219
36	3539	3858	4177	4496	4814	5133	5451	5769	6086	6403
37	6721	7037	7354	7671	7987	8303	8618	8934	9249	9564
38	13 9879	*0194	*0508	*0822	*1136	*1450	*1763	*2076	*2389	*2702
39	14 3015	3327	3639	3951	4263	4574	4885	5196	5507	5818
140	6128	6438	6748	7058	7367	7676	7985	8294	8603	8911
41	14 9219	9527	9835	*0142	*0449	*0756	*1063	*1370	*1676	*1982
42	15 2288	2594	2900	3205	3510	3815	4120	4424	4728	5032
43	5336	5640	5943	6246	6549	6852	7154	7457	7759	8061
44	15 8362	8664	8965	9266	9567	9868	*0168	*0469	*0769	*1068
45	16 1368	1667	1967	2266	2564	2863	3161	3460	3758	4055
46	4353	4650	4947	5244	5541	5838	6134	6430	6726	7022
47	16 7317	7613	7908	8203	8497	8792	9086	9380	9674	9968
48	17 0262	0555	0848	1141	1434	1726	2019	2311	2603	2895
49	3186	3478	3769	4060	4351	4641	4932	5222	5512	5802
150	17 6091	6381	6670	6959	7248	7536	7825	8113	8401	8689
N	0	1	2	3	4	5	6	7	8	9

SIX-PLACE MANTISSAS FOR COMMON LOGARITHMS (Continued)

N	0	1	2	3	4	5	6	7	8	9
150	17 6091	6381	6670	6959	7248	7536	7825	8113	8401	8689
51	17 8977	9264	9552	9839	*0126	*0413	*0699	*0986	*1272	*1558
52	18 1844	2129	2415	2700	2985	3270	3555	3839	4123	4407
53	4691	4975	5259	5542	5825	6108	6391	6674	6956	7239
54	18 7521	7803	8084	8366	8647	8928	9209	9490	9771	*0051
55	19 0332	0612	0892	1171	1451	1730	2010	2289	2567	2846
56	3125	3403	3681	3959	4237	4514	4792	5069	5346	5623
57	5900	6176	6453	6729	7005	7281	7556	7832	8107	8382
58	19 8657	8932	9206	9481	9755	*0029	*0303	*0577	*0850	*1124
59	20 1397	1670	1943	2216	2488	2761	3033	3305	3577	3848
160	4120	4391	4663	4934	5204	5475	5746	6016	6286	6556
61	6826	7096	7365	7634	7904	8173	8441	8710	8979	9247
62	20 9515	9783	*0051	*0319	*0586	*0853	*1121	*1388	*1654	*1921
63	21 2188	2454	2720	2986	3252	3518	3783	4049	4314	4579
64	4844	5109	5373	5638	5902	6166	6430	6694	6957	7221
65	21 7484	7747	8010	8273	8536	8798	9060	9323	9585	9846
66	22 0108	0370	0631	0892	1153	1414	1675	1936	2196	2456
67	2716	2976	3236	3496	3755	4015	4274	4533	4792	5051
68	5309	5568	5826	6084	6342	6600	6858	7115	7372	7630
69	22 7887	8144	8400	8657	8913	9170	9426	9682	9938	*0193
170	23 0449	0704	0960	1215	1470	1724	1979	2234	2488	2742
71	2996	3250	3504	3757	4011	4264	4517	4770	5023	5276
72	5528	5781	6033	6285	6537	6789	7041	7292	7544	7795
73	23 8046	8297	8548	8799	9049	9299	9550	9800	*0050	*0300
74	24 0549	0799	1048	1297	1546	1795	2044	2293	2541	2790
75	3038	3286	3534	3782	4030	4277	4525	4772	5019	5266
76	5513	5759	6006	6252	6499	6745	6991	7237	7482	7728
77	24 7973	8219	8464	8709	8954	9198	9443	9687	9932	*0176
78	25 0420	0664	0908	1151	1395	1638	1881	2125	2368	2610
79	2853	3096	3338	3580	3822	4064	4306	4548	4790	5031
180	5273	5514	5755	5996	6237	6477	6718	6958	7198	7439
81	25 7679	7918	8158	8398	8637	8877	9116	9355	9594	9833
82	26 0071	0310	0548	0787	1025	1263	1501	1739	1976	2214
83	2451	2688	2925	3162	3399	3636	3873	4109	4346	4582
84	4818	5054	5290	5525	5761	5996	6232	6467	6702	6937
85	7172	7406	7641	7875	8110	8344	8578	8812	9046	9279
86	26 9513	9746	9980	*0213	*0446	*0679	*0912	*1144	*1377	*1609
87	27 1842	2074	2306	2538	2770	3001	3233	3464	3696	3927
88	4158	4389	4620	4850	5081	5311	5542	5772	6002	6232
89	6462	6692	6921	7151	7380	7609	7838	8067	8296	8525
190	27 8754	8982	9211	9439	9667	9895	*0123	*0351	*0578	*0806
91	28 1033	1261	1488	1715	1942	2169	2396	2622	2849	3075
92	3301	3527	3753	3979	4205	4431	4656	4882	5107	5332
93	5557	5782	6007	6232	6456	6681	6905	7130	7354	7578
94	28 7802	8026	8249	8473	8696	8920	9143	9366	9589	9812
95	29 0035	0257	0480	0702	0925	1147	1369	1591	1813	2034
96	2256	2478	2699	2920	3141	3363	3584	3804	4025	4246
97	4466	4687	4907	5127	5347	5567	5787	6007	6226	6446
98	6665	6884	7104	7323	7542	7761	7979	8198	8416	8635
99	29 8853	9071	9289	9507	9725	9943	*0161	*0378	*0595	*0813
200	30 1030	1247	1464	1681	1898	2114	2331	2547	2764	2980
N	0	1	2	3	4	5	6	7	8	9

SIX-PLACE MANTISSAS FOR COMMON LOGARITHMS (Continued)

N	0	1	2	3	4	5	6	7	8	9
200	30 1030	1247	1464	1681	1898	2114	2331	2547	2764	2980
01	3196	3412	3628	3844	4059	4275	4491	4706	4921	5136
02	5351	5566	5781	5996	6211	6425	6639	6854	7068	7282
03	7496	7710	7924	8137	8351	8564	8778	8991	9204	9417
04	30 9630	9843	*0056	*0268	*0481	*0693	*0906	*1118	*1330	*1542
05	31 1754	1966	2177	2389	2600	2812	3023	3234	3445	3656
06	3867	4078	4289	4499	4710	4920	5130	5340	5551	5760
07	5970	6180	6390	6599	6809	7018	7227	7436	7646	7854
08	31 8063	8272	8481	8689	8898	9106	9314	9522	9730	9938
09	32 0146	0354	0562	0769	0977	1184	1391	1598	1805	2012
210	2219	2426	2633	2839	3046	3252	3458	3665	3871	4077
11	4282	4488	4694	4899	5105	5310	5516	5721	5926	6131
12	6336	6541	6745	6950	7155	7359	7563	7767	7972	8176
13	32 8380	8583	8787	8991	9194	9398	9601	9805	*0008	*0211
14	33 0414	0617	0819	1022	1225	1427	1630	1832	2034	2236
15	2438	2640	2842	3044	3246	3447	3649	3850	4051	4253
16	4454	4655	4856	5057	5257	5458	5658	5859	6059	6260
17	6460	6660	6860	7060	7260	7459	7659	7858	8058	8257
18	33 8456	8656	8855	9054	9253	9451	9650	9849	*0047	*0246
19	34 0444	0642	0841	1039	1237	1435	1632	1830	2028	2225
220	2423	2620	2817	3014	3212	3409	3606	3802	3999	4196
21	4392	4589	4785	4981	5178	5374	5570	5766	5962	6157
22	6353	6549	6744	6939	7135	7330	7525	7720	7915	8110
23	34 8305	8500	8694	8889	9083	9278	9472	9666	9860	*0054
24	35 0248	0442	0636	0829	1023	1216	1410	1603	1796	1989
25	2183	2375	2568	2761	2954	3147	3339	3532	3724	3916
26	4108	4301	4493	4685	4876	5068	5260	5452	5643	5834
27	6026	6217	6408	6599	6790	6981	7172	7363	7554	7744
28	7935	8125	8316	8506	8696	8886	9076	9266	9456	9646
29	35 9835	*0025	*0215	*0404	*0593	*0783	*0972	*1161	*1350	*1539
230	36 1728	1917	2105	2294	2482	2671	2859	3048	3236	3424
31	3612	3800	3988	4176	4363	4551	4739	4926	5113	5301
32	5488	5675	5862	6049	6236	6423	6610	6796	6983	7169
33	7356	7542	7729	7915	8101	8287	8473	8659	8845	9030
34	36 9216	9401	9587	9772	9958	*0143	*0328	*0513	*0698	*0883
35	37 1068	1253	1437	1622	1806	1991	2175	2360	2544	2728
36	2912	3096	3280	3464	3647	3831	4015	4198	4382	4565
37	4748	4932	5115	5298	5481	5664	5846	6029	6212	6394
38	6577	6759	6942	7124	7306	7488	7670	7852	8034	8216
39	37 8398	8580	8761	8943	9124	9306	9487	9668	9849	*0030
240	38 0211	0392	0573	0754	0934	1115	1296	1476	1656	1837
41	2017	2197	2377	2557	2737	2917	3097	3277	3456	3636
42	3815	3995	4174	4353	4533	4712	4891	5070	5249	5428
43	5606	5785	5964	6142	6321	6499	6677	6856	7034	7212
44	7390	7568	7746	7924	8101	8279	8456	8634	8811	8989
45	38 9166	9343	9520	9698	9875	*0051	*0228	*0405	*0582	*0759
46	39 0935	1112	1288	1464	1641	1817	1993	2169	2345	2521
47	2697	2873	3048	3224	3400	3575	3751	3926	4101	4277
48	4452	4627	4802	4977	5152	5326	5501	5676	5850	6025
49	6199	6374	6548	6722	6896	7071	7245	7419	7592	7766
250	39 7940	8114	8287	8461	8634	8808	8981	9154	9328	9501
N	0	1	2	3	4	5	6	7	8	9

SIX-PLACE MANTISSAS FOR COMMON LOGARITHMS (Continued)

N	0	1	2	3	4	5	6	7	8	9
250	39 7940	8114	8287	8461	8634	8808	8981	9154	9328	9501
51	39 9674	9847	*0020	*0192	*0365	*0538	*0711	*0883	*1056	*1228
52	40 1401	1573	1745	1917	2089	2261	2433	2605	2777	2949
53	3121	3292	3464	3635	3807	3978	4149	4320	4492	4663
54	4834	5005	5176	5346	5517	5688	5858	6029	6199	6370
55	6540	6710	6881	7051	7221	7391	7561	7731	7901	8070
56	8240	8410	8579	8749	8918	9087	9257	9426	9595	9764
57	40 9933	*0102	*0271	*0440	*0609	*0777	*0946	*1114	*1283	*1451
58	41 1620	1788	1956	2124	2293	2461	2629	2796	2964	3132
59	3300	3467	3635	3803	3970	4137	4305	4472	4639	4806
260	4973	5140	5307 ·	5474	5641	5808	5974	6141	6308	6474
61	6641	6807	6973	7139	7306	7472	7638	7804	7970	8135
62	8301	8467	8633	8798	8964	9129	9295	9460	9625	9791
63	41 9956	*0121	*0286	*0451	*0616	*0781	*0945	*1110	*1275	*1439
64	42 1604	1768	1933	2097	2261	2426	2590	2754	2918	3082
65	3246	3410	3574	3737	3901	4065	4228	4392	4555	4718
66	4882	5045	5208	5371	5534	5697	5860	6023	6186	6349
67	6511	6674	6836	6999	7161	7324	7486	7648	7811	7973
68	8135	8297	8459	8621	8783	8944	9106	9268	9429	9591
69	42 9752	9914	*0075	*0236	*0398	*0559	*0720	*0881	*1042	*1203
270	43 1364	1525	1685	1846	2007	2167	2328	2488	2649	2809
71	2969	3130	3290	3450	3610	3770	3930	4090	4249	4409
72	4569	4729	4888	5048	5207	5367	5526	5685	5844	6004
73	6163	6322	6481	6640	6799	6957	7116	7275	7433	7592
74	7751	7909	8067	8226	8384	8542	8701	8859	9017	9175
75	43 9333	9491	9648	9806	9964	*0122	*0279	*0437	*0594	*0752
76	44 0909	1066	1224	1381	1538	1695	1852	2009	2166	2323
77	2480	2637	2793	2950	3106	3263	3419	3576	3732	3889
78	4045	4201	4357	4513	4669	4825	4981	5137	5293	5449
79	5604	5760	5915	6071	6226	6382	6537	6692	6848	7003
280	7158	7313	7468	7623	7778	7933	8088	8242	8397	8552
81	44 8706	8861	9015	9170	9324	9478	9633	9787	9941	*0095
82	45 0249	0403	0557	0711	0865	1018	1172	1326	1479	1633
83	1786	1940	2093	2247	2400	2553	2706	2859	3012	3165
84	3318	3471	3624	3777	3930	4082	4235	4387	4540	4692
85	4845	4997	5150	5302	5454	5606	5758	5910	6062	6214
86	6366	6518	6670	6821	6973	7125	7276	7428	7579	7731
87	7882	8033	8184	8336	8487	8638	8789	8940	9091	9242
88	45 9392	9543	9694	9845	9995	*0146	*0296	*0447	*0597	*0748
89	46 0898	1048	1198	1348	1499	1649	1799	1948	2098	2248
290	2398	2548	2697	2847	2997	3146	3296	3445	3594	3744
91	3893	4042	4191	4340	4490	4639	4788	4936	5085	5234
92	5383	5532	5680	5829	5977	6126	6274	6423	6571	6719
93	6868	7016	7164	7312	7460	7608	7756	7904	8052	8200
94	8347	8495	8643	8790	8938	9085	9233	9380	9527	9675
95	46 9822	9969	*0116	*0263	*0410	*0557	*0704	*0851	*0998	*1145
96	47 1292	1438	1585	1732	1878	2025	2171	2318	2464	2610
97	2756	2903	3049	3195	3341	3487	3633	3779	3925	4071
98	4216	4362	4508	4653	4799	4944	5090	5235	5381	5526
99	5671	5816	5962	6107	6252	6397	6542	6687	6832	6976
300	47 7121	7266	7411	7555	7700	7844	7989	8133	8278	8422
N	0	1	2	3	4	5	6	7	8	9

SIX-PLACE MANTISSAS FOR COMMON LOGARITHMS (Continued)

N	0	1	2	3	4	5	6	7	8	9
300	47 7121	7266	7411	7555	7700	7844	7989	8133	8278	8422
01	47 8566	8711	8855	8999	9143	9287	9431	9575	9719	9863
02	48 0007	0151	0294	0438	0582	0725	0869	1012	1156	1299
03	1443	1586	1729	1872	2016	2159	2302	2445	2588	2731
04	2874	3016	3159	3302	3445	3587	3730	3872	4015	4157
05	4300	4442	4585	4727	4869	5011	5153	5295	5437	5579
06	5721	5863	6005	6147	6289	6430	6572	6714	6855	6997
07	7138	7280	7421	7563	7704	7845	7986	8127	8269	8410
08	8551	8692	8833	8974	9114	9255	9396	9537	9677	9818
09	48 9958	*0099	*0239	*0380	*0520	*0661	*0801	*0941	*1081	*1222
310	49 1362	1502	1642	1782	1922	2062	2201	2341	2481	2621
11	2760	2900	3040	3179	3319	3458	3507	3737	3870	4015
12	4155	4294	4433	4572	4711	4850	4989	5128	5267	5406
13	5544	5683	5822	5960	6099	6238	6376	6515	6653	6791
14	6930	7068	7206	7344	7483	7621	7759	7897	8035	8173
15	8311	8448	8586	8724	8862	8999	9137	9275	9412	9550
16	49 9687	9824	9962	*0099	*0236	*0374	*0511	*0648	*0785	*0922
17	50 1059	1196	1333	1470	1607	1744	1880	2017	2154	2291
18	2427	2564	2700	2837	2973	3109	3246	3382	3518	3655
19	3791	3927	4063	4199	4335	4471	4607	4743	4878	5014
320	5150	5286	5421	5557	5693	5828	5964	6099	6234	6370
21	6505	6640	6776	6911	7046	7181	7316	7451	7586	7721
22	7856	7991	8126	8260	8395	8530	8664	8799	8934	9068
23	50 9203	9337	9471	9606	9740	9874	*0009	*0143	*0277	*0411
24	51 0545	0679	0813	0947	1081	1215	1349	1482	1616	1750
25	1883	2017	2151	2284	2418	2551	2684	2818	2951	3084
26	3218	3351	3484	3617	3750	3883	4016	4149	4282	4415
27	4548	4681	4813	4946	5079	5211	5344	5476	5009	5741
28	5874	0006	6139	6271	6403	6535	6668	6800	6932	7064
29	7196	7328	7460	7592	7724	7855	7987	8119	8251	8382
330	8514	8646	8777	8909	9040	9171	9303	9434	9566	9697
31	51 9828	9959	*0090	*0221	*0353	*0484	*0615	*0745	*0876	*1007
32	52 1138	1269	1400	1530	1661	1792	1922	2053	2183	2314
33	2444	2575	2705	2835	2966	3096	3226	3356	3486	3616
34	3746	3876	4006	4136	4266	4396	4526	4656	4785	4915
35	5045	5174	5304	5434	5563	5693	5822	5951	6081	6210
36	6339	6469	6598	6727	6856	6985	7114	7243	7372	7501
37	7630	7759	7888	8016	8145	8274	8402	8531	8660	8788
38	52 8917	9045	9174	9302	9430	9559	9687	9815	9943	*0072
39	53 0200	0328	0456	0584	0712	0840	0968	1096	1223	1351
340	1479	1607	1734	1862	1990	2117	2245	2372	2500	2627
41	2754	2882	3009	3136	3264	3391	3518	3645	3772	3899
42	4026	4153	4280	4407	4534	4661	4787	4914	5041	5167
43	5294	5421	5547	5674	5800	5927	6053	6180	6306	6432
44	6558	6685	6811	6937	7063	7189	7315	7441	7567	7693
45	7819	7945	8071	8197	8322	8448	8574	8699	8825	8951
46	53 9076	9202	9327	9452	9578	9703	9829	9954	*0079	*0204
47	54 0329	0455	0580	0705	0830	0955	1080	1205	1330	1454
48	1579	1704	1829	1953	2078	2203	2327	2452	2576	2701
49	2825	2950	3074	3199	3323	3447	3571	3696	3820	3944
350	54 4068	4192	4316	4440	4564	4688	4812	4936	5060	5183
N	0	1	2	3	4	5	6	7	8	9

SIX-PLACE MANTISSAS FOR COMMON LOGARITHMS (Continued)

N	0	1	2	3	4	5	6	7	8	9
350	54 4068	4192	4316	4440	4564	4688	4812	4936	5060	5183
51	5307	5431	5555	5678	5802	5925	6049	6172	6296	6419
52	6543	6666	6789	6913	7036	7159	7282	7405	7529	7652
53	7775	7898	8021	8144	8267	8389	8512	8635	8758	8881
54	54 9003	9126	9249	9371	9494	9616	9739	9861	9984	*0106
55	55 0228	0351	0473	0595	0717	0840	0962	1084	1206	1328
56	1450	1572	1694	1816	1938	2060	2181	2303	2425	2547
57	2668	2790	2911	3033	3155	3276	3398	3519	3640	3762
58	3883	4004	4126	4247	4368	4489	4610	4731	4852	4973
59	5094	5215	5336	5457	5578	5699	5820	5940	6061	6182
360	6303	6423	6544	6664	6785	6905	7026	7146	7267	7387
61	7507	7627	7748	7868	7988	8108	8228	8349	8469	8589
62	8709	8829	8948	9068	9188	9308	9428	9548	9667	9787
63	55 9907	*0026	*0146	*0265	*0385	*0504	*0624	*0743	*0863	*0982
64	56 1101	1221	1340	1459	1578	1698	1817	1936	2055	2174
65	2293	2412	2531	2650	2769	2887	3006	3125	3244	3362
66	3481	3600	3718	3837	3955	4074	4192	4311	4429	4548
67	4666	4784	4903	5021	5139	5257	5376	5494	5612	5730
68	5848	5966	6084	6202	6320	6437	6555	6673	6791	6909
69	7026	7144	7262	7379	7497	7614	7732	7849	7967	8084
370	8202	8319	8436	8554	8671	8788	8905	9023	9140	9257
71	56 9374	9491	9608	9725	9842	9959	*0076	*0193	*0309	*0426
72	57 0543	0660	0776	0893	1010	1126	1243	1359	1476	1592
73	1709	1825	1942	2058	2174	2291	2407	2523	2639	2755
74	2872	2988	3104	3220	3336	3452	3568	3684	3800	3915
75	4031	4147	4263	4379	4494	4610	4726	4841	4957	5072
76	5188	5303	5419	5534	5650	5765	5880	5996	6111	6226
77	6341	6457	6572	6687	6802	6917	7032	7147	7262	7377
78	7492	7607	7722	7836	7951	8066	8181	8295	8410	8525
79	8639	8754	8868	8983	9097	9212	9326	9441	9555	9669
380	57 9784	9898	*0012	*0126	*0241	*0355	*0469	*0583	*0697	*0811
81	58 0925	1039	1153	1267	1381	1495	1608	1722	1836	1950
82	2063	2177	2291	2404	2518	2631	2745	2858	2972	3085
83	3199	3312	3426	3539	3652	3765	3879	3992	4105	4218
84	4331	4444	4557	4670	4783	4896	5009	5122	5235	5348
85	5461	5574	5686	5799	5912	6024	6137	6250	6362	6475
86	6587	6700	6812	6925	7037	7149	7262	7374	7486	7599
87	7711	7823	7935	8047	8160	8272	8384	8496	8608	8720
88	8832	8944	9056	9167	9279	9391	9503	9615	9726	9838
89	58 9950	*0061	*0173	*0284	*0396	*0507	*0619	*0730	*0842	*0953
390	59 1065	1176	1287	1399	1510	1621	1732	1843	1955	2066
91	2177	2288	2399	2510	2621	2732	2843	2954	3064	3175
92	3286	3397	3508	3618	3729	3840	3950	4061	4171	4282
93	4393	4503	4614	4724	4834	4945	5055	5165	5276	5386
94	5496	5606	5717	5827	5937	6047	6157	6267	6377	6487
95	6597	6707	6817	6927	7037	7146	7256	7366	7476	7586
96	7695	7805	7914	8024	8134	8243	8353	8462	8572	8681
97	8791	8900	9009	9119	9228	9337	9446	9556	9665	9774
98	9883	9992	*0101	*0210	*0319	*0428	*0537	*0646	*0755	*0864
99	60 0973	1082	1191	1299	1408	1517	1625	1734	1843	1951
400	60 2060	2169	2277	2386	2494	2603	2711	2819	2928	3036
N	0	1	2	3	4	5	6	7	8	9

SIX-PLACE MANTISSAS FOR COMMON LOGARITHMS (Continued)

N	0	1	2	3	4	5	6	7	8	9
400	60 2060	2169	2277	2386	2494	2603	2711	2819	2928	3036
01	3144	3253	3361	3469	3577	3686	3794	3902	4010	4118
02	4226	4334	4442	4550	4658	4766	4874	4982	5089	5197
03	5305	5413	5521	5628	5736	5844	5951	6059	6166	6274
04	6381	6489	6596	6704	6811	6919	7026	7133	7241	7348
05	7455	7562	7669	7777	7884	7991	8098	8205	8312	8419
06	8526	8633	8740	8847	8954	9061	9167	9274	9381	9488
07	60 9594	9701	9808	9914	*0021	*0128	*0234	*0341	*0447	*0554
08	61 0660	0767	0873	0979	1086	1192	1298	1405	1511	1617
09	1723	1829	1936	2042	2148	2254	2360	2466	2572	2678
410	2784	2890	2996	3102	3207	3313	3419	3525	3630	3736
11	3842	3947	4053	4159	4264	4370	4475	4581	4686	4792
12	4897	5003	5108	5213	5319	5424	5529	5634	5740	5845
13	5950	6055	6160	6265	6370	6476	6581	6686	6790	6895
14	7000	7105	7210	7315	7420	7525	7629	7734	7839	7943
15	8048	8153	8257	8362	8466	8571	8676	8780	8884	8989
16	61 9093	9198	9302	9406	9511	9615	9719	9824	9928	*0032
17	62 0136	0240	0344	0448	0552	0656	0760	0864	0968	1072
18	1176	1280	1384	1488	1592	1695	1799	1903	2007	2110
19	2214	2318	2421	2525	2628	2732	2835	2939	3042	3146
420	3249	3353	3456	3559	3663	3766	3869	3973	4076	4179
21	4282	4385	4488	4591	4695	4798	4901	5004	5107	5210
22	5312	5415	5518	5621	5724	5827	5929	6032	6135	6238
23	6340	6443	6546	6648	6751	6853	6956	7058	7161	7263
24	7366	7468	7571	7673	7775	7878	7980	8082	8185	8287
25	8389	8491	8593	8695	8797	8900	9002	9104	9206	9308
26	62 9410	9512	9613	9715	9817	9919	*0021	*0123	*0224	*0326
27	63 0428	0530	0631	0733	0835	0936	1038	1139	1241	1342
28	1444	1545	1647	1748	1849	1951	2052	2153	2255	2356
29	2457	2559	2660	2761	2862	2963	3064	3165	3266	3367
430	3468	3569	3670	3771	3872	3973	4074	4175	4276	4376
31	4477	4578	4679	4779	4880	4981	5081	5182	5283	5383
32	5484	5584	5685	5785	5886	5986	6087	6187	6287	6388
33	6488	6588	6688	6789	6889	6989	7089	7189	7290	7390
34	7490	7590	7690	7790	7890	7990	8090	8190	8290	8389
35	8489	8589	8689	8789	8888	8988	9088	9188	9287	9387
36	63 9486	9586	9686	9785	9885	9984	*0084	*0183	*0283	*0382
37	64 0481	0581	0680	0779	0879	0978	1077	1177	1276	1375
38	1474	1573	1672	1771	1871	1970	2069	2168	2267	2366
39	2465	2563	2662	2761	2860	2959	3058	3156	3255	3354
440	3453	3551	3650	3749	3847	3946	4044	4143	4242	4340
41	4439	4537	4636	4734	4832	4931	5029	5127	5226	5324
42	5422	5521	5619	5717	5815	5913	6011	6110	6208	6306
43	6404	6502	6600	6698	6796	6894	6992	7089	7187	7285
44	7383	7481	7579	7676	7774	7872	7969	8067	8165	8262
45	8360	8458	8555	8653	8750	8848	8945	9043	9140	9237
46	64 9335	9432	9530	9627	9724	9821	9919	*0016	*0113	*0210
47	65 0308	0405	0502	0599	0696	0793	0890	0987	1084	1181
48	1278	1375	1472	1569	1666	1762	1859	1956	2053	2150
49	2246	2343	2440	2536	2633	2730	2826	2923	3019	3116
450	65 3213	3309	3405	3502	3598	3695	3791	3888	3984	4080
N	0	1	2	3	4	5	6	7	8	9

SIX-PLACE MANTISSAS FOR COMMON LOGARITHMS (Continued)

N	0	1	2	3	4	5	6	7	8	9
450	65 3213	3309	3405	3502	3598	3695	3791	3888	3984	4080
51	4177	4273	4369	4465	4562	4658	4754	4850	4946	5042
52	5138	5235	5331	5427	5523	5619	5715	5810	5906	6002
53	6098	6194	6290	6386	6482	6577	6673	6769	6864	6960
54	7056	7152	7247	7343	7438	7534	7629	7725	7820	7916
55	8011	8107	8202	8298	8393	8488	8584	8679	8774	8870
56	8965	9060	9155	9250	9346	9441	9536	9631	9726	9821
57	65 9916	*0011	*0106	*0201	*0296	*0391	*0486	*0581	*0676	*0771
58	66 0865	0960	1055	1150	1245	1339	1434	1529	1623	1718
59	1813	1907	2002	2096	2191	2286	2380	2475	2569	2663
460	2758	2852	2947	3041	3135	3230	3324	3418	3512	3607
61	3701	3795	3889	3983	4078	4172	4266	4360	4454	4548
62	4642	4736	4830	4924	5018	5112	5206	5299	5393	5487
63	5581	5675	5769	5862	5956	6050	6143	6237	6331	6424
64	6518	6612	6705	6799	6892	6986	7079	7173	7266	7360
65	7453	7546	7640	7733	7826	7920	8013	8106	8199	8293
66	8386	8479	8572	8665	8759	8852	8945	9038	9131	9224
67	66 9317	9410	9503	9596	9689	9782	9875	9967	*0060	*0153
68	67 0246	0339	0431	0524	0617	0710	0802	0895	0988	1080
69	1173	1265	1358	1451	1543	1636	1728	1821	1913	2005
470	2098	2190	2283	2375	2467	2560	2652	2744	2836	2929
71	3021	3113	3205	3297	3390	3482	3574	3666	3758	3850
72	3942	4034	4126	4218	4310	4402	4494	4586	4677	4769
73	4861	4953	5045	5137	5228	5320	5412	5503	5595	5687
74	5778	5870	5962	6053	6145	6236	6328	6419	6511	6602
75	6694	6785	6876	6968	7059	7151	7242	7333	7424	7516
76	7607	7698	7789	7881	7972	8063	8154	8245	8336	8427
77	8518	8609	8700	8791	8882	8973	9064	9155	9246	9337
78	67 9428	9519	9610	9700	9791	9882	9973	*0063	*0154	*0245
79	68 0336	0426	0517	0607	0698	0789	0879	0970	1060	1151
480	1241	1332	1422	1513	1603	1693	1784	1874	1964	2055
81	2145	2235	2326	2416	2506	2596	2686	2777	2867	2957
82	3047	3137	3227	3317	3407	3497	3587	3677	3767	3857
83	3947	4037	4127	4217	4307	4396	4486	4576	4666	4756
84	4845	4935	5025	5114	5204	5294	5383	5473	5563	5652
85	5742	5831	5921	6010	6100	6189	6279	6368	6458	6547
86	6636	6726	6815	6904	6994	7083	7172	7261	7351	7440
87	7529	7618	7707	7796	7886	7975	8064	8153	8242	8331
88	8420	8509	8598	8687	8776	8865	8953	9042	9131	9220
89	68 9309	9398	9486	9575	9664	9753	9841	9930	*0019	*0107
490	69 0196	0285	0373	0462	0550	0639	0728	0816	0905	0993
91	1081	1170	1258	1347	1435	1524	1612	1700	1789	1877
92	1965	2053	2142	2230	2318	2406	2494	2583	2671	2759
93	2847	2935	3023	3111	3199	3287	3375	3463	3551	3639
94	3727	3815	3903	3991	4078	4166	4254	4342	4430	4517
95	4605	4693	4781	4868	4956	5044	5131	5219	5307	5394
96	5482	5569	5657	5744	5832	5919	6007	6094	6182	6269
97	6356	6444	6531	6618	6706	6793	6880	6968	7055	7142
98	7229	7317	7404	7491	7578	7665	7752	7839	7926	8014
99	8101	8188	8275	8362	8449	8535	8622	8709	8796	8883
500	69 8970	9057	9144	9231	9317	9404	9491	9578	9664	9751
N	0	1	2	3	4	5	6	7	8	9

SIX-PLACE MANTISSAS FOR COMMON LOGARITHMS (Continued)

N	0	1	2	3	4	5	6	7	8	9
500	69 8970	9057	9144	9231	9317	9404	9491	9578	9664	9751
01	69 9838	9924	*0011	*0098	*0184	*0271	*0358	*0444	*0531	*0617
02	70 0704	0790	0877	0963	1050	1136	1222	1309	1395	1482
03	1568	1654	1741	1827	1913	1999	2086	2172	2258	2344
04	2431	2517	2603	2689	2775	2861	2947	3033	3119	3205
05	3291	3377	3463	3549	3635	3721	3807	3893	3979	4065
06	4151	4236	4322	4408	4494	4579	4665	4751	4837	4922
07	5008	5094	5179	5265	5350	5463	5522	5607	5693	5778
08	5864	5949	6035	6120	6206	6291	6376	6462	6547	6632
09	6718	6803	6888	6974	7059	7144	7229	7315	7400	7485
510	7570	7655	7740	7826	7911	7996	8081	8166	8251	8336
11	8421	8506	8591	8676	8761	8846	8931	9015	9100	9185
12	70 9270	9355	9440	9524	9609	9694	9779	9863	9948	*0033
13	71 0117	0202	0287	0371	0456	0540	0625	0710	0794	0879
14	0963	1048	1132	1217	1301	1385	1470	1554	1639	1723
15	1807	1892	1976	2060	2144	2229	2313	2397	2481	2566
16	2650	2734	2818	2902	2986	3070	3154	3238	3323	3407
17	3491	3575	3659	3742	3826	3910	3994	4078	4162	4246
18	4330	4414	4497	4581	4665	4749	4833	4916	5000	5084
19	5167	5251	5335	5418	5502	5586	5669	5753	5836	5920
520	6003	6087	6170	6254	6337	6421	6504	6588	6671	6754
21	6838	6921	7004	7088	7171	7254	7338	7421	7504	7587
22	7671	7754	7837	7920	8003	8086	8169	8253	8336	8419
23	8502	8585	8668	8751	8834	8917	9000	9083	9165	9248
24	71 9331	9414	9497	9580	9663	9745	9828	9911	9994	*0077
25	72 0159	0242	0325	0407	0490	0573	0655	0738	0821	0903
26	0986	1068	1151	1233	1316	1398	1481	1563	1646	1728
27	1811	1893	1975	2058	2140	2222	2305	2387	2469	2552
28	2634	2716	2798	2881	2963	3045	3127	3209	3291	3374
29	3456	3538	3620	3702	3784	3866	3948	4030	4112	4194
530	4276	4358	4440	4522	4604	4685	4767	4849	4931	5013
31	5095	5176	5258	5340	5422	5503	5585	5667	5748	5830
32	5912	5993	6075	6156	6238	6320	6401	6483	6564	6646
33	6727	6809	6890	6972	7053	7134	7216	7297	7379	7460
34	7541	7623	7704	7785	7866	7948	8029	8110	8191	8273
35	8354	8435	8516	8597	8678	8759	8841	8922	9003	9084
36	9165	9246	9327	9408	9489	9570	9651	9732	9813	9893
37	72 9974	*0055	*0136	*0217	*0298	*0378	*0459	*0540	*0621	*0702
38	73 0782	0863	0944	1024	1105	1186	1266	1347	1428	1508
39	1589	1669	1750	1830	1911	1991	2072	2152	2233	2313
540	2394	2474	2555	2635	2715	2796	2876	2956	3037	3117
41	3197	3278	3358	3438	3518	3598	3679	3759	3839	3919
42	3999	4079	4160	4240	4320	4400	4480	4560	4640	4720
43	4800	4880	4960	5040	5120	5200	5279	5359	5439	5519
44	5599	5679	5759	5838	5918	5998	6078	6157	6237	6317
45	6397	6476	6556	6635	6715	6795	6874	6954	7034	7113
46	7193	7272	7352	7431	7511	7590	7670	7749	7829	7908
47	7987	8067	8146	8225	8305	8384	8463	8543	8622	8701
48	8781	8860	8939	9018	9097	9177	9256	9335	9414	9493
49	73 9572	9651	9731	9810	9889	9968	*0047	*0126	*0205	*0284
550	74 0363	0442	0521	0600	0678	0757	0836	0915	0994	1073
N	0	1	2	3	4	5	6	7	8	9

SIX-PLACE MANTISSAS FOR COMMON LOGARITHMS (Continued)

N	0	1	2	3	4	5	6	7	8	9
550	74 0363	0442	0521	0600	0678	0757	0836	0915	0994	1073
51	1152	1230	1309	1388	1467	1546	1624	1703	1782	1860
52	1939	2018	2096	2175	2254	2332	2411	2489	2568	2647
53	2725	2804	2882	2961	3039	3118	3196	3275	3353	3431
54	3510	3588	3667	3745	3823	3902	3980	4058	4136	4215
55	4293	4371	4449	4528	4606	4684	4762	4840	4919	4997
56	5075	5153	5231	5309	5387	5465	5543	5621	5699	5777
57	5855	5933	6011	6089	6167	6245	6323	6401	6479	6556
58	6634	6712	6790	6868	6945	7023	7101	7179	7256	7334
59	7412	7489	7567	7645	7722	7800	7878	7955	8033	8110
560	8188	8266	8343	8421	8498	8576	8653	8731	8808	8885
61	8963	9040	9118	9195	9272	9350	9427	9504	9582	9659
62	74 9736	9814	9891	9968	*0045	*0123	*0200	*0277	*0354	*0431
63	75 0508	0586	0663	0740	0817	0894	0971	1048	1125	1202
64	1279	1356	1433	1510	1587	1664	1741	1818	1895	1972
65	2048	2125	2202	2279	2356	2433	2509	2586	2663	2740
66	2816	2893	2970	3047	3123	3200	3277	3353	3430	3506
67	3583	3660	3736	3813	3889	3966	4042	4119	4195	4272
68	4348	4425	4501	4578	4654	4730	4807	4883	4960	5036
69	5112	5189	5265	5341	5417	5494	5570	5646	5722	5799
570	5875	5951	6027	6103	6180	6256	6332	6408	6484	6560
71	6636	6712	6788	6864	6940	7016	7092	7168	7244	7320
72	7396	7472	7548	7624	7700	7775	7851	7927	8003	8079
73	8155	8230	8306	8382	8458	8533	8609	8685	8761	8836
74	8912	8988	9063	9139	9214	9290	9366	9441	9517	9592
75	75 9668	9743	9819	9894	9970	*0045	*0121	*0196	*0272	*0347
76	76 0422	0498	0573	0649	0724	0799	0875	0950	1025	1101
77	1176	1251	1326	1402	1477	1552	1627	1702	1778	1853
78	1928	2003	2078	2153	2228	2303	2378	2453	2529	2604
79	2679	2754	2829	2904	2978	3053	3128	3203	3278	3353
580	3428	3503	3578	3653	3727	3802	3877	3952	4027	4101
81	4176	4251	4326	4400	4475	4550	4624	4699	4774	4848
82	4923	4998	5072	5147	5221	5296	5370	5445	5520	5594
83	5669	5743	5818	5892	5966	6041	6115	6190	6264	6338
84	6413	6487	6562	6636	6710	6785	6859	6933	7007	7082
85	7156	7230	7304	7379	7453	7527	7601	7675	7749	7823
86	7898	7972	8046	8120	8194	8268	8342	8416	8490	8564
87	8638	8712	8786	8860	8934	9008	9082	9156	9230	9303
88	76 9377	9451	9525	9599	9673	9746	9820	9894	9968	*0042
89	77 0115	0189	0263	0336	0410	0484	0557	0631	0705	0778
590	0852	0926	0999	1073	1146	1220	1293	1367	1440	1514
91	1587	1661	1734	1808	1881	1955	2028	2102	2175	2248
92	2322	2395	2468	2542	2615	2688	2762	2835	2908	2981
93	3055	3128	3201	3274	3348	3421	3494	3567	3640	3713
94	3786	3860	3933	4006	4079	4152	4225	4298	4371	4444
95	4517	4590	4663	4736	4809	4882	4955	5028	5100	5173
96	5246	5319	5392	5465	5538	5610	5683	5756	5829	5902
97	5974	6047	6120	6193	6265	6338	6411	6483	6556	6629
98	6701	6774	6846	6919	6992	7064	7137	7209	7282	7354
99	7427	7499	7572	7644	7717	7789	7862	7934	8006	8079
600	77 8151	8224	8296	8368	8441	8513	8585	8658	8730	8802
N	0	1	2	3	4	5	6	7	8	9

SIX-PLACE MANTISSAS FOR COMMON LOGARITHMS (Continued)

N	0	1	2	3	4	5	6	7	8	9
600	77 8151	8224	8296	8368	8441	8513	8585	8658	8730	8802
01	8874	8947	9019	9091	9163	9236	9308	9380	9452	9524
02	77 9596	9669	9741	9813	9885	9957	*0029	*0101	*0173	*0245
03	78 0317	0389	0461	0533	0605	0677	0749	0821	0893	0965
04	1037	1109	1181	1253	1324	1396	1468	1540	1612	1684
05	1755	1827	1899	1971	2042	2114	2186	2258	2329	2401
06	2473	2544	2616	2688	2759	2831	2902	2974	3046	3117
07	3189	3260	3332	3403	3475	3546	3618	3689	3761	3832
08	3904	3975	4046	4118	4189	4261	4332	4403	4475	4546
09	4617	4689	4760	4831	4902	4974	5045	5116	5187	5259
610	5330	5401	5472	5543	5615	5686	5757	5828	5899	5970
11	6041	6112	6183	6254	6325	6396	6467	6538	6609	6680
12	6751	6822	6893	6964	7035	7106	7177	7248	7319	7390
13	7460	7531	7602	7673	7744	7815	7885	7956	8027	8098
14	8168	8239	8310	8381	8451	8522	8593	8663	8734	8804
15	8875	8946	9016	9087	9157	9228	9299	9369	9440	9510
16	78 9581	9651	9722	9762	9863	9933	*0004	*0074	*0144	*0215
17	79 0285	0356	0426	0496	0567	0637	0707	0778	0848	0918
18	0988	1059	1129	1199	1269	1340	1410	1480	1550	1620
19	1691	1761	1831	1901	1971	2041	2111	2181	2252	2322
620	2392	2462	2532	2602	2672	2742	2812	2882	2952	3022
21	3092	3162	3231	3301	3371	3441	3511	3581	3651	3721
22	3790	3860	3930	4000	4070	4139	4209	4279	4349	4418
23	4488	4558	4627	4697	4767	4836	4906	4976	5045	5115
24	5185	5254	5324	5393	5463	5532	5602	5672	5741	5811
25	5880	5949	6019	6088	6158	6227	6297	6366	6436	6505
26	6574	6644	6713	6782	6852	6921	6990	7060	7129	7198
27	7268	7337	7406	7475	7545	7614	7683	7752	7821	7890
28	7960	8029	8098	8167	8236	8305	8374	8443	8513	8582
29	8651	8720	8789	8858	8927	8996	9065	9134	9203	9272
630	79 9341	9409	9478	9547	9616	9685	9754	9823	9892	9961
31	80 0029	0098	0167	0236	0305	0373	0442	0511	0580	0648
32	0717	0786	0854	0923	0992	1061	1129	1198	1266	1335
33	1404	1472	1541	1609	1678	1747	1815	1884	1952	2021
34	2089	2158	2226	2295	2363	2432	2500	2568	2637	2705
35	2774	2842	2910	2979	3047	3116	3184	3252	3321	3389
36	3457	3525	3594	3662	3730	3798	3867	3935	4003	4071
37	4139	4208	4276	4344	4412	4480	4548	4616	4685	4753
38	4821	4889	4957	5025	5093	5161	5229	5297	5365	5433
39	5501	5569	5637	5705	5773	5841	5908	5976	6044	6112
640	6180	6248	6316	6384	6451	6519	6587	6655	6723	6790
41	6858	6926	6994	7061	7129	7197	7264	7332	7400	7467
42	7535	7603	7670	7738	7806	7873	7941	8008	8076	8143
43	8211	8279	8346	8414	8481	8549	8616	8684	8751	8818
44	8886	8953	9021	9088	9156	9223	9290	9358	9425	9492
45	80 9560	9627	9694	9762	9829	9896	9964	*0031	*0098	*0165
46	81 0233	0300	0367	0434	0501	0569	0636	0703	0770	0837
47	0904	0971	1039	1106	1173	1240	1307	1374	1441	1508
48	1575	1642	1709	1776	1843	1910	1977	2044	2111	2178
49	2245	2312	2379	2445	2512	2579	2646	2713	2780	2847
650	81 2913	2980	3047	3114	3181	3247	3314	3381	3448	3514
N	0	1	2	3	4	5	6	7	8	9

SIX-PLACE MANTISSAS FOR COMMON LOGARITHMS (Continued)

N	0	1	2	3	4	5	6	7	8	9
650	81 2913	2980	3047	3114	3181	3247	3314	3381	3448	3514
51	3581	3648	3714	3781	3848	3914	3981	4048	4114	4181
52	4248	4314	4381	4447	4514	4581	4647	4714	4780	4847
53	4913	4980	5046	5113	5179	5246	5312	5378	5445	5511
54	5578	5644	5711	5777	5843	5910	5976	6042	6109	6175
55	6241	6308	6374	6440	6506	6573	6639	6705	6771	6838
56	6904	6970	7036	7102	7169	7235	7301	7367	7433	7499
57	7565	7631	7698	7764	7830	7896	7962	8028	8094	8160
58	8226	8292	8358	8424	8490	8556	8622	8688	8754	8820
59	8885	8951	9017	9083	9149	9215	9281	9346	9412	9478
660	81 9544	9610	9676	9741	9807	9873	9939	*0004	*0070	*0136
61	82 0201	0267	0333	0399	0464	0530	0595	0661	0727	0792
62	0858	0924	0989	1055	1120	1186	1251	1317	1382	1448
63	1514	1579	1645	1710	1775	1841	1906	1972	2037	2103
64	2168	2233	2299	2364	2430	2495	2560	2626	2691	2756
65	2822	2887	2952	3018	3083	3148	3213	3279	3344	3409
66	3474	3539	3605	3670	3735	3800	3865	3930	3996	4061
67	4126	4191	4256	4321	4386	4451	4516	4581	4646	4711
68	4776	4841	4906	4971	5036	5101	5166	5231	5296	5361
69	5426	5491	5556	5621	5686	5751	5815	5880	5945	6010
670	6075	6140	6204	6269	6334	6399	6464	6528	6593	6658
71	6723	6787	6852	6917	6981	7046	7111	7175	7240	7305
72	7369	7434	7499	7563	7628	7692	7757	7821	7886	7951
73	8015	8080	8144	8209	8273	8338	8402	8467	8531	8595
74	8660	8724	8789	8853	8918	8982	9046	9111	9175	9239
75	9304	9368	9432	9497	9561	9625	9690	9754	9818	9882
76	82 9947	*0011	*0075	*0139	*0204	*0268	*0332	*0396	*0460	*0525
77	83 0589	0653	0717	0781	0845	0909	0973	1037	1102	1166
78	1230	1294	1358	1422	1486	1550	1614	1678	1742	1806
79	1870	1934	1998	2062	2126	2189	2253	2317	2381	2445
680	2509	2573	2637	2700	2764	2828	2892	2956	3020	3083
81	3147	3211	3275	3338	3402	3466	3530	3593	3657	3721
82	3784	3848	3912	3975	4039	4103	4166	4230	4294	4357
83	4421	4484	4548	4611	4675	4739	4802	4866	4929	4993
84	5056	5120	5183	5247	5310	5373	5437	5500	5564	5627
85	5691	5754	5817	5881	5944	6007	6071	6134	6197	6261
86	6324	6387	6451	6514	6577	6641	6704	6767	6830	6894
87	6957	7020	7083	7146	7210	7273	7336	7399	7462	7525
88	7588	7652	7715	7778	7841	7904	7967	8030	8093	8156
89	8219	8282	8345	8408	8471	8534	8597	8660	8723	8786
690	8849	8912	8975	9038	9101	9164	9227	9289	9352	9415
91	83 9478	9541	9604	9667	9729	9792	9855	9918	9981	*0043
92	84 0106	0169	0232	0294	0357	0420	0482	0545	0608	0671
93	0733	0796	0859	0921	0984	1046	1109	1172	1234	1297
94	1359	1422	1485	1547	1610	1672	1735	1797	1860	1922
95	1985	2047	2110	2172	2235	2297	2360	2422	2484	2547
96	2609	2672	2734	2796	2859	2921	2983	3046	3108	3170
97	3233	3295	3357	3420	3482	3544	3606	3669	3731	3793
98	3855	3918	3980	4042	4104	4166	4229	4291	4353	4415
99	4477	4539	4601	4664	4726	4788	4850	4912	4974	5036
700	84 5098	5160	5222	5284	5346	5408	5470	5532	5594	5656
N	0	1	2	3	4	5	6	7	8	9

SIX-PLACE MANTISSAS FOR COMMON LOGARITHMS (Continued)

N	0	1	2	3	4	5	6	7	8	9
700	84 5098	5160	5222	5284	5346	5408	5470	5532	5594	5656
01	5718	5780	5842	5904	5966	6028	6090	6151	6213	6275
02	6337	6399	6461	6523	6585	6646	6708	6770	6832	6894
03	6955	7017	7079	7141	7202	7264	7326	7388	7449	7511
04	7573	7634	7696	7758	7819	7881	7943	8004	8066	8128
05	8189	8251	8312	8374	8435	8497	8559	8620	8682	8743
06	8805	8866	8928	8989	9051	9112	9174	9235	9297	9358
07	84 9419	9481	9542	9604	9665	9726	9788	9849	9911	9972
08	85 0033	0095	0156	0217	0279	0340	0401	0462	0524	0585
09	0646	0707	0769	0830	0891	0952	1014	1075	1136	1197
710	1258	1320	1381	1442	1503	1564	1625	1686	1747	1809
11	1870	1931	1992	2053	2114	2175	2236	2297	2358	2419
12	2480	2541	2602	2663	2724	2785	2846	2907	2968	3029
13	3090	3150	3211	3272	3333	3394	3455	3516	3577	3637
14	3698	3759	3820	3881	3941	4002	4063	4124	4185	4245
15	4306	4367	4428	4488	4549	4610	4670	4731	4792	4852
16	4913	4974	5034	5095	5156	5216	5277	5337	5398	5459
17	5519	5580	5640	5701	5761	5822	5882	5943	6003	6064
18	6124	6185	6245	6306	6366	6427	6487	6548	6608	6668
19	6729	6789	6850	6910	6970	7031	7091	7152	7212	7272
720	7332	7393	7453	7513	7574	7634	7694	7755	7815	7875
21	7935	7995	8056	8116	8176	8236	8297	8357	8417	8477
22	8537	8597	8657	8718	8778	8838	8898	8958	9018	9078
23	9138	9198	9258	9318	9379	9439	9499	9559	9619	9679
24	85 9739	9799	9859	9918	9978	*0038	*0098	*0158	*0218	*0278
25	86 0338	0398	0458	0518	0578	0637	0697	0757	0817	0877
26	0937	0996	1056	1116	1176	1236	1295	1355	1415	1475
27	1534	1594	1654	1714	1773	1833	1893	1952	2012	2072
28	2131	2191	2251	2310	2370	2430	2489	2549	2608	2668
29	2728	2787	2847	2906	2966	3025	3085	3144	3204	3263
730	3323	3382	3442	3501	3561	3620	3680	3739	3799	3858
31	3917	3977	4036	4096	4155	4214	4274	4333	4392	4452
32	4511	4570	4630	4689	4748	4808	4867	4926	4985	5045
33	5104	5163	5222	5282	5341	5400	5459	5519	5578	5637
34	5696	5755	5814	5874	5933	5992	6051	6110	6169	6228
35	6287	6346	6405	6465	6524	6583	6642	6701	6760	6819
36	6878	6937	6996	7055	7114	7173	7232	7291	7350	7409
37	7467	7526	7585	7644	7703	7762	7821	7880	7939	7998
38	8056	8115	8174	8233	8292	8350	8409	8468	8527	8586
39	8644	8703	8762	8821	8879	8938	8997	9056	9114	9173
740	9232	9290	9349	9408	9466	9525	9584	9642	9701	9760
41	86 9818	9877	9935	9994	*0053	*0111	*0170	*0228	*0287	*0345
42	87 0404	0462	0521	0579	0638	0696	0755	0813	0872	0930
43	0989	1047	1106	1164	1223	1281	1339	1398	1456	1515
44	1573	1631	1690	1748	1806	1865	1923	1981	2040	2098
45	2156	2215	2273	2331	2389	2448	2506	2564	2622	2681
46	2739	2797	2855	2913	2972	3030	3088	3146	3204	3262
47	3321	3379	3437	3495	3553	3611	3669	3727	3785	3844
48	3902	3960	4018	4076	4134	4192	4250	4308	4366	4424
49	4482	4540	4598	4656	4714	4772	4830	4888	4945	5003
750	87 5061	5119	5177	5235	5293	5351	5409	5466	5524	5582
N	0	1	2	3	4	5	6	7	8	9

SIX-PLACE MANTISSAS FOR COMMON LOGARITHMS (Continued)

N	0	1	2	3	4	5	6	7	8	9
750	87 5061	5119	5177	5235	5293	5351	5409	5466	5524	5582
51	5640	5698	5756	5813	5871	5929	5987	6045	6102	6160
52	6218	6276	6333	6391	6449	6507	6564	6622	6680	6737
53	6795	6853	6910	6968	7026	7083	7141	7199	7256	7314
54	7371	7429	7487	7544	7602	7659	7717	7774	7832	7889
55	7947	8004	8062	8119	8177	8234	8292	8349	8407	8464
56	8522	8579	8637	8694	8752	8809	8866	8924	8981	9039
57	9096	9153	9211	9268	9325	9383	9440	9497	9555	9612
58	87 9669	9726	9784	9841	9898	9956	*0013	*0070	*0127	*0185
59	88 0242	0299	0356	0413	0417	0528	0585	0642	0699	0756
760	0814	0871	0928	0985	1042	1099	1156	1213	1271	1328
61	1385	1442	1499	1556	1613	1670	1727	1784	1841	1898
62	1955	2012	2069	2126	2183	2240	2297	2354	2411	2468
63	2525	2581	2638	2695	2752	2809	2866	2923	2980	3037
64	3093	3150	3207	3264	3321	3377	3434	3491	3548	3605
65	3661	3718	3775	3832	3888	3945	4002	4059	4115	4172
66	4229	4285	4342	4399	4455	4512	4569	4625	4682	4739
67	4795	4852	4909	4965	5022	5078	5135	5192	5248	5305
68	5361	5418	5474	5531	5587	5644	5700	5757	5813	5870
69	5926	5983	6039	6096	6152	6209	6265	6321	6378	6434
770	6491	6547	6604	6660	6716	6773	6829	6885	6942	6998
71	7054	7111	7167	7223	7280	7336	7392	7449	7505	7561
72	7617	7674	7730	7786	7842	7898	7955	8011	8067	8123
73	8179	8236	8292	8348	8404	8460	8516	8573	8629	8685
74	8741	8797	8853	8909	8965	9021	9077	9134	9190	9246
75	9302	9358	9414	9470	9526	9582	9638	9694	9750	9806
76	88 9862	9918	9974	*0030	*0086	*0141	*0197	*0253	*0309	*0365
77	89 0421	0477	0533	0589	0645	0700	0756	0812	0868	0924
78	0980	1035	1091	1147	1203	1259	1314	1370	1426	1482
79	1537	1593	1649	1705	1760	1816	1872	1928	1983	2039
780	2095	2150	2206	2262	2317	2373	2429	2484	2540	2595
81	2651	2707	2762	2818	2873	2929	2985	3040	3096	3151
82	3207	3262	3318	3373	3429	3484	3540	3595	3651	3706
83	3762	3817	3873	3928	3984	4039	4094	4150	4205	4261
84	4316	4371	4427	4482	4538	4593	4648	4704	4759	4814
85	4870	4925	4980	5036	5091	5146	5201	5257	5312	5367
86	5423	5478	5533	5588	5644	5699	5754	5809	5864	5920
87	5975	6030	6085	6140	6195	6251	6306	6361	6416	6471
88	6526	6581	6636	6692	6747	6802	6857	6912	6967	7022
89	7077	7132	7187	7242	7297	7352	7407	7462	7517	7572
790	7627	7682	7737	7792	7847	7902	7957	8012	8067	8122
91	8176	8231	8286	8341	8396	8451	8506	8561	8615	8670
92	8725	8780	8835	8890	8944	8999	9054	9109	9164	9218
93	9273	9328	9383	9437	9492	9547	9602	9656	9711	9766
94	89 9821	9875	9930	9985	*0039	*0094	*0149	*0203	*0258	*0312
95	90 0367	0422	0476	0531	0586	0640	0695	0749	0804	0859
96	0913	0968	1022	1077	1131	1186	1240	1295	1349	1404
97	1458	1513	1567	1622	1676	1731	1785	1840	1894	1948
98	2003	2057	2112	2166	2221	2275	2329	2384	2438	2492
99	2547	2601	2655	2710	2764	2818	2873	2927	2981	3036
800	90 3090	3144	3199	3253	3307	3361	3416	3470	3524	3578
N	0	1	2	3	4	5	6	7	8	9

SIX-PLACE MANTISSAS FOR COMMON LOGARITHMS (Continued)

N	0	1	2	3	4	5	6	7	8	9
800	90 3090	3144	3199	3253	3307	3361	3416	3470	3524	3578
01	3633	3687	3741	3795	3849	3904	3958	4012	4066	4120
02	4174	4229	4283	4337	4391	4445	4499	4553	4607	4661
03	4716	4770	4824	4878	4932	4986	5040	5094	5148	5202
04	5256	5310	5364	5418	5472	5526	5580	5634	5688	5742
05	5796	5850	5904	5958	6012	6066	6119	6173	6227	6281
06	6335	6389	6443	6497	6551	6604	6658	6712	6766	6820
07	6874	6927	6981	7035	7089	7143	7196	7250	7304	7358
08	7411	7465	7519	7573	7626	7680	7734	7787	7841	7895
09	7949	8002	8056	8110	8163	8217	8270	8324	8378	8431
810	8485	8539	8592	8646	8699	8753	8807	8860	8914	8967
11	9021	9074	9128	9181	9235	9289	9342	9396	9449	9503
12	90 9556	9610	9663	9716	9770	9823	9877	9930	9984	*0037
13	91 0091	0144	0197	0251	0304	0358	0411	0464	0518	0571
14	0624	0678	0731	0784	0838	0891	0944	0998	1051	1104
15	1158	1211	1264	1317	1371	1424	1477	1530	1584	1637
16	1690	1743	1797	1850	1903	1956	2009	2063	2116	2169
17	2222	2275	2328	2381	2435	2488	2541	2594	2647	2700
18	2753	2806	2859	2913	2966	3019	3072	3125	3178	3231
19	3284	3337	3390	3443	3496	3549	3602	3655	3708	3761
820	3814	3867	3920	3973	4026	4079	4132	4184	4237	4290
21	4343	4396	4449	4502	4555	4608	4660	4713	4766	4819
22	4872	4925	4977	5030	5083	5136	5189	5241	5294	5347
23	5400	5453	5505	5558	5611	5664	5716	5769	5822	5875
24	5927	5980	6033	6085	6138	6191	6243	6296	6349	6401
25	6454	6507	6559	6612	6664	6717	6770	6822	6875	6927
26	6980	7033	7085	7138	7190	7243	7295	7348	7400	7453
27	7506	7558	7611	7663	7716	7768	7820	7873	7925	7978
28	8030	8083	8135	8188	8240	8293	8345	8397	8450	8502
29	8555	8607	8659	8712	8764	8816	8869	8921	8973	9026
830	9078	9130	9183	9235	9287	9340	9392	9444	9496	9549
31	91 9601	9653	9706	9758	9810	9862	9914	9967	*0019	*0071
32	92 0123	0176	0228	0280	0332	0384	0436	0489	0541	0593
33	0645	0697	0749	0801	0853	0906	0958	1010	1062	1114
34	1166	1218	1270	1322	1374	1426	1478	1530	1582	1634
35	1686	1738	1790	1842	1894	1946	1998	2050	2102	2154
36	2206	2258	2310	2362	2414	2466	2518	2570	2622	2674
37	2725	2777	2829	2881	2933	2985	3037	3089	3140	3192
38	3244	3296	3348	3399	3451	3503	3555	3607	3658	3710
39	3762	3814	3865	3917	3969	4021	4072	4124	4176	4228
840	4279	4331	4383	4434	4486	4538	4589	4641	4693	4744
41	4796	4848	4899	4951	5003	5054	5106	5157	5209	5261
42	5312	5364	5415	5467	5518	5570	5621	5673	5725	5776
43	5828	5879	5931	5982	6034	6085	6137	6188	6240	6291
44	6342	6394	6445	6497	6548	6600	6651	6702	6754	6805
45	6857	6908	6959	7011	7062	7114	7165	7216	7268	7319
46	7370	7422	7473	7524	7576	7627	7678	7730	7781	7832
47	7883	7935	7986	8037	8088	8140	8191	8242	8293	8345
48	8396	8447	8498	8549	8601	8652	8703	8754	8805	8857
49	8908	8959	9010	9061	9112	9163	9215	9266	9317	9368
850	92 9419	9470	9521	9572	9623	9674	9725	9776	9827	9879
N	0	1	2	3	4	5	6	7	8	9

SIX-PLACE MANTISSAS FOR COMMON LOGARITHMS (Continued)

N	0	1	2	3	4	5	6	7	8	9
850	92 9419	9470	9521	9572	9623	9674	9725	9776	9827	9879
51	92 9930	9981	*0032	*0083	*0134	*0185	*0236	*0287	*0338	*0389
52	93 0440	0491	0542	0592	0643	0694	0745	0796	0847	0898
53	0949	1000	1051	1102	1153	1204	1254	1305	1356	1407
54	1458	1509	1560	1610	1661	1712	1763	1814	1865	1915
55	1966	2017	2068	2118	2169	2220	2271	2322	2372	2423
56	2474	2524	2575	2626	2677	2727	2778	2829	2879	2930
57	2981	3031	3082	3133	3183	3234	3285	3335	3386	3437
58	3487	3538	3589	3639	3690	3740	3791	3841	3892	3943
59	3993	4044	4094	4145	4195	4246	4296	4347	4397	4448
860	4498	4549	4599	4650	4700	4751	4801	4852	4902	4953
61	5003	5054	5104	5154	5205	5255	5306	5356	5406	5457
62	5507	5558	5608	5658	5709	5759	5809	5860	5910	5960
63	6011	6061	6111	6162	6212	6262	6313	6363	6413	6463
64	6514	6564	6614	6665	6715	6765	6815	6865	6916	6966
65	7016	7066	7166	7167	7217	7267	7317	7367	7418	7468
66	7518	7568	7618	7668	7718	7769	7819	7869	7919	7969
67	8019	8069	8119	8169	8219	8269	8320	8370	8420	8470
68	8520	8570	8620	8670	8720	8770	8820	8870	8920	8970
69	9020	9070	9120	9170	9220	9270	9320	9369	9419	9469
870	93 9519	9569	9619	9669	9719	9769	9819	9869	9918	9968
71	94 0018	0068	0118	0168	0218	0267	0317	0367	0417	0467
72	0516	0566	0616	0666	0716	0765	0815	0865	0915	0964
73	1014	1064	1114	1163	1213	1263	1313	1362	1412	1462
74	1511	1561	1611	1660	1710	1760	1809	1859	1909	1958
75	2008	2058	2107	2157	2207	2256	2306	2355	2405	2455
76	2504	2554	2603	2653	2702	2752	2801	2851	2901	2950
77	3000	3049	3099	3148	3198	3247	3297	3346	3396	3445
78	3495	3544	3593	3643	3692	3742	3791	3841	3890	3939
79	3989	4038	4088	4137	4186	4236	4285	4335	4384	4433
880	4483	4532	4581	4631	4680	4729	4779	4828	4877	4927
81	4976	5025	5074	5124	5173	5222	5272	5321	5370	5419
82	5469	5518	5567	5616	5665	5715	5764	5813	5862	5912
83	5961	6010	6059	6108	6157	6207	6256	6305	6354	6403
84	6452	6501	6551	6600	6649	6698	6747	6796	6845	6894
85	6943	6992	7041	7090	7139	7189	7238	7287	7336	7385
86	7434	7483	7532	7581	7630	7679	7728	7777	7826	7875
87	7924	7973	8022	8070	8119	8168	8217	8266	8315	8364
88	8413	8462	8511	8560	8608	8657	8706	8755	8804	8853
89	8902	8951	8999	9048	9097	9146	9195	9244	9292	9341
890	9390	9439	9488	9536	9585	9634	9683	9731	9780	9829
91	94 9878	9926	9975	*0024	*0073	*0121	*0170	*0219	*0267	*0316
92	95 0365	0414	0462	0511	0560	0608	0657	0706	0754	0803
93	0851	0900	0949	0997	1046	1095	1143	1192	1240	1289
94	1338	1386	1435	1483	1532	1580	1629	1677	1726	1775
95	1823	1872	1920	1969	2017	2066	2114	2163	2211	2260
96	2308	2356	2405	2453	2502	2550	2599	2647	2696	2744
97	2792	2841	2889	2938	2986	3034	3083	3131	3180	3228
98	3276	3325	3373	3421	3470	3518	3566	3615	3663	3711
99	3760	3808	3856	3905	3953	4001	4049	4098	4146	4194
900	95 4243	4291	4339	4387	4435	4484	4532	4580	4628	4677
N	0	1	2	3	4	5	6	7	8	9

SIX-PLACE MANTISSAS FOR COMMON LOGARITHMS (Continued)

N	0	1	2	3	4	5	6	7	8	9
900	95 4243	4291	4339	4387	4435	4484	4532	4580	4628	4677
01	4725	4773	4821	4869	4918	4966	5014	5062	5110	5158
02	5207	5255	5303	5351	5399	5447	5495	5543	5592	5640
03	5688	5736	5784	5832	5880	5928	5976	6024	6072	6120
04	6168	6216	6265	6313	6361	6409	6457	6505	6553	6601
05	6649	6697	6745	6793	6840	6888	6936	6984	7032	7080
06	7128	7176	7224	7272	7320	7368	7416	7464	7512	7559
07	7607	7655	7703	7751	7799	7847	7894	7942	7990	8038
08	8086	8134	8181	8229	8277	8325	8373	8421	8468	8516
09	8564	8612	8659	8707	8755	8803	8850	8898	8946	8994
910	9041	9089	9137	9185	9232	9280	9328	9375	9423	9471
11	9518	9566	9614	9661	9709	9757	9804	9852	9900	9947
12	95 9995	*0042	*0090	*0138	*0185	*0233	*0280	*0328	*0376	*0423
13	96 0471	0518	0566	0613	0661	0709	0756	0804	0851	0899
14	0946	0994	1041	1089	1136	1184	1231	1279	1326	1374
15	1421	1469	1516	1563	1611	1658	1706	1753	1801	1848
16	1895	1943	1990	2038	2085	2132	2180	2227	2275	2322
17	2369	2417	2464	2511	2559	2606	2653	2701	2748	2795
18	2843	2890	2937	2985	3032	3079	3126	3174	3221	3268
19	3316	3363	3410	3457	3504	3552	3599	3646	3693	3741
920	3788	3835	3882	3929	3977	4024	4071	4118	4165	4212
21	4260	4307	4354	4401	4448	4495	4542	4590	4637	4684
22	4731	4778	4825	4872	4919	4966	5013	5061	5108	5155
23	5202	5249	5296	5343	5390	5437	5484	5531	5578	5625
24	5672	5719	5766	5813	5860	5907	5954	6001	6048	6095
25	6142	6189	6236	6283	6329	6376	6423	6470	6517	6564
26	6611	6658	6705	6752	6799	6845	6892	6939	6986	7033
27	7080	7127	7173	7220	7267	7314	7361	7408	7454	7501
28	7548	7595	7642	7688	7735	7782	7829	7875	7922	7969
29	8016	8062	8109	8156	8203	8249	8296	8343	8390	8436
930	8483	8530	8576	8623	8670	8716	8763	8810	8856	8903
31	8950	8996	9043	9090	9136	9183	9229	9276	9323	9369
32	9416	9463	9509	9556	9602	9649	9695	9742	9789	9835
33	96 9882	9928	9975	*0021	*0068	*0114	*0161	*0207	*0254	*0300
34	97 0347	0393	0440	0486	0533	0579	0626	0672	0719	0765
35	0812	0858	0904	0951	0997	1044	1090	1137	1183	1229
36	1276	1322	1369	1415	1461	1508	1554	1601	1647	1693
37	1740	1786	1832	1879	1925	1971	2018	2064	2110	2157
38	2203	2249	2295	2342	2388	2434	2481	2527	2573	2619
39	2666	2712	2758	2804	2851	2897	2943	2989	3035	3082
940	3128	3174	3220	3266	3313	3359	3405	3451	3497	3543
41	3590	3636	3682	3728	3774	3820	3866	3913	3959	4005
42	4051	4097	4143	4189	4235	4281	4327	4374	4420	4466
43	4512	4558	4604	4650	4696	4742	4788	4834	4880	4926
44	4972	5018	5064	5110	5156	5202	5248	5294	5340	5386
45	5432	5478	5524	5570	5616	5662	5707	5753	5799	5845
46	5891	5937	5983	6029	6075	6121	6167	6212	6258	6304
47	6350	6396	6442	6488	6533	6579	6625	6671	6717	6763
48	6808	6854	6900	6946	6992	7037	7083	7129	7175	7220
49	7266	7312	7358	7403	7449	7495	7541	7586	7632	7678
950	97 7724	7769	7815	7861	7906	7952	7998	8043	8089	8135
N	0	1	2	3	4	5	6	7	8	9

SIX-PLACE MANTISSAS FOR COMMON LOGARITHMS (Continued)

N	0	1	2	3	4	5	6	7	8	9
950	97 7724	7769	7815	7861	7906	7952	7998	8043	8089	8135
51	8181	8226	8272	8317	8363	8409	8454	8500	8546	8591
52	8637	8683	8728	8774	8819	8865	8911	8956	9002	9047
53	9093	9138	9184	9230	9275	9321	9366	9412	9457	9503
54	97 9548	9594	9639	9685	9730	9776	9821	9867	9912	9958
55	98 0003	0049	0094	0140	0185	0231	0276	0322	0367	0412
56	0458	0503	0549	0594	0640	0685	0730 ·	0776	0821	0867
57	0912	0957	1003	1048	1093	1139	1184	1229	1275	1320
58	1366	1411	1456	1501	1547	1592	1637	1683	1728	1773
59	1819	1864	1909	1954	2000	2045	2090	2135	2181	2226
960	2271	2316	2362	2407	2452	2497	2543	2588	2633	2678
61	2723	2769	2814	2859	2904	2949	2994	3040	3085	3130
62	3175	3220	3265	3310	3356	3401	3446	3491	3536	3581
63	3626	3671	3716	3762	3807	3852	3897	3942	3987	4032
64	4077	4122	4167	4212	4257	4302	4347	4392	4437	4482
65	4527	4572	4617	4662	4707	4752	4797	4842	4887	4932
66	4977	5022	5067	5112	5157	5202	5247	5292	5337	5382
67	5426	5471	5516	5561	5606	5651	5696	5741	5786	5830
68	5875	5920	5965	6010	6055	6100	6144	6189	6234	6279
69	6324	6369	6413	6458	6503	6548	6593	6637	6682	6727
970	6772	6817	6861	6906	6951	6996	7040	7085	7130	7175
71	7219	7264	7309	7353	7398	7443	7488	7532	7577	7622
72	7666	7711	7756	7800	7845	7890	7934	7979	8024	8068
73	8113	8157	8202	8247	8291	8336	8381	8425	8470	8514
74	8559	8604	8648	8693	8737	8782	8826	8871	8916	8960
75	9005	9049	9094	9138	9183	9227	9272	9316	9361	9405
76	9450	9494	9539	9583	9628	9672	9717	9761	9806	9850
77	98 9895	9939	9983	*0028	*0072	*0117	*0161	*0206	*0250	*0294
78	99 0339	0383	0428	0472	0516	0561	0605	0650	0694	0738
79	0783	0827	0871	0916	0960	1004	1049	1093	1137	1182
980	1226	1270	1315	1359	1403	1448	1492	1536	1580	1625
81	1669	1713	1758	1802	1846	1890	1935	1979	2023	2067
82	2111	2156	2200	2244	2288	2333	2377	2421	2465	2509
83	2554	2598	2642	2686	2730	2774	2819	2863	2907	2951
84	2995	3039	3083	3127	3172	3216	3260	3304	3348	3392
85	3436	3480	3524	3568	3613	3657	3701	3745	3789	3833
86	3877	3921	3965	4009	4053	4097	4141	4185	4229	4273
87	4317	4361	4405	4449	4493	4537	4581	4625	4669	4713
88	4757	4801	4845	4889	4933	4977	5021	5065	5108	5152
89	5196	5240	5284	5328	5372	5416	5460	5504	5547	5591
990	5635	5679	5723	5767	5811	5854	5898	5942	5986	6030
91	6074	6117	6161	6205	6249	6293	6337	6380	6424	6468
92	6512	6555	6599	6643	6687	6731	6774	6818	6862	6906
93	6949	6993	7037	7080	7124	7168	7212	7255	7299	7343
94	7386	7430	7474	7517	7561	7605	7648	7692	7736	7779
95	7823	7867	7910	7954	7998	8041	8085	8129	8172	8216
96	8259	8303	8347	8390	8434	8477	8521	8564	8608	8652
97	8695	8739	8782	8826	8869	8913	8956	9000	9043	9087
98	9131	9174	9218	9261	9305	9348	9392	9435	9479	9522
99	99 9565	9609	9652	9696	9739	9783	9826	9870	9913	9957
1000	00 0000	0043	0087	0130	0174	0217	0260	0304	0347	0391
N	0	1	2	3	4	5	6	7	8	9

NATURAL OR NAPERIAN LOGARITHMS
0.000–0.499

N	0	1	2	3	4	5	6	7	8	9
0.00	− ∞	−6‡ .90776	−6 .21461	−5 .80914	−5 .52146	−5 .29832	−5 .11600	−4 .96185	−4 .82831	−4 .71053
.01	−4.60517	.50986	.42285	.34281	.26870	.19971	.13517	.07454	.01738	*.96332
.02	−3.91202	.86323	.81671	.77226	.72970	.68888	.64966	.61192	.57555	.54046
.03	.50656	.47377	.44202	.41125	.38139	.35241	.32424	.29684	.27017	.24419
.04	.21888	.19418	.17009	.14656	.12357	.10109	.07911	.05761	.03655	.01593
.05	−2.99573	.97593	.95651	.93746	.91877	.90042	.88240	.86470	.84731	.83022
.06	.81341	.79688	.78062	.76462	.74887	.73337	.71810	.70306	.68825	.67365
.07	.65926	.64508	.63109	.61730	.60369	.59027	.57702	.56395	.55105	.53831
.08	.52573	.51331	.50104	.48891	.47694	.46510	45341	.44185	.43042	.41912
.09	.40795	39690	.38597	.37516	.36446	.35388	.34341	.33304	.32279	.31264
0.10	−2.30259	.29263	.28278	.27303	.26336	.25379	.24432	.23493	.22562	.21641
.11	.20727	.19823	.18926	.18037	.17156	.16282	.15417	.14558	.13707	.12863
.12	.12026	.11196	.10373	.09557	.08747	.07944	.07147	.06357	.05573	.04794
.13	.04022	.03256	.02495	.01741	.00992	.00248	*.99510	*.98777	*.98050	*.97328
.14	−1.96611	.95900	.95193	.94491	.93794	.93102	.92415	.91732	.91054	.90381
.15	.89712	.89048	.88387	.87732	.87080	.86433	.85790	.85151	.84516	.83885
.16	.83258	.82635	.82016	.81401	.80789	.80181	.79577	.78976	.78379	.77786
.17	.77196	.76609	.76026	.75446	.74870	.74297	.73727	.73161	.72597	.72037
.18	.71480	.70926	.70375	.69827	.69282	.68740	.68201	.67665	.67131	.66601
.19	.66073	.65548	.65026	.64507	.63990	.63476	.62964	.62455	.61949	.61445
0.20	−1.60944	.60445	.59949	.59455	.58964	.58475	.57988	.57504	.57022	.56542
.21	.56065	.55590	.55117	.54646	.54178	.53712	.53248	.52786	.52326	.51868
.22	.51413	.50959	.50508	.50058	.49611	.49165	.48722	.48281	.47841	.47403
.23	.46968	.46534	.46102	.45672	.45243	.44817	.44392	.43970	.43548	.43129
.24	.42712	.42296	.41882	.41469	.41059	.40650	.40242	.39837	.39433	.39030
25	.38629	.38230	.37833	.37437	.37042	.36649	.36258	.35868	.35480	.35093
.26	.34707	.34323	.33941	.33560	.33181	.32803	.32426	.32051	.31677	.31304
.27	.30933	.30564	.30195	.29828	.29463	.29098	.28735	.28374	.28013	.27654
.28	.27297	.26940	.26585	.26231	.25878	.25527	.25176	.24827	.24479	.24133
.29	.23787	.23443	.23100	.22758	.22418	.22078	.21740	.21402	.21066	.20731
0.30	−1.20397	.20065	.19733	.19402	.19073	.18744	.18417	.18091	.17766	.17441
.31	.17118	.16796	.16475	.16155	.15836	.15518	.15201	.14885	.14570	.14256
.32	.13943	.13631	.13320	.13010	.12701	.12393	.12086	.11780	.11474	.11170
.33	.10866	.10564	.10262	.09961	.09661	.09362	.09064	.08767	.08471	.08176
.34	.07881	.07587	.07294	.07002	.06711	.06421	.06132	.05843	.05555	.05268
.35	−1.04982	.04697	.04412	.04129	.03846	.03564	.03282	.03002	.02722	.02443
.36	.02165	.01888	.01611	.01335	.01060	.00786	.00512	.00239	*.99967	*.99696
.37	−0.99425	.99155	.98886	.98618	.98350	.98083	.97817	.97551	.97286	.97022
.38	.96758	.96496	.96233	.95972	.95711	.95451	.95192	.94933	.94675	.94418
.39	.94161	.93905	.93649	.93395	.93140	.92887	.92634	.92382	.92130	.91879
0.40	−0.91629	.91379	.91130	.90882	.90634	.90387	.90140	.89894	.89649	.89404
.41	.89160	.88916	.88673	.88431	.88189	.87948	.87707	.87467	.87227	.86988
.42	.86750	.86512	.86275	.86038	.85802	.85567	.85332	.85097	.84863	.84630
.43	.84397	.84165	.83933	.83702	.83471	.83241	.83011	.82782	.82554	.82326
.44	.82098	.81871	.81645	.81419	.81193	.80968	.80744	.80520	.80296	.80073
.45	.79851	.79629	.79407	.79186	.78966	.78746	.78526	.78307	.78089	.77871
.46	.77653	.77436	.77219	.77003	.76787	.76572	.76357	.76143	.75929	.75715
.47	.75502	.75290	.75078	.74866	.74655	.74444	.74234	.74024	.73814	.73605
.48	.73397	.73189	.72981	.72774	.72567	.72361	.72155	.71949	.71744	.71539
.49	.71335	.71131	.70928	.70725	.70522	.70320	.70118	.69917	.69716	.69515

‡ Note that the whole number values are given above the decimal values for the first line. In the second and following lines they are given at the left. All decimal values are negative on this page.

NATURAL OR NAPERIAN LOGARITHMS (Continued)
0.500–0.999

N	0	1	2	3	4	5	6	7	8	9
0.50	−0.69315	.69115	.68916	.68717	.68518	.68320	.68122	.67924	.67727	.67531
.51	.67334	.67139	.66943	.66748	.66553	.66359	.66165	.65971	.65778	.65585
.52	.65393	.65201	.65009	.64817	.64626	.64436	.64245	.64055	.63866	.63677
.53	.63488	.63299	.63111	.62923	.62736	.62549	.62362	.62176	.61990	.61804
.54	.61619	.61434	.61249	.61065	.60881	.60697	.60514	.60331	.60148	.59966
.55	.59784	.59602	.59421	.59240	.59059	.58879	.58699	.58519	.58340	.58161
.56	.57982	.57803	.57625	.57448	.57270	.57093	.56916	.56740	.56563	.56387
.57	.56212	.56037	.55862	.55687	.55513	.55339	.55165	.54991	.54818	.54645
.58	.54473	.54300	.54128	.53957	.53785	.53614	.53444	.53273	.53103	.52933
.59	.52763	.52594	.52425	.52256	.52088	.51919	.51751	.51584	.51416	.51249
0.60	−0.51083	.50916	.50750	.50584	.50418	.50253	.50088	.49923	.49758	.49594
.61	.49430	.49266	.49102	.48939	.48776	.48613	.48451	.48289	.48127	.47965
.62	.47804	.47642	.47482	.47321	.47160	.47000	.46840	.46681	.46522	.46362
.63	.46204	.46045	.45887	.45728	.45571	.45413	.45256	.45099	.44942	.44785
.64	.44629	.44473	.44317	.44161	.44006	.43850	.43696	.43541	.43386	.43232
.65	.43078	.42925	.42771	.42618	.42465	.42312	.42159	.42007	.41855	.41703
.66	.41552	.41400	.41249	.41098	.40947	.40797	.40647	.40497	.40347	.40197
.67	.40048	.39899	.39750	.39601	.39453	.39304	.39156	.39008	.38861	.38713
.68	.38566	.38419	.38273	.38126	.37980	.37834	.37688	.37542	.37397	.37251
.69	.37106	.36962	.36817	.36673	.36528	.36384	.36241	.36097	.35954	.35810
0.70	−0.35667	.35525	.35382	.35240	.35098	.34956	.34814	.34672	.34531	.34390
.71	.34249	.34108	.33968	.33827	.33687	.33547	.33408	.33268	.33129	.32989
.72	.32850	.32712	.32573	.32435	.32296	.32158	.32021	.31883	.31745	.31608
.73	.31471	.31334	.31197	.31061	.30925	.30788	.30653	.30517	.30381	.30246
.74	.30111	.29975	.29841	.29706	.29571	.29437	.29303	.29169	.29035	.28902
.75	.28768	.28635	.28502	.28369	.28236	.28104	.27971	.27839	.27707	.27575
.76	.27444	.27312	.27181	.27050	.26919	.26788	.26657	.26527	.26397	.26266
.77	.26136	.26007	.25877	.25748	.25618	.25489	.25360	.25231	.25103	.24974
.78	.24846	.24718	.24590	.24462	.24335	.24207	.24080	.23953	.23826	.23699
.79	.23572	.23446	.23319	.23193	.23067	.22941	.22816	.22690	.22565	.22439
0.80	−0.22314	.22189	.22065	.21940	.21816	.21691	.21567	.21433	.21319	.21196
.81	.21072	.20949	.20825	.20702	.20579	.20457	.20334	.20212	.20089	.19967
.82	.19845	.19723	.19601	.19480	.19358	.19237	.19116	.18995	.18874	.18754
.83	.18633	.18513	.18392	.18272	.18152	.18032	.17913	.17793	.17674	.17554
.84	.17435	.17316	.17198	.17079	.16960	.16842	.16724	.16605	.16487	.16370
.85	−0.16252	.16134	.16017	.15900	.15782	.15665	.15548	.15432	.15315	.15199
.86	.15082	.14966	.14850	.14734	.14618	.14503	.14387	.14272	.14156	.14041
.87	.13926	.13811	.13697	.13582	.13467	.13353	.13239	.13125	.13011	.12897
.88	.12783	.12670	.12556	.12443	.12330	.12217	.12104	.11991	.11878	.11766
.89	.11653	.11541	.11429	.11317	.11205	.11093	.10981	.10870	.10759	.10647
0.90	−0.10536	.10425	.10314	.10203	.10093	.09982	.09872	.09761	.09651	.09541
.91	.09431	.09321	.09212	.09102	.08992	.08883	.08774	.08665	.08556	.08447
.92	.08338	.08230	.08121	.08013	.07904	.07796	.07688	.07580	.07472	.07365
.93	.07257	.07150	.07042	.06935	.06828	.06721	.06614	.06507	.06401	.06294
.94	.06188	.06081	.05975	.05869	.05763	.05657	.05551	.05446	.05340	.05235
.95	.05129	.05024	.04919	.04814	.04709	.04604	.04500	.04395	.04291	.04186
.96	.04082	.03978	.03874	.03770	.03666	.03563	.03459	.03356	.03252	.03149
.97	.03046	.02943	.02840	.02737	.02634	.02532	.02429	.02327	.02225	.02122
.98	.02020	.01918	.01816	.01715	.01613	.01511	.01410	.01309	.01207	.01106
.99	.01005	.00904	.00803	.00702	.00602	.00501	.00401	.00300	.00200	.00100

NATURAL OR NAPERIAN LOGARITHMS (Continued)

To find the natural logarithm of a number which is $\frac{1}{10}$, $\frac{1}{100}$, $\frac{1}{1000}$, etc. of a number whose logarithm is given, subtract from the given logarithm $\log_e 10$, $2 \log_e 10$, $3 \log_e 10$, etc.

To find the natural logarithm of a number which is 10, 100, 1000, etc. times a number whose logarithm is given, add to the given logarithm $\log_e 10$, $2 \log_e 10$, $3 \log_e 10$, etc.

$\log_e 10 = 2.30258\ 50930$	$6 \log_e 10 = 13.81551\ 05580$
$2 \log_e 10 = 4.60517\ 01860$	$7 \log_e 10 = 16.11809\ 56510$
$3 \log_e 10 = 6.90775\ 52790$	$8 \log_e 10 = 18.42068\ 07440$
$4 \log_e 10 = 9.21034\ 03720$	$9 \log_e 10 = 20.72326\ 58369$
$5 \log_e 10 = 11.51292\ 54650$	$10 \log_e 10 = 23.02585\ 09299$

See preceding table for logarithms for numbers between 0.000 and 0.999.

1.00–4.99

N	0	1	2	3	4	5	6	7	8	9
1.0	0.00000	.00995	.01980	.02956	.03922	.04879	.05827	.06766	.07696	.08618
.1	.09531	.10436	.11333	.12222	.13103	.13976	.14842	.15700	.16551	.17395
.2	.18232	.19062	.19885	.20701	.21511	.22314	.23111	.23902	.24686	.25464
.3	.26236	.27003	.27763	.28518	.29267	.30010	.30748	.31481	.32208	.32930
.4	.33647	.34359	.35066	.35767	.36464	.37156	.37844	.38526	.39204	.39878
.5	.40547	.41211	.41871	.42527	.43178	.43825	.44469	.45108	.45742	.46373
.6	.47000	.47623	.48243	.48858	.49470	.50078	.50682	.51282	.51879	.52473
.7	.53063	.53649	.54232	.54812	.55389	.55962	.56531	.57098	.57661	.58222
.8	.58779	.59333	.59884	.60432	.60977	.61519	.62058	.62594	.63127	.63658
.9	.64185	.64710	.65233	.65752	.66269	.66783	.67294	.67803	.68310	.68813
2.0	0.69315	.69813	.70310	.70804	.71295	.71784	.72271	.72755	.73237	.73716
.1	.74194	.74669	.75142	.75612	.76081	.76547	.77011	.77473	.77932	.78390
.2	.78846	.79299	.79751	.80200	.80648	.81093	.81536	.81978	.82418	.82855
.3	.83291	.83725	.84157	.84587	.85015	.85442	.85866	.86289	.86710	.87129
.4	.87547	.87963	.88377	.88789	.89200	.89609	.90016	.90422	.90826	.91228
.5	.91629	.92028	.92426	.92822	.93216	.93609	.94001	.94391	.94779	.95166
.6	.95551	.95935	.96317	.96698	.97078	.97456	.97833	.98208	.98582	.98954
.7	.99325	.99695	*.00063	*.00430	*.00796	*.01160	*.01523	*.01885	*.02245	*.02604
.8	1.02962	.03318	.03674	.04028	.04380	.04732	.05082	.05431	.05779	.06126
.9	.06471	.06815	.07158	.07500	.07841	.08181	.08519	.08856	.09192	.09527
3.0	1.09861	.10194	.10526	.10856	.11186	.11514	.11841	.12168	.12493	.12817
.1	.13140	.13462	.13783	.14103	.14422	.14740	.15057	.15373	.15688	.16002
.2	.16315	.16627	.16938	.17248	.17557	.17865	.18173	.18479	.18784	.19089
.3	.19392	.19695	.19996	.20297	.20597	.20896	.21194	.21491	.21788	.22083
.4	.22378	.22671	.22964	.23256	.23547	.23837	.24127	.24415	.24703	.24990
.5	.25276	.25562	.25846	.26130	.26413	.26695	.26976	.27257	.27536	.27815
.6	.28093	.28371	.28647	.28923	.29198	.29473	.29746	.30019	.30291	.30563
.7	.30833	.31103	.31372	.31641	.31909	.32176	.32442	.32708	.32972	.33237
.8	.33500	.33763	.34025	.34286	.34547	.34807	.35067	.35325	.35584	.35841
.9	.36098	.36354	.36609	.36864	.37118	.37372	.37624	.37877	.38128	.38379
4.0	1.38629	.38879	.39128	.39377	.39624	.39872	.40118	.40364	.40610	.40854
.1	.41099	.41342	.41585	.41828	.42070	.42311	.42552	.42792	.43031	.43270
.2	.43508	.43746	.43984	.44220	.44456	.44692	.44927	.45161	.45395	.45629
.3	.45862	.46094	.46326	.46557	.46787	.47018	.47247	.47476	.47705	.47933
.4	.48160	.48387	.48614	.48840	.49065	.49290	.49515	.49739	.49962	.50185
.5	.50408	.50630	.50851	.51072	.51293	.51513	.51732	.51951	.52170	.52388
.6	.52606	.52823	.53039	.53256	.53471	.53687	.53902	.54116	.54330	.54543
.7	.54756	.54969	.55181	.55393	.55604	.55814	.56025	.56235	.56444	.56653
.8	.56862	.57070	.57277	.57485	.57691	.57898	.58104	.58309	.58515	.58719
.9	.58924	.59127	.59331	.59534	.59737	.59939	.60141	.60342	.60543	.60744

NATURAL OR NAPERIAN LOGARITHMS (Continued)
5.00–9.99

N	0	1	2	3	4	5	6	7	8	9
5.0	1.60944	.61144	.61343	.61542	.61741	.61939	.62137	.62334	.62531	.62728
.1	.62924	.63120	.63315	.63511	.63705	.63900	.64094	.64287	.64481	.64673
.2	.64866	.65058	.65250	.65441	.65632	.65823	.66013	.66203	.66393	.66582
.3	.66771	.66959	.67147	.67335	.67523	.67710	.67896	.68083	.68269	.68455
.4	.68640	.68825	.69010	.69194	.69378	.69562	.69745	.69928	.70111	.70293
.5	.70475	.70656	.70838	.71019	.71199	.71380	.71560	.71740	.71919	.72098
.6	.72277	.72455	.72633	.72811	.72988	.73166	.73342	.73519	.73695	.73871
.7	.74047	.74222	.74397	.74572	.74746	.74920	.75094	.75267	.75440	.75613
.8	.75786	.75958	.76130	.76302	.76473	.76644	.76815	.76985	.77156	.77326
.9	.77495	.77665	.77834	.78002	.78171	.78339	.78507	.78675	.78842	.79009
6.0	1.79176	.79342	.79509	.79675	.79840	.80006	.80171	.80336	.80500	.80665
.1	.80829	.80993	.81156	.81319	.81482	.81645	.81808	.81970	.82132	.82294
.2	.82455	.82616	.82777	.82938	.83098	.83258	.83418	.83578	.83737	.83896
.3	.84055	.84214	.84372	.84530	.84688	.84845	.85003	.85160	.85317	.85473
.4	.85630	.85786	.85942	.86097	.86253	.86408	.86563	.86718	.86872	.87026
.5	.87180	.87334	.87487	.87641	.87794	.87947	.88099	.88251	.88403	.88555
.6	.88707	.88858	.89010	.89160	.89311	.89462	.89612	.89762	.89912	.90061
.7	.90211	.90360	.90509	.90658	.90806	.90954	.91102	.91250	.91398	.91545
.8	.91692	.91839	.91986	.92132	.92279	.92425	.92571	.92716	.92862	.93007
.9	.93152	.93297	.93442	.93586	.93730	.93874	.94018	.94162	.94305	.94448
7.0	1.94591	.94734	.94876	.95019	.95161	.95303	.95445	.95586	.95727	.95869
.1	.96009	.96150	.96291	.96431	.96571	.96711	.96851	.96991	.97130	.97269
.2	.97408	.97547	.97685	.97824	.97962	.98100	.98238	.98376	.98513	.98650
.3	.98787	.98924	.99061	.99198	.99334	.99470	.99606	.99742	.99877	*.00013
.4	2.00148	.00283	.00418	.00553	.00687	.00821	.00956	.01089	.01223	.01357
.5	.01490	.01624	.01757	.01890	.02022	.02155	.02287	.02419	.02551	.02683
.6	.02815	.02946	.03078	.03209	.03340	.03471	.03601	.03732	.03862	.03992
.7	.04122	.04252	.04381	.04511	.04640	.04769	.04898	.05027	.05156	.05284
.8	.05412	.05540	.05668	.05796	.05924	.06051	.06179	.06306	.06433	.06560
.9	.06686	.06813	.06939	.07065	.07191	.07317	.07443	.07568	.07694	.07819
8.0	2.07944	.08069	.08194	.08318	.08443	.08567	.08691	.08815	.08939	.09063
.1	.09186	.09310	.09433	.09556	.09679	.09802	.09924	.10047	.10169	.10291
.2	.10413	.10535	.10657	.10779	.10900	.11021	.11142	.11263	.11384	.11505
.3	.11626	.11746	.11866	.11986	.12106	.12226	.12346	.12465	.12585	.12704
.4	.12823	.12942	.13061	.13180	.13298	.13417	.13535	.13653	.13771	.13889
.5	.14007	.14124	.14242	.14359	.14476	.14593	.14710	.14827	.14943	.15060
.6	.15176	.15292	.15409	.15524	.15640	.15756	.15871	.15987	.16102	.16217
.7	.16332	.16447	.16562	.16677	.16791	.16905	.17020	.17134	.17248	.17361
.8	.17475	.17589	.17702	.17816	.17929	.18042	.18155	.18267	.18380	.18493
.9	.18605	.18717	.18830	.18942	.19054	.19165	.19277	.19389	.19500	.19611
9.0	2.19722	.19834	.19944	.20055	.20166	.20276	.20387	.20497	.20607	.20717
.1	.20827	.20937	.21047	.21157	.21266	.21375	.21485	.21594	.21703	.21812
.2	.21920	.22029	.22138	.22246	.22354	.22462	.22570	.22678	.22786	.22894
.3	.23001	.23109	.23216	.23324	.23431	.23538	.23645	.23751	.23858	.23965
.4	.24071	.24177	.24284	.24390	.24496	.24601	.24707	.24813	.24918	.25024
.5	.25129	.25234	.25339	.25444	.25549	.25654	.25759	.25863	.25968	.26072
.6	.26176	.26280	.26384	.26488	.26592	.26696	.26799	.26903	.27006	.27109
.7	.27213	.27316	.27419	.27521	.27624	.27727	.27829	.27932	.28034	.28136
.8	.28238	.28340	.28442	.28544	.28646	.28747	.28849	.28950	.29051	.29152
.9	.29253	.29354	.29455	.29556	.29657	.29757	.29858	.29958	.30058	.30158

NATURAL OR NAPERIAN LOGARITHMS (Continued)

Constants

$\log_e 10 = 2.30258\ 50930$	$6 \log_e 10 = 13.81551\ 05580$
$2 \log_e 10 = 4.60517\ 01860$	$7 \log_e 10 = 16.11809\ 56510$
$3 \log_e 10 = 6.90775\ 52790$	$8 \log_e 10 = 18.42068\ 07440$
$4 \log_e 10 = 9.21034\ 03720$	$9 \log_e 10 = 20.72326\ 58369$
$5 \log_e 10 = 11.51292\ 54650$	$10 \log_e 10 = 23.02585\ 09299$

10.0–49.9

N	0	1	2	3	4	5	6	7	8	9
10.	2.30259	.31254	.32239	.33214	.34181	.35138	36085	.37024	.37955	.38876
11.	.30700	.40095	.41591	.42480	.43361	.44235	.45101	.45959	.46810	.47654
12.	.48491	.49321	.50144	.50960	.51770	.52573	.53370	.54160	.54945	.55723
13.	.56495	.57261	.58022	.58776	.59525	.60269	.61007	.61740	.62467	.63189
14.	.63906	.64617	.65324	.66026	.66723	.67415	.68102	.68785	.69463	.70136
15.	.70805	.71469	.72130	.72785	.73437	.74084	.74727	.75366	.76001	.76632
16.	.77259	.77882	.78501	.79117	.79728	.80336	.80940	.81541	.82138	.82731
17.	.83321	.83908	.84491	.85071	.85647	.86220	.86790	.87356	.87920	.88480
18.	.89037	.89591	.90142	.90690	.91235	.91777	.92316	.92852	.93386	.93916
19.	.94444	.94969	.95491	.96011	.96527	.97041	.97553	.98062	.98568	.99072
20.	2.99573	*.00072	*.00568	*.01062	*.01553	*.02042	*.02529	*.03013	*.03495	*.03975
21.	3.04452	.04927	.05400	.05871	.06339	.06805	.07269	.07731	.08191	.08649
22.	.09104	.09558	.10009	.10459	.10906	.11352	.11795	.12236	.12676	.13114
23.	.13549	.13983	.14415	.14845	.15274	.15700	.16125	.16548	.16969	.17388
24.	.17805	.18221	.18635	.19048	.19458	.19867	.20275	.20680	.21084	.21487
25.	.21888	.22287	.22684	.23080	.23475	.23868	.24259	.24649	.25037	.25424
26.	.25810	.26194	.26576	.26957	.27336	.27714	.28091	.28466	.28840	.29213
27.	.29584	.29953	.30322	.30689	.31054	.31419	.31782	.32143	.32504	.32863
28.	.33220	.33577	.33932	.34286	.34639	.34990	.35341	.35690	.36038	.36384
29.	.36730	.37074	.37417	.37759	.38099	.38439	.38777	.39115	.39451	.39786
30.	3.40120	.40453	.40784	.41115	.41444	.41773	.42100	.42426	.42751	.43076
31.	.43399	.43721	.44042	.44362	.44681	.44999	.45316	.45632	.45947	.46261
32.	.46574	.46886	.47197	.47507	.47816	.48124	.48431	.48738	.49043	.49347
33.	.49651	.49953	.50255	.50556	.50856	.51155	.51453	.51750	.52046	.52342
34.	.52636	.52930	.53223	.53515	.53806	.54096	.54385	.54674	.54962	.55249
35.	.55535	.55820	.56105	.56388	.56671	.56953	.57235	.57515	.57795	.58074
36.	.58352	.58629	.58906	.59182	.59457	.59731	.60005	.60278	.60550	.60821
37.	.61092	.61362	.61631	.61899	.62167	.62434	.62700	.62966	.63231	.63495
38.	.63759	.64021	.64284	.64545	.64806	.65066	.65325	.65584	.65842	.66099
39.	.66356	.66612	.66868	.67122	.67377	.67630	.67883	.68135	.68387	.68638
40.	3.68888	.69138	.69387	.69635	.69883	.70130	.70377	.70623	.70868	.71113
41.	.71357	.71601	.71844	.72086	.72328	.72569	.72810	.73050	.73290	.73529
42.	.73767	.74005	.74242	.74479	.74715	.74950	.75185	.75420	.75654	.75887
43.	.76120	.76352	.76584	.76815	.77046	.77276	.77506	.77735	.77963	.78191
44.	.78419	.78646	.78872	.79098	.79324	.79549	.79773	.79997	.80221	.80444
45.	.80666	.80888	.81110	.81331	.81551	.81771	.81991	.82210	.82428	.82647
46.	.82864	.83081	.83298	.83514	.83730	.83945	.84160	.84374	.84588	.84802
47.	.85015	.85227	.85439	.85651	.85862	.86073	.86283	.86493	.86703	.86912
48.	.87120	.87328	.87536	.87743	.87950	.88156	.88362	.88568	.88773	.88978
49.	.89182	.89386	.89589	.89792	.89995	.90197	.90399	.90600	.90801	.91002

NATURAL OR NAPERIAN LOGARITHMS (Continued)
50.0–99.9

N	0	1	2	3	4	5	6	7	8	9
50.	3.91202	.91402	.91602	.91801	.91999	.92197	.92395	.92593	.92790	.92986
51.	.93183	.93378	.93574	.93769	.93964	.94158	.94352	.94546	.94739	.94932
52.	.95124	.95316	.95508	.95700	.95891	.96081	.96272	.96462	.96651	.96840
53.	.97029	.97218	.97406	.97594	.97781	.97968	.98155	.98341	.98527	.98713
54.	.98898	.99083	.99268	.99452	.99636	.99820	*.00003	*.00186	*.00369	*.00551
55.	4.00733	.00915	.01096	.01277	.01458	.01638	.01818	.01998	.02177	.02356
56.	.02535	.02714	.02892	.03069	.03247	.03424	.03601	.03777	.03954	.04130
57.	.04305	.04480	.04655	.04830	.05004	.05178	.05352	.05526	.05699	.05872
58.	.06044	.06217	.06389	.06560	.06732	.06903	.07073	.07244	.07414	.07584
59.	.07754	.07923	.08092	.08261	.08429	.08598	.08766	.08933	.09101	.09268
60.	4.09434	.09601	.09767	.09933	.10099	.10264	.10429	.10594	.10759	.10923
61.	.11087	.11251	.11415	.11578	.11741	.11904	.12066	.12228	.12390	.12552
62.	.12713	.12875	.13036	.13196	.13357	.13517	.13677	.13836	.13996	.14155
63.	.14313	.14472	.14630	.14789	.14946	.15104	.15261	.15418	.15575	.15732
64.	.15888	.16044	.16200	.16356	.16511	.16667	.16821	.16976	.17131	.17285
65.	.17439	.17592	.17746	.17899	.18052	.18205	.18358	.18510	.18662	.18814
66.	.18965	.19117	.19268	.19419	.19570	.19720	.19870	.20020	.20170	.20320
67.	.20469	.20618	.20767	.20916	.21065	.21213	.21361	.21509	.21656	.21804
68.	.21951	.22098	.22244	.22391	.22537	.22683	.22829	.22975	.23120	.23266
69.	.23411	.23555	.23700	.23844	.23989	.24133	.24276	.24420	.24563	.24707
70.	4.24850	.24992	.25135	.25277	.25419	.25561	.25703	.25845	.25986	.26127
71.	.26268	.26409	.26549	.26690	.26830	.26970	.27110	.27249	.27388	.27528
72.	.27667	.27805	.27944	.28082	.28221	.28359	.28496	.28634	.28772	.28909
73.	.29046	.29183	.29320	.29456	.29592	.29729	.29865	.30000	.30136	.30271
74.	.30407	.30542	.30676	.30811	.30946	.31080	.31214	.31348	.31482	.31615
75.	.31749	.31882	.32015	.32149	.32281	.32413	.32546	.32678	.32810	.32942
76.	.33073	.33205	.33336	.33467	.33598	.33729	.33860	.33990	.34120	.34251
77.	.34381	.34510	.34640	.34769	.34899	.35028	.35157	.35286	.35414	.35543
78.	.35671	.35800	.35927	.36055	.36182	.36310	.36437	.36564	.36691	.36818
79.	.36945	.37071	.37198	.37324	.37450	.37576	.37701	.37827	.37952	.38078
80.	4.38203	.38328	.38452	.38577	.38701	.38826	.38950	.39074	.39198	.39321
81.	.39445	.39568	.39692	.39815	.39938	.40060	.40183	.40305	.40428	.40550
82.	.40672	.40794	.40916	.41037	.41159	.41280	.41401	.41522	.41643	.41764
83.	.41884	.42004	.42125	.42245	.42365	.42485	.42604	.42724	.42843	.42963
84.	.43082	.43201	.43319	.43438	.43557	.43675	.43793	.43912	.44030	.44147
85.	.44265	.44383	.44500	.44617	.44735	.44852	.44969	.45085	.45202	.45318
86.	.45435	.45551	.45667	.45783	.45899	.46014	.46130	.46245	.46361	.46476
87.	.46591	.46706	.46820	.46935	.47050	.47164	.47278	.47392	.47506	.47620
88.	.47734	.47847	.47961	.48074	.48187	.48300	.48413	.48526	.48639	.48751
89.	.48864	.48976	.49088	.49200	.49312	.49424	.49536	.49647	.49758	.49870
90.	4.49981	.50092	.50203	.50314	.50424	.50535	.50645	.50756	.50866	.50976
91.	.51086	.51196	.51305	.51415	.51525	.51634	.51743	.51852	.51961	.52070
92.	.52179	.52287	.52396	.52504	.52613	.52721	.52829	.52937	.53045	.53152
93.	.53260	.53367	.53475	.53582	.53689	.53796	.53903	.54010	.54116	.54223
94.	.54329	.54436	.54542	.54648	.54754	.54860	.54966	.55071	.55177	.55282
95.	.55388	.55493	.55598	.55703	.55808	.55913	.56017	.56122	.56226	.56331
96.	.56435	.56539	.56643	.56747	.56851	.56954	.57058	.57161	.57265	.57368
97.	.57471	.57574	.57677	.57780	.57883	.57985	.58088	.58190	.58292	.58395
98.	.58497	.58599	.58701	.58802	.58904	.59006	.59107	.59208	.59310	.59411
99.	.59512	.59613	.59714	.59815	.59915	.60016	.60116	.60217	.60317	.60417

NATURAL OR NAPERIAN LOGARITHMS (Continued)
0–499

N	0	1	2	3	4	5	6	7	8	9
0	− ∞	0.00000	0.69315	1.09861	.38629	.60944	.79176	.94591	*.07944	*.19722
1	2.30259	.39790	.48491	.56495	.63906	.70805	.77259	.83321	.89037	.94444
2	.99573	*.04452	*.09104	*.13549	*.17805	*.21888	*.25810	*.29584	*.33220	*.36730
3	3.40120	.43399	.46574	.49651	.52636	.55535	.58352	.61092	.63759	.66356
4	.68888	.71357	.73767	.76120	.78419	.80666	.82864	.85015	.87120	.89182
5	.91202	.93183	.95124	.97029	.98898	*.00733	*02535	*.04305	*.06044	*.07754
6	4.09434	.11087	.12713	.14313	.15888	.17439	.18965	.20469	.21951	.23411
7	.24850	.26268	.27667	.29046	.30407	.31749	.33073	.34381	.35671	.36945
8	.38203	.39445	.40672	.41884	.43082	.44265	.45435	.46591	.47734	.48864
9	.49981	.51086	.52179	.53260	.54329	.55388	.56435	.57471	.58497	.59512
10	4.60517	.61512	.62407	.63473	.64439	.65396	.66344	.67283	.68213	.69135
11	.70048	.70953	.71850	.72739	.73620	.74493	.75359	.76217	.77068	.77912
12	.78749	.79579	.80402	.81218	.82028	.82831	.83628	.84419	.85203	.85981
13	.86753	.87520	.88280	.89035	.89784	.90527	.91265	.91998	.92725	.93447
14	.94164	.94876	.95583	.96284	.96981	.97673	.98361	.99043	.99721	* 00395
15	5.01064	.01728	.02388	.03044	.03695	.04343	.04986	.05625	.06260	.06890
16	.07517	.08140	.08760	.09375	.09987	.10595	.11199	.11799	.12396	.12990
17	.13580	.14166	.14749	.15329	.15906	.16479	.17048	.17615	.18178	.18739
18	.19296	.19850	.20401	.20949	.21494	.22036	.22575	.23111	.23644	.24175
19	.24702	.25227	.25750	.26269	.26786	.27300	.27811	.28320	.28827	.29330
20	5.29832	.30330	.30827	.31321	.31812	.32301	.32788	.33272	.33754	.34233
21	.34711	.35186	.35659	.36129	.36598	.37064	.37528	.37990	.38450.	.38907
22	.39363	.39816	.40268	.40717	.41165	.41610	.42053	.42495	.42935	.43372
23	.43808	.44242	.44674	.45104	.45532	.45959	.46383	.46806	.47227	.47646
24	.48064	.48480	.48894	.49306	.49717	.50126	.50533	.50939	.51343	.51745
25	.52146	.52545	.52943	.53339	53733	.54126	.54518	.54908	.55296	.55683
26	.56068	.56452	.56834	.57215	.57595	.57973	.58350	.58725	.59099	.59471
27	.59842	.60212	.60580	.60947	.61313	.61677	.62040	.62402	.62762	.63121
28	.63479	.63835	.64191	.64545	.64897	.65249	.65599	.65948	.66296	.66643
29	.66988	.67332	.67675	.68017	.68358	.68698	.69036	.69373	.69709	.70044
30	5.70378	.70711	.71043	.71373	.71703	.72031	.72359	.72685	.73010	.73334
31	.73657	.73979	.74300	.74620	.74939	.75257	.75574	.75890	.76205	.76519
32	.76832	.77144	.77455	.77765	.78074	.78383	.78690	.78996	.79301	.79606
33	.79909	.80212	.80513	.80814	.81114	.81413	.81711	.82008	.82305	.82600
34	.82895	.83188	.83481	.83773	.84064	.84354	.84644	.84932	.85220	.85507
35	.85793	.86079	.86363	.86647	.86930	.87212	.87493	.87774	.88053	.88332
36	.88610	.88888	.89164	.89440	.89715	.89990	.90263	.90536	.90808	.91080
37	.91350	.91620	.91889	.92158	.92426	.92693	.92959	.93225	.93489	.93754
38	.94017	.94280	.94542	.94803	.95064	.95324	.95584	.95842	.96101	.96358
39	.96615	.96871	.97126	.97381	.97635	.97889	.98141	.98394	.98645	.98896
40	5.99146	.99396	.99645	.99894	*.00141	*.00389	*.00635	*.00881	*.01127	*.01372
41	6.01616	.01859	.02102	.02345	.02587	.02828	.03069	.03309	.03548	.03787
42	.04025	.04263	.04501	.04737	.04973	.05209	.05444	.05678	.05912	.06146
43	.06379	.06611	.06843	.07074	.07304	.07535	.07764	.07993	.08222	.08450
44	.08677	.08904	.09131	.09357	.09582	.09807	.10032	.10256	.10479	.10702
45	.10925	.11147	.11368	.11589	.11810	.12030	.12249	.12468	.12687	.12905
46	.13123	.13340	.13556	.13773	.13988	.14204	.14419	.14633	.14847	.15060
47	.15273	.15486	.15698	.15910	.16121	.16331	.16542	.16752	.16961	.17170
48	.17379	.17587	.17794	.18002	.18208	.18415	.18621	.18826	.19032	.19236
49	.19441	.19644	.19848	.20051	.20254	.20456	.20658	.20859	.21060	.21261

NATURAL OR NAPERIAN LOGARITHMS (Continued)

500–999

N	0	1	2	3	4	5	6	7	8	9
50	6.21461	.21661	.21860	.22059	.22258	.22456	.22654	.22851	.23048	.23245
51	.23441	.23637	.23832	.24028	.24222	.24417	.24611	.24804	.24998	.25190
52	.25383	.25575	.25767	.25958	.26149	.26340	.26530	.26720	.26910	.27099
53	.27288	.27476	.27664	.27852	.28040	.28227	.28413	.28600	.28786	.28972
54	.29157	.29342	.29527	.29711	.29895	.30079	.30262	.30445	.30628	.30810
55	.30992	.31173	.31355	.31536	.31716	.31897	.32077	.32257	.32436	.32615
56	.32794	.32972	.33150	.33328	.33505	.33683	.33859	.34036	.34212	.34388
57	.34564	.34739	.34914	.35089	.35263	.35437	.35611	.35784	.35957	.36130
58	.36303	.36475	.36647	.36819	.36990	.37161	.37332	.37502	.37673	.37843
59	.38012	.38182	.38351	.38519	.38688	.38856	.39024	.39192	.39359	.39526
60	6.39693	.39859	.40026	.40192	.40357	.40523	.40688	.40853	.41017	.41182
61	.41346	.41510	.41673	.41836	.41999	.42162	.42325	.42487	.42649	.42811
62	.42972	.43133	.43294	.43455	.43615	.43775	.43935	.44095	.44254	.44413
63	.44572	.44731	.44889	.45047	.45205	.45362	.45520	.45677	.45834	.45990
64	.46147	.46303	.46459	.46614	.46770	.46925	.47080	.47235	.47389	.47543
65	.47697	.47851	.48004	.48158	.48311	.48464	.48616	.48768	.48920	.49072
66	.49224	.49375	.49527	.49677	.49828	.49979	.50129	.50279	.50429	.50578
67	.50728	.50877	.51026	.51175	.51323	.51471	.51619	.51767	.51915	.52062
68	.52209	.52356	.52503	.52649	.52796	.52942	.53088	.53233	.53379	.53524
69	.53669	.53814	.53959	.54103	.54247	.54391	.54535	.54679	.54822	.54965
70	6.55108	.55251	.55393	.55536	.55678	.55820	.55962	.56103	.56244	.56386
71	.56526	.56667	.56808	.56948	.57088	.57228	.57368	.57508	.57647	.57786
72	.57925	.58064	.58203	.58341	.58479	.58617	.58755	.58893	.59030	.59167
73	.59304	.59441	.59578	.59715	.59851	.59987	.60123	.60259	.60394	.60530
74	.60665	.60800	.60935	.61070	.61204	.61338	.61473	.61607	.61740	.61874
75	.62007	.62141	.62274	.62407	.62539	.62672	.62804	.62936	.63068	.63200
76	.63332	.63463	.63595	.63726	.63857	.63988	.64118	.64249	.64379	.64509
77	.64639	.64769	.64898	.65028	.65157	.65286	.65415	.65544	.65673	.65801
78	.65929	.66058	.66185	.66313	.66441	.66568	.66696	.66823	.66950	.67077
79	.67203	.67330	.67456	.67582	.67708	.67834	.67960	.68085	.68211	.68336
80	6.68461	.68586	.68711	.68835	.68960	.69084	.69208	.69332	.69456	.69580
81	.69703	.69827	.69950	.70073	.70196	.70319	.70441	.70564	.70686	.70808
82	.70930	.71052	.71174	.71296	.71417	.71538	.71659	.71780	.71901	.72022
83	.72143	.72263	.72383	.72503	.72623	.72743	.72863	.72982	.73102	.73221
84	.73340	.73459	.73578	.73697	.73815	.73934	.74052	.74170	.74288	.74406
85	.74524	.74641	.74759	.74876	.74993	.75110	.75227	.75344	.75460	.75577
86	.75693	.75809	.75926	.76041	.76157	.76273	.76388	.76504	.76619	.76734
87	.76849	.76964	.77079	.77194	.77308	.77422	.77537	.77651	.77765	.77878
88	.77992	.78106	.78219	.78333	.78446	.78559	.78672	.78784	.78897	.79010
89	.79122	.79234	.79347	.79459	.79571	.79682	.79794	.79906	.80017	.80128
90	6.80239	.80351	.80461	.80572	.80683	.80793	.80904	.81014	.81124	.81235
91	.81344	.81454	.81564	.81674	.81783	.81892	.82002	.82111	.82220	.82329
92	.82437	.82546	.82655	.82763	.82871	.82979	.83087	.83195	.83303	.83411
93	.83518	.83626	.83733	.83841	.83948	.84055	.84162	.84268	.84375	.84482
94	.84588	.84694	.84801	.84907	.85013	.85118	.85224	.85330	.85435	.85541
95	.85646	.85751	.85857	.85961	.86066	.86171	.86276	.86380	.86485	.86589
96	.86693	.86797	.86901	.87005	.87109	.87213	.87316	.87420	.87523	.87626
97	.87730	.87833	.87936	.88038	.88141	.88244	.88346	.88449	.88551	.88653
98	.88755	.88857	.88959	.89061	.89163	.89264	.89366	.89467	.89568	.89669
99	.89770	.89871	.89972	.90073	.90174	.90274	.90375	.90475	.90575	.90675

RADIX TABLE OF NATURAL LOGARITHMS

x	n	$\log[1 + x(10^{-n})]$	$-\log[1 - x(10^{-n})]$
1	10	0.00000 00000 99999 99999 50000	0.00000 00001 00000 00000 50000
2	10	0.00000 00001 99999 99998 00000	0.00000 00002 00000 00002 00000
3	10	0.00000 00002 99999 99995 50000	0.00000 00003 00000 00004 50000
4	10	0.00000 00003 99999 99992 00000	0.00000 00004 00000 00008 00000
5	10	0.00000 00004 99999 99987 50000	0.00000 00005 00000 00012 50000
6	10	0.00000 00005 99999 99982 00000	0.00000 00006 00000 00018 00000
7	10	0.00000 00006 99999 99975 50000	0.00000 00007 00000 00024 50000
8	10	0.00000 00007 99999 99968 00000	0.00000 00008 00000 00032 00000
9	10	0.00000 00008 99999 99959 50000	0.00000 00009 00000 00040 50000
1	9	0.00000 00009 99999 99950 00000	0.00000 00010 00000 00050 00000
2	9	0.00000 00019 99999 99800 00000	0.00000 00020 00000 00200 00000
3	9	0.00000 00029 99999 99550 00000	0.00000 00030 00000 00450 00000
4	9	0.00000 00039 99999 99200 00000	0.00000 00040 00000 00800 00000
5	9	0.00000 00049 99999 98750 00000	0.00000 00050 00000 01250 00000
6	9	0.00000 00059 99999 98200 00001	0.00000 00060 00000 01800 00001
7	9	0.00000 00069 99999 97550 00001	0.00000 00070 00000 02450 00001
8	9	0.00000 00079 99999 96800 00002	0.00000 00080 00000 03200 00002
9	9	0.00000 00089 99999 95950 00002	0.00000 00090 00000 04050 00002
1	8	0.00000 00099 99999 95000 00003	0.00000 00100 00000 05000 00003
2	8	0.00000 00199 99999 80000 00027	0.00000 00200 00000 20000 00027
3	8	0.00000 00299 99999 55000 00090	0.00000 00300 00000 45000 00090
4	8	0.00000 00399 99999 20000 00213	0.00000 00400 00000 80000 00213
5	8	0.00000 00499 99998 75000 00417	0.00000 00500 00001 25000 00417
6	8	0.00000 00599 99998 20000 00720	0.00000 00600 00001 80000 00720
7	8	0.00000 00699 99997 55000 01143	0.00000 00700 00002 45000 01143
8	8	0.00000 00799 99996 80000 01707	0.00000 00800 00003 20000 01707
9	8	0.00000 00899 99995 95000 02430	0.00000 00900 00004 05000 02430
1	7	0.00000 00999 99995 00000 03333	0.00000 01000 00005 00000 03333
2	7	0.00000 01999 99980 00000 26667	0.00000 02000 00020 00000 26667
3	7	0.00000 02999 99955 00000 90000	0.00000 03000 00045 00000 90000
4	7	0.00000 03999 99920 00002 13333	0.00000 04000 00080 00002 13333
5	7	0.00000 04999 99875 00004 16667	0.00000 05000 00125 00004 16667
6	7	0.00000 05999 99820 00007 20000	0.00000 06000 00180 00007 20000
7	7	0.00000 06999 99755 00011 43333	0.00000 07000 00245 00011 43334
8	7	0.00000 07999 99680 00017 06666	0.00000 08000 00320 00017 06668
9	7	0.00000 08999 99595 00024 29998	0.00000 09000 00405 00024 30002
1	6	0.00000 09999 99500 00033 33331	0.00000 10000 00500 00033 33336
2	6	0.00000 19999 98000 00266 66627	0.00000 20000 02000 00266 66707
3	6	0.00000 29999 95500 00899 99798	0.00000 30000 04500 00900 00203
4	6	0.00000 39999 92000 02133 32693	0.00000 40000 08000 02133 33973
5	6	0.00000 49999 87500 04166 65104	0.00000 50000 12500 04166 68229
6	6	0.00000 59999 82000 07199 96760	0.00000 60000 18000 07200 03240
7	6	0.00000 69999 75500 11433 27331	0.00000 70000 24500 11433 39336
8	6	0.00000 79999 68000 17066 56427	0.00000 80000 32000 17066 76907
9	6	0.00000 89999 59500 24299 83598	0.00000 90000 40500 24300 16403

For $n > 10$, $\log[1 \pm x(10^{-n})] = \pm x(10^{-n}) - \frac{1}{2}x^2(10^{-2n})$ to 25 decimals.

RADIX TABLE OF NATURAL LOGARITHMS (Continued)

x	n	$\log[1 + x(10^{-n})]$	$-\log[1 - x)10^{-n})]$
1	5	0.00000 99999 50000 33333 08334	0.00001 00000 50000 33333 58334
2	5	0.00001 99998 00002 66662 66673	0.00002 00002 00002 66670 66673
3	5	0.00002 99995 50008 99979 75049	0.00003 00004 50009 00020 25049
4	5	0.00003 99992 00021 33269 33538	0.00004 00008 00021 33397 33538
5	5	0.00004 99987 50041 66510 42292	0.00005 00012 50041 66822 92292
6	5	0.00005 99982 00071 99676 01555	0.00006 00018 00072 00324 01555
7	5	0.00006 99975 50114 32733 11695	0.00007 00024 50114 33933 61695
8	5	0.00007 99968 00170 65642 73220	0.00008 00032 00170 67690 73221
9	5	0.00008 99959 50242 98359 86809	0.00009 00040 50243 01640 36811
1	4	0.00009 99950 00333 30833 53332	0.00010 00050 00333 35833 53335
2	4	0.00019 99800 02666 26673 06560	0.00020 00200 02667 06673 06773
3	4	0.00029 99550 08997 97548 58785	0.00030 00450 09002 02548 61215
4	4	0.00039 99200 21326 93538 06509	0.00040 00800 21339 73538 20162
5	4	0.00049 98750 41651 04791 40636	0.00050 01250 41682 29791 92719
6	4	0.00059 98200 71967 61554 42280	0.00060 01800 72032 41555 97800
7	4	0.00069 97551 14273 34192 77369	0.00070 02451 14393 39196 69533
8	4	0.00079 96801 70564 33215 90059	0.00080 03201 70769 13224 63873
9	4	0.00089 95952 42836 09300 94948	0.00090 04052 43164 14318 66419
1	3	0.00099 95003 33083 53316 68094	0.00100 05003 33583 53350 01430
2	3	0.00199 80026 62673 05601 82538	0.00200 20026 70673 07735 16511
3	3	0.00299 55089 79798 47881 16106	0.00300 45090 20298 72181 32509
4	3	0.00399 20212 69537 45299 90751	0.00400 80213 97538 81834 87927
5	3	0.00498 75415 11039 07361 21022	0.00501 25418 23544 28204 30937
6	3	0.00598 20716 77547 46378 20189	0.00601 80723 25563 01620 19350
7	3	0.00697 56137 36425 24209 95222	0.00702 46149 36964 45987 41123
8	3	0.00796 81696 49176 87351 07973	0.00803 21716 97264 25903 86494
9	3	0.00895 97413 71471 90444 31465	0.00904 07446 52149 06220 55241
1	2	0.00995 03308 53168 08284 82154	0.01005 03358 53501 44118 35489
2	2	0.01980 26272 96179 71302 60291	0.02020 27073 17519 44840 80453
3	2	0.02955 88022 41544 40273 26194	0.03045 92074 84708 54591 92613
4	2	0.03922 07131 53281 29626 92009	0.04082 19945 20255 12955 45771
5	2	0.04879 01641 69432 00306 53744	0.05129 32943 87550 53342 61961
6	2	0.05826 89081 23975 77552 57184	0.06187 54037 18087 47179 78001
7	2	0.06765 86484 73814 80526 84159	0.07257 06928 34835 43071 15733
8	2	0.07696 10411 36128 32498 42170	0.08338 16089 39051 05839 47658
9	2	0.08617 76962 41052 33234 13335	0.09431 06794 71241 32687 71427
1	1	0.09531 01798 04324 86004 39521	0.10536 05156 57826 30122 75010
2	1	0.18232 15567 93954 62621 17180	0.22314 35513 14209 75576 62951
3	1	0.26236 42644 67491 05203 54960	0.35667 49439 38732 37891 26387
4	1	0.33647 22366 21212 93050 45934	0.51082 56237 65990 68320 55141
5	1	0.40546 51081 08164 38197 80131	0.69314 71805 59945 30941 72321
6	1	0.47000 36292 45735 55365 09370	0.91629 07318 74155 06518 35272
7	1	0.53062 82510 62170 39623 15432	1.20397 28043 25935 99262 27462
8	1	0.58778 66649 02119 00818 97311	1.60943 79124 34100 37460 07593
9	1	0.64185 38861 72394 77599 10360	2.30258 50929 94045 68401 79915
1	0	0.69314 71805 59945 30941 72321	∞

For $n > 10$, $\log[1 \pm x(10^{-n})] = \pm x(10^{-n}) - \frac{1}{2} x^2 (10^{-2n})$ to 25 decimals.

EXPONENTIAL FUNCTIONS

Values of e^x, $\log e^x$ and e^{-x} where e is the base of the natural system of logarithms 2.71828... and x has values from 0 to 10. Facilitating the solution of exponential equations, these tables also serve as a table of natural or Naperian antilogarithms. For instance, if the logarithm or exponent $x = 3.26$, the corresponding number or value of e^x is 26.050. Its reciprocal e^{-x} is .038388.

x	e^x	$\text{Log}_{10}(e^x)$	e^{-x}	x	e^x	$\text{Log}_{10}(e^x)$	e^{-x}
0.00	1.0000	0.00000	1.000000	**0.50**	1.6487	0.21715	0.606531
0.01	1.0101	.00434	0.990050	0.51	1.6653	.22149	.600496
0.02	1.0202	.00869	.980199	0.52	1.6820	.22583	.594521
0.03	1.0305	.01303	.970446	0.53	1.6989	.23018	.588605
0.04	1.0408	.01737	.960789	0.54	1.7160	.23452	.582748
0.05	1.0513	0.02171	0.951229	**0.55**	1.7333	0.23886	0.576950
0.06	1.0618	.02606	.941765	0.56	1.7507	.24320	.571209
0.07	1.0725	.03040	.932394	0.57	1.7683	.24755	.565525
0.08	1.0833	.03474	.923116	0.58	1.7860	.25189	.559898
0.09	1.0942	.03909	.913931	0.59	1.8040	.25623	.554327
0.10	1.1052	0.04343	0.904837	**0.60**	1.8221	0.26058	0.548812
0.11	1.1163	.04777	.895834	0.61	1.8404	.26492	.543351
0.12	1.1275	.05212	.886920	0.62	1.8589	.26926	.537944
0.13	1.1388	.05646	.878095	0.63	1.8776	.27361	.532592
0.14	1.1503	.06080	.869358	0.64	1.8965	.27795	.527292
0.15	1.1618	0.06514	0.860708	**0.65**	1.9155	0.28229	0.522046
0.16	1.1735	.06949	.852144	0.66	1.9348	.28663	.516851
0.17	1.1853	.07383	.843665	0.67	1.9542	.29098	.511709
0.18	1.1972	.07817	.835270	0.68	1.9739	.29532	.506617
0.19	1.2092	.08252	.826959	0.69	1.9937	.29966	.501576
0.20	1.2214	0.08686	0.818731	**0.70**	2.0138	0.30401	0.496585
0.21	1.2337	.09120	.810584	0.71	2.0340	.30835	.491644
0.22	1.2461	.09554	.802519	0.72	2.0544	.31269	.486752
0.23	1.2586	.09989	.794534	0.73	2.0751	.31703	.481909
0.24	1.2712	.10423	.786628	0.74	2.0959	.32138	.477114
0.25	1.2840	0.10857	0.778801	**0.75**	2.1170	0.32572	0.472367
0.26	1.2969	.11292	.771052	0.76	2.1383	.33006	.467666
0.27	1.3100	.11726	.763379	0.77	2.1598	.33441	.463013
0.28	1.3231	.12160	.755784	0.78	2.1815	.33875	.458406
0.29	1.3364	.12595	.748264	0.79	2.2034	.34309	.453845
0.30	1.3499	0.13029	0.740818	**0.80**	2.2255	0.34744	0.449329
0.31	1.3634	.13463	.733447	0.81	2.2479	.35178	.444858
0.32	1.3771	.13897	.726149	0.82	2.2705	.35612	.440432
0.33	1.3910	.14332	.718924	0.83	2.2933	.36046	.436049
0.34	1.4049	.14766	.711770	0.84	2.3164	.36481	.431711
0.35	1.4191	0.15200	0.704688	**0.85**	2.3396	0.36915	0.427415
0.36	1.4333	.15635	.697676	0.86	2.3632	.37349	.423162
0.37	1.4477	.16069	.690734	0.87	2.3869	.37784	.418952
0.38	1.4623	.16503	.683861	0.88	2.4109	.38218	.414783
0.39	1.4770	.16937	.677057	0.89	2.4351	.38652	.410656
0.40	1.4918	0.17372	0.670320	**0.90**	2.4596	0.39087	0.406570
0.41	1.5068	.17806	.663650	0.91	2.4843	.39521	.402524
0.42	1.5220	.18240	.657047	0.92	2.5093	.39955	.398519
0.43	1.5373	.18675	.650509	0.93	2.5345	.40389	.394554
0.44	1.5527	.19109	.644036	0.94	2.5600	.40824	.390628
0.45	1.5683	0.19543	0.637628	**0.95**	2.5857	0.41258	0.386741
0.46	1.5841	.19978	.631284	0.96	2.6117	.41692	.382893
0.47	1.6000	.20412	.625002	0.97	2.6379	.42127	.379083
0.48	1.6161	.20846	.618783	0.98	2.6645	.42561	.375311
0.49	1.6323	.21280	.612626	0.99	2.6912	.42995	.371577
0.50	1.6487	0.21715	0.606531	**1.00**	2.7183	0.43429	0.367879

EXPONENTIAL FUNCTIONS (Continued)

x	e^x	$\text{Log}_{10}(e^x)$	e^{-x}	x	e^x	$\text{Log}_{10}(e^x)$	e^{-x}
1.00	2.7183	0.43429	0.367879	**1.50**	4.4817	0.65144	0.223130
1.01	2.7456	.43864	.364219	1.51	4.5267	.65578	.220910
1.02	2.7732	.44298	.360595	1.52	4.5722	.66013	.218712
1.03	2.8011	.44732	.357007	1.53	4.6182	.66447	.216536
1.04	2.8292	.45167	.353455	1.54	4.6646	.66881	.214381
1.05	2.8577	0.45601	0.349938	**1.55**	4.7115	0.67316	0.212248
1.06	2.8864	.46035	.346456	1.56	4.7588	.67750	.210136
1.07	2.9154	.46470	.343009	1.57	4.8066	.68184	.208045
1.08	2.9447	.46904	.339596	1.58	4.8550	.68619	.205975
1.09	2.9743	.47338	.336216	1.59	4.9037	.69053	.203926
1.10	3.0042	0.47772	0.332871	**1.60**	4.9530	0.69487	0.201897
1.11	3.0344	.48207	.329559	1.61	5.0028	.69921	.199888
1.12	3.0649	.48641	.326280	1.62	5.0531	.70356	.197899
1.13	3.0957	.49075	.323033	1.63	5.1039	.70790	.195930
1.14	3.1268	.49510	.319819	1.64	5.1552	.71224	.193980
1.15	3.1582	0.49944	0.316637	**1.65**	5.2070	0.71659	0.192050
1.16	3.1899	.50378	.313486	1.66	5.2593	.72093	.190139
1.17	3.2220	.50812	.310367	1.67	5.3122	.72527	.188247
1.18	3.2544	.51247	.307279	1.68	5.3656	.72961	.186374
1.19	3.2871	.51681	.304221	1.69	5.4195	.73396	.184520
1.20	3.3201	0.52115	0.301194	**1.70**	5.4739	0.73830	0.182684
1.21	3.3535	.52550	.298197	1.71	5.5290	.74264	.180866
1.22	3.3872	.52984	.295230	1.72	5.5845	.74699	.179066
1.23	3.4212	.53418	.292293	1.73	5.6407	.75133	.177284
1.24	3.4556	.53853	.289384	1.74	5.6973	.75567	.175520
1.25	3.4903	0.54287	0.286505	**1.75**	5.7546	0.76002	0.173774
1.26	3.5254	.54721	.283654	1.76	5.8124	.76436	.172045
1.27	3.5609	.55155	.280832	1.77	5.8709	.76870	.170333
1.28	3.5966	.55590	.278037	1.78	5.9299	.77304	.168638
1.29	3.6328	.56024	.275271	1.79	5.9895	.77739	.166960
1.30	3.6693	0.56458	0.272532	**1.80**	6.0496	0.78173	0.165299
1.31	3.7062	.56893	.269820	1.81	6.1104	.78607	.163654
1.32	3.7434	.57327	.267135	1.82	6.1719	.79042	.162026
1.33	3.7810	.57761	.264477	1.83	6.2339	.79476	.160414
1.34	3.8190	.58195	.261846	1.84	6.2965	.79910	.158817
1.35	3.8574	0.58630	0.259240	**1.85**	6.3598	0.80344	0.157237
1.36	3.8962	.59064	.256661	1.86	6.4237	.80779	.155673
1.37	3.9354	.59498	.254107	1.87	6.4883	.81213	.154124
1.38	3.9749	.59933	.251579	1.88	6.5535	.81647	.152590
1.39	4.0149	.60367	.249075	1.89	6.6194	.82082	.151072
1.40	4.0552	0.60801	0.246597	**1.90**	6.6859	0.82516	0.149569
1.41	4.0960	.61236	.244143	1.91	6.7531	.82950	.148080
1.42	4.1371	.61670	.241714	1.92	6.8210	.83385	.146607
1.43	4.1787	.62104	.239309	1.93	6.8895	.83819	.145148
1.44	4.2207	.62538	.236928	1.94	6.9588	.84253	.143704
1.45	4.2631	0.62973	0.234570	**1.95**	7.0287	0.84687	0.142274
1.46	4.3060	.63407	.232236	1.96	7.0993	.85122	.140858
1.47	4.3492	.63841	.229925	1.97	7.1707	.85556	.139457
1.48	4.3929	.64276	.227638	1.98	7.2427	.85990	.138069
1.49	4.4371	.64710	.225373	1.99	7.3155	.86425	.136695
1.50	4.4817	0.65144	0.223130	**2.00**	7.3891	0.86859	0.135335

EXPONENTIAL FUNCTIONS (Continued)

x	e^x	$Log_{10}(e^x)$	e^{-x}	x	e^x	$Log_{10}(e^x)$	e^{-x}
2.00	7.3891	0.86859	0.135335	2.50	12.182	1.08574	0.082085
2.01	7.4633	.87293	.133989	2.51	12.305	1.09008	.081268
2.02	7.5383	.87727	.132655	2.52	12.429	1.09442	.080460
2.03	7.6141	.88162	.131336	2.53	12.554	1.09877	.079659
2.04	7.6906	.88596	.130029	2.54	12.680	1.10311	.078866
2.05	7.7679	0.89030	0.128735	2.55	12.807	1.10745	0.078082
2.06	7.8460	.89465	.127454	2.56	12.936	1.11179	.077305
2.07	7.9248	.89899	.126186	2.57	13.066	1.11614	.076536
2.08	8.0045	.90333	.124930	2.58	13.197	1.12048	.075774
2.09	8.0849	.90768	.123687	2.59	13.330	1.12482	.075020
2.10	8.1662	0.91202	0.122456	2.60	13.464	1.12917	0.074274
2.11	8.2482	.91636	.121238	2.61	13.599	1.13351	.073535
2.12	8.3311	.92070	.120032	2.62	13.736	1.13785	.072803
2.13	8.4149	.92505	.118837	2.63	13.874	1.14219	.072078
2.14	8.4994	.92939	.117655	2.64	14.013	1.14654	.071361
2.15	8.5849	0.93373	0.116484	2.65	14.154	1.15088	0.070651
2.16	8.6711	.93808	.115325	2.66	14.296	1.15522	.069948
2.17	8.7583	.94242	.114178	2.67	14.440	1.15957	.069252
2.18	8.8463	.94676	.113042	2.68	14.585	1.16391	.068563
2.19	8.9352	.95110	.111917	2.69	14.732	1.16825	.067881
2.20	9.0250	0.95545	0.110803	2.70	14.880	1.17260	0.067206
2.21	9.1157	.95979	.109701	2.71	15.029	1.17694	.066537
2.22	9.2073	.96413	.108609	2.72	15.180	1.18128	.065875
2.23	9.2999	.96848	.107528	2.73	15.333	1.18562	.065219
2.24	9.3933	.97282	.106459	2.74	15.487	1.18997	.064570
2.25	9.4877	0.97716	0.105399	2.75	15.643	1.19431	0.063928
2.26	9.5831	.98151	.104350	2.76	15.800	1.19865	.063292
2.27	9.6794	.98585	.103312	2.77	15.959	1.20300	.062662
2.28	9.7767	.99019	.102284	2.78	16.119	1.20734	.062039
2.29	9.8749	.99453	.101266	2.79	16.281	1.21168	.061421
2.30	9.9742	0.99888	0.100259	2.80	16.445	1.21602	0.060810
2.31	10.074	1.00322	.099261	2.81	16.610	1.22037	.060205
2.32	10.176	1.00756	.098274	2.82	16.777	1.22471	.059606
2.33	10.278	1.01191	.097296	2.83	16.945	1.22905	.059013
2.34	10.381	1.01625	.096328	2.84	17.116	1.23340	.058426
2.35	10.486	1.02059	0.095369	2.85	17.288	1.23774	0.057844
2.36	10.591	1.02493	.094420	2.86	17.462	1.24208	.057269
2.37	10.697	1.02928	.093481	2.87	17.637	1.24643	.056699
2.38	10.805	1.03362	.092551	2.88	17.814	1.25077	.056135
2.39	10.913	1.03796	.091630	2.89	17.993	1.25511	.055576
2.40	11.023	1.04231	0.090718	2.90	18.174	1.25945	0.055023
2.41	11.134	1.04665	.089815	2.91	18.357	1.26380	.054476
2.42	11.246	1.05099	.088922	2.92	18.541	1.26814	.053934
2.43	11.359	1.05534	.088037	2.93	18.728	1.27248	.053397
2.44	11.473	1.05968	.087161	2.94	18.916	1.27683	.052866
2.45	11.588	1.06402	0.086294	2.95	19.106	1.28117	0.052340
2.46	11.705	1.06836	.085435	2.96	19.298	1.28551	.051819
2.47	11.822	1.07271	.084585	2.97	19.492	1.28985	.051303
2.48	11.941	1.07705	.083743	2.98	19.688	1.29420	.050793
2.49	12.061	1.08139	.082910	2.99	19.886	1.29854	.050287
2.50	12.182	1.08574	0.082085	3.00	20.086	1.30288	0.049787

EXPONENTIAL FUNCTIONS (Continued)

x	e^x	$Log_{10}(e^x)$	e^{-x}	x	e^x	$Log_{10}(e^x)$	e^{-x}
3.00	20.086	1.30288	0.049787	**3.50**	33.115	1.52003	0.030197
3.01	20.287	1.30723	.049292	3.51	33.448	1.52437	.029897
3.02	20.491	1.31157	.048801	3.52	33.784	1.52872	.029599
3.03	20.697	1.31591	.048316	3.53	34.124	1.53306	.029305
3.04	20.905	1.32026	.047835	3.54	34.467	1.53740	.029013
3.05	21.115	1.32460	0.047359	**3.55**	34.813	1.54175	0.028725
3.06	21.328	1.32894	.046888	3.56	35.163	1.54609	.028439
3.07	21.542	1.33328	.046421	3.57	35.517	1.55043	.028156
3.08	21.758	1.33763	.045959	3.58	35.874	1.55477	.027876
3.09	21.977	1.34197	.045502	3.59	36.234	1.55912	.027598
3.10	22.198	1.34631	0.045049	**3.60**	36.598	1.56346	0.027324
3.11	22.421	1.35066	.044601	3.61	36.966	1.56780	.027052
3.12	22.646	1.35500	.044157	3.62	37.338	1.57215	.026783
3.13	22.874	1.35934	.043718	3.63	37.713	1.57649	.026516
3.14	23.104	1.36368	.043283	3.64	38.092	1.58083	.026252
3.15	23.336	1.36803	0.042852	**3.65**	38.475	1.58517	0.025991
3.16	23.571	1.37237	.042426	3.66	38.861	1.58952	.025733
3.17	23.807	1.37671	.042004	3.67	39.252	1.59386	.025476
3.18	24.047	1.38106	.041586	3.68	39.646	1.59820	.025223
3.19	24.288	1.38540	.041172	3.69	40.045	1.60255	.024972
3.20	24.533	1.38974	0.040762	**3.70**	40.447	1.60689	0.024724
3.21	24.779	1.39409	.040357	3.71	40.854	1.61123	.024478
3.22	25.028	1.39843	.039955	3.72	41.264	1.61558	.024234
3.23	25.280	1.40277	.039557	3.73	41.679	1.61992	.023993
3.24	25.534	1.40711	.039164	3.74	42.098	1.62426	.023754
3.25	25.790	1.41146	0.038774	**3.75**	42.521	1.62860	0.023518
3.26	26.050	1.41580	.038388	3.76	42.948	1.63295	.023284
3.27	26.311	1.42014	.038006	3.77	43.380	1.63729	.023052
3.28	26.576	1.42449	.037628	3.78	43.816	1.64163	.022823
3.29	26.843	1.42883	.037254	3.79	44.256	1.64598	.022596
3.30	27.113	1.43317	0.036883	**3.80**	44.701	1.65032	0.022371
3.31	27.385	1.43751	.036516	3.81	45.150	1.65466	.022148
3.32	27.660	1.44186	.036153	3.82	45.604	1.65900	.021928
3.33	27.938	1.44620	.035793	3.83	46.063	1.66335	.021710
3.34	28.219	1.45054	.035437	3.84	46.525	1.66769	.021494
3.35	28.503	1.45489	0.035084	**3.85**	46.993	1.67203	0.021280
3.36	28.789	1.45923	.034735	3.86	47.465	1.67638	.021068
3.37	29.079	1.46357	.034390	3.87	47.942	1.68072	.020858
3.38	29.371	1.46792	.034047	3.88	48.424	1.68506	.020651
3.39	29.666	1.47226	.033709	3.89	48.911	1.68941	.020445
3.40	29.964	1.47660	0.033373	**3.90**	49.402	1.69375	0.020242
3.41	30.265	1.48094	.033041	3.91	49.899	1.69809	.020041
3.42	30.569	1.48529	.032712	3.92	50.400	1.70243	.019841
3.43	30.877	1.48963	.032387	3.93	50.907	1.70678	.019644
3.44	31.187	1.49397	.032065	3.94	51.419	1.71112	.019448
3.45	31.500	1.49832	0.031746	**3.95**	51.935	1.71546	0.019255
3.46	31.817	1.50266	.031430	3.96	52.457	1.71981	.019063
3.47	32.137	1.50700	.031117	3.97	52.985	1.72415	.018873
3.48	32.460	1.51134	.030807	3.98	53.517	1.72849	.018686
3.49	32.786	1.51569	.030501	3.99	54.055	1.73283	.018500
3.50	33.115	1.52003	0.030197	**4.00**	54.598	1.73718	0.018316

EXPONENTIAL FUNCTIONS (Continued)

x	e^x	$\text{Log}_{10}(e^x)$	e^{-x}	x	e^x	$\text{Log}_{10}(e^x)$	e^{-x}
4.00	54.598	1.73718	0.018316	**4.50**	90.017	1.95433	0.011109
4.01	55.147	1.74152	.018133	4.51	90.922	1.95867	.010998
4.02	55.701	1.74586	.017953	4.52	91.836	1.96301	.010889
4.03	56.261	1.75021	.017774	4.53	92.759	1.96735	.010781
4.04	56.826	1.75455	.017597	4.54	93.691	1.97170	.010673
4.05	57.397	1.75889	0.017422	**4.55**	94.632	1.97604	0.010567
4.06	57.974	1.76324	.017249	4.56	95.583	1.98038	.010462
4.07	58.557	1.76758	.017077	4.57	96.544	1.98473	.010358
4.08	59.145	1.77192	.016907	4.58	97.514	1.98907	.010255
4.09	59.740	1.77626	.016739	4.59	98.494	1.99341	.010153
4.10	60.340	1.78061	0.016573	**4.60**	99.484	1.99775	0.010052
4.11	60.947	1.78495	.016408	4.61	100.48	2.00210	.009952
4.12	61.559	1.78929	.016245	4.62	101.49	2.00644	.009853
4.13	62.178	1.79364	.016083	4.63	102.51	2.01078	.009755
4.14	62.803	1.79798	.015923	4.64	103.54	2.01513	.009658
4.15	63.434	1.80232	0.015764	**4.65**	104.58	2.01947	0.009562
4.16	64.072	1.80667	.015608	4.66	105.64	2.02381	.009466
4.17	64.715	1.81101	.015452	4.67	106.70	2.02816	.009372
4.18	65.366	1.81535	.015299	4.68	107.77	2.03250	.009279
4.19	66.023	1.81969	.015146	4.69	108.85	2.03684	.009187
4.20	66.686	1.82404	0.014996	**4.70**	109.95	2.04118	0.009095
4.21	67.357	1.82838	.014846	4.71	111.05	2.04553	.009005
4.22	68.033	1.83272	.014699	4.72	112.17	2.04987	.008915
4.23	68.717	1.83707	.014552	4.73	113.30	2.05421	.008826
4.24	69.408	1.84141	.014408	4.74	114.43	2.05856	.008739
4.25	70.105	1.84575	0.014264	**4.75**	115.58	2.06290	0.008652
4.26	70.810	1.85009	.014122	4.76	116.75	2.06724	.008566
4.27	71.522	1.85444	.013982	4.77	117.92	2.07158	.008480
4.28	72.240	1.85878	.013843	4.78	119.10	2.07593	.008396
4.29	72.966	1.86312	.013705	4.79	120.30	2.08027	.008312
4.30	73.700	1.86747	0.013569	**4.80**	121.51	2.08461	0.008230
4.31	74.440	1.87181	.013434	4.81	122.73	2.08896	.008148
4.32	75.189	1.87615	.013300	4.82	123.97	2.09330	.008067
4.33	75.944	1.88050	.013168	4.83	125.21	2.09764	.007987
4.34	76.708	1.88484	.013037	4.84	126.47	2.10199	.007907
4.35	77.478	1.88918	0.012907	**4.85**	127.74	2.10633	0.007828
4.36	78.257	1.89352	.012778	4.86	129.02	2.11067	.007750
4.37	79.044	1.89787	.012651	4.87	130.32	2.11501	.007673
4.38	79.838	1.90221	.012525	4.88	131.63	2.11936	.007597
4.39	80.640	1.90655	.012401	4.89	132.95	2.12370	.007521
4.40	81.451	1.91090	0.012277	**4.90**	134.29	2.12804	0.007447
4.41	82.269	1.91524	.012155	4.91	135.64	2.13239	.007372
4.42	83.096	1.91958	.012034	4.92	137.00	2.13673	.007299
4.43	83.931	1.92392	.011914	4.93	138.38	2.14107	.007227
4.44	84.775	1.92827	.011796	4.94	139.77	2.14541	.007155
4.45	85.627	1.93261	0.011679	**4.95**	141.17	2.14976	0.007083
4.46	86.488	1.93695	.011562	4.96	142.59	2.15410	.007013
4.47	87.357	1.94130	.011447	4.97	144.03	2.15844	.006943
4.48	88.235	1.94564	.011333	4.98	145.47	2.16279	.006874
4.49	89.121	1.94998	.011221	4.99	146.94	2.16713	.006806
4.50	90.017	1.95433	0.011109	**5.00**	148.41	2.17147	0.006738

EXPONENTIAL FUNCTIONS (Continued)

x	e^x	$Log_{10}(e^x)$	e^{-x}	x	e^x	$Log_{10}(e^x)$	e^{-x}
5.00	148.41	2.17147	0.006738	**5.50**	244.69	2.38862	0.0040868
5.01	149.90	2.17582	.006671	5.55	257.24	2.41033	.0038875
5.02	151.41	2.18016	.006605	5.60	270.43	2.43205	.0036979
5.03	152.93	2.18450	.006539	5.65	284.29	2.45376	.0035175
5.04	154.47	2.18884	.006474	5.70	298.87	2.47548	.0033460
5.05	156.02	2.19319	0.006409	**5.75**	314.19	2.49719	0.0031828
5.06	157.59	2.19753	.006346	5.80	330.30	2.51891	.0030276
5.07	159.17	2.20187	.006282	5.85	347.23	2.54062	.0028799
5.08	160.77	2.20622	.006220	5.90	365.04	2.56234	.0027394
5.09	162.39	2.21056	.006158	5.95	383.75	2.58405	.0026058
5.10	164.02	2.21490	0.006097	**6.00**	403.43	2.60577	0.0024788
5.11	165.67	2.21924	.006036	6.05	424.11	2.62748	.0023579
5.12	167.34	2.22359	.005976	6.10	445.86	2.64920	.0022429
5.13	169.02	2.22793	.005917	6.15	468.72	2.67091	.0021335
5.14	170.72	2.23227	.005858	6.20	492.75	2.69263	.0020294
5.15	172.43	2.23662	0.005799	**6.25**	518.01	2.71434	0.0019305
5.16	174.16	2.24096	.005742	6.30	544.57	2.73606	.0018363
5.17	175.91	2.24530	.005685	6.35	572.49	2.75777	.0017467
5.18	177.68	2.24965	.005628	6.40	601.85	2.77948	.0016616
5.19	179.47	2.25399	.005572	6.45	632.70	2.80120	.0015805
5.20	181.27	2.25833	0.005517	**6.50**	665.14	2.82291	0.0015034
5.21	183.09	2.26267	.005462	6.55	699.24	2.84463	.0014301
5.22	184.93	2.26702	.005407	6.60	735.10	2.86634	.0013604
5.23	186.79	2.27136	.005354	6.65	772.78	2.88806	.0012940
5.24	188.67	2.27570	.005300	6.70	812.41	2.90977	.0012309
5.25	190.57	2.28005	0.005248	**6.75**	854.06	2.93149	0.0011709
5.26	192.48	2.28439	.005195	6.80	897.85	2.95320	.0011138
5.27	194.42	2.28873	.005144	6.85	943.88	2.97492	.0010595
5.28	196.37	2.29307	.005092	6.90	992.27	2.99663	.0010078
5.29	198.34	2.29742	.005042	6.95	1043.1	3.01835	.0009586
5.30	200.34	2.30176	0.004992	**7.00**	1096.6	3.04006	0.0009119
5.31	202.35	2.30610	.004942	7.05	1152.9	3.06178	.0008674
5.32	204.38	2.31045	.004893	7.10	1212.0	3.08349	.0008251
5.33	206.44	2.31479	.004844	7.15	1274.1	3.10521	.0007849
5.34	208.51	2.31913	.004796	7.20	1339.4	3.12692	.0007466
5.35	210.61	2.32348	0.004748	**7.25**	1408.1	3.14863	0.0007102
5.36	212.72	2.32782	.004701	7.30	1480.3	3.17035	.0006755
5.37	214.86	2.33216	.004654	7.35	1556.2	3.19206	.0006426
5.38	217.02	2.33650	.004608	7.40	1636.0	3.21378	.0006113
5.39	219.20	2.34085	.004562	7.45	1719.9	3.23549	.0005814
5.40	221.41	2.34519	0.004517	**7.50**	1808.0	3.25721	0.0005531
5.41	223.63	2.34953	.004472	7.55	1900.7	3.27892	.0005261
5.42	225.88	2.35388	.004427	7.60	1998.2	3.30064	.0005005
5.43	228.15	2.35822	.004383	7.65	2100.6	3.32235	.0004760
5.44	230.44	2.36256	.004339	7.70	2208.3	3.34407	.0004528
5.45	232.76	2.36690	0.004296	**7.75**	2321.6	3.36578	0.0004307
5.46	235.10	2.37125	.004254	7.80	2440.6	3.38750	.0004097
5.47	237.46	2.37559	.004211	7.85	2565.7	3.40921	.0003898
5.48	239.85	2.37993	.004169	7.90	2697.3	3.43093	.0003707
5.49	242.26	2.38428	.004128	7.95	2835.6	3.45264	.0003527
5.50	244.69	2.38862	0.004087	**8.00**	2981.0	3.47436	0.0003355

EXPONENTIAL FUNCTIONS (Continued)

x	e^x	$\text{Log}_{10}(e^x)$	e^{-x}
8.00	2981.0	3.47436	0.0003355
8.05	3133.8	3.49607	.0003191
8.10	3294.5	3.51779	.0003035
8.15	3463.4	3.53950	.0002887
8.20	3641.0	3.56121	.0002747
8.25	3827.6	3.58293	0.0002613
8.30	4023.9	3.60464	.0002485
8.35	4230.2	3.62636	.0002364
8.40	4447.1	3.64807	.0002249
8.45	4675.1	3.66979	.0002139
8.50	4914.8	3.69150	0.0002035
8.55	5166.8	3.71322	.0001935
8.00	5431.7	3.73493	.0001841
8.65	5710.1	3.75665	.0001751
8.70	6002.9	3.77836	.0001666
8.75	6310.7	3.80008	0.0001585
8.80	6634.2	3.82179	.0001507
8.85	6974.4	3.84351	.0001434
8.90	7332.0	3.86522	.0001364
8.95	7707.9	3.88694	.0001297
9.00	8103.1	3.90865	0.0001234
9.05	8518.5	3.93037	.0001174
9.10	8955.3	3.95208	.0001117
9.15	9414.4	3.97379	.0001062
9.20	9897.1	3.99551	.0001010
9.25	10405	4.01722	0.0000961
9.30	10938	4.03894	.0000914
9.35	11499	4.06065	.0000870
9.40	12088	4.08237	.0000827
9.45	12708	4.10408	.0000787
9.50	13360	4.12580	0.0000749
9.55	14045	4.14751	.0000712
9.60	14765	4.16923	.0000677
9.65	15522	4.19094	.0000644
9.70	16318	4.21266	.0000613
9.75	17154	4.23437	0.0000583
9.80	18034	4.25609	.0000555
9.85	18958	4.27780	.0000527
9.90	19930	4.29952	.0000502
9.95	20952	4.32123	0.0000477
10.00	22026	4.34294	0.0000454

RADIX TABLE OF THE EXPONENTIAL FUNCTION

x	n	$e^{[x(10^{-n})]}$	$e^{-[x(10^{-n})]}$
1	10	1.00000 00001 00000 00000 50000	0.99999 99999 00000 00000 50000
2	10	1.00000 00002 00000 00002 00000	0.99999 99998 00000 00002 00000
3	10	1.00000 00003 00000 00004 50000	0.99999 99997 00000 00004 50000
4	10	1.00000 00004 00000 00008 00000	0.99999 99996 00000 00008 00000
5	10	1.00000 00005 00000 00012 50000	0.99999 99995 00000 00012 50000
6	10	1.00000 00006 00000 00018 00000	0.99999 99994 00000 00018 00000
7	10	1.00000 00007 00000 00024 50000	0.99999 99993 00000 00024 50000
8	10	1.00000 00008 00000 00032 00000	0.99999 99992 00000 00032 00000
9	10	1.00000 00009 00000 00040 50000	0.99999 99991 00000 00040 50000
1	9	1.00000 00010 00000 00050 00000	0.99999 99990 00000 00050 00000
2	9	1.00000 00020 00000 00200 00000	0.99999 99980 00000 00200 00000
3	9	1.00000 00030 00000 00450 00000	0.99999 99970 00000 00450 00000
4	9	1.00000 00040 00000 00800 00000	0.99999 99960 00000 00800 00000
5	9	1.00000 00050 00000 01250 00000	0.99999 99950 00000 01250 00000
6	9	1.00000 00060 00000 01800 00000	0.99999 99940 00000 01800 00000
7	9	1.00000 00070 00000 02450 00001	0.99999 99930 00000 02449 99999
8	9	1.00000 00080 00000 03200 00001	0.99999 99920 00000 03199 99999
9	9	1.00000 00090 00000 04050 00001	0.99999 99910 00000 04049 99999
1	8	1.00000 00100 00000 05000 00002	0.99999 99900 00000 04999 99998
2	8	1.00000 00200 00000 20000 00013	0.99999 99800 00000 19999 99987
3	8	1.00000 00300 00000 45000 00045	0.99999 99700 00000 44999 99955
4	8	1.00000 00400 00000 80000 00107	0.99999 99600 00000 79999 99893
5	8	1.00000 00500 00001 25000 00208	0.99999 99500 00001 24999 99792
6	8	1.00000 00600 00001 80000 00360	0.99999 99400 00001 79999 99640
7	8	1.00000 00700 00002 45000 00572	0.99999 99300 00002 44999 99428
8	8	1.00000 00800 00003 20000 00853	0.99999 99200 00003 19999 99147
9	8	1.00000 00900 00004 05000 01215	0.99999 99100 00004 04999 98785
1	7	1.00000 01000 00005 00000 01667	0.99999 99000 00004 99999 98333
2	7	1.00000 02000 00020 00000 13333	0.99999 98000 00019 99999 86667
3	7	1.00000 03000 00045 00000 45000	0.99999 97000 00044 99999 55000
4	7	1.00000 04000 00080 00001 06667	0.99999 96000 00079 99998 93333
5	7	1.00000 05000 00125 00002 08333	0.99999 95000 00124 99997 91667
6	7	1.00000 06000 00180 00003 60000	0.99999 94000 00179 99996 40000
7	7	1.00000 07000 00245 00005 71667	0.99999 93000 00244 99994 28333
8	7	1.00000 08000 00320 00008 53334	0.99999 92000 00319 99991 46667
9	7	1.00000 09000 00405 00012 15000	0.99999 91000 00404 99987 85000
1	6	1.00000 10000 00500 00016 66667	0.99999 90000 00499 99983 33334
2	6	1.00000 20000 02000 00133 33340	0.99999 80000 01999 99866 66673
3	6	1.00000 30000 04500 00450 00034	0.99999 70000 04499 99550 00034
4	6	1.00000 40000 08000 01066 66773	0.99999 60000 07999 98933 33440
5	6	1.00000 50000 12500 02083 33594	0.99999 50000 12499 97916 66927
6	6	1.00000 60000 18000 03600 00540	0.99999 40000 17999 96400 00540
7	6	1.00000 70000 24500 05716 67667	0.99999 30000 24499 94283 34334
8	6	1.00000 80000 32000 08533 35040	0.99999 20000 31999 91466 68373
9	6	1.00000 90000 40500 12150 02734	0.99999 10000 40499 87850 02734

For $n > 10$, $e^{\pm[x(10^{-n})]} = 1 \pm x(10^{-n}) + \frac{1}{2}x^2(10^{-2n})$ to 25 decimals.

RADIX TABLE OF THE EXPONENTIAL FUNCTION

x	n	$e^{[x(10^{-n})]}$	$e^{-[x(10^{-n})]}$
1	5	1.00001 00000 50000 16666 70833	0.99999 00000 49999 83333 37500
2	5	1.00002 00002 00001 33334 00000	0.99998 00001 99998 66667 33333
3	5	1.00003 00004 50004 50003 37502	0.99997 00004 49995 50003 37498
4	5	1.00004 00008 00010 66677 33342	0.99996 00007 99989 33343 99991
5	5	1.00005 00012 50020 83359 37526	0.99995 00012 49979 16692 70807
6	5	1.00006 00018 00036 00054 00065	0.99994 00017 99964 00053 99935
7	5	1.00007 00024 50057 16766 70973	0.99993 00024 49942 83433 37360
8	5	1.00008 00032 00085 33504 00273	0.99992 00031 99914 66837 33060
9	5	1.00009 00040 50121 50273 37992	0.99991 00040 49878 50273 37008
1	4	1.00010 00050 00166 67083 34167	0.99990 00049 99833 33749 99167
2	4	1.00020 00200 01333 40000 26668	0.99980 00199 98666 73333 06668
3	4	1.00030 00450 04500 33752 02510	0.99970 00449 95500 33747 97510
4	4	1.00040 00800 10667 73341 86724	0.99960 00799 89334 39991 46724
5	4	1.00050 01250 20835 93776 04384	0.99950 01249 79169 27057 29384
6	4	1.00060 01800 36005 40064 80648	0.99940 01799 64005 39935 20648
7	4	1.00070 02450 57176 67223 40801	0.99930 02449 42843 33609 95801
8	4	1.00080 03200 85350 40273 10308	0.99920 03199 14683 73060 30307
9	4	1.00090 04051 21527 34242 14882	0.99910 04048 78527 33257 99880
1	3	1.00100 05001 66708 34166 80558	0.99900 04998 33374 99166 80554
2	3	1.00200 20013 34000 26675 55810	0.99800 19986 67333 06675 55302
3	3	1.00300 45045 03377 02601 29341	0.99700 44955 03372 97601 20662
4	3	1.00400 80106 77341 87235 88080	0.99600 79893 43991 47235 23064
5	3	1.00501 25208 59401 06338 35662	0.99501 24791 92682 31335 25642
6	3	1.00601 80360 54064 86485 55845	0.99401 79640 53935 26474 44988
7	3	1.00702 45572 66848 55523 16000	0.99302 44429 33235 10490 47970
8	3	1.00803 20855 04273 43117 20736	0.99203 19148 37060 63033 98697
9	3	1.00904 06217 73867 81406 25705	0.99104 03787 72883 66216 45648
1	2	1.01005 01670 84168 05754 21655	0.99004 98337 49168 05357 39060
2	2	1.02020 13400 26755 81016 01439	0.98019 86733 06755 30222 08141
3	2	1.03045 45339 53516 85561 24400	0.97044 55335 48508 17693 25284
4	2	1.04081 07741 92388 22675 70448	0.96078 94391 52323 20943 92107
5	2	1.05127 10963 76024 03969 75176	0.95122 94245 00714 00909 14253
6	2	1.06183 65465 45359 62222 46849	0.94176 45335 84248 70953 71528
7	2	1.07250 81812 54216 47905 31039	0.93239 38199 05948 22885 79726
8	2	1.08328 70676 74958 55443 59878	0.92311 63463 86635 78291 07598
9	2	1.09417 42837 05210 35787 28976	0.91393 11852 71228 18674 73535
1	1	1.10517 09180 75647 62481 17078	0.90483 74180 35959 57316 42491
2	1	1.22140 27581 60169 83392 10720	0.81873 07530 77981 85866 99355
3	1	1.34985 88075 76003 10398 37443	0.74081 82206 81717 86606 68738
4	1	1.49182 46976 41270 31782 48530	0.67032 00460 35639 30074 44329
5	1	1.64872 12707 00128 14684 86508	0.60653 06597 12633 42360 37995
6	1	1.82211 88003 90508 97487 53677	0.54881 16360 94026 43262 84589
7	1	2.01375 27074 70476 52162 45494	0.49658 53037 91409 51470 48001
8	1	2.22554 09284 92467 60457 95375	0.44932 89641 17221 59143 01024
9	1	2.45960 31111 56949 66380 01266	0.40656 96597 40599 11188 34542
1	0	2.71828 18284 59045 23536 02875	0.36787 94411 71442 32159 55238

For $n > 10$, $e^{\pm[x(10^{-n})]} = 1 \pm x(10^{-n}) + \frac{1}{2}x^2(10^{-2n})$ to 25 decimals.

HYPERBOLIC AND RELATED FUNCTIONS

Dr. Madhu S. Gupta

HYPERBOLIC FUNCTIONS

Geometrical Definition

Let O be the center, A the vertex, and P any point with coordinates (x,y) on the branch B′AB of the rectangular hyperbola $X^2 - Y^2 = a^2$. Set OM = x, MP = y and OA = a. The shaded area shown in the figure is given by

$$\text{Area OPAP}' = a^2 \log_e \frac{(x + y)}{a}$$

If the angle POP′ in hyperbolic radians is denoted by u,

$$u = \frac{\text{area OPAP}'}{a^2} \text{ hyperbolic radians.}$$

The hyperbolic functions are defined by

hyperbolic sine of u = sinh u = y/a
hyperbolic cosine of u = cosh u = x/a

The approximate length of the hyperbolic curve is given by

$$\text{arc AP} = \frac{3}{2} y - \frac{1}{2} \tan^{-1} y$$

while the straight line distance is

$$\text{line AP} = \sqrt{\sinh^2 u + (\cosh u - 1)^2}$$

Exponential Definition

$$\text{hyperbolic sine of } u = \sinh u = \frac{1}{2}(e^u - e^{-u}) \qquad \qquad \text{csch } u = \frac{1}{\sinh u},$$

$$\text{hyperbolic cosine of } u = \cosh u = \frac{1}{2}(e^u + e^{-u}) \qquad \qquad \text{sech } u = \frac{1}{\cosh u},$$

$$\text{hyperbolic tangent of } u = \tanh u = \frac{\sinh u}{\cosh u} = \frac{e^u - e^{-u}}{e^u + e^{-u}} \qquad \text{coth } u = \frac{1}{\tanh u},$$

Domain and Range for Real Argument

Function	Domain (interval of u)	Range (interval of function)	Remarks
sinh u	$(-\infty, +\infty)$	$(-\infty, +\infty)$	
cosh u	$(-\infty, +\infty)$	$[1, +\infty)$	
tanh u	$(-\infty, +\infty)$	$(-1, +1)$	
cosech u	$(-\infty, 0)$	$(0, -\infty)$	Two branches, pole
	$(0, +\infty)$	$(+\infty, 0)$	at $u = 0$.
sech u	$(-\infty, +\infty)$	$(0, 1]$	
coth u	$(-\infty, 0)$	$(-1, -\infty)$	Two branches, pole
	$(0, +\infty)$	$(+\infty, 1)$	at $u = 0$

Hyperbolic Functions In Terms of One Another

Function	sinh x	cosh x	tanh x
sinh $x =$	sinh x	$\pm \sqrt{\cosh^2 x - 1}$	$\dfrac{\tanh x}{\sqrt{1 - \tanh^2 x}}$
cosh $x =$	$\sqrt{1 + \sinh^2 x}$	cosh x	$\dfrac{1}{\sqrt{1 - \tanh^2 x}}$
tanh $x =$	$\dfrac{\sinh x}{\sqrt{1 + \sinh^2 x}}$	$\pm \dfrac{\sqrt{\cosh^2 x - 1}}{\cosh x}$	tanh x
cosech $x =$	$\dfrac{1}{\sinh x}$	$\pm \dfrac{1}{\sqrt{\cosh^2 x - 1}}$	$\dfrac{\sqrt{1 - \tanh^2 x}}{\tanh x}$
sech $x =$	$\dfrac{1}{\sqrt{1 + \sinh^2 x}}$	$\dfrac{1}{\cosh x}$	$\sqrt{1 - \tanh^2 x}$
coth $x =$	$\dfrac{\sqrt{1 + \sinh^2 x}}{\sinh x}$	$\dfrac{\pm \cosh x}{\sqrt{\cosh^2 x - 1}}$	$\dfrac{1}{\tanh x}$

Function	cosech x	sech x	coth x
sinh $x =$	$\dfrac{1}{\cosech x}$	$\pm \dfrac{\sqrt{1 - \sech^2 x}}{\sech x}$	$\dfrac{\pm 1}{\sqrt{\coth^2 x - 1}}$
cosh $x =$	$\pm \dfrac{\sqrt{\cosech^2 x + 1}}{\cosech x}$	$\dfrac{1}{\sech x}$	$\pm \dfrac{\coth x}{\sqrt{\coth^2 x - 1}}$
tanh $x =$	$\dfrac{1}{\sqrt{\cosech^2 x + 1}}$	$\pm \sqrt{1 - \sech^2 x}$	$\dfrac{1}{\coth x}$
cosech $x =$	cosech x	$\pm \dfrac{\sech x}{\sqrt{1 - \sech^2 x}}$	$\pm \dfrac{\sqrt{\coth^2 x - 1}}{1}$
sech $x =$	$\pm \dfrac{\cosec x}{\sqrt{\cosech^2 x + 1}}$	sech x	$\pm \dfrac{\sqrt{\coth^2 x - 1}}{\coth x}$
coth $x =$	$\sqrt{\cosech^2 x + 1}$	$\pm \dfrac{1}{\sqrt{1 - \sech^2 x}}$	coth x

Whenever two signs are shown, choose $+$ sign if x is positive, $-$ sign if x is negative.

Special Values of Hyperbolic Functions

x	0	$\dfrac{\pi}{2} i$	πi	$\dfrac{3\pi}{2} i$	∞
sinh x	0	i	0	$-i$	∞
cosh x	1	0	-1	0	∞
tanh x	0	∞i	0	$-\infty i$	1
csch x	∞	$-i$	∞	i	0
sech x	1	∞	-1	∞	0
coth x	∞	0	∞	0	1

Symmetry and Periodicity

$$\sinh (-u) = -\sinh u, \qquad \csch (-u) = -\csch u$$
$$\cosh (-u) = \cosh u, \qquad \sech (-u) = \sech u$$
$$\tanh (-u) = -\tanh u, \qquad \coth (-u) = -\coth u$$

sinh u, cosh u, cosech u and sech u are periodic with a period $2\pi i$; tanh u and coth u are periodic with a period πi.

Fundamental Identities

Reciprocal Relations

$$\text{csch } u = \frac{1}{\sinh u}, \quad \text{sech } u = \frac{1}{\cosh u}, \quad \coth u = \frac{1}{\tanh u}$$

Product Relations

$$\sinh u = \tanh u \cosh u \qquad \cosh u = \coth u \sinh u$$
$$\tanh u = \sinh u \text{ sech } u \qquad \coth u = \cosh u \text{ cosech } u$$
$$\text{sech } u = \text{cosech } u \tanh u \qquad \text{cosech } u = \text{sech } u \coth u$$

Quotient Relations

$$\sinh u = \frac{\tanh u}{\text{sech } u} \qquad \cosh u = \frac{\coth u}{\text{cosech } u} \qquad \tanh u = \frac{\sinh u}{\cosh u}$$

$$\text{cosech } u = \frac{\text{sech } u}{\tanh u} \qquad \text{sech } u = \frac{\text{cosech } u}{\coth u} \qquad \coth u = \frac{\cosh u}{\sinh u}$$

Relations Between Squares of Functions

$$\cosh^2 u - \sinh^2 u = 1, \qquad \tanh^2 u + \text{sech}^2 u = 1$$
$$\coth^2 u - \text{csch}^2 u = 1, \qquad \text{csch}^2 u - \text{sech}^2 u = \text{csch}^2 u \text{ sech}^2 u$$

Angle-Sum and Angle-Difference Relations

$$\sinh (u + v) = \sinh u \cosh v + \cosh u \sinh v$$
$$\sinh (u - v) = \sinh u \cosh v - \cosh u \sinh v$$
$$\cosh (u + v) = \cosh u \cosh v + \sinh u \sinh v$$
$$\cosh (u - v) = \cosh u \cosh v - \sinh u \sinh v$$

$$\tanh (u + v) = \frac{\tanh u + \tanh v}{1 + \tanh u \tanh v} = \frac{\sinh 2u + \sinh 2v}{\cosh 2u + \cosh 2v}$$

$$\tanh (u - v) = \frac{\tanh u - \tanh v}{1 - \tanh u \tanh v} = \frac{\sinh 2u - \sinh 2v}{\cosh 2u - \cosh 2v}$$

$$\coth (u + v) = \frac{1 + \coth u \coth v}{\coth u + \coth v} = \frac{\sinh 2u - \sinh 2v}{\cosh 2u - \cosh 2v}$$

$$\coth (u - v) = \frac{1 - \coth u \coth v}{\coth u - \coth v} = \frac{\sinh 2u + \sinh 2v}{\cosh 2u - \cosh 2v}$$

Multiple Angle Relations

$$\sinh 2u = 2 \sinh u \cosh u = \frac{2 \tanh u}{1 - \tanh^2 u}$$

$$\cosh 2u = \cosh^2 u + \sinh^2 u = 2 \cosh^2 u - 1 = 1 + 2 \sinh^2 u = \frac{1 + \tanh^2 u}{1 - \tanh^2 u}$$

$$\tanh 2u = \frac{2 \tanh u}{1 + \tanh^2 u}$$

$$\coth 2u = \frac{\coth^2 u + 1}{2 \coth u}$$

$$\sinh 3u = 3 \sinh u + 4 \sinh^3 u = \sinh u \, (4 \cosh^2 u - 1)$$
$$\cosh 3u = 4 \cosh^3 u - 3 \cosh u = \cosh u \, (1 + 4 \sinh^2 u)$$

$$\tanh 3u = \frac{3 \tanh u + \tanh^3 u}{1 + 3 \tanh^2 u}$$

$$\coth 3u = \frac{3 \coth u + \coth^3 u}{1 + 3 \coth^2 u}$$

$$\sinh 4u = 4 \sinh u \cosh u(2 \cosh^2 u - 1)$$
$$= 4 \sinh u \cosh u(1 + 2 \sinh^2 u)$$
$$= 4 \sinh u \cosh u(\cosh^2 u + \sinh^2 u)$$

$$\cosh 4u = 1 + 8 \cosh^2 u(\cosh^2 u - 1)$$
$$= 1 + 8 \sinh^2 u(\sinh^2 u + 1)$$
$$= \cosh^4 u + 6 \sinh^2 u \cosh^2 u + \sinh^4 u$$

$$\tanh 4u = \frac{4 \tanh u(1 + \tanh^2 u)}{1 + 6 \tanh^2 u + \tanh^4 u}$$

$$\coth 4u = \frac{\coth^4 u + 6 \coth^2 u + 1}{4 \coth u(\coth^2 u + 1)}$$

$$\sinh 5u = \sinh u(16 \sinh^4 u + 20 \sinh^2 u + 5)$$
$$= \sinh u(16 \cosh^4 u - 12 \cosh^2 u + 1)$$

$$\cosh 5u = \cosh u(16 \cosh^4 u - 20 \cosh^2 u + 5)$$
$$= \cosh u(16 \sinh^4 u + 12 \sinh^2 u + 1)$$

$$\sinh 6u = 2 \sinh u \cosh u(16 \cosh^4 u - 16 \cosh^2 u + 3)$$
$$= 2 \sinh u \cosh u(16 \sinh^4 u + 16 \sinh^2 u + 3)$$

$$\cosh 6u = 32 \cosh^6 u - 48 \cosh^4 u + 18 \cosh^2 u - 1$$
$$= 32 \sinh^6 u + 48 \sinh^4 u + 18 \sinh^2 u + 1$$

$$\sinh nu = \sinh u \left[(2 \cosh u)^{n-1} - \frac{(n-2)}{1!} \cdot (2 \cosh u)^{n-3} + \frac{(n-3)(n-4)}{2!} \right.$$
$$\left. \cdot (2 \cosh u)^{n-5} - \frac{(n-4)(n-5)(n-6)}{3!} (2 \cosh u)^{n-7} + \cdots \right].$$

$$\cosh nu = \frac{1}{2} \left[(2 \cosh u)^n - \frac{n}{1!}(2 \cosh u)^{n-2} + \frac{n(n-3)}{2!}(2 \cosh u)^{n-4} \right.$$
$$\left. - \frac{n(n-4)(n-5)}{3!}(2 \cosh u)^{n-6} + \cdots \right].$$

Half Angle Relations

$$\sinh \tfrac{1}{2}u = \pm \sqrt{\tfrac{1}{2}(\cosh u - 1)}$$
$$\cosh \tfrac{1}{2}u = \sqrt{\tfrac{1}{2}(\cosh u + 1)}$$

$$\tanh \frac{u}{2} = \frac{\sinh u}{1 + \cosh u} = \frac{\cosh u - 1}{\sinh u} = \pm \sqrt{\frac{\cosh u - 1}{\cosh u + 1}}$$

$$\coth \frac{u}{2} = \frac{1 + \cosh u}{\sinh u} = \frac{\sinh u}{\cosh u - 1} = \pm \sqrt{\frac{\cosh u + 1}{\cosh u - 1}}$$

Choose $+$ sign if u is positive, otherwise choose the $-$ sign.

Function Sum and Function Difference Relations

$$\sinh u + \sinh v = 2 \sinh \tfrac{1}{2}(u + v) \cosh \tfrac{1}{2}(u - v)$$
$$\sinh u - \sinh v = 2 \cosh \tfrac{1}{2}(u + v) \sinh \tfrac{1}{2}(u - v)$$
$$\cosh u + \cosh v = 2 \cosh \tfrac{1}{2}(u + v) \cosh \tfrac{1}{2}(u - v)$$
$$\cosh u - \cosh v = 2 \sinh \tfrac{1}{2}(u + v) \sinh \tfrac{1}{2}(u - v)$$

$$\tanh u + \tanh v = (1 + \tanh u \tanh v) \tanh (u + v) = \frac{\sinh (u + v)}{\cosh u \cosh v}$$

$$\tanh u - \tanh v = (1 - \tanh u \tanh v) \tanh (u - v) = \frac{\sinh (u - v)}{\cosh u \cosh v}$$

$$\coth u + \coth v = \frac{1 + \coth u \coth v}{\coth (u + v)} = \frac{\sinh (u + v)}{\sinh u \sinh v}$$

$$\coth u - \coth v = \frac{1 - \coth u \coth v}{\coth (u - v)} = \frac{\sinh (u - v)}{\sinh u \sinh v}$$

$$\sinh u + \cosh u = \frac{1 + \tanh \frac{1}{2}u}{1 - \tanh \frac{1}{2}u} = e^u \qquad \cosh u - \sinh u = \frac{1 - \tanh \frac{1}{2}u}{1 + \tanh \frac{1}{2}u} = e^{-u}$$

Function Product Relations

$\sinh u \cosh v = \frac{1}{2} \sinh (u + v) + \frac{1}{2} \sinh (u - v)$

$\cosh u \sinh v = \frac{1}{2} \sinh (u + v) - \frac{1}{2} \sinh (u - v)$

$\cosh u \cosh v = \frac{1}{2} \cosh (u + v) + \frac{1}{2} \cosh (u - v)$

$\sinh u \sinh v = \frac{1}{2} \cosh (u + v) - \frac{1}{2} \cosh (u - v)$

$\sinh (u + v) \sinh (u - v) = \sinh^2 u - \sinh^2 v = \cosh^2 u - \cosh^2 v$

$\cosh (u + v) \cosh (u - v) = \sinh^2 u + \cosh^2 v = \cosh^2 u + \sinh^2 v$

Power Relations

$\sinh^2 u = \frac{1}{2}(\cosh 2u - 1)$

$\cosh^2 u = \frac{1}{2}(\cosh 2u + 1)$

$\sinh^3 u = \frac{1}{4}(-3 \sinh u + \sinh 3u)$

$\cosh^3 u = \frac{1}{4}(3 \cosh u + \cosh 3u)$

$\sinh^4 u = \frac{1}{8}(3 - 4 \cosh 2u + \cosh 4u)$

$\cosh^4 u = \frac{1}{8}(3 + 4 \cosh 2u + \cosh 4u)$

$\sinh^5 u = \frac{1}{16}(10 \sinh u - 5 \sinh 3u + \sinh 5u)$

$\cosh^5 u = \frac{1}{16}(10 \cosh u + 5 \cosh 3u + \cosh 5u)$

$\sinh^6 u = \frac{1}{32}(-10 + 15 \cosh 2u - 6 \cosh 4u + \cosh 6u)$

$\cosh^6 u = \frac{1}{32}(10 + 15 \cosh 2u + 6 \cosh 4u + \cosh 6u)$

$(\cosh u \pm \sinh u)^n = \cosh nu \pm \sinh nu$

Relations with Circular Functions

$\sinh iu = i \sin u, \qquad\qquad \sinh u = -i \sin iu$

$\cosh iu = \cos u, \qquad\qquad \cosh u = \cos iu$

$\tanh iu = i \tan u, \qquad\qquad \tanh u = -i \tan iu$

$\operatorname{cosech} iu = -i \operatorname{cosec} u \qquad \operatorname{cosech} u = i \operatorname{cosec} iu$

$\operatorname{sech} iu = \sec u \qquad\qquad \operatorname{sech} u = \sec iu$

$\coth iu = -i \cot u \qquad\qquad \coth u = i \coth iu$

Hyperbolic Functions of Complex Argument

$\sinh (u + iv) = \sinh u \cos v + i \cosh u \sin v$

$\sinh (u - iv) = \sinh u \cos v - i \cosh u \sin v$

$\cosh (u + iv) = \cosh u \cos v + i \sinh u \sin v$

$\cosh (u - iv) = \cosh u \cos v - i \sinh u \sin v$

$$\tanh (u + iv) = \frac{\sinh 2u + i \sin 2v}{\cosh 2u + \cos 2v}$$

$$\tanh (u - iv) = \frac{\sinh 2u - i \sin 2v}{\cosh 2u + \cos 2v}$$

$$\coth (u + iv) = \frac{\sinh 2u - i \sin 2v}{\cosh 2u - \cos 2v}$$

$$\coth (u - iv) = \frac{\sinh 2u + i \sin 2v}{\cosh 2u - \cos 2v}$$

$$\sinh (u + \tfrac{1}{2}\pi i) = i \cosh u, \qquad \cosh (u + \tfrac{1}{2}\pi i) = i \sinh u$$
$$\sinh (u + \pi i) = -\sinh u, \qquad \cosh (u + \pi i) = -\cosh u$$
$$\sinh (u + 2\pi i) = \sinh u, \qquad \cosh (u + 2\pi i) = \cosh u$$

Series for Hyperbolic Functions

(see series expansions for $\sinh nu$ and $\cosh nu$ under multiple angle relations).

$$\sinh x = x + \frac{x^3}{3!} + \frac{x^5}{5!} + \frac{x^7}{7!} + \cdots + \frac{x^{2n+1}}{(2n+1)!} + \cdots \qquad |x| < \infty$$

$$\sinh ax = \frac{2}{\pi} \sinh \pi a \left[\frac{\sin x}{a^2 + 1^2} - \frac{2 \sin 2x}{a^2 + 2^2} + \frac{3 \sin 3x}{a^2 + 3^2} - + \cdots \right]$$

$$= \frac{2}{\pi} \sinh \pi a \sum_{n=1}^{\infty} (-1)^{n+1} \frac{n \sin nx}{n^2 + a^2}, \qquad |x| < \pi$$

$$\cosh x = 1 + \frac{x^2}{2!} + \frac{x^4}{4!} + \frac{x^6}{6!} + \cdots + \frac{x^{2n}}{(2n)!} + \cdots \qquad |x| < \infty$$

$$\cosh ax = \frac{2a}{\pi} \sinh \pi a \left[\frac{1}{2a^2} - \frac{\cos x}{a^2 + 1^2} + \frac{\cos 2x}{a^2 + 2^2} - \frac{\cos 3x}{a^2 + 3^2} + - \cdots \right]$$

$$= \frac{\sinh \pi a}{a\pi} + \frac{2a}{\pi} \sinh \pi a \sum_{n=1}^{\infty} (-1)^n \frac{\cos nx}{n^2 + a^2}, \qquad |x| < \pi$$

$$\tanh x = x - \frac{1}{3} x^3 + \frac{2}{15} x^5 - \frac{17}{315} x^7 + \frac{62}{2835} x^9 - \cdots$$
$$+ \frac{2^{2n}(2^{2n} - 1) B_{2n} x^{2n-1}}{(2n)!} \pm \cdots \quad \text{(1)} \qquad |x| < \frac{\pi}{2}$$

$$\tanh x = 1 - 2e^{-2x} + 2e^{-4x} - 2e^{-6x} + - \cdots$$

$$= 1 + 2 \sum_{n=1}^{\infty} (-1)^n e^{-2nx}, \qquad \text{Re} (x) > 0$$

$$\tanh x = 2x \left[\frac{1}{\left(\frac{\pi}{2}\right)^2 + x^2} + \frac{1}{\left(\frac{3\pi}{2}\right)^2 + x^2} + \frac{1}{\left(\frac{5\pi}{2}\right)^2 + x^2} + \cdots \right]$$

$$= 2x \sum_{n=0}^{\infty} \frac{1}{\left(n + \frac{1}{2}\right)^2 \pi^2 + x^2}$$

$$\coth x = \frac{1}{x} + \frac{x}{3} - \frac{x^3}{45} + \frac{2x^5}{945} - \frac{x^7}{4725} + \cdots + \frac{2^{2n}}{(2n)!} B_{2n} x^{2n-1} \pm \cdots \text{ (1)}$$
$$0 < |x| < \pi$$

$$\coth x = 1 + 2e^{-2x} + 2e^{-4x} + 2e^{-6x} + \cdots$$

$$= 1 + 2 \sum_{n=1}^{\infty} e^{-2nx} \qquad \text{Re} (x) > 0$$

$$\coth x = \frac{1}{x} + 2x \left[\frac{1}{\pi^2 + x^2} + \frac{1}{(2\pi)^2 + x^2} + \frac{1}{(3\pi)^2 + x^2} + \cdots \right]$$

$$= \frac{1}{x} + 2x \sum_{n=1}^{\infty} \frac{1}{(n\pi)^2 + x^2}$$

(1) B_{2n} denotes Bernoulli numbers.

$$\text{sech } x = 1 - \frac{1}{2!} x^2 + \frac{5}{4!} x^4 - \frac{61}{6!} x^6 + \frac{1385}{8!} x^8 - \cdots + \frac{E_{2n}x^{2n}}{(2n)!} \pm \cdots \quad (2)$$

$$|x| < \frac{\pi}{2}$$

$$\text{sech } x = 2e^{-x} - 2e^{-3x} + 2e^{-5x} - 2e^{-7x} + - \cdots$$

$$= 2 \sum_{n=0}^{\infty} (-1)^n e^{-(2n+1)x}, \text{ Re }(x) > 0$$

$$\text{sech } x = 4\pi \left[\frac{1}{\pi^2 + 4x^2} - \frac{3}{(3\pi)^2 + 4x^2} + \frac{5}{(5\pi)^2 + 4x^2} - + \cdots \right]$$

$$= 4\pi \sum_{n=0}^{\infty} \frac{(-1)^n (2n+1)}{(2n+1)^2 \pi^2 + 4x^2}$$

$$\text{cosech } x = \frac{1}{x} - \frac{x}{6} + \frac{7x^3}{360} - \frac{31x^5}{15,120} + \cdots - \frac{2(2^{2n-1} - 1)}{(2n)!} B_{2n} x^{2n-1} + \cdots \quad (1)$$

$$0 < |x| < \pi$$

$$\text{cosech } x = 2e^{-x} + 2e^{-3x} + 2e^{-5x} + \cdots$$

$$= 2 \sum_{n=0}^{\infty} e^{-(2n+1)x}, \text{ Re }(x) > 0$$

$$\text{cosech } x = \frac{1}{x} - \frac{2x}{\pi^2 + x^2} + \frac{2x}{(2\pi)^2 + x^2} - \frac{2x}{(3\pi)^2 + x^2} + - \cdots$$

$$= \frac{1}{x} + 2x \sum_{n=1}^{\infty} \frac{(-1)^n}{(n\pi)^2 + x^2}$$

(2) E_{2n} denotes Euler numbers.

INVERSE HYPERBOLIC FUNCTIONS

Definitions

If $x = \sinh y$, then $y = \sinh^{-1} x$.   Other inverse functions are denned similarly.

Domain and Range

Function	Domain	Range	Remarks
$\sinh^{-1} x$	$(-\infty, +\infty)$	$(-\infty, +\infty)$	odd function
$\cosh^{-1} x$	$[1, +\infty)$	$(-\infty, +\infty)$	even function double valued
$\tanh^{-1} x$	$(-1, 1)$	$(-\infty, +\infty)$	odd function
$\operatorname{cosech}^{-1} x$	$(-\infty, 0), (0, \infty)$	$(0, -\infty), (\infty, 0)$	odd function two branches, pole at $x = 0$
$\operatorname{sech}^{-1} x$	$(0, 1]$	$(-\infty, +\infty)$	double valued
$\coth^{-1} x$	$(-\infty, -1), (1, \infty)$	$(0, -\infty), (\infty, 0)$	odd function two branches

Graph of the Inverse Functions

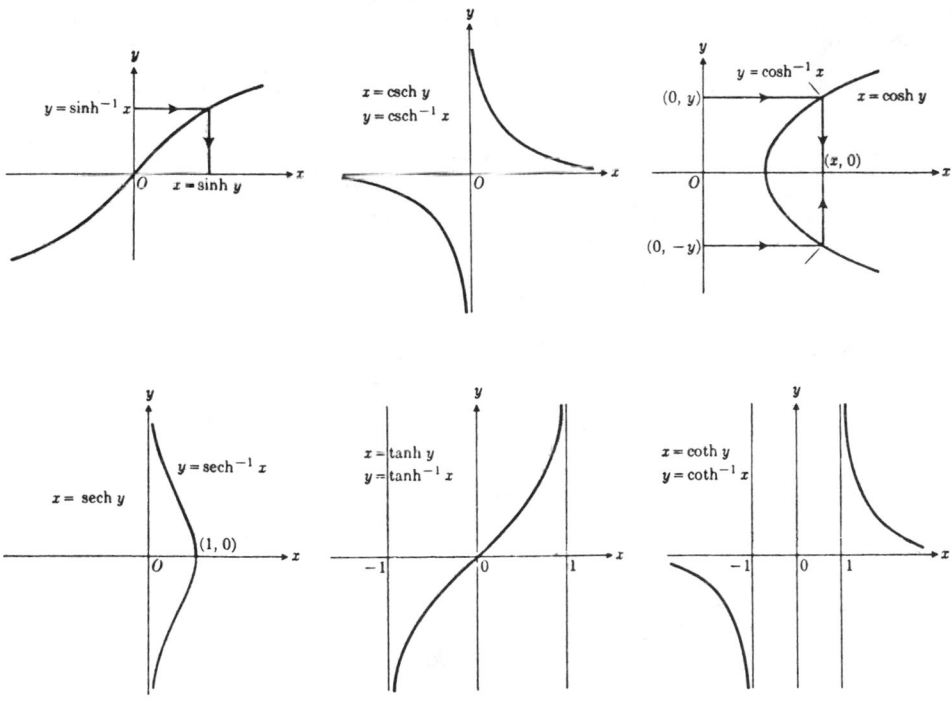

Inverse Hyperbolic Functions In Terms of One Another

Function	$\sinh^{-1} x$	$\cosh^{-1} x$ $+$ if $x > 0$ $-$ if $x < 0$	$\tanh^{-1} x$
$\sinh^{-1} x =$	$\sinh^{-1} x$	$\pm \cosh^{-1} \sqrt{x^2 + 1}$	$\tanh^{-1} \dfrac{x}{\sqrt{1 + x^2}}$
$\cosh^{-1} x =$	$\pm \sinh^{-1} \sqrt{x^2 - 1}$	$\cosh^{-1} x$	$\pm \tanh^{-1} \dfrac{\sqrt{x^2 - 1}}{x}$
$\tanh^{-1} x =$	$\sinh^{-1} \dfrac{x}{\sqrt{1 - x^2}}$	$\pm \cosh^{-1} \dfrac{1}{\sqrt{1 - x^2}}$	$\tanh^{-1} x$
$\operatorname{cosech}^{-1} x =$	$\sinh^{-1} \dfrac{1}{x}$	$\pm \cosh^{-1} \dfrac{\sqrt{x^2 + 1}}{x}$	$\tanh^{-1} \dfrac{1}{\sqrt{1 + x^2}}$
$\operatorname{sech}^{-1} x =$	$\pm \sinh^{-1} \dfrac{\sqrt{1 - x^2}}{x}$	$\cosh^{-1} \dfrac{1}{x}$	$\pm \tanh^{-1} \sqrt{1 - x^2}$
$\coth^{-1} x =$	$\sinh^{-1} \dfrac{1}{\sqrt{x^2 - 1}}$	$\pm \cosh^{-1} \dfrac{x}{\sqrt{x^2 - 1}}$	$\tanh^{-1} \dfrac{1}{x}$

Function	$\operatorname{cosech}^{-1} x$	$\operatorname{sech}^{-1} x$ $+$ if $x > 0$ $-$ if $x < 0$	$\coth^{-1} x$
$\sinh^{-1} x =$	$\operatorname{cosech}^{-1} \dfrac{1}{x}$	$\pm \operatorname{sech}^{-1} \dfrac{1}{\sqrt{1 + x^2}}$	$\coth^{-1} \dfrac{\sqrt{1 + x^2}}{x}$
$\cosh^{-1} x =$	$\pm \operatorname{cosech}^{-1} \dfrac{1}{\sqrt{x^2 - 1}}$	$\operatorname{sech}^{-1} \dfrac{1}{x}$	$\pm \coth^{-1} \dfrac{x}{\sqrt{x^2 - 1}}$
$\tanh^{-1} x =$	$\operatorname{cosech}^{-1} \dfrac{\sqrt{1 - x^2}}{x}$	$\pm \operatorname{sech}^{-1} \sqrt{1 - x^2}$	$\coth^{-1} \dfrac{1}{x}$
$\operatorname{cosech}^{-1} x =$	$\operatorname{cosech}^{-1} x$	$\pm \operatorname{sech}^{-1} \dfrac{x}{\sqrt{x^2 + 1}}$	$\coth^{-1} \sqrt{1 + x^2}$
$\operatorname{sech}^{-1} x =$	$\pm \operatorname{cosech}^{-1} \dfrac{x}{\sqrt{1 - x^2}}$	$\operatorname{sech}^{-1} x$	$\pm \coth^{-1} \dfrac{1}{\sqrt{1 - x^2}}$
$\coth^{-1} x =$	$\operatorname{cosech}^{-1} \sqrt{x^2 - 1}$	$\operatorname{sech}^{-1} \dfrac{\sqrt{x^2 - 1}}{x}$	$\coth^{-1} x$

Fundamental Identities

Relations with Logarithmic Functions

$$\sinh^{-1} x = \log_e (x + \sqrt{x^2 + 1})$$

$$\cosh^{-1} x = \log_e (x \pm \sqrt{x^2 - 1}), \quad x \geqq 1. \quad \text{The plus sign is used for the principal value.}$$

$$\tanh^{-1} x = \tfrac{1}{2} \log_e \left(\frac{1 + x}{1 - x}\right), \quad x^2 < 1$$

$$\operatorname{csch}^{-1} x = \log_e \left(\frac{1 \pm \sqrt{1 + x^2}}{x}\right). \quad \text{The plus sign is used if } x > 0, \text{ the minus sign if } x < 0.$$

$$\operatorname{sech}^{-1} x = \log_e \left(\frac{1 \pm \sqrt{1 - x^2}}{x}\right), \quad 0 < x \leq 1. \quad \text{The plus sign is used for the principal values.}$$

$$\coth^{-1} x = \tfrac{1}{2} \log_e \left(\frac{x + 1}{x - 1}\right), \quad x^2 > 1$$

Relations with Circular Functions

$$\sinh^{-1} x = -i \sin^{-1} (ix) \qquad \sinh^{-1} (ix) = i \sin^{-1} x$$
$$\cosh^{-1} x = \pm i \cos^{-1} x \qquad \cosh^{-1} (ix) = \pm i \cos^{-1} (ix)$$
$$\tanh^{-1} x = -i \tan^{-1} (ix) \qquad \tanh^{-1} (ix) = i \tan^{-1} x$$
$$\operatorname{cosech}^{-1} x = i \operatorname{cosec}^{-1} (ix) \qquad \operatorname{cosech}^{-1} (ix) = -i \operatorname{cosec}^{-1} x$$
$$\operatorname{sech}^{-1} x = \pm i \sec^{-1} x \qquad \operatorname{sech}^{-1} (ix) = \pm i \sec^{-1} (ix)$$
$$\coth^{-1} x = i \coth^{-1} (ix) \qquad \coth^{-1} (ix) = -i \cot^{-1} (x)$$

Function Sum and Function Difference Relations

$$\sinh^{-1} x + \sinh^{-1} y = \sinh^{-1} (x \sqrt{1 + y^2} + y \sqrt{1 + x^2})$$

$$\sinh^{-1} x - \sinh^{-1} y = \sinh^{-1} (x \sqrt{1 + y^2} - y \sqrt{1 + x^2})$$

$$\cosh^{-1} x + \cosh^{-1} y = \cosh^{-1} (xy + \sqrt{(x^2 - 1)(y^2 - 1)})$$

$$\cosh^{-1} x - \cosh^{-1} y = \cosh^{-1} (xy - \sqrt{(x^2 - 1)(y^2 - 1)})$$

$$\tanh^{-1} x + \tanh^{-1} y = \tanh^{-1} \left(\frac{x + y}{1 + xy}\right)$$

$$\tanh^{-1} x - \tanh^{-1} y = \tanh^{-1} \left(\frac{x - y}{1 - xy}\right)$$

$$\sinh^{-1} x + \cosh^{-1} y = \sinh^{-1} (xy + \sqrt{(1 + x^2)(y^2 - 1)})$$
$$= \cosh^{-1} (y \sqrt{1 + x^2} + x \sqrt{y^2 - 1})$$

$$\sinh^{-1} x - \cosh^{-1} y = \sinh^{-1} (xy - \sqrt{(1 + x^2)(y^2 - 1)})$$
$$= \cosh^{-1} (y \sqrt{1 + x^2} - x \sqrt{y^2 - 1})$$

$$\tanh^{-1} x + \coth^{-1} y = \tanh^{-1} \left(\frac{xy + 1}{y + x}\right)$$
$$= \coth^{-1} \left(\frac{y + x}{xy + 1}\right)$$

$$\tanh^{-1} x - \coth^{-1} y = \tanh^{-1} \left(\frac{xy - 1}{y - x}\right)$$
$$= \coth^{-1} \left(\frac{y - x}{xy - 1}\right)$$

Series Expansions

$$\sinh^{-1} x = x - \frac{1}{2 \cdot 3} x^3 + \frac{1 \cdot 3}{2 \cdot 4 \cdot 5} x^5 - \frac{1 \cdot 3 \cdot 5}{2 \cdot 4 \cdot 6 \cdot 7} x^7 + \cdots$$

$$+ (-1)^n \cdot \frac{1 \cdot 3 \cdot 5 \ldots (2n-1)}{2 \cdot 4 \cdot 6 \ldots 2n(2n+1)} x^{2n+1} \pm \cdots \qquad |x| < 1$$

$$\sinh^{-1} x = \ln(2x) + \frac{1}{2} \cdot \frac{1}{2x^2} - \frac{1 \cdot 3}{2 \cdot 4} \frac{1}{4x^4} + \frac{1 \cdot 3 \cdot 5}{2 \cdot 4 \cdot 6} \cdot \frac{1}{6x^6} - \cdots$$

$$= \ln(2x) + \sum_{n=1}^{\infty} (-1)^{n+1} \frac{(2n)! x^{-2n}}{2^{2n}(n!)^2 2n}, \qquad |x| > 1$$

$$\cosh^{-1} x = \pm \left[\ln(2x) - \frac{1}{2 \cdot 2x^2} - \frac{1 \cdot 3}{2 \cdot 4 \cdot 4x^4} - \frac{1 \cdot 3 \cdot 5}{2 \cdot 4 \cdot 6 \cdot 6x^6} - \cdots \right] \qquad x > 1$$

$$\operatorname{cosech}^{-1} x = \frac{1}{x} - \frac{1}{2} \cdot \frac{1}{3x^3} + \frac{1 \cdot 3}{2 \cdot 4} \cdot \frac{1}{5x^5} - \frac{1 \cdot 3 \cdot 5}{2 \cdot 4 \cdot 6} \cdot \frac{1}{7x^7} + - \cdots$$

$$= \sum_{n=0}^{\infty} (-1)^n \frac{(2n)! x^{-2n-1}}{2^{2n}(n!)^2(2n+1)}, \qquad |x| > 1$$

$$\operatorname{cosech}^{-1} x = \ln\frac{2}{x} + \frac{1}{2} \cdot \frac{x^2}{2} - \frac{1 \cdot 3}{2 \cdot 4} \cdot \frac{x^4}{4} + \frac{1 \cdot 3 \cdot 5}{2 \cdot 4 \cdot 6} \cdot \frac{x^6}{6} - + \cdots$$

$$= \ln\frac{2}{x} + \sum_{n=1}^{\infty} \frac{(-1)^{n+1}(2n)! x^{2n}}{2^{2n}(n!)^2 2n}, \qquad 0 < x < 1$$

$$\operatorname{sech}^{-1} x = \ln\frac{2}{x} - \frac{1}{2} \frac{x^2}{2} - \frac{1 \cdot 3}{2 \cdot 4} \cdot \frac{x^4}{4} - \frac{1 \cdot 3 \cdot 5}{2 \cdot 4 \cdot 6} \cdot \frac{x^6}{6} - - \cdots$$

$$= \ln\frac{2}{x} - \sum_{n=1}^{\infty} \frac{(2n)! x^{2n}}{2^{2n}(n!)^2 2n}, \qquad 0 < x < 1$$

$$* \ \tanh^{-1} x = x + \frac{x^3}{3} + \frac{x^5}{5} + \frac{x^7}{7} + \cdots + \frac{x^{2n+1}}{2n+1} + \cdots \qquad |x| < 1$$

$$* \ \coth^{-1} x = \frac{1}{x} + \frac{1}{3x^3} + \frac{1}{5x^5} + \frac{1}{7x^7} + \cdots + \frac{1}{(2n+1)x^{2n+1}} + \cdots \qquad |x| > 1$$

GUDERMANNIAN FUNCTION

This function is useful because it relates circular and hyperbolic functions without the use of functions of imaginary argument.

Definition

$$\gamma = \text{the gudermannian of } x, \text{ written as } \mathbf{gd}\ x$$

$$= 2 \tan^{-1} e^x - \frac{\pi}{2}$$

$$x = \mathbf{gd}^{-1}\ \gamma$$

$$= \ln \tan\left(\frac{\pi}{4} + \frac{\gamma}{2}\right)$$

$$= \ln(\sec \gamma + \tan \gamma)$$

* Can also be written as arg tanh x and arg coth x respectively.

Special Values of Gudermannian Function

$$
\begin{array}{cccc}
x & 0 & \infty & -\infty \\
\text{gd } x & 0 & \tfrac{1}{2}\pi & -\tfrac{1}{2}\pi
\end{array}
$$

Relations with Hyperbolic and Circular Functions

$$\sinh x = \tan (\text{gd } x) \qquad \text{cosech } x = \cot (\text{gd } x)$$
$$\cosh x = \sec (\text{gd } x) \qquad \text{sech } (x) = \cos (\text{gd } x)$$
$$\tanh x = \sin (\text{gd } x) \qquad \coth (x) = \text{cosec } (\text{gd } x)$$

Derivatives

$$\frac{d}{dx} (\text{gd } x) = \text{sech } x$$

$$\frac{d}{dx} (\text{gd}^{-1} \gamma) = \sec \gamma$$

Fundamental Identities

$$\tanh (\tfrac{1}{2}x) = \tan (\tfrac{1}{2}\text{gd } x)$$
$$e^x = \cosh x + \sinh x = \sec (\text{gd } x) + \tan (\text{gd } x)$$
$$= \tan \left(\frac{\pi}{4} + \frac{1}{2} \text{gd } x \right) = \frac{1 + \sin (\text{gd } x)}{\cos (\text{gd } x)} = \frac{1 + \tan (\tfrac{1}{2}\text{gd } x)}{1 - \tan (\tfrac{1}{2}\text{gd } x)}$$
$$\text{gd } x = 2 \tan^{-1} \left(\tanh \frac{1}{2} x \right) = \int_0^x \frac{dt}{\cosh t}$$
$$\text{gd}^{-1} \gamma = \int_0^\gamma \frac{dt}{\cos t}$$
$$i \, \text{gd}^{-1} \gamma = \text{gd } i\gamma, \text{ where } i = \sqrt{-1}.$$

If

$$\alpha + i\beta = \text{gd}(x + iy)$$

then

$$\tan \alpha = \frac{\sinh x}{\cos y} \qquad \tanh \beta = \frac{\sin y}{\cosh x}$$

$$\tanh x = \frac{\sin \alpha}{\cosh \beta} \qquad \tan y = \frac{\sinh \beta}{\cosh \alpha}$$

Series Expansions

$$\text{gd } x = x - \frac{1}{6} x^3 + \frac{1}{24} x^5 - \frac{61}{5040} x^7 + - \cdots$$

$$= \sum_{n=0}^{\infty} \frac{E_{2n}}{(2n + 1)!} x^{2n+1}, \quad |x| < 1 *$$

$$\text{gd } x = \frac{\pi}{2} - \text{sech } x - \frac{1}{2} \frac{\text{sech}^3 x}{3} - \frac{1 \cdot 3}{2 \cdot 4} \frac{\text{sech}^5 x}{5} + - \cdots$$

$$= \frac{\pi}{2} - \sum_{n=0}^{\infty} \frac{(2n)!}{2^{2n}(n!)^2} \frac{\text{sech}^{2n+1} x}{(2n + 1)} \qquad x \text{ large}$$

*E_n denotes Euler numbers.

$$\text{gd } x = \frac{2}{1} \tanh \frac{x}{2} - \frac{2}{3} \tanh^3 \frac{x}{2} + \frac{2}{5} \tanh^5 \frac{x}{2} - \frac{2}{7} \tanh^7 \frac{x}{2} + - \cdots$$

$$= 2 \sum_{n=0}^{\infty} \frac{(-1)^n}{2n+1} \tanh^{2n+1} \frac{x}{2}$$

$$\text{gd}^{-1} \gamma = \gamma + \frac{1}{6} \gamma^3 + \frac{1}{24} \gamma^5 + \frac{61}{5040} \gamma^7 + \cdots$$

$$= \sum_{n=0}^{\infty} \frac{(-1)^n E_{2n} \gamma^{2n+1}}{(2n+1)!}, \qquad |\gamma| < \frac{\pi}{2} \ ^*$$

$$\text{gd}^{-1} \gamma = \frac{2}{1} \tan \frac{\gamma}{2} + \frac{2}{3} \tan^3 \frac{\gamma}{2} + \frac{2}{5} \tan^5 \frac{\gamma}{2} + \cdots$$

$$= 2 \sum_{n=0}^{\infty} \frac{1}{2n+1} \tan^{2n+1} \frac{\gamma}{2}$$

*E_n are Euler's numbers.

References

1. For extensive tables of inverse hyperbolic functions: Tables of Inverse Hyperbolic Functions, Harvard University Computation Laboratory (Cambridge, Harvard University Press, 1949).
2. For tables of hyperbolic functions with complex argument: Tables of Complex Hyperbolic and Circular Functions, A. E. Kennelly (Cambridge, Harvard University Press, 1914).

HYPERBOLIC FUNCTIONS AND THEIR COMMON LOGARITHMS

The logarithms given below show the mantissa only. The proper characteristic must be added.

x	Sinh x Value	Sinh x $\log_{10}$	Cosh x Value	Cosh x $\log_{10}$	Tanh x Value	Tanh x $\log_{10}$	Coth x Value	Coth x $\log_{10}$
0.00	0.00000	$-\infty$	1.00000	.00000	0.00000	$-\infty$	∞	∞
0.01	.01000	.00001	1.00005	.00002	.01000	.99999	100.003	.00001
0.02	.02000	.30106	1.00020	.00009	.02000	.30097	50.007	.69903
0.03	.03000	.47719	1.00045	.00020	.02999	.47699	33.343	.52301
0.04	.04001	.60218	1.00080	.00035	.03998	.60183	25.013	.39817
0.05	0.05002	.69915	1.00125	.00054	0.04996	.69861	20.017	.30139
0.06	.06004	.77841	1.00180	.00078	.05993	.77763	16.687	.22237
0.07	.07006	.84545	1.00245	.00106	.06989	.84439	14.309	.15561
0.08	.08009	.90355	1.00320	.00139	.07983	.90216	12.527	.09784
0.09	.09012	.95483	1.00405	.00176	.08976	.95307	11.141	.04693
0.10	0.10017	.00072	1.00500	.00217	0.09967	.99856	10.0333	.00144
0.11	.11022	.04227	1.00606	.00262	.10956	.03965	9.1275	.96035
0.12	.12029	.08022	1.00721	.00312	.11943	.07710	8.3733	.92290
0.13	.13037	.11517	1.00846	.00366	.12927	.11151	7.7356	.88849
0.14	.14046	.14755	1.00982	.00424	.13909	.14330	7.1895	.85670
0.15	0.15056	.17772	1.01127	.00487	0.14889	.17285	6.7166	.82715
0.16	.16068	.20597	1.01283	.00554	.15865	.20044	6.3032	.79956
0.17	.17082	.23254	1.01448	.00625	.16838	.22629	5.9389	.77371
0.18	.18097	.25762	1.01624	.00700	.17808	.25062	5.6154	.74938
0.19	.19115	.28136	1.01810	.00779	.18775	.27357	5.3263	.72643
0.20	0.20134	.30392	1.02007	.00863	0.19738	.29529	5.0665	.70471
0.21	.21155	.32541	1.02213	.00951	.20697	.31590	4.8317	.68410
0.22	.22178	.34592	1.02430	.01043	.21652	.33540	4.6180	.66451
0.23	.23203	.36555	1.02657	.01139	.22603	.35416	4.4242	.64584
0.24	.24231	.38437	1.02894	.01239	.23550	.37198	4.2464	.62802
0.25	0.25261	.40245	1.03141	.01343	0.24492	.38902	4.0830	.61098
0.26	.26294	.41986	1.03399	.01452	.25430	.40534	3.9324	.59466
0.27	.27329	.43663	1.03667	.01564	.26362	.42099	3.7933	.57901
0.28	.28367	.45282	1.03946	.01681	.27291	.43601	3.6643	.56399
0.29	.29408	.46847	1.04235	.01801	.28213	.45046	3.5444	.54954
0.30	0.30452	.48362	1.04534	.01926	0.29131	.46436	3.4327	.53564
0.31	.31499	.49830	1.04844	.02054	.30044	.47775	3.3285	.52225
0.32	.32549	.51254	1.05164	.02187	.30951	.49067	3.2309	.50933
0.33	.33602	.52637	1.05495	.02323	.31852	.50314	3.1395	.49686
0.34	.34659	.53981	1.05836	.02463	.32748	.51518	3.0536	.48482
0.35	0.35719	.55290	1.06188	.02607	0.33638	.52682	2.9729	.47318
0.36	.36783	.56564	1.06550	.02755	.34521	.53809	2.8968	.46191
0.37	.37850	.57807	1.06923	.02907	.35399	.54899	2.8249	.45101
0.38	.38921	.59019	1.07307	.03063	.36271	.55956	2.7570	.44044
0.39	.39996	.60202	1.07702	.03222	.37136	.56980	2.6928	.43020
0.40	0.41075	.61358	1.08107	.03385	0.37995	.57973	2.6319	.42027
0.41	.42158	.62488	1.08523	.03552	.38847	.58936	2.5742	.41064
0.42	.43246	.63594	1.08950	.03723	.39693	.59871	2.5193	.40129
0.43	.44337	.64677	1.09388	.03897	.40532	.60780	2.4672	.39220
0.44	.45434	.65738	1.09837	.04075	.41364	.61663	2.4175	.38337
0.45	0.46534	.66777	1.10297	.04256	0.42190	.62521	2.3702	.37479
0.46	.47640	.67797	1.10768	.04441	.43008	.63355	2.3251	.36645
0.47	.48750	.68797	1.11250	.04630	.43820	.64167	2.2821	.35833
0.48	.49865	.69779	1.11743	.04822	.44624	.64957	2.2409	.35043
0.49	.50984	.70744	1.12247	.05018	.45422	.65726	2.2016	.34274

HYPERBOLIC FUNCTIONS AND THEIR COMMON LOGARITHMS (Continued)

x	Sinh x Value	Sinh x $\log_{10}$	Cosh x Value	Cosh x $\log_{10}$	Tanh x Value	Tanh x $\log_{10}$	Coth x Value	Coth x $\log_{10}$
0.50	0.52110	.71692	1.12763	.05217	0.46212	.66475	2.1640	.33525
0.51	.53240	.72624	1.13289	.05419	.46995	.67205	2.1279	.32795
0.52	.54375	.73540	1.13827	.05625	.47770	.67916	2.0934	.32084
0.53	.55516	.74442	1.14377	.05834	.48538	.68608	2.0602	.31392
0.54	.56663	.75330	1.14938	.06046	.49299	.69284	2.0284	.30716
0.55	0.57815	.76204	1.15510	.06262	0.50052	.69942	1.9979	.30058
0.56	.58973	.77065	1.16094	.06481	.50798	.70584	1.9686	.29416
0.57	.60137	.77914	1.16690	.06703	.51536	.71211	1.9404	.28789
0.58	.61307	.78751	1.17297	.06929	.52267	.71822	1.9133	.28178
0.59	.62483	.79576	1.17916	.07157	.52990	.72419	1.8872	.27581
0.60	0.63665	.80390	1.18547	.07389	0.53705	.73001	1.8620	.26999
0.61	.64854	.81194	1.19189	.07624	.54413	.73570	1.8378	.26430
0.62	.66049	.81987	1.19844	.07861	.55113	.74125	1.8145	.25875
0.63	.67251	.82770	1.20510	.08102	.55805	.74667	1.7919	.25333
0.64	.68459	.83543	1.21189	.08346	.56490	.75197	1.7702	.24803
0.65	0.69675	.84308	1.21879	.08593	0.57167	.75715	1.7493	.24285
0.66	.70897	.85063	1.22582	.08843	.57836	.76220	1.7290	.23780
0.67	.72126	.85809	1.23297	.09095	.58498	.76714	1.7095	.23286
0.68	.73363	.86548	1.24025	.09351	.59152	.77197	1.6906	.22803
0.69	.74607	.87278	1.24765	.09609	.59798	.77669	1.6723	.22331
0.70	0.75858	.88000	1.25517	.09870	0.60437	.78130	1.6546	.21870
0.71	.77117	.88715	1.26282	.10134	.61068	.78581	1.6375	.21419
0.72	.78384	.89423	1.27059	.10401	.61691	.79022	1.6210	.20978
0.73	.79659	.90123	1.27849	.10670	.62307	.79453	1.6050	.20547
0.74	.80941	.90817	1.28652	.10942	.62915	.79875	1.5895	.20125
0.75	0.82232	.91504	1.29468	.11216	0.63515	.80288	1.5744	.19712
0.76	.83530	.92185	1.30297	.11493	.64108	.80691	1.5599	.19309
0.77	.84838	.92859	1.31139	.11773	.64693	.81086	1.5458	.18914
0.78	.86153	.93527	1.31994	.12055	.65271	.81472	1.5321	.18528
0.79	.87478	.94190	1.32862	.12340	.65841	.81850	1.5188	.18150
0.80	0.88811	.94846	1.33743	.12627	0.66404	.82219	1.5059	.17781
0.81	.90152	.95498	1.34638	.12917	.66959	.82581	1.4935	.17419
0.82	.91503	.96144	1.35547	.13209	.67507	.82935	1.4813	.17065
0.83	.92863	.96784	1.36468	.13503	.68048	.83281	1.4696	.16719
0.84	.94233	.97420	1.37404	.13800	.68581	.83620	1.4581	.16380
0.85	0.95612	.98051	1.38353	.14099	0.69107	.83952	1.4470	.16048
0.86	.97000	.98677	1.39316	.14400	.69626	.84277	1.4362	.15723
0.87	.98398	.99299	1.40293	.14704	.70137	.84595	1.4258	.15405
0.88	.99806	.99916	1.41284	.15009	.70642	.84906	1.4156	.15094
0.89	1.01224	.00528	1.42289	.15317	.71139	.85211	1.4057	.14789
0.90	1.02652	.01137	1.43309	.15627	0.71630	.85509	1.3961	.14491
0.91	1.04090	.01741	1.44342	.15939	.72113	.85801	1.3867	.14199
0.92	1.05539	.02341	1.45390	.16254	.72590	.86088	1.3776	.13912
0.93	1.06998	.02937	1.46453	.16570	.73059	.86368	1.3687	.13632
0.94	1.08468	.03530	1.47530	.16888	.73522	.86642	1.3601	.13358
0.95	1.09948	.04119	1.48623	.17208	0.73978	.86910	1.3517	.13090
0.96	1.11440	.04704	1.49729	.17531	.74428	.87173	1.3436	.12827
0.97	1.12943	.05286	1.50851	.17855	.74870	.87431	1.3356	.12569
0.98	1.14457	.05864	1.51988	.18181	.75307	.87683	1.3279	.12317
0.99	1.15983	.06439	1.53141	.18509	.75736	.87930	1.3204	.12070

HYPERBOLIC FUNCTIONS AND THEIR COMMON LOGARITHMS (Continued)

x	Sinh x Value	Sinh x log$_{10}$	Cosh x Value	Cosh x log$_{10}$	Tanh x Value	Tanh x log$_{10}$	Coth x Value	Coth x log$_{10}$
1.00	1.17520	.07011	1.54308	.18839	0.76159	.88172	1.3130	.11828
1.01	1.19069	.07580	1.55491	.19171	.76576	.88409	1.3059	.11591
1.02	1.20630	.08146	1.56689	.19504	.76987	.88642	1.2989	.11358
1.03	1.22203	.08708	1.57904	.19839	.77391	.88869	1.2921	.11131
1.04	1.23788	.09268	1.59134	.20176	.77789	.89092	1.2855	.10908
1.05	1.25386	.09825	1.60379	.20515	0.78181	.89310	1.2791	.10690
1.06	1.26996	.10379	1.61641	.20855	.78566	.89524	1.2728	.10476
1.07	1.28619	.10930	1.62919	.21197	.78946	.89733	1.2667	.10267
1.08	1.30254	.11479	1.64214	.21541	.79320	.89938	1.2607	.10062
1.09	1.31903	.12025	1.65525	.21886	.79688	.90139	1.2549	.09861
1.10	1.33565	.12569	1.66852	.22233	0.80050	.90336	1.2492	.09664
1.11	1.35240	.13111	1.68196	.22582	.80406	.90529	1.2437	.09471
1.12	1.36929	.13649	1.69557	.22931	.80757	.90718	1.2383	.09282
1.13	1.38631	.14186	1.70934	.23283	.81102	.90903	1.2330	.09097
1.14	1.40347	.14720	1.72329	.23636	.81441	.91085	1.2279	.08915
1.15	1.42078	.15253	1.73741	.23990	0.81775	.91262	1.2229	.08738
1.16	1.43822	.15783	1.75171	.24346	.82104	.91436	1.2180	.08564
1.17	1.45581	.16311	1.76618	.24703	.82427	.91607	1.2132	.08393
1.18	1.47355	.16836	1.78083	.25062	.82745	.91774	1.2085	.08226
1.19	1.49143	.17360	1.79565	.25422	.83058	.91938	1.2040	.08062
1.20	1.50946	.17882	1.81066	.25784	0.83365	.92099	1.1995	.07901
1.21	1.52764	.18402	1.82584	.26146	.83668	.92256	1.1952	.07744
1.22	1.54598	.18920	1.84121	.26510	.83965	.92410	1.1910	.07590
1.23	1.56447	.19437	1.85676	.26876	.84258	.92561	1.1868	.07439
1.24	1.58311	.19951	1.87250	.27242	.84546	.92709	1.1828	.07291
1.25	1.60192	.20464	1.88842	.27610	0.84828	.92854	1.1789	.07146
1.26	1.62088	.20975	1.90454	.27979	.85106	.92996	1.1750	.07004
1.27	1.64001	.21485	1.92084	.28349	.85380	.93135	1.1712	.06865
1.28	1.65930	.21993	1.93734	.28721	.85648	.93272	1.1676	.06728
1.29	1.67876	.22499	1.95403	.29093	.85913	.93406	1.1640	.06594
1.30	1.69838	.23004	1.97091	.29467	0.86172	.93537	1.1605	.06463
1.31	1.71818	.23507	1.98800	.29842	.86428	.93665	1.1570	.06335
1.32	1.73814	.24009	2.00528	.30217	.86678	.93791	1.1537	.06209
1.33	1.75828	.24509	2.02276	.30594	.86925	.93914	1.1504	.06086
1.34	1.77860	.25008	2.04044	.30972	.87167	.94035	1.1472	.05965
1.35	1.79909	.25505	2.05833	.31352	0.87405	.94154	1.1441	.05846
1.36	1.81977	.26002	2.07643	.31732	.87639	.94270	1.1410	.05730
1.37	1.84062	.26496	2.09473	.32113	.87869	.94384	1.1381	.05616
1.38	1.86166	.26990	2.11324	.32495	.88095	.94495	1.1351	.05505
1.39	1.88289	.27482	2.13196	.32878	.88317	.94604	1.1323	.05396
1.40	1.90430	.27974	2.15090	.33262	0.88535	.94712	1.1295	.05288
1.41	1.92591	.28464	2.17005	.33647	.88749	.94817	1.1268	.05183
1.42	1.94770	.28952	2.18942	.34033	.88960	.94919	1.1241	.05081
1.43	1.96970	.29440	2.20900	.34420	.89167	.95020	1.1215	.04980
1.44	1.99188	.29926	2.22881	.34807	.89370	.95119	1.1189	.04881
1.45	2.01427	.30412	2.24884	.35196	0.89569	.95216	1.1165	.04784
1.46	2.03686	.30896	2.26910	.35585	.89765	.95311	1.1140	.04689
1.47	2.05965	.31379	2.28958	.35976	.89958	.95404	1.1116	.04596
1.48	2.08265	.31862	2.31029	.36367	.90147	.95495	1.1093	.04505
1.49	2.10586	.32343	2.33123	.36759	.90332	.95584	1.1070	.04416

HYPERBOLIC FUNCTIONS AND THEIR COMMON LOGARITHMS (Continued)

x	Sinh x Value	Sinh x $\log_{10}$	Cosh x Value	Cosh x $\log_{10}$	Tanh x Value	Tanh x $\log_{10}$	Coth x Value	Coth x $\log_{10}$
1.50	2.12928	.32823	2.35241	.37151	0.90515	.95672	1.1048	.04328
1.51	2.15291	.33303	2.37382	.37545	.90694	.95758	1.1026	.04242
1.52	2.17676	.33781	2.39547	.37939	.90870	.95842	1.1005	.04158
1.53	2.20082	.34258	2.41736	.38334	.91042	.95924	1.0984	.04076
1.54	2.22510	.34735	2.43949	.38730	.91212	.96005	1.0963	.03995
1.55	2.24961	.35211	2.46186	.39126	0.91379	.96084	1.0943	.03916
1.56	2.27434	.35686	2.48448	.39524	.91542	.96162	1.0924	.03838
1.57	2.29930	.36160	2.50735	.39921	.91703	.96238	1.0905	.03762
1.58	2.32449	.36633	2.53047	.40320	.91860	.96313	1.0886	.03687
1.59	2.34991	.37105	2.55384	.40719	.92015	.96386	1.0868	.03614
1.60	2.37557	.37577	2.57746	.41119	0.92167	.96457	1.0850	.03543
1.61	2.40146	.38048	2.60135	.41520	.92316	.96528	1.0832	.03472
1.62	2.42760	.38518	2.62549	.41921	.92462	.96597	1.0815	.03403
1.63	2.45397	.38987	2.64990	.42323	.92606	.96664	1.0798	.03336
1.64	2.48059	.39456	2.67457	.42725	.92747	.96730	1.0782	.03270
1.65	2.50746	.39923	2.69951	.43129	0.92886	.96795	1.0766	.03205
1.66	2.53459	.40391	2.72472	.43532	.93022	.96858	1.0750	.03142
1.67	2.56196	.40857	2.75021	.43937	.93155	.96921	1.0735	.03079
1.68	2.58959	.41323	2.77596	.44341	.93286	.96982	1.0720	.03018
1.69	2.61748	.41788	2.80200	.44747	.93415	.97042	1.0705	.02958
1.70	2.64563	.42253	2.82832	.45153	.93541	.97100	1.0691	.02900
1.71	2.67405	.42717	2.85491	.45559	.93665	.97158	1.0676	.02842
1.72	2.70273	.43180	2.88180	.45966	.93786	.97214	1.0663	.02786
1.73	2.73168	.43643	2.90897	.46374	.93906	.97269	1.0649	.02731
1.74	2.76091	.44105	2.93643	.46782	.94023	.97323	1.0636	.02677
1.75	2.79041	.44567	2.96419	.47191	0.94138	.97376	1.0623	.02624
1.76	2.82020	.45028	2.99224	.47600	.94250	.97428	1.0610	.02572
1.77	2.85026	.45488	3.02059	.48009	.94361	.97479	1.0598	.02521
1.78	2.88061	.45948	3.04925	.48419	.94470	.97529	1.0585	.02471
1.79	2.91125	.46408	3.07821	.48830	.94576	.97578	1.0574	.02422
1.80	2.94217	.46867	3.10747	.49241	0.94681	.97626	1.0562	.02374
1.81	2.97340	.47325	3.13705	.49652	.94783	.97673	1.0550	.02327
1.82	3.00492	.47783	3.16694	.50064	.94884	.97719	1.0539	.02281
1.83	3.03674	.48241	3.19715	.50476	.94983	.97764	1.0528	.02236
1.84	3.06886	.48698	3.22768	.50889	.95080	.97809	1.0518	.02191
1.85	3.10129	.49154	3.25853	.51302	0.95175	.97852	1.0507	.02148
1.86	3.13403	.49610	3.28970	.51716	.95268	.97895	1.0497	.02105
1.87	3.16709	.50066	3.32121	.52130	.95359	.97936	1.0487	.02064
1.88	3.20046	.50521	3.35305	.52544	.95449	.97977	1.0477	.02023
1.89	3.23415	.50976	3.38522	.52959	.95537	.98017	1.0467	.01983
1.90	3.26816	.51430	3.41773	.53374	0.95624	.98057	1.0458	.01943
1.91	3.30250	.51884	3.45058	.53789	.95709	.98095	1.0448	.01905
1.92	3.33718	.52338	3.48378	.54205	.95792	.98133	1.0439	.01867
1.93	3.37218	.52791	3.51733	.54621	.95873	.98170	1.0430	.01830
1.94	3.40752	.53244	3.55123	.55038	.95953	.98206	1.0422	.01794
1.95	3.44321	.53696	3.58548	.55455	0.96032	.98242	1.0413	.01758
1.96	3.47923	.54148	3.62009	.55872	.96109	.98276	1.0405	.01724
1.97	3.51561	.54600	3.65507	.56290	.96185	.98311	1.0397	.01689
1.98	3.55234	.55051	3.69041	.56707	.96259	.98344	1.0389	.01656
1.99	3.58942	.55502	3.72611	.57126	.96331	.98377	1.0381	.01623

HYPERBOLIC FUNCTIONS AND THEIR COMMON LOGARITHMS (Continued)

x	Sinh x Value	Sinh x $\log_{10}$	Cosh x Value	Cosh x $\log_{10}$	Tanh x Value	Tanh x $\log_{10}$	Coth x Value	Coth x $\log_{10}$
2.00	3.62686	.55953	3.76220	.57544	0.96403	.98409	1.0373	.01591
2.01	3.66466	.56403	3.79865	.57963	.96473	.98440	1.0366	.01560
2.02	3.70283	.56853	3.83549	.58382	.96541	.98471	1.0358	.01529
2.03	3.74138	.57303	3.87271	.58802	.96609	.98502	1.0351	.01498
2.04	3.78029	.57753	3.91032	.59221	.96675	.98531	1.0344	.01469
2.05	3.81958	.58202	3.94832	.59641	0.96740	.98560	1.0337	.01440
2.06	3.85926	.58650	3.98671	.60061	.96803	.98589	1.0330	.01411
2.07	3.89932	.59099	4.02550	.60482	.96865	.98617	1.0324	.01383
2.08	3.93977	.59547	4.06470	.60903	.96926	.98644	1.0317	.01356
2.09	3.98061	.59995	4.10430	.61324	.96986	.98671	1.0311	.01329
2.10	4.02186	.60443	4.14431	.61745	0.97045	.98697	1.0304	.01303
2.11	4.06350	.60890	4.18474	.62167	.97103	.98723	1.0298	.01277
2.12	4.10555	.61337	4.22558	.62589	.97159	.98748	1.0292	.01252
2.13	4.14801	.61784	4.26685	.63011	.97215	.98773	1.0286	.01227
2.14	4.19089	.62231	4.30855	.63433	.97269	.98798	1.0281	.01202
2.15	4.23419	.62677	4.35067	.63856	0.97323	.98821	1.0275	.01179
2.16	4.27791	.63123	4.39323	.64278	.97375	.98845	1.0270	.01155
2.17	4.32205	.63569	4.43623	.64701	.97426	.98868	1.0264	.01132
2.18	4.36663	.64015	4.47967	.65125	.97477	.98890	1.0259	.01110
2.19	4.41165	.64460	4.52356	.65548	.97526	.98912	1.0254	.01088
2.20	4.45711	.64905	4.56791	.65972	0.97574	.98934	1.0249	.01066
2.21	4.50301	.65350	4.61271	.66396	.97622	.98955	1.0244	.01045
2.22	4.54936	.65795	4.65797	.66820	.97668	.98975	1.0239	.01025
2.23	4.59617	.66240	4.70370	.67244	.97714	.98996	1.0234	.01004
2.24	4.64344	.66684	4.74989	.67668	.97759	.99016	1.0229	.00984
2.25	4.69117	.67128	4.79657	.68093	0.97803	.99035	1.0225	.00965
2.26	4.73937	.67572	4.84372	.68518	.97846	.99054	1.0220	.00946
2.27	4.78804	.68016	4.89136	.68943	.97888	.99073	1.0216	.00927
2.28	4.83720	.68459	4.93948	.69368	.97929	.99091	1.0211	.00909
2.29	4.88684	.68903	4.98810	.69794	.97970	.99109	1.0207	.00891
2.30	4.93696	.69346	5.03722	.70219	0.98010	.99127	1.0203	.00873
2.31	4.98758	.69789	5.08684	.70645	.98049	.99144	1.0199	.00856
2.32	5.03870	.70232	5.13697	.71071	.98087	.99161	1.0195	00839
2.33	5.09032	.70675	5.18762	.71497	.98124	.99178	1.0191	.00822
2.34	5.14245	.71117	5.23878	.71923	.98161	.99194	1.0187	.00806
2.35	5.19510	.71559	5.29047	.72349	0.98197	.99210	1.0184	.00790
2.36	5.24827	.72002	5.34269	.72776	.98233	.99226	1.0180	.00774
2.37	5.30196	.72444	5.39544	.73203	.98267	.99241	1.0176	.00759
2.38	5.35618	.72885	5.44873	.73630	.98301	.99256	1.0173	.00744
2.39	5.41093	.73327	5.50256	.74056	.98335	.99271	1.0169	.00729
2.40	5.46623	.73769	5.55695	.74484	0.98367	.99285	1.0166	.00715
2.41	5.52207	.74210	5.61189	.74911	.98400	.99299	1.0163	.00701
2.42	5.57847	.74652	5.66739	.75338	.98431	.99313	1.0159	.00687
2.43	5.63542	.75093	5.72346	.75766	.98462	.99327	1.0156	.00673
2.44	5.69294	.75534	5.78010	.76194	.98492	.99340	1.0153	.00660
2.45	5.75103	.75975	5.83732	.76621	0.98522	.99353	1.0150	.00647
2.46	5.80969	.75415	5.89512	.77049	.98551	.99366	1.0147	.00634
2.47	5.86893	.76856	5.95352	.77477	.98579	.99379	1.0144	.00621
2.48	5.92876	.77296	6.01250	.77906	.98607	.99391	1.0141	.00609
2.49	5.98918	.77737	6.07209	.78334	.98635	.99403	1.0138	.00597

HYPERBOLIC FUNCTIONS AND THEIR COMMON LOGARITHMS (Continued)

x	Sinh x Value	Sinh x log$_{10}$	Cosh x Value	Cosh x log$_{10}$	Tanh x Value	Tanh x log$_{10}$	Coth x Value	Coth x log$_{10}$
2.50	6.05020	.78177	6.13229	.78762	0.98661	.99415	1.0136	.00585
2.51	6.11183	.78617	6.19310	.79191	.98688	.99426	1.0133	.00574
2.52	6.17407	.79057	6.25453	.79619	.98714	.99438	1.0130	.00562
2.53	6.23692	.79497	6.31658	.80048	.98739	.99449	1.0128	.00551
2.54	6.30040	.79937	6.37927	.80477	.98764	.99460	1.0125	.00540
2.55	6.36451	.80377	6.44259	.80906	0.98788	.99470	1.0123	.00530
2.56	6.42926	.80816	6.50656	.81335	.98812	.99481	1.0120	.00519
2.57	6.49464	.81256	6.57118	.81764	.98835	.99491	1.0118	.00509
2.58	6.56068	.81695	6.63646	.82194	.98858	.99501	1.0115	.00499
2.59	6.62738	.82134	6.70240	.82623	.98881	.99511	1.0113	.00489
2.60	6.69473	.82573	6.76901	.83052	0.98903	.99521	1.0111	.00479
2.61	6.76276	.83012	6.83629	.83482	.98924	.99530	1.0109	.00470
2.62	6.83146	.83451	6.90426	.83912	.98946	.99540	1.0107	.00460
2.63	6.90085	.83890	6.97292	.84341	.98966	.99549	1.0104	.00451
2.64	6.97092	.84329	7.04228	.84771	.98987	.99558	1.0102	.00442
2.65	7.04169	.84768	7.11234	.85201	0.99007	.99566	1.0100	.00434
2.66	7.11317	.85206	7.18312	.85631	.99026	.99575	1.0098	.00425
2.67	7.18536	.85645	7.25461	.86061	.99045	.99583	1.0096	.00417
2.68	7.25827	.86083	7.32683	.86492	.99064	.99592	1.0094	.00408
2.69	7.33190	.86522	7.39978	.86922	.99083	.99600	1.0093	.00400
2.70	7.40626	.86960	7.47347	.87352	0.99101	.99608	1.0091	.00392
2.71	7.48137	.87398	7.54791	.87783	.99118	.99615	1.0089	.00385
2.72	7.55722	.87836	7.62310	.88213	.99136	.99623	1.0087	.00377
2.73	7.63383	.88274	7.69905	.88644	.99153	.99631	1.0085	.00369
2.74	7.71121	.88712	7.77578	.89074	.99170	.99638	1.0084	.00362
2.75	7.78935	.89150	7.85328	.89505	0.99186	.99645	1.0082	.00355
2.76	7.86828	.89588	7.93157	.89936	.99202	.99652	1.0080	.00348
2.77	7.94799	.90026	8.01065	.90367	.99218	.99659	1.0079	.00341
2.78	8.02849	.90463	8.09053	.90798	.99233	.99666	1.0077	.00334
2.79	8.10980	.90901	8.17122	.91229	.99248	.99672	1.0076	.00328
2.80	8.19192	.91339	8.25273	.91660	0.99263	.99679	1.0074	.00321
2.81	8.27486	.91776	8.33506	.92091	.99278	.99685	1.0073	.00315
2.82	8.35862	.92213	8.41823	.92522	.99292	.99691	1.0071	.00309
2.83	8.44322	.92651	8.50224	.92953	.99306	.99698	1.0070	.00302
2.84	8.52867	.93088	8.58710	.93385	.99320	.99704	1.0069	.00296
2.85	8.61497	.93525	8.67281	.93816	0.99333	.99709	1.0067	.00291
2.86	8.70213	.93963	8.75940	.94247	.99346	.99715	1.0066	.00285
2.87	8.79016	.94400	8.84686	.94679	.99359	.99721	1.0065	.00279
2.88	8.87907	.94837	8.93520	.95110	.99372	.99726	1.0063	.00274
2.89	8.96887	.95274	9.02444	.95542	.99384	.99732	1.0062	.00268
2.90	9.05956	.95711	9.11458	.95974	0.99396	.99737	1.0061	.00263
2.91	9.15116	.96148	9.20564	.96405	.99408	.99742	1.0060	.00258
2.92	9.24368	.96584	9.29761	.96837	.99420	.99747	1.0058	.00253
2.93	9.33712	.97021	9.39051	.97269	.99431	.99752	1.0057	.00248
2.94	9.43149	.97458	9.48436	.97701	.99443	.99757	1.0056	.00243
2.95	9.52681	.97895	9.57915	.98133	0.99454	.99762	1.0055	.00238
2.96	9.62308	.98331	9.67490	.98565	.99464	.99767	1.0054	.00233
2.97	9.72031	.98768	9.77161	.98997	.99475	.99771	1.0053	.00229
2.98	9.81851	.99205	9.86930	.99429	.99485	.99776	1.0052	.00224
2.99	9.91770	.99641	9.96798	.99861	.99496	.99780	1.0051	.00220

HYPERBOLIC FUNCTIONS AND THEIR COMMON LOGARITHMS (Continued)

x	Sinh x		Cosh x		Tanh x		Coth x	
	Value	$\log_{10}$	Value	$\log_{10}$	Value	$\log_{10}$	Value	$\log_{10}$
3.0	10.0179	.00078	10.0677	.00293	0.99505	.99785	1.0050	.00215
3.1	11.0765	.04440	11.1215	.04616	.99595	.99824	1.0041	.00176
3.2	12.2459	.08799	12.2866	.08943	.99668	.99856	1.0033	.00144
3.3	13.5379	.13155	13.5748	.13273	.99728	.99882	1.0027	.00118
3.4	14.9654	.17509	14.9987	.17605	.99777	.99903	1.0022	.00097
3.5	16.5426	.21860	16.5728	.21940	0.99818	.99921	1.0018	.00079
3.6	18.2855	.26211	18.3128	.26275	.99851	.99935	1.0015	.00065
3.7	20.2113	.30559	20.2360	.30612	.99878	.99947	1.0012	.00053
3.8	22.3394	.34907	22.3618	.34951	.99900	.99957	1.0010	.00043
3.9	24.6011	.39254	24.7113	.39290	.99918	.99964	1.0008	.00036
4.0	27.2899	.43600	27.3082	.43629	0.99933	.99971	1.0007	.00029
4.1	30.1619	.47946	30.1784	.47970	.99945	.99976	1.0005	.00024
4.2	33.3357	.52291	33.3507	.52310	.99955	.99980	1.0004	.00020
4.3	36.8431	.56636	36.8567	.56652	.99963	.99984	1.0004	.00016
4.4	40.7193	.60980	40.7316	.60993	.99970	.99987	1.0003	.00013
4.5	45.0030	.65324	45.0141	.65335	0.99975	.99989	1.0002	.00011
4.6	49.7371	.69668	49.7472	.69677	.99980	.99991	1.0002	.00009
4.7	54.9690	.74012	54.9781	.74019	.99983	.99993	1.0002	.00007
4.8	60.7511	.78355	60.7593	.78361	.99986	.99994	1.0001	.00006
4.9	67.1412	.82699	67.1486	.82704	.99989	.99995	1.0001	.00005
5.0	74.2032	.87042	74.2099	.87046	0.99991	.99996	1.0001	.00004
5.1	82.008	.91386	82.014	.91389	.99993	.99997	1.0001	.00003
5.2	90.633	.95729	90.639	.95731	.99994	.99997	1.0001	.00003
5.3	100.17	.00072	100.17	.00074	.99995	.99998	1.0000	.00002
5.4	110.70	.04415	110.71	.04417	.99996	.99998	1.0000	.00002
5.5	122.34	.08758	122.35	.08760	0.99997	.99999	1.0000	.00001
5.6	135.21	.13101	135.22	.13103	.99997	.99999	1.0000	.00001
5.7	149.43	.17444	149.44	.17445	.99998	.99999	1.0000	.00001
5.8	165.15	.21787	165.15	.21788	.99998	.99999	1.0000	.00001
5.9	182.52	.26130	182.52	.26131	.99998	.99999	1.0000	.00001
6.0	201.71	.30473	201.72	.30474	0.99999	.00000	1.0000	.00000
6.1	222.93	.34817	222.93	.34817	.99999	.00000	1.0000	.00000
6.2	246.37	.39159	246.38	.39161	.99999	.00000	1.0000	.00000
6.3	272.29	.43503	272.29	.43503	.99999	.00000	1.0000	.00000
6.4	300.92	.47845	300.92	.47845	.99999	.00000	1.0000	.00000
6.5	332.57	.52188	332.57	.52188	1.0000	.00000	1.0000	.00000
6.6	367.55	.56532	367.55	.56532	1.0000	.00000	1.0000	.00000
6.7	406.20	.60874	406.20	.60874	1.0000	.00000	1.0000	.00000
6.8	448.92	.65217	448.92	.65217	1.0000	.00000	1.0000	.00000
6.9	496.14	.69560	496.14	.69560	1.0000	.00000	1.0000	.00000
7.0	548.32	.73903	548.32	.73903	1.0000	.00000	1.0000	.00000
7.1	605.98	.78246	605.98	.78246	1.0000	.00000	1.0000	.00000
7.2	669.72	.82589	669.72	.82589	1.0000	.00000	1.0000	.00000
7.3	740.15	.86932	740.15	.86932	1.0000	.00000	1.0000	.00000
7.4	817.99	.91275	817.99	.91275	1.0000	.00000	1.0000	.00000
7.5	904.02	.95618	904.02	.95618	1.0000	.00000	1.0000	.00000
7.6	999.10	.99961	999.10	.99961	1.0000	.00000	1.0000	.00000
7.7	1104.2	.04305	1104.2	.04305	1.0000	.00000	1.0000	.00000
7.8	1220.3	.08647	1220.3	.08647	1.0000	.00000	1.0000	.00000
7.9	1348.6	.12988	1348.6	.12988	1.0000	.00000	1.0000	.00000

HYPERBOLIC FUNCTIONS AND THEIR COMMON LOGARITHMS (Continued)

x	Sinh x Value	Sinh x log₁₀	Cosh x Value	Cosh x log₁₀	Tanh x Value	Tanh x log₁₀	Coth x Value	Coth x log₁₀
8.0	1490.5	.17333	1490.5	.17333	1.0000	.00000	1.0000	.00000
8.1	1647.2	.21675	1647.2	.21675	1.0000	.00000	1.0000	.00000
8.2	1820.5	.26019	1820.5	.26019	1.0000	.00000	1.0000	.00000
8.3	2011.9	.30360	2011.9	.30360	1.0000	.00000	1.0000	.00000
8.4	2223.5	.34704	2223.5	.34704	1.0000	.00000	1.0000	.00000
8.5	2457.4	.39048	2457.4	.39048	1.0000	.00000	1.0000	.00000
8.6	2715.8	.43390	2715.8	.43390	1.0000	.00000	1.0000	.00000
8.7	3001.5	.47734	3001.5	.47734	1.0000	.00000	1.0000	.00000
8.8	3317.1	.52076	3317.1	.52076	1.0000	.00000	1.0000	.00000
8.9	3666.0	.56419	3666.0	.56419	1.0000	.00000	1.0000	.00000
9.0	4051.5	.60762	4051.5	.60762	1.0000	.00000	1.0000	.00000
9.1	4477.6	.65105	4477.6	.65105	1.0000	.00000	1.0000	.00000
9.2	4948.6	.69448	4948.6	.69448	1.0000	.00000	1.0000	.00000
9.3	5469.0	.73791	5469.0	.73791	1.0000	.00000	1.0000	.00000
9.4	6044.2	.78134	6044.2	.78134	1.0000	.00000	1.0000	.00000
9.5	6679.9	.82477	6679.9	.82477	1.0000	.00000	1.0000	.00000
9.6	7382.4	.86820	7382.4	.86820	1.0000	.00000	1.0000	.00000
9.7	8158.8	.91163	8158.8	.91163	1.0000	.00000	1.0000	.00000
9.8	9016.9	.95506	9016.9	.95506	1.0000	.00000	1.0000	.00000
9.9	9965.2	.99849	9965.2	.99849	1.0000	.00000	1.0000	.00000
10.0	11013.2	.04191	11013.2	.04191	1.0000	.00000	1.0000	.00000

INVERSE HYPERBOLIC FUNCTIONS

x	$\sinh^{-1} x$	$\tanh^{-1} x$	$\operatorname{cosech}^{-1} x$	$\operatorname{sech}^{-1} x$
0.0	0.00000	0.00000	∞	∞
0.01	0.01000	0.01000	5.29834	5.29829
0.02	0.02000	0.02000	4.60527	4.60507
0.03	0.03000	0.03001	4.19993	4.19948
0.04	0.03999	0.04002	3.91242	3.91162
0.05	0.04998	0.05004	3.68950	3.68825
0.06	0.05996	0.06007	3.50746	3.50566
0.07	0.06994	0.07011	3.35363	3.35118
0.08	0.07991	0.08017	3.22047	3.21727
0.09	0.08988	0.09024	3.10311	3.09906
0.10	0.09983	0.10034	2.99822	2.99322
0.11	0.10978	0.11045	2.90343	2.89738
0.12	0.11971	0.12058	2.81699	2.80979
0.13	0.12964	0.13074	2.73757	2.72912
0.14	0.13955	0.14093	2.66412	2.65432
0.15	0.14944	0.15114	2.59585	2.58459
0.16	0.15933	0.16139	2.53207	2.51927
0.17	0.16919	0.17167	2.47225	2.45780
0.18	0.17904	0.18198	2.41595	2.39975
0.19	0.18888	0.19234	2.36278	2.34473
0.20	0.19869	0.20273	2.31244	2.29243
0.21	0.20849	0.21317	2.26464	2.24258
0.22	0.21826	0.22366	2.21916	2.19495
0.23	0.22802	0.23419	2.17579	2.14933
0.24	0.23775	0.24477	2.13436	2.10554
0.25	0.24747	0.25541	2.09471	2.06344
0.26	0.25716	0.26611	2.05671	2.02288
0.27	0.26682	0.27686	2.02023	1.98374
0.28	0.27646	0.28768	1.98516	1.94591
0.29	0.28608	0.29857	1.95141	1.90930
0.30	0.29567	0.30952	1.91890	1.87382
0.31	0.30524	0.32055	1.88753	1.83939
0.32	0.31478	0.33165	1.85725	1.80594
0.33	0.32429	0.34283	1.82799	1.77340
0.34	0.33377	0.35409	1.79968	1.74172
0.35	0.34322	0.36544	1.77228	1.71083
0.36	0.35265	0.37689	1.74573	1.68070
0.37	0.36204	0.38842	1.71999	1.65127
0.38	0.37140	0.40006	1.69502	1.62250
0.39	0.38073	0.41180	1.67078	1.59436
0.40	0.39004	0.42365	1.64723	1.56680
0.41	0.39930	0.43561	1.62434	1.53979
0.42	0.40854	0.44769	1.60209	1.51331
0.43	0.41774	0.45990	1.58043	1.48731
0.44	0.42691	0.47223	1.55935	1.46178
0.45	0.43605	0.48470	1.53882	1.43669
0.46	0.44515	0.49731	1.51881	1.41200
0.47	0.45422	0.51007	1.49931	1.38771
0.48	0.46325	0.52298	1.48029	1.36379
0.49	0.47225	0.53606	1.46174	1.34021
0.50	0.48121	0.54931	1.44364	1.31696

INVERSE HYPERBOLIC FUNCTIONS (Continued)

x	$\sinh^{-1} x$	$\tanh^{-1} x$	$\operatorname{cosech}^{-1} x$	$\operatorname{sech}^{-1} x$
0.50	0.48121	0.54931	1.44364	1.31696
0.51	0.49014	0.56273	1.42596	1.29401
0.52	0.49903	0.57634	1.40870	1.27136
0.53	0.50788	0.59015	1.39183	1.24898
0.54	0.51670	0.60416	1.37535	1.22686
0.55	0.52548	0.61838	1.35924	1.20497
0.56	0.53422	0.63283	1.34348	1.18331
0.57	0.54293	0.64752	1.32807	1.16186
0.58	0.55160	0.66246	1.31299	1.14060
0.59	0.56023	0.67767	1.29824	1.11952
0.60	0.56882	0.69315	1.28380	1.09861
0.61	0.57738	0.70892	1.26965	1.07785
0.62	0.58590	0.72501	1.25580	1.05723
0.63	0.59438	0.74142	1.24223	1.03673
0.64	0.60282	0.75817	1.22894	1.01635
0.65	0.61122	0.77530	1.21591	0.99606
0.66	0.61959	0.79281	1.20314	0.97585
0.67	0.62792	0.81074	1.19062	0.95572
0.68	0.63620	0.82911	1.17834	0.93564
0.69	0.64446	0.84796	1.16629	0.91560
0.70	0.65267	0.86730	1.15448	0.89559
0.71	0.66084	0.88718	1.14288	0.87559
0.72	0.66897	0.90764	1.13151	0.85558
0.73	0.67707	0.92873	1.12034	0.83555
0.74	0.68513	0.95048	1.10938	0.81549
0.75	0.69315	0.97296	1.09861	0.79537
0.76	0.70113	0.99622	1.08804	0.77517
0.77	0.70907	1.02033	1.07766	0.75487
0.78	0.71697	1.04537	1.06746	0.73445
0.79	0.72484	1.07143	1.05744	0.71388
0.80	0.73267	1.09861	1.04759	0.69315
0.81	0.74046	1.12703	1.03792	0.67221
0.82	0.74821	1.15682	1.02840	0.65103
0.83	0.75592	1.18814	1.01905	0.62958
0.84	0.76360	1.22117	1.00986	0.60781
0.85	0.77124	1.25615	1.00082	0.58568
0.86	0.77884	1.29334	0.99193	0.56313
0.87	0.78640	1.33308	0.98319	0.54008
0.88	0.79393	1.37577	0.97459	0.51647
0.89	0.80142	1.42193	0.96613	0.49220
0.90	0.80887	1.47222	0.95780	0.46715
0.91	0.81628	1.52752	0.94961	0.44116
0.92	0.82366	1.58903	0.94154	0.41406
0.93	0.83100	1.65839	0.93361	0.38560
0.94	0.83830	1.73805	0.92580	0.35542
0.95	0.84557	1.83178	0.91810	0.32304
0.96	0.85280	1.94591	0.91053	0.28768
0.97	0.86000	2.09230	0.90307	0.24807
0.98	0.86716	2.29756	0.89573	0.20169
0.99	0.87428	2.64665	0.88850	0.14201
1.00	0.88137	∞	0.88137	0.00000

INVERSE HYPERBOLIC FUNCTIONS (Continued)

x	$\sinh^{-1} x$	$\cosh^{-1} x$	$\operatorname{cosech}^{-1} x$	$\coth^{-1} x$
1.00	0.88137	0.00000	0.88137	∞
1.01	0.88843	0.14130	0.87436	2.65165
1.02	0.89545	0.19967	0.86744	2.30756
1.03	0.90243	0.24434	0.86063	2.10730
1.04	0.90938	0.28191	0.85391	1.96591
1.05	0.91629	0.31492	0.84730	1.85679
1.06	0.92317	0.34470	0.84078	1.76806
1.07	0.93002	0.37202	0.83435	1.69340
1.08	0.93683	0.39738	0.82801	1.62905
1.09	0.94360	0.42114	0.82177	1.57255
1.10	0.95035	0.44357	0.81561	1.52226
1.11	0.95706	0.46485	0.80954	1.47698
1.12	0.96373	0.48513	0.80355	1.43584
1.13	0.97038	0.50453	0.79764	1.39817
1.14	0.97699	0.52316	0.79182	1.36346
1.15	0.98357	0.54110	0.78607	1.33129
1.16	0.99011	0.55840	0.78041	1.30134
1.17	0.99663	0.57514	0.77482	1.27334
1.18	1.00311	0.59135	0.76930	1.24706
1.19	1.00956	0.60708	0.76386	1.22232
1.20	1.01597	0.62236	0.75849	1.19895
1.21	1.02236	0.63724	0.75319	1.17682
1.22	1.02871	0.65173	0.74796	1.15582
1.23	1.03504	0.66586	0.74279	1.13584
1.24	1.04133	0.67966	0.73770	1.11680
1.25	1.04759	0.69315	0.73267	1.09861
1.26	1.05382	0.70634	0.72770	1.08122
1.27	1.06003	0.71924	0.72280	1.06456
1.28	1.06620	0.73189	0.71796	1.04857
1.29	1.07234	0.74428	0.71318	1.03321
1.30	1.07845	0.75643	0.70846	1.01844
1.31	1.08453	0.76836	0.70380	1.00422
1.32	1.09059	0.78007	0.69920	0.99050
1.33	1.09661	0.79157	0.69465	0.97727
1.34	1.10261	0.80288	0.69016	0.96448
1.35	1.10857	0.81400	0.68572	0.95212
1.36	1.11451	0.82494	0.68134	0.94016
1.37	1.12042	0.83570	0.67701	0.92857
1.38	1.12630	0.84630	0.67273	0.91734
1.39	1.13216	0.85673	0.66851	0.90645
1.40	1.13798	0.86701	0.66433	0.89588
1.41	1.14378	0.87715	0.66020	0.88561
1.42	1.14955	0.88714	0.65612	0.87563
1.43	1.15530	0.89699	0.65209	0.86593
1.44	1.16101	0.90670	0.64811	0.85649
1.45	1.16670	0.91629	0.64417	0.84730
1.46	1.17237	0.92575	0.64028	0.83835
1.47	1.17801	0.93509	0.63643	0.82962
1.48	1.18362	0.94432	0.63263	0.82111
1.49	1.18920	0.95343	0.62886	0.81282
1.50	1.19476	0.96242	0.62515	0.80472

INVERSE HYPERBOLIC FUNCTIONS (Continued)

x	$\sinh^{-1} x$	$\cosh^{-1} x$	$\operatorname{cosech}^{-1} x$	$\coth^{-1} x$
1.50	1.19476	0.96242	0.62515	0.80472
1.51	1.20030	0.97131	0.62147	0.79681
1.52	1.20581	0.98010	0.61783	0.78909
1.53	1.21129	0.98879	0.61424	0.78155
1.54	1.21675	0.99737	0.61068	0.77418
1.55	1.22218	1.00587	0.60716	0.76697
1.56	1.22759	1.01426	0.60368	0.75991
1.57	1.23298	1.02257	0.60024	0.75301
1.58	1.23834	1.03079	0.59684	0.74626
1.59	1.24367	1.03892	0.59347	0.73965
1.60	1.24898	1.04697	0.59014	0.73317
1.61	1.25427	1.05493	0.58685	0.72682
1.62	1.25954	1.06282	0.58359	0.72061
1.63	1.26478	1.07063	0.58036	0.71451
1.64	1.26999	1.07836	0.57717	0.70853
1.65	1.27519	1.08601	0.57401	0.70267
1.66	1.28036	1.09360	0.57089	0.69692
1.67	1.28551	1.10111	0.56780	0.69128
1.68	1.29064	1.10855	0.56474	0.68574
1.69	1.29574	1.11592	0.56171	0.68030
1.70	1.30082	1.12323	0.55871	0.67496
1.71	1.30588	1.13047	0.55574	0.66972
1.72	1.31092	1.13765	0.55281	0.66457
1.73	1.31593	1.14476	0.54990	0.65951
1.74	1.32093	1.15182	0.54702	0.65453
1.75	1.32590	1.15881	0.54417	0.64964
1.76	1.33085	1.16574	0.54135	0.64483
1.77	1.33578	1.17262	0.53856	0.64011
1.78	1.34069	1.17944	0.53579	0.63546
1.79	1.34557	1.18620	0.53305	0.63088
1.80	1.35044	1.19291	0.53034	0.62638
1.81	1.35529	1.19957	0.52766	0.62195
1.82	1.36011	1.20617	0.52500	0.61759
1.83	1.36492	1.21272	0.52237	0.61330
1.84	1.36970	1.21922	0.51976	0.60908
1.85	1.37447	1.22567	0.51718	0.60492
1.86	1.37921	1.23207	0.51462	0.60082
1.87	1.38394	1.23842	0.51208	0.59679
1.88	1.38864	1.24473	0.50957	0.59281
1.89	1.39333	1.25098	0.50709	0.58890
1.90	1.39800	1.25720	0.50462	0.58504
1.91	1.40265	1.26336	0.50218	0.58123
1.92	1.40728	1.26949	0.49977	0.57748
1.93	1.41188	1.27557	0.49737	0.57379
1.94	1.41648	1.28160	0.49500	0.57014
1.95	1.42105	1.28760	0.49265	0.56655
1.96	1.42560	1.29355	0.49032	0.56301
1.97	1.43014	1.29946	0.48801	0.55951
1.98	1.43466	1.30533	0.48572	0.55606
1.99	1.43915	1.31117	0.48346	0.55266
2.00	1.44364	1.31696	0.48121	0.54931

INVERSE HYPERBOLIC FUNCTIONS (Continued)

x	$\sinh^{-1} x$	$\cosh^{-1} x$	$\operatorname{cosech}^{-1} x$	$\coth^{-1} x$
2.00	1.44364	1.31696	0.48121	0.54931
2.10	1.48748	1.37286	0.45982	0.51805
2.20	1.52966	1.42542	0.44019	0.49041
2.30	1.57028	1.47504	0.42213	0.46578
2.40	1.60944	1.52208	0.40547	0.44365
2.50	1.64723	1.56680	0.39004	0.42365
2.60	1.68374	1.60944	0.37571	0.40547
2.70	1.71905	1.65019	0.36239	0.38885
2.80	1.75323	1.68924	0.34996	0.37361
2.90	1.78634	1.72671	0.33834	0.35956
3.00	1.81845	1.76275	0.32745	0.34657
3.10	1.84960	1.79746	0.31723	0.33452
3.20	1.87986	1.83094	0.30763	0.32331
3.30	1.90927	1.86328	0.29857	0.31285
3.40	1.93788	1.89456	0.29003	0.30307
3.50	1.96572	1.92485	0.28196	0.29389
3.60	1.99284	1.95421	0.27432	0.28527
3.70	2.01926	1.98270	0.26708	0.27716
3.80	2.04503	2.01037	0.26021	0.26950
3.90	2.07017	2.03727	0.25368	0.26226
4.00	2.09471	2.06344	0.24747	0.25541
4.10	2.11869	2.08892	0.24155	0.24892
4.20	2.14211	2.11375	0.23590	0.24275
4.30	2.16502	2.13796	0.23051	0.23689
4.40	2.18742	2.16158	0.22536	0.23131
4.50	2.20935	2.18464	0.22043	0.22599
4.60	2.23081	2.20717	0.21571	0.22092
4.70	2.25184	2.22920	0.21119	0.21607
4.80	2.27244	2.25073	0.20685	0.21143
4.90	2.29264	2.27180	0.20269	0.20699
5.00	2.31244	2.29243	0.19869	0.20273
5.10	2.33186	2.31263	0.19484	0.19865
5.20	2.35093	2.33243	0.19114	0.19473
5.30	2.36964	2.35183	0.18758	0.19097
5.40	2.38801	2.37086	0.18414	0.18735
5.50	2.40606	2.38953	0.18083	0.18386
5.60	2.42379	2.40784	0.17764	0.18051
5.70	2.44122	2.42583	0.17455	0.17727
5.80	2.45836	2.44349	0.17157	0.17415
5.90	2.47521	2.46084	0.16869	0.17114
6.00	2.49178	2.47789	0.16590	0.16824
6.10	2.50809	2.49465	0.16321	0.16543
6.20	2.52414	2.51113	0.16060	0.16271
6.30	2.53994	2.52734	0.15807	0.16008
6.40	2.55549	2.54329	0.15562	0.15754
6.50	2.57081	2.55898	0.15325	0.15508
6.60	2.58591	2.57443	0.15094	0.15269
6.70	2.60078	2.58964	0.14871	0.15038
6.80	2.61543	2.60462	0.14653	0.14813
6.90	2.62988	2.61938	0.14442	0.14596
7.00	2.64412	2.63392	0.14238	0.14384

GUDERMANNIAN FUNCTION

x	0	1	2	3	4	5	6	7	8	9
0.0	0.00000	0.01000	0.02000	0.03000	0.03999	0.04998	0.05996	0.06994	0.07991	0.08988
0.1	0.09983	0.10978	0.11971	0.12964	0.13954	0.14944	0.15932	0.16919	0.17904	0.18887
0.2	0.19868	0.20847	0.21825	0.22800	0.23773	0.24744	0.25712	0.26678	0.27641	0.28602
0.3	0.29560	0.30515	0.31467	0.32417	0.33363	0.34307	0.35247	0.36184	0.37117	0.38047
0.4	0.38974	0.39897	0.40817	0.41733	0.42645	0.43554	0.44459	0.45359	0.46256	0.47149
0.5	0.48038	0.48923	0.49803	0.50680	0.51552	0.52420	0.53284	0.54143	0.54997	0.55848
0.6	0.56694	0.57535	0.58372	0.59204	0.60031	0.60854	0.61672	0.62486	0.63294	0.64098
0.7	0.64897	0.65692	0.66481	0.67266	0.68045	0.68820	0.69590	0.70355	0.71115	0.71870
0.8	0.72620	0.73366	0.74106	0.74841	0.75571	0.76297	0.77017	0.77732	0.78443	0.79148
0.9	0.79848	0.80544	0.81234	0.81919	0.82599	0.83275	0.83945	0.84611	0.85271	0.85926
1.0	0.86577	0.87223	0.87863	0.88499	0.89130	0.89756	0.90377	0.90993	0.91604	0.92211
1.1	0.92813	0.93410	0.94002	0.94589	0.95172	0.95750	0.96323	0.96892	0.97455	0.98015
1.2	0.98569	0.99119	0.99665	1.00205	1.00742	1.01274	1.01801	1.02324	1.02842	1.03356
1.3	1.03866	1.04371	1.04872	1.05368	1.05860	1.06348	1.06832	1.07312	1.07787	1.08258
1.4	1.08725	1.09188	1.09647	1.10101	1.10552	1.10999	1.11441	1.11880	1.12315	1.12746
1.5	1.13173	1.13596	1.14015	1.14431	1.14843	1.15251	1.15655	1.16056	1.16453	1.16846
1.6	1.17236	1.17622	1.18005	1.18384	1.18760	1.19132	1.19500	1.19866	1.20228	1.20586
1.7	1.20941	1.21293	1.21642	1.21987	1.22330	1.22668	1.23004	1.23337	1.23666	1.23993
1.8	1.24316	1.24636	1.24954	1.25268	1.25579	1.25888	1.26193	1.26496	1.26795	1.27092
1.9	1.27386	1.27677	1.27966	1.28251	1.28534	1.28815	1.29092	1.29367	1.29639	1.29909
2.0	1.30176	1.30441	1.30703	1.30962	1.31219	1.31473	1.31726	1.31975	1.32222	1.32467
2.1	1.32710	1.32950	1.33188	1.33423	1.33656	1.33887	1.34116	1.34343	1.34567	1.34789
2.2	1.35009	1.35227	1.35443	1.35656	1.35868	1.36077	1.36285	1.36490	1.36694	1.36895
2.3	1.37095	1.37292	1.37488	1.37682	1.37873	1.38063	1.38251	1.38438	1.38622	1.38805
2.4	1.38986	1.39165	1.39342	1.39518	1.39691	1.39864	1.40034	1.40203	1.40370	1.40535
2.5	1.40699	1.40862	1.41022	1.41181	1.41339	1.41495	1.41649	1.41802	1.41954	1.42104
2.6	1.42252	1.42399	1.42545	1.42689	1.42832	1.42973	1.43113	1.43251	1.43388	1.43524
2.7	1.43659	1.43792	1.43924	1.44054	1.44183	1.44311	1.44438	1.44564	1.44688	1.44811
2.8	1.44933	1.45053	1.45173	1.45291	1.45408	1.45524	1.45638	1.45752	1.45864	1.45976
2.9	1.46086	1.46195	1.46303	1.46410	1.46516	1.46621	1.46725	1.46828	1.46930	1.47031
3.0	1.47130	1.47229	1.47327	1.47424	1.47520	1.47615	1.47709	1.47802	1.47894	1.47986
3.1	1.48076	1.48165	1.48254	1.48342	1.48428	1.48514	1.48600	1.48684	1.48767	1.48850
3.2	1.48932	1.49013	1.49093	1.49172	1.49251	1.49329	1.49406	1.49482	1.49558	1.49632
3.3	1.49706	1.49780	1.49852	1.49924	1.49995	1.50066	1.50135	1.50204	1.50273	1.50340
3.4	1.50407	1.50474	1.50539	1.50605	1.50669	1.50733	1.50796	1.50858	1.50920	1.50981
3.5	1.51042	1.51102	1.51161	1.51220	1.51279	1.51336	1.51393	1.51450	1.51506	1.51561
3.6	1.51616	1.51671	1.51724	1.51778	1.51830	1.51883	1.51934	1.51985	1.52036	1.52086
3.7	1.52136	1.52185	1.52234	1.52282	1.52330	1.52377	1.52424	1.52470	1.52516	1.52561
3.8	1.52606	1.52651	1.52695	1.52738	1.52782	1.52824	1.52867	1.52909	1.52950	1.52991
3.9	1.53032	1.53072	1.53112	1.53151	1.53190	1.53229	1.53267	1.53305	1.53343	1.53380
4.0	1.53417	1.53453	1.53489	1.53525	1.53561	1.53596	1.53630	1.53664	1.53698	1.53732
4.1	1.53765	1.53798	1.53831	1.53863	1.53895	1.53927	1.53958	1.53989	1.54020	1.54051
4.2	1.54081	1.54111	1.54140	1.54169	1.54198	1.54227	1.54255	1.54283	1.54311	1.54339
4.3	1.54366	1.54393	1.54420	1.54446	1.54472	1.54498	1.54524	1.54550	1.54575	1.54600
4.4	1.54624	1.54649	1.54673	1.54697	1.54721	1.54744	1.54767	1.54790	1.54813	1.54836
4.5	1.54858	1.54880	1.54902	1.54924	1.54945	1.54966	1.54987	1.55008	1.55029	1.55049
4.6	1.55069	1.55089	1.55109	1.55129	1.55148	1.55167	1.55186	1.55205	1.55224	1.55242
4.7	1.55261	1.55279	1.55297	1.55314	1.55332	1.55349	1.55367	1.55384	1.55400	1.55417
4.8	1.55434	1.55450	1.55466	1.55482	1.55498	1.55514	1.55530	1.55545	1.55560	1.55575
4.9	1.55590	1.55605	1.55620	1.55634	1.55649	1.55663	1.55677	1.55691	1.55705	1.55719

GUDERMANNIAN FUNCTION (Continued)

x	0	1	2	3	4	5	6	7	8	9
5.0	1.55732	1.55745	1.55759	1.55772	1.55785	1.55798	1.55811	1.55823	1.55836	1.55848
5.1	1.55860	1.55872	1.55884	1.55896	1 55908	1.55920	1.55931	1.55943	1.55954	1.55965
5.2	1.55976	1.55987	1.55998	1.56009	1.56020	1.56030	1.56041	1.56051	1.56061	1.56071
5.3	1.56081	1.56091	1.56101	1.56111	1.56120	1.56130	1.56139	1.56149	1.56158	1.56167
5.4	1.56176	1.56185	1.56194	1.56203	1.56212	1.56220	1.56229	1.56237	1.56246	1.56254
5.5	1.56262	1.56270	1.56278	1.56286	1.56294	1.56302	1.56310	1.56318	1.56325	1.56333
5.6	1.56340	1.56347	1.56355	1.56362	1.56369	1.56376	1.56383	1.56390	1.56397	1.56404
5.7	1.56410	1.56417	1.56424	1.56430	1.56437	1.56443	1.56449	1.56456	1.56462	1.56468
5.8	1.56474	1.56480	1.56486	1.56492	1.56498	1.56504	1.56509	1.56515	1.56521	1.56526
5.9	1.56532	1.56537	1.56543	1.56548	1.56553	1.56558	1.56564	1.56569	1.56574	1.56579
6.0	1.56584	1.56589	1.56594	1.56599	1.56603	1.56608	1.56613	1.56617	1.56622	1.56627
6.1	1.56631	1.56636	1.56640	1.56644	1.56649	1.56653	1.56657	1.56661	1.56666	1.56670
6.2	1.56674	1.56678	1.56682	1.56686	1.56690	1.56694	1.56697	1.56701	1.56705	1.56709
6.3	1.56712	1.56716	1.56720	1.56723	1.56727	1.56730	1.56734	1.56737	1.56741	1.56744
6.4	1.56747	1.56751	1.56754	1.56757	1.56760	1.56764	1.56767	1.56770	1.56773	1.56776
6.5	1.56779	1.56782	1.56785	1.56788	1.56791	1.56794	1.56796	1.56799	1.56802	1.56805
6.6	1.56808	1.56810	1.56813	1.56816	1.56818	1.56821	1.56823	1.56826	1.56828	1.56831
6.7	1.56833	1.56836	1.56838	1.56841	1.56843	1.56845	1.56848	1.56850	1.56852	1.56855
6.8	1.56857	1.56859	1.56861	1.56863	1.56866	1.56868	1.56870	1.56872	1.56874	1.56876
6.9	1.56878	1.56880	1.56882	1.56884	1.56886	1.56888	1.56890	1.56892	1.56894	1.56895
7.0	1.56897	1.56899	1.56901	1.56903	1.56904	1.56906	1.56908	1.56910	1.56911	1.56913
7.1	1.56915	1.56916	1.56918	1.56919	1.56921	1.56923	1.56924	1.56926	1.56927	1.56929
7.2	1.56930	1.56932	1.56933	1.56935	1.56936	1.56938	1.56939	1.56940	1.56942	1.56943
7.3	1.56945	1.56946	1.56947	1.56949	1.56950	1.56951	1.56952	1.56954	1.56955	1.56956
7.4	1.56957	1.56959	1.56960	1.56961	1.56962	1.56963	1.56965	1.56966	1.56967	1.56968
7.5	1.56969	1.56970	1.56971	1.56972	1.56973	1.56974	1.56975	1.56976	1.56978	1.56979
7.6	1.56980	1.56981	1.56982	1.56983	1.56983	1.56984	1.56985	1.56986	1.56987	1.56988
7.7	1.56989	1.56990	1.56991	1.56992	1.56993	1.56993	1.56994	1.56995	1.56996	1.56997
7.8	1.56998	1.56999	1.56999	1.57000	1.57001	1.57002	1.57002	1.57003	1.57004	1.57005
7.9	1.57005	1.57006	1.57007	1.57008	1.57008	1.57008	1.57010	1.57010	1.57011	1.57012
8.0	1.57013	1.57013	1.57014	1.57015	1 57015	1.57016	1.57016	1.57017	1.57018	1.57018
8.1	1.57019	1.57020	1.57020	1.57021	1.57021	1.57022	1.57022	1.57023	1.57024	1.57024
8.2	1.57025	1.57025	1.57026	1.57026	1.57027	1.57027	1.57028	1.57028	1.57029	1.57029
8.3	1.57030	1.57030	1.57031	1.57031	1.57032	1.57032	1.57033	1.57033	1.57034	1.57034
8.4	1.57035	1.57035	1.57036	1.57036	1.57036	1.57037	1.57037	1.57038	1.57038	1.57039
8.5	1.57039	1.57039	1.57040	1.57040	1.57041	1.57041	1.57041	1.57042	1.57042	1.57042
8.6	1.57043	1.57043	1.57044	1.57044	1.57044	1.57045	1.57045	1.57045	1.57046	1.57046
8.7	1.57046	1.57047	1.57047	1.57047	1.57048	1.57048	1.57048	1.57049	1.57049	1.57049
8.8	1.57049	1.57050	1.57050	1.57050	1.57051	1.57051	1.57051	1.57052	1.57052	1.57052
8.9	1.57052	1.57053	1.57053	1.57053	1.57053	1.57054	1.57054	1.57054	1.57054	1.57055
9.0	1.57055	1.57055	1.57055	1.57056	1.57056	1.57056	1.57056	1.57057	1.57057	1.57057
9.1	1.57057	1.57058	1.57058	1.57058	1.57058	1.57058	1.57059	1.57059	1.57059	1.57059
9.2	1.57059	1.57060	1.57060	1.57060	1.57060	1.57060	1.57061	1.57061	1.57061	1.57061
9.3	1.57061	1.57062	1.57062	1.57062	1.57062	1.57062	1.57062	1.57063	1.57063	1.57063
9.4	1.57063	1.57063	1.57063	1.57064	1.57064	1.57064	1.57064	1.57064	1.57064	1.57065
9.5	1.57065	1.57065	1.57065	1.57065	1.57065	1.57065	1.57066	1.57066	1.57066	1.57066
9.6	1.57066	1.57066	1.57066	1.57066	1.57067	1.57067	1.57067	1.57067	1.57067	1.57067
9.7	1.57067	1.57067	1.57068	1.57068	1.57068	1.57068	1.57068	1.57068	1.57068	1.57068
9.8	1.57069	1.57069	1.57069	1.57069	1.57069	1.57069	1.57069	1.57069	1.57069	1.57069
9.9	1.57070	1.57070	1.57070	1.57070	1.57070	1.57070	1.57070	1.57070	1.57070	1.57070

INVERSE GUDERMANNIAN FUNCTION

x	0	1	2	3	4	5	6	7	8	9
0.0	0.00000	0.01000	0.02000	0.03000	0.04001	0.05002	0.06004	0.07006	0.08009	0.09012
0.1	0.10017	0.11022	0.12029	0.13037	0.14046	0.15057	0.16069	0.17082	0.18098	0.19115
0.2	0.20135	0.21156	0.22180	0.23206	0.24234	0.25265	0.26298	0.27334	0.28373	0.29415
0.3	0.30460	0.31509	0.32561	0.33616	0.34675	0.35737	0.36804	0.37874	0.38949	0.40028
0.4	0.41111	0.42199	0.43292	0.44390	0.45493	0.46600	0.47714	0.48833	0.49957	0.51087
0.5	0.52224	0.53366	0.54515	0.55671	0.56834	0.58003	0.59180	0.60364	0.61555	0.62755
0.6	0.63962	0.65178	0.66402	0.67636	0.68878	0.70129	0.71390	0.72661	0.73942	0.75233
0.7	0.76535	0.77848	0.79172	0.80508	0.81856	0.83217	0.84590	0.85976	0.87376	0.88790
0.8	0.90218	0.91660	0.93118	0.94592	0.96082	0.97589	0.99113	1.00654	1.02215	1.03794
0.9	1.05392	1.07011	1.08651	1.10313	1.11997	1.13704	1.15435	1.17192	1.18974	1.20783
1.0	1.22619	1.24485	1.26380	1.28306	1.30265	1.32258	1.34285	1.36349	1.38451	1.40593
1.1	1.42776	1.45003	1.47275	1.49594	1.51963	1.54384	1.56860	1.59394	1.61987	1.64645
1.2	1.67370	1.70166	1.73037	1.75987	1.79022	1.82147	1.85367	1.88689	1.92120	1.95667
1.3	1.99340	2.03147	2.07100	2.11210	2.15491	2.19959	2.24630	2.29524	2.34666	2.40080
1.4	2.45800	2.51861	2.58307	2.65193	2.72583	2.80558	2.89219	2.98695	3.09160	3.20843
1.5	3.34068	3.49307	3.67286	3.89217	4.17343	4.56609	5.22169	7.82865		

VII. ANALYTIC GEOMETRY

ANALYTIC GEOMETRY

Dr. Howard Eves

RECTANGULAR COORDINATES IN A PLANE

Rectangular (Cartesian) Coordinates

Let $X'X$ (called the *x-axis*) and $Y'Y$ (called the *y-axis*) be two perpendicular lines (here taken horizontally and vertically, respectively) intersecting in point O (called the *origin*). Then any point P in the plane of the axes is located by the distance x (called the *abscissa*) and the distance y (called the *ordinate*) from $Y'Y$ and $X'X$, respectively, to P, where x is taken as positive to the right and negative to the left of $Y'Y$, and y is taken as positive above and negative below $X'X$. The ordered pair of numbers, (x,y), are called *rectangular coordinates* of the point P.

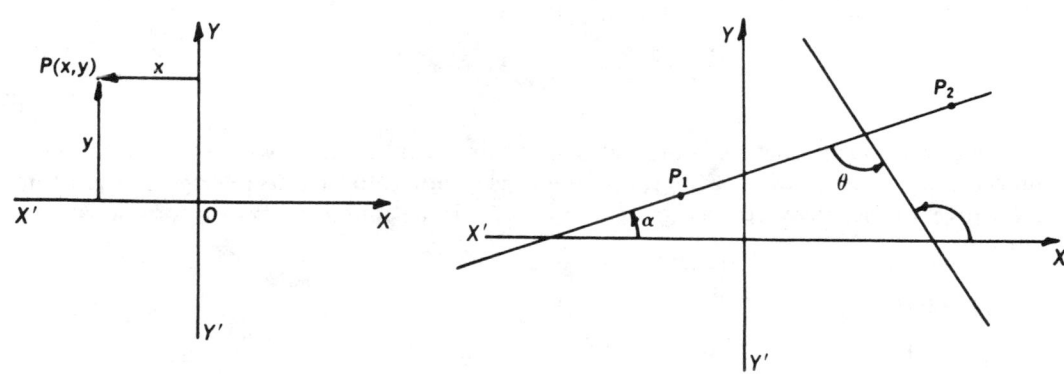

Points, Slopes, Angles

Let $P_1(x_1,y_1)$ and $P_2(x_2,y_2)$ be any two points and let α be the angle measured counterclockwise from $X'X$ to $P_1 P_2$.

Distance between P_1 and P_2:
$$\sqrt{(x_2 - x_1)^2 + (y_2 - y_1)^2}$$

Point dividing $P_1 P_2$ in ratio $\dfrac{r}{s}$:
$$\left(\frac{rx_2 + sx_1}{r + s}, \frac{ry_2 + sy_1}{r + s} \right)$$

Midpoint of $P_1 P_2$:
$$\left(\frac{x_1 + x_2}{2}, \frac{y_1 + y_2}{2} \right)$$

Slope m of $P_1 P_2$:
$$m = \tan \alpha = \frac{y_2 - y_1}{x_2 - x_1}$$

Angle θ between two lines of slopes m_1 and m_2:
$$\tan \theta = \frac{m_2 - m_1}{1 + m_1 m_2}$$

For parallel lines:
$$m_1 = m_2$$

For perpendicular lines:
$$m_1 m_2 = -1$$

Points P_1, P_2, P_3 are collinear if and only if
$$\begin{vmatrix} x_1 & y_1 & 1 \\ x_2 & y_2 & 1 \\ x_3 & y_3 & 1 \end{vmatrix} = 0.$$

Formulas for Use in Analytic Geometry

Polygonal Areas

Area of triangle $P_1 P_2 P_3$:

$$\frac{1}{2}\begin{vmatrix} x_1 & y_1 & 1 \\ x_2 & y_2 & 1 \\ x_3 & y_3 & 1 \end{vmatrix} = \tfrac{1}{2}(x_1 y_2 + x_2 y_3 + x_3 y_1 - y_1 x_2 - y_2 x_3 - y_3 x_1)$$

Area of polygon $P_1 P_2 \cdots P_n$:

$$\tfrac{1}{2}(x_1 y_2 + x_2 y_3 + \cdots + x_{n-1} y_n + x_n y_1 - y_1 x_2 - y_2 x_3 - \cdots - y_{n-1} x_n - y_n x_1)$$

Note. The parenthesis in the last formula is remembered by the device

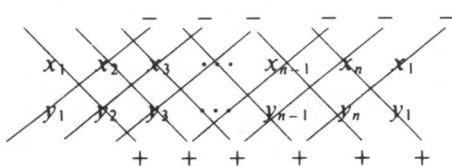

Here one adds the products of coordinates on the lines slanting downward to the right and subtracts the products of coordinates on the lines slanting upward to the right. The area is positive or negative according as $P_1 P_2 \cdots P_n$ is a counterclockwise or clockwise polygon.

Straight Lines

Line parallel to y-axis: $x = a$
Line parallel to x-axis: $y = b$
Slope y-intercept form: $y = mx + b$

Intercept form: $\dfrac{x}{a} + \dfrac{y}{b} = 1$

Point-slope form: $y - y_1 = m(x - x_1)$

Two-point form: $\dfrac{y - y_1}{x - x_1} = \dfrac{y_2 - y_1}{x_2 - x_1}$

or

$$\begin{vmatrix} x & y & 1 \\ x_1 & y_1 & 1 \\ x_2 & y_2 & 1 \end{vmatrix} = 0$$

Normal form: $x \cos \omega + y \sin \omega = p$
General form: $Ax + By + C = 0$

Slope: $m = -\dfrac{A}{B}$

Intercepts: $a = -\dfrac{C}{A}, \quad b = -\dfrac{C}{B}$

To reduce $Ax + By + C = 0$ to normal form, divide by $\pm \sqrt{A^2 + B^2}$, where the sign of the radical is chosen opposite to the sign of C when $C \neq 0$ and the same as the sign of B when $C = 0$.

Distance from $Ax + By + C = 0$ to P_1: $\dfrac{A x_1 + B y_1 + C}{\pm \sqrt{A^2 + B^2}}$

Formulas for Use in Analytic Geometry

Angle θ between lines $A_1x + B_1y + C_1 = 0$
and $A_2x + B_2y + C_2 = 0$

$$\tan \theta = \frac{A_1B_2 - A_2B_1}{A_1A_2 + B_1B_2}$$

Lines parallel: $\qquad A_1B_2 = A_2B_1$

Lines perpendicular: $\qquad A_1A_2 = -B_1B_2$

Lines $A_1x + B_1y + C_1 = 0, A_2x + B_2y + C_2 = 0, A_3x + B_3y + C_3 = 0$ are concurrent if and only if

$$\begin{vmatrix} A_1 & B_1 & C_1 \\ A_2 & B_2 & C_2 \\ A_3 & B_3 & C_3 \end{vmatrix} = 0.$$

Line of Best Fit

In seeking the straight line which best fits a given set of n points $P_1(x_1, y_1)$, $P_2(x_2, y_2)$, $\cdots$, $P_n(x_n, y_n)$, calculate

$$\bar{x} = \frac{x_1 + x_2 + \cdots + x_n}{n}, \quad \bar{y} = \frac{y_1 + y_2 + \cdots + y_n}{n},$$

$$m = \frac{(x_1y_1 + x_2y_2 + \cdots + x_ny_n) - n\bar{x}\bar{y}}{(x_1^2 + x_2^2 + \cdots + x_n^2) - n\bar{x}^2}.$$

Then the sought line is given by

$$y - \bar{y} = m(x - \bar{x}).$$

Circles

Center at origin, radius r: $\qquad x^2 + y^2 = r^2$

Center at (h, k), radius r: $\qquad (x - h)^2 + (y - k)^2 = r^2$

General form: $\qquad \begin{cases} Ax^2 + Ay^2 + Dx + Ey + F = 0, A \neq 0 \\ x^2 + y^2 + 2dx + 2ey + f = 0 \end{cases}$

Center: $\qquad (-d, -e)$

Radius: $\qquad r = \sqrt{d^2 + e^2 - f}$

Circle on P_1P_2 as diameter: $\qquad (x - x_1)(x - x_2) + (y - y_1)(y - y_2) = 0$

Three-point form: $\qquad \begin{vmatrix} x^2 + y^2 & x & y & 1 \\ x_1^2 + y_1^2 & x_1 & y_1 & 1 \\ x_2^2 + y_2^2 & x_2 & y_2 & 1 \\ x_3^2 + y_3^2 & x_3 & y_3 & 1 \end{vmatrix} = 0$

Conic Sections

A *conic section* is the locus of a point P that moves in the plane of a fixed point F (called. a *focus*) and a fixed line d (called a *directrix*), F not on d, such that the ratio of the distance of P from F to its distance from d is a constant e (called the *eccentricity*).

If $e = 1$, the conic is a *parabola*; if $e < 1$, an *ellipse*; if $e > 1$, a *hyperbola*.

Focus, $(0,0)$; directrix, $x = -a$: $\quad x^2 + y^2 = e^2(x + a)^2$

Formulas for Use in Analytic Geometry

Parabolas (e = 1)

Let p = distance from the vertex to the focus, e = eccentricity.

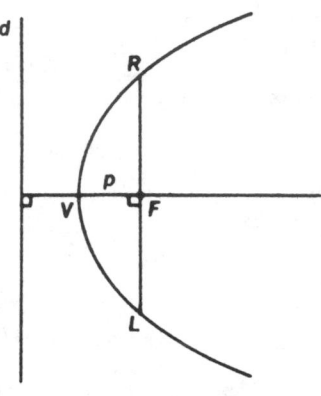

V: vertex
F: focus
d: directrix
LR: latus rectum
line VF: axis

Latus rectum:	$4p$
Distance from vertex to directrix:	p
Vertex at origin, focus at $(p,0)$:	$y^2 = 4px$
Vertex at origin, focus at $(-p,0)$:	$y^2 = -4px$
Vertex at origin, focus at $(0,p)$:	$x^2 = 4py$
Vertex at origin, focus at $(0,-p)$:	$x^2 = -4py$
Vertex, (h,k); focus, $(h + p,k)$:	$(y - k)^2 = 4p(x - h)$
Vertex, (h,k); focus, $(h - p,k)$:	$(y - k)^2 = -4p(x - h)$
Vertex, (h,k); focus, $(h,k + p)$:	$(x - h)^2 = 4p(y - k)$
Vertex, (h,k); focus, $(h,k - p)$:	$(x - h)^2 = -4p(y - k)$
General form, axis parallel to $X'X$:	$Cy^2 + Dx + Ey + F = 0$
General form, axis parallel to $Y'Y$:	$\begin{cases} Ax^2 + Dx + Ey + F = 0 \\ y = ax^2 + bx + c \end{cases}$
General form, axis oblique to coordinate axes:	$Ax^2 + Bxy + Cy^2 + Dx + Ey + F = 0,$ $B^2 - 4AC = 0$

Ellipses (e < 1)

Let $2a$ = major axis, $2b$ = minor axis, e = eccentricity.

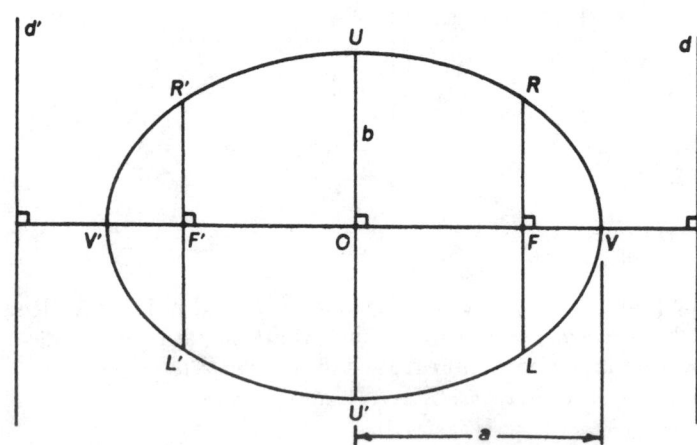

O: center
V, V': vertices
$V'V$: major axis = $2a$
$U'U$: minor axis = $2b$
F, F': foci
d, d': directrices
$LR, L'R'$: latera recta

Formulas for Use in Analytic Geometry

Eccentricity: $\qquad e = \dfrac{\sqrt{a^2 - b^2}}{a}$

Latus rectum: $\qquad \dfrac{2b^2}{a}$

Distance from center to either focus: $\qquad \sqrt{a^2 - b^2}$

Distance from center to either directrix: $\qquad \dfrac{a}{e}$

Sum of distances from any point on ellipse to the foci: $\qquad 2a$

Center at origin, foci on $X'X$: $\qquad \dfrac{x^2}{a^2} + \dfrac{y^2}{b^2} = 1$

Center at origin, foci on $Y'Y$: $\qquad \dfrac{x^2}{b^2} + \dfrac{y^2}{a^2} = 1$

Center at (h,k); major axis parallel to $X'X$: $\qquad \dfrac{(x - h)^2}{a^2} + \dfrac{(y - k)^2}{b^2} = 1$

Center at (h,k), major axis parallel to $Y'Y$: $\qquad \dfrac{(x - h)^2}{b^2} + \dfrac{(y - k)^2}{a^2} = 1$

General form, axes parallel to coordinate axes: $\qquad Ax^2 + Cy^2 + Dx + Ey + F = 0, AC > 0$

General form, axes oblique to coordinate axes: $\qquad Ax^2 + Bxy + Cy^2 + Dx + Ey + F = 0,$
$\qquad\qquad\qquad B^2 - 4AC < 0$

For a *circle*: $a = b$, $e = 0$, foci coincide at the center of the circle, directrices are at infinity.

Hyperbolas ($e > 1$)

Let $2a$ = transverse axis, $2b$ = conjugate axis, e = eccentricity.

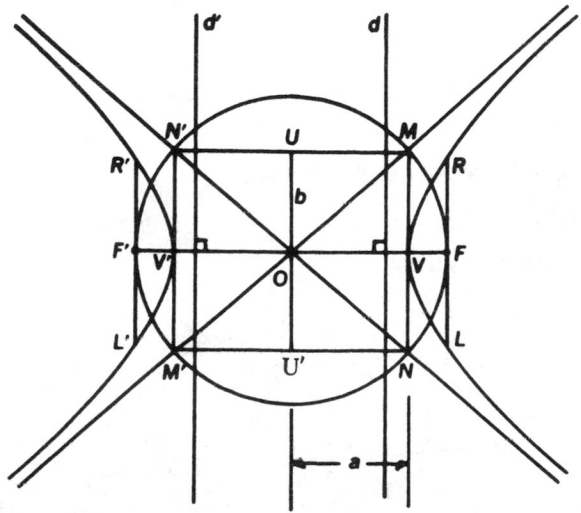

O: center
V, V': vertices
$V'V$: transverse axis = $2a$
$U'U$: conjugate axis = $2b$
F, F': foci
d, d': directrices
$LR, L'R'$: latera recta
lines $M'M$ and $N'N$: asymptotes

Formulas for Use in Analytic Geometry

Eccentricity: $\qquad e = \dfrac{\sqrt{a^2 + b^2}}{a}$

Latus rectum: $\qquad \dfrac{2b^2}{a}$

Distance from center to either focus: $\qquad \sqrt{a^2 + b^2}$

Distance from center to either directrix: $\qquad \dfrac{a}{e}$

Difference of distances of any point on hyperbola from foci: $\qquad 2a$

Center at origin, foci on $X'X$: $\qquad \dfrac{x^2}{a^2} - \dfrac{y^2}{b^2} = 1$

   Slopes of asymptotes: $\qquad \pm \dfrac{b}{a}$

Center at origin, foci on $Y'Y$: $\qquad \dfrac{y^2}{a^2} - \dfrac{x^2}{b^2} = 1$

   Slopes of asymptotes: $\qquad \pm \dfrac{a}{b}$

Center at (h,k), transverse axis parallel to $X'X$: $\qquad \dfrac{(x - h)^2}{a^2} - \dfrac{(y - k)^2}{b^2} = 1$

   Slopes of asymptotes: $\qquad \pm \dfrac{b}{a}$

Center at (h, k), transverse axis parallel to $Y'Y$: $\qquad \dfrac{(y - k)^2}{a^2} - \dfrac{(x - h)^2}{b^2} = 1$

   Slopes of asymptotes: $\qquad \pm \dfrac{a}{b}$

Center at origin, $X'X$ and $Y'Y$ for asymptotes: $\qquad xy = c$

Center at (h,k), asymptotes parallel to $X'X$ and $Y'Y$: $\qquad (x - h)(y - k) = c$

General form, axes parallel to coordinate axes: $\qquad Ax^2 + Cy^2 + Dx + Ey + F = 0, AC < 0$

General form, axes oblique to coordinate axes: $\qquad Ax^2 + Bxy + Cy^2 + Dx + Ey + F = 0,$
$\qquad\qquad B^2 - 4AC > 0$

For a *rectangular hyperbola*: $a = b, e = \sqrt{2}$, asymptotes are perpendicular.

General Equation of Second Degree

The nature of the graph of the general quadratic equation in x and y,

$$ax^2 + 2hxy + by^2 + 2gx + 2fy + c = 0,$$

is described in the following table in terms of the values of

$$\triangle = \begin{vmatrix} a & h & g \\ h & b & f \\ g & f & c \end{vmatrix}, \quad J = \begin{vmatrix} a & h \\ h & b \end{vmatrix},$$

$$I = a + b, \quad K = \begin{vmatrix} a & g \\ g & c \end{vmatrix} + \begin{vmatrix} b & f \\ f & c \end{vmatrix}.$$

Formulas for Use in Analytic Geometry

Case	$\triangle$	J	$\triangle/I$	K	Conic
1	$\neq 0$	> 0	< 0		real ellipse
2	$\neq 0$	> 0	> 0		imaginary ellipse
3	$\neq 0$	< 0			hyperbola
4	$\neq 0$	0			parabola
5	0	< 0			real intersecting lines
6	0	> 0			conjugate complex intersecting lines
7	0	0		< 0	real distinct parallel lines
8	0	0		> 0	conjugate complex parallel lines
9	0	0		0	coincident lines

In cases 1, 2, and 3, the center (x_0, y_0) of the conic is given by the simultaneous solution of the equations

$$ax + hy + g = 0, \quad hx + by + f = 0.$$

The equations of the axes of the conic are

$$y - y_0 = m(x - x_0), \quad y - y_0 = -\frac{1}{m}(x - x_0),$$

where m is the positive root of

$$hm^2 + (a - b)m - h = 0.$$

Transformation of Coordinates

To transform an equation of a curve from an old system of rectangular coordinates (x, y) to a new system of rectangular coordinates (x', y'), substitute for each old variable in the equation of the curve its expression in terms of the new variables.

Translation: $\begin{cases} x = x' + h \\ y = y' + k \end{cases}$ The new axes are parallel to the old axes and the coordinates of the new origin in terms of the old system are (h, k).

Rotation: $\begin{cases} x = x' \cos\theta - y' \sin\theta \\ y = x' \sin\theta + y' \cos\theta \end{cases}$ The new origin is coincident with the old origin and the new axes make an angle θ with the old axes.

To remove the xy-term from the equation

$$ax^2 + 2hxy + by^2 + 2gx + 2fy + c = 0,$$

rotate the coordinate axes about the origin through the acute angle $\theta = \arctan m$, where m is the positive root of

$$hm^2 + (a - b)m - h = 0.$$

OBLIQUE COORDINATES IN A PLANE

Oblique (Cartesian) Coordinates

Let $X'X$ (called the *x-axis*, here taken horizontally) and $Y'Y$ (called the *y-axis*) be two lines intersecting in point O (called the *origin*), and denote by ω the counterclockwise angle from $X'X$ to $Y'Y$. Then any point P in the plane of the axes is located by the distance x (called the *abscissa*) measured parallel to the x-axis and the distance y (called

Formulas for Use in Analytic Geometry

the *ordinate*) measured parallel to the *y*-axis from $Y'Y$ and $X'X$, respectively, to *P*, where *x* is taken as positive to the right and negative to the left of $Y'Y$, and *y* is taken as positive above and negative below $X'X$. The ordered pair of numbers, (x,y), are called *oblique coordinates* of the point *P*. If $\omega = 90°$, this coordinate system becomes a rectangular (Cartesian) coordinate system.

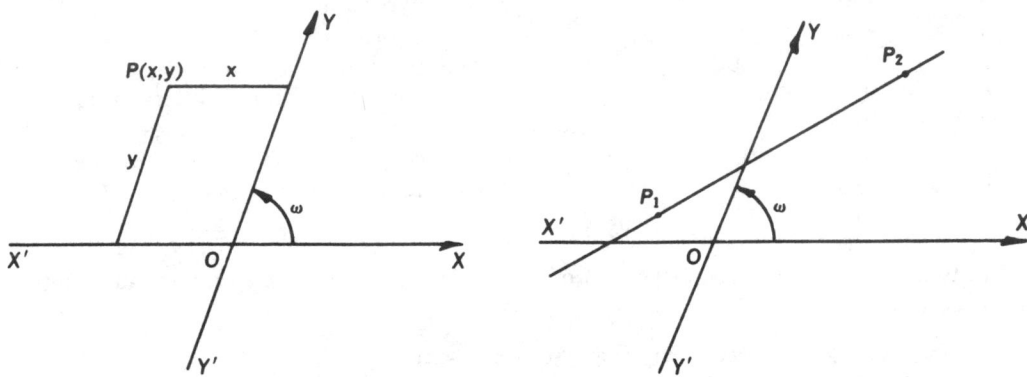

Points

Let $P_1(x_1,y_1)$ and $P_2(x_2,y_2)$ be any two points.
Distance between P_1 and P_2:

$$\sqrt{(x_2 - x_1)^2 + (y_2 - y_1)^2 + 2(x_2 - x_1)(y_2 - y_1)\cos \omega}$$

Point dividing P_1P_2 in ratio $\dfrac{r}{s}$: $\left(\dfrac{rx_2 + sx_1}{r + s}, \dfrac{ry_2 + sy_1}{r + s}\right)$

Midpoint of P_1P_2: $\left(\dfrac{x_1 + x_2}{2}, \dfrac{y_1 + y_2}{2}\right)$

Points P_1, P_2, P_3 are collinear if and only if $\begin{vmatrix} x_1 & y_1 & 1 \\ x_2 & y_2 & 1 \\ x_3 & y_3 & 1 \end{vmatrix} = 0.$

Polygonal Areas

Area of triangle $P_1P_2P_3$:

$$\sin \omega \begin{vmatrix} x_1 & y_1 & 1 \\ x_2 & y_2 & 1 \\ x_3 & y_3 & 1 \end{vmatrix} = \tfrac{1}{2}(\sin \omega)(x_1y_2 + x_2y_3 + x_3y_1 - y_1x_2 - y_2x_3 - y_3x_1)$$

Area of polygon $P_1P_2\cdots P_n$:

$$\tfrac{1}{2}(\sin \omega)(x_1y_2 + x_2y_3 + \cdots + x_{n-1}y_n + x_ny_1 - y_1x_2 - y_2x_3 - \cdots - y_{n-1}x_n - y_nx_1)$$

The area is positive or negative according as $P_1P_2\cdots P_n$ is a counterclockwise or clockwise polygon.

Straight Lines

Line parallel to *y*-axis: $x = a$

Line parallel to *x*-axis: $y = b$

Formulas for Use in Analytic Geometry

Intercept form:

$$\frac{x}{a} + \frac{y}{b} = 1$$

Two-point form:

$$\frac{y - y_1}{x - x_1} = \frac{y_2 - y_1}{x_2 - x_1}, \quad \text{or} \quad \begin{vmatrix} x & y & 1 \\ x_1 & y_1 & 1 \\ x_2 & y_2 & 1 \end{vmatrix} = 0$$

General form: $Ax + By + C = 0$

Intercepts: $a = -\dfrac{C}{A}, \quad b = -\dfrac{C}{B}$

Distance from $Ax + By + C = 0$ **to** P_1:

$$\frac{(Ax_1 + By_1 + C)\sin \omega}{\pm \sqrt{A^2 + B^2 - 2AB\cos \omega}}$$

Angle θ **between lines** $A_1 x + B_1 y + C_1 = 0$
and $A_2 x + B_2 y + C_2 = 0$:

$$\tan \theta = \frac{(A_1 B_2 - A_2 B_1)\sin \omega}{A_1 A_2 + B_1 B_2 - (A_1 B_2 + A_2 B_1)\cos \omega}$$

Lines parallel: $A_1 B_2 = A_2 B_1$

Lines perpendicular: $A_1 A_2 + B_1 B_2 = (A_1 B_2 + A_2 B_1)\cos \omega$

Lines $A_1 x + B_1 y + C_1 = 0$, $A_2 x + B_2 y + C_2 = 0$, $A_3 x + B_3 y + C_3 = 0$ **are concurrent if and only if**

$$\begin{vmatrix} A_1 & B_1 & C_1 \\ A_2 & B_2 & C_2 \\ A_3 & B_3 & C_3 \end{vmatrix} = 0.$$

Circles

Center at (h, k), radius r: $(x - h)^2 + (y - k)^2 + 2(x - h)(y - k)\cos \omega = r^2$

Transformation of Coordinates

Translation: $\begin{cases} x = x' + h \\ y = y' + k \end{cases}$ The new axes are parallel to the old axes and the coordinates of the new origin in terms of the old system are (h, k).

From one oblique system to another, origin fixed:

$$\begin{cases} x = \dfrac{x'\sin(\omega - \theta) + y'\sin(\omega - \omega' - \theta)}{\sin \omega} \\ y = \dfrac{x'\sin \theta + y'\sin(\omega' + \theta)}{\sin \omega} \end{cases}$$

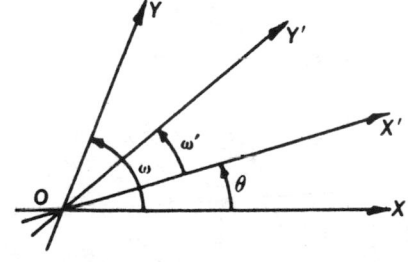

The old and new origins coincide; the old axes intersect at angle ω; the new axes intersect at angle ω'; the counterclockwise angle from x-axis to x'-axis is θ.

Formulas for Use in Analytic Geometry

POLAR COORDINATES IN A PLANE

Polar Coordinates

In a plane, let OX (called the *initial line*) be a fixed ray radiating from point O (called the *pole* or *origin*). Then any point P, other than O, in the plane is located by angle θ (called the *vectorial angle*) measured from OX to the line determined by O and P and the distance r (called the *radius vector*) from O to P, where θ is taken as positive if measured counterclockwise and negative if measured clockwise, and r is taken as positive if measured along the terminal side of angle θ and negative if measured along the terminal side of θ produced through the pole. Such an ordered pair of numbers, (r, θ), are called *polar coordinates* of the point P. The polar coordinates of the pole O are taken as $(0, \theta)$, where θ is arbitrary. It follows that, for a given initial line and pole, each point of the plane has infinitely many polar coordinates, but each pair of coordinates corresponds to only one point.

Example

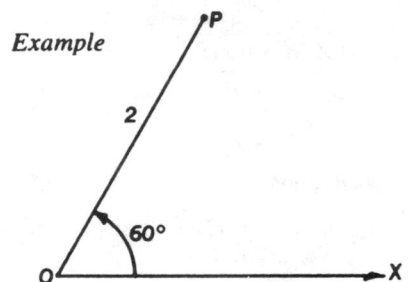

Some polar coordinates of P are: $(2, 60°)$, $(2, 420°)$, $(2, -300°)$, $(-2, 240°)$, $(-2, -120°)$.

Points

Distance between P_1 and P_2: $\sqrt{r_1^2 + r_2^2 - 2r_1 r_2 \cos(\theta_1 - \theta_2)}$

Points P_1, P_2, P_3 are collinear if and only if

$$r_2 r_3 \sin(\theta_3 - \theta_2) + r_3 r_1 \sin(\theta_1 - \theta_3) + r_1 r_2 \sin(\theta_2 - \theta_1) = 0.$$

Polygonal Areas

Area of triangle $P_1 P_2 P_3$:

$$\tfrac{1}{2}[r_1 r_2 \sin(\theta_2 - \theta_1) + r_2 r_3 \sin(\theta_3 - \theta_2) + r_3 r_1 \sin(\theta_1 - \theta_3)]$$

Area of polygon $P_1 P_2 \cdots P_n$:

$$\tfrac{1}{2}[r_1 r_2 \sin(\theta_2 - \theta_1) + r_2 r_3 \sin(\theta_3 - \theta_2) + \cdots + r_{n-1} r_n \sin(\theta_n - \theta_{n-1}) + r_n r_1 \sin(\theta_1 - \theta_n)]$$

The area is positive or negative according as $P_1 P_2 \cdots P_n$ is a counterclockwise or clockwise polygon.

Straight Lines

Let p = distance of line from O, ω = counterclockwise angle from OX to the perpendicular through O to the line.

Normal form: $r \cos(\theta - \omega) = p$

Two-point form: $r[r_1 \sin(\theta - \theta_1) - r_2 \sin(\theta - \theta_2)] = r_1 r_2 \sin(\theta_2 - \theta_1)$

Circles

Center at pole, radius a: $r = a$
Center at $(a, 0)$ and passing
    through the pole: $r = 2a \cos \theta$

Formulas for Use in Analytic Geometry

Center at $\left(a, \dfrac{\pi}{2}\right)$ and passing

through the pole: $r = 2a \sin \theta$

Center (h, α), radius a: $r^2 - 2hr \cos(\theta - \alpha) + h^2 - a^2 = 0$

Conics

Let $2p$ = distance from directrix to focus, e = eccentricity.

Focus at pole, directrix to left of pole: $r = \dfrac{2ep}{1 - e \cos \theta}$

Focus at pole, directrix to right of pole: $r = \dfrac{2ep}{1 + e \cos \theta}$

Focus at pole, directrix below pole: $r = \dfrac{2ep}{1 - e \sin \theta}$

Focus at pole, directrix above pole: $r = \dfrac{2ep}{1 + e \sin \theta}$

Parabola with vertex at pole,
  directrix to left of pole: $r = \dfrac{4p \cos \theta}{\sin^2 \theta}$

Ellipse with center at pole, semiaxes a and b horizontal and vertical, respectively: $r^2 = \dfrac{a^2 b^2}{a^2 \sin^2 \theta + b^2 \cos^2 \theta}$

Hyperbola with center at pole, semiaxes a and b horizontal and vertical, respectively: $r^2 = \dfrac{a^2 b^2}{a^2 \sin^2 \theta - b^2 \cos^2 \theta}$

Relations Between Rectangular and Polar Coordinates

Let the positive x-axis coincide with the initial line and let r be nonnegative.

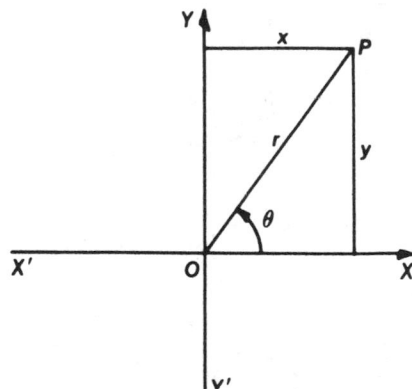

$x = r \cos \theta, \quad y = r \sin \theta,$

$r = \sqrt{x^2 + y^2}, \quad \theta = \arctan \dfrac{y}{x},$

$\sin \theta = \dfrac{y}{\sqrt{x^2 + y^2}}, \quad \cos \theta = \dfrac{x}{\sqrt{x^2 + y^2}}$

RECTANGULAR COORDINATES IN SPACE

Rectangular (Cartesian) Coordinates

Let $X'X$, $Y'Y$, $Z'Z$ (called the *x-axis*, the *y-axis*, and the *z-axis*, respectively) be three mutually perpendicular lines in space intersecting in a point O (called the *origin*), forming in this way three mutually perpendicular planes XOY, XOZ, YOZ (called the *xy-*

Formulas for Use in Analytic Geometry

plane, the *xz-plane*, and the *yz-plane*, respectively). Then any point *P* **of space is** located by its signed distances *x*, *y*, *z* from the *yz*-plane, the *xz*-plane, and the *xy*-plane, respectively, where *x* and *y* are the rectangular coordinates with respect to the axes *X'X* and *Y'Y* of the orthogonal projection *P'* of *P* on the *xy*-plane (here taken horizontally) and *z* is taken as positive above and negative below the *xy*-plane. The ordered triple of numbers, (x, y, z), are called *rectangular coordinates* of the point *P*.

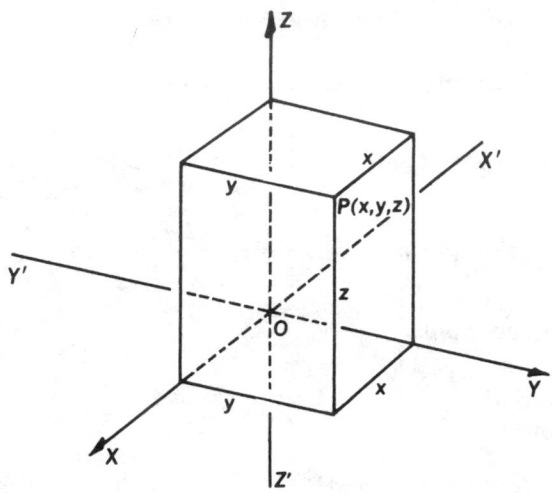

Points

Let $P_1(x_1, y_1, z_1)$ and $P_2(x_2, y_2, z_2)$ be any two points.

Distance between P_1 and P_2: $\sqrt{(x_2 - x_1)^2 + (y_2 - y_1)^2 + (z_2 - z_1)^2}$

Point dividing $P_1 P_2$ in ratio $\dfrac{r}{s}$: $\left(\dfrac{rx_2 + sx_1}{r + s}, \quad \dfrac{ry_2 + sy_1}{r + s}, \quad \dfrac{rz_2 + sz_1}{r + s} \right)$

Midpoint of $P_1 P_2$: $\left(\dfrac{x_1 + x_2}{2}, \quad \dfrac{y_1 + y_2}{2} \quad \dfrac{z_1 + z_2}{2} \right)$

Points P_1, P_2, P_3 are collinear if and only if

$$x_2 - x_1 : y_2 - y_1 : z_2 - z_1 = x_3 - x_1 : y_3 - y_1 : z_3 - z_1.$$

Points P_1, P_2, P_3, P_4 are coplanar if and only if
$$\begin{vmatrix} x_1 & y_1 & z_1 & 1 \\ x_2 & y_2 & z_2 & 1 \\ x_3 & y_3 & z_3 & 1 \\ x_4 & y_4 & z_4 & 1 \end{vmatrix} = 0.$$

Area of triangle $P_1 P_2 P_3$:

$$\frac{1}{2} \sqrt{ \begin{vmatrix} y_1 & z_1 & 1 \\ y_2 & z_2 & 1 \\ y_3 & z_3 & 1 \end{vmatrix}^2 + \begin{vmatrix} z_1 & x_1 & 1 \\ z_2 & x_2 & 1 \\ z_3 & x_3 & 1 \end{vmatrix}^2 + \begin{vmatrix} x_1 & y_1 & 1 \\ x_2 & y_2 & 1 \\ x_3 & y_3 & 1 \end{vmatrix}^2 }$$

Formulas for Use in Analytic Geometry

Volume of tetrahedron $P_1 P_2 P_3 P_4$: $\frac{1}{6} \begin{vmatrix} x_1 & y_1 & z_1 & 1 \\ x_2 & y_2 & z_2 & 1 \\ x_3 & y_3 & z_3 & 1 \\ x_4 & y_4 & z_4 & 1 \end{vmatrix}$

Direction Numbers and Direction Cosines

Let α, β, γ (called *direction angles*) be the angles that $P_1 P_2$, or any line parallel to $P_1 P_2$, makes with the x-, y-, and z-axis, respectively. Let d = distance between P_1 and P_2.

Direction cosines of $P_1 P_2$:

$$\cos \alpha = \frac{x_2 - x_1}{d}, \quad \cos \beta = \frac{y_2 - y_1}{d}, \quad \cos \gamma = \frac{z_2 - z_1}{d}$$

$$\cos^2 \alpha + \cos^2 \beta + \cos^2 \gamma = 1$$

If a, b, c are direction numbers of $P_1 P_2$, then:

$$a : b : c = x_2 - x_1 : y_2 - y_1 : z_2 - z_1$$
$$= \cos \alpha : \cos \beta : \cos \gamma$$

$$\cos \alpha = \frac{a}{\pm \sqrt{a^2 + b^2 + c^2}}, \quad \cos \beta = \frac{b}{\pm \sqrt{a^2 + b^2 + c^2}},$$

$$\cos \gamma = \frac{c}{\pm \sqrt{a^2 + b^2 + c^2}}$$

Angle between two lines with direction angles $\alpha_1, \beta_1, \gamma_1$ and $\alpha_2, \beta_2, \gamma_2$:

$$\cos \theta = \cos \alpha_1 \cos \alpha_2 + \cos \beta_1 \cos \beta_2 + \cos \gamma_1 \cos \gamma_2$$

For parallel lines: $\alpha_1 = \alpha_2, \ \beta_1 = \beta_2, \ \gamma_1 = \gamma_2$

For perpendicular lines:

$$\cos \alpha_1 \cos \alpha_2 + \cos \beta_1 \cos \beta_2 + \cos \gamma_1 \cos \gamma_2 = 0$$

Angle between two lines with directions (a_1, b_1, c_1) and (a_2, b_2, c_2):

$$\cos \theta = \frac{a_1 a_2 + b_1 b_2 + c_1 c_2}{\sqrt{a_1^2 + b_1^2 + c_1^2} \ \sqrt{a_2^2 + b_2^2 + c_2^2}}$$

$$\sin \theta = \frac{\sqrt{(b_1 c_2 - c_1 b_2)^2 + (c_1 a_2 - a_1 c_2)^2 + (a_1 b_2 - b_1 a_2)^2}}{\sqrt{a_1^2 + b_1^2 + c_1^2} \ \sqrt{a_2^2 + b_2^2 + c_2^2}}$$

For parallel lines: $a_1 : b_1 : c_1 = a_2 : b_2 : c_2$

For perpendicular lines: $a_1 a_2 + b_1 b_2 + c_1 c_2 = 0$

The direction

$$(b_1 c_2 - c_1 b_2, c_1 a_2 - a_1 c_2, a_1 b_2 - b_1 a_2)$$

is perpendicular to both directions (a_1, b_1, c_1) and (a_2, b_2, c_2).

The directions (a_1, b_1, c_1), (a_2, b_2, c_2), (a_3, b_3, c_3) are parallel to a common plane if and only if

$$\begin{vmatrix} a_1 & b_1 & c_1 \\ a_2 & b_2 & c_2 \\ a_3 & b_3 & c_3 \end{vmatrix} = 0.$$

Formulas for Use in Analytic Geometry

Straight Lines

Point-direction form: $\dfrac{x - x_1}{a} = \dfrac{y - y_1}{b} = \dfrac{z - z_1}{c}$

Two-point form: $\dfrac{x - x_1}{x_2 - x_1} = \dfrac{y - y_1}{y_2 - y_1} = \dfrac{z - z_1}{z_2 - z_1}$

Parametric form: $x = x_1 + ta, \ y = y_1 + tb, \ z = z_1 + tc$

General form: $\begin{cases} A_1 x + B_1 y + C_1 z + D_1 = 0 \\ A_2 x + B_2 y + C_2 z + D_2 = 0 \end{cases}$

Direction of line: $(B_1 C_2 - C_1 B_2, C_1 A_2 - A_1 C_2, A_1 B_2 - B_1 A_2)$

Projection of segment $P_1 P_2$ on any line having the direction (a, b, c):

$$\frac{(x_2 - x_1)a + (y_2 - y_1)b + (z_2 - z_1)c}{\sqrt{a^2 + b^2 + c^2}}$$

Distance from point P_0 to line through P_1 in direction (a, b, c):

$$\sqrt{\frac{\begin{vmatrix} y_0 - y_1 & z_0 - z_1 \\ b & c \end{vmatrix}^2 + \begin{vmatrix} z_0 - z_1 & x_0 - x_1 \\ c & a \end{vmatrix}^2 + \begin{vmatrix} x_0 - x_1 & y_0 - y_1 \\ a & b \end{vmatrix}^2}{a^2 + b^2 + c^2}}$$

Distance between line through P_1 in direction (a_1, b_1, c_1) and line through P_2 in direction (a_2, b_2, c_2):

$$\pm \frac{\begin{vmatrix} x_2 - x_1 & y_2 - y_1 & z_2 - z_1 \\ a_1 & b_1 & c_1 \\ a_2 & b_2 & c_2 \end{vmatrix}}{\sqrt{\begin{vmatrix} b_1 & c_1 \\ b_2 & c_2 \end{vmatrix}^2 + \begin{vmatrix} c_1 & a_1 \\ c_2 & a_2 \end{vmatrix}^2 + \begin{vmatrix} a_1 & b_1 \\ a_2 & b_2 \end{vmatrix}^2}}$$

The line through P_1 in direction (a_1, b_1, c_1) and the line through P_2 in direction (a_2, b_2, c_2) intersect if and only if

$$\begin{vmatrix} x_2 - x_1 & y_2 - y_1 & z_2 - z_1 \\ a_1 & b_1 & c_1 \\ a_2 & b_2 & c_2 \end{vmatrix} = 0.$$

Planes

General form: $Ax + By + Cz + D = 0$

Direction of normal: (A, B, C)

Perpendicular to yz-plane: $By + Cz + D = 0$

Perpendicular to xz-plane: $Ax + Cz + D = 0$

Perpendicular to xy-plane: $Ax + By + D = 0$

Perpendicular to x-axis: $Ax + D = 0$

Perpendicular to y-axis: $By + D = 0$

Perpendicular to z-axis: $Cz + D = 0$

Intercept form: $\dfrac{x}{a} + \dfrac{y}{b} + \dfrac{z}{c} = 1$

Plane through point P_1 and perpendicular to direction (a, b, c):

$$a(x - x_1) + b(y - y_1) + c(z - z_1) = 0$$

Formulas for Use in Analytic Geometry

Plane through point P_1 and parallel to directions (a_1,b_1,c_1) and (a_2,b_2,c_2):

$$\begin{vmatrix} x - x_1 & y - y_1 & z - z_1 \\ a_1 & b_1 & c_1 \\ a_2 & b_2 & c_2 \end{vmatrix} = 0$$

Plane through points P_1 and P_2 parallel to direction (a,b,c):

$$\begin{vmatrix} x - x_1 & y - y_1 & z - z_1 \\ x_2 - x_1 & y_2 - y_1 & z_2 - z_1 \\ a & b & c \end{vmatrix} = 0$$

Three-point form:

$$\begin{vmatrix} x & y & z & 1 \\ x_1 & y_1 & z_1 & 1 \\ x_2 & y_2 & z_2 & 1 \\ x_3 & y_3 & z_3 & 1 \end{vmatrix} = 0 \quad \text{or} \quad \begin{vmatrix} x - x_1 & y - y_1 & z - z_1 \\ x_2 - x_1 & y_2 - y_1 & z_2 - z_1 \\ x_3 - x_1 & y_3 - y_1 & z_3 - z_1 \end{vmatrix} = 0$$

Normal form (p = distance from origin to plane; α, β, γ are direction angles of perpendicular to plane from origin):

$$x \cos \alpha + y \cos \beta + z \cos \gamma = p$$

To reduce $Ax + By + Cz + D = 0$ to normal form, divide by $\pm \sqrt{A^2 + B^2 + C^2}$, where the sign of the radical is chosen opposite to the sign of D when $D \neq 0$, the same as the sign of C when $D = 0$ and $C \neq 0$, the same as the sign of B when $C = D = 0$.

Distance from point P_1 to plane $Ax + By + Cz + D = 0$:

$$\frac{Ax_1 + By_1 + Cz_1 + D}{\pm \sqrt{A^2 + B^2 + C^2}}$$

Angle θ between planes $A_1x + B_1y + C_1z + D_1 = 0$ and $A_2x + B_2y + C_2z + D_2 = 0$:

$$\cos \theta = \frac{A_1A_2 + B_1B_2 + C_1C_2}{\sqrt{A_1^2 + B_1^2 + C_1^2} \sqrt{A_2^2 + B_2^2 + C_2^2}}$$

Planes parallel: $A_1 : B_1 : C_1 = A_2 : B_2 : C_2$

Planes perpendicular: $A_1A_2 + B_1B_2 + C_1C_2 = 0$

Spheres

Center at origin, radius r: $x^2 + y^2 + z^2 = r^2$

Center at (g,h,k), radius r: $(x - g)^2 + (y - h)^2 + (z - k)^2 = r^2$

General form: $\begin{cases} Ax^2 + Ay^2 + Az^2 + Dx + Ey + Fz + M = 0, \quad A \neq 0 \\ x^2 + y^2 + z^2 + 2dx + 2ey + 2fz + m = 0 \end{cases}$

Center: $(-d, -e, -f)$

Radius: $r = \sqrt{d^2 + e^2 + f^2 - m}$

Formulas for Use in Analytic Geometry

Sphere on $P_1 P_2$ as diameter:

$$(x - x_1)(x - x_2) + (y - y_1)(y - y_2) + (z - z_1)(z - z_2) = 0$$

Four-point form:

$$\begin{vmatrix} x^2 + y^2 + z^2 & x & y & z & 1 \\ x_1^2 + y_1^2 + z_1^2 & x_1 & y_1 & z_1 & 1 \\ x_2^2 + y_2^2 + z_2^2 & x_2 & y_2 & z_2 & 1 \\ x_3^2 + y_3^2 + z_3^2 & x_3 & y_3 & z_3 & 1 \\ x_4^2 + y_4^2 + z_4^2 & x_4 & y_4 & z_4 & 1 \end{vmatrix} = 0$$

The Seventeen Quadric Surfaces in Standard Form

1. Real ellipsoid: $x^2/a^2 + y^2/b^2 + z^2/c^2 = 1$
2. Imaginary ellipsoid: $x^2/a^2 + y^2/b^2 + z^2/c^2 = -1$
3. Hyperboloid of one sheet: $x^2/a^2 + y^2/b^2 - z^2/c^2 = 1$
4. Hyperboloid of two sheets: $x^2/a^2 + y^2/b^2 - z^2/c^2 = -1$
5. Real quadric cone: $x^2/a^2 + y^2/b^2 - z^2/c^2 = 0$
6. Imaginary quadric cone: $x^2/a^2 + y^2/b^2 + z^2/c^2 = 0$
7. Elliptic paraboloid: $x^2/a^2 + y^2/b^2 + 2z = 0$
8. Hyperbolic paraboloid: $x^2/a^2 - y^2/b^2 + 2z = 0$
9. Real elliptic cylinder: $x^2/a^2 + y^2/b^2 = 1$
10. Imaginary elliptic cylinder: $x^2/a^2 + y^2/b^2 = -1$
11. Hyperbolic cylinder: $x^2/a^2 - y^2/b^2 = -1$
12. Real intersecting planes: $x^2/a^2 - y^2/b^2 = 0$
13. Imaginary intersecting planes: $x^2/a^2 + y^2/b^2 = 0$
14. Parabolic cylinder: $x^2 + 2rz = 0$
15. Real parallel planes: $x^2 = a^2$
16. Imaginary parallel planes: $x^2 = -a^2$
17. Coincident planes: $x^2 = 0$

General Equation of Second Degree

The nature of the graph of the general quadratic equation in x, y, z,

$$ax^2 + by^2 + cz^2 + 2fyz + 2gzx + 2hxy + 2px + 2qy + 2rz + d = 0,$$

is described in the following table in terms of $\rho_3, \rho_4, \Delta, k_1, k_2, k_3$, where

$$e = \begin{bmatrix} a & h & g \\ h & b & f \\ g & f & c \end{bmatrix}, \quad E = \begin{bmatrix} a & h & g & p \\ h & b & f & q \\ g & f & c & r \\ p & q & r & d \end{bmatrix},$$

$$\rho_3 = \text{rank } e, \quad \rho_4 = \text{rank } E,$$
$$\Delta = \text{determinant of } E,$$

k_1, k_2, k_3 are the roots of $\begin{vmatrix} a - x & h & g \\ h & b - x & f \\ g & f & c - x \end{vmatrix} = 0.$

Formulas for Use in Analytic Geometry

Case	ρ_3	ρ_4	Sign of Δ	Nonzero k's same sign?	Quadric Surface
1	3	4	$-$	yes	Real ellipsoid
2	3	4	$+$	yes	Imaginary ellipsoid
3	3	4	$+$	no	Hyperboloid of one sheet
4	3	4	$-$	no	Hyperboloid of two sheets
5	3	3		no	Real quadric cone
6	3	3		yes	Imaginary quadric cone
7	2	4	$-$	yes	Elliptic paraboloid
8	2	4	$+$	no	Hyperbolic paraboloid
9	2	3		yes	Real elliptic cylinder
10	2	3		yes	Imaginary elliptic cylinder
11	2	3		no	Hyperbolic cylinder
12	2	2		no	Real intersecting planes
13	2	2		yes	Imaginary intersecting planes
14	1	3			Parabolic cylinder
15	1	2			Real parallel planes
16	1	2			Imaginary parallel planes
17	1	1			Coincident planes

Cylindrical and Conical Surfaces

Any equation in just two of the variables x, y, z represents a *cylindrical surface* whose elements are parallel to the axis of the missing variable.

Any equation homogeneous in the variables x, y, z represents a *conical surface* whose vertex is at the origin.

Transformation of Coordinates

To transform an equation of a surface from an old system of rectangular coordinates (x, y, z) to a new system of rectangular coordinates (x', y', z'), substitute for each old variable in the equation of the surface its expression in terms of the new variables.

Translation:
$$x = x' + h$$
$$y = y' + k$$
$$z = z' + l$$
The new axes are parallel to the old axes and the coordinates of the new origin in terms of the old system are (h, k, l).

Rotation about the origin:
$$x = \lambda_1 x' + \lambda_2 y' + \lambda_3 z'$$
$$y = \mu_1 x' + \mu_2 y' + \mu_3 z'$$
$$z = \nu_1 x' + \nu_2 y' + \nu_3 z'$$
The new origin is coincident with the old origin and the x'-axis, y'-axis, z'-axis have direction cosines $(\lambda_1, \mu_1, \nu_1)$, $(\lambda_2, \mu_2, \nu_2)$, $(\lambda_3, \mu_3, \nu_3)$, respectively, with respect to the old system of axes.

$$x' = \lambda_1 x + \mu_1 y + \nu_1 z$$
$$y' = \lambda_2 x + \mu_2 y + \nu_2 z$$
$$z' = \lambda_3 x + \mu_3 y + \nu_3 z$$

Formulas for Use in Analytic Geometry

Cylindrical Coordinates

If (r, θ, z) are the cylindrical co-ordinates and (x, y, z) the rectangular coordinates of a point P, then

$$x = r \cos \theta, \quad r = \sqrt{x^2 + y^2},$$

$$y = r \sin \theta, \quad \theta = \arctan \frac{y}{x},$$

$$z = z, \qquad z = z.$$

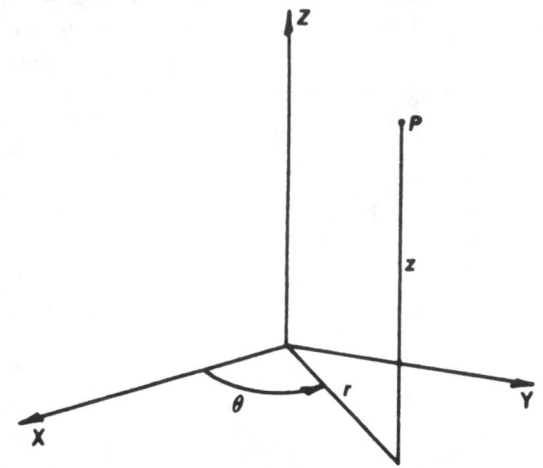

Spherical Coordinates

If (ρ, θ, ϕ) are the spherical co-ordinates and (x, y, z) the rectangular coordinates of a point P, then

$$x = \rho \cos \theta \sin \phi$$

$$y = \rho \sin \theta \sin \phi$$

$$z = \rho \cos \phi$$

$$\phi = \arccos \frac{z}{\sqrt{x^2 + y^2 + z^2}},$$

$$\theta = \arctan \frac{y}{x}.$$

$$p^2 = x^2 + y^2 + z^2$$

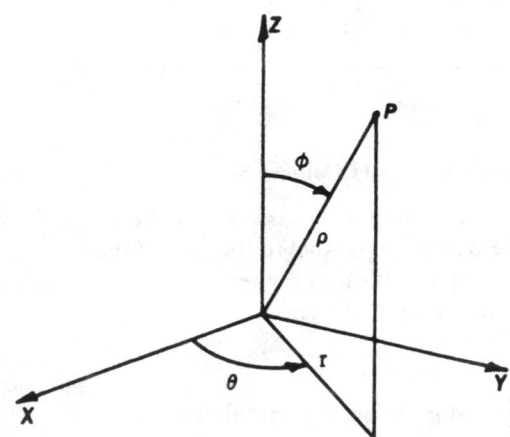

CURVES AND SURFACES

Dr. Howard Eves

The curves and surfaces collected here for reference appear frequently in mathematical literature. The equations most generally associated with each figure are given. The equation of a plane curve when placed otherwise on the coordinate frame of reference may often be found from the given equation by the following rules.

RECTANGULAR COORDINATES

1. If a given curve is reflected in the x-axis, the new equation is obtained from the old by replacing y by $-y$.

2. If a given curve is reflected in the y-axis, the new equation is obtained from the old by replacing x by $-x$.

3. If a given curve is reflected in the origin, the new equation is obtained from the old by replacing x by $-x$ and y by $-y$.

4. If a given curve is reflected in the line $y = x$, the new equation is obtained from the old by interchanging x and y.

5. If a given curve is rotated about the origin through 90°, the new equation is obtained from the old by replacing x by y and y by $-x$.

6. If a given curve is rotated about the origin through −90°, the new equation is obtained from the old by replacing x by $-y$ and y by x.

7. If a given curve is translated a distance h in the x-direction, the new equation is obtained from the old by replacing x by $x - h$.

8. If a given curve is translated a distance k in the y-direction, the new equation is obtained from the old by replacing y by $y - k$.

9. If a given curve is altered by multiplying all the abscissas by a, the new equation is obtained from the old by replacing x by x/a.

10. If a given curve is altered by multiplying all the ordinates by b, the new equation is obtained from the old by replacing y by y/b.

POLAR COORDINATES

1. If a given curve is reflected in the polar axis, the new equation is obtained from the old by replacing θ by $-\theta$, or by replacing r by $-r$ and θ by $\pi - \theta$.

2. If a given curve is reflected in the 90° axis, the new equation is obtained from the old by replacing θ by $\pi - \theta$, or by replacing r by $-r$ and θ by $-\theta$.

3. If a given curve is reflected in the pole, the new equation is obtained from the old by replacing θ by $\pi + \theta$, or by replacing r by $-r$.

4. If a given curve is rotated about the pole through an angle α, the new equation is obtained from the old by replacing θ by $\theta - \alpha$.

PLANE CURVES

Archimedean spiral
See: Spiral of Archimedes

Astroid
See: Hypocycloid of four cusps

Curves and Surfaces

Bifolium

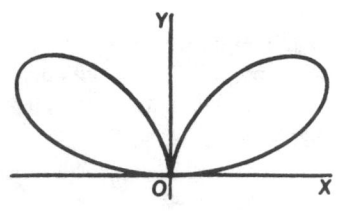

$$(x^2 + y^2)^2 = ax^2y$$
$$r = a \sin \theta \cos^2 \theta$$

Circle

(a)

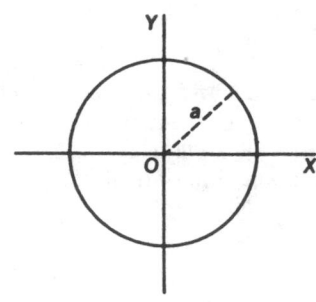

$$x^2 + y^2 = a^2$$
$$r = a$$

Cardioid

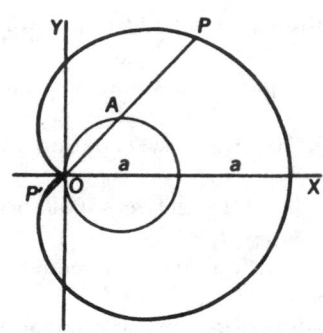

$$(x^2 + y^2 - ax)^2 = a^2(x^2 + y^2)$$
$$r = a(\cos \theta + 1)$$
or
$$r = a(\cos \theta - 1)$$
$$[P'A = AP = a]$$

(b)

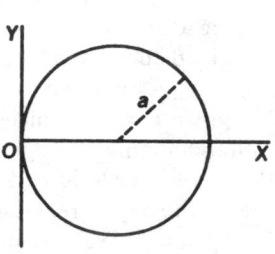

$$x^2 + y^2 = 2ax$$
$$r = 2a \cos \theta$$

(c)

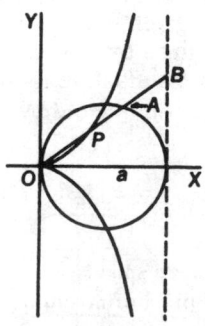

$$x^2 + y^2 = ax + by$$
$$r = a \cos \theta + b \sin \theta$$

Cassinian curves
  See: Ovals of Cassini

Catenary, Hyperbolic cosine

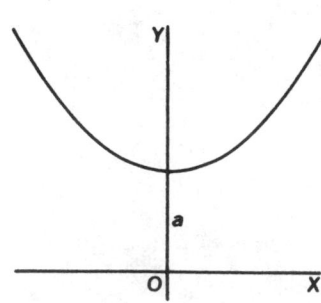

$$y = \frac{a}{2}(e^{x/a} + e^{-x/a}) = a \cosh \frac{x}{a}$$

Cissoid of Diocles

$$y^2(a - x) = x^3$$
$$r = a \sin \theta \tan \theta$$
$$[OP = AB]$$

Curves and Surfaces

Cochleoid, Oui-ja board curve

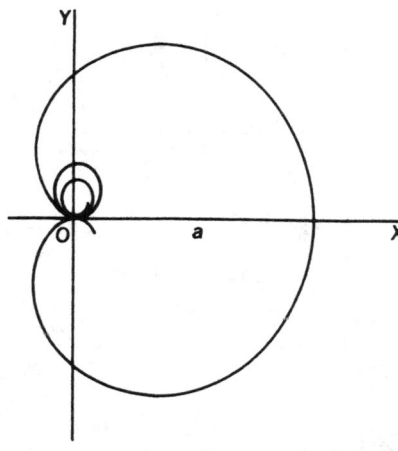

$$(x^2 + y^2)\tan^{-1}(y/x) = ay$$
$$r\theta = a\sin\theta$$

Companion to the cycloid

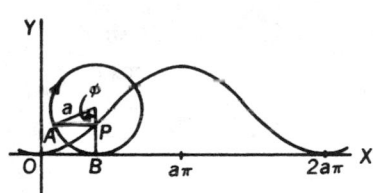

$$\begin{cases} x = a\phi \\ y = a(1 - \cos\phi) \end{cases}$$
$$[OB = \widehat{AB}]$$

(This is a sinusoid)

Conchoid of Nicomedes
(a) $a < b$

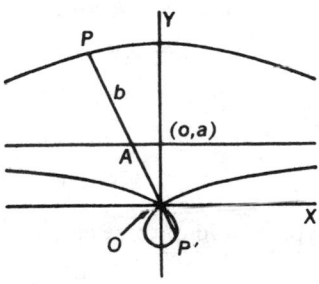

(b) $a > b$

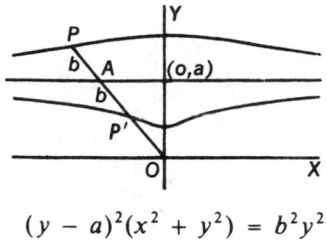

$$(y - a)^2(x^2 + y^2) = b^2y^2$$
$$r = a\csc\theta \pm b$$
$$[P'A = AP = b]$$

Conic sections
See: Circle; Ellipse; Hyperbola; Parabola

Cosecant curve

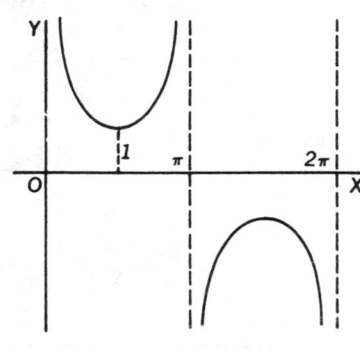

$$y = \csc x$$

Cosine curve

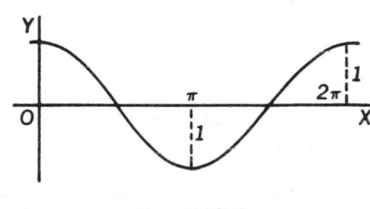

$$y = \cos x$$

Cotangent curve

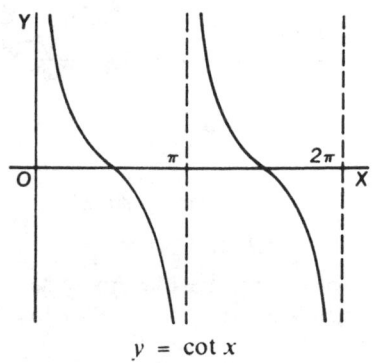

$$y = \cot x$$

Curves and Surfaces

Cubical parabola (special)

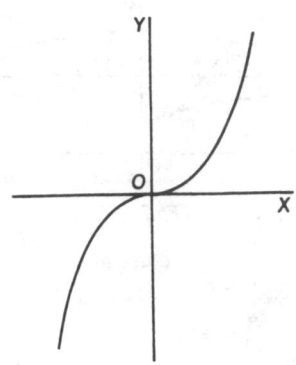

$$y = ax^3, \quad a > 0$$

$$r^2 = \frac{1}{a} \sec^2 \theta \tan \theta, \quad a > 0$$

Cubical parabola (general)

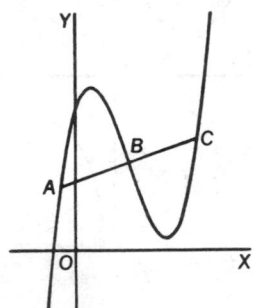

$$y = ax^3 + bx^2 + cx + d, \quad a > 0$$

$$[AB = BC]$$

(abscissa of $B = -b/3a$)

Curtate cycloid, Trochoids
  See: Cycloid, curtate

Cycloid (cusp at origin)

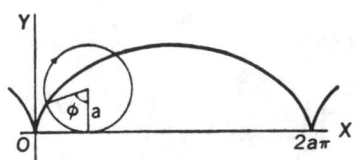

$$x = a \arccos \frac{a - y}{a} \mp \sqrt{2ay - y^2}$$

$$\begin{cases} x = a(\phi - \sin \phi) \\ y = a(1 - \cos \phi) \end{cases}$$

(For one arch: arc length $= 8a$;
area $= 3\pi a^2$)

Cycloid (vertex at origin)

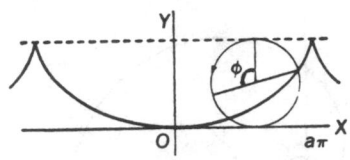

$$x = 2a \arcsin \sqrt{y/2a} + \sqrt{2ay - y^2}$$

$$\begin{cases} x = a(\phi + \sin \phi) \\ y = a(1 - \cos \phi) \end{cases}$$

Cycloid, curtate

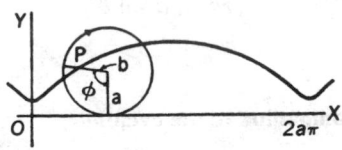

$$\begin{cases} x = a\phi - b \sin \phi \\ y = a - b \cos \phi \end{cases}$$

$$a > b$$

Cycloid, prolate

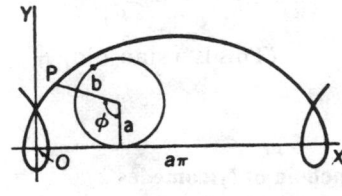

$$\begin{cases} x = a\phi - b \sin \phi \\ y = a - b \cos \phi \end{cases}$$

$$a < b$$

Deltoid
  See: Hypocycloid of three cusps

Curves and Surfaces

Ellipse

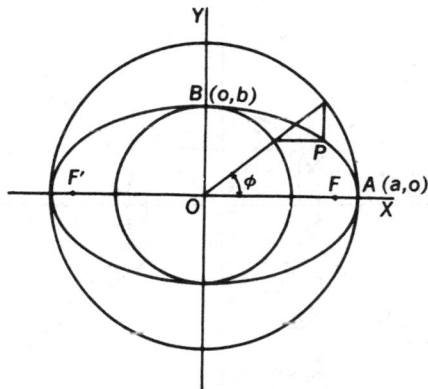

$$x^2/a^2 + y^2/b^2 = 1$$

$$\begin{cases} x = a \cos \phi \\ y = b \sin \phi \end{cases}$$

$$[BF' = BF = a, \quad PF' + PF = 2a]$$

Epicycloid

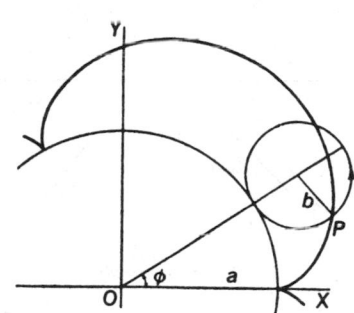

$$\begin{cases} x = (a + b) \cos \phi - b \cos\left(\dfrac{a + b}{b} \phi\right) \\ y = (a + b) \sin \phi - b \sin\left(\dfrac{a + b}{b} \phi\right) \end{cases}$$

Equiangular spiral

See: Spiral, logarithmic or equiangular

Equilateral hyperbola

See: Hyperbola, equilateral or rectangular

Evolute of ellipse

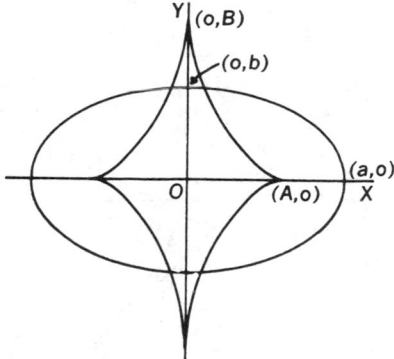

$$(ax)^{2/3} + (by)^{2/3} = (a^2 - b^2)^{2/3}$$

$$\begin{cases} x = A \cos^3 \phi \\ y = B \sin^3 \phi \end{cases}$$

$$[A = (a^2 - b^2)/a, \quad B = (a^2 - b^2)/b]$$

Exponential curve

(1) $a > 0$

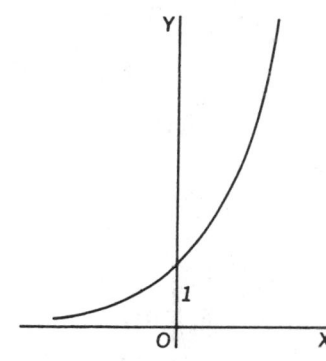

(2) $a < 0$

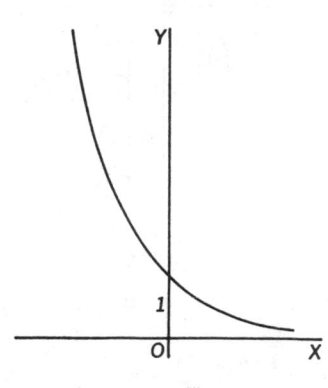

$$y = e^{ax}$$

Curves and Surfaces

Folium of Descartes

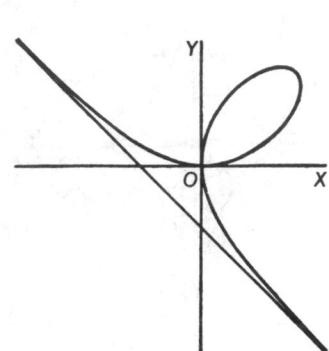

$$x^3 + y^3 - 3axy = 0$$

$$\begin{cases} x = 3a\phi/(1 + \phi^3) \\ y = 3a\phi^2/(1 + \phi^3) \end{cases}$$

$$r = \frac{3a \sin \theta \cos \theta}{\sin^3\theta + \cos^3\theta}$$

[asymptote: $x + y + a = 0$]

Gamma function

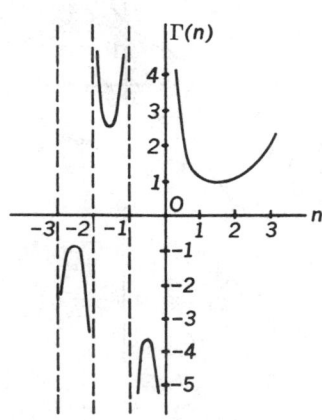

$$\Gamma(n) = \int_0^\infty x^{n-1} e^{-x} dx \quad (n > 0)$$

$$\Gamma(n) = \frac{\Gamma(n + 1)}{n} \quad (0 > n \neq -1, -2, -3, \ldots)$$

Hyperbola

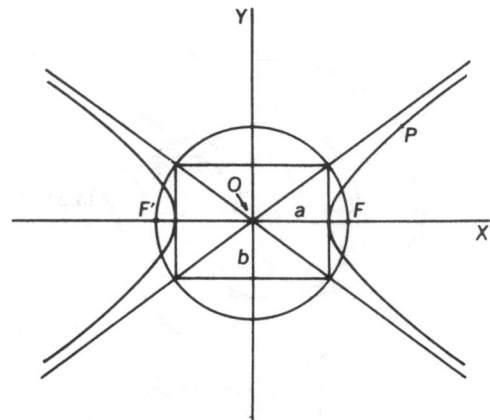

$$x^2/a^2 - y^2/b^2 = 1$$
$$[F'P - FP = 2a]$$

Hyperbola, equilateral or rectangular
(1)

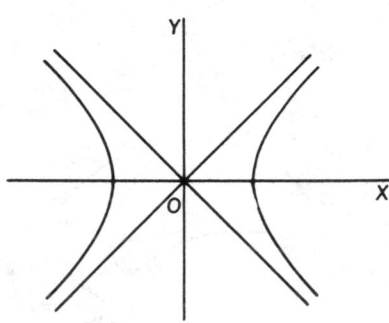

$$x^2 - y^2 = a^2$$

(2)

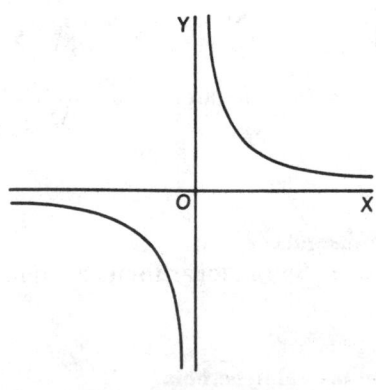

$$xy = k, \quad k > 0$$

Curves and Surfaces

(3)

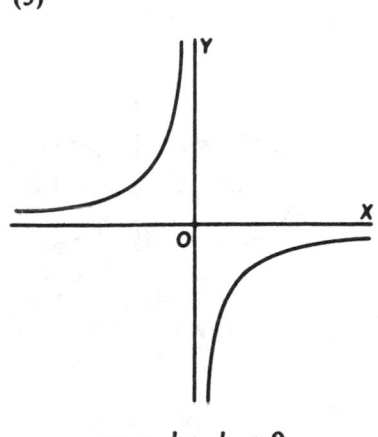

$$xy = k, \quad k < 0$$

Hypocycloid of three cusps, Deltoid

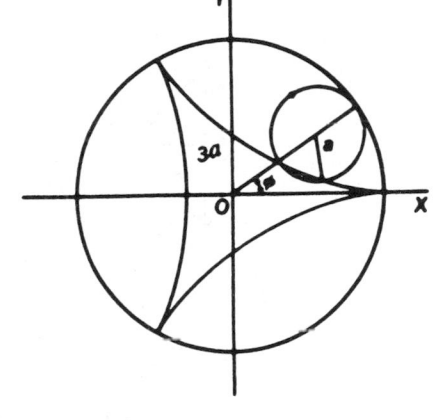

$$\begin{cases} x = 2a \cos \phi + a \cos 2\phi \\ y = 2a \sin \phi - a \sin 2\phi \end{cases}$$

Hyperbolic functions*

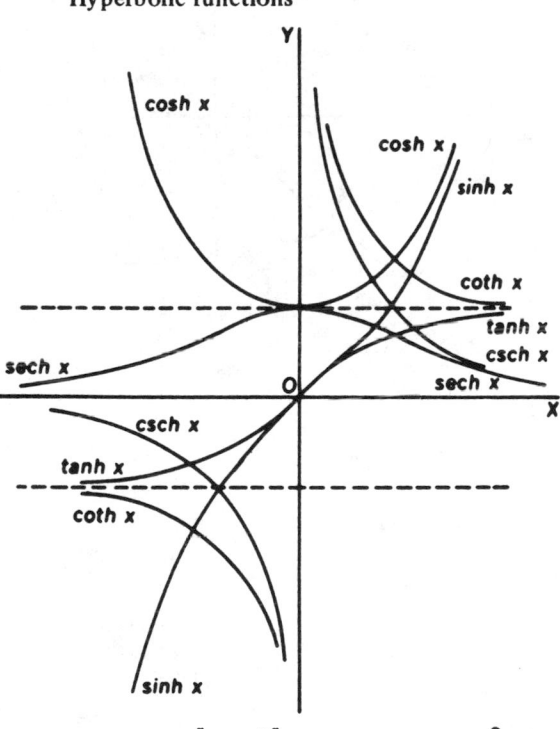

$$\sinh x = \frac{e^x - e^{-x}}{2} \qquad \operatorname{csch} x = \frac{2}{e^x - e^{-x}}$$

$$\cosh x = \frac{e^x + e^{-x}}{2} \qquad \operatorname{sech} x = \frac{2}{e^x + e^{-x}}$$

$$\tanh x = \frac{e^x - e^{-x}}{e^x + e^{-x}} \qquad \coth x = \frac{e^x + e^{-x}}{e^x - e^{-x}}$$

Hyperbolic spiral

   See: Spiral, hyperbolic or reciprocal

*See page 279 for inverse hyperbolic functions.

Hypocycloid of four cusps, Astroid

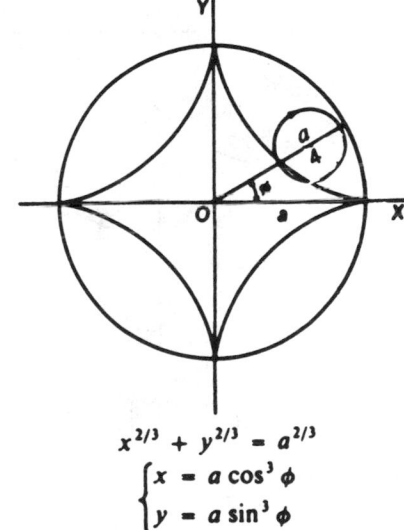

$$x^{2/3} + y^{2/3} = a^{2/3}$$
$$\begin{cases} x = a \cos^3 \phi \\ y = a \sin^3 \phi \end{cases}$$

Inverse cosine curve

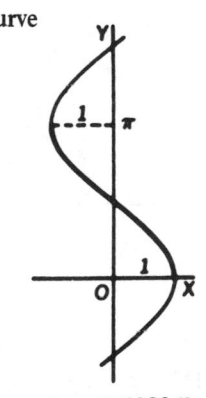

$$y = \arccos x$$

Curves and Surfaces

Inverse sine curve

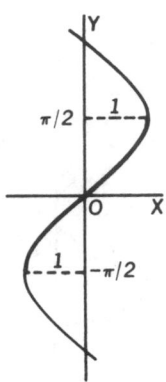

$$y = \arcsin x$$

Inverse tangent curve

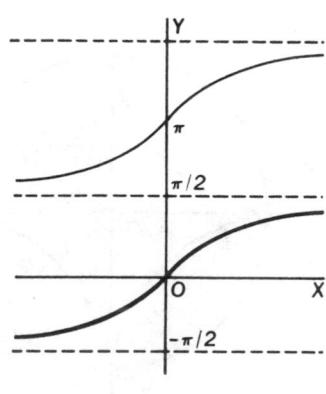

$$y = \arctan x$$

Involute of circle

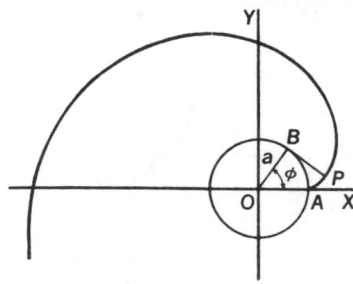

$$\begin{cases} x = a \cos \phi + a\phi \sin \phi \\ y = a \sin \phi - a\phi \cos \phi \end{cases}$$

$$[BP = \widehat{BA}]$$

Lemniscate of Bernoulli, Two-leaved rose

(a)

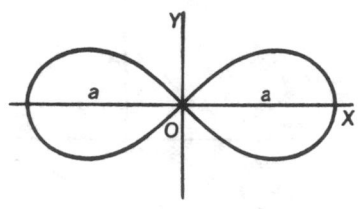

$$(x^2 + y^2)^2 = a^2(x^2 - y^2)$$
$$r^2 = a^2 \cos 2\theta$$

(b)

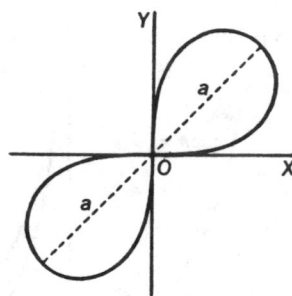

$$(x^2 + y^2)^2 = 2a^2xy$$
$$r^2 = a^2 \sin 2\theta$$

Limacon of Pascal

(1) $a > b$

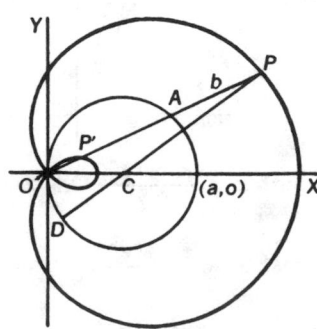

[If $a = 2b$, the curve is called the *trisectrix*, since then $\angle OPD = \frac{1}{3} \angle OCD$.]

(2) $a = b$
   See: Cardioid

Curves and Surfaces

(3) $a < b$

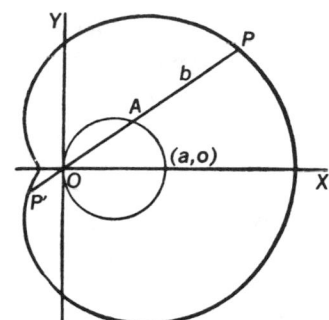

$$(x^2 + y^2 - ax)^2 = b^2(x^2 + y^2)$$
$$r = b + a \cos \theta$$
$$[P'A = AP = b]$$

Lituus

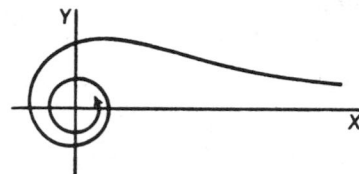

$$r^2 \theta = a^2$$

Logarithmic curve

(1) $a > 1$

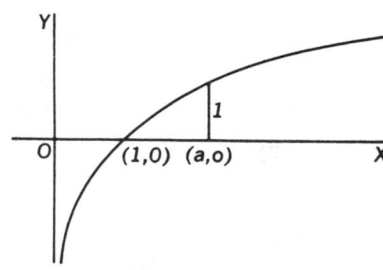

(2) $0 < a < 1$

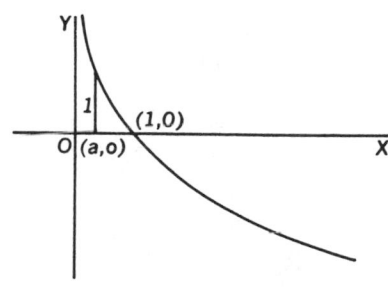

$$y = \log_a x$$

Logarithmic spiral
See: Spiral, logarithmic or equiangular

Nephroid

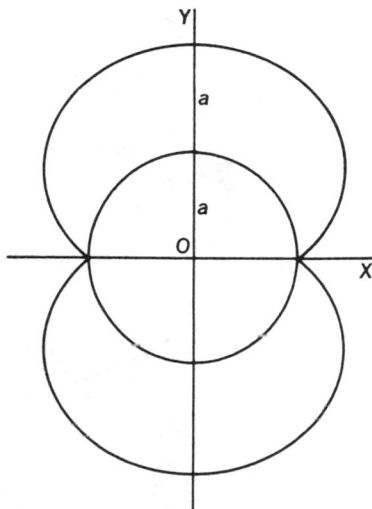

$$\begin{cases} x = \tfrac{1}{2}a(3 \cos \phi - \cos 3\phi) \\ y = \tfrac{1}{2}a(3 \sin \phi - \sin 3\phi) \end{cases}$$

[The nephroid is a 2-cusped epicycloid.]

Oui-ja board curve
See: Cochleoid

Ovals of Cassini

(1) $b > k$

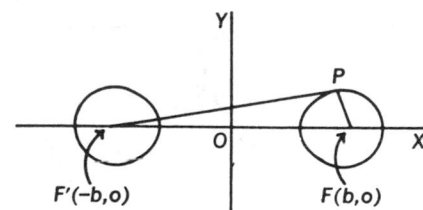

(2) $b = k$
See: Lemniscate of Bernoulli

(3) $b < k$

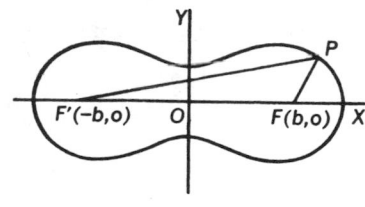

$$(x^2 + y^2 + b^2)^2 - 4b^2x^2 = k^4$$
$$r^4 + b^4 - 2r^2b^2 \cos 2\theta = k^4$$
$$[F'P \cdot FP = k^2]$$

[These curves are sections of a torus on planes parallel to the axis of the torus.]

Curves and Surfaces

Parabola

(1)

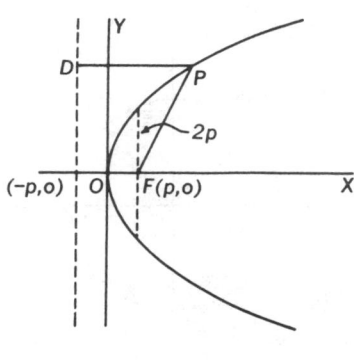

$$y^2 = 4px$$
$$[DP = FP]$$

(2)

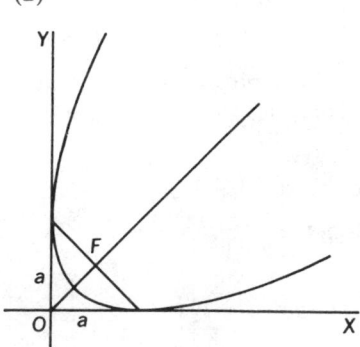

$$\pm x^{1/2} \pm y^{1/2} = a^{1/2}$$
$$(x - y)^2 - 2a(x + y) + a^2 = 0$$

(3)

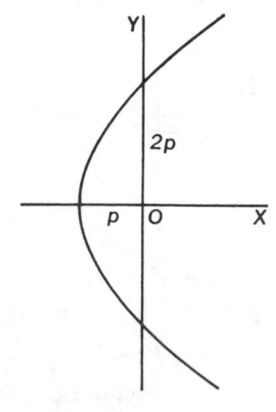

$$r = 2p/(1 - \cos\theta)$$

(4)

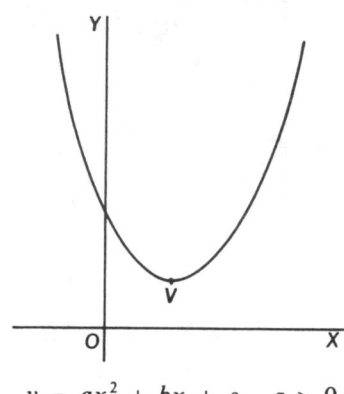

$$y = ax^2 + bx + c, \quad a > 0$$
$$[\text{abscissa of vertex} = -b/2a]$$

Parabolic spiral
 See: Spiral, parabolic

Power functions

(1)

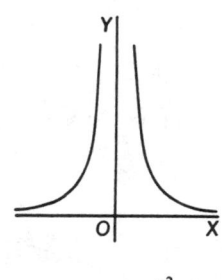

$$y = x^{-2}$$

(2) Equilateral hyperbola

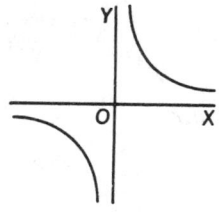

$$y = x^{-1}$$

(3)

$$y = x^{-1/2}$$

Curves and Surfaces

(4) Cubical parabola

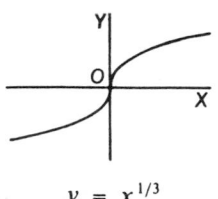

$$y = x^{1/3}$$

(5) Half of a parabola

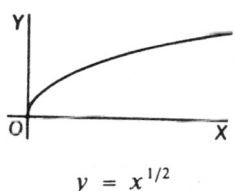

$$y = x^{1/2}$$

(6) Semicubical parabola

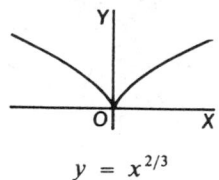

$$y = x^{2/3}$$

(7) Half of semicubical parabola

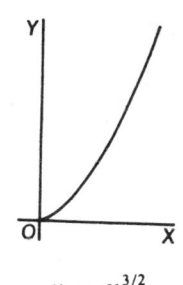

$$y = x^{3/2}$$

(8) Parabola

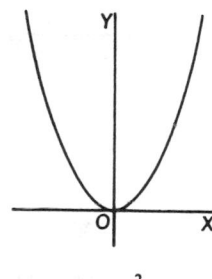

$$y = x^2$$

(9) Cubical parabola

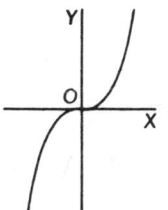

$$y = x^3$$

Probability curve

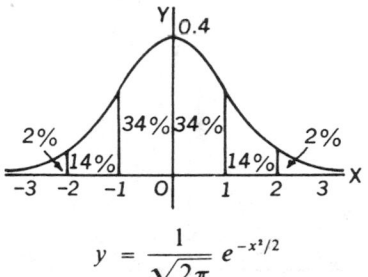

$$y = \frac{1}{\sqrt{2\pi}} \, e^{-x^2/2}$$

Prolate cycloid
  See: Cycloid, prolate

Pursuit curve
  See: Tractrix

Quadratrix of Hippias

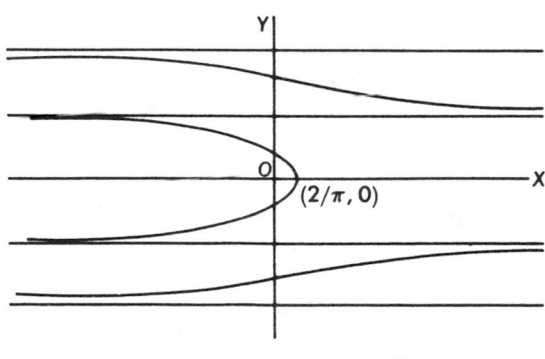

$$y = x \tan(\pi y/2)$$

Reciprocal spiral
  See: Spiral, hyperbolic or reciprocal

Rectangular hyperbola
  See: Hyperbola, equilateral or rectangular

Rose curves
  (1) Two-leaved
    See: Lemniscate of Bernoulli, Two-leaved rose

Curves and Surfaces

(2) Three-leaved

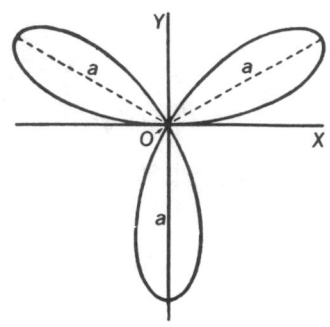

$$r = a \sin 3\theta$$

(3) Three-leaved

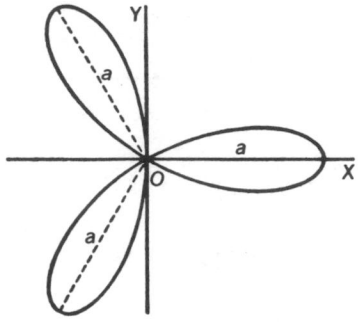

$$r = a \cos 3\theta$$

(4) Four-leaved

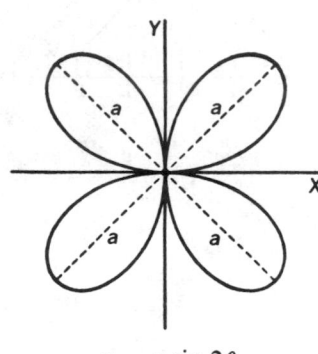

$$r = a \sin 2\theta$$

(5) Four-leaved

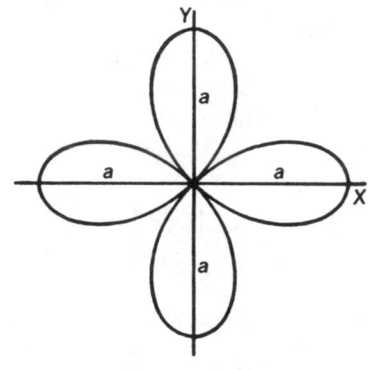

$$r = a \cos 2\theta$$

(6) *n*-leaved

The roses $r = a \sin n\theta$ and $r = a \cos n\theta$, have, for n an even integer, $2n$ leaves; for n an odd integer, n leaves. The roses $r^2 = a \sin n\theta$ and $r^2 = a \cos n\theta$, have, for n an even integer, n leaves; for n an odd integer, $2n$ leaves.

Secant curve

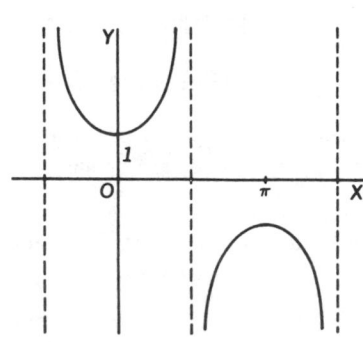

$$y = \sec x$$

Semicubical parabola

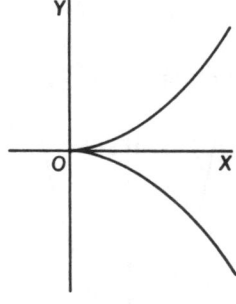

$$y^2 = ax^3$$

$$r = \frac{1}{a} \tan^2 \theta \sec \theta$$

Curves and Surfaces

Serpentine curve

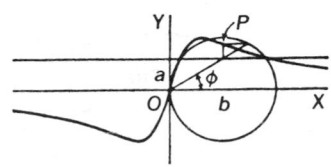

$$(a^2 + x^2)y = abx$$

$$\begin{cases} x = a \cot \phi \\ y = b \sin \phi \cos \phi \end{cases}$$

Sine curve

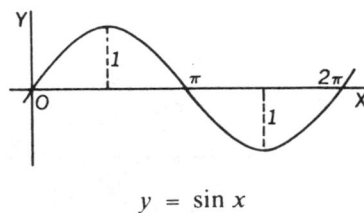

$$y = \sin x$$

Sinusoid

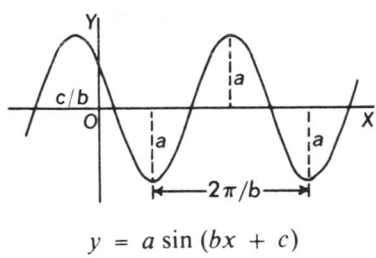

$$y = a \sin (bx + c)$$

Spiral of Archimedes

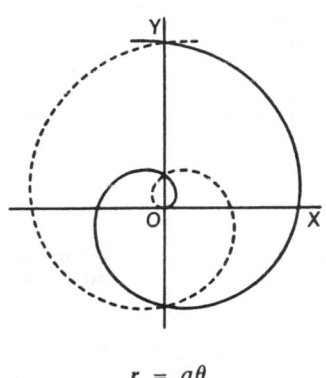

$$r = a\theta$$

Spiral, hyperbolic or reciprocal

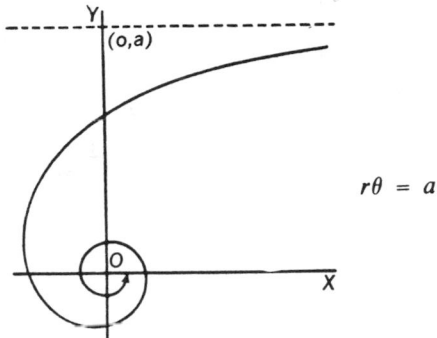

$$r\theta = a$$

Spiral, logarithmic or equiangular

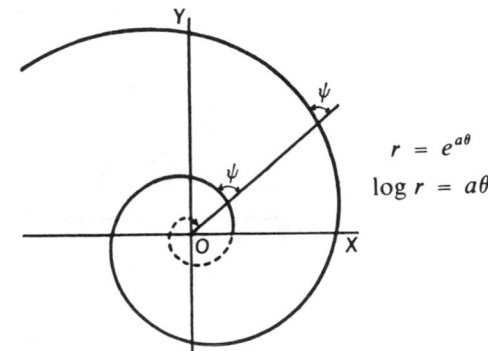

$$r = e^{a\theta}$$
$$\log r = a\theta$$

Spiral, parabolic

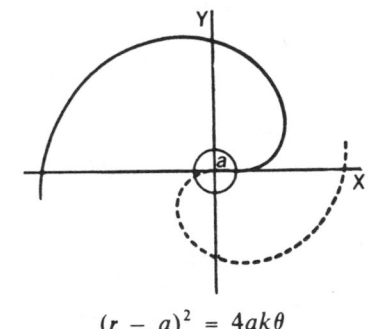

$$(r - a)^2 = 4ak\theta$$

Strophoid

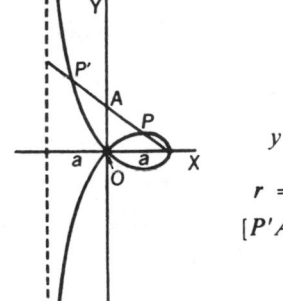

$$y^2 = x^2 \frac{a - x}{a + x}$$

$$r = a \cos 2\theta \sec \theta$$

$$[P'A = AP = OA]$$

Curves and Surfaces

Tangent curve

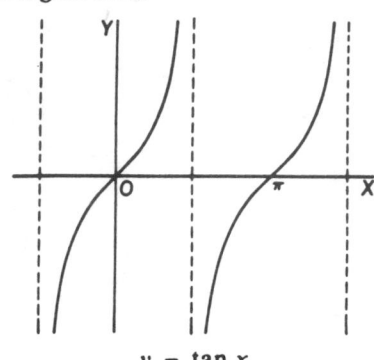

$$y = \tan x$$

Tractrix, Pursuit curve

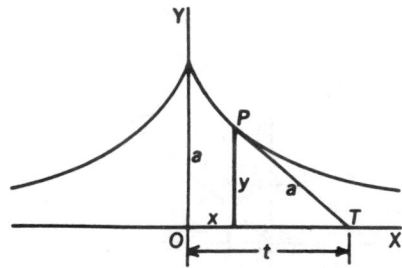

$$x = a \operatorname{sech}^{-1}(y/a) - \sqrt{a^2 - y^2}$$

$$\begin{cases} x = t - a \tanh(t/a) \\ y = a \operatorname{sech}(t/a) \end{cases}$$

$$[PT = a]$$

Trajectory (a parabola)

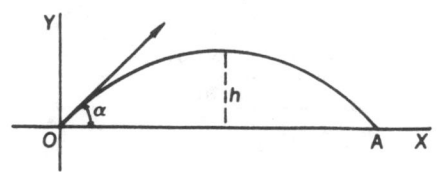

$$y = x \tan \alpha - gx^2/(2v_0^2 \cos^2 \alpha)$$
$$x = (v_0 \cos \alpha)t$$
$$y = (v_0 \sin \alpha)t - gt^2/2$$

Trigonometric functions
   See: Cosecant curve; Cosine curve; Cotangent curve; Secant curve Sine curve; Tangent curve

Trisectrix
   See: Limaçon of Pascal (1)

Witch of Agnesi

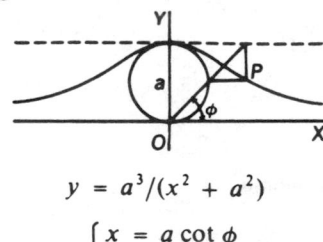

$$y = a^3/(x^2 + a^2)$$

$$\begin{cases} x = a \cot \phi \\ y = a \sin^2 \phi \end{cases}$$

QUADRIC SURFACES*

Ellipsoid

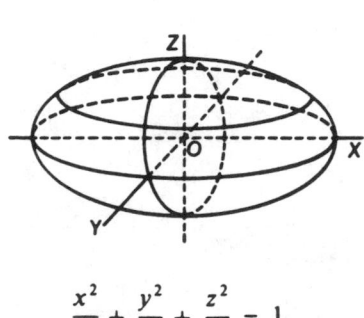

$$\frac{x^2}{a^2} + \frac{y^2}{b^2} + \frac{z^2}{c^2} = 1$$

Elliptic cone

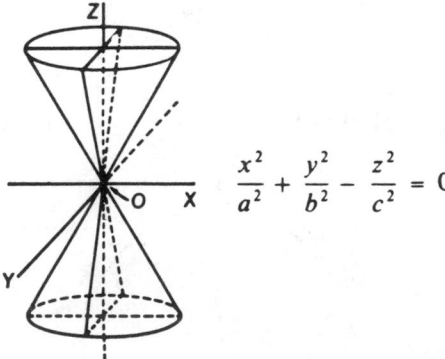

$$\frac{x^2}{a^2} + \frac{y^2}{b^2} - \frac{z^2}{c^2} = 0$$

*Each of the equations is given for the case where the origin is located at $(0, 0, 0)$, the center of the quadric surface. If, however, the center of the surface is at (h, k, l), replace x by $x - h$, y by $y - k$, and z by $z - l$, and the particular standardized form will be that of the surface with center at (h, k, l). For example, the elliptic paraboloid would be

$$\frac{(x - h)^2}{a^2} + \frac{(y - k)^2}{b^2} = c(z - l).$$

Curves and Surfaces

Elliptic cylinder

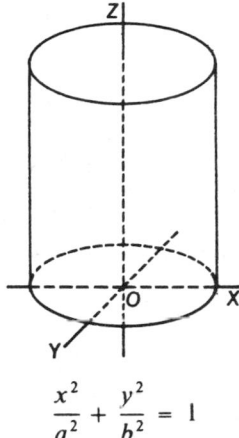

$$\frac{x^2}{a^2} + \frac{y^2}{b^2} = 1$$

Hyperboloid of one sheet

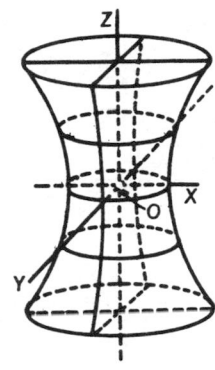

$$\frac{x^2}{a^2} + \frac{y^2}{b^2} - \frac{z^2}{c^2} = 1$$

Elliptic paraboloid

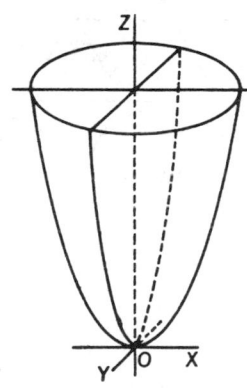

$$\frac{x^2}{a^2} + \frac{y^2}{b^2} = cz$$

Hyperboloid of two sheets

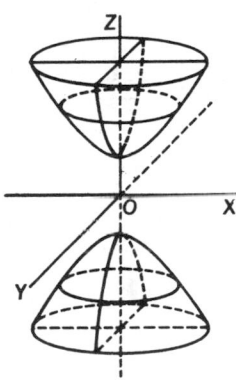

$$\frac{z^2}{c^2} - \frac{x^2}{a^2} - \frac{y^2}{b^2} = 1$$

Sphere

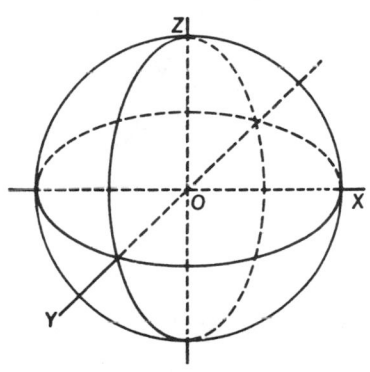

$$x^2 + y^2 + z^2 = a^2$$

Hyperbolic paraboloid

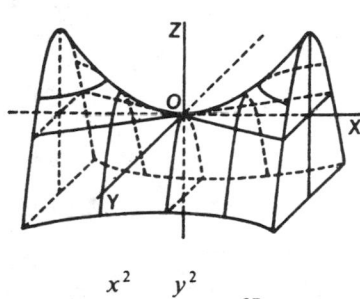

$$\frac{x^2}{a^2} - \frac{y^2}{b^2} = cz$$

Curves and Surfaces

PATTERNS OF REGULAR POLYHEDRA

Tetrahedron

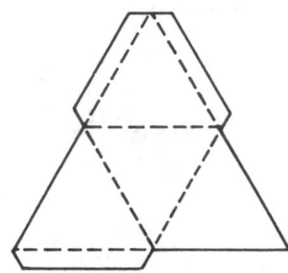

Octahedron

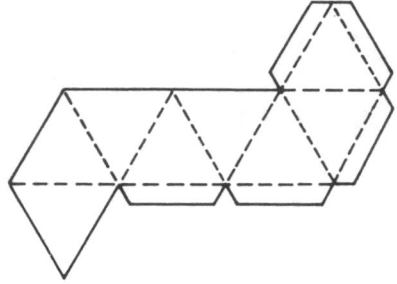

Hexahedron, or Cube

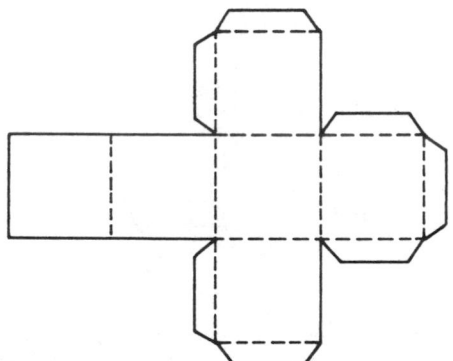

Icosahedron

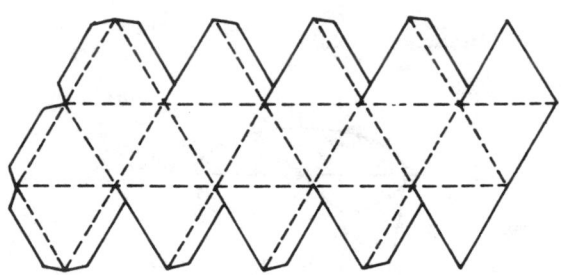

Dodecahedron

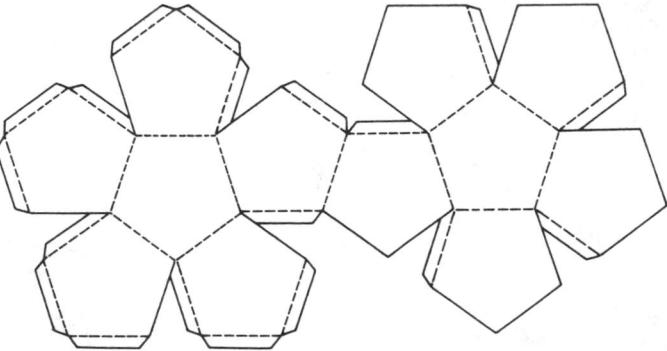

VIII. CALCULUS

Derivatives *

In the following formulas u, v, w represent functions of x, while a, c, n represent fixed real numbers. All arguments in the trigonometric functions are measured in radians, and all inverse trigonometric and hyperbolic functions represent principal values.

1. $\dfrac{d}{dx}(a) = 0$

2. $\dfrac{d}{dx}(x) = 1$

3. $\dfrac{d}{dx}(au) = a\dfrac{du}{dx}$

4. $\dfrac{d}{dx}(u + v - w) = \dfrac{du}{dx} + \dfrac{dv}{dx} - \dfrac{dw}{dx}$

5. $\dfrac{d}{dx}(uv) = u\dfrac{dv}{dx} + v\dfrac{du}{dx}$

6. $\dfrac{d}{dx}(uvw) = uv\dfrac{dw}{dx} + vw\dfrac{du}{dx} + uw\dfrac{dv}{dx}$ and so on to n factors

7. $\dfrac{d}{dx}\left(\dfrac{u}{v}\right) = \dfrac{v\dfrac{du}{dx} - u\dfrac{dv}{dx}}{v^2} = \dfrac{1}{v}\dfrac{du}{dx} - \dfrac{u}{v^2}\dfrac{dv}{dx}$

8. $\dfrac{d}{dx}(u^n) = nu^{n-1}\dfrac{du}{dx}$

9. $\dfrac{d}{dx}(\sqrt{u}) = \dfrac{1}{2\sqrt{u}}\dfrac{du}{dx}$

10. $\dfrac{d}{dx}\left(\dfrac{1}{u}\right) = -\dfrac{1}{u^2}\dfrac{du}{dx}$

11. $\dfrac{d}{dx}\left(\dfrac{1}{u^n}\right) = -\dfrac{n}{u^{n+1}}\dfrac{du}{dx}$

12. $\dfrac{d}{dx}\left(\dfrac{u^n}{v^m}\right) = \dfrac{u^{n-1}}{v^{m+1}}\left(nv\dfrac{du}{dx} - mu\dfrac{dv}{dx}\right)$

---

*Let $y = f(x)$ and $\dfrac{dy}{dx} = \dfrac{d[f(x)]}{dx} = f'(x)$ define respectively a function and its derivative for any value x in their common domain. The differential for the function at such a value x is accordingly defined as

$$dy = d[f(x)] = \dfrac{dy}{dx}\,dx = \dfrac{d[f(x)]}{dx}\,dx = f'(x)dx$$

Each derivative formula has an associated differential formula. For example, formula 6 above has the differential formula

$$d(uvw) = uv\,dw + vw\,du + uw\,dv$$

DERIVATIVES (Continued)

13. $\dfrac{d}{dx}(u^n v^m) = u^{n-1} v^{m-1}\left(nv\dfrac{du}{dx} + mu\dfrac{dv}{dx}\right)$

14. $\dfrac{d}{dx}[f(u)] = \dfrac{d}{du}[f(u)] \cdot \dfrac{du}{dx}$

15. $\dfrac{d^2}{dx^2}[f(u)] = \dfrac{df(u)}{du} \cdot \dfrac{d^2 u}{dx^2} + \dfrac{d^2 f(u)}{du^2} \cdot \left(\dfrac{du}{dx}\right)^2$

16. $\dfrac{d^n}{dx^n}[uv] = \dbinom{n}{0} v\dfrac{d^n u}{dx^n} + \dbinom{n}{1}\dfrac{dv}{dx}\dfrac{d^{n-1} u}{dx^{n-1}} + \dbinom{n}{2}\dfrac{d^2 v}{dx^2}\dfrac{d^{n-2} u}{dx^{n-2}}$

$$+ \cdots + \dbinom{n}{k}\dfrac{d^k v}{dx^k}\dfrac{d^{n-k} u}{dx^{n-k}} + \cdots + \dbinom{n}{n} u\dfrac{d^n v}{dx^n}$$

where $\dbinom{n}{r} = \dfrac{n!}{r!(n-r)!}$ the binomial coefficient, n non-negative integer and $\dbinom{n}{0} = 1$.

17. $\dfrac{du}{dx} = \dfrac{1}{\dfrac{dx}{du}}$ if $\dfrac{dx}{du} \neq 0$

18. $\dfrac{d}{dx}(\log_a u) = (\log_a e)\dfrac{1}{u}\dfrac{du}{dx}$

19. $\dfrac{d}{dx}(\log_e u) = \dfrac{1}{u}\dfrac{du}{dx}$

20. $\dfrac{d}{dx}(a^u) = a^u(\log_e a)\dfrac{du}{dx}$

21. $\dfrac{d}{dx}(e^u) = e^u\dfrac{du}{dx}$

22. $\dfrac{d}{dx}(u^v) = vu^{v-1}\dfrac{du}{dx} + (\log_e u)u^v\dfrac{dv}{dx}$

23. $\dfrac{d}{dx}(\sin u) = \dfrac{du}{dx}(\cos u)$

24. $\dfrac{d}{dx}(\cos u) = -\dfrac{du}{dx}(\sin u)$

25. $\dfrac{d}{dx}(\tan u) = \dfrac{du}{dx}(\sec^2 u)$

26. $\dfrac{d}{dx}(\cot u) = -\dfrac{du}{dx}(\csc^2 u)$

27. $\dfrac{d}{dx}(\sec u) = \dfrac{du}{dx}\sec u \cdot \tan u$

28. $\dfrac{d}{dx}(\csc u) = -\dfrac{du}{dx}\csc u \cdot \cot u$

DERIVATIVES (Continued)

29. $\dfrac{d}{dx}(\text{vers } u) = \dfrac{du}{dx}\sin u$

30. $\dfrac{d}{dx}(\text{arc sin } u) = \dfrac{1}{\sqrt{1-u^2}}\dfrac{du}{dx}, \qquad \left(-\dfrac{\pi}{2} \le \text{arc sin } u \le \dfrac{\pi}{2}\right)$

31. $\dfrac{d}{dx}(\text{arc cos } u) = -\dfrac{1}{\sqrt{1-u^2}}\dfrac{du}{dx}, \qquad (0 \le \text{arc cos } u \le \pi)$

32. $\dfrac{d}{dx}(\text{arc tan } u) = \dfrac{1}{1+u^2}\dfrac{du}{dx}, \qquad \left(-\dfrac{\pi}{2} < \text{arc tan } u < \dfrac{\pi}{2}\right)$

33. $\dfrac{d}{dx}(\text{arc cot } u) = -\dfrac{1}{1+u^2}\dfrac{du}{dx}, \qquad (0 \le \text{arc cot } u \le \pi)$

34. $\dfrac{d}{dx}(\text{arc sec } u) = \dfrac{1}{u\sqrt{u^2-1}}\dfrac{du}{dx}, \qquad \left(0 \le \text{arc sec } u < \dfrac{\pi}{2}, -\pi \le \text{arc sec } u < -\dfrac{\pi}{2}\right)$

35. $\dfrac{d}{dx}(\text{arc csc } u) = -\dfrac{1}{u\sqrt{u^2-1}}\dfrac{du}{dx}, \qquad \left(0 < \text{arc csc } u \le \dfrac{\pi}{2}, -\pi < \text{arc csc } u \le -\dfrac{\pi}{2}\right)$

36. $\dfrac{d}{dx}(\text{arc vers } u) = \dfrac{1}{\sqrt{2u-u^2}}\dfrac{du}{dx}, \qquad (0 \le \text{arc vers } u \le \pi)$

37. $\dfrac{d}{dx}(\sinh u) = \dfrac{du}{dx}(\cosh u)$

38. $\dfrac{d}{dx}(\cosh u) = \dfrac{du}{dx}(\sinh u)$

39. $\dfrac{d}{dx}(\tanh u) = \dfrac{du}{dx}(\text{sech}^2 u)$

40. $\dfrac{d}{dx}(\coth u) = -\dfrac{du}{dx}(\text{csch}^2 u)$

41. $\dfrac{d}{dx}(\text{sech } u) = \dfrac{du}{dx}(\text{sech } u \cdot \tanh u)$

42. $\dfrac{d}{dx}(\text{csch } u) = -\dfrac{du}{dx}(\text{csch } u \cdot \coth u)$

43. $\dfrac{d}{dx}(\sinh^{-1} u) = \dfrac{d}{dx}[\log(u + \sqrt{u^2+1})] = \dfrac{1}{\sqrt{u^2+1}}\dfrac{du}{dx}$

44. $\dfrac{d}{dx}(\cosh^{-1} u) = \dfrac{d}{dx}[\log(u + \sqrt{u^2-1})] = \dfrac{1}{\sqrt{u^2-1}}\dfrac{du}{dx}, \qquad (u > 1, \cosh^{-1} u > 0)$

45. $\dfrac{d}{dx}(\tanh^{-1} u) = \dfrac{d}{dx}\left[\dfrac{1}{2}\log\dfrac{1+u}{1-u}\right] = \dfrac{1}{1-u^2}\dfrac{du}{dx}, \qquad (u^2 < 1)$

46. $\dfrac{d}{dx}(\coth^{-1} u) = \dfrac{d}{dx}\left[\dfrac{1}{2}\log\dfrac{u+1}{u-1}\right] = \dfrac{1}{1-u^2}\dfrac{du}{dx}, \qquad (u^2 > 1)$

47. $\dfrac{d}{dx}(\text{sech}^{-1} u) = \dfrac{d}{dx}\left[\log\dfrac{1+\sqrt{1-u^2}}{u}\right] = -\dfrac{1}{u\sqrt{1-u^2}}\dfrac{du}{dx}, \qquad (0 < u < 1, \text{sech}^{-1} u > 0)$

DERIVATIVES (Continued)

48. $\dfrac{d}{dx}(\operatorname{csch}^{-1} u) = \dfrac{d}{dx}\left[\log\dfrac{1 + \sqrt{1 + u^2}}{u}\right] = -\dfrac{1}{|u|\sqrt{1 + u^2}}\dfrac{du}{dx}$

49. $\dfrac{d}{dq}\displaystyle\int_{p}^{q} f(x)\,dx = f(q), \qquad [p \text{ constant}]$

50. $\dfrac{d}{dp}\displaystyle\int_{p}^{q} f(x)\,dx = -f(p), \qquad [q \text{ constant}]$

51. $\dfrac{d}{da}\displaystyle\int_{p}^{q} f(x, a)\,dx = \int_{p}^{q}\dfrac{\partial}{\partial a}[f(x, a)]\,dx + f(q, a)\dfrac{dq}{da} - f(p, a)\dfrac{dp}{da}$

INTEGRATION

The following is a brief discussion of some integration techniques. A more complete discussion can be found in a number of good text books. However, the purpose of this introduction is simply to discuss a few of the important techniques which may be used, in conjunction with the integral table which follows, to integrate particular functions.

No matter how extensive the integral table, it is a fairly uncommon occurrence to find in the table the exact integral desired. Usually some form of transformation will have to be made. The simplest type of transformation, and yet the most general, is substitution. Simple forms of substitution, such as $y = ax$, are employed almost unconsciously by experienced users of integral tables. Other substitutions may require more thought. In some sections of the tables, appropriate substitutions are suggested for integrals which are similar to, but not exactly like, integrals in the table. Finding the right substitution is largely a matter of intuition and experience.

Several precautions must be observed when using substitutions:

1. Be sure to make the substitution in the dx term, as well as everywhere else in the integral.
2. Be sure that the function substituted is one-to-one and continuous. If this is not the case, the integral must be restricted in such a way as to make it true. See the example following.
3. With definite integrals, the limits should also be expressed in terms of the new dependent variable. With indefinite integrals, it is necessary to perform the reverse substitution to obtain the answer in terms of the original independent variable. This may also be done for definite integrals, but it is usually easier to change the limits.

Example:

$$\int \frac{x^4}{\sqrt{a^2 - x^2}}\, dx$$

Here we make the substitution $x = |a| \sin \theta$. Then $dx = |a| \cos \theta\, d\theta$, and

$$\sqrt{a^2 - x^2} = \sqrt{a^2 - a^2 \sin^2 \theta} = |a|\sqrt{1 - \sin^2 \theta} = |a \cos \theta|$$

Notice the absolute value signs. It is very important to keep in mind that a square root radical always denotes the positive square root, and to assure the sign is always kept positive. Thus $\sqrt{x^2} = |x|$. Failure to observe this is a common cause of errors in integration.

Notice also that the indicated substitution is not a one-to-one function, that is, it does not have a unique inverse. Thus we must restrict the range of θ in such a way as to make the function one-to-one. Fortunately, this is easily done by solving for θ

$$\theta = \sin^{-1} \frac{x}{|a|}$$

and restricting the inverse sine to the principal values, $-\dfrac{\pi}{2} \leq \theta \leq \dfrac{\pi}{2}$.

Thus the integral becomes

$$\int \frac{a^4 \sin^4 \theta |a| \cos \theta\, d\theta}{|a| |\cos \theta|}$$

Now, however, in the range of values chosen for θ, $\cos \theta$ is always positive. Thus we may remove the absolute value signs from $\cos \theta$ in the denominator. (This is one of the reasons that the principal values of the inverse trigonometric functions are defined as they are.)

Then the $\cos \theta$ terms cancel, and the integral becomes

$$a^4 \int \sin^4 \theta \, d\theta$$

By application of integral formulas 299 and 296, we integrate this to

$$-a^4 \frac{\sin^3 \theta \cos \theta}{4} - \frac{3a^4}{8} \cos \theta \sin \theta + \frac{3a^4}{8} \theta + C$$

We now must perform the inverse substitution to get the result in terms of x. We have

$$\theta = \sin^{-1} \frac{x}{|a|}$$

$$\sin \theta = \frac{x}{|a|}$$

Then

$$\cos \theta = \pm \sqrt{1 - \sin^2 \theta} = \pm \sqrt{1 - \frac{x^2}{a^2}} = \pm \frac{\sqrt{a^2 - x^2}}{|a|}.$$

Because of the previously mentioned fact that $\cos \theta$ is positive, we may omit the $\pm$ sign. The reverse substitution then produces the final answer

$$\int \frac{x^4}{\sqrt{a^2 - x^2}} \, dx = -\tfrac{1}{4} x^3 \sqrt{a^2 - x^2} - \tfrac{3}{8} a^2 x \sqrt{a^2 - x^2} + \frac{3a^4}{8} \sin^{-1} \frac{x}{|a|} + C.$$

Any rational function of x may be integrated, if the denominator is factored into linear and irreducible quadratic factors. The function may then be broken into partial fractions, and the individual partial fractions integrated by use of the appropriate formula from the integral table. See the section on partial fractions for further information.

Many integrals may be reduced to rational functions by proper substitutions. For example,

$$z = \tan \frac{x}{2}$$

will reduce any rational function of the six trigonometric functions of x to a rational function of z. (Frequently there are other substitutions which are simpler to use, but this one will always work. See integral formula number 484.)

Any rational function of x and $\sqrt{ax + b}$ may be reduced to a rational function of z by making the substitution

$$z = \sqrt{ax + b}.$$

Other likely substitutions will be suggested by looking at the form of the integrand.

The other main method of transforming integrals is integration by parts. This involves applying formula number 5 or 6 in the accompanying integral table. The critical factor in this method is the choice of the functions u and v. In order for the method to be successful, $v = \int dv$ and $\int v \, du$ must be easier to integrate than the original integral. Again, this choice is largely a matter of intuition and experience.

Example:

$$\int x \sin x \, dx$$

Two obvious choices are $u = x$, $dv = \sin x \, dx$, or $u = \sin x$, $dv = x \, dx$. Since a preliminary mental calculation indicates that $\int v \, du$ in the second choice would be more, rather than less, complicated than the original integral (it would contain x^2), we use the first choice.

$$u = x \qquad\qquad du = dx$$
$$dv = \sin x\,dx \qquad\qquad v = -\cos x$$
$$\int x \sin x\,dx = \int u\,dv = uv - \int v\,du = -x \cos x + \int \cos x\,dx$$
$$= \sin x - x \cos x$$

Of course, this result could have been obtained directly from the integral table, but it provides a simple example of the method. In more complicated examples the choice of u and v may not be so obvious, and several different choices may have to be tried. Of course, there is no guarantee that any of them will work.

Integration by parts may be applied more than once, or combined with substitution. A fairly common case is illustrated by the following example.

Example:

$$\int e^x \sin x\,dx$$

Let

$$u = e^x \qquad \text{Then} \quad du = e^x\,dx$$
$$dv = \sin x\,dx \qquad\qquad v = -\cos x$$
$$\int e^x \sin x\,dx = \int u\,dv = uv - \int v\,du = -e^x \cos x + \int e^x \cos x\,dx$$

In this latter integral,

$$\text{let} \quad u = e^x \qquad \text{Then} \quad du = e^x\,dx$$
$$dv = \cos x\,dx \qquad\qquad v = \sin x$$
$$\int e^x \sin x\,dx = -e^x \cos x + \int e^x \cos x\,dx = -e^x \cos x + \int u\,dv$$
$$= -e^x \cos x + uv - \int v\,du$$
$$= -e^x \cos x + e^x \sin x - \int e^x \sin x\,dx$$

This looks as if a circular transformation has taken place, since we are back at the same integral we started from. However, the above equation can be solved algebraically for the required integral:

$$\int e^x \sin x\,dx = \tfrac{1}{2}(e^x \sin x - e^x \cos x)$$

In the second integration by parts, if the parts had been chosen as $u = \cos x$, $dv = e^x\,dx$, we would indeed have made a circular transformation, and returned to the starting place. In general, when doing repeated integration by parts, one should never choose the function u at any stage to be the same as the function v at the previous stage, or a constant times the previous v.

The following rule is called the extended rule for integration by parts. It is the result of $n + 1$ successive applications of integration by parts.

If

$$g_1(x) = \int g(x)\,dx, \qquad g_2(x) = \int g_1(x)\,dx,$$

$$g_3(x) = \int g_2(x)\,dx, \ldots, g_m(x) = \int g_{m-1}(x)\,dx, \ldots,$$

then

$$\int f(x)\cdot g(x)\,dx = f(x)\cdot g_1(x) - f'(x)\cdot g_2(x) + f''(x)\cdot g_3(x) - + \cdots$$
$$+ (-1)^n f^{(n)}(x)g_{n+1}(x) + (-1)^{n+1}\int f^{(n+1)}(x)g_{n+1}(x)\,dx.$$

A useful special case of the above rule is when $f(x)$ is a polynomial of degree n. Then $f^{(n+1)}(x) = 0$, and

$$\int f(x)\cdot g(x)\,dx = f(x)\cdot g_1(x) - f'(x)\cdot g_2(x) + f''(x)\cdot g_3(x) - + \cdots + (-1)^n f^{(n)}(x)g_{n+1}(x) + C$$

Example:

If $f(x) = x^2$, $g(x) = \sin x$

$$\int x^2 \sin x\,dx = -x^2 \cos x + 2x \sin x + 2\cos x + C$$

Another application of this formula occurs if

$$f''(x) = af(x) \quad \text{and} \quad g''(x) = bg(x),$$

where a and b are unequal constants. In this case, by a process similar to that used in the above example for $\int e^x \sin x\,dx$, we get the formula

$$\int f(x)g(x)\,dx = \frac{f(x)\cdot g'(x) - f'(x)\cdot g(x)}{b - a} + C$$

This formula could have been used in the example mentioned. Here is another example.

Example:

If $f(x) = e^{2x}$, $g(x) = \sin 3x$, then $a = 4, b = -9$, and

$$\int e^{2x} \sin 3x\,dx = \frac{3\,e^{2x}\cos 3x - 2\,e^{2x}\sin 3x}{-9 - 4} + C = \frac{e^{2x}}{13}(2\sin 3x - 3\cos 3x) + C$$

The following additional points should be observed when using this table.

1. A constant of integration is to be supplied with the answers for indefinite integrals.
2. Logarithmic expressions are to base $e = 2.71828\ldots$, unless otherwise specified, and are to be evaluated for the absolute value of the arguments involved therein.
3. All angles are measured in radians, and inverse trigonometric and hyperbolic functions represent principal values, unless otherwise indicated.
4. If the application of a formula produces either a zero denominator or the square root of a negative number in the result, there is usually available another form of the answer which avoids this difficulty. In many of the results, the excluded values are specified, but when such are omitted it is presumed that one can tell what these should be, especially when difficulties of the type herein mentioned are obtained.
5. When inverse trigonometric functions occur in the integrals, be sure that any replacements made for them are strictly in accordance with the rules for such functions. This causes

little difficulty when the argument of the inverse trigonometric function is positive, since then all angles involved are in the first quadrant. However, if the argument is negative, special care must be used. Thus if $u > 0$,

$$\sin^{-1} u = \cos^{-1}\sqrt{1 - u^2} = \csc^{-1}\frac{1}{u}, \text{ etc.}$$

However, if $u < 0$,

$$\sin^{-1} u = -\cos^{-1}\sqrt{1 - u^2} = -\pi - \csc^{-1}\frac{1}{u}, \text{ etc.}$$

See the section on inverse trigonometric functions for a full treatment of the allowable substitutions.

6. In integrals 340–345 and some others, the right side includes expressions of the form

$$A \tan^{-1}[B + C \tan f(x)].$$

In these formulas, the $\tan^{-1}$ does not necessarily represent the principal value. Instead of always employing the principal branch of the inverse tangent function, one must instead use that branch of the inverse tangent function upon which $f(x)$ lies for any particular choice of x.

Example:

$$\int_0^{4\pi} \frac{dx}{2 + \sin x} = \frac{2}{\sqrt{3}}\tan^{-1}\frac{2\tan\frac{x}{2} + 1}{\sqrt{3}}\Bigg]_0^{4\pi}$$

$$= \frac{2}{\sqrt{3}}\left[\tan^{-1}\frac{2\tan 2\pi + 1}{\sqrt{3}} - \tan^{-1}\frac{2\tan 0 + 1}{\sqrt{3}}\right]$$

$$= \frac{2}{\sqrt{3}}\left[\frac{13\pi}{6} - \frac{\pi}{6}\right] = \frac{4\pi}{\sqrt{3}} = \frac{4\sqrt{3}\pi}{3}$$

Here

$$\tan^{-1}\frac{2\tan 2\pi + 1}{\sqrt{3}} = \tan^{-1}\frac{1}{\sqrt{3}} = \frac{13\pi}{6},$$

since $f(x) = 2\pi$; and

$$\tan^{-1}\frac{2\tan 0 + 1}{\sqrt{3}} = \tan^{-1}\frac{1}{\sqrt{3}} = \frac{\pi}{6},$$

since $f(x) = 0$.

7. B_n and E_n where used in Integrals represents the Bernoulli and Euler numbers as defined in the tables contained from pages 520–526.

INTEGRALS

ELEMENTARY FORMS

1. $\displaystyle\int a\,dx = ax$

2. $\displaystyle\int a \cdot f(x)\,dx = a\int f(x)\,dx$

3. $\displaystyle\int \phi(y)\,dx = \int \frac{\phi(y)}{y'}\,dy,$ where $y' = \dfrac{dy}{dx}$

4. $\displaystyle\int (u + v)\,dx = \int u\,dx + \int v\,dx,$ where u and v are any functions of x

5. $\displaystyle\int u\,dv = u\int dv - \int v\,du = uv - \int v\,du$

6. $\displaystyle\int u\frac{dv}{dx}\,dx = uv - \int v\frac{du}{dx}\,dx$

7. $\displaystyle\int x^n\,dx = \frac{x^{n+1}}{n+1},$ except $n = -1$

8. $\displaystyle\int \frac{f'(x)\,dx}{f(x)} = \log f(x),$ $(df(x) = f'(x)\,dx)$

9. $\displaystyle\int \frac{dx}{x} = \log x$

10. $\displaystyle\int \frac{f'(x)\,dx}{2\sqrt{f(x)}} = \sqrt{f(x)},$ $(df(x) = f'(x)\,dx)$

11. $\displaystyle\int e^x\,dx = e^x$

12. $\displaystyle\int e^{ax}\,dx = e^{ax}/a$

13. $\displaystyle\int b^{ax}\,dx = \frac{b^{ax}}{a\log b},$ $(b > 0)$

14. $\displaystyle\int \log x\,dx = x\log x - x$

15. $\displaystyle\int a^x \log a\,dx = a^x,$ $(a > 0)$

16. $\displaystyle\int \frac{dx}{a^2 + x^2} = \frac{1}{a}\tan^{-1}\frac{x}{a}$

INTEGRALS (Continued)

17. $\displaystyle\int \frac{dx}{a^2 - x^2} = \begin{cases} \dfrac{1}{a}\tanh^{-1}\dfrac{x}{a} \\[4pt] \quad\text{or} \\[4pt] \dfrac{1}{2a}\log\dfrac{a+x}{a-x}, \end{cases} \quad (a^2 > x^2)$

18. $\displaystyle\int \frac{dx}{x^2 - a^2} = \begin{cases} -\dfrac{1}{a}\coth^{-1}\dfrac{x}{a} \\[4pt] \quad\text{or} \\[4pt] \dfrac{1}{2a}\log\dfrac{x-a}{x+a}, \end{cases} \quad (x^2 > a^2)$

19. $\displaystyle\int \frac{dx}{\sqrt{a^2 - x^2}} = \begin{cases} \sin^{-1}\dfrac{x}{|a|} \\[4pt] \quad\text{or} \\[4pt] -\cos^{-1}\dfrac{x}{|a|}, \end{cases} \quad (a^2 > x^2)$

20. $\displaystyle\int \frac{dx}{\sqrt{x^2 \pm a^2}} = \log\left(x + \sqrt{x^2 \pm a^2}\right)$

21. $\displaystyle\int \frac{dx}{x\sqrt{x^2 - a^2}} = \frac{1}{|a|}\sec^{-1}\frac{x}{a}$

22. $\displaystyle\int \frac{dx}{x\sqrt{a^2 \pm x^2}} = -\frac{1}{a}\log\left(\frac{a + \sqrt{a^2 \pm x^2}}{x}\right)$

FORMS CONTAINING $(a + bx)$

For forms containing $a + bx$, but not listed in the table, the substitution $u = \dfrac{a + bx}{x}$ may prove helpful.

23. $\displaystyle\int (a + bx)^n\, dx = \frac{(a + bx)^{n+1}}{(n + 1)b}, \quad (n \neq -1)$

24. $\displaystyle\int x(a + bx)^n\, dx$

$$= \frac{1}{b^2(n + 2)}(a + bx)^{n+2} - \frac{a}{b^2(n + 1)}(a + bx)^{n+1}, \quad (n \neq -1, -2)$$

25. $\displaystyle\int x^2(a + bx)^n\, dx = \frac{1}{b^3}\left[\frac{(a + bx)^{n+3}}{n + 3} - 2a\frac{(a + bx)^{n+2}}{n + 2} + a^2\frac{(a + bx)^{n+1}}{n + 1}\right]$

INTEGRALS (Continued)

$$26. \int x^m (a + bx)^n \, dx = \begin{cases} \dfrac{x^{m+1}(a + bx)^n}{m + n + 1} + \dfrac{an}{m + n + 1} \int x^m (a + bx)^{n-1} \, dx \\[6pt] \text{or} \\[6pt] \dfrac{1}{a(n + 1)} \left[-x^{m+1}(a + bx)^{n+1} \right. \\[12pt] \qquad\qquad \left. + (m + n + 2) \int x^m (a + bx)^{n+1} \, dx \right] \\[6pt] \text{or} \\[6pt] \dfrac{1}{b(m + n + 1)} \left[x^m (a + bx)^{n+1} - ma \int x^{m-1}(a + bx)^n \, dx \right] \end{cases}$$

$$27. \int \frac{dx}{a + bx} = \frac{1}{b} \log (a + bx)$$

$$28. \int \frac{dx}{(a + bx)^2} = -\frac{1}{b(a + bx)}$$

$$29. \int \frac{dx}{(a + bx)^3} = -\frac{1}{2b(a + bx)^2}$$

$$30. \int \frac{x \, dx}{a + bx} = \begin{cases} \dfrac{1}{b^2}[a + bx - a \log (a + bx)] \\[6pt] \text{or} \\[6pt] \dfrac{x}{b} - \dfrac{a}{b^2} \log (a + bx) \end{cases}$$

$$31. \int \frac{x \, dx}{(a + bx)^2} = \frac{1}{b^2} \left[\log (a + bx) + \frac{a}{a + bx} \right]$$

$$32. \int \frac{x \, dx}{(a + bx)^n} = \frac{1}{b^2} \left[\frac{-1}{(n - 2)(a + bx)^{n-2}} + \frac{a}{(n - 1)(a + bx)^{n-1}} \right], \qquad n \neq 1, 2$$

$$33. \int \frac{x^2 \, dx}{a + bx} = \frac{1}{b^3} \left[\frac{1}{2}(a + bx)^2 - 2a(a + bx) + a^2 \log (a + bx) \right]$$

$$34. \int \frac{x^2 \, dx}{(a + bx)^2} = \frac{1}{b^3} \left[a + bx - 2a \log (a + bx) - \frac{a^2}{a + bx} \right]$$

$$35. \int \frac{x^2 \, dx}{(a + bx)^3} = \frac{1}{b^3} \left[\log (a + bx) + \frac{2a}{a + bx} - \frac{a^2}{2(a + bx)^2} \right]$$

$$36. \int \frac{x^2 \, dx}{(a + bx)^n} = \frac{1}{b^3} \left[\frac{-1}{(n - 3)(a + bx)^{n-3}} \right. $$
$$\left. + \frac{2a}{(n - 2)(a + bx)^{n-2}} - \frac{a^2}{(n - 1)(a + bx)^{n-1}} \right], \qquad n \neq 1, 2, 3$$

INTEGRALS (Continued)

37. $\displaystyle\int \frac{dx}{x(a + bx)} = -\frac{1}{a}\log\frac{a + bx}{x}$

38. $\displaystyle\int \frac{dx}{x(a + bx)^2} = \frac{1}{a(a + bx)} - \frac{1}{a^2}\log\frac{a + bx}{x}$

39. $\displaystyle\int \frac{dx}{x(a + bx)^3} = \frac{1}{a^3}\left[\frac{1}{2}\left(\frac{2a + bx}{a + bx}\right)^2 + \log\frac{x}{a + bx}\right]$

40. $\displaystyle\int \frac{dx}{x^2(a + bx)} = -\frac{1}{ax} + \frac{b}{a^2}\log\frac{a + bx}{x}$

41. $\displaystyle\int \frac{dx}{x^3(a + bx)} = \frac{2bx - a}{2a^2x^2} + \frac{b^2}{a^3}\log\frac{x}{a + bx}$

42. $\displaystyle\int \frac{dx}{x^2(a + bx)^2} = -\frac{a + 2bx}{a^2x(a + bx)} + \frac{2b}{a^3}\log\frac{a + bx}{x}$

FORMS CONTAINING $c^2 \pm x^2$, $x^2 - c^2$

43. $\displaystyle\int \frac{dx}{c^2 + x^2} = \frac{1}{c}\tan^{-1}\frac{x}{c}$

44. $\displaystyle\int \frac{dx}{c^2 - x^2} = \frac{1}{2c}\log\frac{c + x}{c - x}, \qquad (c^2 > x^2)$

45. $\displaystyle\int \frac{dx}{x^2 - c^2} = \frac{1}{2c}\log\frac{x - c}{x + c}, \qquad (x^2 > c^2)$

46. $\displaystyle\int \frac{x\,dx}{c^2 \pm x^2} = \pm\frac{1}{2}\log(c^2 \pm x^2)$

47. $\displaystyle\int \frac{x\,dx}{(c^2 \pm x^2)^{n+1}} = \mp\frac{1}{2n(c^2 \pm x^2)^n}$

48. $\displaystyle\int \frac{dx}{(c^2 \pm x^2)^n} = \frac{1}{2c^2(n - 1)}\left[\frac{x}{(c^2 \pm x^2)^{n-1}} + (2n - 3)\int \frac{dx}{(c^2 \pm x^2)^{n-1}}\right]$

49. $\displaystyle\int \frac{dx}{(x^2 - c^2)^n} = \frac{1}{2c^2(n - 1)}\left[-\frac{x}{(x^2 - c^2)^{n-1}} - (2n - 3)\int \frac{dx}{(x^2 - c^2)^{n-1}}\right]$

50. $\displaystyle\int \frac{x\,dx}{x^2 - c^2} = \frac{1}{2}\log(x^2 - c^2)$

51. $\displaystyle\int \frac{x\,dx}{(x^2 - c^2)^{n+1}} = -\frac{1}{2n(x^2 - c^2)^n}$

INTEGRALS (Continued)

FORMS CONTAINING $a + bx$ and $c + dx$

$$u = a + bx, \qquad v = c + dx, \qquad k = ad - bc$$

If $k = 0$, then $v = \dfrac{c}{a} u$

52. $\displaystyle \int \frac{dx}{u \cdot v} = \frac{1}{k} \cdot \log\left(\frac{v}{u}\right)$

53. $\displaystyle \int \frac{x\,dx}{u \cdot v} = \frac{1}{k}\left[\frac{a}{b}\log(u) - \frac{c}{d}\log(v)\right]$

54. $\displaystyle \int \frac{dx}{u^2 \cdot v} = \frac{1}{k}\left(\frac{1}{u} + \frac{d}{k}\log\frac{v}{u}\right)$

55. $\displaystyle \int \frac{x\,dx}{u^2 \cdot v} = \frac{-a}{bku} - \frac{c}{k^2}\log\frac{v}{u}$

56. $\displaystyle \int \frac{x^2\,dx}{u^2 \cdot v} = \frac{a^2}{b^2 ku} + \frac{1}{k^2}\left[\frac{c^2}{d}\log(v) + \frac{a(k - bc)}{b^2}\log(u)\right]$

57. $\displaystyle \int \frac{dx}{u^n \cdot v^m} = \frac{1}{k(m - 1)}\left[\frac{-1}{u^{n-1} \cdot v^{m-1}} - (m + n - 2)b\int\frac{dx}{u^n \cdot v^{m-1}}\right]$

58. $\displaystyle \int \frac{u}{v}\,dx = \frac{bx}{d} + \frac{k}{d^2}\log(v)$

59. $\displaystyle \int \frac{u^m\,dx}{v^n} = \begin{cases} \dfrac{-1}{k(n - 1)}\left[\dfrac{u^{m+1}}{v^{n-1}} + b(n - m - 2)\displaystyle\int\frac{u^m}{v^{n-1}}\,dx\right] \\ \text{or} \\ \dfrac{-1}{d(n - m - 1)}\left[\dfrac{u^m}{v^{n-1}} + mk\displaystyle\int\frac{u^{m-1}}{v^n}\,dx\right] \\ \text{or} \\ \dfrac{-1}{d(n - 1)}\left[\dfrac{u^m}{v^{n-1}} - mb\displaystyle\int\frac{u^{m-1}}{v^{n-1}}\,dx\right] \end{cases}$

FORMS CONTAINING $(a + bx^n)$

60. $\displaystyle \int \frac{dx}{a + bx^2} = \frac{1}{\sqrt{ab}}\tan^{-1}\frac{x\sqrt{ab}}{a}, \qquad (ab > 0)$

61. $\displaystyle \int \frac{dx}{a + bx^2} = \begin{cases} \dfrac{1}{2\sqrt{-ab}}\log\dfrac{a + x\sqrt{-ab}}{a - x\sqrt{-ab}}, \qquad (ab < 0) \\ \text{or} \\ \dfrac{1}{\sqrt{-ab}}\tanh^{-1}\dfrac{x\sqrt{-ab}}{a}, \qquad (ab < 0) \end{cases}$

INTEGRALS (Continued)

62. $\displaystyle\int \frac{dx}{a^2 + b^2 x^2} = \frac{1}{ab} \tan^{-1} \frac{bx}{a}$

63. $\displaystyle\int \frac{x\, dx}{a + bx^2} = \frac{1}{2b} \log (a + bx^2)$

64. $\displaystyle\int \frac{x^2\, dx}{a + bx^2} = \frac{x}{b} - \frac{a}{b} \int \frac{dx}{a + bx^2}$

65. $\displaystyle\int \frac{dx}{(a + bx^2)^2} = \frac{x}{2a(a + bx^2)} + \frac{1}{2a} \int \frac{dx}{a + bx^2}$

66. $\displaystyle\int \frac{dx}{a^2 - b^2 x^2} = \frac{1}{2ab} \log \frac{a + bx}{a - bx}$

67. $\displaystyle\int \frac{dx}{(a + bx^2)^{m+1}} = \begin{cases} \dfrac{1}{2ma} \dfrac{x}{(a + bx^2)^m} + \dfrac{2m - 1}{2ma} \displaystyle\int \dfrac{dx}{(a + bx^2)^m} \\[2mm] \qquad\qquad\qquad \text{or} \\[2mm] \dfrac{(2m)!}{(m!)^2} \left[\dfrac{x}{2a} \displaystyle\sum_{r=1}^{m} \dfrac{r!(r-1)!}{(4a)^{m-r}(2r)!(a + bx^2)^r} + \dfrac{1}{(4a)^m} \displaystyle\int \dfrac{dx}{a + bx^2} \right] \end{cases}$

68. $\displaystyle\int \frac{x\, dx}{(a + bx^2)^{m+1}} = -\frac{1}{2bm(a + bx^2)^m}$

69. $\displaystyle\int \frac{x^2\, dx}{(a + bx^2)^{m+1}} = \frac{-x}{2mb(a + bx^2)^m} + \frac{1}{2mb} \int \frac{dx}{(a + bx^2)^m}$

70. $\displaystyle\int \frac{dx}{x(a + bx^2)} = \frac{1}{2a} \log \frac{x^2}{a + bx^2}$

71. $\displaystyle\int \frac{dx}{x^2(a + bx^2)} = -\frac{1}{ax} - \frac{b}{a} \int \frac{dx}{a + bx^2}$

72. $\displaystyle\int \frac{dx}{x(a + bx^2)^{m+1}} = \begin{cases} \dfrac{1}{2am(a + bx^2)^m} + \dfrac{1}{a} \displaystyle\int \dfrac{dx}{x(a + bx^2)^m} \\[2mm] \qquad\qquad\qquad \text{or} \\[2mm] \dfrac{1}{2a^{m+1}} \left[\displaystyle\sum_{r=1}^{m} \dfrac{a^r}{r(a + bx^2)^r} + \log \dfrac{x^2}{a + bx^2} \right] \end{cases}$

73. $\displaystyle\int \frac{dx}{x^2(a + bx^2)^{m+1}} = \frac{1}{a} \int \frac{dx}{x^2(a + bx^2)^m} - \frac{b}{a} \int \frac{dx}{(a + bx^2)^{m+1}}$

74. $\displaystyle\int \frac{dx}{a + bx^3} = \frac{k}{3a} \left[\frac{1}{2} \log \frac{(k + x)^3}{a + bx^3} + \sqrt{3} \tan^{-1} \frac{2x - k}{k\sqrt{3}} \right], \qquad \left(k = \sqrt[3]{\frac{a}{b}} \right)$

75. $\displaystyle\int \frac{x\, dx}{a + bx^3} = \frac{1}{3bk} \left[\frac{1}{2} \log \frac{a + bx^3}{(k + x)^3} + \sqrt{3} \tan^{-1} \frac{2x - k}{k\sqrt{3}} \right], \qquad \left(k = \sqrt[3]{\frac{a}{b}} \right)$

INTEGRALS (Continued)

76. $\int \dfrac{x^2\,dx}{a + bx^3} = \dfrac{1}{3b}\log(a + bx^3)$

77. $\int \dfrac{dx}{a + bx^4} = \dfrac{k}{2a}\left[\dfrac{1}{2}\log\dfrac{x^2 + 2kx + 2k^2}{x^2 - 2kx + 2k^2} + \tan^{-1}\dfrac{2kx}{2k^2 - x^2}\right],$

$$\left(ab > 0, k = \sqrt[4]{\dfrac{a}{4b}}\right)$$

78. $\int \dfrac{dx}{a + bx^4} = \dfrac{k}{2a}\left[\dfrac{1}{2}\log\dfrac{x + k}{x - k} + \tan^{-1}\dfrac{x}{k}\right], \qquad \left(ab < 0, k = \sqrt[4]{-\dfrac{a}{b}}\right)$

79. $\int \dfrac{x\,dx}{a + bx^4} = \dfrac{1}{2bk}\tan^{-1}\dfrac{x^2}{k}, \qquad \left(ab > 0, k = \sqrt{\dfrac{a}{b}}\right)$

80. $\int \dfrac{x\,dx}{a + bx^4} = \dfrac{1}{4bk}\log\dfrac{x^2 - k}{x^2 + k}, \qquad \left(ab < 0, k = \sqrt{-\dfrac{a}{b}}\right)$

81. $\int \dfrac{x^2\,dx}{a + bx^4} = \dfrac{1}{4bk}\left[\dfrac{1}{2}\log\dfrac{x^2 - 2kx + 2k^2}{x^2 + 2kx + 2k^2} + \tan^{-1}\dfrac{2kx}{2k^2 - x^2}\right],$

$$\left(ab > 0, k = \sqrt[4]{\dfrac{a}{4b}}\right)$$

82. $\int \dfrac{x^2\,dx}{a + bx^4} = \dfrac{1}{4bk}\left[\log\dfrac{x - k}{x + k} + 2\tan^{-1}\dfrac{x}{k}\right], \qquad \left(ab < 0, k = \sqrt[4]{-\dfrac{a}{b}}\right)$

83. $\int \dfrac{x^3\,dx}{a + bx^4} = \dfrac{1}{4b}\log(a + bx^4)$

84. $\int \dfrac{dx}{x(a + bx^n)} = \dfrac{1}{an}\log\dfrac{x^n}{a + bx^n}$

85. $\int \dfrac{dx}{(a + bx^n)^{m+1}} = \dfrac{1}{a}\int \dfrac{dx}{(a + bx^n)^m} - \dfrac{b}{a}\int \dfrac{x^n\,dx}{(a + bx^n)^{m+1}}$

86. $\int \dfrac{x^m\,dx}{(a + bx^n)^{p+1}} = \dfrac{1}{b}\int \dfrac{x^{m-n}\,dx}{(a + bx^n)^p} - \dfrac{a}{b}\int \dfrac{x^{m-n}\,dx}{(a + bx^n)^{p+1}}$

87. $\int \dfrac{dx}{x^m(a + bx^n)^{p+1}} = \dfrac{1}{a}\int \dfrac{dx}{x^m(a + bx^n)^p} - \dfrac{b}{a}\int \dfrac{dx}{x^{m-n}(a + bx^n)^{p+1}}$

INTEGRALS (Continued)

88. $\int x^m(a + bx^n)^p \, dx = \begin{cases} \end{cases}$

$$\frac{1}{b(np + m + 1)}\left[x^{m-n+1}(a + bx^n)^{p+1} \right.$$

$$\left. - a(m - n + 1)\int x^{m-n}(a + bx^n)^p \, dx \right]$$

or

$$\frac{1}{np + m + 1}\left[x^{m+1}(a + bx^n)^p \right.$$

$$\left. + anp\int x^m(a + bx^n)^{p-1} \, dx \right]$$

or

$$\frac{1}{a(m + 1)}\left[x^{m+1}(a + bx^n)^{p+1} \right.$$

$$\left. - (m + 1 + np + n)b\int x^{m+n}(a + bx^n)^p \, dx \right]$$

or

$$\frac{1}{an(p + 1)}\left[-x^{m+1}(a + bx^n)^{p+1} \right.$$

$$\left. + (m + 1 + np + n)\int x^m(a + bx^n)^{p+1} \, dx \right]$$

FORMS CONTAINING $c^3 \pm x^3$

89. $\displaystyle\int \frac{dx}{c^3 \pm x^3} = \pm \frac{1}{6c^2} \log \frac{(c \pm x)^3}{c^3 \pm x^3} + \frac{1}{c^2\sqrt{3}} \tan^{-1} \frac{2x \mp c}{c\sqrt{3}}$

90. $\displaystyle\int \frac{dx}{(c^3 \pm x^3)^2} = \frac{x}{3c^3(c^3 \pm x^3)} + \frac{2}{3c^3}\int \frac{dx}{c^3 \pm x^3}$

91. $\displaystyle\int \frac{dx}{(c^3 \pm x^3)^{n+1}} = \frac{1}{3nc^3}\left[\frac{x}{(c^3 \pm x^3)^n} + (3n - 1)\int \frac{dx}{(c^3 \pm x^3)^n} \right]$

92. $\displaystyle\int \frac{x \, dx}{c^3 \pm x^3} = \frac{1}{6c} \log \frac{c^3 \pm x^3}{(c \pm x)^3} \pm \frac{1}{c\sqrt{3}} \tan^{-1} \frac{2x \mp c}{c\sqrt{3}}$

93. $\displaystyle\int \frac{x \, dx}{(c^3 \pm x^3)^2} = \frac{x^2}{3c^3(c^3 \pm x^3)} + \frac{1}{3c^3}\int \frac{x \, dx}{c^3 \pm x^3}$

94. $\displaystyle\int \frac{x \, dx}{(c^3 \pm x^3)^{n+1}} = \frac{1}{3nc^3}\left[\frac{x^2}{(c^3 \pm x^3)^n} + (3n - 2)\int \frac{x \, dx}{(c^3 \pm x^3)^n} \right]$

95. $\displaystyle\int \frac{x^2 \, dx}{c^3 \pm x^3} = \pm \frac{1}{3} \log (c^3 \pm x^3)$

INTEGRALS (Continued)

96. $\displaystyle\int \frac{x^2\,dx}{(c^3 \pm x^3)^{n+1}} = \mp \frac{1}{3n(c^3 \pm x^3)^n}$

97. $\displaystyle\int \frac{dx}{x(c^3 \pm x^3)} = \frac{1}{3c^3}\log\frac{x^3}{c^3 \pm x^3}$

98. $\displaystyle\int \frac{dx}{x(c^3 \pm x^3)^2} = \frac{1}{3c^3(c^3 \pm x^3)} + \frac{1}{3c^6}\log\frac{x^3}{c^3 \pm x^3}$

99. $\displaystyle\int \frac{dx}{x(c^3 \pm x^3)^{n+1}} = \frac{1}{3nc^3(c^3 \pm x^3)^n} + \frac{1}{c^3}\int \frac{dx}{x(c^3 \pm x^3)^n}$

100. $\displaystyle\int \frac{dx}{x^2(c^3 \pm x^3)} = -\frac{1}{c^3 x} \mp \frac{1}{c^3}\int \frac{x\,dx}{c^3 \pm x^3}$

101. $\displaystyle\int \frac{dx}{x^2(c^3 \pm x^3)^{n+1}} = \frac{1}{c^3}\int \frac{dx}{x^2(c^3 \pm x^3)^n} \mp \frac{1}{c^3}\int \frac{x\,dx}{(c^3 \pm x^3)^{n+1}}$

FORMS CONTAINING $c^4 \pm x^4$

102. $\displaystyle\int \frac{dx}{c^4 + x^4} = \frac{1}{2c^3\sqrt{2}}\left[\frac{1}{2}\log\frac{x^2 + cx\sqrt{2} + c^2}{x^2 - cx\sqrt{2} + c^2} + \tan^{-1}\frac{cx\sqrt{2}}{c^2 - x^2}\right]$

103. $\displaystyle\int \frac{dx}{c^4 - x^4} = \frac{1}{2c^3}\left[\frac{1}{2}\log\frac{c + x}{c - x} + \tan^{-1}\frac{x}{c}\right]$

104. $\displaystyle\int \frac{x\,dx}{c^4 + x^4} = \frac{1}{2c^2}\tan^{-1}\frac{x^2}{c^2}$

105. $\displaystyle\int \frac{x\,dx}{c^4 - x^4} = \frac{1}{4c^2}\log\frac{c^2 + x^2}{c^2 - x^2}$

106. $\displaystyle\int \frac{x^2\,dx}{c^4 + x^4} = \frac{1}{2c\sqrt{2}}\left[\frac{1}{2}\log\frac{x^2 - cx\sqrt{2} + c^2}{x^2 + cx\sqrt{2} + c^2} + \tan^{-1}\frac{cx\sqrt{2}}{c^2 - x^2}\right]$

107. $\displaystyle\int \frac{x^2\,dx}{c^4 - x^4} = \frac{1}{2c}\left[\frac{1}{2}\log\frac{c + x}{c - x} - \tan^{-1}\frac{x}{c}\right]$

108. $\displaystyle\int \frac{x^3\,dx}{c^4 \pm x^4} = \pm\frac{1}{4}\log(c^4 \pm x^4)$

FORMS CONTAINING $(a + bx + cx^2)$

$$X = a + bx + cx^2 \text{ and } q = 4ac - b^2$$

If $q = 0$, then $X = c\left(x + \dfrac{b}{2c}\right)^2$, and formulas starting with 23 should be used in place of these.

109. $\displaystyle\int \frac{dx}{X} = \frac{2}{\sqrt{q}}\tan^{-1}\frac{2cx + b}{\sqrt{q}}, \qquad (q > 0)$

INTEGRALS (Continued)

110. $\displaystyle\int \frac{dx}{X} = \begin{cases} \dfrac{-2}{\sqrt{-q}} \tanh^{-1} \dfrac{2cx + b}{\sqrt{-q}} \\ \qquad\qquad \text{or} \\ \dfrac{1}{\sqrt{-q}} \log \dfrac{2cx + b - \sqrt{-q}}{2cx + b + \sqrt{-q}}, \qquad (q < 0) \end{cases}$

111. $\displaystyle\int \frac{dx}{X^2} = \frac{2cx + b}{qX} + \frac{2c}{q} \int \frac{dx}{X}$

112. $\displaystyle\int \frac{dx}{X^3} = \frac{2cx + b}{q} \left(\frac{1}{2X^2} + \frac{3c}{qX} \right) + \frac{6c^2}{q^2} \int \frac{dx}{X}$

113. $\displaystyle\int \frac{dx}{X^{n+1}} = \begin{cases} \dfrac{2cx + b}{nqX^n} + \dfrac{2(2n-1)c}{qn} \int \dfrac{dx}{X^n} \\ \qquad\qquad \text{or} \\ \dfrac{(2n)!}{(n!)^2} \left(\dfrac{c}{q}\right)^n \left[\dfrac{2cx + b}{q} \sum_{r=1}^{n} \left(\dfrac{q}{cX}\right)^r \left(\dfrac{(r-1)!r!}{(2r)!}\right) + \int \dfrac{dx}{X} \right] \end{cases}$

114. $\displaystyle\int \frac{x\,dx}{X} = \frac{1}{2c} \log X - \frac{b}{2c} \int \frac{dx}{X}$

115. $\displaystyle\int \frac{x\,dx}{X^2} = -\frac{bx + 2a}{qX} - \frac{b}{q} \int \frac{dx}{X}$

116. $\displaystyle\int \frac{x\,dx}{X^{n+1}} = -\frac{2a + bx}{nqX^n} - \frac{b(2n-1)}{nq} \int \frac{dx}{X^n}$

117. $\displaystyle\int \frac{x^2}{X}\,dx = \frac{x}{c} - \frac{b}{2c^2} \log X + \frac{b^2 - 2ac}{2c^2} \int \frac{dx}{X}$

118. $\displaystyle\int \frac{x^2}{X^2}\,dx = \frac{(b^2 - 2ac)x + ab}{cqX} + \frac{2a}{q} \int \frac{dx}{X}$

119. $\displaystyle\int \frac{x^m\,dx}{X^{n+1}} = -\frac{x^{m-1}}{(2n - m + 1)cX^n} - \frac{n - m + 1}{2n - m + 1} \cdot \frac{b}{c} \int \frac{x^{m-1}\,dx}{X^{n+1}}$
$$+ \frac{m - 1}{2n - m + 1} \cdot \frac{a}{c} \int \frac{x^{m-2}\,dx}{X^{n+1}}$$

120. $\displaystyle\int \frac{dx}{xX} = \frac{1}{2a} \log \frac{x^2}{X} - \frac{b}{2a} \int \frac{dx}{X}$

121. $\displaystyle\int \frac{dx}{x^2 X} = \frac{b}{2a^2} \log \frac{X}{x^2} - \frac{1}{ax} + \left(\frac{b^2}{2a^2} - \frac{c}{a} \right) \int \frac{dx}{X}$

122. $\displaystyle\int \frac{dx}{xX^n} = \frac{1}{2a(n-1)X^{n-1}} - \frac{b}{2a} \int \frac{dx}{X^n} + \frac{1}{a} \int \frac{dx}{xX^{n-1}}$

INTEGRALS (Continued)

123. $\displaystyle\int \frac{dx}{x^m X^{n+1}} = -\frac{1}{(m-1)ax^{m-1}X^n} - \frac{n+m-1}{m-1}\cdot\frac{b}{a}\int \frac{dx}{x^{m-1}X^{n+1}}$

$$-\frac{2n+m-1}{m-1}\cdot\frac{c}{a}\int \frac{dx}{x^{m-2}X^{n+1}}$$

FORMS CONTAINING $\sqrt{a+bx}$

124. $\displaystyle\int \sqrt{a+bx}\,dx = \frac{2}{3b}\sqrt{(a+bx)^3}$

125. $\displaystyle\int x\sqrt{a+bx}\,dx = -\frac{2(2a-3bx)\sqrt{(a+bx)^3}}{15b^2}$

126. $\displaystyle\int x^2\sqrt{a+bx}\,dx = \frac{2(8a^2-12abx+15b^2x^2)\sqrt{(a+bx)^3}}{105b^3}$

127. $\displaystyle\int x^m\sqrt{a+bx}\,dx = \begin{cases} \dfrac{2}{b(2m+3)}\left[x^m\sqrt{(a+bx)^3} - ma\displaystyle\int x^{m-1}\sqrt{a+bx}\,dx\right] \\ \qquad\text{or} \\ \dfrac{2}{b^{m+1}}\sqrt{a+bx}\displaystyle\sum_{r=0}^{m}\frac{m!(-a)^{m-r}}{r!(m-r)!(2r+3)}(a+bx)^{r+1} \end{cases}$

128. $\displaystyle\int \frac{\sqrt{a+bx}}{x}\,dx = 2\sqrt{a+bx} + a\int \frac{dx}{x\sqrt{a+bx}}$

129. $\displaystyle\int \frac{\sqrt{a+bx}}{x^2}\,dx = -\frac{\sqrt{a+bx}}{x} + \frac{b}{2}\int \frac{dx}{x\sqrt{a+bx}}$

130. $\displaystyle\int \frac{\sqrt{a+bx}}{x^m}\,dx = -\frac{1}{(m-1)a}\left[\frac{\sqrt{(a+bx)^3}}{x^{m-1}} + \frac{(2m-5)b}{2}\int \frac{\sqrt{a+bx}}{x^{m-1}}\,dx\right]$

131. $\displaystyle\int \frac{dx}{\sqrt{a+bx}} = \frac{2\sqrt{a+bx}}{b}$

132. $\displaystyle\int \frac{x\,dx}{\sqrt{a+bx}} = -\frac{2(2a-bx)}{3b^2}\sqrt{a+bx}$

133. $\displaystyle\int \frac{x^2\,dx}{\sqrt{a+bx}} = \frac{2(8a^2-4abx+3b^2x^2)}{15b^3}\sqrt{a+bx}$

INTEGRALS (Continued)

134. $\displaystyle\int \frac{x^m \, dx}{\sqrt{a + bx}} = \begin{cases} \dfrac{2}{(2m + 1)b}\left[x^m\sqrt{a + bx} - ma\displaystyle\int \frac{x^{m-1} \, dx}{\sqrt{a + bx}}\right] \\[2ex] \text{or} \\[2ex] \dfrac{2(-a)^m\sqrt{a + bx}}{b^{m+1}} \displaystyle\sum_{r=0}^{m} \frac{(-1)^r m!(a + bx)^r}{(2r + 1)r!(m - r)!a^r} \end{cases}$

135. $\displaystyle\int \frac{dx}{x\sqrt{a + bx}} = \frac{1}{\sqrt{a}} \log\left(\frac{\sqrt{a + bx} - \sqrt{a}}{\sqrt{a + bx} + \sqrt{a}}\right), \qquad (a > 0)$

136. $\displaystyle\int \frac{dx}{x\sqrt{a + bx}} = \frac{2}{\sqrt{-a}} \tan^{-1}\sqrt{\frac{a + bx}{-a}}, \qquad (a < 0)$

137. $\displaystyle\int \frac{dx}{x^2\sqrt{a + bx}} = -\frac{\sqrt{a + bx}}{ax} - \frac{b}{2a}\int \frac{dx}{x\sqrt{a + bx}}$

138. $\displaystyle\int \frac{dx}{x^n\sqrt{a + bx}} = \begin{cases} -\dfrac{\sqrt{a + bx}}{(n - 1)ax^{n-1}} - \dfrac{(2n - 3)b}{(2n - 2)a}\displaystyle\int \frac{dx}{x^{n-1}\sqrt{a + bx}} \\[2ex] \text{or} \\[2ex] \dfrac{(2n - 2)!}{[(n - 1)!]^2}\left[-\dfrac{\sqrt{a + bx}}{u}\displaystyle\sum_{r=1}^{n-1} \frac{r!(r - 1)!}{x^r(2r)!}\left(-\frac{b}{4a}\right)^{n-r-1}\right. \\[2ex] \qquad\qquad\qquad \left. + \left(-\dfrac{b}{4a}\right)^{n-1}\displaystyle\int \frac{dx}{x\sqrt{a + bx}}\right] \end{cases}$

139. $\displaystyle\int (a + bx)^{\pm\frac{n}{2}} \, dx = \frac{2(a + bx)^{\frac{2\pm n}{2}}}{b(2 \pm n)}$

140. $\displaystyle\int x(a + bx)^{\pm\frac{n}{2}} \, dx = \frac{2}{b^2}\left[\frac{(a + bx)^{\frac{4\pm n}{2}}}{4 \pm n} - \frac{a(a + bx)^{\frac{2\pm n}{2}}}{2 \pm n}\right]$

141. $\displaystyle\int \frac{dx}{x(a + bx)^{\frac{m}{2}}} = \frac{1}{a}\int \frac{dx}{x(a + bx)^{\frac{m-2}{2}}} - \frac{b}{a}\int \frac{dx}{(a + bx)^{\frac{m}{2}}}$

142. $\displaystyle\int \frac{(a + bx)^{\frac{n}{2}} \, dx}{x} = b\int (a + bx)^{\frac{n-2}{2}} \, dx + a\int \frac{(a + bx)^{\frac{n-2}{2}}}{x} \, dx$

143. $\displaystyle\int f(x, \sqrt{a + bx}) \, dx = \frac{2}{b}\int f\left(\frac{z^2 - a}{b}, z\right) z \, dz, \qquad (z = \sqrt{a + bx})$

INTEGRALS (Continued)

FORMS CONTAINING $\sqrt{a + bx}$ and $\sqrt{c + dx}$

$$u = a + bx \qquad v = c + dx \qquad k = ad - bc$$

If $k = 0$, then $v = \dfrac{c}{a}u$, and formulas starting with 124 should be used in place of these.

144. $\displaystyle\int \frac{dx}{\sqrt{uv}} = \begin{cases} \dfrac{2}{\sqrt{bd}} \tanh^{-1} \dfrac{\sqrt{bduv}}{bv}, & bd > o, k < o \\[2mm] \text{or} \\[2mm] \dfrac{2}{\sqrt{bd}} \tanh^{-1} \dfrac{\sqrt{bduv}}{du}, & bd > o, k > o. \\[2mm] \text{or} \\[2mm] \dfrac{1}{\sqrt{bd}} \log \dfrac{(bv + \sqrt{bduv})^2}{v}, & (bd > 0) \end{cases}$

145. $\displaystyle\int \frac{dx}{\sqrt{uv}} = \begin{cases} \dfrac{2}{\sqrt{-bd}} \tan^{-1} \dfrac{\sqrt{-bduv}}{bv} \\[2mm] \text{or} \\[2mm] -\dfrac{1}{\sqrt{-bd}} \sin^{-1}\left(\dfrac{2bdx + ad + bc}{|k|} \right), & (bd < 0) \end{cases}$

146. $\displaystyle\int \sqrt{uv}\, dx = \frac{k + 2bv}{4bd} \sqrt{uv} - \frac{k^2}{8bd} \int \frac{dx}{\sqrt{uv}}$

147. $\displaystyle\int \frac{dx}{v\sqrt{u}} = \begin{cases} \dfrac{1}{\sqrt{kd}} \log \dfrac{d\sqrt{u} - \sqrt{kd}}{d\sqrt{u} + \sqrt{kd}} \\[2mm] \text{or} \\[2mm] \dfrac{1}{\sqrt{kd}} \log \dfrac{(d\sqrt{u} - \sqrt{kd})^2}{v}, & (kd > 0) \end{cases}$

148. $\displaystyle\int \frac{dx}{v\sqrt{u}} = \frac{2}{\sqrt{-kd}} \tan^{-1} \frac{d\sqrt{u}}{\sqrt{-kd}}, \qquad (kd < 0)$

149. $\displaystyle\int \frac{x\, dx}{\sqrt{uv}} = \frac{\sqrt{uv}}{bd} - \frac{ad + bc}{2bd} \int \frac{dx}{\sqrt{uv}}$

150. $\displaystyle\int \frac{dx}{v\sqrt{uv}} = \frac{-2\sqrt{uv}}{kv}$

INTEGRALS (Continued)

151. $\displaystyle\int \frac{v\,dx}{\sqrt{uv}} = \frac{\sqrt{uv}}{b} - \frac{k}{2b}\int \frac{dx}{\sqrt{uv}}$

152. $\displaystyle\int \sqrt{\frac{v}{u}}\,dx = \frac{v}{|v|}\int \frac{v\,dx}{\sqrt{uv}}$

153. $\displaystyle\int v^m\sqrt{u}\,dx = \frac{1}{(2m+3)d}\left(2v^{m+1}\sqrt{u} + k\int \frac{v^m\,dx}{\sqrt{u}}\right)$

154. $\displaystyle\int \frac{dx}{v^m\sqrt{u}} = -\frac{1}{(m-1)k}\left(\frac{\sqrt{u}}{v^{m-1}} + \left(m-\frac{3}{2}\right)b\int \frac{dx}{v^{m-1}\sqrt{u}}\right)$

155. $\displaystyle\int \frac{v^m\,dx}{\sqrt{u}} = \begin{cases} \dfrac{2}{b(2m+1)}\left[v^m\sqrt{u} - mk\displaystyle\int \frac{v^{m-1}}{\sqrt{u}}\,dx\right] \\ \text{or} \\ \dfrac{2(m!)^2\sqrt{u}}{b(2m+1)!}\displaystyle\sum_{r=0}^{m}\left(-\frac{4k}{b}\right)^{m-r}\frac{(2r)!}{(r!)^2}v^r \end{cases}$

FORMS CONTAINING $\sqrt{x^2 \pm a^2}$

156. $\displaystyle\int \sqrt{x^2 \pm a^2}\,dx = \tfrac{1}{2}[x\sqrt{x^2 \pm a^2} \pm a^2 \log(x + \sqrt{x^2 \pm a^2})]$

157. $\displaystyle\int \frac{dx}{\sqrt{x^2 \pm a^2}} = \log(x + \sqrt{x^2 \pm a^2})$

158. $\displaystyle\int \frac{dx}{x\sqrt{x^2 - a^2}} = \frac{1}{|a|}\sec^{-1}\frac{x}{a}$

159. $\displaystyle\int \frac{dx}{x\sqrt{x^2 + a^2}} = -\frac{1}{a}\log\left(\frac{a + \sqrt{x^2 + a^2}}{x}\right)$

160. $\displaystyle\int \frac{\sqrt{x^2 + a^2}}{x}\,dx = \sqrt{x^2 + a^2} - a\log\left(\frac{a + \sqrt{x^2 + a^2}}{x}\right)$

161. $\displaystyle\int \frac{\sqrt{x^2 - a^2}}{x}\,dx = \sqrt{x^2 - a^2} - |a|\sec^{-1}\frac{x}{a}$

162. $\displaystyle\int \frac{x\,dx}{\sqrt{x^2 \pm a^2}} = \sqrt{x^2 \pm a^2}$

163. $\displaystyle\int x\sqrt{x^2 \pm a^2}\,dx = \tfrac{1}{3}\sqrt{(x^2 \pm a^2)^3}$

INTEGRALS (Continued)

164. $\displaystyle\int \sqrt{(x^2 \pm a^2)^3}\, dx = \frac{1}{4}\Bigg[x\sqrt{(x^2 \pm a^2)^3} \pm \frac{3a^2 x}{2}\sqrt{x^2 \pm a^2}$

$$+ \frac{3a^4}{2}\log\left(x + \sqrt{x^2 \pm a^2}\right)\Bigg]$$

165. $\displaystyle\int \frac{dx}{\sqrt{(x^2 \pm a^2)^3}} = \frac{\pm x}{a^2\sqrt{x^2 \pm a^2}}$

166. $\displaystyle\int \frac{x\, dx}{\sqrt{(x^2 \pm a^2)^3}} = \frac{-1}{\sqrt{x^2 \pm a^2}}$

167. $\displaystyle\int x\sqrt{(x^2 \pm a^2)^3}\, dx = \frac{1}{5}\sqrt{(x^2 \pm a^2)^5}$

168. $\displaystyle\int x^2\sqrt{x^2 \pm a^2}\, dx = \frac{x}{4}\sqrt{(x^2 \pm a^2)^3} \mp \frac{a^2}{8}x\sqrt{x^2 \pm a^2} - \frac{a^4}{8}\log\left(x + \sqrt{x^2 \pm a^2}\right)$

169. $\displaystyle\int x^3\sqrt{x^2 + a^2}\, dx = \left(\frac{1}{5}x^2 - \frac{2}{15}a^2\right)\sqrt{(a^2 + x^2)^3}$

170. $\displaystyle\int x^3\sqrt{x^2 - a^2}\, dx = \frac{1}{5}\sqrt{(x^2 - a^2)^5} + \frac{a^2}{3}\sqrt{(x^2 - a^2)^3}$

171. $\displaystyle\int \frac{x^2\, dx}{\sqrt{x^2 \pm a^2}} = \frac{x}{2}\sqrt{x^2 \pm a^2} \mp \frac{a^2}{2}\log\left(x + \sqrt{x^2 \pm a^2}\right)$

172. $\displaystyle\int \frac{x^3\, dx}{\sqrt{x^2 \pm a^2}} = \frac{1}{3}\sqrt{(x^2 \pm a^2)^3} \mp a^2\sqrt{x^2 \pm a^2}$

173. $\displaystyle\int \frac{dx}{x^2\sqrt{x^2 \pm a^2}} = \mp\frac{\sqrt{x^2 \pm a^2}}{a^2 x}$

174. $\displaystyle\int \frac{dx}{x^3\sqrt{x^2 + a^2}} = -\frac{\sqrt{x^2 + a^2}}{2a^2 x^2} + \frac{1}{2a^3}\log\frac{a + \sqrt{x^2 + a^2}}{x}$

175. $\displaystyle\int \frac{dx}{x^3\sqrt{x^2 - a^2}} = \frac{\sqrt{x^2 - a^2}}{2a^2 x^2} + \frac{1}{2|a^3|}\sec^{-1}\frac{x}{a}$

176. $\displaystyle\int x^2\sqrt{(x^2 \pm a^2)^3}\, dx = \frac{x}{6}\sqrt{(x^2 \pm a^2)^5} \mp \frac{a^2 x}{24}\sqrt{(x^2 \pm a^2)^3} - \frac{a^4 x}{16}\sqrt{x^2 \pm a^2}$

$$\mp \frac{a^6}{16}\log\left(x + \sqrt{x^2 \pm a^2}\right)$$

177. $\displaystyle\int x^3\sqrt{(x^2 \pm a^2)^3}\, dx = \frac{1}{7}\sqrt{(x^2 \pm a^2)^7} \mp \frac{a^2}{5}\sqrt{(x^2 \pm a^2)^5}$

INTEGRALS (Continued)

178. $\displaystyle\int \frac{\sqrt{x^2 \pm a^2}\, dx}{x^2} = -\frac{\sqrt{x^2 \pm a^2}}{x} + \log(x + \sqrt{x^2 \pm a^2})$

179. $\displaystyle\int \frac{\sqrt{x^2 + a^2}}{x^3}\, dx = -\frac{\sqrt{x^2 + a^2}}{2x^2} - \frac{1}{2a}\log\frac{a + \sqrt{x^2 + a^2}}{x}$

180. $\displaystyle\int \frac{\sqrt{x^2 - a^2}}{x^3}\, dx = -\frac{\sqrt{x^2 - a^2}}{2x^2} + \frac{1}{2|a|}\sec^{-1}\frac{x}{a}$

181. $\displaystyle\int \frac{\sqrt{x^2 \pm a^2}}{x^4}\, dx = \mp\frac{\sqrt{(x^2 \pm a^2)^3}}{3a^2 x^3}$

182. $\displaystyle\int \frac{x^2\, dx}{\sqrt{(x^2 \pm a^2)^3}} = \frac{-x}{\sqrt{x^2 \pm a^2}} + \log(x + \sqrt{x^2 \pm a^2})$

183. $\displaystyle\int \frac{x^3\, dx}{\sqrt{(x^2 \pm a^2)^3}} = \sqrt{x^2 \pm a^2} \pm \frac{a^2}{\sqrt{x^2 \pm a^2}}$

184. $\displaystyle\int \frac{dx}{x\sqrt{(x^2 + a^2)^3}} = \frac{1}{a^2\sqrt{x^2 + a^2}} - \frac{1}{a^3}\log\frac{a + \sqrt{x^2 + a^2}}{x}$

185. $\displaystyle\int \frac{dx}{x\sqrt{(x^2 - a^2)^3}} = -\frac{1}{a^2\sqrt{x^2 - a^2}} - \frac{1}{|a^3|}\sec^{-1}\frac{x}{a}$

186. $\displaystyle\int \frac{dx}{x^2\sqrt{(x^2 \pm a^2)^3}} = -\frac{1}{a^4}\left[\frac{\sqrt{x^2 \pm a^2}}{x} + \frac{x}{\sqrt{x^2 \pm a^2}}\right]$

187. $\displaystyle\int \frac{dx}{x^3\sqrt{(x^2 + a^2)^3}} = -\frac{1}{2a^2 x^2\sqrt{x^2 + a^2}} - \frac{3}{2a^4\sqrt{x^2 + a^2}}$

$$+ \frac{3}{2a^5}\log\frac{a + \sqrt{x^2 + a^2}}{x}$$

188. $\displaystyle\int \frac{dx}{x^3\sqrt{(x^2 - a^2)^3}} = \frac{1}{2a^2 x^2\sqrt{x^2 - a^2}} - \frac{3}{2a^4\sqrt{x^2 - a^2}} - \frac{3}{2|a^5|}\sec^{-1}\frac{x}{a}$

189. $\displaystyle\int \frac{x^m}{\sqrt{x^2 \pm a^2}}\, dx = \frac{1}{m}x^{m-1}\sqrt{x^2 \pm a^2} \mp \frac{m-1}{m}a^2 \int \frac{x^{m-2}}{\sqrt{x^2 \pm a^2}}\, dx$

190. $\displaystyle\int \frac{x^{2m}}{\sqrt{x^2 \pm a^2}}\, dx = \frac{(2m)!}{2^{2m}(m!)^2}\left[\sqrt{x^2 \pm a^2}\sum_{r=1}^{m} \frac{r!(r-1)!}{(2r)!}(\mp a^2)^{m-r}(2x)^{2r-1}\right.$

$$\left. + (\mp a^2)^m \log(x + \sqrt{x^2 \pm a^2})\right]$$

191. $\displaystyle\int \frac{x^{2m+1}}{\sqrt{x^2 \pm a^2}}\, dx = \sqrt{x^2 \pm a^2}\sum_{r=0}^{m} \frac{(2r)!(m!)^2}{(2m+1)!(r!)^2}(\mp 4a^2)^{m-r}x^{2r}$

INTEGRALS (Continued)

192. $\displaystyle \int \frac{dx}{x^m\sqrt{x^2 \pm a^2}} = \mp \frac{\sqrt{x^2 \pm a^2}}{(m-1)a^2 x^{m-1}} \mp \frac{(m-2)}{(m-1)a^2} \int \frac{dx}{x^{m-2}\sqrt{x^2 \pm a^2}}$

193. $\displaystyle \int \frac{dx}{x^{2m}\sqrt{x^2 \pm a^2}} = \sqrt{x^2 \pm a^2} \sum_{r=0}^{m-1} \frac{(m-1)!\,m!\,(2r)!\,2^{2m-2r-1}}{(r!)^2(2m)!(\mp a^2)^{m-r}x^{2r+1}}$

194. $\displaystyle \int \frac{dx}{x^{2m+1}\sqrt{x^2 + a^2}} = \frac{(2m)!}{(m!)^2}\left[\frac{\sqrt{x^2 + a^2}}{a^2} \sum_{r=1}^{m} (-1)^{m-r+1} \frac{r!(r-1)!}{2(2r)!(4a^2)^{m-r}x^{2r}} \right.$

$$\left. + \frac{(-1)^{m+1}}{2^{2m}a^{2m+1}} \log \frac{\sqrt{x^2 + a^2} + a}{x} \right]$$

195. $\displaystyle \int \frac{dx}{x^{2m+1}\sqrt{x^2 - a^2}} = \frac{(2m)!}{(m!)^2}\left[\frac{\sqrt{x^2 - a^2}}{a^2} \sum_{r=1}^{m} \frac{r!(r-1)!}{2(2r)!(4a^2)^{m-r}x^{2r}} \right.$

$$\left. + \frac{1}{2^{2m}|a|^{2m+1}} \sec^{-1}\frac{x}{a} \right]$$

196. $\displaystyle \int \frac{dx}{(x-a)\sqrt{x^2 - a^2}} = -\frac{\sqrt{x^2 - a^2}}{a(x-a)}$

197. $\displaystyle \int \frac{dx}{(x+a)\sqrt{x^2 - a^2}} = \frac{\sqrt{x^2 - a^2}}{a(x+a)}$

198. $\displaystyle \int f(x, \sqrt{x^2 + a^2})\, dx = a \int f(a\tan u, a\sec u)\sec^2 u\, du, \qquad \left(u = \tan^{-1}\frac{x}{a}, a > 0 \right)$

199. $\displaystyle \int f(x, \sqrt{x^2 - a^2})\, dx = a \int f(a\sec u, a\tan u)\sec u\tan u\, du, \qquad \left(u = \sec^{-1}\frac{x}{a}, \right.$

$$\left. a > 0 \right)$$

FORMS CONTAINING $\sqrt{a^2 - x^2}$

200. $\displaystyle \int \sqrt{a^2 - x^2}\, dx = \frac{1}{2}\left[x\sqrt{a^2 - x^2} + a^2 \sin^{-1}\frac{x}{|a|} \right]$

201. $\displaystyle \int \frac{dx}{\sqrt{a^2 - x^2}} = \begin{cases} \sin^{-1}\dfrac{x}{|a|} \\[2mm] \text{or} \\[2mm] -\cos^{-1}\dfrac{x}{|a|} \end{cases}$

202. $\displaystyle \int \frac{dx}{x\sqrt{a^2 - x^2}} = -\frac{1}{a}\log\left(\frac{a + \sqrt{a^2 - x^2}}{x} \right)$

INTEGRALS (Continued)

203. $\int \dfrac{\sqrt{a^2 - x^2}}{x} \, dx = \sqrt{a^2 - x^2} - a \log \left(\dfrac{a + \sqrt{a^2 - x^2}}{x} \right)$

204. $\int \dfrac{x \, dx}{\sqrt{a^2 - x^2}} = - \sqrt{a^2 - x^2}$

205. $\int x \sqrt{a^2 - x^2} \, dx = -\frac{1}{3} \sqrt{(a^2 - x^2)^3}$

206. $\int \sqrt{(a^2 - x^2)^3} \, dx = \dfrac{1}{4} \left[x \sqrt{(a^2 - x^2)^3} + \dfrac{3a^2 x}{2} \sqrt{a^2 - x^2} + \dfrac{3a^4}{2} \sin^{-1} \dfrac{x}{|a|} \right]$

207. $\int \dfrac{dx}{\sqrt{(a^2 - x^2)^3}} = \dfrac{x}{a^2 \sqrt{a^2 - x^2}}$

208. $\int \dfrac{x \, dx}{\sqrt{(a^2 - x^2)^3}} = \dfrac{1}{\sqrt{a^2 - x^2}}$

209. $\int x \sqrt{(a^2 - x^2)^3} \, dx = -\frac{1}{5} \sqrt{(a^2 - x^2)^5}$

210. $\int x^2 \sqrt{a^2 - x^2} \, dx = -\dfrac{x}{4} \sqrt{(a^2 - x^2)^3} + \dfrac{a^2}{8} \left(x \sqrt{a^2 - x^2} + a^2 \sin^{-1} \dfrac{x}{|a|} \right)$

211. $\int x^3 \sqrt{a^2 - x^2} \, dx = (-\frac{1}{5} x^2 - \frac{2}{15} a^2) \sqrt{(a^2 - x^2)^3}$

212. $\int x^2 \sqrt{(a^2 - x^2)^3} \, dx = -\dfrac{1}{6} x \sqrt{(a^2 - x^2)^5} + \dfrac{a^2 x}{24} \sqrt{(a^2 - x^2)^3}$

$$+ \dfrac{a^4 x}{16} \sqrt{a^2 - x^2} + \dfrac{a^6}{16} \sin^{-1} \dfrac{x}{|a|}$$

213. $\int x^3 \sqrt{(a^2 - x^2)^3} \, dx = \dfrac{1}{7} \sqrt{(a^2 - x^2)^7} - \dfrac{a^2}{5} \sqrt{(a^2 - x^2)^5}$

214. $\int \dfrac{x^2 \, dx}{\sqrt{a^2 - x^2}} = -\dfrac{x}{2} \sqrt{a^2 - x^2} + \dfrac{a^2}{2} \sin^{-1} \dfrac{x}{|a|}$

215. $\int \dfrac{dx}{x^2 \sqrt{a^2 - x^2}} = -\dfrac{\sqrt{a^2 - x^2}}{a^2 x}$

216. $\int \dfrac{\sqrt{a^2 - x^2}}{x^2} \, dx = -\dfrac{\sqrt{a^2 - x^2}}{x} - \sin^{-1} \dfrac{x}{|a|}$

217. $\int \dfrac{\sqrt{a^2 - x^2}}{x^3} \, dx = -\dfrac{\sqrt{a^2 - x^2}}{2x^2} + \dfrac{1}{2a} \log \dfrac{a + \sqrt{a^2 - x^2}}{x}$

218. $\int \dfrac{\sqrt{a^2 - x^2}}{x^4} \, dx = -\dfrac{\sqrt{(a^2 - x^2)^3}}{3a^2 x^3}$

INTEGRALS (Continued)

219. $\displaystyle \int \frac{x^2\, dx}{\sqrt{(a^2 - x^2)^3}} = \frac{x}{\sqrt{a^2 - x^2}} - \sin^{-1}\frac{x}{|a|}$

220. $\displaystyle \int \frac{x^3\, dx}{\sqrt{a^2 - x^2}} = -\frac{2}{3}(a^2 - x^2)^{\frac{1}{2}} - x^2(a^2 - x^2)^{\frac{1}{2}} = -\frac{1}{3}\sqrt{a^2 - x^2}(x^2 + 2a^2)$

221. $\displaystyle \int \frac{x^3\, dx}{\sqrt{(a^2 - x^2)^3}} = 2(a^2 - x^2)^{\frac{1}{2}} + \frac{x^2}{(a^2 - x^2)^{\frac{1}{2}}} = -\frac{a^2}{\sqrt{a^2 - x^2}} + \sqrt{a^2 - x^2}$

222. $\displaystyle \int \frac{dx}{x^3\sqrt{a^2 - x^2}} = -\frac{\sqrt{a^2 - x^2}}{2a^2x^2} - \frac{1}{2a^3}\log\frac{a + \sqrt{a^2 - x^2}}{x}$

223. $\displaystyle \int \frac{dx}{x\sqrt{(a^2 - x^2)^3}} = \frac{1}{a^2\sqrt{a^2 - x^2}} - \frac{1}{a^3}\log\frac{a + \sqrt{a^2 - x^2}}{x}$

224. $\displaystyle \int \frac{dx}{x^2\sqrt{(a^2 - x^2)^3}} = \frac{1}{a^4}\left[-\frac{\sqrt{a^2 - x^2}}{x} + \frac{x}{\sqrt{a^2 - x^2}} \right]$

225. $\displaystyle \int \frac{dx}{x^3\sqrt{(a^2 - x^2)^3}} = -\frac{1}{2a^2x^2\sqrt{a^2 - x^2}} + \frac{3}{2a^4\sqrt{a^2 - x^2}}$

$$-\frac{3}{2a^5}\log\frac{a + \sqrt{a^2 - x^2}}{x}$$

226. $\displaystyle \int \frac{x^m}{\sqrt{a^2 - x^2}}\, dx = -\frac{x^{m-1}\sqrt{a^2 - x^2}}{m} + \frac{(m - 1)a^2}{m}\int \frac{x^{m-2}}{\sqrt{a^2 - x^2}}\, dx$

227. $\displaystyle \int \frac{x^{2m}}{\sqrt{a^2 - x^2}}\, dx = \frac{(2m)!}{(m!)^2}\left[-\sqrt{a^2 - x^2}\sum_{r=1}^{m}\frac{r!(r - 1)!}{2^{2m-2r+1}(2r)!}a^{2m-2r}x^{2r-1} \right.$

$$\left. + \frac{a^{2m}}{2^{2m}}\sin^{-1}\frac{x}{|a|} \right]$$

228. $\displaystyle \int \frac{x^{2m+1}}{\sqrt{a^2 - x^2}}\, dx = -\sqrt{a^2 - x^2}\sum_{r=0}^{m}\frac{(2r)!(m!)^2}{(2m + 1)!(r!)^2}(4a^2)^{m-r}x^{2r}$

229. $\displaystyle \int \frac{dx}{x^m\sqrt{a^2 - x^2}} = -\frac{\sqrt{a^2 - x^2}}{(m - 1)a^2x^{m-1}} + \frac{m - 2}{(m - 1)a^2}\int \frac{dx}{x^{m-2}\sqrt{a^2 - x^2}}$

230. $\displaystyle \int \frac{ax}{x^{2m}\sqrt{a^2 - x^2}} = -\sqrt{a^2 - x^2}\sum_{r=0}^{m-1}\frac{(m - 1)!m!(2r)!2^{2m-2r-1}}{(r!)^2(2m)!a^{2m-2r}x^{2r+1}}$

231. $\displaystyle \int \frac{dx}{x^{2m+1}\sqrt{a^2 - x^2}} = \frac{(2m)!}{(m!)^2}\left[-\frac{\sqrt{a^2 - x^2}}{a^2}\sum_{r=1}^{m}\frac{r!(r - 1)!}{2(2r)!(4a^2)^{m-r}x^{2r}} \right.$

$$\left. + \frac{1}{2^{2m}a^{2m+1}}\log\frac{a - \sqrt{a^2 - x^2}}{x} \right]$$

INTEGRALS (Continued)

232. $\displaystyle\int \frac{dx}{(b^2 - x^2)\sqrt{a^2 - x^2}} = \frac{1}{2b\sqrt{a^2 - b^2}} \log \frac{(b\sqrt{a^2 - x^2} + x\sqrt{a^2 - b^2})^2}{b^2 - x^2},$

$$(a^2 > b^2)$$

233. $\displaystyle\int \frac{dx}{(b^2 - x^2)\sqrt{a^2 - x^2}} = \frac{1}{b\sqrt{b^2 - a^2}} \tan^{-1} \frac{x\sqrt{b^2 - a^2}}{b\sqrt{a^2 - x^2}}, \qquad (b^2 > a^2)$

234. $\displaystyle\int \frac{dx}{(b^2 + x^2)\sqrt{a^2 - x^2}} = \frac{1}{b\sqrt{a^2 + b^2}} \tan^{-1} \frac{x\sqrt{a^2 + b^2}}{b\sqrt{a^2 - x^2}}$

235. $\displaystyle\int \frac{\sqrt{a^2 - x^2}}{b^2 + x^2} dx = \frac{\sqrt{a^2 + b^2}}{|b|} \sin^{-1} \frac{x\sqrt{a^2 + b^2}}{|a|\sqrt{x^2 + b^2}} - \sin^{-1} \frac{x}{|a|}$

236. $\displaystyle\int f(x, \sqrt{a^2 - x^2}) dx = a \int f(a \sin u, a \cos u) \cos u \, du, \qquad \left(u = \sin^{-1} \frac{x}{a}, a > 0\right)$

FORMS CONTAINING $\sqrt{a + bx + cx^2}$

$$X = a + bx + cx^2, q = 4ac - b^2, \text{ and } k = \frac{4c}{q}$$

If $q = 0$, then $\sqrt{X} = \sqrt{c} \left|x + \dfrac{b}{2c}\right|$

237. $\displaystyle\int \frac{dx}{\sqrt{X}} = \begin{cases} \dfrac{1}{\sqrt{c}} \log \dfrac{2\sqrt{cX} + 2cx + b}{\sqrt{q}} \\ \quad\text{or} \\ \dfrac{1}{\sqrt{c}} \sinh^{-1} \dfrac{2cx + b}{\sqrt{q}}, \qquad (c > 0) \end{cases}$

238. $\displaystyle\int \frac{dx}{\sqrt{X}} = -\frac{1}{\sqrt{-c}} \sin^{-1} \frac{2cx + b}{\sqrt{-q}}, \qquad (c < 0)$

239. $\displaystyle\int \frac{dx}{X\sqrt{X}} = \frac{2(2cx + b)}{q\sqrt{X}}$

240. $\displaystyle\int \frac{dx}{X^2\sqrt{X}} = \frac{2(2cx + b)}{3q\sqrt{X}}\left(\frac{1}{X} + 2k\right)$

241. $\displaystyle\int \frac{dx}{X^n\sqrt{X}} = \begin{cases} \dfrac{2(2cx + b)\sqrt{X}}{(2n - 1)qX^n} + \dfrac{2k(n - 1)}{2n - 1} \displaystyle\int \dfrac{dx}{X^{n-1}\sqrt{X}} \\ \quad\text{or} \\ \dfrac{(2cx + b)(n!)(n - 1)!4^n k^{n-1}}{q[(2n)!]\sqrt{X}} \displaystyle\sum_{r=0}^{n-1} \dfrac{(2r)!}{(4kX)^r(r!)^2} \end{cases}$

INTEGRALS (Continued)

242. $\int \sqrt{X}\, dx = \dfrac{(2cx + b)\sqrt{X}}{4c} + \dfrac{1}{2k} \int \dfrac{dx}{\sqrt{X}}$

243. $\int X\sqrt{X}\, dx = \dfrac{(2cx + b)\sqrt{X}}{8c}\left(X + \dfrac{3}{2k}\right) + \dfrac{3}{8k^2} \int \dfrac{dx}{\sqrt{X}}$

244. $\int X^2\sqrt{X}\, dx = \dfrac{(2cx + b)\sqrt{X}}{12c}\left(X^2 + \dfrac{5X}{4k} + \dfrac{15}{8k^2}\right) + \dfrac{5}{16k^3} \int \dfrac{dx}{\sqrt{X}}$

245. $\int X^n\sqrt{X}\, dx = \begin{cases} \dfrac{(2cx + b)X^n\sqrt{X}}{4(n + 1)c} + \dfrac{2n + 1}{2(n + 1)k} \displaystyle\int X^{n-1}\sqrt{X}\, dx \\[2mm] \qquad\qquad\text{or} \\[2mm] \dfrac{(2n + 2)!}{[(n + 1)!]^2(4k)^{n+1}}\left[\dfrac{k(2cx + b)\sqrt{X}}{c}\displaystyle\sum_{r=0}^{n}\dfrac{r!(r + 1)!(4kX)^r}{(2r + 2)!} \right. \\[4mm] \left. \qquad\qquad\qquad\qquad\qquad\qquad + \displaystyle\int \dfrac{dx}{\sqrt{X}}\right] \end{cases}$

246. $\int \dfrac{x\, dx}{\sqrt{X}} = \dfrac{\sqrt{X}}{c} - \dfrac{b}{2c} \int \dfrac{dx}{\sqrt{X}}$

247. $\int \dfrac{x\, dx}{X\sqrt{X}} = -\dfrac{2(bx + 2a)}{q\sqrt{X}}$

248. $\int \dfrac{x\, dx}{X^n\sqrt{X}} = -\dfrac{\sqrt{X}}{(2n - 1)cX^n} - \dfrac{b}{2c} \int \dfrac{dx}{X^n\sqrt{X}}$

249. $\int \dfrac{x^2\, dx}{\sqrt{X}} = \left(\dfrac{x}{2c} - \dfrac{3b}{4c^2}\right)\sqrt{X} + \dfrac{3b^2 - 4ac}{8c^2} \int \dfrac{dx}{\sqrt{X}}$

250. $\int \dfrac{x^2\, dx}{X\sqrt{X}} = \dfrac{(2b^2 - 4ac)x + 2ab}{cq\sqrt{X}} + \dfrac{1}{c} \int \dfrac{dx}{\sqrt{X}}$

251. $\int \dfrac{x^2\, dx}{X^n\sqrt{X}} = \dfrac{(2b^2 - 4ac)x + 2ab}{(2n - 1)cqX^{n-1}\sqrt{X}} + \dfrac{4ac + (2n - 3)b^2}{(2n - 1)cq} \int \dfrac{dx}{X^{n-1}\sqrt{X}}$

252. $\int \dfrac{x^3\, dx}{\sqrt{X}} = \left(\dfrac{x^2}{3c} - \dfrac{5bx}{12c^2} + \dfrac{5b^2}{8c^3} - \dfrac{2a}{3c^2}\right)\sqrt{X} + \left(\dfrac{3ab}{4c^2} - \dfrac{5b^3}{16c^3}\right) \int \dfrac{dx}{\sqrt{X}}$

253. $\int \dfrac{x^n\, dx}{\sqrt{X}} = \dfrac{1}{nc}x^{n-1}\sqrt{X} - \dfrac{(2n - 1)b}{2nc} \int \dfrac{x^{n-1}\, dx}{\sqrt{X}} - \dfrac{(n - 1)a}{nc} \int \dfrac{x^{n-2}\, dx}{\sqrt{X}}$

INTEGRALS (Continued)

254. $\displaystyle \int x\sqrt{X}\,dx = \frac{X\sqrt{X}}{3c} - \frac{b(2cx + b)}{8c^2}\sqrt{X} - \frac{b}{4ck}\int \frac{dx}{\sqrt{X}}$

255. $\displaystyle \int xX\sqrt{X}\,dx = \frac{X^2\sqrt{X}}{5c} - \frac{b}{2c}\int X\sqrt{X}\,dx$

256. $\displaystyle \int xX^n\sqrt{X}\,dx = \frac{X^{n+1}\sqrt{X}}{(2n + 3)c} - \frac{b}{2c}\int X^n\sqrt{X}\,dx$

257. $\displaystyle \int x^2\sqrt{X}\,dx = \left(x - \frac{5b}{6c}\right)\frac{X\sqrt{X}}{4c} + \frac{5b^2 - 4ac}{16c^2}\int \sqrt{X}\,dx$

258. $\displaystyle \int \frac{dx}{x\sqrt{X}} = -\frac{1}{\sqrt{a}}\log\frac{2\sqrt{aX} + bx + 2a}{x}, \qquad (a > 0)$

259. $\displaystyle \int \frac{dx}{x\sqrt{X}} = \frac{1}{\sqrt{-a}}\sin^{-1}\left(\frac{bx + 2a}{|x|\sqrt{-q}}\right), \qquad (a < 0)$

260. $\displaystyle \int \frac{dx}{x\sqrt{X}} = -\frac{2\sqrt{X}}{bx}, \qquad (a = 0)$

261. $\displaystyle \int \frac{dx}{x^2\sqrt{X}} = -\frac{\sqrt{X}}{ax} - \frac{b}{2a}\int \frac{dx}{x\sqrt{X}}$

262. $\displaystyle \int \frac{\sqrt{X}\,dx}{x} = \sqrt{X} + \frac{b}{2}\int \frac{dx}{\sqrt{X}} + a\int \frac{dx}{x\sqrt{X}}$

263. $\displaystyle \int \frac{\sqrt{X}\,dx}{x^2} = -\frac{\sqrt{X}}{x} + \frac{b}{2}\int \frac{dx}{x\sqrt{X}} + c\int \frac{dx}{\sqrt{X}}$

FORMS INVOLVING $\sqrt{2ax - x^2}$

264. $\displaystyle \int \sqrt{2ax - x^2}\,dx = \frac{1}{2}\left[(x - a)\sqrt{2ax - x^2} + a^2\sin^{-1}\frac{x - a}{|a|}\right]$

265. $\displaystyle \int \frac{dx}{\sqrt{2ax - x^2}} = \begin{cases} \cos^{-1}\dfrac{a - x}{|a|} \\[2mm] \text{or} \\[2mm] \sin^{-1}\dfrac{x - a}{|a|} \end{cases}$

INTEGRALS (Continued)

266. $\displaystyle\int x^n\sqrt{2ax-x^2}\,dx = \begin{cases} -\dfrac{x^{n-1}(2ax-x^2)^{\frac{3}{2}}}{n+2} + \dfrac{(2n+1)a}{n+2}\displaystyle\int x^{n-1}\sqrt{2ax-x^2}\,dx \\[2mm] \quad\text{or} \\[2mm] \sqrt{2ax-x^2}\left[\dfrac{x^{n+1}}{n+2} - \displaystyle\sum_{r=0}^{n}\dfrac{(2n+1)!(r!)^2 a^{n-r+1}}{2^{n-r}(2r+1)!(n+2)!n!}x^r\right] \\[4mm] \qquad\qquad + \dfrac{(2n+1)!a^{n+2}}{2^n n!(n+2)!}\sin^{-1}\dfrac{x-a}{|a|} \end{cases}$

267. $\displaystyle\int \frac{\sqrt{2ax-x^2}}{x^n}\,dx = \frac{(2ax-x^2)^{\frac{3}{2}}}{(3-2n)ax^n} + \frac{n-3}{(2n-3)a}\int \frac{\sqrt{2ax-x^2}}{x^{n-1}}\,dx$

268. $\displaystyle\int \frac{x^n\,dx}{\sqrt{2ax-x^2}} = \begin{cases} \dfrac{-x^{n-1}\sqrt{2ax-x^2}}{n} + \dfrac{a(2n-1)}{n}\displaystyle\int \frac{x^{n-1}}{\sqrt{2ax-x^2}}\,dx \\[2mm] \quad\text{or} \\[2mm] -\sqrt{2ax-x^2}\displaystyle\sum_{r=1}^{n}\dfrac{(2n)!r!(r-1)!a^{n-r}}{2^{n-r}(2r)!(n!)^2}x^{r-1} \\[4mm] \qquad\qquad + \dfrac{(2n)!a^n}{2^n(n!)^2}\sin^{-1}\dfrac{x-a}{|a|} \end{cases}$

269. $\displaystyle\int \frac{dx}{x^n\sqrt{2ax-x^2}} = \begin{cases} \dfrac{\sqrt{2ax-x^2}}{a(1-2n)x^n} + \dfrac{n-1}{(2n-1)a}\displaystyle\int \frac{dx}{x^{n-1}\sqrt{2ax-x^2}} \\[2mm] \quad\text{or} \\[2mm] -\sqrt{2ax-x^2}\displaystyle\sum_{r=0}^{n-1}\dfrac{2^{n-r}(n-1)!n!(2r)!}{(2n)!(r!)^2 a^{n-r}x^{r+1}} \end{cases}$

270. $\displaystyle\int \frac{dx}{(2ax-x^2)^{\frac{3}{2}}} = \frac{x-a}{a^2\sqrt{2ax-x^2}}$

271. $\displaystyle\int \frac{x\,dx}{(2ax-x^2)^{\frac{3}{2}}} = \frac{x}{a\sqrt{2ax-x^2}}$

MISCELLANEOUS ALGEBRAIC FORMS

272. $\displaystyle\int \frac{dx}{\sqrt{2ax+x^2}} = \log\left(x+a+\sqrt{2ax+x^2}\right)$

273. $\displaystyle\int \sqrt{ax^2+c}\,dx = \frac{x}{2}\sqrt{ax^2+c} + \frac{c}{2\sqrt{a}}\log\left(x\sqrt{a}+\sqrt{ax^2+c}\right), \qquad (a>0)$

274. $\displaystyle\int \sqrt{ax^2+c}\,dx = \frac{x}{2}\sqrt{ax^2+c} + \frac{c}{2\sqrt{-a}}\sin^{-1}\left(x\sqrt{-\frac{a}{c}}\right), \qquad (a<0)$

INTEGRALS (Continued)

275. $\displaystyle\int\sqrt{\frac{1+x}{1-x}}\,dx = \sin^{-1}x - \sqrt{1-x^2}$

276. $\displaystyle\int\frac{dx}{x\sqrt{ax^n + c}} = \begin{cases} \dfrac{1}{n\sqrt{c}}\log\dfrac{\sqrt{ax^n + c} - \sqrt{c}}{\sqrt{ax^n + c} + \sqrt{c}} \\[2ex] \quad\text{or} \\[2ex] \dfrac{2}{n\sqrt{c}}\log\dfrac{\sqrt{ax^n + c} - \sqrt{c}}{\sqrt{x^n}}, \qquad (c > 0) \end{cases}$

277. $\displaystyle\int\frac{dx}{x\sqrt{ax^n + c}} = \frac{2}{n\sqrt{-c}}\sec^{-1}\sqrt{-\frac{ax^n}{c}}, \qquad (c < 0)$

278. $\displaystyle\int\frac{dx}{\sqrt{ax^2 + c}} = \frac{1}{\sqrt{a}}\log(x\sqrt{a} + \sqrt{ax^2 + c}), \qquad (a > 0)$

279. $\displaystyle\int\frac{dx}{\sqrt{ax^2 + c}} = \frac{1}{\sqrt{-a}}\sin^{-1}\left(x\sqrt{-\frac{a}{c}}\right), \qquad (a < 0)$

280. $\displaystyle\int(ax^2 + c)^{m+\frac{1}{2}}\,dx = \begin{cases} \dfrac{x(ax^2+c)^{m+\frac{1}{2}}}{2(m+1)} + \dfrac{(2m+1)c}{2(m+1)}\displaystyle\int(ax^2+c)^{m-\frac{1}{2}}\,dx \\[2ex] \quad\text{or} \\[2ex] x\sqrt{ax^2+c}\displaystyle\sum_{r=0}^{m}\dfrac{(2m+1)!(r!)^2 c^{m-r}}{2^{2m-2r+1}m!(m+1)!(2r+1)!}(ax^2+c)^r \\[2ex] \quad + \dfrac{(2m+1)!c^{m+1}}{2^{2m+1}m!(m+1)!}\displaystyle\int\dfrac{dx}{\sqrt{ax^2+c}} \end{cases}$

281. $\displaystyle\int x(ax^2 + c)^{m+\frac{1}{2}}\,dx = \frac{(ax^2+c)^{m+\frac{3}{2}}}{(2m+3)a}$

282. $\displaystyle\int\frac{(ax^2 + c)^{m+\frac{1}{2}}}{x}\,dx = \begin{cases} \dfrac{(ax^2+c)^{m+\frac{1}{2}}}{2m+1} + c\displaystyle\int\dfrac{(ax^2+c)^{m-\frac{1}{2}}}{x}\,dx \\[2ex] \quad\text{or} \\[2ex] \sqrt{ax^2+c}\displaystyle\sum_{r=0}^{m}\dfrac{c^{m-r}(ax^2+c)^r}{2r+1} + c^{m+1}\displaystyle\int\dfrac{dx}{x\sqrt{ax^2+c}} \end{cases}$

283. $\displaystyle\int\frac{dx}{(ax^2 + c)^{m+\frac{1}{2}}} = \begin{cases} \dfrac{x}{(2m-1)c(ax^2+c)^{m-\frac{1}{2}}} + \dfrac{2m-2}{(2m-1)c}\displaystyle\int\dfrac{dx}{(ax^2+c)^{m-\frac{1}{2}}} \\[2ex] \quad\text{or} \\[2ex] \dfrac{x}{\sqrt{ax^2+c}}\displaystyle\sum_{r=0}^{m-1}\dfrac{2^{2m-2r-1}(m-1)!m!(2r)!}{(2m)!(r!)^2 c^{m-r}(ax^2+c)^r} \end{cases}$

INTEGRALS (Continued)

284. $\displaystyle\int \frac{dx}{x^m\sqrt{ax^2 + c}} = -\frac{\sqrt{ax^2 + c}}{(m-1)cx^{m-1}} - \frac{(m-2)a}{(m-1)c}\int \frac{dx}{x^{m-2}\sqrt{ax^2 + c}}$

285. $\displaystyle\int \frac{1 + x^2}{(1 - x^2)\sqrt{1 + x^4}}\,dx = \frac{1}{\sqrt{2}}\log \frac{x\sqrt{2} + \sqrt{1 + x^4}}{1 - x^2}$

286. $\displaystyle\int \frac{1 - x^2}{(1 + x^2)\sqrt{1 + x^4}}\,dx = \frac{1}{\sqrt{2}}\tan^{-1}\frac{x\sqrt{2}}{\sqrt{1 + x^4}}$

287. $\displaystyle\int \frac{dx}{x\sqrt{x^n + a^2}} = -\frac{2}{na}\log \frac{a + \sqrt{x^n + a^2}}{\sqrt{x^n}}$

288. $\displaystyle\int \frac{dx}{x\sqrt{x^n - a^2}} = -\frac{2}{na}\sin^{-1}\frac{a}{\sqrt{x^n}}$

289. $\displaystyle\int \sqrt{\frac{x}{a^3 - x^3}}\,dx = \frac{2}{3}\sin^{-1}\left(\frac{x}{a}\right)^{\frac{3}{2}}$

FORMS INVOLVING TRIGONOMETRIC FUNCTIONS

290. $\displaystyle\int (\sin ax)\,dx = -\frac{1}{a}\cos ax$

291. $\displaystyle\int (\cos ax)\,dx = \frac{1}{a}\sin ax$

292. $\displaystyle\int (\tan ax)\,dx = -\frac{1}{a}\log \cos ax = \frac{1}{a}\log \sec ax$

293. $\displaystyle\int (\cot ax)\,dx = \frac{1}{a}\log \sin ax = -\frac{1}{a}\log \csc ax$

294. $\displaystyle\int (\sec ax)\,dx = \frac{1}{a}\log (\sec ax + \tan ax) = \frac{1}{a}\log \tan \left(\frac{\pi}{4} + \frac{ax}{2}\right)$

295. $\displaystyle\int (\csc ax)\,dx = \frac{1}{a}\log (\csc ax - \cot ax) = \frac{1}{a}\log \tan \frac{ax}{2}$

296. $\displaystyle\int (\sin^2 ax)\,dx = -\frac{1}{2a}\cos ax \sin ax + \frac{1}{2}x = \frac{1}{2}x - \frac{1}{4a}\sin 2ax$

297. $\displaystyle\int (\sin^3 ax)\,dx = -\frac{1}{3a}(\cos ax)(\sin^2 ax + 2)$

298. $\displaystyle\int (\sin^4 ax)\,dx = \frac{3x}{8} - \frac{\sin 2ax}{4a} + \frac{\sin 4ax}{32a}$

299. $\displaystyle\int (\sin^n ax)\,dx = -\frac{\sin^{n-1} ax \cos ax}{na} + \frac{n-1}{n}\int (\sin^{n-2} ax)\,dx$

INTEGRALS (Continued)

300. $\displaystyle\int (\sin^{2m} ax)\, dx = -\frac{\cos ax}{a} \sum_{r=0}^{m-1} \frac{(2m)!(r!)^2}{2^{2m-2r}(2r+1)!(m!)^2} \sin^{2r+1} ax + \frac{(2m)!}{2^{2m}(m!)^2} x$

301. $\displaystyle\int (\sin^{2m+1} ax)\, dx = -\frac{\cos ax}{a} \sum_{r=0}^{m} \frac{2^{2m-2r}(m!)^2(2r)!}{(2m+1)!(r!)^2} \sin^{2r} ax$

302. $\displaystyle\int (\cos^2 ax)\, dx = \frac{1}{2a} \sin ax \cos ax + \frac{1}{2}x = \frac{1}{2}x + \frac{1}{4a} \sin 2ax$

303. $\displaystyle\int (\cos^3 ax)\, dx = \frac{1}{3a}(\sin ax)(\cos^2 ax + 2)$

304. $\displaystyle\int (\cos^4 ax)\, dx = \frac{3x}{8} + \frac{\sin 2ax}{4a} + \frac{\sin 4ax}{32a}$

305. $\displaystyle\int (\cos^n ax)\, dx = \frac{1}{na} \cos^{n-1} ax \sin ax + \frac{n-1}{n} \int (\cos^{n-2} ax)\, dx$

306. $\displaystyle\int (\cos^{2m} ax)\, dx = \frac{\sin ax}{a} \sum_{r=0}^{m-1} \frac{(2m)!(r!)^2}{2^{2m-2r}(2r+1)!(m!)^2} \cos^{2r+1} ax + \frac{(2m)!}{2^{2m}(m!)^2} x$

307. $\displaystyle\int (\cos^{2m+1} ax)\, dx = \frac{\sin ax}{a} \sum_{r=0}^{m} \frac{2^{2m-2r}(m!)^2(2r)!}{(2m+1)!(r!)^2} \cos^{2r} ax$

308. $\displaystyle\int \frac{dx}{\sin^2 ax} = \int (\csc^2 ax)\, dx = -\frac{1}{a} \cot ax$

309. $\displaystyle\int \frac{dx}{\sin^m ax} = \int (\csc^m ax)\, dx = -\frac{1}{(m-1)a} \cdot \frac{\cos ax}{\sin^{m-1} ax} + \frac{m-2}{m-1} \int \frac{dx}{\sin^{m-2} ax}$

310. $\displaystyle\int \frac{dx}{\sin^{2m} ax} = \int (\csc^{2m} ax)\, dx = -\frac{1}{a} \cos ax \sum_{r=0}^{m-1} \frac{2^{2m-2r-1}(m-1)!\,m!\,(2r)!}{(2m)!(r!)^2 \sin^{2r+1} ax}$

311. $\displaystyle\int \frac{dx}{\sin^{2m+1} ax} = \int (\csc^{2m+1} ax)\, dx =$

$\displaystyle\qquad -\frac{1}{a} \cos ax \sum_{r=0}^{m-1} \frac{(2m)!(r!)^2}{2^{2m-2r}(m!)^2(2r+1)!\,\sin^{2r+2} ax} + \frac{1}{a} \cdot \frac{(2m)!}{2^{2m}(m!)^2} \log \tan \frac{ax}{2}$

312. $\displaystyle\int \frac{dx}{\cos^2 ax} = \int (\sec^2 ax)\, dx = \frac{1}{a} \tan ax$

313. $\displaystyle\int \frac{dx}{\cos^n ax} = \int (\sec^n ax)\, dx = \frac{1}{(n-1)a} \cdot \frac{\sin ax}{\cos^{n-1} ax} + \frac{n-2}{n-1} \int \frac{dx}{\cos^{n-2} ax}$

314. $\displaystyle\int \frac{dx}{\cos^{2m} ax} = \int (\sec^{2m} ax)\, dx = \frac{1}{a} \sin ax \sum_{r=0}^{m-1} \frac{2^{2m-2r-1}(m-1)!\,m!\,(2r)!}{(2m)!(r!)^2 \cos^{2r+1} ax}$

INTEGRALS (Continued)

315. $\displaystyle\int \frac{dx}{\cos^{2m+1} ax} = \int (\sec^{2m+1} ax)\, dx =$

$$\frac{1}{a} \sin ax \sum_{r=0}^{m-1} \frac{(2m)!(r!)^2}{2^{2m-2r}(m!)^2(2r+1)!\cos^{2r+2} ax}$$

$$+ \frac{1}{a} \cdot \frac{(2m)!}{2^{2m}(m!)^2} \log(\sec ax + \tan ax)$$

316. $\displaystyle\int (\sin mx)(\sin nx)\, dx = \frac{\sin(m-n)x}{2(m-n)} - \frac{\sin(m+n)x}{2(m+n)}, \qquad (m^2 \neq n^2)$

317. $\displaystyle\int (\cos mx)(\cos nx)\, dx = \frac{\sin(m-n)x}{2(m-n)} + \frac{\sin(m+n)x}{2(m+n)}, \qquad (m^2 \neq n^2)$

318. $\displaystyle\int (\sin ax)(\cos ax)\, dx = \frac{1}{2a}\sin^2 ax$

319. $\displaystyle\int (\sin mx)(\cos nx)\, dx = -\frac{\cos(m-n)x}{2(m-n)} - \frac{\cos(m+n)x}{2(m+n)}, \qquad (m^2 \neq n^2)$

320. $\displaystyle\int (\sin^2 ax)(\cos^2 ax)\, dx = -\frac{1}{32a}\sin 4ax + \frac{x}{8}$

321. $\displaystyle\int (\sin ax)(\cos^m ax)\, dx = -\frac{\cos^{m+1} ax}{(m+1)a}$

322. $\displaystyle\int (\sin^m ax)(\cos ax)\, dx = \frac{\sin^{m+1} ax}{(m+1)a}$

323. $\displaystyle\int (\cos^m ax)(\sin^n ax)\, dx = \begin{cases} \dfrac{\cos^{m-1} ax \, \sin^{n+1} ax}{(m+n)a} \\[2mm] \qquad + \dfrac{m-1}{m+n}\displaystyle\int (\cos^{m-2} ax)(\sin^n ax)\, dx \\[2mm] \text{or} \\[2mm] -\dfrac{\sin^{n-1} ax \, \cos^{m+1} ax}{(m+n)a} \\[2mm] \qquad + \dfrac{n-1}{m+n}\displaystyle\int (\cos^m ax)(\sin^{n-2} ax)\, dx \end{cases}$

324. $\displaystyle\int \frac{\cos^m ax}{\sin^n ax}\, dx = \begin{cases} -\dfrac{\cos^{m+1} ax}{(n-1)a\,\sin^{n-1} ax} - \dfrac{m-n+2}{n-1}\displaystyle\int \frac{\cos^m ax}{\sin^{n-2} ax}\, dx \\[2mm] \text{or} \\[2mm] \dfrac{\cos^{m-1} ax}{a(m-n)\sin^{n-1} ax} + \dfrac{m-1}{m-n}\displaystyle\int \frac{\cos^{m-2} ax}{\sin^n ax}\, dx \end{cases}$

INTEGRALS (Continued)

325. $\displaystyle\int \frac{\sin^m ax}{\cos^n ax}\, dx = \begin{cases} \dfrac{\sin^{m+1} ax}{a(n-1)\cos^{n-1} ax} - \dfrac{m-n+2}{n-1}\displaystyle\int \dfrac{\sin^m ax}{\cos^{n-2} ax}\, dx \\[2ex] \text{or} \\[2ex] -\dfrac{\sin^{m-1} ax}{a(m-n)\cos^{n-1} ax} + \dfrac{m-1}{m-n}\displaystyle\int \dfrac{\sin^{m-2} ax}{\cos^n ax}\, dx \end{cases}$

326. $\displaystyle\int \frac{\sin ax}{\cos^2 ax}\, dx = \frac{1}{a\cos ax} = \frac{\sec ax}{a}$

327. $\displaystyle\int \frac{\sin^2 ax}{\cos ax}\, dx = -\frac{1}{a}\sin ax + \frac{1}{a}\log \tan\left(\frac{\pi}{4} + \frac{ax}{2}\right)$

328. $\displaystyle\int \frac{\cos ax}{\sin^2 ax}\, dx = -\frac{1}{a\sin ax} = -\frac{\csc ax}{a}$

329. $\displaystyle\int \frac{dx}{(\sin ax)(\cos ax)} = \frac{1}{a}\log \tan ax$

330. $\displaystyle\int \frac{dx}{(\sin ax)(\cos^2 ax)} = \frac{1}{a}\left(\sec ax + \log \tan\frac{ax}{2}\right)$

331. $\displaystyle\int \frac{dx}{(\sin ax)(\cos^n ax)} = \frac{1}{a(n-1)\cos^{n-1} ax} + \int \frac{dx}{(\sin ax)(\cos^{n-2} ax)}$

332. $\displaystyle\int \frac{dx}{(\sin^2 ax)(\cos ax)} = -\frac{1}{a}\csc ax + \frac{1}{a}\log \tan\left(\frac{\pi}{4} + \frac{ax}{2}\right)$

333. $\displaystyle\int \frac{dx}{(\sin^2 ax)(\cos^2 ax)} = -\frac{2}{a}\cot 2ax$

334. $\displaystyle\int \frac{dx}{\sin^m ax \cos^n ax} = \begin{cases} -\dfrac{1}{a(m-1)(\sin^{m-1} ax)(\cos^{n-1} ax)} \\[2ex] \qquad\qquad + \dfrac{m+n-2}{m-1}\displaystyle\int \dfrac{dx}{(\sin^{m-2} ax)(\cos^n ax)} \\[2ex] \text{or} \\[2ex] \dfrac{1}{a(n-1)\sin^{m-1} ax \cos^{n-1} ax} \\[2ex] \qquad\qquad - \dfrac{m+n-2}{n-1}\displaystyle\int \dfrac{dx}{\sin^m ax \cos^{n-2} ax} \end{cases}$

335. $\displaystyle\int \sin(a + bx)\, dx = -\frac{1}{b}\cos(a + bx)$

336. $\displaystyle\int \cos(a + bx)\, dx = \frac{1}{b}\sin(a + bx)$

337. $\displaystyle\int \frac{dx}{1 \pm \sin ax} = \mp\frac{1}{a}\tan\left(\frac{\pi}{4} \mp \frac{ax}{2}\right)$

INTEGRALS (Continued)

338. $\displaystyle\int \frac{dx}{1 + \cos ax} = \frac{1}{a}\tan\frac{ax}{2}$

339. $\displaystyle\int \frac{dx}{1 - \cos ax} = -\frac{1}{a}\cot\frac{ax}{2}$

*340. $\displaystyle\int \frac{dx}{a + b\sin x} = \begin{cases} \dfrac{2}{\sqrt{a^2 - b^2}}\tan^{-1}\dfrac{a\tan\frac{x}{2} + b}{\sqrt{a^2 - b^2}} \\[2mm] \qquad\text{or} \\[2mm] \dfrac{1}{\sqrt{b^2 - a^2}}\log\dfrac{a\tan\frac{x}{2} + b - \sqrt{b^2 - a^2}}{a\tan\frac{x}{2} + b + \sqrt{b^2 - a^2}} \end{cases}$

*341. $\displaystyle\int \frac{dx}{a + b\cos x} = \begin{cases} \dfrac{2}{\sqrt{a^2 - b^2}}\tan^{-1}\dfrac{\sqrt{a^2 - b^2}\,\tan\frac{x}{2}}{a + b} \\[2mm] \qquad\text{or} \\[2mm] \dfrac{1}{\sqrt{b^2 - a^2}}\log\left(-\dfrac{\sqrt{b^2 - a^2}\,\tan\frac{x}{2} + a + b}{\sqrt{b^2 - a^2}\,\tan\frac{x}{2} - a - b}\right) \end{cases}$

*342. $\displaystyle\int \frac{dx}{a + b\sin x + c\cos x}$

$= \begin{cases} \dfrac{1}{\sqrt{b^2 + c^2 - a^2}}\log\dfrac{b - \sqrt{b^2 + c^2 - a^2} + (a - c)\tan\frac{x}{2}}{b + \sqrt{b^2 + c^2 - a^2} + (a - c)\tan\frac{x}{2}}, & \text{if } a^2 < b^2 + c^2, a \neq c \\[4mm] \qquad\text{or} \\[4mm] \dfrac{2}{\sqrt{a^2 - b^2 - c^2}}\tan^{-1}\dfrac{b + (a - c)\tan\frac{x}{2}}{\sqrt{a^2 - b^2 - c^2}}, & \text{if } a^2 > b^2 + c^2 \\[4mm] \qquad\text{or} \\[4mm] \dfrac{1}{a}\left[\dfrac{a - (b + c)\cos x - (b - c)\sin x}{a - (b - c)\cos x + (b + c)\sin x}\right], & \text{if } a^2 = b^2 + c^2, a \neq c. \end{cases}$

*343. $\displaystyle\int \frac{\sin^2 x\, dx}{a + b\cos^2 x} = \frac{1}{b}\sqrt{\frac{a + b}{a}}\tan^{-1}\left(\sqrt{\frac{a}{a + b}}\tan x\right) - \frac{x}{b}, \qquad (ab > 0, \text{ or } |a| > |b|)$

*See note 6 — page 329.

INTEGRALS (Continued)

*344. $\displaystyle\int \frac{dx}{a^2 \cos^2 x + b^2 \sin^2 x} = \frac{1}{ab} \tan^{-1}\left(\frac{b \tan x}{a}\right)$

*345. $\displaystyle\int \frac{\cos^2 cx}{a^2 + b^2 \sin^2 cx} dx = \frac{\sqrt{a^2 + b^2}}{ab^2 c} \tan^{-1} \frac{\sqrt{a^2 + b^2}\, \tan cx}{a} - \frac{x}{b^2}$

346. $\displaystyle\int \frac{\sin cx \cos cx}{a \cos^2 cx + b \sin^2 cx} dx = \frac{1}{2c(b-a)} \log\left(a \cos^2 cx + b \sin^2 cx\right)$

347. $\displaystyle\int \frac{\cos cx}{a \cos cx + b \sin cx} dx = \int \frac{dx}{a + b \tan cx} =$

$$\frac{1}{c(a^2 + b^2)}[acx + b \log (a \cos cx + b \, \text{si}$$

348. $\displaystyle\int \frac{\sin cx}{a \sin cx + b \cos cx} dx = \int \frac{dx}{a + b \cot cx} =$

$$\frac{1}{c(a^2 + b^2)}[acx - b \log (a \sin cx + b \, \text{cc}$$

*349. $\displaystyle\int \frac{dx}{a \cos^2 x + 2b \cos x \sin x + c \sin^2 x} =$

$$\begin{cases} \dfrac{1}{2\sqrt{b^2 - ac}} \log \dfrac{c \tan x + b - \sqrt{b^2}}{c \tan x + b + \sqrt{b^2}} \\ \qquad\qquad\qquad\qquad\qquad (b^2 \\[4pt] \text{or} \\[4pt] \dfrac{1}{\sqrt{ac - b^2}} \tan^{-1} \dfrac{c \tan x + b}{\sqrt{ac - b^2}}, \quad (b^2 \\[4pt] \text{or} \\[4pt] -\dfrac{1}{c \tan x + b}, \qquad (b^2 = ac) \end{cases}$$

350. $\displaystyle\int \frac{\sin ax}{1 \pm \sin ax} dx = \pm x + \frac{1}{a} \tan\left(\frac{\pi}{4} \mp \frac{ax}{2}\right)$

351. $\displaystyle\int \frac{dx}{(\sin ax)(1 \pm \sin ax)} = \frac{1}{a} \tan\left(\frac{\pi}{4} \mp \frac{ax}{2}\right) + \frac{1}{a} \log \tan \frac{ax}{2}$

352. $\displaystyle\int \frac{dx}{(1 + \sin ax)^2} = -\frac{1}{2a} \tan\left(\frac{\pi}{4} - \frac{ax}{2}\right) - \frac{1}{6a} \tan^3\left(\frac{\pi}{4} - \frac{ax}{2}\right)$

353. $\displaystyle\int \frac{dx}{(1 - \sin ax)^2} = \frac{1}{2a} \cot\left(\frac{\pi}{4} - \frac{ax}{2}\right) + \frac{1}{6a} \cot^3\left(\frac{\pi}{4} - \frac{ax}{2}\right)$

354. $\displaystyle\int \frac{\sin ax}{(1 + \sin ax)^2} dx = -\frac{1}{2a} \tan\left(\frac{\pi}{4} - \frac{ax}{2}\right) + \frac{1}{6a} \tan^3\left(\frac{\pi}{4} - \frac{ax}{2}\right)$

*See note 6 – page 329.

INTEGRALS (Continued)

355. $\displaystyle\int \frac{\sin ax}{(1 - \sin ax)^2}\,dx = -\frac{1}{2a}\cot\left(\frac{\pi}{4} - \frac{ax}{2}\right) + \frac{1}{6a}\cot^3\left(\frac{\pi}{4} - \frac{ax}{2}\right)$

356. $\displaystyle\int \frac{\sin x\,dx}{a + b\sin x} = \frac{x}{b} - \frac{a}{b}\int \frac{dx}{a + b\sin x}$

357. $\displaystyle\int \frac{dx}{(\sin x)(a + b\sin x)} = \frac{1}{a}\log\tan\frac{x}{2} - \frac{b}{a}\int \frac{dx}{a + b\sin x}$

358. $\displaystyle\int \frac{dx}{(a + b\sin x)^2} = \frac{b\cos x}{(a^2 - b^2)(a + b\sin x)} + \frac{a}{a^2 - b^2}\int \frac{dx}{a + b\sin x}$

359. $\displaystyle\int \frac{\sin x\,dx}{(a + b\sin x)^2} = \frac{a\cos x}{(b^2 - a^2)(a + b\sin x)} + \frac{b}{b^2 - a^2}\int \frac{dx}{a + b\sin x}$

*360. $\displaystyle\int \frac{dx}{a^2 + b^2\sin^2 cx} = \frac{1}{ac\sqrt{a^2 + b^2}}\tan^{-1}\frac{\sqrt{a^2 + b^2}\,\tan cx}{a}$

*361. $\displaystyle\int \frac{dx}{a^2 - b^2\sin^2 cx} = \begin{cases} \dfrac{1}{ac\sqrt{a^2 - b^2}}\tan^{-1}\dfrac{\sqrt{a^2 - b^2}\,\tan cx}{a}, & (a^2 > b^2) \\[2ex] \text{or} \\[2ex] \dfrac{1}{2ac\sqrt{b^2 - a^2}}\log\dfrac{\sqrt{b^2 - a^2}\,\tan cx + a}{\sqrt{b^2 - a^2}\,\tan cx - a}, & (a^2 < b^2) \end{cases}$

362. $\displaystyle\int \frac{\cos ax}{1 + \cos ax}\,dx = x - \frac{1}{a}\tan\frac{ax}{2}$

363. $\displaystyle\int \frac{\cos ax}{1 - \cos ax}\,dx = -x - \frac{1}{a}\cot\frac{ax}{2}$

364. $\displaystyle\int \frac{dx}{(\cos ax)(1 + \cos ax)} = \frac{1}{a}\log\tan\left(\frac{\pi}{4} + \frac{ax}{2}\right) - \frac{1}{a}\tan\frac{ax}{2}$

365. $\displaystyle\int \frac{dx}{(\cos ax)(1 - \cos ax)} = \frac{1}{a}\log\tan\left(\frac{\pi}{4} + \frac{ax}{2}\right) - \frac{1}{a}\cot\frac{ax}{2}$

366. $\displaystyle\int \frac{dx}{(1 + \cos ax)^2} = \frac{1}{2a}\tan\frac{ax}{2} + \frac{1}{6a}\tan^3\frac{ax}{2}$

367. $\displaystyle\int \frac{dx}{(1 - \cos ax)^2} = -\frac{1}{2a}\cot\frac{ax}{2} - \frac{1}{6a}\cot^3\frac{ax}{2}$

368. $\displaystyle\int \frac{\cos ax}{(1 + \cos ax)^2}\,dx = \frac{1}{2a}\tan\frac{ax}{2} - \frac{1}{6a}\tan^3\frac{ax}{2}$

369. $\displaystyle\int \frac{\cos ax}{(1 - \cos ax)^2}\,dx = \frac{1}{2a}\cot\frac{ax}{2} - \frac{1}{6a}\cot^3\frac{ax}{2}$

*See note 6—page 329.

INTEGRALS (Continued)

370. $\displaystyle\int \frac{\cos x\, dx}{a + b\cos x} = \frac{x}{b} - \frac{a}{b}\int \frac{dx}{a + b\cos x}$

371. $\displaystyle\int \frac{dx}{(\cos x)(a + b\cos x)} = \frac{1}{a}\log\tan\left(\frac{x}{2} + \frac{\pi}{4}\right) - \frac{b}{a}\int \frac{dx}{a + b\cos x}$

372. $\displaystyle\int \frac{dx}{(a + b\cos x)^2} = \frac{b\sin x}{(b^2 - a^2)(a + b\cos x)} - \frac{a}{b^2 - a^2}\int \frac{dx}{a + b\cos x}$

373. $\displaystyle\int \frac{\cos x}{(a + b\cos x)^2}\,dx = \frac{a\sin x}{(a^2 - b^2)(a + b\cos x)} - \frac{b}{a^2 - b^2}\int \frac{dx}{a + b\cos x}$

*374. $\displaystyle\int \frac{dx}{a^2 + b^2 - 2ab\cos cx} = \frac{2}{c(a^2 - b^2)}\tan^{-1}\left(\frac{a + b}{a - b}\tan\frac{cx}{2}\right)$

*375. $\displaystyle\int \frac{dx}{a^2 + b^2\cos^2 cx} = \frac{1}{ac\sqrt{a^2 + b^2}}\tan^{-1}\frac{a\tan cx}{\sqrt{a^2 + b^2}}$

*376. $\displaystyle\int \frac{dx}{a^2 - b^2\cos^2 cx} = \begin{cases} \dfrac{1}{ac\sqrt{a^2 - b^2}}\tan^{-1}\dfrac{a\tan cx}{\sqrt{a^2 - b^2}}, & (a^2 > b^2) \\[2ex] \text{or} \\[2ex] \dfrac{1}{2ac\sqrt{b^2 - a^2}}\log\dfrac{a\tan cx - \sqrt{b^2 - a^2}}{a\tan cx + \sqrt{b^2 - a^2}}, & (b^2 > a^2) \end{cases}$

377. $\displaystyle\int \frac{\sin ax}{1 \pm \cos ax}\,dx = \mp \frac{1}{a}\log(1 \pm \cos ax)$

378. $\displaystyle\int \frac{\cos ax}{1 \pm \sin ax}\,dx = \pm \frac{1}{a}\log(1 \pm \sin ax)$

379. $\displaystyle\int \frac{dx}{(\sin ax)(1 \pm \cos ax)} = \pm \frac{1}{2a(1 \pm \cos ax)} + \frac{1}{2a}\log\tan\frac{ax}{2}$

380. $\displaystyle\int \frac{dx}{(\cos ax)(1 \pm \sin ax)} = \mp \frac{1}{2a(1 \pm \sin ax)} + \frac{1}{2a}\log\tan\left(\frac{\pi}{4} + \frac{ax}{2}\right)$

381. $\displaystyle\int \frac{\sin ax}{(\cos ax)(1 \pm \cos ax)}\,dx = \frac{1}{a}\log(\sec ax \pm 1)$

382. $\displaystyle\int \frac{\cos ax}{(\sin ax)(1 \pm \sin ax)}\,dx = -\frac{1}{a}\log(\csc ax \pm 1)$

383. $\displaystyle\int \frac{\sin ax}{(\cos ax)(1 \pm \sin ax)}\,dx = \frac{1}{2a(1 \pm \sin ax)} \pm \frac{1}{2a}\log\tan\left(\frac{\pi}{4} + \frac{ax}{2}\right)$

384. $\displaystyle\int \frac{\cos ax}{(\sin ax)(1 \pm \cos ax)}\,dx = -\frac{1}{2a(1 \pm \cos ax)} \pm \frac{1}{2a}\log\tan\frac{ax}{2}$

*See note 6 – page 329.

INTEGRALS (Continued)

385. $\displaystyle\int \frac{dx}{\sin ax \pm \cos ax} = \frac{1}{a\sqrt{2}} \log \tan\left(\frac{ax}{2} \pm \frac{\pi}{8}\right)$

386. $\displaystyle\int \frac{dx}{(\sin ax \pm \cos ax)^2} = \frac{1}{2a} \tan\left(ax \mp \frac{\pi}{4}\right)$

387. $\displaystyle\int \frac{dx}{1 + \cos ax \pm \sin ax} = \pm \frac{1}{a} \log\left(1 \pm \tan \frac{ax}{2}\right)$

388. $\displaystyle\int \frac{dx}{a^2 \cos^2 cx - b^2 \sin^2 cx} = \frac{1}{2abc} \log \frac{b \tan cx + a}{b \tan cx - a}$

389. $\displaystyle\int x(\sin ax)\, dx = \frac{1}{a^2} \sin ax - \frac{x}{a} \cos ax$

390. $\displaystyle\int x^2(\sin ax)\, dx = \frac{2x}{a^2} \sin ax - \frac{a^2 x^2 - 2}{a^3} \cos ax$

391. $\displaystyle\int x^3(\sin ax)\, dx = \frac{3a^2 x^2 - 6}{a^4} \sin ax - \frac{a^2 x^3 - 6x}{a^3} \cos ax$

392. $\displaystyle\int x^m \sin ax\, dx = \begin{cases} -\dfrac{1}{a} x^m \cos ax + \dfrac{m}{a} \displaystyle\int x^{m-1} \cos ax\, dx \\[2mm] \quad\text{or} \\[2mm] \cos ax \displaystyle\sum_{r=0}^{\left[\frac{m}{2}\right]} (-1)^{r+1} \dfrac{m!}{(m-2r)!} \cdot \dfrac{x^{m-2r}}{a^{2r+1}} \\[4mm] + \sin ax \displaystyle\sum_{r=0}^{\left[\frac{m-1}{2}\right]} (-1)^r \dfrac{m!}{(m-2r-1)!} \cdot \dfrac{x^{m-2r-1}}{a^{2r+2}} \end{cases}$

Note: $[s]$ means greatest integer $\le s$; $[3\frac{1}{2}] = 3$, $[\frac{1}{2}] = 0$, etc.

393. $\displaystyle\int x(\cos ax)\, dx = \frac{1}{a^2} \cos ax + \frac{x}{a} \sin ax$

394. $\displaystyle\int x^2(\cos ax)\, dx = \frac{2x \cos ax}{a^2} + \frac{a^2 x^2 - 2}{a^3} \sin ax$

395. $\displaystyle\int x^3(\cos ax)\, dx = \frac{3a^2 x^2 - 6}{a^4} \cos ax + \frac{a^2 x^3 - 6x}{a^3} \sin ax$

396. $\displaystyle\int x^m(\cos ax)\, dx = \begin{cases} \dfrac{x^m \sin ax}{a} - \dfrac{m}{a} \displaystyle\int x^{m-1} \sin ax\, dx \\[2mm] \quad\text{or} \\[2mm] \sin ax \displaystyle\sum_{r=0}^{\left[\frac{m}{2}\right]} (-1)^r \dfrac{m!}{(m-2r)!} \cdot \dfrac{x^{m-2r}}{a^{2r+1}} \\[4mm] + \cos ax \displaystyle\sum_{r=0}^{\left[\frac{m-1}{2}\right]} (-1)^r \dfrac{m!}{(m-2r-1)!} \cdot \dfrac{x^{m-2r-1}}{a^{2r+2}} \end{cases}$

See note integral 392.

INTEGRALS (Continued)

397. $\int \dfrac{\sin ax}{x}\, dx = \displaystyle\sum_{n=0}^{\infty} (-1)^n \dfrac{(ax)^{2n+1}}{(2n+1)(2n+1)!}$

398. $\int \dfrac{\cos ax}{x}\, dx = \log x + \displaystyle\sum_{n=1}^{\infty} (-1)^n \dfrac{(ax)^{2n}}{2n(2n)!}$

399. $\int x(\sin^2 ax)\, dx = \dfrac{x^2}{4} - \dfrac{x \sin 2ax}{4a} - \dfrac{\cos 2ax}{8a^2}$

400. $\int x^2(\sin^2 ax)\, dx = \dfrac{x^3}{6} - \left(\dfrac{x^2}{4a} - \dfrac{1}{8a^3}\right)\sin 2ax - \dfrac{x \cos 2ax}{4a^2}$

401. $\int x(\sin^3 ax)\, dx = \dfrac{x \cos 3ax}{12a} - \dfrac{\sin 3ax}{36a^2} - \dfrac{3x \cos ax}{4a} + \dfrac{3 \sin ax}{4a^2}$

402. $\int x(\cos^2 ax)\, dx = \dfrac{x^2}{4} + \dfrac{x \sin 2ax}{4a} + \dfrac{\cos 2ax}{8a^2}$

403. $\int x^2(\cos^2 ax)\, dx = \dfrac{x^3}{6} + \left(\dfrac{x^2}{4a} - \dfrac{1}{8a^3}\right)\sin 2ax + \dfrac{x \cos 2ax}{4a^2}$

404. $\int x(\cos^3 ax)\, dx = \dfrac{x \sin 3ax}{12a} + \dfrac{\cos 3ax}{36a^2} + \dfrac{3x \sin ax}{4a} + \dfrac{3 \cos ax}{4a^2}$

405. $\int \dfrac{\sin ax}{x^m}\, dx = -\dfrac{\sin ax}{(m-1)x^{m-1}} + \dfrac{a}{m-1}\int \dfrac{\cos ax}{x^{m-1}}\, dx$

406. $\int \dfrac{\cos ax}{x^m}\, dx = -\dfrac{\cos ax}{(m-1)x^{m-1}} - \dfrac{a}{m-1}\int \dfrac{\sin ax}{x^{m-1}}\, dx$

407. $\int \dfrac{x}{1 \pm \sin ax}\, dx = \mp \dfrac{x \cos ax}{a(1 \pm \sin ax)} + \dfrac{1}{a^2}\log(1 \pm \sin ax)$

408. $\int \dfrac{x}{1 + \cos ax}\, dx = \dfrac{x}{a}\tan\dfrac{ax}{2} + \dfrac{2}{a^2}\log\cos\dfrac{ax}{2}$

409. $\int \dfrac{x}{1 - \cos ax}\, dx = -\dfrac{x}{a}\cot\dfrac{ax}{2} + \dfrac{2}{a^2}\log\sin\dfrac{ax}{2}$

410. $\int \dfrac{x + \sin x}{1 + \cos x}\, dx = x \tan\dfrac{x}{2}$

411. $\int \dfrac{x - \sin x}{1 - \cos x}\, dx = -x \cot\dfrac{x}{2}$

412. $\int \sqrt{1 - \cos ax}\, dx = -\dfrac{2 \sin ax}{a\sqrt{1 - \cos ax}} = -\dfrac{2\sqrt{2}}{a}\cos\left(\dfrac{ax}{2}\right)$

413. $\int \sqrt{1 + \cos ax}\, dx = \dfrac{2 \sin ax}{a\sqrt{1 + \cos ax}} = \dfrac{2\sqrt{2}}{a}\sin\left(\dfrac{ax}{2}\right)$

414. $\int \sqrt{1 + \sin x}\, dx = \pm 2 \left(\sin \dfrac{x}{2} - \cos \dfrac{x}{2} \right),$

[use $+$ if $(8k - 1)\dfrac{\pi}{2} < x \le (8k + 3)\dfrac{\pi}{2}$, otherwise $-$; k an integer]

415. $\int \sqrt{1 - \sin x}\, dx = \pm 2 \left(\sin \dfrac{x}{2} + \cos \dfrac{x}{2} \right),$

[use $+$ if $(8k - 3)\dfrac{\pi}{2} < x \le (8k + 1)\dfrac{\pi}{2}$, otherwise $-$; k an integer]

416. $\int \dfrac{dx}{\sqrt{1 - \cos x}} = \pm \sqrt{2} \log \tan \dfrac{x}{4},$

[use $+$ if $4k\pi < x < (4k + 2)\pi$, otherwise $-$; k an integer]

417. $\int \dfrac{dx}{\sqrt{1 + \cos x}} = \pm \sqrt{2} \log \tan \left(\dfrac{x + \pi}{4} \right),$

[use $+$ if $(4k - 1)\pi < x < (4k + 1)\pi$, otherwise $-$; k an integer]

418. $\int \dfrac{dx}{\sqrt{1 - \sin x}} = \pm \sqrt{2} \log \tan \left(\dfrac{x}{4} - \dfrac{\pi}{8} \right),$

[use $+$ if $(8k + 1)\dfrac{\pi}{2} < x < (8k + 5)\dfrac{\pi}{2}$, otherwise $-$; k an integer]

419. $\int \dfrac{dx}{\sqrt{1 + \sin x}} = \pm \sqrt{2} \log \tan \left(\dfrac{x}{4} + \dfrac{\pi}{8} \right),$

[use $+$ if $(8k - 1)\dfrac{\pi}{2} < x < (8k + 3)\dfrac{\pi}{2}$, otherwise $-$; k an integer]

420. $\int (\tan^2 ax)\, dx = \dfrac{1}{a} \tan ax - x$

421. $\int (\tan^3 ax)\, dx = \dfrac{1}{2a} \tan^2 ax + \dfrac{1}{a} \log \cos ax$

422. $\int (\tan^4 ax)\, dx = \dfrac{\tan^3 ax}{3a} - \dfrac{1}{a} \tan x + x$

423. $\int (\tan^n ax)\, dx = \dfrac{\tan^{n-1} ax}{a(n - 1)} - \int (\tan^{n-2} ax)\, dx$

424. $\int (\cot^2 ax)\, dx = -\dfrac{1}{a} \cot ax - x$

425. $\int (\cot^3 ax)\, dx = -\dfrac{1}{2a} \cot^2 ax - \dfrac{1}{a} \log \sin ax$

426. $\int (\cot^4 ax)\, dx = -\dfrac{1}{3a} \cot^3 ax + \dfrac{1}{a} \cot ax + x$

INTEGRALS (Continued)

427. $\displaystyle\int (\cot^n ax)\,dx = -\frac{\cot^{n-1} ax}{a(n-1)} - \int (\cot^{n-2} ax)\,dx$

428. $\displaystyle\int \frac{x}{\sin^2 ax}\,dx = \int x(\csc^2 ax)\,dx = -\frac{x\cot ax}{a} + \frac{1}{a^2}\log\sin ax$

429. $\displaystyle\int \frac{x}{\sin^n ax}\,dx = \int x(\csc^n ax)\,dx = -\frac{x\cos ax}{a(n-1)\sin^{n-1} ax}$

$$-\frac{1}{a^2(n-1)(n-2)\sin^{n-2} ax} + \frac{(n-2)}{(n-1)}\int \frac{x}{\sin^{n-2} ax}\,dx$$

430. $\displaystyle\int \frac{x}{\cos^2 ax}\,dx = \int x(\sec^2 ax)\,dx = \frac{1}{a}x\tan ax + \frac{1}{a^2}\log\cos ax$

431. $\displaystyle\int \frac{x}{\cos^n ax}\,dx = \int x(\sec^n ax)\,dx = \frac{x\sin ax}{a(n-1)\cos^{n-1} ax}$

$$-\frac{1}{a^2(n-1)(n-2)\cos^{n-2} ax} + \frac{n-2}{n-1}\int \frac{x}{\cos^{n-2} ax}\,dx$$

432. $\displaystyle\int \frac{\sin ax}{\sqrt{1+b^2\sin^2 ax}}\,dx = -\frac{1}{ab}\sin^{-1}\frac{b\cos ax}{\sqrt{1+b^2}}$

433. $\displaystyle\int \frac{\sin ax}{\sqrt{1-b^2\sin^2 ax}}\,dx = -\frac{1}{ab}\log\left(b\cos ax + \sqrt{1-b^2\sin^2 ax}\right)$

434. $\displaystyle\int (\sin ax)\sqrt{1+b^2\sin^2 ax}\,dx = -\frac{\cos ax}{2a}\sqrt{1+b^2\sin^2 ax}$

$$-\frac{1+b^2}{2ab}\sin^{-1}\frac{b\cos ax}{\sqrt{1+b^2}}$$

435. $\displaystyle\int (\sin ax)\sqrt{1-b^2\sin^2 ax}\,dx = -\frac{\cos ax}{2a}\sqrt{1-b^2\sin^2 ax}$

$$-\frac{1-b^2}{2ab}\log\left(b\cos ax + \sqrt{1-b^2\sin^2 ax}\right)$$

436. $\displaystyle\int \frac{\cos ax}{\sqrt{1+b^2\sin^2 ax}}\,dx = \frac{1}{ab}\log\left(b\sin ax + \sqrt{1+b^2\sin^2 ax}\right)$

437. $\displaystyle\int \frac{\cos ax}{\sqrt{1-b^2\sin^2 ax}}\,dx = \frac{1}{ab}\sin^{-1}(b\sin ax)$

438. $\displaystyle\int (\cos ax)\sqrt{1+b^2\sin^2 ax}\,dx = \frac{\sin ax}{2a}\sqrt{1+b^2\sin^2 ax}$

$$+\frac{1}{2ab}\log\left(b\sin ax + \sqrt{1+b^2\sin^2 ax}\right)$$

INTEGRALS (Continued)

439. $\int (\cos ax) \sqrt{1 - b^2 \sin^2 ax} \, dx = \dfrac{\sin ax}{2a} \sqrt{1 - b^2 \sin^2 ax} + \dfrac{1}{2ab} \sin^{-1} (b \sin ax)$

440. $\int \dfrac{dx}{\sqrt{a + b \tan^2 cx}} = \dfrac{\pm 1}{c\sqrt{a - b}} \sin^{-1} \left(\sqrt{\dfrac{a - b}{a}} \sin cx \right), \qquad (a > |b|)$

$$\left[\text{use } + \text{ if } (2k - 1)\frac{\pi}{2} < x \le (2k + 1)\frac{\pi}{2}, \text{ otherwise } -; \, k \text{ an integer} \right]$$

FORMS INVOLVING INVERSE TRIGONOMETRIC FUNCTIONS

441. $\int (\sin^{-1} ax) \, dx = x \sin^{-1} ax + \dfrac{\sqrt{1 - a^2 x^2}}{a}$

442. $\int (\cos^{-1} ax) \, dx = x \cos^{-1} ax - \dfrac{\sqrt{1 - a^2 x^2}}{a}$

443. $\int (\tan^{-1} ax) \, dx = x \tan^{-1} ax - \dfrac{1}{2a} \log (1 + a^2 x^2)$

444. $\int (\cot^{-1} ax) \, dx = x \cot^{-1} ax + \dfrac{1}{2a} \log (1 + a^2 x^2)$

445. $\int (\sec^{-1} ax) \, dx = x \sec^{-1} ax - \dfrac{1}{a} \log (ax + \sqrt{a^2 x^2 - 1})$

446. $\int (\csc^{-1} ax) \, dx = x \csc^{-1} ax + \dfrac{1}{a} \log (ax + \sqrt{a^2 x^2 - 1})$

447. $\int \left(\sin^{-1} \dfrac{x}{a} \right) dx = x \sin^{-1} \dfrac{x}{a} + \sqrt{a^2 - x^2}, \qquad (a > 0)$

448. $\int \left(\cos^{-1} \dfrac{x}{a} \right) dx = x \cos^{-1} \dfrac{x}{a} - \sqrt{a^2 - x^2}, \qquad (a > 0)$

449. $\int \left(\tan^{-1} \dfrac{x}{a} \right) dx = x \tan^{-1} \dfrac{x}{a} - \dfrac{a}{2} \log (a^2 + x^2)$

450. $\int \left(\cot^{-1} \dfrac{x}{a} \right) dx = x \cot^{-1} \dfrac{x}{a} + \dfrac{a}{2} \log (a^2 + x^2)$

451. $\int x[\sin^{-1}(ax)] \, dx = \dfrac{1}{4a^2}[(2a^2 x^2 - 1) \sin^{-1}(ax) + ax\sqrt{1 - a^2 x^2}]$

452. $\int x[\cos^{-1}(ax)] \, dx = \dfrac{1}{4a^2}[(2a^2 x^2 - 1) \cos^{-1}(ax) - ax\sqrt{1 - a^2 x^2}]$

INTEGRALS (Continued)

453. $\displaystyle\int x^n[\sin^{-1}(ax)]\,dx = \frac{x^{n+1}}{n+1}\sin^{-1}(ax) - \frac{a}{n+1}\int \frac{x^{n+1}\,dx}{\sqrt{1-a^2x^2}}, \qquad (n \neq -1)$

454. $\displaystyle\int x^n[\cos^{-1}(ax)]\,dx = \frac{x^{n+1}}{n+1}\cos^{-1}(ax) + \frac{a}{n+1}\int \frac{x^{n+1}\,dx}{\sqrt{1-a^2x^2}}, \qquad (n \neq -1)$

455. $\displaystyle\int x(\tan^{-1} ax)\,dx = \frac{1+a^2x^2}{2a^2}\tan^{-1} ax - \frac{x}{2a}$

456. $\displaystyle\int x^n(\tan^{-1} ax)\,dx = \frac{x^{n+1}}{n+1}\tan^{-1} ax - \frac{a}{n+1}\int \frac{x^{n+1}}{1+a^2x^2}\,dx$

457. $\displaystyle\int x(\cot^{-1} ax)\,dx = \frac{1+a^2x^2}{2a^2}\cot^{-1} ax + \frac{x}{2a}$

458. $\displaystyle\int x^n(\cot^{-1} ax)\,dx = \frac{x^{n+1}}{n+1}\cot^{-1} ax + \frac{a}{n+1}\int \frac{x^{n+1}}{1+a^2x^2}\,dx$

459. $\displaystyle\int \frac{\sin^{-1}(ax)}{x^2}\,dx = a\log\left(\frac{1-\sqrt{1-a^2x^2}}{x}\right) - \frac{\sin^{-1}(ax)}{x}$

460. $\displaystyle\int \frac{\cos^{-1}(ax)\,dx}{x^2} = -\frac{1}{x}\cos^{-1}(ax) + a\log\frac{1+\sqrt{1-a^2x^2}}{x}$

461. $\displaystyle\int \frac{\tan^{-1}(ax)\,dx}{x^2} = -\frac{1}{x}\tan^{-1}(ax) - \frac{a}{2}\log\frac{1+a^2x^2}{x^2}$

462. $\displaystyle\int \frac{\cot^{-1} ax}{x^2}\,dx = -\frac{1}{x}\cot^{-1} ax - \frac{a}{2}\log\frac{x^2}{a^2x^2+1}$

463. $\displaystyle\int (\sin^{-1} ax)^2\,dx = x(\sin^{-1} ax)^2 - 2x + \frac{2\sqrt{1-a^2x^2}}{a}\sin^{-1} ax$

464. $\displaystyle\int (\cos^{-1} ax)^2\,dx = x(\cos^{-1} ax)^2 - 2x - \frac{2\sqrt{1-a^2x^2}}{a}\cos^{-1} ax$

465. $\displaystyle\int (\sin^{-1} ax)^n\,dx = \begin{cases} x(\sin^{-1} ax)^n + \dfrac{n\sqrt{1-a^2x^2}}{a}(\sin^{-1} ax)^{n-1} \\[2mm] \qquad\qquad\qquad -n(n-1)\displaystyle\int (\sin^{-1} ax)^{n-2}\,dx \\[4mm] \text{or} \\[2mm] \displaystyle\sum_{r=0}^{\left[\frac{n}{2}\right]} (-1)^r \frac{n!}{(n-2r)!} x(\sin^{-1} ax)^{n-2r} \\[4mm] \qquad + \displaystyle\sum_{r=0}^{\left[\frac{n-1}{2}\right]} (-1)^r \frac{n!\sqrt{1-a^2x^2}}{(n-2r-1)!a}(\sin^{-1} ax)^{n-2r-1} \end{cases}$

Note: [s] means greatest integer $\leq s$. Thus [3.5] means 3; [5] $= 5$, [$\frac{1}{2}$] $= 0$.

INTEGRALS (Continued)

466. $\displaystyle\int (\cos^{-1} ax)^n \, dx = \begin{cases} x(\cos^{-1} ax)^n - \dfrac{n\sqrt{1 - a^2x^2}}{a}(\cos^{-1} ax)^{n-1} \\ \qquad\qquad -n(n-1)\displaystyle\int (\cos^{-1} ax)^{n-2}\, dx \\ \qquad\qquad \text{or} \\ \displaystyle\sum_{r=0}^{\left[\frac{n}{2}\right]} (-1)^r \dfrac{n!}{(n-2r)!} x(\cos^{-1} ax)^{n-2r} \\ \qquad -\displaystyle\sum_{r=0}^{\left[\frac{n-1}{2}\right]} (-1)^r \dfrac{n!\sqrt{1-a^2x^2}}{(n-2r-1)!\,a}(\cos^{-1} ax)^{n-2r-1} \end{cases}$

467. $\displaystyle\int \frac{1}{\sqrt{1 - a^2x^2}}(\sin^{-1} ax)\, dx = \frac{1}{2a}(\sin^{-1} ax)^2$

468. $\displaystyle\int \frac{x^n}{\sqrt{1 - a^2x^2}}(\sin^{-1} ax)\, dx = -\frac{x^{n-1}}{na^2}\sqrt{1 - a^2x^2}\,\sin^{-1} ax + \frac{x^n}{n^2 a}$

$$+ \frac{n-1}{na^2}\int \frac{x^{n-2}}{\sqrt{1 - a^2x^2}}\sin^{-1} ax\, dx$$

469. $\displaystyle\int \frac{1}{\sqrt{1 - a^2x^2}}(\cos^{-1} ax)\, dx = -\frac{1}{2a}(\cos^{-1} ax)^2$

470. $\displaystyle\int \frac{x^n}{\sqrt{1 - a^2x^2}}(\cos^{-1} ax)\, dx = -\frac{x^{n-1}}{na^2}\sqrt{1 - a^2x^2}\,\cos^{-1} ax - \frac{x^n}{n^2 a}$

$$+ \frac{n-1}{na^2}\int \frac{x^{n-2}}{\sqrt{1 - a^2x^2}}\cos^{-1} ax\, dx$$

471. $\displaystyle\int \frac{\tan^{-1} ax}{a^2x^2 + 1}\, dx = \frac{1}{2a}(\tan^{-1} ax)^2$

472. $\displaystyle\int \frac{\cot^{-1} ax}{a^2x^2 + 1}\, dx = -\frac{1}{2a}(\cot^{-1} ax)^2$

473. $\displaystyle\int x \sec^{-1} ax \, dx = \frac{x^2}{2}\sec^{-1} ax - \frac{1}{2a^2}\sqrt{a^2x^2 - 1}$

474. $\displaystyle\int x^n \sec^{-1} ax \, dx = \frac{x^{n+1}}{n+1}\sec^{-1} ax - \frac{1}{n+1}\int \frac{x^n\, dx}{\sqrt{a^2x^2 - 1}}$

475. $\displaystyle\int \frac{\sec^{-1} ax}{x^2}\, dx = -\frac{\sec^{-1} ax}{x} + \frac{\sqrt{a^2x^2 - 1}}{x}$

476. $\displaystyle\int x \csc^{-1} ax \, dx = \frac{x^2}{2}\csc^{-1} ax + \frac{1}{2a^2}\sqrt{a^2x^2 - 1}$

477. $\displaystyle\int x^n \csc^{-1} ax \, dx = \frac{x^{n+1}}{n+1}\csc^{-1} ax + \frac{1}{n+1}\int \frac{x^n\, dx}{\sqrt{a^2x^2 - 1}}$

INTEGRALS (Continued)

478. $\displaystyle\int \frac{\csc^{-1} ax}{x^2} \, dx = -\frac{\csc^{-1} ax}{x} - \frac{\sqrt{a^2 x^2 - 1}}{x}$

FORMS INVOLVING TRIGONOMETRIC SUBSTITUTIONS

479. $\displaystyle\int f(\sin x) \, dx = 2 \int f\left(\frac{2z}{1 + z^2}\right) \frac{dz}{1 + z^2}, \qquad \left(z = \tan\frac{x}{2}\right)$

480. $\displaystyle\int f(\cos x) \, dx = 2 \int f\left(\frac{1 - z^2}{1 + z^2}\right) \frac{dz}{1 + z^2}, \qquad \left(z = \tan\frac{x}{2}\right)$

*481. $\displaystyle\int f(\sin x) \, dx = \int f(u) \frac{du}{\sqrt{1 - u^2}}, \qquad (u = \sin x)$

*482. $\displaystyle\int f(\cos x) \, dx = -\int f(u) \frac{du}{\sqrt{1 - u^2}}, \qquad (u = \cos x)$

*483. $\displaystyle\int f(\sin x, \cos x) \, dx = \int f(u, \sqrt{1 - u^2}) \frac{du}{\sqrt{1 - u^2}}, \qquad (u = \sin x)$

484. $\displaystyle\int f(\sin x, \cos x) \, dx = 2 \int f\left(\frac{2z}{1 + z^2}, \frac{1 - z^2}{1 + z^2}\right) \frac{dz}{1 + z^2}, \qquad \left(z = \tan\frac{x}{2}\right)$

LOGARITHMIC FORMS

485. $\displaystyle\int (\log x) \, dx = x \log x - x$

486. $\displaystyle\int x(\log x) \, dx = \frac{x^2}{2} \log x - \frac{x^2}{4}$

487. $\displaystyle\int x^2(\log x) \, dx = \frac{x^3}{3} \log x - \frac{x^3}{9}$

488. $\displaystyle\int x^n(\log ax) \, dx = \frac{x^{n+1}}{n+1} \log ax - \frac{x^{n+1}}{(n+1)^2}$

489. $\displaystyle\int (\log x)^2 \, dx = x(\log x)^2 - 2x \log x + 2x$

490. $\displaystyle\int (\log x)^n \, dx = \begin{cases} x(\log x)^n - n \displaystyle\int (\log x)^{n-1} \, dx, & (n \neq -1) \\ \text{or} \\ (-1)^n n! \, x \displaystyle\sum_{r=0}^{n} \frac{(-\log x)^r}{r!} \end{cases}$

*The square roots appearing in these formulas may be plus or minus, depending on the quadrant of x. Care must be used to give them the proper sign.

INTEGRALS (Continued)

491. $\displaystyle \int \frac{(\log x)^n}{x}\,dx = \frac{1}{n+1}(\log x)^{n+1}$

492. $\displaystyle \int \frac{dx}{\log x} = \log(\log x) + \log x + \frac{(\log x)^2}{2\cdot 2!} + \frac{(\log x)^3}{3\cdot 3!} + \cdots$

493. $\displaystyle \int \frac{dx}{x \log x} = \log(\log x)$

494. $\displaystyle \int \frac{dx}{x(\log x)^n} = -\frac{1}{(n-1)(\log x)^{n-1}}$

495. $\displaystyle \int \frac{x^m\,dx}{(\log x)^n} = -\frac{x^{m+1}}{(n-1)(\log x)^{n-1}} + \frac{m+1}{n-1}\int \frac{x^m\,dx}{(\log x)^{n-1}}$

496. $\displaystyle \int x^m(\log x)^n\,dx = \begin{cases} \dfrac{x^{m+1}(\log x)^n}{m+1} - \dfrac{n}{m+1}\displaystyle\int x^m(\log x)^{n-1}\,dx \\[2mm] \text{or} \\[2mm] (-1)^n\dfrac{n!}{m+1}x^{m+1}\displaystyle\sum_{r=0}^{n}\frac{(-\log x)^r}{r!(m+1)^{n-r}} \end{cases}$

497. $\displaystyle \int x^p \cos(b\,ln\,x)\,dx = \frac{x^{p+1}}{(p+1)^2+b^2}\cdot[b\sin(b\,ln\,x)+(p+1)\cos(b\,ln\,x)] + c$

498. $\displaystyle \int x^p \sin(b\,ln\,x)\,dx = \frac{x^{p+1}}{(p+1)^2+b^2}\cdot[(p+1)\sin(b\,ln\,x)-b\cos(b\,ln\,x)] + c$

499. $\displaystyle \int [\log(ax+b)]\,dx = \frac{ax+b}{a}\log(ax+b) - x$

500. $\displaystyle \int \frac{\log(ax+b)}{x^2}\,dx = \frac{a}{b}\log x - \frac{ax+b}{bx}\log(ax+b)$

501. $\displaystyle \int x^m[\log(ax+b)]\,dx = \frac{1}{m+1}\left[x^{m+1}-\left(-\frac{b}{a}\right)^{m+1}\right]\log(ax+b)$

$$-\frac{1}{m+1}\left(-\frac{b}{a}\right)^{m+1}\sum_{r=1}^{m+1}\frac{1}{r}\left(-\frac{ax}{b}\right)^r$$

502. $\displaystyle \int \frac{\log(ax+b)}{x^m}\,dx = -\frac{1}{m-1}\frac{\log(ax+b)}{x^{m-1}} + \frac{1}{m-1}\left(-\frac{a}{b}\right)^{m-1}\log\frac{ax+b}{x}$

$$+\frac{1}{m-1}\left(-\frac{a}{b}\right)^{m-1}\sum_{r=1}^{m-2}\frac{1}{r}\left(-\frac{b}{ax}\right)^r,\ (m>2)$$

503. $\displaystyle \int \left[\log\frac{x+a}{x-a}\right]dx = (x+a)\log(x+a) - (x-a)\log(x-a)$

504. $\displaystyle \int x^m\left[\log\frac{x+a}{x-a}\right]dx = \frac{x^{m+1}-(-a)^{m+1}}{m+1}\log(x+a) - \frac{x^{m+1}-a^{m+1}}{m+1}\log(x-a)$

$$+\frac{2a^{m+1}}{m+1}\sum_{r=1}^{\left[\frac{m+1}{2}\right]}\frac{1}{m-2r+2}\left(\frac{x}{a}\right)^{m-2r+2}$$

See note integral 392.

INTEGRALS (Continued)

505. $\displaystyle\int \frac{1}{x^2}\left[\log\frac{x+a}{x-a}\right]dx = \frac{1}{x}\log\frac{x-a}{x+a} - \frac{1}{a}\log\frac{x^2-a^2}{x^2}$

506. $\displaystyle\int (\log X)\,dx = \begin{cases} \left(x+\dfrac{b}{2c}\right)\log X - 2x + \dfrac{\sqrt{4ac-b^2}}{c}\tan^{-1}\dfrac{2cx+b}{\sqrt{4ac-b^2}}, \\ \hspace{6cm}(b^2-4ac<0) \\[4pt] \text{or} \\[4pt] \left(x+\dfrac{b}{2c}\right)\log X - 2x + \dfrac{\sqrt{b^2-4ac}}{c}\tanh^{-1}\dfrac{2cx+b}{\sqrt{b^2-4ac}}, \\ \hspace{6cm}(b^2-4ac>0) \\[4pt] \text{where} \\[4pt] X = a + bx + cx^2 \end{cases}$

507. $\displaystyle\int x^n(\log X)\,dx = \frac{x^{n+1}}{n+1}\log X - \frac{2c}{n+1}\int\frac{x^{n+2}}{X}\,dx - \frac{b}{n+1}\int\frac{x^{n+1}}{X}\,dx$

$$\text{where } X = a + bx + cx^2$$

508. $\displaystyle\int [\log(x^2+a^2)]\,dx = x\log(x^2+a^2) - 2x + 2a\tan^{-1}\frac{x}{a}$

509. $\displaystyle\int [\log(x^2-a^2)]\,dx = x\log(x^2-a^2) - 2x + a\log\frac{x+a}{x-a}$

510. $\displaystyle\int x[\log(x^2\pm a^2)]\,dx = \tfrac{1}{2}(x^2\pm a^2)\log(x^2\pm a^2) - \tfrac{1}{2}x^2$

511. $\displaystyle\int [\log(x+\sqrt{x^2\pm a^2})]\,dx = x\log(x+\sqrt{x^2\pm a^2}) - \sqrt{x^2\pm a^2}$

512. $\displaystyle\int x[\log(x+\sqrt{x^2\pm a^2})]\,dx = \left(\frac{x^2}{2}\pm\frac{a^2}{4}\right)\log(x+\sqrt{x^2\pm a^2}) - \frac{x\sqrt{x^2\pm a^2}}{4}$

513. $\displaystyle\int x^m[\log(x+\sqrt{x^2\pm a^2})]\,dx = \frac{x^{m+1}}{m+1}\log(x+\sqrt{x^2\pm a^2})$

$$-\frac{1}{m+1}\int\frac{x^{m+1}}{\sqrt{x^2\pm a^2}}\,dx$$

514. $\displaystyle\int \frac{\log(x+\sqrt{x^2+a^2})}{x^2}\,dx = -\frac{\log(x+\sqrt{x^2+a^2})}{x} - \frac{1}{a}\log\frac{a+\sqrt{x^2+a^2}}{x}$

515. $\displaystyle\int \frac{\log(x+\sqrt{x^2-a^2})}{x^2}\,dx = -\frac{\log(x+\sqrt{x^2-a^2})}{x} + \frac{1}{|a|}\sec^{-1}\frac{x}{a}$

516. $\displaystyle\int x^n \log(x^2 - a^2)\,dx = \frac{1}{n+1}\bigg[x^{n+1}\log(x^2 - a^2) - a^{n+1}\log(x-a)$

See note integral 392. $\displaystyle - (-a)^{n+1}\log(x+a) - 2\sum_{r=0}^{\left[\frac{n}{2}\right]} \frac{a^{2r}x^{n-2r+1}}{n-2r+1}\bigg]$

EXPONENTIAL FORMS

517. $\displaystyle\int e^x\,dx = e^x$

518. $\displaystyle\int e^{-x}\,dx = -e^{-x}$

519. $\displaystyle\int e^{ax}\,dx = \frac{e^{ax}}{a}$

520. $\displaystyle\int x\,e^{ax}\,dx = \frac{e^{ax}}{a^2}(ax-1)$

521. $\displaystyle\int x^m e^{ax}\,dx = \begin{cases} \dfrac{x^m e^{ax}}{a} - \dfrac{m}{a}\displaystyle\int x^{m-1} e^{ax}\,dx \\ \qquad\qquad \text{or} \\ e^{ax}\displaystyle\sum_{r=0}^{m}(-1)^r \dfrac{m!\,x^{m-r}}{(m-r)!\,a^{r+1}} \end{cases}$

522. $\displaystyle\int \frac{e^{ax}\,dx}{x} = \log x + \frac{ax}{1!} + \frac{a^2x^2}{2\cdot 2!} + \frac{a^3x^3}{3\cdot 3!} + \cdots$

523. $\displaystyle\int \frac{e^{ax}}{x^m}\,dx = -\frac{1}{m-1}\frac{e^{ax}}{x^{m-1}} + \frac{a}{m-1}\int \frac{e^{ax}}{x^{m-1}}\,dx$

524. $\displaystyle\int e^{ax}\log x\,dx = \frac{e^{ax}\log x}{a} - \frac{1}{a}\int \frac{e^{ax}}{x}\,dx$

525. $\displaystyle\int \frac{dx}{1+e^x} = x - \log(1+e^x) = \log\frac{e^x}{1+e^x}$

526. $\displaystyle\int \frac{dx}{a+be^{px}} = \frac{x}{a} - \frac{1}{ap}\log(a+be^{px})$

527. $\displaystyle\int \frac{dx}{ae^{mx}+be^{-mx}} = \frac{1}{m\sqrt{ab}}\tan^{-1}\left(e^{mx}\sqrt{\frac{a}{b}}\right), \qquad (a>0, b>0)$

528. $\displaystyle\int \frac{dx}{ae^{mx}-be^{-mx}} = \begin{cases} \dfrac{1}{2m\sqrt{ab}}\log\dfrac{\sqrt{a}\,e^{mx}-\sqrt{b}}{\sqrt{a}\,e^{mx}+\sqrt{b}} \\ \qquad\qquad \text{or} \\ \dfrac{-1}{m\sqrt{ab}}\tanh^{-1}\left(\sqrt{\dfrac{a}{b}}\,e^{mx}\right), \qquad (a>0, b>0) \end{cases}$

INTEGRALS (Continued)

529. $\int (a^x - a^{-x})\, dx = \dfrac{a^x + a^{-x}}{\log a}$

530. $\int \dfrac{e^{ax}}{b + ce^{ax}}\, dx = \dfrac{1}{ac} \log (b + ce^{ax})$

531. $\int \dfrac{x\, e^{ax}}{(1 + ax)^2}\, dx = \dfrac{e^{ax}}{a^2(1 + ax)}$

532. $\int x\, e^{-x^2}\, dx = -\tfrac{1}{2} e^{-x^2}$

533. $\int e^{ax} [\sin (bx)]\, dx = \dfrac{e^{ax}[a \sin (bx) - b \cos (bx)]}{a^2 + b^2}$

534. $\int e^{ax} [\sin (bx)][\sin (cx)]\, dx = \dfrac{e^{ax}[(b - c) \sin (b - c)x + a \cos (b - c)x]}{2[a^2 + (b - c)^2]}$

$$- \dfrac{e^{ax}[(b + c) \sin (b + c)x + a \cos (b + c)x]}{2[a^2 + (b + c)^2]}$$

535. $\int e^{ax}[\sin (bx)][\cos (cx)]\, dx = \begin{cases} \dfrac{e^{ax}[a \sin (b - c)x - (b - c) \cos (b - c)x]}{2[a^2 + (b - c)^2]} \\ \quad + \dfrac{e^{ax}[a \sin (b + c)x - (b + c) \cos (b + c)x]}{2[a^2 + (b + c)^2]} \\ \qquad\qquad \text{or} \\ \dfrac{e^{ax}}{\rho}[(a \sin bx - b \cos bx)[\cos (cx - \alpha)] \\ \qquad\qquad -c(\sin bx) \sin (cx - \alpha)] \\ \text{where} \\ \rho = \sqrt{(a^2 + b^2 - c^2)^2 + 4a^2 c^2}, \\ \quad \rho \cos \alpha = a^2 + b^2 - c^2, \qquad \rho \sin \alpha = 2ac \end{cases}$

536. $\int e^{ax}[\sin (bx)][\sin (bx + c)]\, dx$

$$= \dfrac{e^{ax} \cos c}{2a} - \dfrac{e^{ax}[a \cos (2bx + c) + 2b \sin (2bx + c)]}{2(a^2 + 4b^2)}$$

537. $\int e^{ax}[\sin (bx)][\cos (bx + c)]\, dx$

$$= \dfrac{-e^{ax} \sin c}{2a} + \dfrac{e^{ax}[a \sin (2bx + c) - 2b \cos (2bx + c)]}{2(a^2 + 4b^2)}$$

538. $\int e^{ax}[\cos (bx)]\, dx = \dfrac{e^{ax}}{a^2 + b^2}[a \cos (bx) + b \sin (bx)]$

INTEGRALS (Continued)

539. $\int e^{ax}[\cos(bx)][\cos(cx)]\,dx = \dfrac{e^{ax}[(b-c)\sin(b-c)x + a\cos(b-c)x]}{2[a^2+(b-c)^2]}$

$$+\dfrac{e^{ax}[(b+c)\sin(b+c)x + a\cos(b+c)x]}{2[a^2+(b+c)^2]}$$

540. $\int e^{ax}[\cos(bx)][\cos(bx+c)]\,dx$

$$= \dfrac{e^{ax}\cos c}{2a} + \dfrac{e^{ax}[a\cos(2bx+c) + 2b\sin(2bx+c)]}{2(a^2+4b^2)}$$

541. $\int e^{ax}[\cos(bx)][\sin(bx+c)]\,dx$

$$= \dfrac{e^{ax}\sin c}{2a} + \dfrac{e^{ax}[a\sin(2bx+c) - 2b\cos(2bx+c)]}{2(a^2+4b^2)}$$

542. $\int e^{ax}[\sin^n bx]\,dx = \dfrac{1}{a^2+n^2b^2}\Bigg[(a\sin bx - nb\cos bx)\,e^{ax}\sin^{n-1}bx$

$$+\,n(n-1)b^2\int e^{ax}[\sin^{n-2}bx]\,dx\Bigg]$$

543. $\int e^{ax}[\cos^n bx]\,dx = \dfrac{1}{a^2+n^2b^2}\Bigg[(a\cos bx + nb\sin bx)\,e^{ax}\cos^{n-1}bx$

$$+\,n(n-1)b^2\int e^{ax}[\cos^{n-2}bx]\,dx\Bigg]$$

544. $\int x^m e^x \sin x\,dx = \dfrac{1}{2}x^m e^x(\sin x - \cos x) - \dfrac{m}{2}\int x^{m-1} e^x \sin x\,dx$

$$+\dfrac{m}{2}\int x^{m-1} e^x \cos x\,dx$$

545. $\int x^m e^{ax}[\sin bx]\,dx = \begin{cases} x^m e^{ax}\dfrac{a\sin bx - b\cos bx}{a^2+b^2} \\[2mm] \qquad -\dfrac{m}{a^2+b^2}\int x^{m-1} e^{ax}(a\sin bx - b\cos bx)\,dx \\[2mm] \qquad\qquad \text{or} \\[2mm] e^{ax}\displaystyle\sum_{r=0}^{m}\dfrac{(-1)^r m!\,x^{m-r}}{\rho^{r+1}(m-r)!}\sin[bx - (r+1)\alpha] \\[2mm] \qquad\qquad \text{where} \\[2mm] \rho = \sqrt{a^2+b^2}, \qquad \rho\cos\alpha = a, \qquad \rho\sin\alpha = b \end{cases}$

546. $\int x^m e^x \cos x\,dx = \dfrac{1}{2}x^m e^x(\sin x + \cos x)$

$$-\dfrac{m}{2}\int x^{m-1} e^x \sin x\,dx - \dfrac{m}{2}\int x^{m-1} e^x \cos x\,dx$$

INTEGRALS (Continued)

547. $\displaystyle\int x^m\, e^{ax}\cos bx\, dx =$

$$\begin{cases}
x^m\, e^{ax}\,\dfrac{a\cos bx + b\sin bx}{a^2 + b^2} \\[2mm]
\qquad\qquad - \dfrac{m}{a^2 + b^2}\displaystyle\int x^{m-1}\,e^{ax}(a\cos bx + b\sin bx)\,dx \\[3mm]
\qquad\text{or} \\[2mm]
e^{ax}\displaystyle\sum_{r=0}^{m}\dfrac{(-1)^r m!\,x^{m-r}}{\rho^{r+1}(m-r)!}\cos[bx - (r+1)\alpha] \\[3mm]
\qquad\text{where} \\[2mm]
\rho = \sqrt{a^2 + b^2}, \qquad \rho\cos\alpha = a, \qquad \rho\sin\alpha = b
\end{cases}$$

548. $\displaystyle\int e^{ax}(\cos^m x)(\sin^n x)\,dx =$

$$\begin{cases}
\dfrac{e^{ax}\cos^{m-1}x\,\sin^n x\,[a\cos x + (m+n)\sin x]}{(m+n)^2 + a^2} \\[3mm]
\qquad - \dfrac{na}{(m+n)^2 + a^2}\displaystyle\int e^{ax}(\cos^{m-1}x)(\sin^{n-1}x)\,dx \\[3mm]
\qquad + \dfrac{(m-1)(m+n)}{(m+n)^2 + a^2}\displaystyle\int e^{ax}(\cos^{m-2}x)(\sin^n x)\,dx \\[3mm]
\qquad\text{or} \\[2mm]
\dfrac{e^{ax}\cos^m x\,\sin^{n-1}x\,[a\sin x - (m+n)\cos x]}{(m+n)^2 + a^2} \\[3mm]
\qquad + \dfrac{ma}{(m+n)^2 + a^2}\displaystyle\int e^{ax}(\cos^{m-1}x)(\sin^{n-1}x)\,dx \\[3mm]
\qquad + \dfrac{(n-1)(m+n)}{(m+n)^2 + a^2}\displaystyle\int e^{ax}(\cos^m x)(\sin^{n-2}x)\,dx \\[3mm]
\qquad\text{or} \\[2mm]
\dfrac{e^{ax}(\cos^{m-1}x)(\sin^{n-1}x)(a\sin x\cos x + m\sin^2 x - n\cos^2 x)}{(m+n)^2 + a^2} \\[3mm]
\qquad + \dfrac{m(m-1)}{(m+n)^2 + a^2}\displaystyle\int e^{ax}(\cos^{m-2}x)(\sin^n x)\,dx \\[3mm]
\qquad + \dfrac{n(n-1)}{(m+n)^2 + a^2}\displaystyle\int e^{ax}(\cos^m x)(\sin^{n-2}x)\,dx \\[3mm]
\qquad\text{or} \\[2mm]
\dfrac{e^{ax}(\cos^{m-1}x)(\sin^{n-1}x)(a\cos x\sin x + m\sin^2 x - n\cos^2 x)}{(m+n)^2 + a^2} \\[3mm]
\qquad + \dfrac{m(m-1)}{(m+n)^2 + a^2}\displaystyle\int e^{ax}(\cos^{m-2}x)(\sin^{n-2}x)\,dx \\[3mm]
\qquad + \dfrac{(n-m)(n+m-1)}{(m+n)^2 + a^2}\displaystyle\int e^{ax}(\cos^m x)(\sin^{n-2}x)\,dx
\end{cases}$$

INTEGRALS (Continued)

549. $\displaystyle\int x\,e^{ax}(\sin bx)\,dx = \frac{x\,e^{ax}}{a^2+b^2}(a\sin bx - b\cos bx)$

$$-\frac{e^{ax}}{(a^2+b^2)^2}[(a^2-b^2)\sin bx - 2ab\cos bx]$$

550. $\displaystyle\int x\,e^{ax}(\cos bx)\,dx = \frac{x\,e^{ax}}{a^2+b^2}(a\cos bx + b\sin bx)$

$$-\frac{e^{ax}}{(a^2+b^2)^2}[(a^2-b^2)\cos bx + 2ab\sin bx]$$

551. $\displaystyle\int \frac{e^{ax}}{\sin^n x}\,dx = -\frac{e^{ax}[a\sin x + (n-2)\cos x]}{(n-1)(n-2)\sin^{n-1} x} + \frac{a^2+(n-2)^2}{(n-1)(n-2)}\int \frac{e^{ax}}{\sin^{n-2} x}\,dx$

552. $\displaystyle\int \frac{e^{ax}}{\cos^n x}\,dx = -\frac{e^{ax}[a\cos x - (n-2)\sin x]}{(n-1)(n-2)\cos^{n-1} x} + \frac{a^2+(n-2)^2}{(n-1)(n-2)}\int \frac{e^{ax}}{\cos^{n-2} x}\,dx$

553. $\displaystyle\int e^{ax}\tan^n x\,dx = e^{ax}\frac{\tan^{n-1} x}{n-1} - \frac{a}{n-1}\int e^{ax}\tan^{n-1} x\,dx - \int e^{ax}\tan^{n-2} x\,dx$

HYPERBOLIC FORMS

554. $\displaystyle\int (\sinh x)\,dx = \cosh x$

555. $\displaystyle\int (\cosh x)\,dx = \sinh x$

556. $\displaystyle\int (\tanh x)\,dx = \log\cosh x$

557. $\displaystyle\int (\coth x)\,dx = \log\sinh x$

558. $\displaystyle\int (\operatorname{sech} x)\,dx = \tan^{-1}(\sinh x)$

559. $\displaystyle\int \operatorname{csch} x\,dx = \log\tanh\left(\frac{x}{2}\right)$

560. $\displaystyle\int x(\sinh x)\,dx = x\cosh x - \sinh x$

561. $\displaystyle\int x^n(\sinh x)\,dx = x^n\cosh x - n\int x^{n-1}(\cosh x)\,dx$

562. $\displaystyle\int x(\cosh x)\,dx = x\sinh x - \cosh x$

563. $\displaystyle\int x^n(\cosh x)\,dx = x^n\sinh x - n\int x^{n-1}(\sinh x)\,dx$

INTEGRALS (Continued)

564. $\displaystyle\int (\operatorname{sech} x)(\tanh x)\, dx = -\operatorname{sech} x$

565. $\displaystyle\int (\operatorname{csch} x)(\coth x)\, dx = -\operatorname{csch} x$

566. $\displaystyle\int (\sinh^2 x)\, dx = \dfrac{\sinh 2x}{4} - \dfrac{x}{2}$

567. $\displaystyle\int (\sinh^m x)(\cosh^n x)\, dx =$
$$
\begin{cases}
\dfrac{1}{m+n}(\sinh^{m+1} x)(\cosh^{n-1} x) \\[2mm]
\qquad + \dfrac{n-1}{m+n}\displaystyle\int (\sinh^m x)(\cosh^{n-2} x)\, dx \\[4mm]
\qquad\qquad \text{or} \\[2mm]
\dfrac{1}{m+n}\sinh^{m-1} x \cosh^{n+1} x \\[2mm]
\qquad - \dfrac{m-1}{m+n}\displaystyle\int (\sinh^{m-2} x)(\cosh^n x)\, dx, \qquad (m+n \ne 0)
\end{cases}
$$

568. $\displaystyle\int \dfrac{dx}{(\sinh^m x)(\cosh^n x)} =$
$$
\begin{cases}
-\dfrac{1}{(m-1)(\sinh^{m-1} x)(\cosh^{n-1} x)} \\[2mm]
\qquad - \dfrac{m+n-2}{m-1}\displaystyle\int \dfrac{dx}{(\sinh^{m-2} x)(\cosh^n x)}, \qquad (m \ne 1) \\[4mm]
\qquad\qquad \text{or} \\[2mm]
\dfrac{1}{(n-1)\sinh^{m-1} x \cosh^{n-1} x} \\[2mm]
\qquad + \dfrac{m+n-2}{n-1}\displaystyle\int \dfrac{dx}{(\sinh^m x)(\cosh^{n-2} x)}, \qquad (n \ne 1)
\end{cases}
$$

569. $\displaystyle\int (\tanh^2 x)\, dx = x - \tanh x$

570. $\displaystyle\int (\tanh^n x)\, dx = -\dfrac{\tanh^{n-1} x}{n-1} + \int (\tanh^{n-2} x)\, dx, \qquad (n \ne 1)$

571. $\displaystyle\int (\operatorname{sech}^2 x)\, dx = \tanh x$

572. $\displaystyle\int (\cosh^2 x)\, dx = \dfrac{\sinh 2x}{4} + \dfrac{x}{2}$

573. $\displaystyle\int (\coth^2 x)\, dx = x - \coth x$

574. $\displaystyle\int (\coth^n x)\, dx = -\dfrac{\coth^{n-1} x}{n-1} + \int \coth^{n-2} x\, dx, \qquad (n \ne 1)$

INTEGRALS (Continued)

575. $\displaystyle\int (\operatorname{csch}^2 x)\, dx = -\operatorname{ctnh} x$

576. $\displaystyle\int (\sinh mx)(\sinh nx)\, dx = \frac{\sinh (m+n)x}{2(m+n)} - \frac{\sinh (m-n)x}{2(m-n)},$ $(m^2 \neq n^2)$

577. $\displaystyle\int (\cosh mx)(\cosh nx)\, dx = \frac{\sinh (m+n)x}{2(m+n)} + \frac{\sinh (m-n)x}{2(m-n)},$ $(m^2 \neq n^2)$

578. $\displaystyle\int (\sinh mx)(\cosh nx)\, dx = \frac{\cosh (m+n)x}{2(m+n)} + \frac{\cosh (m-n)x}{2(m-n)},$ $(m^2 \neq n^2)$

579. $\displaystyle\int \left(\sinh^{-1}\frac{x}{a}\right) dx = x \sinh^{-1}\frac{x}{a} - \sqrt{x^2 + a^2},$ $(a > 0)$

580. $\displaystyle\int x\left(\sinh^{-1}\frac{x}{a}\right) dx = \left(\frac{x^2}{2} + \frac{a^2}{4}\right) \sinh^{-1}\frac{x}{a} - \frac{x}{4}\sqrt{x^2 + a^2},$ $(a > 0)$

581. $\displaystyle\int x^n(\sinh^{-1} x)\, dx = \frac{x^{n+1}}{n+1}\sinh^{-1} x - \frac{1}{n+1}\int \frac{x^{n+1}}{(1+x^2)^{\frac{1}{2}}}\, dx,$ $(n \neq -1)$

582. $\displaystyle\int \left(\cosh^{-1}\frac{x}{a}\right) dx = \begin{cases} x\cosh^{-1}\dfrac{x}{a} - \sqrt{x^2 - a^2}, & \left(\cosh^{-1}\dfrac{x}{a} > 0\right) \\[2mm] \qquad\text{or} \\[2mm] x\cosh^{-1}\dfrac{x}{a} + \sqrt{x^2 - a^2}, & \left(\cosh^{-1}\dfrac{x}{a} < 0\right), \end{cases}$ $(a > 0)$

583. $\displaystyle\int x\left(\cosh^{-1}\frac{x}{a}\right) dx = \frac{2x^2 - a^2}{4}\cosh^{-1}\frac{x}{a} - \frac{x}{4}(x^2 - a^2)^{\frac{1}{2}}$

584. $\displaystyle\int x^n(\cosh^{-1} x)\, dx = \frac{x^{n+1}}{n+1}\cosh^{-1} x - \frac{1}{n+1}\int \frac{x^{n+1}}{(x^2 - 1)^{\frac{1}{2}}}\, dx,$ $(n \neq -1)$

585. $\displaystyle\int \left(\tanh^{-1}\frac{x}{a}\right) dx = x \tanh^{-1}\frac{x}{a} + \frac{a}{2}\log (a^2 - x^2),$ $\left(\left|\frac{x}{a}\right| < 1\right)$

586. $\displaystyle\int \left(\coth^{-1}\frac{x}{a}\right) dx = x \coth^{-1}\frac{x}{a} + \frac{a}{2}\log (x^2 - a^2),$ $\left(\left|\frac{x}{a}\right| > 1\right)$

587. $\displaystyle\int x\left(\tanh^{-1}\frac{x}{a}\right) dx = \frac{x^2 - a^2}{2}\tanh^{-1}\frac{x}{a} + \frac{ax}{2},$ $\left(\left|\frac{x}{a}\right| < 1\right)$

588. $\displaystyle\int x^n\left(\tanh^{-1} x\right) dx = \frac{x^{n+1}}{n+1}\tanh^{-1} x - \frac{1}{n+1}\int \frac{x^{n+1}}{1 - x^2}\, dx,$ $(n \neq -1)$

589. $\displaystyle\int x\left(\coth^{-1}\frac{x}{a}\right) dx = \frac{x^2 - a^2}{2}\coth^{-1}\frac{x}{a} + \frac{ax}{2},$ $\left(\left|\frac{x}{a}\right| > 1\right)$

590. $\displaystyle\int x^n(\coth^{-1} x)\, dx = \frac{x^{n+1}}{n+1}\coth^{-1} x + \frac{1}{n+1}\int \frac{x^{n+1}}{x^2 - 1}\, dx,$ $(n \neq -1)$

INTEGRALS (Continued)

591. $\displaystyle\int (\text{sech}^{-1} x)\, dx = x\, \text{sech}^{-1} x + \sin^{-1} x$

592. $\displaystyle\int x\, \text{sech}^{-1} x\, dx = \frac{x^2}{2}\, \text{sech}^{-1} x - \frac{1}{2}\sqrt{1 - x^2}$

593. $\displaystyle\int x^n\, \text{sech}^{-1} x\, dx = \frac{x^{n+1}}{n+1}\, \text{sech}^{-1} x + \frac{1}{n+1}\int \frac{x^n}{(1 - x^2)^{\frac{1}{2}}}\, dx, \qquad (n \neq -1)$

594. $\displaystyle\int \text{csch}^{-1} x\, dx = x\, \text{csch}^{-1} x + \frac{x}{|x|}\, \sinh^{-1} x$

595. $\displaystyle\int x\, \text{csch}^{-1} x\, dx = \frac{x^2}{2}\, \text{csch}^{-1} x + \frac{1}{2}\frac{x}{|x|}\sqrt{1 + x^2}$

596. $\displaystyle\int x^n\, \text{csch}^{-1} x\, dx = \frac{x^{n+1}}{n+1}\, \text{csch}^{-1} x + \frac{1}{n+1}\frac{x}{|x|}\int \frac{x^n}{(x^2 + 1)^{\frac{1}{2}}}\, dx, \qquad (n \neq -1)$

DEFINITE INTEGRALS

597. $\displaystyle\int_0^\infty x^{n-1} e^{-x}\, dx = \int_0^1 \left(\log \frac{1}{x}\right)^{n-1} dx = \frac{1}{n}\prod_{m=1}^\infty \frac{\left(1 + \dfrac{1}{m}\right)^n}{1 + \dfrac{n}{m}}$

$$= \Gamma(n), n \neq 0, -1, -2, -3, \ldots \qquad \text{(Gamma Function)}$$

598. $\displaystyle\int_0^\infty t^n p^{-t}\, dt = \frac{n!}{(\log p)^{n+1}}, \qquad (n = 0, 1, 2, 3, \ldots \text{ and } p > 0)$

599. $\displaystyle\int_0^\infty t^{n-1} e^{-(a+1)t}\, dt = \frac{\Gamma(n)}{(a+1)^n}, \qquad (n > 0, a > -1)$

600. $\displaystyle\int_0^1 x^m \left(\log \frac{1}{x}\right)^n dx = \frac{\Gamma(n+1)}{(m+1)^{n+1}}, \qquad (m > -1, n > -1)$

601. $\Gamma(n)$ is finite if $n > 0$, $\Gamma(n + 1) = n\Gamma(n)$

602. $\displaystyle \Gamma(n) \cdot \Gamma(1 - n) = \frac{\pi}{\sin n\pi}$

603. $\Gamma(n) = (n - 1)!$ if $n = $ integer > 0

604. $\displaystyle \Gamma(\tfrac{1}{2}) = 2\int_0^\infty e^{-t^2}\, dt = \sqrt{\pi} = 1.7724538509 \cdots = (-\tfrac{1}{2})!$

605. $\displaystyle \Gamma(n + \tfrac{1}{2}) = \frac{1 \cdot 3 \cdot 5 \ldots (2n - 1)}{2^n}\sqrt{\pi} \qquad n = 1, 2, 3, \ldots$

606. $\displaystyle \Gamma(-n + \tfrac{1}{2}) = \frac{(-1)^n 2^n \sqrt{\pi}}{1 \cdot 3 \cdot 5 \ldots (2n - 1)} \qquad n = 1, 2, 3, \ldots$

DEFINITE INTEGRALS (Continued)

607. $\displaystyle\int_0^1 x^{m-1}(1-x)^{n-1}\,dx = \int_0^\infty \frac{x^{m-1}}{(1+x)^{m+n}}\,dx = \frac{\Gamma(m)\Gamma(n)}{\Gamma(m+n)} = B(m,n)$

(Beta function)

608. $B(m,n) = B(n,m) = \dfrac{\Gamma(m)\Gamma(n)}{\Gamma(m+n)}$, where m and n are any positive real numbers.

609. $\displaystyle\int_a^b (x-a)^m(b-x)^n\,dx = (b-a)^{m+n+1}\frac{\Gamma(m+1)\cdot\Gamma(n+1)}{\Gamma(m+n+2)}$,

$(m > -1, n > -1, b > a)$

610. $\displaystyle\int_1^\infty \frac{dx}{x^m} = \frac{1}{m-1},\qquad [m > 1]$

611. $\displaystyle\int_0^\infty \frac{dx}{(1+x)x^p} = \pi\csc p\pi,\qquad [p < 1]$

612. $\displaystyle\int_0^\infty \frac{dx}{(1-x)x^p} = -\pi\cot p\pi,\qquad [p < 1]$

613. $\displaystyle\int_0^\infty \frac{x^{p-1}\,dx}{1+x} = \frac{\pi}{\sin p\pi}$

$= B(p, 1-p) = \Gamma(p)\Gamma(1-p),\qquad [0 < p < 1]$

614. $\displaystyle\int_0^\infty \frac{x^{m-1}\,dx}{1+x^n} = \frac{\pi}{n\sin\dfrac{m\pi}{n}},\qquad [0 < m < n]$

615. $\displaystyle\int_0^\infty \frac{x^a\,dx}{(m+x^b)^c} = \frac{m^{\frac{a+1-bc}{b}}}{b}\left[\frac{\Gamma\!\left(\dfrac{a+1}{b}\right)\Gamma\!\left(c-\dfrac{a+1}{b}\right)}{\Gamma(c)}\right]$

$\left(a > -1, b > 0, m > 0, c > \dfrac{a+1}{b}\right)$

616. $\displaystyle\int_0^\infty \frac{dx}{(1+x)\sqrt{x}} = \pi$

617. $\displaystyle\int_0^\infty \frac{a\,dx}{a^2+x^2} = \frac{\pi}{2}$, if $a > 0$; 0, if $a = 0$; $-\dfrac{\pi}{2}$, if $a < 0$

618. $\displaystyle\int_0^a (a^2-x^2)^{\frac{n}{2}}\,dx = \frac{1}{2}\int_{-a}^a (a^2-x^2)^{\frac{n}{2}}\,dx = \frac{1\cdot3\cdot5\ldots n}{2\cdot4\cdot6\ldots(n+1)}\cdot\frac{\pi}{2}\cdot a^{n+1}$ $(n \text{ odd})$

619. $\displaystyle\int_0^a x^m(a^2-x^2)^{\frac{n}{2}}\,dx = \begin{cases} \dfrac{1}{2}a^{m+n+1}B\!\left(\dfrac{m+1}{2},\dfrac{n+2}{2}\right) \\[4pt] \qquad\qquad\text{or} \\[4pt] \dfrac{1}{2}a^{m+n+1}\dfrac{\Gamma\!\left(\dfrac{m+1}{2}\right)\Gamma\!\left(\dfrac{n+2}{2}\right)}{\Gamma\!\left(\dfrac{m+n+3}{2}\right)} \end{cases}$

DEFINITE INTEGRALS (Continued)

620. $\displaystyle\int_0^{\pi/2} (\sin^n x)\, dx = \begin{cases} \displaystyle\int_0^{\pi/2} (\cos^n x)\, dx \\[2mm] \text{or} \\[2mm] \dfrac{1\cdot 3\cdot 5\cdot 7\ldots(n-1)}{2\cdot 4\cdot 6\cdot 8\ldots(n)}\dfrac{\pi}{2}, \quad (n \text{ an even integer, } n \neq 0) \\[2mm] \text{or} \\[2mm] \dfrac{2\cdot 4\cdot 6\cdot 8\ldots(n-1)}{1\cdot 3\cdot 5\cdot 7\ldots(n)}, \quad (n \text{ an odd integer, } n \neq 1) \\[2mm] \text{or} \\[2mm] \dfrac{\sqrt{\pi}}{2}\dfrac{\Gamma\left(\dfrac{n+1}{2}\right)}{\Gamma\left(\dfrac{n}{2}+1\right)}, \quad (n > -1) \end{cases}$

621. $\displaystyle\int_0^\infty \frac{\sin mx\, dx}{x} = \frac{\pi}{2}, \text{ if } m > 0; 0, \text{ if } m = 0; -\frac{\pi}{2}, \text{ if } m < 0$

622. $\displaystyle\int_0^\infty \frac{\cos x\, dx}{x} = \infty$

623. $\displaystyle\int_0^\infty \frac{\tan x\, dx}{x} = \frac{\pi}{2}$

624. $\displaystyle\int_0^\pi \sin ax \cdot \sin bx\, dx = \int_0^\pi \cos ax \cdot \cos bx\, dx = 0, \quad (a \neq b; a, b \text{ integers})$

625. $\displaystyle\int_0^{\pi/a} [\sin(ax)][\cos(ax)]\, dx = \int_0^\pi [\sin(ax)][\cos(ax)]\, dx = 0$

626. $\displaystyle\int_0^\pi [\sin(ax)][\cos(bx)]\, dx = \frac{2a}{a^2 - b^2}, \text{ if } a - b \text{ is odd, or } 0 \text{ if } a - b \text{ is even}$

627. $\displaystyle\int_0^\infty \frac{\sin x \cos mx\, dx}{x}$

$= 0, \text{ if } m < -1 \text{ or } m > 1; \frac{\pi}{4}, \text{ if } m = \pm 1; \frac{\pi}{2}, \text{ if } m^2 < 1$

628. $\displaystyle\int_0^\infty \frac{\sin ax \sin bx}{x^2}\, dx = \frac{\pi a}{2}, \quad (a \leq b)$

629. $\displaystyle\int_0^\pi \sin^2 mx\, dx = \int_0^\pi \cos^2 mx\, dx = \frac{\pi}{2}$

630. $\displaystyle\int_0^\infty \frac{\sin^2(px)}{x^2}\, dx = \frac{\pi p}{2}$

DEFINITE INTEGRALS (Continued)

631. $\displaystyle\int_0^\infty \frac{\sin x}{x^p}\,dx = \frac{\pi}{2\Gamma(p)\sin(p\pi/2)}, \qquad 0 < p < 1$

632. $\displaystyle\int_0^\infty \frac{\cos x}{x^p}\,dx = \frac{\pi}{2\Gamma(p)\cos(p\pi/2)}, \qquad 0 < p < 1$

633. $\displaystyle\int_0^\infty \frac{1 - \cos px}{x^2}\,dx = \frac{\pi p}{2}$

634. $\displaystyle\int_0^\infty \frac{\sin px \cos qx}{x}\,dx = \left\{0,\quad q > p > 0;\quad \frac{\pi}{2},\ \ p > q > 0;\quad \frac{\pi}{4},\ \ p = q > 0\right\}$

635. $\displaystyle\int_0^\infty \frac{\cos(mx)}{x^2 + a^2}\,dx = \frac{\pi}{2|a|}\,e^{-|ma|}$

636. $\displaystyle\int_0^\infty \cos(x^2)\,dx = \int_0^\infty \sin(x^2)\,dx = \frac{1}{2}\sqrt{\frac{\pi}{2}}$

637. $\displaystyle\int_0^\infty \sin ax^n\,dx = \frac{1}{na^{1/n}}\,\Gamma(1/n)\sin\frac{\pi}{2n}, \qquad n > 1$

638. $\displaystyle\int_0^\infty \cos ax^n\,dx = \frac{1}{na^{1/n}}\,\Gamma(1/n)\cos\frac{\pi}{2n}, \qquad n > 1$

639. $\displaystyle\int_0^\infty \frac{\sin x}{\sqrt{x}}\,dx = \int_0^\infty \frac{\cos x}{\sqrt{x}}\,dx = \sqrt{\frac{\pi}{2}}$

640. (a) $\displaystyle\int_0^\infty \frac{\sin^3 x}{x}\,dx = \frac{\pi}{4}$ (b) $\displaystyle\int_0^\infty \frac{\sin^3 x}{x^2}\,dx\ \frac{3}{4}\log 3$

641. $\displaystyle\int_0^\infty \frac{\sin^3 x}{x^3}\,dx = \frac{3\pi}{8}$

642. $\displaystyle\int_0^\infty \frac{\sin^4 x}{x^4}\,dx = \frac{\pi}{3}$

643. $\displaystyle\int_0^{\pi/2} \frac{dx}{1 + a\cos x} = \frac{\cos^{-1} a}{\sqrt{1 - a^2}}, \qquad (a < 1)$

644. $\displaystyle\int_0^\pi \frac{dx}{a + b\cos x} = \frac{\pi}{\sqrt{a^2 - b^2}}, \qquad (a > b \geq 0)$

645. $\displaystyle\int_0^{2\pi} \frac{dx}{1 + a\cos x} = \frac{2\pi}{\sqrt{1 - a^2}}, \qquad (a^2 < 1)$

646. $\displaystyle\int_0^\infty \frac{\cos ax - \cos bx}{x}\,dx = \log\frac{b}{a}$

647. $\displaystyle\int_0^{\pi/2} \frac{dx}{a^2\sin^2 x + b^2\cos^2 x} = \frac{\pi}{2ab}$

DEFINITE INTEGRALS (Continued)

648. $\displaystyle\int_0^{\pi/2} \frac{dx}{(a^2 \sin^2 x + b^2 \cos^2 x)^2} = \frac{\pi(a^2 + b^2)}{4a^3 b^3}$, $(a, b > 0)$

649. $\displaystyle\int_0^{\pi/2} \sin^{n-1} x \cos^{m-1} x \, dx = \frac{1}{2} B\left(\frac{n}{2}, \frac{m}{2}\right)$, m and n positive integers

650. $\displaystyle\int_0^{\pi/2} (\sin^{2n+1} \theta) \, d\theta = \frac{2 \cdot 4 \cdot 6 \ldots (2n)}{1 \cdot 3 \cdot 5 \ldots (2n + 1)}$, $(n = 1, 2, 3 \ldots)$

651. $\displaystyle\int_0^{\pi/2} (\sin^{2n} \theta) \, d\theta = \frac{1 \cdot 3 \cdot 5 \ldots (2n - 1)}{2 \cdot 4 \ldots (2n)}\left(\frac{\pi}{2}\right)$, $(n = 1, 2, 3 \ldots)$

652. $\displaystyle\int_0^{\pi/2} \frac{x}{\sin x} \, dx = 2\left\{\frac{1}{1^2} - \frac{1}{3^2} + \frac{1}{5^2} - \frac{1}{7^2} + \cdots\right\}$

653. $\displaystyle\int_0^{\pi/2} \frac{dx}{1 + \tan^m x} = \frac{\pi}{4}$

654. $\displaystyle\int_0^{\pi/2} \sqrt{\cos \theta} \, d\theta = \frac{(2\pi)^{\frac{1}{2}}}{[\Gamma(\frac{1}{4})]^2}$

655. $\displaystyle\int_0^{\pi/2} (\tan^h \theta) \, d\theta = \frac{\pi}{2 \cos\left(\dfrac{h\pi}{2}\right)}$, $(0 < h < 1)$

656. $\displaystyle\int_0^{\infty} \frac{\tan^{-1}(ax) - \tan^{-1}(bx)}{x} \, dx = \frac{\pi}{2} \log \frac{a}{b}$, $(a, b > 0)$

657. The area enclosed by a curve defined through the equation $x^{\frac{b}{c}} + y^{\frac{b}{c}} = a^{\frac{b}{c}}$ where $a > 0$, c a positive odd integer and b a positive even integer is given by

$$\frac{\left[\Gamma\left(\dfrac{c}{b}\right)\right]^2}{\Gamma\left(\dfrac{2c}{b}\right)} \left(\frac{2ca^2}{b}\right)$$

658. $\displaystyle I = \iiint_R x^{h-1} y^{m-1} z^{n-1} \, dv$, where R denotes the region of space bounded by

the co-ordinate planes and that portion of the surface $\left(\dfrac{x}{a}\right)^p + \left(\dfrac{y}{b}\right)^q + \left(\dfrac{z}{c}\right)^k = 1$,

which lies in the first octant and where $h, m, n, p, q, k, a, b, c$, denote positive real numbers is given by

$$\int_0^a x^{h-1} \, dx \int_0^{b\left[1-\left(\frac{x}{a}\right)^p\right]^{\frac{1}{q}}} y^m \, dy \int_0^{c\left[1-\left(\frac{x}{a}\right)^p-\left(\frac{y}{b}\right)^q\right]^{\frac{1}{k}}} z^{n-1} \, dz$$

$$= \frac{a^h b^m c^n}{pqk} \frac{\Gamma\left(\dfrac{h}{p}\right)\Gamma\left(\dfrac{m}{q}\right)\Gamma\left(\dfrac{n}{k}\right)}{\Gamma\left(\dfrac{h}{p} + \dfrac{m}{q} + \dfrac{n}{k} + 1\right)}$$

DEFINITE INTEGRALS (Continued)

659. $\displaystyle\int_0^\infty e^{-ax}\,dx = \frac{1}{a}, \qquad (a > 0)$

660. $\displaystyle\int_0^\infty \frac{e^{-ax} - e^{-bx}}{x}\,dx = \log\frac{b}{a}, \qquad (a, b > 0)$

661. $\displaystyle\int_0^\infty x^n e^{-ax}\,dx = \begin{cases} \dfrac{\Gamma(n+1)}{a^{n+1}}, & (n > -1, a > 0) \\[3mm] \qquad\text{or} \\[3mm] \dfrac{n!}{a^{n+1}}, & (a > 0, n \text{ positive integer}) \end{cases}$

662. $\displaystyle\int_0^\infty x^n \exp(-ax^p)\,dx = \frac{\Gamma(k)}{pa^k}, \qquad \left(n > -1, p > 0, a > 0, k = \frac{n+1}{p}\right)$

663. $\displaystyle\int_0^\infty e^{-a^2x^2}\,dx = \frac{1}{2a}\sqrt{\pi} = \frac{1}{2a}\Gamma\left(\frac{1}{2}\right), \qquad (a > 0)$

663a. $\displaystyle\int_0^b e^{-ax^2}\,dx = \frac{1}{2}\sqrt{\frac{\pi}{a}}\ \operatorname{erf}(b\sqrt{a}) \qquad \text{Error Function (see page 442)}$

663b. $\displaystyle\int_0^\infty e^{-ax^2}\,dx \overset{=}{=} \frac{1}{2}\sqrt{\frac{\pi}{a}}\ \operatorname{erf}(b\sqrt{a}) \qquad \text{Complimentary Error Function (see page 442)}$

664. $\displaystyle\int_0^\infty x e^{-x^2}\,dx = \tfrac{1}{2}$

665. $\displaystyle\int_0^\infty x^2 e^{-x^2}\,dx = \frac{\sqrt{\pi}}{4}$

666. $\displaystyle\int_0^\infty x^{2n} e^{-ax^2}\,dx = \frac{1 \cdot 3 \cdot 5 \ldots (2n-1)}{2^{n+1}a^n}\sqrt{\frac{\pi}{a}}$

667. $\displaystyle\int_0^\infty x^{2n+1} e^{-ax^2}\,dx = \frac{n!}{2a^{n+1}}, \qquad (a > 0)$

668. $\displaystyle\int_0^1 x^m e^{-ax}\,dx = \frac{m!}{a^{m+1}}\left[1 - e^{-a}\sum_{r=0}^m \frac{a^r}{r!}\right]$

669. $\displaystyle\int_0^\infty e^{\left(-x^2 - \frac{a^2}{x^2}\right)}\,dx = \frac{e^{-2a}\sqrt{\pi}}{2}, \qquad (a \geq 0)$

670. $\displaystyle\int_0^\infty e^{-nx}\sqrt{x}\,dx = \frac{1}{2n}\sqrt{\frac{\pi}{n}}$

671. $\displaystyle\int_0^\infty \frac{e^{-nx}}{\sqrt{x}}\,dx = \sqrt{\frac{\pi}{n}}$

DEFINITE INTEGRALS (Continued)

672. $\displaystyle\int_0^\infty e^{-ax}(\cos mx)\,dx = \frac{a}{a^2 + m^2}, \qquad (a > 0)$

673. $\displaystyle\int_0^\infty e^{-ax}(\sin mx)\,dx = \frac{m}{a^2 + m^2}, \qquad (a > 0)$

674. $\displaystyle\int_0^\infty x\,e^{-ax}[\sin (bx)]\,dx = \frac{2ab}{(a^2 + b^2)^2}, \qquad (a > 0)$

675. $\displaystyle\int_0^\infty x\,e^{-ax}[\cos (bx)]\,dx = \frac{a^2 - b^2}{(a^2 + b^2)^2}, \qquad (a > 0)$

676. $\displaystyle\int_0^\infty x^n\,e^{-ax}[\sin (bx)]\,dx = \frac{n![(a + ib)^{n+1} - (a - ib)^{n+1}]}{2i(a^2 + b^2)^{n+1}}, \qquad (i^2 = -1, a > 0)$

677. $\displaystyle\int_0^\infty x^n\,e^{-ax}[\cos (bx)]\,dx = \frac{n![(a - ib)^{n+1} + (a + ib)^{n+1}]}{2(a^2 + b^2)^{n+1}}, \qquad (i^2 = -1, a > 0)$

678. $\displaystyle\int_0^\infty \frac{e^{-ax}\sin x}{x}\,dx = \cot^{-1} a, \qquad (a > 0)$

679. $\displaystyle\int_0^\infty e^{-a^2x^2}\cos bx\,dx = \frac{\sqrt{\pi}}{2a}\exp\left(-\frac{b^2}{4a^2}\right), \qquad (ab \neq 0)$

680. $\displaystyle\int_0^\infty e^{-t\cos\phi}\,t^{b-1}\sin (t\sin\phi)\,dt = [\Gamma(b)]\sin (b\phi), \qquad \left(b > 0, -\frac{\pi}{2} < \phi < \frac{\pi}{2}\right)$

681. $\displaystyle\int_0^\infty e^{-t\cos\phi}\,t^{b-1}[\cos (t\sin\phi)]\,dt = [\Gamma(b)]\cos (b\phi), \qquad \left(b > 0, -\frac{\pi}{2} < \phi < \frac{\pi}{2}\right)$

682. $\displaystyle\int_0^\infty t^{b-1}\cos t\,dt = [\Gamma(b)]\cos\left(\frac{b\pi}{2}\right), \qquad (0 < b < 1)$

683. $\displaystyle\int_0^\infty t^{b-1}(\sin t)\,dt = [\Gamma(b)]\sin\left(\frac{b\pi}{2}\right), \qquad (0 < b < 1)$

684. $\displaystyle\int_0^1 (\log x)^n\,dx = (-1)^n \cdot n!$

685. $\displaystyle\int_0^1 \left(\log\frac{1}{x}\right)^{\frac{1}{2}}\,dx = \frac{\sqrt{\pi}}{2}$

686. $\displaystyle\int_0^1 \left(\log\frac{1}{x}\right)^{-\frac{1}{2}}\,dx = \sqrt{\pi}$

687. $\displaystyle\int_0^1 \left(\log\frac{1}{x}\right)^n\,dx = n!$

688. $\displaystyle\int_0^1 x\log (1 - x)\,dx = -\tfrac{3}{4}$

689. $\displaystyle\int_0^1 x\log (1 + x)\,dx = \tfrac{1}{4}$

DEFINITE INTEGRALS (Continued)

690. $\displaystyle\int_0^1 x^m(\log x)^n\,dx = \frac{(-1)^n n!}{(m+1)^{n+1}}, \qquad m > -1, n = 0, 1, 2, \ldots$

If $n \neq 0, 1, 2, \ldots$ replace $n!$ by $\Gamma(n+1)$.

691. $\displaystyle\int_0^1 \frac{\log x}{1+x}\,dx = -\frac{\pi^2}{12}$

692. $\displaystyle\int_0^1 \frac{\log x}{1-x}\,dx = -\frac{\pi^2}{6}$

693. $\displaystyle\int_0^1 \frac{\log(1+x)}{x}\,dx = \frac{\pi^2}{12}$

694. $\displaystyle\int_0^1 \frac{\log(1-x)}{x}\,dx = -\frac{\pi^2}{6}$

695. $\displaystyle\int_0^1 (\log x)[\log(1+x)]\,dx = 2 - 2\log 2 - \frac{\pi^2}{12}$

696. $\displaystyle\int_0^1 (\log x)[\log(1-x)]\,dx = 2 - \frac{\pi^2}{6}$

697. $\displaystyle\int_0^1 \frac{\log x}{1-x^2}\,dx = -\frac{\pi^2}{8}$

698. $\displaystyle\int_0^1 \log\left(\frac{1+x}{1-x}\right) \cdot \frac{dx}{x} = \frac{\pi^2}{4}$

699. $\displaystyle\int_0^1 \frac{\log x\,dx}{\sqrt{1-x^2}} = -\frac{\pi}{2}\log 2$

700. $\displaystyle\int_0^1 x^m \left[\log\left(\frac{1}{x}\right)\right]^n dx = \frac{\Gamma(n+1)}{(m+1)^{n+1}}, \qquad \text{if } m+1 > 0, n+1 > 0$

701. $\displaystyle\int_0^1 \frac{(x^p - x^q)\,dx}{\log x} = \log\left(\frac{p+1}{q+1}\right), \qquad (p+1 > 0, q+1 > 0)$

702. $\displaystyle\int_0^1 \frac{dx}{\sqrt{\log\left(\frac{1}{x}\right)}} = \sqrt{\pi}\ , \text{(same as integral 686)}$

703. $\displaystyle\int_0^\infty \log\left(\frac{e^x + 1}{e^x - 1}\right)dx = \frac{\pi^2}{4}$

704. $\displaystyle\int_0^{\pi/2} (\log \sin x)\,dx = \int_0^{\pi/2} \log \cos x\,dx = -\frac{\pi}{2}\log 2$

705. $\displaystyle\int_0^{\pi/2} (\log \sec x)\,dx = \int_0^{\pi/2} \log \csc x\,dx = \frac{\pi}{2}\log 2$

706. $\displaystyle\int_0^\pi x(\log \sin x)\,dx = -\frac{\pi^2}{2}\log 2$

DEFINITE INTEGRALS (Continued)

707. $\displaystyle\int_0^{\pi/2} (\sin x)(\log \sin x)\, dx = \log 2 - 1$

708. $\displaystyle\int_0^{\pi/2} (\log \tan x)\, dx = 0$

709. $\displaystyle\int_0^{\pi} \log (a \pm b \cos x)\, dx = \pi \log \left(\frac{a + \sqrt{a^2 - b^2}}{2} \right),\qquad (a \geq b)$

710. $\displaystyle\int_0^{\pi} \log (a^2 - 2ab \cos x + b^2)\, dx = \begin{cases} 2\pi \log a, & a \geq b > 0 \\ 2\pi \log b, & b \geq a > 0 \end{cases}$

711. $\displaystyle\int_0^{\infty} \frac{\sin ax}{\sinh bx}\, dx = \frac{\pi}{2b} \tanh \frac{a\pi}{2b}$

712. $\displaystyle\int_0^{\infty} \frac{\cos ax}{\cosh bx}\, dx = \frac{\pi}{2b} \operatorname{sech} \frac{a\pi}{2b}$

713. $\displaystyle\int_0^{\infty} \frac{dx}{\cosh ax} = \frac{\pi}{2a}$

714. $\displaystyle\int_0^{\infty} \frac{x\, dx}{\sinh ax} = \frac{\pi^2}{4a^2}$

715. $\displaystyle\int_0^{\infty} e^{-ax}(\cosh bx)\, dx = \frac{a}{a^2 - b^2},\qquad (0 \leq |b| < a)$

716. $\displaystyle\int_0^{\infty} e^{-ax}(\sinh bx)\, dx = \frac{b}{a^2 - b^2},\qquad (0 \leq |b| < a)$

717. $\displaystyle\int_0^{\infty} \frac{\sinh ax}{e^{bx} + 1}\, dx = \frac{\pi}{2b} \csc \frac{a\pi}{b} - \frac{1}{2a}$

718. $\displaystyle\int_0^{\infty} \frac{\sinh ax}{e^{bx} - 1}\, dx = \frac{1}{2a} - \frac{\pi}{2b} \cot \frac{a\pi}{b}$

719. $\displaystyle\int_0^{\pi/2} \frac{dx}{\sqrt{1 - k^2 \sin^2 x}} = \frac{\pi}{2} \left[1 + \left(\frac{1}{2}\right)^2 k^2 + \left(\frac{1 \cdot 3}{2 \cdot 4}\right)^2 k^4 \right.$

$$\left. + \left(\frac{1 \cdot 3 \cdot 5}{2 \cdot 4 \cdot 6}\right)^2 k^6 + \cdots \right], \text{if } k^2 < 1$$

719a. $\displaystyle\int_0^{\frac{\pi}{2}} \frac{dx}{(1 - k^2 \sin^2 x)^{3/2}} = \frac{\pi}{2} \left[1 + \left(\frac{1}{2}\right)^2 \cdot 3k^2 + \left(\frac{1 \cdot 3}{2 \cdot 4}\right)^2 \cdot 5k^4 + \right.$

$$\left. \left(\frac{1 \cdot 3 \cdot 5}{2 \cdot 4 \cdot 6}\right)^2 \cdot 7k^6 + \cdots \right], \text{if } k^2 < 1$$

DEFINITE INTEGRALS (Continued)

720. $\displaystyle\int_0^{\pi/2} \sqrt{1 - k^2 \sin^2 x}\, dx = \frac{\pi}{2}\left[1 - \left(\frac{1}{2}\right)^2 k^2 - \left(\frac{1\cdot 3}{2\cdot 4}\right)^2 \frac{k^4}{3} \right.$

$$\left. - \left(\frac{1\cdot 3\cdot 5}{2\cdot 4\cdot 6}\right)^2 \frac{k^6}{5} - \cdots \right], \text{ if } k^2 < 1$$

721. $\displaystyle\int_0^\infty e^{-x} \log x \, dx = -\gamma = -0.5772157\ldots$

722. $\displaystyle\int_0^\infty e^{-x^2} \log x \, dx = -\frac{\sqrt{\pi}}{4}(\gamma + 2\log 2)$

723. $\displaystyle\int_0^\infty \left(\frac{1}{1 - e^{-x}} - \frac{1}{x}\right) e^{-x}\, dx = \gamma = 0.5772157\ldots$ [Euler's Constant]

724. $\displaystyle\int_0^\infty \frac{1}{x}\left(\frac{1}{1 + x} - e^{-x}\right) dx = \gamma = 0.5772157\ldots$

For n even:

725. $\displaystyle\int\cos^n x\, dx \;=\; \frac{1}{2}n^{-1} \sum_{k=0}^{\frac{n}{2}-1} \binom{n}{k}\, \frac{\sin(n-2k)\,x}{(n-2k)} \;+\; \frac{1}{2^n}\binom{n}{\frac{n}{2}}\, x$

726. $\displaystyle\int\sin^n x\, dx \;=\; \frac{1}{2^{n-1}} \sum_{k=0}^{\frac{n}{2}-1} \binom{n}{k}\, \frac{\sin\left[(n-2k)\left(\frac{\pi}{2} - x\right)\right]}{2k - n} \;+\; \frac{1}{2^n}\binom{n}{\frac{n}{2}}\, x$

For n odd:

727. $\displaystyle\int\cos^n x\, dx \;=\; \frac{1}{2^{n-1}} \sum_{k=0}^{\frac{n-1}{2}} \binom{n}{k}\, \frac{\sin(n-2k)\,x}{(n-2k)}$

728. $\displaystyle\int\sin^n x\, dx \;=\; \frac{1}{2^{n-1}} \sum_{k=0}^{\frac{n-1}{2}} \binom{n}{k}\, \frac{\sin\left[n-2k)\left(\frac{\pi}{2} - x\right)\right]}{2k - n}$

SERIES EXPANSION

The expression in parentheses following certain of the series indicates the region of convergence. If not otherwise indicated it is to be understood that the series converges for all finite values of x.

BINOMIAL

$$(x + y)^n = x^n + nx^{n-1}y + \frac{n(n-1)}{2!} x^{n-2}y^2$$

$$+ \frac{n(n-1)(n-2)}{3!} x^{n-3}y^3 + \cdots \quad (y^2 < x^2)$$

$$(1 \pm x)^n = 1 \pm nx + \frac{n(n-1)x^2}{2!} \pm \frac{n(n-1)(n-2)x^3}{3!} + \cdots \text{ etc.} \quad (x^2 < 1)$$

$$(1 \pm x)^{-n} = 1 \mp nx + \frac{n(n+1)x^2}{2!} \mp \frac{n(n+1)(n+2)x^3}{3!} + \cdots \text{ etc.} \quad (x^2 < 1)$$

$$(1 \pm x)^{-1} = 1 \mp x + x^2 \mp x^3 + x^4 \mp x^5 + \cdots \qquad (x^2 < 1)$$

$$(1 \pm x)^{-2} = 1 \mp 2x + 3x^2 \mp 4x^3 + 5x^4 \mp 6x^5 + \cdots \qquad (x^2 < 1)$$

REVERSION OF SERIES

Let a series be represented by

$$y = a_1 x + a_2 x^2 + a_3 x^3 + a_4 x^4 + a_5 x^5 + a_6 x^6 + \cdots \quad (a_1 \neq 0)$$

to find the coefficients of the series

$$x = A_1 y + A_2 y^2 + A_3 y^3 + A_4 y^4 + \cdots$$

$$A_1 = \frac{1}{a_1} \qquad A_2 = -\frac{a_2}{a_1^3} \qquad A_3 = \frac{1}{a_1^5}(2a_2^2 - a_1 a_3)$$

$$A_4 = \frac{1}{a_1^7}(5a_1 a_2 a_3 - a_1^2 a_4 - 5a_2^3)$$

$$A_5 = \frac{1}{a_1^9}(6a_1^2 a_2 a_4 + 3a_1^2 a_3^2 + 14a_2^4 - a_1^3 a_5 - 21a_1 a_2^2 a_3)$$

$$A_6 = \frac{1}{a_1^{11}}(7a_1^3 a_2 a_5 + 7a_1^3 a_3 a_4 + 84a_1 a_2^3 a_3 - a_1^4 a_6 - 28a_1^2 a_2^2 a_4 - 28a_1^2 a_2 a_3^2 - 42a_2^5)$$

$$A_7 = \frac{1}{a_1^{13}}(8a_1^4 a_2 a_6 + 8a_1^4 a_3 a_5 + 4a_1^4 a_4^2 + 120a_1^2 a_2^3 a_4$$

$$+ 180a_1^2 a_2^2 a_3^2 + 132a_2^6 - a_1^5 a_7$$
$$- 36a_1^3 a_2^2 a_5 - 72a_1^3 a_2 a_3 a_4 - 12a_1^3 a_3^3 - 330a_1 a_2^4 a_3)$$

TAYLOR

1. $f(x) = f(a) + (x - a)f'(a) + \dfrac{(x-a)^2}{2!} f''(a) + \dfrac{(x-a)^3}{3!} f'''(a)$

$$+ \cdots + \frac{(x-a)^n}{n!} f^{(n)}(a) + \cdots \text{ (Taylor's Series)}$$

Series

(Increment form)

2. $f(x + h) = f(x) + hf'(x) + \dfrac{h^2}{2!} f''(x) + \dfrac{h^3}{3!} f'''(x) + \cdots$

$$= f(h) + xf'(h) + \frac{x^2}{2!} f''(h) + \frac{x^3}{3!} f'''(h) + \cdots$$

3. If $f(x)$ is a function possessing derivatives of all orders throughout the interval $a \leq x \leq b$, then there is a value X, with $a < X < b$, such that

$$f(b) = f(a) + (b - a)f'(a) + \frac{(b - a)^2}{2!} f''(a) + \cdots$$

$$+ \frac{(b - a)^{n-1}}{(n - 1)!} f^{(n-1)}(a) + \frac{(b - a)^n}{n!} f^{(n)}(X)$$

$$f(a + h) = f(a) + hf'(a) + \frac{h^2}{2!} f''(a) + \cdots + \frac{h^{n-1}}{(n - 1)!} f^{(n-1)}(a)$$

$$+ \frac{h^n}{n!} f^{(n)}(a + \theta h), \quad b = a + h, 0 < \theta < 1.$$

or

$$f(x) = f(a) + (x - a)f'(a) + \frac{(x - a)^2}{2!} f''(a) + \cdots + (x - a)^{n-1} \frac{f^{(n-1)}(a)}{(n - 1)!} + R_n,$$

where

$$R_n = \frac{f^{(n)}[a + \theta \cdot (x - a)]}{n!} (x - a)^n, \quad 0 < \theta < 1.$$

The above forms are known as Taylor's series with the remainder term.

4. *Taylor's series for a function of two variables*

If $\left(h \dfrac{\partial}{\partial x} + k \dfrac{\partial}{\partial y} \right) f(x, y) = h \dfrac{\partial f(x, y)}{\partial x} + k \dfrac{\partial f(x, y)}{\partial y}$;

$$\left(h \frac{\partial}{\partial x} + k \frac{\partial}{\partial y} \right)^2 f(x, y) = h^2 \frac{\partial^2 f(x, y)}{\partial x^2} + 2hk \frac{\partial^2 f(x, y)}{\partial x \, \partial y} + k^2 \frac{\partial^2 f(x, y)}{\partial y^2}$$

etc., and if $\left(h \dfrac{\partial}{\partial x} + k \dfrac{\partial}{\partial y} \right)^n f(x, y) \Big|_{\substack{x = a \\ y = b}}$ with the bar and subscripts means that after differentiation we are to replace x by a and y by b,

$$f(a + h, b + k) = f(a, b) + \left(h \frac{\partial}{\partial x} + k \frac{\partial}{\partial y} \right) f(x, y) \Big|_{\substack{x = a \\ y = b}} + \cdots$$

$$+ \frac{1}{n!} \left(h \frac{\partial}{\partial x} + k \frac{\partial}{\partial y} \right)^n f(x, y) \Big|_{\substack{x = a \\ y = b}} + \cdots$$

MACLAURIN

$$f(x) = f(0) + xf'(0) + \frac{x^2}{2!} f''(0) + \frac{x^3}{3!} f'''(0) + \cdots + x^{n-1} \frac{f^{(n-1)}(0)}{(n - 1)!} + R_n,$$

where

$$R_n = \frac{x^n f^{(n)}(\theta x)}{n!}, \quad 0 < \theta < 1.$$

Series

EXPONENTIAL

$$e = 1 + \frac{1}{1!} + \frac{1}{2!} + \frac{1}{3!} + \frac{1}{4!} + \cdots$$

$$e^x = 1 + x + \frac{x^2}{2!} + \frac{x^3}{3!} + \frac{x^4}{4!} + \cdots \qquad \text{(all real values of } x\text{)}$$

$$a^x = 1 + x \log_e a + \frac{(x \log_e a)^2}{2!} + \frac{(x \log_e a)^3}{3!} + \cdots$$

$$e^x = e^a \left[1 + (x - a) + \frac{(x - a)^2}{2!} + \frac{(x - a)^3}{3!} + \cdots \right]$$

LOGARITHMIC

$$\log_e x = \frac{x - 1}{x} + \frac{1}{2} \left(\frac{x - 1}{x} \right)^2 + \frac{1}{3} \left(\frac{x - 1}{x} \right)^3 + \cdots \qquad (x > \tfrac{1}{2})$$

$$\log_e x = (x - 1) - \tfrac{1}{2}(x - 1)^2 + \tfrac{1}{3}(x - 1)^3 - \cdots \qquad (2 \geq x > 0)$$

$$\log_e x = 2 \left[\frac{x - 1}{x + 1} + \frac{1}{3} \left(\frac{x - 1}{x + 1} \right)^3 + \frac{1}{5} \left(\frac{x - 1}{x + 1} \right)^5 + \cdots \right] \qquad (x > 0)$$

$$\log_e (1 + x) = x - \tfrac{1}{2}x^2 + \tfrac{1}{3}x^3 - \tfrac{1}{4}x^4 + \cdots \qquad (-1 < x \leq 1)$$

$$\log_e (n + 1) - \log_e (n - 1) = 2 \left[\frac{1}{n} + \frac{1}{3n^3} + \frac{1}{5n^5} + \cdots \right]$$

$$\log_e (a + x) = \log_e a + 2 \left[\frac{x}{2a + x} + \frac{1}{3} \left(\frac{x}{2a + x} \right)^3 + \frac{1}{5} \left(\frac{x}{2a + x} \right)^5 + \cdots \right]$$

$$(a > 0,\ -a < x < +\infty)$$

$$\log_e \frac{1 + x}{1 - x} = 2 \left[x + \frac{x^3}{3} + \frac{x^5}{5} + \cdots + \frac{x^{2n-1}}{2n - 1} + \cdots \right], \qquad -1 < x < 1$$

$$\log_e x = \log_e a + \frac{(x - a)}{a} - \frac{(x - a)^2}{2a^2} + \frac{(x - a)^3}{3a^3} - + \cdots, \qquad 0 < x \leq 2a$$

TRIGONOMETRIC

$$\sin x = x - \frac{x^3}{3!} + \frac{x^5}{5!} - \frac{x^7}{7!} + \cdots \qquad \text{(all real values of } x\text{)}$$

$$\cos x = 1 - \frac{x^2}{2!} + \frac{x^4}{4!} - \frac{x^6}{6!} + \cdots \qquad \text{(all real values of } x\text{)}$$

$$\tan x = x + \frac{x^3}{3} + \frac{2x^5}{15} + \frac{17x^7}{315} + \frac{62x^9}{2835} + \cdots + \frac{(-1)^{n-1} 2^{2n} (2^{2n} - 1) B_{2n}}{(2n)!} x^{2n-1} + \cdots,$$

$$\left[x^2 < \frac{\pi^2}{4}, \text{ and } B_n \text{ represents the } n\text{'th Bernoulli number.} \right]$$

$$\cot x = \frac{1}{x} - \frac{x}{3} - \frac{x^3}{45} - \frac{2x^5}{945} - \frac{x^7}{4725} - \cdots - \frac{(-1)^{n+1} 2^{2n}}{(2n)!} B_{2n} x^{2n-1} - \cdots,$$

$$[x^2 < \pi^2, \text{ and } B_n \text{ represents the } n\text{'th Bernoulli number.}]$$

Series

$$\sec x = 1 + \frac{x^2}{2} + \frac{5}{24} x^4 + \frac{61}{720} x^6 + \frac{277}{8064} x^8 + \cdots + \frac{(-1)^n}{(2n)!} E_{2n} x^{2n} + \cdots,$$

$$\left[x^2 < \frac{\pi^2}{4}, \text{ and } E_n \text{ represents the } n\text{'th Euler number.} \right]$$

$$\csc x = \frac{1}{x} + \frac{x}{6} + \frac{7}{360} x^3 + \frac{31}{15,120} x^5 + \frac{127}{604,800} x^7 + \cdots$$

$$+ \frac{(-1)^{n+1} 2 (2^{2n-1} - 1)}{(2n)!} B_{2n} x^{2n-1} + \cdots,$$

$$[x^2 < \pi^2, \text{ and } B_n \text{ represents } n\text{'th Bernoulli number.}]$$

$$\sin x = x \left(1 - \frac{x^2}{\pi^2} \right) \left(1 - \frac{x^2}{2^2 \pi^2} \right) \left(1 - \frac{x^2}{3^2 \pi^2} \right) \cdots \qquad (x^2 < \infty)$$

$$\cos x = \left(1 - \frac{4x^2}{\pi^2} \right) \left(1 - \frac{4x^2}{3^2 \pi^2} \right) \left(1 - \frac{4x^2}{5^2 \pi^2} \right) \cdots \qquad (x^2 < \infty)$$

$$\sin^{-1} x = x + \frac{x^3}{2 \cdot 3} + \frac{1 \cdot 3}{2 \cdot 4 \cdot 5} x^5 + \frac{1 \cdot 3 \cdot 5}{2 \cdot 4 \cdot 6 \cdot 7} x^7 + \cdots \quad \left(x^2 < 1, -\frac{\pi}{2} < \sin^{-1} x < \frac{\pi}{2} \right)$$

$$\cos^{-1} x = \frac{\pi}{2} - \left(x + \frac{x^3}{2 \cdot 3} + \frac{1 \cdot 3}{2 \cdot 4 \cdot 5} x^5 + \frac{1 \cdot 3 \cdot 5 x^7}{2 \cdot 4 \cdot 6 \cdot 7} + \cdots \right) \quad (x^2 < 1, 0 < \cos^{-1} x < \pi)$$

$$\tan^{-1} x = x - \frac{x^3}{3} + \frac{x^5}{5} - \frac{x^7}{7} + \cdots \qquad (x^2 < 1)$$

$$\tan^{-1} x = \frac{\pi}{2} - \frac{1}{x} + \frac{1}{3x^3} - \frac{1}{5x^5} + \frac{1}{7x^7} - \cdots \qquad (x > 1)$$

$$\tan^{-1} x = -\frac{\pi}{2} - \frac{1}{x} + \frac{1}{3x^3} - \frac{1}{5x^5} + \frac{1}{7x^7} - \cdots \qquad (x < -1)$$

$$\cot^{-1} x = \frac{\pi}{2} - x + \frac{x^3}{3} - \frac{x^5}{5} + \frac{x^7}{7} - \cdots \qquad (x^2 < 1)$$

$$\log_e \sin x = \log_e x - \frac{x^2}{6} - \frac{x^4}{180} - \frac{x^6}{2835} - \cdots \qquad (x^2 < \pi^2)$$

$$\log_e \cos x = -\frac{x^2}{2} - \frac{x^4}{12} - \frac{x^6}{45} - \frac{17x^8}{2520} - \cdots \qquad \left(x^2 < \frac{\pi^2}{4} \right)$$

$$\log_e \tan x = \log_e x + \frac{x^2}{3} + \frac{7x^4}{90} + \frac{62x^6}{2835} + \cdots \qquad \left(x^2 < \frac{\pi^2}{4} \right)$$

$$e^{\sin x} = 1 + x + \frac{x^2}{2!} - \frac{3x^4}{4!} - \frac{8x^5}{5!} - \frac{3x^6}{6!} + \frac{56x^7}{7!} + \cdots$$

$$e^{\cos x} = e \left(1 - \frac{x^2}{2!} + \frac{4x^4}{4!} - \frac{31x^6}{6!} + \cdots \right)$$

$$e^{\tan x} = 1 + x + \frac{x^2}{2!} + \frac{3x^3}{3!} + \frac{9x^4}{4!} + \frac{37x^5}{5!} + \cdots \qquad \left(x^2 < \frac{\pi^2}{4} \right)$$

$$\sin x = \sin a + (x - a) \cos a - \frac{(x - a)^2}{2!} \sin a$$

$$- \frac{(x - a)^3}{3!} \cos a + \frac{(x - a)^4}{4!} \sin a + \cdots$$

VECTOR ANALYSIS

Definitions

Any quantity which is completely determined by its magnitude is called a *scalar*. Examples of such are mass, density, temperature, etc. Any quantity which is completely determined by its magnitude and direction is called a *vector*. Examples of such are velocity, acceleration, force, etc. A vector quantity is represented by a directed line segment, the length of which represents the magnitude of the vector. A vector quantity is usually represented by a boldfaced letter such as $\mathbf{V}$. Two vectors $\mathbf{V}_1$ and $\mathbf{V}_2$ are equal to one another if they have equal magnitudes and are acting in the same directions. A negative vector, written as $-\mathbf{V}$, is one which acts in the opposite direction to $\mathbf{V}$, but is of equal magnitude to it. If we represent the magnitude of $\mathbf{V}$ by v, we write $|\mathbf{V}| = v$. A vector parallel to $\mathbf{V}$, but equal to the reciprocal of its magnitude is written as $\mathbf{V}^{-1}$ or as $\dfrac{1}{\mathbf{V}}$.

The *unit vector* $\dfrac{\mathbf{V}}{v}$ $(v \neq 0)$ is that vector which has the same direction as $\mathbf{V}$, but has a magnitude of unity (sometimes represented as $\mathbf{V}_0$ or $\hat{\mathbf{v}}$).

Vector Algebra

The vector sum of $\mathbf{V}_1$ and $\mathbf{V}_2$ is represented by $\mathbf{V}_1 + \mathbf{V}_2$. The vector sum of $\mathbf{V}_1$ and $-\mathbf{V}_2$, or the difference of the vector $\mathbf{V}_2$ from $\mathbf{V}_1$ is represented by $\mathbf{V}_1 - \mathbf{V}_2$.

If r is a scalar, then $r\mathbf{V} = \mathbf{V}r$, and represents a vector r times the magnitude of $\mathbf{V}$, in the same direction as $\mathbf{V}$ if r is positive, and in the opposite direction if r is negative. If r and s are scalars, $\mathbf{V}_1$, $\mathbf{V}_2$, $\mathbf{V}_3$, vectors, then the following rules of scalars and vectors hold:

$$\mathbf{V}_1 + \mathbf{V}_2 = \mathbf{V}_2 + \mathbf{V}_1$$

$$(r + s)\mathbf{V}_1 = r\mathbf{V}_1 + s\mathbf{V}_1; \qquad r(\mathbf{V}_1 + \mathbf{V}_2) = r\mathbf{V}_1 + r\mathbf{V}_2$$

$$\mathbf{V}_1 + (\mathbf{V}_2 + \mathbf{V}_3) = (\mathbf{V}_1 + \mathbf{V}_2) + \mathbf{V}_3 = \mathbf{V}_1 + \mathbf{V}_2 + \mathbf{V}_3$$

Vectors in Space

A plane is described by two distinct vectors $\mathbf{V}_1$ and $\mathbf{V}_2$. Should these vectors not intersect each other, then one is displaced parallel to itself until they do (fig. 1.) Any other vector $\mathbf{V}$ lying in this plane is given by

$$\mathbf{V} = r\mathbf{V}_1 + s\mathbf{V}_2$$

VECTOR ANALYSIS (Continued)

A *position vector* specifies the position in space of a point relative to a fixed origin. If therefore V_1 and V_2 are the position vectors of the points A and B, relative to the origin O, then any point P on the line AB has a position vector V given by

$$V = rV_1 + (1 - r)V_2$$

The scalar "r" can be taken as the parametric representation of P since $r = 0$ implies $P = B$ and $r = 1$ implies $P = A$. (fig. 2). If P divides the line AB in the ratio $r:s$ then

$$V = \left(\frac{r}{r + s}\right) V_1 + \left(\frac{s}{r + s}\right) V_2$$

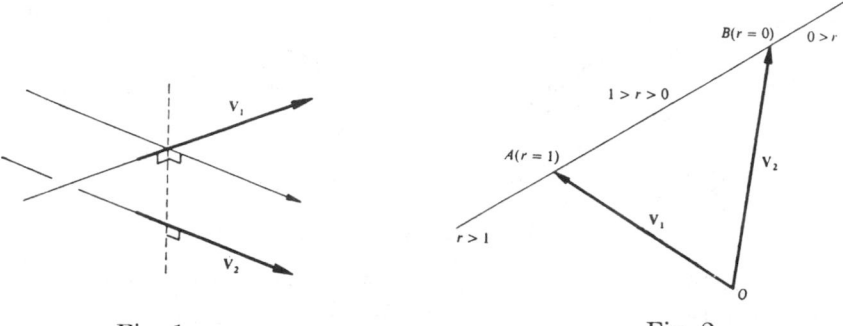

Fig. 1.   Fig. 2.

The vectors V_1, V_2, V_3, . . . , V_n are said to be *linearly dependent* if there exist scalars r_1, r_2, r_3, . . . , r_n, not all zero, such that

$$r_1V_1 + r_2V_2 + \cdots + r_nV_n = 0$$

A vector V is linearly dependent upon the set of vectors V_1, V_2, V_3, . . . , V_n if

$$V = r_1V_1 + r_2V_2 + r_3V_3 + \cdots + r_nV_n$$

Three vectors are linearly dependent if and only if they are co-planar.

All points in space can be uniquely determined by linear dependence upon three *base vectors* i.e. three vectors any one of which is linearly independent of the other two. The simplest set of base vectors are the unit vectors along the coordinate Ox, Oy and Oz axes. These are usually designated by i, j and k respectively.

If V is a vector in space, and a, b and c are the respective magnitudes of the projections of the vector along the axes then

$$V = ai + bj + ck$$

and
$$v = \sqrt{a^2 + b^2 + c^2}$$

and the direction cosines of V are

$$\cos \alpha = a/v, \quad \cos \beta = b/v, \quad \cos \gamma = c/v.$$

The law of addition yields

$$V_1 + V_2 = (a_1 + a_2)i + (b_1 + b_2)j + (c_1 + c_2)k$$

The Scalar, Dot, or Inner Product of Two Vectors V_1 and V_2

This product is represented as $V_1 \cdot V_2$ and is defined to be equal to $v_1 v_2 \cos \theta$, where θ is the angle from V_1 to V_2, i.e.,

$$V_1 \cdot V_2 = v_1 v_2 \cos \theta$$

The following rules apply for this product:

$$V_1 \cdot V_2 = a_1 a_2 + b_1 b_2 + c_1 c_2 = V_2 \cdot V_1$$

It should be noted that this verifies that scalar multiplication is commutative.

$$(V_1 + V_2) \cdot V_3 = V_1 \cdot V_3 + V_2 \cdot V_3$$
$$V_1 \cdot (V_2 + V_3) = V_1 \cdot V_2 + V_1 \cdot V_3$$

If V_1 is perpendicular to V_2 then $V_1 \cdot V_2 = 0$, and if V_1 is parallel to V_2 then $V_1 \cdot V_2 = v_1 v_2 = rw_1{}^2$

In particular

$$i \cdot i = j \cdot j = k \cdot k = 1,$$

and

$$i \cdot j = j \cdot k = k \cdot i = 0$$

The Vector or Cross Product of Vectors V_1 and V_2

This product is represented as $V_1 \times V_2$ and is defined to be equal to $v_1 v_2 (\sin \theta) 1$, where θ is the angle from V_1 to V_2 and 1 is a unit vector perpendicular to the plane of V_1 and V_2 and so directed that a right-handed screw driven in the direction of 1 would carry V_1 into V_2, i.e.,

$$V_1 \times V_2 = v_1 v_2 (\sin \theta) 1$$

and

$$\tan \theta = \frac{|V_1 \times V_2|}{V_1 \cdot V_2}$$

The following rules apply for vector products:

$$V_1 \times V_2 = -V_2 \times V_1$$
$$V_1 \times (V_2 + V_3) = V_1 \times V_2 + V_1 \times V_3$$
$$(V_1 + V_2) \times V_3 = V_1 \times V_3 + V_2 \times V_3$$
$$V_1 \times (V_2 \times V_3) = V_2(V_3 \cdot V_1) - V_3(V_1 \cdot V_2)$$
$$i \times i = j \times j = k \times k = 0.1 \text{ (zero vector)}$$
$$= 0$$

$$i \times j = k, \qquad j \times k = i, \qquad k \times i = j$$

If $V_1 = a_1 i + b_1 j + c_1 k$, $\qquad V_2 = a_2 i + b_2 j + c_2 k$, $\qquad V_3 = a_3 i + b_3 j + c_3 k$,

then

$$V_1 \times V_2 = \begin{vmatrix} i & j & k \\ a_1 & b_1 & c_1 \\ a_2 & b_2 & c_2 \end{vmatrix} = (b_1 c_2 - b_2 c_1)i + (c_1 a_2 - c_2 a_1)j + (a_1 b_2 - a_2 b_1)k$$

It should be noted that, since $V_1 \times V_2 = -V_2 \times V_1$, the vector product is not commutative.

Scalar Triple Product

There is only one possible interpretation of the expression $\mathbf{V}_1 \cdot \mathbf{V}_2 \times \mathbf{V}_3$ and that is $\mathbf{V}_1 \cdot (\mathbf{V}_2 \times \mathbf{V}_3)$ which is obviously a scalar.

Further $\mathbf{V}_1 \cdot (\mathbf{V}_2 \times \mathbf{V}_3) = (\mathbf{V}_1 \times \mathbf{V}_2) \cdot \mathbf{V}_3 = \mathbf{V}_2 \cdot (\mathbf{V}_3 \times \mathbf{V}_1)$

$$= \begin{vmatrix} a_1 & b_1 & c_1 \\ a_2 & b_2 & c_2 \\ a_3 & b_3 & c_3 \end{vmatrix}$$

$$= v_1 v_2 v_3 \cos \phi \sin \theta,$$

Where θ is the angle between $\mathbf{V}_2$ and $\mathbf{V}_3$ and ϕ is the angle between $\mathbf{V}_1$ and the normal to the plane of $\mathbf{V}_2$ and $\mathbf{V}_3$.

This product is called the *scalar triple product* and is written as $[\mathbf{V}_1\mathbf{V}_2\mathbf{V}_3]$.

The determinant indicates that it can be considered as the volume of the parallelepiped whose three determining edges are $\mathbf{V}_1$, $\mathbf{V}_2$ and $\mathbf{V}_3$.

It also follows that cyclic permutation of the subscripts does not change the value of the scalar triple product so that

$$[\mathbf{V}_1\mathbf{V}_2\mathbf{V}_3] = [\mathbf{V}_2\mathbf{V}_3\mathbf{V}_1] = [\mathbf{V}_3\mathbf{V}_1\mathbf{V}_2]$$

but $\quad [\mathbf{V}_1\mathbf{V}_2\mathbf{V}_3] = -[\mathbf{V}_2\mathbf{V}_1\mathbf{V}_3] \quad$ etc. $\qquad$ and $\quad [\mathbf{V}_1\mathbf{V}_1\mathbf{V}_2] \equiv 0$ etc.

Given three non-coplanar reference vectors $\mathbf{V}_1$, $\mathbf{V}_2$ and $\mathbf{V}_3$, the *reciprocal system is* given by $\mathbf{V}_1^*$, $\mathbf{V}_2^*$ and $\mathbf{V}_3^*$, where

$$1 = v_1 v_1^* = v_2 v_2^* = v_3 v_3^*$$

$$0 = v_1 v_2^* = v_1 v_3^* = v_2 v_1^* \quad \text{etc.}$$

$$\mathbf{V}_1^* = \frac{\mathbf{V}_2 \times \mathbf{V}_3}{[\mathbf{V}_1\mathbf{V}_2\mathbf{V}_3]}, \qquad \mathbf{V}_2^* = \frac{\mathbf{V}_3 \times \mathbf{V}_1}{[\mathbf{V}_1\mathbf{V}_2\mathbf{V}_3]}, \qquad \mathbf{V}_3^* = \frac{\mathbf{V}_1 \times \mathbf{V}_2}{[\mathbf{V}_1\mathbf{V}_2\mathbf{V}_3]}$$

The system $\mathbf{i}, \mathbf{j}, \mathbf{k}$ is its own reciprocal.

Vector Triple Product

The product $\mathbf{V}_1 \times (\mathbf{V}_2 \times \mathbf{V}_3)$ defines the *vector triple product*. Obviously, in this case, the brackets are vital to the definition.

$$\mathbf{V}_1 \times (\mathbf{V}_2 \times \mathbf{V}_3) = (\mathbf{V}_1 \cdot \mathbf{V}_3)\mathbf{V}_2 - (\mathbf{V}_1 \cdot \mathbf{V}_2)\mathbf{V}_3$$

$$= \begin{vmatrix} \mathbf{i} & \mathbf{j} & \mathbf{k} \\ a_1 & b_1 & c_1 \\ \begin{vmatrix} b_2 & c_2 \\ b_3 & c_3 \end{vmatrix} & \begin{vmatrix} c_2 & a_2 \\ c_3 & a_3 \end{vmatrix} & \begin{vmatrix} a_2 & b_2 \\ a_3 & b_3 \end{vmatrix} \end{vmatrix}$$

i.e. it is a vector, perpendicular to $\mathbf{V}_1$, lying in the plane of $\mathbf{V}_2$, $\mathbf{V}_3$.

Similarly $\qquad (\mathbf{V}_1 \times \mathbf{V}_2) \times \mathbf{V}_3 = \begin{vmatrix} \mathbf{i} & \mathbf{j} & \mathbf{k} \\ \begin{vmatrix} b_1 & c_1 \\ b_2 & c_2 \end{vmatrix} & \begin{vmatrix} c_1 & a_1 \\ c_2 & a_2 \end{vmatrix} & \begin{vmatrix} a_1 & b_1 \\ a_2 & b_2 \end{vmatrix} \\ a_3 & b_3 & c_3 \end{vmatrix}$

$$\mathbf{V}_1 \times (\mathbf{V}_2 \times \mathbf{V}_3) + \mathbf{V}_2 \times (\mathbf{V}_3 \times \mathbf{V}_1) + \mathbf{V}_3 \times (\mathbf{V}_1 + \mathbf{V}_2) \equiv 0$$

If $\mathbf{V}_1 \times (\mathbf{V}_2 \times \mathbf{V}_3) = (\mathbf{V}_1 \times \mathbf{V}_2) \times \mathbf{V}_3$ then $\mathbf{V}_1, \mathbf{V}_2, \mathbf{V}_3$ form an *orthogonal set*. Thus $\mathbf{i}, \mathbf{j}, \mathbf{k}$ form an orthogonal set.

Geometry of the Plane, Straight Line and Sphere

The position vectors of the fixed points A, B, C, D relative to O are $\mathbf{V}_1, \mathbf{V}_2, \mathbf{V}_3, \mathbf{V}_4$ and the position vector of the variable point P is $\mathbf{V}$.

The vector form of the equation of the straight line through A parallel to $\mathbf{V}_2$ is

$$\mathbf{V} = \mathbf{V}_1 + r\mathbf{V}_2$$

or $\quad (\mathbf{V} - \mathbf{V}_1) = r\mathbf{V}_2$

or $\quad (\mathbf{V} - \mathbf{V}_1) \times \mathbf{V}_2 = 0$

while that of the plane through A perpendicular to $\mathbf{V}_2$ is

$$(\mathbf{V} - \mathbf{V}_1) \cdot \mathbf{V}_2 = 0$$

The equation of the line AB is

$$\mathbf{V} = r\mathbf{V}_1 + (1 - r)\mathbf{V}_2$$

and those of the bisectors of the angles between $\mathbf{V}_1$ and $\mathbf{V}_2$ are

$$\mathbf{V} = r\left(\frac{\mathbf{V}_1}{v} \pm \frac{\mathbf{V}_2}{v_2}\right)$$

or $\quad \mathbf{V} = r(\hat{\mathbf{v}}_1 \pm \hat{\mathbf{v}}_2)$

The perpendicular from C to the line through A parallel to $\mathbf{V}_2$ has as its equation

$$\mathbf{V} = \mathbf{V}_1 - \mathbf{V}_3 - \hat{\mathbf{v}}_2 \cdot (\mathbf{V}_1 - \mathbf{V}_3)\hat{\mathbf{v}}_2.$$

The condition for the intersection of the two lines,

$$\mathbf{V} = \mathbf{V}_1 + r\mathbf{V}_3$$

and $\quad \mathbf{V} = \mathbf{V}_2 + s\mathbf{V}_4$

is $\quad [(\mathbf{V}_1 - \mathbf{V}_2)\mathbf{V}_3\mathbf{V}_4] = 0.$

The common perpendicular to the above two lines is the line of intersection of the two planes

$$[(\mathbf{V} - \mathbf{V}_1)\mathbf{V}_3(\mathbf{V}_3 \times \mathbf{V}_4)] = 0$$

and $\quad [(\mathbf{V} - \mathbf{V}_2)\mathbf{V}_4(\mathbf{V}_3 \times \mathbf{V}_4)] = 0$

and the length of this perpendicular is

$$\frac{[(\mathbf{V}_1 - \mathbf{V}_2)\mathbf{V}_3\mathbf{V}_4]}{|\mathbf{V}_3 \times \mathbf{V}_4|}.$$

The equation of the line perpendicular to the plane ABC is

$$\mathbf{V} = \mathbf{V}_1 \times \mathbf{V}_2 + \mathbf{V}_2 \times \mathbf{V}_3 + \mathbf{V}_3 \times \mathbf{V}_1$$

and the distance of the plane from the origin is

$$\frac{[\mathbf{V}_1\mathbf{V}_2\mathbf{V}_3]}{|(\mathbf{V}_2 - \mathbf{V}_1) \times (\mathbf{V}_3 - \mathbf{V}_1)|}.$$

In general the vector equation

$$\mathbf{V} \cdot \mathbf{V}_2 = r$$

defines the plane which is perpendicular to $\mathbf{V}_2$, and the perpendicular distance from A to this plane is

$$\frac{r - \mathbf{V}_1 \cdot \mathbf{V}_2}{v_2}.$$

The distance from A, measured along a line parallel to $\mathbf{V}_3$, is

$$\frac{r - \mathbf{V}_1 \cdot \mathbf{V}_2}{\mathbf{V}_2 \cdot \hat{\mathbf{v}}_3} \quad \text{or} \quad \frac{r - \mathbf{V}_1 \cdot \mathbf{V}_2}{v_2 \cos \theta}$$

where θ is the angle beween $\mathbf{V}_2$ and $\mathbf{V}_3$.
(If this plane contains the point C then $r = \mathbf{V}_3 \cdot \mathbf{V}_2$ and if it passes through the origin then $r = 0$.)

Given two planes

$$\mathbf{V} \cdot \mathbf{V}_1 = r$$

$$\mathbf{V} \cdot \mathbf{V}_2 = s$$

then any plane through the line of intersection of these two planes is given by

$$\mathbf{V} \cdot (\mathbf{V}_1 + \lambda \mathbf{V}_2) = r + \lambda s$$

where λ is a scalar parameter. In particular $\lambda = \pm v_1/v_2$ yields the equation of the two planes bisecting the angle between the given planes.
The plane through A parallel to the plane of $\mathbf{V}_2$, $\mathbf{V}_3$ is

$$\mathbf{V} = \mathbf{V}_1 + r\mathbf{V}_2 + s\mathbf{V}_3$$

$$\text{or} \quad (\mathbf{V} - \mathbf{V}_1) \cdot \mathbf{V}_2 \times \mathbf{V}_3 = 0$$

$$\text{or} \quad [\mathbf{V}\mathbf{V}_2\mathbf{V}_3] - [\mathbf{V}_1\mathbf{V}_2\mathbf{V}_3] = 0$$

so that the expansion in rectangular Cartesian coordinates yields

$$\begin{vmatrix} (x - a_1) & (y - b_1) & (z - c_1) \\ a_2 & b_2 & c_2 \\ a_3 & b_3 & c_3 \end{vmatrix} = 0 \qquad (\mathbf{V} \equiv x\mathbf{i} + y\mathbf{j} + z\mathbf{k})$$

which is obviously the usual linear equation in x, y and z.
The plane through AB parallel to $\mathbf{V}_3$ is given by

$$[(\mathbf{V} - \mathbf{V}_1)(\mathbf{V}_1 - \mathbf{V}_2)\mathbf{V}_3] = 0$$

$$\text{or} \quad [\mathbf{V}\mathbf{V}_2\mathbf{V}_3] - [\mathbf{V}\mathbf{V}_1\mathbf{V}_3] - [\mathbf{V}_1\mathbf{V}_2\mathbf{V}_3] = 0$$

The plane through the three points A, B and C is

$$\mathbf{V} = \mathbf{V}_1 + s(\mathbf{V}_2 - \mathbf{V}_1) + t(\mathbf{V}_3 - \mathbf{V}_1)$$

$$\text{or} \quad \mathbf{V} = r\mathbf{V}_1 + s\mathbf{V}_2 + t\mathbf{V}_3 \qquad (r + s + t \equiv 1)$$

$$\text{or} \quad [(\mathbf{V} - \mathbf{V}_1)(\mathbf{V}_1 - \mathbf{V}_2)(\mathbf{V}_2 - \mathbf{V}_3)] = 0$$

$$\text{or} \quad [\mathbf{V}\mathbf{V}_1\mathbf{V}_2] + [\mathbf{V}\mathbf{V}_2\mathbf{V}_3] + [\mathbf{V}\mathbf{V}_3\mathbf{V}_1] - [\mathbf{V}_1\mathbf{V}_2\mathbf{V}_3] = 0$$

For four points A, B, C, D to be coplanar, then

$$r\mathbf{Y}_1 + s\mathbf{V}_2 + t\mathbf{V}_3 + u\mathbf{V}_4 \equiv 0 \equiv r + s + t + u$$

The following formulae relate to a sphere when the vectors are taken to lie in three dimensional space and to a circle when the space is two dimensional. For a circle in three dimensions take the intersection of the sphere with a plane.

The equation of a sphere with center O and radius OA is

$$\mathbf{V} \cdot \mathbf{V} = v_1^2 \qquad \text{(not } \mathbf{V} = \mathbf{V}_1\text{)}$$

or
$$(\mathbf{V} - \mathbf{V}_1) \cdot (\mathbf{V} + \mathbf{V}_1) = 0$$

while that of a sphere with center B radius v_1 is

$$(\mathbf{V} - \mathbf{V}_2) \cdot (\mathbf{V} - \mathbf{V}_2) = v_1^2$$

or
$$\mathbf{V} \cdot (\mathbf{V} - 2\mathbf{V}_2) = v_1^2 - v_2^2$$

If the above sphere passes through the origin then

$$\mathbf{V} \cdot (\mathbf{V} - 2\mathbf{V}_2) = 0$$

(note that in two dimensional polar coordinates this is simply)

$$r = 2a \cdot \cos \theta$$

while in three dimensional Cartesian coordinates it is

$$x^2 + y^2 + z^2 - 2(a_2 x + b_2 y + c_2 x) = 0.$$

The equation of a sphere having the points A and B as the extremities of a diameter is

$$(\mathbf{V} - \mathbf{V}_1) \cdot (\mathbf{V} - \mathbf{V}_2) = 0.$$

The square of the length of the tangent from C to the sphere with center B and radius v_1 is given by

$$(\mathbf{V}_3 - \mathbf{V}_2) \cdot (\mathbf{V}_3 - \mathbf{V}_2) = v_1^2$$

The condition that the plane $\mathbf{V} \cdot \mathbf{V}_3 = s$ is tangential to the sphere $(\mathbf{V} - \mathbf{V}_2) \cdot (\mathbf{V} - \mathbf{V}_2) = v_1^2$ is

$$(s - \mathbf{V}_3 \cdot \mathbf{V}_2) \cdot (s - \mathbf{V}_3 \cdot \mathbf{V}_2) = v_1^2 v_3^2.$$

The equation of the tangent plane at D, on the surface of sphere $(\mathbf{V} - \mathbf{V}_2) \cdot (\mathbf{V} - \mathbf{V}_2) = v_1^2$, is

$$(\mathbf{V} - \mathbf{V}_4) \cdot (\mathbf{V}_4 - \mathbf{V}_2) = 0$$

or
$$\mathbf{V} \cdot \mathbf{V}_4 - \mathbf{V}_2 \cdot (\mathbf{V} + \mathbf{V}_4) = v_1^2 - v_2^2$$

The condition that the two circles $(\mathbf{V} - \mathbf{V}_2) \cdot (\mathbf{V} - \mathbf{V}_2) = v_1^2$ and $(\mathbf{V} - \mathbf{V}_4) \cdot (\mathbf{V} - \mathbf{V}_4) = v_3^2$ intersect orthogonally is clearly

$$(\mathbf{V}_2 - \mathbf{V}_4) \cdot (\mathbf{V}_2 - \mathbf{V}_4) = v_1^2 + v_3^2$$

The polar plane of D with respect to the circle

$$(\mathbf{V} - \mathbf{V}_2) \cdot (\mathbf{V} - \mathbf{V}_2) = v_1^2 \quad \text{is}$$

$$\mathbf{V} \cdot \mathbf{V}_4 - \mathbf{V}_2 \cdot (\mathbf{V} + \mathbf{V}_4) = v_1^2 - v_2^2$$

Any sphere through the intersection of the two spheres $(\mathbf{V} - \mathbf{V}_2) \cdot (\mathbf{V} - \mathbf{V}_2) = v_1^2$ and $(\mathbf{V} - \mathbf{V}_4) \cdot (\mathbf{V} - \mathbf{V}_4) = v_3^2$ is given by

$$(\mathbf{V} - \mathbf{V}_2) \cdot (\mathbf{V} - \mathbf{V}_2) + \lambda(\mathbf{V} - \mathbf{V}_4) \cdot (\mathbf{V} - \mathbf{V}_4) = v_1^2 + \lambda v_3^2$$

while the radical plane of two such spheres is

$$\mathbf{V} \cdot (\mathbf{V}_2 - \mathbf{V}_4) = -\tfrac{1}{2}(v_1^2 - r_2^2 - v_3^2 + v_4^2)$$

Differentiation of Vectors

If $\mathbf{V}_1 = a_1\mathbf{i} + b_1\mathbf{j} + c_1\mathbf{k}$, and $\mathbf{V}_2 = a_2\mathbf{i} + b_2\mathbf{j} + c_2\mathbf{k}$, and if $\mathbf{V}_1$ and $\mathbf{V}_2$ are functions of the scalar t, then

$$\frac{d}{dt}(\mathbf{V}_1 + \mathbf{V}_2 + \cdots) = \frac{d\mathbf{V}_1}{dt} + \frac{d\mathbf{V}_2}{dt} + \cdots,$$

$$\text{where } \frac{d\mathbf{V}_1}{dt} = \frac{da_1}{dt}\mathbf{i} + \frac{db_1}{dt}\mathbf{j} + \frac{dc_1}{dt}\mathbf{k}, \text{ etc.}$$

$$\frac{d}{dt}(\mathbf{V}_1 \cdot \mathbf{V}_2) = \frac{d\mathbf{V}_1}{dt} \cdot \mathbf{V}_2 + \mathbf{V}_1 \cdot \frac{d\mathbf{V}_2}{dt}$$

$$\frac{d}{dt}(\mathbf{V}_1 \times \mathbf{V}_2) = \frac{d\mathbf{V}_1}{dt} \times \mathbf{V}_2 + \mathbf{V}_1 \times \frac{d\mathbf{V}_2}{dt}$$

$$\mathbf{V} \cdot \frac{d\mathbf{V}}{dt} = v \cdot \frac{dv}{dt}$$

In particular, if $\mathbf{V}$ is a vector of constant length then the right hand side of the last equation is identically zero showing that $\mathbf{V}$ is perpendicular to its derivative.

The derivatives of the triple products are

$$\frac{d}{dt}[\mathbf{V}_1\mathbf{V}_2\mathbf{V}_3] = \left[\left(\frac{d\mathbf{V}_1}{dt}\right)\mathbf{V}_2\mathbf{V}_3\right] + \left[\mathbf{V}_1\left(\frac{d\mathbf{V}_2}{dt}\right)\mathbf{V}_3\right] + \left[\mathbf{V}_1\mathbf{V}_2\left(\frac{d\mathbf{V}_3}{dt}\right)\right]$$

and $$\frac{d}{dt}\{\mathbf{V}_1 \times (\mathbf{V}_2 \times \mathbf{V}_3)\} = \left(\frac{d\mathbf{V}_1}{dt}\right) \times (\mathbf{V}_2 \times \mathbf{V}_3) + \mathbf{V}_1$$

$$\times \left(\left(\frac{d\mathbf{V}_2}{dt}\right) \times \mathbf{V}_3\right) + \mathbf{V}_1 \times \left(\mathbf{V}_2 \times \left(\frac{d\mathbf{V}_3}{dt}\right)\right)$$

Geometry of Curves in Space

s = the *length of arc*, measured from some fixed point on the curve (fig. 3).

$\mathbf{V}_1$ = the position vector of the point A on the curve

$\mathbf{V}_1 + \delta\mathbf{V}_1$ = the position vector of the point P in the neighborhood of A

$\hat{\mathbf{t}}$ = the *unit tangent* to the curve at the point A, measured in the direction of s increasing.

The *normal plane* is that plane which is perpendicular to the unit tangent. The principal normal is defined as the intersection of the normal plane with the plane defined by $\mathbf{V}_1$ and $\mathbf{V}_1 + \delta\mathbf{V}_1$ in the limit as $\delta\mathbf{V}_1 - 0$.

$\hat{\mathbf{n}}$ = the *unit normal* (principal) at the point A. The plane defined by $\hat{\mathbf{t}}$ and $\hat{\mathbf{n}}$ is called the *osculating plane* (alternatively plane of curvature or local plane).

ρ = the radius of curvature at A

$\delta\theta$ = the angle subtended at the origin by $\delta\mathbf{V}_1$.

$$\kappa = \frac{d\theta}{ds} = \frac{1}{\rho}$$

$\hat{\mathbf{b}}$ = the *unit binormal* i.e. the unit vector which is parallel to $\hat{\mathbf{t}} \times \hat{\mathbf{n}}$ at the point A :
λ = the *torsion* of the curve at A

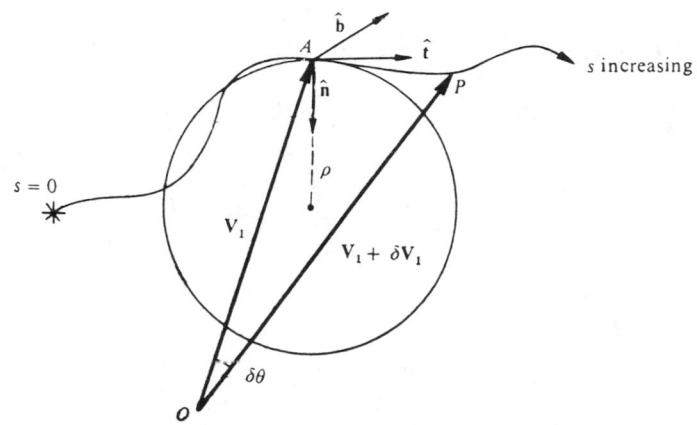

Figure 3.

Frenet's Formulae:

$$\frac{d\hat{\mathbf{t}}}{ds} = \kappa\hat{\mathbf{n}}$$

$$\frac{d\hat{\mathbf{n}}}{ds} = -\kappa\hat{\mathbf{t}} \qquad + \lambda\hat{\mathbf{b}}$$

$$\frac{d\hat{\mathbf{b}}}{ds} = \qquad -\lambda\hat{\mathbf{n}}$$

The following formulae are also applicable:

Unit tangent $\qquad\qquad\qquad\qquad \hat{\mathbf{t}} = \dfrac{d\mathbf{V}_1}{ds}$

Equation of the tangent $\qquad\qquad (\mathbf{V} - \mathbf{V}_1) \times \hat{\mathbf{t}} = 0$

$\qquad\qquad\qquad$ or $\qquad \mathbf{V} = \mathbf{V}_1 + q\hat{\mathbf{t}}$

Unit normal $\qquad\qquad\qquad\qquad \hat{\mathbf{n}} = \dfrac{1}{\kappa}\dfrac{d^2\mathbf{V}_1}{ds^2}$

Equation of the normal plane $\qquad (\mathbf{V} - \mathbf{V}_1) \cdot \hat{\mathbf{t}} = 0$

Equation of the normal $\qquad\qquad (\mathbf{V} - \mathbf{V}_1) \times \hat{\mathbf{n}} = 0$

$\qquad\qquad\qquad$ or $\qquad \mathbf{V} = \mathbf{V}_1 + r\hat{\mathbf{n}}$

Unit binormal $\qquad\qquad \hat{\mathbf{b}} = \hat{\mathbf{t}} \times \hat{\mathbf{n}}$

Equation of the binormal $\qquad\quad (\mathbf{V} - \mathbf{V}_1) \times \hat{\mathbf{b}} = 0$

$\qquad\qquad$ or $\qquad\qquad \mathbf{V} = \mathbf{V}_1 + u\hat{\mathbf{b}}$

$\qquad\qquad$ or $\qquad\qquad \mathbf{V} = \mathbf{V}_1 + w\dfrac{d\mathbf{V}_1}{ds} \times \dfrac{d^2\mathbf{V}_1}{ds^2}$

Equation of the osculating plane:

$$[(\mathbf{V} - \mathbf{V}_1)\hat{\mathbf{t}}\hat{\mathbf{n}}] = 0$$

$$\text{or} \qquad \left[(\mathbf{V} - \mathbf{V}_1) \left(\frac{d\mathbf{V}_1}{ds}\right) \left(\frac{d^2\mathbf{V}_1}{ds^2}\right) \right] = 0$$

A *geodetic line* on a surface is a curve, the osculating plane of which is everywhere normal to the surface.

The differential equation of the geodetic is

$$[\hat{\mathbf{n}}\, d\mathbf{V}_1\, d^2\mathbf{V}_1] = 0$$

Differential Operators—Rectangular Coordinates

$$dS = \frac{\partial S}{\partial x} \cdot dx + \frac{\partial S}{\partial y} \cdot dy + \frac{\partial S}{\partial z} \cdot dz$$

By definition

$$\nabla \equiv \text{del} \equiv \mathbf{i}\, \frac{\partial}{\partial x} + \mathbf{j}\, \frac{\partial}{\partial y} + \mathbf{k}\, \frac{\partial}{\partial z}$$

$$\nabla^2 \equiv \text{Laplacian} \equiv \frac{\partial^2}{\partial x^2} + \frac{\partial^2}{\partial y^2} + \frac{\partial^2}{\partial z^2}$$

If S is a scalar function, then

$$\nabla S \equiv \text{grad } S \equiv \frac{\partial S}{dx}\, \mathbf{i} + \frac{\partial S}{dy}\, \mathbf{j} + \frac{\partial S}{dz}\, \mathbf{k}$$

Grad S defines both the direction and magnitude of the maximum rate of increase of S at any point. Hence the name *gradient* and also its vectorial nature. ∇S is independent of the choice of rectangular coordinates.

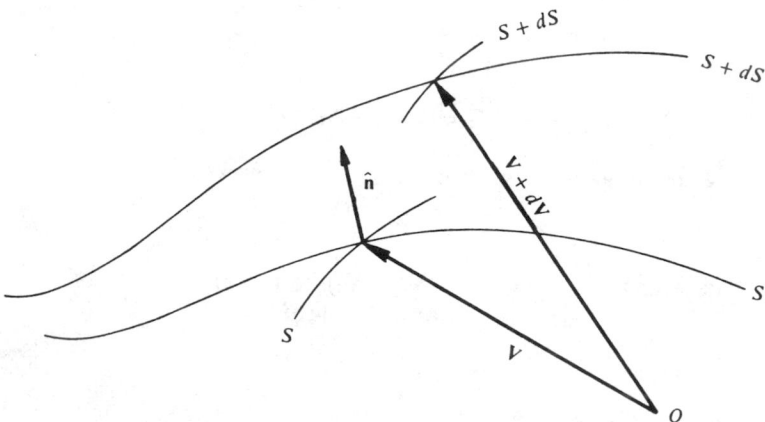

Figure 4

$$\nabla S = \frac{\partial S}{\partial n}\, \hat{\mathbf{n}}$$

where $\hat{\mathbf{n}}$ is the unit normal to the surface $S = $ constant, in the direction of S increasing. The total derivative of S at a point having the position vector $\mathbf{V}$ is given by (fig. 4)

$$dS = \frac{\partial S}{\partial n}\, \hat{\mathbf{n}} \cdot d\mathbf{V}$$

$$= d\mathbf{V} \cdot \nabla S$$

and the directional derivative of S in the direction of $\mathbf{U}$ is

$$\mathbf{U} \cdot \nabla S = \mathbf{U} \cdot (\nabla S) = (\mathbf{U} \cdot \nabla)S$$

Similarly the directional derivative of the vector $\mathbf{V}$ in the direction of $\mathbf{U}$ is

$$(\mathbf{U} \cdot \nabla)\mathbf{V}$$

The *distributive* law holds for finding a gradient. Thus if S and T are scalar functions

$$\nabla(S + T) = \nabla S + \nabla T$$

The *associative* law becomes the rule for differentiating a product:

$$\nabla(ST) = S\nabla T + T\nabla S$$

If **V** is a vector function with the magnitudes of the components parallel to the three coordinate axes V_x, V_y, V_z, then

$$\nabla \cdot \mathbf{V} \equiv \text{div } \mathbf{V} \equiv \frac{\partial V_x}{\partial x} + \frac{\partial V_y}{\partial y} + \frac{\partial V_z}{\partial z}$$

The divergence obeys the distributive law. Thus, if **V** and **U** are vectors functions, then

$$\nabla \cdot (\mathbf{V} + \mathbf{U}) = \nabla \cdot \mathbf{V} + \nabla \cdot \mathbf{U}$$
$$\nabla \cdot (S\mathbf{V}) = (\nabla S) \cdot \mathbf{V} + S(\nabla \cdot \mathbf{V})$$
$$\nabla \cdot (\mathbf{U} \times \mathbf{V}) = \mathbf{V} \cdot (\nabla \times \mathbf{U}) - \mathbf{U} \cdot (\nabla \times \mathbf{V})$$

As with the gradient of a scalar, the divergence of a vector is invariant under a transformation from one set of rectangular coordinates to another.

$$\nabla \times \mathbf{V} \equiv \text{curl } \mathbf{V} \quad (\text{sometimes } \nabla \wedge \mathbf{V} \text{ or rot } \mathbf{V})$$

$$\equiv \left(\frac{\partial \mathbf{V}_z}{\partial y} - \frac{\partial \mathbf{V}_y}{\partial z}\right)\mathbf{i} + \left(\frac{\partial \mathbf{V}_x}{\partial z} - \frac{\partial \mathbf{V}_z}{\partial x}\right)\mathbf{j} + \left(\frac{\partial \mathbf{V}_y}{\partial x} - \frac{\partial \mathbf{V}_x}{\partial y}\right)\mathbf{k}$$

$$= \begin{vmatrix} \mathbf{i} & \mathbf{j} & \mathbf{k} \\ \frac{\partial}{\partial x} & \frac{\partial}{\partial y} & \frac{\partial}{\partial z} \\ V_x & V_y & V_z \end{vmatrix}$$

The *curl* (or *rotation*) of a vector is a vector which is invariant under a transformation from one set of rectangular coordinates to another.

$$\nabla \times (\mathbf{U} + \mathbf{V}) = \nabla \times \mathbf{U} + \nabla \times \mathbf{V}$$
$$\nabla \times (S\mathbf{V}) = (\nabla S) \times \mathbf{V} + S(\nabla \times \mathbf{V})$$
$$\nabla \times (\mathbf{U} \times \mathbf{V}) = (\mathbf{V} \cdot \nabla)\mathbf{U} - (\mathbf{U} \cdot \nabla)\mathbf{V} + \mathbf{U}(\nabla \cdot \mathbf{V}) - \mathbf{V}(\nabla \cdot \mathbf{U})$$

$$\text{grad } (\mathbf{U} \cdot \mathbf{V}) = \nabla(\mathbf{U} \cdot \mathbf{V})$$
$$= (\mathbf{V} \cdot \nabla)\mathbf{U} + (\mathbf{U} \cdot \nabla)\mathbf{V} + \mathbf{V} \times (\nabla \times \mathbf{U}) + \mathbf{U} \times (\nabla \times \mathbf{V})$$

If
$$\mathbf{V} = V_x\mathbf{i} + V_y\mathbf{j} + V_z\mathbf{k}$$
$$\nabla \cdot \mathbf{V} = \nabla V_x \cdot \mathbf{i} + \nabla V_y \cdot \mathbf{j} + \nabla V_z \cdot \mathbf{k}$$
and
$$\nabla \times \mathbf{V} = \nabla V_x \times \mathbf{i} + \nabla V_y \times \mathbf{j} + \nabla V_z \times \mathbf{k}$$

The operator ∇ can be used more than once. The number of possibilities where ∇ is used twice are

$$\nabla \cdot (\nabla\theta) \equiv \text{div grad } \theta$$
$$\nabla \times (\nabla\theta) \equiv \text{curl grad } \theta$$
$$\nabla(\nabla \cdot \mathbf{V}) \equiv \text{grad div } \mathbf{V}$$
$$\nabla \cdot (\nabla \times \mathbf{V}) \equiv \text{div curl } \mathbf{V}$$
$$\nabla \times (\nabla \times \mathbf{V}) \equiv \text{curl curl } \mathbf{V}$$

Thus: div grad $S \equiv \nabla \cdot (\nabla S) \equiv$ Laplacian $S \equiv \nabla^2 S$

$$\equiv \frac{\partial^2 S}{\partial x^2} + \frac{\partial^2 S}{\partial y^2} + \frac{\partial^2 S}{\partial z^2}$$

curl grad $S \equiv 0$;   curl curl $\mathbf{V} \equiv$ grad div $\mathbf{V} - \nabla^2 \mathbf{V}$;

$$\text{div curl } \mathbf{V} \equiv \qquad 0$$

Taylor's expansion in three dimensions can be written

$$f(\mathbf{V} + \boldsymbol{\varepsilon}) = e^{\boldsymbol{\varepsilon} \cdot \nabla} f(\mathbf{V})$$

$$\text{where} \qquad \mathbf{V} = x\mathbf{i} + y\mathbf{j} + z\mathbf{k}$$

$$\text{and} \qquad \boldsymbol{\varepsilon} = h\mathbf{i} + l\mathbf{j} + m\mathbf{k}$$

(note the analogy with $f_p = e^{phD}f_0$ in finite difference methods).

Orthogonal Curvilinear Coordinates

If at a point P there exist three uniform point functions u, v and w so that the surfaces $u = $ const., $v = $ const., and $w = $ const., intersect in three distinct curves through P then the surfaces are called the *coordinate surfaces* through P. The three lines of intersection are referred to as the *coordinate lines* and their tangents a, b, and c as the *coordinate axes*. When the coordinate axes form an orthogonal set the system is said to define *orthogonal curvilinear coordinates* at P.

Consider an infinitesimal volume enclosed by the surfaces u, v, w, $u + du$, $v + dv$, and $w + dw$ (fig. 5).

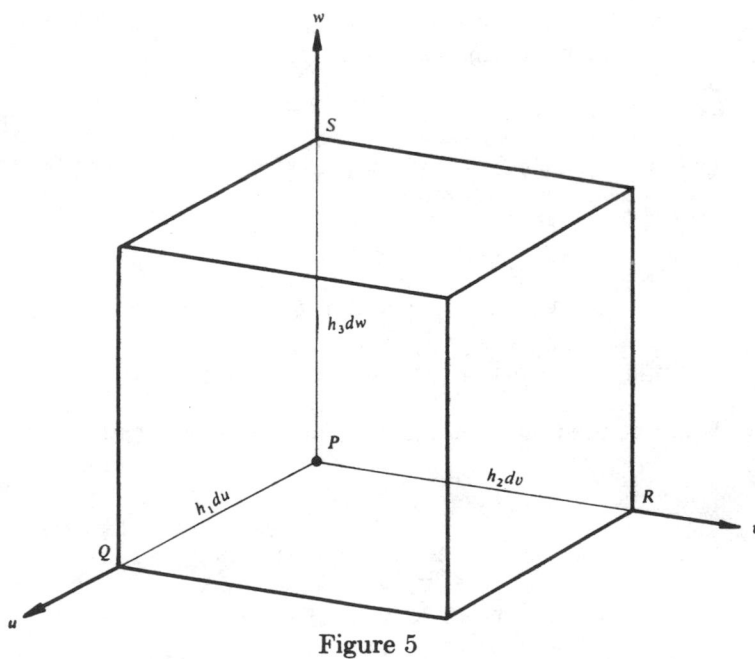

Figure 5

The surface $PRS \equiv u = $ const., and the face of the curvilinear figure immediately opposite this is $u + du = $ const. etc.

In terms of these surface constants

$$P = P(u,v,w)$$
$$Q = Q(u + du,v,w) \qquad \text{and} \qquad PQ = h_1 du$$
$$R = R(u,v + dv,w) \qquad\qquad\qquad PR = h_2 dv$$
$$S = S(u,v,w + dw) \qquad\qquad\qquad PS = h_3 dw$$

where h_1, h_2, and h_3 are functions of u, v, and w.

In rectangular Cartesians $\mathbf{i}$, $\mathbf{j}$, $\mathbf{k}$

$$h_1 = 1, \qquad h_2 = 1, \qquad h_3 = 1.$$

$$\frac{\hat{\mathbf{a}}}{h_1}\frac{\partial}{\partial u} = \mathbf{i}\frac{\partial}{\partial x}, \qquad \frac{\hat{\mathbf{b}}}{h_2}\frac{\partial}{\partial v} = \mathbf{j}\frac{\partial}{\partial y}, \qquad \frac{\hat{\mathbf{c}}}{h_3}\frac{\partial}{\partial w} = \mathbf{k}\frac{\partial}{\partial z}.$$

In cylindrical coordinates $\hat{\mathbf{r}}$, $\hat{\boldsymbol{\phi}}$, $\hat{\mathbf{k}}$

$$h_1 = 1, \qquad h_2 = r \qquad h_3 = 1.$$

$$\frac{\hat{\mathbf{a}}}{h_1}\frac{\partial}{\partial u} = \hat{\mathbf{r}}\frac{\partial}{\partial r}, \qquad \frac{\hat{\mathbf{b}}}{h_2}\frac{\partial}{\partial v} = \frac{\hat{\boldsymbol{\phi}}}{r}\frac{\partial}{\partial \phi} \qquad \frac{\hat{\mathbf{c}}}{h_3}\frac{\partial}{\partial w} = \hat{\mathbf{k}}\frac{\partial}{\partial z}$$

In spherical coordinates $\hat{\mathbf{r}}$, $\hat{\boldsymbol{\theta}}$, $\hat{\boldsymbol{\phi}}$

$$h_1 = 1, \qquad h_2 = r, \qquad h_3 = r\sin\theta$$

$$\frac{\hat{\mathbf{a}}}{h_1}\frac{\partial}{\partial u} = \hat{\mathbf{r}}\frac{\partial}{\partial r}, \qquad \frac{\mathbf{b}}{h_2}\frac{\partial}{\partial v} = \frac{\hat{\boldsymbol{\phi}}}{r}\frac{\partial}{\partial \theta}, \qquad \frac{\hat{\mathbf{c}}}{h_3}\frac{\partial}{\partial w} = \frac{\hat{\boldsymbol{\phi}}}{r\sin\theta}\frac{\partial}{\partial \phi}$$

The general expressions for grad, div and curl together with those for ∇^2 and the directional derivative are, in orthogonal curvilinear coordinates, given by

$$\nabla S = \frac{\hat{\mathbf{a}}}{h_1}\frac{\partial S}{\partial u} + \frac{\hat{\mathbf{b}}}{h_2}\frac{\partial S}{\partial v} + \frac{\hat{\mathbf{c}}}{h_3}\frac{\partial S}{\partial w}$$

$$(\mathbf{V}\cdot\nabla)S = \frac{V_1}{h_1}\frac{\partial S}{\partial u} + \frac{V_2}{h_2}\frac{\partial S}{\partial v} + \frac{V_3}{h_3}\frac{\partial S}{\partial w}$$

$$\nabla\cdot\mathbf{V} = \frac{1}{h_1 h_2 h_3}\left\{\frac{\partial}{\partial u}(h_2 h_3 V_1) + \frac{\partial}{\partial v}(h_3 h_1 V_2) + \frac{\partial}{\partial w}(h_1 h_2 V_3)\right.$$

$$\nabla\times\mathbf{V} = \frac{\hat{\mathbf{a}}}{h_2 h_3}\left\{\frac{\partial}{\partial v}(h_3 V_3) - \frac{\partial}{\partial w}(h_2 V_2)\right\} + \frac{\hat{\mathbf{b}}}{h_3 h_1}\left\{\frac{\partial}{\partial w}(h_1 V_1) - \frac{\partial}{\partial u}(h_3 V_3)\right\}$$

$$+ \frac{\hat{\mathbf{c}}}{h_1 h_2}\left\{\frac{\partial}{\partial u}(h_2 V_2) - \frac{\partial}{\partial v}(h_1 V_1)\right\}$$

$$\nabla^2 S = \frac{1}{h_1 h_2 h_3}\left\{\frac{\partial}{\partial u}\left(\frac{h_2 h_3}{h_1}\frac{\partial S}{\partial u}\right) + \frac{\partial}{\partial v}\left(\frac{h_3 h_1}{h_2}\frac{\partial S}{\partial v}\right) + \frac{\partial}{\partial w}\left(\frac{h_1 h_2}{h_3}\frac{\partial S}{\partial w}\right)\right\}$$

FORMULAS OF VECTOR ANALYSIS

	Rectangular coordinates	Cylindrical coordinates	Spherical coordinates
Conversion to rectangular coordinates		$x = r\cos\varphi \quad y = r\sin\varphi \quad z = z$	$x = r\cos\varphi\sin\theta \quad y = r\sin\varphi\sin\theta$ $z = r\cos\theta$
Gradient	$\nabla\phi = \dfrac{\partial\phi}{\partial x}\mathbf{i} + \dfrac{\partial\phi}{\partial y}\mathbf{j} + \dfrac{\partial\phi}{\partial z}\mathbf{k}$	$\nabla\phi = \dfrac{\partial\phi}{\partial r}\mathbf{r} + \dfrac{1}{r}\dfrac{\partial\phi}{\partial\varphi}\boldsymbol{\phi} + \dfrac{\partial\phi}{\partial z}\mathbf{k}$	$\nabla\phi = \dfrac{\partial\phi}{\partial r}\mathbf{r} + \dfrac{1}{r}\dfrac{\partial\phi}{\partial\theta}\boldsymbol{\theta} + \dfrac{1}{r\sin\theta}\dfrac{\partial\phi}{\partial\varphi}\boldsymbol{\phi}$
Divergence	$\nabla\cdot\mathbf{A} = \dfrac{\partial A_x}{\partial x} + \dfrac{\partial A_y}{\partial y} + \dfrac{\partial A_z}{\partial z}$	$\nabla\cdot\mathbf{A} = \dfrac{1}{r}\dfrac{\partial(rA_r)}{\partial r} + \dfrac{1}{r}\dfrac{\partial A_\varphi}{\partial\varphi} + \dfrac{\partial A_z}{\partial z}$	$\nabla\cdot\mathbf{A} = \dfrac{1}{r^2}\dfrac{\partial(r^2A_r)}{\partial r} + \dfrac{1}{r\sin\theta}\dfrac{\partial(A_\theta\sin\theta)}{\partial\theta} + \dfrac{1}{r\sin\theta}\dfrac{\partial A_\varphi}{\partial\varphi}$
Curl	$\nabla\times\mathbf{A} = \begin{vmatrix} \mathbf{i} & \mathbf{j} & \mathbf{k} \\ \dfrac{\partial}{\partial x} & \dfrac{\partial}{\partial y} & \dfrac{\partial}{\partial z} \\ A_x & A_y & A_z \end{vmatrix}$	$\nabla\times\mathbf{A} = \begin{vmatrix} \dfrac{1}{r}\mathbf{r} & \boldsymbol{\phi} & \dfrac{1}{r}\mathbf{k} \\ \dfrac{\partial}{\partial r} & \dfrac{\partial}{\partial\varphi} & \dfrac{\partial}{\partial z} \\ A_r & rA_\varphi & A_z \end{vmatrix}$	$\nabla\times\mathbf{A} = \begin{vmatrix} \dfrac{\mathbf{r}}{r^2\sin\theta} & \dfrac{\boldsymbol{\theta}}{r\sin\theta} & \dfrac{\boldsymbol{\phi}}{r} \\ \dfrac{\partial}{\partial r} & \dfrac{\partial}{\partial\theta} & \dfrac{\partial}{\partial\varphi} \\ A_r & rA_\theta & rA_\varphi\sin\theta \end{vmatrix}$
Laplacian	$\nabla^2\phi = \dfrac{\partial^2\phi}{\partial x^2} + \dfrac{\partial^2\phi}{\partial y^2} + \dfrac{\partial^2\phi}{\partial z^2}$	$\nabla^2\phi = \dfrac{1}{r}\dfrac{\partial}{\partial r}\left(r\dfrac{\partial\phi}{\partial r}\right) + \dfrac{1}{r^2}\dfrac{\partial^2\phi}{\partial\varphi^2} + \dfrac{\partial^2\phi}{\partial z^2}$	$\nabla^2\phi = \dfrac{1}{r^2}\dfrac{\partial}{\partial r}\left(r^2\dfrac{\partial\phi}{\partial r}\right) + \dfrac{1}{r^2\sin\theta}\dfrac{\partial}{\partial\theta}\left(\sin\theta\dfrac{\partial\phi}{\partial\theta}\right) + \dfrac{1}{r^2\sin^2\theta}\dfrac{\partial^2\phi}{\partial\varphi^2}$

Transformation of Integrals

s = the distance along some curve "C" in space and is measured from some fixed point.

S = a surface area

V = a volume contained by a specified surface

$\hat{\mathbf{t}}$ = the unit tangent to C at the point P

$\hat{\mathbf{n}}$ = the unit outward pointing normal

F = some vector function

ds = the vector element of curve ($= \hat{\mathbf{t}}\, ds$)

dS = the vector element of surface ($= \hat{\mathbf{n}}\, dS$)

Then
$$\int_{(c)} \mathbf{F} \cdot \hat{\mathbf{t}}\, ds = \int_{(c)} \mathbf{F} \cdot d\mathbf{s}$$

and when
$$\mathbf{F} = \nabla\phi$$

$$\int_{(c)} (\nabla\phi) \cdot \hat{\mathbf{t}}\, ds = \int_{(c)} d\phi$$

Gauss' Theorem (Green's Theorem)

When S defines a closed region having a volume V

$$\iiint_{(v)} (\nabla \cdot \mathbf{F})\, dV = \iint_{(s)} (\mathbf{F} \cdot \hat{\mathbf{n}})\, dS = \iint_{(s)} \mathbf{F} \cdot d\mathbf{S}$$

also
$$\iiint_{(v)} (\nabla\phi)\, dV = \iint_{(s)} \phi\hat{\mathbf{n}}\, dS$$

and
$$\iiint_{(v)} (\nabla \times \mathbf{F})\, dV = \iint_{(s)} (\hat{\mathbf{n}} \times \mathbf{F})\, dS$$

Stokes' Theorem

When C is closed and bounds the open surface S.

$$\iint_{(s)} \hat{\mathbf{n}} \cdot (\nabla \times \mathbf{F})\, dS = \int_{(c)} \mathbf{F} \cdot d\mathbf{s}$$

also
$$\iint_{(s)} (\hat{\mathbf{n}} \times \nabla\phi)\, dS = \int_{(c)} \phi\, d\mathbf{s}$$

Green's Theorem

$$\iint_{(s)} (\nabla\phi \cdot \nabla\theta)\, dS = \iint_{(s)} \phi\hat{\mathbf{n}} \cdot (\nabla\theta)\, dS = \iiint_{(v)} \phi(\nabla^2\theta)\, dV$$

$$= \iint_{(s)} \theta \cdot \hat{\mathbf{n}}(\nabla\phi)\, dS = \iiint_{(v)} \theta(\nabla^2\phi)\, dV$$

MOMENT OF INERTIA FOR VARIOUS BODIES OF MASS

The mass of the body is indicated by m.

Body	Axis	Moment of inertia
Uniform thin rod	Normal to the length, at one end	$m\dfrac{l^2}{3}$
Uniform thin rod	Normal to the length, at the center	$m\dfrac{l^2}{12}$
Thin rectangular sheet, sides a and b	Through the center parallel to b	$m\dfrac{a^2}{12}$
Thin rectangular sheet, sides a and b	Through the center perpendicular to the sheet	$m\dfrac{a^2 + b^2}{12}$
Thin circular sheet of radius r	Normal to the plate through the center	$m\dfrac{r^2}{2}$
Thin circular sheet of radius r	Along any diameter	$m\dfrac{r^2}{4}$
Thin circular ring. Radii r_1 and r_2	Through center normal to plane of ring	$m\dfrac{r_1^2 + r_2^2}{2}$
Thin circular ring. Radii r_1 and r_2	Any diameter	$m\dfrac{r_1^2 + r_2^2}{4}$
Rectangular parallelopiped, edges a, b, and c	Through center perpendicular to face ab, (parallel to edge c)	$m\dfrac{a^2 + b^2}{12}$
Sphere, radius r	Any diameter	$m\dfrac{2}{5}r^2$
Spherical shell, external radius, r_1, internal radius r_2	Any diameter	$m\dfrac{2}{5}\dfrac{(r_1^5 - r_2^5)}{(r_1^3 - r_2^3)}$
Spherical shell, very thin, mean radius, r	Any diameter	$m\dfrac{2}{3}r^2$
Right circular cylinder of radius r, length l	The longitudinal axis of the solid	$m\dfrac{r^2}{2}$
Right circular cylinder of radius r, length l	Transverse diameter	$m\left(\dfrac{r^2}{4} + \dfrac{l^2}{12}\right)$
Hollow circular cylinder, length l, radii r_1 and r_2	The longitudinal axis of the figure	$m\dfrac{(r_1^2 + r_2^2)}{2}$
Thin cylindrical shell, length l, mean radius, r	The longitudinal axis of the figure	mr^2
Hollow circular cylinder, length l, radii r_1 and r_2	Transverse diameter	$m\left[\dfrac{r_1^2 + r_2^2}{4} + \dfrac{l^2}{12}\right]$
Hollow circular cylinder, length l, very thin, mean radius	Transverse diameter	$m\left(\dfrac{r^2}{2} + \dfrac{l^2}{12}\right)$
Elliptic cylinder, length l, transverse semiaxes a and b	Longitudinal axis	$m\left(\dfrac{a^2 + b^2}{4}\right)$
Right cone, altitude h, radius of base r	Axis of the figure	$m\dfrac{3}{10}r^2$
Spheroid of revolution, equatorial radius r	Polar axis	$m\dfrac{2r^2}{5}$
Ellipsoid, axes $2a$, $2b$, $2c$	Axis $2a$	$m\dfrac{(b^2 + c^2)}{5}$

IX. DIFFERENTIAL EQUATIONS

Differential equation	Method of solution
Separation of variables $f_1(x) g_1(y)\, dx + f_2(x) g_2(y)\, dy = 0$	$$\int \frac{f_1(x)}{f_2(x)}\, dx + \int \frac{g_2(y)}{g_1(y)}\, dy = c$$
Exact equation $M(x,y)dx + N(x,y)dy = 0$ where $\partial M/\partial y = \partial N/\partial x$	$$\int M\, \partial x + \int \left(N - \frac{\partial}{\partial y} \int M\, \partial x \right) dy = c$$ where ∂x indicates that the integration is to be performed with respect to x keeping y constant.
Linear first order equation $\frac{dy}{dx} + P(x)y = Q(x)$	$$ye^{\int P dx} = \int Q e^{\int P dx}\, dx + c$$
Bernoulli's equation $\frac{dy}{dx} + P(x)y = Q(x)y^n$	$$ve^{(1-n)\int P dx} = (1-n)\int Q e^{(1-n)\int P dx}\, dx + c$$ where $v = y^{1-n}$. If $n = 1$, the solution is $$\ln y = \int (Q - P)\, dx + c$$
Homogeneous equation $\frac{dy}{dx} = F\left(\frac{y}{x}\right)$	$$\ln x = \int \frac{dv}{F(v) - v} + c$$ where $v = y/x$. If $F(v) = v$, the solution is $y = cx$
Reducible to homogeneous $(a_1 x + b_1 y + c_1)\, dx + (a_2 x + b_2 y + c_2)\, dy = 0$ $\frac{a_1}{a_2} \neq \frac{b_1}{b_2}$	Set $u = a_1 x + b_1 y + c_1$. $v = a_2 x + b_2 y + c_2$ Eliminate x and y and the equation becomes homogenous
Reducible to separable $(a_1 x + b_1 y + c_1)\, dx + (a_2 x + b_2 y + c_2)\, dy = 0$ $\frac{a_1}{a_2} = \frac{b_1}{b_2}$	Set $u = a_1 x + b_1 y$ Eliminate x or y and equation becomes separable

$y\,F(xy)\,dx + x\,G(xy)dy = 0$	$\ln x = \displaystyle\int \frac{G(v)\,dv}{v\{G(v) - F(v)\}} + c$ where $v = xy$. If $G(v) = F(v)$, the solution is $xy = c$.
Linear, homogeneous second order equation $\dfrac{d^2 y}{dx^2} + b\,\dfrac{dy}{dx} + cy = 0$ b, c are real constants	Let m_1, m_2 be the roots of $m^2 + bm + c = 0$. Then there are 3 cases: Case 1. m_1, m_2 real and distinct: $$y = c_1 e^{m_1 x} + c_2 e^{m_2 x}$$ Case 2. m_1, m_2 real and equal: $$y = c_1 e^{m_1 x} + c_2 x e^{m_1 x}$$ Case 3. $m_1 = p + qi,\ m_2 = p - qi$: $$y = e^{px}(c_1 \cos qx + c_2 \sin qx)$$ where $p = -b/2,\ q = \sqrt{4c - b^2}/2$
Linear, nonhomogeneous second order equation $\dfrac{d^2 y}{dx^2} + b\,\dfrac{dy}{dx} + cy = R(x)$ b, c are real constants	There are 3 cases corresponding to those immediately above: Case 1. $$y = c_1 e^{m_1 x} + c_2 e^{m_2 x}$$ $$+ \frac{e^{m_1 x}}{m_1 - m_2} \int e^{-m_1 x} R(x)\,dx$$ $$+ \frac{e^{m_2 x}}{m_2 - m_1} \int e^{-m_2 x} R(x)\,dx$$ Case 2. $$y = c_1 e^{m_1 x} + c_2 x e^{m_1 x}$$ $$+ x e^{m_1 x} \int e^{-m_1 x} R(x)\,dx$$ $$- e^{m_1 x} \int x e^{-m_1 x} R(x)\,dx$$ Case 3. $$y = e^{px}(c_1 \cos qx + c_2 \sin qx)$$ $$+ \frac{e^{px} \sin qx}{q} \int e^{-px} R(x) \cos qx\,dx$$ $$- \frac{e^{px} \cos qx}{q} \int e^{-px} R(x) \sin qx\,dx$$

Euler or Cauchy equation	Putting $x = e^t$, the equation becomes
$$x^2 \frac{d^2 y}{dx^2} + bx \frac{dy}{dx} + cy = S(x)$$	$$\frac{d^2 y}{dt^2} + (b-1) \frac{dy}{dt} + cy = S(e^t)$$ and can then be solved as a linear second order equation.
Bessel's equation $$x^2 \frac{d^2 y}{dx^2} + x \frac{dy}{dx} + (\lambda^2 x^2 - n^2)y = 0$$	$$y = c_1 J_n(\lambda x) + c_2 Y_n(\lambda x)$$ See pages 500–534.
Transformed Bessel's equation $$x^2 \frac{d^2 y}{dx^2} + (2p+1)x \frac{dy}{dx} + (\alpha^2 x^{2r} + \beta^2)y = 0$$	$$y = x^{-p} \left\{ c_1 J_{q/r}\left(\frac{\alpha}{r} x^r\right) + c_2 Y_{q/r}\left(\frac{\alpha}{r} x^r\right) \right\}$$ where $q = \sqrt{p^2 - \beta^2}$.
Legendre's equation $$(1 - x^2)\frac{d^2 y}{dx^2} - 2x \frac{dy}{dx} + n(n+1)y = 0$$	$$y = c_1 P_n(x) + c_2 Q_n(x)$$ See pages 557, 561, 564–576.

SPECIAL FORMULAS

Certain types of differential equations occur sufficiently often to justify the use of formulas for the corresponding particular solutions. The following set of tables I to XIV covers all first, second and nth order ordinary linear differential equations with constant coefficients for which the right members are of the form $P(x)e^{rx}\sin sx$ or $P(x)e^{rx}\cos sx$, where r and s are constants and $P(x)$ is a polynomial of degree n.

When the right member of a reducible linear partial differential equation with constant coefficients is not zero, particular solutions for certain types of right members are contained in tables XV to XXI. In these tables both F and P are used to denote polynomials, and it is assumed that no denominator is zero. In any formula the roles of x and y may be reversed throughout, changing a formula in which x dominates to one in which y dominates. Tables XIX, XX, XXI are applicable whether the equations are reducible or not. The symbol $\binom{m}{n}$ stands for $\dfrac{m!}{(m-n)!n!}$ and is the $n+1$ st coefficient in the expansion of $(a+b)^m$. Also $0! = 1$ by definition.

The tables as herewith given are those contained in the text *Differential Equations* by Ginn and Company (1955) and are published with their kind permission and that of the author, Professor Frederick H. Steen.

Solution of Linear Differential Equations with Constant Coefficients

Any linear differential equation with constant coefficients may be written in the form

$$p(D)y = R(x),$$

where D is the differential operator

$$Dy = \frac{dy}{dx},$$

$p(D)$ is a polynomial in D,
y is the dependent variable,
x is the independent variable,
$R(x)$ is an arbitrary function of x.

A power of D represents repeated differentiation, that is

$$D^n y = \frac{d^n y}{dx^n}.$$

For such an equation, the general solution may be written in the form

$$y = y_c + y_p,$$

where y_p is any particular solution, and y_c is called the *complementary function*. This complementary function is defined as the general solution of the *homogeneous equation*, which is the original differential equation with the right side replaced by zero, i.e.

$$p(D)y = 0$$

The complementary function y_c may be determined as follows:

1. Factor the polynomial $p(D)$ into real and complex linear factors, just as if D were a variable instead of an operator.

2. For each non-repeated linear factor of the form $(D - a)$, where a is real, write down a term of the form

$$ce^{ax},$$

where c is an arbitrary constant.

3. For each repeated real linear factor of the form $(D - a)^n$, write down n terms of the form

$$c_1 e^{ax} + c_2 x e^{ax} + c_3 x^2 e^{ax} + \cdots + c_n x^{n-1} e^{ax},$$

where the c_i's are arbitrary constants.

4. For each non-repeated conjugate complex pair of factors of the form $(D - a + ib)(D - a - ib)$, write down 2 terms of the form

$$c_1 e^{ax} \cos bx + c_2 e^{ax} \sin bx$$

5. For each repeated conjugate complex pair of factors of the form $(D - a + ib)^n (D - a - ib)^n$, write down $2n$ terms of the form

$$c_1 e^{ax} \cos bx + c_2 e^{ax} \sin bx + c_3 x e^{ax} \cos bx + c_4 x e^{ax} \sin bx + \cdots$$
$$+ c_{2n-1} x^{n-1} e^{ax} \cos bx + c_{2n} x^{n-1} e^{ax} \sin bx$$

6. The sum of all the terms thus written down is the complementary function y_c.

To find the particular solution y_p, use the following tables, as shown in the examples. For cases not shown in the tables, there are various methods of finding y_p. The most general method is called *variation of parameters*. The following example illustrates the method:

Find y_p for $(D^2 - 4) y = e^x$.

This example can be solved most easily by use of equation 63 in the tables following. However it is given here as an example of the method of variation of parameters.

The complementary function is

$$y_c = c_1 e^{2x} + c_2 e^{2x}$$

To find y_p, replace the constants in the complementary function with unknown functions,

$$y_p = ue^{2x} + ve^{-2x}.$$

We now prepare to substitute this assumed solution into the original equation. We begin by taking all the necessary derivatives:

$$y_p = ue^{2x} + ve^{-2x}$$
$$y_p' = 2ue^{2x} - 2ve^{-2x} + u'e^{2x} + v'e^{-2x}$$

For each derivative of y_p except the highest, we set the sum of all the terms containing u' and v' to 0. Thus the above equation becomes

$$u'e^{2x} + v'e^{-2x} = 0 \quad \text{and} \quad y_p' = 2ue^{2x} - 2ve^{-2x}$$

Continuing to differentiate, we have

$$y_p'' = 4ue^{2x} + 4ve^{-2x} + 2u'e^{2x} - 2v'e^{-2x}$$

When we substitute into the original equation, all the terms not containing u' or v' cancel out. This is a consequence of the method by which y_p was set up.

Thus all that is necessary is to write down the terms containing u' or v' in the highest order derivative of y_p, multiply by the constant coefficient of the highest power of D in $p(D)$, and set it equal to $R(x)$. Together with the previous terms in u' and v' which were set equal to 0, this gives us as many linear equations in the first derivatives of the unknown functions as there are unknown functions. The first derivatives may then be solved for by algebra, and the unknown functions found by integration. In the present example, this becomes

$$u'e^{2x} + v'e^{-2x} = 0$$
$$2u'e^{2x} - 2v'e^{-2x} = e^x.$$

We eliminate v' and u' separately, getting

$$4u'e^{2x} = e^x$$
$$4v'e^{-2x} = -e^x.$$

Thus

$$u' = \tfrac{1}{4}e^{-x}$$
$$v' = -\tfrac{1}{4}e^{3x}.$$

Therefore, by integrating

$$u = -\tfrac{1}{4}e^{-x}$$
$$v = -\tfrac{1}{12}e^{3x}.$$

A constant of integration is not needed, since we need only one particular solution. Thus

$$y_p = ue^{2x} + ve^{-2x} = -\tfrac{1}{4}e^{-x}e^{2x} - \tfrac{1}{12}e^{3x}e^{-2x}$$
$$= -\tfrac{1}{4}e^x - \tfrac{1}{12}e^x = -\tfrac{1}{3}e^x,$$

and the general solution is

$$y = y_c + y_p = c_1e^{2x} + c_2e^{-2x} - \tfrac{1}{3}e^x.$$

The following examples illustrate the use of the tables.

Example 1.   Solve $(D^2 - 4)y = \sin 3x$.

Substitution of $q = -4, s = 3$ in formula 24 gives

$$y_p = \frac{\sin 3x}{-9 - 4},$$

wherefore the general solution is

$$y = c_1e^{2x} + c_2e^{-2x} - \frac{\sin 3x}{13}.$$

Example 2.   Obtain a particular solution of $(D^2 - 4D + 5)y = x^2e^{3x}\sin x$.

Applying formula 40 with $a = 2, b = 1, r = 3, s = 1, P(x) = x^2, s + b = 2, s - b = 0, a - r = -1, (a - r)^2 + (s + b)^2 = 5, (a - r)^2 + (s - b)^2 = 1$, we have

$$y_p = \frac{e^{3x}\sin x}{2}\left[\left(\frac{2}{5} - \frac{0}{1}\right)x^2 + \left(\frac{2(-1)2}{25} - \frac{2(-1)0}{1}\right)2x\right.$$
$$\left. + \left(\frac{3\cdot1\cdot2 - 2^3}{125} - \frac{3\cdot1\cdot0 - 0}{1}\right)2\right]$$
$$- \frac{e^{3x}\cos x}{2}\left[\left(\frac{-1}{5} - \frac{-1}{1}\right)x^2 + \left(\frac{1 - 4}{25} - \frac{1 - 0}{1}\right)2x\right.$$
$$\left. + \left(\frac{-1 - 3(-1)4}{125} - \frac{-1 - 3(-1)0}{1}\right)2\right]$$
$$= (\tfrac{1}{5}x^2 - \tfrac{4}{25}x - \tfrac{2}{125})e^{3x}\sin x + (-\tfrac{2}{5}x^2 + \tfrac{28}{25}x - \tfrac{136}{125})e^{3x}\cos x.$$

The special formulas effect a very considerable saving of time in problems of this type.

Example 3. Obtain a particular solution of $(D^2 - 4D + 5)y = x^2 e^{2x} \cos x$. (Compare with Example 2.)

Formula 40 is not applicable here since for this equation $r = a$, $s = b$, wherefore the denominator $(a - r)^2 + (s - b)^2 = 0$. We turn instead to formula 44. Substituting $a = 2, b = 1, P(x) = x^2$ and replacing sin by cos, cos by $-\sin$, we obtain

$$y_p = \frac{e^{2x} \cos x}{4} \left(x^2 - \tfrac{2}{4}\right) + \frac{e^{2x} \sin x}{2} \int \left(x^2 - \tfrac{1}{2}\right) dx$$

$$= \left(\frac{x^2}{4} - \frac{1}{8}\right) e^{2x} \cos x + \left(\frac{x^3}{6} - \frac{x}{4}\right) e^{2x} \sin x,$$

which is the required solution.

Example 4. Find z_p for $(D_x - 3D_y)z = \ln(y + 3x)$.

Referring to Table XV we note that formula 69 (not 68) is applicable. This gives

$$z_p = x \ln(y + 3x).$$

It is easily seen that $-\dfrac{y}{3} \ln(y + 3x)$ would serve equally well.

Example 5. Solve $(D_x + 2D_y - 4)z = y \cos(y - 2x)$.

Since R in formula 76 contains a polynomial in x, not y, we rewrite the given equation in the form

$$(D_y + \tfrac{1}{2}D_x - 2)z = \tfrac{1}{2}y \cos(y - 2x).$$

Then

$$z_c = e^{2y} F(x - \tfrac{1}{2}y) = e^{2y} f(2x - y),$$

and by the formula

$$z_p = -\tfrac{1}{2} \cos(y - 2x) \cdot \left(\frac{y}{2} + \frac{\tfrac{1}{2}}{2}\right)$$

$$= -\tfrac{1}{8}(2y + 1) \cos(y - 2x).$$

Example 6. Find z_p for $(D_x + 4D_y)^3 z = (2x - y)^2$.

Using formula 79, we obtain

$$z_p = \frac{\int \int \int u^2 du^3}{[2 + 4(-1)]^3} = \frac{u^5}{5 \cdot 4 \cdot 3 \cdot (-8)} = -\frac{(2x - y)^5}{480}.$$

Example 7. Find z_p for $(D_x^3 + 5D_x^2 D_y - 7D_x + 4)z = e^{2x+3y}$.

By formula 87

$$z_p = \frac{e^{2x+3y}}{2^3 + 5 \cdot 2^2 \cdot 3 - 7 \cdot 2 + 4} = \frac{e^{2x+3y}}{58}.$$

Example 8. Find z_p for

$$(D_x^4 + 6D_x^3 D_y + D_x D_y + D_y^2 + 9)z = \sin(3x + 4y).$$

Since every term in the left member is of *even* degree in the two operators D_x and D_y, formula 90 is applicable.

It gives

$$z_p = \frac{\sin(3x + 4y)}{(-9)^2 + 6(-9)(-12) + (-12) + (-16) + 9}$$

$$= \frac{\sin(3x + 4y)}{710}$$

DIFFERENTIAL EQUATIONS

TABLE I: $(D - a)y = R$

R	y_p
1. e^{rx}	$\dfrac{e^{rx}}{r-a}$
2. $\sin sx$*	$-\dfrac{a\sin sx + s\cos sx}{a^2+s^2} = -\dfrac{1}{\sqrt{a^2+s^2}}\sin\left(sx + \tan^{-1}\dfrac{s}{a}\right)$
3. $P(x)$	$-\dfrac{1}{a}\left[P(x) + \dfrac{P'(x)}{a} + \dfrac{P''(x)}{a^2} + \cdots + \dfrac{P^{(n)}(x)}{a^n}\right]$
4. $e^{rx}\sin sx$*	Replace a by $a - r$ in formula 2 and multiply by e^{rx}.
5. $P(x)e^{rx}$	Replace a by $a - r$ in formula 3 and multiply by e^{rx}.
6. $P(x)\sin sx$*	$-\sin sx\left[\dfrac{a}{a^2+s^2}P(x) + \dfrac{a^2-s^2}{(a^2+s^2)^2}P'(x) + \dfrac{a^3-3as^2}{(a^2+s^2)^3}P''(x) + \cdots\right.$ $\left. + \dfrac{a^k - \binom{k}{2}a^{k-2}s^2 + \binom{k}{4}a^{k-4}s^4 - \cdots}{(a^2+s^2)^k}P^{(k-1)}(x) + \cdots\right]$ $-\cos sx\left[\dfrac{s}{a^2+s^2}P(x) + \dfrac{2as}{(a^2+s^2)^2}P'(x) + \dfrac{3a^2s-s^3}{(a^2+s^2)^3}P''(x) + \cdots\right.$ $\left. + \dfrac{\binom{k}{1}a^{k-1}s - \binom{k}{3}a^{k-3}s^3 + \cdots}{(a^2+s^2)^k}P^{(k-1)}(x) + \cdots\right]$
7. $P(x)e^{rx}\sin sx$*	Replace a by $a - r$ in formula 6 and multiply by e^{rx}.
8. e^{ax}	xe^{ax}
9. $e^{ax}\sin sx$*	$-\dfrac{e^{ax}\cos sx}{s}$
10. $P(x)e^{ax}$	$e^{ax}\displaystyle\int\int P(x)\,dx$
11. $P(x)e^{ax}\sin sx$*	$\dfrac{e^{ax}\sin sx}{s}\left[\dfrac{P'(x)}{s} - \dfrac{P'''(x)}{s^3} + \dfrac{P^{v}(x)}{s^5} - \cdots\right] - \dfrac{e^{ax}\cos sx}{s}\left[P(x) - \dfrac{P''(x)}{s^2} + \dfrac{P^{iv}(x)}{s^4} - \cdots\right]$

*For cos sx in R replace "sin" by "cos" and "cos" by "$-$ sin" in y_p.

$$D^n = \frac{d^n}{dx^n} \qquad \binom{m}{n} = \frac{m!}{(m-n)!n!} \qquad 0! = 1$$

DIFFERENTIAL EQUATIONS (Continued)

TABLE II: $(D-a)^2 y = R$

R	y_p
12. e^{rz}	$\dfrac{e^{rz}}{(r-a)^2}$
13. $\sin sx$*	$\dfrac{1}{(a^2+s^2)^2}[(a^2-s^2)\sin sx + 2as\cos sx] = \dfrac{1}{a^2+s^2}\sin\left(sx+\tan^{-1}\dfrac{2as}{a^2-s^2}\right)$
14. $P(x)$	$\dfrac{1}{a^2}\left[P(x) + \dfrac{2P'(x)}{a} + \dfrac{3P''(x)}{a^2} + \cdots + \dfrac{(n+1)P^{(n)}(x)}{a^n}\right]$
15. $e^{rz}\sin sx$*	Replace a by $a-r$ in formula 13 and multiply by e^{rz}.
16. $P(x)e^{rz}$	Replace a by $a-r$ in formula 14 and multiply by e^{rz}.
17. $P(x)\sin sx$*	$\sin sx\left[\dfrac{a^2-s^2}{(a^2+s^2)^2}P(x) + 2\dfrac{a^3-3as^2}{(a^2+s^2)^3}P'(x) + 3\dfrac{a^4-6a^2s^2+s^4}{(a^2+s^2)^4}P''(x) + \cdots\right.$ $\left. + (k-1)\dfrac{a^k - \binom{k}{2}a^{k-2}s^2 + \binom{k}{4}a^{k-4}s^4 - \cdots}{(a^2+s^2)^k}P^{(k-2)}(x) + \cdots\right]$ $+ \cos sx\left[\dfrac{2as}{(a^2+s^2)^2}P(x) + 2\dfrac{3a^2s-s^3}{(a^2+s^2)^3}P'(x) + 3\dfrac{4a^3s-4as^3}{(a^2+s^2)^4}P''(x) + \cdots\right.$ $\left. + (k-1)\dfrac{\binom{k}{1}a^{k-1}s - \binom{k}{3}a^{k-3}s^3 + \cdots}{(a^2+s^2)^k}P^{(k-2)}(x) + \cdots\right]$
18. $P(x)e^{rz}\sin sx$*	Replace a by $a-r$ in formula 17 and multiply by e^{rz}.
19. e^{az}	$\tfrac{1}{2}x^2 e^{az}$
20. $e^{az}\sin sx$*	$-\dfrac{e^{az}\sin sx}{s^2}$
21. $P(x)e^{az}$	$e^{az}\displaystyle\iint P(x)\,dx\,dx$
22. $P(x)e^{az}\sin sx$*	$-\dfrac{e^{az}\sin sx}{s^2}\left[P(x) - \dfrac{3P''(x)}{s^2} + \dfrac{5P^{iv}(x)}{s^4} - \dfrac{7P^{vi}(x)}{s^6} + \cdots\right]$ $-\dfrac{e^{az}\cos sx}{s^2}\left[\dfrac{2P'(\cdot)}{s} - \dfrac{4P'''(x)}{s^3} + \dfrac{6P^{v}(x)}{s^5} - \cdots\right]$

* For $\cos sx$ in R replace "sin" by "cos" and "cos" by "$-$ sin" in y_p.

DIFFERENTIAL EQUATIONS (Continued)

TABLE III: $(D^2 + q)y = R$

R	y_p
23. e^{rx}	$\dfrac{e^{rx}}{r^2+q}$
24. $\sin sx$*	$\dfrac{\sin sx}{-s^2+q}$
25. $P(x)$	$\dfrac{1}{q}\left[P(x) - \dfrac{P''(x)}{q} + \dfrac{P^{iv}(x)}{q^2} - \cdots + (-1)^k\dfrac{P^{(2k)}(x)}{q^k} \cdots\right]$
26. $e^{rx}\sin sx$*	$\dfrac{(r^2-s^2+q)e^{rx}\sin sx - 2\,rse^{rx}\cos sx}{(r^2-s^2+q)^2+(2rs)^2} = \dfrac{e^{rx}}{\sqrt{(r^2-s^2+q)^2+(2rs)^2}}\sin\left[sx - \tan^{-1}\dfrac{2rs}{r^2-s^2+q}\right]$
27. $P(x)e^{rx}$	$\dfrac{e^{rx}}{r^2+q}\left[P(x) - \dfrac{2r}{r^2+q}P'(x) + \dfrac{3r^2-q}{(r^2+q)^2}P''(x) - \dfrac{4r^3-4qr}{(r^2+q)^3}P'''(x)\right.$
	$\left. + \cdots + (-1)^{k-1}\dfrac{\binom{k}{1}r^{k-1} - \binom{k}{3}r^{k-3}q + \binom{k}{5}r^{k-5}q^2 - \cdots}{(r^2+q)^{k-1}}P^{(k-1)}(x) + \cdots\right]$
28. $P(x)\sin sx$*	$\dfrac{\sin sx}{(-s^2+q)}\left[P(x) - \dfrac{3s^2+q}{(-s^2+q)^2}P''(x) + \dfrac{5s^4+10s^2q+q^2}{(-s^2+q)^4}P^{iv}(x) + \cdots\right.$
	$\left. + (-1)^k\dfrac{\binom{2k+1}{1}s^{2k} + \binom{2k+1}{3}s^{2k-2}q + \binom{2k+1}{5}s^{2k-4}q^2 + \cdots}{(-s^2+q)^{2k}}P^{(2k)}(x) + \cdots\right]$
	$-\dfrac{s\cos sx}{(-s^2+q)}\left[\dfrac{2P'(x)}{(-s^2+q)} - \dfrac{4s^2+4q}{(-s^2+q)^3}P'''(x) + \cdots\right.$
	$\left. + (-1)^{k+1}\dfrac{\binom{2k}{1}s^{2k-2} + \binom{2k}{3}s^{2k-4}q + \cdots}{(-s^2+q)^{2k-1}}P^{(2k-1)}(x) + \cdots\right]$

TABLE IV: $(D^2 + b^2)y = R$

R	y_p
29. $\sin bx$*	$-\dfrac{x\cos bx}{2b}$
30. $P(x)\sin bx$*	$\dfrac{\sin bx}{(2b)^2}\left[P(x) - \dfrac{P''(x)}{(2b)^2} + \dfrac{P^{iv}(x)}{(2b)^4} - \cdots\right] - \dfrac{\cos bx}{2b}\int\left[P(x) - \dfrac{P''(x)}{(2b)^2} + \cdots\right]dx$

* For $\cos sx$ in R replace "sin" by "cos" and "cos" by "$-$sin" in y_p.

DIFFERENTIAL EQUATIONS (Continued)

TABLE V: $(D^2 + pD + q)y = R$

R	y_p
31. e^{rx}	$\dfrac{e^{rx}}{r^2 + pr + q}$
32. $\sin sx$*	$\dfrac{(q - s^2)\sin sx - ps\cos sx}{(q - s^2)^2 + (ps)^2} = \dfrac{1}{\sqrt{(q - s^2)^2 + (ps)^2}}\sin\left(sx - \tan^{-1}\dfrac{ps}{q - s^2}\right)$
33. $P(x)$	$\dfrac{1}{q}\left[P(x) - \dfrac{p}{q}P'(x) + \dfrac{p^2 - q}{q^2}P''(x) - \dfrac{p^3 - 2pq}{q^3}P'''(x) + \cdots \right.$ $\left. + (-1)^n\dfrac{p^n - \binom{n-1}{1}p^{n-2}q + \binom{n-2}{2}p^{n-4}q^2 - \cdots}{q^n}P^{(n)}(x)\right]$
34. $e^{rx}\sin sx$*	Replace p by $p + 2r$, q by $q + pr + r^2$ in formula 32 and multiply by e^{rx}.
35. $P(x)e^{rx}$	Replace p by $p + 2r$, q by $q + pr + r^2$ in formula 33 and multiply by e^{rx}.

TABLE VI: $(D - b)(D - a)y = R$

36. $P(x)\sin sx$*	$\dfrac{\sin sx}{b - a}\left[\left(\dfrac{a}{a^2 + s^2} - \dfrac{b}{b^2 + s^2}\right)P(x) + \left(\dfrac{a^2 - s^2}{(a^2 + s^2)^2} - \dfrac{b^2 - s^2}{(b^2 + s^2)^2}\right)P'(x)\right.$ $\left. + \left(\dfrac{a^3 - 3as^2}{(a^2 + s^2)^3} - \dfrac{b^3 - 3bs^2}{(b^2 + s^2)^3}\right)P''(x) + \cdots\right]$ $+ \dfrac{\cos sx}{b - a}\left[\left(\dfrac{s}{a^2 + s^2} - \dfrac{s}{b^2 + s^2}\right)P(x) + \left(\dfrac{2as}{(a^2 + s^2)^2} - \dfrac{2bs}{(b^2 + s^2)^2}\right)P'(x)\right.$ $\left. + \left(\dfrac{3a^2s - s^3}{(a^2 + s^2)^3} - \dfrac{3b^2s - s^3}{(b^2 + s^2)^3}\right)P''(x) + \cdots\right]$†
37. $P(x)e^{rx}\sin sx$*	Replace a by $a - r$, b by $b - r$ in formula 36 and multiply by e^{rx}.
38. $P(x)e^{ax}$	$\dfrac{e^{ax}}{a - b}\left[\displaystyle\int P(x)\,dx + \dfrac{P(x)}{(b - a)} + \dfrac{P'(x)}{(b - a)^2} + \dfrac{P''(x)}{(b - a)^3} + \cdots + \dfrac{P^{(n)}(x)}{(b - a)^{n+1}}\right]$

* For cos sx in R replace "sin" by "cos" and "cos" by "− sin" in y_p.

† For additional terms, compare with formula 6.

DIFFERENTIAL EQUATIONS (Continued)

TABLE VII: $(D^2 - 2aD + a^2 + b^2)y = R$

R	y_p

39. $P(x) \sin sx$*

$$\frac{\sin sx}{2b} \left[\left(\frac{s+b}{a^2 + (s+b)^2} - \frac{s-b}{a^2 + (s-b)^2} \right) P(x) + \left(\frac{2a(s+b)}{[a^2+(s+b)^2]^2} - \frac{2a(s-b)}{[a^2+(s-b)^2]^2} \right) P'(x) \right.$$

$$+ \left(\frac{3a^2(s+b) - (s+b)^3}{[a^2+(s+b)^2]^3} - \frac{3a^2(s-b)-(s-b)^3}{[a^2+(s-b)^2]^3} \right) P''(x) + \cdots \right]$$

$$- \frac{\cos sx}{2b} \left[\left(\frac{a}{a^2 + (s+b)^2} - \frac{a}{a^2 + (s-b)^2} \right) P(x) + \left(\frac{a^2 - (s+b)^2}{[a^2+(s+b)^2]^2} - \frac{a^2 - (s-b)^2}{[a^2+(s-b)^2]^2} \right) P'(x) \right.$$

$$+ \left(\frac{a^3 - 3a(s+b)^2}{[a^2+(s+b)^2]^3} - \frac{a^3 - 3a(s-b)^2}{[a^2+(s-b)^2]^3} \right) P''(x) + \cdots \right]^\dagger$$

40. $P(x)e^{rx} \sin sx$*   Replace a by $a - r$ in formula 39 and multiply by e^{rx}.

41. $P(x)e^{ax}$ $\dfrac{e^{ax}}{b^2} \left[P(x) - \dfrac{P''(x)}{b^2} + \dfrac{P^{iv}(x)}{b^4} - \cdots \right]$

42. $e^{ax} \sin sx$* $\dfrac{e^{ax} \sin sx}{-s^2 + b^2}$

43. $e^{ax} \sin bx$* $-\dfrac{xe^{ax} \cos bx}{2b}$

44. $P(x)e^{ax} \sin bx$* $\dfrac{e^{ax} \sin bx}{(2b)^2} \left[P(x) - \dfrac{P''(x)}{(2b)^2} + \dfrac{P^{iv}(x)}{(2b)^4} - \cdots \right] - \dfrac{e^{ax} \cos bx}{2b} \int \int \left[P(x) - \dfrac{P''(x)}{(2b)^2} + \dfrac{P^{iv}(x)}{(2b)^4} - \cdots \right] dx$

* For $\cos sx$ in R replace "sin" by "cos" and "cos" by "$-\sin$" in y_p.

† For additional terms, compare with formula 6.

DIFFERENTIAL EQUATIONS (Continued)

TABLE VIII: $f(D)y = [D^n + a_{n-1}D^{n-1} + \cdots + a_1 D + a_0]y = R$

R	y_p
45. e^{rx}	$\dfrac{e^{rx}}{f(r)}$
46. $\sin sx$*	$\dfrac{[a_0 - a_2 s^2 + a_4 s^4 - \cdots]\sin sx - [a_1 s - a_3 s^3 + a_5 s^5 + \cdots]\cos sx}{[a_0 - a_2 s^2 + a_4 s^4 - \cdots]^2 + [a_1 s - a_3 s^3 + a_5 s^5 - \cdots]^2}$

TABLE IX: $f(D^2)y = R$

47. $\sin sx$*	$\dfrac{\sin sx}{f(-s^2)} = \dfrac{\sin sx}{a_0 - a_2 s^2 + \cdots \pm s^{2n}}$

TABLE X: $(D - a)^n y = R$

48. e^{rx}	$\dfrac{e^{rx}}{(r - a)^n}$
49. $\sin sx$*	$\dfrac{(-1)^n}{(a^2 + s^2)^n}\{[(a^n - \binom{n}{2}a^{n-2}s^2 + \binom{n}{4}a^{n-4}s^4 - \cdots]\sin sx + [\binom{n}{1}a^{n-1}s - \binom{n}{3}a^{n-3}s^3 + \cdots]\cos sx\}$
50. $P(x)$	$\dfrac{(-1)^n}{a^n}\left[P(x) + \binom{n}{1}\dfrac{P'(x)}{a} + \binom{n+1}{2}\dfrac{P''(x)}{a^2} + \binom{n+2}{3}\dfrac{P'''(x)}{a^3} + \cdots \right]$
51. $e^{rx}\sin sx$*	Replace a by $a - r$ in formula 49 and multiply by e^{rx}.
52. $e^{rx}P(x)$	Replace a by $a - r$ in formula 50 and multiply by e^{rx}.

* For $\cos sx$ in R replace "sin" by "cos" and "cos" by "$-$ sin" in y_p.

DIFFERENTIAL EQUATIONS (Continued)

53. $P(x)\sin sx$*

$$(-1)^n \sin sx[A_n P(x) + \tbinom{n}{1}A_{n+1}P'(x) + \tbinom{n+1}{2}A_{n+2}P''(x) + \tbinom{n+2}{3}A_{n+3}P'''(x) + \cdots]$$
$$+ (-1)^n \cos sx[B_n P(x) + \tbinom{n}{1}B_{n+1}P'(x) + \tbinom{n+1}{2}B_{n+2}P''(x) + \tbinom{n+2}{3}B_{n+3}P'''(x) + \cdots]$$

$$A_1 = \frac{a}{a^2+s^2}, \quad A_2 = \frac{a^2-s^2}{(a^2+s^2)^2}, \quad \cdots, \quad A_k = \frac{a^k - \binom{k}{2}a^{k-2}s^2 + \binom{k}{4}a^{k-4}s^4 - \cdots}{(a^2+s^2)^k}$$

$$B_1 = \frac{s}{a^2+s^2}, \quad B_2 = \frac{2as}{(a^2+s^2)^2}, \quad \cdots, \quad B_k = \frac{\binom{k}{1}a^{k-1}s - \binom{k}{3}a^{k-3}s^3 + \cdots}{(a^2+s^2)^k}$$

54. $P(x)e^{rx}\sin sx$*   Replace a by $a-r$ in formula 53 and multiply by e^{rx}.

55. $e^{az}P(x)$ $\displaystyle e^{az}\underbrace{\int\int\cdots\int}_{n-1} P(x)dx^n$

56. $P(x)e^{az}\sin sx$*

$$\frac{(-1)^{\frac{n-1}{2}}e^{az}\sin sx}{s^n}\left[\binom{n}{n-1}\frac{P'(x)}{s} - \binom{n+2}{n-1}\frac{P'''(x)}{s^3} + \binom{n+4}{n-1}\frac{P^v(x)}{s^5} - \cdots\right]$$
$$+ \frac{(-1)^{\frac{n+1}{2}}e^{az}\cos sx}{s^n}\left[\binom{n-1}{n-1}P(x) - \binom{n+1}{n-1}\frac{P''(x)}{s^2} + \binom{n+3}{n-1}\frac{P^{iv}(x)}{s^4} - \cdots\right] \quad (n\ \text{odd})$$

$$\frac{(-1)^{\frac{n}{2}}e^{az}\sin sx}{s^n}\left[\binom{n-1}{n-1}P(x) - \binom{n+1}{n-1}\frac{P''(x)}{s^2} + \binom{n+3}{n-1}\frac{P^{iv}(x)}{s^4} - \cdots\right]$$
$$+ \frac{(-1)^{\frac{n}{2}}e^{az}\cos sx}{s^n}\left[\binom{n}{n-1}\frac{P'(x)}{s} - \binom{n+2}{n-1}\frac{P'''(x)}{s^3} + \binom{n+4}{n-1}\frac{P^v(x)}{s^5} - \cdots\right] \quad (n\ \text{even})$$

TABLE XI: $(D-a)^n f(D)y = R$

57. e^{az} $\displaystyle \frac{x^n}{n!}\cdot\frac{e^{az}}{f(a)}$

* For cos sx in R replace "sin" by "cos" and "cos" by "− sin" in y_p.

DIFFERENTIAL EQUATIONS (Continued)

TABLE XII: $(D^2 + q)^n y = R$

R	y_p
58. e^{rx}	$e^{rx}/(r^2 + q)^n$
59. $\sin sx*$	$\sin sx/(q - s^2)^n$
60. $P(x)$	$\dfrac{1}{q^n}\left[P(x) - \binom{n}{1}\dfrac{P''(x)}{q} + \binom{n+1}{2}\dfrac{P^{iv}(x)}{q^2} - \binom{n+2}{3}\dfrac{P^{vi}(x)}{q^3} + \cdots\right]$
61. $e^{rx}\sin sx*$	$\dfrac{e^{rx}}{(A^2 + B^2)^n}\left\{[A^n - \binom{n}{2}A^{n-2}B^2 + \binom{n}{4}A^{n-4}B^4 - \cdots]\sin sx - [\binom{n}{1}A^{n-1}B - \binom{n}{3}A^{n-3}B^3 + \cdots]\cos sx\right\}$

$$A = r^2 - s^2 + q, \quad B = 2rs$$

TABLE XIII: $(D^2 + b^2)^n y = R$

62. $\sin bx*$	$(-1)^{\frac{n+1}{2}}\dfrac{x^n \cos bx}{n!(2b)^n}$ (n odd), $\quad (-1)^{\frac{n}{2}}\dfrac{x^n \sin bx}{n!(2b)^n}$ (n even)

TABLE XIV: $(D^n - q)y = R$

63. e^{rx}	$e^{rx}/(r^n - q)$
64. $P(x)$	$-\dfrac{1}{q}\left[P(x) + \dfrac{P^{(n)}(x)}{q} + \dfrac{P^{(2n)}(x)}{q^2} + \cdots\right]$
65. $\sin sx*$	$-\dfrac{q\sin sx + (-1)^{\frac{n-1}{2}}s^n \cos sx}{q^2 + s^{2n}}$ (n odd), $\quad \dfrac{\sin sx}{(-s^2)^{n/2} - q}$ (n even)
66. $e^{rx}\sin sx*$	$\dfrac{Ae^{rx}\sin sr - Be^{rx}\cos sx}{A^2 + B^2} = \dfrac{e^{rx}}{\sqrt{A^2 + B^2}}\sin\left(sx - \tan^{-1}\dfrac{B}{A}\right)] - q$

$$A = [r^n - \binom{n}{2}r^{n-2}s^2 + \binom{n}{4}r^{n-4}s^4 - \cdots] - q, \quad B = [\binom{n}{1}r^{n-1}s - \binom{n}{3}r^{n-3}s^3 + \cdots]$$

*For cos sx in R replace "sin" by "cos" and "cos" by "$-$ sin" in y_p.

DIFFERENTIAL EQUATIONS (Continued)

TABLE XV: $(D_x + mD_y)z = R$

R	z_p
67. e^{ax+by}	$\dfrac{e^{ax+by}}{a+mb}$
68. $f(ax+by)$	$\dfrac{\int f(u)du}{a+mb}$, $u = ax+by$
69. $f(y-mx)$	$xf(y-mx)$
70. $\phi(x,y)f(y-mx)$	$f(y-mx)\int \phi(x,\,a+mx)dx$ $(a = y - mx$ after integration$)$

TABLE XVI: $(D_x + mD_y - k)z = R$

71. e^{ax+by}	$\dfrac{e^{ax+by}}{a+mb-k}$
72. $\sin(ax+by)$*	$-\dfrac{(a+bm)\cos(ax+by) + k\sin(ax+by)}{(a+bm)^2 + k^2}$
73. $e^{\alpha x+\beta y}\sin(ax+by)$*	Replace k in 72 by $k - \alpha - m\beta$ and multiply by $e^{\alpha x+\beta y}$
74. $e^{kx}f(ax+by)$	$\dfrac{e^{kx}\int f(u)du}{a+mb}$, $u = ax+by$
75. $f(y-mx)$	$-\dfrac{f(y-mx)}{k}$
76. $P(x)f(y-mx)$	$-\dfrac{1}{k}f(y-mx)\left[P(x) + \dfrac{P'(x)}{k} + \dfrac{P''(x)}{k^2} + \cdots + \dfrac{P^{(n)}(x)}{k^n}\right]$
77. $e^{kx}f(y-mx)$	$xe^{kx}f(y-mx)$

*For $\cos(ax+by)$ replace "sin" by "cos," and "cos" by "$-\sin$" in z_p.

$$D_x = \frac{\partial}{\partial x}; \qquad D_y = \frac{\partial}{\partial y}; \qquad D_x^k D_y^r = \frac{\partial^{k+r}}{\partial_x^k \partial_y^r}$$

DIFFERENTIAL EQUATIONS (Continued)

TABLE XVII: $(D_x + mD_y)^n z = R$

R	z_p
78. e^{ax+by}	$\dfrac{e^{ax+by}}{(a+mb)^n}$
79. $f(ax+by)$	$\dfrac{\int\int\cdots\int f(u)du^n}{(a+mb)^n}$, $u = ax + by$
80. $f(y-mx)$	$\dfrac{x^n}{n!}f(y-mx)$
81. $\phi(x,y)f(y+mx)$	$f(y-mx)\int\int\int\cdots\int\phi(x, a+mx)dx^n$ $\quad(a = y - mx$ after integration$)$

TABLE XVIII: $(D_x + mD_y - k)^n z = R$

R	
82. e^{ax+by}	$\dfrac{e^{ax+by}}{(a+mb-k)^n}$
83. $f(y-mx)$	$\dfrac{(-1)^n f(y-mx)}{k^n}$
84. $P(x)f(y-mx)$	$\dfrac{(-1)^n}{k^n}f(y-mx)\left[P(x) + \binom{n}{1}\dfrac{P'(x)}{k} + \binom{n+1}{2}\dfrac{P''(x)}{k^2} + \binom{n+2}{3}\dfrac{P'''(x)}{k^3} + \cdots\right]$
85. $e^{kx}f(ax+by)$	$\dfrac{e^{kx}\int\int\cdots\int f(u)du^n}{(a+mb)^n}$, $u = ax + by$
86. $e^{kx}f(y-mx)$	$\dfrac{x^n}{n!}e^{kx}f(y-mx)$

DIFFERENTIAL EQUATIONS (Continued)

TABLE XIX: $[D_x^n + a_1 D_x^{n-1} D_y + a_2 D_x^{n-2} D_y^2 + \cdots + a^n D_y^n] z = R$

87. e^{ax+by}

$$\dfrac{e^{ax+by}}{a + a_1 a^{n-1} b + a_2 a^{n-2} b^2 + \cdots + a_n b^n}$$

88. $f(ax + by)$

$$\dfrac{\int\!\int \cdots \int f(u) du^n}{a^n + a_1 a^{n-1} b + a_2 a^{n-2} b^2 + \cdots + a^n b^n}, \quad (u = ax + by)$$

TABLE XX: $F(D_x, D_y) z = R$

89. e^{ax+by}

$$\dfrac{e^{ax+by}}{F(a, b)}$$

TABLE XXI: $F(D_x^2, D_x D_y, D_y^2) z = R$

90. $\sin(ax + by)$*

$$\dfrac{\sin(ax + by)}{F(-a^2, -ab, -b^2)}$$

* For $\cos(ax + by)$ replace "sin" by "cos", and "cos" by "$-\sin$" in z_p.

X. SPECIAL FUNCTIONS

The Gamma Function

Definition: $\Gamma(n) = \displaystyle\int_0^\infty t^{n-1} e^{-t}\, dt \quad n > 0$

Recursion Formula: $\Gamma(n + 1) = n\Gamma(n)$

$\Gamma(n + 1) = n!$ if $n = 0, 1, 2, \ldots$ where $0! = 1$

For $n < 0$ the gamma function can be defined by using

$$\Gamma(n) = \frac{\Gamma(n + 1)}{n}$$

Graph:

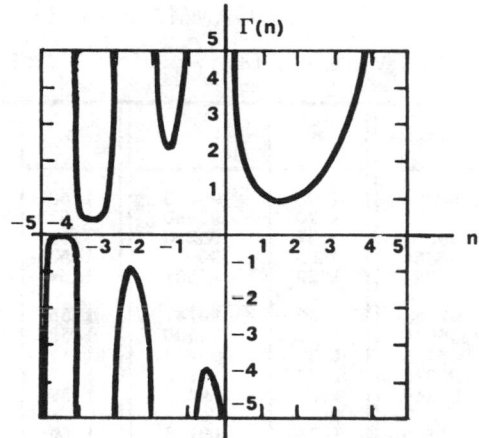

Special Values: $\qquad\qquad \Gamma(\tfrac{1}{2}) = \sqrt{\pi}$

$$\Gamma(m + \tfrac{1}{2}) = \frac{1 \cdot 3 \cdot 5 \cdots (2m - 1)}{2^m} \sqrt{\pi} \qquad m = 1, 2, 3, \ldots$$

$$\Gamma(-m + \tfrac{1}{2}) = \frac{(-1)^m 2^m \sqrt{\pi}}{1 \cdot 3 \cdot 5 \cdots (2m - 1)} \qquad m = 1, 2, 3, \ldots$$

Definition: $\quad \Gamma(x + 1) = \displaystyle\lim_{k \to \infty} \frac{1 \cdot 2 \cdot 3 \cdots k}{(x + 1)(x + 2) \cdots (x + k)} k^x$

$$\frac{1}{\Gamma(x)} = x e^{\gamma x} \prod_{m=1}^{\infty} \left\{ \left(1 + \frac{x}{m}\right) e^{-x/m} \right\}$$

This is an infinite product representation for the gamma function where γ is Euler's constant.

Properties:

$$\Gamma'(1) = \int_0^\infty e^{-x} \ln x \, dx = -\gamma$$

$$\frac{\Gamma'(x)}{\Gamma(x)} = -\gamma + \left(\frac{1}{1} - \frac{1}{x}\right) + \left(\frac{1}{2} - \frac{1}{x+1}\right) + \cdots + \left(\frac{1}{n} - \frac{1}{x+n-1}\right) + \cdots$$

$$\Gamma(x+1) = \sqrt{2\pi x} \; x^x e^{-x} \left\{1 + \frac{1}{12x} + \frac{1}{288x^2} - \frac{139}{51,840x^3} + \cdots\right\}$$

This is called *Stirling's asymptotic series.*

If we let $x = n$ a positive integer, then a useful approximation for $n!$ where n is large [e.g. $n > 10$] is given by *Stirling's formula*

$$n! \approx \sqrt{2\pi n} \; n^n e^{-n}$$

GAMMA FUNCTION*

Values of $\Gamma(n) = \int_0^\infty e^{-x} x^{n-1} \, dx$; $\Gamma(n+1) = n\Gamma(n)$

n	$\Gamma(n)$	n	$\Gamma(n)$	n	$\Gamma(n)$	n	$\Gamma(n)$
1.00	1.00000	1.25	.90640	1.50	.88623	1.75	.91906
1.01	.99433	1.26	.90440	1.51	.88659	1.76	.92137
1.02	.98884	1.27	.90250	1.52	.88704	1.77	.92376
1.03	.98355	1.28	.90072	1.53	.88757	1.78	.92623
1.04	.97844	1.29	.89904	1.54	.88818	1.79	.92877
1.05	.97350	1.30	.89747	1.55	.88887	1.80	.93138
1.06	.96874	1.31	.89600	1.56	.88964	1.81	.93408
1.07	.96415	1.32	.89464	1.57	.89049	1.82	.93685
1.08	.95973	1.33	.89338	1.58	.89142	1.83	.93969
1.09	.95546	1.34	.89222	1.59	.89243	1.84	.94261
1.10	.95135	1.35	.89115	1.60	.89352	1.85	.94561
1.11	.94740	1.36	.89018	1.61	.89468	1.86	.94869
1.12	.94359	1.37	.88931	1.62	.89592	1.87	.95184
1.13	.93993	1.38	.88854	1.63	.89724	1.88	.95507
1.14	.93642	1.39	.88785	1.64	.89864	1.89	.95838
1.15	.93304	1.40	.88726	1.65	.90012	1.90	.96177
1.16	.92980	1.41	.88676	1.66	.90167	1.91	.96523
1.17	.92670	1.42	.88636	1.67	.90330	1.92	.96877
1.18	.92373	1.43	.88604	1.68	.90500	1.93	.97240
1.19	.92089	1.44	.88581	1.69	.90678	1.94	.97610
1.20	.91817	1.45	.88566	1.70	.90864	1.95	.97988
1.21	.91558	1.46	.88560	1.71	.91057	1.96	.98374
1.22	.91311	1.47	.88563	1.72	.91258	1.97	.98768
1.23	.91075	1.48	.88575	1.73	.91466	1.98	.99171
1.24	.90852	1.49	.88595	1.74	.91683	1.99	.99581
						2.00	1.00000

* For large positive values of z, $\Gamma(z)$ approximates Stirling's asymptotic series

$$z^z e^{-z} \sqrt{\frac{2\pi}{z}} \left[1 + \frac{1}{12z} + \frac{1}{288z^2} - \frac{139}{51840z^3} - \frac{571}{2488320z^4} + \cdots\right].$$

GAMMA FUNCTION AND ITS NATURAL LOGARITHM

n	$\Gamma(n)$	$\log_e \Gamma(n)$		n	$\Gamma(n)$	$\log_e \Gamma(n)$	
1.000	1.00000 00	0.00000 00	0.000	1.250	0.90640 25	−0.09827 18	0.250
1.005	0.99713 85	−0.00286 56	0.005	1.255	0.90538 58	−0.09939 42	0.255
1.010	0.99432 59	−0.00569 03	0.010	1.260	0.90439 71	−0.10048 67	0.260
1.015	0.99156 13	−0.00847 45	0.015	1.265	0.90343 63	−0.10154 97	0.265
1.020	0.98884 42	−0.01121 85	0.020	1.270	0.90250 31	−0.10258 32	0.270
1.025	0.98617 40	−0.01392 25	0.025	1.275	0.90159 72	−0.10358 74	0.275
1.030	0.98355 00	−0.01658 69	0.030	1.280	0.90071 85	−0.10456 25	0.280
1.035	0.98097 16	−0.01921 18	0.035	1.285	0.89986 67	−0.10550 87	0.285
1.040	0.97843 82	−0.02179 77	0.040	1.290	0.89904 16	−0.10642 60	0.290
1.045	0.97594 93	−0.02434 46	0.045	1.295	0.89824 30	−0.10731 47	0.295
1.050	0.97350 43	−0.02685 31	0.050	1.300	0.89747 07	−0.10817 48	0.300
1.055	0.97110 26	−0.02932 32	0.055	1.305	0.89672 45	−0.10900 66	0.305
1.060	0.96874 36	−0.03175 53	0.060	1.310	0.89600 42	−0.10981 02	0.310
1.065	0.96642 70	0.03414 95	0.065	1.315	0.89530 96	−0.11058 57	0.315
1.070	0.96415 20	−0.03650 63	0.070	1.320	0.89464 05	−0.11133 34	0.320
1.075	0.96191 83	−0.03882 57	0.075	1.325	0.89399 67	−0.11205 32	0.325
1.080	0.95972 53	−0.04110 82	0.080	1.330	0.89337 81	−0.11274 54	0.330
1.085	0.95757 25	−0.04335 38	0.085	1.335	0.89278 44	−0.11341 02	0.335
1.090	0.95545 95	−0.04556 29	0.090	1.340	0.89221 55	−0.11404 76	0.340
1.095	0.95338 57	−0.04773 57	0.095	1.345	0.89167 12	−0.11465 78	0.345
1.100	0.95135 08	−0.04987 24	0.100	1.350	0.89115 14	−0.11524 09	0.350
1.105	0.94935 42	−0.05197 33	0.105	1.355	0.89065 59	−0.11579 71	0.355
1.110	0.94739 55	−0.05403 86	0.110	1.360	0.89018 45	−0.11632 65	0.360
1.115	0.94547 43	−0.05606 86	0.115	1.365	0.88973 71	−0.11682 92	0.365
1.120	0.94359 02	−0.05806 33	0.120	1.370	0.88931 35	−0.11730 55	0.370
1.125	0.94174 27	−0.06002 32	0.125	1.375	0.88891 36	−0.11775 53	0.375
1.130	0.93993 14	−0.06194 83	0.130	1.380	0.88853 71	−0.11817 88	0.380
1.135	0.93815 60	−0.06383 90	0.135	1.385	0.88818 41	−0.11857 62	0.385
1.140	0.93641 61	−0.06569 54	0.140	1.390	0.88785 43	−0.11894 76	0.390
1.145	0.93471 12	−0.06751 77	0.145	1.395	0.88754 76	−0.11929 32	0.395
1.150	0.93304 09	−0.06930 62	0.150	1.400	0.88726 38	−0.11961 29	0.400
1.155	0.93140 50	−0.07106 11	0.155	1.405	0.88700 29	−0.11990 70	0.405
1.160	0.92980 31	−0.07278 25	0.160	1.410	0.88676 47	−0.12017 57	0.410
1.165	0.92823 47	−0.07447 07	0.165	1.415	0.88654 90	−0.12041 89	0.415
1.170	0.92669 96	−0.07612 58	0.170	1.420	0.88635 58	−0.12063 68	0 420
1.175	0.92519 74	−0.07774 81	0.175	1.425	0.88618 49	−0.12082 97	0.425
1.180	0.92372 78	−0.07933 78	0.180	1.430	0.88603 62	−0.12099 74	0.430
1.185	0.92229 05	−0.08089 51	0.185	1.435	0.88590 97	−0.12114 03	0.435
1.190	0.92088 50	−0.08242 01	0.190	1.440	0.88580 51	−0.12125 84	0.440
1.195	0.91951 12	−0.08391 30	0.195	1.445	0.88572 23	−0.12135 18	0.445
1.200	0.91816 87	−0.08537 41	0.200	1.450	0.88566 14	−0.12142 06	0.450
1.205	0.91685 73	−0.08680 35	0.205	1.455	0.88562 21	−0.12146 50	0.455
1.210	0.91557 65	−0.08820 14	0.210	1.460	0.88560 43	−0.12148 50	0.460
1.215	0.91432 62	−0.08956 79	0.215	1.465	0.88560 80	−0.12148 08	0.465
1.220	0.91310 59	−0.09090 34	0.220	1.470	0.88563 31	−0.12145 25	0.470
1.225	0.91191 56	−0.09220 78	0.225	1.475	0.88567 95	−0.12140 02	0.475
1.230	0.91075 49	−0.09348 15	0.230	1.480	0.88574 70	−0.12132 40	0.480
1.235	0.90962 34	−0.09472 46	0.235	1.485	0.88583 56	−0.12122 40	0.485
1.240	0.90852 11	−0.09593 72	0.240	1.490	0.88594 51	−0.12110 03	0.490
1.245	0.90744 75	−0.09711 96	0.245	1.495	0.88607 56	−0.12095 30	0.495
	$y!$	$\log_e y!$	y		$y!$	$\log_e y!$	y

GAMMA FUNCTION AND ITS NATURAL LOGARITHM (Continued)

n	$\Gamma(n)$	$\log_e \Gamma(n)$		n	$\Gamma(n)$	$\log_e \Gamma(n)$	
1.500	0.88622 69	−0.12078 22	0.500	1.750	0.91906 25	−0.08440 11	0.750
1.505	0.88639 90	−0.12058 81	0.505	1.755	0.92020 92	−0.08315 42	0.755
1.510	0.88659 17	−0.12037 07	0.510	1.760	0.92137 49	−0.08188 83	0.760
1.515	0.88680 50	−0.12013 02	0.515	1.765	0.92255 95	−0.08060 34	0.765
1.520	0.88703 88	−0.11986 66	0.520	1.770	0.92376 31	−0.07929 96	0.770
1.525	0.88729 30	−0.11958 00	0.525	1.775	0.92498 57	−0.07797 70	0.775
1.530	0.88756 76	−0.11927 06	0.530	1.780	0.92622 73	−0.07663 56	0.780
1.535	0.88786 25	−0.11893 84	0.535	1.785	0.92748 79	−0.07527 55	0.785
1 540	0.88817 77	−0.11858 35	0.540	1.790	0.92876 75	−0.07389 69	0.790
1.545	0.88851 30	−0.11820 61	0.545	1.795	0.93006 61	−0.07249 96	0.795
1.550	0.88886 83	−0.11780 61	0.550	1.800	0.93138 38	−0.07108 39	0.800
1.555	0.88924 38	−0.11738 39	0.555	1.805	0.93272 05	−0.06964 97	0.805
1.560	0.88963 92	−0.11693 93	0.560	1.810	0.93407 63	−0.06819 72	0.810
1.565	0.89005 45	−0.11647 25	0.565	1.815	0.93545 11	−0.06672 64	0.815
1.570	0.89048 97	−0.11598 37	0.570	1.820	0.93684 51	−0.06523 73	0.820
1.575	0.89094 48	−0.11547 28	0.575	1.825	0.93825 82	−0.06373 01	0.825
1.580	0.89141 96	−0.11494 01	0.580	1.830	0.93969 04	−0.06220 48	0.830
1.585	0.89191 41	−0.11438 55	0.585	1.835	0.94114 18	−0.06066 15	0.835
1.590	0.89242 82	−0.11380 92	0.590	1.840	0.94261 24	−0.05910 01	0.840
1.595	0.89296 20	−0.11321 13	0.595	1.845	0.94410 22	−0.05752 09	0.845
1.600	0.89351 53	−0.11259 18	0.600	1.850	0.94561 12	−0.05592 38	0.850
1.605	0.89408 82	−0.11195 08	0.605	1.855	0.94713 95	−0.05430 89	0.855
1.610	0.89468 06	−0.11128 85	0.610	1.860	0.94868 70	−0.05267 63	0.860
1.615	0.89529 24	−0.11060 49	0.615	1.865	0.95025 39	−0.05102 60	0.865
1.620	0.89592 37	−0.10990 01	0.620	1.870	0.95184 02	−0.04935 81	0.870
1.625	0.89657 43	−0.10917 41	0.625	1.875	0.95344 58	−0.04767 27	0.875
1.630	0.89724 42	−0.10842 72	0.630	1.880	0.95507 09	−0.04596 97	0.880
1.635	0.89793 35	−0.10765 93	0.635	1.885	0.95671 53	−0.04424 94	0.885
1.640	0.89864 20	−0.10687 05	0.640	1.890	0.95837 93	−0.04251 16	0.890
1.645	0.89936 98	−0.10606 10	0.645	1.895	0.96006 28	−0.04075 66	0.895
1.650	0.90011 68	−0.10523 07	0.650	1.900	0.96176 58	−0.03898 43	0.900
1.655	0.90088 30	−0.10437 99	0.655	1.905	0.96348 85	−0.03719 48	0.905
1.660	0.90166 84	−0.10350 85	0.660	1.910	0.96523 07	−0.03538 81	0.910
1.665	0.90247 29	−0.10261 66	0.665	1.915	0.96699 27	−0.03356 44	0.915
1.670	0.90329 65	−0.10170 44	0.670	1.920	0.96877 43	−0.03172 36	0.920
1.675	0.90413 92	−0.10077 19	0.675	1.925	0.97057 57	−0.02986 59	0.925
1.680	0.90500 10	−0.09981 92	0.680	1.930	0.97239 69	−0.02799 12	0.930
1.685	0.90588 19	−0.09884 63	0.685	1.935	0.97423 80	−0.02609 97	0.935
1.690	0.90678 18	−0.09785 34	0.690	1.940	0.97609 89	−0.02419 14	0.940
1.695	0.90770 08	−0.09684 05	0.695	1.945	0.97797 98	−0.02226 63	0.945
1.700	0.90863 87	−0.09580 77	0.700	1.950	0.97988 07	−0.02032 45	0.950
1.705	0.90959 57	−0.09475 51	0.705	1.955	0.98180 16	−0.01836 61	0.955
1.710	0.91057 17	−0.09368 27	G.710	1.960	0.98374 25	−0.01639 11	0.960
1.715	0.91156 66	−0.09259 06	0.715	1.965	0.98570 37	−0.01439 95	0.965
1.720	0.91258 06	−0.09147 89	0.720	1.970	0.98768 50	−0.01239 15	0.970
1.725	0.91361 35	−0.09034 77	0.725	1.975	0.98968 65	−0.01036 70	0.975
1.730	0.91466 54	−0.08919 70	0.730	1.980	0.99170 84	−0.00832 62	0.980
1.735	0.91573 62	−0.08802 69	0.735	1.985	0.99375 06	−0.00626 90	0.985
1.740	0.91682 60	−0.08683 75	0.740	1.990	0.99581 33	−0.00419 55	0.990
1.745	0.91793 48	−0.08562 89	0.745	1.995	0.99789 64	−0.00210 59	0.995
				2.000	1.00000 00	0.00000 00	1.000
	$y!$	$\log_e y!$	y		$y!$	$\log_e y!$	y

GAMMA FUNCTION FOR COMPLEX ARGUMENTS
$x = 1.0$

y	$\Re \ln \Gamma(z) = \log_e \lvert \Gamma(z) \rvert$	$\Im \ln \Gamma(z)$	y	$\Re \ln \Gamma(z) = \log_e \lvert \Gamma(z) \rvert$	$\Im \ln \Gamma(z)$
0.0	0.00000	0.00000	5.0	− 6.13032	3.81590
0.1	− 0.00820	− 0.05732	5.1	− 6.27750	3.97816
0.2	− 0.03248	− 0.11230	5.2	− 6.42487	4.14238
0.3	− 0.07195	− 0.16282	5.3	− 6.57243	4.30850
0.4	− 0.12529	− 0.20716	5.4	− 6.72016	4.47650
0.5	− 0.19095	− 0.24406	5.5	− 6.86807	4.64634
0.6	− 0.26729	− 0.27274	5.6	− 7.01614	4.81799
0.7	− 0.35277	− 0.29283	5.7	− 7.16437	4.99142
0.8	− 0.44598	− 0.30423	5.8	− 7.31275	5.16659
0.9	− 0.54571	− 0.30707	5.9	− 7.46128	5.34348
1.0	− 0.65092	− 0.30164	6.0	− 7.60996	5.52205
1.1	− 0.76078	− 0.28827	6.1	− 7.75877	5.70229
1.2	− 0.87459	− 0.26733	6.2	− 7.90772	5.88415
1.3	− 0.99177	− 0.23922	6.3	− 8.05680	6.06762
1.4	− 1.11186	− 0.20430	6.4	− 8.20601	6.25267
1.5	− 1.23448	− 0.16294	6.5	− 8.35534	6.43928
1.6	− 1.35931	− 0.11547	6.6	− 8.50478	6.62742
1.7	− 1.48609	− 0.06220	6.7	− 8.65434	6.81707
1.8	− 1.61460	− 0.00342	6.8	− 8.80402	7.00821
1.9	− 1.74464	+0.06061	6.9	− 8.95380	7.20081
2.0	− 1.87608	0.12965	7.0	− 9.10368	7.39486
2.1	− 2.00876	0.20346	7.1	− 9.25367	7.59033
2.2	− 2.14258	0.28185	7.2	− 9.40375	7.78720
2.3	− 2.27744	0.36461	7.3	− 9.55394	7.98546
2.4	− 2.41324	0.45159	7.4	− 9.70421	8.18508
2.5	− 2.54991	0.54260	7.5	− 9.85458	8.38605
2.6	− 2.68738	0.63751	7.6	− 10.00504	8.58835
2.7	− 2.82259	0.73617	7.7	− 10.15558	8.79197
2.8	− 2.96448	0.83844	7.8	− 10.30621	8.99687
2.9	− 3.10402	0.94421	7.9	− 10.45692	9.20306
3.0	− 3.24414	1.05335	8.0	− 10.60771	9.41051
3.1	− 3.38483	1.16577	8.1	− 10.75858	9.61920
3.2	− 3.52603	1.28135	8.2	− 10.90952	9.82913
3.3	− 3.66773	1.40001	8.3	− 11.06054	10.04027
3.4	− 3.80988	1.52165	8.4	− 11.21163	10.25262
3.5	− 3.95247	1.64619	8.5	− 11.36280	10.46615
3.6	− 4.09546	1.77355	8.6	− 11.51403	10.68086
3.7	− 4.23884	1.90365	8.7	− 11.66533	10.89673
3.8	− 4.38259	2.03642	8.8	− 11.81669	11.11374
3.9	− 4.52668	2.17179	8.9	− 11.96812	11.33189
4.0	− 4.67110	2.30970	9.0	− 12.11962	11.55116
4.1	− 4.81583	2.45008	9.1	− 12.27117	11.77153
4.2	− 4.96086	2.59287	9.2	− 12.42279	11.99301
4.3	− 5.10618	2.73803	9.3	− 12.57446	12.21557
4.4	− 5.25176	2.88549	9.4	− 12.72619	12.43920
4.5	− 5.39761	3.03520	9.5	− 12.87798	12.66390
4.6	− 5.54370	3.18711	9.6	− 13.02982	12.88964
4.7	− 5.69002	3.34118	9.7	− 13.18172	13.11643
4.8	− 5.83658	3.49737	9.8	− 13.33367	13.34424
4.9	− 5.98335	3.65562	9.9	− 13.48568	13.57307
5.0	− 6.13032	3.81590	10.0	− 13.63773	13.80291

Linear interpolation will yield about three figures; eight-point interpolation will yield about eight figures.

$\Re \ln \Gamma(z) = \log_e \lvert \Gamma(z) \rvert$ $\Im \ln \Gamma(z) = \arg \Gamma(z)$

GAMMA FUNCTION FOR COMPLEX ARGUMENTS
$x = 1.1$

y	$\Re \ln \Gamma(z) = \log_e \lvert \Gamma(z) \rvert$	$\Im \ln \Gamma(z)$	y	$\Re \ln \Gamma(z) = l_e \lvert \Gamma(z) \rvert$	$\Im \ln \Gamma(z)$
0.0	−0.04987	0.00000	5.0	− 5.96894	3.96199
0.1	−0.05702	−0.04207	5.1	− 6.11415	4.12447
0.2	−0.07824	−0.08231	5.2	− 6.25960	4.28889
0.3	−0.11291	−0.11905	5.3	− 6.40527	4.45521
0.4	−0.16008	−0.15087	5.4	− 6.55114	4.62340
0.5	−0.21859	−0.17666	5.5	− 6.69723	4.79343
0.6	−0.28719	−0.19566	5.6	− 6.84351	4.96526
0.7	−0.36464	−0.20740	5.7	− 6.98998	5.13886
0.8	−0.44979	−0.21167	5.8	− 7.13664	5.31419
0.9	−0.54158	−0.20844	5.9	− 7.28347	5.49124
1.0	−0.63909	−0.19782	6.0	− 7.43048	5.66997
1.1	−0.74154	−0.18001	6.1	− 7.57765	5.85035
1.2	−0.84826	−0.15525	6.2	− 7.72498	6.03236
1.3	−0.95869	−0.12384	6.3	− 7.87247	6.25198
1.4	−1.07235	−0.08605	6.4	− 8.02011	6.40116
1.5	−1.18886	−0.04217	6.5	− 8.16790	6.58790
1.6	−1.30787	+0.00752	6.6	− 8.31582	6.77617
1.7	−1.42911	0.06276	6.7	− 8.46389	6.96595
1.8	−1.55234	0.12330	6.8	− 8.61208	7.15720
1.9	−1.67734	0.18890	6.9	− 8.76041	7.34992
2.0	−1.80396	0.25936	7.0	− 8.90886	7.54408
2.1	−1.93203	0.33446	7.1	− 9.05744	7.73966
2.2	−2.06143	0.41402	7.2	− 9.20613	7.93664
2.3	−2.19204	0.49787	7.3	− 9.35494	8.13501
2.4	−2.32376	0.58583	7.4	− 9.50387	8.33473
2.5	−2.45650	0.67775	7.5	− 9.65290	8.53580
2.6	−2.59018	0.77350	7.6	− 9.80203	8.73820
2.7	−2.72474	0.87293	7.7	− 9.95128	8.94191
2.8	−2.86010	0.97592	7.8	−10.10062	9.14690
2.9	−2.99623	1.08236	7.9	−10.25006	9.35318
3.0	−3.13305	1.19214	8.0	−10.39960	9.56071
3.1	−3.27054	1.30514	8.1	−10.54923	9.76950
3.2	−3.40864	1.42128	8.2	−10.69895	9.97950
3.3	−3.54732	1.54045	8.3	−10.84876	10.19073
3.4	−3.68655	1.66258	8.4	−10.99866	10.40315
3.5	−3.82629	1.78758	8.5	−11.14864	10.61676
3.6	−3.96651	1.91537	8.6	−11.29870	10.83154
3.7	−4.10720	2.04589	8.7	−11.44885	11.04749
3.8	−4.24832	2.17905	8.8	−11.59908	11.26457
3.9	−4.38985	2.31479	8.9	−11.74938	11.48279
4.0	−4.53178	2.45304	9.0	−11.89976	11.70213
4.1	−4.67408	2.59376	9.1	−12.05021	11.92257
4.2	−4.81673	2.73687	9.2	−12.20074	12.14411
4.3	−4.95972	2.88233	9.3	−12.35133	12.36673
4.4	−5.10303	3.03008	9.4	−12.50200	12.59043
4.5	−5.24665	3.18007	9.5	−12.65273	12.81519
4.6	−5.39057	3.33225	9.6	−12.80353	13.04099
4.7	−5.53477	3.48657	9.7	−12.95439	13.26784
4.8	−5.67924	3.64300	9.8	−13.10532	13.49571
4.9	−5.82396	3.80148	9.9	−13.25631	13.72460
5.0	−5.96894	3.96199	10.0	−13.40736	13.95449

Linear interpolation will yield about three figures; eight-point interpolation will yield about eight figures.

$$\Re \ln \Gamma(z) = \log_e \lvert \Gamma(z) \rvert \qquad \Im \ln \Gamma(z) = \arg \Gamma(z)$$

GAMMA FUNCTION FOR COMPLEX ARGUMENTS
x = 1.2

y	$\Re \ln \Gamma(z) = \log_e \| \Gamma(z) \|$	$\Im \ln \Gamma(z)$	y	$\Re \ln \Gamma(z) = \log_e \| \Gamma(z) \|$	$\Im \ln \Gamma(z)$
0.0	−0.08537	0.00000	5.0	− 5.80732	4.10610
0.1	−0.09170	−0.02866	5.1	− 5.95058	4.26883
0.2	−0.11051	−0.05586	5.2	− 6.09410	4.43349
0.3	−0.14135	−0.08026	5.3	− 6.23789	4.60005
0.4	−0.18352	−0.10066	5.4	− 6.38192	4.76847
0.5	−0.23614	−0.11611	5.5	− 6.52619	4.93871
0.6	−0.29825	−0.12588	5.6	− 6.67069	5.11075
0.7	−0.36885	−0.12949	5.7	− 6.81541	5.28455
0.8	−0.44698	−0.12664	5.8	− 6.96035	5.46009
0.9	−0.53174	−0.11721	5.9	− 7.10549	5.63732
1.0	−0.62233	−0.10119	6.0	− 7.25083	5.81623
1.1	−0.71804	−0.07869	6.1	− 7.39636	5.99679
1.2	−0.81823	−0.04984	6.2	− 7.54209	6.17898
1.3	−0.92238	−0.01484	6.3	− 7.68799	6.36275
1.4	−1.03001	+0.02611	6.4	− 7.83407	6.54810
1.5	−1.14074	0.07278	6.5	− 7.98031	6.73500
1.6	−1.25421	0.12496	6.6	− 8.12673	6.92342
1.7	−1.37015	0.18241	6.7	− 8.27330	7.11334
1.8	−1.48830	0.24494	6.8	− 8.42002	7.30474
1.9	−1.60844	0.31234	6.9	− 8.56690	7.49759
2.0	−1.73039	0.38443	7.0	− 8.71393	7.69189
2.1	−1.85398	0.46101	7.1	− 8.86109	7.88760
2.2	−1.97907	0.54192	7.2	− 9.00840	8.08471
2.3	−2.10553	0.62700	7.3	− 9.15584	8.28319
2.4	−2.23326	0.71610	7.4	− 9.30341	8.48304
2.5	−2.36215	0.80908	7.5	− 9.45110	8.68422
2.6	−2.49211	0.90579	7.6	− 9.59892	8.88673
2.7	−2.62308	1.00613	7.7	− 9.74687	9.09055
2.8	−2.75497	1.10996	7.8	− 9.89493	9.29566
2.9	−2.88773	1.21718	7.9	−10.04310	9.50204
3.0	−3.02130	1.32769	8.0	−10.19139	9.70968
3.1	−3.15563	1.44138	8.1	−10.33978	9.91856
3.2	−3.29066	1.55816	8.2	−10.48828	10.12866
3.3	−3.42637	1.67794	8.3	−10.63689	10.33998
3.4	−3.56270	1.80064	8.4	−10.78559	10.55250
3.5	−3.69962	1.92618	8.5	−10.93440	10.76620
3.6	−3.83711	2.05448	8.6	−11.08330	10.98107
3.7	−3.97513	2.18547	8.7	−11.23229	11.19710
3.8	−4.11364	2.31909	8.8	−11.38138	11.41427
3.9	−4.25264	2.45526	8.9	−11.53056	11.63257
4.0	−4.39209	2.59393	9.0	−11.67983	11.85199
4.1	−4.53197	2.73504	9.1	−11.82918	12.07251
4.2	−4.67226	2.87853	9.2	−11.97861	12.29413
4.3	−4.81294	3.02434	9.3	−12.12813	12.51683
4.4	−4.95399	3.17243	9.4	−12.27773	12.74060
4.5	−5.09541	3.32274	9.5	−12.42741	12.96543
4.6	−5.23716	3.47523	9.6	−12.57717	13.19130
4.7	−5.37924	3.62986	9.7	−12.72700	13.41822
4.8	−5.52164	3.78657	9.8	−12.87691	13.64616
4.9	−5.66433	3.94533	9.9	−13.02689	13.87511
5.0	−5.80732	4.10610	10.0	−13.17694	14.10508

Linear interpolation will yield about three figures; eight-point interpolation will yield about eight figures.

$$\Re \ln \Gamma(z) = \log_e | \Gamma(z) | \qquad\qquad \Im \ln \Gamma(z) = \arg \Gamma(z)$$

GAMMA FUNCTION FOR COMPLEX ARGUMENTS
$x = 1.3$

y	$\Re \ln \Gamma(z) = \log_e |\Gamma(z)|$	$\mathcal{I} \ln \Gamma(z)$	y	$\Re \ln \Gamma(z) = \log_e |\Gamma(z)|$	$\mathcal{I} \ln \Gamma(z)$
0.0	−0.10817	−0.00000	5.0	− 5.64541	4.24824
0.1	−0.11384	−0.01672	5.1	− 5.78673	4.41126
0.2	−0.13070	−0.03226	5.2	− 5.92835	4.57621
0.3	−0.15843	−0.04550	5.3	− 6.07027	4.74303
0.4	−0.19649	−0.05545	5.4	− 6.21246	4.91171
0.5	−0.24421	−0.06127	5.5	− 6.35492	5.08220
0.6	−0.30082	−0.06230	5.6	− 6.49765	5.25448
0.7	−0.36553	−0.05805	5.7	− 6.64063	5.42852
0.8	−0.43755	−0.04821	5.8	− 6.78385	5.60428
0.9	−0.15610	−0.03257	5.9	− 6.92730	5.78173
1.0	−0.60048	−0.01108	6.0	− 7.07099	5.96085
1.1	−0.69007	+0.01628	6.1	− 7.21489	6.14161
1.2	−0.78427	0.04942	6.2	− 7.35901	6.32399
1.3	−0.88259	0.08822	6.3	− 7.50333	6.50796
1.4	−0.98459	0.13255	6.4	− 7.64785	6.69349
1.5	−1.08987	0.18224	6.5	− 7.79257	6.88057
1.6	−1.19810	0.23711	6.6	− 7.93747	7.06916
1.7	−1.30899	0.29700	6.7	− 8.08255	7.25925
1.8	−1.42227	0.36172	6.8	− 8.22781	7.45081
1.9	−1.53773	0.43111	6.9	− 8.37325	7.64383
2.0	−1.65518	0.50500	7.0	− 8.51885	7.83828
2.1	−1.77443	0.58323	7.1	− 8.66461	8.03414
2.2	−1.89533	0.66565	7.2	− 8.81053	8.23139
2.3	−2.01776	0.75212	7.3	− 8.95660	8.43002
2.4	−2.14159	0.84250	7.4	− 9.10282	8.63000
2.5	−2.26672	0.93666	7.5	− 9.24919	8.83132
2.6	−2.39305	1.03448	7.6	− 9.39569	9.03396
2.7	−2.52049	1.13583	7.7	− 9.54234	9.23791
2.8	−2.64898	1.24062	7.8	− 9.68912	9.44314
2.9	−2.77843	1.34873	7.9	− 9.83603	9.64964
3.0	−2.90879	1.46006	8.0	− 9.98307	9.85740
3.1	−3.04001	1.57453	8.1	−10.13023	10.06639
3.2	−3.17202	1.69204	8.2	−10.27751	10.27661
3.3	−3.30478	1.81251	8.3	−10.42491	10.48804
3.4	−3.43826	1.93586	8.4	−10.57243	10.70067
3.5	−3.57240	2.06201	8.5	−10.72006	10.91447
3.6	−3.70717	2.19090	8.6	−10.86780	11.12944
3.7	−3.84255	2.32244	8.7	−11.01564	11.34557
3.8	−3.97849	2.45658	8.8	−11.16360	11.56284
3.9	−4.11497	2.59324	8.9	−11.31165	11.78123
4.0	−4.25196	2.73239	9.0	−11.45981	12.00075
4.1	−4.38945	2.87394	9.1	−11.60806	12.22136
4.2	−4.52739	3.01786	9.2	−11.75641	12.44307
4.3	−4.66578	3.16408	9.3	−11.90485	12.66585
4.4	−4.80460	3.31256	9.4	−12.05339	12.88971
4.5	−4.94382	3.46324	9.5	−12.20202	13.11462
4.6	−5.08342	3.61609	9.6	−12.35073	13.34058
4.7	−5.22340	3.77106	9.7	−12.49953	13.56757
4.8	−5.36374	3.92810	9.8	−12.64842	13.79559
4.9	−5.50441	4.08717	9.9	−12.79739	14.02463
5.0	−5.64541	4.24824	10.0	−12.94644	14.25466

Linear interpolation will yield about three figures; eight-point interpolation will yield about eight figures.

$$\Re \ln \Gamma(z) = \log_e |\Gamma(z)| \qquad\qquad \mathcal{I} \ln \Gamma(z) = \arg \Gamma(z)$$

GAMMA FUNCTION FOR COMPLEX ARGUMENTS
$x = 1.4$

y	$\Re \ln \Gamma(z) = \log_e |\Gamma(z)|$	$\Im \ln \Gamma(z)$	y	$\Re \ln \Gamma(z) = \log_e |\Gamma(z)|$	$\Im \ln \Gamma(z)$
0.0	−0.11961	0.00000	5.0	−5.48320	4.38843
0.1	−0.12473	−0.00597	5.1	−5.62259	4.55178
0.2	−0.14000	−0.01097	5.2	−5.76231	4.71704
0.3	−0.16516	−0.01406	5.3	−5.90236	4.88417
0.4	−0.19979	−0.01439	5.4	−6.04273	5.05314
0.5	−0.24337	−0.01125	5.5	−6.18340	5.22391
0.6	−0.29530	−0.00402	5.6	−6.32436	5.39646
0.7	−0.35492	+0.00776	5.7	−6.46560	5.57076
0.8	−0.42158	0.02442	5.8	−6.60712	5.74677
0.9	−0.49463	0.04618	5.9	−6.74889	5.92447
1.0	−0.57345	0.07318	6.0	−6.89093	6.10383
1.1	−0.65748	0.10546	6.1	−7.03321	6.28482
1.2	−0.74620	0.14300	6.2	−7.17572	6.46742
1.3	−0.83914	0.18576	6.3	−7.31847	6.65160
1.4	−0.93588	0.23363	6.4	−7.46144	6.83734
1.5	−1.03606	0.28650	6.5	−7.60463	7.02462
1.6	−1.13934	0.34426	6.6	−7.74803	7.21341
1.7	−1.24543	0.40674	6.7	−7.89163	7.40369
1.8	−1.35407	0.47383	6.8	−8.03543	7.59543
1.9	−1.46505	0.54537	6.9	−8.17943	7.78863
2.0	−1.57816	0.62123	7.0	−8.32360	7.98325
2.1	−1.69322	0.70126	7.1	−8.46797	8.17928
2.2	−1.81008	0.78534	7.2	−8.61250	8.37670
2.3	−1.92859	0.87334	7.3	−8.75721	8.57549
2.4	−2.04863	0.96513	7.4	−8.90209	8.77563
2.5	−2.17009	1.06059	7.5	−9.04713	8.97710
2.6	−2.29287	1.15962	7.6	−9.19233	9.17989
2.7	−2.41687	1.26211	7.7	−9.33768	9.38398
2.8	−2.54201	1.36795	7.8	−9.48318	9.58935
2.9	−2.66822	1.47704	7.9	−9.62883	9.79599
3.0	−2.79544	1.58930	8.0	−9.77462	10.00388
3.1	−2.92359	1.70464	8.1	−9.92056	10.21300
3.2	−3.05262	1.82297	8.2	−10.06662	10.42335
3.3	−3.18249	1.94421	8.3	−10.21282	10.63490
3.4	−3.31315	2.06828	8.4	−10.35915	10.84765
3.5	−3.44454	2.19512	8.5	−10.50561	11.06157
3.6	−3.57664	2.32465	8.6	−10.65219	11.27666
3.7	−3.70940	2.45681	8.7	−10.79889	11.49290
3.8	−3.84280	2.59153	8.8	−10.94571	11.71028
3.9	−3.97679	2.72876	8.9	−11.09264	11.92878
4.0	−4.11135	2.86842	9.0	−11.23969	12.14840
4.1	−4.24646	3.01048	9.1	−11.38685	12.36912
4.2	−4.38209	3.15488	9.2	−11.53411	12.59093
4.3	−4.51821	3.30156	9.3	−11.68148	12.81381
4.4	−4.65480	3.45047	9.4	−11.82896	13.03776
4.5	−4.79184	3.60158	9.5	−11.97653	13.26277
4.6	−4.92932	3.75483	9.6	−12.12421	13.48882
4.7	−5.06721	3.91018	9.7	−12.27198	13.71591
4.8	−5.20549	4.06759	9.8	−12.41985	13.94401
4.9	−5.34416	4.22702	9.9	−12.56781	14.17313
5.0	−5.48320	4.38843	10.0	−12.71586	14.40326

Linear interpolation will yield about three figures; eight-point interpolation will yield about eight figures.

$$\Re \ln \Gamma(z) = \log_e |\Gamma(z)| \qquad \Im \ln \Gamma(z) = \arg \Gamma(z)$$

GAMMA FUNCTION FOR COMPLEX ARGUMENTS
$x = 1.5$

y	$\Re \ln \Gamma(z) = \log_e \lvert \Gamma(z) \rvert$	$\Im \ln \Gamma(z)$	y	$\Re \ln \Gamma(z) = \log_e \lvert \Gamma(z) \rvert$	$\Im \ln \Gamma(z)$
0.0	−0.12078	0.00000	5.0	−5.32063	4.52667
0.1	−0.12545	0.00379	5.1	−5.45810	4.69038
0.2	−0.13939	0.00839	5.2	−5.59594	4.85599
0.3	−0.16238	0.01461	5.3	−5.73414	5.02346
0.4	−0.19412	0.02315	5.4	−5.87269	5.19275
0.5	−0.23419	0.03467	5.5	−6.01158	5.36384
0.6	−0.28208	0.04969	5.6	−6.15078	5.53669
0.7	−0.33728	0.06867	5.7	−6.29030	5.71128
0.8	−0.39924	0.09192	5.8	−6.43012	5.88757
0.9	−0.46739	0.11969	5.9	−6.57023	6.06555
1.0	−0.54122	0.15214	6.0	−6.71062	6.24517
1.1	−0.62022	0.18936	6.1	−6.85128	6.42641
1.2	−0.70392	0.23137	6.2	−6.99221	6.60926
1.3	−0.79189	0.27817	6.3	−7.13339	6.79368
1.4	−0.88376	0.32970	6.4	−7.27482	6.97966
1.5	−0.97915	0.38589	6.5	−7.41649	7.16716
1.6	−1.07776	0.44666	6.6	−7.55839	7.35616
1.7	−1.17932	0.51190	6.7	−7.70051	7.54666
1.8	−1.28355	0.58149	6.8	−7.84286	7.73861
1.9	−1.39024	0.65532	6.9	−7.98542	7.93200
2.0	−1.49920	0.73328	7.0	−8.12818	8.12682
2.1	−1.61023	0.81525	7.1	−8.27115	8.32304
2.2	−1.72317	0.90111	7.2	−8.41431	8.52064
2.3	−1.83790	0.99076	7.3	−8.55766	8.71961
2.4	−1.95426	1.08407	7.4	−8.70120	8.91992
2.5	−2.07215	1.18096	7.5	−8.84491	9.12156
2.6	−2.19146	1.28131	7.6	−8.98881	9.32452
2.7	−2.31210	1.38503	7.7	−9.13287	9.52877
2.8	−2.43398	1.49202	7.8	−9.27710	9.73429
2.9	−2.55701	1.60219	7.9	−9.42149	9.94108
3.0	−2.68114	1.71547	8.0	−9.56604	10.14912
3.1	−2.80629	1.83176	8.1	−9.71075	10.35839
3.2	−2.93240	1.95098	8.2	−9.85560	10.56888
3.3	−3.05942	2.07307	8.3	−10.00060	10.78057
3.4	−3.18730	2.19794	8.4	−10.14575	10.99345
3.5	−3.31598	2.32553	8.5	−10.29104	11.20751
3.6	−3.44544	2.45578	8.6	−10.43646	11.42273
3.7	−3.57563	2.58862	8.7	−10.58202	11.63909
3.8	−3.70650	2.72398	8.8	−10.72771	11.85659
3.9	−3.83804	2.86182	8.9	−10.87352	12.07522
4.0	−3.97020	3.00207	9.0	−11.01946	12.29495
4.1	−4.10296	3.14469	9.1	−11.16553	12.51579
4.2	−4.23629	3.28962	9.2	−11.31171	12.73771
4.3	−4.37016	3.43680	9.3	−11.45801	12.96070
4.4	−4.50455	3.58620	9.4	−11.60442	13.18476
4.5	−4.63943	3.73777	9.5	−11.75095	13.40987
4.6	−4.77480	3.89147	9.6	−11.89759	13.63603
4.7	−4.91061	4.04724	9.7	−12.04433	13.86321
4.8	−5.04687	4.20506	9.8	−12.19118	14.09142
4.9	−5.18355	4.36488	9.9	−12.33814	14.32064
5.0	−5.32063	4.52667	10.0	−12.48519	14.55086

Linear interpolation will yield about three figures; eight-point interpolation will yield about eight figures.

$\Re \ln \Gamma(z) = \log_e \lvert \Gamma(z) \rvert$ \qquad $\Im \ln \Gamma(z) = \arg \Gamma(z)$

GAMMA FUNCTION FOR COMPLEX ARGUMENTS
$x = 1.6$

y	$\Re \ln \Gamma(z) = \log_e |\Gamma(z)|$	$\Im \ln \Gamma(z)$	y	$\Re \ln \Gamma(z) = \log_e |\Gamma(z)|$	$\Im \ln \Gamma(z)$
0.0	−0.11259	0.00000	5.0	−5.15767	4.66299
0.1	−0.11688	0.01272	5.1	−5.29324	4.82710
0.2	−0.12969	0.02614	5.2	−5.42921	4.99309
0.3	−0.15085	0.04093	5.3	−5.56558	5.16093
0.4	−0.18012	0.05771	5.4	−5.70233	5.33058
0.5	−0.21716	0.07706	5.5	−5.83944	5.50201
0.6	−0.26156	0.09944	5.6	−5.97690	5.67519
0.7	−0.31289	0.12528	5.7	−6.11470	5.85010
0.8	−0.37069	0.15489	5.8	−6.25283	6.02670
0.9	−0.43449	0.18851	5.9	−6.39128	6.20497
1.0	−0.50382	0.22633	6.0	−6.53004	6.38488
1.1	−0.57826	0.26845	6.1	−6.66910	6.56641
1.2	−0.65737	0.31495	6.2	−6.80844	6.74953
1.3	−0.74077	0.36584	6.3	−6.94806	6.93422
1.4	−0.82810	0.42111	6.4	−7.08795	7.12044
1.5	−0.91903	0.48071	6.5	−7.22811	7.30819
1.6	−1.01326	0.54460	6.6	−7.36852	7.49744
1.7	−1.11052	0.61269	6.7	−7.50917	7.68816
1.8	−1.21057	0.68490	6.8	−7.65007	7.88034
1.9	−1.31318	0.76114	6.9	−7.79120	8.07396
2.0	−1.41816	0.84132	7.0	−7.93256	8.26899
2.1	−1.52532	0.92534	7.1	−8.07413	8.46541
2.2	−1.63450	1.01310	7.2	−8.21592	8.66322
2.3	−1.74556	1.10450	7.3	−8.35792	8.86238
2.4	−1.85836	1.19946	7.4	−8.50012	9.06288
2.5	−1.97279	1.29786	7.5	8.64252	9.26471
2.6	−2.08874	1.39963	7.6	−8.78511	9.46785
2.7	−2.20610	1.50467	7.7	−8.92789	9.67227
2.8	−2.32478	1.61291	7.8	−9.07085	9.87797
2.9	−2.44472	1.72425	7.9	−9.21399	10.08493
3.0	−2.56582	1.83862	8.0	−9.35730	10.29314
3.1	−2.68802	1.95594	8.1	−9.50079	10.50257
3.2	−2.81127	2.07613	8.2	−9.64443	10.71321
3.3	−2.93549	2.19914	8.3	−9.78824	10.92505
3.4	−3.06063	2.32488	8.4	−9.93221	11.13808
3.5	−3.18666	2.45329	8.5	−10.07633	11.35228
3.6	−3.31351	2.58432	8.6	−10.22060	11.56764
3.7	−3.44116	2.71790	8.7	−10.36501	11.78415
3.8	−3.56955	2.85396	8.8	−10.50958	12.00178
3.9	−3.69866	2.99247	8.9	−10.65428	12.22054
4.0	−3.82845	3.13336	9.0	−10.79911	12.44041
4.1	−3.95889	3.27659	9.1	−10.94409	12.66137
4.2	−4.08994	3.42209	9.2	−11.08919	12.88341
4.3	−4.22159	3.56983	9.3	−11.23442	13.10653
4.4	−4.35380	3.71977	9.4	−11.37978	13.33071
4.5	−4.48655	3.87184	9.5	−11.52526	13.55594
4.6	−4.61982	4.02602	9.6	−11.67086	13.78220
4.7	−4.75359	4.18227	9.7	−11.81658	14.00950
4.8	−4.88783	4.34053	9.8	−11.96242	14.23782
4.9	−5.02253	4.50079	9.9	−12.10836	14.46714
5.0	−5.15767	4.66299	10.0	−12.25442	14.69747

Linear interpolation will yield about three figures; eight-point interpolation will yield about eight figures

$$\Re\ln\Gamma(z) = \log_e|\Gamma(z)| \qquad\qquad \Im\ln\Gamma(z) = \arg\Gamma(z)$$

GAMMA FUNCTION FOR COMPLEX ARGUMENTS
$x = 1.7$

y	$\Re \ln \Gamma(z) = \log_e \mid \Gamma(z) \mid$	$\Im \ln \Gamma(z)$	y	$\Re \ln \Gamma(z) = \log_e \mid \Gamma(z) \mid$	$\Im \ln \Gamma(z)$
0.0	−0.09581	0.00000	5.0	−4.99429	4.79739
0.1	−0.09977	0.02096	5.1	−5.12797	4.96193
0.2	−0.11161	0.04251	5.2	−5.26209	5.12834
0.3	−0.13121	0.06524	5.3	−5.39664	5.29658
0.4	−0.15835	0.08971	5.4	−5.53159	5.46662
0.5	−0.19275	0.11639	5.5	−5.66694	5.63842
0.6	−0.23410	0.14573	5.6	−5.80267	5.81197
0.7	−0.28203	0.17811	5.7	−5.93877	5.98722
0.8	−0.33614	0.21382	5.8	−6.07523	6.16416
0.9	−0.39604	0.25313	5.9	−6.21203	6.34276
1.0	−0.46133	0.29620	6.0	−6.34916	6.52299
1.1	−0.53162	0.34317	6.1	−6.48662	6.70482
1.2	−0.60653	0.39413	6.2	−6.62439	6.88823
1.3	−0.68572	0.44913	6.3	−6.76246	7.07320
1.4	−0.76885	0.50817	6.4	−6.90083	7.25971
1.5	−0.85561	0.57125	6.5	−7.03948	7.44773
1.6	−0.94574	0.63833	6.6	−7.17840	7.63724
1.7	−1.03895	0.70936	6.7	−7.31760	7.82821
1.8	−1.13503	0.78428	6.8	−7.45705	8.02064
1.9	−1.23376	0.86303	6.9	−7.59676	8.21449
2.0	−1.33493	0.94553	7.0	−7.73671	8.40976
2.1	−1.43838	1.03169	7.1	−7.87690	8.60641
2.2	−1.54395	1.12145	7.2	−8.01733	8.80443
2.3	−1.65148	1.21470	7.3	−8.15798	9.00381
2.4	−1.76084	1.31138	7.4	−8.29885	9.20452
2.5	−1.87192	1.41140	7.5	−8.43994	9.40656
2.6	−1.98459	1.51468	7.6	−8.58123	9.60989
2.7	−2.09877	1.62113	7.7	−8.72273	9.81451
2.8	−2.21435	1.73069	7.8	−8.86443	10.02040
2.9	−2.33125	1.84328	7.9	−9.00632	10.22754
3.0	−2.44940	1.95882	8.0	−9.14840	10.43592
3.1	−2.56872	2.07724	8.1	−9.29066	10.64553
3.2	−2.68915	2.19848	8.2	−9.43310	10.85634
3.3	−2.81063	2.32247	8.3	−9.57572	11.06835
3.4	−2.93310	2.44914	8.4	−9.71851	11.28154
3.5	−3.05650	2.57844	8.5	−9.86147	11.49590
3.6	−3.18080	2.71031	8.6	−10.00459	11.71142
3.7	−3.30594	2.84468	8.7	−10.14787	11.92807
3.8	−3.43189	2.98151	8.8	−10.29130	12.14586
3.9	−3.55861	3.12074	8.9	−10.43489	12.36476
4.0	−3.68605	3.26232	9.0	−10.57863	12.58476
4.1	−3.81420	3.40620	9.1	−10.72251	12.80586
4.2	−3.94301	3.55233	9.2	−10.86654	13.02804
4.3	−4.07245	3.70068	9.3	−11.01071	13.25129
4.4	−4.20251	3.85118	9.4	−11.15501	13.47560
4.5	−4.33315	4.00381	9.5	−11.29945	13.70096
4.6	−4.46434	4.15852	9.6	−11.44402	13.92735
4.7	−4.59608	4.31527	9.7	−11.58871	14.15477
4.8	−4.72833	4.47402	9.8	−11.73354	14.38321
4.9	−4.86107	4.63474	9.9	−11.87848	14.61265
5.0	−4.99429	4.79739	10.0	−12.02355	14.84309

Linear interpolation will yield about three figures; eight-point interpolation will yield about eight figures.

$\Re \ln \Gamma(z) = \log_e \mid \Gamma(z) \mid$ $\qquad\qquad$ $\Im \ln \Gamma(z) = \arg \Gamma(z)$

GAMMA FUNCTION FOR COMPLEX ARGUMENTS
$x = 1.8$

y	$\Re \ln \Gamma(z) = \log_e |\Gamma(z)|$	$\Im \ln \Gamma(z)$	y	$\Re \ln \Gamma(z) = \log_e |\Gamma(z)|$	$\Im \ln \Gamma(z)$
0.0	− 0.07108	0.00000	5.0	− 4.83046	4.92990
0.1	− 0.07477	0.02859	5.1	− 4.96227	5.09491
0.2	− 0.08578	0.05769	5.2	− 5.09455	5.26177
0.3	− 0.10401	0.08783	5.3	− 5.22729	5.43044
0.4	− 0.12929	0.11946	5.4	− 5.36046	5.60089
0.5	− 0.16140	0.15305	5.5	− 5.49407	5.77310
0.6	− 0.20007	0.18897	5.6	− 5.62808	5.94703
0.7	− 0.24498	0.22758	5.7	− 5.76249	6.12266
0.8	− 0.29581	0.26917	5.8	− 5.89728	6.29997
0.9	− 0.35222	0.31396	5.9	− 6.03244	6.47891
1.0	− 0.41385	0.36216	6.0	− 6.16796	6.65948
1.1	− 0.48036	0.41390	6.1	− 6.30383	6.84164
1.2	− 0.55143	0.46928	6.2	− 6.44004	7.02538
1.3	− 0.62673	0.52837	6.3	− 6.57657	7.21066
1.4	− 0.70597	0.59120	6.4	− 6.71341	7.39746
1.5	− 0.78885	0.65778	6.5	− 6.85057	7.58577
1.6	− 0.87511	0.72810	6.6	− 6.98802	7.77556
1.7	− 0.96452	0.80213	6.7	− 7.12576	7.96682
1.8	− 1.05685	0.87984	6.8	− 7.26378	8.15951
1.9	− 1.15188	0.96117	6.9	− 7.40207	8.35362
2.0	− 1.24944	1.04606	7.0	− 7.54063	8.54914
2.1	− 1.34934	1.13446	7.1	− 7.67944	8.74603
2.2	− 1.45143	1.22629	7.2	− 7.81851	8.94430
2.3	− 1.55557	1.32148	7.3	− 7.95782	9.14391
2.4	− 1.66161	1.41997	7.4	− 8.09737	9.34485
2.5	− 1.76944	1.52168	7.5	− 8.23715	9.54710
2.6	− 1.87895	1.62655	7.6	− 8.37715	9.75064
2.7	− 1.99003	1.73449	7.7	− 8.51737	9.95547
2.8	− 1.10259	1.84545	7.8	− 8.65781	10.16156
2.9	− 2.21654	1.95935	7.9	− 8.79846	10.36891
3.0	− 2.33181	2.07613	8.0	− 8.93931	10.57748
3.1	− 2.44832	2.19573	8.1	− 9.08036	10.78728
3.2	− 2.56599	2.31808	8.2	− 9.22160	10.99827
3.3	− 2.68478	2.44311	8.3	− 9.36303	11.21046
3.4	− 2.80462	2.57078	8.4	− 9.50465	11.42383
3.5	− 2.92546	2.70103	8.5	− 9.64645	11.63836
3.6	− 3.04724	2.83379	8.6	− 9.78842	11.85405
3.7	− 3.16992	2.96901	8.7	− 9.93057	12.07087
3.8	− 3.29346	3.10665	8.8	− 10.07288	12.28881
3.9	−- 3.41782	3.24666	8.9	− 10.21536	12.50787
4.0	− 3.54296	3.38897	9.0	− 10.35800	12.72803
4.1	− 3.66884	3.53356	9.1	− 10.50080	12.94928
4.2	− 3.79543	3.68037	9.2	− 10.64375	13.17161
4.3	− 3.92271	3.82935	9.3	− 10.78686	13.39500
4.4	− 4.05063	3.98048	9.4	− 10.93011	13.61945
4.5	− 4.17918	4.13370	9.5	− 11.07350	13.84494
4.6	− 4.30833	4.28897	9.6	− 11.21704	14.07147
4.7	− 4.43806	4.44627	9.7	− 11.36072	14.29902
4.8	− 4.56833	4.60554	9.8	− 11.50453	14.52759
4.9	− 4.66914	4.76676	9.9	− 11.64848	14.75716
5.0	− 4.83046	4.92990	10.0	− 11.79255	14.98772

Linear interpolation will yield about three figures; eight-point interpolation will yield about eight figures.

$$\Re \ln \Gamma(z) = \log_e |\Gamma(z)| \qquad\qquad \Im \ln \Gamma(z) = \arg \Gamma(z)$$

GAMMA FUNCTION FOR COMPLETE ARGUMENTS
$x = 1.9$

y	$\Re \ln \Gamma(z) = \log_e |\Gamma(z)|$	$\mathcal{I} \ln \Gamma(z)$	y	$\Re \ln \Gamma(z) = \log_e |\Gamma(z)|$	$\mathcal{I} \ln \Gamma(z)$
0.0	−0.03898	0.00000	5.0	− 4.66613	5.06053
0.1	−0.04242	0.03569	5.1	− 4.79608	5.22604
0.2	−0.05270	0.07184	5.2	− 4.92655	5.39337
0.3	−0.06975	0.10890	5.3	− 5.05749	5.56251
0.4	−0.09340	0.14727	5.4	− 5.18891	5.73341
0.5	−0.12349	0.18736	5.5	− 5.32078	5.90605
0.6	−0.15978	0.22952	5.6	− 5.45309	6.08040
0.7	−0.20201	0.27408	5.7	− 5.58582	6.25643
0.8	−0.24990	0.32129	5.8	− 5.71896	6.43413
0.9	−0.30316	0.37139	5.9	− 5.85250	6.61345
1.0	−0.36148	0.42457	6.0	− 5.98642	6.79438
1.1	−0.42456	0.48098	6.1	− 6.12071	6.97690
1.2	−0.49210	0.54071	6.2	− 6.25536	7.16097
1.3	−0.56382	0.60386	6.3	− 6.39036	7.34659
1.4	−0.63944	0.67047	6.4	− 6.52570	7.53372
1.5	−0.71870	0.74056	6.5	− 6.66136	7.72234
1.6	−0.80136	0.81416	6.6	− 6.79735	7.92143
1.7	−0.88718	0.89124	6.7	− 6.93364	8.10398
1.8	−0.97596	0.97178	6.8	− 7.07023	8.29696
1.9	−1.06749	1.05574	6.9	− 7.20712	8.49135
2.0	−1.16160	1.14310	7.0	− 7.34429	8.68713
2.1	−1.25812	1.23379	7.1	− 7.48173	8.88429
2.2	−1.35688	1.32777	7.2	− 7.61945	9.08281
2.3	−1.45775	1.42497	7.3	− 7.75742	9.28267
2.4	−1.56060	1.52534	7.4	− 7.89565	9.48385
2.5	−1.66529	1.62881	7.5	− 8.03413	9.68634
2.6	−1.77174	1.73533	7.6	− 8.17285	9.89012
2.7	−1.87982	1.84483	7.7	− 8.31180	10.09517
2.8	−1.98944	1.95726	7.8	− 8.45099	10.30148
2.9	−2.10052	2.07255	7.9	− 8.59039	10.50904
3.0	−2.21298	2.19064	8.0	− 8.73002	10.71782
3.1	−2.32674	2.31147	8.1	− 8.86986	10.92782
3.2	−2.44173	2.43498	8.2	− 9.00991	11.13902
3.3	−2.55788	2.56113	8.3	− 9.15016	11.35140
3.4	−2.67515	2.68985	8.4	− 9.29060	11.56496
3.5	−2.79346	2.82109	8.5	− 9.43125	11.77967
3.6	−2.91278	2.95480	8.6	− 9.57208	11.99554
3.7	−3.03304	3.09093	8.7	− 9.71309	12.21253
3.8	−3.15422	3.22944	8.8	− 9.85429	12.43065
3.9	−3.27626	3.37026	8.9	− 9.99567	12.64988
4.0	−3.39912	3.51336	9.0	−10.13721	12.87021
4.1	−3.52277	3.65870	9.1	−10.27893	13.09162
4.2	−3.64718	3.80622	9.2	−10.42081	13.31410
4.3	−3.77231	3.95589	9.3	−10.56286	13.53765
4.4	−3.89814	4.10768	9.4	−10.70506	13.76225
4.5	−4.02462	4.26153	9.5	−10.84742	13.98789
4.6	−4.15175	4.41741	9.6	−10.98993	14.21457
4.7	−4.27948	4.57528	9.7	−11.13259	14.44226
4.8	−4.40781	4.73512	9.8	−11.27539	14.67097
4.9	−4.53670	4.89688	9.9	−11.41834	14.90067
5.0	−4.66613	5.06053	10.0	−11.56143	15.13137

Linear interpolation will yield about three figures; eight-point interpolation will yield about eight figures.

$\Re \ln \Gamma(z) = \log_e |\Gamma(z)|$ $\mathcal{I} \ln \Gamma(z) = \arg \Gamma(z)$

GAMMA FUNCTION FOR COMPLEX ARGUMENTS
$x = 2.0$

y	$\Re \ln \Gamma(z) = \log_e |\Gamma(z)|$	$\mathscr{I} \ln \Gamma(z)$	y	$\Re \ln \Gamma(z) = \log_e |\Gamma(z)|$	$\mathscr{I} \ln \Gamma(z)$
0.0	−0.00000	0.00000	5.0	− 4.50128	5.18930
0.1	−0.00322	0.04235	5.1	− 4.62940	5.35534
0.2	−0.01287	0.08509	5.2	− 4.75806	5.52319
0.3	−0.02886	0.12864	5.3	− 4.88723	5.69281
0.4	−0.05108	0.17335	5.4	− 5.01690	5.86419
0.5	−0.07937	0.21959	5.5	− 5.14706	6.03729
0.6	−0.11355	0.26768	5.6	− 5.27768	6.21208
0.7	−0.15338	0.31790	5.7	− 5.40874	6.38855
0.8	−0.19863	0.37052	5.8	− 5.54025	6.56665
0.9	−0.24904	0.42574	5.9	− 5.67217	6.74638
1.0	−0.30435	0.48376	6.0	− 5.80450	6.92770
1.1	−0.36429	0.54471	6.1	− 5.93723	7.11059
1.2	−0.42859	0.60873	6.2	− 6.07033	7.29503
1.3	−0.49700	0.67588	6.3	− 6.20381	7.48100
1.4	−0.56927	0.74625	6.4	− 6.33765	7.66847
1.5	−0.64516	0.81985	6.5	− 6.47184	7.85743
1.6	−0.72443	0.89673	6.6	− 6.60636	8.04785
1.7	−0.80689	0.97687	6.7	− 6.74122	8.23971
1.8	−0.89231	1.06028	6.8	− 6.87639	8.43299
1.9	−0.98053	1.14693	6.9	− 7.01188	8.62768
2.0	−1.07136	1.23680	7.0	− 7.14767	8.82376
2.1	−1.16464	1.32984	7.1	− 7.28375	9.02120
2.2	−1.26022	1.42601	7.2	− 7.42012	9.21999
2.3	−1.35796	1.52528	7.3	− 7.55677	9.42012
2.4	−1.45773	1.62759	7.4	− 7.69369	9.62156
2.5	−1.55941	1.73289	7.5	− 7.83087	9.82430
2.6	−1.66288	1.84113	7.6	− 7.96831	10.02832
2.7	−1.76806	1.95226	7.7	− 8.10600	10.23362
2.8	−1.87484	2.06621	7.8	− 8.24394	10.44016
2.9	−1.98313	2.18294	7.9	− 8.38211	10.64794
3.0	−2.09285	2.30240	8.0	− 8.52052	10.85695
3.1	−2.20393	2.42452	8.1	− 8.65915	11.06716
3.2	−2.31629	2.54926	8.2	− 8.79801	11.27857
3.3	−2.42988	2.67657	8.3	− 8.93708	11.49117
3.4	−2.54462	2.80640	8.4	− 9.07637	11.70493
3.5	−2.66047	2.93869	8.5	− 9.21586	11.91984
3.6	−2.77736	3.07340	8.6	− 9.35555	12.13590
3.7	−2.89526	3.21048	8.7	− 9.49544	12.35308
3.8	−3.01411	3.34989	8.8	− 9.63553	12.57138
3.9	−3.13386	3.49159	8.9	− 9.77580	12.79079
4.0	−3.25449	3.63552	9.0	− 9.91626	13.01130
4.1	−3.37595	3.78164	9.1	−10.05689	13.23288
4.2	−3.49821	3.92993	9.2	−10.19771	13.45553
4.3	−3.62123	4.08033	9.3	−10.33870	13.67925
4.4	−3.74498	4.23281	9.4	−10.47986	13.90401
4.5	−3.86943	4.38732	9.5	−10.62118	14.12981
4.6	−3.99455	4.54385	9.6	−10.76267	14.35664
4.7	−4.12032	4.70234	9.7	−10.90431	14.58449
4.8	−4.24672	4.86277	9.8	−11.04611	14.81335
4.9	−4.37371	5.02510	9.9	−11.18807	15.04320
5.0	−4.50128	5.18930	10.0	−11.33017	15.27404

Linear interpolation will yield about three figures; eight-point interpolation will yield about eight figures.

$\Re \ln \Gamma(z) = \log_e |\Gamma(z)|$ $\mathscr{I} \ln \Gamma(z) = \arg \Gamma(z)$

The Beta Function

Definition: $B(m,n) = \displaystyle\int_0^1 t^{m-1}(1-t)^{n-1}\, dt \quad m > 0, n > 0$

Relationship with Gamma Function $B(m,n) = \dfrac{\Gamma(m)\Gamma(n)}{\Gamma(m+n)}$

Properties:

$$B(m,n) = B(n,m)$$

$$B(m,n) = 2\int_0^{\pi/2} \sin^{2m-1}\theta \, \cos^{2n-1}\theta \, d\theta$$

$$B(m,n) = \int_0^\infty \frac{t^{m-1}}{(1+t)^{m+n}}\, dt$$

$$B(m,n) = r^n(r+1)^m \int_0^1 \frac{t^{m-1}(1-t)^{n-1}}{(r+t)^{m+n}}\, dt$$

The Error Function

Definition: $\operatorname{erf} x = \dfrac{2}{\sqrt{\pi}}\displaystyle\int_0^x e^{-t^2}\, dt$

Series: $\operatorname{erf} x = \dfrac{2}{\sqrt{\pi}}\left(x - \dfrac{x^3}{3} + \dfrac{1}{2!}\dfrac{x^5}{5} - \dfrac{1}{3!}\dfrac{x^7}{7} + \ldots\right)$

Property: $\operatorname{erf} x = -\operatorname{erf}(-x)$

Relationship with Normal Probability Function $f(t)$: $\displaystyle\int_0^x f(t)\, dt = \dfrac{1}{2}\operatorname{erf}\left(\dfrac{x}{\sqrt{2}}\right)$

To evaluate $\operatorname{erf}(2.3)$, one proceeds as follows: Since $\dfrac{x}{\sqrt{2}} = 2.3$, one finds $x = (2.3)(\sqrt{2}) = 3.25$. In the normal probability function table (page 753), one finds the entry 0.4994 opposite the value 3.25. Thus $\operatorname{erf}(2.3) = 2(0.4994) = 0.9988$.

$$\operatorname{erfc} z = 1 - \operatorname{erf} z = \frac{2}{\sqrt{\pi}}\int_z^\infty e^{-t^2}\, dt$$

is known as the complementary error function.

BESSEL FUNCTIONS

1. **Bessel's** differential equation for a real variable x is

$$x^2 \frac{d^2y}{dx^2} + x \frac{dy}{dx} + (x^2 - n^2)y = 0$$

2. When n is not an integer, two independent solutions of the equation are $J_n(x)$ and $J_{-n}(x)$, where

$$J_n(x) = \sum_{k=0}^{\infty} \frac{(-1)^k}{k!\Gamma(n + k + 1)} \left(\frac{x}{2}\right)^{n+2k}$$

3. If n is an integer $J_{-n}(x) = (-1)^n J_n(x)$, where

$$J_n(x) = \frac{x^n}{2^n n!}\left\{1 - \frac{x^2}{2^2 \cdot 1!(n+1)} + \frac{x^4}{2^4 \cdot 2!(n+1)(n+2)}\right.$$
$$\left. - \frac{x^6}{2^6 \cdot 3!(n+1)(n+2)(n+3)} + \cdots\right\}$$

4. For $n = 0$ and $n = 1$, this formula becomes

$$J_0(x) = 1 - \frac{x^2}{2^2(1!)^2} + \frac{x^4}{2^4(2!)^2} - \frac{x^6}{2^6(3!)^2} + \frac{x^8}{2^8(4!)^2} - \cdots$$

$$J_1(x) = \frac{x}{2} - \frac{x^3}{2^3 \cdot 1!2!} + \frac{x^5}{2^5 \cdot 2!3!} - \frac{x^7}{2^7 \cdot 3!4!} + \frac{x^9}{2^9 \cdot 4!5!} - \cdots$$

5. When x is large and positive, the following asymptotic series may be used

$$J_0(x) = \left(\frac{2}{\pi x}\right)^{\frac{1}{2}}\left\{P_0(x) \cos\left(x - \frac{\pi}{4}\right) - Q_0(x) \sin\left(x - \frac{\pi}{4}\right)\right\}$$

$$J_1(x) = \left(\frac{2}{\pi x}\right)^{\frac{1}{2}}\left\{P_1(x) \cos\left(x - \frac{3\pi}{4}\right) - Q_1(x) \sin\left(x - \frac{3\pi}{4}\right)\right\},$$

where

$$P_0(x) \sim 1 - \frac{1^2 \cdot 3^2}{2!(8x)^2} + \frac{1^2 \cdot 3^2 \cdot 5^2 \cdot 7^2}{4!(8x)^4} - \frac{1^2 \cdot 3^2 \cdot 5^2 \cdot 7^2 \cdot 9^2 \cdot 11^2}{6!(8x)^6} + \cdots$$

$$Q_0(x) \sim -\frac{1^2}{1!8x} + \frac{1^2 \cdot 3^2 \cdot 5^2}{3!(8x)^3} - \frac{1^2 \cdot 3^2 \cdot 5^2 \cdot 7^2 \cdot 9^2}{5!(8x)^5} + - \cdots$$

$$P_1(x) \sim 1 + \frac{1^2 \cdot 3 \cdot 5}{2!(8x)^2} - \frac{1^2 \cdot 3^2 \cdot 5^2 \cdot 7 \cdot 9}{4!(8x)^4} + \frac{1^2 \cdot 3^2 \cdot 5^2 \cdot 7^2 \cdot 9^2 \cdot 11 \cdot 13}{6!(8x)^6} - + \cdots$$

$$Q_1(x) \sim \frac{1 \cdot 3}{1!8x} - \frac{1^2 \cdot 3^2 \cdot 5 \cdot 7}{3!(8x)^3} + \frac{1^2 \cdot 3^2 \cdot 5^2 \cdot 7^2 \cdot 9 \cdot 11}{5!(8x)^5} - \cdots$$

[In $P_1(x)$ the signs alternate from + to − after the first term]

6. If $x > 25$, it is convenient to use the formulas

$$J_0(x) = A_0(x) \sin x + B_0(x) \cos x$$
$$J_1(x) = B_\top(x) \sin x - A_1(x) \cos x,$$

where

$$A_0(x) = \frac{P_0(x) - Q_0(x)}{(\pi x)^{\frac{1}{2}}} \quad \text{and} \quad A_1(x) = \frac{P_1(x) - Q_1(x)}{(\pi x)^{\frac{1}{2}}}$$

$$B_0(x) = \frac{P_0(x) + Q_0(x)}{(\pi x)^{\frac{1}{2}}} \quad \text{and} \quad B_1(x) = \frac{P_1(x) + Q_1(x)}{(\pi x)^{\frac{1}{2}}}$$

7. The zeros of $J_0(x)$ and $J_1(x)$

If $j_{0,s}$ and $j_{1,s}$ are the s'th zeros of $J_0(x)$ and $J_1(x)$ respectively, and if $a = 4s - 1$, $b = 4s + 1$

$$j_{0,s} \sim \frac{1}{4} \pi a \left\{ 1 + \frac{2}{\pi^2 a^2} - \frac{62}{3\pi^4 a^4} + \frac{15{,}116}{15\pi^6 a^6} - \frac{12{,}554{,}474}{105\pi^8 a^8} + \frac{8{,}368{,}654{,}292}{315\pi^{10} a^{10}} - + \cdots \right\}$$

$$j_{1,s} \sim \frac{1}{4} \pi b \left\{ 1 - \frac{6}{\pi^2 b^2} + \frac{6}{\pi^4 b^4} - \frac{4716}{5\pi^6 b^6} + \frac{3{,}902{,}418}{35\pi^8 b^8} - \frac{895{,}167{,}324}{35\pi^{10} b^{10}} + \cdots \right\}$$

$$J_1(j_{0,s}) \sim \frac{(-1)^{s+1} 2^{\frac{3}{2}}}{\pi a^{\frac{1}{2}}} \left\{ 1 - \frac{56}{3\pi^4 a^4} + \frac{9664}{5\pi^6 a^6} - \frac{7{,}381{,}280}{21\pi^8 a^8} + \cdots \right\}$$

$$J_0(j_{1,s}) \sim \frac{(-1)^s 2^{\frac{3}{2}}}{\pi b^{\frac{1}{2}}} \left\{ 1 + \frac{24}{\pi^4 b^4} - \frac{19{,}584}{10\pi^6 b^6} + \frac{2{,}466{,}720}{7\pi^8 b^8} - \cdots \right\}$$

8. Table of zeros for $J_0(x)$ and $J_1(x)$

$$J_0(\alpha_n) = 0 \qquad J_1(\beta_n) = 0$$

Roots α_n	$J_1(\alpha_n)$	Roots β_n	$J_0(\beta_n)$
2.4048	0.5191	0.0000	1.0000
5.5201	−0.3403	3.8317	−0.4028
8.6537	0.2715	7.0156	0.3001
11.7915	−0.2325	10.1735	−0.2497
14.9309	0.2065	13.3237	0.2184
18.0711	−0.1877	16.4706	−0.1965
21.2116	0.1733	19.6159	0.1801

9. Recurrence formulas

$$J_{n-1}(x) + J_{n+1}(x) = \frac{2n}{x} J_n(x) \qquad\qquad nJ_n(x) + xJ_n'(x) = xJ_{n-1}(x)$$

$$J_{n-1}(x) - J_{n+1}(x) = 2J_n'(x) \qquad\qquad nJ_n(x) - xJ_n'(x) = xJ_{n+1}(x)$$

10. If J_n is written for $J_n(x)$ and $J_n^{(k)}$ is written for $\dfrac{d^k}{dx^k}\{J_n(x)\}$, then the following derivative relationships are important

$$J_0^{(r)} = -J_1^{(r-1)}$$

$$J_0^{(2)} = -J_0 + \frac{1}{x} J_1 = \frac{1}{2} (J_2 - J_0)$$

$$J_0^{(3)} = \frac{1}{x} J_0 + \left(1 - \frac{2}{x^2}\right) J_1 = \frac{1}{4} (-J_3 + 3J_1)$$

$$J_0^{(4)} = \left(1 - \frac{3}{x^2}\right) J_0 - \left(\frac{2}{x} - \frac{6}{x^3}\right) J_1 = \frac{1}{8} (J_4 - 4J_2 + 3J_0), \text{ etc.}$$

11. Half order Bessel functions

$$J_{\frac{1}{2}}(x) = \sqrt{\frac{2}{\pi x}} \sin x$$

$$J_{-\frac{1}{2}}(x) = \sqrt{\frac{2}{\pi x}} \cos x$$

$$J_{n+\frac{3}{2}}(x) = -x^{n+\frac{1}{2}} \frac{d}{dx} \{x^{-(n+\frac{1}{2})} J_{n+\frac{1}{2}}(x)\}$$

$$J_{n-\frac{1}{2}}(x) = x^{-(n+\frac{1}{2})} \frac{d}{dx} \{x^{n+\frac{1}{2}} J_{n+\frac{1}{2}}(x)\}$$

n	$\left(\frac{\pi x}{2}\right)^{\frac{1}{2}} J_{n+\frac{1}{2}}(x)$	$\left(\frac{\pi x}{2}\right)^{\frac{1}{2}} J_{-(n+\frac{1}{2})}(x)$
0	$\sin x$	$\cos x$
1	$\dfrac{\sin x}{x} - \cos x$	$-\dfrac{\cos x}{x} - \sin x$
2	$\left(\dfrac{3}{x^2} - 1\right) \sin x - \dfrac{3}{x} \cos x$	$\left(\dfrac{3}{x^2} - 1\right) \cos x + \dfrac{3}{x} \sin x$
3	$\left(\dfrac{15}{x^3} - \dfrac{6}{x}\right) \sin x - \left(\dfrac{15}{x^2} - 1\right) \cos x$	$-\left(\dfrac{15}{x^3} - \dfrac{6}{x}\right) \cos x - \left(\dfrac{15}{x^2} - 1\right) \sin x$
	etc.	

12. Additional solutions to Bessel's equation are

$Y_n(x)$ (also called Weber's function, and sometimes denoted by $N_n(x)$)

$H_n^{(1)}(x)$ and $H_n^{(2)}(x)$ (also called Hankel functions)

These solutions are defined as follows

$$Y_n(x) = \begin{cases} \dfrac{J_n(x) \cos(n\pi) - J_{-n}(x)}{\sin(n\pi)} & n \text{ not an integer} \\[2em] \lim\limits_{v \to n} \dfrac{J_v(x) \cos(v\pi) - J_{-v}(x)}{\sin(v\pi)} & n \text{ an integer} \end{cases}$$

$$H_n^{(1)}(x) = J_n(x) + iY_n(x)$$
$$H_n^{(2)}(x) = J_n(x) - iY_n(x)$$

The additional properties of these functions may all be derived from the above relations and the known properties of $J_n(x)$.

13. Complete solutions to Bessel's equation may be written as

$$c_1 J_n(x) + c_2 J_{-n}(x) \qquad \text{if } n \text{ is not an integer,}$$

or

$$c_1 J_n(x) + c_2 Y_n(x)$$

or

$$c_1 H_n^{(1)}(x) + c_2 H_n^{(2)}(x)$$ for any value of n

14. The modified (or hyperbolic) Bessel's differential equation is

$$x^2 \frac{d^2 y}{dx^2} + x \frac{dy}{dx} - (x^2 + n^2) y = 0$$

15. When n is not an integer, two independent solutions of the equation are $I_n(x)$ and $I_{-n}(x)$, where

$$I_n(x) = \sum_{k=0}^{\infty} \frac{1}{k! \Gamma(n + k + 1)} \left(\frac{x}{2}\right)^{n+2k}$$

16. If n is an integer,

$$I_n(x) = I_{-n}(x) = \frac{x^n}{2^n n!} \left\{ 1 + \frac{x^2}{2^2 \cdot 1!(n + 1)} + \frac{x^4}{2^4 \cdot 2!(n + 1)(n + 2)} \right.$$
$$\left. + \frac{x^6}{2^6 \cdot 3!(n + 1)(n + 2)(n + 3)} + \cdots \right\}$$

17. For $n = 0$ and $n = 1$, this formula becomes

$$I_0(x) = 1 + \frac{x^2}{2^2(1!)^2} + \frac{x^4}{2^4(2!)^2} + \frac{x^6}{2^6(3!)^2} + \frac{x^8}{2^8(4!)^2} + \cdots$$

$$I_1(x) = \frac{x}{2} + \frac{x^3}{2^3 \cdot 1!2!} + \frac{x^5}{2^5 \cdot 2!3!} + \frac{x^7}{2^7 \cdot 3!4!} + \frac{x^9}{2^9 \cdot 4!5!} + \cdots$$

18. Another solution to the modified Bessel's equation is

$$K_n(x) = \begin{cases} \dfrac{1}{2} \pi \dfrac{I_{-n}(x) - I_n(x)}{\sin(n\pi)} & n \text{ not an integer} \\ \\ \lim_{v \to n} \dfrac{1}{2} \pi \dfrac{I_{-v}(x) - I_v(x)}{\sin(v\pi)} & n \text{ an integer} \end{cases}$$

This function is linearly independent of $I_n(x)$ for all values of n. Thus the complete solution to the modified Bessel's equation may be written as

$$c_1 I_n(x) + c_2 I_{-n}(x) \qquad n \text{ not an integer}$$

or

$$c_1 I_n(x) + c_2 K_n(x) \qquad \text{any } n$$

19. The following relations hold among the various Bessel functions:

$$I_n(z) = i^{-m} J_m(iz)$$

$$Y_n(iz) = (i)^{n+1} I_n(z) - \frac{2}{\pi} i^{-n} K_n(z)$$

Most of the properties of the modified Bessel function may be deduced from the known properties of $J_n(x)$ by use of these relations and those previously given.

20. Recurrence formulas

$$I_{n-1}(x) - I_{n+1}(x) = \frac{2n}{x} I_n(x) \qquad I_{n-1}(x) + I_{n+1}(x) = 2I_n'(x)$$

$$I_{n-1}(x) - \frac{n}{x} I_n(x) = I_n'(x) \qquad I_n'(x) = I_{n+1}(x) + \frac{n}{x} I_n(x)$$

BESSEL FUNCTIONS FOR SPHERICAL COORDINATES

$$j_n(x) = \sqrt{\frac{\pi}{2x}} \, J_{(n+\frac{1}{2})}(x), \; y_n(x) = \sqrt{\frac{\pi}{2x}} \, Y_{(n+\frac{1}{2})}(x) = (-1)^{n+1} \sqrt{\frac{\pi}{2x}} \, J_{-(n+\frac{1}{2})}(x)$$

x	$j_0(x)$	$y_0(x)$	$j_1(x)$	$y_1(x)$	$j_2(x)$	$y_2(x)$
0.0	1.0000	$-\infty$	0.0000	$-\infty$	0.0000	$-\infty$
0.1	0.9983	-9.9500	0.0333	-100.50	0.0007	-3005.0
0.2	0.9933	-4.9003	0.0664	-25.495	0.0027	-377.52
0.4	0.9735	-2.3027	0.1312	-6.7302	0.0105	-48.174
0.6	0.9411	-1.3756	0.1929	-3.2337	0.0234	-14.793
0.8	0.8967	-0.8709	0.2500	-1.9853	0.0408	-6.5740
1.0	0.8415	-0.5403	0.3012	-1.3818	0.0620	-3.6050
1.2	0.7767	-0.3020	0.3453	-1.0283	0.0865	-2.2689
1.4	0.7039	-0.1214	0.3814	-0.7906	0.1133	-1.5728
1.6	0.6247	$+0.0182$	0.4087	-0.6133	0.1416	-1.1682
1.8	0.5410	0.1262	0.4268	-0.4709	0.1703	-0.9111
2.0	0.4546	0.2081	0.4354	-0.3506	0.1984	-0.7340
2.2	0.3675	0.2675	0.4345	-0.2459	0.2251	-0.6028
2.4	0.2814	0.3072	0.4245	-0.1534	0.2492	-0.4990
2.6	0.1983	0.3296	0.4058	-0.0715	0.2700	-0.4121
2.8	0.1196	0.3365	0.3792	$+0.0005$	0.2867	-0.3359
3.0	$+0.0470$	0.3300	0.3457	0.0630	0.2986	-0.2670
3.2	-0.0182	0.3120	0.3063	0.1157	0.3054	-0.2035
3.4	-0.0752	0.2844	0.2622	0.1588	0.3066	-0.1442
3.6	-0.1229	0.2491	0.2150	0.1921	0.3021	-0.0890
3.8	-0.1610	0.2082	0.1658	0.2158	0.2919	-0.0378
4.0	-0.1892	0.1634	0.1161	0.2301	0.2763	$+0.0091$
4.2	-0.2075	0.1167	0.0673	0.2353	0.2556	0.0514
4.4	-0.2163	0.0698	$+0.0207$	0.2321	0.2304	0.0884
4.6	-0.2160	$+0.0244$	-0.0226	0.2213	0.2013	0.1200
4.8	-0.2075	-0.0182	-0.0615	0.2037	0.1691	0.1456
5.0	-0.1918	-0.0567	-0.0951	0.1804	0.1347	0.1650
5.2	-0.1699	-0.0901	-0.1228	0.1526	0.0991	0.1781
5.4	-0.1431	-0.1175	-0.1440	0.1213	0.0631	0.1850
5.6	-0.1127	-0.1385	-0.1586	0.0880	$+0.0277$	0.1856
5.8	-0.0801	-0.1527	-0.1665	0.0538	-0.0060	0.1805
6.0	-0.0466	-0.1600	-0.1678	$+0.0199$	-0.0373	0.1700
6.2	-0.0134	-0.1607	-0.1629	-0.0125	-0.0654	0.1547
6.4	$+0.0182$	-0.1552	-0.1523	-0.0425	-0.0896	0.1353
6.6	0.0472	-0.1440	-0.1368	-0.0690	-0.1094	0.1126
6.8	0.0727	-0.1278	-0.1172	-0.0915	-0.1243	0.0875
7.0	0.0939	-0.1077	-0.0943	-0.1092	-0.1343	0.0609
7.2	0.1102	-0.0845	-0.0692	-0.1220	-0.1391	0.0337
7.4	0.1215	-0.0593	-0.0429	-0.1295	-0.1388	$+0.0068$
7.6	0.1274	-0.0331	-0.0163	-0.1317	-0.1338	-0.0189
7.8	0.1280	-0.0069	$+0.0095$	-0.1289	-0.1244	-0.0427
8.0	0.1237	$+0.0182$	0.0336	-0.1214	-0.1111	-0.0637

Taken from Vibration and Sound with the permission of Philip Morse, author, and McGraw-Hill Book Company, Inc., publisher.

BESSEL FUNCTIONS $J_0(x)$ AND $J_1(x)$

x	$J_0(x)$	$J_1(x)$	x	$J_0(x)$	$J_1(x)$	x	$J_0(x)$	$J_1(x)$
0.0	1.0000	.0000	**5.0**	−.1776	−.3276	**10.0**	−.2459	.0435
0.1	.9975	.0499	5.1	−.1443	−.3371	10.1	−.2490	.0184
0.2	.9900	.0995	5.2	−.1103	−.3432	10.2	−.2496	−.0066
0.3	.9776	.1483	5.3	−.0758	−.3460	10.3	−.2477	−.0313
0.4	.9604	.1960	5.4	−.0412	−.3453	10.4	−.2434	−.0555
0.5	.9385	.2423	**5.5**	−.0068	−.3414	**10.5**	−.2366	−.0789
0.6	.9120	.2867	5.6	.0270	−.3343	10.6	−.2276	−.1012
0.7	.8812	.3290	5.7	.0599	−.3241	10.7	−.2164	−.1224
0.8	.8463	.3688	5.8	.0917	−.3110	10.8	−.2032	−.1422
0.9	.8075	.4059	5.9	.1220	−.2951	10.9	−.1881	−.1603
1.0	.7652	.4401	**6.0**	.1506	−.2767	**11.0**	−.1712	−.1768
1.1	.7196	.4709	6.1	.1773	−.2559	11.1	−.1528	−.1913
1.2	.6711	.4983	6.2	.2017	−.2329	11.2	−.1330	−.2039
1.3	.6201	.5220	6.3	.2238	−.2081	11.3	−.1121	−.2143
1.4	.5669	.5419	6.4	.2433	−.1816	11.4	−.0902	−.2225
1.5	.5118	.5579	**6.5**	.2601	−.1538	**11.5**	−.0677	−.2284
1.6	.4554	.5699	6.6	.2740	−.1250	11.6	−.0446	−.2320
1.7	.3980	.5778	6.7	.2851	−.0953	11.7	−.0213	−.2333
1.8	.3400	.5815	6.8	.2931	−.0652	11.8	.0020	−.2323
1.9	.2818	.5812	6.9	.2981	−.0349	11.9	.0250	−.2290
2.0	.2239	.5767	**7.0**	.3001	−.0047	**12.0**	.0477	−.2234
2.1	.1666	.5683	7.1	.2991	.0252	12.1	.0697	−.2157
2.2	.1104	.5560	7.2	.2951	.0543	12.2	.0908	−.2060
2.3	.0555	.5399	7.3	.2882	.0826	12.3	.1108	−.1943
2.4	.0025	.5202	7.4	.2786	.1096	12.4	.1296	−.1807
2.5	−.0484	.4971	**7.5**	.2663	.1352	**12.5**	.1469	−.1655
2.6	−.0968	.4708	7.6	.2516	.1592	12.6	.1626	−.1487
2.7	−.1424	.4416	7.7	.2346	.1813	12.7	.1766	−.1307
2.8	−.1850	.4097	7.8	.2154	.2014	12.8	.1887	−.1114
2.9	−.2243	.3754	7.9	.1944	.2192	12.9	.1988	−.0912
3.0	−.2601	.3391	**8.0**	.1717	.2346	**13.0**	.2069	−.0703
3.1	−.2921	.3009	8.1	.1475	.2476	13.1	.2129	−.0489
3.2	−.3202	.2613	8.2	.1222	.2580	13.2	.2167	−.0271
3.3	−.3443	.2207	8.3	.0960	.2657	13.3	.2183	−.0052
3.4	−.3643	.1792	8.4	.0692	.2708	13.4	.2177	.0166
3.5	−.3801	.1374	**8.5**	.0419	.2731	**13.5**	.2150	.0380
3.6	−.3918	.0955	8.6	.0146	.2728	13.6	.2101	.0590
3.7	−.3992	.0538	8.7	−.0125	.2697	13.7	.2032	.0791
3.8	−.4026	.0128	8.8	−.0392	.2641	13.8	.1943	.0984
3.9	−.4018	−.0272	8.9	−.0653	.2559	13.9	.1836	.1165
4.0	−.3971	−.0660	**9.0**	−.0903	.2453	**14.0**	.1711	.1334
4.1	−.3887	−.1033	9.1	−.1142	.2324	14.1	.1570	.1488
4.2	−.3766	−.1386	9.2	−.1367	.2174	14.2	.1414	.1626
4.3	−.3610	−.1719	9.3	−.1577	.2004	14.3	.1245	.1747
4.4	−.3423	−.2028	9.4	−.1768	.1816	14.4	.1065	.1850
4.5	−.3205	−.2311	**9.5**	−.1939	.1613	**14.5**	.0875	.1934
4.6	−.2961	−.2566	9.6	−.2090	.1395	14.6	.0679	.1999
4.7	−.2693	−.2791	9.7	−.2218	.1166	14.7	.0476	.2043
4.8	−.2404	−.2985	9.8	−.2323	.0928	14.8	.0271	.2066
4.9	−.2097	−.3147	9.9	−.2403	.0684	14.9	.0064	.2069

BESSEL FUNCTION J_0

T	.00	.01	.02	.03	.04
0.0	1.0000000	.9999750	.9999000	.9997750	.9996000
.1	.9975016	.9969773	.9964032	.9957795	.9951060
.2	.9900250	.9890054	.9879366	.9868187	.9856518
.3	.9776262	.9761189	.9745634	.9729597	.9713081
.4	.9603982	.9584145	.9563838	.9543065	.9521825
.5	.9384698	.9360245	.9335339	.9309983	.9284179
.6	.9120049	.9091162	.9061843	.9032094	.9001918
.7	.8812009	.8778904	.8745391	.8711471	.8677147
.8	.8462874	.8425797	.8388338	.8350500	.8312284
.9	.8075238	.8034465	.7993339	.7951863	.7910039
1.0	.7651977	.7607810	.7563321	.7518513	.7473390
1.1	.7196220	.7148985	.7101461	.7053653	.7005564
1.2	.6711327	.6661371	.6611163	.6560706	.6510004
1.3	.6200860	.6148549	.6096023	.6043287	.5990343
1.4	.5668551	.5614267	.5559807	.5505176	.5450376
1.5	.5118277	.5062414	.5006415	.4950285	.4894026
1.6	.4554022	.4496983	.4439850	.4382625	.4325313
1.7	.3979849	.3922044	.3864185	.3806276	.3748321
1.8	.3399864	.3341705	.3283532	.3225351	.3167166
1.9	.2818186	.2760083	.2702008	.2643965	.2585959
2.0	.2238908	.2181268	.2123697	.2066198	.2008776
2.1	.1666070	.1609293	.1552625	.1496068	.1439626
2.2	.1103623	.1048098	.0992720	.0937491	.0882416
2.3	.0555398	.0501501	.0447787	.0394259	.0340921
2.4	.0025077	−.0026834	−.0078527	−.0129999	−.0181247
2.5	−.0483838	−.0533423	−.0582758	−.0631839	−.0680664
2.6	−.0968050	−.1014992	−.1061654	−.1108031	−.1154123
2.7	−.1424494	−.1468500	−.1512198	−.1555585	−.1598658
2.8	−.1850360	−.1891165	−.1931636	−.1971771	−.2011568
2.9	−.2243115	−.2280481	−.2317491	−.2354142	−.2390434
3.0	−.2600520	−.2634239	−.2667583	−.2700551	−.2733140
3.1	−.2920643	−.2950541	−.2980048	−.3009162	−.3037884
3.2	−.3201882	−.3227815	−.3253345	−.3278471	−.3303193
3.3	−.3442963	−.3464823	−.3486272	−.3507308	−.3527931
3.4	−.3642956	−.3660670	−.3677967	−.3694845	−.3711306
3.5	−.3801277	−.3814805	−.3827914	−.3840603	−.3852873
3.6	−.3917690	−.3927027	−.3935947	−.3944449	−.3952533
3.7	−.3992302	−.3997479	−.4002242	−.4006593	−.4010532
3.8	−.4025564	−.4026643	−.4027318	−.4027588	−.4027456
3.9	−.4018260	−.4015339	−.4012023	−.4008316	−.4004218
4.0	−.3971498	−.3964704	−.3957530	−.3949979	−.3942053
4.1	−.3886697	−.3876188	−.3865318	−.3854088	−.3842500
4.2	−.3765571	−.3751534	−.3737157	−.3722440	−.3707386
4.3	−.3610111	−.3592761	−.3575093	−.3557108	−.3538810
4.4	−.3422568	−.3402143	−.3381424	−.3360414	−.3339116
4.5	−.3205425	−.3182185	−.3158678	−.3134908	−.3110877
4.6	−.2961378	−.2935603	−.2909591	−.2883344	−.2856866
4.7	−.2693308	−.2665295	−.2637076	−.2608653	−.2580029
4.8	−.2404253	−.2374315	−.2344201	−.2313916	−.2283462
4.9	−.2097383	−.2065842	−.2034158	−.2002335	−.1970377

BESSEL FUNCTION J_0

T	.05	.06	.07	.08	.09
0.0	.9993751	.9991002	.9987754	.9984006	.9979760
.1	.9943829	.9936102	.9927880	.9919164	.9909953
.2	.9844359	.9831713	.9818579	.9804958	.9790853
.3	.9696087	.9678615	.9660667	.9642245	.9623350
.4	.9500121	.9477955	.9455328	.9432242	.9408698
.5	.9257928	.9231233	.9204096	.9176518	.9148501
.6	.8971316	.8940292	.8908846	.8876982	.8844702
.7	.8642423	.8607300	.8571780	.8535868	.8499565
.8	.8273695	.8234734	.8195405	.8155711	.8115654
.9	.7867871	.7825361	.7782514	.7739332	.7695819
1.0	.7427956	.7382212	.7336163	.7289813	.7243164
1.1	.6957198	.6908557	.6859646	.6810469	.6761028
1.2	.6459061	.6407880	.6356466	.6304822	.6252952
1.3	.5937196	.5883850	.5830309	.5776576	.5722655
1.4	.5395413	.5340289	.5285010	.5229579	.5174000
1.5	.4837644	.4781143	.4724526	.4667797	.4610961
1.6	.4267919	.4210446	.4152898	.4095280	.4037595
1.7	.3690325	.3632292	.3574225	.3516128	.3458007
1.8	.3108980	.3050797	.2992623	.2934460	.2876313
1.9	.2527992	.2470071	.2412197	.2354376	.2296612
2.0	.1951434	.1894177	.1837008	.1779931	.1722950
2.1	.1383305	.1327106	.1271035	.1215095	.1159290
2.2	.0827499	.0772742	.0718150	.0663726	.0609474
2.3	.0287776	.0234828	.0182081	.0129538	.0077202
2.4	−.0232267	−.0283057	−.0333611	−.0383929	−.0434005
2.5	−.0729229	−.0777531	−.0825567	−.0873334	−.0920829
2.6	−.1199924	−.1245434	−.1290648	−.1335565	−.1380181
2.7	−.1641414	−.1683852	−.1725967	−.1767759	−.1809224
2.8	−.2051024	−.2090137	−.2128905	−.2167325	−.2205396
2.9	−.2426364	−.2461931	−.2497131	−.2531964	−.2566427
3.0	−.2765350	−.2797178	−.2828623	−.2859683	−.2890357
3.1	−.3066211	−.3094142	−.3121675	−.3148811	−.3175547
3.2	−.3327508	−.3351416	−.3374917	−.3398009	−.3420691
3.3	−.3548140	−.3567934	−.3587314	−.3606277	−.3624825
3.4	−.3727349	−.3742972	−.3758177	−.3772963	−.3787330
3.5	−.3864724	−.3876155	−.3887167	−.3897760	−.3907934
3.6	−.3960201	−.3967452	−.3974287	−.3980707	−.3986712
3.7	−.4014061	−.4017178	−.4019887	−.4022187	−.4024079
3.8	−.4026921	−.4025986	−.4024651	−.4022918	−.4020787
3.9	−.3999730	−.3994854	−.3989591	−.3983943	−.3977912
4.0	−.3933752	−.3925079	−.3916035	−.3906622	−.3896842
4.1	−.3830556	−.3818259	−.3805609	−.3792610	−.3779263
4.2	−.3691998	−.3676276	−.3660225	−.3643845	−.3627140
4.3	−.3520200	−.3501281	−.3482056	−.3462527	−.3442697
4.4	−.3317533	−.3295666	−.3273519	−.3251095	−.3228396
4.5	−.3086589	−.3062045	−.3037249	−.3012204	−.2986913
4.6	−.2830159	−.2803228	−.2776073	−.2748700	−.2721110
4.7	−.2551208	−.2522193	−.2492987	−.2463592	−.2434014
4.8	−.2252843	−.2222062	−.2191122	−.2160027	−.2128779
4.9	−.1938286	−.1906067	−.1873722	−.1841255	−.1808669

BESSEL FUNCTION J_0

T	.00	.01	.02	.03	.04
5.0	−.1775968	−.1743154	−.1710232	−.1677205	−.1644075
5.1	−.1443347	−.1409599	−.1375776	−.1341882	−.1307919
5.2	−.1102904	−.1068561	−.1034176	−.0999753	−.0965297
5.3	−.0758031	−.0723430	−.0688822	−.0654211	−.0619598
5.4	−.0412101	−.0377578	−.0343082	−.0308615	−.0274180
5.5	−.0068439	−.0034323	−.0000266	.0033730	.0067661
5.6	.0269709	.0303098	.0336398	.0369605	.0402716
5.7	.0599200	.0631556	.0663792	.0695906	.0727894
5.8	.0917026	.0948055	.0978937	.1009668	.1040245
5.9	.1220334	.1249761	.1279015	.1308091	.1336987
6.0	.1506453	.1534022	.1561393	.1588562	.1615527
6.1	.1772914	.1798391	.1823646	.1848678	.1873484
6.2	.2017472	.2040644	.2063574	.2086262	.2108705
6.3	.2238120	.2258800	.2279222	.2299383	.2319283
6.4	.2433106	.2451134	.2468888	.2486369	.2503573
6.5	.2600946	.2616188	.2631145	.2645817	.2660201
6.6	.2740434	.2752785	.2764843	.2776606	.2788074
6.7	.2850647	.2860032	.2869117	.2877901	.2886385
6.8	.2930956	.2937327	.2943394	.2949159	.2954620
6.9	.2981020	.2984359	.2987395	.2990127	.2992557
7.0	.3000793	.3001111	.3001128	.3000846	.3000264
7.1	.2990514	.2987851	.2984893	.2981641	.2978096
7.2	.2950707	.2945131	.2939268	.2933119	.2926686
7.3	.2882169	.2873774	.2865103	.2856158	.2846939
7.4	.2785962	.2774868	.2763512	.2751894	.2740018
7.5	.2663397	.2649748	.2635853	.2621712	.2607329
7.6	.2516018	.2499982	.2483717	.2467225	.2450508
7.7	.2345591	.2327355	.2308910	.2290257	.2271400
7.8	.2154078	.2133848	.2113430	.2092828	.2072042
7.9	.1943618	.1921618	.1899452	.1877126	.1854639
8.0	.1716508	.1692974	.1669299	.1645488	.1621542
8.1	.1475175	.1450356	.1425423	.1400378	.1375223
8.2	.1222153	.1196308	.1170375	.1144357	.1118256
8.3	.0960061	.0933456	.0906789	.0880063	.0853282
8.4	.0691573	.0664476	.0637345	.0610183	.0582992
8.5	.0419393	.0392076	.0364752	.0337424	.0310094
8.6	.0146230	.0118963	.0091717	.0064492	.0037293
8.7	−.0125227	−.0152177	−.0179081	−.0205935	−.0232739
8.8	−.0392338	−.0418710	−.0445011	−.0471237	−.0497387
8.9	−.0652532	−.0678075	−.0703522	−.0728869	−.0754116
9.0	−.0903336	−.0927808	−.0952160	−.0976390	−.1000496
9.1	−.1142392	−.1165565	−.1188596	−.1211483	−.1234224
9.2	−.1367484	−.1389144	−.1410642	−.1431976	−.1453143
9.3	−.1576552	−.1596503	−.1616274	−.1635862	−.1655265
9.4	−.1767716	−.1785781	−.1803648	−.1821316	−.1838783
9.5	−.1939287	−.1955308	−.1971117	−.1986712	−.2002092
9.6	−.2089787	−.2103628	−.2117244	−.2130634	−.2143797
9.7	−.2217955	−.2229502	−.2240814	−.2251890	−.2262730
9.8	−.2322760	−.2331923	−.2340844	−.2349521	−.2357955
9.9	−.2403411	−.2410124	−.2416590	−.2422808	−.2428777

BESSEL FUNCTION J_0

T	.05	.06	.07	.08	.09
5.0	−.1610847	−.1577524	−.1544109	−.1510606	−.1477018
5.1	−.1273892	−.1239803	−.1205657	−.1171456	−.1137204
5.2	−.0930810	−.0896295	−.0861757	−.0827198	−.0792621
5.3	−.0584989	−.0550386	−.0515792	−.0481211	−.0446646
5.4	−.0239781	−.0205422	−.0171104	−.0136833	−.0102610
5.5	.0101524	.0135315	.0169033	.0202673	.0236233
5.6	.0435728	.0468638	.0501444	.0534141	.0566727
5.7	.0759753	.0791482	.0823076	.0854533	.0885851
5.8	.1070666	.1100928	.1131028	.1160964	.1190734
5.9	.1365701	.1394230	.1422573	.1450725	.1478686
6.0	.1642286	.1668837	.1695178	.1721306	.1747218
6.1	.1898062	.1922411	.1946529	.1970413	.1994061
6.2	.2130901	.2152848	.2174546	.2195991	.2217183
6.3	.2338920	.2358292	.2377398	.2396237	.2414807
6.4	.2520501	.2537151	.2553522	.2569612	.2585420
6.5	.2674298	.2688106	.2701625	.2714853	.2727789
6.6	.2799246	.2810122	.2820700	.2830981	.2840964
6.7	.2894568	.2902449	.2910029	.2917307	.2924283
6.8	.2959779	.2964633	.2969185	.2973434	.2977379
6.9	.2994685	.2996510	.2998033	.2999254	.3000174
7.0	.2999383	.2998204	.2996727	.2994953	.2992881
7.1	.2974258	.2970128	.2965707	.2960996	.2955996
7.2	.2919969	.2912970	.2905689	.2898128	.2890288
7.3	.2837448	.2827687	.2817656	.2807358	.2796793
7.4	.2727883	.2715492	.2702846	.2689947	.2676797
7.5	.2592704	.2577839	.2562736	.2547397	.2531824
7.6	.2433568	.2416407	.2399026	.2381429	.2363617
7.7	.2252341	.2233081	.2213622	.2193967	.2174119
7.8	.2051076	.2029932	.2008612	.1987118	.1965453
7.9	.1831996	.1809198	.1786247	.1763147	.1739900
8.0	.1597463	.1573255	.1548919	.1524459	.1499876
8.1	.1349963	.1324598	.1299132	.1273568	.1247907
8.2	.1092075	.1065816	.1039483	.1013077	.0986602
8.3	.0826448	.0799563	.0772630	.0745652	.0718632
8.4	.0555775	.0528534	.0501273	.0473994	.0446699
8.5	.0282765	.0255440	.0228121	.0200812	.0173513
8.6	.0010122	−.0017019	−.0044128	−.0071200	−.0098234
8.7	−.0259489	−.0286182	−.0312816	−.0339388	−.0365896
8.8	−.0523457	−.0549445	−.0575350	−.0601167	−.0626896
8.9	−.0779258	−.0804295	−.0829224	−.0854042	−.0878747
9.0	−.1024475	−.1048325	−.1072044	−.1095629	−.1119080
9.1	−.1256816	−.1279258	−.1301548	−.1323684	−.1345663
9.2	−.1474143	−.1494972	−.1515629	−.1536113	−.1556421
9.3	−.1674482	−.1693511	−.1712351	−.1730999	−.1749455
9.4	−.1856048	−.1873109	−.1889965	−.1906615	−.1923056
9.5	−.2017255	−.2032202	−.2046929	−.2061437	−.2075723
9.6	−.2156732	−.2169439	−.2181915	−.2194161	−.2206174
9.7	−.2273333	−.2283698	−.2293823	−.2303710	−.2313356
9.8	−.2366145	−.2374090	−.2381789	−.2389243	−.2396451
9.9	−.2434497	−.2439968	−.2445190	−.2450163	−.2454885

BESSEL FUNCTION J_0

T	.00	.01	.02	.03	.04
10.0	− .2459358	− .2463580	− .2467551	− .2471272	− .2474743
10.1	− .2490297	− .2492011	− .2493474	− .2494687	− .2495649
10.2	− .2496171	− .2495385	− .2494350	− .2493067	− .2491536
10.3	− .2477168	− .2473914	− .2470416	− .2466674	− .2462690
10.4	− .2433718	− .2428051	− .2422148	− .2416008	− .2409633
10.5	− .2366482	− .2358483	− .2350255	− .2341800	− .2333120
10.6	− .2276350	− .2266119	− .2255670	− .2245006	− .2234127
10.7	− .2164427	− .2152085	− .2139539	− .2126791	− .2113843
10.8	− .2032020	− .2017708	− .2003208	− .1988522	− .1973650
10.9	− .1880622	− .1864501	− .1848208	− .1831745	− .1815115
11.0	− .1711903	− .1694147	− .1676238	− .1658178	− .1639968
11.1	− .1527683	− .1508483	− .1489149	− .1469684	− .1450089
11.2	− .1329919	− .1309477	− .1288922	− .1268256	− .1247483
11.3	− .1120685	− .1099213	− .1077650	− .1055999	− .1034261
11.4	− .0902145	− .0879865	− .0857517	− .0835102	− .0812623
11.5	− .0676539	− .0653678	− .0630771	− .0607821	− .0584830
11.6	− .0446157	− .0422945	− .0399711	− .0376456	− .0353184
11.7	− .0213313	− .0189982	− .0166653	− .0143327	− .0120006
11.8	.0019672	.0042889	.0066082	.0089249	.0112388
11.9	.0250494	.0273370	.0296200	.0318980	.0341710
12.0	.0476893	.0499204	.0521447	.0543619	.0565718
12.1	.0696668	.0718199	.0739640	.0760989	.0782245
12.2	.0907701	.0928245	.0948680	.0969003	.0989212
12.3	.1107980	.1127342	.1146576	.1165679	.1184651
12.4	.1295610	.1313609	.1331462	.1349167	.1366724
12.5	.1468841	.1485309	.1501615	.1517758	.1533737
12.6	.1626073	.1640860	.1655471	.1669905	.1684160
12.7	.1765879	.1778851	.1791636	.1804231	.1816637
12.8	.1887014	.1898058	.1908904	.1919550	.1929997
12.9	.1988424	.1997446	.2006261	.2014869	.2023269
13.0	.2069261	.2076187	.2082899	.2089399	.2095684
13.1	.2128882	.2133659	.2138219	.2142562	.2146687
13.2	.2166859	.2169457	.2171835	.2173995	.2175935
13.3	.2182981	.2183389	.2183579	.2183551	.2183304
13.4	.2177252	.2175484	.2173499	.2171299	.2168884
13.5	.2149892	.2145981	.2141858	.2137525	.2132981
13.6	.2101332	.2095332	.2089128	.2082720	.2076107
13.7	.2032208	.2024195	.2015986	.2007581	.1998982
13.8	.1943356	.1933424	.1923305	.1913002	.1902515
13.9	.1835799	.1824059	.1812145	.1800059	.1787801
14.0	.1710735	.1697317	.1683739	.1670002	.1656108
14.1	.1569529	.1554577	.1539481	.1524242	.1508861
14.2	.1413694	.1397368	.1380914	.1364334	.1347629
14.3	.1244877	.1227348	.1209709	.1191961	.1174107
14.4	.1064841	.1046292	.1027650	.1008919	.0990100
14.5	.0875449	.0856069	.0836617	.0817095	.0797504
14.6	.0678641	.0658629	.0638565	.0618450	.0598288
14.7	.0476418	.0455977	.0435503	.0415000	.0394470
14.8	.0270823	.0250157	.0229481	.0208795	.0188102
14.9	.0063915	.0043232	.0022558	.0001896	− .0018753

BESSEL FUNCTION J_0

T	.05	.06	.07	.08	.09
10.0	−.2477962	−.2480931	−.2483649	−.2486116	−.2488332
10.1	−.2496361	−.2496822	−.2497034	−.2496996	−.2496708
10.2	−.2489758	−.2487732	−.2485460	−.2482942	−.2480177
10.3	−.2458463	−.2453994	−.2449285	−.2444335	−.2439146
10.4	−.2403022	−.2396178	−.2389101	−.2381792	−.2374252
10.5	−.2324214	−.2315085	−.2305732	−.2296158	−.2286364
10.6	−.2223036	−.2211732	−.2200218	−.2188495	−.2176565
10.7	−.2100695	−.2087349	−.2073807	−.2060071	−.2046141
10.8	−.1958594	−.1943357	−.1927939	−.1912343	−.1896570
10.9	−.1798318	−.1781356	−.1764232	−.1746947	−.1729504
11.0	−.1621611	−.1603109	−.1584463	−.1565675	−.1546748
11.1	−.1430367	−.1410520	−.1390549	−.1370458	−.1350247
11.2	−.1226602	−.1205618	−.1184532	−.1163346	−.1142063
11.3	−.1012439	−.0990535	−.0968552	−.0946491	−.0924354
11.4	−.0790083	−.0767484	−.0744828	−.0722117	−.0699353
11.5	−.0561801	−.0538735	−.0515636	−.0492505	−.0469344
11.6	−.0329897	−.0306597	−.0283286	−.0259967	−.0236642
11.7	−.0096694	−.0073391	−.0050101	−.0026825	−.0003567
11.8	.0135496	.0158571	.0181610	.0204612	.0227574
11.9	.0364386	.0387007	.0409570	.0432074	.0454516
12.0	.0587743	.0609690	.0631559	.0653346	.0675049
12.1	.0803406	.0824468	.0845431	.0866292	.0887050
12.2	.1009306	.1029283	.1049140	.1068877	.1088491
12.3	.1203489	.1222191	.1240756	.1259182	.1277467
12.4	.1384129	.1401382	.1418482	.1435426	.1452212
12.5	.1549550	.1565195	.1580671	.1595977	.1611111
12.6	.1698236	.1712131	.1725845	.1739374	.1752719
12.7	.1828851	.1840872	.1852700	.1864334	.1875772
12.8	.1940242	.1950286	.1960127	.1969764	.1979197
12.9	.2031460	.2039441	.2047213	.2054773	.2062123
13.0	.2101755	.2107612	.2113253	.2118679	.2123888
13.1	.2150594	.2154284	.2157755	.2161009	.2164043
13.2	.2177657	.2179159	.2180443	.2181508	.2182354
13.3	.2182839	.2182156	.2181256	.2180138	.2178803
13.4	.2166254	.2163409	.2160349	.2157076	.2153590
13.5	.2128227	.2123263	.2118092	.2112712	.2107125
13.6	.2069293	.2062276	.2055059	.2047641	.2040024
13.7	.1990189	.1981203	.1972026	.1962659	.1953102
13.8	.1891845	.1880993	.1869962	.1858751	.1847363
13.9	.1775373	.1762777	.1750014	.1737085	.1723991
14.0	.1642058	.1627855	.1613498	.1598991	.1584334
14.1	.1493340	.1477681	.1461885	.1445954	.1429890
14.2	.1330800	.1313851	.1296782	.1279596	.1262293
14.3	.1156147	.1138085	.1119921	.1101658	.1083297
14.4	.0971195	.0952206	.0933135	.0913984	.0894754
14.5	.0777848	.0758127	.0738344	.0718500	.0698599
14.6	.0578079	.0557827	.0537533	.0517198	.0496826
14.7	.0373914	.0353334	.0332733	.0312113	.0291476
14.8	.0167404	.0146704	.0126003	.0105303	.0084606
14.9	−.0039386	−.0060002	−.0080597	−.0101171	−.0121721

BESSEL FUNCTION J_0

T	.00	.01	.02	.03	.04
15.0	− .0142245	− .0162741	− .0183207	− .0203641	− .0224042
15.1	− .0345619	− .0365725	− .0385782	− .0405787	− .0425738
15.2	− .0544208	− .0563729	− .0583180	− .0602560	− .0621868
15.3	− .0736075	− .0754820	− .0773477	− .0792045	− .0810521
15.4	− .0919362	− .0937150	− .0954833	− .0972409	− .0989876
15.5	− .1092307	− .1108968	− .1125507	− .1141923	− .1158215
15.6	− .1253260	− .1268636	− .1283875	− .1298977	− .1313938
15.7	− .1400702	− .1414649	− .1428446	− .1442091	− .1455583
15.8	− .1533257	− .1545646	− .1557872	− .1569934	− .1581832
15.9	− .1649705	− .1660422	− .1670966	− .1681336	− .1691532
16.0	− .1748991	− .1757940	− .1766708	− .1775294	− .1783697
16.1	− .1830237	− .1837341	− .1844257	− .1850984	− .1857523
16.2	− .1892749	− .1897949	− .1902956	− .1907770	− .1912390
16.3	− .1936024	− .1939279	− .1942339	− .1945203	− .1947870
16.4	− .1959748	− .1961039	− .1962134	− .1963031	− .1963732
16.5	− .1963807	− .1963133	− .1962262	− .1961196	− .1959935
16.6	− .1948279	− .1945657	− .1942843	− .1939836	− .1936637
16.7	− .1913435	− .1908905	− .1904186	− .1899280	− .1894187
16.8	− .1859739	− .1853355	− .1846791	− .1840045	− .1833120
16.9	− .1787834	− .1779672	− .1771337	− .1762830	− .1754152
17.0	− .1698543	− .1688694	− .1678682	− .1668508	− .1658174
17.1	− .1592853	− .1581425	− .1569846	− .1558116	− .1546238
17.2	− .1471911	− .1459027	− .1446004	− .1432844	− .1419548
17.3	− .1337006	− .1322802	− .1308473	− .1294021	− .1279449
17.4	− .1189559	− .1174182	− .1158697	− .1143105	− .1127408
17.5	− .1031104	− .1014715	− .0998235	− .0981664	− .0965004
17.6	− .0863279	− .0846047	− .0828740	− .0811360	− .0793908
17.7	− .0687804	− .0669904	− .0651947	− .0633935	− .0615870
17.8	− .0506464	− .0488078	− .0469654	− .0451192	− .0432696
17.9	− .0321095	− .0302408	− .0283701	− .0264976	− .0246235
18.0	− .0133558	− .0114757	− .0095956	− .0077155	− .0058357
18.1	.0054270	.0072997	.0091706	.0110396	.0129064
18.2	.0240523	.0258990	.0277422	.0295815	.0314169
18.3	.0423358	.0441384	.0459356	.0477272	.0495131
18.4	.0600979	.0618387	.0635723	.0652987	.0670175
18.5	.0771648	.0788268	.0804800	.0821243	.0837595
18.6	.0933708	.0949380	.0964948	.0980411	.0995768
18.7	.1085595	.1100167	.1114621	.1128956	.1143171
18.8	.1225853	.1239187	.1252390	.1265461	.1278398
18.9	.1353152	.1365122	.1376949	.1388632	.1400170
19.0	.1466294	.1476788	.1487129	.1497316	.1507347
19.1	.1564231	.1573152	.1581912	.1590508	.1598942
19.2	.1646067	.1653335	.1660435	.1667364	.1674124
19.3	.1711073	.1716625	.1722003	.1727206	.1732233
19.4	.1758692	.1762481	.1766092	.1769525	.1772779
19.5	.1788538	.1790536	.1792354	.1793991	.1795449
19.6	.1800407	.1800603	.1800618	.1800454	.1800109
19.7	.1794272	.1792673	.1790895	.1788939	.1786806
19.8	.1770286	.1766917	.1763373	.1759654	.1755761
19.9	.1728777	.1723681	.1718414	.1712978	.1707374

BESSEL FUNCTION J_0

T	.05	.06	.07	.08	.09
15.0	−.0244406	−.0264732	−.0285019	−.0305263	−.0325464
15.1	−.0445634	−.0465472	−.0485250	−.0504967	−.0524620
15.2	−.0641100	−.0660256	−.0679333	−.0698330	−.0717245
15.3	−.0828904	−.0847192	−.0865384	−.0883477	−.0901471
15.4	−.1007233	−.1024478	−.1041610	−.1058626	−.1075525
15.5	−.1174380	−.1190418	−.1206326	−.1222103	−.1237748
15.6	−.1328759	−.1343438	−.1357973	−.1372363	−.1386606
15.7	···.1468921	−.1482104	−.1495130	−.1507998	−.1520708
15.8	−.1593565	−.1605130	−.1616528	−.1627757	−.1638816
15.9	−.1701552	−.1711396	−.1721063	−.1730551	−.1739861
16.0	−.1791917	−.1799952	−.1807802	−.1815467	−.1822945
16.1	−.1863871	−.1870029	−.1875996	−.1881773	−.1887357
16.2	−.1916815	−.1921047	−.1925084	−.1928926	−.1932572
16.3	−.1950342	−.1952616	−.1954694	−.1956576	−.1958260
16.4	−.1964235	−.1964543	−.1964653	−.1964567	−.1964285
16.5	−.1958479	−.1956827	−.1954981	−.1952941	−.1950707
16.6	−.1933247	−.1929665	−.1925892	−.1921930	−.1917777
16.7	−.1888908	−.1883442	−.1877792	−.1871958	−.1865940
16.8	−.1826015	−.1818732	−.1811272	−.1803635	−.1795822
16.9	−.1745303	−.1736286	−.1727100	−.1717747	−.1708227
17.0	−.1647680	−.1637027	−.1626217	−.1615250	−.1604129
17.1	−.1534212	−.1522039	−.1509721	−.1497260	−.1484656
17.2	−.1406118	−.1392556	−.1378862	−.1365038	−.1351086
17.3	−.1264757	−.1249947	−.1235021	−.1219979	−.1204825
17.4	−.1111607	−.1095704	−.1079700	−.1063598	−.1047399
17.5	−.0948257	−.0931425	−.0914510	−.0897512	−.0880435
17.6	−.0776387	−.0758799	−.0741145	−.0723426	−.0705645
17.7	−.0597754	−.0579587	−.0561374	−.0543114	−.0524810
17.8	−.0414167	−.0395607	−.0377018	−.0358402	−.0339760
17.9	−.0227480	−.0208713	−.0189935	−.0171149	−.0152356
18.0	−.0039563	−.0020776	−.0001997	.0016771	.0035528
18.1	.0147709	.0166330	.0184923	.0203487	.0222021
18.2	.0332481	.0350750	.0368974	.0387152	.0405280
18.3	.0512931	.0530669	.0548345	.0565956	.0583501
18.4	.0687288	.0704323	.0721278	.0738151	.0754942
18.5	.0853855	.0870020	.0886089	.0902061	.0917935
18.6	.1011017	.1026157	.1041186	.1056103	.1070906
18.7	.1157264	.1171234	.1185079	.1198798	.1212390
18.8	.1291201	.1303868	.1316397	.1328789	.1341041
18.9	.1411562	.1422807	.1433904	.1444851	.1455648
19.0	.1517223	.1526942	.1536503	.1545905	.1555148
19.1	.1607211	.1615315	.1623253	.1631025	.1638630
19.2	.1680713	.1687130	.1693375	.1699448	.1705347
19.3	.1737084	.1741760	.1746258	.1750580	.1754725
19.4	.1775854	.1778749	.1781466	.1784003	.1786361
19.5	.1796726	.1797823	.1798739	.1799475	.1800031
19.6	.1799585	.1798881	.1797998	.1796935	.1795693
19.7	.1784494	.1782006	.1779340	.1776498	.1773480
19.8	.1751695	.1747456	.1743044	.1738459	.1733704
19.9	.1701602	.1695663	.1689557	.1683285	.1676848

BESSEL FUNCTION J_0

T	.00	.01	.02	.03	.04
20.0	.1670247	.1663482	.1656554	.1649464	.1642212
20.1	.1595361	.1587003	.1578491	.1569825	.1561007
20.2	.1504946	.1495086	.1485082	.1474935	.1464645
20.3	.1399977	.1388722	.1377333	.1365812	.1354160
20.4	.1281571	.1269038	.1256384	.1243611	.1230720
20.5	.1150970	.1137290	.1123504	.1109612	.1095615
20.6	.1009532	.0994848	.0980071	.0965203	.0950246
20.7	.0858717	.0843178	.0827563	.0811872	.0796108
20.8	.0700069	.0683835	.0667540	.0651187	.0634776
20.9	.0535204	.0518439	.0501631	.0484780	.0467889
21.0	.0365791	.0348665	.0331512	.0314334	.0297133
21.1	.0193536	.0176221	.0158896	.0141564	.0124226
21.2	.0020167	.0002836	−.0014488	−.0031802	−.0049105
21.3	−.0152587	−.0169762	−.0186913	−.0204036	−.0221131
21.4	−.0323011	−.0339860	−.0356667	−.0373431	−.0390149
21.5	−.0489420	−.0505777	−.0522076	−.0538314	−.0554492
21.6	−.0650179	−.0665883	−.0681513	−.0697068	−.0712545
21.7	−.0803714	−.0818612	−.0833421	−.0848140	−.0862767
21.8	−.0948530	−.0962478	−.0976323	−.0990063	−.1003699
21.9	−.1083229	−.1096091	−.1108838	−.1121469	−.1133981
22.0	−.1206515	−.1218169	−.1229697	−.1241096	−.1252366
22.1	−.1317214	−.1327550	−.1337749	−.1347809	−.1357731
22.2	−.1414282	−.1423203	−.1431977	−.1440605	−.1449085
22.3	−.1496814	−.1504237	−.1511507	−.1518622	−.1525582
22.4	−.1564055	−.1569913	−.1575612	−.1581151	−.1586529
22.5	.1615403	−.1619646	−.1623724	−.1627639	−.1631389
22.6	−.1650419	−.1653010	−.1655435	−.1657694	−.1659785
22.7	−.1668827	−.1669748	−.1670502	−.1671089	−.1671509
22.8	−.1670515	−.1669765	−.1668849	−.1667766	−.1666517
22.9	−.1655542	−.1653135	−.1650564	−.1647829	−.1644931
23.0	−.1624128	−.1620096	−.1615903	−.1611551	−.1607040
23.1	−.1576658	−.1571047	−.1565282	−.1559363	−.1553290
23.2	−.1513673	−.1506547	−.1499273	−.1491852	−.1484286
23.3	−.1435868	−.1427303	−.1418600	−.1409758	−.1400779
23.4	−.1344080	−.1334168	−.1324128	−.1313959	−.1303663
23.5	−.1239282	−.1228128	−.1216856	−.1205467	−.1193963
23.6	−.1122573	−.1110293	−.1097907	−.1085417	−.1072823
23.7	−.0995165	−.0981886	−.0968514	−.0955051	−.0941498
23.8	−.0858371	−.0844229	−.0830008	−.0815710	−.0801337
23.9	−.0713592	−.0698732	−.0683807	−.0668820	−.0653773
24.0	−.0562303	−.0546874	−.0531398	−.0515874	−.0500306
24.1	−.0406036	−.0390195	−.0374323	−.0358420	−.0342487
24.2	−.0246367	−.0230274	−.0214166	−.0198042	−.0181905
24.3	−.0084900	−.0068717	−.0052533	−.0036351	−.0020171
24.4	.0076750	.0092863	.0108961	.0125040	.0141101
24.5	.0236974	.0252857	.0268708	.0284525	.0300307
24.6	.0394183	.0409677	.0425125	.0440524	.0455872
24.7	.0546823	.0561777	.0576669	.0591498	.0606261
24.8	.0693393	.0707661	.0721851	.0735964	.0749998
24.9	.0832460	.0845901	.0859252	.0872512	.0885679

BESSEL FUNCTION J_0

T	.05	.06	.07	.08	.09
20.0	.1634800	.1627228	.1619498	.1611609	.1603563
20.1	.1552037	.1542916	.1533645	.1524226	.1514659
20.2	.1454214	.1443643	.1432932	.1422083	.1411098
20.3	.1342379	.1330469	.1318432	.1306269	.1293982
20.4	.1217712	.1204589	.1191352	.1178002	.1164541
20.5	.1081516	.1067316	.1053016	.1038618	.1024123
20.6	.0935202	.0920071	.0904855	.0889557	.0874177
20.7	.0780272	.0764366	.0748391	.0732348	.0716240
20.8	.0618310	.0601790	.0585217	.0568595	.0551923
20.9	.0450960	.0433993	.0416991	.0399956	.0382888
21.0	.0279911	.0262669	.0245408	.0228132	.0210841
21.1	.0106883	.0089538	.0072192	.0054847	.0037505
21.2	−.0066394	−.0083669	−.0100928	−.0118168	−.0135388
21.3	−.0238196	−.0255230	−.0272229	−.0289194	−.0306122
21.4	−.0406821	−.0423444	−.0440018	−.0456539	−.0473007
21.5	−.0570606	−.0586656	−.0602640	−.0618556	−.0634403
21.6	−.0727945	−.0743264	−.0758502	−.0773658	−.0788729
21.7	−.0877302	−.0891742	−.0906086	−.0920333	−.0934482
21.8	−.1017228	−.1030649	−.1043961	−.1057162	−.1070252
21.9	−.1146375	−.1158648	−.1170800	−.1182829	−.1194734
22.0	−.1263506	−.1274514	−.1285390	−.1296133	−.1306741
22.1	−.1367512	−.1377151	−.1386649	−.1396004	−.1405215
22.2	−.1457416	−.1465597	−.1473628	−.1481509	−.1489238
22.3	−.1532387	−.1539035	−.1545527	−.1551861	−.1558037
22.4	−.1591747	−.1596803	−.1601697	−.1606428	−.1610997
22.5	−.1634974	−.1638394	−.1641649	−.1644738	−.1647662
22.6	−.1661710	−.1663467	−.1665058	−.1666481	−.1667738
22.7	−.1671761	−.1671846	−.1671764	−.1671515	−.1671098
22.8	−.1665101	−.1663520	−.1661773	−.1659861	−.1657784
22.9	−.1641870	−.1638645	−.1635258	−.1631710	−.1627999
23.0	−.1602370	−.1597541	−.1592555	−.1587412	−.1582113
23.1	−.1547065	−.1540688	−.1534159	−.1527480	−.1520651
23.2	−.1476574	−.1468718	−.1460719	−.1452576	−.1444293
23.3	−.1391664	−.1382414	−.1373029	−.1363511	−.1353861
23.4	−.1293241	−.1282695	−.1272024	−.1261232	−.1250317
23.5	−.1182343	−.1170611	−.1158767	−.1146811	−.1134747
23.6	−.1060127	−.1047331	−.1034435	−.1021441	−.1008351
23.7	−.0927856	−.0914128	−.0900313	−.0886415	−.0872434
23.8	−.0786889	−.0772369	−.0757778	−.0743117	−.0728388
23.9	−.0638667	−.0623503	−.0608283	−.0593009	−.0577681
24.0	−.0484694	−.0469040	−.0453345	−.0437612	−.0421842
24.1	−.0326527	−.0310540	−.0294530	−.0278496	−.0262441
24.2	−.0165757	−.0149599	−.0133433	−.0117260	−.0101082
24.3	−.0003997	.0012172	.0028332	.0044483	.0060623
24.4	.0157141	.0173158	.0189152	.0205120	.0221062
24.5	.0316054	.0331762	.0347430	.0363057	.0378642
24.6	.0471169	.0486412	.0501601	.0516733	.0531807
24.7	.0620958	.0635587	.0650146	.0664635	.0679051
24.8	.0763951	.0777822	.0791610	.0805313	.0818930
24.9	.0898752	.0911731	.0924613	.0937397	.0950082

BESSEL FUNCTION J_0

T	.00	.01	.02	.03	.04
25.0	.0962668	.0975152	.0987534	.0999812	.1011985
25.1	.1082757	.1094164	.1105457	.1116634	.1127696
25.2	.1191571	.1201791	.1211887	.1221858	.1231703
25.3	.1288072	.1297009	.1305812	.1314482	.1323016
25.4	.1371348	.1378917	.1386345	.1393632	.1400777
25.5	.1440622	.1446753	.1452738	.1458574	.1464263
25.6	.1495258	.1499896	.1504383	.1508718	.1512900
25.7	.1534770	.1537875	.1540825	.1543621	.1546260
25.8	.1558824	.1560371	.1561761	.1562995	.1564071
25.9	.1567239	.1567219	.1567041	.1566707	.1566216
26.0	.1559993	1558411	.1556673	.1554781	.1552734
26.1	.1537218	.1534096	.1530822	.1527396	.1523819
26.2	.1499200	.1494575	.1489803	.1484883	.1479817
26.3	.1446375	.1440300	.1434082	.1427724	.1421225
26.4	.1379327	.1371866	.1364271	.1356543	.1348682
26.5	.1298776	.1290010	.1281119	.1272102	.1262962
26.6	.1205577	.1195597	.1185503	.1175293	.1164970
26.7	.1100704	.1089615	.1078422	.1067125	.1055726
26.8	.0985245	.0973162	.0960986	.0948719	.0936361
26.9	.0860391	.0847437	.0834404	.0821291	.0808102
27.0	.0727419	.0713727	.0699969	.0686145	.0672259
27.1	.0587684	.0573393	.0559049	.0544655	.0530212
27.2	.0442603	.0427857	.0413073	.0398254	.0383400
27.3	.0293640	.0278587	.0263513	.0248417	.0233302
27.4	.0142293	.0127085	.0111870	.0096650	.0081426
27.5	−.0009922	−.0025133	−.0040336	−.0055529	−.0070711
27.6	−.0161486	−.0176548	−.0191587	−.0206602	−.0221590
27.7	−.0310890	−.0325655	−.0340381	−.0355069	−.0369715
27.8	−.0456656	−.0470977	−.0485245	−.0499460	−.0513619
27.9	−.0597344	−.0611079	−.0624749	−.0638351	−.0651885
28.0	−.0731570	−.0744586	−.0757523	−.0770380	−.0783155
28.1	.0858021	−.0870191	−.0882269	−.0894254	−.0906146
28.2	−.0975466	−.0986670	−.0997772	−.1008770	−.1019663
28.3	−.1082765	−.1092896	−.1102915	−.1112819	−.1122609
28.4	−.1178886	−.1187847	−.1196686	−.1205403	−.1213995
28.5	−.1262911	−.1270617	−.1278193	−.1285638	−.1292952
28.6	−.1334046	−.1340423	−.1346665	−.1352769	−.1358737
28.7	−.1391625	−.1396616	−.1401466	−.1406174	−.1410739
28.8	−.1435124	−.1438684	−.1442098	−.1445367	−.1448490
28.9	−.1464158	−.1466256	−.1468206	−.1470008	−.1471663
29.0	−.1478488	−.1479107	−.1479578	−.1479901	−.1480077
29.1	−.1478020	−.1477161	−.1476154	−.1475000	−.1473698
29.2	−.1462813	−.1460487	−.1458017	−.1455402	−.1452642
29.3	−.1433065	−.1429303	−.1425399	−.1421354	−.1417168
29.4	−.1389125	−.1383968	−.1378675	−.1373245	−.1367680
29.5	−.1331479	−.1324983	−.1318357	−.1311601	−.1304717
29.6	−.1260746	−.1252980	−.1245092	−.1237083	−.1228952
29.7	−.1177675	−.1168722	−.1159655	−.1150475	−.1141184
29.8	−.1083137	−.1073089	−.1062938	−.1052683	−.1042327
29.9	−.0978112	−.0967073	−.0955941	−.0944718	−.0933404

BESSEL FUNCTION J_0

T	.05	.06	.07	.08	.09
25.0	.1024052	.1036012	.1047863	.1059606	.1071237
25.1	.1138641	.1149467	.1160175	.1170762	.1181228
25.2	.1241421	.1251011	.1260472	.1269803	.1279003
25.3	.1331415	.1339678	.1347803	.1355790	.1363639
25.4	.1407778	.1414636	.1421350	.1427920	.1434344
25.5	.1469803	.1475194	.1480436	.1485527	.1490468
25.6	.1516929	.1520805	.1524527	.1528096	.1531510
25.7	.1548744	.1551072	.1553244	.1555261	.1557120
25.8	.1564991	.1565755	.1566361	.1566811	.1567103
25.9	.1565569	.1564766	.1563807	.1562691	.1561420
26.0	.1550532	.1548177	.1545667	.1543004	.1540187
26.1	.1520090	.1516211	.1512182	.1508004	.1503676
26.2	.1474605	.1469248	.1463745	.1458099	.1452308
26.3	.1414587	.1407810	.1400895	.1393842	.1386652
26.4	.1340689	.1332565	.1324311	.1315928	.1307416
26.5	.1253699	.1244315	.1234809	.1225183	.1215439
26.6	.1154534	.1143986	.1133328	.1122561	.1111686
26.7	.1044225	.1032625	.1020925	.1009128	.0997234
26.8	.0923915	.0911380	.0898759	.0886053	.0873263
26.9	.0794837	.0781497	.0768084	.0754599	.0741044
27.0	.0658310	.0644300	.0630232	.0616105	.0601922
27.1	.0515721	.0501184	.0486602	.0471977	.0457310
27.2	.0368513	.0353596	.0338648	.0323671	.0308668
27.3	.0218170	.0203021	.0187857	.0172680	.0157491
27.4	.0066199	.0050970	.0035743	.0020517	.0005295
27.5	−.0085880	−.0101035	−.0116175	−.0131298	−.0146402
27.6	−.0236551	−.0251483	−.0266384	−.0281254	−.0296090
27.7	−.0384319	−.0398880	−.0413395	−.0427864	−.0442285
27.8	−.0527722	−.0541768	−.0555754	−.0569680	−.0583543
27.9	−.0665348	−.0678741	−.0692060	−.0705306	−.0718476
28.0	−.0795847	−.0808455	−.0820977	−.0833414	−.0845762
28.1	−.0917943	−.0929644	−.0941248	−.0952754	−.0964160
28.2	−.1030451	−.1041131	−.1051704	−.1062168	−.1072522
28.3	−.1132283	−.1141841	−.1151281	−.1160602	−.1169804
28.4	−.1222464	−.1230807	−.1239024	−.1247114	−.1255077
28.5	−.1300134	−.1307184	−.1314101	−.1320884	−.1327532
28.6	−.1364566	−.1370257	−.1375809	−.1381221	−.1386494
28.7	−.1415162	−.1419442	−.1423579	−.1427572	−.1431420
28.8	−.1451468	−.1454299	−.1456984	−.1459522	−.1461914
28.9	−.1473170	−.1474530	−.1475741	−.1476805	−.1477720
29.0	−.1480104	−.1479983	−.1479714	−.1479297	−.1478733
29.1	−.1472250	−.1470655	−.1468913	−.1467026	−.1464992
29.2	−.1449738	−.1446690	−.1443498	−.1440163	−.1436686
29.3	−.1412842	−.1408377	−.1403771	−.1399028	−.1394145
29.4	−.1361980	−.1356145	−.1350177	−.1344077	−.1337843
29.5	−.1297704	−.1290564	−.1283297	−.1275905	−.1268387
29.6	−.1220701	−.1212331	−.1203842	−.1195236	−.1186514
29.7	−.1131781	−.1122268	−.1112647	−.1102917	−.1093080
29.8	−.1031870	−.1021313	−.1010658	−.0999905	−.0989056
29.9	−.0922000	−.0910508	−.0898929	−.0887264	−.0875514

BESSEL FUNCTION J_0

T	.00	.01	.02	.03	.04
30.0	− .0863680	− .0851764	− .0839766	− .0827689	−.0815533
30.1	− .0741014	− .0728342	− .0715601	− .0702793	− .0689919
30.2	− .0611363	− .0598064	− .0584709	− .0571301	− .0557839
30.3	− .0476042	− .0462250	− .0448417	− .0434543	− .0420631
30.4	− .0336418	− .0322273	− .0308101	− .0293902	− .0279678
30.5	− .0193897	− .0179541	− .0165171	− .0150790	− .0136398
30.6	− .0049909	− .0035485	− .0021062	− .0006641	.0007775
30.7	.0094108	.0108456	.0122788	.0137103	.0151400
30.8	.0236718	.0250846	.0264945	.0279013	.0293049
30.9	.0376503	.0390273	.0403999	.0417680	.0431315
31.0	.0512082	.0525356	.0538574	.0551733	.0564834
31.1	.0642115	.0654763	.0667343	.0679851	.0692287
31.2	.0765325	.0777224	.0789042	.0800776	.0812426
31.3	.0880507	.0891540	.0902480	.0913326	.0924077
31.4	.0986537	.0996597	.1006553	.1016406	.1026154

BESSEL FUNCTION J_0

T	.05	.06	.07	.08	.09
30.0	− .0803300	− .0790990	− .0778605	− .0766147	− .0753616
30.1	− .0676981	− .0663979	− .0650914	− .0637790	− .0624605
30.2	− .0544327	− .0530764	− .0517153	− .0503495	− .0489791
30.3	− .0406681	− .0392695	− .0378674	− .0364620	− .0350535
30.4	− .0265431	− .0251162	− .0236873	− .0222565	− .0208239
30.5	− .0121997	− .0107588	− .0093174	− .0078755	− .0064332
30.6	.0022186	.0036590	.0050986	.0065371	.0079746
30.7	.0165677	.0179934	.0194167	.0208376	.0222560
30.8	.0307051	.0321017	.0334947	.0348839	.0362691
30.9	.0444903	.0458442	.0471930	.0485367	.0498752
31.0	.0577873	.0590850	.0603764	.0616614	.0629398
31.1	.0704650	.0716939	.0729152	.0741288	.0753347
31.2	.0823992	.0835472	.0846864	.0858168	.0869383
31.3	.0934732	.0945292	.0955752	.0966114	.0976377
31.4	.1035796	.1045332	.1054759	.1064079	.1073289

BESSEL FUNCTION J_1

T	.00	.01	.02	.03	.04
0.0	0.0000000	.0049999	.0099995	.0149983	.0199960
.1	.0499375	.0549169	.0598921	.0648628	.0698286
.2	.0995008	.1044223	.1093358	.1142412	.1191381
.3	.1483188	.1531455	.1579607	.1627641	.1675553
.4	.1960266	.2007225	.2054034	.2100689	.2147188
.5	.2422685	.2467987	.2513105	.2558035	.2602774
.6	.2867010	.2910319	.2953412	.2996284	.3038932
.7	.3289957	.3330955	.3371705	.3412203	.3452448
.8	.3688420	.3726806	.3764916	.3802745	.3840292
.9	.4059495	.4094991	.4130184	.4165071	.4199649
1.0	.4400506	.4432858	.4464882	.4496577	.4527939
1.1	.4709024	.4738003	.4766634	.4794913	.4822840
1.2	.4982891	.5008297	.5033336	.5058006	.5082305
1.3	.5220232	.5241895	.5263174	.5284070	.5304580
1.4	.5419477	.5437255	.5454638	.5471625	.5488215
1.5	.5579365	.5593150	.5606532	.5619508	.5632079
1.6	.5698959	.5708676	.5717984	.5726881	.5735368
1.7	.5777652	.5783259	.5788453	.5793235	.5797604
1.8	.5815170	.5816656	.5817731	.5818396	.5818649
1.9	.5811571	.5808962	.5805946	.5802523	.5798695
2.0	.5767248	.5760601	.5753554	.5746109	.5738267
2.1	.5682921	.5672326	.5661342	.5649970	.5638212
2.2	.5559630	.5545208	.5530410	.5515239	.5499696
2.3	.5398725	.5380627	.5362170	.5343358	.5324190
2.4	.5201853	.5180259	.5158327	.5136058	.5113456
2.5	.4970941	.4946060	.4920863	.4895351	.4869528
2.6	.4708183	.4680247	.4652020	.4623503	.4594700
2.7	.4416014	.4385280	.4354281	.4323020	.4291500
2.8	.4097092	.4063837	.4030346	.3996622	.3962667
2.9	.3754275	.3718794	.3683108	.3647218	.3611130
3.0	.3390590	.3353194	.3315626	.3277886	.3239979
3.1	.3009211	.2970226	.2931100	.2891837	.2852440
3.2	.2613432	.2573192	.2532845	.2492394	.2451844
3.3	.2206635	.2165481	.2124255	.2082960	.2041599
3.4	.1792259	.1750538	.1708779	.1666987	.1625163
3.5	.1373775	.1331835	.1289892	.1247949	.1206010
3.6	.0954655	.0912841	.0871059	.0829311	.0787602
3.7	.0538340	.0496993	.0455712	.0414500	.0373359
3.8	.0128210	.0087665	.0047218	.0006872	−.0033369
3.9	−.0272440	−.0311861	−.0351151	−.0390308	−.0429330
4.0	−.0660433	−.0698418	−.0736243	−.0773905	−.0811401
4.1	−.1032733	−.1068987	−.1105054	−.1140928	−.1176609
4.2	−.1386469	−.1420717	−.1454750	−.1488565	−.1522160
4.3	−.1718966	−.1750950	−.1782695	−.1814198	−.1845457
4.4	−.2027755	−.2057242	−.2086467	−.2115429	−.2144125
4.5	−.2310604	−.2337384	−.2363882	−.2390097	−.2416027
4.6	−.2565528	−.2589416	−.2613006	−.2636296	−.2659284
4.7	−.2790807	−.2811647	−.2832174	−.2852387	−.2872286
4.8	−.2984999	−.3002661	−.3019999	−.3037013	−.3053702
4.9	−.3146947	−.3161332	−.3175386	−.3189107	−.3202495

BESSEL FUNCTION J_1

T	.05	.06	.07	.08	.09
0.0	.0249922	.0299865	.0349786	.0399680	.0449545
.1	.0747893	.0797443	.0846933	.0896360	.0945720
.2	.1240260	.1289046	.1337735	.1386325	.1434810
.3	.1723340	.1770997	.1818522	.1865911	.1913160
.4	.2193525	.2239699	.2285705	.2331540	.2377201
.5	.2647318	.2691665	.2735811	.2779752	.2823486
.6	.3081355	.3123547	.3165506	.3207230	.3248715
.7	.3492436	.3532164	.3571629	.3610829	.3649760
.8	.3877554	.3914529	.3951213	.3987603	.4023699
.9	.4233917	.4267871	.4301509	.4334829	.4367829
1.0	.4558968	.4589660	.4620014	.4650027	.4679698
1.1	.4850413	.4877629	.4904486	.4930984	.4957119
1.2	.5106233	.5129786	.5152965	.5175766	.5198189
1.3	.5324703	.5344439	.5363785	.5382741	.5401305
1.4	.5504407	.5520200	.5535593	.5550586	.5565177
1.5	.5644245	.5656003	.5667354	.5678298	.5688833
1.6	.5743443	.5751108	.5758362	.5765204	.5771634
1.7	.5801562	.5805107	.5808241	.5810962	.5813272
1.8	.5818493	.5817926	.5816951	.5815566	.5813772
1.9	.5794463	.5789825	.5784784	.5779341	.5773495
2.0	.5730028	.5721393	.5712364	.5702942	.5693127
2.1	.5626069	.5613543	.5600635	.5587345	.5573677
2.2	.5483784	.5467502	.5450854	.5433841	.5416464
2.3	.5304671	.5284801	.5264582	.5244016	.5223106
2.4	.5090521	.5067256	.5043663	.5019745	.4995503
2.5	.4843396	.4816957	.4790214	.4763168	.4735824
2.6	.4565613	.4536245	.4506598	.4476676	.4446480
2.7	.4259723	.4227693	.4195412	.4162882	.4130109
2.8	.3928485	.3894079	.3859452	.3824607	.3789547
2.9	.3574845	.3538368	.3501700	.3464846	.3427808
3.0	.3201909	.3163677	.3125289	.3086746	.3048052
3.1	.2812912	.2773257	.2733478	.2693579	.2653563
3.2	.2411197	.2370457	.2329627	.2288711	.2247712
3.3	.2000177	.1958696	.1917161	.1875574	.1833938
3.4	.1583313	.1541439	.1499545	.1457634	.1415709
3.5	.1164079	.1122159	.1080253	.1038365	.0996498
3.6	.0745934	.0704312	.0662737	.0621215	.0579748
3.7	.0332293	.0291307	.0250402	.0209582	.0168850
3.8	−.0073502	−.0113524	−.0153432	−.0193223	−.0232894
3.9	−.0468212	−.0506953	−.0545548	−.0583995	−.0622291
4.0	−.0848728	−.0885884	−.0922865	−.0959669	−.0996292
4.1	−.1212093	−.1247378	−.1282461	−.1317339	−.1352009
4.2	−.1555532	−.1588679	−.1621598	−.1654287	−.1686744
4.3	−.1876469	−.1907233	−.1937745	−.1968005	−.1998009
4.4	−.2172552	−.2200710	−.2228596	−.2256209	−.2283545
4.5	−.2441671	−.2467026	−.2492091	−.2516864	−.2541344
4.6	−.2681970	−.2704352	−.2726428	−.2748196	−.2769657
4.7	−.2891868	−.2911133	−.2930080	−.2948707	−.2967014
4.8	−.3070064	−.3086098	−.3101805	−.3117182	−.3132230
4.9	−.3215549	−.3228269	−.3240653	−.3252702	−.3264415

BESSEL FUNCTION J_1

T	.00	.01	.02	.03	.04
5.0	−.3275791	−.3286831	−.3297533	−.3307898	−.3317925
5.1	−.3370972	−.3378626	−.3385940	−.3392915	−.3399550
5.2	−.3432230	−.3436489	−.3440409	−.3443991	−.3447234
5.3	−.3459608	−.3460493	−.3461043	−.3461259	−.3461140
5.4	−.3453448	−.3451009	−.3448242	−.3445147	−.3441725
5.5	−.3414382	−.3408699	−.3402696	−.3396376	−.3389739
5.6	−.3343328	−.3334507	−.3325379	−.3315946	−.3306208
5.7	−.3241477	−.3229651	−.3217534	−.3205126	−.3192429
5.8	−.3110277	−.3095607	−.3080661	−.3065442	−.3049952
5.9	−.2951424	−.2934090	−.2916501	−.2898658	−.2880563
6.0	−.2766839	−.2747045	−.2727017	−.2706758	−.2686269
6.1	−.2558648	−.2536618	−.2514378	−.2491929	−.2469275
6.2	−.2329166	−.2305141	−.2280930	−.2256536	−.2231961
6.3	−.2080869	−.2055105	−.2029180	−.2003100	−.1976865
6.4	−.1816375	−.1789139	−.1761771	−.1734274	−.1706650
6.5	−.1538413	−.1509984	−.1481451	−.1452818	−.1424086
6.6	−.1249802	−.1220465	−.1191054	−.1161571	−.1132019
6.7	−.0953421	−.0923468	−.0893469	−.0863427	−.0833346
6.8	−.0652187	−.0621909	−.0591615	−.0561307	−.0530989
6.9	−.0349021	−.0318710	−.0288412	−.0258130	−.0227866
7.0	−.0046828	−.0016773	.0013241	.0043211	.0073134
7.1	.0251533	.0281050	.0310498	.0339875	.0369177
7.2	.0543274	.0571980	.0600589	.0629100	.0657511
7.3	.0825704	.0853335	.0880844	.0908230	.0935488
7.4	.1096251	.1122557	.1148718	.1174730	.1200593
7.5	.1352484	.1377232	.1401811	.1426220	.1450456
7.6	.1592138	.1615109	.1637892	.1660484	.1682883
7.7	.1813127	.1834125	.1854916	.1875497	.1895868
7.8	.2013569	.2032417	.2051041	.2069439	.2087611
7.9	.2191794	.2208337	.2224642	.2240708	.2256533
8.0	.2346363	.2360471	.2374329	.2387936	.2401291
8.1	.2476078	.2487643	.2498950	.2509996	.2520782
8.2	.2579986	.2588928	.2597605	.2606016	.2614159
8.3	.2657393	.2663657	.2669651	.2675375	.2680829
8.4	.2707863	.2711419	.2714704	.2717718	.2720460
8.5	.2731220	.2732065	.2732640	.2732946	.2732981
8.6	.2727548	.2725706	.2723596	.2721221	.2718580
8.7	.2697190	.2692707	.2687964	.2682961	.2677699
8.8	.2640737	.2633687	.2626384	.2618831	.2611028
8.9	.2559024	.2549502	.2539740	.2529738	.2519497
9.0	.2453118	.2441244	.2429143	.2416816	.2404263
9.1	.2324307	.2310222	.2295925	.2281416	.2266698
9.2	.2174087	.2157950	.2141618	.2125092	.2108375
9.3	.2004139	.1986130	.1967943	.1949582	.1931047
9.4	.1816322	.1796634	.1776789	.1756789	.1736637
9.5	.1612644	.1591486	.1570192	.1548765	.1527208
9.6	.1395248	.1372840	.1350319	.1327688	.1304950
9.7	.1166386	.1142959	.1119443	.1095840	.1072154
9.8	.0928401	.0904193	.0879920	.0855585	.0831189
9.9	.0683698	.0658953	.0634167	.0609343	.0584484

BESSEL FUNCTION J_1

T	.05	.06	.07	.08	.09
5.0	−.3327613	−.3336963	−.3345974	−.3354646	−.3362979
5.1	−.3405846	−.3411802	−.3417418	−.3422695	−.3427632
5.2	−.3450140	−.3452707	−.3454938	−.3456831	−.3458388
5.3	−.3460688	−.3459903	−.3458785	−.3457337	−.3455557
5.4	−.3437977	−.3433905	−.3429508	−.3424788	−.3419746
5.5	−.3382786	−.3375518	−.3367938	−.3360045	−.3351841
5.6	−.3296168	−.3285826	−.3275185	−.3264245	−.3253009
5.7	−.3179445	−.3166176	−.3152623	−.3138787	−.3124672
5.8	−.3034193	−.3018166	−.3001874	−.2985318	−.2968501
5.9	−.2862220	−.2843629	−.2824793	−.2805715	−.2786396
6.0	−.2665553	−.2644612	−.2623449	−.2602066	−.2580464
6.1	−.2446417	−.2423358	−.2400101	−.2376649	−.2353003
6.2	−.2207209	−.2182281	−.2157181	−.2131910	−.2106472
6.3	−.1950479	−.1923944	−.1897264	−.1870440	−.1843476
6.4	−.1678903	−.1651035	−.1623049	−.1594949	−.1566736
6.5	−.1395260	−.1366341	−.1337333	−.1308238	−.1279060
6.6	−.1102401	−.1072720	−.1042978	−.1013179	−.0983326
6.7	−.0803228	−.0773076	−.0742893	−.0712681	−.0682445
6.8	−.0500663	−.0470332	−.0440000	−.0409669	−.0379341
6.9	−.0197623	−.0167404	−.0137213	−.0107051	−.0076922
7.0	.0103007	.0132828	.0162594	.0192302	.0221949
7.1	.0398402	.0427547	.0456609	.0485586	.0514476
7.2	.0685817	.0714017	.0742109	.0770089	.0797955
7.3	.0962619	.0989617	.1016482	.1043211	.1069802
7.4	.1226303	.1251857	.1277255	.1302494	.1327571
7.5	.1474518	.1498404	.1522110	.1545636	.1568979
7.6	.1705088	.1727096	.1748906	.1770516	.1791923
7.7	.1916026	.1935970	.1955697	.1975208	.1994499
7.8	.2105554	.2123267	.2140749	.2157999	.2175014
7.9	.2272116	.2287457	.2302553	.2317403	.2332007
8.0	.2414393	.2427241	.2439835	.2452173	.2464254
8.1	.2531307	.2541570	.2551569	.2561306	.2570778
8.2	.2622036	.2629644	.2636985	.2644056	.2650859
8.3	.2686012	.2690924	.2695566	.2699936	.2704035
8.4	.2722931	.2725131	.2727059	.2728717	.2730104
8.5	.2732747	.2732244	.2731472	.2730432	.2729124
8.6	.2715674	.2712504	.2709069	.2705372	.2701412
8.7	.2672179	.2666402	.2660368	.2654079	.2647535
8.8	.2602976	.2594677	.2586131	.2577339	.2568303
8.9	.2509019	.2498306	.2487357	.2476176	.2464762
9.0	.2391487	.2378489	.2365270	.2351833	.2338178
9.1	.2251772	.2236640	.2221304	.2205765	.2190026
9.2	.2091467	.2074370	.2057087	.2039620	.2021970
9.3	.1912342	.1893468	.1874427	.1855221	.1835852
9.4	.1716335	.1695884	.1675288	.1654548	.1633666
9.5	.1505523	.1483711	.1461775	.1439718	.1417542
9.6	.1282106	.1259159	.1236111	.1212965	.1189722
9.7	.1048385	.1024537	.1000612	.0976613	.0952542
9.8	.0806737	.0782229	.0757669	.0733059	.0708401
9.9	.0559592	.0534670	.0509720	.0484745	.0459746

BESSEL FUNCTION J_1

T	.00	.01	.02	.03	.04
10.0	.0434727	.0409691	.0384638	.0359573	.0334497
10.1	.0183955	.0158874	.0133801	.0108741	.0083694
10.2	− .0066157	− .0091038	− .0115886	− .0140698	− .0165471
10.3	− .0313178	− .0337618	− .0362001	− .0386324	− .0410586
10.4	− .0554728	− .0578492	− .0602176	− .0625779	− .0649296
10.5	− .0788500	− .0811364	− .0834125	− .0856782	− .0879333
10.6	− .1012287	− .1034034	− .1055659	− .1077159	− .1098532
10.7	− .1223994	− .1244424	− .1264711	− .1284855	− .1304852
10.8	− .1421666	− .1440590	− .1459354	− .1477956	− .1496394
10.9	− .1603497	− .1620744	− .1637815	− .1654708	− .1671422
11.0	− .1767853	− .1783270	− .1798496	− .1813530	− .1828371
11.1	− .1913283	− .1926735	− .1939984	− .1953028	− .1965868
11.2	− .2038531	− .2049904	− .2061063	− .2072008	− .2082738
11.3	− .2142550	− .2151751	− .2160729	− .2169486	− .2178019
11.4	− .2224506	− .2231462	− .2238192	− .2244693	− .2250966
11.5	− .2283786	− .2288450	− .2292883	− .2297085	− .2301055
11.6	− .2320005	− .2322350	− .2324463	− .2326344	− .2327992
11.7	− .2333002	− .2333026	− .2332817	− .2332378	− .2331707
11.8	− .2322847	− .2320568	− .2318060	− .2315324	− .2312361
11.9	− .2289832	− .2285292	− .2280528	− .2275541	− .2270334
12.0	− .2234471	− .2227732	− .2220777	− .2213608	− .2206225
12.1	− .2157490	− .2148637	− .2139578	− .2130314	− .2120846
12.2	− .2059820	− .2048957	− .2037900	− .2026649	− .2015206
12.3	− .1942588	− .1929838	− .1916907	− .1903795	− .1890506
12.4	− .1807102	− .1792605	− .1777942	− .1763114	− .1748122
12.5	− .1654838	− .1638750	− .1622513	− .1606127	− .1589594
12.6	− .1487423	− .1469916	− .1452276	− .1434505	− .1416606
12.7	− .1306622	− .1287877	− .1269019	− .1250050	− .1230971
12.8	− .1114316	− .1094528	− .1074646	− .1054674	− .1034612
12.9	− .0912483	− .0891854	− .0871153	− .0850381	− .0829541
13.0	− .0703181	− .0681921	− .0660609	− .0639249	− .0617841
13.1	− .0488525	− .0466847	− .0445140	− .0423405	− .0401645
13.2	− .0270667	− .0248789	− .0226902	− .0205009	− .0183113
13.3	− .0051775	− .0029912	− .0008063	.0013771	.0035587
13.4	.0165990	.0187622	.0209219	.0230780	.0252301
13.5	.0380493	.0401683	.0422817	.0443894	.0464911
13.6	.0589646	.0610188	.0630655	.0651044	.0671353
13.7	.0791428	.0811125	.0830728	.0850233	.0869640
13.8	.0983905	.1002570	.1021121	.1039558	.1057877
13.9	.1165249	.1182704	.1200029	.1217222	.1234282
14.0	.1333752	.1349834	.1365770	.1381560	.1397201
14.1	.1487844	.1502404	.1516805	.1531045	.1545122
14.2	.1626107	.1639013	.1651747	.1664308	.1676695
14.3	.1747291	.1758426	.1769380	.1780149	.1790734
14.4	.1850317	.1859585	.1868661	.1877546	.1886237
14.5	.1934295	.1941616	.1948740	.1955664	.1962389
14.6	.1998527	.2003843	.2008956	.2013866	.2018572
14.7	.2042513	.2045785	.2048851	.2051711	.2054365
14.8	.2065956	.2067165	.2068167	.2068963	.2069553
14.9	.2068762	.2067910	.2066853	.2065590	.2064124

BESSEL FUNCTION J_1

T	.05	.06	.07	.08	.09
10.0	.0309412	.0284322	.0259229	.0234135	.0209043
10.1	.0058663	.0033652	.0008662	−.0016305	−.0041246
10.2	−.0190205	−.0214895	−.0239540	−.0264137	−.0288684
10.3	−.0434783	−.0458914	−.0482976	−.0506967	−.0530885
10.4	−.0672727	−.0696068	−.0719318	−.0742475	−.0765537
10.5	−.0901775	−.0924107	−.0946326	−.0968431	−.0990418
10.6	−.1119776	−.1140889	−.1161870	−.1182715	−.1203424
10.7	−.1324701	−.1344401	−.1363949	−.1383343	−.1402583
10.8	−.1514668	−.1532774	−.1550711	−.1568479	−.1586075
10.9	−.1687954	−.1704305	−.1720471	−.1736452	−.1752247
11.0	−.1843017	−.1857467	−.1871720	−.1885774	−.1899629
11.1	−.1978500	−.1990926	−.2003142	−.2015150	−.2026946
11.2	−.2093252	−.2103549	−.2113628	−.2123488	−.2133129
11.3	−.2186329	−.2194415	−.2202277	−.2209912	−.2217322
11.4	−.2257010	−.2262825	−.2268410	−.2273766	−.2278891
11.5	−.2304793	−.2308300	−.2311575	−.2314617	−.2317427
11.6	−.2329407	−.2330591	−.2331542	−.2332261	−.2332747
11.7	−.2330806	−.2329674	−.2328312	−.2326720	−.2324898
11.8	−.2309170	−.2305754	−.2302111	−.2298243	−.2294150
11.9	−.2264905	−.2259255	−.2253387	−.2247299	−.2240994
12.0	−.2198629	−.2190821	−.2182803	−.2174574	−.2166136
12.1	−.2111175	−.2101303	−.2091230	−.2080958	−.2070488
12.2	−.2003572	−.1991749	−.1979738	−.1967540	−.1955156
12.3	−.1877039	−.1863397	−.1849580	−.1835591	−.1821432
12.4	−.1732969	−.1717656	−.1702185	−.1686557	−.1670774
12.5	−.1572917	−.1556097	−.1539136	−.1522036	−.1504797
12.6	−.1398581	−.1380431	−.1362158	−.1343765	−.1325252
12.7	−.1211786	−.1192494	−.1173100	−.1153604	−.1134008
12.8	−.1014463	−.0994229	−.0973912	−.0953513	−.0933036
12.9	−.0808634	−.0787663	−.0766630	−.0745538	−.0724387
13.0	−.0596388	−.0574892	−.0553356	−.0531781	−.0510170
13.1	−.0379861	−.0358056	−.0336232	−.0314391	−.0292535
13.2	−.0161215	−.0139317	−.0117422	−.0095532	−.0073649
13.3	.0057383	.0079157	.0100907	.0122630	.0144326
13.4	.0273781	.0295218	.0316610	.0337954	.0359249
13.5	.0485866	.0506758	.0527583	.0548341	.0569029
13.6	.0691581	.0711725	.0731784	.0751755	.0771637
13.7	.0888946	.0908150	.0927250	.0946243	.0965129
13.8	.1076079	.1094160	.1112119	.1129955	.1147665
13.9	.1251207	.1267995	.1284645	.1301156	.1317525
14.0	.1412691	.1428030	.1443216	.1458248	.1473124
14.1	.1559036	.1572785	.1586368	.1599783	.1613030
14.2	.1688906	.1700940	.1712797	.1724475	.1735973
14.3	.1801133	.1811346	.1821372	.1831209	.1840858
14.4	.1894735	.1903038	.1911147	.1919059	.1926775
14.5	.1968915	.1975240	.1981364	.1987287	.1993008
14.6	.2023074	.2027371	.2031464	.2035352	.2039035
14.7	.2056813	.2059054	.2061089	.2062918	.2064540
14.8	.2069936	.2070113	.2070084	.2069849	.2069408
14.9	.2062452	.2060577	.2058498	.2056215	.2053729

BESSEL FUNCTION J_1

T	.00	.01	.02	.03	.04
15.0	.2051040	.2048149	.2045057	.2041762	.2038267
15.1	.2013102	.2008214	.2003130	.1997849	.1992373
15.2	.1955454	.1948631	.1941618	.1934416	.1927027
15.3	.1878794	.1870115	.1861255	.1852216	.1842998
15.4	1784003	.1773565	.1762958	.1752183	.1741239
15.5	.1672132	.1660051	.1647812	.1635418	.1622868
15.6	.1544396	.1530801	.1517062	.1503181	.1489160
15.7	.1402157	.1387193	.1372099	.1356878	.1341533
15.8	.1246913	.1230735	.1214444	.1198043	.1181532
15.9	.1080279	.1063054	.1045735	.1028322	.1010818
16.0	.0903972	.0885878	.0867707	.0849461	.0831142
16.1	.0719794	.0701015	.0682178	.0663284	.0644336
16.2	.0529615	.0510340	.0491027	.0471677	.0452291
16.3	.0335351	.0315774	.0296179	.0276565	.0256936
16.4	.0138947	.0119264	.0099581	.0079901	.0060224
16.5	−.0057642	−.0077236	−.0096811	−.0116364	−.0135893
16.6	−.0252471	−.0271783	−.0291056	−.0310289	−.0329479
16.7	−.0443624	−.0462465	−.0481248	−.0499972	−.0518635
16.8	−.0629232	−.0647418	−.0665528	−.0683562	−.0701516
16.9	−.0807493	−.0824847	−.0842110	−.0859278	−.0876351
17.0	−.0976685	−.0993042	−.1009291	−.1025429	−.1041456
17.1	−.1135188	−.1150392	−.1165472	−.1180427	−.1195255
17.2	−.1281497	−.1295403	−.1309172	−.1322802	−.1336293
17.3	−.1414233	−.1426712	−.1439041	−.1451219	−.1463246
17.4	−.1532162	−.1543097	−.1553872	−.1564487	−.1574939
17.5	−.1634200	−.1643493	−.1652617	−.1661571	−.1670354
17.6	−.1719427	−.1726995	−.1734387	−.1741601	−.1748638
17.7	−.1787096	−.1792874	−.1798469	−.1803883	−.1809113
17.8	−.1836635	−.1840575	−.1844329	−.1847898	−.1851280
17.9	−.1867654	−.1869728	−.1871614	−.1873313	−.1874824
18.0	−.1879949	−.1880146	−.1880156	−.1879979	−.1879614
18.1	−.1873502	−.1871831	−.1869975	−.1867933	−.1865706
18.2	−.1848479	−.1844967	−.1841273	−.1837398	−.1833341
18.3	−.1805231	−.1799922	−.1794437	−.1788776	−.1782940
18.4	−.1744283	−.1737241	−.1730029	−.1722648	−.1715100
18.5	−.1666336	−.1657639	−.1648780	−.1639762	−.1630586
18.6	−.1572254	−.1561996	−.1551588	−.1541031	−.1530325
18.7	−.1463053	−.1451345	−.1439499	−.1427515	−.1415396
18.8	−.1339897	−.1326863	−.1313703	−.1300419	−.1287013
18.9	−.1204080	−.1189855	−.1175520	−.1161074	−.1146521
19.0	−.1057014	−.1041747	−.1026383	−.1010925	−.0995375
19.1	−.0900216	−.0884062	−.0867828	−.0851517	−.0835128
19.2	−.0735290	−.0718414	−.0701476	−.0684476	−.0667417
19.3	−.0563913	−.0546486	−.0529015	−.0511499	−.0493942
19.4	−.0387816	−.0370015	−.0352186	−.0334331	−.0316452
19.5	−.0208771	−.0190773	−.0172765	−.0154749	−.0136727
19.6	−.0028566	−.0010551	.0007456	.0025454	.0043439
19.7	.0151006	.0168860	.0186688	.0204488	.0222259
19.8	.0328168	.0345684	.0363156	.0380584	.0397965
19.9	.0501174	.0518181	.0535127	.0552011	.0568832

BESSEL FUNCTION J_1

T	.05	.06	.07	.08	.09
15.0	.2034571	.2030675	.2026580	.2022286	.2017793
15.1	.1986703	.1980838	.1974780	.1968530	.1962088
15.2	.1919450	.1911688	.1903740	.1895608	.1887292
15.3	.1833603	.1824032	.1814285	.1804363	.1794269
15.4	.1730130	.1718855	.1707417	.1695816	.1684054
15.5	.1610165	.1597310	.1584304	.1571149	.1557846
15.6	.1474999	.1460700	.1446265	.1431695	.1416992
15.7	.1326063	.1310471	.1294758	.1278927	.1262978
15.8	.1164915	.1148192	.1131365	.1114436	.1097407
15.9	.0993224	.0975542	.0957774	.0939922	.0921987
16.0	.0812751	.0794291	.0775764	.0757170	.0738513
16.1	.0625336	.0606285	.0587186	.0568040	.0548849
16.2	.0432872	.0413423	.0393944	.0374437	.0354906
16.3	.0237294	.0217640	.0197976	.0178305	.0158628
16.4	.0040553	.0020891	.0001238	−.0018403	−.0038031
16.5	−.0155397	−.0174874	−.0194322	−.0213739	−.0233122
16.6	−.0348625	−.0367725	−.0386776	−.0405778	−.0424728
16.7	−.0537236	−.0555771	−.0574241	−.0592642	−.0610973
16.8	−.0719390	−.0737182	−.0754890	−.0772512	−.0790047
16.9	−.0893326	−.0910202	−.0926978	−.0943651	−.0960221
17.0	−.1057370	−.1073169	−.1088852	−.1104417	−.1119863
17.1	−.1209956	−.1224528	−.1238970	−.1253279	−.1267455
17.2	−.1349642	−.1362850	−.1375914	−.1388833	−.1401607
17.3	−.1475120	−.1486840	−.1498406	−.1509815	−.1521067
17.4	−.1585228	−.1595354	−.1605315	−.1615110	−.1624738
17.5	−.1678966	−.1687405	−.1695672	−.1703765	−.1711684
17.6	−.1755497	−.1762176	−.1768677	−.1774997	−.1781137
17.7	−.1814160	−.1819024	−.1823704	−.1828199	−.1832509
17.8	−.1854476	−.1857486	−.1860308	−.1862944	−.1865392
17.9	−.1876147	−.1877283	−.1878231	−.1878991	−.1879564
18.0	−.1879062	−.1878323	−.1877397	−.1876285	−.1874987
18.1	−.1863295	−.1860699	−.1857919	−.1854956	−.1851809
18.2	−.1829104	−.1824688	−.1820091	−.1815316	−.1810362
18.3	−.1776930	−.1770745	−.1764388	−.1757858	−.1751156
18.4	−.1707385	−.1699503	−.1691457	−.1683247	−.1674873
18.5	−.1621252	−.1611761	−.1602115	−.1592314	−.1582360
18.6	−.1519474	−.1508476	−.1497334	−.1486049	−.1474621
18.7	−.1403142	−.1390754	−.1378235	−.1365584	−.1352805
18.8	−.1273486	−.1259838	−.1246073	−.1232190	−.1218192
18.9	−.1131861	−.1117096	−.1102227	−.1087256	−.1072185
19.0	−.0979733	−.0964002	−.0948183	−.0932278	−.0916289
19.1	−.0818666	−.0802130	−.0785523	−.0768846	−.0752101
19.2	−.0650301	−.0633128	−.0615901	−.0598622	−.0581292
19.3	−.0476344	−.0458709	−.0441036	−.0423329	−.0405588
19.4	−.0298551	−.0280629	−.0262688	−.0244730	−.0226757
19.5	−.0118701	−.0100672	−.0082642	−.0064613	−.0046587
19.6	.0061411	.0079368	.0097307	.0115228	.0133128
19.7	.0239998	.0257705	.0275377	.0293012	.0310610
19.8	.0415298	.0432580	.0449810	.0466988	.0484109
19.9	.0585587	.0602276	.0618897	.0635447	.0651926

BESSEL FUNCTION J_1

T	.00	.01	.02	.03	.04
20.0	.0668331	.0684662	.0700916	.0717092	.0733189
20.1	.0828010	.0843506	.0858911	.0874222	.0889438
20.2	.0978664	.0993176	.1007583	.1021881	.1036071
20.3	.1118844	.1132233	.1145503	.1158652	.1171679
20.4	.1247210	.1259349	.1271356	.1283231	.1294972
20.5	.1362547	.1373321	.1383953	.1394442	.1404787
20.6	.1463774	.1473083	.1482241	.1491247	.1500099
20.7	.1549955	.1557714	.1565314	.1572754	.1580034
20.8	.1620307	.1626447	.1632421	.1638229	.1643871
20.9	.1674209	.1678675	.1682972	.1687099	.1691055
21.0	.1711203	.1713960	.1716544	.1718957	.1721196
21.1	.1731003	.1732031	.1732886	.1733568	.1734076
21.2	.1733493	.1732790	.1731915	.1730868	.1729649
21.3	.1718730	.1716312	.1713724	.1710966	.1708039
21.4	.1686941	.1682839	.1678572	.1674139	.1669541
21.5	.1638521	.1632784	.1626887	.1620831	.1614616
21.6	.1574027	.1566720	.1559260	.1551648	.1543884
21.7	.1494174	.1485376	.1476434	.1467348	.1458121
21.8	.1399825	.1389630	.1379301	.1368840	.1358246
21.9	.1291982	.1280498	.1268892	.1257164	.1245316
22.0	.1171778	.1159125	.1146362	.1133490	.1120511
22.1	.1040461	.1026770	.1012982	.0999099	.0985123
22.2	.0899387	.0884797	.0870126	.0855375	.0840545
22.3	.0749998	.0734659	.0719255	.0703785	.0688253
22.4	.0593815	.0577884	.0561902	.0545871	.0529793
22.5	.0432420	.0416056	.0399658	.0383228	.0366766
22.6	.0267439	.0250807	.0234157	.0217492	.0200812
22.7	.0100524	.0083791	.0067056	.0050322	.0033591
22.8	−.0066657	−.0083325	−.0099978	−.0116614	−.0133231
22.9	−.0232443	−.0248881	−.0265288	−.0281660	−.0297998
23.0	−.0395193	−.0411239	−.0427237	−.0443186	−.0459083
23.1	−.0553305	−.0568801	−.0584233	−.0599601	−.0614902
23.2	−.0705228	−.0720022	−.0734738	−.0749374	−.0763929
23.3	−.0849479	−.0863427	−.0877284	−.0891046	−.0904714
23.4	−.0984658	−.0997626	−.1010489	−.1023245	−.1035894
23.5	−.1109461	−.1121324	−.1133070	−.1144697	−.1156206
23.6	−.1222693	−.1233337	−.1243854	−.1254242	−.1264500
23.7	−.1323277	−.1332602	−.1341790	−.1350840	−.1359752
23.8	−.1410266	−.1418185	−.1425959	−.1433588	−.1441070
23.9	−.1482855	−.1489295	−.1495583	−.1501720	−.1507705
24.0	−.1540381	−.1545284	−.1550031	−.1554621	−.1559054
24.1	−.1582335	−.1585659	−.1588823	−.1591828	−.1594672
24.2	−.1608365	−.1610084	−.1611641	−.1613036	−.1614270
24.3	−.1618278	−.1618381	−.1618321	−.1618101	−.1617718
24.4	−.1612042	−.1610534	−.1608865	−.1607037	−.1605049
24.5	−.1589784	−.1586687	−.1583432	−.1580021	−.1576454
24.6	−.1551791	−.1547142	−.1542340	−.1537387	−.1532282
24.7	−.1498504	−.1492356	−.1486061	−.1479620	−.1473034
24.8	−.1430514	−.1422934	−.1415214	−.1407357	−.1399362
24.9	−.1348557	−.1339625	−.1330564	−.1321373	−.1312054

BESSEL FUNCTION J_1

T	.05	.06	.07	.08	.09
20.0	.0749204	.0765137	.0780985	.0796748	.0812423
20.1	.0904558	.0919581	.0934504	.0949326	.0964047
20.2	.1050150	.1064117	.1077972	.1091712	.1105336
20.3	.1184583	.1197362	.1210015	.1222542	.1234941
20.4	.1306578	.1318048	.1329381	.1340576	.1351631
20.5	.1414986	.1425040	.1434946	.1444704	.1454314
20.6	.1508798	.1517342	.1525730	.1533962	.1542037
20.7	.1587152	.1594109	.1600904	.1607535	.1614003
20.8	.1649347	.1654655	.1659796	.1664769	.1669573
20.9	.1694841	.1698456	.1701900	.1705173	.1708274
21.0	.1723263	.1725157	.1726878	.1728426	.1729801
21.1	.1734412	.1734574	.1734563	.1734379	.1734022
21.2	.1728258	.1726695	.1724960	.1723054	.1720977
21.3	.1704943	.1701678	.1698245	.1694644	.1690876
21.4	.1664779	.1659852	.1654763	.1649511	.1644097
21.5	.1608242	.1601711	.1595023	.1588180	.1581181
21.6	.1535970	.1527906	.1519694	.1511334	.1502827
21.7	.1448752	.1439243	.1429595	.1419808	.1409885
21.8	.1347522	.1336669	.1325687	.1314578	.1303342
21.9	.1233350	.1221266	.1209065	.1196749	.1184320
22.0	.1107426	.1094237	.1080944	.1067550	.1054055
22.1	.0971055	.0956897	.0942649	.0928314	.0913893
22.2	.0825638	.0810655	.0795598	.0780469	.0765268
22.3	.0672658	.0657003	.0641290	.0625520	.0609695
22.4	.0513669	.0497501	.0481290	.0465039	.0448748
22.5	.0350275	.0333756	.0317212	.0300643	.0284051
22.6	.0184119	.0167415	.0150702	.0133982	.0117255
22.7	.0016863	.0000141	−.0016574	−.0033279	−.0049974
22.8	−.0149827	−.0166401	−.0182951	−.0199476	−.0215974
22.9	−.0314299	−.0330561	−.0346783	−.0362964	−.0379101
23.0	−.0474927	−.0490718	−.0506452	−.0522129	−.0537747
23.1	−.0630134	−.0645298	−.0660391	−.0675411	−.0690357
23.2	−.0778402	−.0792791	−.0807094	−.0821311	−.0835440
23.3	−.0918286	−.0931761	−.0945136	−.0958412	−.0971586
23.4	−.1048434	−.1060864	−.1073183	−.1085390	−.1097483
23.5	−.1167594	−.1178861	−.1190005	−.1201026	−.1211922
23.6	−.1274628	−.1284624	−.1294488	−.1304219	−.1313815
23.7	−.1368525	−.1377157	−.1385648	−.1393997	−.1402203
23.8	−.1448406	−.1455593	−.1462633	−.1469523	−.1476264
23.9	−.1513536	−.1519214	−.1524738	−.1530107	−.1535322
24.0	−.1563330	−.1567448	−.1571408	−.1575209	−.1578852
24.1	−.1597356	−.1599880	−.1602243	−.1604445	−.1606486
24.2	−.1615342	−.1616253	−.1617002	−.1617589	−.1618015
24.3	−.1617175	−.1616470	−.1615604	−.1614577	−.1613390
24.4	−.1602902	−.1600595	−.1598130	−.1595506	−.1592724
24.5	−.1572730	−.1568852	−.1564818	−.1560629	−.1556287
24.6	−.1527026	−.1521619	−.1516064	−.1510359	−.1504505
24.7	−.1466304	−.1459431	−.1452414	−.1445255	−.1437955
24.8	−.1391230	−.1382963	−.1374562	−.1366026	−.1357357
24.9	−.1302607	−.1293035	−.1283337	−.1273515	−.1263570

BESSEL FUNCTION J_1

T	.00	.01	.02	.03	.04
25.0	−.1253502	−.1243314	−.1233006	−.1222578	−.1212033
25.1	−.1146348	−.1135009	−.1123561	−.1112005	−.1100344
25.2	−.1028206	−.1015834	−.1003365	−.0990801	−.0978143
25.3	−.0900295	−.0887017	−.0873655	−.0860211	−.0846686
25.4	−.0763926	−.0749876	−.0735757	−.0721571	−.0707318
25.5	−.0620485	−.0605808	−.0591076	−.0576290	−.0561453
25.6	−.0471429	−.0456272	−.0441075	−.0425840	−.0410569
25.7	−.0318259	−.0302775	−.0287266	−.0271735	−.0256183
25.8	−.0162515	−.0146859	−.0131194	−.0115523	−.0099846
25.9	−.0005755	.0009916	.0025581	.0041237	.0056883
26.0	.0150457	.0165989	.0181497	.0196982	.0212441
26.1	.0304572	.0319809	.0335009	.0350169	.0365288
26.2	.0455065	.0469857	.0484597	.0499283	.0513914
26.3	.0600453	.0614655	.0628791	.0642858	.0656857
26.4	.0739309	.0752783	.0766176	.0779488	.0792716
26.5	.0870278	.0882891	.0895412	.0907838	.0920169
26.6	.0992087	.1003718	.1015244	.1026665	.1037979
26.7	.1103559	.1114095	.1124517	.1134822	.1145010
26.8	.1203624	.1212965	.1222182	.1231273	.1240238
26.9	.1291329	.1299386	.1307311	.1315103	.1322760
27.0	.1365847	.1372546	.1379105	.1385524	.1391803
27.1	.1426487	.1431765	.1436899	.1441887	.1446729
27.2	.1472696	.1476507	.1480168	.1483681	.1487043
27.3	.1504068	.1506378	.1508537	.1510544	.1512400
27.4	.1520345	.1521137	.1521776	.1522264	.1522600
27.5	.1521419	.1520691	.1519811	.1518779	.1517597
27.6	.1507335	.1505100	.1502714	.1500180	.1497496
27.7	.1478289	.1474573	.1470712	.1466705	.1462553
27.8	.1434622	.1429469	.1424175	.1418741	.1413166
27.9	.1376822	.1370288	.1363619	.1356817	.1349881
28.0	.1305515	.1297669	.1289697	.1281599	.1273375
28.1	.1221457	.1212383	.1203191	.1193882	.1184457
28.2	.1125531	.1115323	.1105007	.1094584	.1084056
28.3	.1018732	.1007495	.0996162	.0984734	.0973211
28.4	.0902161	.0890012	.0877778	.0865461	.0853062
28.5	.0777014	.0764076	.0751066	.0737985	.0724836
28.6	.0644564	.0630968	.0617315	.0603604	.0589838
28.7	.0506155	.0492040	.0477880	.0463678	.0449434
28.8	.0363185	.0348692	.0334170	.0319619	.0305041
28.9	.0217093	.0202368	.0187628	.0172874	.0158109
29.0	.0069342	.0054533	.0039723	.0024914	.0010108
29.1	−.0078591	−.0093337	−.0108069	−.0122786	−.0137485
29.2	−.0225233	−.0239770	−.0254278	−.0268756	−.0283202
29.3	−.0369130	−.0383314	−.0397454	−.0411550	−.0425600
29.4	−.0508859	−.0522549	−.0536183	−.0549758	−.0563273
29.5	−.0643044	−.0656106	−.0669098	−.0682019	−.0694868
29.6	−.0770368	−.0782674	−.0794898	−.0807039	−.0819095
29.7	−.0889586	−.0901017	−.0912354	−.0923596	−.0934742
29.8	−.0999541	−.1009985	−.1020325	−.1030559	−.1040687
29.9	−.1099168	−.1108525	−.1117768	−.1126896	−.1135909

BESSEL FUNCTION J_1

T	.05	.06	.07	.08	.09
25.0	−.1201371	−.1190593	−.1179701.	−.1168695	−.1157577
25.1	−.1088577	−.1076706	−.1064732	−.1052656	−.1040481
25.2	−.0965392	−.0952550	−.0939618	−.0926597	−.0913489
25.3	−.0833083	−.0819401	−.0805643	−.0791811	−.0777904
25.4	−.0693000	−.0678618	−.0664174	−.0649670	−.0635107
25.5	−.0546566	−.0531630	−.0516647	−.0501618	−.0486544
25.6	−.0395262	−.0379922	−.0364551	−.0349148	−.0333717
25.7	−.0240612	−.0225022	−.0209416	−.0193795	−.0178161
25.8	−.0084164	−.0068481	−.0052797	−.0037114	−.0021433
25.9	.0072517	.0088138	.0103744	.0119333	.0134905
26.0	.0227873	.0243276	.0258649	.0273991	.0289299
26.1	.0380365	.0395398	.0410387	.0425328	.0440221
26.2	.0528487	.0543003	.0557458	.0571853	.0586185
26.3	.0670784	.0684639	.0698420	.0712126	.0725757
26.4	.0805861	.0818920	.0831893	.0844778	.0857573
26.5	.0932404	.0944540	.0956578	.0968516	.0980353
26.6	.1049185	.1060282	.1071269	.1082144	.1092908
26.7	.1155080	.1165030	.1174861	.1184571	.1194159
26.8	.1249076	.1257786	.1266367	.1274818	.1283139
26.9	.1330283	.1337670	.1344920	.1352034	.1359010
27.0	.1397940	.1403935	.1409788	.1415498	.1421065
27.1	.1451425	.1455974	.1460376	.1464631	.1468738
27.2	.1490257	.1493320	.1496233	.1498995	.1501607
27.3	.1514104	.1515656	.1517056	.1518304	.1519401
27.4	.1522783	.1522814	.1522693	.1522420	.1521996
27.5	.1516263	.1514779	.1513143	.1511358	.1509421
27.6	.1494665	.1491684	.1488556	.1485281	.1481858
27.7	.1458256	.1453815	.1449231	.1444504	.1439634
27.8	.1407453	.1401601	.1395612	.1389485	.1383222
27.9	.1342813	.1335613	.1328283	.1320822	.1313233
28.0	.1265027	.1256556	.1247963	.1239248	.1230412
28.1	.1174918	.1165264	.1155498	.1145619	.1135630
28.2	.1073424	.1062688	.1051849	.1040910	.1029870
28.3	.0961595	.0949888	.0938089	.0926201	.0914225
28.4	.0840582	.0828022	.0815384	.0802669	.0789879
28.5	.0711619	.0698335	.0684987	.0671574	.0658099
28.6	.0576018	.0562145	.0548221	.0534247	.0520224
28.7	.0435151	.0420829	.0406470	.0392075	.0377646
28.8	.0290438	.0275811	.0261161	.0246491	.0231801
28.9	.0143332	.0128547	.0113754	.0098954	.0084150
29.0	−.0004694	−.0019490	−.0034280	−.0049060	−.0063831
29.1	−.0152165	−.0166825	−.0181463	−.0196078	−.0210669
29.2	−.0297615	−.0311993	−.0326335	−.0340639	−.0354905
29.3	−.0439603	−.0453557	−.0467461	−.0481314	−.0485113
29.4	−.0576728	−.0590121	−.0603450	−.0616715	−.0629913
29.5	−.0707643	−.0720343	−.0732966	−.0745513	−.0757980
29.6	−.0831065	−.0842948	−.0854742	−.0866448	−.0878063
29.7	−.0945791	−.0956741	−.0967593	−.0978344	−.0988994
29.8	−.1050708	−.1060620	−.1070423	−.1080116	−.1089698
29.9	−.1144805	−.1153584	−.1162245	−.1170787	−.1179209

BESSEL FUNCTION J_1

T	.00	.01	.02	.03	.04
30.0	−.1187511	−.1195691	−.1203749	−.1211684	−.1219495
30.1	−.1263727	−.1270653	−.1277450	−.1284116	−.1290653
30.2	−.1327098	−.1332705	−.1338177	−.1343514	−.1348714
30.3	−.1377037	−.1381273	−.1385370	−.1389328	−.1393145
30.4	−.1413090	−.1415918	−.1418604	−.1421147	−.1423547
30.5	−.1434943	−.1436339	−.1437592	−.1438701	−.1439666
30.6	−.1442426	−.1442382	−.1442194	−.1441861	−.1441385
30.7	−.1435512	−.1434032	−.1432409	−.1430644	−.1428736
30.8	−.1414315	−.1411419	−.1408382	−.1405206	−.1401891
30.9	−.1379094	−.1374814	−.1370399	−.1365848	−.1361162
31.0	−.1330243	−.1324628	−.1318882	−.1313006	−.1307001
31.1	−.1268294	−.1261403	−.1254388	−.1247250	−.1239990
31.2	−.1193904	−.1185810	−.1177600	−.1169275	−.1160836
31.3	−.1107855	−.1098642	−.1089323	−.1079897	−.1070368
31.4	−.1011040	−.1000804	−.0990471	−.0980043	−.0969520

BESSEL FUNCTION J_1

T	.05	.06	.07	.08	.09
30.0	−.1227182	−.1234744	−.1242180	−.1249490	−.1256672
30.1	−.1297058	−.1303332	−.1309473	−.1315482	−.1321357
30.2	−.1353778	−.1358705	−.1363495	−.1368148	−.1372661
30.3	−.1396822	−.1400358	−.1403753	−.1407007	−.1410119
30.4	−.1425805	−.1427920	−.1429891	−.1431719	−.1433403
30.5	−.1440486	−.1441162	−.1441695	−.1442083	−.1442327
30.6	−.1440765	−.1440001	−.1439094	−.1438043	−.1436849
30.7	−.1426686	−.1424495	−.1422161	−.1419687	−.1417071
30.8	−.1398436	−.1394843	−.1391112	−.1387243	−.1383237
30.9	−.1356342	−.1351388	−.1346300	−.1341080	−.1335727
31.0	−.1300868	−.1294606	−.1288217	−.1281702	−.1275060
31.1	−.1232608	−.1225106	−.1217483	−.1209742	−.1201882
31.2	−.1152284	−.1143619	−.1134843	−.1125956	−.1116960
31.3	−.1060733	−.1050996	−.1041158	−.1031218	−.1021178
31.4	−.0958903	−.0948194	−.0937394	−.0926504	−.0915524

HYPERBOLIC BESSEL FUNCTIONS

$$I_m(x) = i^{-m}J_m(ix)$$

x	$I_0(x)$	$I_1(x)$	$I_2(x)$
0.0	1.0000	0.0000	0.0000
0.1	1.0025	0.0501	0.0012
0.2	1.0100	0.1005	0.0050
0.4	1.0404	0.2040	0.0203
0.6	1.0920	0.3137	0.0464
0.8	1.1665	0.4329	0.0844
1.0	1.2661	0.5652	0.1357
1.2	1.3937	0.7147	0.2026
1.4	1.5534	0.8861	0.2875
1.6	1.7500	1.0848	0.3940
1.8	1.9896	1.3172	0.5260
2.0	2.2796	1.5906	0.6889
2.2	2.6291	1.9141	0.8891
2.4	3.0493	2.2981	1.1342
2.6	3.5533	2.7554	1.4337
2.8	4.1573	3.3011	1.7994
3.0	4.8808	3.9534	2.2452
3.2	5.7472	4.7343	2.7883
3.4	6.7848	5.6701	3.4495
3.6	8.0277	6.7927	4.2540
3.8	9.5169	8.1404	5.2325
4.0	11.302	9.7595	6.4222
4.2	13.442	11.706	7.8684
4.4	16.010	14.046	9.6258
4.6	19.093	16.863	11.761
4.8	22.794	20.253	14.355
5.0	27.240	24.336	17.506
5.2	32.584	29.254	21.332
5.4	39.009	35.182	25.978
5.6	46.738	42.328	31.620
5.8	56.038	50.946	38.470
6.0	67.234	61.342	46.787
6.2	80.718	73.886	56.884
6.4	96.962	89.026	69.141
6.6	116.54	107.30	84.021
6.8	140.14	129.38	102.08
7.0	168.59	156.04	124.01
7.2	202.92	188.25	150.63
7.4	244.34	227.17	182.94
7.6	294.33	274.22	222.17
7.8	354.68	331.10	269.79
8.0	427.56	399.87	327.60

Taken from Vibration and Sound with the permission of Philip Morse, author, and McGraw-Hill Book Company, Inc., publisher.

ELLIPTIC INTEGRALS OF THE FIRST, SECOND AND THIRD KIND

An elliptic integral has the form $\int R(x, \sqrt{f(x)})dx$, where R represents a rational function and $f(x) = a + bx + cx^2 + dx^3 + ex^4$, an algebraic function of the third or fourth degree.

1. Elliptic integrals of the *first kind* are represented by

$$F(k, \phi) = \int_0^\phi \frac{d\Phi}{\sqrt{1 - k^2 \sin^2 \Phi}}$$

$$= \int_0^x \frac{d\xi}{\sqrt{(1 - \xi^2)(1 - k^2\xi^2)}}, \quad x = \sin \phi, \, k^2 < 1.$$

2. Elliptic integrals of the second kind are represented by

$$E(k, \phi) = \int_0^\phi \sqrt{1 - k^2 \sin^2 \Phi} \, d\Phi$$

$$= \int_0^x \frac{\sqrt{1 - k^2\xi^2}}{\sqrt{1 - \xi^2}} \, d\xi, \quad x = \sin \phi, \, k^2 < 1.$$

3. Elliptic integrals of the third kind are represented as

$$\pi(k, n, \phi) = \int_0^\phi \frac{d\Phi}{(1 + n \sin^2 \Phi) \sqrt{1 - k^2 \sin^2 \Phi}},$$

$$k^2 < 1, \, n \text{ an integer.}$$

Elliptic integrals of the third kind are also presented as

$$\pi_1(k, n, x) = \int_0^x \frac{d\xi}{(1 + n\xi^2) \sqrt{(1 - \xi^2)(1 - k^2\xi^2)}},$$

$$x = \sin \phi, \, k^2 < 1, \, n \text{ an integer.}$$

4. The complete integrals are

$$K = F\left(k, \frac{\pi}{2}\right) = \frac{\pi}{2}\left[1 + \left(\frac{1}{2}\right)^2 k^2 + \left(\frac{3}{2 \cdot 4}\right)^2 k^4 \right.$$

$$\left. + \left(\frac{3 \cdot 5}{2 \cdot 4 \cdot 6}\right)^2 k^6 + \cdots \right]$$

$$E = E\left(k, \frac{\pi}{2}\right) = \frac{\pi}{2}\left[1 - \left(\frac{1}{2^2}\right) k^2 - \left(\frac{3^2}{2^2 \cdot 4^2}\right)\frac{k^4}{3} \right.$$

$$\left. - \left(\frac{3^2 \cdot 5^2}{2^2 \cdot 4^2 \cdot 6^2}\right)\frac{k^6}{5} - \left(\frac{3^2 \cdot 5^2 \cdot 7^2}{2^2 \cdot 4^2 \cdot 6^2 \cdot 8^2}\right)\frac{k^8}{7} - \cdots \right].$$

$$K' = F\left(\sqrt{1 - k^2}, \frac{\pi}{2}\right), \, E' = E\left(\sqrt{1 - k^2}, \frac{\pi}{2}\right)$$

5. The following relation holds between K, K', E, E', namely

$$KE' + EK' - KK' = \frac{\pi}{2} \qquad \text{Legendre's relation}$$

E: see 2 above. $E' = \int_0^{\pi/2} (1 - k'^2 \sin^2 \phi)^{\frac{1}{2}}d\phi.$ $k' = \sqrt{(1 - k^2)}$

$K = \int_0^{\pi/2} (1 - k^2 \sin^2 \phi)^{-\frac{1}{2}}d\phi \qquad K' = \int_0^{\pi/2} (1 - k'^2 \sin^2 \phi)^{-\frac{1}{2}}d\phi$

6. To evaluate elliptic integrals for values outside the range contained in the following tables, these relations are useful

$$F(k, \pi) = 2K;\ E(k, \pi) = 2E$$

$$F(k, \phi + m\pi) = mF(k, \pi) + F(k, \phi) = 2mK + F(k, \phi)$$
$$m = 0, 1, 2, 3, \ldots$$

$$E(k, \phi + m\pi) = mE(k, \pi) + E(k, \phi) = 2mE + E(k, \phi)$$
$$m = 0, 1, 2, 3, \ldots$$

7. If $u = F(k, \phi) = \displaystyle\int_0^\phi \frac{d\Phi}{\sqrt{1 - k^2 \sin^2 \Phi}}$ $(k^2 < 1),$

= elliptic integral of the first kind.

$$u = \int_0^x \frac{dx}{\sqrt{(1 - \xi^2)(1 - k^2 \xi^2)}}, \text{ where } x = \sin \phi.$$

ϕ is called the amplitude of u or am u.
k is called the modulus.

$$k' = \sqrt{1 - k^2} = \text{the complementary modulus.}$$

$\sin \phi = \operatorname{sn} u = x$ $\tan \phi = \operatorname{tn} u = \dfrac{x}{\sqrt{1 - x^2}}.$

$\cos \phi = \operatorname{cn} u = \sqrt{1 - x^2}.$ $\Delta \phi = \operatorname{dn} u = \sqrt{1 - k^2 x^2}.$

am $0 = 0.$ sn $0 = 0.$

cn $0 = 1.$ dn $0 = 1.$

am $(-u) = -$am $u.$ sn $(-u) = -$sn $u.$

cn $(-u) = $ cn $u.$ dn $(-u) = $ dn $u.$

tn $(-u) = -$tn $u.$
$\operatorname{sn}^2 u + \operatorname{cn}^2 u = 1.$
$\operatorname{dn}^2 u + k^2 \operatorname{sn}^2 u = 1.$
$\operatorname{dn}^2 u - k^2 \operatorname{cn}^2 u = 1 - k^2 = k'^2.$

$$\operatorname{sn} u = u - (1 + k^2) \frac{u^3}{3!} + (1 + 14k^2 + k^4) \frac{u^5}{5!}$$
$$- (1 + 135k^2 + 135k^4 + k^6) \frac{u^7}{7!} + \cdots$$

Periods: $4k$ and $2ik'$

$$\operatorname{cn} u = 1 - \frac{u^2}{2!} + (1 + 4k^2) \frac{u^4}{4!} - (1 + 44k^2 + 16k^4) \frac{u^6}{6!} + \cdots$$

Periods: $4k$ and $2k + 2ik'$

$$\operatorname{dn} u = 1 - k^2 \frac{u^2}{2!} + k^2(4 + k^2) \frac{u^4}{4!} - k^2(16 + 44k^2 + k^4) \frac{u^6}{6!} + \cdots$$

Periods: $2k$ and $4ik'$

ELLIPTIC INTEGRALS OF THE FIRST KIND: $F(k, \phi)$*

$$F(k, \phi) = \int_0^\phi \frac{d\Phi}{\sqrt{1 - k^2 \sin^2 \Phi}}, \qquad \theta = \sin^{-1} k$$

θ / ϕ	5°	10°	15°	20°	25°	30°	35°	40°	45°
1°	0.0175	0.0175	0.0175	0.0175	0.0175	0.0175	0.0175	0.0175	0.0175
2°	0.0349	0.0349	0.0349	0.0349	0.0349	0.0349	0.0349	0.0349	0.0349
3°	0.0524	0.0524	0.0524	0.0524	0.0524	0.0524	0.0524	0.0524	0.0524
4°	0.0698	0.0698	0.0698	0.0698	0.0698	0.0698	0.0698	0.0698	0.0698
5°	0.0873	0.0873	0.0873	0.0873	0.0873	0.0873	0.0873	0.0873	0.0873
6°	0.1047	0.1047	0.1047	0.1047	0.1048	0.1048	0.1048	0.1048	0.1048
7°	0.1222	0.1222	0.1222	0.1222	0.1222	0.1222	0.1223	0.1223	0.1223
8°	0.1396	0.1396	0.1397	0.1397	0.1397	0.1397	0.1398	0.1398	0.1399
9°	0.1571	0.1571	0.1571	0.1572	0.1572	0.1572	0.1573	0.1573	0.1574
10°	0.1745	0.1746	0.1746	0.1746	0.1747	0.1748	0.1748	0.1749	0.1750
11°	0.1920	0.1920	0.1921	0.1921	0.1922	0.1923	0.1924	0.1925	0.1926
12°	0.2095	0.2095	0.2095	0.2096	0.2097	0.2098	0.2099	0.2101	0.2102
13°	0.2269	0.2270	0.2270	0.2271	0.2272	0.2274	0.2275	0.2277	0.2279
14°	0.2444	0.2444	0.2445	0.2446	0.2448	0.2450	0.2451	0.2453	0.2456
15°	0.2618	0.2619	0.2620	0.2621	0.2623	0.2625	0.2628	0.2630	0.2633
16°	0.2793	0.2794	0.2795	0.2797	0.2799	0.2802	0.2804	0.2808	0.2811
17°	0.2967	0.2968	0.2970	0.2972	0.2975	0.2978	0.2981	0.2985	0.2989
18°	0.3142	0.3143	0.3145	0.3148	0.3151	0.3154	0.3159	0.3163	0.3167
19°	0.3317	0.3318	0.3320	0.3323	0.3327	0.3331	0.3336	0.3341	0.3347
20°	0.3491	0.3493	0.3495	0.3499	0.3503	0.3508	0.3514	0.3520	0.3526
21°	0.3666	0.3668	0.3671	0.3675	0.3680	0.3685	0.3692	0.3699	0.3706
22°	0.3840	0.3842	0.3846	0.3851	0.3856	0.3863	0.3871	0.3879	0.3887
23°	0.4015	0.4017	0.4021	0.4027	0.4033	0.4041	0.4049	0.4059	0.4068
24°	0.4190	0.4192	0.4197	0.4203	0.4210	0.4219	0.4229	0.4239	0.4250
25°	0.4364	0.4367	0.4372	0.4379	0.4387	0.4397	0.4408	0.4420	0.4433
26°	0.4539	0.4542	0.4548	0.4556	0.4565	0.4576	0.4588	0.4602	0.4616
27°	0.4714	0.4717	0.4724	0.4732	0.4743	0.4755	0.4769	0.4784	0.4800
28°	0.4888	0.4893	0.4899	0.4909	0.4921	0.4934	0.4950	0.4967	0.4985
29°	0.5063	0.5068	0.5075	0.5086	0.5099	0.5114	0.5132	0.5150	0.5170
30°	0.5238	0.5243	0.5251	0.5263	0.5277	0.5294	0.5313	0.5334	0.5356
31°	0.5412	0.5418	0.5427	0.5440	0.5456	0.5475	0.5496	0.5519	0.5543
32°	0.5587	0.5593	0.5603	0.5617	0.5635	0.5656	0.5679	0.5704	0.5731
33°	0.5762	0.5769	0.5780	0.5795	0.5814	0.5837	0.5862	0.5890	0.5920
34°	0.5937	0.5944	0.5956	0.5973	0.5994	0.6018	0.6046	0.6077	0.6109
35°	0.6111	0.6119	0.6133	0.6151	0.6173	0.6200	0.6231	0.6264	0.6300
36°	0.6286	0.6295	0.6309	0.6329	0.6353	0.6383	0.6416	0.6452	0.6491
37°	0.6461	0.6470	0.6486	0.6507	0.6534	0.6565	0.6602	0.6641	0.6684
38°	0.6636	0.6646	0.6662	0.6685	0.6714	0.6749	0.6788	0.6831	0.6877
39°	0.6810	0.6821	0.6839	0.6864	0.6895	0.6932	0.6975	0.7021	0.7071
40°	0.6985	0.6997	0.7016	0.7043	0.7076	0.7116	0.7162	0.7213	0.7267
41°	0.7160	0.7173	0.7193	0.7222	0.7258	0.7301	0.7350	0.7405	0.7463
42°	0.7335	0.7348	0.7370	0.7401	0.7440	0.7486	0.7539	0.7598	0.7661
43°	0.7510	0.7524	0.7548	0.7580	0.7622	0.7671	0.7728	0.7791	0.7859
44°	0.7685	0.7700	0.7725	0.7760	0.7804	0.7857	0.7918	0.7986	0.8059
45°	0.7859	0.7876	0.7903	0.7940	0.7987	0.8044	0.8109	0.8181	0.8260

* For useful information about these tables see preceding page.

ELLIPTIC INTEGRALS OF THE FIRST KIND: $F(k, \phi)$ (Continued)

$$F(k, \phi) = \int_0^\phi \frac{d\Phi}{\sqrt{1 - k^2 \sin^2 \Phi}}, \qquad \theta = \sin^{-1} k$$

θ / ϕ	50°	55°	60°	65°	70°	75°	80°	85°	90°
1°	0.0175	0.0175	0.0175	0.0175	0.0175	0.0175	0.0175	0.0175	0.0175
2°	0.0349	0.0349	0.0349	0.0349	0.0349	0.0349	0.0349	0.0349	0.0349
3°	0.0524	0.0524	0.0524	0.0524	0.0524	0.0524	0.0524	0.0524	0.0524
4°	0.0698	0.0699	0.0699	0.0699	0.0699	0.0699	0.0699	0.0699	0.0699
5°	0.0873	0.0873	0.0873	0.0874	0.0874	0.0874	0.0874	0.0874	0.0874
6°	0.1048	0.1048	0.1049	0.1049	0.1049	0.1049	0.1049	0.1049	0.1049
7°	0.1224	0.1224	0.1224	0.1224	0.1224	0.1225	0.1225	0.1225	0.1225
8°	0.1399	0.1399	0.1400	0.1400	0.1400	0.1401	0.1401	0.1401	0.1401
9°	0.1575	0.1575	0.1576	0.1576	0.1577	0.1577	0.1577	0.1577	0.1577
10°	0.1751	0.1751	0.1752	0.1753	0.1753	0.1754	0.1754	0.1754	0.1754
11°	0.1927	0.1928	0.1929	0.1930	0.1930	0.1931	0.1931	0.1932	0.1932
12°	0.2103	0.2105	0.2106	0.2107	0.2108	0.2109	0.2109	0.2110	0.2110
13°	0.2280	0.2282	0.2284	0.2285	0.2286	0.2287	0.2288	0.2288	0.2289
14°	0.2458	0.2460	0.2462	0.2464	0.2465	0.2466	0.2467	0.2468	0.2468
15°	0.2636	0.2638	0.2641	0.2643	0.2645	0.2646	0.2647	0.2648	0.2648
16°	0.2814	0.2817	0.2820	0.2823	0.2825	0.2827	0.2828	0.2829	0.2830
17°	0.2993	0.2997	0.3000	0.3003	0.3006	0.3008	0.3010	0.3011	0.3012
18°	0.3172	0.3177	0.3181	0.3185	0.3188	0.3191	0.3193	0.3194	0.3195
19°	0.3352	0.3357	0.3362	0.3367	0.3371	0.3374	0.3377	0.3378	0.3379
20°	0.3533	0.3539	0.3545	0.3550	0.3555	0.3559	0.3561	0.3563	0.3564
21°	0.3714	0.3721	0.3728	0.3734	0.3740	0.3744	0.3747	0.3749	0.3750
22°	0.3896	0.3904	0.3912	0.3919	0.3926	0.3931	0.3935	0.3937	0.3938
23°	0.4078	0.4088	0.4097	0.4105	0.4113	0.4119	0.4123	0.4126	0.4127
24°	0.4261	0.4272	0.4283	0.4292	0.4301	0.4308	0.4313	0.4316	0.4317
25°	0.4446	0.4458	0.4470	0.4481	0.4490	0.4498	0.4504	0.4508	0.4509
26°	0.4630	0.4645	0.4658	0.4670	0.4681	0.4690	0.4697	0.4701	0.4702
27°	0.4816	0.4832	0.4847	0.4861	0.4873	0.4884	0.4891	0.4896	0.4897
28°	0.5003	0.5021	0.5038	0.5053	0.5067	0.5079	0.5087	0.5092	0.5094
29°	0.5190	0.5210	0.5229	0.5247	0.5262	0.5275	0.5285	0.5291	0.5293
30°	0.5379	0.5401	0.5422	0.5442	0.5459	0.5474	0.5484	0.5491	0.5493
31°	0.5568	0.5593	0.5617	0.5639	0.5658	0.5674	0.5686	0.5693	0.5696
32°	0.5759	0.5786	0.5812	0.5837	0.5858	0.5876	0.5889	0.5898	0.5900
33°	0.5950	0.5980	0.6010	0.6037	0.6060	0.6080	0.6095	0.6104	0.6107
34°	0.6143	0.6176	0.6208	0.6238	0.6265	0.6287	0.6303	0.6313	0.6317
35°	0.6336	0.6373	0.6408	0.6441	0.6471	0.6495	0.6513	0.6525	0.6528
36°	0.6531	0.6571	0.6610	0.6647	0.6679	0.6706	0.6726	0.6739	0.6743
37°	0.6727	0.6771	0.6814	0.6854	0.6890	0.6919	0.6941	0.6955	0.6960
38°	0.6925	0.6973	0.7019	0.7063	0.7102	0.7135	0.7159	0.7175	0.7180
39°	0.7123	0.7176	0.7227	0.7275	0.7318	0.7353	0.7380	0.7397	0.7403
40°	0.7323	0.7380	0.7436	0.7488	0.7535	0.7575	0.7604	0.7623	0.7629
41°	0.7524	0.7586	0.7647	0.7704	0.7756	0.7799	0.7831	0.7852	0.7859
42°	0.7727	0.7794	0.7860	0.7922	0.7979	0.8026	0.8062	0.8084	0.8092
43°	0.7931	0.8004	0.8075	0.8143	0.8204	0.8256	0.8295	0.8320	0.8328
44°	0.8136	0.8215	0.8293	0.8367	0.8433	0.8490	0.8533	0.8560	0.8569
45°	0.8343	0.8428	0.8512	0.8592	0.8665	0.8727	0.8774	0.8804	0.8814

ELLIPTIC INTEGRALS OF THE FIRST KIND: $F(k, \phi)$ (Continued)

$$F(k, \phi) = \int_0^\phi \frac{d\Phi}{\sqrt{1 - k^2 \sin^2 \Phi}}, \qquad \theta = \sin^{-1} k$$

θ / ϕ	5°	10°	15°	20°	25°	30°	35°	40°	45°
46°	0.8034	0.8052	0.8080	0.8120	0.8170	0.8230	0.8300	0.8378	0.8462
47°	0.8209	0.8227	0.8258	0.8300	0.8353	0.8418	0.8492	0.8575	0.8666
48°	0.8384	0.8403	0.8436	0.8480	0.8537	0.8606	0.8685	0.8773	0.8870
49°	0.8559	0.8579	0.8614	0.8661	0.8721	0.8794	0.8878	0.8972	0.9076
50°	0.8734	0.8756	0.8792	0.8842	0.8905	0.8982	0.9072	0.9173	0.9283
51°	0.8909	0.8932	0.8970	0.9023	0.9090	0.9172	0.9267	0.9374	0.9491
52°	0.9084	0.9108	0.9148	0.9204	0.9275	0.9361	0.9462	0.9575	0.9701
53°	0.9259	0.9284	0.9326	0.9385	0.9460	0.9551	0.9658	0.9778	0.9912
54°	0.9434	0.9460	0.9505	0.9567	0.9646	0.9742	0.9855	0.9982	1.0124
55°	0.9609	0.9637	0.9683	0.9748	0.9832	0.9933	1.0052	1.0187	1.0337
56°	0.9784	0.9813	0.9862	0.9930	1.0018	1.0125	1.0250	1.0393	1.0552
57°	0.9959	0.9989	1.0041	1.0112	1.0204	1.0317	1.0449	1.0600	1.0768
58°	1.0134	1.0166	1.0219	1.0295	1.0391	1.0509	1.0648	1.0807	1.0985
59°	1.0309	1.0342	1.0398	1.0477	1.0578	1.0702	1.0848	1.1016	1.1204
60°	1.0484	1.0519	1.0577	1.0660	1.0766	1.0896	1.1049	1.1226	1.1424
61°	1.0659	1.0695	1.0757	1.0843	1.0953	1.1089	1.1250	1.1436	1.1646
62°	1.0834	1.0872	1.0936	1.1026	1.1141	1.1284	1.1452	1.1648	1.1868
63°	1.1009	1.1049	1.1115	1.1209	1.1330	1.1478	1.1655	1.1860	1.2093
64°	1.1184	1.1225	1.1295	1.1392	1.1518	1.1674	1.1859	1.2073	1.2318
65°	1.1359	1.1402	1.1474	1.1575	1.1707	1.1869	1.2063	1.2288	1.2545
66°	1.1534	1.1579	1.1654	1.1759	1.1896	1.2065	1.2267	1.2503	1.2773
67°	1.1709	1.1756	1.1833	1.1943	1.2085	1.2262	1.2472	1.2719	1.3002
68°	1.1884	1.1932	1.2013	1.2127	1.2275	1.2458	1.2678	1.2936	1.3232
69°	1.2059	1.2109	1.2193	1.2311	1.2465	1.2655	1.2885	1.3154	1.3464
70°	1.2234	1.2286	1.2373	1.2495	1.2655	1.2853	1.3092	1.3372	1.3697
71°	1.2410	1.2463	1.2553	1.2680	1.2845	1.3051	1.3299	1.3592	1.3931
72°	1.2585	1.2640	1.2733	1.2864	1.3036	1.3249	1.3507	1.3812	1.4167
73°	1.2760	1.2817	1.2913	1.3049	1.3226	1.3448	1.3715	1.4033	1.4403
74°	1.2935	1.2994	1.3093	1.3234	1.3417	1.3647	1.3924	1.4254	1.4640
75°	1.3110	1.3171	1.3273	1.3418	1.3608	1.3846	1.4134	1.4477	1.4879
76°	1.3285	1.3348	1.3454	1.3603	1.3800	1.4045	1.4344	1.4700	1.5118
77°	1.3460	1.3525	1.3634	1.3788	1.3991	1.4245	1.4554	1.4923	1.5359
78°	1.3636	1.3702	1.3814	1.3974	1.4183	1.4445	1.4765	1.5147	1.5600
79°	1.3811	1.3879	1.3995	1.4159	1.4374	1.4645	1.4976	1.5372	1.5842
80°	1.3986	1.4056	1.4175	1.4344	1.4566	1.4846	1.5187	1.5597	1.6085
81°	1.4161	1.4234	1.4356	1.4530	1.4758	1.5046	1.5399	1.5823	1.6328
82°	1.4336	1.4411	1.4536	1.4715	1.4950	1.5247	1.5611	1.6049	1.6572
83°	1.4512	1.4588	1.4717	1.4901	1.5143	1.5448	1.5823	1.6276	1.6817
84°	1.4687	1.4765	1.4897	1.5086	1.5335	1.5649	1.6035	1.6502	1.7062
85°	1.4862	1.4942	1.5078	1.5272	1.5527	1.5850	1.6248	1.6730	1.7308
86°	1.5037	1.5120	1.5259	1.5457	1.5720	1.6052	1.6461	1.6957	1.7554
87°	1.5212	1.5297	1.5439	1.5643	1.5912	1.6253	1.6673	1.7184	1.7801
88°	1.5388	1.5474	1.5620	1.5829	1.6105	1.6454	1.6886	1.7412	1.8047
89°	1.5563	1.5651	1.5801	1.6015	1.6297	1.6656	1.7099	1.7640	1.8294
90°	1.5738	1.5828	1.5981	1.6200	1.6490	1.6858	1.7312	1.7863	1.8541

ELLIPTIC INTEGRALS OF THE FIRST KIND: $F(k, \phi)$ (Continued)

$$F(k, \phi) = \int_0^\phi \frac{d\Phi}{\sqrt{1 - k^2 \sin^2 \Phi}}, \qquad \theta = \sin^{-1} k$$

θ \ ϕ	50°	55°	60°	65°	70°	75°	80°	85°	90°
46°	0.8552	0.8643	0.8734	0.8821	0.8900	0.8968	0.9019	0.9052	0.9063
47°	0.8761	0.8860	0.8958	0.9053	0.9139	0.9212	0.9269	0.9304	0.9316
48°	0.8973	0.9079	0.9185	0.9287	0.9381	0.9461	0.9523	0.9561	0.9575
49°	0.9186	0.9300	0.9415	0.9525	0.9627	0.9714	0.9781	0.9824	0.9838
50°	0.9401	0.9523	0.9647	0.9766	0.9876	0.9971	1.0044	1.0091	1.0107
51°	0.9617	0.9748	0.9881	1.0010	1.0130	1.0233	1.0313	1.0364	1.0381
52°	0.9835	0.9976	1.0118	1.0258	1.0387	1.0499	1.0587	1.0642	1.0662
53°	1.0055	1.0205	1.0359	1.0509	1.0649	1.0771	1.0866	1.0927	1.0948
54°	1.0277	1.0437	1.0602	1.0764	1.0915	1.1048	1.1152	1.1219	1.1242
55°	1.0500	1.0672	1.0848	1.1022	1.1186	1.1331	1.1444	1.1517	1.1542
56°	1.0725	1.0908	1.1097	1.1285	1.1462	1.1619	1.1743	1.1823	1.1851
57°	1.0952	1.1147	1.1349	1.1551	1.1743	1.1914	1.2049	1.2136	1.2167
58°	1.1180	1.1389	1.1605	1.1822	1.2030	1.2215	1.2362	1.2458	1.2492
59°	1.1411	1.1632	1.1864	1.2097	1.2321	1.2522	1.2684	1.2789	1.2826
60°	1.1643	1.1879	1.2125	1.2376	1.2619	1.2837	1.3014	1.3129	1.3170
61°	1.1877	1.2128	1.2392	1.2660	1.2922	1.3159	1.3352	1.3480	1.3524
62°	1.2113	1.2379	1.2661	1.2949	1.3231	1.3490	1.3701	1.3841	1.3890
63°	1.2351	1.2633	1.2933	1.3242	1.3547	1.3828	1.4059	1.4214	1.4268
64°	1.2591	1.2890	1.3209	1.3541	1.3870	1.4175	1.4429	1.4599	1.4659
65°	1.2833	1.3149	1.3489	1.3844	1.4199	1.4532	1.4810	1.4998	1.5065
66°	1.3076	1.3411	1.3773	1.4153	1.4536	1.4898	1.5203	1.5411	1.5485
67°	1.3321	1.3675	1.4060	1.4467	1.4880	1.5274	1.5610	1.5840	1.5923
68°	1.3568	1.3942	1.4351	1.4786	1.5232	1.5661	1.6030	1.6287	1.6379
69°	1.3817	1.4212	1.4646	1.5111	1.5591	1.6059	1.6466	1.6752	1.6856
70°	1.4068	1.4484	1.4944	1.5441	1.5959	1.6468	1.6918	1.7237	1.7354
71°	1.4320	1.4759	1.5246	1.5777	1.6335	1.6891	1.7388	1.7745	1.7877
72°	1.4574	1.5036	1.5552	1.6118	1.6720	1.7326	1.7876	1.8277	1.8427
73°	1.4830	1.5315	1.5862	1.6465	1.7113	1.7774	1.8384	1.8837	1.9008
74°	1.5087	1.5597	1.6175	1.6818	1.7516	1.8237	1.8915	1.9427	1.9623
75°	1.5345	1.5882	1.6492	1.7176	1.7927	1.8715	1.9468	2.0050	2.0276
76°	1.5606	1.6168	1.6812	1.7540	1.8347	1.9207	2.0047	2.0711	2.0973
77°	1.5867	1.6457	1.7136	1.7909	1.8777	1.9716	2.0653	2.1414	2.1721
78°	1.6130	1.6748	1.7462	1.8284	1.9215	2.0240	2.1288	2.2164	2.2528
79°	1.6394	1.7040	1.7792	1.8664	1.9663	2.0781	2.1954	2.2969	2.3404
80°	1.6660	1.7335	1.8125	1.9048	2.0119	2.1339	2.2653	2.3836	2.4362
81°	1.6926	1.7631	1.8461	1.9438	2.0584	2.1913	2.3387	2.4775	2.5421
82°	1.7193	1.7929	1.8799	1.9831	2.1057	2.2504	2.4157	2.5795	2.6603
83°	1.7462	1.8228	1.9140	2.0229	2.1537	2.3110	2.4965	2.6911	2.7942
84°	1.7731	1.8528	1.9482	2.0630	2.2024	2.3731	2.5811	2.8136	2.9487
85°	1.8001	1.8830	1.9826	2.1035	2.2518	2.4366	2.6694	2.9487	3.1313
86°	1.8271	1.9132	2.0172	2.1442	2.3017	2.5013	2.7612	3.0978	3.3547
87°	1.8542	1.9435	2.0519	2.1852	2.3520	2.5670	2.8561	3.2620	3.6425
88°	1.8813	1.9739	2.0867	2.2263	2.4026	2.6336	2.9537	3.4412	4.0481
89°	1.9084	2.0043	2.1216	2.2675	2.4535	2.7007	3.0530	3.6328	4.7413
90°	1.9356	2.0347	2.1565	2.3088	2.5046	2.7681	3.1534	3.8317	———

ELLIPTIC INTEGRALS OF THE SECOND KIND: $E(k, \phi)$

$$E(k, \phi) = \int_0^\phi \sqrt{(1 - k^2 \sin^2 \Phi)}d\Phi, \quad \theta = \sin^{-1} k$$

θ / ϕ	5°	10°	15°	20°	25°	30°	35°	40°	45°
1°	0.0175	0.0175	0.0175	0.0175	0.0175	0.0175	0.0175	0.0175	0.0175
2°	0.0349	0.0349	0.0349	0.0349	0.0349	0.0349	0.0349	0.0349	0.0349
3°	0.0524	0.0524	0.0524	0.0524	0.0524	0.0524	0.0524	0.0523	0.0523
4°	0.0698	0.0698	0.0698	0.0698	0.0698	0.0698	0.0698	0.0698	0.0698
5°	0.0873	0.0873	0.0873	0.0873	0.0872	0.0872	0.0872	0.0872	0.0872
6°	0.1047	0.1047	0.1047	0.1047	0.1047	0.1047	0.1047	0.1046	0.1046
7°	0.1222	0.1222	0.1222	0.1221	0.1221	0.1221	0.1221	0.1220	0.1220
8°	0.1396	0.1396	0.1396	0.1396	0.1395	0.1395	0.1395	0.1394	0.1394
9°	0.1571	0.1571	0.1570	0.1570	0.1570	0.1569	0.1569	0.1568	0.1568
10°	0.1745	0.1745	0.1745	0.1744	0.1744	0.1743	0.1742	0.1742	0.1741
11°	0.1920	0.1920	0.1919	0.1918	0.1918	0.1917	0.1916	0.1915	0.1914
12°	0.2094	0.2094	0.2093	0.2093	0.2092	0.2091	0.2089	0.2088	0.2087
13°	0.2269	0.2268	0.2268	0.2267	0.2265	0.2264	0.2263	0.2261	0.2259
14°	0.2443	0.2443	0.2442	0.2441	0.2439	0.2437	0.2436	0.2433	0.2431
15°	0.2618	0.2617	0.2616	0.2615	0.2613	0.2611	0.2608	0.2606	0.2603
16°	0.2792	0.2791	0.2790	0.2788	0.2786	0.2784	0.2781	0.2778	0.2775
17°	0.2967	0.2966	0.2964	0.2962	0.2959	0.2956	0.2953	0.2949	0.2946
18°	0.3141	0.3140	0.3138	0.3136	0.3133	0.3129	0.3125	0.3121	0.3116
19°	0.3316	0.3314	0.3312	0.3309	0.3305	0.3301	0.3296	0.3291	0.3286
20°	0.3490	0.3489	0.3486	0.3483	0.3478	0.3473	0.3468	0.3462	0.3456
21°	0.3665	0.3663	0.3660	0.3656	0.3651	0.3645	0.3639	0.3632	0.3625
22°	0.3839	0.3837	0.3834	0.3829	0.3823	0.3817	0.3809	0.3802	0.3793
23°	0.4013	0.4011	0.4007	0.4002	0.3996	0.3988	0.3980	0.3971	0.3961
24°	0.4188	0.4185	0.4181	0.4175	0.4168	0.4159	0.4150	0.4139	0.4129
25°	0.4362	0.4359	0.4354	0.4348	0.4339	0.4330	0.4319	0.4308	0.4296
26°	0.4537	0.4533	0.4528	0.4520	0.4511	0.4500	0.4488	0.4475	0.4462
27°	0.4711	0.4707	0.4701	0.4693	0.4682	0.4670	0.4657	0.4643	0.4628
28°	0.4886	0.4881	0.4874	0.4865	0.4854	0.4840	0.4825	0.4809	0.4793
29°	0.5060	0.5055	0.5048	0.5037	0.5025	0.5010	0.4993	0.4975	0.4957
30°	0.5234	0.5229	0.5221	0.5209	0.5195	0.5179	0.5161	0.5141	0.5120
31°	0.5409	0.5403	0.5394	0.5381	0.5366	0.5348	0.5327	0.5306	0.5283
32°	0.5583	0.5577	0.5567	0.5553	0.5536	0.5516	0.5494	0.5470	0.5446
33°	0.5757	0.5751	0.5740	0.5725	0.5706	0.5684	0.5660	0.5634	0.5607
34°	0.5932	0.5924	0.5912	0.5896	0.5876	0.5852	0.5826	0.5797	0.5768
35°	0.6106	0.6098	0.6085	0.6067	0.6045	0.6019	0.5991	0.5960	0.5928
36°	0.6280	0.6272	0.6258	0.6238	0.6214	0.6186	0.6155	0.6122	0.6087
37°	0.6455	0.6445	0.6430	0.6409	0.6383	0.6353	0.6319	0.6283	0.6245
38°	0.6629	0.6619	0.6602	0.6580	0.6552	0.6519	0.6483	0.6444	0.6403
39°	0.6803	0.6792	0.6775	0.6750	0.6720	0.6685	0.6646	0.6604	0.6559
40°	0.6977	0.6966	0.6947	0.6921	0.6888	0.6851	0.6808	0.6763	0.6715
41°	0.7152	0.7139	0.7119	0.7091	0.7056	0.7016	0.6970	0.6921	0.6870
42°	0.7326	0.7313	0.7291	0.7261	0.7224	0.7180	0.7132	0.7079	0.7024
43°	0.7500	0.7486	0.7463	0.7431	0.7391	0.7345	0.7293	0.7237	0.7178
44°	0.7674	0.7659	0.7634	0.7600	0.7558	0.7508	0.7453	0.7393	0.7330
45°	0.7849	0.7832	0.7806	0.7770	0.7725	0.7672	0.7613	0.7549	0.7482

ELLIPTIC INTEGRALS OF THE SECOND KIND: $E(k, \phi)$ (Continued)

$$E(k, \phi) = \int_0^{\phi} \sqrt{(1 - k^2 \sin^2 \Phi)}\, d\Phi, \quad \theta = \sin^{-1} k$$

ϕ θ	50°	55°	60°	65°	70°	75°	80°	85°	90°
1°	0.0175	0.0175	0.0175	0.0175	0.0175	0.0175	0.0175	0.0175	0.0175
2°	0.0349	0.0349	0.0349	0.0349	0.0349	0.0349	0.0349	0.0349	0.0349
3°	0.0523	0.0523	0.0523	0.0523	0.0523	0.0523	0.0523	0.0523	0.0523
4°	0.0698	0.0698	0.0698	0.0698	0.0698	0.0698	0.0698	0.0698	0.0698
5°	0.0872	0.0872	0.0872	0.0872	0.0872	0.0872	0.0872	0.0872	0.0872
6°	0.1046	0.1046	0.1046	0.1046	0.1046	0.1045	0.1045	0.1045	0.1045
7°	0.1220	0.1220	0.1219	0.1219	0.1219	0.1219	0.1219	0.1219	0.1219
8°	0.1394	0.1393	0.1393	0.1393	0.1392	0.1392	0.1392	0.1392	0.1392
9°	0.1567	0.1566	0.1566	0.1566	0.1565	0.1565	0.1565	0.1564	0.1564
10°	0.1740	0.1739	0.1739	0.1738	0.1738	0.1737	0.1737	0.1737	0.1736
11°	0.1913	0.1912	0.1911	0.1910	0.1909	0.1909	0.1908	0.1908	0.1908
12°	0.2085	0.2084	0.2083	0.2082	0.2081	0.2080	0.2080	0.2079	0.2079
13°	0.2258	0.2256	0.2254	0.2253	0.2252	0.2251	0.2250	0.2250	0.2250
14°	0.2429	0.2427	0.2425	0.2424	0.2422	0.2421	0.2420	0.2419	0.2419
15°	0.2601	0.2598	0.2596	0.2594	0.2592	0.2590	0.2589	0.2588	0.2588
16°	0.2771	0.2768	0.2765	0.2763	0.2761	0.2759	0.2757	0.2757	0.2756
17°	0.2942	0.2938	0.2935	0.2932	0.2929	0.2927	0.2925	0.2924	0.2924
18°	0.3112	0.3107	0.3103	0.3099	0.3096	0.3094	0.3092	0.3091	0.3090
19°	0.3281	0.3276	0.3271	0.3267	0.3263	0.3260	0.3258	0.3256	0.3256
20°	0.3450	0.3444	0.3438	0.3433	0.3429	0.3425	0.3422	0.3421	0.3420
21°	0.3618	0.3611	0.3604	0.3598	0.3593	0.3589	0.3586	0.3584	0.3584
22°	0.3785	0.3777	0.3770	0.3763	0.3757	0.3752	0.3749	0.3747	0.3746
23°	0.3952	0.3943	0.3935	0.3927	0.3920	0.3915	0.3911	0.3908	0.3907
24°	0.4118	0.4108	0.4098	0.4090	0.4082	0.4076	0.4071	0.4068	0.4067
25°	0.4284	0.4272	0.4261	0.4251	0.4243	0.4236	0.4230	0.4227	0.4226
26°	0.4449	0.4436	0.4423	0.4412	0.4402	0.4394	0.4389	0.4385	0.4384
27°	0.4613	0.4598	0.4584	0.4572	0.4561	0.4552	0.4545	0.4541	0.4540
28°	0.4776	0.4760	0.4744	0.4730	0.4718	0.4708	0.4701	0.4696	0.4695
29°	0.4938	0.4920	0.4903	0.4887	0.4874	0.4863	0.4855	0.4850	0.4848
30°	0.5100	0.5080	0.5061	0.5044	0.5029	0.5016	0.5007	0.5002	0.5000
31°	0.5261	0.5239	0.5218	0.5199	0.5182	0.5169	0.5159	0.5152	0.5150
32°	0.5421	0.5396	0.5373	0.5352	0.5334	0.5319	0.5308	0.5301	0.5299
33°	0.5580	0.5553	0.5528	0.5505	0.5485	0.5468	0.5456	0.5449	0.5446
34°	0.5738	0.5709	0.5681	0.5656	0.5634	0.5616	0.5603	0.5595	0.5592
35°	0.5895	0.5863	0.5833	0.5806	0.5782	0.5762	0.5748	0.5739	0.5736
36°	0.6051	0.6017	0.5984	0.5954	0.5928	0.5907	0.5891	0.5881	0.5878
37°	0.6207	0.6169	0.6134	0.6101	0.6073	0.6050	0.6032	0.6022	0.6018
38°	0.6361	0.6321	0.6282	0.6247	0.6216	0.6191	0.6172	0.6160	0.6157
39°	0.6515	0.6471	0.6429	0.6391	0.6357	0.6330	0.6310	0.6297	0.6293
40°	0.6667	0.6620	0.6575	0.6533	0.6497	0.6468	0.6446	0.6432	0.6428
41°	0.6818	0.6767	0.6719	0.6674	0.6636	0.6604	0.6580	0.6566	0.6561
42°	0.6969	0.6914	0.6862	0.6814	0.6772	0.6738	0.6712	0.6697	0.6691
43°	0.7118	0.7059	0.7003	0.6952	0.6907	0.6870	0.6843	0.6826	0.6820
44°	0.7266	0.7204	0.7144	0.7088	0.7040	0.7000	0.6971	0.6953	0.6947
45°	0.7414	0.7346	0.7282	0.7223	0.7171	0.7129	0.7097	0.7078	0.7071

ELLIPTIC INTEGRALS OF THE SECOND KIND: $E(k, \phi)$ (Continued)

$$E(k, \phi) = \int_0^\phi \sqrt{(1 - k^2 \sin^2 \Phi)}d\Phi, \quad \theta = \sin^{-1} k$$

ϕ \ θ	5°	10°	15°	20°	25°	30°	35°	40°	45°
46°	0.8023	0.8006	0.7977	0.7939	0.7891	0.7835	0.7772	0.7704	0.7633
47°	0.8197	0.8179	0.8149	0.8108	0.8057	0.7998	0.7931	0.7858	0.7782
48°	0.8371	0.8352	0.8320	0.8277	0.8223	0.8160	0.8089	0.8012	0.7931
49°	0.8545	0.8525	0.8491	0.8446	0.8389	0.8322	0.8247	0.8165	0.8079
50°	0.8719	0.8698	0.8663	0.8614	0.8554	0.8483	0.8404	0.8317	0.8227
51°	0.8894	0.8871	0.8834	0.8783	0.8719	0.8644	0.8560	0.8469	0.8373
52°	0.9068	0.9044	0.9004	0.8951	0.8884	0.8805	0.8716	0.8620	0.8518
53°	0.9242	0.9217	0.9175	0.9119	0.9048	0.8965	0.8872	0.8770	0.8663
54°	0.9416	0.9389	0.9345	0.9287	0.9212	0.9125	0.9026	0.8919	0.8806
55°	0.9590	0.9562	0.9517	0.9454	0.9376	0.9284	0.9181	0.9068	0.8949
56°	0.9764	0.9735	0.9687	0.9622	0.9540	0.9443	0.9335	0.9216	0.9091
57°	0.9938	0.9908	0.9858	0.9789	0.9703	0.9602	0.9488	0.9363	0.9232
58°	1.0112	1.0080	1.0028	0.9956	0.9866	0.9760	0.9641	0.9510	0.9372
59°	1.0286	1.0253	1.0198	1.0123	1.0029	1.9918	0.9793	0.9656	0.9511
60°	1.0460	1.0426	1.0368	1.0290	1.0191	1.0076	0.9945	0.9801	0.9650
61°	1.0634	1.0598	1.0538	1.0456	1.0354	1.0233	1.0096	0.9946	0.9787
62°	1.0808	1.0771	1.0708	1.0623	1.0516	1.0389	1.0246	1 0090	0.9924
63°	1.0982	1.0943	1.0878	1.0789	1.0678	1.0546	1.0397	1.0233	1.0060
64°	1.1156	1.1115	1.1048	1.0955	1.0839	1.0702	1.0547	1.0376	1.0195
65°	1.1330	1.1288	1.1218	1.1121	1.1001	1.0858	1.0696	1.0518	1.0329
66°	1.1504	1.1460	1.1387	1.1287	1.1162	1.1013	1.0845	1.0660	1.0463
67°	1.1678	1.1632	1.1557	1.1453	1.1323	1.1168	1.0993	1.0801	1.0596
68°	1.1852	1.1805	1.1726	1.1618	1.1483	1.1323	1.1141	1.0941	1.0728
69°	1.2026	1.1977	1.1896	1.1784	1.1644	1.1478	1.1289	1.1081	1.0859
70°	1.2200	1.2149	1.2065	1.1949	1.1804	1.1632	1.1436	1.1221	1.0990
71°	1.2374	1.2321	1.2234	1.2114	1.1964	1.1786	1.1583	1.1359	1.1120
72°	1.2548	1.2493	1.2403	1.2280	1.2124	1.1939	1.1729	1.1498	1.1250
73°	1.2722	1.2666	1.2573	1.2445	1.2284	1.2093	1.1875	1.1636	1.1379
74°	1.2896	1.2838	1.2742	1.2609	1.2443	1.2246	1.2021	1.1773	1.1507
75°	1.3070	1.3010	1.2911	1.2774	1.2603	1.2399	1.2167	1.1910	1.1635
76°	1.3244	1.3182	1.3080	1.2939	1.2762	1.2552	1.2312	1.2047	1.1762
77°	1.3418	1.3354	1.3249	1.3104	1.2921	1.2704	1.2457	1.2183	1.1889
78°	1.3592	1.3526	1.3417	1.3268	1.3080	1.2856	1.2601	1.2319	1.2015
79°	1.3765	1.3698	1.3586	1.3432	1.3239	1.3009	1.2746	1.2454	1.2141
80°	1.3939	1.3870	1.3755	1.3597	1.3398	1.3161	1.2890	1.2590	1.2266
81°	1.4113	1.4042	1.3924	1.3761	1.3556	1.3312	1.3034	1.2725	1.2391
82°	1.4287	1.4214	1.4093	1.3925	1.3715	1.3464	1.3177	1.2859	1.2516
83°	1.4461	1.4386	1.4261	1.4090	1.3873	1.3616	1.3321	1.2994	1.2640
84°	1.4635	1.4558	1.4430	1.4254	1.4032	1.3767	1.3464	1.3128	1.2765
85°	1.4809	1.4729	1.4598	1.4418	1.4190	1.3919	1.3608	1.3262	1.2889
86°	1.4983	1.4901	1.4767	1.4582	1.4348	1.4070	1.3751	1.3396	1.3012
87°	1.5156	1.5073	1.4936	1.4746	1.4507	1.4221	1.3894	1.3530	1.3136
88°	1.5330	1.5245	1.5104	1.4910	1.4665	1.4372	1.4037	1.3664	1.3260
89°	1.5504	1.5417	1.5273	1.5074	1.4823	1.4523	1.4180	1.3798	1.3383
90°	1.5678	1.5589	1.5442	1.5238	1.4981	1.4675	1.4323	1.3931	1.3506

ELLIPTIC INTEGRALS OF THE SECOND KIND: $E(k, \phi)$ (Continued)

$$E(k, \phi) = \int_0^\phi \sqrt{(1 - k^2 \sin^2 \Phi)}\,d\Phi, \quad \theta = \sin^{-1} k$$

ϕ \ θ	50°	55°	60°	65°	70°	75°	80°	85°	90°
46°	0.7560	0.7488	0.7419	0.7356	0.7301	0.7255	0.7221	0.7200	0.7193
47°	0.7705	0.7628	0.7555	0.7488	0.7429	0.7380	0.7344	0.7321	0.7314
48°	0.7849	0.7768	0.7690	0.7618	0.7555	0.7502	0.7464	0.7440	0.7431
49°	0.7992	0.7905	0.7822	0.7746	0.7679	0.7623	0.7581	0.7556	0.7547
50°	0.8134	0.8042	0.7954	0.7872	0.7801	0.7741	0.7697	0.7670	0.7660
51°	0.8275	0.8177	0.8084	0.7997	0.7921	0.7858	0.7811	0.7781	0.7771
52°	0.8414	0.8311	0.8212	0.8120	0.8039	0.7972	0.7922	0.7891	0.7880
53°	0.8553	0.8444	0.8339	0.8241	0.8155	0.8084	0.8031	0.7998	0.7986
54°	0.8690	0.8575	0.8464	0.8361	0.8270	0.8194	0.8137	0.8102	0.8090
55°	0.8827	0.8705	0.8588	0.8479	0.8382	0.8302	0.8242	0.8204	0.8192
56°	0.8962	0.8834	0.8710	0.8595	0.8493	0.8408	0.8344	0.8304	0.8290
57°	0.9096	0.8961	0.8831	0.8709	0.8601	0.8511	0.8443	0.8401	0.8387
58°	0.9230	0.9088	0.8950	0.8822	0.8707	0.8612	0.8540	0.8496	0.8480
59°	0.9362	0.9213	0.9068	0.8932	0.8812	0.8711	0.8635	0.8588	0.8572
60°	0.9493	0.9336	0.9184	0.9042	0.8914	0.8808	0.8728	0.8677	0.8660
61°	0.9623	0.9459	0.9299	0.9149	0.9015	0.8903	0.8817	0.8764	0.8746
62°	0.9752	0.9580	0.9412	0.9254	0.9113	0.8995	0.8905	0.8849	0.8829
63°	0.9880	0.9700	0.9524	0.9358	0.9210	0.9085	0.8990	0.8930	0.8910
64°	1.0007	0.9818	0.9634	0.9460	0.9304	0.9173	0.9072	0.9009	0.8988
65°	1.0133	0.9936	0.9743	0.9561	0.9397	0.9258	0.9152	0.9086	0.9063
66°	1.0258	1.0052	0.9850	0.9659	0.9487	0.9341	0.9230	0.9159	0.9135
67°	1.0383	1.0167	0.9956	0.9756	0.9576	0.9422	0.9305	0.9230	0.9205
68°	1.0506	1.0281	1.0061	0.9852	0.9662	0.9501	0.9377	0.9299	0.9272
69°	1.0628	1.0394	1.0164	0.9946	0.9747	0.9578	0.9447	0.9364	0.9336
70°	1.0750	1.0506	1.0266	1.0038	0.9830	0.9652	0.9514	0.9427	0.9397
71°	1.0871	1.0617	1.0367	1.0129	0.9911	0.9724	0.9579	0.9487	0.9455
72°	1.0991	1.0727	1.0467	1.0218	0.9990	0.9794	0.9642	0.9544	0.9511
73°	1.1110	1.0836	1.0565	1.0306	1.0067	0.9862	0.9702	0.9599	0.9563
74°	1.1228	1.0944	1.0662	1.0392	1.0143	0.9928	0.9759	0.9650	0.9613
75°	1.1346	1.1051	1.0759	1.0477	1.0217	0.9992	0.9814	0.9699	0.9659
76°	1.1463	1.1158	1.0854	1.0561	1.0290	1.0053	0.9867	0.9745	0.9703
77°	1.1580	1.1263	1.0948	1.0643	1.0361	1.0113	0.9917	0.9789	0.9744
78°	1.1695	1.1368	1.1041	1.0724	1.0430	1.0171	0.9965	0.9829	0.9781
79°	1.1811	1.1472	1.1133	1.0805	1.0498	1.0228	1.0011	0.9867	0.9816
80°	1.1926	1.1576	1.1225	1.0884	1.0565	1.0282	1.0054	0.9902	0.9848
81°	1.2040	1.1678	1.1316	1.0962	1.0630	1.0335	1.0096	0.9935	0.9877
82°	1.2154	1.1781	1.1406	1.1040	1.0695	1.0387	1.0135	0.9965	0.9903
83°	1.2267	1.1883	1.1495	1.1116	1.0758	1.0437	1.0173	0.9992	0.9925
84°	1.2381	1.1984	1.1584	1.1192	1.0821	1.0486	1.0209	1.0017	0.9945
85°	1.2493	1.2085	1.1673	1.1267	1.0882	1.0534	1.0244	1.0039	0.9962
86°	1.2606	1.2186	1.1761	1.1342	1.0944	1.0581	1.0277	1.0060	0.9976
87°	1.2719	1.2286	1.1848	1.1417	1.1004	1.0628	1.0309	1.0078	0.9986
88°	1.2831	1.2386	1.1936	1.1491	1.1064	1.0673	1.0340	1.0095	0.9994
89°	1.2943	1.2487	1.2023	1.1565	1.1124	1.0719	1.0371	1.0111	0.9998
90°	1.3055	1.2587	1.2111	1.1638	1.1184	1.0764	1.0401	1.0127	1.0000

COMPLETE ELLIPTIC INTEGRALS

$$K = \int_0^{\pi/2} \frac{d\Phi}{\sqrt{1 - k^2 \sin^2 \Phi}} = F\left(k, \frac{\pi}{2}\right)$$

$\sin^{-1} k$	K	$\log K$	$\sin^{-1} k$	K	$\log K$
0°	1.5708	0.196120	40°	1.7868	0.252068
1	1.5709	0.196153	41	1.7992	0.255085
2	1.5713	0.196252	42	1.8122	0.258197
3	1.5719	0.196418	43	1.8256	0.261406
4	1.5727	0.196649	44	1.8396	0.264716
5	1.5738	0.196947	45	1.8541	0.268127
6	1.5751	0.197312	46	1.8691	0.271644
7	1.5767	0.197743	47	1.8848	0.275267
8	1.5785	0.198241	48	1.9011	0.279001
9	1.5805	0.198806	49	1.9180	0.282848
10	1.5828	0.199438	50	1.9356	0.286811
11	1.5854	0.200137	51	1.9539	0.290895
12	1.5882	0.200904	52	1.9729	0.295101
13	1.5913	0.201740	53	1.9927	0.299435
14	1.5946	0.202643	54	2.0133	0.303901
15	1.5981	0.203615	55	2.0347	0.308504
16	1.6020	0.204657	56	2.0571	0.313247
17	1.6061	0.205768	57	2.0804	0.318138
18	1.6105	0.206948	58	2.1047	0.323182
19	1.6151	0.208200	59	2.1300	0.328384
20	1.6200	0.209522	60	2.1565	0.333753
21	1.6252	0.210916	61	2.1842	0.339295
22	1.6307	0.212382	62	2.2132	0.345020
23	1.6365	0.213921	63	2.2435	0.350936
24	1.6426	0.215533	64	2.2754	0.357053
25	1.6490	0.217219	65	2.3088	0.363384
26	1.6557	0.218981	66	2.3439	0.369940
27	1.6627	0.220818	67	2.3809	0.376736
28	1.6701	0.222732	68	2.4198	0.383787
29	1.6777	0.224723	69	2.4610	0.391112
30	1.6858	0.226793	70	2.5046	0.398730
31	1.6941	0.228943	71	2.5507	0.406665
32	1.7028	0.231173	72	2.5998	0.414943
33	1.7119	0.233485	73	2.6521	0.423596
34	1.7214	0.235880	74	2.7081	0.432660
35	1.7312	0.238359	75	2.7681	0.442176
36	1.7415	0.240923	76	2.8327	0.452196
37	1.7522	0.243575	77	2.9026	0.462782
38	1.7633	0.246315	78	2.9786	0.474008
39	1.7748	0.249146	79	3.0617	0.485967
40	1.7868	0.252068	80	3.1534	0.498777

COMPLETE ELLIPTIC INTEGRALS (Continued)

$$K = \int_0^{\pi/2} \frac{d\Phi}{\sqrt{1 - k^2 \sin^2 \Phi}} = F\left(k, \frac{\pi}{2}\right)$$

$\sin^{-1} k$	K	$\log K$	$\sin^{-1} k$	K	$\log K$
80°	3.1534	0.498777	**85°**	3.8317	0.583396
81	3.2553	0.512591	86	4.0528	0.607751
82	3.3699	0.527613	87	4.3387	0.637355
83	3.5004	0.544120	88	4.7427	0.676027
84	3.6519	0.562514	89	5.4349	0.735192
85	3.8317	0.583396	**90**	∞	∞

Values of K for $\sin^{-1} k = 85°$ to $89°$ by $0.1°$ and $89°$ to $90°$ by minutes

$\sin^{-1} k$	K	$\log K$	$\sin^{-1} k$		K	$\log K$
85.0°	3.832	0.58343	**89°**	**0'**	5.435	0.73520
85.1	3.852	0.58569	89	2	5.469	0.73791
85.2	3.872	0.58794	89	4	5.504	0.74068
85.3	3.893	0.59028	89	6	5.540	0.74351
85.4	3.914	0.59262	89	8	5.578	0.74648
85.5	3.936	0.59506	**89**	**10**	5.617	0.74950
85.6	3.958	0.59748	89	12	5.658	0.75266
85.7	3.981	0.59999	89	14	5.700	0.75587
85.8	4.004	0.60249	89	16	5.745	0.75929
85.9	4.028	0.60509	89	18	5.791	0.76275
86.0	4.053	0.60778	**89**	**20**	5.840	0.76641
86.1	4.078	0.61045	89	22	5.891	0.77019
86.2	4.104	0.61321	89	24	5.946	0.77422
86.3	4.130	0.61595	89	26	6.003	0.77837
86.4	4.157	0.61878	89	28	6.063	0.78269
86.5	4.185	0.62170	**89**	**30**	6.128	0.78732
86.6	4.214	0.62469	89	32	6.197	0.79218
86.7	4.244	0.62778	89	34	6.271	0.79734
86.8	4.274	0.63083	89	36	6.351	0.80284
86.9	4.306	0.63407	89	38	6.438	0.80875
87.0	4.339	0.63739	**89**	**40**	6.533	0.81511
87.1	4.372	0.64068	89	41	6.584	0.81849
87.2	4.407	0.64414	89	42	6.639	0.82210
87.3	4.444	0.64777	89	43	6.696	0.82582
87.4	4.481	0.65137	89	44	6.756	0.82969
87.5	4.520	0.65514	**89**	**45**	6.821	0.83385
87.6	4.561	0.65916	89	46	6.890	0.83822
87.7	4.603	0.66304	89	47	6.964	0.84286
87.8	4.648	0.66727	89	48	7.044	0.84782
87.9	4.694	0.67154	89	49	7.131	0.85315
88.0	4.743	0.67605	**89**	**50**	7.226	0.85890
88.1	4.794	0.68070	89	51	7.332	0.86522
88.2	4.848	0.68556	89	52	7.449	0.87210
88.3	4.905	0.69064	89	53	7.583	0.87984
88.4	4.965	0.69592	89	54	7.737	0.88857
88.5	5.030	0.70157	**89**	**55**	7.919	0.89867
88.6	5.099	0.70749	89	56	8.143	0.91078
88.7	5.173	0.71374	89	57	8.430	0.92583
88.8	5.253	0.72041	89	58	8.836	0.94626
88.9	5.340	0.72754	89	59	9.529	0.97905
89.0	5.435	0.73520	**90**	**0**	∞	∞

COMPLETE ELLIPTIC INTEGRALS (Continued)

$$E = \int_0^{\pi/2} \sqrt{1 - k^2 \sin^2 \Phi} \cdot d\Phi = E\left(k, \frac{\pi}{2}\right)$$

$\sin^{-1} k$	E	$\log E$	$\sin^{-1} k$	E	$\log E$
0°	1.5708	0.196120	**45°**	1.3506	0.130541
1	1.5707	0.196087	46	1.3418	0.127690
2	1.5703	0.195988	47	1.3329	0.124788
3	1.5697	0.195822	48	1.3238	0.121836
4	1.5689	0.195591	49	1.3147	0.118836
5	1.5678	0.195293	**50**	1.3055	0.115790
6	1.5665	0.194930	51	1.2963	0.112698
7	1.5649	0.194500	52	1.2870	0.109563
8	1.5632	0.194004	53	1.2776	0.106386
9	1.5611	0.193442	54	1.2681	0.103169
10	1.5589	0.192815	**55**	1.2587	0.099915
11	1.5564	0.192121	56	1.2492	0.096626
12	1.5537	0.191362	57	1.2397	0.093303
13	1.5507	0.190537	58	1.2301	0.089950
14	1.5476	0.189646	59	1.2206	0.086569
15	1.5442	0.188690	**60**	1.2111	0.083164
16	1.5405	0.187668	61	1.2015	0.079738
17	1.5367	0.186581	62	1.1920	0.076293
18	1.5326	0.185428	63	1.1826	0.072834
19	1.5283	0.184210	64	1.1732	0.069364
20	1.5238	0.182928	**65**	1.1638	0.065889
21	1.5191	0.181580	66	1.1545	0.062412
22	1.5141	0.180168	67	1.1453	0.058937
23	1.5090	0.178691	68	1.1362	0.055472
24	1.5037	0.177150	69	1.1272	0.052020
25	1.4981	0.175545	**70**	1.1184	0.048589
26	1.4924	0.173876	71	1.1096	0.045183
27	1.4864	0.172144	72	1.1011	0.041812
28	1.4803	0.170348	73	1.0927	0.038481
29	1.4740	0.168489	74	1.0844	0.035200
30	1.4675	0.166567	**75**	1.0764	0.031976
31	1.4608	0.164583	76	1.0686	0.028819
32	1.4539	0.162537	77	1.0611	0.025740
33	1.4469	0.160429	78	1.0538	0.022749
34	1.4397	0.158261	79	1.0468	0.019858
35	1.4323	0.156031	**80**	1.0401	0.017081
36	1.4248	0.153742	81	1.0338	0.014432
37	1.4171	0.151393	82	1.0278	0.011927
38	1.4092	0.148985	83	1.0223	0.009584
39	1.4013	0.146519	84	1.0172	0.007422
40	1.3931	0.143995	**85**	1.0127	0.005465
41	1.3849	0.141414	86	1.0086	0.003740
42	1.3765	0.138778	87	1.0053	0.002278
43	1.3680	0.136086	88	1.0026	0.001121
44	1.3594	0.133340	89	1.0008	0.000326
45	1.3506	0.130541	**90**	1.0000	0.000000

SINE, COSINE, AND EXPONENTIAL INTEGRALS

$$Si(x) = \int_0^x \frac{\sin v}{v}\,dv; \qquad Ci(x) = \int_\infty^x \frac{\cos v}{v}\,dv;$$

$$Ei(x) = \int_{-\infty}^x \frac{e^v}{v}\,dv; \qquad E_1(x) = -Ei(-x) = \int_x^\infty \frac{e^{-v}}{v}\,dv$$

Interpolation can be by iterative linear interpolation or by means of Everett's Formula using throwback to second differences. With this in view the modified second differences have been tabulated where relevant.

Everett's Formula.

$$f_p = (1 - p)f_0 + pf_1 + E_2\delta_{m0}^2 + F_2\delta_{m1}^2$$

SINE, COSINE, AND EXPONENTIAL INTEGRALS

x	$Si(x)$	δ_m^2	$Ci(x)$	δ_m^2	$Ei(x)$	δ_m^2	$E_1(x) = -Ei(-x)$	δ_m^2
.00	.00000		$-\infty$		$-\infty$		$+\infty$	
.01	.01000	0	-4.02798		-4.01790		4.03790	
.02	.02000	0	-3.33482		-3.31476		3.35476	
.03	.03000	0	-2.92957		-2.89912		2.95912	
.04	.04000	-2	-2.64206		-2.60126		2.68126	
.05	.04999	2	-2.41914		-2.36788		2.46790	
.06	.05999	-2	-2.23709		-2.17528		2.29531	
.07	.06998	0	-2.08327		-2.01080		2.15084	
.08	.07997	0	-1.95011		-1.86688		2.02694	
.09	.08996	-2	-1.83275		-1.73866		1.91874	
.10	.09994	2	-1.72787	-997	-1.62281	-989	1.82292	990
.11	.10993	-3	-1.63308	-829	-1.51696	-816	1.73711	816
.12	.11990	2	-1.54665	-695	-1.41935	-687	1.65954	688
.13	.12988	-1	-1.46723	-594	-1.32866	-582	1.58890	585
.14	.13985	-1	-1.39379	-514	-1.24384	-505	1.52415	503
.15	.14981	0	-1.32552	-449	-1.16409	-436	1.46446	441
.16	.15977	0	-1.26176	-394	-1.08873	-384	1.40919	384
.17	.16973	-1	-1.20196	-349	-1.01723	-341	1.35778	342
.18	.17968	-1	-1.14567	-314	$-.94915$	-302	1.30980	303
.19	.18962	0	-1.09253	-281	$-.88410$	-270	1.26486	272
.20	.19956	-1	-1.04221	-254	$-.82176$	-244	1.22265	245
.21	.20949	-1	$-.99444$	-231	$-.76187$	-222	1.18290	223
.22	.21941	1	$-.94899$	-212	$-.70420$	-199	1.14538	201
.23	.22933	-3	$-.90566$	-194	$-.64853$	-184	1.10988	186
.24	.23923	0	$-.86427$	-177	$-.59470$	-166	1.07624	167
.25	.24913	0	$-.82466$	-166	$-.54254$	-155	1.04428	158
.26	.25903	-3	$-.78671$	-153	$-.49193$	-142	1.01389	142
.27	.26891	-1	$-.75029$	-142	$-.44274$	-131	.98493	134
.28	.27878	1	$-.71529$	-132	$-.39486$	-122	.95731	122
.29	.28865	-3	$-.68161$	-124	$-.34820$	-113	.93092	115
.30	.29850	1	$-.64917$	-117	$-.30267$	-105	.90568	107
.31	.30835	-1	$-.61790$	-107	$-.25819$	-96	.88151	100
.32	.31819	-2	$-.58771$	-103	$-.21468$	-94	.85834	92
.33	.32801	-1	$-.55855$	-97	$-.17210$	-83	.83610	90
.34	.33782	1	$-.53036$	-91	$-.13036$	-81	.81475	82
.35	.34763	-3	$-.50308$	-86	$-.08943$	-76	.79422	77
.36	.35742	-1	$-.47666$	-84	$-.04926$	-69	.77446	74
.37	.36720	-3	$-.45107$	-76	$-.00979$	-67	.75544	69
.38	.37696	1	$-.42625$	-76	.02901	-63	.73711	66
.39	.38672	-3	$-.40218$	-70	.06718	-57	.71944	60
.40	.39646	-1	$-.37881$	-67	.10477	-58	.70238	59
.41	.40619	-1	$-.35611$	-65	.14179	-53	.68591	56
.42	.41591	-2	$-.33406$	-60	.17828	-50	.67000	51
.43	.42561	-2	$-.31262$	-61	.21427	-47	.65461	51
.44	.43529	1	$-.29178$	-54	.24979	-45	.63973	48
.45	.44497	-2	$-.27149$	-56	.28486	-43	.62533	46
.46	.45463	-2	$-.25175$	-52	.31950	-40	.61139	43
.47	.46427	-1	$-.23253$	-48	.35374	-39	.59788	41
.48	.47390	-2	$-.21380$	-50	.38759	-36	.58478	42
.49	.48351	-1	.19556	-46	.42108	-35	.57209	36
.50	.49311	-2	$-.17778$	-45	.45422	-33	.55977	37

SINE, COSINE, AND EXPONENTIAL INTEGRALS (Continued)

x	$Si(x)$	δ_m^2	$Ci(x)$	δ_m^2	$Ei(x)$	δ_m^2	$E_1(x) = -Ei(-x)$	δ_m^2
.50	.49311	− 2	− .17778	− 45	.45422	− 33	.55977	37
.51	.50269	− 2	− .16045	− 43	.48703	− 31	.54782	35
.52	.51225	− 1	− .14355	− 42	.51953	− 30	.53622	33
.53	.52180	− 2	− .12707	− 40	.55173	− 28	.52495	32
.54	.53133	− 2	− .11099	− 39	.58365	− 29	.51400	31
.55	.54084	− 2	− .09530	− 38	.61529	− 24	.50336	30
.56	.55033	− 1	− .07999	− 36	.64668	− 25	.49302	28
.57	.55981	− 2	− .06504	− 35	.67782	− 23	.48296	27
.58	.56927	− 2	− .05044	− 35	.70873	− 23	.47317	27
.59	.57871	− 2	− .03619	− 33	.73941	− 21	.46365	25
.60	.58813	− 2	− .02227	− 34	.76988	− 20	.45438	24
.61	.59753	− 2	− .00868	− 29	.80015	− 19	.44535	24
.62	.60691	− 2	.00461	− 33	.83023	− 19	.43656	23
.63	.61627	− 2	.01758	− 28	.86012	− 17	.42800	20
.64	.62561	0	.03026	− 29	.88984	− 17	.41965	23
.65	.63494	− 3	.04265	− 28	.91939	− 16	.41152	20
.66	.64424	− 3	.05476	− 28	.94878	− 15	.40359	19
.67	.65351	0	.06659	− 25	.97802	− 14	.39585	21
.68	.66277	− 2	.07816	− 28	1.00712	− 14	.38831	18
.69	.67201	− 3	.08946	− 25	1.03608	− 13	.38095	18
.70	.68122	− 2	.10051	− 24	1.06491	− 13	.37377	17
.71	.69041	− 2	.11132	− 25	1.09361	− 11	.36676	17
.72	.69958	− 2	.12188	− 24	1.12220	− 11	.35992	16
.73	.70873	− 3	.13220	− 21	1.15068	− 10	.35324	15
.74	.71785	− 2	.14230	− 25	1.17906	− 12	.34671	16
.75	.72695	− 2	.15216	− 20	1.20733	− 9	.34034	15
.76	.73603	− 3	.16181	− 22	1.23551	− 9	.33412	12
.77	.74508	− 2	.17124	− 21	1.26360	− 8	.32803	16
.78	.75411	− 2	.18046	− 21	1.29161	− 8	.32209	13
.79	.76312	− 3	.18947	− 20	1.31954	− 7	.31628	13
.80	.77210	− 3	.19828	− 20	1.34740	− 9	.31060	12
.81	.78105	− 2	.20689	− 20	1.37518	− 6	.30504	13
.82	.78998	− 3	.21530	− 17	1.40290	− 6	.29961	12
.83	.79888	− 2	.22353	− 19	1.43056	− 6	.29430	11
.84	.80776	− 3	.23157	− 19	1.45816	− 5	.28910	12
.85	.81661	− 2	.23942	− 16	1.48571	− 3	.28402	11
.86	.82544	− 3	.24710	− 18	1.51322	− 7	.27905	10
.87	.83424	− 3	.25460	− 18	1.54067	− 2	.27418	10
.88	.84301	− 3	.26192	− 15	1.56809	− 4	.26941	12
.89	.85175	− 2	.26908	− 17	1.59547	− 4	.26475	8
.90	.86047	− 3	.27607	− 17	1.62281	− 1	.26018	10
.91	.86916	− 3	.28289	14	1.65013	− 5	.25571	10
.92	.87782	− 1	.28956	− 18	1.67741	0	.25134	7
.93	.88646	− 5	.29606	− 13	1.70468	− 4	.24705	9
.94	.89506	− 1	.30242	− 18	1.73192	0	.24285	9
.95	.90364	− 3	.30861	− 13	1.75915	− 2	.23874	8
.96	.91219	− 5	.31466	− 15	1.78636	− 1	.23471	8
.97	.92070	− 1	.32056	− 14	1.81356	− 1	.23076	8
.98	.92919	− 3	.32632	− 15	1.84075	− 1	.22689	8
.99	.93765	− 3	.33193	− 14	1.86793	2	.22310	7
1.00	.94608	− 3	.33740	− 13	1.89512	− 2	.21938	8

SINE, COSINE, AND EXPONENTIAL INTEGRALS (Continued)

x	$Si(x)$	δ_m^2	$Ci(x)$	δ_m^2	$Ei(x)$	$E_1(x) = -Ei(-x)$
1.0	.94608	−301	.33740	−1376	1.89512	.21938
1.1	1.02869	−324	.38487	−1181	2.16738	.18599
1.2	1.10805	−345	.42046	−1025	2.44209	.15841
1.3	1.18396	−365	.44574	−897	2.72140	.13545
1.4	1.25623	−382	.46201	−789	3.00721	.11622
1.5	1.32468	−396	.47036	−695	3.30129	.10002
1.6	1.38918	−409	.47173	−612	3.60532	.08631
1.7	1.44959	−419	.46697	−538	3.92096	.07465
1.8	1.50582	−427	.45681	−470	4.24987	.06471
1.9	1.55778	−433	.44194	−408	4.59371	.05620
2.0	1.60541	−436	.42298	−350	4.95423	.04890
2.1	1.64870	−436	.40051	−296	5.33324	.04261
2.2	1.68762	−435	.37507	−246	5.73261	.03719
2.3	1.72221	−431	.34718	−198	6.15438	.03250
2.4	1.75249	−425	.31729	−153	6.60067	.02844
2.5	1.77852	−416	.28587	−111	7.07377	.02491
2.6	1.80039	−406	.25334	−71	7.57611	.02185
2.7	1.81821	−394	.22008	−34	8.11035	.01918
2.8	1.83210	−379	.18649	1	8.67930	.01686
2.9	1.84220	−363	.15290	33	9.28602	.01482
3.0	1.84865	−346	.11963	63	9.93383	.01305
3.1	1.85166	−327	.08699	91	10.6263	.01149
3.2	1.85140	−306	.05526	116	11.3673	.01013
3.3	1.84808	−285	.02468	139	12.1610	.00894
3.4	1.84191	−262	−.00452	159	13.0121	.00789
3.5	1.83313	−239	−.03213	177	13.9254	.00697
3.6	1.82195	−215	−.05797	192	14.9063	.00616
3.7	1.80862	−191	−.08190	205	15.9606	.00545
3.8	1.79339	−166	−.10378	216	17.0948	.00482
3.9	1.77650	−141	−.12350	224	18.3157	.00427
4.0	1.75820	−116	−.14098	230	19.6309	.00378
4.1	1.73874	−91	−.15617	234	21.0485	.00335
4.2	1.71837	−67	−.16901	236	22.5774	.00297
4.3	1.69732	−44	−.17951	235	24.2274	.00263
4.4	1.67583	−21	−.18766	232	26.0090	.00234
4.5	1.65414	2	−.19349	228	27.9337	.00207
4.6	1.63246	23	−.19705	221	30.0141	.00184
4.7	1.61101	43	−.19839	213	32.2639	.00164
4.8	1.58998	62	−.19760	204	34.6979	.00145
4.9	1.56956	79	−.19478	193	37.3325	.00129
5.0	1.54993	95	−.19003	181	40.1853	.00115
5.1	1.53125	110	−.18348	167	43.2757	.00102
5.2	1.51367	123	−.17525	153	46.6249	.00091
5.3	1.49732	134	−.16551	137	50.2557	.00081
5.4	1.48230	144	−.15439	121	54.1935	.00072
5.5	1.46872	152	−.14205	105	58.4655	.00064
5.6	1.45667	159	−.12867	88	63.1018	.00057
5.7	1.44620	164	−.11441	71	68.1350	.00051
5.8	1.43736	167	−.09944	54	73.6008	.00045
5.9	1.43018	168	−.08393	37	79.5382	.00040

SINE, COSINE, AND EXPONENTIAL INTEGRALS (Continued)

x	$Si(x)$	δ_m^2	$Ci(x)$	δ_m^2	$Ei(x)$	$E_1(x) = -Ei(-x)$
6.0	1.42469	168	−.06806	20	85.9898	.00036
6.1	1.42087	166	−.05198	3	93.0020	.00032
6.2	1.41871	163	−.03587	−13	100.626	.00029
6.3	1.41817	158	−.01989	−28	108.916	.00026
6.4	1.41922	152	−.00418	−43	117.935	.00023
6.5	1.42179	145	.01110	−56	127.747	.00020
6.6	1.42582	137	.02582	−69	138.426	.00018
6.7	1.43121	128	.03986	−81	150.050	.00016
6.8	1.43787	117	.05308	−92	162.707	.00014
6.9	1.44570	106	.06539	−101	176.491	.00013
7.0	1.45460	94	.07670	−109	191.505	.00012
7.1	1.46443	82	.08691	−116	207.863	.00010
7.2	1.47509	69	.09596	−122	225.688	.00009
7.3	1.48644	56	.10379	−127	245.116	.00008
7.4	1.49834	43	.11036	−130	266.296	.00007
7.5	1.51068	30	.11563	−131	289.388	.00007
7.6	1.52331	16	.11960	−132	314.572	.00006
7.7	1.53611	3	.12225	−131	342.040	.00005
7.8	1.54894	−10	.12359	−129	372.006	.00005
7.9	1.56167	−22	.12364	−126	404.701	.00004
8.0	1.57419	−34	.12	−122	440.380	.00004
8.1	1.58637	−45	.12002	−116	479.322	.00003
8.2	1.59810	−55	.11644	−110	521.831	.00003
8.3	1.60928	−65	.11177	−103	568.242	.00003
8.4	1.61981	−74	.10607	−94	618.919	.00002
8.5	1.62960	−82	.09943	−86	674.264	.00002
8.6	1.63857	−89	.09194	−76	734.714	.00002
8.7	1.64665	−95	.08368	−66	800.749	.00002
8.8	1.65379	−100	.07476	−56	872.895	.00002
8.9	1.65993	−104	.06528	−45	951.728	.00001
9.0	1.66504	−106	.05535	−35	1037.88	.00001
9.1	1.66908	−108	.04507	−24	1132.04	.00001
9.2	1.67205	−109	.03455	−13	1234.96	.00001
9.3	1.67393	−108	.02391	−2	1347.48	.00001
9.4	1.67473	−107	.01325	9	1470.51	.00001
9.5	1.67446	−104	.00268	19	1605.03	.00001
9.6	1.67316	−101	−.00771	29	1752.14	.00001
9.7	1.67084	−96	−.01780	38	1913.05	.00001
9.8	1.66757	−91	−.02752	47	2089.05	.00001
9.9	1.66338	−85	−.03676	55	2281.58	.00000
10.0	1.65835	−78	−.04546	63	2492.23	.00000

SINE, COSINE, AND EXPONENTIAL INTEGRALS (Continued)

x	$Si(x)$	δ_m^2	$Ci(x)$	δ_m^2
10.0	1.65835	−2001	−.04546	1616
10.5	1.62294	−944	−.07828	2255
11.0	1.57831	229	−.08956	2323
11.5	1.53571	1252	−.07857	1849
12.0	1.50497	1898	−.04978	987
12.5	1.49234	2051	−.01141	−35
13.0	1.49936	1720	.02676	−968
13.5	1.52291	1007	.05576	−1610
14.0	1.55621	116	.06940	−1828
14.5	1.59072	−745	.06554	−1604
15.0	1.61819	−1371	.04628	−1019
15.5	1.63258	−1638	.01719	−231
16.0	1.63130	−1500	−.01420	559
16.5	1.61563	−1020	−.04031	1173
17.0	1.59014	−326	−.05524	1471
17.5	1.56146	407	−.05610	1406
18.0	1.53661	999	−.04348	1014
18.5	1.52128	1327	−.02111	400
19.0	1.51863	1320	.00515	−276
19.5	1.52863	1000	.02883	−849
20.0	1.54824	462	.04442	−1195
20.5	1.57232	−163	.04859	−1240
21.0	1.59489	−717	.04089	−985
21.5	1.61063	−1077	.02373	−511
22.0	1.61608	−1161	.00164	67
22.5	1.61041	−967	−.01986	599
23.0	1.59546	−550	−.03566	969
23.5	1.57521	−19	−.04221	1090
24.0	1.55474	493	−.03833	946
24.5	1.53896	870	−.02538	579
25.0	1.53148	1020	−.00685	96
25.5	1.53376	921	.01262	−399
26.0	1.54487	605	.02830	−776
26.5	1.56180	161	.03657	−954
27.0	1.58029	−314	.03572	−894
27.5	1.59581	−700	.02630	−626
28.0	1.60475	−781	.01087	−214
28.5	1.60581	−1062	−.00664	232
29.0	1.59731	−530	−.02195	610
29.5	1.58331	−275	−.03144	826
30.0	1.56676	160	−.03303	837
30.5	1.55172	534	−.02660	649
31.0	1.54177	765	−.01395	307
31.5	1.53914	808	.00166	−94
32.0	1.54424	653	.01639	−464
32.5	1.55560	345	.02670	−705
33.0	1.57028	−33	.03026	−777
33.5	1.58466	−397	.02639	−653
34.0	1.59526	−651	.01626	−377
34.5	1.59964	−742	.00250	−22

SINE, COSINE, AND EXPONENTIAL INTEGRALS (Continued)

x	$Si(x)$	δ_m^2	$Ci(x)$	δ_m^2
35.0	1.59692	-650	$-.01148$	335
35.5	1.58797	-408	$-.02228$	592
36.0	1.57511	-70	$-.02741$	709
36.5	1.56156	273	$-.02576$	646
37.0	1.55061	541	$-.01792$	429
37.5	1.54483	673	$-.00596$	117
38.0	1.54549	638	.00713	-216
38.5	1.55226	449	.01816	-490
39.0	1.56334	157	.02451	-638
39.5	1.57594	-164	.02476	-625
40.0	1.58698	-437	.01902	-467
40.5	1.59384	-603	.00881	-195
41.0	1.59494	-611	$-.00328$	114
41.5	1.59018	-479	$-.01429$	391
42.0	1.58083	-230	$-.02157$	563
42.5	1.56927	68	$-.02346$	599
43.0	1.55835	343	$-.01962$	486
43.5	1.55070	527	$-.01112$	262
44.0	1.54809	581	$-.00011$	-26
44.5	1.55104	494	.01067	-295
45.0	1.55872	286	.01863	-491
45.5	1.56915	18	.02190	-562
46.0	1.57976	-254	.01979	-496
46.5	1.58796	-451	.01293	-313
47.0	1.59184	-544	.00307	-52
47.5	1.59052	-496	$-.00731$	207
48.0	1.58445	-331	$-.01571$	416
48.5	1.57520	-93	$-.02013$	521
49.0	1.56506	168	$-.01957$	494
49.5	1.55652	381	$-.01428$	351
50.0	1.55162	495	$-.00563$	125
50.5	1.55146	491	.00422	-128
51.0	1.55600	365	.01285	-345
51.5	1.56404	154	.01819	-471
52.0	1.57356	-88	.01902	-484
52.5	1.58224	-310	.01522	-379
53.0	1.58797	-446	.00779	-182
53.5	1.58943	-477	$-.00139$	54
54.0	1.58633	-385	$-.01006$	272
54.5	1.57953	-210	$-.01613$	422
55.0	1.57072	21	$-.01817$	463
55.5	1.56210	236	$-.01577$	396
56.0	1.55574	393	$-.00958$	232
56.5	1.55314	454	$-.00117$	12
57.0	1.55488	401	.00737	-202
57.5	1.56046	251	.01398	-369
58.0	1.56845	44	.01707	-437
58.5	1.57687	-168	.01597	-403
59.0	1.58368	-340	.01101	-272
59.5	1.58724	-423	.00345	-72

SINE, COSINE, AND EXPONENTIAL INTEGRALS (Continued)

x	$Si(x)$	δ_m^2	$Ci(x)$	δ_m^2
60.0	1.58675	−404	−.00481	136
60.5	1.58239	−287	−.01177	311
61.0	1.57528	−99	−.01576	408
61.5	1.56721	103	−.01585	402
62.0	1.56013	284	−.01209	301
62.5	1.55576	389	−.00545	127
63.0	1.55511	401	.00241	−75
63.5	1.55829	311	.00956	−256
64.0	1.56445	151	.01427	−368
64.5	1.57206	−44	.01545	−395
65.0	1.57925	−227	.01285	−321
65.5	1.58427	−350	.00717	−173
66.0	1.58594	−388	−.00017	16
66.5	1.58390	−329	−.00736	199
67.0	1.57871	−193	−.01265	331
67.5	1.57167	−11	−.01478	378
68.0	1.56452	171	−.01329	335
68.5	1.55900	309	−.00859	211
69.0	1.55643	369	−.00187	37
69.5	1.55739	338	.00521	−144
70.0	1.56159	229	.01092	−285
70.5	1.56798	61	.01390	−359
71.0	1.57496	−116	.01345	−340
71.5	1.58083	−263	.00974	−242
72.0	1.58418	−346	.00371	−85
72.5	1.58422	−340	−.00314	90
73.0	1.58100	−253	−.00913	240
73.5	1.57535	−109	−.01282	332
74.0	1.56866	64	−.01334	339
74.5	1.56258	218	−.01061	267
75.0	1.55858	317	−.00533	127
75.5	1.55761	336	.00117	−39
76.0	1.55986	275	.00730	−193
76.5	1.56474	147	.01158	−302
77.0	1.57103	−15	.01298	−331
77.5	1.57718	−169	.01121	−283
78.0	1.58171	−287	.00673	−165
78.5	1.58351	−323	.00067	−9
79.0	1.58221	−290	−.00548	149
79.5	1.57814	−182	−.01021	265
80.0	1.57233	−31	−.01240	317
80.5	1.56622	126	−.01155	294
81.0	1.56131	246	−.00789	195
81.5	1.55876	309	−.00236	53
82.0	1.55917	296	.00368	−102
82.5	1.56241	209	.00875	−229
83.0	1.56765	73	.01163	−300
83.5	1.57359	−77	.01164	−296
84.0	1.57879	−212	.00882	−220
84.5	1.58197	−287	.00389	−93

SINE, COSINE, AND EXPONENTIAL INTEGRALS (Continued)

x	$Si(x)$	δ_m^2	$Ci(x)$	δ_m^2
85.0	1.58240	−296	−.00193	55
85.5	1.58000	−232	−.00722	192
86.0	1.57538	−110	−.01068	274
86.5	1.56970	36	−.01151	295
87.0	1.56436	169	−.00952	239
87.5	1.56064	264	−.00524	129
88.0	1.55944	291	.00027	−14
88.5	1.56102	246	.00565	−149
89.0	1.56496	144	.00960	−250
89.5	1.57028	5	.01116	−285
90.0	1.57566	−128	.00999	−254
90.5	1.57981	−236	.00639	−157
91.0	1.58171	−280	.00128	−26
91.5	1.58093	−255	−.00408	108
92.0	1.57770	−174	−.00840	220
92.5	1.57281	−44	−.01062	273
93.0	1.56749	90	−.01023	259
93.5	1.56303	203	−.00736	185
94.0	1.56051	264	−.00272	63
94.5	1.56052	263	.00253	−70
95.0	1.56304	192	.00711	−186
95.5	1.56741	83	.00991	−256
96.0	1.57257	−52	.01026	−261
96.5	1.57724	−170	.00811	−203
97.0	1.58029	−244	.00401	−98
97.5	1.58100	−262	−.00103	33
98.0	1.57921	−210	−.00576	150
98.5	1.57540	−114	−.00905	234
99.0	1.57050	13	−.01010	260
99.5	1.56572	135	−.00867	218
100.0	1.56223	222	−.00515	127

ORTHOGONAL POLYNOMIALS

I

Name: Legendre   *Symbol*: $P_n(x)$ *Interval*: $[-1, 1]$

Differential Equation: $(1 - x^2)y'' - 2xy' + n(n + 1)y = 0$

$$y = P_n(x)$$

Explicit Expression: $P_n(x) = \dfrac{1}{2^n} \displaystyle\sum_{m=0}^{[n/2]} (-1)^m \binom{n}{m}\binom{2n - 2m}{n} x^{n-2m}$

Recurrence Relation: $(n + 1)P_{n+1}(x) = (2n + 1)xP_n(x) - nP_{n-1}(x)$

Weight: 1   *Standardization*: $P_n(1) = 1$

Norm: $\displaystyle\int_{-1}^{+1} [P_n(x)]^2 dx = \dfrac{2}{2n + 1}$

Rodrigues' Formula: $P_n(x) = \dfrac{(-1)^n}{2^n n!} \dfrac{d^n}{dx^n} \{(1 - x^2)^n\}$

Generating Function: $R^{-1} = \displaystyle\sum_{n=0}^{\infty} P_n(x)z^n$; $-1 < x < 1$, $|z| < 1$,

$$R = \sqrt{1 - 2xz + z^2}.$$

Inequality: $|P_n(x)| \le 1$, $-1 \le x \le 1$.

II

Name: Tschebysheff, First Kind   *Symbol*: $T_n(x)$ *Interval*: $[-1, 1]$

Differential Equation: $(1 - x^2)y'' - xy' + n^2 y = 0$

$$y = T_n(x)$$

Explicit Expression: $\dfrac{n}{2} \displaystyle\sum_{m=0}^{[n/2]} (-1)^m \dfrac{(n - m - 1)!}{m!(n - 2m)!} (2x)^{n-2m} = \cos(n \arccos x) = T_n(x)$

Recurrence Relation: $T_{n+1}(x) = 2xT_n(x) - T_{n-1}(x)$

Weight: $(1 - x^2)^{-1/2}$ *Standardization*: $T_n(1) = 1$

Norm: $\displaystyle\int_{-1}^{+1} (1 - x^2)^{-1/2}[T_n(x)]^2 dx = \begin{cases} \pi/2, & n \ne 0 \\ \pi, & n = 0 \end{cases}$

Rodrigues' Formula: $\dfrac{(-1)^n(1 - x^2)^{1/2}\sqrt{\pi}}{2^n \Gamma(n + \frac{1}{2})} \dfrac{d^n}{dx^n} \{(1 - x^2)^{n-(1/2)}\} = T_n(x)$

Generating Function: $\dfrac{1 - xz}{1 - 2xz + z^2} = \displaystyle\sum_{n=0}^{\infty} T_n(x)z^n$, $-1 < x < 1$, $|z| < 1$.

Inequality: $|T_n(x)| \le 1$, $-1 \le x \le 1$.

III

Name: Tschebysheff, Second Kind   *Symbol*: $U_n(x)$ *Interval*: $[-1, 1]$

Differential Equation: $(1 - x^2)y'' - 3xy' + n(n + 2)y = 0$

$$y = U_n(x)$$

Explicit Expression:
$$U_n(x) = \sum_{m=0}^{[n/2]} (-1)^m \frac{(m - n)!}{m!(n - 2m)!} (2x)^{n-2m}$$

$$U_n(\cos \theta) = \frac{\sin[(n + 1)\theta]}{\sin \theta}$$

Recurrence Relation: $U_{n+1}(x) = 2x U_n(x) - U_{n-1}(x)$

Weight: $(1 - x^2)^{1/2}$ *Standardization*: $U_n(1) = n + 1$

Norm: $\int_{-1}^{+1} (1 - x^2)^{1/2} [U_n(x)]^2 dx = \frac{\pi}{2}$

Rodrigues' Formula: $U_n(x) = \dfrac{(-1)^n (n + 1) \sqrt{\pi}}{(1 - x^2)^{1/2} 2^{n+1} \Gamma(n + \frac{3}{2})} \dfrac{d^n}{dx^n} \{(1 - x^2)^{n+(1/2)}\}$

Generating Function: $\dfrac{1}{1 - 2xz + z^2} = \sum_{n=0}^{\infty} U_n(x) z^n, \quad -1 < x < 1, \ |z| < 1.$

Inequality: $|U_n(x)| \le n + 1, \ -1 \le x \le 1.$

IV

Name: Jacobi   *Symbol*: $P_n^{(\alpha,\beta)}(x)$ *Interval*: $[-1, 1]$

Differential Equation:

$$(1 - x^2)y'' + [\beta - \alpha - (\alpha + \beta + 2)x]y' + n(n + \alpha + \beta + 1)y = 0$$

$$y = P_n^{(\alpha,\beta)}(x)$$

Explicit Expression: $P_n^{(\alpha,\beta)}(x) = \dfrac{1}{2^n} \sum_{m=0}^{n} \binom{n + \alpha}{m} \binom{n + \beta}{n - m} (x - 1)^{n-m}(x + 1)^m$

Recurrence Relation: $2(n + 1)(n + \alpha + \beta + 1)(2n + \alpha + \beta) P_{n+1}^{(\alpha,\beta)}(x)$
$$= (2n + \alpha + \beta + 1)[(\alpha^2 - \beta^2) + (2n + \alpha + \beta + 2)$$
$$\times (2n + \alpha + \beta)x] P_n^{(\alpha,\beta)}(x)$$
$$- 2(n + \alpha)(n + \beta)(2n + \alpha + \beta + 2) P_{n-1}^{(\alpha,\beta)}(x)$$

Weight: $(1 - x)^\alpha (1 + x)^\beta; \ \alpha, \beta > 1$ *Standardization*: $P_n^{(\alpha,\beta)}(x) = \binom{n + \alpha}{n}$

Norm: $\int_{-1}^{+1} (1 - x)^\alpha (1 + x)^\beta [P_n^{(\alpha,\beta)}(x)]^2 dx = \dfrac{2^{\alpha+\beta+1} \Gamma(n + \alpha + 1) \Gamma(n + \beta + 1)}{(2n + \alpha + \beta + 1)n! \Gamma(n + \alpha + \beta + 1)}$

Rodrigues' Formula: $P_n^{(\alpha,\beta)}(x) = \dfrac{(-1)^n}{2^n n!(1 - x)^\alpha (1 + x)^\beta} \dfrac{d^n}{dx^n} \{(1 - x)^{n+\alpha}(1 + x)^{n+\beta}\}$

IV (Continued)

Generating Function: $R^{-1}(1 - z + R)^{-\alpha}(1 + z + R)^{-\beta} = \sum_{n=0}^{\infty} 2^{-\alpha-\beta} P_n^{(\alpha,\beta)}(x) z^n,$

$$R = \sqrt{1 - 2xz + z^2}, \; |z| < 1$$

Inequality: $\max_{-1 \leq x \leq 1} |P_n^{(\alpha,\beta)}(x)| = \begin{cases} \binom{n+q}{n} \sim n^q \text{ if } q = \max(\alpha,\beta) \geq -\frac{1}{2} \\ |P_n^{(\alpha,\beta)}(x')| \sim n^{-1/2} \text{ if } q < -\frac{1}{2} \\ x' \text{ is one of the two maximum points nearest} \\ \dfrac{\beta - \alpha}{\alpha + \beta + 1} \end{cases}$

V

Name: Generalized Laguerre   *Symbol*: $L_n^{(\alpha)}(x)$ *Interval*: $[0, \infty]$

Differential Equation: $xy'' + (\alpha + 1 - x)y' + ny = 0$

$$y = L_n^{(\alpha)}(x)$$

Explicit Expression: $L_n^{(\alpha)}(x) = \sum_{m=0}^{n} (-1)^m \binom{n+\alpha}{n-m} \frac{1}{m!} x^m$

Recurrence Relation: $(n + 1) L_{n+1}^{(\alpha)}(x) = [(2n + \alpha + 1) - x] L_n^{(\alpha)}(x) - (n + \alpha) L_{n-1}^{(\alpha)}(x)$

Weight: $x^\alpha e^{-x}, \; \alpha > -1$ *Standardization*: $L_n^{(\alpha)}(x) = \dfrac{(-1)^n}{n!} x^n + \cdots$

Norm: $\displaystyle\int_0^\infty x^\alpha e^{-x} [L_n^{(\alpha)}(x)]^2 \, dx = \dfrac{\Gamma(n + \alpha + 1)}{n!}$

Rodrigues' Formula: $L_n^{(\alpha)}(x) = \dfrac{1}{n! x^\alpha e^{-x}} \dfrac{d^n}{dx^n} \{x^{n+\alpha} e^{-x}\}$

Generating Function: $(1 - z)^{-\alpha-1} \exp\left(\dfrac{xz}{z-1}\right) = \sum_{n=0}^{\infty} L_n^{(\alpha)}(x) z^n$

Inequality: $|L_n^{(\alpha)}(x)| \leq \dfrac{\Gamma(n + \alpha + 1)}{n! \Gamma(\alpha + 1)} e^{x/2}; \quad \begin{array}{l} x \geq 0 \\ \alpha > 0 \end{array}$

$$|L_n^{(\alpha)}(x)| \leq \left[2 - \dfrac{\Gamma(\alpha + n + 1)}{n! \Gamma(\alpha + 1)}\right] e^{x/2}; \quad \begin{array}{l} x \geq 0 \\ -1 < \alpha < 0 \end{array}$$

Orthogonal Polynomials

Name: Hermite *Symbol:* $H_n(x)$ *Interval:* $[-\infty, \infty]$

Differential Equation: $y'' - 2xy' + 2ny = 0$

Explicit Expression: $H_n(x) = \sum_{m=0}^{[n/2]} \dfrac{(-1)^m \, n! \, (2x)^{n-2m}}{m! \, (n-2m)!}$

Recurrence Relation: $H_{n+1}(x) = 2x \, H_n(x) - 2n H_{n-1}(x)$

Weight: e^{-x^2} *Standardization:* $H_n(1) = 2^n x^n + \ldots$

Norm: $\displaystyle\int_{-\infty}^{\infty} e^{-x^2} \, [H_n(x)]^2 \, dx = 2^n \, n! \, \sqrt{\pi}$

Rodriques' Formula: $H_n(x) = (-1)^n \, e^{x^2} \, \dfrac{d^n}{dx^n} \, (e^{-x^2})$

Generating Function: $e^{-z^2 + 2zx} = \displaystyle\sum_{n=0}^{\infty} H_n(x) \, \dfrac{z^n}{n!}$

Inequality: $|H_n(x)| < e^{\frac{x^2}{2}} \, k \, 2^{n/2} \, \sqrt{n!} \quad k \approx 1.086435$

Coefficients for Orthogonal Polynomials, and for x^n in Terms of Orthogonal Polynomials*

I. Legendre Polynomials: $P_n(x) = a_n^{-1} \sum_{m=0}^{n} c_m x^m \qquad x^n = b_n^{-1} \sum_{m=0}^{n} d_m P_m(x)$

	a_n	x^0	x^1	x^2	x^3	x^4	x^5	x^6	x^7
b_n		1	1	3	5	35	63	231	429
P_0	1	1 1		1		7		33	
P_1	1		1 1		3		27		143
P_2	2	−1		3 2		20		110	
P_3	2		−3		5 2		28		182
P_4	8	3		−30		35 8		72	
P_5	8		15		−70		63 8		88
P_6	16	−5		105		−315		231 16	
P_7	16		−35		315		−693		429 16

$$P_6(x) = \frac{1}{16}[231x^6 - 315x^4 + 105x^2 - 5] \qquad x^6 = \frac{1}{231}[33P_0 + 110P_2 + 72P_4 + 16P_6]$$

II. Tschebysheff Polynomials: $T_n(x) = \sum_{m=0}^{n} c_m x^m \qquad x^n = b_n^{-1} \sum_{m=0}^{n} d_m T_m(x)$

	x^0	x^1	x^2	x^3	x^4	x^5	x^6	x^7
b_n	1	1	2	4	8	16	32	64
T_0	1 1		1		3		10	
T_1		1 1		3		10		35
T_2	−1		2 1		4		15	
T_3		−3		4 1		5		21
T_4	1		−8		8 1		6	
T_5		5		−20		16 1		7
T_6	−1		18		−48		32 1	
T_7		−7		56		−112		64 1

$$T_6(x) = 32x^6 - 48x^4 + 18x^2 - 1 \qquad x^6 = \frac{1}{32}[10T_0 + 15T_2 + 6T_4 + T_6]$$

III. Tschebysheff Polynomials: $U_n(x) = \sum_{m=0}^{n} c_m x^m \qquad x^n = b_n^{-1} \sum_{m=0}^{n} d_m U_m(x)$

	x^0	x^1	x^2	x^3	x^4	x^5	x^6	x^7
b_n	1	2	4	8	16	32	64	128
U_0	1 1		1		2		5	
U_1		2 1		2		5		14
U_2	−1		4 1		3		9	
U_3		−4		8 1		4		14
U_4	1		−12		16 1		5	
U_5		6		−32		32 1		6
U_6	−1		24		−80		64 1	
U_7		−8		80		−192		128 1

$$U_6(x) = 64x^6 - 80x^4 + 24x^3 - 1 \qquad x^6 = \frac{1}{64}[5U_0 + 9U_2 + 5U_4 + U_6]$$

Coefficients for Orthogonal Polynomials, and for x^n in Terms of Orthogonal Polynomials* (continued)

IV. Jacobi Polynomials $P_n^{(\alpha,\beta)}(x) = a_n^{-1} \sum_{m=0}^{n} c_m (x-1)^m$

	α_n	$(x-1)^0$	$(x-1)^1$	$(x-1)^2$	$(x-1)^3$	$(x-1)^4$	$(x-1)^5$	$(x-1)^6$
$P_0(\alpha,\beta)$	1	1						
$P_1(\alpha,\beta)$	2	$2(\alpha+1)$	$\alpha+\beta+2$					
$P_2(\alpha,\beta)$	8	$4(\alpha+1)_2$	$4(\alpha+\beta+3)(\alpha+2)$	$(\alpha+\beta+3)_2$				
$P_3(\alpha,\beta)$	48	$8(\alpha+1)_3$	$12(\alpha+\beta+4)(\alpha+2)_2$	$6(\alpha+\beta+4)_2(\alpha+3)$	$(\alpha+\beta+4)_3$			
$P_4(\alpha,\beta)$	384	$16(\alpha+1)_4$	$32(\alpha+\beta+5)(\alpha+2)_3$	$24(\alpha+\beta+5)_2(\alpha+3)_2$	$8(\alpha+\beta+5)_3(\alpha+4)$	$(\alpha+\beta+5)_4$		
$P_5(\alpha,\beta)$	3840	$32(\alpha+1)_5$	$80(\alpha+\beta+6)(\alpha+2)_4$	$80(\alpha+\beta+6)_2(\alpha+3)_3$	$40(\alpha+\beta+6)_3(\alpha+4)_2$	$10(\alpha+\beta+6)_4(\alpha+5)$	$(\alpha+\beta+6)_5$	
$P_6(\alpha,\beta)$	46080	$64(\alpha+1)_6$	$192(\alpha+\beta+7)(\alpha+2)_5$	$240(\alpha+\beta+7)_2(\alpha+3)_4$	$160(\alpha+\beta+7)_3(\alpha+4)_3$	$60(\alpha+\beta+7)_4(\alpha+5)_2$	$12(\alpha+\beta+7)_5(\alpha+6)$	$(\alpha+\beta+7)_6$

$$(m)_n = m(m+1)(m+2)\cdots(m+n-1)$$

$$P_5^{(1,1)}(x) = \frac{1}{3840}\left[(8)_5(x-1)^5+10(8)_4(6)(x-1)^4+40(8)_3(5)_2(x-1)^3+80(8)_2(4)_3(x-1)^2+80(8)(3)_4(x-1)+32(2)_5\right]$$

$$P_5^{(1,1)}(x) = \frac{1}{3840}\left[95040(x-1)^5+475200(x-1)^4+864000(x-1)^3+691200(x-1)^2+230400(x-1)+23040\right]$$

Coefficients for Orthogonal Polynomials, and for x^n in Terms of Orthogonal Polynomials (continued)

V. Laguerre Polynomials: $L_n(x) = \sum_{m=0}^{n} c_m x^m \qquad x^n = b_n^{-1} \sum_{m=0}^{n} d_m L_m(x)$

	x^0	x^1	x^2	x^3	x^4	x^5	x^6	x^7
b_m	1	1	2	6	24	120	720	5040
L_0	1 1	1	2	6	24	120	720	5040
L_1	1	−1 −1	−4	−18	−96	−600	−4320	−35280
L_2	2	−4	1 2	18	144	1200	10800	105840
L_3	6	−18	9	−1 −6	−96	−1200	−14400	−176400
L_4	24	−96	72	−16	1 24	600	10800	17640
L_5	120	−600	600	−200	25	−1 −120	−4320	−105840
L_6	720	−4320	5400	−2400	450	−36	1 720	35280
L_7	5040	−35280	52920	−29400	7350	−882	49	−1 −5040

$$L_6(x) = x^6 - 36x^5 + 450x^4 - 2400x^3 + 5400x^2 - 4320x + 720$$
$$x^6 = \frac{1}{720}\left[720L_0 - 4320L_1 + 10800L_2 - 14400L_3 + 10800L_4 - 4320L_5 + 720L_6\right]$$

VI. Hermite polynomials: $H_n(x) = \sum_{m=0}^{n} c_m x^m \qquad x^n = b_n^{-1} \sum_{m=0}^{n} d_m H_m(x)$

	x^0	x^1	x^2	x^3	x^4	x^5	x^6	x^7
b_n	1	2	4	8	16	32	64	128
H_0	1 1		2		12		120	
H_1		2 1		6		60		840
H_2	−2		4 1		12		180	
H_3		−12		8 1		20		420
H_4	12		−48		16 1		30	
H_5		120		−160		32 1		42
H_6	−120		720		−480		64 1	
H_7		−1680		3360		−1344		128 1

$$H_6(x) = 64x^6 - 480x^4 + 720x^2 - 120 \qquad x^6 = \frac{1}{64}\left[120H_0 + 180H_2 + 30H_4 + H_6\right]$$

Abridged from Abramowitz, M. and Stegun, I. A., Eds., *Handbook of Mathematical Functions,* National Bureau of Standards, Washington, D. C., 1964.

LEGENDRE FUNCTIONS

In the following, m and n are positive integers.

1. The differential equation $(1 - z^2) \dfrac{d^2w}{dz^2} - 2z \dfrac{dw}{dz} + n(n + 1)z = 0$ is known as Legendre's differential equation. The Legendre polynomials $P_0(z)$, $P_1(z)$, $P_2(z)$, . . . are solutions of this equation, and are referred to as Surface Zonal Harmonics, for which tables are herewith given.

2. The solution of Legendre's equation can be stated as

$$w = AP_n(z) + BQ_n(z) \qquad |z| < 1$$

or

$$w = AP_n(z) + B\mathcal{Q}_n(z) \qquad |z| > 1,$$

where A and B are arbitrary constants, $P_n(z)$ the Legendre polynomials, $Q_n(z)$ and $\mathcal{Q}_n(z)$ Legendre functions of the second kind.

3. The differential equation $(1 - z^2) \dfrac{d^2u}{dz^2} - 2z \dfrac{du}{dz} + \left[n(n + 1) - \dfrac{m^2}{1 - z^2} \right] u = 0$ which is obtained from the Legendre equation in 1. after it has been differentiated m times and u replaced by $(1 - z^2)^{m/2} \left(\dfrac{d^m w}{dz^m} \right)$ is referred to as the "associated Legendre differential equation."

4. The solution for the "associated Legendre differential equation" is given by

$$u = AP_n^m(z) + BQ_n^m(z),$$

where A and B are arbitrary constants and $P_n^m(z)$, $Q_n^m(z)$ are called associated Legendre functions.

5. $P_n(z) = \displaystyle\sum_{1=0}^{m} (-1)^r \dfrac{(2n - 2r)!}{2^n(r!)(n - r_r)!(n - 2r)!} z^{n-2r} \quad m = \tfrac{1}{2}n, \ n \text{ even}$

$$m = \tfrac{1}{2}(n - 1), \ n \text{ odd}$$

$$Q_n(z) = \tfrac{1}{2}P_n(z) \log_e \frac{1 + z}{1 - z} - \sum_{r=1}^{n} \frac{1}{r} P_{r-1}(z)P_{n-r}(z)$$

$$= P_n(z)Q_0(z) - \sum_{r=1}^{n} \frac{1}{r} P_{r-1}(z)P_{n-r}(z)$$

$$Q_0(z) = \tfrac{1}{2} \log_e \frac{1 + z}{1 - z} = \tanh^{-1} z$$

$$\mathcal{Q}_n(z) = \tfrac{1}{2}P_n(z) \log_e \frac{z + 1}{z - 1} - \sum_{r=1}^{n} \frac{1}{r} P_{r-1}(z) P_{n-r}(z)$$

$$= P_n(z)\mathcal{Q}_0(z) - \sum_{r=1}^{n} \frac{1}{r} P_{r-1}(z)P_{n-r}(z)$$

$$\mathcal{Q}_0(z) = \tfrac{1}{2} \log_e \frac{z + .1}{z - 1} = \operatorname{ctnh}^{-1} z$$

LEGENDRE FUNCTIONS (Continued)

6.

n	$P_n(z)$	$Q_n(z)$
0	1	$\dfrac{\log_e}{2} \dfrac{(1+z)}{(1-z)}$
1	z	$zQ_0(z) - 1$
2	$\dfrac{1}{2}(3z^2 - 1)$	$\dfrac{1}{2}(3z^2 - 1)Q_0(z) - \dfrac{3}{2}z$
3	$\dfrac{1}{2}(5z^3 - 3z)$	$\dfrac{1}{2}(5z^3 - 3z)Q_0(z) - \dfrac{5}{2}z^2 + \dfrac{2}{3}$
4	$\dfrac{1}{8}(35z^4 - 30z^2 + 3)$	$\dfrac{1}{8}(35z^4 - 30z^2 + 3)Q_0(z) - \dfrac{35}{8}z^3 + \dfrac{55}{24}z$
5	$\dfrac{1}{8}(63z^5 - 70z^3 + 15z)$	$\dfrac{1}{8}(63z^5 - 70z^3 + 15z)Q_0(z) - \dfrac{63}{8}z^4 + \dfrac{49}{8}z^2 - \dfrac{8}{15}$

7. $\dfrac{1}{(1 - 2xZ + Z^2)^{1/2}} = \sum P_n(x)Z^n$

If $x = \cos\theta = \tfrac{1}{2}(e^{i\theta} + e^{-i\theta})$, $i = \sqrt{-1}$, then

$$[1 - Z(e^{i\theta} + e^{-i\theta}) + Z^2]^{-1/2} = \sum_{n=0}^{\infty} P_n(\cos\theta)Z^n$$

$P_0(\cos\theta) = 1$

$P_1(\cos\theta) = \cos\theta$

$P_2(\cos\theta) = \dfrac{1}{4}(3\cos 2\theta + 1)$

$P_3(\cos\theta) = \dfrac{1}{8}(5\cos 3\theta + 3\cos\theta)$

$P_4(\cos\theta) = \dfrac{1}{64}(35\cos 4\theta + 20\cos 2\theta + 9)$

$$P_n(\cos\theta) = \frac{(2n)!}{2^{2n}(n!)^2}\left[\cos n\theta + \frac{1}{1}\frac{n}{2n-1}\cos(n-2)\theta\right.$$
$$+ \frac{1\cdot 3}{1\cdot 2}\frac{n(n-1)}{(2n-1)(2n-3)}\cos(n-4)\theta$$
$$\left.+ \frac{1\cdot 3\cdot 5}{1\cdot 2\cdot 3}\frac{n(n-1)(n-2)}{(2n-1)(2n-3)(2n-5)}\cos(n-6)\theta + \cdots\right]$$

LEGENDRE FUNCTIONS (Continued)

8. $P_{2n+1}(0) = 0$

$$P_{2n}(0) = (-1)^n \frac{1 \cdot 3 \cdot 5 \cdots (2n-1)}{2 \cdot 4 \cdot 6 \cdot 8 \cdots (2n)}$$

$$P_n(1) = 1$$

$$P_n(-1) = (-1)^n$$

$$P_n(z) = \frac{1}{2^n n!} \frac{d^n}{dz^n} (z^2 - 1)^n \text{ (Rodrigues' formula)}$$

$$P_n^m(z) = (1 - z^2)^{m/2} \frac{d^m[P_n(z)]}{dz^m}$$

$$Q_n^m(z) = (1 - z^2)^{m/2} \frac{d^m[Q_n(z)]}{dz^m}$$

9. Important recurrence and orthogonality relations (note that w_n stands for either P_n or Q_n in the following)

$$(n + 1)[w_{n+1}(z)] - (2n + 1)z[w_n(z)] + n[w_{n-1}(z)] = 0$$

$$z \frac{d[w_n(z)]}{dz} - \frac{d[w_{n-1}(z)]}{dz} = n[w_n(z)]$$

$$\frac{d[w_{n+1}(z)]}{dz} - z \frac{d[w_n(z)]}{dz} = (n + 1)[w_n(z)]$$

$$\frac{d[w_{n+1}(z)]}{dz} - \frac{d}{dz}[w_{n-1}(z)] = (2n + 1)[w_n(z)]$$

$$(z^2 - 1) \frac{d}{dz}[w_n(z)] = nz[w_n(z)] - n[w_{n-1}(z)]$$

$$\int_{-1}^{1} [P_m(z)][P_n(z)]dz \begin{cases} = 0 & m \neq n \\ = \dfrac{2}{2n + 1} & m = n \end{cases}$$

$$\int_{-1}^{1} [P_n^m(z)][P_r^m(z)]dz \begin{cases} = 0 & r \neq n \\ = \dfrac{2}{2n + 1} \dfrac{(n + m)!}{(n - m)!} & r = n > m. \end{cases}$$

SURFACE ZONAL HARMONICS

$$P_n(x) = \frac{1}{2^n n!} \frac{d^n}{dx^n} (x^2 - 1)^n$$

x	$P_1(x)$	$P_2(x)$	$P_3(x)$	$P_4(x)$	$P_5(x)$
0	0	− .5	0	.375	0
.01	.01	− .49985	− .0149975	.37463	.018741
.02	.02	− .4994	− .02998	.37350	.037430
.03	.03	− .49865	− .0449325	.37163	.056014
.04	.04	− .4976	− .05984	.36901	.074441
.05	.05	− .49625	− .0746875	.36565	.092659
.06	.06	− .4946	− .08946	.36156	.11062
.07	.07	− .49265	− .10414	.35673	.12826
.08	.08	− .4904	− .11872	.35118	.14555
.09	.09	− .48785	− .13318	.34491	.16242
.10	.10	− .485	− .1475	.33794	.17883
.11	.11	− .48185	− .16167	.33027	.19473
.12	.12	− .4784	− .17568	.32191	.21008
.13	.13	− .47465	− .18951	.31287	.22482
.14	.14	− .4706	− .20314	.30318	.23891
.15	.15	− .46625	− .21656	.29284	.25232
.16	.16	− .4616	− .22976	.28187	.26499
.17	.17	− .45665	− .24272	.27028	.27688
.18	.18	− .4514	− .25542	.25809	.28796
.19	.19	− .44585	− .26785	.24533	.29818
.20	.20	− .44	− .28	.232	.30752
.21	.21	− .43385	− .29185	.21813	.31593
.22	.22	− .4274	− .30338	.20375	.32339
.23	.23	− .42065	− .31458	.18887	.32986
.24	.24	− .4136	− .32544	.17352	.33531
.25	.25	− .40625	− .33594	.15771	.33972
.26	.26	− .3986	− .34606	.14149	.34307
.27	.27	− .39065	− .35579	.12488	.34532
.28	.28	− .3824	− .36512	.10789	.34647
.29	.29	− .37385	− .37403	.09057	.34650
.30	.30	− .365	− .3825	.07294	.34539
.31	.31	− .35585	− .39052	.05503	.34312
.32	.32	− .3464	− .39808	.03688	.33970
.33	.33	− .33665	− .40516	.01851	.33512
.34	.34	− .3266	− .41174	− .00004	.32937
.35	.35	− .31625	− .41781	− .01872	.32245
.36	.36	− .3056	− .42336	− .03752	.31438
.37	.37	− .29465	− .42837	− .05638	.30514
.38	.38	− .2834	− .43282	− .07528	.29477
.39	.39	− .27185	− .43670	− .09416	.28326
.40	.40	− .26	− .44	− .113	.27064
.41	.41	− .24785	− .44270	− .13175	.25693
.42	.42	− .2354	− .44478	− .15036	.24215
.43	.43	− .22265	− .44623	− .16880	.22633
.44	.44	− .2096	− .44704	− .18702	.20951
.45	.45	− .19625	− .44719	− .20497	.19172
.46	.46	− .1826	− .44666	− .22261	.17301
.47	.47	− .16865	− .44544	− .23989	.15341
.48	.48	− .1544	− .44352	− .25676	.13298
.49	.49	− .13985	− .44088	− .27316	.11177

SURFACE ZONAL HARMONICS (Continued)

x	$P_6(x)$	$P_7(x)$	$P_8(x)$	$P_9(x)$	$P_{10}(x)$
0	$-.3125$	-0	$.2734$	0	$-.2461$
.01	$-.31184$	$-.021855$	$.2725$	$.0246$	$-.2447$
.02	$-.30988$	$-.043593$	$.2695$	$.0489$	$-.2407$
.03	$-.30661$	$-.065094$	$.2646$	$.0729$	$-.2340$
.04	$-.30205$	$-.086244$	$.2578$	$.0961$	$-.2247$
.05	$-.29622$	$-.10693$	$.2492$	$.1186$	$-.2130$
.06	$-.28913$	$-.12703$	$.2387$	$.1400$	$-.1989$
.07	$-.28081$	$-.14644$	$.2265$	$.1601$	$-.1826$
.08	$-.27130$	$-.16506$	$.2126$	$.1789$	$-.1642$
.09	$-.26063$	$-.18278$	$.1972$	$.1960$	$-.1440$
.10	$-.24883$	$-.19949$	$.1803$	$.2114$	$-.1221$
.11	$-.23595$	$-.21511$	$.1621$	$.2249$	$-.0989$
.12	$-.22204$	$-.22955$	$.1426$	$.2364$	$-.0745$
.13	$-.20715$	$-.24271$	$.1221$	$.2457$	$-.0492$
.14	$-.19133$	$-.25453$	$.1006$	$.2529$	$-.0233$
.15	$-.17465$	$-.26492$	$.0783$	$.2577$	$.0020$
.16	$-.15716$	$-.27383$	$.0554$	$.2601$	$.0293$
.17	$-.13894$	$-.28119$	$.0319$	$.2602$	$.0553$
18	$-.12005$	$-.28695$	$.0082$	$.2579$	$.0808$
.19	$-.10057$	$-.29107$	$-.0157$	$.2531$	$.1055$
.20	$-.080576$	$-.29352$	$-.0396$	$.2460$	$.1291$
.21	$-.060144$	$-.29426$	$-.0632$	$.2365$	$.1513$
.22	$-.039357$	$-.29327$	$-.0865$	$.2247$	$.1718$
.23	$-.018300$	$-.29055$	$-.1093$	$.2108$	$.1905$
.24	$.002941$	$-.28610$	$-.1313$	$.1948$	$.2070$
.25	$.024277$	$-.27992$	$-.1525$	$.1768$	$.2212$
.26	$.045618$	$-.27203$	$-.1725$	$.1571$	$.2329$
.27	$.066872$	$-.26246$	$-.1914$	$.1357$	$.2419$
.28	$.087947$	$-.25124$	$-.2089$	$.1129$	$.2480$
.29	$.10875$	$-.23843$	$-.2248$	$.0888$	$.2513$
.30	$.12918$	$-.22407$	$-.2391$	$.0637$	$.2515$
.31	$.14915$	$-.20824$	$-.2515$	$.0378$	$.2487$
.32	$.16856$	$-.19100$	$-.2621$	$.0114$	$.2428$
.33	$.18732$	$-.17244$	$-.2706$	$-.0154$	$.2339$
.34	$.20534$	$-.15266$	$-.2770$	$-.0422$	$.2220$
.35	$.22251$	$-.13176$	$-.2812$	$-.0688$	$.2073$
.36	$.23875$	$-.10984$	$-.2831$	$-.0948$	$.1899$
.37	$.25397$	$-.087036$	$-.2826$	$-.1201$	$.1699$
.38	$.26808$	$-.063467$	$-.2798$	$-.1444$	$.1476$
.39	$.28100$	$-.039271$	$-.2746$	$-.1674$	$.1231$
.40	$.29264$	$-.014590$	$-.2670$	$-.1888$	$.0968$
.41	$.30291$	$.010424$	$-.2570$	$-.2083$	$.0690$
.42	$.31176$	$.035614$	$-.2447$	$-.2258$	$.0401$
.43	$.31909$	$.060820$	$-.2302$	$-.2410$	$.0103$
.44	32486	$.085873$	$-.2134$	$-.2537$	$-.0200$
.45	$.32898$	$.11060$	$-.1945$	$-.2637$	$-.0503$
.46	$.33141$	$.13483$	$-.1737$	$-.2708$	$-.0803$
.47	$.33209$	$.15838$	$-.1510$	$-.2749$	$-.1095$
.48	$.33098$	$.18107$	$-.1267$	$-.2758$	$-.1375$
.49	$.32804$	$.20272$	$-.1008$	$-.2735$	$-.1639$

SURFACE ZONAL HARMONICS (Continued)

x	$P_1(x)$	$P_2(x)$	$P_3(x)$	$P_4(x)$	$P_5(x)$
0.50	0.50	−0.125	−0.4375	−0.28906	0.08984
.51	.51	−.10985	−.43337	−.30440	.06726
.52	.52	−.0944	−.42848	−.31912	.04409
.53	53	−.07865	−.42281	−.33317	.02041
.54	.54	−.0626	−.41634	−.34649	−.00372
55	.55	−.04625	−.40906	−.35904	−.02819
.56	.56	−.0296	−.40096	−.37074	−.05294
.57	.57	−.01265	−.39202	−.38155	−.07786
.58	.58	.0046	−.38222	−.39140	−.10285
.59	.59	.02215	−.37155	−.40024	−.12781
.60	.60	.04	−.36	−.408	−.15264
.61	.61	.05815	−.34755	−.41462	−.17721
.62	.62	.0766	−.33418	−.42004	−.20142
.63	.63	.09535	−.31988	−.42418	−.22512
.64	.64	.1144	−.30464	−.42700	−.24819
.65	.65	.13375	−.28844	−.42841	−.27049
.66	.66	.1534	−.27126	−.42836	−.29188
.67	.67	.17335	−.25309	−.42676	−.31220
.68	.68	.1936	−.23392	−.42356	−.33131
.69	.69	.21415	−.21373	−.41869	−.34903
.70	.70	.235	−.1925	−.41206	−.36520
.71	.71	.25615	−.17022	−.40361	−.37964
.72	.72	.2776	−.14688	−.39327	−.39217
.73	.73	.29935	−.12246	−.38095	−.40260
.74	.74	.3214	−.09694	−.36659	−.41074
.75	.75	.34375	−.07031	−.35010	−.41638
.76	.76	.3664	−.04256	−.33140	−.41931
.77	.77	.38935	−.01367	−.31043	−.41932
.78	.78	.4126	.01638	−.28709	−.41618
.79	.79	.43615	.04760	−.26131	−.40966
.80	.80	.46	.08	−.233	−.39952
.81	.81	.48415	.11360	−.20208	−.38552
.82	.82	.5086	.14842	−.16847	−.36739
.83	.83	.53335	.18447	−.13207	−.34489
.84	.84	.5584	.22176	−.09281	−.31774
.85	.85	.58375	.26031	−.05060	−.28566
.86	.86	.6094	.30014	−.00534	−.24838
.87	.87	.63535	.34126	.04305	−.20559
.88	.88	.6616	.38368	.09467	−.15699
.89	.89	.68815	.42742	.14960	−.10228
.90	.90	.715	.4725	.20794	−.04114
.91	.91	.74215	.51893	.26978	0.02676
.92	.92	.7696	.56672	.33522	.10175
.93	.93	.79735	.61589	.40435	.18417
.94	.94	.8254	.66646	.47728	.27438
.95	.95	.85373	.71844	.55409	.37274
.96	.96	.8824	.77184	.63489	.47962
.97	.97	.91135	.82668	.71978	.59539
.98	.98	.9406	.88298	.80886	.72045
.99	.99	.97015	.94075	.90223	.85518
100	1	1	1	1	1

SURFACE ZONAL HARMONICS (Continued)

x	$P_6(x)$	$P_7(x)$	$P_8(x)$	$P_9(x)$	$P_{10}(x)$
.50	0.32324	0.22314	−0.0736	−0.2679	−0.1882
.51	.31655	.24217	−.0454	−.2590	−.2101
.52	.30796	.25961	−.0163	−.2468	−.2291
.53	.29747	.27530	.0133	−.2314	−.2450
.54	.28506	.28906	.0432	−.2128	−.2573
.55	.27077	.30074	.0732	−.1913	−.2658
.56	.25460	.31016	.1029	−.1669	−.2701
.57	.23660	.31719	.1320	−.1399	−.2702
.58	.21681	.32169	.1601	−.1105	−.2659
.59	.19528	.32353	.1870	−.0791	−.2570
.60	.17210	.32260	.2133	−.0450	−.2433
.61	.14733	.31880	.2357	−.0118	−.2258
.63	.12109	.31207	.2568	.0234	−.2036
.63	.093475	.30232	.2753	.0589	−.1773
.64	.064623	.28954	.2909	.0943	−.1471
.65	.034675	.27371	.3032	.1290	−.1136
.66	.003790	.25483	.3120	.1625	−.0771
.67	−.027853	.23295	.3170	.1941	−.0382
.68	−.060059	.20813	.3179	.2233	.0024
.69	−.092615	.18049	.3145	.2495	.0440
.70	−.12529	.15016	.3067	.2721	.0858
.71	−.15782	.11731	.2943	.2904	.1269
.72	−.18994	.082166	.2771	.3039	.1663
.73	−.22136	.044990	.2553	.3120	.2030
.74	−.25175	.006087	.2287	.3143	.2360
.75	−.28078	−.034184	.1976	.3103	.2644
.76	−.30807	−.075411	.1621	.2997	.2869
.77	−.33325	−.11713	.1225	.2823	.3027
.78	−.35589	−.15881	.0791	.2578	.3108
.79	−.37557	−.19987	.0326	.2263	.3103
.80	−.39180	−.23965	−.0167	.1879	.3005
.81	−.40409	−.27743	−.0678	.1429	.2810
.82	−.41193	−.31240	−.1199	.0920	.2513
.83	−.41475	−.34368	−.1720	.0359	.1942
.84	−.41198	−.37033	−.2228	−.0243	.1617
.85	−.40300	−.39130	−.2710	−.0873	.1029
.86	−.38716	−.40545	−.3150	−.1513	.0362
.87	−.36379	−.41156	−.3530	−.2143	−.0366
.88	−.33217	−.40829	−.3830	−.2738	−.1130
.89	−.29156	−.39423	−.4028	−.3267	−.1899
.90	−.24116	−.36782	−.4097	−.3695	−.2631
.91	−.18018	−.32743	−.4010	−.3983	−.3277
.92	−.10774	−.27129	−.3737	−.4083	−.3773
.93	−.02295	−.19749	−.3243	−.3941	−.4046
.94	.07512	−.10404	−.2491	−.3498	−.4006
.95	.18745	.01123	−.1440	−.2684	−.3549
.96	.31506	.15060	−.0046	−.1422	−.2552
.97	.45899	.31650	.1740	.0375	−.0875
.98	.62035	.51151	.3971	.2804	.1647
.99	.80029	.73838	.6704	.5973	.5201
1.00	1	1	1	1	1

SURFACE ZONAL HARMONICS (Continued)

θ deg.	$P_1(\cos\theta)$	$P_2(\cos\theta)$	$P_3(\cos\theta)$	$P_4(\cos\theta)$	$P_5(\cos\theta)$
0	1	1	1	1	1
1	0.99985	0.99954	0.99909	0.99848	0.99772
2	.99939	.99817	.99635	.99392	.99088
3	.99863	.99589	.99179	.98634	.97954
4	.99756	.99270	.98543	.97577	.96377
5	.99619	.98861	.97728	.96227	.94368
6	.99452	.98361	.96736	.94589	.91939
7	.99255	.97772	.95569	.92670	.89108
8	.99027	.97095	.94232	.90480	.85893
9	.98769	.96329	.92726	.88026	.82315
10	.98481	.95477	.91057	.85321	.78399
11	.98163	.94539	.89228	.82376	.74170
12	.97815	.93516	.87244	.79204	.69656
13	.97437	.92410	.85111	.75819	.64888
14	.97030	.91221	.82833	.72235	.59895
15	.96593	.89952	.80416	.68470	.54713
16	.96126	.88604	.77868	.64537	.49373
17	.95630	.87178	.75194	.60456	.43911
18	.95106	.85676	.72401	.56244	.38363
19	.94552	.84101	.69497	.51918	.32763
20	.93969	.82453	.66488	.47498	.27149
21	.93358	.80736	.63384	.43002	.21556
22	.92718	.78950	.60190	.38450	.16019
23	.92050	.77099	.56917	.33862	.10573
24	.91355	.75185	.53572	.29256	.05252
25	.90631	.73209	.50163	.24653	.00088
26	.89879	.71175	.46699	.20072	− .04887
27	.89101	.69084	.43190	.15531	− .09642
28	.88295	.66939	.39644	.11051	− .14151
29	.87462	.64744	.36069	.06649	− .18388
30	.86603	.62500	.32476	.02344	− .22327
31	.85717	.60210	.28873	− .01847	− .25949
32	.84805	.57878	.25269	− .05907	− .29233
33	.83867	.55505	.21673	− .09820	− .32163
34	.82904	.53095	.18094	− .13570	− .34726
35	.81915	.50652	.14542	− .17142	− .36910
36	.80902	.48176	.11025	− .20524	− .38707
37	.79864	.45673	.07551	− .23701	− .40113
38	.78801	.43144	.04129	− .26664	− .41124
39	.77715	.40593	.00769	− .29400	− .41741
40	.76604	.38024	− .02523	− .31900	− .41968
41	.75471	.35438	− .05738	− .34157	− .41811
42	.74314	.32840	− .08869	− .36163	− .41279
43	.73135	.30232	− .11907	− .37913	− .40385
44	.71934	.27617	− .14845	− .39401	− .39141
45	.70711	.25000	− .17678	− .40625	− .37565
46	.69466	.22383	− .20397	− .41582	− .35677
47	.68200	.19768	− .22997	− .42273	− .33496
48	.66913	.17160	− .25471	− .42696	− .31048
49	.65606	.14562	− .27815	− .42856	− .28357

For $P_n(\cos\theta)$, use the definition of $P_n(x)$ given on the first page of this table, and put $x = \cos\theta$.

SURFACE ZONAL HARMONICS (Continued)

$\theta°$	$P_6(\cos\theta)$	$P_7(\cos\theta)$	$P_8(\cos\theta)$	$P_9(\cos\theta)$	$P_{10}(\cos\theta)$
0	1	1	1	1	1
1	0.99680	0.99574	0.9945	0.9932	0.9916
2	.98725	.98301	.9782	.9728	.9668
3	.97142	.96198	.9513	.9393	.9260
4	.94947	.93291	.9142	.8933	.8704
5	.92160	.89616	.8675	.8358	.8012
6	.88808	.85220	.8121	.7680	.7203
7	.84922	.80158	.7487	.6911	.6296
8	.80538	.74493	.6784	.6069	.5312
9	.75698	.68296	.6024	.5168	.4277
10	.70447	.61644	.5218	.4228	.3214
11	.64833	.54619	.4380	.3266	.2150
12	.58909	.47307	.3522	.2302	.1108
13	.52729	.39798	.2657	.1353	.0113
14	.46350	.32183	.1799	.0437	− .0813
15	.39831	.24554	.0962	− .0428	− .1651
16	.33229	.17001	.0157	− .1227	− .2381
17	.26606	.09614	− .0604	− .1946	− .2992
18	.20020	.02477	− .1310	− .2573	− .3471
19	.13529	− .04327	− .1951	− .3100	− .3813
20	.07190	− .10723	− .2518	− .3517	− .4013
21	.01059	− .16640	− .3005	− .3821	− .4073
22	− .04813	− .22017	− .3407	− .4009	− .3997
23	− .10376	− .26800	− .3718	− .4082	− .3793
24	− .15585	− .30942	− .3936	− .4042	− .3473
25	− .20398	− .34408	− .4062	− .3896	− .3052
26	− .24779	− .37172	− .4096	− .3650	− .2547
27	− .28694	− .39216	− .4041	− .3515	− .1975
28	− .32117	− .40534	− .3900	− .2902	− .1358
29	− .35025	− .41130	− .3680	− .2424	− .0716
30	− .37402	− .41018	− .3388	− .1896	− .0070
31	− .39238	− .40221	− .3031	− .1332	.0559
32	− .40527	− .38771	− .2619	− .0749	.1150
33	− .41269	− .36710	− .2162	− .0161	.1689
34	− .41471	− .34086	− .1670	.0415	.2157
35	− .41145	− .30956	− .1154	.0966	.2542
36	− .40307	− .27382	− .0627	.1476	.2833
37	− .38980	− .23432	− .0098	.1935	.3024
38	− .37191	− .19178	.0421	.2331	.3111
39	− .34972	− .14695	.0919	.2655	.3093
40	− .32357	− .10060	.1386	.2900	.2973
41	− .29387	− .05351	.1814	.3062	.2758
42	− .26104	− .00645	.2194	.3137	.2455
43	− .22554	.03982	.2519	.3127	.2077
44	− .18784	.08455	.2784	.3031	.1637
45	− .14844	.12706	.2983	.2855	.1151
46	− .10783	.16668	.3115	.2605	.0635
47	− .06654	.20283	.3176	.2288	.0107
48	− .02508	.23497	.3167	.1915	− .0416
49	.01606	.26263	.3090	.1495	− .0918

SURFACE ZONAL HARMONICS (Continued)

$\theta°$	$P_1(\cos\theta)$	$P_2(\cos\theta)$	$P_3(\cos\theta)$	$P_4(\cos\theta)$	$P_5(\cos\theta)$
50	0.64279	0.11976	−0.30022	−0.42753	−0.25449
51	.62932	.09407	−.32088	−.42394	−.22353
52	.61566	.06856	−.34009	−.41784	−.19097
53	.60182	.04327	−.35781	−.40929	−.15712
54	.58779	.01824	−.37399	−.39837	−.12229
55	.57358	−.00652	−.38861	−.38519	−.08679
56	.55919	−.03095	−.40164	−.36983	−.05093
57	.54464	−.05505	−.41307	−.35241	−.01503
58	.52992	−.07878	−.42286	−.33306	.02060
59	.51504	−.10210	−.43100	−.31189	.05566
60	.50000	−.12500	−.43750	−.28906	.08984
61	.48481	−.14744	−.44234	−.26471	.12287
62	.46947	−.16939	−.44552	−.23899	.15446
63	.45399	−.19084	−.44706	−.21205	.18436
64	.43837	−.21175	−.44695	−.18407	.21232
65	.42262	−.23209	−.44522	−.15521	.23811
66	.40674	−.25185	−.44188	−.12564	.26152
67	.39073	−.27099	−.43696	−.09554	.28238
68	.37461	−.28950	−.43049	−.06508	.30051
69	.35837	−.30736	−.42249	−.03444	.31577
70	.34202	−.32453	−.41301	−.00380	.32807
71	.32557	−.34101	−.40208	.02667	.33730
72	.30902	−.35676	−.38975	.05680	.34340
73	.29237	−.37178	−.37608	.08641	.34634
74	.27564	−.38604	−.36110	.11534	.34611
75	.25882	−.39952	−.34488	.14343	.34273
76	.24192	−.41221	−.32749	.17051	.33624
77	.22495	−.42410	−.30897	.19644	.32672
78	.20791	−.43516	−.28940	.22107	.31425
79	.19081	−.44539	−.26885	.24427	.29897
80	.17365	−.45477	−.24738	.26590	.28102
81	.15643	−.46329	−.22508	.28585	.26056
82	.13917	−.47095	−.20202	.30401	.23777
83	.12187	−.47772	−.17828	.32027	.21288
84	.10453	−.48361	−.15394	.33455	.18610
85	.08716	−.48861	−.12908	.34677	.15766
86	.06976	−.49270	−.10379	.35686	.12784
87	.05234	−.49589	−.07815	.36476	.09688
88	.03490	−.49817	−.05224	.37044	.06507
89	.01745	−.49954	−.02617	.37386	.03268
90	0	−.50000	0	.37500	0

SURFACE ZONAL HARMONICS (Continued)

$\theta°$	$P_6(\cos\theta)$	$P_7(\cos\theta)$	$P_8(\cos\theta)$	$P_9(\cos\theta)$	$P_{10}(\cos\theta)$
50	0.05638	0.28543	0.2947	0.1041	−0.1381
51	.09539	.30308	.2742	.0565	−.1792
52	.13265	.31535	.2480	.0080	−.2137
53	.16772	.32213	.2167	−.0400	−.2407
54	.20020	.32336	.1812	−.0863	−.2594
55	.22972	.31910	.1422	−.1296	−.2692
56	.25597	.30949	.1005	−.1689	−.2700
57	.27866	.29475	.0572	−.2032	−.2617
58	.29756	.27518	.0131	−.2315	−.2449
59	.31246	.25117	−.0309	−.2533	−.2201
60	.32324	.22314	−.0736	−.2679	−.1882
61	.32980	.19162	−.1144	−.2751	−.1504
62	.33210	.15715	−.1523	−.2747	−.1080
63	.33016	.12034	−.1865	−.2669	−.0624
64	.32403	.08181	−.2163	−.2518	−.0151
65	.31383	.04222	−.2411	−.2300	.0323
66	.29971	.00223	−.2605	−.2022	.0783
67	.28189	−.03748	−.2741	−.1690	.1213
68	.26062	−.07627	−.2816	−.1315	.1599
69	.23617	−.11348	−.2829	−.0906	.1929
70	.20888	−.14853	−.2780	−.0476	.2193
71	.17910	−.18082	−.2671	−.0035	.2382
72	.14721	−.20986	−.2504	.0404	.2491
73	.11363	−.23516	−.2283	.0829	.2516
74	.07878	−.25634	−.2014	.1230	.2457
75	.04310	−.27305	−.1702	.1595	.2316
76	.00704	−.28504	−.1355	.1915	.2099
77	−.02896	−.29214	−.0070	.2181	.1813
78	−.06444	−.29424	−.0583	.2387	.1468
79	−.09897	−.29133	−.0176	.2526	.1074
80	−.13212	−.28348	.0233	.2596	.0647
81	−.16348	−.27083	.0636	.2595	.0199
82	−.19267	−.25360	.1024	.2523	−.0254
83	−.21933	−.23211	.1389	.2383	−.0698
84	−.24313	.20671	.1722	.2177	−.1118
85	−.26378	−.17784	.2017	.1913	−.1499
86	−.28103	−.14598	.2268	.1597	−.1830
87	−.29467	−.11168	.2469	.1237	−.2099
88	−.30454	−.07551	.2615	.0844	−.2298
89	−.31050	−.03807	.2704	.0428	−.2420
90	−.31250	0	.2734	0	−.2461

For larger tables, see "Report of the British Association for the Advancement of Science," 1879, pp. 54–57, Philosophical Magazine, Dec., 1891, "Tafeln der Besselschen, Theta-, Kugel- und anderer Funktionen" by K. Hayashi and "Fünfstellige Funktionentafeln" by K. Hayashi.

SURFACE ZONAL HARMONICS—FIRST DERIVATIVES

$$\frac{d}{d\theta} P_n(\cos\theta)$$

$\theta°$	$\frac{d}{d\theta} P_1$	$\frac{d}{d\theta} P_2$	$\frac{d}{d\theta} P_3$	$\frac{d}{d\theta} P_4$	$\frac{d}{d\theta} P_5$	$\frac{d}{d\theta} P_6$	$\frac{d}{d\theta} P_7$
0	0	0	0	0	0	0	0
1	−0.0175	−0.0523	−0.1047	−0.1744	−0.2615	−0.3659	−0.488
2	−.0349	−.1046	−.2091	−.3480	−.5213	−.7284	−.969
3	−.0523	−.1568	−.3129	−.5201	−.7775	−1.0841	−1.438
4	−.0698	−.2088	−.4160	−.6901	−1.0286	−1.4295	−1.890
5	−.0872	−.2605	−.5180	−.8570	−1.2728	−1.7614	−2.317
6	−.1045	−.3119	−.6186	−1.0197	−1.5085	−2.0768	−2.715
7	−.1219	−.3629	−.7176	−1.1781	−1.7341	−2.3727	−3.080
8	−.1392	−.4135	−.8148	−1.3315	−1.9481	−2.6465	−3.405
9	−.1564	−.4635	−.9099	−1.4789	−2.1492	−2.8954	−3.689
10	−.1736	−.5130	−1.0026	−1.6199	−2.3360	−3.1174	−3.926
11	−.1908	−.5619	−1.0928	−1.7537	−2.5074	−3.3104	−4.116
12	−.2079	−.6101	−1.1801	−1.8799	−2.6621	−3.4729	−4.254
13	−.2250	−.6576	−1.2643	−1.9978	−2.7993	−3.6034	−4.341
14	−.2419	−.7042	−1.3453	−2.1069	−2.9181	−3.7008	−4.376
15	−.2588	−.7500	−1.4229	−2.2069	−3.0178	−3.7646	−4.358
16	−.2756	−.7949	−1.4968	−2.2973	−3.0978	−3.7943	−4.288
17	−.2924	−.8388	−1.5668	−2.3777	−3.1576	−3.7899	−4.169
18	−.3090	−.8817	−1.6328	−2.4478	−3.1970	−3.7518	−4.001
19	−.3256	−.9235	−1.6946	−2.5073	−3.2158	−3.6806	−3.788
20	−.3420	−.9642	−1.7520	−2.5560	−3.2141	−3.5774	−3.534
21	−.3584	−1.0037	−1.8050	−2.5937	−3.1920	−3.4435	−3.241
22	−.3746	−1.0420	−1.8534	−2.6203	−3.1497	−3.2804	−2.915
23	−.3907	−1.0790	−1.8970	−2.6357	−3.0877	−3.0917	−2.561
24	−.4067	−1.1147	−1.9357	−2.6400	−3.0067	−2.8749	−2.183
25	−.4226	−1.1491	−1.9696	−2.6330	−2.9073	−2.6371	−1.787
26	−.4384	−1.1820	−1.9984	−2.6150	−2.7903	−2.3794	−1.378
27	−.4540	−1.2135	−2.0222	−2.5861	−2.6568	−2.1046	−.963
28	−.4695	−1.2436	−2.0408	−2.5464	−2.5077	−1.8156	−.548
29	−.4848	−1.2721	−2.0542	−2.4961	−2.3443	−1.5155	−.136
30	−.5000	−1.2990	−2.0625	−2.4357	−2.1680	−1.2077	.263
31	−.5150	−1.3244	−2.0654	−2.3654	−1.9799	−.8953	.647
32	−.5299	−1.3482	−2.0635	−2.2855	−1.7817	−.5815	1.010
33	−.5446	−1.3703	−2.0562	−2.1966	−1.5748	−.2697	1.347
34	−.5592	−1.3908	−2.0437	−2.0991	−1.3608	.0370	1.654
35	−.5736	−1.4095	−2.0262	−1.9934	−1.1413	.3354	1.927
36	−.5878	−1.4266	−2.0036	−1.8802	−.9179	.6225	2.162
37	−.6018	−1.4419	−1.9761	−1.7600	−.6924	.8955	2.357
38	−.6157	−1.4554	−1.9438	−1.6334	−.4664	1.1516	2.510
39	−.6293	−1.4672	−1.9066	−1.5011	−.2415	1.3885	2.620
40	−.6428	−1.4772	−1.8648	−1.3637	−.0194	1.6038	2.684
41	−.6561	−1.4854	−1.8185	−1.2219	.1983	1.7955	2.705
43	−.6691	−1.4918	−1.7678	−1.0764	.4100	1.9620	2.681
43	−.6820	−1.4963	−1.7129	−.9279	.6142	2.1017	2.614
44	−.6947	−1.4991	−1.6539	−.7772	.8095	2.2136	2.506
45	−.7071	−1.5000	−1.5910	−.6250	.9943	2.2969	2.359
46	−.7193	−1.4991	−1.5244	−.4720	1.1677	2.3510	2.176
47	−.7314	−1.4963	−1.4542	−.3190	1.3282	2.3757	1.961
48	−.7431	−1.4918	−1.3808	−.1668	1.4749	2.3715	1.717
49	−.7547	−1.4854	−1.3042	−.0160	1.6067	2.3382	1.449

SURFACE ZONAL HARMONICS—FIRST DERIVATIVES (Continued)

$$\frac{d}{d\theta} P_n(\cos \theta)$$

$\theta°$	$\frac{d}{d\theta} P_1$	$\frac{d}{d\theta} P_2$	$\frac{d}{d\theta} P_3$	$\frac{d}{d\theta} P_4$	$\frac{d}{d\theta} P_5$	$\frac{d}{d\theta} P_6$	$\frac{d}{d\theta} P_7$
50	−0.7660	−1.4772	−1.2248	0.1327	1.7229	2.2772	1.161
51	−.7771	−1.4672	−1.1427	.2784	1.8225	2.1892	.859
52	−.7880	−1.4554	−1.0581	.4205	1.9053	2.0760	.546
53	−.7986	−1.4419	−.9714	.5584	1.9704	1.9387	.230
54	−.8090	−1.4266	−.8828	.6914	2.0173	1.7797	−.088
55	−.8192	−1.4095	−.7925	.8188	2.0474	1.6009	−.399
56	−.8290	−1.3908	−.7007	.9401	2.0587	1.4046	−.700
57	−.8387	−1.3703	−.6078	1.0546	2.0521	1.1936	−.986
58	−.8480	−1.3482	−.5140	1.1620	2.0281	.9099	−1.252
59	−.8572	−1.3244	−.4196	1.2617	1.9866	.7369	−1.495
60	−.8660	−1.2990	−.3248	1.3532	1.9283	.4973	−1.711
61	−.8746	−1.2721	−.2299	1.4361	1.8538	.2540	−1.896
62	−.8829	−1.2436	−.1351	1.5101	1.7640	.0099	−2.048
63	−.8910	−1.2135	−.0409	1.5748	1.6596	−.2321	−2.165
64	−.8988	−1.1820	.0527	1.6306	1.5420	−.4691	−2.244
65	−.9063	−1.1491	.1454	1.6755	1.4114	−.6983	−2.286
66	−.9135	−1.1147	.2368	1.7111	1.2698	−.9170	−2.290
67	−.9205	−1.0790	.3268	1.7366	1.1187	−1.1226	−2.255
68	−.9272	−1.0420	.4149	1.7520	.9580	−1.3129	−2.183
69	−.9336	−1.0037	.5011	1.7574	.7906	−1.4855	−2.076
70	−.9397	−.9642	.5851	1.7525	.6173	−1.6386	−1.984
71	−.9455	−.9235	.6666	1.7376	.4397	−1.7704	−1.762
72	−.9511	−.8817	.7455	1.7131	.2594	−1.8794	−1.561
73	−.9563	−.8388	.8214	1.6787	.0776	−1.9646	−1.334
74	−.9613	−.7949	.8941	1.6349	−.1037	−2.0248	−1.088
75	−.9659	−.7500	.9636	1.5819	−.2833	−2.0596	−.824
76	−.9703	−.7042	1.0295	1.5200	−.4595	−2.0687	−.548
77	−.9744	−.6576	1.0918	1.4498	−.6309	−2.0520	−.264
78	−.9781	−.6101	1.1501	1.3714	−.7961	−2.0098	.023
79	−.9816	−.5619	1.2044	1.2854	−.9537	−1.9428	.309
80	−.9848	−.5130	1.2545	1.1923	−1.1023	−1.8519	.589
81	−.9877	−.4635	1.3003	1.0296	−1.2407	−1.7382	.858
82	−.9903	−.4135	1.3416	.9870	−1.3679	−1.6032	1.112
83	−.9925	−.3629	1.3783	.8758	−1.4831	−1.4487	1.348
84	−.9945	−.3119	1.4103	.7598	−1.5841	−1.2760	1.559
85	−.9962	−.2605	1.4376	.6396	−1.6715	−1.0881	1.745
86	−.9976	−.2088	1.4599	.5160	−1.7440	−.8868	1.901
87	−.9986	−.1568	1.4774	.3895	−1.8009	−.6747	2.024
88	−.9994	−.1046	1.4899	.2608	−1.8420	−.4544	2.115
89	−.9998	−.0523	1.4975	.1308	−1.8667	−.2286	2.169
90	−1.0000	0	1.5000	0	−1.8750	0	2.187

Note that

$$\frac{d}{dx} P_n(x) = -\frac{1}{\sin \theta} \frac{d}{d\theta} P_n(\cos \theta)$$

where

$$x = \cos \theta$$

BERNOULLI AND EULER NUMBERS—POLYNOMIALS

There are numerous sets of defined numbers and polynomials, among which the more important ones are those classified under the names of Bernoulli and Euler.

The *Bernoulli Polynomials* are generated by the defining condition

$$\frac{te^{tx}}{e^t - 1} = B_0(x) + B_1(x)t + B_2(x)\frac{t^2}{2!} + B_3(x)\frac{t^3}{3!} + \cdots$$

$$B_0(x) = 1, \quad B_1(x) = x - \tfrac{1}{2}, \quad B_2(x) = x^2 - x + \tfrac{1}{6}$$

$$B_3(x) = x^3 - \frac{3}{2}x^2 + \frac{x}{2}, \quad B_4(x) = x^4 - 2x^3 + x^2 - \tfrac{1}{30}, \ldots$$

The following useful relations should be noted

$$B_n'(x) = nB_{n-1}(x), \quad \int_a^x B_n(x)\,dx = \frac{1}{n+1}[B_{n+1}(x) - B_{n+1}(a)]$$

The Bernoulli numbers $B_0, B_1, B_2, \ldots$, can be obtained by putting $x = 0$ in the respective polynomials. A simpler method is to use the generating function

$$\frac{x}{e^x - 1} = \sum_{n=0}^{\infty} \frac{B_n x^n}{n!}$$

and read the coefficients from the expansion. Here,

$$B_0 = 1, \quad B_1 = -\tfrac{1}{2}, \quad B_2 = \tfrac{1}{6}, \quad B_4 = -\tfrac{1}{30},$$

$$B_6 = \tfrac{1}{42}, \quad B_8 = -\tfrac{1}{30}, \ldots B_{2n+1} = 0 \quad (n \geq 1)$$

An important application for the Bernoulli coefficients is in its use involved for the Euler-Maclaurin Sum Formula:

$$\sum_{x=1}^{n-1} f(x) = \int_1^n f(x)\,dx + \left[\sum_{i=1}^{\infty} \frac{B_i}{i!} f^{(i-1)}(x)\right]_{x=1}^{x=n}$$

$$= \left[\int f(x)\,dx - \frac{1}{2}f(x) + \frac{1}{12}f'(x) - \frac{1}{720}f'''(x)\right.$$

$$\left. + \frac{1}{30,240}f^{(V)}(x) - \frac{1}{1,209,600}f^{(VII)}(x)\right]_{x=1}^{x=n}$$

Examples:

$$\sum_{x=1}^{n-1} \sqrt{x} = \sqrt{n}\left\{\frac{2}{3}n - \frac{1}{2} + \frac{1}{24n} - \frac{1}{1,920n^3}\right.$$

$$\left. + \frac{1}{9,216n^5} - \frac{11}{163,840n^7}\right\} - 0.207,886,224,977,355$$

correct to 12 places for $n \geq 10$.

$$\log_e(x!) = \log_e \Gamma(x+1)$$

$$= \left(x + \frac{1}{2}\right)\log_e x - x + \frac{1}{12x} - \frac{1}{360x^3} + \frac{1}{1,260x^5} - \frac{1}{1,680x^7}$$

$$+ 0.918,938,533,205$$

accurate to 12 places for $x \geq 10$.

$$f(x) \, (\text{digamma function}) = \frac{d \log \Gamma(x)}{dx}$$

$$= 1 + \frac{1}{2} + \frac{1}{3} + \cdots + \frac{1}{x-1} - \gamma$$

(Euler Constant) for x integer.

By Euler-Maclaurin

$$f(x) = \log_e x - \frac{1}{2x} - \frac{1}{12x^2} + \frac{1}{120x^4} - \frac{1}{252x^6} + \frac{1}{240x^8} - \frac{5}{660x^{10}} + \frac{691}{32,760x^{12}}$$

correct to 12 places for $x \geq 10$.

The Euler numbers together with their respective polynomials can be generated from

$$\frac{2e^{tx}}{e^t + 1} = \sum_{i=0}^{\infty} E_i(x) \frac{t^i}{i!}$$

and the relation

$$x^n = \frac{1}{2} \left[E_n(x+1) + E_n(x) \right]$$

The Euler polynomials are

$$E_0(x) = 1, \quad E_1(x) = x - \frac{1}{2}, \quad E_2(x) = x^2 - x$$

$$E_3(x) = x^3 - \frac{3}{2}x^2 + \frac{1}{4}, \quad E_4(x) = x^4 - 2x^3 + x$$

$$E_5(x) = x^5 - \frac{5}{2}x^4 + \frac{5}{2}x^2 - \frac{1}{2}$$

Tables which follow record the first fifteen polynomials of $B_k(x)$ and $E_k(x)$. The first sixty Bernoulli and Euler numbers are given in a separate table. The value of $x^n/n!$ is also important and this is given in a succeeding table for values of x from 1 to 9 and n from 1 to 50.

COEFFICIENTS b_k OF THE BERNOULLI POLYNOMIALS $B_n(x) = \sum_{k=0}^{n} b_k x^k$

$n\backslash k$	0	1	2	3	4	5	6	7	8	9	10	11	12	13	14	15
0	1															
1	$-\frac{1}{2}$	1														
2	$\frac{1}{6}$	-1	1													
3	0	$\frac{1}{2}$	$-\frac{3}{2}$	1												
4	$-\frac{1}{30}$	0	1	-2	1											
5	0	$-\frac{1}{6}$	0	$\frac{5}{3}$	$-\frac{5}{2}$	1										
6	$\frac{1}{42}$	0	$-\frac{1}{2}$	0	$\frac{5}{2}$	-3	1									
7	0	$\frac{1}{6}$	0	$-\frac{7}{6}$	0	$\frac{7}{2}$	$-\frac{7}{2}$	1								
8	$-\frac{1}{30}$	0	$\frac{2}{3}$	0	$-\frac{7}{3}$	0	$\frac{14}{3}$	-4	1							
9	0	$-\frac{3}{10}$	0	2	0	$-\frac{21}{5}$	0	6	$-\frac{9}{2}$	1						
10	$\frac{5}{66}$	0	$-\frac{3}{2}$	0	5	0	-7	0	$\frac{15}{2}$	-5	1					
11	0	$\frac{5}{6}$	0	$-\frac{11}{2}$	0	11	0	-11	0	$\frac{55}{6}$	$-\frac{11}{2}$	1				
12	$-\frac{691}{2730}$	0	5	0	$-\frac{33}{2}$	0	22	0	$-\frac{33}{2}$	0	11	-6	1			
13	0	$-\frac{691}{210}$	0	$\frac{65}{3}$	0	$-\frac{429}{10}$	0	$\frac{286}{7}$	0	$-\frac{143}{6}$	0	13	$-\frac{13}{2}$	1		
14	$\frac{7}{6}$	0	$-\frac{691}{30}$	0	$\frac{455}{6}$	0	$-\frac{1001}{10}$	0	$\frac{143}{2}$	0	$-\frac{1001}{30}$	0	$\frac{91}{6}$	-7	1	
15	0	$\frac{35}{2}$	0	$-\frac{691}{6}$	0	$\frac{455}{2}$	0	$-\frac{429}{2}$	0	$\frac{715}{6}$	0	$-\frac{91}{2}$	0	$\frac{35}{2}$	$-\frac{15}{2}$	1

COEFFICIENTS e_k OF THE EULER POLYNOMIALS $E_n(x) = \sum_{k=0}^{n} e_k x^k$

$n\backslash k$	0	1	2	3	4	5	6	7	8	9	10	11	12	13	14	15
0	1															
1	$-\frac{1}{2}$	1														
2	0	-1	1													
3	$\frac{1}{4}$	0	$-\frac{3}{2}$	1												
4	0	1	0	-2	1											
5	$-\frac{1}{2}$	0	$\frac{5}{2}$	0	$-\frac{5}{2}$	1										
6	0	-3	0	5	0	-3	1									
7	$\frac{17}{8}$	0	$-\frac{21}{2}$	0	$\frac{35}{4}$	0	$-\frac{7}{2}$	1								
8	0	17	0	-28	0	14	0	-4	1							
9	$-\frac{31}{2}$	0	$\frac{153}{2}$	0	-63	0	21	0	$-\frac{9}{2}$	1						
10	0	-155	0	255	0	-126	0	30	0	-5	1					
11	$\frac{691}{4}$	0	$-\frac{1705}{2}$	0	$\frac{2805}{4}$	0	-231	0	$\frac{165}{2}$	0	$-\frac{11}{2}$	1				
12	0	2073	0	-3410	0	1683	0	-396	0	55	0	-6	1			
13	$-\frac{5461}{2}$	0	$\frac{26949}{2}$	0	$-\frac{22165}{2}$	0	$\frac{7293}{2}$	0	$-\frac{1287}{2}$	0	$\frac{143}{2}$	0	$-\frac{13}{2}$	1		
14	0	-38227	0	62881	0	-31031	0	7293	0	-1001	0	91	0	-7	1	
15	$\frac{929569}{16}$	0	$-\frac{573405}{2}$	0	$\frac{943215}{4}$	0	$-\frac{155155}{2}$	0	$\frac{109395}{8}$	0	-3003	0	$\frac{455}{4}$	0	$-\frac{15}{2}$	1

BERNOULLI NUMBERS

$$B_n = N/D$$

n	N	D	B_n
0	1	1	(0) 1.0000 00000
1	−1	2	(−1)−5.0000 00000*
2	1	6	(−1) 1.6666 66667
4	−1	30	(−2)−3.3333 33333
6	1	42	(−2) 2.3809 52381
8	−1	30	(−2)−3.3333 33333
10	5	66	(−2) 7.5757 57576
12	−691	2730	(−1)−2.5311 35531
14	7	6	(0) 1.1666 66667
16	−3617	510	(0)−7.0921 56863
18	43867	798	(1) 5.4971 17794
20	−1 74611	330	(2)−5.2912 42424
22	8 54513	138	(3) 6.1921 23188
24	−2363 64091	2730	(4)−8.6580 25311
26	85 53103	6	(6) 1.4255 17167
28	−2 37494 61029	870	(7)−2.7298 23107
30	861 58412 76005	14322	(8) 6.0158 08739
32	−770 93210 41217	510	(10)−1.5116 31577
34	257 76878 58367	6	(11) 4.2961 46431
36	−26315 27155 30534 77373	19 19190	(13)−1.3711 65521
38	2 92999 39138 41559	6	(14) 4.8833 23190
40	−2 61082 71849 64491 22051	13530	(16)−1.9296 57934
42	15 20097 64391 80708 02691	1806	(17) 8.4169 30476
44	−278 33269 57930 10242 35023	690	(19)−4.0338 07185
46	5964 51111 59391 21632 77961	282	(21) 2.1150 74864
48	−560 94033 68997 81768 62491 27547	46410	(23)−1.2086 62652
50	49 50572 05241 07964 82124 77525	66	(24) 7.5008 66746
52	−80116 57181 35489 95734 79249 91853	1590	(26)−5.0387 78101
54	29 14996 36348 84862 42141 81238 12691	798	(28) 3.6528 77648
56	−2479 39292 93132 26753 68541 57396 63229	870	(30)−2.8498 76930
58	84483 61334 88800 41862 04677 59940 36021	354	(32) 2.3865 42750
60	−121 52331 40483 75557 20403 04994 07982 02460 41491	567 86730	(34)−2.1399 94926

*The floating decimal point notation is used here. For example for $n = 1$, $B_1 = -\frac{1}{2} = -.500000$ $= (-5.00000)(10^{-1}) = (-1) - 5.0000,0000.$

EULER NUMBERS

n	En
0	1
2	−1
4	5
6	−61
8	1385
10	−50521
12	27 02765
14	−1993 60981
16	1 93915 12145
18	−240 48796 75441
20	37037 11882 37525
22	−69 34887 43931 37901
24	15514 53416 35570 86905
26	−40 87072 50929 31238 92361
28	12522 59641 40362 98654 68285
30	−44 15438 93249 02310 45536 82821
32	17751 93915 79539 28943 66647 89665
34	−80 72329 92358 87898 06216 82474 53281
36	41222 06033 95177 02122 34707 96712 59045
38	−234 89580 52704 31082 52017 82857 61989 47741
40	1 48511 50718 11498 00178 77156 78140 58266 84425
42	−1036 46227 33519 61211 93979 57304 74518 59763 10201
44	7 94757 94225 97592 70360 80405 10088 07061 95192 73805
46	−6667 53751 66855 44977 43502 84747 73748 19752 41076 84661
48	60 96278 64556 85421 58691 68574 28768 43153 97653 90444 35185
50	−60532 85248 18862 18963 14383 78511 16490 88103 49822 51468 15121
52	650 61624 86684 60884 77158 70634 08082 29834 83644 23676 53855 76565
54	−7 54665 99390 08739 09806 14325 65889 73674 42122 40024 71169 98586 45581
56	9420 32189 64202 41204 20228 62376 90583 22720 93888 52599 64600 93949 05945
58	−126 22019 25180 62187 19903 40923 72874 89255 48234 10611 91825 59406 99649 20041
60	181089 11496 57923 04965 45807 74165 21586 88733 48734 92363 14106 00809 54542 31325

BERNOULLI AND EULER POLYNOMIALS, RIEMANN ZETA FUNCTION

$x^n/n!$

$n\backslash x$	2		3		4		5	
1	(0) 2.0000	00000	(0) 3.0000	00000	(0) 4.0000	00000	(0) 5.0000	00000
2	(0) 2.0000	00000	(0) 4.5000	00000	(0) 8.0000	00000	(1) 1.2500	00000
3	(0) 1.3333	33333	(0) 4.5000	00000	(1) 1.0666	66667	(1) 2.0833	33333
4	(− 1) 6.6666	66667	(0) 3.3750	00000	(1) 1.0666	66667	(1) 2.6041	66667
5	(− 1) 2.6666	66667	(0) 2.0250	00000	(0) 8.5333	33333	(1) 2.6041	66667
6	(− 2) 8.8888	88889	(0) 1.0125	00000	(0) 5.6888	88889	(1) 2.1701	38889
7	(− 2) 2.5396	82540	(− 1) 4.3392	85714	(0) 3.2507	93651	(1) 1.5500	99206
8	(− 3) 6.3492	06349	(− 1) 1.6272	32143	(0) 1.6253	96825	(0) 9.6881	20040
9	(− 3) 1.4109	34744	(− 2) 5.4241	07143	(− 1) 7.2239	85891	(0) 5.3822	88911
10	(− 4) 2.8218	69489	(− 2) 1.6272	32144	(− 1) 2.8895	94356	(0) 2.6911	44455
11	(− 5) 5.1306	71797	(− 3) 4.4379	05844	(− 1) 1.0507	61584	(0) 1.2232	47480
12	(− 6) 8.5511	19662	(− 3) 1.1094	76461	(− 2) 3.5025	38614	(− 1) 5.0968	64499
13	(− 6) 1.3155	56871	(− 4) 2.5603	30295	(− 2) 1.0777	04189	(− 1) 1.9603	32500
14	(− 7) 1.8793	66959	(− 5) 5.4864	22060	(− 3) 3.0791	54825	(− 2) 7.0011	87499
15	(− 8) 2.5058	22612	(− 5) 1.0972	84412	(− 4) 8.2110	79534	(− 2) 2.3337	29166
16	(− 9) 3.1322	78264	(− 6) 2.0574	08272	(− 4) 2.0527	69883	(− 3) 7.2929	03644
17	(−10) 3.6850	33252	(− 7) 3.6307	20481	(− 5) 4.8300	46785	(− 3) 2.1449	71660
18	(−11) 4.0944	81391	(− 8) 6.0512	00801	(− 5) 1.0733	43730	(− 4) 5.9582	54611
19	(−12) 4.3099	80412	(− 9) 9.5545	27582	(− 6) 2.2596	71011	(− 4) 1.5679	61740
20	(−13) 4.3099	80413	(− 9) 1.4331	79137	(− 7) 4.5193	42021	(− 5) 3.9199	04350
21	(−14) 4.1047	43250	(−10) 2.0473	98768	(− 8) 8.6082	70516	(− 6) 9.3331	05595
22	(−15) 3.7315	84772	(−11) 2.7919	07410	(− 8) 1.5651	40093	(− 6) 2.1211	60362
23	(−16) 3.2448	56324	(−12) 3.6416	18361	(− 9) 2.7219	82772	(− 7) 4.6112	18179
24	(−17) 2.7040	46937	(−13) 4.5520	22952	(−10) 4.5366	37953	(− 8) 9.6067	04540
25	(−18) 2.1632	37550	(−14) 5.4624	27543	(−11) 7.2586	20726	(− 8) 1.9213	40908
26	(−19) 1.6640	28884	(−15) 6.3028	01010	(−11) 1.1167	10881	(− 9) 3.6948	86362
27	(−20) 1.2326	13988	(−16) 7.0031	12233	(−12) 1.6543	86490	(−10) 6.8423	82151
28	(−22) 8.8043	85630	(−17) 7.5033	34535	(−13) 2.3634	09271	(−10) 1.2218	53956
29	(−23) 6.0719	90089	(−18) 7.7620	70209	(−14) 3.2598	74857	(−11) 2.1066	44751
30	(−24) 4.0479	93393	(−19) 7.7620	70209	(−15) 4.3464	99810	(−12) 3.5110	74585
31	(−25) 2.6116	08641	(−20) 7.5116	80847	(−16) 5.6083	86851	(−13) 5.6630	23524
32	(−26) 1.6322	55401	(−21) 7.0422	00795	(−17) 7.0104	83564	(−14) 8.8484	74257
33	(−28) 9.8924	56972	(−22) 6.4020	00722	(−18) 8.4975	55834	(−14) 1.3406	77918
34	(−29) 5.8190	92337	(−23) 5.6488	24167	(−19) 9.9971	24511	(−15) 1.9715	85173
35	(−30) 3.3251	95620	(−24) 4.8418	49286	(−19) 1.1425	28515	(−16) 2.8165	50246
36	(−31) 1.8473	30900	(−25) 4.0348	74405	(−20) 1.2694	76128	(−17) 3.9118	75343
37	(−33) 9.9855	72436	(−26) 3.2715	19788	(−21) 1.3724	06625	(−18) 5.2863	18032
38	(−34) 5.2555	64439	(−27) 2.5827	78779	(−22) 1.4446	38552	(−19) 6.9556	81619
39	(−35) 2.6951	61251	(−28) 1.9867	52908	(−23) 1.4816	80567	(−20) 8.9175	40539
40	(−36) 1.3475	80626	(−29) 1.4900	64681	(−24) 1.4816	80567	(−20) 1.1146	92567
41	(−38) 6.5735	64028	(−30) 1.0902	91230	(−25) 1.4455	42017	(−21) 1.3593	81180
42	(−39) 3.1302	68584	(−32) 7.7877	94498	(−26) 1.3767	06682	(−22) 1.6183	10928
43	(−40) 1.4559	38876	(−33) 5.4333	44999	(−27) 1.2806	57379	(−23) 1.8817	56893
44	(−42) 6.6179	03983	(−34) 3.7045	53408	(−28) 1.1642	33981	(−24) 2.1383	60106
45	(−43) 2.9412	90659	(−35) 2.4697	02271	(−29) 1.0348	74650	(−25) 2.3759	55673
46	(−44) 1.2788	22026	(−36) 1.6106	75395	(−31) 8.9989	09998	(−26) 2.5825	60514
47	(−46) 5.4417	95855	(−37) 1.0280	90677	(−32) 7.6586	46807	(−27) 2.7474	04803
48	(−47) 2.2674	14940	(−39) 6.4255	66736	(−33) 6.3822	05674	(−28) 2.8618	80003
49	(−49) 9.2547	54855	(−40) 3.9340	20450	(−34) 5.2099	63815	(−29) 2.9202	85717
50	(−50) 3.7019	10942	(−41) 2.3604	12270	(−35) 4.1679	71052	(−30) 2.9202	85717

BERNOULLI AND EULER POLYNOMIALS, RIEMANN ZETA FUNCTION

$$x^n/n!$$

n\x	6		7		8		9	
1	(0) 6.0000	00000	(0) 7.0000	00000	(0) 8.0000	00000	(0) 9.0000	00000
2*	(1) 1.8000	00000	(1) 2.4500	00000	(1) 3.2000	00000	(1) 4.0500	00000
3	(1) 3.6000	00000	(1) 5.7166	66667	(1) 8.5333	33333	(2) 1.2150	00000
4	(1) 5.4000	00000	(2) 1.0004	16667	(2) 1.7066	66667	(2) 2.7337	50000
5	(1) 6.4800	00000	(2) 1.4005	83333	(2) 2.7306	66667	(2) 4.9207	50000
6	(1) 6.4800	00000	(2) 1.6340	13889	(2) 3.6408	88889	(2) 7.3811	25000
7	(1) 5.5542	85714	(2) 1.6340	13889	(2) 4.1610	15873	(2) 9.4900	17857
8	(1) 4.1657	14286	(2) 1.4297	62153	(2) 4.1610	15873	(3) 1.0676	27009
9	(1) 2.7771	42857	(2) 1.1120	37230	(2) 3.6986	80776	(3) 1.0676	27009
10	(1) 1.6662	85714	(1) 7.7842	60610	(2) 2.9589	44621	(2) 9.6086	43080
11	(0) 9.0888	31169	(1) 4.9536	20388	(2) 2.1519	59724	(2) 7.8616	17066
12	(0) 4.5444	15584	(1) 2.8896	11893	(2) 1.4346	39816	(2) 5.8962	12799
13	(0) 2.0974	22577	(1) 1.5559	44865	(1) 8.8285	52715	(2) 4.0819	93476
14	(– 1) 8.9889	53903	(0) 7.7797	24327	(1) 5.0448	87266	(2) 2.6241	38663
15	(– 1) 3.5955	81561	(0) 3.6305	38019	(1) 2.6906	06542	(2) 1.5744	83198
16	(– 1) 1.3483	43085	(0) 1.5883	60383	(1) 1.3453	03271	(1) 8.8564	67988
17	(– 2) 4.7588	57949	(– 1) 6.5403	07461	(0) 6.3308	38921	(1) 4.6887	18347
18	(– 2) 1.5862	85983	(– 1) 2.5434	52902	(0) 2.8137	06187	(1) 2.3443	59173
19	(– 3) 5.0093	24157	(– 2) 9.3706	15954	(0) 1.1847	18395	(1) 1.1104	85924
20	(– 3) 1.5027	97247	(– 2) 3.2797	15584	(– 1) 4.7388	73579	(0) 4.9971	86660
21	(– 4) 4.2937	06421	(– 2) 1.0932	38528	(– 1) 1.8052	85173	(0) 2.1416	51426
22	(– 4) 1.1710	10841	(– 3) 3.4784	86224	(– 2) 6.5646	73356	(– 1) 8.7613	01286
23	(– 5) 3.0548	10892	(– 3) 1.0586	69721	(– 2) 2.2833	64645	(– 1) 3.4283	35286
24	(– 6) 7.6370	27230	(– 4) 3.0877	86685	(– 3) 7.6112	15485	(– 1) 1.2856	25732
25	(– 6) 1.8328	86535	(– 5) 8.6458	02719	(– 3) 2.4355	88956	(– 2) 4.6282	52637
26	(– 7) 4.2297	38158	(– 5) 2.3277	16117	(– 4) 7.4941	19863	(– 2) 1.6020	87451
27	(– 8) 9.3994	18129	(– 6) 6.0348	19562	(– 4) 2.2204	79959	(– 3) 5.3402	91503
28	(– 8) 2.0141	61028	(– 6) 1.5087	04890	(– 5) 6.3442	28454	(– 3) 1.7165	22269
29	(– 9) 4.1672	29712	(– 7) 3.6417	01460	(– 5) 1.7501	31987	(– 4) 5.3271	38075
30	(– 10) 8.3344	59424	(– 8) 8.4973	03406	(– 6) 4.6670	18634	(– 4) 1.5981	41423
31	(– 10) 1.6131	21179	(– 8) 1.9187	45930	(– 6) 1.2043	91905	(– 5) 4.6397	65421
32	(– 11) 3.0246	02211	(– 9) 4.1972	56723	(– 7) 3.0109	79764	(– 5) 1.3049	34025
33	(– 12) 5.4992	76746	(– 10) 8.9032	71836	(– 8) 7.2993	44881	(– 6) 3.5589	10976
34	(– 13) 9.7046	06022	(– 10) 1.8330	26555	(– 8) 1.7174	92913	(– 7) 9.4206	46701
35	(– 13) 1.6636	46746	(– 11) 3.6660	53108	(– 9) 3.9256	98086	(– 7) 2.4224	52008
36	(– 14) 2.7727	44578	(– 12) 7.1284	36600	(– 10) 8.7237	73527	(– 8) 6.0561	30022
37	(– 15) 4.4963	42559	(– 12) 1.3486	23141	(– 10) 1.8862	21303	(– 8) 1.4731	12708
38	(– 16) 7.0994	88250	(– 13) 2.4843	05785	(– 11) 3.9709	92217	(– 9) 3.4889	51151
39	(– 16) 1.0922	28962	(– 14) 4.4590	10384	(– 12) 8.1456	25061	(– 10) 8.0514	25733
40	(– 17) 1.6383	43443	(– 15) 7.8032	68172	(– 12) 1.6291	25012	(– 10) 1.8115	70790
41	(– 18) 2.3975	75770	(– 15) 1.3322	65298	(– 13) 3.1787	80512	(– 11) 3.9766	18807
42	(– 19) 3.4251	08241	(– 16) 2.2204	42162	(– 14) 6.0548	20021	(– 12) 8.5213	26014
43	(– 20) 4.7792	20803	(– 17) 3.6146	73288	(– 14) 1.1264	78144	(– 12) 1.7835	33352
44	(– 21) 6.5171	19276	(– 18) 5.7506	16594	(– 15) 2.0481	42079	(– 13) 3.6481	36401
45	(– 22) 8.6894	92369	(– 19) 8.9454	03592	(– 16) 3.6411	41473	(– 14) 7.2962	72804
46	(– 22) 1.1334	12048	(– 19) 1.3612	57068	(– 17) 6.3324	19955	(– 14) 1.4275	31635
47	(– 23) 1.4469	08998	(– 20) 2.0274	04144	(– 17) 1.0778	58716	(– 15) 2.7335	71217
48	(– 24) 1.8086	36247	(– 21) 2.9566	31045	(– 18) 1.7964	31193	(– 16) 5.1254	46033
49	(– 25) 2.2146	56629	(– 22) 4.2237	58634	(– 19) 2.9329	48887	(– 17) 9.4140	84548
50	(– 26) 2.6575	87955	(– 23) 5.9132	62088	(– 20) 4.6927	18219	(– 17) 1 6945	35219

*The floating decimal point notation is used here. For example (1) 1.8000,000 = (1.8000,000) 10.

STIRLING NUMBERS

Stirling numbers

The Stirling numbers are used for reducing factorials such as $x^{(n)}$ to polynomials in x and vice versa.

Stirling numbers of the first kind

The factorial polynomial $x^{(n)}$ is defined and represented by

$$x^{(n)} = x(x - 1)(x - 2)\cdots(x - n + 1)$$

where $x^{(0)}$ is 1 by definition.

If n is a non-negative integer, then

$$x^{(n)} = s_{n1}x + s_{n2}x^2 + \cdots + s_{nn}x^n.$$

Here the numbers $s_{n1}, s_{n2}, s_{n3}, \ldots$, are Stirling numbers of the first kind. A table listing these numbers with an example using same follows.

STIRLING NUMBERS OF THE FIRST KIND

n	s_{n1}	s_{n2}	s_{n3}	s_{n4}	s_{n5}	s_{n6}	s_{n7}	s_{n8}
1	1	0	0	0	0	0	0	0
2	-1	1	0	0	0	0	0	0
3	2	-3	1	0	0	0	0	0
4	-6	11	-6	1	0	0	0	0
5	24	-50	35	-10	1	0	0	0
6	-120	274	-225	85	-15	1	0	0
7	720	-1 764	1 624	-735	175	-21	1	0
8	-5 040	13 068	-13 132	6 769	-1 960	322	-28	1

The table may be continued by using the recurrence formula

$$s_{ni} = s_{n-1,i-1} - (n - 1)s_{n-1,i}, \quad i = 1, 2, \ldots, n$$

where $s_{n0} = 0$ for all n.

Example

Express $3x^{(3)} + 2x^{(1)}$ using Stirling's numbers of the first kind

$$3x^{(3)} = 3(2x - 3x^2 + x^3)$$
$$= 6x - 9x^2 + 3x^3$$
$$2x^{(1)} = 2x$$
$$\therefore 3x^{(3)} + 2x^{(1)} = 8x - 9x^2 + 3x^3$$

This may be verified by carrying out the indicated operations as follows

$$3x^{(3)} + 2x^{(1)} = 3(x)(x - 1)(x - 2) + 2x$$
$$= 3x^3 - 9x^2 + 6x + 2x$$
$$= 8x - 9x^2 + 3x^3$$

Stirling numbers of the second kind

For every non-negative integer n the function defined by x^n can be expressed as a linear combination of factorial powers of x not higher than the n'th. In other words

$$x^n = t_{n1}x^{(1)} + t_{n2}x^{(2)} + \cdots + t_{nn}x^{(n)}.$$

The numbers $t_{n1}, t_{n2}, \ldots, t_{nn}$ are called Stirling's numbers of the second kind.

A table listing these numbers with an example using same follows.

STIRLING NUMBERS OF THE SECOND KIND

n	t_{n1}	t_{n2}	t_{n3}	t_{n4}	t_{n5}	t_{n6}	t_{n7}	t_{n8}
1	1	0	0	0	0	0	0	0
2	1	1	0	0	0	0	0	0
3	1	3	1	0	0	0	0	0
4	1	7	6	1	0	0	0	0
5	1	15	25	10	1	0	0	0
6	1	31	90	65	15	1	0	0
7	1	63	301	350	140	21	1	0
8	1	127	966	1 701	1 050	266	28	1

This table may be continued by using the recurrence formula

$$t_{ni} = i t_{n-1,i} + t_{n-1,i-1}, \quad i = 1, 2, \ldots, n$$

where $t_{n0} = 0$ for all n.

Example

Express $2x^3 - 3x^2 + x + 2$ by use of factorial powers

$$2x^3 = 2[x^{(1)} + 3x^{(2)} + x^{(3)}] = 2x^{(1)} + 6x^{(2)} + 2x^{(3)}$$
$$-3x^2 = -3[x^{(1)} + x^{(2)}] \qquad = -3x^{(1)} - 3x^{(2)}$$
$$x = x^{(1)} \qquad\qquad\qquad = x^{(1)}$$
$$+2 = +2x^{(0)} \qquad\qquad = 2x^{(0)}$$

$$\therefore 2x^3 - 3x^2 + x + 2 = 2 + 3x^{(2)} + 2x^{(3)}$$

FOURIER SERIES

(Also see Index for Cosine and Sine Transforms)

1. If $f(x)$ is a bounded periodic function of period $2L$ (i.e. $f(x + 2L) = f(x)$), and satisfies the *Dirichlet conditions*:

   a) In any period $f(x)$ is continuous, except possibly for a finite number of jump discontinuities

   b) In any period $f(x)$ has only a finite number of maxima and minima.

Then $f(x)$ may be represented by the *Fourier series*

$$\frac{a_0}{2} + \sum_{n=1}^{\infty} \left(a_n \cos \frac{n\pi x}{L} + b_n \sin \frac{n\pi x}{L} \right),$$

where a_n and b_n are as determined below. This series will converge to $f(x)$ at every point where $f(x)$ is continuous, and to

$$\frac{f(x^+) + f(x^-)}{2}$$

(i.e. the average of the left-hand and right-hand limits) at every point where $f(x)$ has a jump discontinuity.

$$a_n = \frac{1}{L} \int_{-L}^{L} f(x) \cos \frac{n\pi x}{L} \, dx, \quad n = 0, 1, 2, 3, \ldots;$$

$$b_n = \frac{1}{L} \int_{-L}^{L} f(x) \sin \frac{n\pi x}{L} \, dx, \quad n = 1, 2, 3, \ldots$$

We may also write

$$a_n = \frac{1}{L} \int_{\alpha}^{\alpha+2L} f(x) \cos \frac{n\pi x}{L} \, dx \text{ and } b_n = \frac{1}{L} \int_{\alpha}^{\alpha+2L} f(x) \sin \frac{n\pi x}{L} \, dx,$$

where α is any real number. Thus if $\alpha = 0$,

$$a_n = \frac{1}{L} \int_{0}^{2L} f(x) \cos \frac{n\pi x}{L} \, dx, \quad n = 0, 1, 2, 3, \ldots;$$

$$b_n = \frac{1}{L} \int_{0}^{2L} f(x) \sin \frac{n\pi x}{L} \, dx, \quad n = 1, 2, 3, \ldots$$

2. If in addition to the above restrictions, $f(x)$ is even (i.e. $f(-x) = f(x)$), the Fourier series reduces to

$$\frac{a_0}{2} + \sum_{n=1}^{\infty} a_n \cos \frac{n\pi x}{L}.$$

That is, $b_n = 0$. In this case, a simpler formula for a_n is

$$a_n = \frac{2}{L} \int_0^L f(x) \cos \frac{n\pi x}{L} \, dx, \quad n = 0, 1, 2, 3, \ldots$$

3. If in addition to the restrictions in (1), $f(x)$ is an odd function (i.e. $f(-x) = -f(x)$), then the Fourier series reduces to

$$\sum_{n=1}^{\infty} b_n \sin \frac{n\pi x}{L}.$$

That is, $a_n = 0$. In this case, a simpler formula for the b_n is

$$b_n = \frac{2}{L} \int_0^L f(x) \sin \frac{n\pi x}{L} \, dx, \quad n = 1, 2, 3, \ldots$$

4. If in addition to the restrictions in (2) above, $f(x) = -f(L - x)$, then a_n will be 0 for all even values of n, including $n = 0$. Thus in this case, the expansion reduces to

$$\sum_{m=1}^{\infty} a_{2m-1} \cos \frac{(2m - 1)\pi x}{L}.$$

5. If in addition to the restrictions in (3) above, $f(x) = f(L - x)$, then b_n will be 0 for all even values of n. Thus in this case, the expansion reduces to

$$\sum_{m=1}^{\infty} b_{2m-1} \sin \frac{(2m - 1)\pi x}{L}.$$

(The series in (4) and (5) are known as *odd-harmonic series*, since only the odd harmonics appear. Similar rules may be stated for even-harmonic series, but when a series appears in the even-harmonic form, it means that $2L$ has not been taken as the smallest period of $f(x)$. Since any integral multiple of a period is also a period, series obtained in this way will also work, but in general computation is simplified if $2L$ is taken to be the smallest period.)

6. If we write the Euler definitions for $\cos \theta$ and $\sin \theta$, we obtain the complex form of the Fourier Series known either as the "Complex Fourier Series" or the "Exponential Fourier Series" of $f(x)$. It is represented as

$$f(x) = \frac{1}{2} \sum_{n=-\infty}^{n=+\infty} c_n e^{i\omega_n x},$$

where

$$c_n = \frac{1}{L} \int_{-L}^L f(x) \, e^{-i\omega_n x} dx, \quad n = 0, \pm 1, \pm 2, \pm 3, \ldots$$

with $\omega_n = \frac{n\pi}{L}, \quad n = 0, \pm 1, \pm 2, \ldots$

The set of coefficients $\{c_n\}$ is often referred to as the Fourier spectrum.

7. If both sine and cosine terms are present and if $f(x)$ is of period $2L$ and expandable by a Fourier series, it can be represented as

$$f(x) = \frac{a_0}{2} + \sum_{n=1}^{\infty} c_n \cos\left(\frac{n\pi x}{L} + \phi_n\right),$$

where $a_n = c_n \cos \phi_n$, $\quad b_n = -c_n \sin \phi_n$, $\quad c_n = \sqrt{a_n^2 + b_n^2}$, $\quad \phi_n = \text{arc tan}\left(-\frac{b_n}{a_n}\right)$

It can also be represented as

$$f(x) = \frac{a_0}{2} + \sum_{n=1}^{\infty} c_n \sin\left(\frac{n\pi x}{L} + \phi_n\right),$$

where $a_n = c_n \sin \phi_n$, $\quad b_n = c_n \cos \phi_n$, $\quad c_n = \sqrt{a_n^2 + b_n^2}$, $\quad \phi_n = \text{arc tan}\left(\frac{a_n}{b_n}\right)$

where the quadrant of ϕ_n is chosen so as to make the formulas for a_n, b_n, and c_n hold.

8. The following table of trigonometric identities should be helpful for developing Fourier Series.

	n **	n even	n odd	$n/2$ odd	$n/2$ even
$\sin n\pi$	0	0	0	0	0
$\cos n\pi$	$(-1)^n$	$+1$	-1	$+1$	$+1$
$\sin \frac{n\pi}{2}$ *		0	$(-1)^{(n-1)/2}$	0	0
$\cos \frac{n\pi}{2}$ *		$(-1)^{n/2}$	0	-1	$+1$
$\sin \frac{n\pi}{4}$			$\frac{\sqrt{2}}{2}(-1)^{(n^2+4n+11)/8}$	$(-1)^{(n-2)/4}$	0

*A useful formula for $\sin \frac{n\pi}{2}$ and $\cos \frac{n\pi}{2}$ is given by

$$\sin \frac{n\pi}{2} = \frac{(i)^{n+1}}{2}[(-1)^n - 1] \text{ and } \cos \frac{n\pi}{2} = \frac{(i)^n}{2}[(-1)^n + 1], \text{ where } i^2 = -1.$$

** n any integer.

AUXILIARY FORMULAS FOR FOURIER SERIES

$$1 = \frac{4}{\pi}\left[\sin\frac{\pi x}{k} + \frac{1}{3}\sin\frac{3\pi x}{k} + \frac{1}{5}\sin\frac{5\pi x}{k} + \cdots\right] \qquad [0 < x < k]$$

$$x = \frac{2k}{\pi}\left[\sin\frac{\pi x}{k} - \frac{1}{2}\sin\frac{2\pi x}{k} + \frac{1}{3}\sin\frac{3\pi x}{k} - \cdots\right] \qquad [-k < x < k]$$

$$x = \frac{k}{2} - \frac{4k}{\pi^2}\left[\cos\frac{\pi x}{k} + \frac{1}{3^2}\cos\frac{3\pi x}{k} + \frac{1}{5^2}\cos\frac{5\pi x}{k} + \cdots\right] \qquad [0 < x < k]$$

$$x^2 = \frac{2k^2}{\pi^3}\left[\left(\frac{\pi^2}{1} - \frac{4}{1}\right)\sin\frac{\pi x}{k} - \frac{\pi^2}{2}\sin\frac{2\pi x}{k} + \left(\frac{\pi^2}{3} - \frac{4}{3^3}\right)\sin\frac{3\pi x}{k}\right.$$
$$\left. - \frac{\pi^2}{4}\sin\frac{4\pi x}{k} + \left(\frac{\pi^2}{5} - \frac{4}{5^3}\right)\sin\frac{5\pi x}{k} + \cdots\right] \quad [0 < x < k]$$

$$x^2 = \frac{k^2}{3} - \frac{4k^2}{\pi^2}\left[\cos\frac{\pi x}{k} - \frac{1}{2^2}\cos\frac{2\pi x}{k} + \frac{1}{3^2}\cos\frac{3\pi x}{k} - \frac{1}{4^2}\cos\frac{4\pi x}{k} + \cdots\right]$$
$$[-k < x < k]$$

$$1 - \frac{1}{3} + \frac{1}{5} - \frac{1}{7} + \cdots = \frac{\pi}{4}$$

$$1 + \frac{1}{2^2} + \frac{1}{3^2} + \frac{1}{4^2} + \cdots = \frac{\pi^2}{6}$$

$$1 - \frac{1}{2^2} + \frac{1}{3^2} - \frac{1}{4^2} + \cdots = \frac{\pi^2}{12}$$

$$1 + \frac{1}{3^2} + \frac{1}{5^2} + \frac{1}{7^2} + \cdots = \frac{\pi^2}{8}$$

$$\frac{1}{2^2} + \frac{1}{4^2} + \frac{1}{6^2} + \frac{1}{8^2} + \cdots = \frac{\pi^2}{24}$$

FOURIER EXPANSIONS FOR BASIC PERIODIC FUNCTIONS

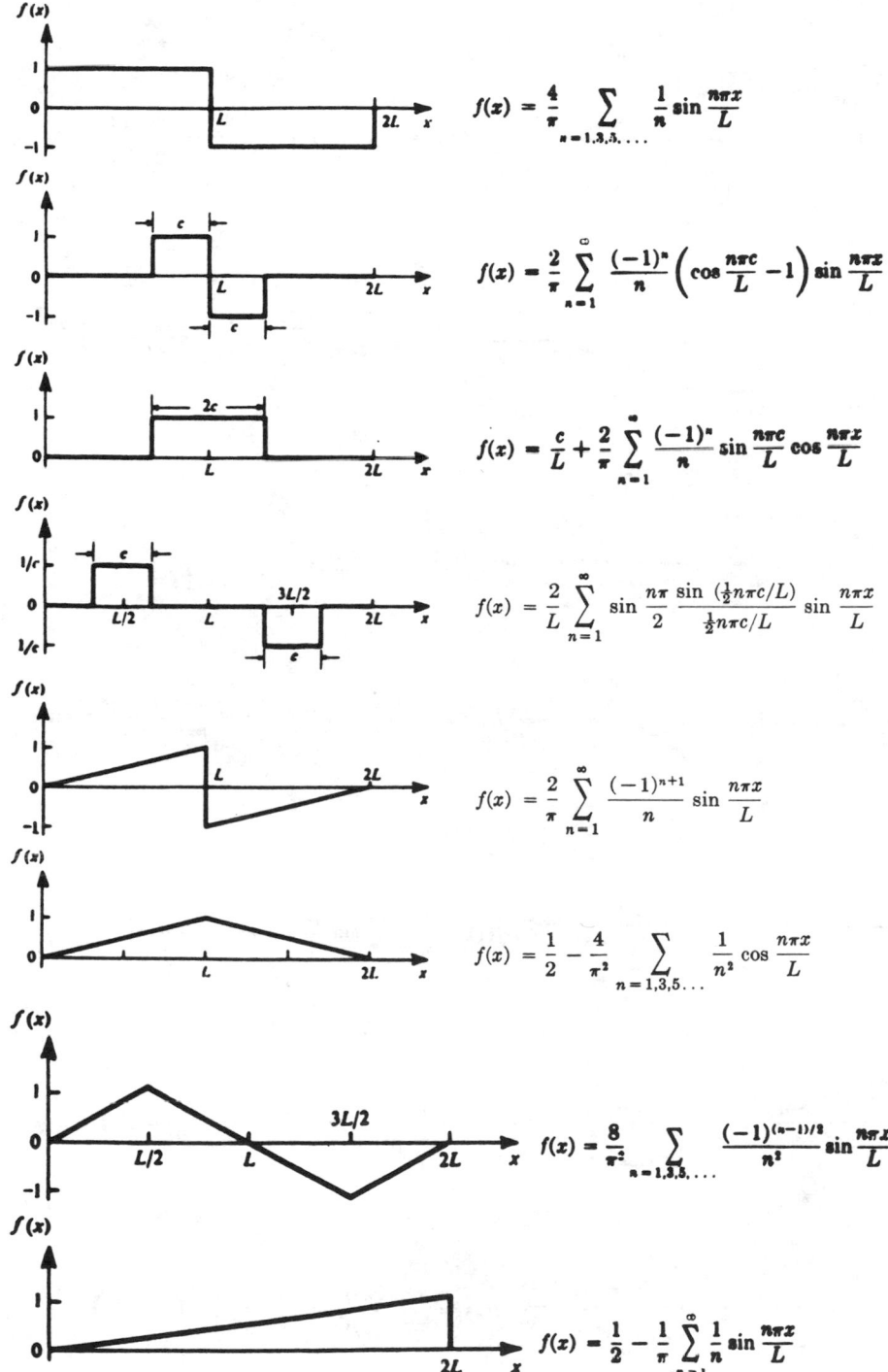

$$f(x) = \frac{4}{\pi} \sum_{n=1,3,5,\dots} \frac{1}{n} \sin \frac{n\pi x}{L}$$

$$f(x) = \frac{2}{\pi} \sum_{n=1}^{\infty} \frac{(-1)^n}{n} \left(\cos \frac{n\pi c}{L} - 1 \right) \sin \frac{n\pi x}{L}$$

$$f(x) = \frac{c}{L} + \frac{2}{\pi} \sum_{n=1}^{\infty} \frac{(-1)^n}{n} \sin \frac{n\pi c}{L} \cos \frac{n\pi x}{L}$$

$$f(x) = \frac{2}{L} \sum_{n=1}^{\infty} \sin \frac{n\pi}{2} \frac{\sin \left(\frac{1}{2} n\pi c / L \right)}{\frac{1}{2} n\pi c / L} \sin \frac{n\pi x}{L}$$

$$f(x) = \frac{2}{\pi} \sum_{n=1}^{\infty} \frac{(-1)^{n+1}}{n} \sin \frac{n\pi x}{L}$$

$$f(x) = \frac{1}{2} - \frac{4}{\pi^2} \sum_{n=1,3,5\dots} \frac{1}{n^2} \cos \frac{n\pi x}{L}$$

$$f(x) = \frac{8}{\pi^2} \sum_{n=1,3,5,\dots} \frac{(-1)^{(n-1)/2}}{n^2} \sin \frac{n\pi x}{L}$$

$$f(x) = \frac{1}{2} - \frac{1}{\pi} \sum_{n=1}^{\infty} \frac{1}{n} \sin \frac{n\pi x}{L}$$

FOURIER EXPANSIONS FOR BASIC PERIODIC FUNCTIONS (Continued)

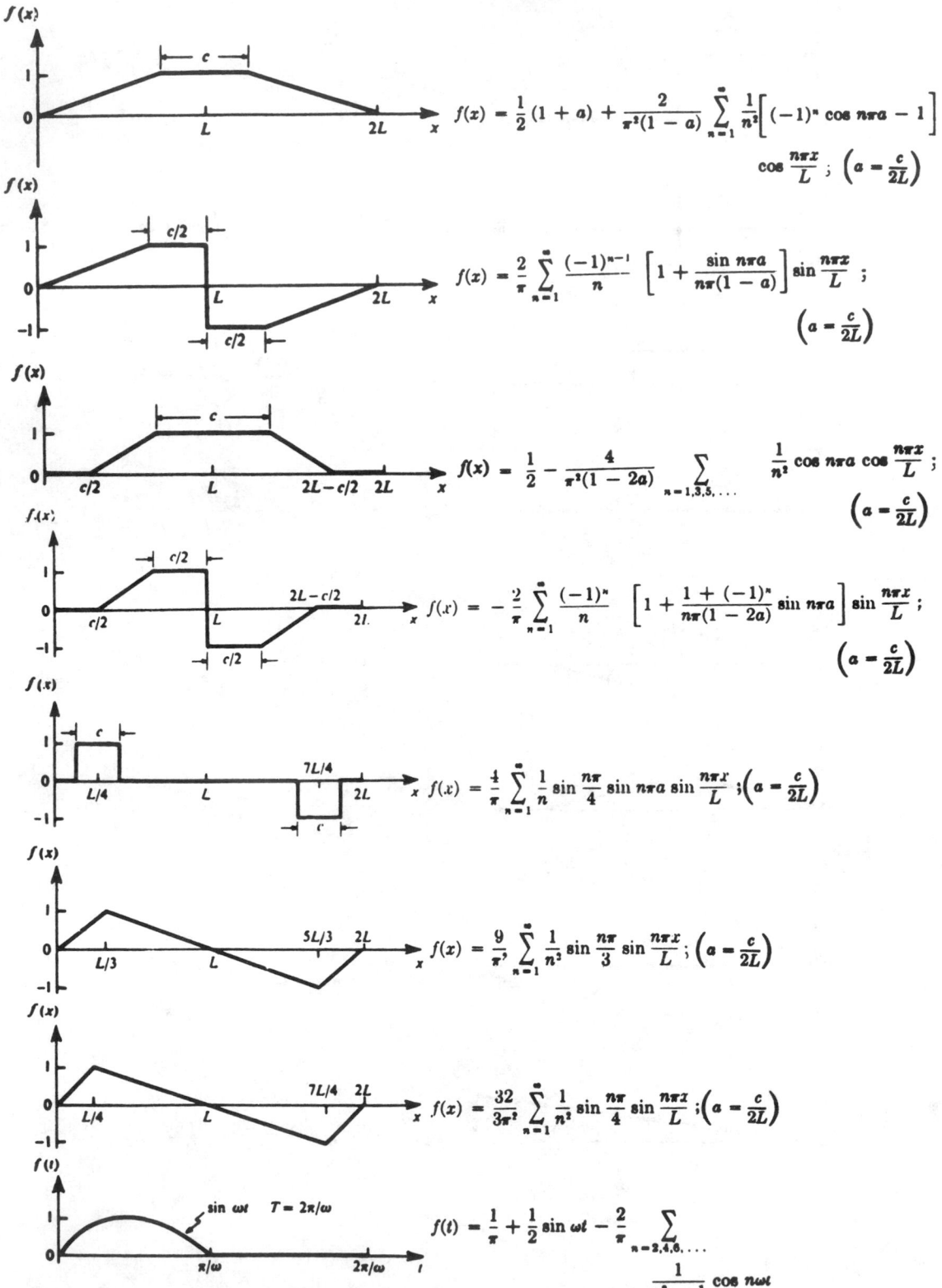

$$f(x) = \frac{1}{2}(1 + a) + \frac{2}{\pi^2(1 - a)}\sum_{n=1}^{\infty}\frac{1}{n^2}\Big[(-1)^n \cos n\pi a - 1\Big] \cos \frac{n\pi x}{L} ; \quad \Big(a = \frac{c}{2L}\Big)$$

$$f(x) = \frac{2}{\pi}\sum_{n=1}^{\infty}\frac{(-1)^{n-1}}{n}\left[1 + \frac{\sin n\pi a}{n\pi(1 - a)}\right]\sin \frac{n\pi x}{L} ; \quad \Big(a = \frac{c}{2L}\Big)$$

$$f(x) = \frac{1}{2} - \frac{4}{\pi^2(1 - 2a)}\sum_{n=1,3,5,\dots}\frac{1}{n^2}\cos n\pi a \cos \frac{n\pi x}{L} ; \quad \Big(a = \frac{c}{2L}\Big)$$

$$f(x) = -\frac{2}{\pi}\sum_{n=1}^{\infty}\frac{(-1)^n}{n}\left[1 + \frac{1 + (-1)^n}{n\pi(1 - 2a)}\sin n\pi a\right]\sin \frac{n\pi x}{L} ; \quad \Big(a = \frac{c}{2L}\Big)$$

$$f(x) = \frac{4}{\pi}\sum_{n=1}^{\infty}\frac{1}{n}\sin \frac{n\pi}{4}\sin n\pi a \sin \frac{n\pi x}{L} ; \Big(a = \frac{c}{2L}\Big)$$

$$f(x) = \frac{9}{\pi^2}\sum_{n=1}^{\infty}\frac{1}{n^2}\sin \frac{n\pi}{3}\sin \frac{n\pi x}{L} ; \Big(a = \frac{c}{2L}\Big)$$

$$f(x) = \frac{32}{3\pi^2}\sum_{n=1}^{\infty}\frac{1}{n^2}\sin \frac{n\pi}{4}\sin \frac{n\pi x}{L} ; \Big(a = \frac{c}{2L}\Big)$$

$$f(t) = \frac{1}{\pi} + \frac{1}{2}\sin \omega t - \frac{2}{\pi}\sum_{n=2,4,6,\dots}\frac{1}{n^2 - 1}\cos n\omega t$$

Extracted from graphs and formulas, pages 372, 373, Differential Equations in Engineering Problems, Salvadori and Schwarz, published by Prentice-Hall, Inc., 1954.

THE FOURIER TRANSFORMS

Dr. R. E. Gaskell

For a piecewise continuous function $F(x)$ over a finite interval $0 \leq x \leq \pi$, the *finite Fourier cosine transform* of $F(x)$ is

$$f_c(n) = \int_0^\pi F(x) \cos nx \, dx \quad (n = 0, 1, 2, \ldots). \tag{1}$$

If x ranges over the interval $0 \leq x \leq L$, the substitution $x' = \pi x / L$ allows the use of this definition, also. The inverse transform is written

$$\bar{F}(x) = \frac{1}{\pi} f_c(0) + \frac{2}{\pi} \sum_{n=1}^\pi f_c(n) \cos nx \quad (0 < x < \pi) \tag{2}$$

where $\bar{F}(x) = \dfrac{[F(x + 0) + F(x - 0)]}{2}$. We observe that $\bar{F}(x) = F(x)$ at points of continuity. The formula

$$f_c^{(2)}(n) = \int_0^\pi F''(x) \cos nx \, dx$$

$$= -n^2 f_c(n) - F'(0) + (-1)^n F'(\pi) \tag{3}$$

makes the finite Fourier cosine transform useful in certain boundary value problems.

Analogously, the *finite Fourier sine transform* of $F(x)$ is

$$f_s(n) = \int_0^\pi F(x) \sin nx \, dx \quad (n = 1, 2, 3, \ldots) \tag{4}$$

and

$$\bar{F}(x) = \frac{2}{\pi} \sum_{n=1}^\pi f_s(n) \sin nx \quad (0 < x < \pi) \tag{5}$$

Corresponding to (6) we have

$$f_s^{(2)}(n) = \int_0^\pi F''(x) \sin nx \, dx$$

$$= -n^2 f_s(n) - nF(0) - n(-1)^n F(\pi). \tag{6}$$

FOURIER TRANSFORMS

If $F(x)$ is defined for $x \geq 0$ and is piecewise continuous over any finite interval, and if

$$\int_0^\infty F(x) \, dx$$

is absolutely convergent, then

$$f_c(\alpha) = \sqrt{\frac{2}{\pi}} \int_0^\infty F(x) \cos (\alpha x) \, dx \tag{7}$$

is the *Fourier cosine transform* of $F(x)$. Furthermore,

$$\bar{F}(x) = \sqrt{\frac{2}{\pi}} \int_0^\infty f_c(\alpha) \cos (\alpha x) \, d\alpha. \tag{8}$$

The Fourier Transforms

If $\lim_{x \to \infty} \dfrac{d^n F}{dx^n} = 0$, an important property of the Fourier cosine transform

$$f_c^{(2r)}(\alpha) = \sqrt{\frac{2}{\pi}} \int_0^\infty \left(\frac{d^{2r} F}{dx^{2r}} \right) \cos(\alpha x) \, dx$$

$$= -\sqrt{\frac{2}{\pi}} \sum_{n=0}^{r-1} (-1)^n a_{2r-2n-1} \alpha^{2n} + (-1)^r \alpha^{2r} f_c(\alpha) \tag{9}$$

where $\lim_{x \to 0} \dfrac{d^r F}{dx^r} = a_r$, makes it useful in the solution of many problems.

Under the same conditions,

$$f_s(\alpha) = \sqrt{\frac{2}{\pi}} \int_0^\infty F(x) \sin(\alpha x) \, dx \tag{10}$$

defines the *Fourier sine transform* of $F(x)$, and

$$\bar{F}(x) = \sqrt{\frac{2}{\pi}} \int_0^\infty f_s(\alpha) \sin(\alpha x) \, d\alpha. \tag{11}$$

Corresponding to (12) we have

$$f_s^{(2r)}(\alpha) = \sqrt{\frac{2}{\pi}} \int_0^\infty \frac{d^{2r} F}{dx^{2r}} \sin(\alpha x) \, dx$$

$$= -\sqrt{\frac{2}{\pi}} \sum_{n=1}^{r} (-1)^n \alpha^{2n-1} a_{2r-2n} + (-1)^{r-1} \alpha^{2r} f_s(\alpha). \tag{12}$$

Similarly, if $F(x)$ is defined for $-\infty < x < \infty$, and if $\displaystyle\int_{-\infty}^\infty F(x) \, dx$ is absolutely convergent, then

$$f(\alpha) = \frac{1}{\sqrt{2\pi}} \int_{-\infty}^\infty F(x) e^{i\alpha x} \, dx \tag{13}$$

is the *Fourier transform* of $F(x)$, and

$$\bar{F}(x) = \frac{1}{\sqrt{2\pi}} \int_{-\infty}^\infty f(\alpha) e^{-i\alpha x} \, d\alpha. \tag{14}$$

Also, if

$$\lim_{|x| \to \infty} \left| \frac{d^n F}{dx^n} \right| = 0 \quad (n = 1, 2, \ldots, r - 1),$$

then

$$f^{(r)}(\alpha) = \frac{1}{\sqrt{2\pi}} \int_{-\infty}^\infty F^{(r)}(x) e^{i\alpha x} \, dx = (-i\alpha)^r f(\alpha). \tag{15}$$

The Fourier Transforms

FINITE SINE TRANSFORMS

	$f_s(n)$	$F(x)$
1	$f_s(n) = \int_0^\pi F(x) \sin nx\, dx \ (n = 1, 2, \cdots)$	$F(x)$
2	$(-1)^{n+1} f_s(n)$	$F(\pi - x)$
3	$\dfrac{1}{n}$	$\dfrac{\pi - x}{\pi}$
4	$\dfrac{(-1)^{n+1}}{n}$	$\dfrac{x}{\pi}$
5	$\dfrac{1 - (-1)^n}{n}$	1
6	$\dfrac{2}{n^2} \sin \dfrac{n\pi}{2}$	$\begin{cases} x & \text{when } 0 < x < \pi/2 \\ \pi - x & \text{when } \pi/2 < x < \pi \end{cases}$
7	$\dfrac{(-1)^{n+1}}{n^3}$	$\dfrac{x(\pi^2 - x^2)}{6\pi}$
8	$\dfrac{1 - (-1)^n}{n^3}$	$\dfrac{x(\pi - x)}{2}$
9	$\dfrac{\pi^2(-1)^{n-1}}{n} - \dfrac{2[1 - (-1)^n]}{n^3}$	x^2
10	$\pi(-1)^n \left(\dfrac{6}{n^3} - \dfrac{\pi^2}{n} \right)$	x^3
11	$\dfrac{n}{n^2 + c^2} [1 - (-1)^n e^{c\pi}]$	e^{cx}
12	$\dfrac{n}{n^2 + c^2}$	$\dfrac{\sinh c(\pi - x)}{\sinh c\pi}$
13	$\dfrac{n}{n^2 - k^2} \ (k \neq 0, 1, 2, \cdots)$	$\dfrac{\sin k(\pi - x)}{\sin k\pi}$
14	$\begin{cases} \dfrac{\pi}{2} & \text{when } n = m \\ 0 & \text{when } n \neq m \end{cases} \quad (m = 1, 2, \cdots)$	$\sin mx$
15	$\dfrac{n}{n^2 - k^2} [1 - (-1)^n \cos k\pi]$ $(k \neq 1, 2, \cdots)$	$\cos kx$
16	$\begin{cases} \dfrac{n}{n^2 - m^2} [1 - (-1)^{n+m}] & \text{when } n \neq m = 1, 2, \cdots \\ 0 & \text{when } n = m \end{cases}$	$\cos mx$
17	$\dfrac{n}{(n^2 - k^2)^2} \ (k \neq 0, 1, 2, \cdots)$	$\dfrac{\pi \sin kx}{2k \sin^2 k\pi} - \dfrac{x \cos k(\pi - x)}{2k \sin k\pi}$
18	$\dfrac{b^n}{n} \ (|b| \leqq 1)$	$\dfrac{2}{\pi} \arctan \dfrac{b \sin x}{1 - b \cos x}$
19	$\dfrac{1 - (-1)^n}{n} b^n \ (|b| \leq 1)$	$\dfrac{2}{\pi} \arctan \dfrac{2b \sin x}{1 - b^2}$

The Fourier Transforms

FINITE COSINE TRANSFORMS

	$f_c(n)$	$F(x)$
1	$f_c(n) = \displaystyle\int_0^\pi F(x) \cos nx\, dx \quad (n = 0, 1, 2, \cdots)$	$F(x)$
2	$(-1)^n f_c(n)$	$F(\pi - x)$
3	0 when $n = 1, 2, \cdots$; $f_c(0) = \pi$	1
4	$\dfrac{2}{n} \sin \dfrac{n\pi}{2}$; $f_c(0) = 0$	$\begin{cases} 1 \text{ when } 0 < x < \pi/2 \\ -1 \text{ when } \pi/2 < x < \pi \end{cases}$
5	$-\dfrac{1 - (-1)^n}{n^2}$; $f_c(0) = \dfrac{\pi^2}{2}$	x
6	$\dfrac{(-1)^n}{n^2}$; $f_c(0) = \dfrac{\pi^2}{6}$	$\dfrac{x^2}{2\pi}$
7	$\dfrac{1}{n^2}$; $f_c(0) = 0$	$\dfrac{(\pi - x)^2}{2\pi} - \dfrac{\pi}{6}$
8	$3\pi^2 \dfrac{(-1)^n}{n^2} - 6 \dfrac{1 - (-1)^n}{n^4}$; $f_c(0) = \dfrac{\pi^4}{4}$	x^3
9	$\dfrac{(-1)^n e^c \pi - 1}{n^2 + c^2}$	$\dfrac{1}{c} e^{cx}$
10	$\dfrac{1}{n^2 + c^2}$	$\dfrac{\cosh c(\pi - x)}{c \sinh c\pi}$
11	$\dfrac{k}{n^2 - k^2} [(-1)^n \cos \pi k - 1]$ $(k \neq 0, 1, 2, \cdots)$	$\sin kx$
12	$\dfrac{(-1)^{n+m} - 1}{n^2 - m^2}$; $f_c(m) = 0 \quad (m = 1, 2, \cdots)$	$\dfrac{1}{m} \sin mx$
13	$\dfrac{1}{n^2 - k^2} \quad (k \neq 0, 1, 2, \cdots)$	$-\dfrac{\cos k(\pi - x)}{k \sin k\pi}$
14	0 when $n = 1, 2, \cdots$; $f_c(m) = \dfrac{\pi}{2} \quad (m = 1, 2, \cdots)$	$\cos mx$

The Fourier Transforms

FOURIER SINE TRANSFORMS**

$F(x)$	$f_s(\alpha)$
1 $\begin{cases} 1 & (0 < x < a) \\ 0 & (x > a) \end{cases}$	$\sqrt{\dfrac{2}{\pi}} \left[\dfrac{1 - \cos \alpha}{\alpha} \right]$
2 $x^{p-1} \ (0 < p < 1)$	$\sqrt{\dfrac{2}{\pi}} \dfrac{\Gamma(p)}{\alpha^p} \sin \dfrac{p\pi}{2}$
3 $\begin{cases} \sin x & (0 < x < a) \\ 0 & (x > a) \end{cases}$	$\dfrac{1}{\sqrt{2\pi}} \left[\dfrac{\sin [a(1 - \alpha)]}{1 - \alpha} - \dfrac{\sin [a(1 + \alpha)]}{1 + \alpha} \right]$
4 e^{-x}	$\sqrt{\dfrac{2}{\pi}} \left[\dfrac{\alpha}{1 + \alpha^2} \right]$
5 $xe^{-x^2/2}$	$\alpha e^{-\alpha^2/2}$
6 $\cos \dfrac{x^2}{2}$	$\sqrt{2} \left[\sin \dfrac{\alpha^2}{2} C\left(\dfrac{\alpha^2}{2}\right) - \cos \dfrac{\alpha^2}{2} S\left(\dfrac{\alpha^2}{2}\right) \right]$ *
7 $\sin \dfrac{x^2}{2}$	$\sqrt{2} \left[\cos \dfrac{\alpha^2}{2} C\left(\dfrac{\alpha^2}{2}\right) + \sin \dfrac{\alpha^2}{2} S\left(\dfrac{\alpha^2}{2}\right) \right]$ *

*$C(y)$ and $S(y)$ are the Fresnel integrals

$$C(y) = \frac{1}{\sqrt{2\pi}} \int_0^y \frac{1}{\sqrt{t}} \cos t \, dt,$$

$$S(y) = \frac{1}{\sqrt{2\pi}} \int_0^y \frac{1}{\sqrt{t}} \sin t \, dt.$$

**More extensive tables of the Fourier sine and cosine transforms can be found in Fritz Oberhettinger, "Tabellen zur-Fourier Transformation," Springer (1957).

FOURIER COSINE TRANSFORMS

$F(x)$	$f_c(\alpha)$
1 $\begin{cases} 1 & (0 < x < a) \\ 0 & (x > a) \end{cases}$	$\sqrt{\dfrac{2}{\pi}} \dfrac{\sin a\alpha}{\alpha}$
2 $x^{p-1} \ (0 < p < 1)$	$\sqrt{\dfrac{2}{\pi}} \dfrac{\Gamma(p)}{\alpha^p} \cos \dfrac{p\pi}{2}$
3 $\begin{cases} \cos x & (0 < x < a) \\ 0 & (x > a) \end{cases}$	$\dfrac{1}{\sqrt{2\pi}} \left[\dfrac{\sin [a(1 - \alpha)]}{1 - \alpha} + \dfrac{\sin [a(1 + \alpha)]}{1 + \alpha} \right]$
4 e^{-x}	$\sqrt{\dfrac{2}{\pi}} \left(\dfrac{1}{1 + \alpha^2} \right)$
5 $e^{-x^2/2}$	$e^{-\alpha^2/2}$
6 $\cos \dfrac{x^2}{2}$	$\cos \left(\dfrac{\alpha^2}{2} - \dfrac{\pi}{4} \right)$
7 $\sin \dfrac{x^2}{2}$	$\cos \left(\dfrac{\alpha^2}{2} + \dfrac{\pi}{4} \right)$

The Fourier Transforms

FOURIER TRANSFORMS*

$F(x)$	$f(\alpha)$
1 $\dfrac{\sin ax}{x}$	$\begin{cases} \sqrt{\dfrac{\pi}{2}} & \|\alpha\| < a \\ 0 & \|\alpha\| > a \end{cases}$
2 $\begin{cases} e^{iwx} & (p < x < q) \\ 0 & (x < p, x > q) \end{cases}$	$\dfrac{i}{\sqrt{2\pi}} \dfrac{e^{ip(w+\alpha)} - e^{iq(w+\alpha)}}{(w + \alpha)}$
3 $\begin{cases} e^{-cx+iwx} & (x > 0) \\ 0 & (x < 0) \end{cases} \quad (c > 0)$	$\dfrac{i}{\sqrt{2\pi}(w + \alpha + ic)}$
4 $e^{-px^2} \quad R(p) > 0$	$\dfrac{1}{\sqrt{2p}} e^{-\alpha^2/4p}$
5 $\cos px^2$	$\dfrac{1}{\sqrt{2p}} \cos\left[\dfrac{\alpha^2}{4p} - \dfrac{\pi}{4}\right]$
6 $\sin px^2$	$\dfrac{1}{\sqrt{2p}} \cos\left[\dfrac{\alpha^2}{4p} + \dfrac{\pi}{4}\right]$
7 $\|x\|^{-p} \quad (0 < p < 1)$	$\sqrt{\dfrac{2}{\pi}} \dfrac{\Gamma(1-p) \sin\dfrac{p\pi}{2}}{\|\alpha\|^{(1-p)}}$
8 $\dfrac{e^{-a\|x\|}}{\sqrt{\|x\|}}$	$\dfrac{\sqrt{\sqrt{(a^2+\alpha^2)} + a}}{\sqrt{a^2+\alpha^2}}$
9 $\dfrac{\cosh ax}{\cosh \pi x} \quad (-\pi < a < \pi)$	$\sqrt{\dfrac{2}{\pi}} \dfrac{\cos\dfrac{a}{2} \cosh\dfrac{\alpha}{2}}{\cosh \alpha + \cos a}$
10 $\dfrac{\sinh ax}{\sinh \pi x} \quad (-\pi < a < \pi)$	$\dfrac{1}{\sqrt{2\pi}} \dfrac{\sin a}{\cosh \alpha + \cos a}$
11 $\begin{cases} \dfrac{1}{\sqrt{a^2 - x^2}} & (\|x\| < a) \\ 0 & (\|x\| > a) \end{cases}$	$\sqrt{\dfrac{\pi}{2}} J_0(a\alpha)$
12 $\dfrac{\sin[b\sqrt{a^2+x^2}]}{\sqrt{a^2+x^2}}$	$\begin{cases} 0 & (\|\alpha\| > b) \\ \sqrt{\dfrac{\pi}{2}} J_0(a\sqrt{b^2 - \alpha^2}) & (\|\alpha\| < b) \end{cases}$
13 $\begin{cases} P_n(x) & (\|x\| < 1) \\ 0 & (\|x\| > 1) \end{cases}$	$\dfrac{i^n}{\sqrt{\alpha}} J_{n+\frac{1}{2}}(\alpha)$
14 $\begin{cases} \dfrac{\cos[b\sqrt{a^2-x^2}]}{\sqrt{a^2-x^2}} & (\|x\| < a) \\ 0 & (\|x\| > a) \end{cases}$	$\sqrt{\dfrac{\pi}{2}} J_0(a\sqrt{a^2 + b^2})$
15 $\begin{cases} \dfrac{\cosh[b\sqrt{a^2-x^2}]}{\sqrt{a^2-x^2}} & (\|x\| < a) \\ 0 & (\|x\| > a) \end{cases}$	$\sqrt{\dfrac{\pi}{2}} J_0(a\sqrt{\alpha^2 - b^2})$

*More extensive tables of Fourier transforms can be found in W. Magnus and F. Oberhettinger, "Formulas and Theorems of the Special Functions of Mathematical Physics," pp. 116–120. Chelsea (1949).

The Fourier Transforms

The following functions appear among the entries of the tables on transforms.

Function	Definition	Name
$Ei(x)$	$\displaystyle\int_{-\infty}^{x} \frac{e^v}{v}\, dv$; or sometimes defined as $$-Ei(-x) = \int_{x}^{\infty} \frac{e^{-v}}{v}\, dv$$	Sine, Cosine, and Exponential Integral tables pages 548–556
$Si(x)$	$\displaystyle\int_{0}^{x} \frac{\sin v}{v}\, dv$	Sine, Cosine, and Exponential Integral tables pages 548–556
$Ci(x)$	$\displaystyle\int_{\infty}^{x} \frac{\cos v}{v}\, dv$; or sometimes defined as negative of this integral	Sine, Cosine, and Exponential Integral tables pages 548–546
$erf(x)$	$\displaystyle\frac{2}{\sqrt{\pi}}\int_{0}^{x} e^{-v^2}\, dv$	Error function
$erfc(x)$	$\displaystyle 1 - erf(x) = \frac{2}{\sqrt{\pi}}\int_{x}^{\infty} e^{-v^2}\, dv$	Complementary function to error function
$L_n(x)$	$\displaystyle\frac{e^x}{n!}\frac{d^n}{dx^n}(x^n e^{-x}),\quad n = 0, 1, \cdots$	Laguerre polynomial of degree n

THE LAPLACE TRANSFORM

Dr. R. E. Gaskell

If $F(t)$ is a piecewise continuous real-valued function of the real variable $t(0 \leq t < \infty)$, and if $F(t)$ is of exponential order, that is if $|F(t)| < Me^{at}(t > T; M, a, T$ positive constants) then the *Laplace transform* of $F(t)$,

$$L\{F(t)\} = f(s) = \int_0^\infty e^{-st} F(t)\,dt, \tag{1}$$

exists in the half-plane of the complex variable s for which the real part of s is greater than some fixed value s_0, i.e., $R(s) \geq s_0$. Furthermore

$$F(t) = \frac{1}{2\pi i} \int_{a-i\infty}^{a+i\infty} e^{st} f(s)\,ds, \tag{2}$$

where $a > s_0$. The important property

$$L\{F^{(r)}(t)\} = \int_0^\infty e^{-st} \left(\frac{d^r F}{dt^r}\right) dt$$
$$= s^r f(s) - \sum_{n=0}^{r-1} s^{r-1-n} F^{(n)}(+0) \tag{3}$$

makes the Laplace transform very useful for solving linear differential equations with constant coefficients, and many boundary value problems.

The Laplace Transform
LAPLACE OPERATIONS*

	$F(t)$	$f(s)$
1	$F(t)$	$\displaystyle\int_0^\infty e^{-st}F(t)\,dt$
2	$AF(t) + BG(t)$	$Af(s) + Bg(s)$
3	$F'(t)$	$sf(s) - F(+0)$
4	$F^{(n)}(t)$	$s^n f(s) - s^{n-1}F(+0) - s^{n-2}F'(+0) - \cdots$ $- F^{(n-1)}(+0)$
5	$\displaystyle\int_0^t F(\tau)\,d\tau$	$\dfrac{1}{s} f(s)$
6	$\displaystyle\int_0^t \int_0^\tau F(\lambda)\,d\lambda\,d\tau$	$\dfrac{1}{s^2} f(s)$
7	$\displaystyle\int_0^t F_1(t - \tau)F_2(\tau)\,d\tau = F_1 * F_2$	$f_1(s)\,f_2(s)$
8	$tF(t)$	$-f'(s)$
9	$t^n F(t)$	$(-1)^n f^{(n)}(s)$
10	$\dfrac{1}{t} F(t)$	$\displaystyle\int_s^\infty f(x)\,dx$
11	$e^{at} F(t)$	$f(s - a)$
12	$F(t - b)$, where $F(t) = 0$ when $t < 0$	$e^{-bs} f(s)$
13	$\dfrac{1}{c} F\left(\dfrac{t}{c}\right)$	$f(cs)$
14	$\dfrac{1}{c} e^{(bt)/c} \; F\left(\dfrac{t}{c}\right)$	$f(cs - b)$
15	$F(t + a) = F(t)$	$\dfrac{\displaystyle\int_0^a e^{-st}F(t)\,dt}{1 - e^{-as}}$
16	$F(t + a) = -F(t)$	$\dfrac{\displaystyle\int_0^a e^{-st}F(t)\,dt}{1 + e^{-as}}$
17	$F_1(t)$, the half-wave rectification of $F(t)$ in No. 16	$\dfrac{f(s)}{1 - e^{-as}}$
18	$F_2(t)$, the full-wave rectification of $F(t)$ in No. 16	$f(s) \coth \dfrac{as}{2}$
19	$\displaystyle\sum_1^m \dfrac{p(a_n)}{q'(a_n)} e^{a_n t}$	$\dfrac{p(s)}{q(s)}, q(s) = (s - a_1)(s - a_2) \cdots (s - a_m)$
20	$\displaystyle e^{at} \sum_{n=1}^r \dfrac{\phi^{(r-n)}(a)}{(r - n)!} \dfrac{t^{n-1}}{(n - 1)!} + \cdots$	$\dfrac{p(s)}{q(s)} = \dfrac{\phi(s)}{(s - a)^r}$

*These tables of Laplace Operations, Laplace Transforms, and Finite Fourier sine and cosine transforms were taken from "Modern Operational Mathematics in Engineering", by permission of the author, R. V. Churchill, and the publisher, McGraw-Hill Book Company, Inc.

The Laplace Transforms

LAPLACE TRANSFORMS

	$f(s)$	$F(t)$
1	$\dfrac{1}{s}$	$\mu(t)$, unit step function
2	$\dfrac{1}{s^2}$	t
3	$\dfrac{1}{s^n}$ $(n = 1, 2, \ldots)$	$\dfrac{t^{n-1}}{(n-1)!}$
4	$\dfrac{1}{\sqrt{s}}$	$\dfrac{1}{\sqrt{\pi t}}$
5	$s^{-3/2}$	$2\sqrt{\dfrac{t}{\pi}}$
6	$s^{-[n+(1/2)]}$ $(n = 1, 2, \ldots)$	$\dfrac{2^n t^{n-(1/2)}}{1 \cdot 3 \cdot 5 \cdots (2n-1)\sqrt{\pi}}$
7	$\dfrac{\Gamma(k)}{s^k}$ $(k > 0)$	t^{k-1}
8	$\dfrac{1}{s-a}$	e^{at}
9	$\dfrac{1}{(s-a)^2}$	te^{at}
10	$\dfrac{1}{(s-a)^n}$ $(n = 1, 2, \ldots)$	$\dfrac{1}{(n-1)!}\, t^{n-1} e^{at}$
11	$\dfrac{\Gamma(k)}{(s-a)^k}$ $(k > 0)$	$t^{k-1} e^{at}$
12*	$\dfrac{1}{(s-a)(s-b)}$	$\dfrac{1}{a-b}(e^{at} - e^{bt})$
13*	$\dfrac{s}{(s-a)(s-b)}$	$\dfrac{1}{a-b}(ae^{at} - be^{bt})$
14*	$\dfrac{1}{(s-a)(s-b)(s-c)}$	$-\dfrac{(b-c)e^{at} + (c-a)e^{bt} + (a-b)e^{ct}}{(a-b)(b-c)(c-a)}$
15	$\dfrac{1}{s^2 + a^2}$	$\dfrac{1}{a}\sin at$
16	$\dfrac{s}{s^2 + a^2}$	$\cos at$
17	$\dfrac{1}{s^2 - a^2}$	$\dfrac{1}{a}\sinh at$
18	$\dfrac{s}{s^2 - a^2}$	$\cosh at$

*Here a, b, and (in 14) c represent distinct constants.

The Laplace Transforms

LAPLACE TRANSFORMS (Continued)

	$f(s)$	$F(t)$
19	$\dfrac{1}{s(s^2 + a^2)}$	$\dfrac{1}{a^2}(1 - \cos at)$
20	$\dfrac{1}{s^2(s^2 + a^2)}$	$\dfrac{1}{a^3}(at - \sin at)$
21	$\dfrac{1}{(s^2 + a^2)^2}$	$\dfrac{1}{2a^3}(\sin at - at \cos at)$
22	$\dfrac{s}{(s^2 + a^2)^2}$	$\dfrac{t}{2a}\sin at$
23	$\dfrac{s^2}{(s^2 + a^2)^2}$	$\dfrac{1}{2a}(\sin at + at \cos at)$
24	$\dfrac{s^2 - a^2}{(s^2 + a^2)^2}$	$t \cos at$
25	$\dfrac{s}{(s^2 + a^2)(s^2 + b^2)}\ (a^2 \neq b^2)$	$\dfrac{\cos at - \cos bt}{b^2 - a^2}$
26	$\dfrac{1}{(s - a)^2 + b^2}$	$\dfrac{1}{b}e^{at}\sin bt$
27	$\dfrac{s - a}{(s - a)^2 + b^2}$	$e^{at}\cos bt$
27.1	$\dfrac{1}{[(s + a)^2 + b^2]^n}$	$\dfrac{-e^{-at}}{4^{n-1}b^{2n}}\sum\limits_{r=1}^{n}\begin{pmatrix} 2n - r - 1 \\ n - 1 \end{pmatrix}(-2t)^{r-1}\dfrac{d^r}{dt^r}[\cos(bt)]$
27.2	$\dfrac{s}{[(s + a)^2 + b^2]^n}$	$\dfrac{e^{-at}}{4^{n-1}b^{2n}}\left\{\sum\limits_{r=1}^{n}\begin{pmatrix} 2n - r - 1 \\ n - 1 \end{pmatrix}\dfrac{1}{(r-1)!}\right.$ $(-2t)^{r-1}\dfrac{d^r}{dt^r}[a\cos(bt) + b\sin(bt)]$ $-2b\sum\limits_{r=1}^{n-1}\dfrac{1}{(r-1)!}\begin{pmatrix} 2n - r - 2 \\ n - 1 \end{pmatrix}$ $\left.(-2t)^{r-1}\dfrac{d^r}{dt^r}[\sin bt]\right\}$
28	$\dfrac{3a^2}{s^3 + a^3}$	$e^{-at} - e^{(at)/2}\left(\cos\dfrac{at\sqrt{3}}{2} - \sqrt{3}\sin\dfrac{at\sqrt{3}}{2}\right)$
29	$\dfrac{4a^3}{s^4 + 4a^4}$	$\sin at \cosh at - \cos at \sinh at$
30	$\dfrac{s}{s^4 + 4a^4}$	$\dfrac{1}{2a^2}\sin at \sinh at$

The Laplace Transforms

LAPLACE TRANSFORMS (Continued)

	$f(s)$	$F(t)$
31	$\dfrac{1}{s^4 - a^4}$	$\dfrac{1}{2a^3}(\sinh at - \sin at)$
32	$\dfrac{s}{s^4 - a^4}$	$\dfrac{1}{2a^2}(\cosh at - \cos at)$
33	$\dfrac{8a^3 s^2}{(s^2 + a^2)^3}$	$(1 + a^2 t^2)\sin at - \cos at$
34*	$\dfrac{1}{s}\left(\dfrac{s-1}{s}\right)^n$	$L_n(t) = \dfrac{e^t}{n!}\dfrac{d^n}{dt^n}(t^n e^{-t})$
35	$\dfrac{s}{(s-a)^{3/2}}$	$\dfrac{1}{\sqrt{\pi t}}e^{at}(1 + 2at)$
36	$\sqrt{s-a} - \sqrt{s-b}$	$\dfrac{1}{2\sqrt{\pi t^3}}(e^{bt} - e^{at})$
37	$\dfrac{1}{\sqrt{s} + a}$	$\dfrac{1}{\sqrt{\pi t}} - ae^{a^2 t}\,\mathrm{erfc}\,(a\sqrt{t})$
38	$\dfrac{\sqrt{s}}{s - a^2}$	$\dfrac{1}{\sqrt{\pi t}} + ae^{a^2 t}\,\mathrm{erf}\,(a\sqrt{t})$
39	$\dfrac{\sqrt{s}}{s + a^2}$	$\dfrac{1}{\sqrt{\pi t}} - \dfrac{2a}{\sqrt{\pi}}e^{-a^2 t}\displaystyle\int_0^{a\sqrt{t}} e^{\lambda^2}\,d\lambda$
40	$\dfrac{1}{\sqrt{s}(s - a^2)}$	$\dfrac{1}{a}e^{a^2 t}\,\mathrm{erf}\,(a\sqrt{t})$
41	$\dfrac{1}{\sqrt{s}(s + a^2)}$	$\dfrac{2}{a\sqrt{\pi}}e^{-a^2 t}\displaystyle\int_0^{a\sqrt{t}} e^{\lambda^2}\,d\lambda$
42	$\dfrac{b^2 - a^2}{(s - a^2)(b + \sqrt{s})}$	$e^{a^2 t}[b - a\,\mathrm{erf}\,(a\sqrt{t})] - be^{b^2 t}\,\mathrm{erfc}\,(b\sqrt{t})$
43	$\dfrac{1}{\sqrt{s}(\sqrt{s} + a)}$	$e^{a^2 t}\,\mathrm{erfc}\,(a\sqrt{t})$
44	$\dfrac{1}{(s + a)\sqrt{s + b}}$	$\dfrac{1}{\sqrt{b-a}}e^{-at}\,\mathrm{erf}\,(\sqrt{b-a}\,\sqrt{t})$
45	$\dfrac{b^2 - a^2}{\sqrt{s}(s - a^2)(\sqrt{s} + b)}$	$e^{a^2 t}\left[\dfrac{b}{a}\,\mathrm{erf}\,(a\sqrt{t}) - 1\right] + e^{b^2 t}\,\mathrm{erfc}\,(b\sqrt{t})$
46†	$\dfrac{(1 - s)^n}{s^{n+(1/2)}}$	$\dfrac{n!}{(2n)!\sqrt{\pi t}}H_{2n}(\sqrt{t})$
47	$\dfrac{(1 - s)^n}{s^{n+(3/2)}}$	$-\dfrac{n!}{\sqrt{\pi}(2n+1)!}H_{2n+1}(\sqrt{t})$

*$L_n(t)$ is the Laguerre polynomial of degree n.

†$H_n(x)$ is the Hermite polynomial, $H_n(x) = e^{x^2}\dfrac{d^n}{dx^n}(e^{-x^2})$.

LAPLACE TRANSFORMS (Continued)

	$f(s)$	$F(t)$
48*	$\dfrac{\sqrt{s+2a}}{\sqrt{s}} - 1$	$ae^{-at}[I_1(at) + I_0(at)]$
49	$\dfrac{1}{\sqrt{s+a}\,\sqrt{s+b}}$	$e^{-(1/2)(a+b)t} I_0\left(\dfrac{a-b}{2}t\right)$
50	$\dfrac{\Gamma(k)}{(s+a)^k(s+b)^k}\ (k \geq 0)$	$\sqrt{\pi}\left(\dfrac{t}{a-b}\right)^{k-(1/2)} e^{-(1/2)(a+b)t} I_{k-(1/2)}\left(\dfrac{a-b}{2}t\right)$
51	$\dfrac{1}{(s+a)^{1/2}(s+b)^{3/2}}$	$te^{-(1/2)(a+b)t}\left[I_0\left(\dfrac{a-b}{2}t\right) + I_1\left(\dfrac{a-b}{2}t\right)\right]$
52	$\dfrac{\sqrt{s+2a} - \sqrt{s}}{\sqrt{s+2a} + \sqrt{s}}$	$\dfrac{1}{t}e^{-at}I_1(at)$
53	$\dfrac{(a-b)^k}{(\sqrt{s+a} + \sqrt{s+b})^{2k}}\ (k > 0)$	$\dfrac{k}{t}e^{-(1/2)(a+b)t}I_k\left(\dfrac{a-b}{2}t\right)$
54	$\dfrac{(\sqrt{s+a} + \sqrt{s})^{-2\nu}}{\sqrt{s}\,\sqrt{s+a}}\ (\nu > -1)$	$\dfrac{1}{a^\nu}e^{-(1/2)(at)}I_\nu\left(\dfrac{1}{2}at\right)$
55	$\dfrac{1}{\sqrt{s^2+a^2}}$	$J_0(at)$
56	$\dfrac{(\sqrt{s^2+a^2} - s)^\nu}{\sqrt{s^2+a^2}}\ (\nu > -1)$	$a^\nu J_\nu(at)$
57	$\dfrac{1}{(s^2+a^2)^k}\ (k > 0)$	$\dfrac{\sqrt{\pi}}{\Gamma(k)}\left(\dfrac{t}{2a}\right)^{k-(1/2)}J_{k-(1/2)}(at)$
58	$(\sqrt{s^2+a^2} - s)^k\ (k > 0)$	$\dfrac{ka^k}{t}J_k(at)$
59	$\dfrac{(s - \sqrt{s^2-a^2})^\nu}{\sqrt{s^2-a^2}}\ (\nu > -1)$	$a^\nu I_\nu(at)$
60	$\dfrac{1}{(s^2-a^2)^k}\ (k > 0)$	$\dfrac{\sqrt{\pi}}{\Gamma(k)}\left(\dfrac{t}{2a}\right)^{k-(1/2)}I_{k-(1/2)}(at)$
61	$\dfrac{e^{-ks}}{s}$	$S_k(t) = \begin{cases} 0 \text{ when } 0 < t < k \\ 1 \text{ when } t > k \end{cases}$
62	$\dfrac{e^{-ks}}{s^2}$	$\begin{cases} 0 \text{ when } 0 < t < k \\ t - k \text{ when } t > k \end{cases}$
63	$\dfrac{e^{-ks}}{s^\mu}\ (\mu > 0)$	$\begin{cases} 0 \quad\text{when } 0 < t < k \\ \dfrac{(t-k)^{\mu-1}}{\Gamma(\mu)} \text{ when } t > k \end{cases}$
64	$\dfrac{1 - e^{-ks}}{s}$	$\begin{cases} 1 \text{ when } 0 < t < k \\ 0 \text{ when } t > k \end{cases}$

*$I_n(x) = i^{-n}J_n(ix)$, where J_n is Bessel's function of the first kind.

The Laplace Transforms

LAPLACE TRANSFORMS (Continued)

	$f(s)$	$F(t)$
65	$\dfrac{1}{s(1 - e^{-ks})} = \dfrac{1 + \coth \frac{1}{2}ks}{2s}$	$S(k, t) = n$ when $\qquad (n - 1)k < t < nk(n = 1, 2, \ldots)$
66	$\dfrac{1}{s(e^{ks} - a)}$	$\begin{cases} 0 \text{ when } 0 < t < k \\ 1 + a + a^2 + \cdots + a^{n-1} \\ \qquad \text{when } nk < t < (n + 1)k(n = 1, 2, \ldots) \end{cases}$
67	$\dfrac{1}{s} \tanh ks$	$M(2k, t) = (-1)^{n-1}$ $\qquad \text{when } 2k(n - 1) < t < 2kn$ $\qquad\qquad\qquad (n = 1, 2, \ldots)$
68	$\dfrac{1}{s(1 + e^{-ks})}$	$\dfrac{1}{2} M(k, t) + \dfrac{1}{2} = \dfrac{1 - (-1)^n}{2}$ $\qquad\qquad \text{when } (n - 1)k < t < nk$
69*	$\dfrac{1}{s^2} \tanh ks$	$H(2k, t)$
70	$\dfrac{1}{s \sinh ks}$	$2S(2k, t + k) - 2 = 2(n - 1)$ $\qquad \text{when } (2n - 3)k < t < (2n - 1)k \;\; (t > 0)$
71	$\dfrac{1}{s \cosh ks}$	$M(2k, t + 3k) + 1 = 1 + (-1)^n$ $\qquad \text{when } (2n - 3)k < t < (2n - 1)k \;\; (t > 0)$
72	$\dfrac{1}{s} \coth ks$	$2S(2k, t) - 1 = 2n - 1$ $\qquad\qquad \text{when } 2k(n - 1) < t < 2kn$
73	$\dfrac{k}{s^2 + k^2} \coth \dfrac{\pi s}{2k}$	$\lvert \sin kt \rvert$
74	$\dfrac{1}{(s^2 + 1)(1 - e^{-\pi s})}$	$\begin{cases} \sin t \text{ when } (2n - 2)\pi < t < (2n - 1)\pi \\ 0 \quad \text{when } (2n - 1)\pi < t < 2n\pi \end{cases}$
75	$\dfrac{1}{s} e^{-k/s}$	$J_0(2\sqrt{kt})$
76	$\dfrac{1}{\sqrt{s}} e^{-k/s}$	$\dfrac{1}{\sqrt{\pi t}} \cos 2\sqrt{kt}$
77	$\dfrac{1}{\sqrt{s}} e^{k/s}$	$\dfrac{1}{\sqrt{\pi t}} \cosh 2\sqrt{kt}$
78	$\dfrac{1}{s^{3/2}} e^{-k/s}$	$\dfrac{1}{\sqrt{\pi k}} \sin 2\sqrt{kt}$
79	$\dfrac{1}{s^{3/2}} e^{k/s}$	$\dfrac{1}{\sqrt{\pi k}} \sinh 2\sqrt{kt}$
80	$\dfrac{1}{s^\mu} e^{-k/s}(\mu > 0)$	$\left(\dfrac{t}{k}\right)^{(\mu - 1)/2} J_{\mu-1}(2\sqrt{kt})$

*$H(2k, t) = k + (r - k)(-1)^n$ where $t = 2kn + r; 0 \le r < 2k; n = 0, 1, 2, \ldots$.

The Laplace Transforms

LAPLACE TRANSFORMS (Continued)

	$f(s)$	$F(t)$
81	$\dfrac{1}{s^\mu}\, e^{k/s}\,(\mu > 0)$	$\left(\dfrac{t}{k}\right)^{(\mu-1)/2} I_{\mu-1}(2\sqrt{kt})$
82	$e^{-k\sqrt{s}}\,(k > 0)$	$\dfrac{k}{2\sqrt{\pi t^3}}\exp\left(-\dfrac{k^2}{4t}\right)$
83	$\dfrac{1}{s}\, e^{-k\sqrt{s}}\,(k \geqq 0)$	$\operatorname{erfc}\left(\dfrac{k}{2\sqrt{t}}\right)$
84	$\dfrac{1}{\sqrt{s}}\, e^{-k\sqrt{s}}\,(k \geqq 0)$	$\dfrac{1}{\sqrt{\pi t}}\exp\left(-\dfrac{k^2}{4t}\right)$
85	$s^{-3/2} e^{-k\sqrt{s}}\,(k \geqq 0)$	$2\sqrt{\dfrac{t}{\pi}}\exp\left(-\dfrac{k^2}{4t}\right) - k\operatorname{erfc}\left(\dfrac{k}{2\sqrt{t}}\right)$
86	$\dfrac{a e^{-k\sqrt{s}}}{s(a+\sqrt{s})}\,(k \geqq 0)$	$-e^{ak}e^{a^2 t}\operatorname{erfc}\left(a\sqrt{t}+\dfrac{k}{2\sqrt{t}}\right) + \operatorname{erfc}\left(\dfrac{k}{2\sqrt{t}}\right)$
87	$\dfrac{e^{-k\sqrt{s}}}{\sqrt{s}(a+\sqrt{s})}\,(k \geqq 0)$	$e^{ak}e^{a^2 t}\operatorname{erfc}\left(a\sqrt{t}+\dfrac{k}{2\sqrt{t}}\right)$
88	$\dfrac{e^{-k\sqrt{s(s+a)}}}{\sqrt{s(s+a)}}$	$\begin{cases} 0 & \text{when } 0 < t < k \\ e^{-(1/2)(at)} I_0(\tfrac{1}{2}a\sqrt{t^2-k^2}) & \text{when } t > k \end{cases}$
89	$\dfrac{e^{-k\sqrt{s^2+a^2}}}{\sqrt{s^2+a^2}}$	$\begin{cases} 0 & \text{when } 0 < t < k \\ J_0(a\sqrt{t^2-k^2}) & \text{when } t > k \end{cases}$
90	$\dfrac{e^{-k\sqrt{s^2-a^2}}}{\sqrt{s^2-a^2}}$	$\begin{cases} 0 & \text{when } 0 < t < k \\ I_0(a\sqrt{t^2-k^2}) & \text{when } t > k \end{cases}$
91	$\dfrac{e^{-k(\sqrt{s^2+a^2}-s)}}{\sqrt{s^2+a^2}}\,(k \geqq 0)$	$J_0(a\sqrt{t^2+2kt})$
92	$e^{-ks} - e^{-k\sqrt{s^2+a^2}}$	$\begin{cases} 0 & \text{when } 0 < t < k \\ \dfrac{ak}{\sqrt{t^2-k^2}} J_1(a\sqrt{t^2-k^2}) & \text{when } t > k \end{cases}$
93	$e^{-k\sqrt{s^2+a^2}} - e^{-ks}$	$\begin{cases} 0 & \text{when } 0 < t < k \\ \dfrac{ak}{\sqrt{t^2-k^2}} I_1(a\sqrt{t^2-k^2}) & \text{when } t > k \end{cases}$
94	$\dfrac{a^\nu e^{-k\sqrt{s^2-a^2}}}{\sqrt{s^2+a^2}(\sqrt{s^2+a^2}+s)^\nu}$ $(\nu > -1)$	$\begin{cases} 0 & \text{when } 0 < t < k \\ \left(\dfrac{t-k}{t+k}\right)^{(1/2)\nu} J_\nu(a\sqrt{t^2-k^2}) & \text{when } t > k \end{cases}$
95	$\dfrac{1}{s}\log s$	$\Gamma'(1) - \log t\ [\Gamma'(1) = -0.5772]$
96	$\dfrac{1}{s^k}\log s\,(k > 0)$	$t^{k-1}\left\{\dfrac{\Gamma'(k)}{[\Gamma(k)]^2} - \dfrac{\log t}{\Gamma(k)}\right\}$
97	$\dfrac{\log s}{s-a}\,(a > 0)$	$e^{at}[\log a - \operatorname{Ei}(-at)]$

The Laplace Transforms

LAPLACE TRANSFORMS (Continued)

	$f(s)$	$F(t)$
98	$\dfrac{\log s}{s^2 + 1}$	$\cos t \, \mathrm{Si}(t) - \sin t \, \mathrm{Ci}(t)$
99	$\dfrac{s \log s}{s^2 + 1}$	$-\sin t \, \mathrm{Si}(t) - \cos t \, \mathrm{Ci}(t)$
100	$\dfrac{1}{s} \log(1 + ks)\,(k > 0)$	$-\mathrm{Ei}\left(-\dfrac{t}{k}\right)$
101	$\log \dfrac{s - a}{s - b}$	$\dfrac{1}{t}(e^{bt} - e^{at})$
102	$\dfrac{1}{s} \log(1 + k^2 s^2)$	$-2\mathrm{Ci}\left(\dfrac{t}{k}\right)$
103	$\dfrac{1}{s} \log(s^2 + a^2)\ \ (a > 0)$	$2 \log a - 2\mathrm{Ci}(at)$
104	$\dfrac{1}{s^2} \log(s^2 + a^2)\ \ (a > 0)$	$\dfrac{2}{a}[at \log a + \sin at - at \, \mathrm{Ci}(at)]$
105	$\log \dfrac{s^2 + a^2}{s^2}$	$\dfrac{2}{t}(1 - \cos at)$
106	$\log \dfrac{s^2 - a^2}{s^2}$	$\dfrac{2}{t}(1 - \cosh at)$
107	$\arctan \dfrac{k}{s}$	$\dfrac{1}{t} \sin kt$
108	$\dfrac{1}{s} \arctan \dfrac{k}{s}$	$\mathrm{Si}(kt)$
109	$e^{k^2 s^2} \mathrm{erfc}\,(ks)\ \ (k > 0)$	$\dfrac{1}{k\sqrt{\pi}} \exp\left(-\dfrac{t^2}{4k^2}\right)$
110	$\dfrac{1}{s} e^{k^2 s^2} \mathrm{erfc}\,(ks)\ \ (k > 0)$	$\mathrm{erf}\left(\dfrac{t}{2k}\right)$
111	$e^{ks} \mathrm{erfc}\,(\sqrt{ks})\ \ (k > 0)$	$\dfrac{\sqrt{k}}{\pi \sqrt{t}(t + k)}$
112	$\dfrac{1}{\sqrt{s}} \mathrm{erfc}\,(\sqrt{ks})$	$\begin{cases} 0 & \text{when } 0 < t < k \\ (\pi t)^{-1/2} & \text{when } t > k \end{cases}$
113	$\dfrac{1}{\sqrt{s}} e^{ks} \mathrm{erfc}\,(\sqrt{ks})\,(k > 0)$	$\dfrac{1}{\sqrt{\pi(t + k)}}$
114	$\mathrm{erf}\left(\dfrac{k}{\sqrt{s}}\right)$	$\dfrac{1}{\pi t} \sin(2k\sqrt{t})$
115	$\dfrac{1}{\sqrt{s}} e^{k^2/s} \mathrm{erfc}\left(\dfrac{k}{\sqrt{s}}\right)$	$\dfrac{1}{\sqrt{\pi t}} e^{-2k\sqrt{t}}$

The Laplace Transforms

LAPLACE TRANSFORMS (Continued)

	$f(s)$	$F(t)$
115.1	$-e^{as} \text{Ei}(-as)$	$\dfrac{1}{t+a}$; $(a > 0)$
115.2	$\dfrac{1}{a} + se^{as} \text{Ei}(-as)$	$\dfrac{1}{(t+a)^2}$; $(a > 0)$
115.3	$\left[\dfrac{\pi}{2} - \text{Si}(s)\right]\cos s + \text{Ci}(s)\sin s$	$\dfrac{1}{t^2 + 1}$
116*	$K_0(ks)$	$\begin{cases} 0 & \text{when } 0 < t < k \\ (t^2 - k^2)^{-1/2} & \text{when } t > k \end{cases}$
117	$K_0(k\sqrt{s})$	$\dfrac{1}{2t}\exp\left(-\dfrac{k^2}{4t}\right)$
118	$\dfrac{1}{s}e^{ks}K_1(ks)$	$\dfrac{1}{k}\sqrt{t(t + 2k)}$
119	$\dfrac{1}{\sqrt{s}}K_1(k\sqrt{s})$	$\dfrac{1}{k}\exp\left(-\dfrac{k^2}{4t}\right)$
120	$\dfrac{1}{\sqrt{s}}e^{k/s}K_0\left(\dfrac{k}{s}\right)$	$\dfrac{2}{\sqrt{\pi t}}K_0(2\sqrt{2kt})$
121	$\pi e^{-ks}I_0(ks)$	$\begin{cases} [t(2k - t)]^{-1/2} & \text{when } 0 < t < 2k \\ 0 & \text{when } t > 2k \end{cases}$
122**	$e^{-ks}I_1(ks)$	$\begin{cases} \dfrac{k - t}{\pi k\sqrt{t(2k - t)}} & \text{when } 0 < t < 2k \\ 0 & \text{when } t > 2k \end{cases}$

*$K_n(x)$ is Bessel's function of the second kind for the imaginary argument.

**Several additional transforms, especially those involving other Bessel functions, can be found in the tables by G. A. Campbell and R. M. Foster, "Fourier Integrals for Practical Applications", or "Vol. 1, Bateman Manuscript Project, Transform Tables, McGraw-Hill, 1955", or N. W. McLachlan and P. Humbert, "Formulaire pour le calcul symbolique". In the tables by Campbell and Foster, only those entries containing the condition $0 < g$ or $k < g$, where g is our t, are Laplace transforms.

THE Z TRANSFORM

B. Girling

When $F(t)$, a continuous function of time, is sampled at regular intervals of period T the usual Laplace transform techniques are modified. The diagramatic form of a simple sampler together with its associated input-output waveforms is shown below

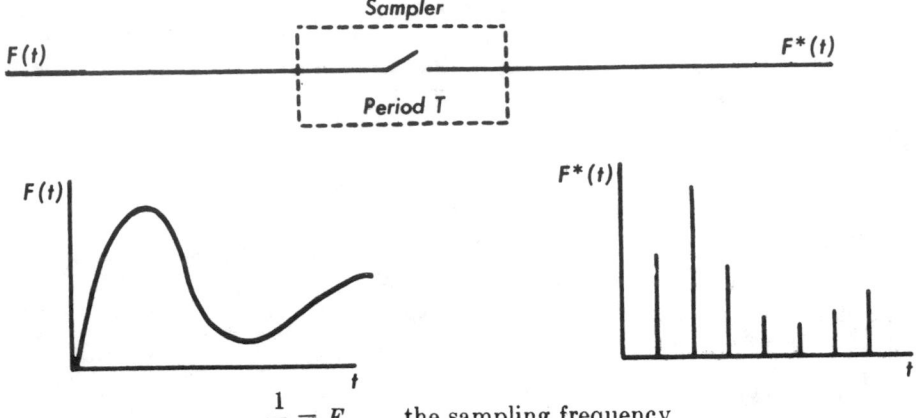

$$\frac{1}{T} \equiv F_s \qquad \text{the sampling frequency}$$

Defining the set of impulse functions $\delta_T(t)$ by

$$\delta_T(t) \equiv \sum_{n=0}^{\infty} \delta(t - nT)$$

the input-output relationship of the sampler becomes

$$F^*(t) = F(t) \cdot \delta_T(t)$$
$$= \sum_{n=0}^{\infty} F(nT) \cdot \delta(t - nT).$$

While for a given $F(t)$ and T the $F^*(t)$ is unique, the converse is not true.

The Laplace transform can be used to define $F^*(s)$ as follows

$$L\{F^*(t)\} \equiv f^*(s)$$
$$= \sum_{n=0}^{\infty} F(nT) \cdot e^{-nTs}.$$

The variable 'z' is introduced by means of the transformation

$$z = e^{Ts}$$

and since any function of s can now be replaced by a corresponding function of z we have

$$f(z) = \sum_{n=0}^{\infty} F(nT) \cdot z^{-n}$$

where $\qquad f^*(s) \equiv f(z)$

and $\qquad s = \frac{1}{T} \ln z$

The Z operator can now be defined in terms of the Laplace operator by the relationship

$$Z\{F(t)\} \equiv L\{F^*(t)\}$$

The Z Transform

THE Z TRANSFORM (Continued)

An alternative definition (quoted without proof) is

$$Z\{F(t)\} = \sum \text{residues of} \left[\left(\frac{1}{1 - e^{Tx}z^{-1}}\right) \cdot f(z)\right]$$

The inverse z transform

$$Z^{-1}\{f(z)\} \equiv F^*(t)$$

$$= \frac{1}{2\pi j} \oint f(z) \cdot z^{n-1} \, dz$$

where the contour of integration encloses all the singularities of the integrand. In the following table Greek letters denote constants.

$F(t)$	$f(z) = Z\{F(t)\}$
$\alpha F(t)$	$\alpha f(z)$
$F(t) + G(t)$	$f(z) + g(z)$
$F(t + T)$	$zf(z) - zF(0)$
$F(t + 2T)$	$z^2 f(z) - z^2 F(0) - zF(T)$
$F(t + mT)$	$z^m f(z) - \displaystyle\sum_{r=0}^{m-1} z^{m-r} F(rT)$
	$= z^m f(z)$ when $F(rT) = 0, 0 \le r \le m - 1$
$F(t - mT)$	$z^{-m} f(z)$
$e^{\alpha t} F(t)$	$f(e^{-\alpha T}z)$
$e^{-\alpha t} F(t)$	$f(e^{\alpha T}z)$
$t \cdot F(t)$	$-Tz \dfrac{d}{dz} f(z)$
$t^{-1} F(t)$	$-\dfrac{1}{T} \displaystyle\int_0^z \frac{f(z)}{z} \, dz$
$\displaystyle\sum_{m=0}^{T/t} F(mT)$	$\left(\dfrac{z}{z - 1}\right) f(z)$

The following limits are also valid

$$\lim_{t \to 0} F(t) = \lim_{z \to \infty} f(z)$$

$$\lim_{t \to \infty} F^*(t) = \lim_{z \to 1} \left[\left(\frac{z - 1}{z}\right) f(z)\right]$$

In the table which follows, the Heavyside unit step function is defined by

$$H(t - nT) \equiv \begin{cases} 1; t \ge nT \\ 0; t < nT. \end{cases}$$

The Z Transform

THE Z TRANSFORM (Continued)

$F(t)$	$f(z)$
$\delta(t)$	1
$\delta(t - mT)$	$\dfrac{1}{z^m}$
$H(t)$	$\dfrac{z}{z - 1}$
$H(t - T)$	$\dfrac{1}{z - 1}$
$H(t - mT)$	$\dfrac{z}{z^m \cdot (z - 1)}$
$H(t) - H(t - T)$	1
$H(t) - H(t - 2T)$	$1 + \dfrac{1}{z}$
$H(t - mT) - H(t - \overline{m + 1}T)$	$\dfrac{1}{z}\, m$
$\dfrac{T}{t} H(t - T)$	$\ln\left(\dfrac{z}{z - 1}\right)$
t	$\dfrac{Tz}{(z - 1)^2}$
t^2	$\dfrac{T^2 z(z + 1)}{(z - 1)^3}$
t^3	$\dfrac{T^3 z(z^2 + 4z + 1)}{(z - 1)^4}$
t^n	$(-1)^n \lim\limits_{\chi \to 0} \dfrac{\partial^n}{\partial \chi^n}\left(\dfrac{z}{z - e^{-\chi T}}\right)$
$1 - a^{\omega t}$	$\dfrac{z(1 - a^{\omega T})}{(z - 1)(z - a^{\omega T})}$
$a^{\omega t}$	$\dfrac{z}{(z - a^{\omega T})}$
$t a^{\omega t}$	$\dfrac{T z a^{\omega T}}{(z - a^{\omega T})^2}$
$t^2 a^{\omega t}$	$\dfrac{T^2 a^{\omega T} z(z + a^{\omega T})}{(z - a^{\omega T})^3}$
$\sin \omega t$	$\dfrac{z \sin \omega T}{z^2 - 2z \cos \omega T + 1}$
$\cos \omega t$	$\dfrac{z(z - \cos \omega T)}{z^2 - 2z \cos \omega T + 1}$
$\sinh \omega t$	$\dfrac{z \sinh \omega T}{z^2 - 2z \cosh \omega T + 1}$
$\cosh \omega t$	$\dfrac{z(z - \cosh \omega T)}{z^2 - 2z \cosh \omega T + 1}$
$e^{-\alpha t} \sin \omega t$	$\dfrac{z e^{-\alpha T} \sin \omega T}{z^2 - 2z e^{-\alpha T} \cos \omega T + e^{-2\alpha T}}$
$e^{-\alpha t} \cos \omega t$	$\dfrac{z(z - e^{-\alpha T} \cos \omega T)}{z^2 - 2z e^{-\alpha T} \cos \omega T + e^{-2\alpha T}}$
$e^{-\alpha t} \sinh \omega t$	$\dfrac{z e^{-\alpha T} \sinh \omega T}{z^2 - 2z e^{-\alpha T} \cosh \omega T + e^{-2\alpha T}}$
$e^{-\alpha t} \cosh \omega t$	$\dfrac{z(z - e^{-\alpha T} \cosh \omega T)}{z^2 - 2z e^{-\alpha T} \cosh \omega T + e^{-2\alpha T}}$

THE Z TRANSFORM (Continued)

$F(t)$	$f(z)$
$-\dfrac{1}{a}\left[\delta(t) - a^{t/T}\right]$	$\dfrac{1}{z-a}$
$\dfrac{1}{(a-b)}\left[a^{\left(\frac{t}{T}-1\right)} - b^{\left(\frac{t}{T}-1\right)}\right]$	$\dfrac{1}{(z-a)(z-b)}$
$\dfrac{1}{(a-b)}\left[a^{\frac{t}{T}} - b^{\frac{t}{T}}\right]$	$\dfrac{z}{(z-a)(z-b)}$
$\dfrac{1}{(a-b)}\left[(a-c)a^{\left(\frac{t}{T}-1\right)} - (b-c)b^{\left(\frac{t}{T}-1\right)}\right]$	$\dfrac{z-c}{(z-a)(z-b)}$
$\dfrac{1}{(a-b)}\left[a^{\left(\frac{t}{T}+1\right)} - b^{\left(\frac{t}{T}+1\right)}\right]$	$\dfrac{z^2}{(z-a)(z-b)}$
$\dfrac{1}{\left(\frac{t}{T}\right)!}$	$e^{1/z}$
$\dfrac{1}{\left(\frac{2t}{T}\right)!}$	$\cosh\left(z^{-\frac{1}{2}}\right)$

Methods of evaluating inverse *z* transforms.

(1) Cauchy's residue theorem.

For $t = nT$,

$$G(nT) = \sum_{\text{all } z_k} \left[\text{residues of } g(z)z^{n-1} \text{ at } z_k\right]$$

where the z_k define all the poles of $g(z)z^{n-1}$.

(2) Partial fractions.

Expand $g(z)/z$ into partial fractions. The product of z with each of the partial fractions will then be recognizable from the standard forms in the table of z transforms. Note however that the continuous functions obtained are only valid at the sampling instants.

(3) Power series expansion by long division using detached coefficients.

$g(z)$ is expanded into a power series in z^{-1} and the coefficient of the term in z^{-n} is the value of $g(nT)$ i.e. the value of $G(t)$ at the nth sampling instant.

The z transform as a means of determining approximately the inverse Laplace transform.

Since

$$z \equiv e^{Ts}$$

$$s^{-1} = \frac{T}{2}\left[\frac{1}{v} - \frac{v}{3} - \frac{4v^3}{45} - \frac{44v^5}{945} - \cdots\right]$$

where

$$v \equiv \frac{1 - z^{-1}}{1 + z^{-1}},$$

the series being very rapid in its convergence. Given $g(s)$, to find its inverse Laplace transform the following operations are carried out:-

(i) Divide the numerator and denominator of $g(s)$ by the highest power of s yielding as an alternative form for $g(s)$ the quotient of two polynomials in s^{-1}.

(ii) Chose as a numerical value of T, that which makes $2\pi/T$ much larger than the imaginary part of the poles of $G(s)$.

(iii) Substitute into the alternative form for $g(s)$ obtained in (i) above the expansion for s^{-n} determined from the following short table of approximations.

The Z Transform

THE Z TRANSFORM (Continued)

Do not at this stage insert the numerical value for T as tabulations with different intervals may be required.

(iv) Divide by T.

(v) Insert the chosen value for T and divide the numerator by the denominator.

(vi) The coefficient of z^{-n} is the required value of the function at $t = nT$.

s^{-n}	z transform (approximate)
s^{-1}	$\dfrac{T}{2}\left[\dfrac{1 + z^{-1}}{1 - z^{-1}}\right]$
s^{-2}	$\dfrac{T^2}{12}\left[\dfrac{1 + 10z^{-1} + z^{-2}}{(1 - z^{-1})^2}\right]$
s^{-3}	$\dfrac{T^3}{3}\left[\dfrac{z^{-1} + z^{-2}}{(1 - z^{-1})^3}\right]$
s^{-4}	$\dfrac{T^4}{144}\left[\dfrac{1 + 20z^{-1} + 102z^{-2} + 20z^{-3} + z^{-4}}{(1 - z^{-1})^4}\right]$
s^{-5}	$\dfrac{T^5}{24}\left[\dfrac{z^{-1} + 11z^{-2} + 11z^{-3} + z^{-4}}{(1 - z^{-1})^5}\right]$
s^{-6}	$\dfrac{T^6}{4}\left[\dfrac{z^{-2} + 2z^{-3} + z^{-4}}{(1 - z^{-1})^6}\right]$
s^{-7}	$\dfrac{T^7}{8}\left[\dfrac{z^{-2} + 3z^{-3} + 3z^{-4} + z^{-5}}{(1 - z^{-1})^7}\right]$

Additional material on Z-transforms can be found in the papers by Boxer, R. and Thaler, S., A simplified method of solving linear and nonlinear systems. *Proc. IEE,* 1956, 89–101, and Boxer, R., A note on numerical transform calculus, *Proc. IEE,* 1957, 1401–1406.

COMPLEX VARIABLES

Complex Numbers

Cartesian Form

The cartesian form of a complex number is $z = x + iy$, where x and y are real numbers and i, called the imaginary unit, has the property that $i^2 = -1$. The real numbers x and y are called the real and imaginary parts of $x + iy$, respectively.

Polar Form

$$z = re^{i\theta} = r(\cos\theta + i\sin\theta)$$

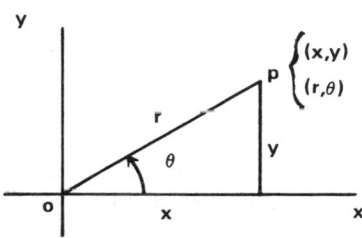

Modulus

$$r = |z| = (x^2 + y^2)^{1/2}$$

Argument

$$\theta = \arg z = \arctan\frac{y}{x}$$

Complex Conjugate

$$\bar{z} = x - iy, \quad |\bar{z}| = |z|, \quad \arg\bar{z} = -\arg z$$

Addition and Subtraction

$$z_1 \pm z_2 = (x_1 + iy_1) \pm (x_2 + iy_2) = (x_1 \pm x_2) + (y_1 \pm y_2)$$

Multiplication

$$z_1 z_2 = (x_1 + iy_1)(x_2 + iy_2) = (x_1 x_2 - y_1 y_2) + i(x_1 y_2 + x_2 y_1)$$

$$|z_1 z_2| = |z_1| \, |z_2|, \quad \arg(z_1 z_2) = \arg z_1 + \arg z_2$$

Division

$$\frac{z_1}{z_2} = \frac{z_1 \bar{z}_2}{z_2 \bar{z}_2} = \frac{(x_1 x_2 + y_1 y_2) + i(x_2 y_1 - x_1 y_2)}{x_2^2 + y_2^2}$$

$$\left|\frac{z_1}{z_2}\right| = \frac{|z_1|}{|z_2|}, \quad \arg\left(\frac{z_1}{z_2}\right) = \arg z_1 - \arg z_2$$

Powers

$$z^n = r^n e^{in\theta} = r^n(\cos n\theta + i\sin n\theta) \quad \text{(DeMoivre's Theorem)}$$

Roots

$$z^{\frac{1}{n}} = r^{\frac{1}{n}} e^{i\theta/n} = r^{\frac{1}{n}} \left[\cos \frac{\theta + 2k\pi}{n} + i \sin \frac{\theta + 2k\pi}{n}, k = 1, 2, \ldots, n-1 \right]$$

(Principal root if $-\pi < \theta < \pi$)

Functions of a Complex Variable

A complex function

$$w = f(z) = u(x, y) + iv(x, y) = |w| e^{i\phi}$$

$$(z = x + iy = |z| e^{i\theta})$$

associates one or more values of the complex dependent variable w with each value of the complex independent variable z in a given domain of definition.

Cauchy-Riemann Equations

$$\frac{\partial u}{\partial x} = \frac{\partial v}{\partial y}, \ \frac{\partial u}{\partial y} = -\frac{\partial v}{\partial x}$$

A function $w = f(z)$ differentiable at each point of a certain neighborhood of a point z_o (i.e., at each point of a circle with center z_o and an arbitrary small radius) is said to be analytic at the point z_o. A function is called analytic in a connected domain if it is analytic at each point in the domain.

A necessary and sufficient condition that the function $f(z) = u(x,y) + iv(x,y)$ is analytic is that it satisfy the Cauchy-Riemann conditions.

Cauchy-Goursat Theorem

If a function f(z) is analytic at all points within and on a simple closed curve C, then

$$\int_c f(z) \, dz = 0$$

Cauchy's Integral Formula

If f(z) is analytic inside and on a simple closed curve C and if z_o is interior to C, then

$$f(z_0) = \frac{1}{2\pi i} \int_c \frac{f(z)}{z - z_0} \, dz$$

Also, the derivatives $f'(z)$, $f''(z)$, ... of all orders exist, and

$$f^n(z_0) = \frac{n!}{2\pi i} \int_c \frac{f(z)}{(z - z_0)^{n+1}} \, dz$$

Taylor-Series Expansion

If f(z) is analytic inside and on a circle C of radius r about $z = z_o$, then there exists a unique and uniformly convergent series expansion in powers of $z - z_o$,

$$f(z) = \sum_{n=0}^{\infty} a_n (z - z_0)^n, \ (|z - z_0| < r, z_0 \neq \infty)$$

where

$$a_n = \frac{1}{n!} \, f^n(z_0) = \frac{1}{2\pi i} \int_C \frac{f(s)}{(s-z_0)^{n+1}} \, ds$$

If M(r) is an upper bound on |f(z)| on C, then

$$|a_n| = \frac{1}{n!} \, |f^{(n)}(z_0)| \leqslant \frac{M(r)}{r^n} \qquad \text{(Cauchy's Inequality)}$$

If Taylor's series is terminated with the term $a_{n-1}(z - z_0)^{n-1}$, the remainder $R_n(z)$ is given by

$$R_n(z) = \frac{(z-z_0)^n}{2\pi i} \int_C \frac{f(s) \, ds}{(s-z_0)^n (s-z)}$$

$$|R_n(z)| \leqslant \left(\frac{|z-z_0|}{r} \right)^n \frac{M(r) \, r}{r - |z-z_0|}$$

Laurent-Series Expansion

If f(z) is analytic throughout the annular region between and on the concentric circles C_1 and C_2 centered at $z = z_0$ and radii r_1 and $r_2 < r_1$, respectively, there exists a unique series expansion in terms of positive and negative powers of $z - z_0$

$$f(z) = \sum_{n=0}^{\infty} a_n (z-z_0)^n + \sum_{n=1}^{\infty} b_n (z-z_0)^{-n}$$

where

$$a_n = \frac{1}{2\pi i} \int_C \frac{f(s)}{(s-z_0)^{n+1}} \, ds, \qquad b_n = \frac{1}{2\pi i} \int_C \frac{f(s)}{(s-z_0)^{-n+1}} \, ds$$

$$(n = 0, 1, 2, \dots) \qquad\qquad\qquad (n = 1, 2, \dots)$$

Zeros and Isolated Singularities

Zeros

The points z for which f(z) = 0 are called the zeros of f(z). A function f(z) analytic at $z = z_0$ has a zero of order m, where m is a positive integer at $z = z_0$ if and only if the first m coefficients $a_0, a_1, \dots, a_{m-1}$ in the Taylor-series expansion about $z = z_0$ vanish.

Singularities

A singular point or singularity of the function f(z) is any point where f(z) is not analytic. An isolated singularity of f(z) at $z = z_0$ is

1. A removable singularity if and only if all coefficients b_n in the Laurent-series expansion of f(z) about $z = z_0$ vanish.
2. A pole of order m (m = 1,2,...) if and only if $(z - z_0)^m f(z)$, but not $(z - z_0)^{m-1} f(z)$ is analytic at $z = z_0$, i.e., if and only if $b_m \neq 0$, $b_{m+1} = b_{m+2} = \dots = 0$ in the Laurent-series expansion of f(z) about $z = z_0$, or if $\frac{1}{f(z)}$ is analytic and has a zero of order m at $z = z_0$.
3. An isolated essential singularity if and only if the Laurent-series expansion of f(z) about $z = z_0$ has an infinite number of terms involving negative powers of $z - z_0$.

Residues

Given a point $z = z_0$, where $f(z)$ is either analytic or has an isolated singularity, the residue of $f(z)$ at $z = z_0$ is the coefficient of $(z - z_0)^{-1}$ in the Laurent-series expansion of $f(z)$ about $z = z_0$, or

$$R_k = b_1 = \frac{1}{2\pi i} \int_C f(z) \, dz$$

If $f(z)$ is either analytic or has a removable singularity at $z = z_0$, then $b_1 = 0$. If $z = z_0$ is a pole of order m, then

$$b_1 = \frac{1}{(m-1)!} \frac{d^{m-1}}{dz^{m-1}} \left[(z - z_0)^m f(z) \right]_{z = z_0}$$

For every simple closed contour C enclosing at most a finite number of singularities $z_1, z_2, \ldots, z_n$ of a single-valued function $f(z)$ continuous on C,

$$\frac{1}{2\pi i} \int_C f(z) \, dz = \sum_{k=1}^{n} R_k$$

where the R_k are the residues of $f(z)$ at the singularities.

Mappings or Transformations

A function $w = f(z)$ maps points of the z-plane into corresponding points of the w-plane. At every point z such that $f(z)$ is analytic and $f'(z) \neq 0$, the mapping is conformal, i.e., the angle between two curves through such a point is reproduced in magnitude and sense by the angle between the corresponding curves in the w-plane.

A table giving real and imaginary parts, zeros, and singularities for frequently used functions of a complex variable and a table illustrating a number of special transformations of interest in various applications are provided.

Real and Imaginary Parts, Zeros, and Singularities for a Number of Frequently Used Functions

$$f(z) = u(x,y) + iv(x,y) \text{ of a Complex Variable } z = x + iy$$

Note: $|f(z)| = \sqrt{u^2 + v^2}$ $\arg f(z) = \arctan \dfrac{v}{u}$

Function $f(z)$	$u(x,y)$	$v(x,y)$	Zeros (order m)	Isolated singularities
z	x	y	$z = 0$ $m = 1$	Pole ($m = 1$) at $z = \infty$
z^2	$x^2 - y^2$	$2xy$	$z = 0$ $m = 2$	Pole ($m = 2$) at $z = \infty$
$\dfrac{1}{z}$	$\dfrac{x}{x^2 + y^2}$	$-\dfrac{y}{x^2 + y^2}$	$z = \infty$ $m = 1$	Pole ($m = 1$) at $z = 0$
$\dfrac{1}{z^2}$	$\dfrac{x^2 - y^2}{(x^2 + y^2)^2}$	$\dfrac{-2xy}{(x^2 + y^2)^2}$	$z = \infty$ $m = 2$	Pole ($m = 2$) at $z = 0$
$\dfrac{1}{z - (a + ib)}$ $(a, b\ \text{real})$	$\dfrac{(x - a)}{(x-a)^2 + (y-b)^2}$	$-\dfrac{(y - b)}{(x-a)^2 + (y-b)^2}$	$z = \infty$ $m = 1$	Pole ($m = 1$) at $z = a + ib$
$\sqrt{z}$	$\pm\left(\dfrac{x + \sqrt{x^2 + y^2}}{2}\right)^{1/2}$	$\pm\left(\dfrac{-x + \sqrt{x^2 + y^2}}{2}\right)^{1/2}$	Zero of order 1 at $z = 0$ (branch point)	Branch point ($m = 1$) at $z = 0$ Branch point ($m = 1$) at $z = \infty$
e^z	$e^x \cos y$	$e^x \sin y$	$\cdots\cdots\cdots\cdots\cdots$	Essential singularity at $z = \infty$
$\sin z$	$\sin x \cosh y$	$\cos x \sinh y$	$z = k\pi$ $(k = 0, \pm 1, \pm 2, \pm \ldots)$ $m = 1$	Essential singularity at $z = \infty$
$\cos z$	$\cos x \cosh y$	$-\sin x \sinh y$	$z = (k + \tfrac{1}{2})\pi$ $(k = 0, \pm 1, \pm 2, \pm \ldots)$ $m = 1$	Essential singularity at $z = \infty$
$\sinh z$	$\sinh x \cos y$	$\cosh x \sin y$	$z = k\pi i$ $(k = 0, \pm 1, \pm 2, \pm \ldots)$ $m = 1$	Essential singularity at $z = \infty$
$\cosh z$	$\cosh x \cos y$	$\sinh x \sin y$	$z = (k + \tfrac{1}{2})\pi i$ $(k = 0, \pm 1, \pm 2, \pm \ldots)$ $m = 1$	Essential singularity at $z = \infty$

Real and Imaginary Parts, Zeros, and Singularities for a Number of Frequently Used Functions

$$f(z) = u\,(x,y) + iv\,(x,y) \text{ of a Complex Variable } z = x + iy$$

Function $f(z)$	$u(x,y)$	$v(x,y)$	Zeros (order m)	Isolated singularities
$\tan z$	$\dfrac{\sin 2x}{\cos 2x + \cosh 2y}$	$\dfrac{\sinh 2y}{\cos 2x + \cosh 2y}$	$z = k\pi \qquad m = 1$ $(k = 0, \pm 1, \pm 2, \pm \dots)$	Essential singularity at $z = \infty$ Poles $(m = 1)$ at $z = (k + \frac{1}{2})\pi$ $(k = 0, \pm 1, \pm 2, \pm \dots)$
$\tanh z$	$\dfrac{\sinh 2x}{\cosh 2x + \cos 2y}$	$\dfrac{\sin 2y}{\cosh 2x + \cos 2y}$	$z = k\pi i \qquad m = 1$ $(k = 0, \pm 1, \pm 2, \pm \dots)$	Essential singularity at $z = \infty$ Poles $(m = 1)$ at $z = (k+\frac{1}{2})\pi i$ $(k = 0, \pm 1, \pm 2, \pm \dots)$
$\log_e z$	$\frac{1}{2} \log_e (x^2 + y^2)$	$\arctan \dfrac{y}{x} + 2k\pi$ $(k = 0, \pm 1, \pm 2, + \dots)$	$z = 1 \qquad m = 1$ (branch corresponding to $k=0$ only)	Branch points of infinite order at $z = 0$, $z = \infty$; both are essential singularities

From Korn, G. A. and Korn, T. M., *Mathematical Handbook for Scientists & Engineers*, 2nd ed., McGraw-Hill, New York, 1968, 189. With permission.

Table of Transformations of Regions

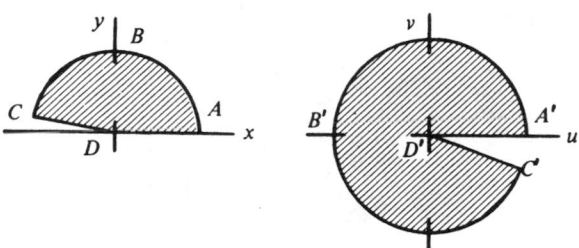

FIGURE 1. $\omega = z^2$.

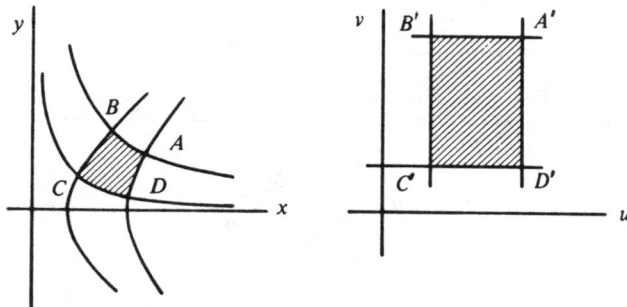

FIGURE 2. $\omega = z^2$.

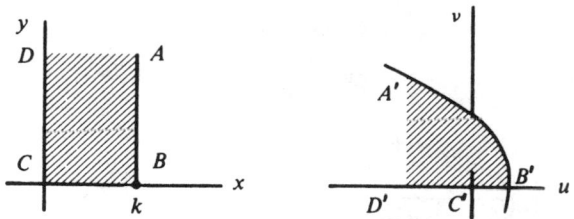

FIGURE 3. $\omega = z^2$; $A'\ B'$ on parabola $\rho = \dfrac{2\,k^2}{1 + \cos\phi}$

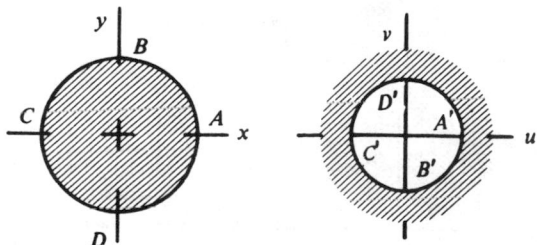

FIGURE 4. $\omega = 1/z$.

(From Churchill, R. V., Brown, J. W., and Verhey, R. F., *Complex Variables and Applications*, 34d ed., McGraw-Hill, New York, 1974, 314 – 324. With permission.)

Table of Transformations of Regions (continued)

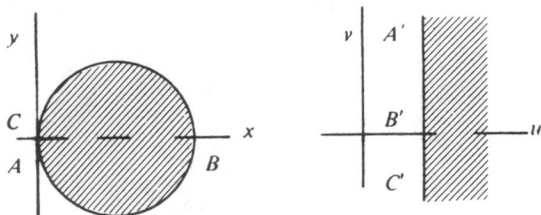

FIGURE 5. $\omega = 1/z$.

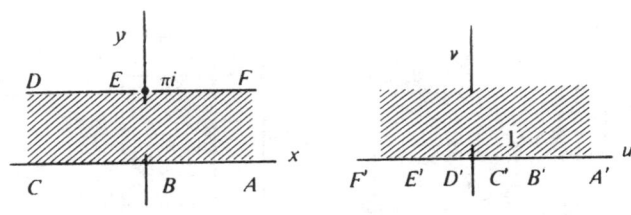

FIGURE 6. $\omega = e^z$.

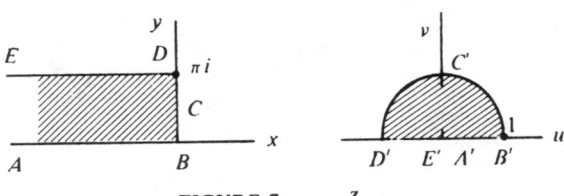

FIGURE 7. $\omega = e^z$.

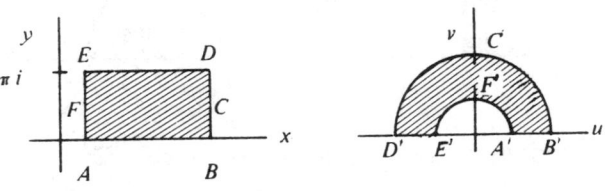

FIGURE 8. $\omega = e^z$.

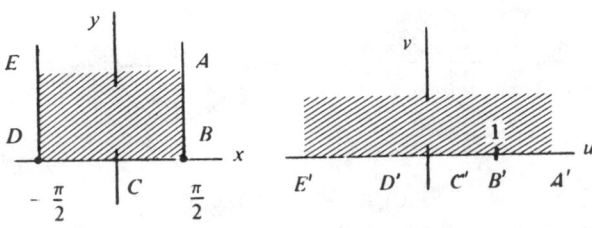

FIGURE 9. $\omega = \sin z$.

Table of Transformations of Regions (continued)

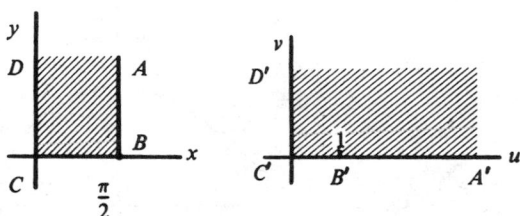

FIGURE 10. $\omega = \sin z$.

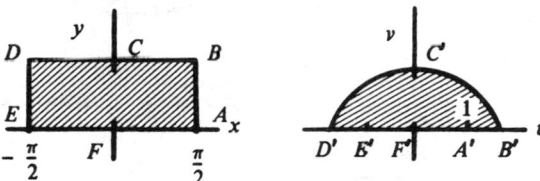

FIGURE 11. $\omega = \sin z$; BCD: $y = k$, $B'\,C'\,D'$ is on ellipse $\left(\dfrac{u}{\cosh k}\right)^2 + \left(\dfrac{v}{\sinh k}\right)^2 = 1$.

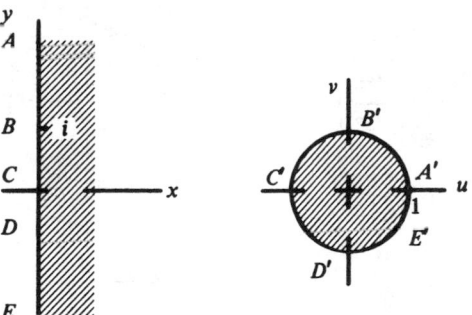

FIGURE 12. $\omega = \dfrac{z-1}{z+1}$.

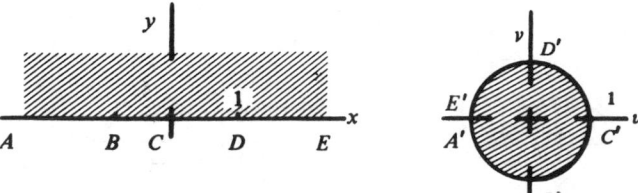

FIGURE 13. $\omega = \dfrac{i-z}{i+z}$.

Table of Transformations of Regions (continued)

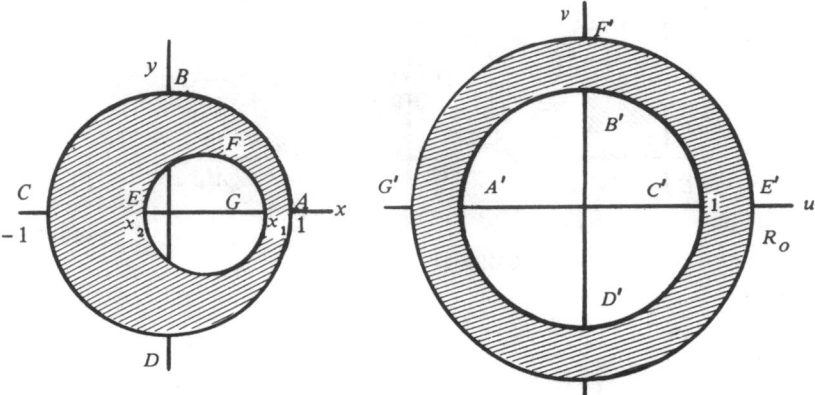

FIGURE 14. $\omega = \dfrac{z-a}{az-1}$; $a = \dfrac{1 + x_1 x_2 + \sqrt{(1 - x_1{}^2)(1 - x_2{}^2)}}{x_1 + x_2}$;

$R_0 = \dfrac{1 - x_1 x_2 + \sqrt{(1 - x_1{}^2)(1 - x_2{}^2)}}{x_1 - x_2}$ (a > 1 and R_0 > 1 when $-1 < x_2 < x_1 < 1$).

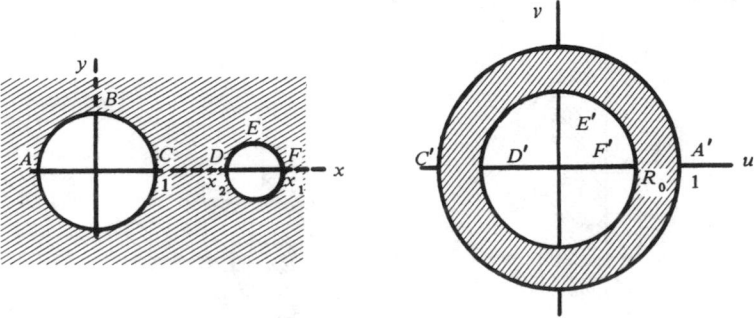

FIGURE 15. $\omega = \dfrac{z-a}{az-1}$; $a = \dfrac{1 + x_1 x_2 + \sqrt{(x_1{}^2 - 1)(x_2{}^2 - 1)}}{x_1 + x_2}$;

$R_0 = \dfrac{x_1 x_2 - 1 - \sqrt{(x_1{}^2 - 1)(x_2{}^2 - 1)}}{x_1 - x_2}$ ($x_2 < a < x_1$ and $0 < R_0 < 1$ when $1 < x_2 < x_1$).

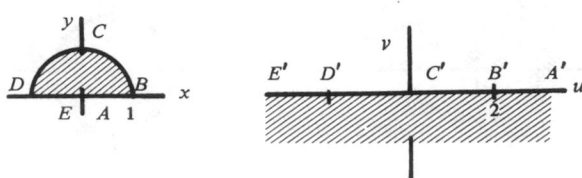

FIGURE 16. $\omega = z + 1/z$.

Table of Transformations of Regions (continued)

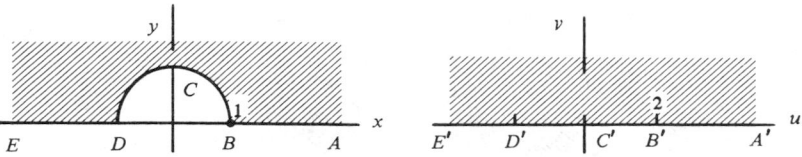

FIGURE 17. $\omega = z + 1/z$.

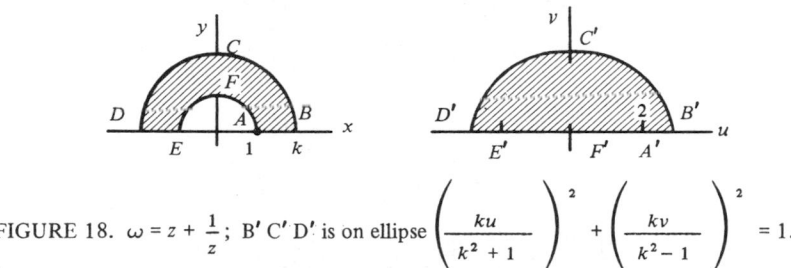

FIGURE 18. $\omega = z + \dfrac{1}{z}$; $B' C' D'$ is on ellipse $\left(\dfrac{ku}{k^2 + 1}\right)^2 + \left(\dfrac{kv}{k^2 - 1}\right)^2 = 1$.

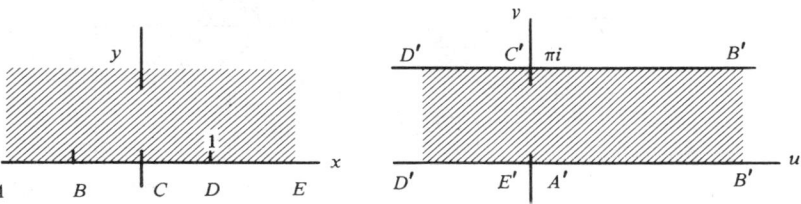

FIGURE 19. $\omega = \log_e \dfrac{z - 1}{z + 1}$; $z = -\coth \dfrac{\omega}{2}$.

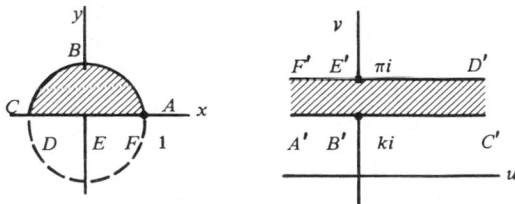

FIGURE 20. $\omega = \log_e \dfrac{z - 1}{z + 1}$; ABC is on circle $x^2 + y^2 - 2y \cot k = 1$.

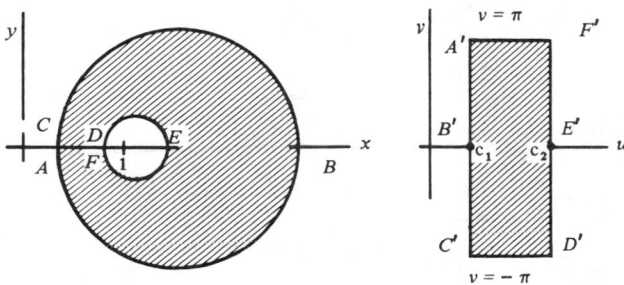

FIGURE 21. $\omega = \log_e \dfrac{z + 1}{z - 1}$; centers of circles at $z = \coth c_n$, radii: $\operatorname{csch} c_n$ ($n = 1, 2$).

Table of Transformations of Regions (continued)

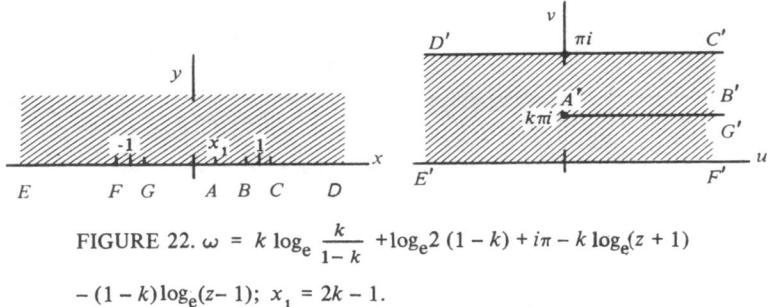

FIGURE 22. $\omega = k \log_e \dfrac{k}{1-k} + \log_e 2(1-k) + i\pi - k \log_e(z+1)$

$-(1-k)\log_e(z-1); \ x_1 = 2k-1.$

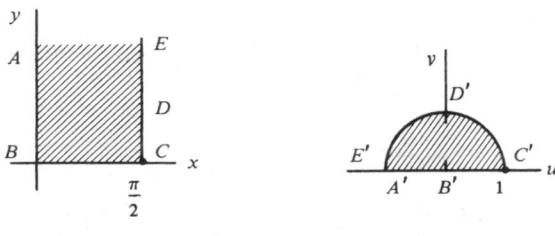

FIGURE 23. $\omega = \tan^2 \dfrac{z}{2} = \dfrac{1-\cos z}{1+\cos z}$

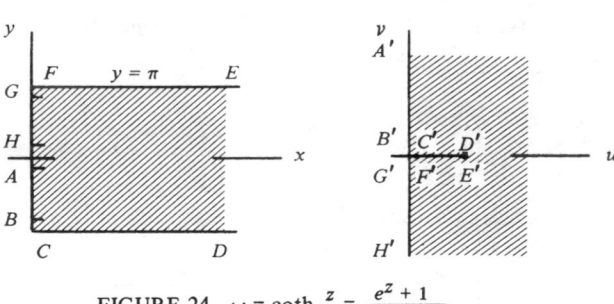

FIGURE 24. $\omega = \coth \dfrac{z}{2} = \dfrac{e^z + 1}{e^z - 1}$.

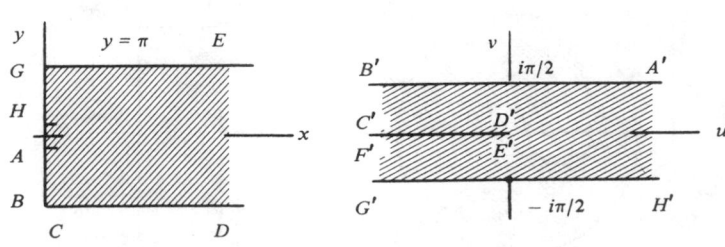

FIGURE 25. $\omega = \log_e \coth \dfrac{z}{2}.$

Table of Transformations of Regions (continued)

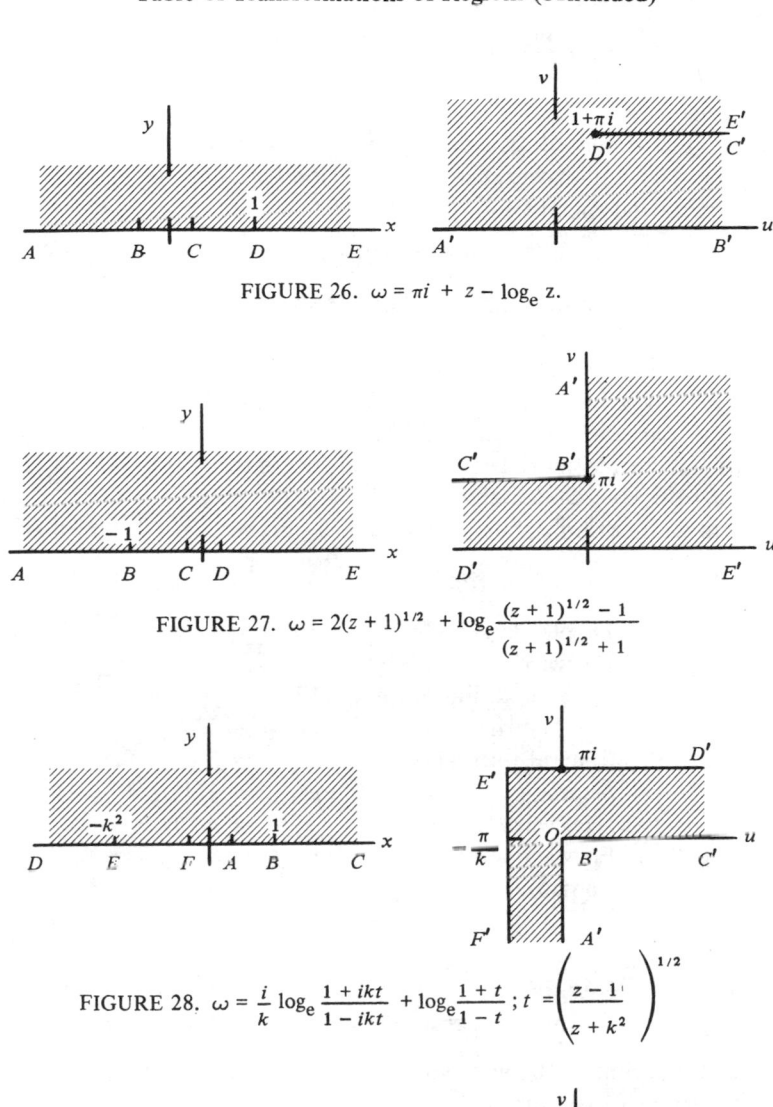

FIGURE 26. $\omega = \pi i + z - \log_e z$.

FIGURE 27. $\omega = 2(z + 1)^{1/2} + \log_e \dfrac{(z + 1)^{1/2} - 1}{(z + 1)^{1/2} + 1}$

FIGURE 28. $\omega = \dfrac{i}{k} \log_e \dfrac{1 + ikt}{1 - ikt} + \log_e \dfrac{1 + t}{1 - t} \; ; \; t = \left(\dfrac{z - 1}{z + k^2}\right)^{1/2}$

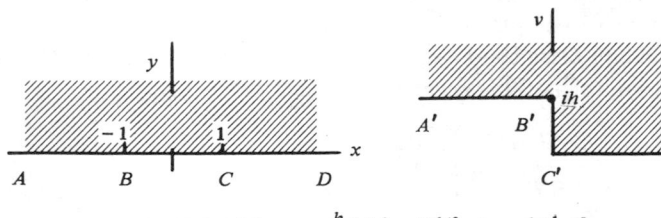

FIGURE 29. $\omega = \dfrac{h}{\pi}\left[(z^2 - 1)^{1/2} + \cosh^{-1} z\right]$.

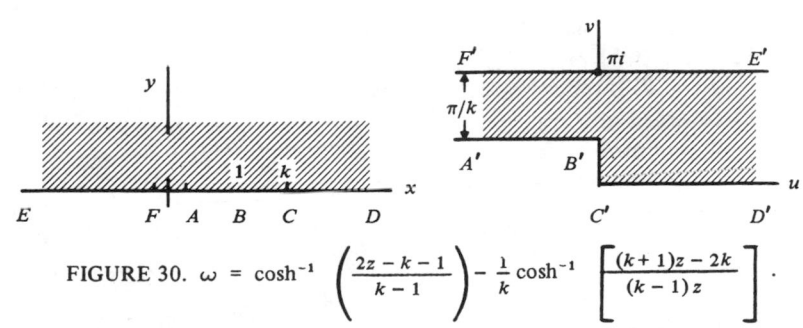

FIGURE 30. $\omega = \cosh^{-1}\left(\dfrac{2z - k - 1}{k - 1}\right) - \dfrac{1}{k} \cosh^{-1}\left[\dfrac{(k + 1)z - 2k}{(k - 1)z}\right]$.

XI. NUMERICAL METHODS
FINITE DIFFERENCES

Uniform interval h.

If a function $f(x)$ is tabulated at a uniform interval h, that is, for arguments given by $x_n = x_0 + nh$, where n is an integer, then the function $f(x)$ may be denoted by f_n. This can be generalized so that for all values of p, and in particular for $0 \leq p \leq 1$,

$$f(x_0 + ph) = f(x_p) = f_p \,,$$

where the argument designated x_0 can be chosen quite arbitrarily.

The notation $x = a(h)b(H)C$ means that the arguments run from a to b inclusive at the uniform interval h and thereafter continue to C at the uniform interval H.

The statement $f(x)$ $4D$ means that the function $f(x)$ is to be tabulated and the answer rounded off to four places of decimals.

The statement $f(x)$ $5S$ means that the function $f(x)$ is to be tabulated with the answer being rounded off to five significant figures.

The following table lists and defines the standard operators used in numerical analysis.

Symbol	Function	Definition
E	Displacement	$Ef_p = f_{p+1}$
Δ	Forward difference	$\Delta f_p = f_{p+1} - f_p$
∇	Backward difference	$\nabla f_p = f_p - f_{p-1}$
$\wedge$	Divided difference	See page 637
δ	Central difference	$\delta f_p = f_{p+\frac{1}{2}} - f_{p-\frac{1}{2}}$
μ	Average	$\mu f_p = \frac{1}{2}(f_{p+\frac{1}{2}} + f_{p-\frac{1}{2}})$
Δ^{-1}	Backward sum	$\Delta^{-1}f_p = \Delta^{-1}f_{p-1} + f_{p-1}$
∇^{-1}	Forward sum	$\nabla^{-1}f_p = \nabla^{-1}f_{p-1} + f_p$
δ^{-1}	Central sum	$\delta^{-1}f_p = \delta^{-1}f_{p-1} + f_{p-\frac{1}{2}}$
D	Differentiation	$Df_p = \dfrac{d}{dx}f(x) = \dfrac{1}{h} \cdot \dfrac{d}{dp}f_p$
$I(=D^{-1})$	Integration	$If_p = \int^{x_p}f(x)dx = h\int^p f_p dp$
$J(=\Delta D^{-1})$	Definite integration	$Jf_p = h\int_p^{p+1}f_p dp$

I, Δ^{-1}, ∇^{-1} and δ^{-1} all imply the existence of an arbitrary constant which is determined by the initial conditions of the problem.

Where no confusion can arise the f can be omitted as, for example in writing Δ_p for Δf_p. Higher differences are formed by successive operations, e.g.,

$$
\begin{aligned}
\Delta^2 f_p &= \Delta_p^2 \\
&= \Delta \cdot \Delta_p \\
&= \Delta(f_{p+1} - f_p) \\
&= \Delta_{p+1} - \Delta_p \\
&= f_{p+2} - f_{p+1} - f_{p+1} + f_p \,.
\end{aligned}
$$

Thus $\Delta_p^2 = f_{p+2} - 2f_{p+1} + f_p$.

Note that $f_p \equiv \Delta_p^0 \equiv \nabla_p^0 \equiv \delta_p^0$.

The disposition of the differences and sums relative to the function values is as shown (the arguments are omitted in these cases in the interests of clarity). In manuscript working double spacing is recommended.

Calculus of Finite Differences

Forward difference scheme						Backward difference scheme			

Forward difference scheme

$$\Delta_{-1}^{-2} \qquad f_{-2} \qquad \Delta_{-3}^{2}$$
$$\Delta_{-1}^{-1} \qquad \Delta_{-2} \qquad \Delta_{-3}^{3}$$
$$\Delta_{0}^{-2} \qquad f_{-1} \qquad \Delta_{-2}^{2} \qquad \Delta_{-2}^{3}$$
$$\Delta_{0}^{-1} \qquad \Delta_{-1} \qquad \Delta_{-2}^{3}$$
$$\Delta_{1}^{-2} \qquad f_{0} \qquad \Delta_{-1}^{2}$$
$$\Delta_{1}^{-1} \qquad \Delta_{0} \qquad \Delta_{-1}^{3}$$
$$\Delta_{2}^{-2} \qquad f_{1} \qquad \Delta_{0}^{2}$$
$$\Delta_{2}^{-1} \qquad \Delta_{1} \qquad \Delta_{0}^{3}$$
$$\Delta_{3}^{-2} \qquad f_{2} \qquad \Delta_{1}^{2}$$

Backward difference scheme

$$\nabla_{-3}^{-2} \qquad f_{-2} \qquad \nabla_{-1}^{2}$$
$$\nabla_{-2}^{-1} \qquad \nabla_{-1} \qquad \nabla_{0}^{3}$$
$$\nabla_{-2}^{-2} \qquad f_{-1} \qquad \nabla_{0}^{2} \qquad \nabla_{1}^{3}$$
$$\nabla_{-1}^{-1} \qquad \nabla_{0} \qquad \nabla_{1}^{3}$$
$$\nabla_{-1}^{-2} \qquad f_{0} \qquad \nabla_{1}^{2}$$
$$\nabla_{0}^{-1} \qquad \nabla_{1} \qquad \nabla_{2}^{3}$$
$$\nabla_{0}^{-2} \qquad f_{1} \qquad \nabla_{2}^{2}$$
$$\nabla_{1}^{-1} \qquad \nabla_{2} \qquad \nabla_{3}^{3}$$
$$\nabla_{1}^{-2} \qquad f_{2} \qquad \nabla_{3}^{2}$$

Central difference scheme

$$\delta_{-2}^{-2} \qquad f_{-2} \qquad \delta_{-2}^{2} \qquad \delta_{-2}^{4}$$
$$\delta_{-1\frac{1}{2}}^{-1} \qquad \delta_{-1\frac{1}{2}} \qquad \delta_{-1\frac{1}{2}}^{3}$$
$$\delta_{-1}^{-2} \qquad f_{-1} \qquad \delta_{-1}^{2} \qquad \delta_{-1}^{4}$$
$$\delta_{-\frac{1}{2}}^{-1} \qquad \delta_{-\frac{1}{2}} \qquad \delta_{-\frac{1}{2}}^{3}$$
$$\delta_{0}^{-2} \qquad f_{0} \qquad \delta_{0}^{2} \qquad \delta_{0}^{4}$$
$$\delta_{\frac{1}{2}}^{-1} \qquad \delta_{\frac{1}{2}} \qquad \delta_{\frac{1}{2}}^{3}$$
$$\delta_{1}^{-2} \qquad f_{1} \qquad \delta_{1}^{2} \qquad \delta_{1}^{4}$$
$$\delta_{1\frac{1}{2}}^{-1} \qquad \delta_{1\frac{1}{2}} \qquad \delta_{1\frac{1}{2}}^{3}$$
$$\delta_{2}^{-2} \qquad f_{2} \qquad \delta_{2}^{2} \qquad \delta_{2}^{4}$$

In the forward difference scheme the subscripts are seen to move forward into the difference table and no fractional subscripts occur. In the backward difference scheme the subscripts lie on diagonals slanting backwards into the table while in the central difference scheme the subscripts maintain their position and the odd order subscripts are fractional.

All three however are merely alternative ways of labeling the same numerical quantities as any difference is the result of subtracting the number diagonally above it in the preceding column from that diagonally below it in the preceding column or, alternatively, it is the sum of the number diagonally above it in the subsequent column with that immediately above it in its own column.

In general $\Delta_{p-\frac{1}{2}n}^{n} \equiv \delta_{p}^{n} \equiv \nabla_{p+\frac{1}{2}n}^{n}$.

If a polynomial of degree r is tabulated exactly i.e., without any round-off errors, then the rth differences are constant.

Tabulate $f(x)3D$ for $x = -.4 \ (.2) \ .8$ where $f(x) = x^3 - .3x^2 + .07x - .123$.

x	f	δ	δ^2	δ^3	δ^4
$-.4$	$-.263$				
		106			
$-.2$	$-.157$		-72		
		34		48	
$+0$	$-.123$		-24		0
		10		48	
$.2$	$-.113$		$+24$		0
		34		48	
$.4$	$-.079$		72		0
		106		48	
$.6$	$+.027$		120		
		226			
$.8$	$.253$				

Calculus of Finite Differences

Had the function been rounded off to two places of decimals then the third differences would not have been constant. The differences are conventionally expressed in units of the least significant decimal as shown. This obviates the need to write down non-significant zeros. In any well tabulated function the differences decrease in magnitude.

The effect of an error term e in any function value is superimposed upon the difference table in a manner shown in the following table:

f	δ	δ^2	δ^3	δ^4	δ^5
0		0		0	
	0		0		$+1e$
0		0		$+1e$	
	0		$+1e$		$-5e$
0		$+1e$		$-4e$	
	$+1e$		$-3e$		$+10e$
$+e$		$-2e$		$+6e$	
	$-1e$		$+3e$		$-10e$
0		$+1e$		$-4e$	
	0		$-1e$		$+5e$
0		0		$+1e$	
	0		0		$-1e$
0		0		0	

In particular since round-off can contribute an error of half a unit (of the least significant decimal) in the function values, reference to the table of binomial coefficients will show that the maximum effect of this upon the twelfth differences (say) is $924(\frac{1}{2}) = 462$. This could give trouble if the problem does not require this to be multiplied by a small constant and is in fact one of the reasons why numerical differentiation is unreliable.

The following table enables the simpler operators to be expressed in terms of the others:

	E	Δ	δ, μ	∇
E	—	$1 + \Delta$	$1 + \mu\delta + \frac{1}{2}\delta^2$	$(1 - \nabla)^{-1}$
Δ	$E - 1$	—	$\mu\delta + \frac{1}{2}\delta^2$	$\nabla(1 - \nabla)^{-1}$
δ	$E^{\frac{1}{2}} - E^{-\frac{1}{2}}$	$\Delta(1 + \Delta)^{-\frac{1}{2}}$	$2(\mu^2 - 1)^{\frac{1}{2}}$	$\nabla(1 - \nabla)^{-\frac{1}{2}}$
∇	$-E^{-1}$	$\Delta(1 + \Delta)^{-1}$	$\mu\delta - \frac{1}{2}\delta^2$	—
μ	$\frac{1}{2}(E^{\frac{1}{2}} + E^{-\frac{1}{2}})$	$\frac{1}{2}(2 + \Delta)(1 + \Delta)^{-\frac{1}{2}}$	$(1 + \frac{1}{4}\delta^2)^{\frac{1}{2}}$	$\frac{1}{2}(2 - \nabla)(1 - \nabla)^{-\frac{1}{2}}$

In addition to the above there are other identities by means of which the above table can be extended, viz.,

$$E = e^{hD} = \Delta\nabla^{-1}$$

$$\mu = E^{-\frac{1}{2}} + \frac{1}{2}\delta = E^{\frac{1}{2}} - \frac{1}{2}\delta = \cosh\left(\frac{1}{2}hD\right)$$

$$\delta = E^{-\frac{1}{2}}\Delta = E^{\frac{1}{2}}\nabla = (\Delta\nabla)^{\frac{1}{2}} = 2\sinh\left(\frac{1}{2}hD\right).$$

Note the emergence of Taylor's series from

$$f_p = E^p f_0$$

$$= e^{phD} f_0$$

$$= f_0 + phDf_0 + \frac{1}{2!}p^2 h^2 D^2 f_0 + \cdots.$$

Calculus of Finite Differences

FUNCTION BUILD-UP FROM DIFFERENCES

Backward differences will be used here since the notation is simpler in this case.

If a function is tabulated as far as f_0 say, then the difference table can be built up. Assuming the differences up to and including ∇_0'' have been formed, if ∇_1'' is known it follows from the definition of backward differences that f_1 can be found from

$$f_1 = f_0 + \nabla_0 + \nabla_0^2 + \nabla_0^3 + \nabla_0^4 + \cdots + \nabla_0^{n-1} + \nabla_1^n .$$

For example, in the cubic tabulated earlier if we take $x_0 = .8$ then $f_0 = .253$ and f_1 can be built-up using the above scheme as follows:

$$f_1 = .253 + (226 + 120 + 48)(10^{-3})$$
$$= .647$$

INTERPOLATION

Finite difference interpolation entails taking a given set of points and fitting a function to them. This function is usualy a polynomial, although it is rarely found explicitly as such. Central differences formulae are in general to be preferred to those involving either forward or backward differences, although the latter may be essential at the extremities of a table where the central differences may not extend far enough.

The most useful of the central differences formulae are those of *Bessel* and *Everett*, the latter particularly so since it only utilizes even order differences. Nevertheless all central difference formulae stopping at the same difference are equivalent.

Notation used:

$$\Delta_{-m_r}^r \equiv \delta_{(r/2)-m_r}^r \equiv \nabla_{r-m_r}^r$$

$$p = \frac{x - x_0}{h} = 1 - q \quad \text{(usually } 0 \leq p \leq 1\text{) .}$$

Any number of finite difference interpolation formulae can be obtained from the following schematic (shown for central differences only, but easily extensible by the above identities):

$$y_p = \sum_{r=0} \binom{p + m_{r-1}}{r} \cdot \delta_{(r/2)-m_r}^r$$

$$= f_{-m_0} + \binom{p + m_0}{1} \delta_{\frac{1}{2}-m_1} + \binom{p + m_1}{2} \delta_{1-m_2}^2 + \binom{p + m_2}{3} \delta_{1\frac{1}{2}-m_3}^3 \cdots ,$$

where the m_r are integers so chosen that the binomial coefficient term $\binom{p + m_{r-1}}{r}$ contains all the linear factors occurring in the preceding term $\binom{p + m_{r-2}}{r - 1}$. Obviously $m_{r+1} = m_r$ or $1 + m_r$.

Further formulae are obtained by taking linear combinations of the formulae obtained from the above schematic, or by using the standard identity transformations upon chosen terms of one of the formulae.

Another way of obtaining interpolation formulae is to apply the standard identity transformations to the operator E in the operational form of the interpolation formula, viz., $f_p = E^p f_0$,

Newton's forward formula

$$f_p = f_0 + p\Delta_0 + \frac{1}{2!} p(p - 1)\Delta_0^2 + \frac{1}{3!} p(p - 1)(p - 2)\Delta_0^3 \cdots \quad 0 \leq p \leq 1$$

Calculus of Finite Differences

Newton's backward formula

$$f_p = f_0 + p\nabla_0 + \frac{1}{2!}p(p+1)\nabla_0^2 + \frac{1}{3!}p(p+1)(p+2)\nabla_0^3 \cdots \quad\quad 0 \leqq p \leqq 1$$

Gauss' forward formula

$$f_p = f_0 + p\delta_{\frac{1}{2}} + G_2\delta_0^2 + G_3\delta_{\frac{1}{2}}^3 + G_4\delta_0^4 + G_5\delta_{\frac{1}{2}}^5 \cdots . \quad\quad 0 \leqq p \leqq 1$$

Gauss' backward formula

$$f_p = f_0 + p\delta_{-\frac{1}{2}} + G_2^*\delta_0^2 + G_3\delta_{-\frac{1}{2}}^3 + G_4^*\delta_0^4 + G_5\delta_{-\frac{1}{2}}^5 \cdots \quad\quad 0 \leqq p \leqq 1$$

In the above $G_{2n} = \begin{pmatrix} p+n-1 \\ 2n \end{pmatrix}$

$$G_{2n}^* = \begin{pmatrix} p+n \\ 2n \end{pmatrix}$$

$$G_{2n+1} = \begin{pmatrix} p+n \\ 2n+1 \end{pmatrix}$$

Stirling's formula

$$f_p = f_0 + \tfrac{1}{2}p(\delta_{\frac{1}{2}} + \delta_{-\frac{1}{2}}) + \tfrac{1}{2}p^2\delta_0^2 + S_3(\delta_{\frac{1}{2}}^3 + \delta_{-\frac{1}{2}}^3) + S_4\delta_0^4 + \cdots . \quad\quad -\tfrac{1}{2} \leqq p \leqq \tfrac{1}{2}$$

Steffenson's formula

$$f_p = f_0 + \tfrac{1}{2}p(p+1)\delta_{\frac{1}{2}} - \tfrac{1}{2}(p-1)p\delta_{-\frac{1}{2}} + (S_3 + S_4)\delta_{\frac{1}{2}}^3 + (S_3 - S_4)\delta_{-\frac{1}{2}}^3 \cdots . \quad -\tfrac{1}{2} \leqq p \leqq \tfrac{1}{2}.$$

In the above $S_{2n+1} = \dfrac{1}{2}\begin{pmatrix} p+n \\ 2n+1 \end{pmatrix}$

$$S_{2n+2} = \frac{p}{2n+2}\begin{pmatrix} p+n \\ 2n+1 \end{pmatrix}$$

$$S_{2n+1} + S_{2n+2} = \begin{pmatrix} p+n+1 \\ 2n+2 \end{pmatrix}$$

$$S_{2n+1} - S_{2n+2} = -\begin{pmatrix} p+n \\ 2n+2 \end{pmatrix}$$

Bessel's formula

$$f_p = f_0 + p\delta_{\frac{1}{2}} + B_2(\delta_0^2 + \delta_1^2) + B_3\delta_{\frac{1}{2}}^3 + B_4(\delta_0^4 + \delta_1^4) + B_5\delta_{\frac{1}{2}}^5 + \cdots \quad\quad 0 \leqq p \leqq 1$$

Everett's formula

$$f_p = (1-p)f_0 + pf_1 + E_2\delta_0^2 + F_2\delta_1^2 + E_4\delta_0^4 + F_4\delta_1^4 + E_6\delta_0^6 + F_6\delta_1^6 + \cdots \quad\quad 0 \leqq p \leqq 1$$

The coefficients in the above two formulae are related to each other and to the coefficients in the Gaussian formulae by the identities

$$\begin{aligned}
B_{2n} &\equiv \tfrac{1}{2}G_{2n} \equiv \tfrac{1}{2}(E_{2n} + F_{2n}) \\
B_{2n+1} &\equiv G_{2n+1} - \tfrac{1}{2}G_{2n} \equiv \tfrac{1}{2}(F_{2n} - E_{2n}) \\
E_{2n} &\equiv G_{2n} - G_{2n+1} \equiv B_{2n} - B_{2n+1} \\
F_{2n} &\equiv G_{2n+1} \equiv B_{2n} + B_{2n+1}
\end{aligned}$$

Also for $q \equiv 1 - p$ the following symmetrical relationships hold:

$$\begin{aligned}
B_{2n}(p) &\equiv B_{2n}(q) \\
B_{2n+1}(p) &\equiv -B_{2n+1}(q) \\
E_{2n}(p) &\equiv F_{2n}(q) \\
F_{2n}(p) &\equiv E_{2n}(q)
\end{aligned}$$

as can be seen from the tables of these coefficients.

p	B_2	B_3	B_4	B_5	B_6	B_7	$(1-p)$
0.00	−0.000000	0.0000000	0.000000	−0.000000	−0.00000	0.00000	1.00
0.01	2475	8085	415	81	8	1	0.99
0.02	4900	15680	825	158	17	2	0.98
0.03	7275	22795	1230	231	25	3	0.97
0.04	9600	29440	1631	300	33	4	0.96
0.05	−0.011875	0.0035625	0.002026	−0.000365	−0.00041	0.00005	0.95
0.06	14100	41360	2416	425	49	6	0.94
0.07	16275	46655	2801	482	57	7	0.93
0.08	18400	51520	3180	534	64	8	0.92
0.09	20475	55965	3552	583	72	8	0.91
0.10	−0.022500	0.0060000	0.003919	−0.000627	−0.00080	0.00009	0.90
0.11	24475	63635	4279	667	87	10	0.89
0.12	26400	66880	4632	704	94	10	0.88
0.13	28275	60745	4979	737	101	11	0.87
0.14	30100	72240	5319	766	109	11	0.86
0.15	−0.031875	0.0074375	0.005651	−0.000791	−0.00115	0.00012	0.85
0.16	33600	76160	5976	813	122	12	0.84
0.17	35275	77605	6294	831	129	12	0.83
0.18	36900	78720	6604	845	135	12	0.82
0.19	38475	79515	6906	856	142	13	0.81
0.20	−0.040000	0.0080000	0.007200	−0.000864	−0.00148	0.00013	0.80
0.21	41475	80185	7486	868	154	13	0.79
0.22	42900	80080	7763	870	160	13	0.78
0.23	44275	79695	8033	868	165	13	0.77
0.24	45600	79040	8293	862	171	13	0.76
0.25	−0.046875	0.0078125	0.008545	−0.000554	−0.00176	0.00013	0.75
0.26	48100	76960	8788	844	181	12	0.74
0.27	49275	75555	9022	830	186	12	0.73
0.28	50400	73920	9247	814	191	12	0.72
0.29	51475	72065	9462	795	196	12	0.71
0.30	−0.052500	0.0070000	0.009669	−0.000773	−0.00200	0.00011	0.70
0.31	53475	67735	9866	750	204	11	0.69
0.32	54400	65280	10053	724	208	11	0.68
0.33	55275	62645	10231	696	212	10	0.67
0.34	56100	59840	10399	666	216	10	0.66
0.35	−0.056875	0.0056875	0.010557	−0.000633	−0.00219	0.00009	0.65
0.36	57600	53760	10706	600	222	9	0.64
0.37	58275	50505	10844	564	225	8	0.63
0.38	58900	47120	10973	527	228	8	0.62
0.39	59475	43615	11092	488	231	7	0.61
0.40	−0.060000	0.0040000	0.011200	−0.000448	−0.00233	0.00007	0.60
0.41	60475	36285	11298	407	235	6	0.59
0.42	60900	32480	11386	364	237	5	0.58
0.43	61275	28595	11464	321	239	5	0.57
0.44	61600	24640	11532	277	240	4	0.56
0.45	−0.061875	0.0020625	0.011589	−0.000232	−0.00241	0.00003	0.55
0.46	62100	16560	11635	186	242	3	0.54
0.47	62275	12455	11672	140	243	2	0.53
0.48	62400	8320	11698	94	244	1	0.52
0.49	62475	4165	11714	47	244	1	0.51
0.50	−0.062500	0.0000000	0.011719	−0.000000	−0.00244	0.00000	0.50
$(1-p)$	B_2	$-B_3$	B_4	$-B_5$	B_6	$-B_7$	p

Bessel's formula (*unmodified*)

(a) $f_p = f_0 + p\delta_{\frac{1}{2}} + B_2(\delta_0^2 + \delta_1^2) + B_3\delta_{\frac{1}{2}}^3 + B_4(\delta_0^4 + \delta_1^4)$
$$+ B_5\delta_{\frac{1}{2}}^5 + B_6(\delta_0^6 + \delta_1^6) + B_7\delta_{\frac{1}{2}}^7 + \cdots .$$

The end terms can be neglected,i.e.,they contribute less than half a unit in the last place of decimals, if the differences are less than the values shown in the following list:

$$
\begin{array}{rll}
\text{Neglect} \quad & B_7\delta_{\frac{1}{2}}^7 \text{ if } & \delta^7 < 3500 \\
B_6(\delta_0 + \delta_1) & & \delta^6 < 100 \\
B_5\delta_{\frac{1}{2}}^5 & & \delta^5 < 500 \\
B_4(\delta_0^4 + \delta_1^4) & & \delta^4 < 20 \\
B_3\delta_{\frac{1}{2}}^3 & & \delta^3 < 60 \\
B_2(\delta_0^2 + \delta_1^2) & & \delta^2 < 4
\end{array}
$$

A powerful simplification is that of throwback. If the higher differences are less than a specified value they can be combined with the lower order ones as follows:

If $\delta^6 < 10,000$ use $\delta_m^4 \equiv \delta^4 - 0.20697\delta^6$

and if $\delta^6 < 1,000$* use $\delta_m^4 \equiv \delta^4 - 0.2\delta^6$
in the modified form of Bessel's formula

(b) $f_p = f_0 + p\delta_{\frac{1}{2}} + B_2(\delta_0^2 + \delta_1^2) + B_3\delta_{\frac{1}{2}}^3 + B_4(\delta_{m0}^4 + \delta_{m1}^4) + B_5\delta_{\frac{1}{2}}^5$.

If $\delta^4 < 1,000$ and $\delta^6 < 1,000$* use $\delta_m^2 \equiv \delta^2 - 0.18393\delta^4 + 0.03808\delta^6$
and if $\delta^4 < 1,000$* use $\delta_m^2 \equiv \delta^2 - 0.18393\delta^4$
in the modified form of Bessel's formula

(c) $f_p = f_0 + p\delta_{\frac{1}{2}} + B_2(\delta_{m0}^2 + \delta_{m1}^2) + B_3\delta_{\frac{1}{2}}^3$.

---

* Higher even order differences are negligible. Throwback involving odd order differences is of little practical use.

p	E_2	F_2	E_4	F_4	E_6	F_6	$(1-p)$
0.00	-0.000000	-0.0000000	0.000000	0.000000	-0.00000	-0.00000	1.00
0.01	32835	16665	496	33	9	7	0.99
0.02	64680	33320	983	67	19	14	0.98
0.03	95545	49955	1461	100	28	21	0.97
0.04	125440	66560	1931	133	37	29	0.96
0.05	-0.0154375	-0.0083125	0.002391	0.001661	-0.00046	-0.00036	0.95
0.06	182360	99640	2842	199	55	43	0.94
0.07	209405	116095	3283	232	64	50	0.93
0.08	235520	132480	3714	264	72	57	0.92
0.09	260715	148785	4135	297	80	64	0.91
0.10	-0.0285000	-0.0165000	0.004546	0.003292	-0.00089	-0.00070	0.90
0.11	308385	181115	4946	361	97	77	0.89
0.12	330880	197120	5336	393	105	84	0.88
0.13	352495	213005	5716	424	112	91	0.87
0.14	373240	228760	6085	455	120	97	0.86
0.15	-0.0393125	-0.0244375	0.006442	0.004860	-0.00127	-0.00104	0.85
0.16	412160	259840	6789	516	134	110	0.84
0.17	430355	275145	7125	546	141	117	0.83
0.18	447720	290280	7449	576	148	123	0.82
0.19	464265	305235	7762	605	154	129	0.81
0.20	-0.0480000	-0.0320000	0.008064	0.006336	-0.00161	-0.00135	0.80
0.21	494935	334565	8354	662	167	141	0.79
0.22	509080	348920	8633	689	172	147	0.78
0.23	522445	363055	8900	716	178	153	0.77
0.24	535040	376960	9156	743	184	158	0.76
0.25	-0.0546875	-0.0390625	0.009399	0.007690	-0.00189	-0.00164	0.75
0.26	557960	404040	9632	794	194	169	0.74
0.27	568305	417195	9852	819	199	174	0.73
0.28	577920	430080	10060	843	203	179	0.72
0.29	586815	442685	10257	867	207	184	0.71
0.30	-0.0595000	-0.0455000	0.010442	0.008895	-0.00212	-0.00189	0.70
0.31	602485	467015	10615	912	215	193	0.69
0.32	609280	478720	10777	933	219	198	0.68
0.33	615395	490105	10927	954	222	202	0.67
0.34	620840	501160	11065	973	226	206	0.66
0.35	-0.0625625	-0.0511875	0.011191	0.009924	-0.00229	-0.00210	0.65
0.36	629760	522240	11305	1011	231	213	0.64
0.37	633255	532245	11408	1028	234	217	0.63
0.38	636120	541880	11500	1045	236	220	0.62
0.39	638365	551135	11580	1060	238	223	0.61
0.40	-0.0640000	-0.0560000	0.011648	0.010752	$-.00240$	-0.00226	0.60
0.41	641035	568465	11705	1089	241	229	0.59
0.42	641480	576520	11751	1102	242	232	0.58
0.43	641345	584155	11785	1114	243	234	0.57
0.44	640640	591360	11808	1125	244	236	0.56
0.45	-0.0639375	-0.0598125	0.011820	0.011357	-0.00245	-0.00238	0.55
0.46	637560	604440	11822	1145	245	240	0.54
0.47	635205	610295	11812	1153	245	241	0.53
0.48	632320	615680	11792	1160	245	242	0.52
0.49	628915	620585	11760	1167	245	243	0.51
0.50	-0.0625000	-0.0625000	0.011719	0.011719	-0.00244	-0.00244	0.50
$(1-p)$	F_2	E_2	F_4	E_4	F_6	E_6	p

Everett's formula (unmodified)

(a) $f_p = (1 - p)f_0 + pf_1 + E_2\delta_0^2 + F_2\delta_1^2 + E_4\delta_0^4 + F_4\delta_1^4 + E_6\delta_0^6 + F_6\delta_1^6 + \cdots$

As with Bessel's formula, end terms can be neglected as follows:

$$\text{Neglect} \quad E_6\delta_0^6 + F_6\delta_1^6 \quad \text{if} \quad \delta^6 < 100$$
$$E_4\delta_0^4 + F_4\delta_1^4 \qquad \delta^4 < 20$$
$$E_2\delta_0^2 + F_2\delta_1^2 \qquad \delta^2 < 4$$

The corresponding throwback formulae are

(b) $f_p = (1 - p)f_0 + pf_1 + E_2\delta_0^2 + F_2\delta_1^2 + E_4\delta_{m0}^4 + F_4\delta_{m1}^4,$

where $\delta_m^4 \equiv \delta^4 - 0.20697\delta^6$ providing $\delta^6 < 10,000*$

or $\delta_m^4 \equiv \delta^4 - 0.2\delta^6$ providing $\delta^6 < 1,000*$;

(c) $f_p = (1 - p)f_0 + pf_1 + E_2\delta_{m0}^4 + F_2\delta_{m1}^2$

where $\delta_m^2 \equiv \delta^2 - 0.18393\delta^4 + 0.03808\delta^6$ if $\delta^4, \delta^6 < 1,000*.$

Generalized Throwback

Using 0.184 instead of the more precise value of 0.18393 we obtain a very simple but nevertheless accurate interpolation formulae

$$f \equiv (1 - p)f_0 + pf_1 + E_2\delta_{m0}^2 + F_2\delta_{m1}^2 + P_4\theta_0^4 + Q_4\theta_1^4,$$

where $\delta_m^2 \equiv \delta^2 - 0.184\delta^4 + 0.03882\delta^6 - 0.0083\delta^8 + 0.0019\delta^{10}$

and $100\theta^4 \equiv \delta^4 - 0.2783\delta^6 + 0.0685\delta^8 - 0.0168\delta^{10}$

and $P_4 \equiv 100(E_4 + 0.184E_2); \quad Q_4 \equiv 100(F_4 + 0.184F_2).$

Symmetric formulae for interpolation to halves (sub-tabulation to halves)

When $p = \frac{1}{2}$ is substituted into Bessel's interpolation formula the odd coefficients B_{2n+1} all become zero and the resulting formula is

$$f_{\frac{1}{2}} = \left(1 - \frac{1}{8}\delta^2 + \frac{3}{128}\delta^4 - \frac{5}{1024}\delta^6 + \frac{35}{32768}\delta^8 - \cdots\right)\mu f_{\frac{1}{2}}.$$

If this is truncated after one, two, three terms etc. the following formulae and error terms are obtained:

Error

$f_{\frac{1}{2}} = \frac{1}{2}(f_0 + f_1)$ $\qquad\qquad\qquad\qquad\qquad\qquad \dfrac{-1}{8}\mu\delta_{\frac{1}{2}}^2$

$f_{\frac{1}{2}} = (-f_{-1} + 9f_0 + 9f_1 - f_2)/16$ $\qquad\qquad\qquad \dfrac{+3}{128}\mu\delta_{\frac{1}{2}}^4$

$f_{\frac{1}{2}} = (3f_{-2} - 25f_{-1} + 150f_0 + 150f_1 - 25f_2 + 3f_3)/256$ $\qquad \dfrac{-5}{1024}\mu\delta_{\frac{1}{2}}^6$

$f_{\frac{1}{2}} = (-5f_{-3} + 49f_{-2} - 245f_{-1} + 1225f_0 + 1225f_1 - 245f_2 + 49f_3 - 5f_4)/2048$ $\qquad \dfrac{+35}{32768}\mu\delta_{\frac{1}{2}}^8$

---

* Higher order differences are negligible.

Similarly, unsymmetric formulae for subtabulation to halves can be obtained from Newton's formula by substituting $p = \frac{1}{2}$ and truncating, e.g.,

Error

$$f_{\frac{1}{2}} = \left(1 + \frac{1}{2}\Delta - \frac{1}{8}\Delta^2 + \frac{1}{16}\Delta^3 - \frac{5}{128}\Delta^4 + \frac{7}{256}\Delta^5 \cdots \right)f_0$$

$$f_{\frac{1}{2}} = (3f_0 + 6f_1 - f_2)/8 \qquad\qquad\qquad \frac{+1}{16}\Delta_0^3$$

$$f_{\frac{1}{2}} = (5f_0 + 15f_1 - 5f_2 + f_3)/16 \qquad\qquad \frac{-5}{128}\Delta_0^4$$

$$f_{\frac{1}{2}} = (35f_0 + 140f_1 - 70f_2 + 28f_3 - 5f_4)/128 \qquad \frac{+7}{256}\Delta_0^5$$

The formula obtained by considering the first two terms only of the series is identical with the first one derived from Bessel's formula.

Calculus of Finite Differences

Interpolation techniques which do not require the function to be tabulated for equal interval of the argument

a) Lagrangian Polynomials

The interpolated value is obtained from a set of points which bridge the required point. The nearer the argument of the required value is to the center of the range of arguments of the function values used the better.

For an odd number of points
$$f(x) = \sum_{r=-n}^{n} L_r f_r \; .$$

For an even number of points
$$f(x) = \sum_{r=-n}^{n+1} L_r f_r \; .$$

The interpolating polynomial is of degree $2n$ in the first case and $2n + 1$ in the second. Further, it must pass through the given points $f_{-n}, f_{1-n}, f_{2-n}, \cdots, f_n$ or f_{1+n} as the case may be.

The coefficient L_r associated with the function value f_r is given by

$$L_r = \frac{(x - x_{-n})(x - x_{1-n}) \cdots (x - x_{r-1})(x - x_{r+1}) \cdots (x - x_{n-1})(x - x_n)}{(x_r - x_{-n})(x_r - x_{1-n}) \cdots (x_r - x_{r-1})(x_r - x_{r+1}) \cdots (x_r - x_{n-1})(x_r - x_n)}$$

for the first case with an additional factor $(x - x_{1+n})/(x_r - x_{1+n})$ in the second.

The error involved is equal to the $(2n + 2)$th. derivative, at some point in the range, multiplied by

$$(x - x_{-n})(x - x_{1-n}) \cdots (x - x_{1-n})/(2n + 2)!$$

in the case of the even number of points, and the $(2n + 1)$th. derivative, again at some point in the range, multiplied by

$$(x - x_{-n})(x - x_{1-n}) \cdots (x - x_n)/(2n + 1)!$$

in the other case.

In both cases the multipliers are functions which oscillate with an amplitude which increases substantially as the argument of the required point departs from the center of the range. When the arguments are spaced at equal intervals h and the point x is nearer to x_0 than to the other points, then $L_r(p)$ can be tabulated for $p = (x - x_0)/h$.

If the degree of the approximating polynomial is known the method is very powerful, otherwise it is best avoided.

b) Divided differences

The layout of a divided difference table is similar to that of an ordinary finite difference table.

x_{-1}	f_{-1}		Δ^2_{-1}		Δ^4_{-1}
		$\Delta_{-\frac{1}{2}}$		$\Delta_{-\frac{1}{2}}$	
x_0	f_0		Δ^2_0		Δ^4_0
		$\Delta_{\frac{1}{2}}$		$\Delta^3_{\frac{1}{2}}$	
x_1	f_1		Δ^2_1		Δ^4_1

Calculus of Finite Differences

where the $\triangle$'s are defined as follows:

$$\triangle_r^0 \equiv f_r, \qquad \triangle_{r+\frac{1}{2}} \equiv (f_{r+1} - f_r)/(x_{r+1} - x_r),$$

and in general

$$\triangle_r^{2n} \equiv (\triangle_{r+\frac{1}{2}}^{2n-1} - \triangle_{r-\frac{1}{2}}^{2n-1})/(x_{r+n} - x_{r-n})$$

and

$$\triangle_{r+\frac{1}{2}}^{2n+1} \equiv (\triangle_{r+1}^{2n} - \triangle_r^{2n})/(x_{r+1+n} - x_{r-n}).$$

Divided differences can with advantage be replaced by adjusted divided differences.

c) Adjusted divided differences

These differences are a modified form of the above and, when the interval of tabulation becomes constant, they reduce to ordinary central differences. Formulae analogous to all the existing formulae involving forward, central and backward differences can be obtained by means of Sheppard's rules which are stated below.

x_{-1}	p_{-1}	f_{-1}		δ_{-1}^2		δ_{-1}^4
			$\delta_{-\frac{1}{2}}$		$\delta_{-\frac{1}{2}}^3$	
x_0	p_0	f_0		δ_0^2		δ_0^4
			$\delta_{\frac{1}{2}}$		$\delta_{\frac{1}{2}}^3$	
x_1	p_1	f_1		δ_1^2		δ_1^4
			$\delta_{1\frac{1}{2}}$		$\delta_{1\frac{1}{2}}^3$	
x_2	p_2	f_2		δ_2^2		δ_2^4

where h is the total range of arguments divided by the number of intervals in that range, the result being rounded to a suitable figure and $x_k \cong p_k \cdot h$ or $(x/h)_k = p_k + w_k$ where the w's are used to shift the origin, if necessary, and to round-off the p's.

Define

$$\delta_r \equiv \frac{1}{(p_{r+\frac{1}{2}} - p_{r-\frac{1}{2}})} (f_{r+\frac{1}{2}} - f_{r-\frac{1}{2}})$$

and

$$\delta_r^2 \equiv \frac{2}{(p_{r+1} - p_{r-1})} (\delta_{r+\frac{1}{2}} - \delta_{r-\frac{1}{2}}).$$

In general

$$\delta_r^{2n+1} \equiv \frac{2n+1}{(p_{r+\frac{1}{2}+n} - p_{r-\frac{1}{2}-n})} (\delta_{r+\frac{1}{2}}^{2n} - \delta_{r-\frac{1}{2}}^{2n}),$$

$$\delta_r^{2n} \equiv \frac{2n}{(p_{r+n} - p_{r-n})} (\delta_{r+\frac{1}{2}}^{2n-1} - \delta_{r+\frac{1}{2}}^{2n-1}).$$

The formula will be in the form

$$f(x) = f_0 + \sum_{r=1}^{N} \left\{ \prod (p - p_j \cdot) \frac{\delta_s}{r!} \right\},$$

where the product part of each term of the formula consists of r factors and the p_j's and s are chosen as follows:

a) When $r = 1$, $j = 0$ and $s = +\frac{1}{2}$, or $-\frac{1}{2}$, thereafter:—

Calculus of Finite Differences

b) r is increased by unity and s is either increased or decreased by $\frac{1}{2}$. Thus the p's, involved in the new divided difference of the series, are all those which are necessary for the computation of the divided difference used in the *preceding term* plus one more added either to the beginning or to the end of the set.

c) The product part of the term consists of factors $(p - p_j)$ such that all the p's necessary for the evaluation of the *previous* divided difference are involved. For example, Gauss' forward difference interpolation formula in terms of adjusted divided differences is obtained by following the path shown below

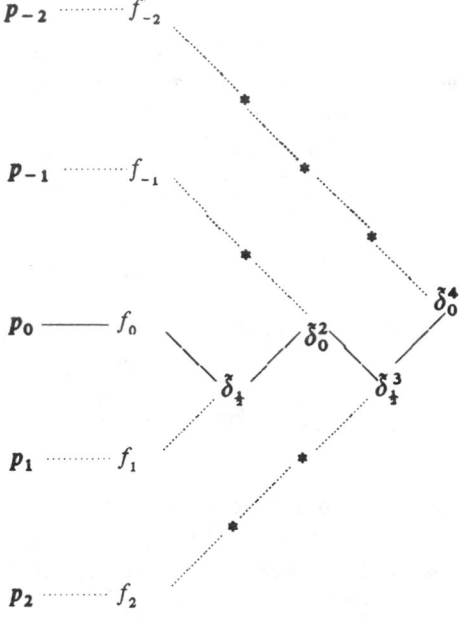

The solid line indicates the path chosen through the difference table.

$$f(x) = f_0 + (p - p_0)\delta_{\frac{1}{2}} + \frac{1}{2!} \cdot (p - p_0)(p - p_1)\delta_0^2 + \frac{1}{3!} \cdot (p - p_0)(p - p_1)(p - p_{-1})\delta_{\frac{1}{2}}^3$$

$$+ \frac{1}{4!} \cdot (p - p_0)(p - p_1)(p - p_{-1})(p - p_2)\delta_0^4 \cdots$$

Similarly Gauss' backward difference interpolation formula is

$$f(x) = f_0 + (p - p_0)\delta_{-\frac{1}{2}} + \frac{1}{2!} \cdot (p - p_0)(p - p_{-1})\delta_0^2 + \frac{1}{3!} \cdot (p - p_0)(p - p_{-1})(p - p_1)\delta_{-\frac{1}{2}}^3$$

$$+ \frac{1}{4!} \cdot (p - p_0)(p - p_{-1})(p - p_1)(p - p_{-2})\delta_0^4 \cdots,$$

and Stirling's formula is

$$f(x) = f_0 + \frac{(p - p_0)}{2}[\delta_{\frac{1}{2}} + \delta_{-\frac{1}{2}}] + \frac{1}{2!} \cdot \frac{(p - p_0)}{2}[(p - p_{-1}) + (p - p_1)]\delta_0^2$$

$$+ \frac{1}{3!} \cdot \frac{(p - p_1)(p - p_0)(p - p_{-1})}{2}[\delta_{\frac{1}{2}}^3 + \delta_{-\frac{1}{2}}^3]$$

$$+ \frac{1}{4!} \cdot \frac{(p - p_1)(p - p_0)(p - p_{-1})}{2}[(p - p_{-2}) + (p - p_2)]\delta_0^4 \cdots$$

Calculus of Finite Differences

Replacing the odd order differences, in Gauss' forward formula, by combinations of the previous even order differences the formula analogous to Everett's is obtained, i.e.,

$$f(x) = \left[1 - \frac{(p - p_0)}{(p_1 - p_0)}\right] f_0 + \frac{!(p - p_0)}{(p_1 - p_0)} f_1 - \frac{1}{2!} \frac{(p - p_0)(p - p_1)(p - p_2)}{(p_2 - p_{-1})} \delta_0^2$$

$$+ \frac{1}{2!} \frac{(p - p_0)(p - p_1)(p - p_{-1})}{(p_2 - p_{-1})} \delta_1^2 - \cdots .$$

d) Iterative linear interpolation

Neville's modification of Aiken's method of iterative linear interpolation is one of the most powerful methods of interpolation when the arguments are unevenly spaced as no prior knowledge of the order of the approximating polynomial is necessary nor is a difference table required.

The values obtained are successive approximations to the required result and the process terminates when there is no appreciable change. These values are of course useless if a new interpolation is required when the procedure must be started afresh.

Defining

$$f_{r,s} \equiv \frac{(x_s - x)f_r - (x_r - x)f_s}{(x_s - x_r)}$$

$$f_{r,s,t} \equiv \frac{(x_t - x)f_{r,s} - (x_r - x)f_{s,t}}{(x_t - x_r)}$$

$$f_{r,s,t,u} \equiv \frac{(x_u - x)f_{r,s,t} - (x_r - x)f_{s,t,u}}{(x_u - x_r)} ,$$

the computation is laid out as follows:

x_{-1}	$(x_{-1} - x)$	f_{-1}			
			$f_{-1,0}$		
x_0	$(x_0 - x)$	f_0		$f_{-1\,0,1}$	
			$f_{0,1}$		$f_{-1\,0\,1,2}$
x_1	$(x_1 - x)$	f_1		$f_{0,1,2}$	
			$f_{1,2}$		
x_2	$(x_2 - x)$	f_2			

As the iterates tend to their limit the common leading figures can be omitted.

e) Gauss' trigonometric interpolation formula

This is of greatest value when the function is periodic, i.e., a Fourier series expansion is possible.

$$f(x) = \sum_{r=0}^{n} C_r f_r ,$$

where $C_r = N_r(x)/N_r(x_r)$ and

$$N_r(x) = \left[\sin \frac{(x - x_0)}{2}\right]\left[\sin \frac{(x - x_1)}{2}\right] \cdots \left[\sin \frac{(x - x_{r-1})}{2}\right]$$

$$\left[\sin \frac{(x - x_{r+1})}{2}\right] \cdots \left[\sin \frac{(x - x_n)}{2}\right] .$$

Calculus of Finite Differences

This is similar to the Lagrangian formula.

f) Reciprocal differences

These are used when the quotient of two polynomials will give a better representation of the interpolating function than a simple polynomial expression.

A convenient layout is as shown below:

$$
\begin{array}{lll}
x_{-1} & f_{-1} \\
 & & \rho_{-\frac{1}{2}} \\
x_0 & f_0 & & \rho_0^2 \\
 & & \rho_{\frac{1}{2}} & & \rho_{\frac{1}{2}}^3 \\
x_1 & f_1 & & \rho_1^2 & & \rho_1^4 \\
 & & \rho_{1\frac{1}{2}} & & \rho_{1\frac{1}{2}}^3 \\
x_2 & f_2 & & \rho_2^2 \\
 & & \rho_{2\frac{1}{2}} \\
x_3 & f_3
\end{array}
$$

where

$$\rho_{r+\frac{1}{2}} \equiv \frac{x_{r+1} - x_r}{f_{r+1} - f_r}$$

and

$$\rho_r^2 \equiv \frac{x_{r+1} - x_{r-1}}{f_{r+\frac{1}{2}} - f_{r-\frac{1}{2}}} + f_r$$

In general

$$\rho_{r+\frac{1}{2}}^{2n+1} \equiv \frac{x_{r+n+1} - x_{r-n}}{\rho_{r+1}^{2n} - \rho_r^{2n}} + \rho_{r+\frac{1}{2}}^{2n-1}$$

$$\rho_r^{2n} \equiv \frac{x_{r+n} - x_{r-n}}{\rho_{r+\frac{1}{2}}^{2n-1} - \rho_{r-\frac{1}{2}}^{2n-1}} + \rho_r^{2n-2}$$

The interpolation formula is expressed in the form of a continued fraction expansion.

The expansion corresponding to Newton's forward difference interpolation formula, in the sense of the differences involved, is

$$f(x) = f_0 + \cfrac{(x - x_0)}{\rho_{\frac{1}{2}} + \cfrac{(x_2 - x_1)}{\rho_1 - f_0 + \cfrac{(x - x_2)}{\rho_{1\frac{1}{2}}^3 - \rho_{\frac{1}{2}} + \cfrac{(x_4 - x_3)}{\rho_2^4 - \rho_1^2 + (x - x_4)}}}}$$
$$\text{etc.}$$

while that corresponding to Gauss' forward formula is

$$f(x) = f_0 + \cfrac{(x - x_0)}{\rho_{\frac{1}{2}} + \cfrac{(x_2 - x_1)}{\rho_0^2 - f_0 + \cfrac{(x_3 - x_{-1})}{\rho_{\frac{1}{2}}^3 - \rho_{\frac{1}{2}} + \cfrac{(x_4 - x_2)}{\rho_0^4 - \rho_0^2 + (x - x_{-2})}}}}$$
$$\text{etc.}$$

Inverse interpolation

Any method of interpolation which does not require the arguments to be evenly spaced will be satisfactory, by simply interchanging the roles of the arguments and the function values.

Alternative (i) Sub-tabulate the function until linear interpolation is adequate and simple ratio and proportion will yield the result.

Alternative (ii) Find an approximate value for p from

$$p \cong (f_p - f_0)/(f_1 - f_0)$$

and then iterate using *any* of the standard interpolation formulae;
e.g., Bessel

$$p = [f_p - f_0 - B_2(\delta_0^2 + \delta_1^2) - B_3\delta_{\frac{1}{2}}^3 - \cdots \text{ etc.}]/\delta_{\frac{1}{2}}$$

where the Bessel's coefficients on the right hand side are evaluated for the approximate p.
Everrett

$$p = [f_p - f_0 - E_2\delta_0^2 - F_2\delta_1^2 - E_4\delta^4 - F_4\delta_0^4 \cdots]/\delta_{\frac{1}{2}}$$

Newton forward

$$p = [f_p - f_0 - \frac{1}{2!}p(p-1)\Delta_0^2 - \frac{1}{3!}p(p-1)(p-2)\Delta_0^3 \cdots]/\Delta_0$$

Newton backward

$$p = [f_p - f_0 - \frac{1}{2!}p(p+1)\nabla_0^2 - \frac{1}{3!}p(p+1)(p+2)\nabla_0^3 \cdots]/\nabla_0 .$$

Note that the divisors are in fact identical.

Lozenge Diagram — Interpolation

The following figure shows a scheme for writing down any of the standard interpolation formulas by associating it with a prescribed path through the diagram, called a lozenge diagram.

y_{-4}　$(u+4)_1$　$\Delta^2 y_{-5}$　$(u+5)_3$　$\Delta^4 y_{-6}$　$(u+6)_5$　$\Delta^6 y_{-7}$　$(u+7)_7$

1　Δy_{-4}　$(u+4)_2$　$\Delta^3 y_{-5}$　$(u+5)_4$　$\Delta^5 y_{-6}$　$(u+6)_6$　$\Delta^7 y_{-7}$

y_{-3}　$(u+3)_1$　$\Delta^2 y_{-4}$　$(u+4)_3$　$\Delta^4 y_{-5}$　$(u+5)_5$　$\Delta^6 y_{-6}$　$(u+6)_7$

1　Δy_{-3}　$(u+3)_2$　$\Delta^3 y_{-4}$　$(u+4)_4$　$\Delta^5 y_{-5}$　$(u+5)_6$　$\Delta^7 y_{-6}$

y_{-2}　$(u+2)_1$　$\Delta^2 y_{-3}$　$(u+3)_3$　$\Delta^4 y_{-4}$　$(u+4)_5$　$\Delta^6 y_{-5}$　$(u+5)_7$

1　Δy_{-2}　$(u+2)_2$　$\Delta^3 y_{-3}$　$(u+3)_4$　$\Delta^5 y_{-4}$　$(u+4)_6$　$\Delta^7 y_{-5}$

y_{-1}　$(u+1)_1$　$\Delta^2 y_{-2}$　$(u+2)_3$　$\Delta^4 y_{-3}$　$(u+3)_5$　$\Delta^6 y_{-4}$　$(u+4)_7$

1　Δy_{-1}　$(u+1)_2$　$\Delta^3 y_{-2}$　$(u+2)_4$　$\Delta^5 y_{-3}$　$(u+3)_6$　$\Delta^7 y_{-4}$

y_0　$(u)_1$　$\Delta^2 y_{-1}$　$(u+1)_3$　$\Delta^4 y_{-2}$　$(u+2)_5$　$\Delta^6 y_{-3}$　$(u+3)_7$

1　Δy_0　$(u)_2$　$\Delta^3 y_{-1}$　$(u+1)_4$　$\Delta^5 y_{-2}$　$(u+2)_6$　$\Delta^7 y_{-3}$

y_1　$(u-1)_1$　$\Delta^2 y_0$　$(u)_3$　$\Delta^4 y_{-1}$　$(u+1)_5$　$\Delta^6 y_{-2}$　$(u+2)_7$

1　Δy_1　$(u-1)_2$　$\Delta^3 y_0$　$(u)_4$　$\Delta^5 y_{-1}$　$(u+1)_6$　$\Delta^7 y_{-2}$

y_2　$(u-2)_1$　$\Delta^2 y_1$　$(u-1)_3$　$\Delta^4 y_0$　$(u)_5$　$\Delta^6 y_{-1}$　$(u+1)_7$

1　Δy_2　$(u-2)_2$　$\Delta^3 y_1$　$(u-1)_4$　$\Delta^5 y_0$　$(u)_6$　$\Delta^7 y_{-1}$

y_3　$(u-3)_1$　$\Delta^2 y_2$　$(u-2)_3$　$\Delta^4 y_1$　$(u-1)_5$　$\Delta^6 y_0$　$(u)_7$

1　Δy_3　$(u-3)_2$　$\Delta^3 y_2$　$(u-2)_4$　$\Delta^5 y_1$　$(u-1)_6$　$\Delta^7 y_0$

y_4　$(u-4)_1$　$\Delta^2 y_3$　$(u-3)_3$　$\Delta^4 y_2$　$(u-2)_5$　$\Delta^6 y_1$　$(u-1)_7$

→—— **Gregory-Newton (Forward)**　　—→— **Stirling**

⇥—— **Gregory-Newton (Backward)**　　···⤑···· **Bessel**

〜〜〜 **Gauss (I)**　　$(u+k)_s \equiv \dfrac{1}{s!}\,(u+k)^{[s]}$

Lozenge diagram showing paths for some of the standard interpolation formulas. (From Kunz, K. S., *Numerical Analysis*, McGraw-Hill, New York, 1957, 75–77, 80, 81. With permission.)

To convert a path through the lozenge to an interpolation formula the following rules are formulated:

1. Each time a difference column is crossed from left to right a term is added.
2. If a path enters a difference column (from the left) at a positive slope, the term added is the product of the difference, say $\Delta^k y_{-p}$, at which the column is crossed and the factorial $(u + p - 1)_k$ lying just below that difference.
3. If a path enters a difference column (from the left) at a positive slope, the term added is the product of the difference, say $\Delta^k y_{-p}$, at which the column is crossed and the factorial $(u + p)_k$ lying just above that difference.
4. If a path enters a difference column horizontally (from the left), the term added is the product of the difference, say $\Delta^k y_{-p}$, at which the column is crossed and the average of the two factorials $(u + p)_k$ and $(u + p-1)_k$ lying, respectively, just above and just below that difference.

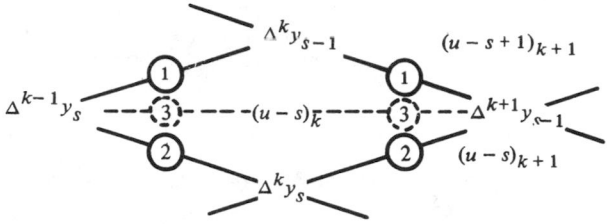

5. If a path crosses a difference column (from left to right) between two differences, say $\Delta^k y_{-(p+1)}$ and $\Delta^k y_{-p}$, then the added term is the product of the average of these two differences and the factorial $(u + p)_k$ at which the column is crossed.
6. Any portion of a path traversed from right to left gives rise to the same terms as would arise from going along this portion from left to right except that the sign of each term is changed.
7. The zero-difference column corresponding to the tabulated values may be treated by the same rules as any other difference columns provided one thinks of the lozenge as being entered just to the left of this column. Thus this column can be crossed by a path making a positive, negative, or zero slope, just as is true of the other columns.

Examples

Write down the interpolation formulas whose paths are shown on the lozenge diagram.

It is clear that, since the lozenge employs descending differences, all formulas will be in terms of descending differences. If one wishes the central-difference formulas to appear as they have been written in the earlier sections of the chapter, one should use central-difference notation for the differences.

Gregory-Newton (Forward) — Applying Rules 1, 3, and 7 to the diagonal path that slopes downward from y_0, one obtains the formula

$$y = y_0 + (u)_1 \, \Delta y_0 + (u)_2 \, \Delta^2 y_0 + (u)_3 \, \Delta^3 y_0 + \ldots + (u)_m \Delta^m y_0$$

Note: $(u)_r = u^{[r]}/r!$

Gregory-Newton (Backward) — Employing Rules 1, 2, and 7 to the diagonal path that slopes upward from y_0, one obtains the formula

$$y = y_0 + (u)_1 \, \Delta y_{-1} + (u + 1)_2 \, \Delta^2 y_{-2} + (u + 2)_3 \, \Delta^3 y_{-3} + \ldots$$

$$+ (u + m - 1)_m \, \Delta^m y_{-m}$$

Gauss (I) — Employing Rules 1, 2, 3, and 7 to the zigzag path shown, one has

$$y = y_0 + (u)_1 \, \Delta y_0 + (u)_2 \, \Delta^2 y_{-1} + (u+1)_3 \, \Delta^3 y_{-1} + (u+1)_4 \, \Delta^4 y_{-2} + \ldots$$

$$+ (u+k-1)_{2k-1} \, \Delta^{2k-1} y_{-k+1} + (u+k-1)_{2k} \, \Delta^{2k} y_{-k} + \ldots$$

where, as before, the series is terminated at any desired difference.

Stirling's — Employing Rules 1, 4, 5, and 7 to the horizontal path through y_0,

$$y = y_0 + (u)_1 \, \frac{\Delta y_{-1} + \Delta y_0}{2} + \frac{(u)_2 + (u+1)_2}{2} \, \Delta^2 y_{-1}$$

$$+ (u+1)_3 \, \frac{\Delta^2 y_{-2} + \Delta^3 y_{-1}}{2} + \ldots$$

$$+ (u+k-1)_{2k-1} \, \frac{\Delta^{2k-1} y_{-k} + \Delta^{2k-1} y_{-k+1}}{2}$$

$$+ \frac{(u+k-1)_{2k} + (u+k)_{2k}}{2} \, \Delta^{2k} y_{-k} + \ldots$$

Now

$$(u+k-1)_{2k-1} = \frac{1}{(2k-1)!} \, u(u^2-1)(u^2-4) \ldots [u^2-(k-1)^2]$$

and

$$\frac{(u+k-1)_{2k} + (u+k)_{2k}}{2} = \frac{1}{2k!} \, u^2(u^2-1)(u^2-4) \ldots [u^2-(k-1)^2]$$

Therefore, one may write

$$y = y_0 + u \, \frac{\Delta y_{-1} + \Delta y_0}{2} + \frac{1}{2!} \, u^2 \, \Delta^2 y_{-1} + \frac{1}{3!} \, u(u^2-1) \, \frac{\Delta^3 y_{-2} + \Delta^3 y_{-1}}{2}$$

$$+ \frac{1}{4!} \, u^2(u^2-1) \, \Delta^4 y_{-2} + \ldots$$

$$+ \frac{1}{(2k-1)!} \, u(u^2-1)(u^2-4) \ldots [u^2-(k-1)^2] \, \frac{\Delta^{2k-1} y_{-k} + \Delta^{2k-1} y_{-k+1}}{2}$$

$$+ \frac{1}{(2k)!} \, u^2(u^2-1)(u^2-4) \ldots [u^2-(k-1)^2] \, \Delta^{2k} y_{-k} + \ldots$$

This is the expression of Stirling's formula when descending differences are employed.

Bessel's — Employing the rules above to the horizontal path through Δy_0, one has

$$y = \frac{y_0 + y_1}{2} + \frac{(u)_1 + (u-1)_1}{2} \, \Delta y_0 + (u)_2 \, \frac{\Delta^2 y_{-1} + \Delta^2 y_0}{2}$$

$$+ \frac{(u+1)_3 + (u)_3}{2} \, \Delta^3 y_{-1} + \ldots$$

$$+ \frac{(u+k-1)_{2k-1} + (u+k-2)_{2k-1}}{2} \, \Delta^{2k-1} y_{-k+1}$$

$$+ (u+k-1)_{2k} \, \frac{\Delta^{2k} y_{-k} + \Delta^{2k} y_{-k+1}}{2}$$

or

$$y = \frac{y_0 + y_1}{2} + (u - \tfrac{1}{2}) \Delta y_0 + \frac{1}{2!} u^{[2]} \frac{\Delta^2 y_{-1} + \Delta^2 y_0}{2}$$

$$+ \frac{1}{3!} (u - \tfrac{1}{2}) u^{[2]} \Delta^3 y_{-1} + \frac{1}{4!} (u + 1)^{[4]} \frac{\Delta^4 y_{-2} + \Delta^4 y_{-1}}{2} + \dots$$

$$+ \frac{1}{(2k)!} (u + k - 1)^{[2k]} \frac{\Delta^{2k} y_{-k} + \Delta^{2k} y_{-k+1}}{2}$$

$$+ \frac{1}{(2k+1)!} (u - \tfrac{1}{2})(u + k - 1)^{[2k]} \Delta^{2k+1} y_{-k} + \dots$$

Remainder Term

The remainder term for a series terminated on a single difference, say the mth difference, is just the (m + 1)th term of the series if the series were extended, with, however, $h^{m+1} f^{m+1} (\xi)$ replacing the (m + 1)th difference (or mean difference). For a formula terminated on the mean of two differences, one must average the error terms for those formulas ending on each of the two differences.

NUMERICAL DIFFERENTIATION FORMULAS

The notation used is defined as follows:

$$h^n \frac{d^n}{dx^n} f(x_0 + ph) \equiv h^n \frac{d^n}{dx^n} f_p \equiv \frac{d^n}{dp^n} f_p$$

Derivatives at tabular points.

$$h f_0^{(1)} = \left(\Delta - \frac{\Delta^2}{2} + \frac{\Delta^3}{3} - \frac{\Delta^4}{4} + \frac{\Delta^5}{5} - \frac{\Delta^6}{6} \cdots \right) f_0$$

$$= \left(\mu\delta - \tfrac{1}{6}\mu\delta^3 + \tfrac{1}{30}\mu\delta^5 - \tfrac{1}{140}\mu\delta^7 \cdots \right) f_0$$

$$= \left(\nabla + \frac{\nabla^2}{2} + \frac{\nabla^3}{3} + \frac{\nabla^4}{4} + \frac{\nabla^5}{5} + \frac{\nabla^6}{6} \cdots \right) f_0$$

$$h^2 f_0^{(2)} = \left(\Delta^2 - \Delta^3 + \tfrac{11}{12}\Delta^4 - \tfrac{5}{6}\Delta^5 + \tfrac{137}{180}\Delta^6 - \tfrac{7}{10}\Delta^7 \cdots \right) f_0$$

$$= \left(\delta^2 - \tfrac{1}{12}\delta^4 + \tfrac{1}{90}\delta^6 - \tfrac{1}{560}\delta^8 \cdots \right) f_0$$

$$= \left(\nabla^2 + \nabla^3 + \tfrac{11}{12}\nabla^4 + \tfrac{5}{6}\nabla^5 + \tfrac{137}{180}\nabla^6 + \tfrac{7}{10}\nabla^7 \cdots \right) f_0$$

$$h^3 f_0^{(3)} = \left(\Delta^3 - \tfrac{3}{2}\Delta^4 + \tfrac{21}{12}\Delta^5 - \tfrac{45}{24}\Delta^6 + \tfrac{348}{180}\Delta^7 - \tfrac{79}{40}\Delta^8 \cdots \right) f_0$$

$$= \left(\mu\delta^3 - \tfrac{1}{4}\mu\delta^5 + \tfrac{7}{120}\mu\delta^7 - \tfrac{41}{3024}\mu\delta^9 \cdots \right) f_0$$

$$= \left(\nabla^3 + \tfrac{3}{2}\nabla^4 + \tfrac{21}{12}\nabla^5 + \tfrac{45}{24}\nabla^6 + \tfrac{348}{180}\nabla^7 + \tfrac{79}{40}\nabla^8 \cdots \right) f_0$$

Higher derivatives can be found by direct multiplication of the above formulae e.g.

$$h^5 f_0^{(5)} = h^3 \frac{d^3}{dp^3} \left(h^2 \frac{d^2}{dp^2} f_0 \right)$$

$$\therefore h^5 f_0^{(5)} = \left(\mu\delta^3 - \tfrac{1}{4}\mu\delta^5 + \tfrac{7}{120}\mu\delta^7 \cdots \right)\left(\delta^2 - \tfrac{1}{12}\delta^4 + \tfrac{1}{90}\delta^6 - \tfrac{1}{560}\delta^8 \cdots \right) f_0$$

$$= \left(\mu\delta^5 - \tfrac{1}{3}\mu\delta^7 + \tfrac{13}{144}\mu\delta^9 \cdots \right) f_0$$

The identity $\mu^2 \equiv 1 + \tfrac{1}{4}\delta^2$ is used to ensure that no powers of 'μ,' other than the first, appear in any formula.

$$h^4 f_0^{(4)} = \left(\delta^4 - \tfrac{1}{6}\delta^6 + \tfrac{7}{240}\delta^8 - \tfrac{41}{7560}\delta^{10} \cdots \right) f_0$$

$$h^6 f_0^{(6)} = \left(\delta^6 - \tfrac{1}{4}\delta^8 + \tfrac{13}{240}\delta^{10} \cdots \right) f_0$$

$$h^7 f_0^{(7)} = \left(\mu\delta^7 - \tfrac{5}{12}\mu\delta^9 + \tfrac{31}{240}\mu\delta^{11} \cdots \right) f_0$$

The preceding formulae can be truncated, as in the examples, and the differences expressed in terms of the function values. The resulting formulae are of the type

$$h^n f_0^{(n)} = \sum_{r=i}^{k} a_r \cdot f_r + \epsilon$$

where

$$\sum_{r=i}^{k} a_r = 0$$

and the principal part of the error 'ϵ' is the first neglected difference. It is still necessary for the differences to be converging.

In the following formulae only the order of the differences constituting the error term are indicated.

$$hf_0^{(1)} \simeq (f_1 - f_0) - \tfrac{1}{2}\Delta^2$$
$$\simeq \tfrac{1}{2}(-f_2 + 4f_1 - 3f_0) + \tfrac{1}{3}\Delta^3$$
$$\simeq \tfrac{1}{6}(2f_3 - 9f_2 + 18f_1 - 11f_0) - \tfrac{1}{4}\Delta^4$$
$$\simeq \tfrac{1}{2}(f_1 - f_{-1}) - \tfrac{1}{6}\mu\delta^3$$
$$\simeq \tfrac{1}{12}(-f_2 + 8f_1 - 8f_{-1} + f_{-2}) + \tfrac{1}{30}\mu\delta^5$$
$$\simeq \tfrac{1}{60}(f_3 - 9f_2 + 45f_1 - 45f_{-1} + 9f_{-2} - f_{-3}) - \tfrac{1}{140}\mu\delta^7$$
$$\simeq (f_0 - f_{-1}) + \tfrac{1}{2}\nabla^2$$
$$\simeq \tfrac{1}{2}(3f_0 - 4f_{-1} + f_{-2}) + \tfrac{1}{3}\nabla^3$$
$$\simeq \tfrac{1}{6}(11f_0 - 18f_{-1} + 9f_{-2} - 2f_{-3}) + \tfrac{1}{4}\nabla^4$$

$$h^2f_0^{(2)} \simeq (f_2 - 2f_1 + f_0) - \Delta^3$$
$$\simeq (-f_3 + 4f_2 - 5f_1 + 2f_0) + \tfrac{11}{12}\Delta^4$$
$$\simeq (11f_4 - 56f_3 + 114f_2 - 104f_1 + 35f_0) - \tfrac{5}{6}\Delta^5$$
$$\simeq (f_1 - 2f_0 + f_{-1}) - \tfrac{1}{12}\delta^4$$
$$\simeq \tfrac{1}{12}(-f_2 + 16f_1 - 30f_0 + 16f_{-1} - f_{-2}) + \tfrac{1}{90}\delta^6$$
$$\simeq \tfrac{1}{180}(2f_3 - 27f_2 + 270f_1 - 490f_0 + 270f_{-1} - 27f_{-2} + 2f_{-3}) - \tfrac{1}{560}\delta^8$$
$$\simeq (f_0 - 2f_{-1} + f_{-2}) + \nabla^3$$
$$\simeq (2f_0 - 5f_{-1} + 4f_{-2} - f_{-3}) + \tfrac{11}{12}\nabla^4$$
$$\simeq (35f_0 - 104f_{-1} + 114f_{-2} - 56f_{-3} + 11f_{-4}) + \tfrac{5}{6}\nabla^5$$

$$h^3f_0^{(3)} \simeq (f_3 - 3f_2 + 3f_1 - f_0) - \tfrac{3}{2}\Delta^4$$
$$\simeq \tfrac{1}{2}(-3f_4 + 14f_3 - 24f_2 + 18f_1 - 5f_0) + \tfrac{21}{12}\Delta^5$$
$$\simeq \tfrac{1}{2}(f_2 - 2f_1 + 2f_{-1} - f_{-2}) - \tfrac{1}{4}\mu\delta^5$$
$$\simeq \tfrac{1}{8}(-f_3 + 8f_2 - 13f_1 + 13f_{-1} - 8f_{-2} + f_{-3}) + \tfrac{7}{120}\mu\delta^7$$
$$\simeq (f_0 - 3f_{-1} + 3f_{-2} - f_{-3}) + \tfrac{3}{2}\nabla^4$$
$$\simeq \tfrac{1}{2}(5f_0 - 18f_{-1} + 24f_{-2} - 14f_{-3} + 3f_{-4}) + \tfrac{21}{12}\nabla^5$$

For derivatives at half-way points there exist analogous formulae. The operators act upon f_0 and f_1, in the case of the forward and backward differences, and upon $f_{1/2}$ in the case of the central differences δ_0^{2n+1} being meaningless.

$$hf_{1/2}^{(1)} = (\Delta - \tfrac{1}{24}\Delta^3 + \tfrac{1}{24}\Delta^4 - \tfrac{71}{1920}\Delta^5 \cdots)f_0$$
$$= (\delta - \tfrac{1}{24}\delta^3 + \tfrac{3}{640}\delta^5 - \tfrac{5}{7168}\delta^7 \cdots)f_{1/2}$$
$$= (\nabla - \nabla^2 + \tfrac{23}{24}\nabla^3 + \tfrac{11}{12}\nabla^4 + \tfrac{563}{840}\nabla^5 \cdots)f_0$$
$$= (\nabla - \tfrac{1}{24}\nabla^3 - \tfrac{1}{24}\nabla^4 - \tfrac{71}{1920}\nabla^5 \cdots)f_1$$

$$h^2f_{1/2}^{(2)} = (\Delta^2 - \tfrac{1}{12}\Delta^4 + \tfrac{1}{12}\Delta^5 - \tfrac{13}{180}\Delta^6 \cdots)f_0$$
$$= (\mu\delta^2 - \tfrac{5}{24}\mu\delta^4 + \tfrac{259}{5760}\mu\delta^6 - \tfrac{3229}{322560}\mu\delta^8 \cdots)f_{1/2}$$
$$= (\nabla^2 - \tfrac{1}{12}\nabla^4 - \tfrac{1}{12}\nabla^5 - \tfrac{13}{180}\nabla^6 \cdots)f_1$$

$$h^3f_{1/2}^{(3)} = (\Delta^3 - \tfrac{1}{8}\Delta^5 + \tfrac{1}{8}\Delta^6 - \tfrac{203}{1920}\Delta^7 \cdots)f_0$$
$$= (\delta^3 - \tfrac{1}{8}\delta^5 + \tfrac{37}{1920}\delta^7 - \tfrac{2229}{967680}\delta^9 \cdots)f_{1/2}$$
$$= (\nabla^3 - \tfrac{1}{8}\nabla^5 - \tfrac{1}{8}\nabla^6 - \tfrac{203}{1920}\nabla^7 \cdots)f_1$$

As for tabular values, these formulae can be used to yield higher derivatives by multiplication of the operator series. Derivative formulae involving function values can also be obtained by truncation of the above series.

$$hf_{1/2}^{(1)} \simeq \tfrac{1}{24}(-f_3 + 3f_2 + 21f_1 - 23f_0) - \tfrac{1}{24}\Delta^4$$
$$\simeq \tfrac{1}{24}(f_4 - 5f_3 + 9f_2 + 13f_1 - 22f_0) - \tfrac{71}{1920}\Delta^5$$
$$\simeq (f_1 - f_0) - \tfrac{1}{24}\delta^3$$
$$\simeq \tfrac{1}{24}(-f_2 + 27f_1 - 27f_0 + f_{-1}) + \tfrac{3}{640}\delta^5$$
$$\simeq \tfrac{1}{1920}(3f_3 - 95f_2 + 2190f_1 - 2190f_0 + 95f_{-1} - 3f_{-2}) - \tfrac{5}{7168}\delta^7$$
$$\simeq (f_0 - f_{-1}) + \nabla^2$$
$$\simeq \tfrac{1}{24}(23f_1 - 21f_0 - 3f_{-1} + f_{-2}) - \tfrac{1}{24}\nabla^4$$
$$\simeq \tfrac{1}{24}(22f_1 - 17f_0 - 9f_{-1} + 5f_{-2} - f_{-3}) - \tfrac{71}{1920}\nabla^5$$

$$h^2 f_{1/2}^{(2)} \simeq (f_2 - 2f_1 + f_0) - \tfrac{1}{12}\Delta^4$$
$$\simeq \tfrac{1}{12}(-f_4 + 4f_3 + 6f_2 - 20f_1 + 11f_0) + \tfrac{1}{12}\Delta^5$$
$$\simeq \tfrac{1}{2}(f_2 - f_1 - f_0 + f_{-1}) - \tfrac{5}{24}\mu\delta^4$$
$$\simeq \tfrac{1}{48}(5f_3 + 33f_2 - 38f_1 - 38f_0 + 33f_{-1} + 5f_{-2}) + \tfrac{259}{5760}\mu\delta^6$$
$$\simeq (f_1 - 2f_0 + f_{-1}) - \tfrac{1}{12}\nabla^4$$
$$\simeq \tfrac{1}{12}(11f_1 - 20f_0 + 6f_{-1} + 4f_{-2} - f_{-3}) - \tfrac{1}{12}\nabla^5$$

$$h^3 f_{1/2}^{(3)} \simeq (f_3 - 3f_2 + 3f_1 - f_0) - \tfrac{1}{8}\Delta^5$$
$$\simeq (f_2 - 3f_1 + 3f_0 - f_{-1}) - \tfrac{1}{8}\delta^5$$
$$\simeq \tfrac{1}{8}(-f_3 + 13f_2 - 34f_1 + 34f_0 - 13f_{-1} + f_{-2}) + \tfrac{37}{1920}\delta^7$$
$$\simeq (f_1 - 3f_0 + 3f_{-1} - f_{-2}) - \tfrac{1}{8}\nabla^5$$

To calculate the derivative at an intermediate point 'p' the values of the derivative at tabular and half-way points can be evaluated and then interpolated, or the following formulae which involve the derivatives of Bessel's coefficients can be used. B_r' denotes the first derivative and B_r'' the second etc. Note that the formula for the fourth derivative involves $B_4'' - \tfrac{1}{24}$ which is the same as B_6^{IV}

$$hf_p' = \delta_{1/2} + \tfrac{1}{2}(p - \tfrac{1}{2})(\delta_0^2 + \delta_1^2) + B_3'\delta_{1/2}^3 + B_4'(\delta_0^4 + \delta_1^4) + B_5'\delta_{1/2}^5$$
$$+ B_6'(\delta_0^6 + \delta_1^6) + B_7'\delta_{1/2}^7 + \cdots$$
$$h^2 f_p'' = \delta_0^2 + p\delta_{1/2}^3 + B_4''(\delta_0^4 + \delta_1^4) + B_5''\delta_{1/2}^5 + B_6''(\delta_0^6 + \delta_1^6) + B_7''\delta_{1/2}^7 + \cdots$$
$$h^3 f_p'' = \delta_{1/2}^3 + \tfrac{1}{2}(p - \tfrac{1}{2})(\delta_0^4 + \delta_1^4) + B_5'''\delta_{1/2}^5 + B_6'''(\delta_0^6 + \delta_1^6) + B_7'''\delta_{1/2}^7 + \cdots$$
$$h^4 f_p = \delta_0^4 + p\delta_{1/2}^5 + (B_4'' - 0.417)(\delta_0^6 + \delta_1^6) + \cdots$$

To find the value of p at which a derivative has a given value a technique of successive approximation using these tables can be employed as an alternative to inverse interpolation in a table of the appropriate derivative.

Starting values for this procedure are given by

$$p \simeq \tfrac{1}{2} + 2(h^n f_p^{(n)} - \delta_{1/2}^n)/(\delta_0^{n+1} + \delta_1^{n+1}) \quad \text{when } n \text{ is odd}$$

and

$$p \simeq (h^n f_p^{(n)} - \delta_0^n)/\delta_{1/2}^{n+1} \quad \text{when } n \text{ is even}$$

DIFFERENTIATION TABLES FOR NON-TABULAR POINTS

p	B'_3	B'_4	B'_5	B'_6	B'_7	$(1-p)$
.00	+.083333	+.041667	−.00833	−.0083	+.0012	1.00
.01	78383	41238	792	83	11	.99
.02	73533	40784	750	82	11	.98
.03	68783	40306	709	81	10	.97
.04	64133	39805	667	81	10	.96
.05	+.059583	+.039281	−.00626	−.0080	+.0009	.95
.06	55133	38735	585	79	9	.94
.07	50783	38166	544	78	8	.93
.08	46533	37576	504	77	7	.92
.09	42383	36965	464	76	7	.91
.10	+.038333	+.036333	−.00425	−.0075	+.0006	.90
.11	34383	35682	385	74	6	.89
.12	30533	35011	347	72	5	.88
.13	26783	34321	309	71	5	.87
.14	23133	33612	271	70	4	.86
.15	+.019583	+.032885	−.00234	−.0068	+.0004	.85
.16	16133	32141	198	67	3	.84
.17	12783	31380	162	66	3	.83
.18	9533	30603	128	64	2	.82
.19	6383	29809	93	63	1	.81
.20	+.003333	+.029000	−.00060	−.0061	+.0001	.80
.21	+ 383	28176	− 27	59	+ 1	.79
.22	− 2467	27337	+ 4	58	− 0	.78
.23	5217	26485	35	56	0	.77
.24	7867	25619	65	54	1	.76
.25	−.010417	+.024740	+.00094	−.0052	−.0001	.75
.26	12867	23848	123	51	2	.74
.27	15217	22944	150	49	2	.73
.28	17467	22029	176	47	3	.72
.29	19617	21103	201	45	3	.71
.30	−.021667	+.020167	+.00225	−.0043	−.0003	.70
.31	23617	19220	249	41	4	.69
.32	25467	18264	271	39	4	.68
.33	27217	17299	292	37	4	.67
.34	28867	16325	311	35	5	.66
.35	−.030417	+.015344	+.00330	−.0033	−.0005	.65
.36	31867	14355	348	31	5	.64
.37	33217	13359	364	29	5	.63
.38	34467	12356	380	27	6	.62
.39	35617	11347	394	24	6	.61
.40	−.036667	+.010333	+.00407	−.0022	−.0006	.60
.41	37617	9314	418	20	6	.59
.42	38467	8291	429	18	6	.58
.43	39217	7263	438	16	7	.57
.44	39867	6232	446	13	7	.56
.45	−.040417	+.005198	+.00453	−.0011	−.0007	.55
.46	40867	4161	459	9	7	.54
.47	41217	3123	463	7	7	.53
.48	41467	2083	466	4	7	.52
.49	41617	1042	468	2	7	.51
.50	−.041667	+.000000	+.00469	−.0000	−.0007	.50
$(1-p)$	B'_3	$-B'_4$	B'_5	$-B'_6$	B'_7	p

DIFFERENTIATION TABLES FOR NON-TABULAR POINTS

p	B_4''	B_5''	B_6''	B_7''	$(1-p)$
.00	$-.041667$	$+.04167$	$+.0056$	$-.0056$	1.00
.01	44142	4164	62	56	.99
.02	46567	4157	68	56	.98
.03	48942	4145	74	56	.97
.04	51267	4128	80	57	.96
.05	$-.053542$	$+.04106$	$+.0086$	$-.0057$	.95
.06	55767	4080	91	57	.94
.07	57942	4050	97	56	.93
.08	60067	4015	103	56	.92
.09	62142	3976	108	56	.91
.10	$-.064167$	$+.03933$	$+.0113$	$-.0056$	.90
.11	66142	3886	119	55	.89
.12	68067	3835	124	55	.88
.13	69942	3781	129	54	.87
.14	71767	3722	134	54	.86
.15	$-.073542$	$+.03660$	$+.0139$	$-.0053$	.85
.16	75267	3595	143	52	.84
.17	76942	3526	148	51	.83
.18	78567	3454	152	51	.82
.19	80142	3378	157	50	.81
.20	$-.081667$	$+.03300$	$+.0161$	$-.0049$	.80
.21	83142	3219	165	48	.79
.22	84567	3134	169	46	.78
.23	85942	3047	173	45	.77
.24	87267	2957	176	44	.76
.25	$-.088542$	$+.02865$	$+.0180$	$-.0043$	.75
.26	89767	2770	184	42	.74
.27	90942	2672	187	40	.73
.28	92067	2573	190	39	.72
.29	93142	2471	193	37	.71
.30	$-.094167$	$+.02367$	$+.0196$	$-.0036$	.70
.31	95142	2261	199	34	.69
.32	96067	2153	201	33	.68
.33	96942	2043	204	31	.67
.34	97767	1932	206	29	.66
.35	$-.098542$	$+.01819$	$+.0209$	$-.0028$	.65
.36	99267	1704	211	26	.64
.37	99942	1588	213	24	.63
.38	100567	1471	214	23	.62
.39	101142	1353	216	21	.61
.40	$-.101667$	$+.01233$	$+.0218$	$-.0019$	.60
.41	102142	1113	219	17	.59
.42	102567	991	220	15	.58
.43	102942	869	221	13	.57
.44	103267	746	222	11	.56
.45	$-.103542$	$+.00623$	$+.0223$	$-.0010$	.55
.46	103767	499	224	8	.54
.47	103942	375	224	6	.53
.48	104067	250	225	4	.52
.49	104142	125	225	2	.51
.50	$-.104167$	$+.00000$	$+.0225$	$-.0000$	.50
$(1-p)$	B_4''	$-B_5''$	B_6''	$-B_7''$	p

DIFFERENTIATION TABLES FOR NON-TABULAR POINTS

p	B_5'''	B_6'''	B_7'''	$(1 - p)$
.00	− .00000	+ .0625	− .0042	1.00
.01	495	617	33	.99
.02	980	608	25	.98
.03	1455	599	17	.97
.04	1920	590	9	.96
.05	− .02375	+ .0580	− .0001	.95
.06	2820	571	+ 7	.94
.07	3255	561	14	.93
.08	3680	551	22	.92
.09	4095	540	29	.91
.10	− .04500	+ .0530	+ .0037	.90
.11	4895	519	44	.89
.12	5280	508	51	.88
.13	5655	497	58	.87
.14	6020	486	65	.86
.15	− .06375	+ .0475	+ .0071	.85
.16	6720	463	78	.84
.17	7055	451	84	.83
.18	7380	439	90	.82
.19	7695	427	96	.81
.20	− .08000	+ .0415	+ .0102	.80
.21	8295	403	108	.79
.22	8580	390	114	.78
.23	8855	377	119	.77
.24	9120	365	124	76
.25	− .09375	+ .0352	+ .0129	.75
.26	9620	338	134	.74
.27	9855	325	139	.73
.28	10080	312	143	.72
.29	10295	299	148	.71
.30	− .10500	+ .0285	+ .0152	.70
.31	10695	271	156	.69
.32	10880	258	159	.68
.33	11055	244	163	.67
.34	11220	230	166	.66
.35	− .11375	+ .0216	.0169	.65
.36	11520	202	172	.64
.37	11655	188	175	.63
.38	11780	174	178	.62
.39	11895	159	180	.61
.40	− .12000	+ .0145	+ .0182	.60
.41	12095	131	184	.59
.42	12180	116	186	.58
.43	12255	102	188	.57
.44	12320	87	189	.56
.45	− .12375	+ .0073	+ .0190	.55
.46	12420	58	191	.54
.47	12455	44	192	.53
.48	12480	29	192	.52
.49	12495	15	193	.51
.50	− .12500	+ .0000	+ .0193	.50
$(1 - p)$	B_5'''	B_6'''	B_7'''	p

Caution

Numerical differentiation is error magnifying and many guarding figures are usually necessary. This is due to two factors namely:

    a) the interval h enters each formula in the form $1/h^n$

    b) each formula starts with the nth order difference.

Both of these combine to make numerical differentiation unreliable. e.g. Suppose the formula $f^{(2)} = (\Delta^2 - \Delta^3 + \frac{11}{12}\Delta^4)/h^2$ is used with $h = 0.1$ and an error of 5×10^{-4} can be tolerated in the derivative. For an error of e in each of the tabulated values

$$5 \times 10^{-4} = \frac{1}{0.01}(2e + 3e + \tfrac{11}{12}6e)$$
$$= 100(\tfrac{21}{2}e)$$
$$= 1050e$$

and therefore $e \simeq 5 \times 10^{-7}$ i.e. the function must be tabulated to *six* places of decimals before the second derivative can be correct to *three*.

If, on the other hand, h had been unity in magnitude then

$$5 \times 10^{-4} = 10.5e$$

i.e. $e \simeq 5 \times 10^{-5}$ and the function would need only one guarding figure to produce the required degree of accuracy in the second derivative.

Obviously this effect is magnified for higher derivatives. In any case, the largest value of h, which permits the function to be well represented by a converging table of differences, should be used.

INTEGRATION FORMULAS

Integration is a smoothing procedure and as such errors in tabulated values tend to be cancelled out.

Notation used:

$$\int_{x_0}^{x_n} f(x) \cdot dx = h \int_0^n f_p \cdot dp$$

where

$$f_p \equiv f(x_0 + ph) \equiv f(x)$$

$$\int_0^n f_p \cdot dp = \tfrac{1}{2} f_0 + f_1 + f_2 + \cdots + f_{n-1} + \tfrac{1}{2} f_n + \tfrac{1}{12}(\Delta_0 - \nabla_n) - \tfrac{1}{24}(\Delta_0^2 + \nabla_n^2)$$
$$+ \tfrac{19}{720}(\Delta_0^3 - \nabla_n^3) - \tfrac{3}{160}(\Delta_0^4 + \nabla_n^4) \cdots$$
$$= \tfrac{1}{2} f_0 + f_1 + f_2 + \cdots + f_{n-1} + \tfrac{1}{2} f_n + \tfrac{1}{12}(\mu\delta_0 - \mu\delta_n)$$
$$- \tfrac{11}{720}(\mu\delta_0^3 - \mu\delta_n^3) + \tfrac{191}{60480}(\mu\delta_0^5 - \mu\delta_n^5) - \cdots$$

In particular

$$\int_0^1 f_p \cdot dp = (1 + \tfrac{1}{2}\Delta - \tfrac{1}{12}\Delta^2 + \tfrac{1}{24}\Delta^3 - \tfrac{19}{720}\Delta^4 + \tfrac{3}{160}\Delta^5 - \tfrac{863}{60480}\Delta^6 + \cdots) f_0$$
$$= (1 + \tfrac{1}{2}\nabla + \tfrac{5}{12}\nabla^2 + \tfrac{3}{8}\nabla^3 + \tfrac{251}{720}\nabla^4 + \tfrac{95}{288}\nabla^5 + \tfrac{19087}{60480}\nabla^6 \cdots) f_0$$
$$= (1 - \tfrac{1}{2}\nabla - \tfrac{1}{12}\nabla^2 - \tfrac{1}{24}\nabla^3 - \tfrac{19}{720}\nabla^4 - \tfrac{3}{160}\nabla^5 - \tfrac{863}{60480}\nabla^6 \cdots) f_1$$
$$= (\mu - \tfrac{1}{12}\mu\delta^2 + \tfrac{11}{720}\mu\delta^4 - \tfrac{191}{60480}\mu\delta^6 + \cdots) f_{1/2}$$

$$\int_0^{n+(1/2)} f_p \cdot dp = \tfrac{1}{2} f_0 + f_1 + f_2 \cdots + f_n + (\tfrac{1}{12}\mu\delta - \tfrac{11}{720}\mu\delta^3 + \tfrac{191}{60480}\mu\delta^5 \cdots) f_0$$
$$+ (\tfrac{1}{24}\delta - \tfrac{17}{5760}\delta^3 + \tfrac{367}{967680}\delta^5 \cdots) f_{n+(1/2)}$$

In particular

$$\int_0^{1/2} f_p \cdot dp = \tfrac{1}{2} f_0 + (\tfrac{1}{8}\delta - \tfrac{1}{24}\mu\delta^2 + \tfrac{1}{384}\delta^3 + \tfrac{11}{1440}\mu\delta^4 - \tfrac{13}{46080}\delta^5 \cdots) f_{1/2}$$

Symmetric range integrals

$$\int_{-1/2}^{1/2} f_p \cdot dp = (1 + \tfrac{1}{24}\delta^2 - \tfrac{17}{5760}\delta^4 + \tfrac{367}{967680}\delta^6 - \tfrac{27859}{464486400}\delta^8 \cdots) f_0$$

$$\int_{-1}^{1} f_p \cdot dp = 2(1 + \tfrac{1}{6}\delta^2 - \tfrac{1}{180}\delta^4 + \tfrac{1}{1512}\delta^6 - \tfrac{23}{226800}\delta^8 + \cdots) f_0$$

$$\int_{-2}^{2} f_p \cdot dp = 4(1 + \tfrac{2}{3}\delta^2 + \tfrac{7}{90}\delta^4 - \tfrac{2}{945}\delta^6 + \tfrac{13}{56700}\delta^8 - \cdots) f_0$$

$$\int_{-3}^{3} f_p \cdot dp = 6(1 + \tfrac{3}{2}\delta^2 + \tfrac{11}{20}\delta^4 + \tfrac{41}{840}\delta^6 - \tfrac{3}{2800}\delta^8 + \cdots) f_0$$

$$\int_{-4}^{4} f_p \cdot dp = 8(1 + \tfrac{8}{3}\delta^2 + \tfrac{217}{90}\delta^4 + \tfrac{92}{189}\delta^6 + \tfrac{989}{28350}\delta^8 + \cdots) f_0$$

The differences in the above formulae can be replaced by their function values, after truncation of the series, if extensive differencing is to be avoided.

When this is done, the formulae take on the forms

$$\sum_m A_m \cdot f_m$$

Some typical formulae of this type, used in the iterative starting of the solution of ordinary differential equations, can be found on pages 680. These are essentially quadrature formulae as are the predictor corrector formulae of pages 687.

When the limits involved in the integrals are not tabular or half-way points, formulae analagous to Bessel's interpolation formula can be used, the corresponding coefficients being designated by B^* etc. Single integration.

$$\int_0^p f_p \cdot dp = \tfrac{1}{2} p(f_0 + f_1)' + B_1^* \delta_{1/2} + B_2^* (\delta_0^2 + \delta_1^2) + B_3^* \delta_{1/2}^3$$

$$+ B_4^* (\delta_0^4 + \delta_1^4) + B_5^* \delta_{1/2}^5 + B_6^* (\delta_0^6 + \delta_1^6)$$

$$\iint_0^p f_p \cdot d^2 p = B_0^{**}(f_0 + f_1) + B_1^{**} \delta_{1/2} + B_2^{**}(\delta_0 + \delta_1) + B_3^{**} \delta_{1/2}$$

$$+ B_4^{**}(\delta_0 + \delta_1) + B_5^{**} \delta_{1/2}$$

INTEGRATION TABLES FOR NON-TABULAR POINTS

p	B_1^*	B_2^*	B_3^*	B_4^*	B_5^*	B_6^*
.00	− .000000	− .000000	+ .000000	+ .00000	− .00000	− .0000
.01	4950	12	4	0	0	0
.02	9800	49	16	1	0	0
.03	14550	110	35	2	0	0
.04	19200	195	61	3	1	0
.05	− .023750	− .000302	+ .000094	+ .00005	− .00001	− .0000
.06	28200	432	133	7	1	0
.07	32550	584	177	10	2	0
.08	36800	757	226	13	2	0
.09	40950	952	279	16	3	0
.10	− .045000	− .001167	+ .000338	+ .00020	− .00003	− .0000
.11	48950	1402	399	24	4	0
.12	52800	1656	465	29	5	1
.13	56550	1929	533	33	6	1
.14	60200	2221	604	39	6	1
.15	− .063750	− .002531	+ .000677	+ .00044	− .00007	− .0001
.16	67200	2859	753	50	8	1
.17	70550	3203	830	56	9	1
.18	73800	3564	908	62	10	1
.19	76950	3941	987	69	10	1
.20	− .080000	− .004333	+ .001067	+ .00076	− .00011	− .0002
.21	82950	4741	1147	84	12	2
.22	85800	5163	1227	91	13	2
.23	88550	5599	1307	99	14	2
.24	91200	6048	1386	107	15	2
.25	− .093750	− .006510	+ .001465	+ .00116	− .00016	− .0002
.26	96200	6985	1542	124	16	3
.27	98550	7472	1619	133	17	3
.28	100800	7971	1693	142	18	3
.29	102950	8480	1766	152	19	3
.30	− .105000	− .009000	+ .001838	+ .00161	− .00020	− .0003
.31	106950	9530	1906	171	20	4
.32	108800	10069	1973	181	21	4
.33	110550	10618	2037	191	22	4
.34	112200	11175	2098	202	23	4
.35	− .113750	− .011740	+ .002157	+ .00212	− .00023	− .0004
.36	115200	12312	2212	223	24	5
.37	116550	12891	2264	233	24	5
.38	117800	13477	2313	244	25	5
.39	118950	14069	2358	255	25	5
.40	− .120000	− .014667	+ .002400	+ .00266	− .00026	− .0005
.41	120950	15269	2438	278	26	6
.42	121800	15876	2473	289	27	6
.43	122550	16487	2503	301	27	6
.44	123200	17101	2530	312	27	6
.45	− .123750	− .017719	+ .002552	+ .00324	− .00028	− .0007
.46	124200	18339	2571	335	28	7
.47	124550	18961	2585	347	28	7
.48	124800	19584	2596	359	28	7
.49	124950	20208	2602	370	28	8
.50	− .125000	− .020833	+ .002604	+ .00382	− .00028	− .0008

INTEGRATION TABLES FOR NON-TABULAR POINTS (Continued)

p	B_1^*	B_2^*	B_3^*	B_4^*	B_5^*	B_6^*
.50	−.125000	−.020833	+.002604	+.00382	−.00028	−.0008
.51	124950	21458	2602	394	28	8
.52	124800	22083	2596	405	28	8
.53	124550	22706	2585	417	28	9
.54	124200	23328	2571	429	28	9
.55	−.123750	−.023948	+.002552	+.00440	−.00028	−.0009
.56	123200	24565	2530	452	27	9
.57	122550	25180	2503	463	27	10
.58	121800	25791	2473	475	27	10
.59	120950	26398	2438	486	26	10
.60	−.120000	−.027000	+.002400	+.00497	−.00026	−.0010
.61	118950	27597	2358	509	25	11
.62	117800	28189	2313	520	25	11
.63	116550	28775	2264	530	24	11
.64	115200	29355	2212	541	24	11
.65	−.113750	−.029927	+.002157	+.00552	−.00023	−.0011
.66	112200	30492	2098	562	23	12
.67	110550	31049	2037	573	22	12
.68	108800	31597	1973	583	21	12
.69	106950	32137	1906	593	20	12
.70	−.105000	−.032667	+.001838	+.00603	−.00020	−.0012
.71	102950	33187	1766	612	19	13
.72	100800	33696	1693	621	18	13
.73	98550	34194	1619	631	17	13
.74	96200	34681	1542	640	16	13
.75	−.093750	−.035156	+.001465	+.00648	−.00016	−.0013
.76	91200	35619	1386	657	15	14
.77	88550	36068	1307	665	14	14
.78	85800	36504	1227	673	13	14
.79	82950	36926	1147	680	12	14
.80	−.080000	−.037333	+.001067	+.00688	−.00011	−.0014
.81	76950	37726	987	695	10	14
.82	73800	38103	908	701	10	15
.83	70550	38464	830	708	9	15
.84	67200	38808	753	714	8	15
.85	−.063750	−.039135	+.000677	+.00720	−.00007	−.0015
.86	60200	39445	604	725	6	15
.87	56550	39737	533	730	6	15
.88	52800	40011	465	735	5	15
.89	48950	40265	399	740	4	15
.90	−.045000	−.040500	+.000337	+.00744	−.00003	−.0015
.91	40950	40715	279	748	3	15
.92	36800	40909	226	751	2	16
.93	32550	41083	177	754	2	16
.94	28200	41235	133	757	1	16
.95	−.023750	−.041365	+.000094	+.00759	−.00001	−.0016
.96	19200	41472	61	761	1	16
.97	14550	41556	35	762	0	16
.98	9800	41617	16	763	0	16
.99	4950	41654	4	764	0	16
1.00	−.000000	−.041667	+.000000	+.00764	−.00000	−.0016

INTEGRATION TABLES FOR NON-TABULAR POINTS

p	B_0^{**}	B_1^{**}	B_2^{**}	B_3^{**}	B_4^{**}	B_5^{**}
.00	+.000000	−.000000	−.000000	+.00000	+.00000	−.0000
.01	25	25	0	0	0	0
.02	100	99	0	0	0	0
.03	225	220	1	0	0	0
.04	400	389	3	0	0	0
.05	+.000625	−.000604	−.000005	+.00000	+.00000	−.0000
.06	900	864	9	0	0	0
.07	1225	1168	14	0	0	0
.08	1600	1515	20	1	0	0
.09	2025	1904	29	1	0	0
.10	+.002500	−.002333	−.000040	+.00001	+.00001	−.0000
.11	3025	2803	52	2	1	0
.12	3600	3312	68	2	1	0
.13	4225	3859	86	2	1	0
.14	4900	4443	106	3	2	0
.15	+.005625	−.005063	−.000130	+.00004	+.00002	−.0000
.16	6400	5717	157	4	3	0
.17	7225	6406	187	5	3	0
.18	8100	7128	221	6	4	0
.19	9025	7882	259	7	4	0
.20	+.010000	−.008667	−.000300	+.00008	+.00005	−.0000
.21	11025	9481	345	9	6	0
.22	12100	10325	395	10	7	0
.23	13225	11197	449	12	8	0
.24	14400	12096	507	13	9	0
.25	+.015625	−.013021	−.000570	+.00014	+.00010	−.0000
.26	16900	13971	637	16	11	0
.27	18225	14945	709	17	12	0
.28	19600	15941	787	19	14	0
.29	21025	16960	869	21	15	0
.30	+.022500	−.018000	−.000956	+.00023	+.00017	−.0000
.31	24025	19060	1049	25	19	0
.32	25600	20139	1147	26	20	0
.33	27225	21235	1250	28	22	0
.34	28900	22349	1359	31	24	0
.35	+.030625	−.023479	−.001474	+.00033	+.00026	−.0000
.36	32400	24624	1594	35	28	0
.37	34225	25783	1720	37	31	0
.38	36100	26955	1852	39	33	0
.39	38025	28138	1990	42	36	0
.40	+.040000	−.029333	−.002133	+.00044	+.00038	−.0000
.41	42025	30538	2283	47	41	0
.42	44100	31752	2439	49	44	1
.43	46225	32974	2601	51	47	1
.44	48400	34203	2768	54	50	1
.45	+.050625	−.035438	−.002943	+.00057	+.00053	−.0001
.46	52900	36677	3123	59	56	1
.47	55225	37921	3309	62	60	1
.48	57600	39168	3502	64	63	1
.49	60025	40417	3701	67	67	1
.50	+.062500	−.041667	−.003906	+.00069	+.00071	−.0001

INTEGRATION TABLES FOR NON-TABULAR POINTS (Continued)

p	B_0^{**}	B_1^{**}	B_2^{**}	B_3^{**}	B_4^{**}	B_5^{**}
.50	+.062500	−.041667	−.003906	+.00069	+.00071	−.0001
.51	65025	42916	4118	72	74	1
.52	67600	44165	4335	75	78	1
.53	70225	45412	4559	77	83	1
.54	72900	46656	4790	80	87	1
.55	+.075625	−.047896	−.005026	+.00082	+.00091	−.0001
.56	78400	49131	5268	85	96	1
.57	81225	50359	5517	87	100	1
.58	84100	51581	5772	90	105	1
.59	87025	52795	6033	92	110	1
.60	+.090000	−.054000	−.006300	+.00095	+.00115	−.0001
.61	93025	55195	6573	97	120	1
.62	96100	56379	6852	100	125	1
.63	99225	57550	7137	102	130	1
.64	102400	58709	7427	104	135	1
.65	+.105625	−.059854	−.007724	+.00106	+.00141	−.0001
.66	108900	60984	8026	108	146	1
.67	112225	62098	8334	110	152	1
.68	115600	63195	8647	112	158	1
.69	119025	64274	8966	114	164	1
.70	+.122500	−.065333	−.009290	+.00116	+.00170	−.0001
.71	126025	66373	9619	118	176	1
.72	129600	67392	9953	120	182	1
.73	133225	68389	10293	121	188	1
.74	136900	69363	10637	123	195	1
.75	+.140625	−.070313	−.010986	+.00125	+.00201	−.0001
.76	144400	71237	11340	126	207	1
.77	148225	72136	11699	127	214	1
.78	152100	73008	12062	129	221	1
.79	156025	73852	12429	130	228	1
.80	+.160000	−.074667	−.012800	+.00131	+.00234	−.0001
.81	164025	75451	13175	132	241	1
.82	168100	76205	13554	133	248	1
.83	172225	76927	13937	134	255	1
.84	176400	77616	14324	134	262	1
.85	+.180625	−.078271	−.014713	+.00135	+.00270	−.0001
.86	184900	78891	15106	136	277	1
.87	189225	79474	15502	136	284	1
.88	193600	80021	15901	137	291	1
.89	198025	80530	16302	137	299	1
.90	+.202500	−.081000	−.016706	+.00138	+.00306	−.0001
.91	207025	81430	17112	138	314	1
.92	211600	81819	17520	138	321	1
.93	216225	82165	17930	138	329	1
.94	220900	82469	18342	139	336	1
.95	+.225625	−.082729	−.018755	+.00139	+.00344	−.0001
.96	230400	82944	19169	139	351	1
.97	235225	83113	19584	139	359	1
.98	240100	83235	20000	139	367	1
.99	245025	83308	20417	139	374	1
1.00	+.250000	−.083333	−.020833	+.00139	+.00382	−.0001

GAUSS-TYPE WEIGHTS ABSCISSAE

Quadrature formulae using unevenly spaced ordinates. Some high accuracy formulae involve the use of abscissae which are unevenly spaced throughout the interval. The interval itself is usually subject to a transformation and various 'weighting functions' are involved.

To change the interval (a, b) in x to that of (θ, ϕ) in t the transformation

$$t = \frac{(a\theta - b\phi)}{(a - b)} + \frac{(\theta - \phi)x}{(a - b)}$$

is used.

Some of the more popular weighting functions and their associated intervals are given below. The general formulae are of two main types derived from

$$\int_a^b W(x) \cdot F(x) \cdot dx = \int_\theta^\phi \omega(t) \cdot f(t) \cdot dt$$

namely

$$\int_\theta^\phi \omega(t) \cdot f(t) \cdot dt = \sum_t \{H_t \cdot f(x_t)\}$$

and

$$\int_\theta^\phi \omega(t) \cdot f(t) \cdot dt = H \sum_t \{f(x_t) \pm f(-x_t)\}$$

$\omega(t)$	(θ, ϕ)	x_t are the zeros of:
1	$(-1, 1)$	$P_n(x)$
$t^{1/2}$	$(0, 1)$	$x^{-1/2} \cdot P_{2n+1}(\sqrt{x})$
$t^{-1/2}$	$(0, 1)$	$P_n(\sqrt{x})$
$(1 - t^2)^{1/2}$	$(-1, 1)$	$S_n(x)$
$(1 - t^2)^{-1/2}$	$(-1, 1)$	$T_n(x)$
e^{-t}	$(0, \infty)$	$L_n(x)$
e^{-t^2}	$(-\infty, \infty)$	$H_n(x)$

Some useful tables of abscissae x_t and their corresponding H_t are as follows.

Gaussian Quadrature.

$$\int_{-1}^{1} f(x) \cdot dx = \sum_{1}^{n} \{H_t \cdot f(x_t)\}$$

The x_t occur in pairs symmetrically placed with respect to the origin.

$\pm x_t$	H_t
n = 2	
0.5773503	1.0000000
n = 3	
0.7745967	0.5555556
0.0000000	0.8888889
n = 4	
0.8611363	0.3478548
0.3399810	0.6521452
n = 5	
0.9061798	0.2369269
0.5384693	0.4786287
0.0000000	0.5688889
n = 6	
0.9324695	0.1713245
0.6612094	0.3607616
0.2386192	0.4679139
n = 7	
0.9491079	0.1294850
0.7415312	0.2797054
0.4058452	0.3818301
0.0000000	0.4179592
n = 8	
0.9602899	0.1012285
0.7966665	0.2223810
0.5255324	0.3137066
0.1834346	0.3626838
n = 9	
0.9681602	0.0812744
0.8360311	0.1806482
0.6133714	0.2606107
0.3242534	0.3123471
0.0000000	0.3302394
n = 10	
0.9739065	0.0666713
0.8650634	0.1494513
0.6794096	0.2190864
0.4333954	0.2692602
0.1488743	0.2955242

$\pm x_t$	H_t
$n = 11$	
0.9782287	0.0556686
0.8870626	0.1255804
0.7301520	0.1862902
0.5190961	0.2331938
0.2695432	0.2628045
0.0000000	0.2729251
$n = 12$	
0.9815606	0.0471753
0.9041173	0.1069393
0.7699027	0.1600783
0.5873180	0.2031674
0.3678315	0.2334925
0.1253334	0.2491470
$n = 13$	
0.9841831	0.0404840
0.9175984	0.0921215
0.8015781	0.1388735
0.6423493	0.1781460
0.4484928	0.2078160
0.2304583	0.2262832
0.0000000	0.2325516
$n = 14$	
0.9862838	0.0351195
0.9284349	0.0801581
0.8272013	0.1215186
0.6872929	0.1572032
0.5152486	0.1855384
0.3191124	0.2051985
0.1080549	0.2152639
$n = 15$	
0.9879925	0.0307532
0.9372734	0.0703660
0.8482066	0.1071592
0.7244177	0.1395707
0.5709722	0.1662692
0.3941513	0.1861610
0.2011941	0.1984315
0.0000000	0.2025782
$n = 16$	
0.9894009	0.0271525
0.9445750	0.0622535
0.8656312	0.0951585
0.7554044	0.1246290
0.6178762	0.1495960
0.4580168	0.1691565
0.2816036	0.1826034
0.0950125	0.1894506

Laguerre Quadrature.

$$\int_0^\infty e^{-x} \cdot f(x) \cdot dx = \sum_1^n \{H_t \cdot f(x_t)\}$$

x_t	H_t
n = 2	
0.5857864	0.8535534
3.4142136	0.1464466
n = 3	
0.4157746	0.7110930
2.2942804	0.2785177
6.2899451	0.0103893
n = 4	
0.3225477	0.6031541
1.7457611	0.3574187
4.5366203	0.0388879
9.3950709	0.0005393
n = 5	
0.2635603	0.5217556
1.4134031	0.3986668
3.5964258	0.0759424
7.0858100	0.0036118
12.6408008	0.0000234
n = 6	
0.2228466	0.4589647
1.1889321	0.4170008
2.9927363	0.1133734
5.7751436	0.0103992
9.8374674	0.0002610
15.9828740	0.0000009

Hermitian Quadrature.

$$e^{-x^2} \cdot f(x) \cdot dx = H_t \cdot f(x_t)$$

The abscissae are symmetrically placed with respect to the origin.

$\pm x_t$	H_t
$n = 2$	
0.7071068	0.8862269
$n = 3$	
0.0000000	1.1816359
1.2247449	0.2954090
$n = 4$	
0.5246476	0.8049141
1.6506801	0.0813128
$n = 5$	
0.0000000	0.9453087
0.9585725	0.3936193
2.0201829	0.0199532
$n = 6$	
0.4360774	0.7246296
1.3358491	0.1570673
2.3506050	0.0045300
$n = 7$	
0.0000000	0.8102646
0.8162879	0.4256073
1.6735516	0.0545156
2.6519614	0.0009718
$n = 8$	
0.3811870	0.6611470
1.1571937	0.2078023
1.9816568	0.0170780
2.9306374	0.0001996
$n = 9$	
0.0000000	0.7202352
0.7235510	0.4326516
1.4685533	0.0884745
2.2665806	0.0049436
3.1909932	0.0000396
$n = 10$	
0.3429013	0.6108626
1.0366108	0.2401386
1.7566836	0.0338744
2.5327317	0.0013436
3.4361591	0.0000076

$\pm x_t$	H_t
n = 11	
0.0000000	0.6547593
0.6568096	0.4293598
1.3265571	0.1172279
2.0259480	0.0119114
2.7832901	0.0003468
3.6684708	0.0000014
n = 12	
0.3142404	0.5701352
0.9477884	0.2604923
1.5976826	0.0516080
2.2795071	0.0039054
3.0206370	0.0000857
3.8897249	0.0000003

Radau Quadrature.

$$\int_{-1}^{1} f(x) \cdot dx = H_1 \cdot f(-1) + H_n \cdot f(1) + \sum_{2}^{n-1} \{H_t \cdot f(x_t)\}$$

$\pm x_t$	H_t
$n = 2$	
1.0000000	1.0000000
$n = 3$	
1.0000000	0.3333333
0.0000000	1.3333333
$n = 4$	
1.0000000	0.1666667
0.4472136	0.8333333
$n = 5$	
1.0000000	0.1000000
0.6546537	0.5444444
0.0000000	0.7111111
$n = 6$	
1.0000000	0.0666667
0.7650553	0.3784750
0.2852315	0.5548584
$n = 7$	
1.0000000	0.0476190
0.8302239	0.2768260
0.4688488	0.4317454
0.0000000	0.4876190
$n = 8$	
1.0000000	0.0357143
0.8717402	0.2107042
0.5917002	0.3411227
0.2092992	0.4124588
$n = 9$	
1.0000000	0.0277778
0.8997580	0.1654954
0.6771863	0.2745387
0.3631175	0.3464285
0.0000000	0.3715193
$n = 10$	
1.0000000	0.0222222
0.9195339	0.1333060
0.7387739	0.2248893
0.4779249	0.2920427
0.1652790	0.3275398

$\pm x_i$	H_i
$n = 11$	
1.0000000	0.0181818
0.9340014	0.1096123
0.7844835	0.1871699
0.5652353	0.2480481
0.2957581	0.2868791
0.0000000	0.3002176

Chebyshev-Radau Quadrature.

$$\int_{-1}^{1} x \cdot f(x) \cdot dx = H \cdot \sum_{1}^{n} \{f(x_t) - f(-x_t)\}$$

x_t	H
$n = 1$	
0.7745967	0.4303315
$n = 2$	
0.5002990	0.2393715
0.8922365	
$n = 3$	
0.4429861	
0.7121545	0.1599145
0.9293066	
$n = 4$	
0.3549416	
0.6433097	
0.7783202	0.1223363
0.9481574	

Chebyshev Quadrature.

$$\int_{-1}^{1} f(x) \cdot dx = \frac{2}{n} \cdot \sum_{1}^{n} f(x_t)$$

The abscissae are skew symmetric with respect to the origin.

x_t	x_t
$n = 2$	$n = 6$
−0.5773503	−0.8662468
+0.5773503	−0.4225187
$n = 3$	−0.2666354
−0.7071068	+0.2666354
0.0000000	+0.4225187
+0.7071068	+0.8662468
$n = 4$	$n = 7$
−0.7946545	−0.8838617
−0.1875925	−0.5296568
+0.1875925	−0.3239118
+0.7946545	0.0000000
$n = 5$	+0.3239118
−0.8324975	+0.5296568
−0.3745414	+0.8838617
0.0000000	
+0.3745414	
+0.8324975	

SOLUTION OF NONLINEAR EQUATIONS

John A. Heminger

INTRODUCTION

F denotes a real N-vector valued-function on R^N, the set of real N-vectors; f_i denotes the ith component function of F and $x \in R^N$ has i^{th} component x_i. In the case $N = 1$, f denotes F and x denotes x.

It is desired to find $x^* \in R^N$ such that $F(x^*) = 0$. The methods described below all generate a sequence $\{x^{(k)}\}$. It is expected that this sequence converges to x^*. Whether or not this actually happens depends on the individual problem being solved.

SOLUTION OF A SINGLE EQUATION

Zero Bracketing Methods

For the following two methods, bisection and false postion, it is assumed f is continuous on [a,b] and $f(a)f(b) < 0$. Then the Intermediate Value Theorem guarantees the existence of $x^* \in (a,b)$ with $f(x^*) = 0$.

Bisection

This method produces a nested sequence of subintervals, $\{[a^{(k)}, b^{(k)}]\}$ each of length $1/2$ of the previous subinterval, such that each subinterval always contains a solution of $f(x) = 0$.

Step o: Set $a^{(1)} = a$, $b^{(1)} = b$

Step k: Set $x^{(k)} = \dfrac{a^{(k)} + b^{(k)}}{2}$

If $f(x^{(k)}) = 0$, then $x^{(k)} = x^*$, so stop. Otherwise there are two cases:

1. If $f(a^{(k)})f(x^{(k)}) < 0$, then set $a^{(k+1)} = a^{(k)}$, $b^{(k+1)} = x^{(k)}$
2. If $f(x^{(k)})f(b^{(k)}) < 0$, then set $a^{(k+1)} = x^{(k)}$, $b^{(k+1)} = b^{(k)}$

False Position (Regula Falsi)

This method produces a nested sequence of subintervals $\{[a^{(k)}, b^{(k)}]\}$, not necessarily with lengths converging to zero, such that each subinterval always contains a solution of $f(x) = 0$. This method uses the linear function which interpolates function values of f at $a^{(k)}$ and $b^{(k)}$, i.e., the function

$$y = f(a^{(k)}) + \frac{f(b^{(k)}) - f(a^{(k)})}{b^{(k)} - a^{(k)}} (x - a^{(k)})$$

The zero of this is taken as the next iterate.

Step o: Set $a^{(1)} = a$, $b^{(1)} = b$

Step k: Set $x^{(k)} = a^{(k)} - \dfrac{b^{(k)} - a^{(k)}}{f(b^{(k)}) - f(a^{(k)})} f(a^{(k)})$

If $f(x^{(k)}) = 0$, then $x^{(k)} = x^*$, so stop. Otherwise there are two cases:

1. If $f(a^{(k)})f(x^{(k)}) < 0$, then set $a^{(k+1)} = a^{(k)}$, $b^{(k+1)} = x^{(k)}$
2. If $f(x^{(k)})f(b^{(k)}) < 0$, then set $a^{(k+1)} = x^{(k)}$, $b^{(k+1)} = b^{(k)}$

Linear Approximations of $f(x)$ and $f^{-1}(x)$

The following three methods, secant, Newton's, and discretized Newton's, are all based on a linear approximation of f. Each is equivalent to a similar linear approximation of f^{-1}.

Interpolation: Secant Method

This method uses the linear function which interpolates function values of f at $x^{(k-1)}$ and $x^{(k-2)}$, i.e., the function

$$y = f(x^{(k-1)}) + \frac{f(x^{(k-1)}) - f(x^{(k-2)})}{x^{(k-1)} - x^{(k-2)}} (x - x^{(k-1)})$$

The zero of this is taken as the next iterate.

Step o: Choose $x^{(-1)}$ and $x^{(0)}$ to be initial estimates of x*.

Step k: Set $x^{(k)} = x^{(k-1)} - \dfrac{x^{(k-1)} - x^{(k-2)}}{f(x^{(k-1)}) - f(x^{(k-2)})} f(x^{(k-1)})$

Taylor Series: Newton's Method

This method uses the linear Taylor polynomial approximation to f at $x^{(k-1)}$, i.e., $y = f(x^{(k-1)}) + f'(x^{(k-1)})(x - x^{(k-1)})$. The zero of this is taken as the next iterate.

Step o: Choose $x^{(0)}$ to be an initial estimate of x*

Step k: Set $x^{(k)} = x^{(k-1)} - \dfrac{f(x^{(k-1)})}{f'(x^{(k-1)})}$

Discretized Newton's Method

This method uses the linear Taylor polynomial approximation to $f(x)$, but approximates $f'(x^{(k-1)})$ using a difference quotient. So the next iterate is chosen to be the zero of the function

$$y = f(x^{(k-1)}) + \frac{f(x^{(k-1)} + \delta) - f(x^{(k-1)})}{\delta} (x - x^{(k-1)})$$

where δ is a suitable number near zero.

Step o: Choose $x^{(0)}$ to be an initial estimate of x*

Step k: Set $x^{(k)} = x^{(k-1)} - \dfrac{\delta}{f(x^{(k-1)} + \delta) - f(x^{(k-1)})} f(x^{(k-1)})$.

Quadratic Approximation of f

Interpolation: Muller's Method

This method uses the quadratic polynomial which interpolates function values of f at $x^{(k-1)}$, $x^{(k-2)}$, and $x^{(k-3)}$. That function is

$$y = f(x^{(k-1)}) + d_1^{(k)}(x - x^{(k-1)}) + d_3^{(k)}(x - x^{(k-1)})(x - x^{(k-2)})$$

where

$$d_1^{(k)} = \frac{f(x^{(k-1)}) - f(x^{(k-2)})}{x^{(k-1)} - x^{(k-2)}}$$

$$d_2^{(k)} = \frac{f(x^{(k-2)}) - f(x^{(k-3)})}{x^{(k-2)} - x^{(k-3)}}$$

$$d_3^{(k)} = \frac{d_1^{(k)} - d_2^{(k)}}{x^{(k-1)} - x^{(k-3)}}$$

This is rewritten as

$$y = d_3^{(k)}(x - x^{(k-1)})^2 + b^{(k)}(x - x^{(k-1)}) + c^{(k)}$$

where

$$b^{(k)} = d_1^{(k)} + d_3^{(k)}(x^{(k-1)} - x^{(k-2)})$$
$$c^{(k)} = f(x^{(k-1)})$$

The zero of this which is closer to $x^{(k-1)}$ is taken as the next iterate.

Step o: Choose $x^{(-2)}$, $x^{(-1)}$, and $x^{(0)}$ to be initial estimates of x^*
Step k: a. Set $d = \max \{o, [b^{(k)}]^2 - 4d_3^{(k)} c^{(k)}\}$
  b. Set $x^{(k)} = x^{(k-1)} + \dfrac{2c^{(k)}}{-b^{(k)} \pm \sqrt{d}}$
  where the sign in the denominator is chosen to yield the larger magnitude denominator.

The use of d in Step k: a is done to avoid complex arithmetic in finding real roots. If f is defined for complex numbers, then Step k: a can be

$$\text{Set } d = [b^{(k)}]^2 - 4d_3^{(k)} c^{(k)}$$

Taylor Series
  This method approximates the function by its quadratic Taylor polynomial at $x^{(k-1)}$ and takes its zero nearer to $x^{(k-1)}$ to be the next iterate.

Step o: Choose $x^{(0)}$ to be an initial estimate of x^*
Step k: a. Set $d = \max \{o, [f'(x^{(k-1)})]^2 - 2f(x^{(k-1)}) f''(x^{(k-1)})\}$
  b. Set $x^{(k)} = x^{(k-1)} + \dfrac{2f(x^{(k-1)})}{-f'(x^{(k-1)}) \pm \sqrt{d}}$
  where the sign in the denominator is chosen to yield the larger magnitude denominator.

The use of d in Step k: a is done to avoid complex arithmetic in finding real roots. If f is defined for complex numbers, then Step k: a can be

$$\text{Set } d = [f'(x^{(k-1)})]^2 - 2f(x^{(k-1)}) f''(x^{(k-1)})$$

Quadratic Approximation of f^{-1}

Interpolation

This method uses the quadratic polynomial which interpolates function values of f^{-1} at $f(x^{(k-1)})$, $f(x^{(k-2)})$, and $f(x^{(k-3)})$. This polynomial in y is

$$x = x^{(k-1)} + d_1^{(k)}(y - f(x^{(k-1)})) + d_3^{(k)}(y - f(x^{(k-1)}))(y - f(x^{(k-2)}))$$

where

$$d_1^{(k)} = \frac{x^{(k-1)} - x^{(k-2)}}{f(x^{(k-1)}) - f(x^{(k-2)})}$$

$$d_2^{(k)} = \frac{x^{(k-2)} - x^{(k-3)}}{f(x^{(k-2)}) - f(x^{(k-3)})}$$

$$d_3^{(k)} = \frac{d_1^{(k)} - d_2^{(k)}}{f(x^{(k-1)}) - f(x^{(k-3)})}$$

Evaluating this at $y = 0$ gives the next iterate.

Step o: Choose $x^{(-2)}$, $x^{(-1)}$, and $x^{(0)}$ to be initial estimates of x^*
Step k: Set $x^{(k)} = x^{(k-1)} + [-d_1^{(k)} + d_3^{(k)} f(x^{(k-2)})] f(x^{(x-1)})$

Taylor Series

This method uses the quadratic Taylor polynomial of f^{-1} at $f(x^{(k-1)})$:

$$x = x^{(k-1)} + [f^{-1}]'(f(x^{(k-1)}))(y - f(x^{(k-1)})) + \tfrac{1}{2}[f^{-1}]'' (f(x^{(k-1)}))(y - f(x^{(k-1)}))^2$$

$$= x^{(k-1)} + \frac{1}{f'(x^{(k-1)})} (y - f(x^{(k-1)})) - \frac{f''(x^{(k-1)})}{2[f'(x^{(k-1)})]^3} (y - f(x^{(k-1)}))^2.$$

The next iterate is the value of this polynomial at $y = 0$.

Step o: Choose $x^{(0)}$ to be an initial estimate of x^*
Step k: Set $x^{(k)} = x^{(k-1)} - \dfrac{2f(x^{(k-1)})[f'(x^{(k-1)})]^2 + f''(x^{(k-1)})[f(x^{(k-1)})]^2}{2[f'(x^{(k-1)})]^3}$

1,1 Rational Approximation of f

Pade Approximation: Halley's Method

This method uses the rational function of the form

$$\frac{a^{(k)}(x - x^{(k-1)}) + b^{(k)}}{c^{(k)}(x - x^{(k-1)}) + 1}$$

whose value, first derivative, and second derivative at $x^{(k-1)}$ agree with $f(x^{(k-1)})$, $f'(x^{(k-1)})$, and $f''(x^{(k-1)})$, respectively. The coefficients are

$$a^{(k)} = \frac{2[f'(x^{(k-1)})]^2 - f(x^{(k-1)}) f''(x^{(k-1)})}{2f'(x^{(k-1)})}$$

$$b^{(k)} = f(x^{(k-1)})$$

$$c^{(k)} = -\frac{f''(x^{(k-1)})}{2f'(x^{(k-1)})}$$

The zero of this function is taken as the next iterate. The method is equivalent to a similar approximation of f^{-1}.

Step o: Choose $x^{(0)}$ to be an initial estimate of $x*$

Step k: Set $x^{(k)} = x^{(k-1)} - \dfrac{2f(x^{(k-1)})f'(x^{(k-1)})}{2[f'(x^{(k-1)})]^2 - f(x^{(k-1)}) \, f''(x^{(k-1)})}$

Continuation

These methods start with a related nonlinear equation for which the solution is known. That equation is changed slightly and the solution of this equation, which is not expected to be too different from the solution of the previous equation, is calculated. After a sequence of changes, the system becomes the desired equation with a corresponding solution. For example, let $h(x,t) = f(x) + (t - 1) f(x^{(0)})$ for some fixed $x^{(0)}$. Then $0 = h(x,0) = f(x) - f(x^{(0)})$ has the solution $x^{(0)}$; changing t changes the solution of $h(x,t) = 0$; and the solution of $0 = h(x,1) = f(x)$ is eventually obtained.

The methods of Numerical Continuation and Continuation by Differentiation for a single equation are special cases of solving systems of equations with n = 1. See below for the discussion.

SOLUTION OF A SYSTEM OF EQUATIONS

Linear Approximation of F and F^{-1}

The following three methods, sequential secant, Newton's, and discretized Newton's, are all based on a linear approximation of **F**. Each is equivalent to a similar linear approximation of F^{-1}. These are N-dimensional analogs of the corresponding one-dimensional methods.

Interpolation: Sequential Secant Method

Let

$$\Delta^{(k)}x = (x^{(k-2)} - x^{(k-1)}, x^{(k-3)} - x^{(k-1)}, \ldots, x^{(k-1-N)} - x^{(k-1)})$$

i.e., the N × N matrix whose j^{th} column is $x^{(k-1-j)} - x^{(x-1)}$. Also let

$$\Delta^{(k)}F = (F(x^{(k-2)}) - F(x^{(k-1)}), \ldots, F(x^{(k-1-N)}) - F(x^{(k-1)}))$$

and

$$L^{(k)}(x) = F(x^{(k-1)}) + (\Delta^{(k)}F)(\Delta^{(k)}x)^{-1}(x - x^{(k-1)})$$

Then $L^{(k)}$ is the affine function on R^N which satisfies

$$L^{(k)}(x^{(j)}) = F(x^{(j)}), \, j = k - 1 - N, k - N, \ldots, k - 2, k - 1$$

The **x** for which $L^{(k)}(x) = 0$ is chosen as $x^{(k)}$, i.e.,

$$0 = F(x^{(k-1)}) + (\Delta^{(k)}F)(\Delta^{(k)}x)^{-1}(x^{(k)} - x^{(k-1)})$$

Letting $\mathbf{z} = (\Delta^{(k)}\mathbf{x})^{-1}(\mathbf{x}^{(k)} - \mathbf{x}^{(k-1)})$, this condition on $\mathbf{x}^{(k)}$ becomes

$$\mathbf{0} = \left(1 - \sum_{i=1}^{N} z_i\right) \mathbf{F}(\mathbf{x}^{(k-1)}) + \sum_{i=1}^{N} z_i \mathbf{F}(\mathbf{x}^{(k-1-i)})$$

With $z_0 = 1 - \sum_{i=1}^{N} z_i$, this becomes

$$1 = \sum_{i=0}^{N} z_i$$

$$0 = \sum_{i=0}^{N} z_i \mathbf{F}(\mathbf{x}^{(k-1-i)})$$

Then from the definition of $\mathbf{z}$,

$$\mathbf{x}^{(k)} = \mathbf{x}^{(k-1)} + (\Delta^{(k)}\mathbf{x})\mathbf{z}$$

$$= (1 - \sum_{i=1}^{N} z_i)\mathbf{x}^{(k-1)} + \sum_{i=1}^{N} z_i\mathbf{x}^{(k-1-i)}$$

So,

$$\mathbf{x}^{(k)} = \sum_{i=0}^{N} z_i\mathbf{x}^{(k-1-i)}$$

Step o: Choose $\mathbf{x}^{(0)}, \mathbf{x}^{(-1)}, \ldots, \mathbf{x}^{(-N)}$ in general position in R^N as initial estimates of $\mathbf{x}^*$

Step k: a. Solve the following system of $N + 1$ linear equations for $z_0, z_1, \ldots, z_N$

$$\begin{cases} 1 = \sum_{i=0}^{N} z_i \\ 0 = \sum_{i=0}^{N} z_i \mathbf{F}(\mathbf{x}^{(k-1-i)}) \end{cases}$$

b. Set $\mathbf{x}^{(k)} = \sum_{i=0}^{N} z_i\mathbf{x}^{(k-1-i)}$

Taylor Series: Newton's Method

This method uses the two term Taylor polynomial approximation to $\mathbf{F}(\mathbf{x})$, i.e.,

$$\mathbf{L}^{(k)}(\mathbf{x}) = \mathbf{F}(\mathbf{x}^{(k-1)}) + \mathbf{F}'(\mathbf{x}^{(k-1)})(\mathbf{x} - \mathbf{x}^{(k-1)})$$

The next iterate $\mathbf{x}^{(k)}$ is chosen as the solution of $\mathbf{L}^{(k)}(\mathbf{x}) = 0$. Here $\mathbf{F}'(\mathbf{x}^{(k-1)})$ denotes the Jacobian matrix of $\mathbf{F}$ at $\mathbf{x}^{(k-1)}$, i.e., the matrix whose i^{th} row, j^{th} column entry is $\frac{\partial f_i}{\partial x_j}(\mathbf{x}^{(k-1)})$.

Step o: Choose $\mathbf{x}^{(0)}$ to be an initial estimate of $\mathbf{x}^*$

Step k: a. Solve the following system of N linear equations for $\mathbf{z}$
$$\mathbf{F}'(\mathbf{x}^{(k-1)})\mathbf{z} = -\mathbf{F}(\mathbf{x}^{(k-1)})$$
b. Set $\mathbf{x}^{(k)} = \mathbf{x}^{(k-1)} + \mathbf{z}$

Discretized Newton's Method

This method uses the two term Taylor polynomial approximation to $\mathbf{F}(\mathbf{x})$, but approximates $\mathbf{F}'(\mathbf{x}^{(k-1)})$ using a difference quotient, i.e.,

$$\mathbf{F}'(\mathbf{x}^{(k-1)}) \simeq \mathbf{D}^{(k)}$$

where the ith row, j^{th} column entry of $\mathbf{D}^{(k)}$ is

$$\frac{f_i(\mathbf{x}^{(k-1)} + \delta_j \mathbf{e}^{(j)}) - f_i(\mathbf{x}^{(k-1)})}{\delta_j}$$

$\mathbf{e}^{(j)}$ has j^{th} component 1 and all others 0, and δ is a suitable vector whose components are all near zero. The next iterate is chosen as the zero of the affine function

$$\mathbf{L}^{(k)}(\mathbf{x}) = \mathbf{F}(\mathbf{x}^{(k-1)}) + \mathbf{D}^{(k)}(\mathbf{x} - \mathbf{x}^{(k-1)})$$

Step o: Choose $\mathbf{x}^{(0)}$ to be an initial estimate of x*
Step k: a. Solve the following system of N linear equations for $\mathbf{z}$:
$$\mathbf{D}^{(k)}\mathbf{z} = -\mathbf{F}(\mathbf{x}^{(k-1)})$$
b. Set $\mathbf{x}^{(k)} = \mathbf{x}^{(k-1)} + \mathbf{z}$

Note: In each of the three previous methods it is necessary to solve a system of linear equations. If an iterative method is used to approximate the solution of the linear equations, then that iteration is called the inner iteration. The inner iteration may use a small number of iterations to obtain only a rough approximation instead of iterating to convergence. When an inner iteration is used with the above three methods, the name of the inner iteration becomes a suffix to the name of the basic method, e.g., sequential secant-SOR, Newton-Gauss-Seidel, discretized Newton-Jacobi.

Sequential Solution of One-Dimensional Nonlinear Equations

These methods sequentially solve the individual component equations for one component of the unknown. In solving $f_i(\mathbf{x}) = 0$, an estimate is known for the solution x* and f_i is treated as a function of x_i alone. Any of the methods given above for solving a single equation can be used here. Then another equation $f_j(\mathbf{x}) = 0$ is selected and a new value for x_j is determined.

Jacobi
Step o: Choose $\mathbf{x}^{(0)}$ to be an initial estimate of x*
Step k: For i = 1, 2, . . . , N solve
$$f_i(x_1^{(k-1)}, \ldots , x_{i-1}^{(k-1)}, x, x_{i+1}^{(k-1)}, \ldots , x_N^{(k-1)}) = 0$$
to obtain $x_i^{(k)}$

Gauss-Seidel
Step o: Choose $\mathbf{x}^{(0)}$ to be an initial estimate of x*
Step k: For i = 1, 2, . . . , N solve
$$f_i(x_1^{(k)}, \ldots , x_{i-1}^{(k)}, x, x_{i+1}^{(k-1)}, \ldots , x_N^{(k-1)}) = 0$$
to obtain $x_i^{(k)}$

Note: The iteration needed to solve the one-dimensional component equation is called the inner iteration. It may use a small number of iterations to obtain only a rough approximation instead of iterating to convergence. When an inner iteration is used with the above two methods, the name of the inner iteration becomes a suffix to the name of that method, e.g., Jacobi-Muller, Gauss-Seidel-5-step-Newton.

Update Methods

These methods are also known as quasi-Newton methods.

The sequential secant method, Newton's method, and the discretized Newton's method, all have the general form

$$\mathbf{x}^{(k)} = \mathbf{x}^{(k-1)} - [\mathbf{B}^{(k-1)}]^{-1}\mathbf{F}(\mathbf{x}^{(k-1)})$$

and were derived from a linear approximation to $\mathbf{F}$ (see above).

Update methods also have this same general form. At each iteration the matrix $\mathbf{B}^{(k-1)}$ is modified, i.e., updated, by a matrix of small rank in order to provide a better linear approximation to $\mathbf{F}$. The update $\mathbf{B}^{(k)} = \mathbf{B}^{(k-1)} + \Delta\mathbf{B}^{(k-1)}$, where $\Delta\mathbf{B}^{(k-1)}$ is a matrix of small rank, is chosen such that

$$\mathbf{F}(\mathbf{x}^{(k)}) = \mathbf{F}(\mathbf{x}^{(k-1)}) + \mathbf{B}^{(k)}(\mathbf{x}^{(k)} - \mathbf{x}^{(k-1)})$$

This equation is known as the secant or finite difference equation. Thus, $\mathbf{B}^{(k)}$ is chosen so the affine approximation

$$\mathbf{L}(\mathbf{x}) = \mathbf{F}(\mathbf{x}^{(k-1)}) + \mathbf{B}^{(k)}(\mathbf{x} - \mathbf{x}^{(k-1)})$$

to $\mathbf{F}$ agrees with it at $\mathbf{x}^{(k-1)}$ and $\mathbf{x}^{(k)}$.

Since $[\mathbf{B}^{(k-1)}]^{-1}$ is desired, the updating is usually done to $\mathbf{H}^{(k-1)} = [\mathbf{B}^{(k-1)}]^{-1}$. In this case, the updated matrix $\mathbf{H}^{(k)} = \mathbf{H}^{(k-1)} + \Delta\mathbf{H}^{(k-1)}$ is chosen such that

$$\mathbf{x}^{(k)} = \mathbf{x}^{(k-1)} + \mathbf{H}^{(k)}(\mathbf{F}(\mathbf{x}^{(k)}) - \mathbf{F}(\mathbf{x}^{(k-1)})).$$

This equation is known as the secant or finite difference condition for the inverse update method.

Broyden Rank-One Update

This method chooses $\Delta\mathbf{B}^{(k-1)}$ to be a matrix of rank one, i.e., $\Delta\mathbf{B}^{(k-1)} = \mathbf{c}^{(k-1)}[\mathbf{d}^{(k-1)}]^{\mathrm{T}}$, where T denotes the transpose. The updating is done such that $\mathbf{B}^{(k)}$ satisfies the secant condition

$$\mathbf{B}^{(k)}(\mathbf{x}^{(k)} - \mathbf{x}^{(k-1)}) = \mathbf{F}(\mathbf{x}^{(k)}) - \mathbf{F}(\mathbf{x}^{(k-1)})$$

and

$$\mathbf{B}^{(k)}\mathbf{x} = \mathbf{B}^{(k-1)}\mathbf{x} \text{ for all } \mathbf{x} \in \{\mathbf{x}^{(k)} - \mathbf{x}^{(k-1)}\}^{\perp}$$

The second condition requires that for any vector x orthogonal to $\mathbf{x}^{(k)} - \mathbf{x}^{(k-1)}$, $\mathbf{B}^{(k)}$ maps x to the same vector $\mathbf{B}^{(k-1)}$ does. From the first requirement, $\mathbf{B}^{(k-1)}$ is modified so that for $\mathbf{x}^{(k)} - \mathbf{x}^{(k-1)}$, $\mathbf{B}^{(k)}$ produces the change $\mathbf{F}(\mathbf{x}^{(k)}) - \mathbf{F}(\mathbf{x}^{(k-1)})$. In this way the affine approximation to $\mathbf{F}$ is modified to agree with $\mathbf{F}$ at $\mathbf{x}^{(k)}$ as well as at $\mathbf{x}^{(k-1)}$.

This update choice requires $\mathbf{d}^{(k-1)} = \alpha^{(k-1)}(\mathbf{x}^{(k)} - \mathbf{x}^{(k-1)})$ for some nonzero scalar $\alpha^{(k-1)}$.

After substituting this into the secant condition, solving for $c^{(k-1)}$, and substituting the result into the equation for $B^{(k)}$, one has

$$\mathbf{B}^{(k)} = \mathbf{B}^{(k-1)} + \frac{\{[\mathbf{F}(\mathbf{x}^{(k)}) - \mathbf{F}(\mathbf{x}^{(k-1)})] - \mathbf{B}^{(k-1)}(\mathbf{x}^{(k)} - \mathbf{x}^{(k-1)})\}(\mathbf{x}^{(k)} - \mathbf{x}^{(k-1)})^{\mathrm{T}}}{(\mathbf{x}^{(k)} - \mathbf{x}^{(k-1)})^{\mathrm{T}}(\mathbf{x}^{(k)} - \mathbf{x}^{(k-1)})}$$

Recall the Sherman-Morrison formula for the inverse of a matrix modified by a rank-one matrix:

$$(\mathbf{A} + \mathbf{u}\mathbf{v}^{\mathrm{T}})^{-1} = \mathbf{A}^{-1} - \frac{\mathbf{A}^{-1}\mathbf{u}\mathbf{v}^{\mathrm{T}}\mathbf{A}^{-1}}{1 + \mathbf{v}^{\mathrm{T}}\mathbf{A}^{-1}\mathbf{u}}$$

Applying this formula to the expression for $B^{(k)}$ one obtains

$$[\mathbf{B}^{(k)}]^{-1} = [\mathbf{B}^{(k-1)}]^{-1}$$

$$+ \frac{\{(\mathbf{x}^{(k)} - \mathbf{x}^{(k-1)}) - [\mathbf{B}^{(k-1)}]^{-1}[\mathbf{F}(\mathbf{x}^{(k)}) - \mathbf{F}(\mathbf{x}^{(k-1)})]\}(\mathbf{x}^{(k)} - \mathbf{x}^{(k-1)})^{\mathrm{T}}[\mathbf{B}^{(k-1)}]^{-1}}{(\mathbf{x}^{(k)} - \mathbf{x}^{(k-1)})^{\mathrm{T}}[\mathbf{B}^{(k-1)}]^{-1}[\mathbf{F}(\mathbf{x}^{(k)}) - \mathbf{F}(\mathbf{x}^{(k-1)})]}$$

Step o: a. Choose $\mathbf{x}^{(0)}$ to be an initial estimate of x^*
         b. Choose $\mathbf{B}^{(0)}$ to be an initial estimate of the Jacobian matrix of $\mathbf{F}$ at $\mathbf{x}^{(0)}$
         c. Calculate $[\mathbf{B}^{(0)}]^{-1}$

Step k: a. Set $\mathbf{x}^{(k)} = \mathbf{x}^{(k-1)} - [\mathbf{B}^{(k-1)}]^{-1}\mathbf{F}(\mathbf{x}^{(k-1)})$
         b. Update $[\mathbf{B}^{(k-1)}]^{-1}$ as shown above to obtain $[\mathbf{B}^{(k)}]^{-1}$

Broyden Rank-One Inverse Update

This method chooses $\Delta\mathbf{H}^{(k-1)}$ to be a matrix of rank one, i.e., $\Delta\mathbf{H}^{(k-1)} = \mathbf{c}^{(k-1)}[\mathbf{d}^{(k-1)}]^{\mathrm{T}}$. The updating is done such that $\mathbf{H}^{(k)}$ satisfies the inverse secant condition

$$\mathbf{H}^{(k)}[\mathbf{F}(\mathbf{x}^{(k)}) - \mathbf{F}(\mathbf{x}^{(k-1)})] = \mathbf{x}^{(k)} - \mathbf{x}^{(k-1)}$$

and

$$\mathbf{H}^{(k)}\mathbf{y} = \mathbf{H}^{(k-1)}\mathbf{y} \text{ for all } \mathbf{y} \in \{\mathbf{F}(\mathbf{x}^{(k)}) - \mathbf{F}(\mathbf{x}^{(k-1)})\}^{\perp}$$

So for any vector $\mathbf{y}$ orthogonal to $\mathbf{F}(\mathbf{x}^{(k)}) - \mathbf{F}(\mathbf{x}^{(k-1)})$, $\mathbf{H}^{(k)}$ maps $\mathbf{y}$ to the same vector $\mathbf{H}^{(k-1)}$ does. And from the first condition, $\mathbf{H}^{(k-1)}$ is modified so that on $\mathbf{F}(\mathbf{x}^{(k)}) - \mathbf{F}(\mathbf{x}^{(k-1)})$, $\mathbf{H}^{(k)}$ produces the change $\mathbf{x}^{(k)} - \mathbf{x}^{(k-1)}$.

This update choice requires $\mathbf{d}^{(k-1)} = \alpha^{(k-1)}[\mathbf{F}(\mathbf{x}^{(k)}) - \mathbf{F}(\mathbf{x}^{(k-1)})]$ for some nonzero scalar $\alpha^{(k-1)}$. After substituting this into the inverse secant condition, solving for $\mathbf{c}^{(k-1)}$, and substituting the result into the equation for $\mathbf{H}^{(k)}$, one has

$$\mathbf{H}^{(k)} = \mathbf{H}^{(k-1)} + \frac{\{(\mathbf{x}^{(k)} - \mathbf{x}^{(k-1)}) - \mathbf{H}^{(k-1)}[\mathbf{F}(\mathbf{x}^{(k)}) - \mathbf{F}(\mathbf{x}^{(k-1)})]\}[\mathbf{F}(\mathbf{x}^{(k)}) - \mathbf{F}(\mathbf{x}^{(k-1)})]^{\mathrm{T}}}{[\mathbf{F}(\mathbf{x}^{(k)}) - \mathbf{F}(\mathbf{x}^{(k-1)})]^{\mathrm{T}}[\mathbf{F}(\mathbf{x}^{(k)}) - \mathbf{F}(\mathbf{x}^{(k-1)})]}$$

Step o: a. Choose $\mathbf{x}^{(0)}$ to be an initial estimate of x^*
         b. Choose $\mathbf{H}^{(0)}$ to be an initial estimate of the inverse of the Jacobian matrix of $\mathbf{F}$ at $\mathbf{x}^{(0)}$

Step k: a. Set $\mathbf{x}^{(k)} = \mathbf{x}^{(k-1)} - \mathbf{H}^{(k-1)}\mathbf{F}(\mathbf{x}^{(k-1)})$
         b. Update $\mathbf{H}^{(k-1)}$ as shown above to obtain $\mathbf{H}^{(k)}$

Continuation

These methods start with a related nonlinear system for which the solution is known. The system is changed slightly and the solution of this system, which is not expected to be too different from the solution of the previous system, is calculated. After a sequence of changes, the system becomes the desired nonlinear system with a corresponding solution. For example, let $\mathbf{H}(\mathbf{x},t) = \mathbf{F}(\mathbf{x}) + (t - 1) \mathbf{F}(\mathbf{x}^{(0)})$ for some fixed $\mathbf{x}^{(0)}$. Then $\mathbf{0} = \mathbf{H}(\mathbf{x},0) = \mathbf{F}(\mathbf{x}) - \mathbf{F}(\mathbf{x}^{(0)})$ has the solution $\mathbf{x}^{(0)}$; changing t changes the solution of $\mathbf{H}(\mathbf{x},t) = \mathbf{0}$; and the solution of $\mathbf{0} = \mathbf{H}(\mathbf{x},1) = \mathbf{F}(\mathbf{x})$ is eventually obtained.

Numerical Continuation

Let $0 = t_0 < t_1 < \ldots < t_M = 1$. These prescribe the sequence of modifications to the system of equations. For each t_i the system $\mathbf{H}(\mathbf{x},t_i) = \mathbf{0}$ is solved using one of the methods described above, e.g., sequential secant, Newton, Broyden rank-one update. This system should require few iterations since a good initial guess is available, namely, the solution of $\mathbf{H}(\mathbf{x},t_{i-1}) = \mathbf{0}$ which was previously determined.

Step o: a. Choose $\mathbf{x}^{(0)}$ as an initial estimate of $\mathbf{x}^*$
   b. Choose $\mathbf{H}(\mathbf{x},t)$ such that $\mathbf{H}(\mathbf{x}^{(0)},0) = \mathbf{0}$ and $\mathbf{H}(\mathbf{x},1) = \mathbf{F}(\mathbf{x})$
   c. Choose $0 = t_0 < t_1 < \ldots < t_M = 1$
Step k: a. Solve the system $\mathbf{H}(\mathbf{x},t_k) = \mathbf{0}$ using an iterative method with the initial guess $\mathbf{x}^{(k-1)}$
   b. Set $\mathbf{x}^{(k)}$ equal to the solution obtained in a

Then $\mathbf{x}^{(M)} = \mathbf{x}^*$.

Continuation by Differentiation

If $\mathbf{H}(\mathbf{x},t) = \mathbf{0}$ has a unique solution for each t from 0 to 1, then the solution can be treated as a function of t. That is, $\mathbf{x}^*(t)$ satisfies $\mathbf{H}(\mathbf{x}^*(t),t) = \mathbf{0}$ for $0 \leq t \leq 1$, $\mathbf{x}^*(0) = \mathbf{x}^{(0)}$, and $\mathbf{x}^*(1)$ is the desired solution. To find $\mathbf{x}^*(1)$, one follows the curve $\mathbf{x}^*(t)$, $0 \leq t \leq 1$, from $t = 0$ to $t = 1$.

This can be accomplished by solving an initial value problem. Assuming the derivatives exist and differentiating $\mathbf{H}(\mathbf{x}(t),t) = \mathbf{0}$ with respect to t, one obtains

$$D_1\mathbf{H}(\mathbf{x}(t),t) \frac{d\mathbf{x}(t)}{dt} + D_2\mathbf{H}(\mathbf{x}(t),t) = \mathbf{0}$$

where $D_i\mathbf{H}$ denotes the derivative of $\mathbf{H}$ with respect to its ith argument, $i = 1, 2$. If $D_1\mathbf{H}(\mathbf{x}(t),t)$ is nonsingular, then one solves the intial value problem

$$\begin{cases} \dfrac{d\mathbf{x}(t)}{dt} = -[D_1\mathbf{H}(\mathbf{x}(t),t)]^{-1} D_2\mathbf{H}(\mathbf{x}(t),t), \ t \in [0,1] \\ \\ \mathbf{x}(0) = \mathbf{x}^{(0)} \end{cases}$$

Various numerical methods exist for solving such problems, e.g., Runge-Kutta, Adams-Bashforth.

In the case $\mathbf{H}(\mathbf{x}(t),t) = \mathbf{F}(\mathbf{x}(t)) + (t - 1)\mathbf{F}(\mathbf{x}^{(0)})$, one obtains by differentiation

$$\mathbf{F}'(\mathbf{x}(t)) \frac{d\mathbf{x}(t)}{dt} + \mathbf{F}(\mathbf{x}^{(0)}) = \mathbf{0}$$

where F' denotes the Jacobian matrix of F. Then the initial value problem to be solved is

$$\begin{cases} \dfrac{d\mathbf{x}(t)}{dt} = -[\mathbf{F}'(\mathbf{x}(t))]^{-1}\,\mathbf{F}(\mathbf{x}^{(0)}), & t \in [0,1] \\ \mathbf{x}(0) = \mathbf{x}^{(0)} \end{cases}$$

THE NUMERICAL SOLUTION OF DIFFERENTIAL EQUATIONS

Introduction

The aim is to produce a numerical approximation to the solution of the differential equation, correct to a specified number of decimals.

The principal factors affecting accuracy are
  a) the form of the equation,
  b) the interval of tabulation,
  c) the formula chosen to effect the integration.

The main types of equation considered here are classified as follows

TABLE I

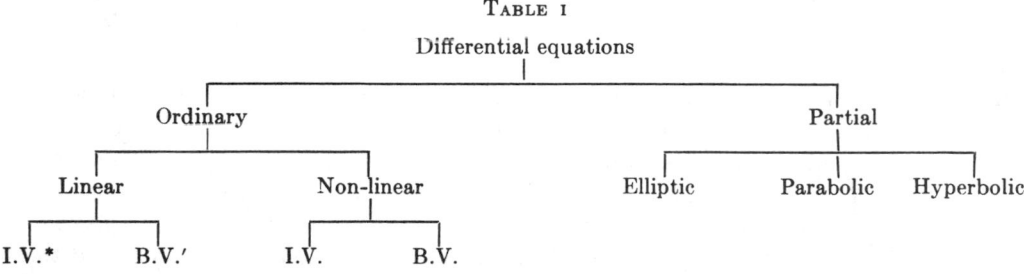

I.V.* = Initial value problems, sometimes referred to as Marching problems.
B.V.' = Boundary value problems, sometimes referred to as Jury problems.

Ordinary differential equations

With initial value problems, all the conditions required to determine uniquely the constants arising from the integration procedure are given at one point. With boundary value problems these conditions are given at two or more points.

For ordinary boundary value problems the solution can usually be based upon initial value techniques as follows. One of the boundary points is chosen as the principal point and the variables which are given at the other boundary point(s) are estimated at this point. This is repeated for a number of different estimates and a combination of the resulting solutions is chosen so that the numerical values satisfy both the differential equation and also all the original boundary conditions.

The principal methods of solving ordinary initial value differential equations to be considered are classified as follows. Purely analytical techniques will not be dealt with.

In the direct methods the value at any point is obtained from the value at the preceeding point without any iteration. In the case of the Taylor's series approach it is a simple matter to include a running check on previously computed values. With the Runge-Kutta and Chebyshev techniques checking is more difficult. For linear differential equations, or any for which repeated differentiation is practical, the Taylors series method can be used with a

THE NUMERICAL SOLUTION OF DIFFERENTIAL EQUATIONS (Continued)

TABLE II

Ordinary Initial value differential equations

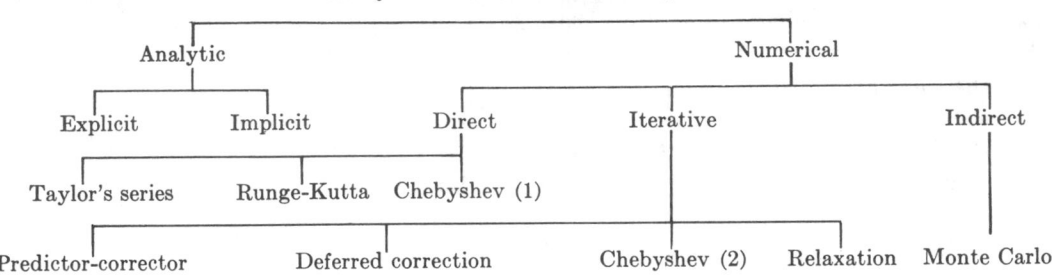

large step length. The basic disadvantages of direct methods are that the amount of calculation can be heavy and previously computed values are ignored. In addition a prior knowledge of the order of the approximation is necessary for the most efficient use of the basic Runga-Kutta approach and the direct Chebyshev method requires normalization of the range of the argument to either $(-1, +1)$ or $(0, +1)$.

With the iterative methods to be described, initial approximations are made and these are then refined until successive approximations agree to the required number of decimals. This is carried out either a point at a time or simultaneously over the whole range. They have the advantage of being self correcting in the sense that small computational errors are eliminated in subsequent iterations.

In this group of methods the predictor-corrector methods are particularly suitable for desk machine operation, especially with the more complicated non-linear differential equation. Their accuracy is controlled by the corrector formula which has associated with it an indication of the maximum interval to ensure convergence. These formulae are expressed either in terms of ordinates or backward and or central differences.

The deferred corrector methods are particularly suitable for desk machine work and are highly recommended. Relaxation methods are seldom used for initial value problems but they are naturally applicable to boundary value problems and in fact an initial value problem has to be converted to a boundary value one if relaxation is to be used.

Some predictor-corrector methods can be modified to be non-iterative but all will require special starting techniques to get under way.

General observations applicable to all methods

There will always be an optimum interval "h" of the argument (independent variable) and the method chosen should be able to take into account the need for changing this as the solution proceeds.

Build-up error due to rounding-off of differences can cause trouble. Two or three guarding figures should be held and there should always be running and terminal checks incorporated in any solution.

The stability of the method i.e. the convergence to the wanted solution, depends upon the differential equation as well as the method chosen. Starting with the last computed value and integrating in the reverse direction back to the original origin will check this.

The truncation error due to using a finite approximation to an infinite series will reduce the number of significant decimals obtained. However, this can be corrected at a later stage.

THE NUMERICAL SOLUTION OF DIFFERENTIAL EQUATIONS (Continued)

Any order differential equation can be reduced to a number of simultaneous differential equations of the first order and so the solution of first order differential equations is considered in detail. In some cases there will however be a loss of convenience if such a reduction is employed.

$$\begin{aligned}
\text{e.g.} \quad \text{Given} \quad & y'' = xy - \sin x & \text{with} \quad & y(0) = 1, \quad y'(0) = 0.\\
\text{This reduces to} \quad & y' = w & \text{with} \quad & y(0) = 1\\
& z' = v & & z(0) = 0\\
& v' = -z & & v(0) = 1\\
& w' = xy - z & & w(0) = 0
\end{aligned}$$

DIRECT METHODS

Taylor's series

Applicable to all linear equations and those non-linear equations whose higher derivatives can be easily found.

$$\text{e.g.} \qquad y' = f(x,y) \qquad \text{i.e.} \quad y^{(1)} = f(x,y^{(0)})$$

$$\text{then} \qquad y^{(n+1)} = \frac{\partial y^{(n)}}{\partial x} + y^{(1)} \frac{\partial y^{(n)}}{\partial y}$$

(since $y^{(n)}$ can be reduced to a function of x and y only.)

$$\text{Finally} \qquad x_1 = x_0 + h$$

$$\text{and} \qquad y_1 = y_0 + \frac{hy_0^{(1)}}{1!} + \frac{h^2 y_0^{(2)}}{2!} + \frac{h^3 y_0^{(3)}}{3!} + \frac{h^4 y_0^{(4)}}{4!} + \cdots$$

In practice the computational procedure is simplified by introducing the reduced derivatives defined by

$$\tau^{(n)} \equiv \frac{h^n y^{(n)}}{n!} \qquad\qquad (\tau^{(0)} \equiv y)$$

from which it follows that
$$\frac{d\tau^{(n)}}{dx} = \frac{(n+1)\tau^{(n+1)}}{h}$$

For linear equations a recurrence relationship is obtained by means of Leibnitz's theorem to generate any particular $\tau^{(n)}$ from the preceeding τ's. Thus, given the differential equation and the initial conditions, the sequence $\tau^{(0)}, \tau^{(1)} \tau^{(2)}, \tau^{(3)}, \ldots, \tau^{(n)}$ is computed until the terms cease to have any significance with regard to the number of places of decimals required.

If the Taylor's series is re-arranged then not only can the new function value be computed but a running check on the previously computed value be obtained with a minimum of effort. This can also be applied to as many of the derivatives as are desired.

$$\begin{aligned}
y_1 &= (\tau_0^{(0)} + \tau_0^{(2)} + \tau_0^{(4)} + \tau_0^{(6)} \cdots) + (\tau_0^{(1)} + \tau_0^{(3)} + \tau_0^{(5)} + \cdots)\\
\tau_1^{(1)} &= (2\tau_0^{(2)} + 4\tau_0^{(4)} + 6\tau_0^{(6)} + \cdots) + (\tau_0^{(1)} + 3\tau_0^{(3)} + 5\tau_0^{(5)} + \cdots)
\end{aligned}$$

The checking formulae corresponding to the above pair are

$$\begin{aligned}
y_{-1} &= (\tau_0^{(0)} + \tau_0^{(2)} + \tau_0^{(4)} + \tau_0^{(6)} \cdots) - (\tau_0^{(1)} + \tau_0^{(3)} + \tau_0^{(5)} \cdots)\\
\tau_{-1}^{(1)} &= (2\tau_0^{(2)} + 4\tau_0^{(4)} + 6\tau_0^{(6)} + \cdots) - (\tau_0^{(1)} + 3\tau_0^{(3)} + 5\tau_0^{(5)} + \cdots)
\end{aligned}$$

THE NUMERICAL SOLUTION OF DIFFERENTIAL EQUATIONS (Continued)

It will be seen that the sums inside the brackets which are used to evaluate the y_1 and the $\tau_1^{(1)}$ are also used to evaluate the y_{-1} and the $\tau_{-1}^{(1)}$. A shift of origin from (x_0, y_0) to (x_1, y_1) is then made, in accordance with the usual procedure for advancing the solution of a differential equation, and the process repeated. If too many terms are necessary, before the τ's assume negligible proportions, then the interval of tabulation "h" should be reduced. Conversely, if only a few terms have significance, then "h" should be increased.

Computational layout for one such stage

$$x_0 \qquad\qquad\qquad\qquad\qquad\qquad\qquad\qquad y_0$$

$$\tau_0^{(0)} \qquad\qquad \tau_0^{(1)}$$

$$\tau_0^{(2)} \qquad\qquad \tau_0^{(3)}$$

$$\tau_0^{(4)} \qquad\qquad \tau_0^{(5)}$$

$$\cdots \qquad\qquad \cdots\cdots$$

$$\tau_0^{(2n)} \qquad\qquad \tau_0^{(2n+1)}$$

$$x_1 \qquad \sum_{r=0}^{n} \tau_0^{(2n)} \qquad \sum_{r=0}^{n} \tau_0^{(2n+1)} \qquad y_1 \qquad y_{-1}$$

$$\sum_{r=0}^{n} 2r\tau_0^{(2r)} \qquad \sum_{r=0}^{n} (2r+1)\tau_0^{(2r+1)} \qquad \tau_1^{(1)} \qquad \tau_{-1}^{(1)}$$

Method of Runge-Kutta

An nth order Runge-Kutta process has an error of $0(h^{n+1})$. If $y' = f(x,y)$ then $y_1 = y_0 + K$, where K is the weighted mean of "n" evaluations of $h \cdot y'$ taken at specified points. There are an infinity of such formulae and the weights, and the constants necessary to determine the specified points, are tabulated for some simple cases only.

At any stage of the numerical solution, only the function value at the beginning of the stage need be known. This means that no special starting technique is necessary, but against this must be set the fact that the function must be evaluated "n" times per step. Further, previously computed values are not used to check the computation. Error analysis is not easy and the usual check is simply one of repeating the calculations at a different interval (one simple formula involving an error check is given at the end). Alternatively a finite difference check can be used.

If the functions are not smooth or if a frequent change in step length, particularly when such a change is expected to be a decrease, is envisaged then this process is of particular importance.

Second order systems $\qquad\qquad$ Error $= 0(h^3)$

Given $y' = f(x,y)$ and y_0

Then $x_1 = x_0 + h$ and $y_1 = y_0 + K$

where $K = \omega_1 k_1 + \omega_2 k_2 \qquad (\omega_1 + \omega_2 = 1)$

and $k_r = h \cdot f(x_0 + a_r h, y_0 + b_r k_1)$

THE NUMERICAL SOLUTION OF DIFFERENTIAL EQUATIONS (Continued)

Six typical systems are listed

(i)

r	a_r	b_r	ω_r
1	0	0	-1
2	$\frac{1}{4}$	$\frac{1}{4}$	2

(ii)

r	a_r	b_r	ω_r
1	0	0	0
2	$\frac{1}{2}$	$\frac{1}{2}$	1

(iii)

r	a_r	b_r	ω_r
1	0	0	$\frac{2}{3}$
2	$\frac{9}{4}$	$\frac{9}{4}$	$\frac{1}{3}$

(iv)

r	a_r	b_r	ω_r
1	0	0	$\frac{1}{2}$
2	1	1	$\frac{1}{2}$

(v)

r	a_r	b_r	ω_r
1	0	0	$\frac{1}{4}$
2	$\frac{2}{3}$	$\frac{2}{3}$	$\frac{3}{4}$

(vi)

r	a_r	b_r	ω_r
1	0	0	$-\frac{1}{2}$
2	$\frac{1}{3}$	$\frac{1}{3}$	$1\frac{1}{2}$

Third order systems   Error $= 0(h^4)$

Given $y' = f(x,y)$ and y_0
Then $x_1 = x_0 + h$ and $y_1 = y_0 + K$
where $K = \omega_1 k_1 + \omega_2 k_2 + \omega_3 k_3$ $(\omega_1 + \omega_2 + \omega_3 = 1)$
and $k_r = h \cdot f(x_0 + a_r h, y_0 + b_r k_1 + c_r k_2)$
Four typical systems are listed

(i)

r	a_r	b_r	c_r	ω_r
1	0	0	0	$\frac{2}{3}$
2	$\frac{1}{4}$	$\frac{1}{4}$	0	$-\frac{4}{3}$
3	$\frac{1}{2}$	$\frac{1}{10}$	$\frac{2}{5}$	$\frac{5}{3}$

(ii)

r	a_r	b_r	c_r	ω_r
1	0	0	0	$\frac{1}{4}$
2	$\frac{1}{3}$	$\frac{1}{3}$	0	0
3	$\frac{2}{3}$	0	$\frac{2}{3}$	$\frac{3}{4}$

(iii)

r	a_r	b_r	c_r	ω_r
1	0	0	0	$\frac{1}{6}$
2	$\frac{1}{2}$	$\frac{1}{2}$	0	$\frac{2}{3}$
3	1	-1	2	$\frac{1}{6}$

(iv)

r	a_r	b_r	c_r	ω_r
1	0	0	0	$\frac{1}{4}$
2	$\frac{2}{3}$	$\frac{2}{3}$	0	$\frac{1}{2}$
3	$\frac{2}{3}$	$-\frac{1}{3}$	1	$\frac{1}{4}$

Fourth order systems   Error $= (h^5)$

Given $y' = f(x,y)$ and y_0
Then $x_1 = x_0 + h$ and $y_1 = y_0 + K$
where $K = \omega_1 k_1 + \omega_2 k_2 + \omega_3 k_3 + \omega_4 k_4$ $(\omega_1 + \omega_2 + \omega_3 + \omega_4 = 1)$
and $k_r = h \cdot f(x_0 + a_r, y_0 + b_r k_1 + c_r k_2 + d_r k_3)$

THE NUMERICAL SOLUTION OF DIFFERENTIAL EQUATIONS (Continued)

Four typical systems are listed. The third is of particular importance both from the point of view of simplicity of numerical constants and of and the sequential relationships between the various k's.

(i)

r	a_r	b_r	c_r	d_r	ω_r
1	0	0	0	0	$\frac{1}{8}$
2	$\frac{1}{3}$	$\frac{1}{3}$	0	0	$\frac{3}{8}$
3	$\frac{2}{3}$	$-\frac{1}{3}$	1	0	$\frac{3}{8}$
4	1	1	-1	1	$\frac{1}{8}$

(ii)

r	a_r	b_r	c_r	d_r	ω_r
1	0	0	0	0	$\frac{1}{6}$
2	1	1	0	0	0
3	$\frac{1}{2}$	$\frac{3}{8}$	$\frac{1}{8}$	0	$\frac{2}{3}$
4	1	$-\frac{1}{2}$	$-\frac{1}{2}$	2	$\frac{1}{6}$

(iii)

r	a_r	b_r	c_r	d_r	ω_r
1	0	0	0	0	$\frac{1}{6}$
2	$\frac{1}{2}$	$\frac{1}{2}$	0	0	$\frac{1}{3}$
3	$\frac{1}{2}$	0	$\frac{1}{2}$	0	$\frac{1}{3}$
4	1	0	0	1	$\frac{1}{6}$

(iv)

r	a_r	b_r	c_r	d_r	ω_r
1	0	0	0	0	0
2	$\frac{1}{2}$	$\frac{1}{2}$	0	0	$\frac{2}{3}$
3	0	$-\frac{1}{2}$	$\frac{1}{2}$	0	$\frac{1}{6}$
4	1	$-\frac{1}{2}$	$1\frac{1}{2}$	1	$\frac{1}{6}$

Fifth order systems   Error $= 0(h^6)$

The number of terms required to form the weighted mean "K" increases, for orders greater than the fourth and, in addition, the coefficients are much less easy to handle.

The following system requires one additional term.

Given $y' = f(x,y)$ and y_0

Then $x_1 = x_0 + h$, and $y_1 = y_0 + K$

where $K = \omega_1 k_1 + \omega_2 k_2 + \omega_3 k_3 + \omega_4 k_4 + \omega_5 k_5 + \omega_6 k_6$

$$(\omega_1 + \omega_2 + \omega_3 + \omega_4 + \omega_5 + \omega_6 = 1)$$

and $k_r = h \cdot f(x_0 + a_r h, y_0 + b_r k_1 + c_r k_2 + d_r k_3 + e_r k_4 + f_r k_5)$

r	a_r	b_r	c_r	d_r	e_r	f_r	ω_r
1	0	0	0	0	0	0	23/192
2	$\frac{1}{3}$	$\frac{1}{3}$	0	0	0	0	0
3	$\frac{2}{5}$	$\frac{4}{25}$	$\frac{6}{25}$	0	0	0	125/192
4	1	$\frac{1}{4}$	-3	$\frac{15}{4}$	0	0	0
5	$\frac{2}{3}$	$\frac{2}{27}$	$\frac{10}{9}$	$-\frac{50}{81}$	$\frac{8}{81}$	0	$-27/64$
6	$\frac{4}{5}$	$\frac{2}{25}$	$\frac{12}{25}$	$\frac{2}{15}$	$\frac{8}{75}$	0	125/192

THE NUMERICAL SOLUTION OF DIFFERENTIAL EQUATIONS (Continued)

An example of computational layout

Consider formula (iii) of the fourth order systems. On substituting the values from the table the resulting formulae are as follows

$$y' = f(x,y)$$
$$x_1 = x_0 + h$$
$$k_1 = h \cdot f(x_0, y_0)$$
$$k_2 = h \cdot f(x_0 + \tfrac{1}{2}h, \; y_0 + \tfrac{1}{2}k_1)$$
$$k_3 = h \cdot f(x_0 + \tfrac{1}{2}h, \; y_0 + \tfrac{1}{2}k_2)$$
$$k_4 = h \cdot f(x_0 + h, \; y_0 + k_3)$$
$$y_1 = y_0 + \tfrac{1}{6}(k_1 + 2k_2 + 2k_3 + k_4)$$

and the layout of one state of the calculation is

x	y	$k_r = hy'$	K
$\mathbf{x}_0$	$\mathbf{y}_0$	k_1	
$x_0 + \tfrac{1}{2}h$	$y_0 + \tfrac{1}{2}k_1$	k_2	
$x_0 + \tfrac{1}{2}h$	$y_0 + \tfrac{1}{2}k_2$	k_3	
$x_0 + h$	$y_0 + k_3$	k_4	
			$(k_1 + 2k_2 + 2k_3 + k_4)$
$x_1 = x_0 + h$	$y_1 = y_0 + K$		

The starting values for this stage are underlined.

Extension of the Runge-Kutta approach to simultaneous differential equations

Consider again table (iii) of the fourth order systems. It can be extended to cover any number of simultaneous D.E.'s. The extension to two is shown in detail and the general case can be easily deduced from it.

$$\text{Given} \quad y' = f(x,y,z)$$
$$\text{and} \quad z' = g(x,y,z)$$
$$x_1 = x_0 + h$$
$$\text{we obtain} \quad y_1 = y_0 + K$$
$$\text{and} \quad z_1 = z_0 + M$$
$$\text{where} \quad K = \omega_1 k_1 + \omega_2 k_2 + \omega_3 k_3 + \omega_4 k_4$$
$$M = \omega_1 m_1 + \omega_2 m_2 + \omega_3 m_3 + \omega_4 m_4$$
$$\text{and} \quad k_1 = h \cdot f(x_0, \; y_0, \; z_0)$$
$$k_2 = h \cdot f(x_0 + \tfrac{1}{2}h, \; y_0 + \tfrac{1}{2}k_1, \; z_0 + \tfrac{1}{2}m_1)$$
$$k_3 = h \cdot f(x_0 + \tfrac{1}{2}h, \; y_0 + \tfrac{1}{2}k_2, \; z_0 + \tfrac{1}{2}m_2)$$
$$k_4 = h \cdot f(x_0 + h, \; y_0 + k_3, \; z_0 + m_3)$$

THE NUMERICAL SOLUTION OF DIFFERENTIAL EQUATIONS (Continued)

and $m_1 = h \cdot g(x_0, y_0, z_0)$

$$m_2 = h \cdot g(x_0 + \tfrac{1}{2}h, y_0 + \tfrac{1}{2}k_1, z_0 + \tfrac{1}{2}m_1)$$

$$m_3 = h \cdot g(x_0 + \tfrac{1}{2}h, y_0 + \tfrac{1}{2}k_2, z_0 + \tfrac{1}{2}m_2)$$

$$m_4 = h \cdot g(x_0 + h, y_0 + k_3, z_0 + m_3)$$

and the computational layout will be similar to the previous one

x	y	z	$k_r = hf$ $K = \Sigma\omega_r k_r$	$m_r = hg$ $M = \Sigma\omega_r m_r$
x_0	y_0	z_0	k_1	m_1
$x_0 + \tfrac{1}{2}h$	$y_0 + \tfrac{1}{2}k_1$	$z_0 + \tfrac{1}{2}m_1$	k_2	m_2
$x_0 + \tfrac{1}{2}h$	$y_0 + \tfrac{1}{2}k_2$	$z_0 + \tfrac{1}{2}m_2$	k_3	m_3
$x_0 + h$	$y_0 + k_3$	$z_0 + m_3$	k_4	m_4
			"K"	"M"
$x_1 = x_0 + h$	$y_1 = y_0 + K$	$z_1 = z_0 + M$		

The use of Runge-Kutta techniques for higher order differential equations

A schematic for the general second order differential equation is shown.

Given $y'' = f(x, y, y')$ and y_0, y_0' the reduced derivatives can be used i.e.

$$\tau^{(n)} \equiv \frac{h^n}{n!} \cdot y^{(n)}$$

Then $y'' = f\left(x, y, \dfrac{\tau^{(1)}}{h}\right)$

and $x_1 = x_0 + h$

$$y_1 = y_0 + \tau_0^{(1)} + K$$

$$\tau_1^{(1)} = \tau_0^{(1)} + K'$$

where $K = \displaystyle\sum_{r=1}^{4} \omega_r k_r$ $\left(1 = \displaystyle\sum_{r=1}^{4} \omega_r\right)$

and $K' = \displaystyle\sum_{r=1}^{4} \omega_r' k_r$ $\left(1 = \displaystyle\sum_{r=1}^{4} \omega_r'\right)$

$$k_r = \frac{h^2}{2} \cdot f(x_0 + a_r h, y_0 + b_r \tau_0^{(1)} + c_r k_1 + d_r k_2 + e_r k_3, \{\tau_0^{(1)} + g_r k_{r-1}\}/h)$$

THE NUMERICAL SOLUTION OF DIFFERENTIAL EQUATIONS (Continued)

r	a_r	b_r	c_r	d_r	e_r	g_r	ω_r	ω_r'
1	0	0	0	0	0	0	0	$\frac{1}{3}$
2	$\frac{1}{2}$	$\frac{1}{2}$	$\frac{1}{4}$	0	0	1	$\frac{1}{3}$	$\frac{2}{3}$
3	$\frac{1}{2}$	$\frac{1}{2}$	$\frac{1}{4}$	0	0	1	$\frac{1}{3}$	$\frac{3}{2}$
4	1	1	0	0	1	2	$\frac{1}{3}$	$\frac{1}{3}$

The computational layout being as follows

x	y	τ	k, K, K'
x_0	y_0	$\tau_0^{(1)}$	k_1
$x_0 + \frac{1}{2}h$	$y_0 + \frac{1}{2}\tau_0^{(1)} + \frac{1}{4}k_1$	$\tau_0^{(1)} + k_1$	k_2
$x_0 + \frac{1}{2}h$	$y_0 + \frac{1}{2}\tau_0^{(1)} + \frac{1}{4}k_1$	$\tau_0^{(1)} + k_2$	k_3
$x_0 + h$	$y_0 + \tau_0^{(1)} + k_3$	$\tau_0^{(1)} + 2k_3$	k_4
			K
			K'

$$x_1 = x_0 + h \qquad y_1 = y_0 + \tau_0^{(1)} + K \tau_1^{(1)} = \tau_0^{(1)} + K'$$

where
$$K = \omega_1 k_1 + \omega_2 k_2 + \omega_3 k_3 + \omega_4 k_4$$
$$K' = \omega_1' k_1 + \omega_2' k_2 + \omega_3' k_3 + \omega_4' k_4$$

Chebyshev Polynomials (1)

The Chebyshev polynomial of degree r in x where $-1 \leq x \leq 1$ is denoted by $T_r \equiv T_r(x)$.

$$T_r(x) \equiv \cos (r \cdot \cos^{-1} x)$$
$$\text{and } 2x \cdot T_r = T_{r+1} + T_{r-1}$$

The shifted Chebyshev polynomials of degree r in x where $0 \leq x \leq 1$ are denoted by $T_r^* \equiv T_r^*(x)$.

$$T_r^*(x) \equiv \cos (r \cdot \cos^{-1} (2x - 1))$$
$$\text{and} \quad 2(2x - 1) \cdot T_r^* = T_{r+1}^* + T_{r-1}^*$$

N.B. $T_r^*(x^2) = T_{2r}(x)$

Both types of polynomials can be used for linear differential equations over a finite interval. The equation is assumed to have polynomial coefficients—if not then suitable polynomial approximations should be inserted. The finite range must first be normalized to either $(-1, 1)$ or $(0, 1)$.

a) Chebyshev polynomials

$$T_0 = 1$$
$$T_r(1) = 1$$
$$T_r(-1) = (-1)^r$$
$$T_{2r}(0) = (-1)^r$$
$$T_{2r+1}(0) = 0$$

THE NUMERICAL SOLUTION OF DIFFERENTIAL EQUATIONS (Continued)

The solution of the equation $y^{(1)} = f(x,y)$ is assumed to have the form

$$y = \tfrac{1}{2}a_0 + a_1 T_1 + a_2 T_2 + a_3 T_3 + a_4 T_4 + \cdots$$

where $a_r = 0$ for all $r > N$

i.e.
$$y = \tfrac{1}{2}a_0 + a_1 T_1 + a_2 T_2 + \cdots + a_N T_N.$$

similarly
$$y^{(1)} = \tfrac{1}{2}a_0^{(1)} + \sum_{r=1}^{N} a_r^{(1)} T_r \text{ etc.}$$

Writing $C_r(y^{(s)}) \equiv$ the coefficient of T_r in the expansion of $y^{(s)}$ (the s th derivative of y with respect to x)

$$C_r(y^{(s)}) = a_r^{(s)} \tag{1}$$

Further
$$C_r(xy) = \tfrac{1}{2}(a_{r+1} + a_{r-1})$$

and
$$C_r(x^2 y) = \tfrac{1}{4}(a_{r+2} + 2a_r + a_{r-2})$$

in general
$$C_r(x^p y^{(s)}) = 2^{-p} \sum_{i=0}^{p} \binom{p}{i} a_{|r-p+2i|}^{(s)} \qquad \left(\begin{matrix} r = 0,1,2 \cdots \\ s = 1,2, \ldots \end{matrix} \right) \tag{2}$$

In addition the Chebyshev recurrence relationship itself yields

$$a_{r-1}^{(s)} = 2r \cdot a_r^{(s-1)} + a_{r+1}^{(s)} \tag{3}$$

Method of solution

(i) Obtain a relationship between the a's and the $a^{(1)}$'s from $C_r\{y^{(1)} - f(x,y)\} = 0$, e.g. If $(x^2 + 1)y^{(1)} - 4xy = 0$.

Then

$$a_{r-2}^{(1)} + 6a_r^{(1)} + a_{r+2}^{(1)} - 8a_{r-1} - 8a_{r+1} = 0 \tag{4}$$

since

$$C_r\{x^2 y^{(1)}\} + C_r\{y^{(1)}\} - 4C_r\{xy\} = 0$$

(ii) From (3) with $s = 1$.

$$a_{r-1}^{(1)} = 2ra_r + a_{r+1}^{(1)} \tag{5}$$

(iii) Substitute (5) into (4) in such a manner that the suffix of one of the a's is *less* than all the other suffices. The above example yields

$$2(r - 5)a_{r-1} = 8a_{r+1} - 7a_r^{(1)} - a_{r+2}^{(1)} \tag{4a}$$

(iv) Equations (4a) and (5) are then used to generate the a's and $a^{(1)}$'s by recurrence *backwards* from some arbitrarily chosen values of a_N and $a_N^{(1)}$.

e.g.   Set $a_N = 1$, $a_N^{(1)} = 1$ and $a_r = a_r^{(1)} = 0$, all $r > N$

The values found will however be far too large, in general, so a scale factor has to be determined.

(v) The initial conditions will be given either at the origin or at the extremities of the range.

THE NUMERICAL SOLUTION OF DIFFERENTIAL EQUATIONS (Continued)

Noting that

$$y(0) \quad = \tfrac{1}{2}a_0 - a_2 + a_4 - a_6 + a_8 \cdots$$

$$y(1) \quad = \tfrac{1}{2}a_0 + a_1 + a_2 + a_3 + a_4 \cdots$$

$$y(-1) = \tfrac{1}{2}a_0 - a_1 + a_2 - a_3 + a_4 \cdots$$

If $y(0)$ is given then the scale factor is

$$K = y(0)/(\tfrac{1}{2}a_0 - a_1 + a_2 - a_3 + a_4 \cdots)$$

where the values inside the brackets are the values just computed.

(vi) The values for $a_r^{(1)}$ just computed, when multiplied by K, should satisfy the given equation at the origin

$$\text{i.e.} \qquad y^{(1)} = f(x,y) \qquad \text{at } x = 0, \quad y = y(0),$$

$$\text{or} \qquad y^{(1)}(0) = K(\tfrac{1}{2}a_0^{(1)} - a_2^{(1)} + a_4^{(1)} - a_6^{(1)} + \cdots)$$

This in general will not be the case and so the computation is repeated with different starting values for the a_N and the $a_N^{(1)}$. A linear combination of the two sets is then chosen so that both the a's *and* the $a^{(1)}$'s satisfy the given conditions.

The method can be extended to equations of higher degree if the necessary $a_r^{(s)}$ are tabulated. This will, of course, necessitate a linear combination of $s + 1$ sets of results.

(b) Shifted Chebyshev polynomials
   Fundamental values

$$T_0^* = 1$$

$$T_r^*(1) = 1$$

$$T_{2r}^*(\tfrac{1}{2}) = (-1)^r$$

$$T_{2r+1}^*(\tfrac{1}{2}) = 0$$

$$T_{2r}^*(0) = 1$$

$$T_{2r+1}^*(0) = -1$$

And correspondingly

$$y = \tfrac{1}{2}a_0 + a_1 T_1^* + a_2 T_2^* + \cdots + a_N T_N^*$$

$$C_r^*(y^{(s)}) = a_r^{(s)}$$

$$C_r^*(x^p y^{(s)}) = 2^{-2p} \sum_{i=0}^{2p} \binom{2p}{i} a_{|r-p+i|}^{(s)} \qquad \left(\begin{matrix} r = 0,1,2 \cdots \\ s = 1,2, \cdots \end{matrix} \right)$$

also

$$a_{r-1}^{(s)} = 4r \ a_r^{(s-1)} + a_{r+1}^{(s)}$$

In addition the initial values yield

$$y(0) = \tfrac{1}{2}a_0 - a_1 + a_2 - a_3 + \cdots$$

$$y(\tfrac{1}{2}) = \tfrac{1}{2}a_0 - a_2 + a_4 - a_6 + \cdots$$

$$y(1) = \tfrac{1}{2}a_0 + a_1 + a_2 + a_3 + \cdots$$

The computational procedure being identical with that of the preceding section.

THE NUMERICAL SOLUTION OF DIFFERENTIAL EQUATIONS (Continued)

ITERATIVE METHODS

With the iterative methods of solution of differential equations, two fundamental approaches exist

a) Point iteration.

b) Block iteration.

In the former, the solution at each point is refined as far as is necessary before the next point is considered. The latter, on the other hand, involves successive sweeps over the whole range of integration until no further improvement is detected.

With either group of methods the number of iterations should be small, i.e. if more than three or four applications of the iterative formula in question are necessary then the interval of tabulation should be reduced—in practice this is usually taken to mean halved.

The restriction on the number of iterations does not apply to the Chebyshev method since here the coefficients of an approximating polynomial are being determined and thus the iterations are continued until the required number of significant places is obtained. The solution of the differential equation is then obtained by direct substitution into the polynomial.

When only one starting value is given, as is usual at the beginning of the tabulation of a first order equation, then special starting techniques are necessary for the point iteration approach as a knowledge of the function at several consecutive points is necessary before it can be evaluated at the next point. Either one of the direct methods previously tabulated can be invoked or an iterative starting formula used.

With the iterative predictor-corrector methods, the starting formulae are of the same type.

Doubling the interval of tabulation does not give any trouble with the point iteration schemes whereas halving the interval requires the use of an interpolation routine.

The block iteration schemes require a fresh start to be made if the interval is to be changed.

Iterative Starting Formulae

Notation used: $x_r \equiv x_0 + rh$

$y_r \equiv y(x_r)$

$f_r \equiv f(x_r, y_r) \equiv hy'_r$

$E_r \equiv$ The truncation error in the expression for y_r.

The superscript (ν) denotes the νth iterate.

Since the iterative starting formulae require initial values to be inserted the **first** iterative values are obtained from

$$y_r^{(1)} \rightleftharpoons y_0 + rf_0 \qquad\qquad r = 1(1)n - 1$$

This enables "n" starting values to be determined, including the given value y_0, for use with the following groups of formulae.

It will be seen that the values of y_0 and f_0 are unaltered as they will be the given values; f_0 being simply a function of h, x_0 and y_0 determined by the differential equation in question. Further, as each of the other values are refined, they are used in the succeeding equations.

E.g., anticipating the table associated with $n = 3$ in group 1, we obtain, on writing out the formulae in full

ITERATIVE STARTING FORMULAE (Continued)

$$y_1^{(\nu+1)} = \left(y_0 + \frac{5}{12}f_0\right) + \frac{(8f_1^{(\nu)} - f_2^{(\nu)})}{12} + \frac{\delta^3}{24}$$

$$y_2^{(\nu+1)} = \left(y_0 + \frac{1}{3}f_0\right) + \frac{4}{3}f_1^{(\nu+1)} + \frac{1}{3}f_2^{(\nu)} - \frac{\delta^4}{90}$$

The worst error term in the above is $\delta^3/24$ and thus for the contribution from this source to be less than 1/2 unit in the last significant figure $\delta^3/24 < 1/2$ i.e. $\delta^3 < 12$. If this is not so then the interval h must be reduced.

The coefficients A_{rs} (and A_{rs}^*) can be expressed as rational fractions in the form $A_{rs} =$ Numerator of $A_{rs}/D_r \equiv$ Num. A_{rs}/D_r. The tabulation will therefore be of

Num. A_{rs}	D_r	E_r
$s = 0(1)n - 1$ $r = 1(1)n - 1$	$r = 1(1)n - 1$	$r = 1(1)n - 1$

where E_r indicates the principal error term associated with f_r.

Group 1

$$y_1^{(\nu+1)} = y_0 + A_{10}f_0 + \sum_{s=1}^{n-1} A_{1s}f_s^{(\nu)} + E_1$$

$$y_r^{(\nu+1)} = y_0 + A_{r0}f_0 + \sum_{s=1}^{r-1} A_{rs}f_s^{(\nu+1)} + \sum_{s=r}^{n-1} A_{rs}f_s^{(\nu)} + E_r, \quad r = 2(1)n - 1$$

$n = 2$

r \ s	Num. A_{rs} 0	1	D_r	E_r
1	1	1	2	$-\dfrac{1}{12}\delta^2$

Requires $\delta^2 < 6$

$n = 3$

r \ s	Num. A_{rs}, 0	1	2	D_r	E_r
1	5	8	-1	12	$\dfrac{1}{24}\delta^3$
2	1	4	1	3	$-\dfrac{1}{90}\delta^4$

Requires $\delta^3 < 12$

$n = 4$

r \ s	Num. A_{rs} 0	1	2	3	D_r	E_r
1	9	19	-5	1	24	$-\dfrac{19}{720}\delta_4$
2	1	4	1	0	3	$-\dfrac{1}{90}\delta^4$
3	3	9	9	3	8	$-\dfrac{3}{80}\delta^4$

Requires $\delta^4 < 13$

ITERATIVE STARTING FORMULAE (Continued)

$n = 5$

		Num. A_{rs}				D_r	E_r
r \ s	0	1	2	3	4		
1	251	646	-264	106	-19	720	$\dfrac{3}{160}\delta^5$
2	29	124	24	4	-1	90	$\dfrac{1}{90}\delta^5$
3	27	102	72	42	-3	80	$\dfrac{3}{160}\delta^5$
4	14	64	24	64	14	45	$-\dfrac{1}{105}\delta^6$

Requires $\delta^5 < 27$

$n = 6$

		Num. A_{rs}					D_r	E_r
r \ s	0	1	2	3	4	5		
1	493	1337	-618	302	-83	9	1440	$-\dfrac{14}{695}\delta^6$
2	28	129	14	14	-6	1	90	$-\dfrac{1}{105}\delta^6$
3	51	219	114	114	-21	3	160	$-\dfrac{3}{224}\delta^6$
4	28	128	48	128	28	0	90	$-\dfrac{1}{105}\delta^6$
5	95	375	250	250	375	95	288	$-\dfrac{1}{42}\delta^6$

Requires $\delta^6 < 21$

Group 2

The formulae of this group involve the function values as well as the derivatives and consequently tend to have smaller truncation errors.

$$y_r^{(\nu+1)} = \sum_{s=0}^{r-1} A_{rs}^* y_s^{(\nu+1)} + \sum_{s=0}^{r-1} A_{rs} f_s^{(\nu+1)} + \sum_{s=r}^{n-1} A_{rs} f_s^{(\nu)} + E_r, \qquad r = 1(1)n - 1$$

$n = 4$

	Num. A_{rs}			Num. A_{rs}				D_r	E_r
r \ s	0	1	2	0	1	2	3		
1	24	0	0	9	19	-5	1	24	$-\dfrac{19}{720}\delta^4$
2	33	24	0	10	57	24	-1	57	$\dfrac{1}{342}\delta^5$
3	11	27	-27	3	27	27	3	11	$-\dfrac{1}{513}\delta^6$

Requires $\delta^4 < 18$

ITERATIVE STARTING FORMULAE (Continued)

$n=5$	Num. A_{rs}^*				Num. A_{rs}					D_r	E_r
r \ s	0	1	2	3	0	1	2	3	4		
1	720	0	0	0	251	646	-264	106	-19	720	$\dfrac{3}{160}\delta^5$
2	990	1440	0	0	281	2056	1176	-104	11	2430	$-\dfrac{1}{729}\delta^6$
3	550	2160	-1350	0	141	1656	2376	456	-9	1360	$\dfrac{1}{2539}\delta^7$
4	25	160	0	-160	6	96	216	96	6	25	$-\dfrac{1}{2625}\delta^8$

Requires $\delta^5 < 27$

$n=6$	Num. A_{rs}^*				
r \ s	0	1	2	3	4
1	1440	0	0	0	0
2	24390	53280	0	0	0
3	9550	51570	-17550	0	0
4	3425	29560	14040	-35960	0
5	137	1625	2000	-2000	-1625

	Num. A_{rs}						D_r	E_r
r \ s	0	1	2	3	4	5		
1	475	1427	-798	482	-173	27	1440	$-\dfrac{1}{70}\delta^6$
2	6589	58528	41608	-5752	1223	-136	77670	$\dfrac{31}{290}\delta^7$
3	2337	33687	62352	16512	-693	45	43570	$-\dfrac{1}{9072}\delta^8$
4	786	15540	45720	29280	3210	-36	11065	$\dfrac{1}{16800}\delta^9$
5	30	750	3000	3000	750	30	137	$-\dfrac{1}{12659}\delta^{10}$

Requires $\delta^6 < 35$

Group 3

In this group neither y_r nor f_r appears on the right hand side i.e. $A_{ss}^* \equiv A_{ss} \equiv 0$.

$$y_r^{(\nu+1)} = \sum_{s=0}^{r-1} A_{rs}^* y_s^{(\nu+1)} + \sum_{s=r+1}^{n-1} A_{rs}^* y^{(\nu)} + \sum_{s=0}^{r-1} A_{rs} f^{(\nu+1)} + \sum_{s=r+1}^{n-1} A_{rs} f^{(\nu)} + E_r$$

ITERATIVE STARTING FORMULAE (Continued)

$n = 3$

r \ s	Num. A_{rs}^* 0	1	2	Num. A_{rs} 0	1	2	D_r	E_r
1	2	0	2	1	0	-1	4	$\dfrac{1}{24}\delta^3$
2	5	-4	0	2	4	0	1	$\dfrac{1}{12}\delta^3$

Requires $\delta^3 < 6$

$n = 4$

r \ s	Num. A_{rs}^* 0	1	2	3	Num. A_{rs} 0	1	2	3	D_r	E_r
1	8	0	0	19	3	0	-27	-6	27	$\dfrac{1}{180}\delta^5$
2	19	0	0	8	6	27	0	-3	27	$\dfrac{1}{180}\delta^5$
3	10	9	-18	0	3	18	9	0	1	$\dfrac{1}{20}\delta^5$

Requires $\delta^5 < 10$

$n = 5$

r \ s	Num. A_{rs}^* 0	1	2	3	4	Num. A_{rs} 0	1	2	3	4	D_r	E_r
1	19	0	-216	224	69	6	0	-216	-192	-18	96	$\dfrac{1}{1120}\delta^7$
2	11	16	0	16	11	3	24	0	-24	-3	54	$\dfrac{1}{2520}\delta^7$
3	69	224	-216	0	19	18	192	216	0	-6	96	$\dfrac{1}{805}\delta^7$
4	47	192	-108	-128	0	12	144	216	48	0	3	$\dfrac{1}{70}\delta^7$

Requires $\delta^7 < 35$

Group 4

In this group the starting values are symmetrically placed with respect to the origin. These formulae usually have a better truncation error and their coefficients exhibit a certain degree of skew symmetry.

Assuming that y_r, and hence f_r, are found before y_{-r} and f_{-r} the formulae used are, for $N \equiv (n - 1)/2$ where n denotes the odd number of points involved, as follows

$$y_r^{(\nu+1)} = y_0 + \sum_{s=-N}^{-r} A_{rs}f_s^{(\nu)} + \sum_{s=-r+1}^{r-1} A_{rs}f_s^{(\nu+1)} + \sum_{s=r}^{N} A_{rs}f_s^{(\nu)} + E_r$$

which is followed immediately by

$$y_{-r}^{(\nu+1)} = y_0 + \sum_{s=-N}^{-r} A_{rs}f_s^{(\nu)} + \sum_{s=-r+1}^{r} A_{rs}f_s^{(\nu+1)} + \sum_{s=r+1}^{N} A_{rs}f_s^{(r)} + E_r$$

ITERATIVE STARTING FORMULAE (Continued)

The formulae are tabulated in the order in which they are to be used i.e. $r = 1, -1,$ $2, -2, 3, -3.$

$n = 3$

	Num. A_{rs}			D_r	E_r
s	-1	0	1		
r					
1	-1	8	5	12	$\dfrac{1}{24} \delta^3$
-1	-5	-8	1	12	$-\dfrac{1}{24} \delta^3$

Requires $\delta^3 < 12$

$n = 5$

	Num. A_{rs}					D_r	E_r
s	-2	-1	0	1	2		
r							
1	11	-74	456	346	-19	720	$\dfrac{1}{144} \delta^5$
1	19	-346	-456	74	-11	720	$\dfrac{1}{144} \delta^5$
2	-1	4	24	124	29	90	$-\dfrac{1}{90} \delta^5$
-2	-29	-124	-24	-4	1	90	$-\dfrac{1}{90} \delta^5$

Requires $\delta^5 < 45$

$n = 7$

	Num. A_{rs}							D_r	E_r
s	-3	-2	-1	0	1	2	3		
r									
1	-191	1608	-6771	37504	30819	-2760	271	60480	$\dfrac{1}{633} \delta^7$
-1	-271	2760	-30819	-37504	6771	-1608	191	60480	$\dfrac{1}{633} \delta^7$
2	5	-30	33	1328	4863	1398	-37	3780	$\dfrac{1}{8172} \delta^7$
-2	37	-1398	-4863	-1328	-33	30	-5	3780	$\dfrac{1}{8172} \delta^7$
3	-29	216	-729	2176	1161	3240	685	2240	$\dfrac{9}{896} \delta^7$
-3	-685	-3240	-1161	-2176	729	-216	29	2240	$\dfrac{9}{896} \delta^7$

Requires $\delta^7 < 50$

Group 5

As group 4, with the addition of function values. The same observations apply. y_0 and f_0 do not alter from cycle to cycle. Defining N the same as for group 4, the formulae used are:

ITERATIVE STARTING FORMULAE (Continued)

$$y_r^{(\nu+1)} = \sum_{s=-N}^{-r} A_{rs}^* y_s^{(\nu)} + \sum_{s=-r+1}^{r-1} A_{rs}^* y_s^{(\nu+1)} + \sum_{s=r}^{N} A_{rs}^* y_s^{(\nu)}$$

$$+ \sum_{s=-N}^{-r} A_{rs} f_s^{(\nu)} + \sum_{s=-r+1}^{r-1} A_{rs} f_s^{(\nu+1)} + \sum_{s=r}^{N} A_{rs} f_s^{(\nu)} + E_r$$

$n = 5$ Num. A_{rs}^*

r \ s	−2	−1	0	1	2
1	−85	0	540	0	−135
−1	−135	0	540	0	−85
2	0	1360	1350	−2160	0
−2	0	−2160	1350	1360	0

r \ s	Num. A_{rs} −2	−1	0	1	2	D_r	E_r
1	−24	−144	216	336	36	320	$-\dfrac{1}{1643}\delta^7$
−1	−36	−336	−216	144	24	320	$-\dfrac{1}{1643}\delta^7$
2	−9	456	2376	1656	141	550	$-\dfrac{1}{1067}\delta^7$
−2	−141	−1656	−2376	−456	9	550	$-\dfrac{1}{1067}\delta^7$

Requires $\delta^7 < 533$

The coefficients become unwieldy for higher values of n.

Continuing the solution

Two formulae are used to continue the solution once sufficient starting values have been obtained. The first, the "predictor," is used to obtain an estimate of the next value. It is used once only and can be appreciably less accurate than the second, the "corrector," which determines the precision.

The corrector is used iteratively until sufficient significant figures are obtained.

Predictors and correctors can involve either backward differences, central differences or function values.

Predictor-corrector formulae involving backward differences

A typical predictor formula is

$$y_1 = y_0 + \left(1 + \frac{1}{2}\nabla + \frac{5}{12}\nabla^2 + \frac{3}{8}\nabla^3 + \frac{251}{720}\nabla^4 + \frac{95}{288}\nabla^5 + \frac{19087}{60480}\nabla^6 + \cdots\right)f_0$$

this is followed by

$$f_1 = h \cdot y_1',$$

where y' is defined as a function of x and y.

PREDICTOR-CORRECTOR METHODS

This predicted value of f_1 is used to generate all the backward differences ∇_1^r for $r = 1(1)6$.

These are then used with a corrector formula such as

$$y_1 = y_0 + \left(1 - \frac{1}{2}\nabla - \frac{1}{12}\nabla^2 - \frac{1}{24}\nabla^3 - \frac{19}{720}\nabla^4 - \frac{3}{160}\nabla^5 - \frac{863}{60480}\nabla^6 - \cdots\right)f_1$$

to recompute y_1 and thus f_1. Note that all the backward differences ∇_1^r must then be recomputed before the corrector is applied once more.

The numerical coefficients tend to be rather high and round-off errors can become troublesome if differences beyond ∇^6 are significant.

An alternative method of *prediction* when one particular column of differences is nearly constant (say for example $\nabla^6 \sim k$) is to use the recurrence formula

$$\nabla_1^{r-1} = \nabla_1^r + \nabla_0^{r-1}$$

to generate the required differences. The overall effect is as if

$$f_1 = (1 + \nabla + \nabla^2 + \nabla^3 + \nabla^4 + \nabla^5)f_0 + k$$

had been used as a predictor.

Even when ∇^6 is not constant it can usually be estimated with sufficient accuracy from the preceding entries in the column, i.e., it is extrapolated from a knowledge of the previous differences, and is then used in the predictor

$$f_1 = (1 + \nabla + \nabla^2 + \nabla^3 + \nabla^4 + \nabla^5)f_0 + \nabla^6 f_1$$

The corrector quoted in the example is the first of a group of such formulae having the general form

$$y_1 = y_r + \sum_{s=0} A_{rs}\nabla_s f_1 \qquad r < 0.$$

Truncating the infinite series at $s = 6$, we have

$$A_{rs}$$

r \ s	0	1	2	3	4	5	6
0	1	$-\dfrac{1}{2}$	$-\dfrac{1}{12}$	$-\dfrac{1}{24}$	$-\dfrac{19}{720}$	$-\dfrac{3}{160}$	$-\dfrac{863}{60480}$
-1	2	-2	$\dfrac{1}{3}$	0	$-\dfrac{1}{90}$	$-\dfrac{1}{90}$	$-\dfrac{37}{3780}$
-2	3	$-\dfrac{9}{2}$	$\dfrac{9}{4}$	$-\dfrac{3}{8}$	$-\dfrac{3}{80}$	$-\dfrac{3}{160}$	$-\dfrac{29}{2240}$
-3	4	-8	$\dfrac{20}{3}$	$-\dfrac{8}{3}$	$\dfrac{14}{45}$	0	$-\dfrac{8}{945}$
-4	5	$-\dfrac{25}{2}$	$\dfrac{175}{12}$	$-\dfrac{225}{24}$	$\dfrac{425}{144}$	$-\dfrac{95}{288}$	$-\dfrac{275}{12096}$
-5	6	-18	27	-24	$\dfrac{123}{10}$	$-\dfrac{33}{10}$	$\dfrac{41}{140}$

PREDICTOR-CORRECTOR METHODS (Continued)

Predictor-corrector formulae involving central differences

In these formulae, the predictors and the correctors have the same form. In fact the corrector has the same coefficients as the predictor on the previous line of the table with the addition of one more term, together with an estimate of the error.

The formulae involved are

$$\text{Predictor} \qquad y_1 = y_{1-2r} + \sum_{s=0}^{r-1} A_{rs}\delta^{2s}f_{1-r}$$

$$\text{Corrector} \qquad y_1 = y_{3-2r} + \sum_{s=0}^{r-1} B_{rs}\delta^{2s}f_{2-r} + \text{Error} \ (\equiv B_{rr}\delta^{2r})$$

A_{rs}

s \ r	0	1	2	3	4
2	4	$\frac{8}{3}$			
3	6	9	$\frac{33}{10}$		
4	8	$\frac{64}{3}$	$\frac{868}{45}$	$\frac{736}{189}$	
5	10	$\frac{125}{3}$	$\frac{875}{18}$	$\frac{17225}{756}$	$\frac{20225}{4536}$

B_{rs}

s \ r	0	1	2	3	4	5
2	2	$\frac{1}{3}$	$\frac{-1}{90}$			
3	4	$\frac{8}{3}$	$\frac{14}{45}$	$\frac{-8}{945}$		
4	6	9	$\frac{33}{10}$	$\frac{41}{140}$	$\frac{-9}{1400}$	
5	8	$\frac{64}{3}$	$\frac{868}{45}$	$\frac{736}{189}$	$\frac{3956}{14175}$	$\frac{-2368}{467775}$

e.g. Suppose that enough starting values have been obtained to indicate that the sixth differences are of the order of 50 (say).

The corrector corresponding to $r = 3$ has an error term of magnitude $8\delta^6/945$ which is less than $\frac{1}{2}$ when $\delta^6 = 50$.

Therefore label the last computed value y_0 and use the predictor

$$y_1 = y_{-5} + 6f_{-2} + 9\delta^2_{-2} + \frac{33}{10}\delta^4_{-2} \quad (r = 3 \text{ in the above table}).$$

From this value of y_1, generate $f_1 \ (= hy'_1)$ and carry the differences back to the sixth differences column.

PREDICTOR-CORRECTOR METHODS (Continued)

The corrector to use is then

$$y_1 = y_{-3} + 4f_{-1} + \frac{8}{3}\delta^2_{-1} + \frac{14}{45}\delta^4_{-1} \qquad (r = 3 \text{ above}).$$

Note that the error term should strictly be expressed in terms of δ^6_{-1} and so an adequate check cannot be carried out until the next term has been computed.

Note also that, in common with all predictor-corrector methods, any form of predictor can be used, as the refinement depends only upon the corrector.

Predictor-corrector formulae involving function values

The general form of the simplest type of predictor is

$$y_1 = y_r + \sum_{s=r}^{0} A_{rs}f_s + E_r \qquad < 0$$

r \ s	0	−1	−2	−3	−4	−5	E_r
0	1						$\frac{1}{2}\delta$
−1	2	0					$\frac{1}{3}\delta^2$
−2	$\frac{9}{4}$	0	$\frac{3}{4}$				$\frac{3}{8}\delta^3$
−3	$\frac{8}{3}$	$-\frac{4}{3}$	$\frac{8}{3}$	0			$\frac{14}{45}\delta^4$
−4	$\frac{55}{24}$	$\frac{5}{24}$	$\frac{5}{24}$	$\frac{55}{24}$	0		$\frac{95}{144}\delta^5$
−5	$\frac{33}{10}$	$-\frac{42}{10}$	$\frac{78}{10}$	$-\frac{42}{10}$	$\frac{33}{10}$	0	$\frac{41}{140}\delta^6$

An alternative group of predictor formulae have the form

$$y_1 = y_0 + \sum_{s=r}^{0} A_{rs}f_s + E_r \qquad (r < 0)$$

r \ s	0	−1	−2	−3	−4	E_r
−1	$\frac{3}{2}$	$-\frac{1}{2}$				$\frac{5}{12}\delta^2$
−2	$\frac{23}{12}$	$-\frac{4}{3}$	$\frac{5}{12}$			$\frac{3}{8}\delta^3$
−3	$\frac{55}{24}$	$-\frac{59}{24}$	$\frac{37}{24}$	$-\frac{3}{8}$		$\frac{25}{72}\delta^4$
−4	$\frac{1901}{720}$	$-\frac{1387}{360}$	$\frac{327}{90}$	$-\frac{637}{360}$	$\frac{251}{720}$	$\frac{95}{288}\delta^5$

The correctors fall roughly into main groups. The first group is similar to the predictors tabulated above in that the emphasis is upon a weighted mean of the derivatives rather than the function values.

PREDICTOR-CORRECTOR METHODS (Continued)

Group 1

$$y_1 = y_r + \sum_{s=r}^{0} A_{rs}f_s + E_r \qquad (r < 0)$$

				A_{rs}				E_r
s	1	0	-1	-2	-3	-4	-5	
r								
0	$\dfrac{1}{2}$	$\dfrac{1}{2}$						$\dfrac{-1}{12}\delta^2$
-1	$\dfrac{1}{3}$	$\dfrac{4}{3}$	$\dfrac{1}{3}$					$\dfrac{-1}{90}\delta^4$
-2	$\dfrac{3}{8}$	$\dfrac{9}{8}$	$\dfrac{9}{8}$	$\dfrac{3}{8}$				$\dfrac{-3}{80}\delta^4$
-3	$\dfrac{14}{45}$	$\dfrac{64}{45}$	$\dfrac{24}{45}$	$\dfrac{64}{45}$	$\dfrac{14}{45}$			$\dfrac{-1}{105}\delta^6$
-4	$\dfrac{95}{288}$	$\dfrac{375}{288}$	$\dfrac{125}{144}$	$\dfrac{125}{144}$	$\dfrac{375}{288}$	$\dfrac{95}{288}$		$\dfrac{-1}{44}\delta^6$
-5	$\dfrac{41}{140}$	$\dfrac{54}{35}$	$\dfrac{27}{140}$	$\dfrac{68}{35}$	$\dfrac{27}{140}$	$\dfrac{54}{35}$	$\dfrac{41}{140}$	$\dfrac{-1}{140}\delta^8$
-6	$\dfrac{3}{10}$	$\dfrac{3}{2}$	$\dfrac{3}{10}$	$\dfrac{9}{5}$	$\dfrac{3}{10}$	$\dfrac{3}{2}$	$\dfrac{3}{10}$	$\dfrac{-1}{140}\delta^6$

The second group involves the computation of a weighted mean of the function value as well as the derivatives.

Group 2

$$y_1 = \sum_{s=-4}^{0} A_{rs}^* y_s + \sum_{s=-4}^{1} A_{rs}f_s + E_r$$

			A_{rs}^*		
s	-4	-3	-2	-1	0
r					
1	0	0	0	-1	2
2	0	0	1	$\dfrac{27}{11}$	$\dfrac{-27}{11}$
3	0	1	$\dfrac{32}{5}$	0	$\dfrac{-32}{5}$
4	1	$\dfrac{1625}{137}$	$\dfrac{2000}{137}$	$\dfrac{-2000}{137}$	$\dfrac{-1625}{137}$
5	0	0	0	$\dfrac{1}{5}$	$\dfrac{4}{5}$
6	0	0	0	$\dfrac{-19}{13}$	$\dfrac{32}{13}$
7	0	0	$\dfrac{32}{5}$	$\dfrac{-27}{5}$	0
8	0	0	$\dfrac{136}{55}$	$\dfrac{27}{11}$	$\dfrac{-216}{55}$

PREDICTOR-CORRECTOR METHODS (Continued)

s	-4	-3	-2	-1	0	1	E_r
r			A_{rs}				
1	0	0	0	$\dfrac{1}{2}$	0	$\dfrac{-1}{2}$	$\dfrac{-1}{12}\delta^3$
2	0	0	$\dfrac{3}{11}$	$\dfrac{27}{11}$	$\dfrac{27}{11}$	$\dfrac{3}{11}$	$\dfrac{-1}{513}\delta^6$
3	0	$\dfrac{6}{25}$	$\dfrac{96}{25}$	$\dfrac{216}{25}$	$\dfrac{96}{25}$	$\dfrac{6}{25}$	$\dfrac{-1}{2625}\delta^8$
4	$\dfrac{30}{137}$	$\dfrac{750}{137}$	$\dfrac{3000}{137}$	$\dfrac{3000}{137}$	$\dfrac{750}{137}$	$\dfrac{30}{137}$	$\dfrac{-1}{12659}\delta^{10}$
5	0	0	0	0	$\dfrac{4}{5}$	$\dfrac{2}{5}$	$\dfrac{-1}{30}\delta^3$
6	0	0	$\dfrac{4}{39}$	-1	0	$\dfrac{17}{39}$	$\dfrac{-19}{390}\delta^4$
7	0	0	$\dfrac{12}{5}$	$\dfrac{27}{5}$	0	$\dfrac{3}{5}$	$\dfrac{-1}{30}\delta^4$
8	0	$\dfrac{-9}{550}$	$\dfrac{228}{275}$	$\dfrac{1188}{275}$	$\dfrac{828}{275}$	$\dfrac{141}{550}$	$\dfrac{-1}{1067}\delta^7$

Iterative methods

Should it become necessary to halve the interval, because the error term is becoming significant or the number of iterations required is becoming excessive, then interpolation becomes necessary.

This interpolation, actually subtabulation to halves, is based upon previously computed values and the formulae used can take several forms. One such form is as follows

$$y_k = \frac{1}{D_r}\sum_{s=-n}^{0} A_{rs}y_s + E_r, \qquad (r = 1(1)n)$$

where $k \equiv (1 - 2r)/2$, and E_r denotes the principal error term.

$n = 2$

s	0	-1	-2	D_r	E_r
r		A_{rs}			
1	3	6	-1	8	$\dfrac{-1}{16}\nabla_0^3$
2	-1	6	3	8	$\dfrac{1}{16}\nabla_0^3$

$n = 3$

s	0	-1	-2	-3	D_r	E_r
r		A_{rs}				
1	5	15	-5	1	16	$\dfrac{-5}{128}\nabla_0^4$
2	-1	9	9	-1	16	$\dfrac{3}{128}\nabla_0^4$
3	1	-5	15	5	16	$\dfrac{-5}{128}\nabla_0^4$

DEFERRED-CORRECTOR METHODS

$n = 4$ $\quad$ r	s $\quad$ 0	-1	A_{rs} -2	-3	-4	D_r	E_r
1	35	140	-70	28	-5	128	$\dfrac{-7}{256} \nabla_0^5$
2	-5	60	90	-20	3	128	$\dfrac{3}{256} \nabla_0^5$
3	3	-20	90	60	-5	128	$\dfrac{-3}{256} \nabla_0^5$
4	-5	28	-70	140	35	128	$\dfrac{7}{256} \nabla_0^5$

For a given "n" there will be $2n + 1$ values available at the new interval to continue the solution.

Deferred-correction methods

One of the most powerful of the iterative methods is that of "deferred correction." This is a block iteration method, a complete set of function values being obtained from a recurrence relationship, their differences formed and corrections applied from these differences.

Central difference corrections are the most useful. Let the first order differential equation be

$$y' + g(x) \cdot y = k(x)$$

The first recurrence relation, assuming an initial y only given, is

$$y_1 \left(1 + \frac{1}{2} hg_1\right) - y_0 \left(1 - \frac{1}{2} hg_0\right) = \frac{1}{2} h(k_0 + k_1)$$

This is used to generate y_r for both positive and negative values of r if required. When the required range of values of y_r has been covered, with sufficient additional terms to generate all the differences required, a difference table is formed and the correction terms $C(y_{\frac{1}{2}})$ tabulated from

$$C(y_{\frac{1}{2}}) \equiv \frac{1}{12} \delta_{\frac{1}{2}}^3 - \frac{1}{120} \delta_{\frac{1}{2}}^5 + \frac{1}{840} \delta_{\frac{1}{2}}^7 - \frac{1}{5040} \delta_{\frac{1}{2}}^9 \cdots$$

The correction terms $C(y_{\frac{1}{2}})$ are corrections to the recurrence relation. The corrections to the function values y_r are η_r i.e. True value = Approximate value + η

where $\qquad$ $$\eta_1 \left(1 + \frac{1}{2} hg_1\right) - \eta_0 \left(1 - \frac{1}{2} hg_0\right) + C(y_{\frac{1}{2}}) = 0$$

In this formula the initial η_0 is zero and the recurrence relation can be used in either direction.

New $C(y_{\frac{1}{2}})$ are then computed and the correction process repeated until there is no change in the values of $C(y_{\frac{1}{2}})$. At this point the computed values and the differences derived there-from should be checked for consistency with the differential equation. The method of deferred correction is particularly useful for linear differential equations as its convergence is extremely rapid, even for very large intervals. If more starting values are known formulae which have even faster convergence can be employed. Two such formulae are as follows: Two starting values given

DEFERRED-CORRECTOR METHODS (Continued)

Predictor $\qquad y_1\left(1+\dfrac{1}{3}hg_1\right)+\dfrac{4}{3}hg_0y_0-y_{-1}\left(1-\dfrac{1}{3}hg_{-1}\right)=\dfrac{1}{3}h(k_1+4k_0+k_{-1})$

$$C(y_0)\equiv\frac{1}{180}(\delta_{-\frac{1}{2}}^5+\delta_{\frac{1}{2}}^5)-\frac{1}{630}(\delta_{-\frac{1}{2}}^7+\delta_{\frac{1}{2}}^7)+\frac{1}{2510}(\delta_{-\frac{1}{2}}^9+\delta_{\frac{1}{2}}^9)\cdots$$

Corrector $\qquad \eta_1\left(1+\dfrac{1}{3}hg_1\right)+\dfrac{4}{3}hg_0\eta_0-\eta_{-1}\left(1-\dfrac{1}{3}hg_{-1}\right)+C(y_0)=0.$

Three starting values given

Predictor $\qquad y_1+y_0(9+6hg_0)-y_{-1}(9-6hg_{-1})-y_{-2}=6h(k_0+k_{-1})$

$$C(y_{-\frac{1}{2}})\equiv-\frac{1}{10}\delta_{-\frac{1}{2}}^5+\frac{1}{70}\delta_{-\frac{1}{2}}^7-\frac{1}{420}\delta_{-\frac{1}{2}}^9+\cdots$$

Corrector $\qquad \eta_1+\eta_0(9+6hg_0)-\eta_{-1}(9-6hg_{-1})-\eta_{-2}+C(y_{-\frac{1}{2}})=0$

The use of the deferred correction technique for higher order differential equations.
  Second order differential equations.
  Two starting values given e.g. $y''+m\cdot y'+g\cdot y=k$ where y_0 and y_{-1} are known and $y\equiv y(x)$; $m\equiv m(x)$; $g\equiv g(x)$; $k\equiv k(x)$.

Predictor $\qquad y_1\left(1+\dfrac{1}{2}hm_0\right)-y_0(2-h^2g_0)+y_{-1}\left(1-\dfrac{1}{2}hm_0\right)=h^2k_0$

$$C(y_0)\equiv\left[-\frac{1}{12}\delta_0^4+\frac{1}{90}\delta_0^6-\frac{1}{560}\delta_0^8+\frac{1}{3150}\delta_0^{10}\right]$$

$$+\,hg_0\left[-\frac{1}{12}(\delta_{\frac{1}{2}}^3+\delta_{-\frac{1}{2}}^3)+\frac{1}{60}(\delta_{\frac{1}{2}}^5+\delta_{-\frac{1}{2}}^5)-\frac{1}{280}(\delta_{\frac{1}{2}}^7+\delta_{-\frac{1}{2}}^7)+\frac{1}{1260}(\delta_{\frac{1}{2}}^9+\delta_{-\frac{1}{2}}^9)\right].$$

Corrector $\qquad \eta_1\left(1+\dfrac{1}{2}hm_0\right)-\eta_0(2-h^2g_0)+\eta_{-1}\left(1-\dfrac{1}{2}hm_0\right)+C(y_0)=0.$

Three starting values require the known functions $m(x)$, $g(x)$ $g(x)$ and $k(x)$ to be tabulated at half way points and the advantage in speed of convergence is relatively slight.

Predictor $\qquad y_1\left(1-\dfrac{1}{12}hm_{-\frac{1}{2}}-\dfrac{1}{8}h^2g_{-\frac{1}{2}}\right)-y_0\left(1-\dfrac{9}{4}hm_{-\frac{1}{2}}-\dfrac{9}{8}h^2g_{-\frac{1}{2}}\right)$

$$-y_{-1}\left(1+\frac{9}{4}hm_{-\frac{1}{2}}-\frac{9}{8}h^2g_{-\frac{1}{2}}\right)+y_{-2}\left(1+\frac{1}{12}hm_{-\frac{1}{2}}-\frac{1}{8}h^2g_{-\frac{1}{2}}\right)=2h^2k_{-\frac{1}{2}}$$

$$C(y_{-\frac{1}{2}})\equiv\left[\frac{-5}{24}(\delta_0^4+\delta_{-1}^4)+\frac{259}{5760}(\delta_0^6+\delta_{-1}^6)-\frac{3229}{322560}(\delta_0^8+\delta_{-1}^8)\right.$$

$$+\frac{117469}{51609600}(\delta_0^{10}+\delta_{-1}^{10})\left]+hm_{-\frac{1}{2}}\left[\frac{3}{320}\delta_{-\frac{1}{2}}^5-\frac{5}{3584}\delta_{-\frac{1}{2}}^7+\frac{35}{147456}\delta_{-\frac{1}{2}}^9\right]\right.$$

$$+\,h^2g_{-\frac{1}{2}}\left[\frac{3}{128}(\delta_0^4+\delta_{-1}^4)-\frac{5}{1024}(\delta_0^6+\delta_{-1}^6)+\frac{35}{32768}(\delta_0^8+\delta_{-1}^8)\right]$$

Corrector $\qquad \eta_1\left(1-\dfrac{1}{12}hm_{-\frac{1}{2}}-\dfrac{1}{8}h^2g_{-\frac{1}{2}}\right)-\eta_0\left(1-\dfrac{9}{4}hm_{-\frac{1}{2}}-\dfrac{9}{8}h^2g_{-\frac{1}{2}}\right)$

$$-\eta_{-1}\left(1+\frac{9}{4}hm_{-\frac{1}{2}}-\frac{9}{8}h^2g_{-\frac{1}{2}}\right)+\eta_{-2}\left(1+\frac{1}{12}hm_{-\frac{1}{2}}-\frac{1}{8}h^2g_{-\frac{1}{2}}\right)+C(y_{-\frac{1}{2}})=0$$

CHEBYSHEV METHODS

In general, for second order differential equations, recurence relations involving an even number of known values are simpler than those involving an odd number of such values. The formulae for four starting values are given below and they are seen to be much simpler than those for three values. Further the correction terms should have more rapid convergence properties.

Predictor
$$y_1(1 + hm_{-1}) - y_0(16 + 8hm_{-1}) + y_{-1}(30 - 12h^2g_{-1})$$
$$+ y_{-2}(16 - 8hm_{-1}) + y_{-3}(1 - hm_{-1}) = -12h^2k_{-1}$$

$$C(y_{-1}) \equiv \left[-\frac{4}{30}\delta^6_{-1} + \frac{3}{140}\delta^8_{-1} - \frac{2}{525}\delta^{10}_1 + \frac{1}{1386}\delta^{12}_{-1} \right]$$
$$+ hm_{-1}\left[-\frac{1}{5}(\delta^5_{-\frac{1}{2}} + \delta^5_{-1\frac{1}{2}}) + \frac{3}{70}(\delta^7_{-\frac{1}{2}} + \delta^7_{-1\frac{1}{2}}) \right.$$
$$\left. -\frac{1}{105}(\delta^9_{-\frac{1}{2}} + \delta^9_{-1\frac{1}{2}}) + \frac{1}{462}(\delta^{11}_{-\frac{1}{2}} + \delta^{11}_{-1\frac{1}{2}}) \right].$$

Corrector
$$\eta_1(1 + hm_{-1}) - \eta_0(16 + 8hm_{-1}) + \eta_{-1}(30 - 12h^2g_{-1})$$
$$+ \eta_{-2}(16 - 8hm_{-1}) + \eta_{-3}(1 - hm_{-1}) + C(y_{-1}) = 0$$

Iterative Chebyshev

The recurrence relationships used are identical with those set up in the direct Chebyshev method. The first value of $a_0^{(s)}$ is obtained from the first term of the series for the initial value of $y^{(s)}$, all the other terms being initially set at zero. e.g. Consider the case when $a_r^{(s)}$ are required for $s = 0,1,2$ and 3. (The initial values of y, y', y'' being known at $x = -1$.) The computational sequence is as follows

Set all $a_r^{(3)} = 0$
Obtain $a_0^{(2)}$ from $y''(-1) = \frac{1}{2}a_0^{(2)}$
Use the Chebyshev recurrence formula to obtain $a_1^{(1)}$ from this value of $a_0^{(2)}$
Obtain $a_0^{(1)}$ from $y'(-1) = \frac{1}{2}a_0^{(1)} - a_1^{(1)}$
Use the Chebyshev formula again to obtain both $a_1^{(0)}$ and $a_2^{(0)}$
Obtain $a_0^{(0)}$ from $y(-1) = \frac{1}{2}a_1^{(0)} - a_1^{(0)} + a_2^{(0)}$
Now utilise the recurrence relation given by the differential equation to generate $a_0^{(3)}$.

The cycle is now repeated each series being increased by one term if desired. The determining factor being the number of significant decimals required in the result. The method is in general slower than the direct Chebyshev process but it has the merit that no predetermined knowledge of the number of terms required is necessary—further computational errors die away as the method is truly iterative, whereas in the direct approach they can nullify all the working.

Direct Chebyshev method

This method is suitable for a differential equation of any order. The range of the arguments must however be normalized to $-1 \leq x \leq 1$ (or $0 \leq x \leq 1$ if the "shifted" Chebyshev polynomials are to be used).

A polynomial approximation to the analytical solution is sought, the degree of the polynomial N being decided beforehand.

CHEBYSHEV METHODS (Continued)

Assume

$$y(x) = \tfrac{1}{2}a_0 + \sum_{r=1}^{N} a_r \cdot T_r(x)$$

$$y'(x) = \tfrac{1}{2}a_0' + \sum_{r=1}^{N} a_r' \cdot T_r(x)$$

$$y''(x) = \tfrac{1}{2}a_0'' + \sum_{r=1}^{N} a_r'' \cdot T_r(x) \qquad \text{etc.}$$

where the a's are constants.

Writing $y^{(0)}$, $y^{(1)}$, $y^{(2)}$ for $y(x)$, $y'(x)$, $y''(x)$ etc., $a_r^{(0)}$, $a_r^{(1)}$, $a_r^{(2)}$ for a_r, a_r', a_r'', etc., and T_r for $T_r(x)$.

The recurrence relationship for the Chebyshev polynomials can be written as

$$2xT_r = T_{|r-1|} + T_{r+1} \qquad (r = 0,1,2, \ldots)$$

further

$$2 \int T_r \cdot dx = \frac{T_{r+1}}{(r+1)} - \frac{T_{r-1}}{(r-1)} \qquad (r > 1)$$

$$= \tfrac{1}{2}T_2 \qquad (r = 1)$$

$$= 2T_1 \qquad (r = 0)$$

Thus
$$2r \cdot a_r^{(s)} = a_{r-1}^{(s+1)} - a_{r+1}^{(s+1)} \qquad (r > 1)$$

If, in addition, the notation $C_r(x^p y^{(s)})$ is introduced to denote the coefficient of T_r in the expansion of $x^p y^{(s)}$ *and twice this coefficient when $r = 0$* then

$$C_r(y^{(0)}) = a_r$$

$$C_r(y^{(s)}) = a_r^{(s)}$$

$$C_r(x^p y^{(s)}) = \frac{1}{2^p} \sum_{k=0}^{p} \binom{p}{k} \cdot a_{|r-p+2k|}^{(s)}$$

Method of use of Chebyshev polynomials

Suppose the differential equation is

$$f(x) \cdot y'' + g(x) \cdot y' + h(x) \cdot y = 0$$

i.e.
$$fy^{(2)} + gy^{(1)} + hy^{(0)} = 0$$

where f, g and h are polynomials in x (or can be approximated to by such polynomials).
Then

$$C_r(fy^{(2)}) + C_r(gy^{(1)}) + C_r(hy^{(0)}) = 0$$

will yield a recurrence relationship based upon the differential equation. The Chebyshev polynomials themselves yield the second recurrence relationship, namely

$$2r \cdot a_r^{(s)} = a_{r-1}^{(s+1)} - a_{r+1}^{(s+1)}$$

To start the procedure, assume that

$$a_r^{(s)} = 0 \qquad r > N$$

and assign arbitrary integer values to $a_N^{(1)}$ and $a_N^{(2)}$. The recurrence relationships are used to generate $a_r^{(s)}$ for $r = N - 1$, $N - 2$, etc.

CHEBYSHEV METHODS (Continued)

The formulae for the initial conditions are

$$y^{(s)}(-1) = \tfrac{1}{2}a_0^{(s)} - a_1^{(s)} + a_2^{(s)} - a_3^{(s)} + \cdots$$

$$y^{(s)}(0) = \tfrac{1}{2}a_0^{(s)} - a_2^{(s)} + a_4^{(s)} - a_6^{(s)} + \cdots$$

$$y^{(s)}(1) = \tfrac{1}{2}a_0^{(s)} + a_1^{(s)} + a_2^{(s)} + a_3^{(s)} + \cdots$$

It will be found that the series obtained do not satisfy the given initial conditions. Therefore a second set of coefficients must be obtained by repeating the process for different arbitrary starting values and a linear combination of the two sets will yield the required solution.

$$T_0 = 1 \qquad\qquad\qquad T_{r+1} - 2xT_r + T_{r-1} = 0$$

$$T_1 = x$$

$$T_2 = 2x^2 - 1$$

$$T_3 = 4x^3 - 3x$$

$$T_4 = 8x^4 - 8x^2 + 1$$

$$T_5 = 16x^5 - 20x^3 + 5x$$

$$T_6 = 32x^6 - 48x^4 + 18x^2 - 1$$

If the "shifted" Chebyshev polynomials T_r^* are to be used then

$$y(x) = \tfrac{1}{2}a_0 + \sum_{r=1}^{N} a_r T_r^*$$

$$4r \cdot a_r^{(s)} = a_{|r-1|}^{(s+1)} - a_{r+1}^{(s+1)}$$

$$C_r(x^p y^{(s)}) = \frac{1}{2^p} \sum_{k=0}^{2p} \binom{2p}{k} a_{|r-p+k|}^{(s)}$$

with

$$y^{(s)}(0) = \tfrac{1}{2}a_0^{(s)} - a_1^{(s)} + a_2^{(s)} - a_3^{(s)} \cdots$$

$$y^{(s)}(\tfrac{1}{2}) = \tfrac{1}{2}a_0^{(s)} - a_2^{(s)} + a_4^{(s)} - a_6^{(s)} \cdots$$

$$y^{(s)}(1) = \tfrac{1}{2}a_0^{(s)} + a_1^{(s)} + a_2^{(s)} + a_3^{(s)} \cdots$$

$$T_r^*(x) = T_r(2x - 1) = \cos\{r \cdot \text{arc cos}\,(2x - 1)\}$$

$$T_0^* = 1 \qquad\qquad (4x - 2)T_r^* = T_{r-1}^* + T_{r+1}^*$$

$$T_1^* = 2x - 1$$

$$T_2^* = 8x^2 - 8x + 1$$

$$T_3^* = 32x^3 - 48x^2 + 18x - 1$$

$$T_4^* = 128x^4 - 256x^3 + 160x^2 - 32x + 1$$

The approach is otherwise the same.

INDIRECT METHODS

MONTE CARLO

Applicable to first order equations i.e. quadrature type, for which an upper bound can be set for the function.   e.g. $y' = f(x); a \leq x \leq b; 0 \leq f(x) < e$.

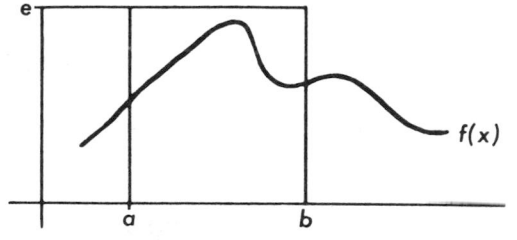 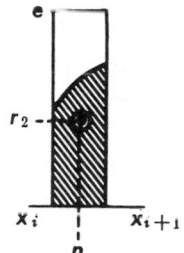

The range $[a,b]$ is subdivided into n intervals, not necessarily equal, such that $x_0 = a$ and $x_n = b$.   For the sub-interval $[x_i, x_{i+1}]$ the procedure is as follows:

Set N_i equal to zero and after each trial increase N_i by unity so that it will always indicate the number of completed trials.

Set T_i equal to zero and after each trial increase T_i by unity *only if the trial was successful* otherwise allow it to retain its previous value.

A trial consists of selecting two random numbers r_1 and r_2.   The first of these is selected from a uniform (rectangular) distribution in the range $[x_i, x_{i+1}]$ and the second from a uniform distribution in the range $[0,e]$.   If $r_2 \leq f(r_1)$ then the trial is deemed to be successful—otherwise it is a failure.

As $N_i \to \infty$,
$$\frac{T_i}{N_i} e(x_{i+1} - x_i) \to \int_{x_i}^{x_{i+1}} f(x) \cdot dx$$

and hence
$$y_{i+1} = y_i + \frac{T_i e(x_{i+1} - x_i)}{N_i} \qquad \text{as } N_i \to \infty.$$

If the function has a negative lower bound such that $-d \leq f(x) \leq e$ then a simple axis transformation is necessary i.e. replace $f(x)$ by $d + f(x)$ and e by $(e + d)$ in the above, giving

$$y_{i+1} = y_i + \frac{T_i(e + d)(x_{i+1} - x_i)}{N_i} \qquad \text{as } N_i \to \infty.$$

In general, the range of values for which a uniform distribution is required will not be standard and a transformation will be necessary.   Such a transformation is obtained as follows:

if r^* is a random number from a uniform distribution in the range $[a,b]$ then r a random number from a uniform distribution in the range $[s,t]$ can be obtained from

$$r = \left(\frac{bs - at}{b - a}\right) - r^*\left(\frac{s - t}{b - a}\right)$$

In particular, if the given range is [0, 1] then

$$r = s + r(t - s)$$

while if the given range is $[-1, -1]$ then

$$r = \tfrac{1}{2}(s + t) + \tfrac{1}{2}r(t - s).$$

In all *Monte Carlo* techniques the number of trials must be large—usually several thousand are required.

PARTIAL DIFFERENTIAL EQUATIONS

Introduction

In the section dealing with finite differences and their applications, some attention has been paid to the derivation of the formulae quoted and a few words of explanation of their use given. Formulae of interest to users of high speed digital computers have been given, as well as those for desk computation.

Physical phenomena frequently give rise to partial differential equations, rather than ordinary ones. The "field" equations of Laplace and Poisson being particular examples. The analytical solution of partial differential equations is frequently tedious, if not impossible, even in those cases where an analytical solution is feasible, a numerical treatment may well be quicker.

Partial differential equations in practice are often reduced to simultaneous equations in two dimensions, one of which is frequently time-like. It is this type of equation which will be considered here.

The classification of partial differential equations is, by analogy with the conic sections, into elliptic, parabolic and hyperbolic equations.

The physical criterion for this grouping is one of boundary conditions. It is also convenient to define the *characteristics* of the equation under consideration. If these are real and distinct then the classification is that of hyperbolic, if real and equal then it is parabolic, while if they are complex it is elliptic.

Consider partial differential equations in two dimensions.

a) Second order equation.
Notation used:

$$p = \frac{\partial u}{\partial x}; \quad q = \frac{\partial u}{\partial y}; \quad r = \frac{\partial^2 u}{\partial x^2}; \quad s = \frac{\partial^2 u}{\partial x \partial y}; \quad t = \frac{\partial^2 u}{\partial y^2}$$

A, B, C and D are functions of u, x, y, p and q.

The physical problem is determined by a second order partial differential equation, the general form of which is

$$A \cdot r + B \cdot s + C \cdot t = D$$

The geometry of the space in which the equation holds yields two more equations:

$$dp = d\left(\frac{\partial u}{\partial x}\right) = r \cdot dx + s \cdot dy; \qquad \text{and similarly, } dq = s \cdot dx + t \cdot dy$$

In matrix form these can be expressed as

$$\begin{bmatrix} A & B & C \\ dx & dy & 0 \\ 0 & dx & dy \end{bmatrix} \begin{bmatrix} r \\ s \\ t \end{bmatrix} = \begin{bmatrix} D \\ dp \\ dq \end{bmatrix}$$

PARTIAL DIFFERENTIAL EQUATIONS (Continued)

At any point at which the determinant of the 3×3 matrix is zero

i.e.
$$\begin{vmatrix} A & B & C \\ dx & dy & 0 \\ 0 & dx & dy \end{vmatrix} = 0,$$

a necessary condition for the equations to have a solution is that the rank of the following matrix is two.

$$\begin{bmatrix} A & B & C & D \\ dx & dy & 0 & dp \\ 0 & dx & dy & dq \end{bmatrix}$$

This condition can be stated in an alternative fashion as follows. The equations have a solution if any of the following three determinental equations are satisfied

$$\begin{vmatrix} A & B & D \\ dx & dy & dp \\ 0 & dx & dq \end{vmatrix} = 0$$

$$\begin{vmatrix} A & D & C \\ dx & dp & 0 \\ 0 & dq & dy \end{vmatrix} = 0$$

$$\begin{vmatrix} D & B & C \\ dp & dy & 0 \\ dq & dx & dy \end{vmatrix} = 0.$$

The equation
$$\begin{vmatrix} A & B & C \\ dx & dy & 0 \\ 0 & dx & dy \end{vmatrix} = 0$$

can be written in the form $\quad A\left(\dfrac{dy}{dx}\right)^2 - B\left(\dfrac{dy}{dx}\right) + C = 0$

and thus for the discriminant Δ defined by

$$\Delta \equiv B^2 - 4AC$$

PARTIAL DIFFERENTIAL EQUATIONS (Continued)

there exist three possibilities

(i) $\Delta > 0$, in which case two real distinct values can be obtained for $\dfrac{dy}{dx}$, corresponding

to real characteristics, and the equation is said to be hyperbolic.

(ii) $\Delta = 0$, when the equation is said to be parabolic.

(iii) $\Delta < 0$, when the equation is said to be elliptic.

Simple examples of these equations are

Hyperbolic
$$\frac{\partial^2 u}{\partial x^2} - \frac{\partial^2 u}{\partial y^2} = 0$$

Parabolic
$$\frac{\partial^2 u}{\partial x^2} - \frac{\partial u}{\partial y} = 0$$

Elliptic
$$\frac{\partial^2 u}{\partial x^2} + \frac{\partial^2 u}{\partial y^2} = 0$$

In the above simple forms A, B, C and D are constants while in fact they can depend upon the solution, and its first derivatives, as well as position in the region. Thus the classification of the equation *can* depend upon the solution itself.

b) Two simultaneous first order equations.

The techniques are similar to those employed in section (*a*) above.

A, B, C, D and E are functions of u, v, x, y, $\dfrac{\partial u}{\partial x}$, $\dfrac{\partial u}{\partial y}$, $\dfrac{\partial v}{\partial x}$, $\dfrac{\partial v}{\partial y}$ and the general form of the

equations are

$$A_1 \frac{\partial u}{\partial x} + B_1 \frac{\partial u}{\partial y} + C_1 \frac{\partial v}{\partial x} + D_1 \frac{\partial v}{\partial y} = E_1$$

$$A_2 \frac{\partial u}{\partial x} + B_2 \frac{\partial u}{\partial y} + C_2 \frac{\partial v}{\partial x} + D_2 \frac{\partial v}{\partial y} = E_2.$$

The geometry of the space yields, as before

$$du = \frac{\partial u}{\partial x} \cdot dx + \frac{\partial u}{\partial y} \cdot dy$$

$$dv = \frac{\partial v}{\partial x} \cdot dx + \frac{\partial v}{\partial y} \cdot dy$$

The discriminant Δ now being given by

$$\Delta \equiv (A_1 D_2 - A_2 D_1 + B_1 C_2 - B_2 C_1) - 4(A_1 C_2 - A_2 C_1)(B_1 D_2 - B_2 D_1).$$

Δ decides the class into which the two equations fall in the same manner as in the previous case.

i.e. Hyperbolic   if $\Delta > 0$
     Parabolic   if $\Delta = 0$
     Elliptic   if $\Delta < 0$.

An elegant method for the solution of hyperbolic equations, based on the determination of the equations of the characteristics, exists but will not be discussed further here.

PARTIAL DIFFERENTIAL EQUATIONS (Continued)

Elliptic equations are best treated numerically by iteration. This enables both the interval to be adjusted arbitrarily and also the powerful convergence techniques of numerical analysis to be invoked.

Parabolic and hyperbolic equations on the other hand can be covered by step-by-step methods. These latter fall into two main groups. On the one hand, there are those methods which yield a direct solution, and, on the other, those which yield the solution implicitly.

In general, however, elliptic equations arise when the boundary conditions are specified on a contour enclosing a region, which need not be simply connected, i.e. elliptic equations are of the boundary value type.

The hyperbolic equations are of the initial value type, and require the solution over a domain having an open boundary.

For parabolic equations the boundary is also open, but in the main only in one direction, the boundary conditions themselves being mixed.

Elliptic Partial Differential Equations

A square mesh of side "h" is superimposed upon the region to enable the points to be labelled with respect to an arbitrary origin.

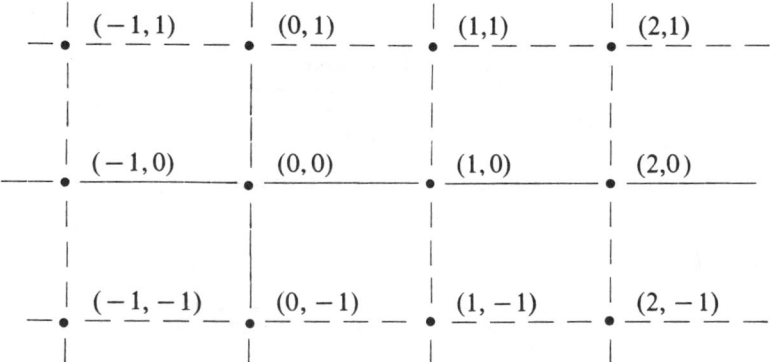

This presupposes that the boundary is formed from the sides of such a mesh. If such is not the case, the procedure must be modified at the boundary. This will be discussed at the end of this section.

The emphasis here will be on the equations of Poisson and Laplace, although the methods involved can easily be extended to any elliptic partial differential equation. In essence, the Taylor's series expansion for a number of mesh points is used to express the partial derivatives occurring in the equation at the origin, in terms of the function values at neighboring points, and higher partial derivatives at the origin. These function values are used to replace the differential equation by a finite difference equation, which is cast in the form of an iteration formula, the higher derivatives consituting the truncation error term.

Consider the equation

$$\nabla^2 u(x,y) = f(x,y)$$

where

$$\nabla^2 \equiv \frac{\partial^2}{\partial x^2} + \frac{\partial^2}{\partial y^2}$$

ELLIPTIC DIFFERENTIAL EQUATIONS

The differential equation is to be replaced by the iteration formula having the form

$$A_{00}u_{00} = \sum_{s=-n}^{n} \sum_{r=-n}^{n}{}^{*} A_{rs}u_{rs} - Bh^2 f_{00} + \text{error term}$$

(the asterisk attached to the double summation indicating that the term involving u_{00} is to be omitted from the sum). In practice n will be restricted to 1 or 2.

Notation used:

$$x_r \ = x_0 + rh$$
$$y_r \ = y_0 + sh$$
$$(r,s) \ = (x_r, y_s)$$
$$f_{00} \ = f(x_0, y_0)$$
$$u_{rs} \ = u(x_r, y_s)$$
$$A_{rs} \ = \text{Integer weight associated with } u_{rs}$$
$$B \ \ = \text{Integer constant.}$$

The principal part of the error term will be of the form

$$h^4(\theta \nabla^4 + \phi \mathfrak{D}^4)u_{00}$$

where
$$\mathfrak{D}^4 \equiv \frac{\partial^4}{\partial x^2 \partial y^2} \text{ and } \theta, \phi \text{ are constants.}$$

A_{rs}, B, θ and ϕ are tabulated below.

Restricting the values of the weights to the doubly symmetric case, i.e. $A_{rs} = A_{sr} = A_{|r||s|}$, and assuming the validity of the Taylor's expansion in the neighborhood of u_{00}, integer values can be obtained for A_{rs} and B, and rational fractions for θ and ϕ. However, the assumptions of the existence of the Taylor's expansion implies that $u(x,y)$ can be adequately represented by a polynomial in x and y over the region in question. Thus if a square region, centered on an arbitrary origin, is considered, the terms involving odd powers of either variable will contribute nothing to the sum providing the coordinates involved are specially selected.

The significant part of the expansion is therefore

$$u_{rs} = \left\{ 1 + \frac{h^2}{2!}\left(r^2 \frac{\partial^2}{\partial x^2} + s^2 \frac{\partial^2}{\partial y^2} \right) + \frac{h^4}{4!}\left(r^4 \frac{\partial^4}{\partial x^4} + 6r^2 s^2 \frac{\partial^4}{\partial x^2 \partial y^2} + s^4 \frac{\partial^4}{\partial y^4} \right) \cdots \right\} u_{00}$$

The coordinate groupings that are used are as follows:

(i) $(0,1)$, $(1,0)$, $(0,-1)$ and $(-1,0)$
(ii) $(1,1)$, $(1,-1)$, $(-1,1)$ and $(-1,-1)$
(iii) $(0,2)$, $(2,0)$, $(0,-2)$ and $(-2,0)$
(iv) $(2,2)$, $(2,-2)$, $(-2,2)$ and $(-2,-2)$
(v) $(2,1)$, $(1,2)$, $(-1,2)$, $(-2,1)$, $(-2,-1)$, $(-1,-2)$, $(1,-2)$ and $(2,-1)$.

ELLIPTIC DIFFERENTIAL EQUATIONS (Continued)

Group 1.   The general form is

A_{rs}

r \ s	-1	0	1
1	1	α	1
0	α	$4(1+\alpha)$	α
-1	1	α	1

$B = (\alpha + 2)$

$\theta = -(\alpha + 2)/12$

$\phi = (\alpha - 4)/6$

in particular

A_{rs}

r \ s	-1	0	1
1	1	0	1
0	0	4	0
-1	1	0	1

$B = 2$

$\theta = -1/6$

$\phi = -2/3$

A_{rs}

r \ s	-1	0	1
1	1	2	1
0	2	12	2
-1	1	2	1

$B = 4$

$\theta = -1/3$

$\phi = -1/3$

A_{rs}

r \ s	-1	0	1
1	1	3	1
0	3	16	3
-1	1	3	1

$B = 5$

$\theta = -5/12$

$\phi = -1/6$

A_{rs}

r \ s	-1	0	1
1	1	4	1
0	4	20	4
-1	1	4	1

$B = 6$

$\theta = -1/2$

$\phi = 0$

A_{rs}

r \ s	-1	0	1
1	1	6	1
0	6	28	6
-1	1	6	1

$B = 8$

$\theta = -2/3$

$\phi = 1/3$

ELLIPTIC DIFFERENTIAL EQUATIONS (Continued)

Group 2.   The general form is

A_{rs}			
s \\ r	-1	0	1
1	α	1	α
0	1	$4(1+\alpha)$	1
-1	α	1	α

$B = (1 + 2\alpha)$

$\theta = -(1 + 2\alpha)/12$

$\phi = (1 - 4\alpha)/6$

In particular

A_{rs}			
s \\ r	-1	0	1
1	0	1	0
0	1	4	1
-1	0	1	0

$B = 1$

$\theta = -1/2$

$\phi = 1/6$

A_{rs}			
s \\ r	-1	0	1
1	1	1	1
0	1	8	1
-1	1	1	1

$B = 3$

$\theta = -1/4$

$\phi = -1/2$

A_{rs}			
s \\ r	-1	0	1
1	2	1	2
0	1	12	1
-1	2	1	2

$B = 5$

$\theta = -5/12$

$\phi = -7/6$

A_{rs}			
s \\ r	-1	0	1
1	4	1	4
0	1	20	1
-1	4	1	4

$B = 9$

$\theta = -3/4$

$\phi = -5/3$

A_{rs}			
s \\ r	-1	0	1
1	8	1	8
0	1	36	1
-1	8	1	8

$B = 17$

$\theta = -17/12$

$\phi = -31/6$

ELLIPTIC DIFFERENTIAL EQUATIONS (Continued)

Group 3.   The general form is

A_{rs}

r \ s	1	0	−1
1	β	α	β
0	α	$4(\alpha + \beta)$	α
−1	β	α	β

$B = \alpha + 2\beta$

$\theta = -(\alpha + 2\beta)/12$

$\phi = (\alpha - 4\beta)/6$

In particular

A_{rs}

r \ s	−1	0	1
1	2	3	2
0	3	20	3
−1	2	3	2

$B = 7$

$\theta = -7/2$

$\phi = -5/6$

A_{rs}

r \ s	−1	0	1
1	2	5	2
0	5	28	5
−1	2	5	2

$B = 9$

$\theta = -3/4$

$\phi = -1/2$

A_{rs}

r \ s	−1	0	1
1	3	4	3
0	4	28	4
−1	3	4	3

$B = 10$

$\theta = -5/6$

$\phi = -4/3$

A_{rs}

r \ s	−1	0	1
1	3	5	3
0	5	32	5
−1	3	5	3

$B = 11$

$\theta = -11/12$

$\phi = -7/3$

A_{rs}

r \ s	−1	0	1
1	4	5	4
0	5	36	5
−1	4	5	4

$B = 13$

$\theta = -13/12$

$\phi = -11/6$

ELLIPTIC DIFFERENTIAL EQUATIONS (Continued)

While negative weights can be used in all groups care must be taken that the subtraction of comparable terms does not reduce the number of significant decimals left in the result.

The above formulae are of the nine-point type. To increase the accuracy, twenty-five-point formulae can be used. For these the error term is taken to be

$$\{h^4(\theta\nabla^4 + \phi\mathfrak{D}^4) + h^6(\eta\nabla^6 + \gamma\mathfrak{D}^4\nabla^2)\}u_{00}$$

The twenty-five-point formulae which form the basic set are

$$A_{rs}$$

r \ s	−2	−1	0	1	2	
2	0	0	0	0	0	$B = 1$
1	0	0	1	0	0	$\theta = -1/12$
0	0	1	4	1	0	$\phi = 1/6$
−1	0	0	1	0	0	$\eta = -1/360$
−2	0	0	0	0	0	$\gamma = 1/120$

$$A_{rs}$$

r \ s	−2	−1	0	1	2	
2	0	0	0	0	0	$B = 2$
1	0	1	0	1	0	$\theta = -1/6$
0	0	0	4	0	0	$\phi = -2/3$
−1	0	1	0	1	0	$\eta = -1/180$
−2	0	0	0	0	0	$\gamma = -1/15$

$$A_{rs}$$

r \ s	−2	−1	0	1	2	
2	0	1	0	1	0	$B = 10$
1	1	0	0	0	1	$\theta = -17/6$
0	0	0	8	0	0	$\phi = -7/3$
−1	1	0	0	0	1	$\eta = -13/36$
−2	0	1	0	1	0	$\gamma = -7/12$

ELLIPTIC DIFFERENTIAL EQUATIONS (Continued)

A_{rs}

r \ s	−2	−1	0	1	2	
2	0	0	1	0	0	$B = 4$
1	0	0	0	0	0	$\theta = -4/3$
0	1	0	4	0	1	$\phi = 8/3$
−1	0	0	0	0	0	$\eta = -8/45$
−2	0	0	1	0	0	$\gamma = 8/15$

A_{rs}

r \ s	−2	−1	0	1	2	
2	1	0	0	0	1	$B = 8$
1	0	0	0	0	0	$\theta = -8/3$
0	0	0	4	0	0	$\phi = -32/3$
−1	0	0	0	0	0	$\eta = -16/45$
−2	1	0	0	0	1	$\gamma = -64/15$

Any linear combination of the above five sets can be used.

Group 4   The general form is

A_{rs}

r \ s	−2	−1	0	1	2	
2	μ	κ	ρ	κ	μ	$B = \alpha + 2\beta + 4\rho + 8\mu + 10\kappa$
1	κ	β	α	β	κ	$\theta = -(\alpha + 2\beta + 16\rho + 32\mu + 34\kappa)/12$
0	ρ	α	$4(\alpha + \beta + \rho + \mu + 2\kappa)$	α	ρ	$\phi = (\alpha - 4\beta + 16\rho - 64\mu - 14\kappa)/6$
−1	κ	β	α	β	κ	$\eta = -(\alpha + 2\beta + 64\rho + 128\mu + 130\kappa)/360$
−2	μ	κ	ρ	κ	μ	$\gamma = (\alpha - 8\beta + 64\rho - 512\mu - 70\kappa)/120$

In particular θ and ϕ can be reduced to zero simultaneously if either *a*) $\rho = -1$, $\alpha = 16$, $\mu = 0 = \beta = \kappa$.

A_{rs}

r \ s	−2	−1	0	1	2	
2	0	0	−1	0	0	$B = 12$
1	0	0	16	0	0	$\theta = 0$
0	−1	16	60	16	−1	$\phi = 0$
−1	0	0	16	0	0	$\eta = 2/15$
−2	0	0	−1	0	0	$\gamma = -2/5$

ELLIPTIC DIFFERENTIAL EQUATIONS (Continued)

or   b) $\mu = -1$, $\beta = 16$, $\alpha = 0 = \rho = \kappa$.

A_{rs}

r \ s	-2	-1	0	1	2
2	-1	0	0	0	-1
1	0	16	0	16	0
0	0	0	60	0	0
-1	0	16	0	16	0
-2	-1	0	0	0	-1

$B = 24$

$\theta = 0$

$\phi = 0$

$\eta = 4/15$

$\gamma = 16/5$

or   c) $\rho = -1 = \mu$, $\alpha = 16 = \beta$.

A_{rs}

r \ s	-2	-1	0	1	2
2	-1	0	-1	0	-1
1	0	16	16	16	0
0	-1	16	120	16	-1
-1	0	16	16	16	0
-2	-1	0	-1	0	-1

$B = 36$

$\theta = 0$

$\phi = 0$

$\eta = 3/5$

$\gamma = 14/5$

When working with digital computers it is frequently advantageous to have, as weights, powers of two. The following sets of weights yield this, and in addition the overall divisor A_{00} is also a power of two.

A_{rs}

r \ s	-2	-1	0	1	2
2	0	1	2	1	0
1	1	0	0	0	1
0	2	0	16	0	2
-1	1	0	0	0	1
-2	1	1	2	1	0

$B = 14$

$\theta = -11/2$

$\phi = 3$

$\eta = -43/60$

$\gamma = 29/60$

A_{rs}

r \ s	-2	-1	0	1	2
2	0	1	2	1	0
1	1	4	0	4	1
0	2	0	32	0	2
-1	1	4	0	4	1
-2	0	1	2	1	0

$B = 22$

$\theta = -37/6$

$\phi = 1/3$

$\eta = -133/180$

$\gamma = 13/60$

ELLIPTIC DIFFERENTIAL EQUATIONS (Continued)

A_{rs}

r \ s	−2	−1	0	1	2
2	0	1	2	1	0
1	1	0	4	0	1
0	2	4	32	4	2
−1	1	0	4	0	1
−2	0	1	2	1	0

$B = 18$

$\theta = -35/6$

$\phi = 11/3$

$\eta = -131/180$

$\gamma = 31/60$

A_{rs}

r \ s	−2	−1	0	1	2
2	0	1	2	1	0
1	1	4	8	4	1
0	2	8	64	8	2
−1	1	4	8	4	1
−2	0	1	2	1	0

$B = 34$

$\theta = -41/3$

$\phi = 5/3$

$\eta = -137/180$

$\gamma = 17/60$

A_{rs}

r \ s	−2	−1	0	1	2
2	0	0	1	0	0
1	0	1	2	1	0
0	1	2	16	2	1
−1	0	1	2	1	0
−2	1	0	1	0	0

$B = 8$

$\theta = -5/3$

$\phi = 7/3$

$\eta = -17/90$

$\gamma = 29/60$

A_{rs}

r \ s	−2	−1	0	1	2
2	0	1	0	1	0
1	1	1	1	1	1
0	0	1	16	1	0
−1	1	1	1	1	1
−2	0	1	0	1	0

$B = 13$

$\theta = -37/12$

$\phi = -17/6$

$\eta = -133/360$

$\gamma = -77/120$

ELLIPTIC DIFFERENTIAL EQUATIONS (Continued)

For desk machine working the value of A_{00} should be chosen so as to simplify the process of division. Some typical sets of weights in which A_{00} is 100 are

A_{rs}

r \ s	−2	−1	0	1	2	
2	1	7	10	7	1	$B = 118$
1	7	0	0	0	7	$\theta = -215/6$
0	10	0	100	0	10	$\phi = -1/6$
−1	7	0	0	0	7	$\eta = -839/180$
−2	1	7	10	7	1	$\gamma = -181/60$

A_{rs}

r \ s	−2	−1	0	1	2	
2	0	1	5	1	0	$B = 56$
1	1	8	10	8	1	$\theta = -35/3$
0	5	10	100	10	5	$\phi = 22/3$
−1	1	8	10	8	1	$\eta = -119/90$
−2	0	1	5	1	0	$\gamma = 49/30$

A_{rs}

r \ s	−2	−1	0	1	2	
2	0	5	0	5	0	$B = 70$
1	5	5	10	5	5	$\theta = -95/6$
0	0	10	100	10	0	$\phi = -40/3$
−1	5	5	10	5	5	$\eta = -67/36$
−2	0	5	0	5	0	$\gamma = -19/6$

Starting the solution

Any rough estimate can be used to start the solution. In the absence of such an estimate all the interior points can be set at zero. In the case of Laplace's equation, straight lines can be drawn through any point to cut the boundary. The estimated value of the function at that point is then obtained by simple ratio and proportion.

Continuing the solution.

A systematic sweep through the region is essential, although the actual direction of travel chosen is immaterial for all practical purposes. It is important however that once the value at a specific point has been recomputed the new value should immediately replace the old.

Refining the mesh.

Once all the values at the mesh points have converged to the degree of accuracy permitted by the interval "h," the mesh can be subdivided in the following fashion and convergence to more accurate values obtained.

ELLIPTIC DIFFERENTIAL EQUATIONS (Continued)

Fig. 1 shows the mesh of side h, near a boundary where it is required to produce a mesh of side $\frac{1}{2}h$

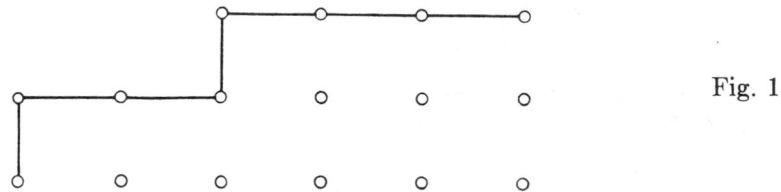

Fig. 1

The points marked with a $+$ in Fig. 2 can be inserted by means of the first formula of Group 1.

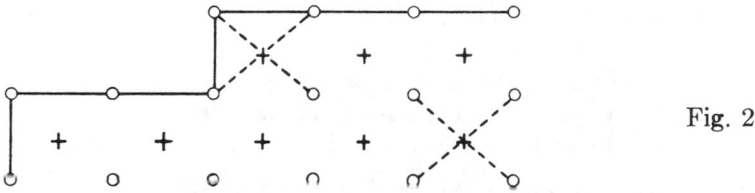

Fig. 2

The points marked with a $\bullet$ in Fig. 3 can then be inserted by means of the first formula of Group 2.

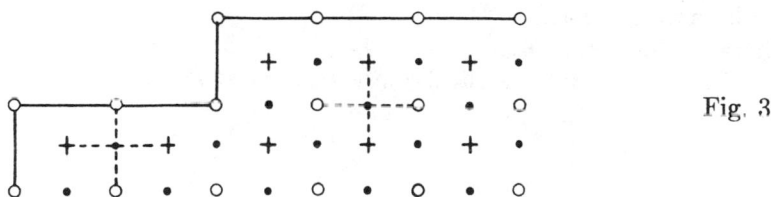

Fig. 3

Finally the boundary conditions enable the values at the new mesh points on the boundary to be inserted. These are indicated by $\bullet$ in Fig. 4.

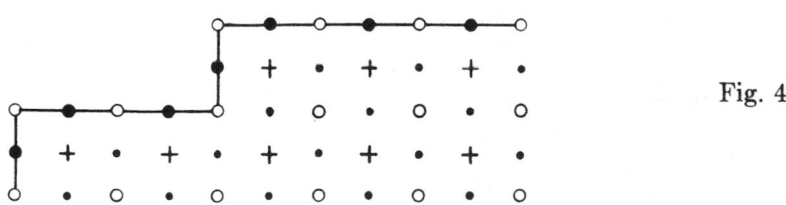

Fig. 4

Improving the numerical solution

(i) Aitken's δ^2 process.

If $u^{(1)}$, $u^{(2)}$ and $u^{(3)}$ are three successive approximations for a given h to u, then a better approximation is given by

$$u = u^{(3)} - \frac{[u^{(3} - u^{(2)}]^2}{u^{(3)} - 2u^{(2)} + u^{(1)}}$$

(ii) Richardson's deferred approach to the limit.

ELLIPTIC DIFFERENTIAL EQUATIONS (Continued)

If $u_{(1)}$ is the value obtained for a mesh size h and $u_{(2)}$ is the value when the mesh size is $\frac{1}{2}h$ then a better value for u is given by

$$u = u_{(2)} + \frac{1}{3}[u_{(2)} - u_{(1)}]$$

If, in addition, $u_{(3)}$ is the value obtained for a mesh size $\frac{1}{4}h$ then the improved value is given by

$$u = u_{(3)} + \frac{4}{9}[u_{(3)} - u_{(2)}] - \frac{1}{45}[u_{(3)} - u_{(1)}]$$

It is not necessary to refine the mesh over the entire region. In most cases only specially selected areas will need refining.

(iii) Difference correction.

The technique here is the same as is described in the section on ordinary differential equations. The derivatives in the error term are replaced by their finite difference equivalents, in both the x and y directions, and these constitute the difference correction.

Non-rectangular boundaries.

In some cases a transformation of coordinates will produce a boundary which will fit a rectangular mesh at the expense of increased complexity of the partial differential equation.

In others a mesh which is not rectangular can be employed, the usual alternatives being triangular and hexagonal.

Otherwise the rectangular mesh can be preserved for the mean body of the calculation and the iterative formula modified on the boundary. Thus two basic forms will be needed at a curved boundary.

Case 1.

$$A_0 u_0 = -Bf_0 + \sum_{r=1}^{4} A_r u_r + 0(h^3)$$

$$h \geq H_i \quad (i = 1,2,3,4)$$
$$h_i = H_i/h$$

$$A_0 = (h_1 + h_3)(h_2 + h_4)(h_1 h_3 + h_2 h_4)$$
$$B = \tfrac{1}{2}h_1 h_2 h_3 h_4(h_1 + h_3)(h_2 + h_4)$$
$$A_1 = h_2 h_3 h_4(h_2 + h_4)$$
$$A_2 = h_3 h_4 h_1(h_3 + h_1)$$
$$A_3 = h_4 h_1 h_2(h_4 + h_2)$$
$$A_4 = h_1 h_2 h_3(h_1 + h_3)$$

ELLIPTIC DIFFERENTIAL EQUATIONS (Continued)

Case 2

The formulae are the same as in Case 1; it is only the orientation of the mesh which is changed

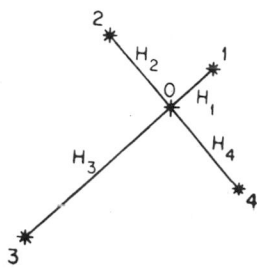

Triangular mesh formulae for Poisson's equation
The points are numbered as shown

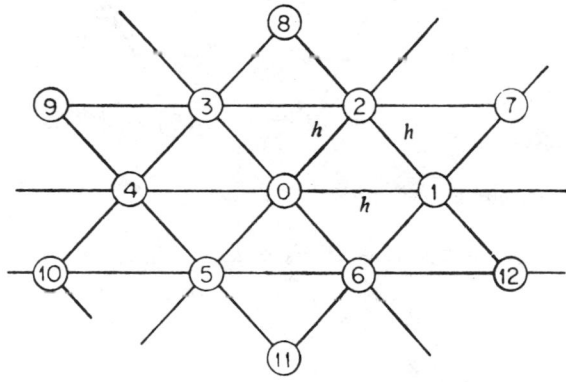

The iteration formula for the above mesh is

$$A_0 = \Sigma A_r u_r - B h^2 f_0 + \epsilon$$

a) Summation extends over $r = 1(1)6$

$$A_0 = 6, \qquad A_r = 1, \qquad B = 3/2$$

$$\epsilon = -\frac{32}{32} h^4 \nabla_0^4$$

b) Summation extends over $r = 7(1)12$

$$A_0 = 6, \qquad A_r = 1, \qquad B = 9/2$$

$$\epsilon = -\frac{27}{32} h^4 \nabla_0^4$$

c) Summation extends over $r = 1(1)12$

$$A_0 = 48, \qquad A_r = 9 \qquad (r = 1(1)6)$$
$$B = 8 \qquad A_r = -1 \qquad (r = 7(1)12)$$
$$\epsilon = -\frac{h^6}{3840}\left(11\frac{\partial^6}{\partial x^6} + 15\frac{\partial^6}{\partial x^4 \partial y^2} + 45\frac{\partial^6}{\partial x^2 \partial y^4} + 9\frac{\partial^6}{\partial y^6}\right)u_0$$

ELLIPTIC DIFFERENTIAL EQUATIONS (Continued)

Hexagonal mesh formulae for Poisson's equation

The points are numbered as shown

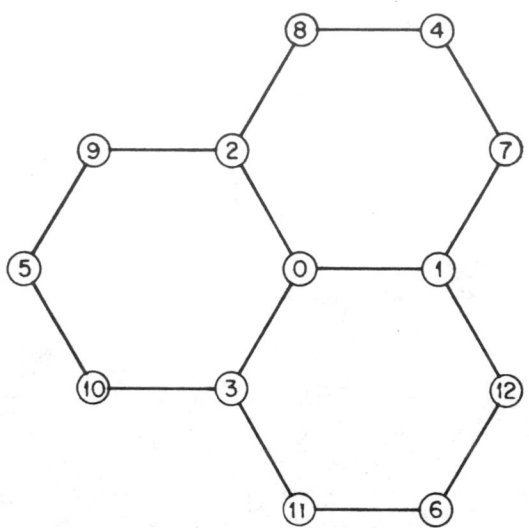

The iteration formula for the above mesh being

$$A_0 = \Sigma A_r u_r - Bh^2 f_0 + \epsilon$$

a) Summation extends over $r = 1(1)3$

$$A_0 = 3, \qquad A_r = 1, \qquad B = 3/4$$

$$\epsilon = -\frac{h^3}{8}\left(\frac{\partial^3}{\partial x^3} - 3\frac{\partial^3}{\partial x \partial y^2}\right)u_0$$

b) Summation extends over $r = 4(1)6$

$$A_0 = 3, \qquad A_r = 1, \qquad B = 3$$

$$\epsilon = -h^3\left(\frac{\partial^3}{\partial x^3} - 3\frac{\partial^3}{\partial x \partial y^2}\right)u_0$$

c) Summation extends over $r = 1(1)6$

$$A_0 = 21, \qquad A_r = 8 \qquad (r = 1(1)3)$$

$$B = 3, \qquad A_r = -1 \qquad (r = 4(1)6)$$

$$\epsilon = \frac{h^4}{8}\left(3\frac{\partial^4}{\partial x^4} + \frac{\partial^4}{\partial x^2 \partial y^2} + 3\frac{\partial^4}{\partial y^4}\right)u_0$$

PARABOLIC DIFFERENTIAL EQUATIONS

Parabolic and Hyperbolic Partial Differential Equations

While this section will be restricted to partial differential equations with two independent variables, the techniques involved can easily be extended to those with three or more.

For simplicity, a rectangular mesh will be superimposed upon the region of integration. The pivotal points will have co-ordinates (x_s, t_r) where, for the intervals "h" and "k,"

$$x_s = x_0 + sh$$

$$t_r = t_0 + rk$$

$$f_{s,r} = f(x_s, t_r).$$

If the derivatives, with respect to either x or t, are replaced by their finite difference approximations, the result will be a set of simultaneous ordinary differential equations.

If, however, all the derivatives are replaced by their relevant approximations, a difference equation corresponding to the original differential equations is obtained. It should be noted, of course, that the approximation chosen will not be unique. Some typical examples are as follows,

$$\left(\frac{\partial f}{\partial t}\right)_{s,r} = (f_{s,r+1} - f_{s,r})/k$$

or $$= (f_{s,r} - f_{s,r-1})/k$$

or $$= (f_{s,r+1} - f_{s,r-1})/2k$$

$$\left(\frac{\partial^2 f}{\partial x^2}\right)_{s,r} = (f_{s+1,r} - 2f_{s,r} + f_{s-1,r})/h^2$$

or $$= (2f_{s,r} - 5f_{s-1,r} + 4f_{s-2,r} - f_{s-3,r})/h^2$$

or $$= (11f_{s+1,r} - 20f_{s,r} + 6f_{s-1,r} + 4f_{s-2,r} - f_{s-3,r})/12h^2$$

and so forth.

Parabolic Partial Differential Equations

The examples in this section will be concerned in the main, with the simple but nevertheless typical one-dimensional heat equation given by:

$$\frac{\partial f}{\partial t} = \frac{\partial^2 f}{\partial x^2}$$

with the initial conditions $\qquad f = I(x) \qquad\qquad [x_0 \leqslant x \leqslant x_n,\, t = 0]$

and the boundary conditions $\qquad f = B(t) \qquad\qquad [t > 0, \quad x = x_0]$

$$f = B^*(t) \qquad\qquad [t > 0, \quad x = x_n]$$

PARABOLIC DIFFERENTIAL EQUATIONS (Continued)

Without loss of generality, x_0 can be replaced by 0 so that

$$x_s = sh, \quad t_r = rk,$$

and
$$I(x_s) = f_{s,0}, \quad B(t_r) = f_{0,r}, \quad B^*(t_r) = f_{n,r}.$$

Two decisions have now to be made.

(*a*) Is the relationship between the values of "*f*" at the pivotal points to be an implicit one or an explicit one? That is to say, if the solution is being determined in the direction of t increasing, is the value at $f_{s,r}$ going to involve other values for which $t = rk$, or is it only going to involve those points for which $t < rk$?

(*b*) How many points are going to be involved at any one time?

A small number of points will mean less computational labor, but generally at the expense of the truncation error.

The explicit formulae are more direct than the implicit ones, but may require the mesh size to be unacceptably small in order to ensure the stability of the solution. In some cases the explicit form is unstable for all meshes. On the other hand, the implicit formulae are stable but they require the solution of large numbers of simultaneous equations.

Some typical explicit formulae

i.
$$f_{s,r+1} = f_{s,r} + (k/h^2)(f_{s-1,r} - 2f_{s,r} + f_{s+1,r})$$

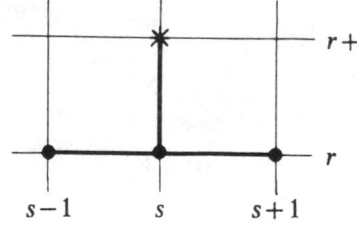

Stable if $0 < k/h^2 \leqslant 1/2$

Truncation error $= 0(k) + 0(h^2)$

In particular when $k/h^2 = 1/2$ the point pattern is simplified by the omission of the pivot $f_{s,r}$.

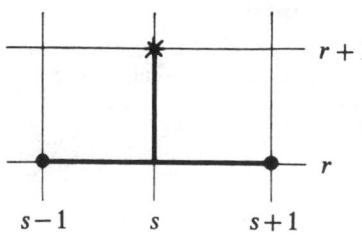

$$f_{s,r+1} = \tfrac{1}{2}(f_{s-1,r} + f_{s+1,r})$$

Milne has shown that an improved truncation error results when $k/h^2 = 1/6$

$$f_{s,r+1} = (1/6)(f_{s-1,r} + 4f_{s,r} + f_{s+1,r})$$

Truncation error $= 0(k^2) + 0(h^4)$

PARABOLIC DIFFERENTIAL EQUATIONS (Continued)

ii.
$$f_{s,r+1} = \frac{(h^2 - 2k)f_{s,r-1} + 2k(f_{s-1,r} + f_{s+1,r})}{(h^2 + 2k)}$$

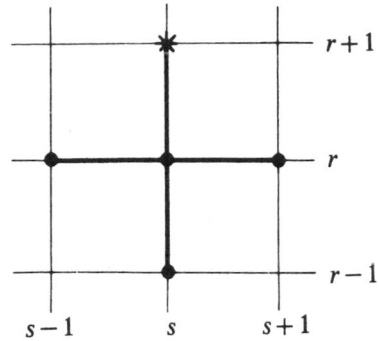

Always stable, but $f_{s,r-1}$ is required.

Truncation error $= 0(k) + 0(h^2) + 0(k/h)$

iii.
$$f_{s,r+1} = f_{s,r} - (1/12)(k/h^2)(f_{s-2,r} - 16f_{s-1,r} + 30f_{s,r} - 16f_{s+1,r} + f_{s+2,r})$$

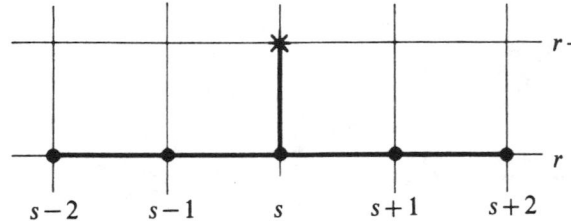

Stable if $0 < (k/h^2) \leq 3/8$

Truncation error $= 0(k) + 0(h^4)$

If the above mesh is used for the main body of the computation, then unsymmetric formulae are required at the ends of the range of s. The formula for the left hand side is shown below. (The formula for the right hand side is the mirror image.)

$$f_{s,r+1} = f_{s,r} + (1/12)(k/h^2)(11f_{s-1,r} - 20f_{s,r} + 6f_{s+1,r} + 4f_{s+2,r} - f_{s+3,r})$$

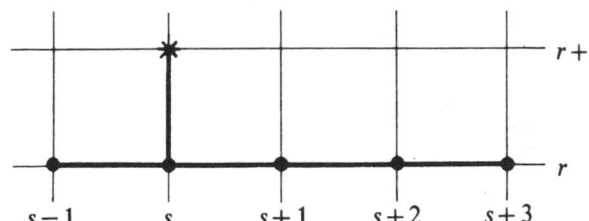

Stable if $0 < (k/h^2) \leq 2/7$

Truncation error $= 0(h^3) + 0(k)$

The truncation error of the unsymmetrical case will of course determine the overall truncation error if all three formulae are to be used in the one problem. The same observation will hold for the stability requirements.

One way of reducing the truncation error is to use additional, previously computed values; i.e., instead of using only those for which $t = rk$, increase the number of pivotal values involved by incorporating those for which $t = (r - 1)k, (r - 2)k$ etc. This method was used in example ii. However, it is generally unwise as such formulae tend toward high instability.

Some typical implicit formulae

These have the merit of stability but require the solution of a number of simultaneous equations. However, when matrix methods are used, often only one inversion is required. This simplifies the problem considerably.

PARABOLIC DIFFERENTIAL EQUATIONS (Continued)

i. $(k/h^2)(f_{s+1,r+1} + f_{s-1,r+1} + f_{s+1,r} + f_{s-1,r})$

$$- 2[1 + (k/h^2)]f_{s,r+1} + 2[1 - (k/h^2)]f_{s,r} = 0$$

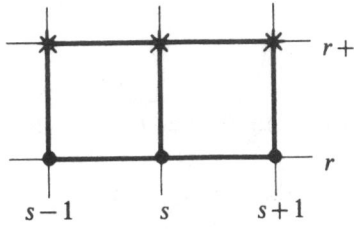

Always stable

Truncation error $= 0(k^2) + 0(h^2)$

ii. $$f_{s,r+1} - f_{s,r} = (k/h^2)(f_{s-1,r+1} - 2f_{s,r+1} + f_{s+1,r+1})$$

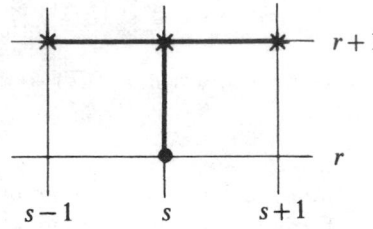

Always stable

Truncation error $= 0(k) + 0(h^2))$

iii. $12(f_{s,r+1} - f_{s,r}) + (k/h^2)(f_{s-2,r+1} - 16f_{s-1,r+1} + 30f_{s,r+1}$

$$- 16f_{s+1,r+1} + f_{s+2,r+1}) = 0$$

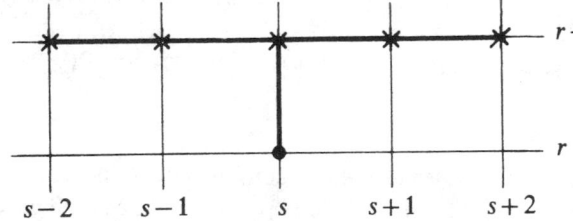

Always stable

Truncation error $= 0(k) + 0(h^4)$

As in the explicit case, unsymmetrical forms of the above formula are required at the extremities of the range. Only one will be quoted here (the other is merely a mirror image):

$$12(f_{s,r+1} - f_{s,r}) = (k/h^2)(11f_{s-1,r+1} - 20f_{s,r+1} + 6f_{s+1,r+1} + 4f_{s+2,r+1} - f_{s+3,r+1})$$

Always stable

Truncation error $= 0(k) + 0(h^3)$

HYPERBOLIC DIFFERENTIAL EQUATIONS (Continued)

iv. $[18(k/h^2) - 1](f_{s,r+1} - f_{s,r}) - [6(k/h^2) - 1](f_{s,r} - f_{s,r-1})$

$$= 12(k^2/h^4)(f_{s+1,r+1} - 2f_{s,r+1} + f_{s-1,r+1})$$

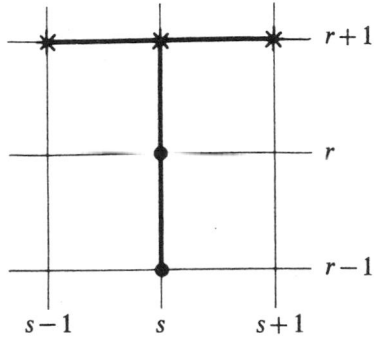

Always stable

Truncation error $= 0(k^2) + 0(h^4)$

v. An implicit method in its own right which can be used as a check on (iv) above is given by

$(3f_{s+1,r+1} - 4f_{s+1,r} + f_{s+1,r-1}) + (30f_{s,r+1} - 40f_{s,r} + 10f_{s,r-1})$

$$+ (3f_{s-1,r+1} - 4f_{s,r+1} + f_{s-1,r-1}) = 24(k/h^2)(f_{s+1,r+1} - 2f_{s,r+1} + f_{s-1,r+1})$$

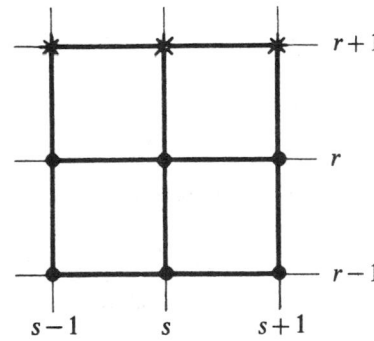

Always stable

Truncation error $= 0(k^2) + 0(h^4)$

Hyperbolic Partial Differential Equations

The methods used in the parabolic case carry over into the hyperbolic case; therefore, only one example of each will be quoted.

Consider

$$(\partial^2 f/\partial x^2) = (\partial^2 f/\partial t^2)$$

with appropriate boundary conditions.

An explicit formula:

$$f_{s,r+1} = 2f_{s,r} - f_{s,r-1} + (k^2/h)(f_{s-1,r} - 2f_{s,r} + f_{s+1,r})$$

PARABOLIC AND HYPERBOLIC DIFFERENTIAL EQUATIONS (Continued)

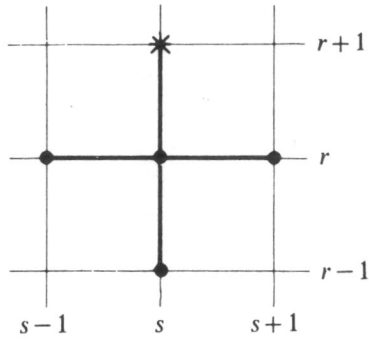

Stable if $0 < (k/h)^2 \leq 1$

Truncation error $= 0(h^4) + 0(k^4)$

An implicit formula:

$$(f_{s,r+1} - 2f_{s,r} + f_{s,r-1}) = (k/h)^2(f_{s-1,r+1} - 2f_{s,r+1} + f_{s+1,r+1})$$

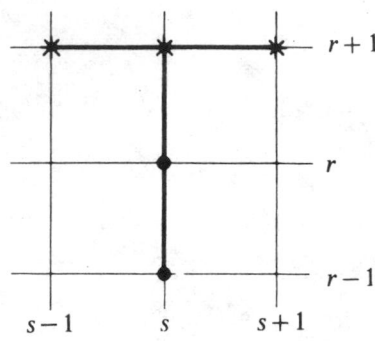

Always stable

Truncation error $= 0(h^4) + 0(k^4)$

Mixed Boundary Conditions for Parabolic and Hyperbolic Differential Equations

When the boundary conditions involve a derivative some slight modifications are necessary to take this condition into account.

Example 1 $\dfrac{\partial f}{\partial t} = \dfrac{\partial^2 f}{\partial x^2};$ $f = I(x),$ $[t = 0,\ \ 0 \leq x \leq nh]$

$f = B(t),$ $[t > 0,\ \ x = 0]$

$f + A(\partial f/\partial x) = B^*(t),$ $[t > 0,\ \ x = nh]$

The derivative can be replaced by a finite difference approximation as follows

$$f_{n,r} + \frac{A(f_{n,r} - f_{n-1,r})}{k} = B^*(t_r)$$

which can be rewritten as $f_{n,r} = \dfrac{(k/A)B^*(t_r) + f_{n-1,r}}{(1 + k/A)}$

This enables the end points to be computed. A more accurate approximation to the derivative can be used of course, e.g.,

$$f_{n,r} + \frac{A(6f_{n,r} - 10f_{n-1,r} + 5f_{n-2,r} - f_{n-3,r})}{6k} = B^*(t_r)$$

PARABOLIC AND HYPERBOLIC DIFFERENTIAL EQUATIONS (Continued)

Example 2. $\qquad \dfrac{\partial f}{\partial t} = \dfrac{\partial^2 f}{\partial x^2};$ $\qquad f + A\,\dfrac{\partial f}{\partial t} = I(x),$ $\qquad\qquad\qquad$ $[t = 0, \, 0 \le x \le nh]$

$$f = B(t), \qquad\qquad\qquad\qquad [t > 0, \quad x = 0]$$

$$f = B^*(t), \qquad\qquad\qquad\qquad [t > 0, \quad x = nh]$$

In this example the unconditionally stable form (ii) of the explicit set can be used. This is rewritten to give

$$f_{s,-1} = \frac{(h^2 + 2k)}{(h^2 - 2k)} \left(f_{s,1} - \frac{2k}{(h^2 + 2k)}\,(f_{s-1,0} + f_{s+1,0}) \right)$$

The initial conditions yield

$$f_{s,0} + (A/2k)(f_{s,1} - f_{s,-1}) = I(x_s)$$

or

$$f_{s,-1} = (2k/A)[f_{s,0} - I(x_s)] + f_{s,1}$$

Eliminating $f_{s,-1}$ from these two equations will yield an expression for $f_{s,1}[1 \le s \le (n-1)]$ and thereafter explicit formula (ii) can be used without modification.

MONTE CARLO TECHNIQUES FOR PARTIAL DIFFERENTIAL EQUATIONS

(*i*) *Elliptic Equations.*

The two dimensional case will be discussed here—the extension to more than two dimensions follows by increasing the number of the directions of the random walks by two for each added dimension.

The finite difference approximations

$$\frac{\partial^2 u}{\partial x^2} = \frac{u_{i+1,j} - 2u_{i,j} + u_{i-1,j}}{h_1^2}$$

and

$$\frac{\partial^2 u}{\partial y^2} = \frac{u_{i,j+1} - 2u_{i,j} + u_{i,j-1}}{h_2^2}$$

when substituted into $\nabla^2 u = 0$, yield

$$u_{i,j} = \frac{h_2^2 u_{i+1,j}}{2(h_1^2 + h_2^2)} + \frac{h_2^2 u_{i-1,j}}{2(h_1^2 + h_2^2)} + \frac{h_1^2 u_{i,j+1}}{2(h_1^2 + h_2^2)} + \frac{h_1^2 u_{i,j-1}}{2(h_1^2 + h_2^2)}$$

and, defining

$$p_1 = p_2 = \frac{h_2^2}{2(h_1^2 + h_2^2)}$$

$$p_3 = p_4 = \frac{h_1^2}{2(h_1^2 + h_2^2)}$$

It can be seen that $\sum_{i=1}^{4} p_i = 1$, which is the condition for carrying out a random walk along the four directions x increasing, x decreasing, y increasing and y decreasing. Thus to determine $u_{i,j}$ the value of u at the point with coordinates (x_i, y_j) the procedure is as follows; Set both T and N to zero initially.

For each walk completed add 1 to N.

A walk is completed when the boundary is reached and the value of u *at that point on the boundary* has been added to T. As the number of completed walks increases, $N \rightarrow \infty$ and $\frac{T}{N} \rightarrow u_{i,j}$.

For the equation $\nabla^2 u = f$ use the relationship

$$\frac{h_1^2 h_2^2}{2(h_1^2 + h_2^2)} f_{i,j} + u_{i,j} = p_1 u_{i+1,j} + p_2 u_{i-1,j} + p_3 u_{i,j+1} + p_4 u_{i,j-1}$$

As before the expression on the left hand side is the limiting value of the ratio $\frac{T}{N}$ as $N \rightarrow \infty$.

(*ii*) *Parabolic equations.*

Formula (ii) on page 717 yields, as a finite difference representation of

$$\frac{\partial f_{s,r}}{\partial t} = \frac{\partial^2 f_{s,r}}{\partial x^2}$$

$$f_{s,r+1} = \left(\frac{h^2 - 2k}{h^2 + 2k}\right) f_{s,r-1} + \left(\frac{2k}{h^2 + 2k}\right) f_{s-1,r} + \left(\frac{2k}{h^2 + 2k}\right) f_{s+1,r}$$

This can be written as

$$f_{s,r+1} = p_1 f_{s-1,r} + p_2 f_{s+1,r} + p_3 f_{s,r-1}$$

where as before $\sum_{r=1}^{3} p_r = 1$

MONTE CARLO TECHNIQUES (Continued)

for
$$p_1 = p_2 = \left(\frac{2k}{h^2 + 2k}\right); \qquad p_3 = \left(\frac{h^2 - 2k}{h^2 + 2k}\right).$$

The three directions are x increasing, x decreasing and t decreasing. In other words progress is restricted to be either towards the boundaries (along the x directions) or back towards the initial conditions along the direction of t decreasing. As before, N will be the total number of completed walks and T will be the sum of the boundary values or initial values reached.

XII. PROBABILITY AND STATISTICS

DESCRIPTIVE STATISTICS

a) Ungrouped Data

The formulas of this section designated as *a*) apply to a random sample of size n, denoted by x_i, $i = 1, 2, \ldots, n$.

b) Grouped Data

The formulas of this section designated as *b*) apply to data grouped into a frequency distribution having class marks x_i, $i = 1, 2, \ldots, k$, and corresponding class frequencies f_i, $i = 1, 2, \ldots, k$. The total number of observations given by

$$n = \sum_{i=1}^{k} f_i$$

In the formulas that follow, c denotes the width of the class interval, x_o denotes one of the class marks taken to be the computing origin, and $u_i = \dfrac{x_i - x_o}{c}$. Then coded class marks are obtained by replacing the original class marks with the integers $\ldots, -3, -2, -1, 0, 1, 2, 3, \ldots$ where 0 corresponds to class mark x_o in the original scale.

Mean (Arithmetic Mean)

a) $\displaystyle \bar{x} = \frac{1}{n} \sum_{i=1}^{n} x_i = \frac{x_1 + x_2 + \cdots + x_n}{n}$

b.1) $\displaystyle \bar{x} = \frac{1}{n} \sum_{i=1}^{k} f_i x_i = \frac{f_1 x_1 + f_2 x_2 + \cdots + f_k x_k}{n}$

If data is coded

b.2) $\displaystyle \bar{x} = x_o + c \frac{\displaystyle\sum_{i=1}^{k} f_i u_i}{n}$

Weighted Mean (Weighted Arithmetic Mean)

If with each value x_i is associated a weighting factor $w_i \geq 0$, then $\displaystyle\sum_{i=1}^{n} w_i$ is the total weight, and

a) $\displaystyle \bar{x} = \frac{\displaystyle\sum_{i=1}^{n} w_i x_i}{\displaystyle\sum_{i=1}^{n} w_i} = \frac{w_1 x_1 + w_2 x_2 + \cdots + w_n x_n}{w_1 + w_2 + \cdots + w_n}$

Geometric Mean

a) $\mathrm{G.M.} = \sqrt[n]{x_1 \cdot x_2 \cdots x_n}$

In logarithmic form

$$\log\,(\text{G.M.}) = \frac{1}{n}\sum_{i=1}^{n}\log x_i = \frac{\log x_1 + \log x_2 + \cdots + \log x_n}{n}$$

b) $\text{G.M.} = \sqrt[n]{x_1^{f_1}\cdot x_2^{f_2}\cdots x_k^{f_k}}$

In logarithmic form

$$\log\,(\text{G.M.}) = \frac{1}{n}\sum_{i=1}^{k}f_i\log x_i = \frac{f_1\log x_1 + f_2\log x_2 + \cdots + f_k\log x_k}{n}$$

Harmonic Mean

a) $\text{H.M.} = \dfrac{n}{\displaystyle\sum_{i=1}^{n}\dfrac{1}{x_i}} = \dfrac{n}{\dfrac{1}{x_1} + \dfrac{1}{x_2} + \cdots + \dfrac{1}{x_n}}$

b) $\text{H.M.} = \dfrac{n}{\displaystyle\sum_{i=1}^{k}\dfrac{f_i}{x_i}} = \dfrac{n}{\dfrac{f_1}{x_1} + \dfrac{f_2}{r_9} + \cdots + \dfrac{f_k}{x_k}}$

Relation Between Arithmetic, Geometric, and Harmonic Mean

$\text{H.M.} \le \text{G.M.} \le \bar{x}$, (Equality sign holds only if all sample values are identical.)

Mode

a) A mode M_o of a sample of size n is a value which occurs with greatest frequency, i.e., it is the most common value. A mode may not exist, and even if it does exist it may not be unique.

b) $M_o = L + c\,\dfrac{\Delta_1}{\Delta_1 + \Delta_2}$,

where L is the lower class boundary of the modal class (class containing the mode),

Δ_1 is the excess of modal frequency over frequency of next lower class,

Δ_2 is the excess of modal frequency over frequency of next higher class.

Median

a) If the sample is arranged in ascending order of magnitude, then the median M_d is given by the $\dfrac{n+1}{2}$ nd value. When n is odd, the median is the middle value of the set of ordered data; when n is even, the median is usually taken as the mean of the two middle values of the set of ordered data.

b) $M_d = L + c\,\dfrac{\dfrac{n}{2} - F_c}{f_m}$,

where L is lower class boundary of median class (class containing the median),

F_c is the sum of the frequencies of all classes lower than the median class,

f_m is the frequency of the median class.

Empirical Relation Between Mean, Median, and Mode

$$\text{Mean} - \text{Mode} = 3\,(\text{Mean} - \text{Median})$$

Quartiles

a) If the data is arranged in ascending order of magnitude, the jth quartile Q_j, $j = 1$, 2, or 3, is given by the $\dfrac{j(n+1)}{4}$ th value. It may be necessary to interpolate between successive values.

b) The jth quartile Q_j, $j = 1$, 2, or 3, is obtained from formula *b)* for the median by counting $\dfrac{jn}{4}$ cases starting at the bottom of the distribution.

Deciles

a) If the sample is arranged in ascending order of magnitude, the jth decile D_j, $j = 1, 2, \ldots$, or 9, is given by the $\dfrac{j(n+1)}{10}$ th value. It may be necessary to interpolate between successive values.

b) The jth decile D_j, $j = 1, 2, \ldots$, or 9, is obtained from formula *b)* for the median by counting $\dfrac{jn}{10}$ cases starting at the bottom of the distribution.

Percentiles

a) If the sample is arranged in ascending order of magnitude, the jth percentile P_j, $j = 1, 2, \ldots$, or 99 is given by the $\dfrac{j(n+1)}{100}$ th value. It may be necessary to interpolate between successive values.

b) The jth percentile P_j, $j = 1, 2, \ldots$, or 99, is obtained from formulas *b)* for the median by counting $\dfrac{jn}{100}$ cases starting at the bottom of the distribution.

Mean Deviation

a)
$$\text{M.D.} = \frac{1}{n} \sum_{i=1}^{n} |x_i - \bar{x}|$$

or

$$\text{M.D.} = \frac{1}{n} \sum_{i=1}^{n} |x_i - M_d|$$

where $\bar{x}$ is the mean and M_d is the median of the sample.

b)
$$\text{M.D.} = \frac{1}{n} \sum_{i=1}^{k} f_i |x_i - \bar{x}|$$

or

$$\text{M.D.} = \frac{1}{n} \sum_{i=1}^{k} f_i |x_i - M_d|$$

where $\bar{x}$ is the mean and M_d the median of the sample.

Standard Deviation

a) $s = \sqrt{\dfrac{\sum\limits_{i=1}^{n} (x_i - \bar{x})^2}{n - 1}}$, where $\bar{x}$ is the mean of the sample.

For computational purposes,

$$s = \sqrt{\dfrac{\sum\limits_{i=1}^{n} x_i^2 - n\bar{x}^2}{n - 1}}$$

$$s = \sqrt{\dfrac{n \sum\limits_{i=1}^{n} x_i^2 - \left(\sum\limits_{i=1}^{n} x_i\right)^2}{n(n - 1)}}$$

b) $s = \sqrt{\dfrac{\sum\limits_{i=1}^{k} f_i(x_i - \bar{x})^2}{n - 1}}$, where $\bar{x}$ is the mean of the sample.

For computational purposes

$$s = \sqrt{\dfrac{\sum\limits_{i=1}^{k} f_i x_i^2 - n\bar{x}^2}{n - 1}}$$

$$s = \sqrt{\dfrac{n \sum\limits_{i=1}^{k} f_i x_i^2 - \left(\sum\limits_{i=1}^{k} f_i x_i\right)^2}{n(n - 1)}}$$

If data is coded,

$$s = c \sqrt{\dfrac{n \sum\limits_{i=1}^{k} f_i u_i^2 - \left(\sum\limits_{i=1}^{k} f_i u_i\right)^2}{n(n - 1)}}$$

Variance

The variance is the square of the standard deviation.

Range

The range of a set of values is the difference between the largest and smallest values in the set.

Root Mean Square

a) $\text{R.M.S.} = \left[\dfrac{1}{n} \sum\limits_{i=1}^{n} x_i^2\right]^{\frac{1}{2}}$

b) $\text{R.M.S.} = \left[\dfrac{1}{n} \sum\limits_{i=1}^{k} f_i x_i^2\right]^{\frac{1}{2}}$

Interquartile Range

$$Q_3 - Q_1,$$

where Q_1 and Q_3 are the first and third quartiles.

Quartile Deviation (Semi-Interquartile Range)

$$\frac{Q_3 - Q_1}{2},$$

where Q_1 and Q_3 are the first and third quartiles.

Coefficient of Variation

$$V = \frac{100s}{\bar{x}},$$

where $\bar{x}$ is the mean and s the standard deviation of the sample.

Coefficient of Quartile Variation

$$V = 100\,\frac{Q_3 - Q_1}{Q_3 + Q_1},$$

where Q_1 and Q_3 are the first and third quartiles.

Standardized Variable (Standard Scores)

$$z = \frac{x_i - \bar{x}}{s},$$

where $\bar{x}$ is the mean and s the standard deviation of the sample.

Moments

a) The rth moment about the origin is given by

$$m'_r = \frac{1}{n}\sum_{i=1}^{n} x_i^{\,r}$$

The rth moment about the mean $\bar{x}$ is given by

$$m_r = \frac{1}{n}\sum_{i=1}^{n} (x_i - \bar{x})^r$$

If $\sum_{i=1}^{n} (x_i - \bar{x})^r$ is expanded by use of the binomial theorem, moments about the mean may be expressed in terms of moments about the origin.

b) The rth moment about the origin is given by

$$m'_r = \frac{1}{n}\sum_{i=1}^{k} f_i x_i^{\,r}$$

The rth moment about the mean $\bar{x}$ is given by

$$m_r = \frac{1}{n}\sum_{i=1}^{k} f_i (x_i - \bar{x})^r$$

If $\sum_{i=1}^{k} f_i(x_i - \bar{x})^r$ is expanded by use of the binomial theorem, moments **about the** mean may be expressed in terms of moments about the origin.

If data is coded

$$m'_r = c^r \frac{\sum_{i=1}^{k} f_i u_i^r}{n}$$

Coefficient of Skewness

$$\alpha_3 = \frac{m_3}{(m_2)^{3/2}},$$

where m_2 and m_3 are the second and third moments about the mean of the sample.

Coefficient of Momental Skewness

$$\frac{\alpha_3}{2} = \frac{m_3}{2(m_2)^{3/2}},$$

where m_2 and m_3 are the second and third moments about the mean of the sample.

Pearson's First Coefficient of Skewness

$$S_{k_1} = \frac{3(\bar{x} - M_o)}{s},$$

where $\bar{x}$ is the mean, M_o the mode, and s the standard deviation of the sample.

Pearson's Second Coefficient of Skewness

$$S_{k_2} = \frac{3(\bar{x} - M_d)}{s},$$

where $\bar{x}$ is the mean, M_d the median, and s the standard deviation of the sample.

Quartile Coefficient of Skewness

$$S_{k_Q} = \frac{Q_3 - 2Q_2 + Q_1}{Q_3 - Q_1},$$

where Q_1, Q_2, and Q_3 are the first, second, and third quartiles.

Coefficient of Kurtosis

$$\alpha_4 = \frac{m_4}{(m_2)^2},$$

where m_2 and m_4 are the second and fourth moments about the mean of the sample.

Coefficient of Excess (Kurtosis)

$$\alpha_4 - 3 = \frac{m_4}{(m_2)^2} - 3,$$

where m_2 and m_4 are the second and fourth moments about the mean of the sample.

Sheppards Corrections for Grouping

Let all class intervals be of equal length c. If the distribution of x has a high order of contact with the x-axis at both tails, (i.e., if the distribution of x has tails which are very

nearly tangent to the x-axis), one may improve the grouped data approximation to the variance by adding Sheppard's correction $-\dfrac{c^2}{12}$. Thus

$$\text{corrected variance} = \text{grouped data variance} - \frac{c^2}{12}$$

Analogous corrections for grouped data sample moments

$$m'_r = \frac{1}{n} \sum_{i=1}^{k} f_i x_i{}^r \qquad \text{and} \qquad m_r = \frac{1}{n} \sum_{i=1}^{k} f_i (x_i - \bar{x})^r$$

yield improved estimates m'_{r_e} and m_{r_e} given by

$$m'_{1_e} = m'_1 \qquad\qquad\qquad m_{1_e} = m_1$$

$$m'_{2_e} = m'_2 - \frac{c^2}{12} \qquad\qquad m_{2_e} = m_2 - \frac{c^2}{12}$$

$$m'_{3_e} = m'_3 - \frac{c^2}{4} m'_1 \qquad\qquad m_{3_e} = m_3$$

$$m'_{4_e} = m'_4 - \frac{c^2}{2} m'_1 + \frac{7c^4}{240} \qquad m_{4_e} = m_4 - \frac{c^2}{2} m_2 + \frac{7c^4}{240}$$

Curve Fitting, Regression, and Correlation

The following formulas apply to a set of n ordered pairs $\{(x_i, y_i)\}$, $i = 1, 2, \ldots, n$. The assumptions of normal regression analysis are that the x's are fixed variables, and the y's are independent random variables having normal distributions with common variance σ^2. The assumptions of normal correlation analysis are that $\{(x_i, y_i)\}$ constitute a random sample from a bivariate normal population.

Curve Fitting

1. Polynomial Function

$$y = b_0 + b_1 x + b_2 x^2 + \cdots + b_m x^m$$

For a polynomial function fit by the method of least squares, the values of $b_0, b_1, \ldots, b_m$ are obtained by solving the system of $m + 1$ normal equations

$$nb_0 + b_1 \Sigma x_i + b_2 \Sigma x_i{}^2 + \cdots + b_m \Sigma x_i{}^m = \Sigma y_i$$
$$b_0 \Sigma x_i + b_1 \Sigma x_i{}^2 + b_2 \Sigma x_i{}^3 + \cdots + b_m \Sigma x_i{}^{m+1} = \Sigma x_i y_i$$
$$\cdots\cdots\cdots\cdots\cdots\cdots\cdots\cdots\cdots\cdots\cdots\cdots$$
$$b_0 \Sigma x_i{}^m + b_1 \Sigma x_i{}^{m+1} + b_2 \Sigma x_i{}^{m+2} + \cdots + b_m \Sigma x_i{}^{2m} = \Sigma x_i{}^m y_i$$

2. Straight Line

$$y = b_0 + b_1 x$$

For a straight line fit by the method of least squares, the values b_0 and b_1 are obtained by solving the normal equations

$$nb_0 + b_1 \Sigma x_i = \Sigma y_i$$
$$b_0 \Sigma x_i + b_1 \Sigma x_i{}^2 = \Sigma x_i y_i$$

The solutions of these normal equations are

$$b_1 = \frac{n \Sigma x_i y_i - (\Sigma x_i)(\Sigma y_i)}{n \Sigma x_i{}^2 - (\Sigma x_i)^2}$$

$$b_0 = \frac{\Sigma y_i}{n} - b_1 \frac{\Sigma x_i}{n} = \bar{y} - b_1 \bar{x}$$

3. Exponential Curve

$$y = ab^x$$

or

$$\log y = \log a + (\log b)x$$

For an exponential curve fit by the method of least squares, the values $\log a$ and $\log b$ are obtained by fitting a straight line to the set of ordered pairs $\{(x_i, \log y_i)\}$.

4. Power Function

$$y = ax^b$$

or

$$\log y = \log a + b \log x$$

For a power function fit by the method of least squares, the values $\log a$ and b are obtained by fitting a straight line to the set of ordered pairs $\{(\log x_i, \log y_i)\}$.

Regression and Correlation

1. Simple Linear Regression.

For a regression of y on x

$$E(y/x) = b_0 + b_1 x,$$

where $E(y/x)$ is the mean of the distribution of y for a given x.

Standard Error of Estimate

$$s_e = \sqrt{\frac{\Sigma[y_i - (b_0 + b_1 x_i)]^2}{n - 2}},$$

where b_0 and b_1 are given by

$$b_1 = \frac{n\Sigma x_i y_i - (\Sigma x_i)(\Sigma y_i)}{n\Sigma x_i^2 - (\Sigma x_i)^2}$$

$$b_0 = \frac{\Sigma y_i}{n} - b_1 \frac{\Sigma x_i}{n} = \bar{y} - b_1 \bar{x}$$

2. Correlation.

An estimate of the population correlation coefficient ρ is given by

$$r = \frac{\Sigma(x_i - \bar{x})(y_i - \bar{y})}{\sqrt{[\Sigma(x_i - \bar{x})^2][\Sigma(y_i - \bar{y})^2]}}$$

or by the computing formula

$$r = \frac{n\Sigma x_i y_i - (\Sigma x_i)(\Sigma y_i)}{\sqrt{[n\Sigma x_i^2 - (\Sigma x_i)^2][n\Sigma y_i^2 - (\Sigma y_i)^2]}}$$

For grouped data

$$r = \frac{n\Sigma f x_i y_i - (\Sigma f_x x_i)(\Sigma f_y y_i)}{\sqrt{[n\Sigma f_x x_i^2 - (\Sigma f_x x_i)^2][n\Sigma f_y y_i^2 - (\Sigma f_y y_i)^2]}}$$

where f_x and f_y denote the frequencies corresponding to the class marks x and y, and f denotes the frequency of the corresponding cell of the correlation table.

If the data is coded

$$r = \frac{n\Sigma fuv - (\Sigma f_u u)(\Sigma f_v v)}{\sqrt{[n\Sigma f_u u^2 - (\Sigma f_u u)^2][n\Sigma f_v v^2 - (\Sigma f_v v)^2]}}$$

where the u's and v's are coded class marks. The frequencies f_u and f_v are defined analogous to f_x and f_y.

PROBABILITY

Definitions

A sample space S associated with an experiment is a set S of elements such that any outcome of the experiment corresponds to one and only one element of the set. An event E is a subset of a sample space S. An element in a sample space is called a sample point or a simple event (Unit subset of S).

Definition of Probability

If an experiment can occur in n mutually exclusive and equally likely ways, and if exactly m of these ways correspond to an event E, then the probability of E is given by

$$P(E) = \frac{m}{n}.$$

If E is a subset of S, and if to each unit subset of S, a non-negative number, called its probability, is assigned, and if E is the union of two or more different simple events, then the probability of E, denoted by $P(E)$, is the sum of the probabilities of those simple events whose union is E.

Marginal and Conditional Probability

Suppose a sample space S is partioned into rs disjoint subsets where the general subset is denoted by $E_i \cap F_j$. Then the marginal probability of E_i is defined as

$$P(E_i) = \sum_{j=1}^{s} P(E_i \cap F_j)$$

and the marginal probability of F_j is defined as

$$P(F_j) = \sum_{i=1}^{r} P(E_i \cap F_j)$$

The conditional probability of E_i, given that F_j has occurred, is defined as

$$P(E_i/F_j) = \frac{P(E_i \cap F_j)}{P(F_j)} , \qquad P(F_j) \neq 0$$

and that of F_j, given that E_i has occurred, is defined as

$$P(F_j/E_i) = \frac{P(E_i \cap F_j)}{P(E_i)} , \qquad P(E_i) \neq 0 .$$

Probability Theorems

1. If ϕ is the null set, $P(\phi) = 0$.
2. If S is the sample space, $P(S) = 1$.
3. If E and F are two events

$$P(E \cup F) = P(E) + P(F) - P(E \cap F).$$

4. If E and F are mutually exclusive events,

$$P(E \cup F) = P(E) + P(F).$$

5. If E and E' are complementary events,

$$P(E) = 1 - P(E').$$

6. The conditional probability of an event E, given an event F, is denoted by $P(E/F)$ and is defined as

$$P(E/F) = \frac{P(E \cap F)}{P(F)},$$

where $P(F) \neq 0$.

7. Two events E and F are said to be independent if and only if

$$P(E \cap F) = P(E) \cdot P(F).$$

E is said to be statistically independent of F if $P(E/F) = P(E)$ and $P(F/E) = P(F)$.

8. The events $E_1, E_2, \ldots, E_n$ are called mutually independent for all combinations if and only if every combination of these events taken any number at a time is independent.

9. *Bayes Theorem.*

If $E_1, E_2, \ldots, E_n$ are n mutually exclusive events whose union is the sample space S, and E is any arbitrary event of S such that $P(E) \neq 0$, then

$$P(E_k/E) = \frac{P(E_k) \cdot P(E/E_k)}{\displaystyle\sum_{j=1}^{n} [P(E_j) \cdot P(E/E_j)]}$$

Random Variable

A function whose domain is a sample space S and whose range is some set of real numbers is called a random variable, denoted by $\mathbf{X}$. The function $\mathbf{X}$ transforms sample points of S into points on the x-axis. $\mathbf{X}$ will be called a discrete random variable if it is a random variable that assumes only a finite or denumerable number of values on the x-axis. $\mathbf{X}$ will be called a continuous random variable if it assumes a continuum of values on the x-axis.

Probability Function (Discrete Case)

The random variable $\mathbf{X}$ will be called a discrete random variable if there exists a function f such that $f(x_i) \geq 0$ and $\sum_i f(x_i) = 1$ for $i = 1, 2, 3, \ldots$ and such that for any event E,

$$P(E) = P[\mathbf{X} \text{ is in } E] = \sum_E f(x)$$

where $\sum_E$ means sum $f(x)$ over those values x_i that are in E and where $f(x) = P[\mathbf{X} = x]$.

The probability that the value of $\mathbf{X}$ is some real number x, is given by $f(x) = P[\mathbf{X} = x]$, where f is called the probability function of the random variable $\mathbf{X}$.

Cumulative Distribution Function (Discrete Case)

The probability that the value of a random variable $\mathbf{X}$ is less than or equal to some real number x is defined as

$$F(x) = P(\mathbf{X} \leq x)$$
$$= \Sigma f(x_i), \quad -\infty < x < \infty,$$

where the summation extends over those values of i such that $x_i \leq x$.

Probability Density (Continuous Case)

The random variable $\mathbf{X}$ will be called a continuous random variable if there exists a function f such that $f(x) \geq 0$ and $\int_{-\infty}^{\infty} f(x)\, dx = 1$ for all x in interval $-\infty < x < \infty$ and such that for any event E

$$P(E) = P(\mathbf{X} \text{ is in } E) = \int_E f(x)\, dx.$$

$f(x)$ is called the probability density of the random variable $\mathbf{X}$. The probability that $\mathbf{X}$ assumes any given value of x is equal to zero and the probability that it assumes a value on the interval from a to b, including or excluding either end point, is equal to

$$\int_a^b f(x)\, dx.$$

Cumulative Distribution Function (Continuous Case)

The probability that the value of a random variable $\mathbf{X}$ is less than or equal to some real number x is defined as

$$F(x) = P(\mathbf{X} \leq x), \quad -\infty < x < \infty$$
$$= \int_{-\infty}^{x} f(x)\, dx.$$

From the cumulative distribution, the density, if it exists, can be found from

$$f(x) = \frac{dF(x)}{dx}.$$

From the cumulative distribution

$$P(a \leq \mathbf{X} \leq b) = P(\mathbf{X} \leq b) - P(\mathbf{X} \leq a)$$
$$= F(b) - F(a)$$

Mathematical Expectation

A. EXPECTED VALUE

Let $\mathbf{X}$ be a random variable with density $f(x)$. Then the expected value of $\mathbf{X}$, $E(\mathbf{X})$, is defined to be

$$E(\mathbf{X}) = \sum_x x f(x)$$

if $\mathbf{X}$ is discrete and

$$E(\mathbf{X}) = \int_{-\infty}^{\infty} x f(x)\, dx$$

if $\mathbf{X}$ is continuous. The expected value of a function g of a random variable $\mathbf{X}$ is defined as

$$E[g(\mathbf{X})] = \sum_x g(x) \cdot f(x)$$

if $\mathbf{X}$ is discrete and

$$E[g(\mathbf{X})] = \int_{-\infty}^{\infty} g(x) \cdot f(x)\, dx$$

if $\mathbf{X}$ is continuous.

Theorems

1. $E[aX + bY] = aE(X) + bE(Y)$
2. $E[X \cdot Y] = E(X) \cdot E(Y)$ if **X** and **Y** are statistically independent.

B. MOMENTS

a. Moments About the Origin. The moments about the origin of a probability distribution are the expected values of the random variable which has the given distribution. The rth moment of **X**, usually denoted by μ'_r, is defined as

$$\mu'_r = E[X^r] = \sum_x x^r f(x)$$

if **X** is discrete and

$$\mu'_r = E[X^r] = \int_{-\infty}^{\infty} x^r f(x) \, dx$$

if **X** is continuous.

The first moment, μ'_1, is called the mean of the random variable **X** and is usually denoted by μ.

b. Moments About the Mean. The rth moment about the mean, usually denoted by μ_r, is defined as

$$\mu_r = E[(X - \mu)^r] = \sum_x (x - \mu)^r f(x)$$

if **X** is discrete and

$$\mu_r = E[(X - \mu)^r] = \int_{-\infty}^{\infty} (x - \mu)^r f(x) \, dx$$

if **X** is continuous.

The second moment about the mean, μ_2, is given by

$$\mu_2 = E[(X - \mu)^2] = \mu'_2 - \mu^2$$

and is called the variance of the random variable **X**, and is denoted by σ^2. The square root of the variance, σ, is called the standard deviation.

Theorems

1. $\sigma^2_{cX} = c^2 \sigma^2_X$
2. $\sigma^2_{c+X} = \sigma^2_X$
3. $\sigma^2_{aX+b} = a^2 \sigma^2_X$

c. Factorial Moments. The rth factorial moment of a probability distribution is defined as

$$\mu'_{(r)} = E[X^{[r]}] = \sum_x x^{[r]} f(x)$$

if **X** is discrete and

$$\mu'_{(r)} = E[X^{[r]}] = \int_{-\infty}^{\infty} x^{[r]} f(x) \, dx$$

if **X** is continuous, where the symbol $x^{[r]}$ denotes the factorial expression

$$x^{[r]} = x(x - 1)(x - 2) \cdots (x - r + 1), \, r = 1, 2, 3, \ldots$$

C. Generating Functions

a. Moment Generating Functions. The moment generating function (m.g.f.) of the random variable $\mathbf{X}$ is defined as

$$m_x(t) = E(e^{t\mathbf{X}}) = \sum_x e^{tx} f(x)$$

if $\mathbf{X}$ is discrete and

$$m_x(t) = E(e^{t\mathbf{X}}) = \int_{-\infty}^{\infty} e^{tx} f(x)\, dx$$

if $\mathbf{X}$ is continuous.

$E(e^{t\mathbf{X}})$ is the expected value of $e^{t\mathbf{X}}$. If $m_x(t)$ and its derivatives exist, $|t| < h^2$, the rth moment about the origin is

$$\mu'_r = m_x^{(r)}(0), \qquad r = 0, 1, 2, \ldots$$

where $m_x^{(r)}(0)$ is the rth derivative of $m_x(t)$ with respect to t, evaluated at $t = 0$. For

$$m_x(t) = E(e^{t\mathbf{X}})$$
$$= E\left[1 + \mathbf{X}t + \frac{(\mathbf{X}t)^2}{2!} + \cdots \right]$$
$$= 1 + \mu'_1 t + \mu'_2 \frac{t^2}{2} + \cdots$$

Thus, the moments μ'_r appear as coefficients of $\dfrac{t^r}{r!}$, and $m_x(t)$ may be regarded as generating the moments ν_r. The moments μ_r may be generated by the generating function

$$M_x(t) = E[e^{t(\mathbf{X}-\mu)}] = e^{-\mu t} E(e^{t\mathbf{X}}) = e^{-\mu t} m_x(t) \ .$$

b. Factorial Moment Generating Function. The factorial moment generating function is defined as

$$E(t^{\mathbf{X}}) = \sum_x t^x f(x) \qquad \text{(probability generation function)}$$

if $\mathbf{X}$ is discrete and

$$E(t^{\mathbf{X}}) = \int_{-\infty}^{\infty} t^x f(x)\, dx$$

if $\mathbf{X}$ is continuous.

The rth factorial moment is obtained from the factorial moment generating function by differentiating it r times with respect to t and then evaluating the result when $t = 1$.

Theorems

1. If c is a constant, the m.g.f. of $c + \mathbf{X}$ is $e^{ct} m_x(t)$.
2. If c is a constant, the m.g.f. of $c\mathbf{X}$ is $m_x(ct)$.
3. If $\mathbf{Y} = \sum_{i=1}^{n} \mathbf{X}_i$, and $m_x(t)$ is the m.g.f. of $\mathbf{X}_i$, where $\mathbf{X}_1, \ldots, \mathbf{X}_n$ is a random sample from $f(x)$, then the m.g.f. of $\mathbf{Y}$ is $[m_x(t)]^n$.

D. Cumulant Generating Function

Let $m_x(t)$ be a m.g.f. If $\ln m_x(t)$ can be expanded in the form

$$c(t) = \ln m_x(t) = \kappa_1 t + \kappa_2 \frac{t^2}{2!} + \kappa_3 \frac{t^3}{3!} + \cdots + \kappa_r \frac{t^r}{r!} + \cdots,$$

then $c(t)$ is called the cumulant generating function (semi-invariant generating function) and κ_r are called the cumulants (semi-invariants) of a distribution.

$$\kappa_r = c^{(r)}(0)$$

where $c^{(r)}(0)$ is the rth derivative of $c(t)$ with respect to t evaluated at $t = 0$.

E. CHARACTERISTIC FUNCTIONS

The characteristic function of a distribution is defined as

$$\phi(t) = E(e^{itX}) = \sum_x e^{itx} \cdot f(x)$$

if **X** is discrete and

$$\phi(t) = E(e^{itX}) = \int_{-\infty}^{\infty} e^{itx} \cdot f(x) \, dx$$

if **X** is continuous.

Here t is a real number, $i^2 = -1$, and $e^{itX} = \cos(tX) + i \sin(tX)$. The characteristic function also generates moments, if they exist for

$$i^r \mu'_r = \phi^{(r)}(0)$$

where $\phi^{(r)}(0)$ is the rth derivative of $\phi(t)$ with respect to t evaluated at $t = 0$.

Multivariate Distributions

A. DISCRETE CASE

The k-dimensional random variable $(X_1, X_2, \ldots, X_k)$ is a k-dimensional discrete random variable if it assumes values only at a finite or denumerable number of points $(x_1, x_2, \ldots, x_k)$. Define

$$P[X_1 = x_1, X_2 = x_2, \ldots, X_k = x_k] = f(x_1, x_2, \ldots, x_k)$$

for every value that the random variable can assume. $f(x_1, x_2, \ldots, x_k)$ is called the joint density of the k-dimensional random variable. If E is any subset of the set of values that the random variable can assume, then

$$P(E) = P[(X_1, X_2, \ldots, X_k) \text{ is in } E] = \sum_E f(x_1, x_2, \ldots, x_k)$$

where the sum is over all those points in E. The cumulative distribution is defined as

$$F(x_1, x_2, \ldots, x_k) = \sum_{x_1} \sum_{x_2} \cdots \sum_{x_k} f(x_1, x_2, \ldots, x_k).$$

B. CONTINUOUS CASE

The k random variables $X_1, X_2, \ldots, X_k$ are said to be jointly distributed if there exists a function f such that $f(x_1, x_2, \ldots, x_k) \geq 0$ for all $-\infty < x_i < \infty$, $i = 1, 2, \ldots, k$ and such that for any event E

$$P(E) = P[(X_1, X_2, \ldots, X_k) \text{ is in } E]$$
$$= \int_E f(x_1, x_2, \ldots, x_k) \, dx_1 \, dx_2 \cdots dx_k.$$

$f(x_1, x_2, \ldots, x_k)$ is called the joint density of the random variables $X_1, X_2, \ldots, X_k$.

The cumulative distribution is defined as

$$F(x_1, x_2, \ldots, x_k) = \int_{-\infty}^{x_1} \int_{-\infty}^{x_2} \cdots \int_{-\infty}^{x_k} f(x_1, x_2, \ldots, x_k) \, dx_k \cdots dx_2 \, dx_1 .$$

Given the cumulative distribution, the density may be found by

$$f(x_1, x_2, \ldots, x_k) = \frac{\partial}{\partial x_1} \cdot \frac{\partial}{\partial x_2} \cdots \frac{\partial}{\partial x_k} F(x_1, x_2, \ldots, x_k) \ .$$

Moments

The rth moment of X_i, say, is defined as

$$E(X_i^r) = \sum_{x_1} \sum_{x_2} \cdots \sum_{x_k} x_i^r f(x_1, x_2, \ldots, x_k)$$

if the X_i are discrete and

$$E(X_i^r) = \int_{-\infty}^{\infty} \int_{-\infty}^{\infty} \cdots \int_{-\infty}^{\infty} x_i^r f(x_1, x_2, \ldots, x_k) \, dx_k \cdots dx_2 \, dx_1$$

if the X_i are continuous.

Joint moments about the origin are defined as

$$E(X_1^{r_1} X_2^{r_2} \cdots X_k^{r_k})$$

where $r_1 + r_2 + \cdots + r_k$ is the order of the moment.

Joint moments about the mean are defined as

$$E[(X_1 - \mu_1)^{r_1} (X_2 - \mu_2)^{r_2} \cdots (X_k - \mu_k)^{r_k}].$$

Marginal and Conditional Distributions

If the random variables $X_1, X_2, \ldots, X_k$ have the joint density function $f(x_1, x_2, \ldots, x_k)$, then the marginal distribution of the subset of the random variables, say, $X_1, X_2, \ldots, X_p$ ($p < k$), is given by

$$g(x_1, x_2, \ldots, x_p) = \sum_{x_{p+1}} \sum_{x_{p+2}} \cdots \sum_{x_k} f(x_1, x_2, \ldots, x_k)$$

if the X's are discrete, and

$$g(x_1, x_2, \ldots, x_p) = \int_{-\infty}^{\infty} \int_{-\infty}^{\infty} \cdots \int_{-\infty}^{\infty} f(x_1, x_2, \ldots, x_k) \, dx_{p+1} \cdots dx_{k-1} \, dx_k$$

if the X's are continuous.

The conditional distribution of a certain subset of the random variables is the joint distribution of this subset under the condition that the remaining variables are given certain values. The conditional distribution of $X_1, X_2, \ldots, X_p$ given $X_{p+1}, X_{p+2}, \ldots, X_k$ is

$$h(x_1, x_2, \ldots, x_p | x_{p+1}, x_{p+2}, \ldots, x_k) = \frac{f(x_1, x_2, \ldots, x_k)}{g(x_{p+1}, x_{p+2}, \ldots, x_k)}$$

if $g(x_{p+1}, x_{p+2}, \ldots, x_k) \neq 0$.

The variance σ_{ii} of X_i and the covariance σ_{ij} of X_i and X_j are given by

$$\sigma_{ii} = \sigma_i^2 = E[(X_i - \mu_i)^2]$$

and

$$\sigma_{ij} = \rho_{ij}\sigma_i\sigma_j = E[(X_i - \mu_i)(X_j - \mu_j)]$$

where ρ_{ij} is the correlation coefficient and σ_i and σ_j are the standard deviations of X_i and X_j.

A joint m.g.f. is defined as

$$m(t_1, t_2, \ldots, t_k) = E[e^{t_1 X_1 + t_2 X_2 + \cdots + t_k X_k}]$$

if it exists for all values of t_i such that $|t_i| < h^2$.

The rth moment of X_i may be obtained by differentiating the m.g.f. r times with respect to t_i and then evaluating the result when all t's are set equal to zero. Similarly, a joint moment would be found by differentiating the m.g.f. r_1 times with respect to t_1, . . . , r_k times with respect to t_k, and then evaluating the result when all t's are set equal to zero.

Probability Distributions

A. DISCRETE CASE

1. *Discrete Uniform Distribution.* If the random variable X has a probability function given by

$$P(X = x) = f(x) = \frac{1}{n}, \qquad x = x_1, x_2, \ldots, x_n,$$

then the variable X is said to possess a discrete uniform distribution.

Properties

When $x_i = i$ for $i = 1, 2, \ldots$, and n

$$\text{Mean} = \mu = \frac{n+1}{2}$$

$$\text{Variance} = \sigma^2 = \frac{n^2 - 1}{12}$$

$$\text{Standard Deviation} = \sigma = \sqrt{\frac{n^2 - 1}{12}}$$

$$\text{Moment Generating Function} = m_x(t) = \frac{e^t(1 - e^{nt})}{n(1 - e^t)}$$

2. *Binomial Distribution.* If the random variable X has a probability function given by

$$P(X = x) = f(x) = \binom{n}{x} \theta^x (1 - \theta)^{n-x}, \qquad x = 0, 1, 2, \ldots, n$$

where

$$\binom{n}{x} = \frac{n!}{x!(n - x)!},$$

then the variable X is said to possess a binomial distribution. $f(x)$ is the general term of the expansion of $[\theta + (1 - \theta)]^n$.

Properties

$$\text{Mean} = \mu = n\theta$$
$$\text{Variance} = \sigma^2 = n\theta(1 - \theta)$$
$$\text{Standard Deviation} = \sigma = \sqrt{n\theta(1 - \theta)}$$
$$\text{Moment Generating Function} = m_x(t) = [\theta e^t + (1 - \theta)]^n$$

3. *Geometric Distribution.* If the random variable X has a probability function given by

$$P(X = x) = f(x) = \theta(1 - \theta)^{x-1}, \qquad x = 1, 2, 3, \ldots,$$

then the variable X is said to possess a geometric distribution.

Properties

$$\text{Mean} = \mu = \frac{1}{\theta}$$

$$\text{Variance} = \sigma^2 = \frac{1 - \theta}{\theta^2}$$

$$\text{Standard Deviation} = \sigma = \sqrt{\frac{1 - \theta}{\theta^2}}$$

$$\text{Moment Generating Function} = m_x(t) = \frac{\theta e^t}{1 - e^t(1 - \theta)}$$

4. *Multinomial Distribution.* If a set of random variables $\mathbf{X}_1$, $\mathbf{X}_2$, . . . , $\mathbf{X}_n$ has a probability function given by

$$P(\mathbf{X}_1 = x_1, \mathbf{X}_2 = x_2, \ldots, \mathbf{X}_n = x_n) = f(x_1, x_2, \ldots, x_n) = \frac{N!}{\prod_{i=1}^{n} x_i!} \prod_{i=1}^{n} \theta_i^{x_i}$$

where x_i are positive integers and each $\theta_i > 0$ for $i = 1, 2, \ldots, n$ and

$$\sum_{i=1}^{n} \theta_i = 1, \qquad \sum_{i=1}^{n} x_i = N,$$

then the joint distribution of $\mathbf{X}_1$, $\mathbf{X}_2$, . . . , $\mathbf{X}_n$ is called the multinomial distribution. $f(x_1, x_2, \ldots, x_n)$ is the general term of the expansion of $(\theta_1 + \theta_2 + \cdots + \theta_n)^N$.

Properties

$$\text{Mean of } \mathbf{X}_i = \mu_i = N\theta_i$$
$$\text{Variance of } \mathbf{X}_i = \sigma_i^2 = N\theta_i(1 - \theta_i)$$
$$\text{Covariance of } \mathbf{X}_i \text{ and } \mathbf{X}_j = \sigma_{ij}^2 = -N\theta_i\theta_j$$
$$\text{Joint Moment Generating Function} = (\theta_1 e^{t_1} + \cdots + \theta_n e^{t_n})^N$$

5. *Poisson Distribution.* If the random variable $\mathbf{X}$ has a probability function given by

$$P(\mathbf{X} = x) = f(x) = \frac{e^{-\lambda}\lambda^x}{x!}, \qquad \lambda > 0, x = 0, 1, \ldots,$$

then the variable $\mathbf{X}$ is said to possess a Poisson distribution.

Properties

$$\text{Mean} = \mu = \lambda$$
$$\text{Variance} = \sigma^2 = \lambda$$
$$\text{Standard Deviation} = \sigma = \sqrt{\lambda}$$
$$\text{Moment Generating Function} = m_x(t) = e^{\lambda(e^t - 1)}$$

6. *Hypergeometric Distribution.* If the random variable $\mathbf{X}$ has a probability function given by

$$P(\mathbf{X} = x) = f(x) = \frac{\binom{k}{x}\binom{N-k}{n-x}}{\binom{N}{n}}, \qquad x = 0, 1, 2, \ldots, [n, k],$$

where $[n, k]$ means the smaller of the two numbers n, k, then the variable $\mathbf{X}$ is said to possess a hypergeometric distribution.

Properties

$$\text{Mean} = \mu = \frac{kn}{N}$$

$$\text{Variance} = \sigma^2 = \frac{k(N - k)n(N - n)}{N^2(N - 1)}$$

$$\text{Standard Deviation} = \sigma = \sqrt{\frac{k(N - k)n(N - n)}{N^2(N - 1)}}$$

7. *Negative Binomial Distribution.* If the random variable **X** has a probability function given by

$$P(\mathbf{X} = x) = f(x) = \binom{x + r - 1}{r - 1} \theta^r (1 - \theta)^x, \qquad x = 0, 1, 2, \ldots;$$

then the variable **X** is said to possess a negative binomial distribution, known also as the Pascal or Pólya distribution.

Properties

$$\text{Mean} = \mu = \frac{r}{\theta}$$

$$\text{Variance} = \sigma^2 = \frac{r}{\theta}\left(\frac{1}{\theta} - 1\right) = \frac{r(1 - \theta)}{\theta^2}$$

$$\text{Standard Deviation} = \sigma = \sqrt{\frac{r}{\theta}\left(\frac{1}{\theta} - 1\right)} = \sqrt{\frac{r(1 - \theta)}{\theta^2}}$$

$$\text{Moment Generating Function} = m_x(t) = e^{tr}\theta^r[1 - (1 - \theta)e^t]^{-r}$$

B. Continuous Case

1. *Uniform Distribution.* A random variable **X** is said to be distributed as the uniform distribution if the density function is given by

$$f(x) = \frac{1}{\beta - \alpha}, \qquad \alpha < x < \beta,$$

where α and β are parameters with $\alpha < \beta$.

Properties

$$\text{Mean} = \mu = \frac{\alpha + \beta}{2}$$

$$\text{Variance} = \sigma^2 = \frac{(\beta - \alpha)^2}{12}$$

$$\text{Standard Deviation} = \sigma = \sqrt{\frac{(\beta - \alpha)^2}{12}}$$

$$\text{Moment Generating Function} = m_x(t) = \frac{2}{(\beta - \alpha)t}\sinh\left[\frac{(\beta - \alpha)t}{2}\right]e^{\frac{\alpha + \beta}{2}t} = \frac{e^{\beta t} - e^{\alpha t}}{(\beta - \alpha)t}$$

2. *Normal Distribution.* A random variable **X** is said to be normally distributed if its density function is given by

$$f(x) = \frac{1}{\sqrt{2\pi}\,\sigma} e^{-(x-\mu)^2/2\sigma^2}, \qquad -\infty < x < \infty$$

where μ and σ are parameters, called the mean and the standard deviation or the random variable **X**, respectively.

Properties

$$\text{Mean} = \mu$$
$$\text{Variance} = \sigma^2$$
$$\text{Standard Deviation} = \sigma$$

$$\text{Moment Generating Function} = m_x(t) = e^{t\mu + \frac{\sigma^2 t^2}{2}}$$

Cumulative Distribution

$$F(x) = \int_{-\infty}^{x} \frac{1}{\sqrt{2\pi}\,\sigma}\, e^{-(x-\mu)^2/2\sigma^2}\, dx$$

Set $y = \dfrac{x - \mu}{\sigma}$ to obtain the cumulative standard normal.

3. *Gamma Distribution.* A random variable **X** is said to be distributed as the **gamma** distribution if the density function is given by

$$f(x) = \frac{1}{\Gamma(\alpha + 1)\beta^{\alpha+1}}\, x^\alpha e^{-x/\beta}, \qquad 0 < x < \infty$$

where α and β are parameters with $\alpha > -1$ and $\beta > 0$.

Properties

$$\text{Mean} = \mu = \beta(\alpha + 1)$$
$$\text{Variance} = \sigma^2 = \beta^2(\alpha + 1)$$
$$\text{Standard Deviation} = \sigma = \beta\sqrt{\alpha + 1}$$
$$\text{Moment Generating Function} = m_x(t) = (1 - \beta t)^{-(\alpha+1)}, \qquad t < \frac{1}{\beta}.$$

4. *Exponential Distribution.* A random variable **X** is said to be distributed as the exponential distribution if the density function is given by

$$f(x) = \frac{1}{\theta}\, e^{-x/\theta}, \qquad x > 0$$

where θ is a parameter and $\theta > 0$.

Properties

$$\text{Mean} = \mu = \theta$$
$$\text{Variance} = \sigma^2 = \theta^2$$
$$\text{Standard Deviation} = \sigma = \sqrt{\theta^2}$$
$$\text{Moment Generating Function} = m_x(t) = (1 - \theta t)^{-1}$$

5. *Beta Distribution.* A random variable **X** is said to be distributed as the beta distribution if the density function is given by

$$f(x) = \frac{\Gamma(\alpha + \beta + 2)}{\Gamma(\alpha + 1)\Gamma(\beta + 1)}\, x^\alpha (1 - x)^\beta, \qquad 0 < x < 1$$

where α and β are parameters with $\alpha > -1$ and $\beta > -1$.

Properties

$$\text{Mean} = \mu = \frac{\alpha + 1}{\alpha + \beta + 2}$$
$$\text{Variance} = \sigma^2 = \frac{(\alpha + 1)(\beta + 1)}{(\alpha + \beta + 2)^2(\alpha + \beta + 3)}$$
$$\text{rth moment about the origin} = \nu_r = \frac{\Gamma(\alpha + \beta + 2)\Gamma(\alpha + r + 1)}{\Gamma(\alpha + \beta + r + 2)\Gamma(\alpha + 1)}.$$

Sampling Distributions

Population—A finite or infinite set of elements of a random variable **X**.

Random Sample—If the random variables **X₁**, **X₂**, . . . , **Xₙ** have a joint density,

$$g(x_1, x_2, \ldots, x_n) = f(x_1)f(x_2) \cdots f(x_n)$$

where the density of each X_i is $f(x)$, then $X_1, X_2, \ldots, X_n$ is said to be a random sample of size n from the population with density $f(x)$.

Sampling Distributions

A random sample is selected from a population in which the form of the probability function is known, and from the joint density of the random variables a distribution, called the sampling distribution, of a function of the random variables is derived.

1. *Chi-Square Distribution.* If $Y_1, Y_2, \ldots, Y_n$ are normally and independently distributed with mean 0 and variance 1, then

$$\chi^2 = \sum_{i=1}^{n} Y_i^2$$

is distributed as Chi-Square (χ^2) with n degrees of freedom. The density function is given by

$$f(\chi^2) = \frac{(\chi^2)^{\frac{1}{2}(n-2)}}{2^{\frac{n}{2}}\Gamma\left(\frac{n}{2}\right)} e^{-\chi^2/2}, \qquad 0 < \chi^2 < \infty.$$

Properties

$$\text{Mean} = \mu = n$$
$$\text{Variance} = \sigma^2 = 2n$$

Reproductive Property of χ^2 - Distribution

If $\chi_1^2, \chi_2^2, \ldots, \chi_k^2$ are independently distributed according to χ^2 - distributions with $n_1, n_2, \ldots, n_k$ degrees of freedom, respectively, then $\sum_{j=1}^{k} \chi_j^2$ is distributed according to a χ^2 - distribution with $n = \sum_{j=1}^{k} n_j$ degrees of freedom.

2. *Snedecor's F-Distribution.* If a random variable X is distributed as χ^2 with m degrees of freedom (χ_m^2) and a random variable Y is distributed as χ^2 with n degrees of freedom (χ_n^2) and if X and Y are independent, then $F = \frac{X/m}{Y/n}$ is distributed as Snedecor's F with m and n degrees of freedom, denoted by $F(m, n)$. The density function of the F-distribution is given by

$$f(F) = \frac{\Gamma\left(\frac{m+n}{2}\right)\left(\frac{m}{n}\right)^{m/2} F^{(m-2)/2}}{\Gamma\left(\frac{m}{2}\right)\Gamma\left(\frac{n}{2}\right)\left(1 + \frac{m}{n}F\right)^{(m+n)/2}}, \qquad 0 < F < \infty.$$

Properties

$$\text{Mean} = \mu = \frac{n}{n-2}, \qquad n > 2$$

$$\text{Variance} = \sigma^2 = \frac{2n^2(m+n-2)}{m(n-2)^2(n-4)}, \qquad n > 4 .$$

3. *Student's t-Distribution.* If a random variable X is normally distributed with mean 0 and variance σ^2, and if Y^2/σ^2 is distributed as χ^2 with n degrees of freedom and if X and Y are independent, then

$$t = \frac{X \sqrt{n}}{Y}$$

is distributed as Student's t with n degrees of freedom. The density function is given by

$$f(t) = \frac{\Gamma\left(\dfrac{n+1}{2}\right)}{\sqrt{n\pi}\,\Gamma\left(\dfrac{n}{2}\right)\left(1+\dfrac{t^2}{n}\right)^{\frac{1}{2}(n+1)}} , \qquad -\infty < t < \infty .$$

Properties

$$\text{Mean} = \mu = 0$$

$$\text{Variance} = \sigma^2 = \frac{n}{n-2}, \qquad n > 2 .$$

The transformation $w = \dfrac{mF/n}{1+\dfrac{mF}{n}}$ transforms the F-density into a Beta density.

SUMMARY OF SIGNIFICANCE TESTS: TESTING FOR THE VALUE OF A SPECIFIED PARAMETER

Hypothesis	Conditions	Test Statistic	Distribution of Test Statistic	Critical Region
1. $\mu = \mu_0$	Known σ	$z = \dfrac{(\bar{x} - \mu_0)\sqrt{n}}{\sigma}$	Normal	$z > z_\alpha$ if we wish to reject when $\mu > \mu_0$ $z < -z_\alpha$ if we wish to reject when $\mu < \mu_0$ $|z| > z_{\alpha/2}$ if we wish to reject when $\mu \neq \mu_0$
2. $\mu = \mu_0$	Unknown σ	$t = \dfrac{(\bar{x} - \mu_0)\sqrt{n}}{s}$	Student's t with $(n - 1)$ d.f.	$t > t_{\alpha;n-1}$ if we wish to reject when $\mu > \mu_0$ $t < -t_{\alpha;n-1}$ if we wish to reject when $\mu < \mu_0$ $|t| > t_{\alpha/2;n-1}$ if we wish to reject when $\mu \neq \mu_0$
3. $\sigma = \sigma_0$		$\chi^2 = \dfrac{(n-1)s^2}{\sigma_0^2}$	χ^2 with $n - 1$ d.f.	$\chi^2 > \chi^2_{\alpha;n-1}$ if we wish to reject when $\sigma > \sigma_0$ $\chi^2 < \chi^2_{1-\alpha;n-1}$ if we wish to reject when $\sigma < \sigma_0$ $\chi^2 < \chi^2_{1-\alpha/2;n-1}$ or $\chi^2 > \chi^2_{\alpha/2;n-1}$ if we wish to reject when $\sigma \neq \sigma_0$
4. $\theta = \theta_0$	Large sample. (For small samples, exact tests are based on tables of binomial probabilities)	$z = \dfrac{\dfrac{x}{n} - \theta_0}{\sqrt{\dfrac{\theta_0(1 - \theta_0)}{n}}}$ Continuity correction: Replace x in numerator of formula with $x - \frac{1}{2}$ or $x + \frac{1}{2}$, whichever makes z numerically smallest.	Normal	$z > z_\alpha$ if we wish to reject when $\theta > \theta_0$ $z < -z_\alpha$ if we wish to reject when $\theta < \theta_0$ $|z| > -z_{\alpha/2}$ if we wish to reject when $\theta \neq \theta_0$

SUMMARY OF SIGNIFICANCE TESTS: COMPARISON OF TWO POPULATIONS

Hypothesis	Conditions	Test Statistic	Distribution of Test Statistic	Critical Region
1. $\mu_x = \mu_y$	Known σ_x and σ_y	$z = \dfrac{\bar{x} - \bar{y}}{\sqrt{\dfrac{\sigma_x^2}{n_x} + \dfrac{\sigma_y^2}{n_y}}}$	Normal	$z > z_\alpha$ if we wish to reject when $\mu_x > \mu_y$ $z < -z_\alpha$ if we wish to reject when $\mu_x < \mu_y$ $|z| > z_{\alpha/2}$ if we wish to reject when $\mu_x \neq \mu_y$
2. $\mu_x = \mu_y$	Unknown σ_x and σ_y $\sigma_x = \sigma_y$	$t = \dfrac{\bar{x} - \bar{y}}{\sqrt{\dfrac{(n_x-1)s_x^2 + (n_y-1)s_y^2}{n_x + n_y - 2}}\sqrt{\dfrac{1}{n_x} + \dfrac{1}{n_y}}}$	Student's t with $n-1$ d.f.	$t > t_{\alpha;n_x+n_y-2}$ if we wish to reject when $\mu_x > \mu_y$ $t < -t_{\alpha;n_x+n_y-2}$ if we wish to reject when $\mu_x < \mu_y$ $|t| > t_{\alpha/2;n_x+n_y-2}$ if we wish to reject when $\mu_x \neq \mu_y$
3. $\mu_x = \mu_y$	Unknown σ_x and σ_y $\sigma_x \neq \sigma_y$	$t = \dfrac{\bar{x} - \bar{y}}{\sqrt{\dfrac{s_x^2}{n_x} + \dfrac{s_y^2}{n_y}}}$	Student's t with ν d.f.	$t > t_{\alpha;\nu}$ if we wish to reject when $\mu_x > \mu_y$ $t < -t_{\alpha;\nu}$ if we wish to reject when $\mu_x < \mu_y$ $|t| > t_{\alpha/2;\nu}$ if we wish to reject when $\mu_x \neq \mu_y$ where d.f. ν is given by closest integer to $\nu = -2 + \dfrac{\left(\dfrac{s_x^2}{n_x} + \dfrac{s_y^2}{n_y}\right)^2}{\dfrac{\left(\dfrac{s_x^2}{n_x}\right)^2}{n_x+1} + \dfrac{\left(\dfrac{s_y^2}{n_y}\right)^2}{n_y+1}}$
4. $\mu_x = \mu_y$	Correlated pairs	$t = \dfrac{\bar{d}\sqrt{n}}{s_d}$ where $d_i = x_i - y_i$	Student's t with $n-1$ d.f.	$t > t_{\alpha;n-1}$ if we wish to reject when $\mu_x > \mu_y$ $t < -t_{\alpha;n-1}$ if we wish to reject when $\mu_x < \mu_y$ $|t| > t_{\alpha/2;n-1}$ if we wish to reject when $\mu_x \neq \mu_y$
5. $\sigma_x^2 = \sigma_y^2$		$F = \dfrac{s_x^2}{s_y^2}$ In a two-sided test, put larger mean square in the numerator	F with $n_x - 1$ and $n_y - 1$ d.f.	$F > F_{\alpha;n_x-1,n_y-1}$ if we wish to reject when $\sigma_x > \sigma_y$ $F > F_{\alpha/2;n_x-1,n_y-1}$ if $s_x^2 > s_y^2$ and we wish to reject when $\sigma_x \neq \sigma_y$ $F > F_{\alpha/2;n_y-1,n_x-1}$ if $s_x^2 < s_y^2$ and we wish to reject when $\sigma_x \neq \sigma_y$
6. $\theta_1 = \theta_2$	Large sample	$z = \dfrac{\dfrac{x_1}{n_1} - \dfrac{x_2}{n_2}}{\sqrt{\dfrac{\dfrac{x_1}{n_1}\left(1 - \dfrac{x_1}{n_1}\right)}{n_1} + \dfrac{\dfrac{x_2}{n_2}\left(1 - \dfrac{x_2}{n_2}\right)}{n_2}}}$ Continuity Correction: Replace x in numerator of formula with $x - \frac{1}{2}$ or $x + \frac{1}{2}$, whichever makes z numerically smallest.	Normal	$z > z_\alpha$ if we wish to reject when $\theta_1 > \theta_2$ $z < -z_\alpha$ if we wish to reject when $\theta_1 < \theta_2$ $|z| > z_{\alpha/2}$ if we wish to reject when $\theta_1 \neq \theta_2$

SUMMARY OF CONFIDENCE INTERVALS

Parameter	Conditions	Point Estimate	Confidence Interval
1. μ	Known σ	$\bar{x}$	$\bar{x} - z_{\alpha/2} \dfrac{\sigma}{\sqrt{n}} < \mu < \bar{x} + z_{\alpha/2} \dfrac{\sigma}{\sqrt{n}}$
2. μ	Unknown σ	$\bar{x}$	$\bar{x} - t_{\alpha/2} \dfrac{s}{\sqrt{n}} < \mu < \bar{x} + t_{\alpha/2} \dfrac{s}{\sqrt{n}}$
3. $\mu_x - \mu_y$	$\sigma_x = \sigma_y$ known	$\bar{x} - \bar{y}$	$\bar{x} - \bar{y} - z_{\alpha/2} \sqrt{\dfrac{\sigma_x{}^2}{n_x} + \dfrac{\sigma_y{}^2}{n_y}} < \mu_x - \mu_y$ $< \bar{x} - \bar{y} + z_{\alpha/2} \sqrt{\dfrac{\sigma_x{}^2}{n_x} + \dfrac{\sigma_y{}^2}{n_y}}$
4. $\mu_x - \mu_y$	$\sigma_x = \sigma_y$ unknown	$\bar{x} - \bar{y}$	$\bar{x} - \bar{y} - t_{\alpha/2} \dfrac{\sqrt{(n_x - 1)s_x{}^2 + (n_y - 1)s_y{}^2}}{\sqrt{\dfrac{n_x n_y (n_x + n_y - 2)}{n_x + n_y}}}$ $< \mu_x - \mu_y < \bar{x} - \bar{y}$ $+ t_{\alpha/2} \dfrac{\sqrt{(n_x - 1)s_x{}^2 + (n_y - 1)s_y{}^2}}{\sqrt{\dfrac{n_x n_y (n_x + n_y - 2)}{n_x + n_y}}}$
5. $\mu_d = \mu_x - \mu_y$	Correlated pairs σ_x and σ_y unknown	$\bar{d} = \bar{x} - \bar{y}$	$\bar{d} - t_{\alpha/2} \dfrac{s_d}{\sqrt{n}} < \mu_d < \bar{d} + t_{\alpha/2} \dfrac{s_d}{\sqrt{n}}$
6. σ		s	$\sqrt{\dfrac{(n - 1)s^2}{\chi^2_{\alpha/2; n-1}}} < \sigma < \sqrt{\dfrac{(n - 1)s^2}{\chi^2_{1-\alpha/2; n-1}}}$
7. $\dfrac{\sigma_x{}^2}{\sigma_y{}^2}$		$\dfrac{s_x{}^2}{s_y{}^2}$	$\dfrac{s_x{}^2}{s_y{}^2} \dfrac{1}{F_{\alpha/2; n_x-1, n_y-1}} < \dfrac{\sigma_x{}^2}{\sigma_y{}^2} < \dfrac{s_x{}^2}{s_y{}^2} \dfrac{1}{F_{1-\alpha/2; n_x-1, n_y-1}}$
8. θ	Large sample	$\dfrac{x}{n}$	$\dfrac{x}{n} - z_{\alpha/2} \sqrt{\dfrac{\dfrac{x}{n}\left(1 - \dfrac{x}{n}\right)}{n}} < \theta < \dfrac{x}{n}$ $+ z_{\alpha/2} \sqrt{\dfrac{\dfrac{x}{n}\left(1 - \dfrac{x}{n}\right)}{n}}$
9. $\theta_1 - \theta_2$	Large sample	$\dfrac{x_1}{n_1} - \dfrac{x_2}{n_2}$	$\dfrac{x_1}{n_1} - \dfrac{x_2}{n_2} - z_{\alpha/2} \sqrt{\dfrac{\dfrac{x_1}{n_1}\left(1 - \dfrac{x_1}{n_1}\right)}{n_1} + \dfrac{\dfrac{x_2}{n_2}\left(1 - \dfrac{x_2}{n_2}\right)}{n_2}}$ $< \theta_1 - \theta_2 < \dfrac{x_1}{n_1} - \dfrac{x_2}{n_2}$ $+ z_{\alpha/2} \sqrt{\dfrac{\dfrac{x_1}{n_1}\left(1 - \dfrac{x_1}{n_1}\right)}{n_1} + \dfrac{\dfrac{x_2}{n_2}\left(1 - \dfrac{x_2}{n_2}\right)}{n_2}}$

ANALYSIS OF VARIANCE (ANOVA) TABLES

The analysis of variance (ANOVA) table containing the sum of squares, degrees of freedom, mean square, expectations, etc., present the initial analysis in a compact form. This kind of tabular representation is customarily used to set out the results of analysis of variance calculations. Appropriate ANOVA tables for various experimental design models are presented here. In the tables, the use of "dot notation" indicates a summing over all observations in the population, i.e., when summing over a suffix, that suffix is replaced by a dot. Small letters refer to observations, whereas capital letters refer to observation totals.

ANALYSIS OF VARIANCE AND EXPECTED MEAN SQUARES FOR THE ONE-WAY CLASSIFICATION

Model: $y_{ij} = \mu + \alpha_i + \epsilon_{ij}$ $(i = 1, 2, \ldots, k; j = 1, 2, \ldots, n_i)$

Source of variation	Degrees of freedom	Sum of squares	Mean square	Test statistic
Between groups	$k - 1$	$S_1 = \sum_i n_i(\bar{y}_i - \bar{y}_{..})^2 = \sum_i \left(\frac{Y_i^2}{n_i}\right) - \frac{Y^2}{n}$	$s_1^2 = \dfrac{S_1}{k-1}$	$F = \dfrac{s_1^2}{s_c^2}$
Within groups	$n - k$	$S_c = \sum_i \sum_j (\bar{y}_{ij} - \bar{y}_i)^2 = \sum_i \sum_j y_{ij}^2 - \sum_i \left(\frac{Y_i^2}{n_i}\right)$	$s_c^2 = \dfrac{S_c}{n-k}$	
Total	$n - 1$	$S = \sum_i \sum_j (y_{ij} - \bar{y})^2 = \sum_i \sum_j y_{ij}^2 - \frac{Y^2}{n}$		

Source of variation	Degrees of freedom	Mean square	Expected mean square for	
			Fixed model	**Random model**
Between groups	$k - 1$	s_1^2	$\sigma^2 + \dfrac{\sum_i n_i \alpha_i^2}{k-1}$	$\sigma^2 + \dfrac{1}{k-1}\left(n - \dfrac{\sum n_i^2}{n}\right)\sigma_\alpha^2$
Within groups	$n - k$	s_c^2	σ^2	σ^2
Total	$n - 1$			

Notation:

$$Y_i = \sum_j y_{ij}; \quad Y = \sum_i \sum_j y_{ij}; \quad \bar{y}_i = \frac{1}{n_i}\sum_j y_{ij} = \frac{1}{n_i}Y_i;$$

$$n = \sum_i n_i; \quad \bar{y} = \frac{1}{n}\sum_i \sum_j y_{ij} = \frac{Y}{n}$$

ANALYSIS OF VARIANCE AND EXPECTED MEAN SQUARES FOR THE TWO-WAY CLASSIFICATION WITH ONE OBSERVATION PER CELL

Model: $y_{ij} = \mu + \alpha_i + \beta_j + \epsilon_{ij}$ $(i = 1, 2, \ldots c; j = 1, 2, \ldots, r)$

Source of variation	Degrees of freedom	Sum of squares	Mean square	Test statistic
Column effects	$c - 1$	$SSC = \dfrac{\sum_i Y_{i.}^2}{r} - \dfrac{Y_{..}^2}{cr}$	$s_1^2 = \dfrac{SSC}{c - 1}$	$\dfrac{s_1^2}{s_e^2}$
Row effects	$r - 1$	$SSR = \dfrac{\sum_j Y_{.j}^2}{c} - \dfrac{Y_{..}^2}{cr}$	$s_2^2 = \dfrac{SSR}{r - 1}$	$\dfrac{s_2^2}{s_e^2}$
Error	$(c - 1)(r - 1)$	$SSE = SST - SSC - SSR$	$s_e^2 = \dfrac{SSE}{(c - 1)(r - 1)}$	
Total	$cr - 1$	$SST = \sum_i \sum_j y_{ij}^2 - \dfrac{Y_{..}^2}{cr}$		

Source of variation	Degrees of freedom	Mean square	Expected mean square for		
			Fixed model	**Mixed model (α)**	**Random model**
Column effects	$c - 1$	s_1^2	$\sigma^2 + r\left(\dfrac{\sum_i \alpha_i^2}{c - 1}\right)$	$\sigma^2 + r\left(\dfrac{\sum_i \alpha_i^2}{c - 1}\right)$	$\sigma^2 + r\sigma_\alpha^2$
Row effects	$r - 1$	s_2^2	$\sigma^2 + c\left(\dfrac{\sum_j \beta_j^2}{r - 1}\right)$	$\sigma^2 + c\sigma_\beta^2$	$\sigma^2 + c\sigma_\beta^2$
Error	$(c - 1)(r - 1)$	s_e^2	σ^2	σ^2	σ^2
Total	$cr - 1$				

ANALYSIS OF VARIANCE AND EXPECTED MEAN SQUARES FOR NESTED CLASSIFICATIONS WITH UNEQUAL SAMPLES

Model: $y_{iju} = \mu + \alpha_i + \delta_{ij} + \epsilon_{iju}$ $(i = 1, 2, \ldots, k; j = 1, 2, \ldots, n_i; u = 1, 2, \ldots, n_{ij})$

Source of variation	Degrees of freedom	Sum of squares	Mean square	Expected mean square for fixed model (α, δ)
Between main groups	$k - 1$	$S_1 = \sum_i \dfrac{Y_{i.}^2}{n_i} - \dfrac{Y_{..}^2}{n_{..}}$	$s_1^2 = \dfrac{S_1}{k - 1}$	$\sigma^2 + \dfrac{\sum_i n_i \alpha_i^2}{k - 1}$
Subgroups within main groups (experimental error)	$\sum_i n_i - k$	$S_2 = \sum_i \sum_j \dfrac{Y_{ij}^2}{n_{ij}} - \sum_i \dfrac{Y_{i.}^2}{n_i}$	$s_2^2 = \dfrac{S_2}{\sum_i n_i - k}$	$\sigma^2 + \dfrac{\sum_i \sum_j n_{ij} \delta_{ij}^2}{\sum_i n_i - k}$
Within subgroups (sampling error)	$n_{..} - \sum_i n_i$	$S_e = \sum_i \sum_j \sum_u y_{iju}^2 - \sum_i \sum_j \dfrac{Y_{ij}^2}{n_{ij}}$	$s_e^2 = \dfrac{S_3}{n_{..} - \sum_i n_i}$	σ^2
Total	$n_{..} - 1$	$S = \sum_i \sum_j \sum_u y_{iju}^2 - \dfrac{Y_{..}^2}{n_{..}}$		

Source of variation	Degrees of freedom	Mean square	Expected mean square for		
			Mixed model (α)	Mixed model (δ)	Random model
Between main groups	$k - 1$	s_1^2	$\sigma^2 + b\sigma_\delta^2 + \dfrac{\sum_i n_i \alpha_i^2}{k - 1}$	$\sigma^2 + c\delta_\alpha^2$	$\sigma^2 + b\sigma_\delta^2 + c\sigma_\alpha^2$
Experimental error	$\sum_i n_i - k$	s_2^2	$\sigma^2 + a\sigma_\delta^2$	$\sigma^2 + \dfrac{\sum_i \sum_j n_{ij} \delta_{ij}^2}{\sum_i n_i - k}$	$\sigma^2 + a\sigma_\delta^2$
Sampling error	$n_{..} - \sum_i n_i$	s_e^2	σ^2	σ^2	σ^2
Total	$n_{..} - 1$				

where

$$a = \frac{n_{..} - \sum_i \dfrac{\sum_j n_{ij}^2}{n_i}}{\sum_i n_i - k}$$

$$b = \frac{\sum_i \dfrac{\sum_j n_{ij}^2}{n_i} - \dfrac{\sum_i \sum_j n_{ij}^2}{n_{..}}}{k - 1}$$

$$c = \frac{n_{..} - \dfrac{\sum_i n_i^2}{n_{..}}}{k - 1}$$

ANALYSIS OF VARIANCE AND EXPECTED MEAN SQUARES FOR NESTED CLASSIFICATIONS WITH EQUAL SAMPLES

Model: $y_{iju} = \mu + \alpha_i + \delta_{ij} + \epsilon_{iju}$ $(i = 1, 2, \ldots, k; j = 1, 2, \ldots, n; u = 1, 2, \ldots, r)$

Source of variation	Degrees of freedom	Sum of squares	Mean square	Expected mean square for fixed model (α, δ)
Between main groups	$k - 1$	$S_1 = \sum_i \dfrac{Y_i^2}{nr} - \dfrac{Y^2}{knr}$	$s_1^2 = \dfrac{S_1}{k - 1}$	$\sigma^2 + nr \dfrac{\sum_i \alpha_i^2}{k - 1}$
Experimental error	$k(n - 1)$	$S_2 = \dfrac{\sum_i \sum_j Y_{ij}^2}{r} - \dfrac{\sum_i Y_i^2}{nr}$	$s_2^2 = \dfrac{S_2}{k(n - 1)}$	$\sigma^2 + r \dfrac{\sum_i \sum_j \delta_{ij}^2}{k(n - 1)}$
Sampling error	$kn(r - 1)$	$S_e = \sum_i \sum_j \sum_u y_{iju}^2 - \dfrac{\sum_i \sum_j Y_{ij}^2}{r}$	$s_e^2 = \dfrac{S_e}{kn(r - 1)}$	σ^2
Total	$kn(r - 1)$	$S = \sum_i \sum_j \sum_u y_{iju}^2 - \dfrac{Y^2}{knr}$		

Source of variation	Degrees of freedom	Mean square	Expected mean square for		
			Mixed model (α)	Mixed model (δ)	Random model
Between main groups	$k - 1$	s_1^2	$\sigma^2 + r\sigma_\delta^2 + nr\left(\dfrac{\sum_i \alpha_i^2}{k - 1}\right)$	$\sigma^2 + nr\sigma_\alpha^2$	$\sigma^2 + r\sigma_\delta^2 + nr\sigma_\alpha^2$
Experimental error	$k(n - 1)$	s_2^2	$\sigma^2 + r\sigma_\delta^2$	$\sigma^2 + \dfrac{r\sum_i \sum_j \delta_{ij}^2}{k(n - 1)}$	$\sigma^2 + r\sigma_\delta^2$
Sampling error	$kn(r - 1)$	s_e^2	σ^2	σ^2	σ^2
Total	$knr - 1$				

where

$$a = b = r$$
$$c = nr$$

ANALYSIS OF VARIANCE AND EXPECTED MEAN SQUARES FOR A FIXED MODEL TWO-FACTOR FACTORIAL EXPERIMENT IN A ONE-WAY CLASSIFICATION DESIGN

Model: $y_{iju} = \mu + \alpha_i + \beta_j + (\alpha\beta)_{ij} + \epsilon_{iju}$
$(i = 1, 2, \ldots, c; j = 1, 2, \ldots, r; u = 1, 2, \ldots, n)$

Source of variation	Degrees of freedom	Sum of squares	Mean square	Expected mean square for fixed model $[\alpha, \beta, (\alpha\beta)]$
Treatment combinations	$cr - 1$	$SSTr$	$s_0^2 = \dfrac{SSTr}{cr - 1}$	$\sigma^2 + n\dfrac{\sum_i^c \sum_j^r (\mu_{ij} - \mu)^2}{cr - 1}$
Factor A	$c - 1$	SSA	$s_1^2 = \dfrac{SSA}{c - 1}$	$\sigma^2 + rn\dfrac{\sum_i^c \alpha_i^2}{c - 1}$
Factor B	$r - 1$	SSB	$s_2^2 = \dfrac{SSB}{r - 1}$	$\sigma^2 + cn\dfrac{\sum_j^r \beta_j^2}{r - 1}$
Interaction	$(c - 1)(r - 1)$	$SSAB = SSTr - SSA - SSB$	$s_3^2 = \dfrac{SSAB}{(c - 1)(r - 1)}$	$\sigma^2 + n\dfrac{\sum_i^c \sum_j^r (\alpha\beta)_{ij}^2}{(c - 1)(r - 1)}$
Within (error)	$cr(n - 1)$	$SSW = SST - SSTr$	$s_e^2 = \dfrac{SSW}{cr(n - 1)}$	σ^2
Total	$crn - 1$	SST		

where

$$SSTr = \frac{\sum_i^c \sum_j^r Y_{ij}^2}{n} - \frac{Y_{\ldots}^2}{crn} \qquad SSA = \frac{\sum_i^c Y_{i\ldots}^2}{rn} - \frac{Y_{\ldots}^2}{crn}$$

$$SSB = \frac{\sum_j^r Y_{.j.}^2}{cn} - \frac{Y_{\ldots}^2}{crn} \qquad SST = \sum_i^c \sum_j^r \sum_u^n y_{iju}^2 - \frac{Y_{\ldots}^2}{crn}$$

$$Y_{ij} = \sum_u^n y_{iju} \qquad Y_{i\ldots} = \sum_j^r \sum_u^n y_{iju} \qquad Y_{.j.} = \sum_i^c \sum_u^n y_{iju}$$

Factor A	$s_1^2 = \dfrac{SSA}{c - 1}$	$\sigma^2 + n\sigma_{\alpha\beta}^2 + rn\sigma_\alpha^2$	$\sigma^2 + n\sigma_{\alpha\beta}^2 + rn\dfrac{\sum_i \alpha_i^2}{c - 1}$
Factor B	$s_2^2 = \dfrac{SSB}{r - 1}$	$\sigma^2 + n\sigma_{\alpha\beta}^2 + cn\sigma_\beta^2$	$\sigma^2 + cn\sigma_\beta^2$
Interaction	$s_3^2 = \dfrac{SSAB}{(c - 1)(r - 1)}$	$\sigma^2 + n\sigma_{\alpha\beta}^2$	$\sigma^2 + n\sigma_{\alpha\beta}^2$
Within (error)	$s_e^2 = \dfrac{SSW}{cr(n - 1)}$	σ^2	σ^2
Total	$s_5^2 = \dfrac{SST}{crn - 1}$		

ANALYSIS OF VARIANCE AND EXPECTED MEAN SQUARES FOR A THREE-FACTOR FACTORIAL EXPERIMENT IN A COMPLETELY RANDOMIZED DESIGN

Model: $y_{ijku} = \mu + \alpha_i + \beta_j + \gamma_k + (\alpha\beta)_{ij} + (\alpha\gamma)_{ik} + (\beta\gamma)_{jk} + (\alpha\beta\gamma)_{ijk} + \epsilon_{ijku}$

$(i = 1, 2, \ldots, c; j = 1, 2, \ldots, r; k = 1, 2, \ldots, l; u = 1, 2, \ldots, n)$

Source of variation	Degrees of freedom	Sum of squares	Mean square	Expected mean square for *Fixed Model*
Factor A	$c - 1$	SSA	s_1^2	$\sigma^2 + rln \dfrac{\sum_i \alpha_i^2}{c - 1}$
Factor B	$r - 1$	SSB	s_2^2	$\sigma^2 + cln \dfrac{\sum_j \beta_j^2}{r - 1}$
Factor C	$l - 1$	SSC	s_3^2	$\sigma^2 + crn \dfrac{\sum_k \gamma_k^2}{l - 1}$
Interaction $A \times B$	$(c - 1)(r - 1)$	SSAB	s_4^2	$\sigma^2 + ln \dfrac{\sum_i \sum_j (\alpha\beta)_{ij}^2}{(c - 1)(r - 1)}$
Interaction $A \times C$	$(c - 1)(l - 1)$	SSAC	s_5^2	$\sigma^2 + rn \dfrac{\sum_i \sum_k (\alpha\beta)_{ik}^2}{(c - 1)(l - 1)}$
Interaction $B \times C$	$(r - 1)(l - 1)$	SSBC	s_6^2	$\sigma^2 + cn \dfrac{\sum_j \sum_k (\beta\gamma)_{jk}^2}{(r - 1)(l - 1)}$
Interaction $A \times B \times C$	$(c - 1)(r - 1)(l - 1)$	SSABC	s_7^2	$\sigma^2 + n \dfrac{\sum_i \sum_j \sum_k (\alpha\beta\gamma)_{ijk}^2}{(c - 1)(r - 1)(l - 1)}$
Within (error)	$crl(n - 1)$	SSE	s_e^2	σ^2
Total	$crln - 1$	SST		

where

$$SST = \sum_i \sum_j \sum_k \sum_u y_{ijku}^2 - \frac{Y_{\ldots}^2}{crln}$$

$$SSA = \frac{\sum_i Y_{i\ldots}^2}{rln} - \frac{Y_{\ldots}^2}{crln}$$

$$SSB = \frac{\sum_j Y_{.j..}^2}{cln} - \frac{Y_{\ldots}^2}{crln}$$

$$SSC = \frac{\sum_k Y_{..k.}^2}{crn} - \frac{Y_{\ldots}^2}{crln}$$

$$SSTr(ABC) = \frac{\sum_i \sum_j \sum_k Y_{ijk.}^2}{n} - \frac{Y_{\ldots}^2}{crln}$$

$$SSTr(AB) = \frac{\sum_i \sum_j Y_{ij..}^2}{ln} - \frac{Y_{\ldots}^2}{crln}$$

ANALYSIS OF VARIANCE AND EXPECTED MEAN SQUARES FOR A THREE-FACTOR FACTORIAL EXPERIMENT IN A COMPLETELY RANDOMIZED DESIGN (continued)

$$SSTr(AC) = \frac{\sum_i \sum_k Y_{i.k}^2}{rn} - \frac{Y_{...}^2}{crln}$$

$$SSTr(BC) = \frac{\sum_j \sum_k Y_{.jk}^2}{cn} - \frac{Y_{...}^2}{crln}$$

$$SSAB = SSTr(AB) - SSA - SSB$$
$$SSAC = SSTr(AC) - SSA - SSC$$
$$SSBC = SSTr(BC) - SSB - SSC$$
$$SSABC = SSTr(ABC) - SSA - SSB - SSC - SSAB - SSAC - SSBC$$
$$SSE = SST - SSTr(ABC)$$

Source of variation	Mean square	Expected mean square for the		
		Random model	Mixed model (α)	Mixed model (α,β)
Factor A	$s_1^2 = \dfrac{SSA}{c-1}$	$\sigma^2 + n\sigma_{\alpha\beta\gamma}^2 + ln\sigma_{\alpha\beta}^2$ $+ rn\sigma_{\alpha\gamma}^2 + rln\sigma_\alpha^2$	$\sigma^2 + n\sigma_{\alpha\beta\gamma}^2 + ln\sigma_{\alpha\beta}^2$ $+ rn\sigma_{\alpha\gamma}^2 + rln\dfrac{\sum_i \alpha_i^2}{c-1}$	$\sigma^2 + rn\sigma_{\alpha\gamma}^2$ $+ rln\dfrac{\sum_i \alpha_i^2}{c-1}$
Factor B	$s_2^2 = \dfrac{SSB}{r-1}$	$\sigma^2 + n\sigma_{\alpha\beta\gamma}^2 + ln\sigma_{\alpha\beta}^2$ $+ cn\sigma_{\beta\gamma}^2 + cln\sigma_\beta^2$	$\sigma^2 + cn\sigma_{\beta\gamma}^2 + cln\sigma_\beta^2$	$\sigma^2 + cn\sigma_{\beta\gamma}^2$ $+ cln\dfrac{\sum_j \beta_j^2}{r-1}$
Factor C	$s_3^2 = \dfrac{SSC}{l-1}$	$\sigma^2 + n\sigma_{\alpha\beta\gamma}^2 + rn\sigma_{\alpha\gamma}^2$ $+ cn\sigma_{\beta\gamma}^2 + crn\sigma_\gamma^2$	$\sigma^2 + cn\sigma_{\beta\gamma}^2 + crn\sigma_\gamma^2$	$\sigma^2 + crn\sigma_\gamma^2$
$A \times B$	$s_4^2 = \dfrac{SSAB}{(c-1)(r-1)}$	$\sigma^2 + n\sigma_{\alpha\beta\gamma}^2 + ln_{\alpha\beta}^2$	$\sigma^2 + n\sigma_{\alpha\beta\gamma}^2 + ln_{\alpha\beta}^2$	$\sigma^2 + n\sigma_{\alpha\beta\gamma}^2$ $+ ln\dfrac{\sum_i \sum_j (\alpha\beta)_{ij}^2}{(c-1)(r-1)}$
$A \times C$	$s_5^2 = \dfrac{SSAC}{(c-1)(l-1)}$	$\sigma^2 + n\sigma_{\alpha\beta\gamma}^2 + rn\sigma_{\alpha\gamma}^2$	$\sigma^2 + n\sigma_{\alpha\beta\gamma}^2 + rn\sigma_{\alpha\gamma}^2$	$\sigma^2 + rn\sigma_{\alpha\gamma}^2$
$B \times C$	$s_6^2 = \dfrac{SSBC}{(r-1)(l-1)}$	$\sigma^2 + n\sigma_{\alpha\beta\gamma}^2 + cn\sigma_{\beta\gamma}^2$	$\sigma^2 + cn\sigma_{\beta\gamma}^2$	$\sigma^2 + cn\sigma_{\beta\gamma}^2$
$A \times B \times C$	$s_7^2 = \dfrac{SSABC}{(c-1)(r-1)(l-1)}$	$\sigma^2 + n\sigma_{\alpha\beta\gamma}^2$	$\sigma^2 + n\sigma_{\alpha\beta\gamma}^2$	$\sigma^2 + n\sigma_{\alpha\beta\gamma}^2$
Within (error)	$s_r^2 = \dfrac{SSE}{crl(n-1)}$	σ^2	σ^2	σ^2
Total	$s_0^2 = \dfrac{SST}{crln-1}$			

ANALYSIS OF VARIANCE AND EXPECTED MEAN SQUARES
FOR A $t \times t$ LATIN SQUARE

Model: $y_{ij(k)} = \mu + \alpha_i + \beta_j + \gamma_{(k)} + \epsilon_{ij(k)}$
$(i = 1, 2, \ldots, t; j = 1, 2, \ldots, t; k = 1, 2, \ldots, t)$

Source of variation	Degrees of freedom	Sum of squares	Mean square	Expected mean square for fixed model
Columns	$t - 1$	$SSC = \dfrac{\sum_i Y^2_{i.}}{t} - \dfrac{Y^2_{..}}{t^2}$	$s^2_1 = \dfrac{SSC}{t - 1}$	$\sigma^2 + t \dfrac{\sum_i \alpha^2_i}{t - 1}$
Rows	$t - 1$	$SSR = \dfrac{\sum_j Y^2_{.j}}{t} - \dfrac{Y^2_{..}}{t^2}$	$s^2_2 = \dfrac{SSR}{t - 1}$	$\sigma^2 + t \dfrac{\sum_j \beta^2_j}{t - 1}$
Treatments	$t - 1$	$SSTr = \dfrac{\sum_k Y^2_{(k)}}{t} - \dfrac{Y^2_{..}}{t^2}$	$s^2_3 = \dfrac{SSTr}{t - 1}$	$\sigma^2 + t \dfrac{\sum_k \gamma^2_k}{t - 1}$
Error	$(t - 1)(t - 2)$	$SSE = SST - SSC - SSR - SSTr$	$s^2_e = \dfrac{SSE}{(t - 1)(t - 2)}$	σ^2
Total	$t^2 - 1$	$SST = \sum_i \sum_j y^2_{ij(k)} - \dfrac{Y^2_{..}}{t^2}$		

Source of variation	Mean square	Expected mean square for Random model	Expected mean square for Mixed model (γ)	Expected mean square for Mixed model (α, γ)
Columns	$s^2_1 = \dfrac{SSC}{t - 1}$	$\sigma^2 + t\sigma^2_\alpha$	$\sigma^2 + t\sigma^2_\alpha$	$\sigma^2 + t \dfrac{\sum_i \alpha^2_i}{t - 1}$
Rows	$s^2_2 = \dfrac{SSR}{t - 1}$	$\sigma^2 + t\sigma^2_\beta$	$\sigma^2 + t\sigma^2_\beta$	$\sigma^2 + t\sigma^2_\beta$
Treatments	$s^2_3 = \dfrac{SSTr}{t - 1}$	$\sigma^2 + t\sigma^2_\gamma$	$\sigma^2 + t \dfrac{\sum_k \gamma^2_k}{t - 1}$	$\sigma^2 + t \dfrac{\sum_k \gamma^2_k}{t - 1}$
Error	$s^2_e = \dfrac{SSE}{(t - 1)(t - 2)}$	σ^2	σ^2	σ^2

ANALYSIS OF VARIANCE FOR A GRAECO-LATIN SQUARE

Model: $y_{ijuk} = \mu + \alpha_i + \beta_j + \gamma_u + \delta_k + \epsilon_{ijuk}$ $(i, j, u, k = 1, 2, \ldots, n)$

Source of variation	Degrees of freedom	Sum of squares	Mean square
Factor I (rows)	$n - 1$	$S_1 = \dfrac{\sum_i Y_i^2}{n} - \dfrac{Y^2}{n^2}$	$s_1^2 = \dfrac{S_1}{n - 1}$
Factor II (columns)	$n - 1$	$S_2 = \dfrac{\sum_j Y_j^2}{n} - \dfrac{Y^2}{n^2}$	$s_2^2 = \dfrac{S_2}{n - 1}$
Factor III (Latin letters)	$n - 1$	$S_3 = \dfrac{\sum_u Y_u^2}{n - 1} - \dfrac{Y^2}{n^2}$	$s_3^2 = \dfrac{S_3}{n - 1}$
Factor IV (Greek letters)	$n - 1$	$S_4 = \dfrac{\sum_k Y_k^2}{n} - \dfrac{Y^2}{n^2}$	$s_4^2 = \dfrac{S_4}{n - 1}$
Residual	$(n - 1)(n - 3)$	$S_c = $ difference	$s_c^2 = \dfrac{S_c}{(n - 1)(n - 3)}$
Total	$n^2 - 1$	$S = \sum_i \sum_j y_{ijuk}^2 - \dfrac{Y^2}{n^2}$	

ANALYSIS OF VARIANCE FOR A YOUDEN SQUARE

Model: $y_{iju} = \mu + \alpha_i + \beta_j + \gamma_u + \epsilon_{iju}$

$(i = 1, 2, \ldots, b; j = 1, 2, \ldots, t(= b); u = 1, 2, \ldots, k(<t))$

Source of variation	Degrees of freedom	Sum of squares	Mean square
Blocks (crude)		$S_1 = \sum_i \dfrac{Y_i^2}{k} - \dfrac{Y^2}{bk}$	
Treatments (adjusted)	$t - 1$	$S_2 = \dfrac{t - 1}{bk^2(k - 1)} \sum_i \left(kY_i^2 - \sum_{u(i)} Y_i^2 \right)$	$s_1^2 = \dfrac{S_2}{t - 1}$
Treatments (crude)		$S_3 = \sum_i \dfrac{Y_i^2}{r} - \dfrac{Y^2}{tr}$	
Blocks (adjusted)	$b - 1$	$S_4 = \dfrac{b - 1}{bk^2(k - 1)} \sum_i \left(rY_i^2 - \sum_{u(i)} Y_i^2 \right)$	$s_2^2 = \dfrac{S_4}{b - 1}$
Factor II (γ)	$k - 1$	$S_5 = \dfrac{\sum_u Y_u^2}{k} - \dfrac{Y^2}{bk}$	$s_3^2 = \dfrac{S_5}{k - 1}$
Residual	$bk - t - b - k + 2$	$\begin{aligned} S_c &= S - (S_1 + S_2 + S_5) \\ &= S - (S_3 + S_4 + S_5) \end{aligned}$	$s_c^2 = \dfrac{S_c}{bk - t - b - k + 2}$
Total	$bk - 1$	$S = \sum_i \sum_j y_{iju}^2 - \dfrac{Y^2}{bk}$	

(Note that $S_1 + S_2 = S_3 + S_4$.)

ANALYSIS OF VARIANCE FOR BALANCED INCOMPLETE BLOCK (BIB)

Model: $y_{iju} = \mu + \alpha_i + \beta_j + \epsilon_{iju}$ $(i = 1, 2, \ldots, b; j = 1, 2, \ldots, t; u = n_{ij})$

Source of variation	Degrees of freedom	Sum of squares	Mean square
Blocks	$b - 1$	$S_1 = \dfrac{\sum_i Y_i^2}{k} - \dfrac{Y^2}{bk}$	$s_1^2 = \dfrac{S_1}{b - 1}$
Treatments (adjusted)	$t - 1$	$S_2 = \dfrac{t - 1}{bk^2(k - 1)} \sum_i \left[kY_i - \sum_{ij} Y_i \right]^2$	$s_2^2 = \dfrac{S_2}{t - 1}$
Residual	$bk - t - b + 1$	$S_r = \text{difference}$	$s_r^2 = \dfrac{S_r}{bk - t - b + 1}$
Total	$bk - 1$	$S = \sum_i \sum_j y_{iju}^2 - \dfrac{Y^2}{bk}$	

where

t = number of treatment levels
b = number of blocks
k = number of treatment levels per block
r = number of replications of each treatment level
λ = number of blocks in which any given pair of treatment levels appear together
$bk = tr$
$r(k - 1) = \lambda(t - 1)$

THE NORMAL PROBABILITY FUNCTION
AND RELATED FUNCTIONS

This table gives values of:

a) $f(x)$ = the probability density of a standardized random variable

$$= \frac{1}{\sqrt{2\pi}} e^{-\frac{1}{2}x^2}$$

For negative values of x, one uses the fact that $f(-x) = f(x)$.

b) $F(x)$ = the cumulative distribution function of a standardized normal random variable

$$= \int_{-\infty}^{x} \frac{1}{\sqrt{2\pi}} e^{-\frac{1}{2}t^2} dt$$

For negative values of x, one uses the relationship $F(-x) = 1 - F(x)$. Values of x corresponding to a few special values of $F(x)$ are given in a separate table following the main table.

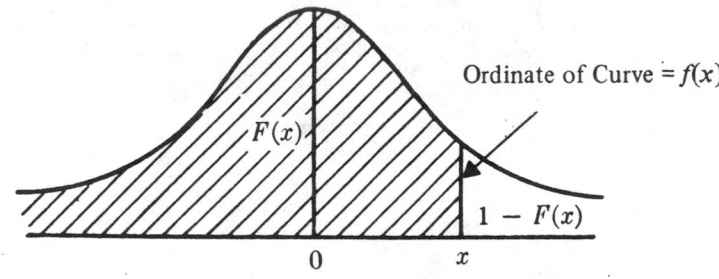

c) $f'(x)$ = the first derivative of $f(x)$ with respect to x

$$= -\frac{x}{\sqrt{2\pi}} e^{-\frac{1}{2}x^2} = -xf(x)$$

d) $f''(x)$ = the second derivative of $f(x)$ with respect to x

$$= \frac{(x^2 - 1)}{\sqrt{2\pi}} e^{-\frac{1}{2}x^2} = (x^2 - 1)f(x)$$

e) $f'''(x)$ = the third derivative of $f(x)$ with respect to x

$$= \frac{3x - x^3}{\sqrt{2\pi}} e^{-\frac{1}{2}x^2} = (3x - x^3)f(x)$$

f) $f^{\mathrm{iv}}(x)$ = the fourth derivative of $f(x)$ with respect to x

$$= \frac{x^4 - 6x^2 + 3}{\sqrt{2\pi}} e^{-\frac{1}{2}x^2} = (x^4 - 6x^2 + 3)f(x)$$

THE NORMAL PROBABILITY FUNCTION AND
RELATED FUNCTIONS (Continued)

It should be noted that other probability integrals can be evaluated by the use of these tables. For example,

$$\int_0^x f(t)dt = \tfrac{1}{2}\operatorname{erf}\left(\frac{x}{\sqrt{2}}\right),$$

where $\operatorname{erf}\left(\dfrac{x}{\sqrt{2}}\right)$ represents the error function associated with the normal curve.

To evaluate erf (2.3) one proceeds as follows: Since $\dfrac{x}{\sqrt{2}} = 2.3$, one finds $x = (2.3)(\sqrt{2}) = 3.25$. In the entry opposite $x = 3.25$, the value 0.9994 is given. Subtracting 0.5000 from the tabular value, one finds the value 0.4994. Thus erf (2.3) = 2(0.4994) = 0.9988.

NORMAL PROBABILITY FUNCTION AND RELATED FUNCTIONS (Continued)

x	$F(x)$	$1 - F(x)$	$f(x)$	$f'(x)$	$f''(x)$	$f'''(x)$	$f^{\mathrm{iv}}(x)$
.00	.5000	.5000	.3989	− .0000	− .3989	.0000	1.1968
.01	.5040	.4960	.3989	− .0040	− .3989	.0120	1.1965
.02	.5080	.4920	.3989	− .0080	− .3987	.0239	1.1956
.03	.5120	.4880	.3988	− .0120	− .3984	.0359	1.1941
.04	.5160	.4840	.3986	− .0159	− .3980	.0478	1.1920
.05	.5199	.4801	.3984	− .0199	− .3975	.0597	1.1894
.06	.5239	.4761	.3982	− .0239	− .3968	.0716	1.1861
.07	.5279	.4721	.3980	− .0279	− .3960	.0834	1.1822
.08	.5319	.4681	.3977	− .0318	− .3951	.0952	1.1778
.09	.5359	.4641	.3973	− .0358	− .3941	.1070	1.1727
.10	.5398	.4602	.3970	− .0397	− .3930	.1187	1.1671
.11	.5438	.4562	.3965	− .0436	− .3917	.1303	1.1609
.12	.5478	.4522	.3961	− .0475	− .3904	.1419	1.1541
.13	.5517	.4483	.3956	− .0514	− .3889	.1534	1.1468
.14	.5557	.4443	.3951	− .0553	− .3873	.1648	1.1389
.15	.5596	.4404	.3945	− .0592	− .3856	.1762	1.1304
.16	.5636	.4364	.3939	− .0630	− .3838	.1874	1.1214
.17	.5675	.4325	.3932	− .0668	− .3819	.1986	1.1118
.18	.5714	.4286	.3925	− .0707	− .3798	.2097	1.1017
.19	.5753	.4247	.3918	− .0744	− .3777	.2206	1.0911
.20	.5793	.4207	.3910	− .0782	− .3754	.2315	1.0799
.21	.5832	.4168	.3902	− .0820	− .3730	.2422	1.0682
.22	.5871	.4129	.3894	− .0857	− .3706	.2529	1.0560
.23	.5910	.4090	.3885	− .0894	− .3680	.2634	1.0434
.24	.5948	.4052	.3876	− .0930	− .3653	.2737	1.0302
.25	.5987	.4013	.3867	− .0967	− .3625	.2840	1.0165
.26	.6026	.3974	.3857	− .1003	− .3596	.2941	1.0024
.27	.6064	.3936	.3847	− .1039	− .3566	.3040	0.9878
.28	.6103	.3897	.3836	− .1074	− .3535	.3138	0.9727
.29	.6141	.3859	.3825	− .1109	− .3504	.3235	0.9572
.30	.6179	.3821	.3814	− .1144	− .3471	.3330	0.9413
.31	.6217	.3783	.3802	− .1179	− .3437	.3423	0.9250
.32	.6255	.3745	.3790	− .1213	− .3402	.3515	0.9082
.33	.6293	.3707	.3778	− .1247	− .3367	.3605	0.8910
.34	.6331	.3669	.3765	− .1280	− .3330	.3693	0.8735
.35	.6368	.3632	.3752	− .1313	− .3293	.3779	0.8556
.36	.6406	.3594	.3739	− .1346	− .3255	.3864	0.8373
.37	.6443	.3557	.3726	− .1378	− .3216	.3947	0.8186
.38	.6480	.3520	.3712	− .1410	− .3176	.4028	0.7996
.39	.6517	.3483	.3697	− .1442	− .3135	.4107	0.7803
.40	.6554	.3446	.3683	− .1473	− .3094	.4184	0.7607
.41	.6591	.3409	.3668	− .1504	− .3051	.4259	0.7408
.42	.6628	.3372	.3653	− .1534	− .3008	.4332	0.7206
.43	.6664	.3336	.3637	− .1564	− .2965	.4403	0.7001
.44	.6700	.3300	.3621	− .1593	− .2920	.4472	0.6793
.45	.6736	.3264	.3605	− .1622	− .2875	.4539	0.6583
.46	.6772	.3228	.3589	− .1651	− .2830	.4603	0.6371
.47	.6808	.3192	.3572	− .1679	− .2783	.4666	0.6156
.48	.6844	.3156	.3555	− .1707	− .2736	.4727	0.5940
.49	.6879	.3121	.3538	− .1734	− .2689	.4785	0.5721
.50	.6915	.3085	.3521	− .1760	− .2641	.4841	0.5501

NORMAL PROBABILITY FUNCTION AND RELATED FUNCTIONS (Continued)

x	$F(x)$	$1 - F(x)$	$f(x)$	$f'(x)$	$f''(x)$	$f'''(x)$	$f^{IV}(x)$
.50	.6915	.3085	.3521	−.1760	−.2641	.4841	.5501
.51	.6950	.3050	.3503	−.1787	−.2592	.4895	.5279
.52	.6985	.3015	.3485	−.1812	−.2543	.4947	.5056
.53	.7019	.2981	.3467	−.1837	−.2493	.4996	.4831
.54	.7054	.2946	.3448	−.1862	−.2443	.5043	.4605
.55	.7088	.2912	.3429	−.1886	−.2392	.5088	.4378
.56	.7123	.2877	.3410	−.1920	−.2341	.5131	.4150
.57	.7157	.2843	.3391	−.1933	−.2289	.5171	.3921
.58	.7190	.2810	.3372	−.1956	−.2238	.5209	.3691
.59	.7224	.2776	.3352	−.1978	−.2185	.5245	.3461
.60	.7257	.2743	.3332	−.1999	−.2133	.5278	.3231
.61	.7291	.2709	.3312	−.2020	−.2080	.5309	.3000
.62	.7324	.2676	.3292	−.2041	−.2027	.5338	.2770
.63	.7357	.2643	.3271	−.2061	−.1973	.5365	.2539
.64	.7389	.2611	.3251	−.2080	−.1919	.5389	.2309
.65	.7422	.2578	.3230	−.2099	−.1865	.5411	.2078
.66	.7454	.2546	.3209	−.2118	−.1811	.5431	.1849
.67	.7486	.2514	.3187	−.2136	−.1757	.5448	.1620
.68	.7517	.2483	.3166	−.2153	−.1702	.5463	.1391
.69	.7549	.2451	.3144	−.2170	−.1647	.5476	.1164
.70	.7580	.2420	.3123	−.2186	−.1593	.5486	.0937
.71	.7611	.2389	.3101	−.2201	−.1538	.5495	.0712
.72	.7642	.2358	.3079	−.2217	−.1483	.5501	.0487
.73	.7673	.2327	.3056	−.2231	−.1428	.5504	.0265
.74	.7704	.2296	.3034	−.2245	−.1373	.5506	.0043
.75	.7734	.2266	.3011	−.2259	−.1318	.5505	−.0176
.76	.7764	.2236	.2989	−.2271	−.1262	.5502	−.0394
.77	.7794	.2206	.2966	−.2284	−.1207	.5497	−.0611
.78	.7823	.2177	.2943	−.2296	−.1153	.5490	−.0825
.79	.7852	.2148	.2920	−.2307	−.1098	.5481	−.1037
.80	.7881	.2119	.2897	−.2318	−.1043	.5469	−.1247
.81	.7910	.2090	.2874	−.2328	−.0988	.5456	−.1455
.82	.7939	.2061	.2850	−.2337	−.0934	.5440	−.1660
.83	.7967	.2033	.2827	−.2346	−.0880	.5423	−.1862
.84	.7995	.2005	.2803	−.2355	−.0825	.5403	−.2063
.85	.8023	.1977	.2780	−.2363	−.0771	.5381	−.2260
.86	.8051	.1949	.2756	−.2370	−.0718	.5358	−.2455
.87	.8079	.1921	.2732	−.2377	−.0664	.5332	−.2646
.88	.8106	.1894	.2709	−.2384	−.0611	.5305	−.2835
.89	.8133	.1867	.2685	−.2389	−.0558	.5276	−.3021
.90	.8159	.1841	.2661	−.2395	−.0506	.5245	−.3203
.91	.8186	.1814	.2637	−.2400	−.0453	.5212	−.3383
.92	.8212	.1788	.2613	−.2404	−.0401	.5177	−.3559
.93	.8238	.1762	.2589	−.2408	−.0350	.5140	−.3731
.94	.8264	.1736	.2565	−.2411	−.0299	.5102	−.3901
.95	.8289	.1711	.2541	−.2414	−.0248	.5062	−.4066
.96	.8315	.1685	.2516	−.2416	−.0197	.5021	−.4228
.97	.8340	.1660	.2492	−.2417	−.0147	.4978	−.4387
.98	.8365	.1635	.2468	−.2419	−.0098	.4933	−.4541
.99	.8389	.1611	.2444	−.2420	−.0049	.4887	−.4692
1.00	.8413	.1587	.2420	−.2420	.0000	.4839	−.4839

NORMAL PROBABILITY FUNCTION AND RELATED FUNCTIONS (Continued)

x	$F(x)$	$1 - F(x)$	$f(x)$	$f'(x)$	$f''(x)$	$f'''(x)$	$f^{\mathrm{iv}}(x)$
1.00	.8413	.1587	.2420	− .2420	.0000	.4839	− .4839
1.01	.8438	.1562	.2396	− .2420	.0048	.4790	− .4983
1.02	.8461	.1539	.2371	− .2419	.0096	.4740	− .5122
1.03	.8485	.1515	.2347	− .2418	.0143	.4688	− .5257
1.04	.8508	.1492	.2323	− .2416	.0190	.4635	− .5389
1.05	.8531	.1469	.2299	− .2414	.0236	.4580	− .5516
1.06	.8554	.1446	.2275	− .2411	.0281	.4524	− .5639
1.07	.8577	.1423	.2251	− .2408	.0326	.4467	− .5758
1.08	.8599	.1401	.2227	− .2405	.0371	.4409	− .5873
1.09	.8621	.1379	.2203	− .2401	.0414	.4350	− .5984
1.10	.8643	.1357	.2179	− .2396	.0458	.4290	− .6091
1.11	.8665	.1335	.2155	− .2392	.0500	.4228	− .6193
1.12	.8686	.1314	.2131	− .2386	.0542	.4166	− .6292
1.13	.8708	.1292	.2107	− .2381	.0583	.4102	− .6386
1.14	.8729	.1271	.2083	− .2375	.0624	.4038	− .6476
1.15	.8749	.1251	.2059	− .2368	.0664	.3973	− .6561
1.16	.8770	.1230	.2036	− .2361	.0704	.3907	− .6643
1.17	.8790	.1210	.2012	− .2354	.0742	.3840	− .6720
1.18	.8810	.1190	.1989	− .2347	.0780	.3772	− .6792
1.19	.8830	.1170	.1965	− .2339	.0818	.3704	− .6861
1.20	.8849	.1151	.1942	− .2330	.0854	.3635	− .6926
1.21	.8869	.1131	.1919	− .2322	.0890	.3566	− .6986
1.22	.8888	.1112	.1895	− .2312	.0926	.3496	− .7042
1.23	.8907	.1093	.1872	− .2303	.0960	.3425	− .7094
1.24	.8925	.1075	.1849	− .2293	.0994	.3354	− .7141
1.25	.8944	.1056	.1826	− .2283	.1027	.3282	− .7185
1.26	.8962	.1038	.1804	− .2273	.1060	.3210	− .7224
1.27	.8980	.1020	.1781	− .2262	.1092	.3138	− .7259
1.28	.8997	.1003	.1758	− .2251	.1123	.3065	− .7291
1.29	.9015	.0985	.1736	− .2240	.1153	.2992	− .7318
1.30	.9032	.0968	.1714	− .2228	.1182	.2918	− .7341
1.31	.9049	.0951	.1691	− .2216	.1211	.2845	− .7361
1.32	.9066	.0934	.1669	− .2204	.1239	.2771	− .7376
1.33	.9082	.0918	.1647	− .2191	.1267	.2697	− .7388
1.34	.9099	.0901	.1626	− .2178	.1293	.2624	− .7395
1.35	.9115	.0885	.1604	− .2165	.1319	.2550	− .7399
1.36	.9131	.0869	.1582	− .2152	.1344	.2476	− .7400
1.37	.9147	.0853	.1561	− .2138	.1369	.2402	− .7396
1.38	.9162	.0838	.1539	− .2125	.1392	.2328	− .7389
1.39	.9177	.0823	.1518	− .2110	.1415	.2254	− .7378
1.40	.9192	.0808	.1497	− .2096	.1437	.2180	− .7364
1.41	.9207	.0793	.1476	− .2082	.1459	.2107	− .7347
1.42	.9222	.0778	.1456	− .2067	.1480	.2033	− .7326
1.43	.9236	.0764	.1435	− .2052	.1500	.1960	− .7301
1.44	.9251	.0749	.1415	− .2037	.1519	.1887	− .7274
1.45	.9265	.0735	.1394	− .2022	.1537	.1815	− .7243
1.46	.9279	.0721	.1374	− .2006	.1555	.1742	− .7209
1.47	.9292	.0708	.1354	− .1991	.1572	.1670	− .7172
1.48	.9306	.0694	.1334	− .1975	.1588	.1599	− .7132
1.49	.9319	.0681	.1315	− .1959	.1604	.1528	− .7089
1.50	.9332	.0668	.1295	− .1943	.1619	.1457	− .7043

NORMAL PROBABILITY FUNCTION AND RELATED FUNCTIONS (Continued)

x	$F(x)$	$1 - F(x)$	$f(x)$	$f'(x)$	$f''(x)$	$f'''(x)$	$f^{\mathrm{iv}}(x)$
1.50	.9332	.0668	.1295	−.1943	.1619	.1457	−.7043
1.51	.9345	.0655	.1276	−.1927	.1633	.1387	−.6994
1.52	.9357	.0643	.1257	−.1910	.1647	.1317	−.6942
1.53	.9370	.0630	.1238	−.1894	.1660	.1248	−.6888
1.54	.9382	.0618	.1219	−.1877	.1672	.1180	−.6831
1.55	.9394	.0606	.1200	−.1860	.1683	.1111	−.6772
1.56	.9406	.0594	.1182	−.1843	.1694	.1044	−.6710
1.57	.9418	.0582	.1163	−.1826	.1704	.0977	−.6646
1.58	.9429	.0571	.1145	−.1809	.1714	.0911	−.6580
1.59	.9441	.0559	.1127	−.1792	.1722	.0846	.6511
1.60	.9452	.0548	.1109	−.1775	.1730	.0781	−.6441
1.61	.9463	.0537	.1092	−.1757	.1738	.0717	−.6368
1.62	.9474	.0526	.1074	−.1740	.1745	.0654	−.6293
1.63	.9484	.0516	.1057	−.1723	.1751	.0591	−.6216
1.64	.9495	.0505	.1040	−.1705	.1757	.0529	−.6138
1.65	.9505	.0495	.1023	−.1687	.1762	.0468	−.6057
1.66	.9515	.0485	.1006	−.1670	.1766	.0408	−.5975
1.67	.9525	.0475	.0989	−.1652	.1770	.0340	−.5891
1.68	.9535	.0465	.0973	−.1634	.1773	.0290	−.5806
1.69	.9545	.0455	.0957	−.1617	.1776	.0233	−.5720
1.70	.9554	.0446	.0940	−.1599	.1778	.0176	−.5632
1.71	.9564	.0436	.0925	−.1581	.1779	.0120	−.5542
1.72	.9573	.0427	.0909	−.1563	.1780	.0065	−.5452
1.73	.9582	.0418	.0893	−.1546	.1780	.0011	−.5360
1.74	.9591	.0409	.0878	−.1528	.1780	−.0042	−.5267
1.75	.9599	.0401	.0863	−.1510	.1780	−.0094	−.5173
1.76	.9608	.0392	.0848	−.1492	.1778	−.0146	−.5079
1.77	.9616	.0384	.0833	−.1474	.1777	−.0196	−.4983
1.78	.9625	.0375	.0818	−.1457	.1774	−.0245	−.4887
1.79	.9633	.0367	.0804	−.1439	.1772	−.0294	−.4789
1.80	.9641	.0359	.0790	−.1421	.1769	−.0341	−.4692
1.81	.9649	.0351	.0775	−.1403	.1765	−.0388	−.4593
1.82	.9656	.0344	.0761	−.1386	.1761	−.0433	−.4494
1.83	.9664	.0336	.0748	−.1368	.1756	−.0477	−.4395
1.84	.9671	.0329	.0734	−.1351	.1751	−.0521	−.4295
1.85	.9678	.0322	.0721	−.1333	.1746	−.0563	−.4195
1.86	.9686	.0314	.0707	−.1316	.1740	−.0605	−.4095
1.87	.9693	.0307	.0694	−.1298	.1734	−.0645	−.3995
1.88	.9699	.0301	.0681	−.1281	.1727	−.0685	−.3894
1.89	.9706	.0294	.0669	−.1264	.1720	−.0723	−.3793
1.90	.9713	.0287	.0656	−.1247	.1713	−.0761	−.3693
1.91	.9719	.0281	.0644	−.1230	.1705	−.0797	−.3592
1.92	.9726	.0274	.0632	−.1213	.1697	−.0832	−.3492
1.93	.9732	.0268	.0620	−.1196	.1688	−.0867	−.3392
1.94	.9738	.0262	.0608	−.1179	.1679	−.0900	−.3292
1.95	.9744	.0256	.0596	−.1162	.1670	−.0933	−.3192
1.96	.9750	.0250	.0584	−.1145	.1661	−.0964	−.3093
1.97	.9756	.0244	.0573	−.1129	.1651	−.0994	−.2994
1.98	.9761	.0239	.0562	−.1112	.1641	−.1024	−.2895
1.99	.9767	.0233	.0551	−.1096	.1630	−.1052	−.2797
2.00	.9772	.0228	.0540	−.1080	.1620	−.1080	−.2700

NORMAL PROBABILITY FUNCTION AND RELATED FUNCTIONS (Continued)

x	$F(x)$	$1 - F(x)$	$f(x)$	$f'(x)$	$f''(x)$	$f'''(x)$	$f^{\mathrm{iv}}(x)$
2.00	.9772	.0228	.0540	−.1080	.1620	−.1080	−.2700
2.01	9778	.0222	.0529	−.1064	.1609	−.1106	−.2603
2.02	.9783	.0217	.0519	−.1048	.1598	−.1132	−.2506
2.03	.9788	.0212	.0508	−.1032	.1586	−.1157	−.2411
2.04	.9793	.0207	.0498	−.1016	.1575	−.1180	−.2316
2.05	.9798	.0202	.0488	−.1000	.1563	−.1203	−.2222
2.06	.9803	.0197	.0478	−.0985	.1550	−.1225	−.2129
2.07	.9808	.0192	.0468	−.0969	.1538	−.1245	−.2036
2.08	.9812	.0188	.0459	−.0954	.1526	−.1265	−.1945
2.09	.9817	.0183	.0449	−.0939	.1513	−.1284	−.1854
2.10	.9821	.0179	.0440	−.0924	.1500	−.1302	−.1765
2.11	.9826	.0174	.0431	−.0909	.1487	−.1320	−.1676
2.12	.9830	.0170	.0422	−.0894	.1474	−.1336	−.1588
2.13	.9834	.0166	.0413	−.0879	.1460	−.1351	−.1502
2.14	.9838	.0162	.0404	−.0865	.1446	−.1366	−.1416
2.15	.9842	.0158	.0396	−.0850	.1433	−.1380	−.1332
2.16	.9846	.0154	.0387	−.0836	.1419	−.1393	−.1249
2.17	.9850	.0150	.0379	−.0822	.1405	−.1405	−.1167
2.18	.9854	.0146	.0371	−.0808	.1391	−.1416	−.1086
2.19	.9857	.0143	.0363	−.0794	.1377	−.1426	−.1006
2.20	.9861	.0139	.0355	−.0780	.1362	−.1436	−.0927
2.21	.9864	.0136	.0347	−.0767	.1348	−.1445	−.0850
2.22	.9868	.0132	.0339	−.0754	.1333	−.1453	−.0774
2.23	.9871	.0129	.0332	−.0740	.1319	−.1460	−.0700
2.24	.9875	.0125	.0325	−.0727	.1304	−.1467	−.0626
2.25	.9878	.0122	.0317	−.0714	.1289	−.1473	−.0554
2.26	.9881	.0119	.0310	−.0701	.1275	−.1478	−.0484
2.27	.9884	.0116	.0303	−.0689	.1260	−.1483	−.0414
2.28	.9887	.0113	.0297	−.0676	.1245	−.1486	−.0346
2.29	.9890	.0110	.0290	−.0664	.1230	−.1490	−.0279
2.30	.9893	.0107	.0283	−.0652	.1215	−.1492	−.0214
2.31	.9896	.0104	.0277	−.0639	.1200	−.1494	−.0150
2.32	.9898	.0102	.0270	−.0628	.1185	−.1495	−.0088
2.33	.9901	.0099	.0264	−.0616	.1170	−.1496	−.0027
2.34	.9904	.0096	.0258	−.0604	.1155	−.1496	.0033
2.35	.9906	.0094	.0252	−.0593	.1141	−.1495	.0092
2.36	.9909	.0091	.0246	−.0581	.1126	−.1494	.0149
2.37	.9911	.0089	.0241	−.0570	.1111	−.1492	.0204
2.38	.9913	.0087	.0235	−.0559	.1096	−.1490	.0258
2.39	.9916	.0084	.0229	−.0548	.1081	−.1487	.0311
2.40	.9918	.0082	.0224	−.0538	.1066	−.1483	.0362
2.41	.9920	.0080	.0219	−.0527	.1051	−.1480	.0412
2.42	.9922	.0078	.0213	−.0516	.1036	−.1475	.0461
2.43	.9925	.0075	.0208	−.0506	.1022	−.1470	.0508
2.44	.9927	.0073	.0203	−.0496	.1007	−.1465	.0554
2.45	.9929	.0071	.0198	−.0486	.0992	−.1459	.0598
2.46	.9931	.0069	.0194	−.0476	.0978	−.1453	.0641
2.47	.9932	.0068	.0189	−.0467	.0963	−.1446	.0683
2.48	.9934	.0066	.0184	−.0457	.0949	−.1439	.0723
2.49	.9936	.0064	.0180	−.0448	.0935	−.1432	.0762
2.50	.9938	.0062	.0175	−.0438	.0920	−.1424	.0800

NORMAL PROBABILITY FUNCTION AND RELATED FUNCTIONS (Continued)

x	$F(x)$	$1 - F(x)$	$f(x)$	$f'(x)$	$f''(x)$	$f'''(x)$	$f^{iv}(x)$
2.50	.9938	.0062	.0175	− .0438	.0920	− .1424	.0800
2.51	.9940	.0060	.0171	− .0429	.0906	− .1416	.0836
2.52	.9941	.0059	.0167	− .0420	.0892	− .1408	.0871
2.53	.9943	.0057	.0163	− .0411	.0878	− .1399	.0905
2.54	.9945	.0055	.0158	− .0403	.0864	− .1389	.0937
2.55	.9946	.0054	.0155	− .0394	.0850	− .1380	.0968
2.56	.9948	.0052	.0151	− .0386	.0836	− .1370	.0998
2.57	.9949	.0051	.0147	− .0377	.0823	− .1360	.1027
2.58	.9951	.0049	.0143	− .0369	.0809	− .1350	.1054
2.59	.9952	.0048	.0139	− .0361	.0796	− .1339	.1080
2.60	.9953	.0047	.0136	− .0353	.0782	− .1328	.1105
2.61	.9955	.0045	.0132	− .0345	.0769	− .1317	.1129
2.62	.9956	.0044	.0129	− .0338	.0756	− .1305	.1152
2.63	.9957	.0043	.0126	− .0330	.0743	− .1294	.1173
2.64	.9959	.0041	.0122	− .0323	.0730	− .1282	.1194
2.65	.9960	.0040	.0119	− .0316	.0717	− .1270	.1213
2.66	.9961	.0039	.0116	− .0309	.0705	− .1258	.1231
2.67	.9962	.0038	.0113	− .0302	.0692	− .1245	.1248
2.68	.9963	.0037	.0110	− .0295	.0680	− .1233	.1264
2.69	.9964	.0036	.0107	− .0288	.0668	− .1220	.1279
2.70	.9965	.0035	.0104	− .0281	.0656	− .1207	.1293
2.71	.9966	.0034	.0101	− .0275	.0644	− .1194	.1306
2.72	.9967	.0033	.0099	− .0269	.0632	− .1181	.1317
2.73	.9968	.0032	.0096	− .0262	.0620	− .1168	.1328
2.74	.9969	.0031	.0093	− .0256	.0608	− .1154	.1338
2.75	.9970	.0030	.0091	− .0250	.0597	− .1141	.1347
2.76	.9971	.0029	.0088	− .0244	.0585	− .1127	.1356
2.77	.9972	.0028	.0086	− .0238	.0574	− .1114	.1363
2.78	.9973	.0027	.0084	− .0233	.0563	− .1100	.1369
2.79	.9974	.0026	.0081	− .0227	.0552	− .1087	.1375
2.80	.9974	.0026	.0079	− .0222	.0541	− .1073	.1379
2.81	.9975	.0025	.0077	− .0216	.0531	− .1059	.1383
2.82	.9976	.0024	.0075	− .0211	.0520	− .1045	.1386
2.83	.9977	.0023	.0073	− .0206	.0510	− .1031	.1389
2.84	.9977	.0023	.0071	− .0201	.0500	− .1017	.1390
2.85	.9978	.0022	.0069	− .0196	.0490	− .1003	.1391
2.86	.9979	.0021	.0067	− .0191	.0480	− .0990	.1391
2.87	.9979	.0021	.0065	− .0186	.0470	− .0976	.1391
2.88	.9980	.0020	.0063	− .0182	.0460	− .0962	.1389
2.89	.9981	.0019	.0061	− .0177	.0451	− .0948	.1388
2.90	.9981	.0019	.0060	− .0173	.0441	− .0934	.1385
2.91	.9982	.0018	.0058	− .0168	.0432	− .0920	.1382
2.92	.9982	.0018	.0056	− .0164	.0423	− .0906	.1378
2.93	.9983	.0017	.0055	− .0160	.0414	− .0893	.1374
2.94	.9984	.0016	.0053	− .0156	.0405	− .0879	.1369
2.95	.9984	.0016	.0051	− .0152	.0396	− .0865	.1364
2.96	.9985	.0015	.0050	− .0148	.0388	− .0852	.1358
2.97	.9985	.0015	.0048	− .0144	.0379	− .0838	.1352
2.98	.9986	.0014	.0047	− .0140	.0371	− .0825	.1345
2.99	.9986	.0014	.0046	− .0137	.0363	− .0811	.1337
3.00	.9987	.0013	.0044	− .0133	.0355	− .0798	.1330

NORMAL PROBABILITY FUNCTION AND RELATED FUNCTIONS (Continued)

x	$F(x)$	$1 - F(x)$	$f(x)$	$f'(x)$	$f''(x)$	$f'''(x)$	$f^{\mathrm{IV}}(x)$
3.00	.9987	.0013	.0044	− .0133	.0355	− .0798	.1330
3.01	.9987	.0013	.0043	− .0130	.0347	− .0785	.1321
3.02	.9987	.0013	.0042	− .0126	.0339	− .0771	.1313
3.03	.9988	.0012	.0040	− .0123	.0331	− .0758	.1304
3.04	.9988	.0012	.0039	− .0119	.0324	− .0745	.1294
3.05	.9989	.0011	.0038	− .0116	.0316	− .0732	.1285
3.06	.9989	.0011	.0037	− .0113	.0720	− .0720	.1275
3.07	.9989	.0011	.0036	− .0110	.0302	− .0707	.1264
3.08	.9990	.0010	.0035	− .0107	.0295	− .0694	.1254
3.09	.9990	.0010	.0034	− .0104	.0288	− .0682	.1243
3.10	.9990	.0010	.0033	− .0101	.0281	− .0669	.1231
3.11	.9991	.0009	.0032	− .0099	.0275	− .0657	.1220
3.12	.9991	.0009	.0031	− .0096	.0268	− .0645	.1208
3.13	.9991	.0009	.0030	− .0093	.0262	− .0633	.1196
3.14	.9992	.0008	.0029	− .0091	.0256	− .0621	.1184
3.15	.9992	.0008	.0028	− .0088	.0249	− .0609	.1171
3.16	.9992	.0008	.0027	− .0086	.0243	− .0598	.1159
3.17	.9992	.0008	.0026	− .0083	.0237	− .0586	.1146
3.18	.9993	.0007	.0025	− .0081	.0232	− .0575	.1133
3.19	.9993	.0007	.0025	− .0079	.0226	− .0564	.1120
3.20	.9993	.0007	.0024	− .0076	.0220	− .0552	.1107
3.21	.9993	.0007	.0023	− .0074	.0215	− .0541	.1093
3.22	.9994	.0006	.0022	− .0072	.0210	− .0531	.1080
3.23	.9994	.0006	.0022	− .0070	.0204	− .0520	.1066
3.24	.9994	.0006	.0021	− .0068	.0199	− .0509	.1053
3.25	.9994	.0006	.0020	− .0066	.0194	− .0499	.1039
3.26	.9994	.0006	.0020	− .0064	.0189	− .0488	.1025
3.27	.9995	.0005	.0019	− .0062	.0184	− .0478	.1011
3.28	.9995	.0005	.0018	− .0060	.0180	− .0468	.0997
3.29	.9995	.0005	.0018	− .0059	.0175	− .0458	.0983
3.30	.9995	.0005	.0017	− .0057	.0170	− .0449	.0969
3.31	.9995	.0005	.0017	− .0055	.0166	− .0439	.0955
3.32	.9995	.0005	.0016	− .0054	.0162	− .0429	.0941
3.33	.9996	.0004	.0016	− .0052	.0157	− .0420	.0927
3.34	.9996	.0004	.0015	− .0050	.0153	− .0411	.0913
3.35	.9996	.0004	.0015	− .0049	.0149	− .0402	.0899
3.36	.9996	.0004	.0014	− .0047	.0145	− .0393	.0885
3.37	.9996	.0004	.0014	− .0046	.0141	− .0384	.0871
3.38	.9996	.0004	.0013	− .0045	.0138	− .0376	.0857
3.39	.9997	.0003	.0013	− .0043	.0134	− .0367	.0843
3.40	.9997	.0003	.0012	− .0042	.0130	− .0359	.0829
3.41	.9997	.0003	.0012	− .0041	.0127	− .0350	.0815
3.42	.9997	.0003	.0012	− .0039	.0123	− .0342	.0801
3.43	.9997	.0003	.0011	− .0038	.0120	− .0334	.0788
3.44	.9997	.0003	.0011	− .0037	.0116	− .0327	.0774
3.45	.9997	.0003	.0010	− .0036	.0113	− .0319	.0761
3.46	.9997	.0003	.0010	− .0035	.0110	− .0311	.0747
3.47	.9997	.0003	.0010	− .0034	.0107	− .0304	.0734
3.48	.9997	.0003	.0009	− .0033	.0104	− .0297	.0721
3.49	.9998	.0002	.0009	− .0032	.0101	− .0290	.0707
3.50	.9998	.0002	.0009	− .0031	.0098	− .0283	.0694

NORMAL PROBABILITY FUNCTION AND RELATED FUNCTIONS (Continued)

x	$F(x)$	$1-F(x)$	$f(x)$	$f'(x)$	$f''(x)$	$f'''(x)$	$f^{iv}(x)$
3.50	.9998	.0002	.0009	− .0031	.0098	− .0283	.0694
3.51	.9998	.0002	.0008	− .0030	.0095	− .0276	.0681
3.52	.9998	.0002	.0008	− .0029	.0093	− .0269	.0669
3.53	.9998	.0002	.0008	− .0028	.0090	− .0262	.0656
3.54	.9998	.0002	.0008	− .0027	.0087	− .0256	.0643
3.55	.9998	.0002	.0007	− .0026	.0085	− .0249	.0631
3.56	.9998	.0002	.0007	− .0025	.0082	− .0243	.0618
3.57	.9998	.0002	.0007	− .0024	.0080	− .0237	.0606
3.58	.9998	.0002	.0007	− .0024	.0078	− .0231	.0594
3.59	.9998	.0002	.0006	− .0023	.0075	− .0225	.0582
3.60	.9998	.0002	.0006	− .0022	.0073	− .0219	.0570
3.61	.9998	.0002	.0006	− .0021	.0071	− .0214	.0559
3.62	.9999	.0001	.0006	− .0021	.0069	− .0208	.0547
3.63	.9999	.0001	.0005	− .0020	.0067	− .0203	.0536
3.64	.9999	.0001	.0005	− .0019	.0065	− .0198	.0524
3.65	.9999	.0001	.0005	− .0019	.0063	− .0192	.0513
3.66	.9999	.0001	.0005	− .0018	.0061	− .0187	.0502
3.67	.9999	.0001	.0005	− .0017	.0059	− .0182	.0492
3.68	.9999	.0001	.0005	− .0017	.0057	− .0177	.0481
3.69	.9999	.0001	.0004	− .0016	.0056	− .0173	.0470
3.70	.9999	.0001	.0004	− .0016	.0054	− .0168	.0460
3.71	.9999	.0001	.0004	− .0015	.0052	− .0164	.0450
3.72	.9999	.0001	.0004	− .0015	.0051	− .0159	.0440
3.73	.9999	.0001	.0004	− .0014	.0049	− .0155	.0430
3.74	.9999	.0001	.0004	− .0014	.0048	− .0150	.0420
3.75	.9999	.0001	.0004	− .0013	.0046	− .0146	.0410
3.76	.9999	.0001	.0003	− .0013	.0045	− .0142	.0401
3.77	.9999	.0001	.0003	− .0012	.0043	− .0138	.0392
3.78	.9999	.0001	.0003	− .0012	.0042	− .0134	.0382
3.79	.9999	.0001	.0003	− .0012	.0041	− .0131	.0373
3.80	.9999	.0001	.0003	− .0011	.0039	− .0127	.0365
3.81	.9999	.0001	.0003	− .0011	.0038	− .0123	.0356
3.82	.9999	.0001	.0003	− .0010	.0037	− .0120	.0347
3.83	.9999	.0001	.0003	− .0010	.0036	− .0116	.0339
3.84	.9999	.0001	.0003	− .0010	.0034	− .0113	.0331
3.85	.9999	.0001	.0002	− .0009	.0033	− .0110	.0323
3.86	.9999	.0001	.0002	− .0009	.0032	− .0107	.0315
3.87	.9999	.0001	.0002	− .0009	.0031	− .0104	.0307
3.88	.9999	.0001	.0002	− .0008	.0030	− .0100	.0299
3.89	1.0000	.0000	.0002	− .0008	.0029	− .0098	.0292
3.90	1.0000	.0000	.0002	− .0008	.0028	− .0095	.0284
3.91	1.0000	.0000	.0002	− .0008	.0027	− .0092	.0277
3.92	1.0000	.0000	.0002	− .0007	.0026	− .0089	.0270
3.93	1.0000	.0000	.0002	− .0007	.0026	− .0086	.0263
3.94	1.0000	.0000	.0002	− .0007	.0025	− .0084	.0256
3.95	1.0000	.0000	.0002	− .0006	.0024	− .0081	.0250
3.96	1.0000	.0000	.0002	− .0006	.0023	− .0079	.0243
3.97	1.0000	.0000	.0002	− .0006	.0022	− .0076	.0237
3.98	1.0000	.0000	.0001	− .0006	.0022	− .0074	.0230
3.99	1.0000	.0000	.0001	− .0006	.0021	− .0072	.0224
4.00	1.0000	.0000	.0001	− .0005	.0020	− .0070	.0218

x	1.282	1.645	1.960	2.326	2.576	3.090
$F(x)$	.90	.95	.975	.99	.995	.999
$2[1-F(x)]$	.20	.10	.05	.02	.01	.002

INDIVIDUAL TERMS, BINOMIAL DISTRIBUTION

The $(x + 1)^{st}$ term in the expansion of the binomial $[\theta + (1 - \theta)]^n$ is given by

$$f(x;n,\theta) = \binom{n}{x} \theta^x (1 - \theta)^{n-x}, \qquad x = 0, 1, 2, \ldots, n.$$

This is the probability of exactly x successes in n independent binomial trials with probability of success on a single trial equal to θ. This table contains the individual terms of $f(x;n,\theta)$ for specified choices of x, n, and θ.

For $\theta > 0.5$, the value of $\binom{n}{x} \theta^x (1 - \theta)^{n-x}$ is found by using the table entry for $\binom{n}{n-x} (1 - \theta)^{n-x} \theta^x$.

INDIVIDUAL TERMS, BINOMIAL DISTRIBUTION (Continued)

n	x	.05	.10	.15	.20	.25	θ .30	.35	.40	.45	.50
1	0	9500	.9000	.8500	.8000	.7500	.7000	.6500	.6000	.5500	.5000
	1	.0500	.1000	.1500	.2000	.2500	.3000	.3500	.4000	.4500	.5000
2	0	.9025	.8100	.7225	.6400	.5625	.4900	.4225	.3600	.3025	.2500
	1	.0950	.1800	.2550	.3200	.3750	.4200	.4550	.4800	.4950	.5000
	2	.0025	.0100	.0225	.0400	.0625	.0900	.1225	.1600	.2025	.2500
3	0	.8574	.7290	.6141	.5120	.4219	.3430	.2746	.2160	.1664	.1250
	1	.1354	.2430	.3251	.3840	.4219	.4410	.4436	.4320	.4084	.3750
	2	.0071	.0270	.0574	.0960	.1406	.1890	.2389	.2880	.3341	.3750
	3	.0001	.0010	.0034	.0080	.0156	.0270	.0429	.0640	.0911	.1250
4	0	.8145	.6561	.5220	.4096	.3164	.2401	.1785	.1296	.0915	.0625
	1	.1715	.2916	.3685	.4096	.4219	.4116	.3845	.3456	.2995	.2500
	2	.0135	.0486	.0975	.1536	.2109	.2646	.3105	.3456	.3675	.3750
	3	.0005	.0036	.0115	.0256	.0469	.0756	.1115	.1536	.2005	.2500
	4	.0000	.0001	.0005	.0016	.0039	.0081	.0150	.0256	.0410	.0625
5	0	.7738	.5905	.4437	.3277	.2373	.1681	.1160	.0778	.0503	.0312
	1	.2036	.3280	.3915	.4096	.3955	.3602	.3124	.2592	.2059	.1562
	2	.0214	.0729	.1382	.2048	.2637	.3087	.3364	.3456	.3369	.3125
	3	.0011	.0081	.0244	.0512	.0879	.1323	.1811	.2304	.2757	.3125
	4	.0000	.0004	.0022	.0064	.0146	.0284	.0488	.0768	.1128	.1562
	5	.0000	.0000	.0001	.0003	.0010	.0024	.0053	.0102	.0185	.0312
6	0	.7351	.5314	.3771	.2621	.1780	.1176	.0754	.0467	.0277	.0156
	1	.2321	.3543	.3993	.3932	.3560	.3025	.2437	.1866	.1359	.0938
	2	.0305	.0984	.1762	.2458	.2966	.3241	.3280	.3110	.2780	.2344
	3	.0021	.0146	.0415	.0819	.1318	.1852	.2355	.2765	.3032	.3125
	4	.0001	.0012	.0055	.0154	.0330	.0595	.0951	.1382	.1861	.2344
	5	.0000	.0001	.0004	.0015	.0044	.0102	.0205	.0369	.0609	.0938
	6	.0000	.0000	.0000	.0001	.0002	.0007	.0018	.0041	.0083	.0156
7	0	.6983	.4783	.3206	.2097	.1335	.0824	.0490	.0280	.0152	.0078
	1	.2573	.3720	.3960	.3670	.3115	.2471	.1848	.1306	.0872	.0547
	2	.0406	.1240	.2097	.2753	.3115	.3177	.2985	.2613	.2140	.1641
	3	.0036	.0230	.0617	.1147	.1730	.2269	.2679	.2903	.2918	.2734
	4	.0002	.0026	.0109	.0287	.0577	.0972	.1442	.1935	.2388	.2734
	5	.0000	.0002	.0012	.0043	.0115	.0250	.0466	.0774	.1172	.1641
	6	.0000	.0000	.0001	.0004	.0013	.0036	.0084	.0172	.0320	.0547
	7	.0000	.0000	.0000	.0000	.0001	.0002	.0006	.0016	.0037	.0078
8	0	.6634	.4305	.2725	.1678	.1001	.0576	.0319	.0168	.0084	.0039
	1	.2793	.3826	.3847	.3355	.2670	.1977	.1373	.0896	.0548	.0312
	2	.0515	.1488	.2376	.2936	.3115	.2965	.2587	.2090	.1569	.1094
	3	.0054	.0331	.0839	.1468	.2076	.2541	.2786	.2787	.2568	.2188
	4	.0004	.0046	.0185	.0459	.0865	.1361	.1875	.2322	.2627	.2734
	5	.0000	.0004	.0026	.0092	.0231	.0467	.0808	.1239	.1719	.2188
	6	.0000	.0000	.0002	.0011	.0038	.0100	.0217	.0413	.0703	.1094
	7	.0000	.0000	.0000	.0001	.0004	.0012	.0033	.0079	.0164	.0312
	8	.0000	.0000	.0000	.0000	.0000	.0001	.0002	.0007	.0017	.0039

Linear interpolations with respect to p will in general be accurate at most to two decimal places.

INDIVIDUAL TERMS, BINOMIAL DISTRIBUTION (Continued)

						θ					
n	x	.05	.10	.15	.20	.25	.30	.35	.40	.45	.50
9	0	.6302	.3874	.2316	.1342	.0751	.0404	.0207	.0101	.0046	.0020
	1	.2985	.3874	.3679	.3020	.2253	.1556	.1004	.0605	.0339	.0176
	2	.0629	.1722	.2597	.3020	.3003	.2668	.2162	.1612	.1110	.0703
	3	.0077	.0446	.1069	.1762	.2336	.2668	.2716	.2508	.2119	.1641
	4	.0006	.0074	.0283	.0661	.1168	.1715	.2194	.2508	.2600	.2461
	5	.0000	.0008	.0050	.0165	.0389	.0735	.1181	.1672	.2128	.2461
	6	.0000	.0001	.0006	.0028	.0087	.0210	.0424	.0743	.1160	.1641
	7	.0000	.0000	.0000	.0003	.0012	.0039	.0098	.0212	.0407	.0703
	8	.0000	.0000	.0000	.0000	.0001	.0004	.0013	.0035	.0083	.0176
	9	.0000	.0000	.0000	.0000	.0000	.0000	.0001	.0003	.0008	.0020
10	0	.5987	.3487	.1969	.1074	.0563	.0282	.0135	.0060	.0025	.0010
	1	.3151	.3874	.3474	.2684	.1877	.1211	.0725	.0403	.0207	.0098
	2	.0746	.1937	.2759	.3020	.2816	.2335	.1757	.1209	.0763	.0439
	3	.0105	.0574	.1298	.2013	.2503	.2668	.2522	.2150	.1665	.1172
	4	.0010	.0112	.0401	.0881	.1460	.2001	.2377	.2508	.2384	.2051
	5	.0001	.0015	.0085	.0264	.0584	.1029	.1536	.2007	.2340	.2461
	6	.0000	.0001	.0012	.0055	.0162	.0368	.0689	.1115	.1596	.2051
	7	.0000	.0000	.0001	.0008	.0031	.0090	.0212	.0425	.0746	.1172
	8	.0000	.0000	.0000	.0001	.0004	.0014	.0043	.0106	.0229	.0439
	9	.0000	.0000	.0000	.0000	.0000	.0001	.0005	.0016	.0042	.0098
	10	.0000	.0000	.0000	.0000	.0000	.0000	.0000	.0001	.0003	.0010
11	0	.5688	.3138	.1673	.0859	.0422	.0198	.0088	.0036	.0014	.0004
	1	.3293	.3835	.3248	.2362	.1549	.0932	.0518	.0266	.0125	.0055
	2	.0867	.2131	.2866	.2953	.2581	.1998	.1395	.0887	.0513	.0269
	3	.0137	.0710	.1517	.2215	.2581	.2568	.2254	.1774	.1259	.0806
	4	.0014	.0158	.0536	.1107	.1721	.2201	.2428	.2365	.2060	.1611
	5	.0001	.0025	.0132	.0388	.0803	.1321	.1830	.2207	.2360	.2256
	6	.0000	.0003	.0023	.0097	.0268	.0566	.0985	.1471	.1931	.2256
	7	.0000	.0000	.0003	.0017	.0064	.0173	.0379	.0701	.1128	.1611
	8	.0000	.0000	.0000	.0002	.0011	.0037	.0102	.0234	.0462	.0806
	9	.0000	.0000	.0000	.0000	.0001	.0005	.0018	.0052	.0126	.0269
	10	.0000	.0000	.0000	.0000	.0000	.0000	.0002	.0007	.0021	.0054
	11	.0000	.0000	.0000	.0000	.0000	.0000	.0000	.0000	.0002	.0005
12	0	.5404	.2824	.1422	.0687	.0317	.0138	.0057	.0022	.0008	.0002
	1	.3413	.3766	.3012	.2062	.1267	.0712	.0368	.0174	.0075	.0029
	2	.0988	.2301	.2924	.2835	.2323	.1678	.1088	.0639	.0339	.0161
	3	.0173	.0852	.1720	.2362	.2581	.2397	.1954	.1419	.0923	.0537
	4	.0021	.0213	.0683	.1329	.1936	.2311	.2367	.2128	.1700	.1208
	5	.0002	.0038	.0193	.0532	.1032	.1585	.2039	.2270	.2225	.1934
	6	.0000	.0005	.0040	.0155	.0401	.0792	.1281	.1766	.2124	.2256
	7	.0000	.0000	.0006	.0033	.0115	.0291	.0591	.1009	.1489	.1934
	8	.0000	.0000	.0001	.0005	.0024	.0078	.0199	.0420	.0762	.1208
	9	.0000	.0000	.0000	.0001	.0004	.0015	.0048	.0125	.0277	.0537
	10	.0000	.0000	.0000	.0000	.0000	.0002	.0008	.0025	.0068	.0161
	11	.0000	.0000	.0000	.0000	.0000	.0000	.0001	.0003	.0010	.0029
	12	.0000	.0000	.0000	.0000	.0000	.0000	.0000	.0000	.0001	.0002

INDIVIDUAL TERMS, BINOMIAL DISTRIBUTION (Continued)

n	x	.05	.10	.15	.20	.25	.30	.35	.40	.45	.50
13	0	.5133	.2542	.1209	.0550	.0238	.0097	.0037	.0013	.0004	.0001
	1	.3512	.3672	.2774	.1787	.1029	.0540	.0259	.0113	.0045	.0016
	2	.1109	.2448	.2937	.2680	.2059	.1388	.0836	.0453	.0220	.0095
	3	.0214	.0997	.1900	.2457	.2517	.2181	.1651	.1107	.0660	.0349
	4	.0028	.0277	.0838	.1535	.2097	.2337	.2222	.1845	.1350	.0873
	5	.0003	.0055	.0266	.0691	.1258	.1803	.2154	.2214	.1989	.1571
	6	.0000	.0008	.0063	.0230	.0559	.1030	.1546	.1968	.2169	.2095
	7	.0000	.0001	.0011	.0058	.0186	.0442	.0833	.1312	.1775	.2095
	8	.0000	.0000	.0001	.0011	.0047	.0142	.0336	.0656	.1089	.1571
	9	.0000	.0000	.0000	.0001	.0009	.0034	.0101	.0243	.0495	.0873
	10	.0000	.0000	.0000	.0000	.0001	.0006	.0022	.0065	.0162	.0349
	11	.0000	.0000	.0000	.0000	.0000	.0001	.0003	.0012	.0036	.0095
	12	.0000	.0000	.0000	.0000	.0000	.0000	.0000	.0001	.0005	.0016
	13	.0000	.0000	.0000	.0000	.0000	.0000	.0000	.0000	.0000	.0001
14	0	.4877	.2288	.1028	.0440	.0178	.0068	.0024	.0008	.0002	.0001
	1	.3593	.3559	.2539	.1539	.0832	.0407	.0181	.0073	.0027	.0009
	2	.1229	.2570	.2912	.2501	.1802	.1134	.0634	.0317	.0141	.0056
	3	.0259	.1142	.2056	.2501	.2402	.1943	.1366	.0845	.0462	.0222
	4	.0037	.0349	.0998	.1720	.2202	.2290	.2022	.1549	.1040	.0611
	5	.0004	.0078	.0352	.0860	.1468	.1963	.2178	.2066	.1701	.1222
	6	.0000	.0013	.0093	.0322	.0734	.1262	.1759	.2066	.2088	.1833
	7	.0000	.0002	.0019	.0092	.0280	.0618	.1082	.1574	.1952	.2095
	8	.0000	.0000	.0003	.0020	.0082	.0232	.0510	.0918	.1398	.1833
	9	.0000	.0000	.0000	.0003	.0018	.0066	.0183	.0408	.0762	.1222
	10	.0000	.0000	.0000	.0000	.0003	.0014	.0049	.0136	.0312	.0611
	11	.0000	.0000	.0000	.0000	.0000	.0002	.0010	.0033	.0093	.0222
	12	.0000	.0000	.0000	.0000	.0000	.0000	.0001	.0005	.0019	.0056
	13	.0000	.0000	.0000	.0000	.0000	.0000	.0000	.0001	.0002	.0009
	14	.0000	.0000	.0000	.0000	.0000	.0000	.0000	.0000	.0000	.0001
15	0	.4633	.2059	.0874	.0352	.0134	.0047	.0016	.0005	.0001	.0000
	1	.3658	.3432	.2312	.1319	.0668	.0305	.0126	.0047	.0016	.0005
	2	.1348	.2669	.2856	.2309	.1559	.0916	.0476	.0219	.0090	.0032
	3	.0307	.1285	.2184	.2501	.2252	.1700	.1110	.0634	.0318	.0139
	4	.0049	.0428	.1156	.1876	.2252	.2186	.1792	.1268	.0780	.0417
	5	.0006	.0105	.0449	.1032	.1651	.2061	.2123	.1859	.1404	.0916
	6	.0000	.0019	.0132	.0430	.0917	.1472	.1906	.2066	.1914	.1527
	7	.0000	.0003	.0030	.0138	.0393	.0811	.1319	.1771	.2013	.1964
	8	.0000	.0000	.0005	.0035	.0131	.0348	.0710	.1181	.1647	.1964
	9	.0000	.0000	.0001	.0007	.0034	.0116	.0298	.0612	.1048	.1527
	10	.0000	.0000	.0000	.0001	.0007	.0030	.0096	.0245	.0515	.0916
	11	.0000	.0000	.0000	.0000	.0001	.0006	.0024	.0074	.0191	.0417
	12	.0000	.0000	.0000	.0000	.0000	.0001	.0004	.0016	.0052	.0139
	13	.0000	.0000	.0000	.0000	.0000	.0000	.0001	.0003	.0010	.0032
	14	.0000	.0000	.0000	.0000	.0000	.0000	.0000	.0000	.0001	.0005
	15	.0000	.0000	.0000	.0000	.0000	.0000	.0000	.0000	.0000	.0000

INDIVIDUAL TERMS, BINOMIAL DISTRIBUTION (Continued)

n	x	.05	.10	.15	.20	θ .25	.30	.35	.40	.45	.50
16	0	.4401	.1853	.0743	.0281	.0100	.0033	.0010	.0003	.0001	.0000
	1	.3706	.3294	.2097	.1126	.0535	.0228	.0087	.0030	.0009	.0002
	2	.1463	.2745	.2775	.2111	.1336	.0732	.0353	.0150	.0056	.0018
	3	.0359	.1423	.2285	.2463	.2079	.1465	.0888	.0468	.0215	.0085
	4	.0061	.0514	.1311	.2001	.2252	.2040	.1553	.1014	.0572	.0278
	5	.0008	.0137	.0555	.1201	.1802	.2099	.2008	.1623	.1123	.0667
	6	.0001	.0028	.0180	.0550	.1101	.1649	.1982	.1983	.1684	.1222
	7	.0000	.0004	.0045	.0197	.0524	.1010	.1524	.1889	.1969	.1746
	8	.0000	.0001	.0009	.0055	.0197	.0487	.0923	.1417	.1812	.1964
	9	.0000	.0000	.0001	.0012	.0058	.0185	.0442	.0840	.1318	.1746
	10	.0000	.0000	.0000	.0002	.0014	.0056	.0167	.0392	.0755	.1222
	11	.0000	.0000	.0000	.0000	.0002	.0013	.0049	.0142	.0337	.0667
	12	.0000	.0000	.0000	.0000	.0000	.0002	.0011	.0040	.0115	.0278
	13	.0000	.0000	.0000	.0000	.0000	.0000	.0002	.0008	.0029	.0085
	14	.0000	.0000	.0000	.0000	.0000	.0000	.0000	.0001	.0005	.0018
	15	.0000	.0000	.0000	.0000	.0000	.0000	.0000	.0000	.0001	.0002
	16	.0000	.0000	.0000	.0000	.0000	.0000	.0000	.0000	.0000	.0000
17	0	.4181	.1668	.0631	.0225	.0075	.0023	.0007	.0002	.0000	.0000
	1	.3741	.3150	.1893	.0957	.0426	.0169	.0060	.0019	.0005	.0001
	2	.1575	.2800	.2673	.1914	.1136	.0581	.0260	.0102	.0035	.0010
	3	.0415	.1556	.2359	.2393	.1893	.1245	.0701	.0341	.0144	.0052
	4	.9076	.0605	.1457	.2093	.2209	.1868	.1320	.0796	.0411	.0182
	5	.0010	.0175	.0668	.1361	.1914	.2081	.1849	.1379	.0875	.0472
	6	.0001	.0039	.0236	.0680	.1276	.1784	.1991	.1839	.1432	.0944
	7	.0000	.0007	.0065	.0267	.0668	.1201	.1685	.1927	.1841	.1484
	8	.0000	.0001	.0014	.0084	.0279	.0644	.1134	.1606	.1883	.1855
	9	.0000	.0000	.0003	.0021	.0093	.0276	.0611	.1070	.1540	.1855
	10	.0000	.0000	.0000	.0004	.0025	.0095	.0263	.0571	.1008	.1484
	11	.0000	.0000	.0000	.0001	.0005	.0026	.0090	.0242	.0525	.0944
	12	.0000	.0000	.0000	.0000	.0001	.0006	.0024	.0081	.0215	.0472
	13	.0000	.0000	.0000	.0000	.0000	.0001	.0005	.0021	.0068	.0182
	14	.0000	.0000	.0000	.0000	.0000	.0000	.0001	.0004	.0016	.0052
	15	.0000	.0000	.0000	.0000	.0000	.0000	.0000	.0001	.0003	.0010
	16	.0000	.0000	.0000	.0000	.0000	.0000	.0000	.0000	.0000	.0001
	17	.0000	.0000	.0000	.0000	.0000	.0000	.0000	.0000	.0000	.0000
18	0	.3972	.1501	.0536	.0180	.0056	.0016	.0004	.0001	.0000	.0000
	1	.3763	.3002	.1704	.0811	.0338	.0126	.0042	.0012	.0003	.0001
	2	.1683	.2835	.2556	.1723	.0958	.0458	.0190	.0069	.0022	.0006
	3	.0473	.1680	.2406	.2297	.1704	.1046	.0547	.0246	.0095	.0031
	4	.0093	.0700	.1592	.2153	.2130	.1681	.1104	.0614	.0291	.0117
	5	.0014	.0218	.0787	.1507	.1988	.2017	.1664	.1146	.0666	.0327
	6	.0002	.0052	.0301	.0816	.1436	.1873	.1941	.1655	.1181	.0708
	7	.0000	.0010	.0091	.0350	.0820	.1376	.1792	.1892	.1657	.1214
	8	.0000	.0002	.0022	.0120	.0376	.0811	.1327	.1734	.1864	.1669
	9	.0000	.0000	.0004	.0033	.0139	.0380	.0794	.1284	.1694	.1855
	10	.0000	.0000	.0001	.0008	.0042	.0149	.0385	.0771	.1248	.1669
	11	.0000	.0000	.0000	.0001	.0010	.0046	.0151	.0374	.0742	.1214

INDIVIDUAL TERMS, BINOMIAL DISTRIBUTION (Continued)

n	x	.05	.10	.15	.20	.25	.30	.35	.40	.45	.50
18	12	.0000	.0000	.0000	.0000	.0002	.0012	.0047	.0145	.0354	.0708
	13	.0000	.0000	.0000	.0000	.0000	.0002	.0012	.0045	.0134	.0327
	14	.0000	.0000	.0000	.0000	.0000	.0000	.0002	.0011	.0039	.0117
	15	.0000	.0000	.0000	.0000	.0000	.0000	.0000	.0002	.0009	.0031
	16	.0000	.0000	.0000	.0000	.0000	.0000	.0000	.0000	.0001	.0006
	17	.0000	.0000	.0000	.0000	.0000	.0000	.0000	.0000	.0000	.0001
	18	.0000	.0000	.0000	.0000	.0000	.0000	.0000	.0000	.0000	.0000
19	0	.3774	.1351	.0456	.0144	.0042	.0011	.0003	.0001	.0000	.0000
	1	.3774	.2852	.1529	.0685	.0268	.0093	.0029	.0008	.0002	.0000
	2	.1787	.2852	.2428	.1540	.0803	.0358	.0138	.0046	.0013	.0003
	3	.0533	.1796	.2428	.2182	.1517	.0869	.0422	.0175	.0062	.0018
	4	.0112	.0798	.1714	.2182	.2023	.1491	.0909	.0467	.0203	.0074
	5	.0018	.0266	.0907	.1636	.2023	.1916	.1468	.0933	.0497	.0222
	6	.0002	.0069	.0374	.0955	.1574	.1916	.1844	.1451	.0949	.0518
	7	.0000	.0014	.0122	.0443	.0974	.1525	.1844	.1797	.1443	.0961
	8	.0000	.0002	.0032	.0166	.0487	.0981	.1489	.1797	.1771	.1442
	9	.0000	.0000	.0007	.0051	.0198	.0514	.0980	.1464	.1771	.1762
	10	.0000	.0000	.0001	.0013	.0066	.0220	.0528	.0976	.1449	.1762
	11	.0000	.0000	.0000	.0003	.0018	.0077	.0233	.0532	.0970	.1442
	12	.0000	.0000	.0000	.0000	.0004	.0022	.0083	.0237	.0529	.0961
	13	.0000	.0000	.0000	.0000	.0001	.0005	.0024	.0085	.0233	.0518
	14	.0000	.0000	.0000	.0000	.0000	.0001	.0006	.0024	.0082	.0222
	15	.0000	.0000	.0000	.0000	.0000	.0000	.0001	.0005	.0022	.0074
	16	.0000	.0000	.0000	.0000	.0000	.0000	.0000	.0001	.0005	.0018
	17	.0000	.0000	.0000	.0000	.0000	.0000	.0000	.0000	.0001	.0003
	18	.0000	.0000	.0000	.0000	.0000	.0000	.0000	.0000	.0000	.0000
	19	.0000	.0000	.0000	.0000	.0000	.0000	.0000	.0000	.0000	.0000
20	0	.3585	.1216	.0388	.0115	.0032	.0008	.0002	.0000	.0000	.0000
	1	.3774	.2702	.1368	.0576	.0211	.0068	.0020	.0005	.0001	.0000
	2	.1887	.2852	.2293	.1369	.0669	.0278	.0100	.0031	.0008	.0002
	3	.0596	.1901	.2428	.2054	.1339	.0716	.0323	.0123	.0040	.0011
	4	.0133	.0898	.1821	.2182	.1897	.1304	.0738	.0350	.0139	.0046
	5	.0022	.0319	.1028	.1746	.2023	.1789	.1272	.0746	.0365	.0148
	6	.0003	.0089	.0454	.1091	.1686	.1916	.1712	.1244	.0746	.0370
	7	.0000	.0020	.0160	.0545	.1124	.1643	.1844	.1659	.1221	.0739
	8	.0000	.0004	.0046	.0222	.0609	.1144	.1614	.1797	.1623	.1201
	9	.0000	.0001	.0011	.0074	.0271	.0654	.1158	.1597	.1771	.1602
	10	.0000	.0000	.0002	.0020	.0099	.0308	.0686	.1171	.1593	.1762
	11	.0000	.0000	.0000	.0005	.0030	.0120	.0336	.0710	.1185	.1602
	12	.0000	.0000	.0000	.0001	.0008	.0039	.0136	.0355	.0727	.1201
	13	.0000	.0000	.0000	.0000	.0002	.0010	.0045	.0146	.0366	.0739
	14	.0000	.0000	.0000	.0000	.0000	.0002	.0012	.0049	.0150	.0370
	15	.0000	.0000	.0000	.0000	.0000	.0000	.0003	.0013	.0049	.0148
	16	.0000	.0000	.0000	.0000	.0000	.0000	.0000	.0003	.0013	.0046
	17	.0000	.0000	.0000	.0000	.0000	.0000	.0000	.0000	.0002	.0011
	18	.0000	.0000	.0000	.0000	.0000	.0000	.0000	.0000	.0000	.0002
	19	.0000	.0000	.0000	.0000	.0000	.0000	.0000	.0000	.0000	.0000
	20	.0000	.0000	.0000	.0000	.0000	.0000	.0000	.0000	.0000	.0000

The column header row is labeled θ centered above the .25 and .30 columns.

CUMULATIVE TERMS, BINOMIAL DISTRIBUTION

For the binomial probability function $f(x;n,\theta)$ the probability of observing x' or more successes is given by

$$\sum_{x=x'}^{n} \binom{n}{x} \theta^x (1-\theta)^{n-x},$$

This table contains the values of $\displaystyle\sum_{x=x'}^{n} \binom{n}{x} \theta^x (1-\theta)^{n-x}$ for specified values of n, x', and

θ. If $\theta > 0.5$, the values for $\displaystyle\sum_{x=x'}^{n} \binom{n}{x} \theta^x (1-\theta)^{n-x}$ are obtained using the corresponding

results obtained from

$$1 - \sum_{x=n-x'+1}^{n} \binom{n}{x} (1-\theta)^x \theta^{n-x}$$

The cumulative binomial distribution is related to the incomplete beta function as follows:

$$\sum_{x=x'}^{n} \binom{n}{x} \theta^x (1-\theta)^{n-x} = I_\theta(x', n - x' + 1),$$

$$\sum_{x=0}^{x'-1} \binom{n}{x} \theta^x (1-\theta)^{n-x} = 1 - I_\theta(x', n - x' + 1)$$

$$= 1 - \int_0^\theta u^{x'-1}(1-u)^{n-x'} \, du \Big/ \int_0^1 u^{x'-1}(1-u)^{n-x'} \, du .$$

The cumulative binomial distribution is related to the cumulative negative binomial distribution as follows:

$$1 - \sum_{x'=0}^{r-1} \binom{x+r}{x'} \theta^{x'}(1-\theta)^{x+r-x'} = \sum_{x'=0}^{x} \binom{x'+r-1}{r-1} \theta^r (1-\theta)^{x'}$$

or

$$\sum_{x'=r}^{x+r} \binom{x+r}{x'} \theta^{x'}(1-\theta)^{x+r-x'} = \sum_{x'=0}^{x} \binom{x'+r-1}{r-1} \theta^r (1-\theta)^{x'} .$$

CUMULATIVE TERMS, BINOMIAL DISTRIBUTION (Continued)

						θ					
n	*x'*	.05	.10	.15	.20	.25	.30	.35	.40	.45	.50
2	1	.0975	.1900	.2775	.3600	.4375	.5100	.5775	.6400	.6975	.7500
	2	.0025	.0100	.0225	.0400	.0625	.0900	.1225	.1600	.2025	.2500
3	1	.1426	.2710	.3859	.4880	.5781	.6570	.7254	.7840	.8336	.8750
	2	.0072	.0280	.0608	.1040	.1562	.2160	.2818	.3520	.4252	.5000
	3	.0001	.0010	.0034	.0080	.0156	.0270	.0429	.0640	.0911	.1250
4	1	.1855	.3439	.4780	.5904	.6836	.7599	.8215	.8704	.9085	.9375
	2	.0140	.0523	.1095	.1808	.2617	.3483	.4370	.5248	.6090	.6875
	3	.0005	.0037	.0120	.0272	.0508	.0837	.1265	.1792	.2415	.3125
	4	.0000	.0001	.0005	.0016	.0039	.0081	.0150	.0256	.0410	.0625
5	1	.2262	.4095	.5563	.6723	.7627	.8319	.8840	.9222	.9497	.9688
	2	.0226	.0815	.1648	.2627	.3672	.4718	.5716	.6630	.7438	.8125
	3	.0012	.0086	.0266	.0579	.1035	.1631	.2352	.3174	.4069	.5000
	4	.0000	.0005	.0022	.0067	.0156	.0308	.0540	.0870	.1312	.1875
	5	.0000	.0000	.0001	.0003	.0010	.0024	.0053	.0102	.0185	.0312
6	1	.2649	.4686	.6229	.7379	.8220	.8824	.9246	.9533	.9723	.9844
	2	.0328	.1143	.2235	.3447	.4661	.5798	.6809	.7667	.8364	.8906
	3	.0022	.0158	.0473	.0989	.1694	.2557	.3529	.4557	.5585	.6562
	4	.0001	.0013	.0059	.0170	.0376	.0705	.1174	.1792	.2553	.3438
	5	.0000	.0001	.0004	.0016	.0046	.0109	.0223	.0410	.0692	.1094
	6	.0000	.0000	.0000	.0001	.0002	.0007	.0018	.0041	.0083	.0156
7	1	.3017	.5217	.6794	.7903	.8665	.9176	.9510	.9720	.9848	.9922
	2	.0444	.1497	.2834	.4233	.5551	.6706	.7662	.8414	.8976	.9375
	3	.0038	.0257	.0738	.1480	.2436	.3529	.4677	.5801	.6836	.7734
	4	.0002	.0027	.0121	.0333	.0706	.1260	.1998	.2898	.3917	.5000
	5	.0000	.0002	.0012	.0047	.0129	.0288	.0556	.0963	.1529	.2266
	6	.0000	.0000	.0001	.0004	.0013	.0038	.0090	.0188	.0357	.0625
	7	.0000	.0000	.0000	.0000	.0001	.0002	.0006	.0016	.0037	.0078
8	1	.3366	.5695	.7275	.8322	.8999	.9424	.9681	.9832	.9916	.9961
	2	.0572	.1869	.3428	.4967	.6329	.7447	.8309	.8936	.9368	.9648
	3	.0058	.0381	.1052	.2031	.3215	.4482	.5722	.6846	.7799	.8555
	4	.0004	.0050	.0214	.0563	.1138	.1941	.2936	.4059	.5230	.6367
	5	.0000	.0004	.0029	.0104	.0273	.0580	.1061	.1737	.2604	.3633
	6	.0000	.0000	.0002	.0012	.0042	.0113	.0253	.0498	.0885	.1445
	7	.0000	.0000	.0000	.0001	.0004	.0013	.0036	.0085	.0181	.0352
	8	.0000	.0000	.0000	.0000	.0000	.0001	.0002	.0007	.0017	.0039
9	1	.3698	.6126	.7684	.8658	.9249	.9596	.9793	.9899	.9954	.9980
	2	.0712	.2252	.4005	.5638	.6997	.8040	.8789	.9295	.9615	.9805
	3	.0084	.0530	.1409	.2618	.3993	.5372	.6627	.7682	.8505	.9102
	4	.0006	.0083	.0339	.0856	.1657	.2703	.3911	.5174	.6386	.7461
	5	.0000	.0009	.0056	.0196	.0489	.0988	.1717	.2666	.3786	.5000
	6	.0000	.0001	.0006	.0031	.0100	.0253	.0536	.0994	.1658	.2539
	7	.0000	.0000	.0000	.0003	.0013	.0043	.0112	.0250	.0498	.0898
	8	.0000	.0000	.0000	.0000	.0001	.0004	.0014	.0038	.0091	.0195
	9	.0000	.0000	.0000	.0000	.0000	.0000	.0001	.0003	.0008	.0020

Linear interpolation will be accurate at most to two decimal places.

CUMULATIVE TERMS, BINOMIAL DISTRIBUTION (Continued)

n	x′	.05	.10	.15	.20	.25	.30	.35	.40	.45	.50
10	1	.4013	.6513	.8031	.8926	.9437	.9718	.9865	.9940	.9975	.9990
	2	.0861	.2639	.4557	.6242	.7560	.8507	.9140	.9536	.9767	.9893
	3	.0115	.0702	.1798	.3222	.4744	.6172	.7384	.8327	.9004	.9453
	4	.0010	.0128	.0500	.1209	.2241	.3504	.4862	.6177	.7340	.8281
	5	.0001	.0016	.0099	.0328	.0781	.1503	.2485	.3669	.4956	.6230
	6	.0000	.0001	.0014	.0064	.0197	.0473	.0949	.1662	.2616	.3770
	7	.0000	.0000	.0001	.0009	.0035	.0106	.0260	.0548	.1020	.1719
	8	.0000	.0000	.0000	.0001	.0004	.0016	.0048	.0123	.0274	.0547
	9	.0000	.0000	.0000	.0000	.0000	.0001	.0005	.0017	.0045	.0107
	10	.0000	.0000	.0000	.0000	.0000	.0000	.0000	.0001	.0003	.0010
11	1	.4312	.6862	.8327	.9141	.9578	.9802	.9912	.9964	.9986	.9995
	2	.1019	.3026	.5078	.6779	.8029	.8870	.9394	.9698	.9861	.9941
	3	.0152	.0896	.2212	.3826	.5448	.6873	.7999	.8811	.9348	.9673
	4	.0016	.0185	.0694	.1611	.2867	.4304	.5744	.7037	.8089	.8867
	5	.0001	.0028	.0159	.0504	.1146	.2103	.3317	.4672	.6029	.7256
	6	.0000	.0003	.0027	.0117	.0343	.0782	.1487	.2465	.3669	.5000
	7	.0000	.0000	.0003	.0020	.0076	.0216	.0501	.0994	.1738	.2744
	8	.0000	.0000	.0000	.0002	.0012	.0043	.0122	.0293	.0610	.1133
	9	.0000	.0000	.0000	.0000	.0001	.0006	.0020	.0059	.0148	.0327
	10	.0000	.0000	.0000	.0000	.0000	.0000	.0002	.0007	.0022	.0059
	11	.0000	.0000	.0000	.0000	.0000	.0000	.0000	.0000	.0002	.0005
12	1	.4596	.7176	.8578	.9313	.9683	.9862	.9943	.9978	.9992	.9998
	2	.1184	.3410	.5565	.7251	.8416	.9150	.9576	.9804	.9917	.9968
	3	.0196	.1109	.2642	.4417	.6093	.7472	.8487	.9166	.9579	.9807
	4	.0022	.0256	.0922	.2054	.3512	.5075	.6533	.7747	.8655	.9270
	5	.0002	.0043	.0239	.0726	.1576	.2763	.4167	.5618	.6956	.8062
	6	.0000	.0005	.0046	.0194	.0544	.1178	.2127	.3348	.4731	.6128
	7	.0000	.0001	.0007	.0039	.0143	.0386	.0846	.1582	.2607	.3872
	8	.0000	.0000	.0001	.0006	.0028	.0095	.0255	.0573	.1117	.1938
	9	.0000	.0000	.0000	.0001	.0004	.0017	.0053	.0153	.0356	.0730
	10	.0000	.0000	.0000	.0000	.0000	.0002	.0008	.0028	.0079	.0193
	11	.0000	.0000	.0000	.0000	.0000	.0000	.0001	.0003	.0011	.0032
	12	.0000	.0000	.0000	.0000	.0000	.0000	.0000	.0000	.0001	.0002
13	1	.4867	.7458	.8791	.9450	.9762	.9903	.9963	.9987	.9996	.9999
	2	.1354	.3787	.6017	.7664	.8733	.9363	.9704	.9874	.9951	.9983
	3	.0245	.1339	.3080	.4983	.6674	.7975	.8868	.9421	.9731	.9888
	4	.0031	.0342	.1180	.2527	.4157	.5794	.7217	.8314	.9071	.9539
	5	.0003	.0065	.0342	.0991	.2060	.3457	.4995	.6470	.7721	.8666
	6	.0000	.0009	.0075	.0300	.0802	.1654	.2841	.4256	.5732	.7095
	7	.0000	.0001	.0013	.0070	.0243	.0624	.1295	.2288	.3563	.5000
	8	.0000	.0000	.0002	.0012	.0056	.0182	.0462	.0977	.1788	.2905
	9	.0000	.0000	.0000	.0002	.0010	.0040	.0126	.0321	.0698	.1334
	10	.0000	.0000	.0000	.0000	.0001	.0007	.0025	.0078	.0203	.0461
	11	.0000	.0000	.0000	.0000	.0000	.0001	.0003	.0013	.0041	.0112
	12	.0000	.0000	.0000	.0000	.0000	.0000	.0000	.0001	.0005	.0017
	13	.0000	.0000	.0000	.0000	.0000	.0000	.0000	.0000	.0000	.0001

CUMULATIVE TERMS, BINOMIAL DISTRIBUTION (Continued)

n	x'	.05	.10	.15	.20	.25 θ	.30	.35	.40	.45	.50
14	1	.5123	.7712	.8972	.9560	.9822	.9932	.9976	.9992	.9998	.9999
	2	.1530	.4154	.6433	.8021	.8990	.9525	.9795	.9919	.9971	.9991
	3	.0301	.1584	.3521	.5519	.7189	.8392	.9161	.9602	.9830	.9935
	4	.0042	.0441	.1465	.3018	.4787	.6448	.7795	.8757	.9368	.9713
	5	.0004	.0092	.0467	.1298	.2585	.4158	.5773	.7207	.8328	.9102
	6	.0000	.0015	.0115	.0439	.1117	.2195	.3595	.5141	.6627	.7880
	7	.0000	.0002	.0022	.0116	.0383	.0933	.1836	.3075	.4539	.6047
	8	.0000	.0000	.0003	.0024	.0103	.0315	.0753	.1501	.2586	.3953
	9	.0000	.0000	.0000	.0004	.0022	.0083	.0243	.0583	.1189	.2120
	10	.0000	.0000	.0000	.0000	.0003	.0017	.0060	.0175	.0426	.0898
	11	.0000	.0000	.0000	.0000	.0000	.0002	.0011	.0039	.0114	.0287
	12	.0000	.0000	.0000	.0000	.0000	.0000	.0001	.0006	.0022	.0065
	13	.0000	.0000	.0000	.0000	.0000	.0000	.0000	.0001	.0003	.0009
	14	.0000	.0000	.0000	.0000	.0000	.0000	.0000	.0000	.0000	.0001
15	1	.5367	.7941	.9126	.9648	.9866	.9953	.9984	.9995	.9999	1.0000
	2	.1710	.4510	.6814	.8329	.9198	.9647	.9858	.9948	.9983	.9995
	3	.0362	.1841	.3958	.6020	.7639	.8732	.9383	.9729	.9893	.9963
	4	.0055	.0556	.1773	.3518	.5387	.7031	.8273	.9095	.9576	.9824
	5	.0006	.0127	.0617	.1642	.3135	.4845	.6481	.7827	.8796	.9408
	6	.0001	.0022	.0168	.0611	.1484	.2784	.4357	.5968	.7392	.8491
	7	.0000	.0003	.0036	.0181	.0566	.1311	.2452	.3902	.5478	.6964
	8	.0000	.0000	.0006	.0042	.0173	.0500	.1132	.2131	.3465	.5000
	9	.0000	.0000	.0001	.0008	.0042	.0152	.0422	.0950	.1818	.3036
	10	.0000	.0000	.0000	.0001	.0008	.0037	.0124	.0338	.0769	.1509
	11	.0000	.0000	.0000	.0000	.0001	.0007	.0028	.0093	.0255	.0592
	12	.0000	.0000	.0000	.0000	.0000	.0001	.0005	.0019	.0063	.0176
	13	.0000	.0000	.0000	.0000	.0000	.0000	.0001	.0003	.0011	.0037
	14	.0000	.0000	.0000	.0000	.0000	.0000	.0000	.0000	.0001	.0005
	15	.0000	.0000	.0000	.0000	.0000	.0000	.0000	.0000	.0000	.0000
16	1	.5599	.8147	.9257	.9719	.9900	.9967	.9990	.9997	.9999	1.0000
	2	.1892	.4853	.7161	.8593	.9365	.9739	.9902	.9967	.9990	.9997
	3	.0429	.2108	.4386	.6482	.8029	.9006	.9549	.9817	.9934	.9979
	4	.0070	.0684	.2101	.4019	.5950	.7541	.8661	.9349	.9719	.9894
	5	.0009	.0170	.0791	.2018	.3698	.5501	.7108	.8334	.9147	.9616
	6	.0001	.0033	.0235	.0817	.1897	.3402	.5100	.6712	.8024	.8949
	7	.0000	.0005	.0056	.0267	.0796	.1753	.3119	.4728	.6340	.7228
	8	.0000	.0001	.0011	.0070	.0271	.0744	.1594	.2839	.4371	.5982
	9	.0000	.0000	.0002	.0015	.0075	.0257	.0671	.1423	.2559	.4018
	10	.0000	.0000	.0000	.0002	.0016	.0071	.0229	.0583	.1241	.2272
	11	.0000	.0000	.0000	.0000	.0003	.0016	.0062	.0191	.0486	.1051
	12	.0000	.0000	.0000	.0000	.0000	.0003	.0013	.0049	.0149	.0384
	13	.0000	.0000	.0000	.0000	.0000	.0000	.0002	.0009	.0035	.0106
	14	.0000	.0000	.0000	.0000	.0000	.0000	.0000	.0001	.0006	.0021
	15	.0000	.0000	.0000	.0000	.0000	.0000	.0000	.0000	.0001	.0003
	16	.0000	.0000	.0000	.0000	.0000	.0000	.0000	.0000	.0000	.0000

CUMULATIVE TERMS, BINOMIAL DISTRIBUTION (Continued)

n	x'	.05	.10	.15	.20	.25 θ	.30	.35	.40	.45	.50
17	1	.5819	.8332	.9369	.9775	.9925	.9977	.9993	.9998	1.0000	1.0000
	2	.2078	.5182	.7475	.8818	.9499	.9807	.9933	.9979	.9994	.9999
	3	.0503	.2382	.4802	.6904	.8363	.9226	.9673	.9877	.9959	.9988
	4	.0088	.0826	.2444	.4511	.6470	.7981	.8972	.9536	.9816	.9936
	5	.0012	.0221	.0987	.2418	.4261	.6113	.7652	.8740	.9404	.9755
	6	.0001	.0047	.0319	.1057	.2347	.4032	.5803	.7361	.8529	.9283
	7	.0000	.0008	.0083	.0377	.1071	.2248	.3812	.5522	.7098	.8338
	8	.0000	.0001	.0017	.0109	.0402	.1046	.2128	.3595	.5257	.6855
	9	.0000	.0000	.0003	.0026	.0124	.0403	.0994	.1989	.3374	.5000
	10	.0000	.0000	.0000	.0005	.0031	.0127	.0383	.0919	.1834	.3145
	11	.0000	.0000	.0000	.0001	.0006	.0032	.0120	.0348	.0826	.1662
	12	.0000	.0000	.0000	.0000	.0001	.0007	.0030	.0106	.0301	.0717
	13	.0000	.0000	.0000	.0000	.0000	.0001	.0006	.0025	.0086	.0245
	14	.0000	.0000	.0000	.0000	.0000	.0000	.0001	.0005	.0019	.0064
	15	.0000	.0000	.0000	.0000	.0000	.0000	.0000	.0001	.0003	.0012
	16	.0000	.0000	.0000	.0000	.0000	.0000	.0000	.0000	.0000	.0001
	17	.0000	.0000	.0000	.0000	.0000	.0000	.0000	.0000	.0000	.0000
18	1	.6028	.8499	.9464	.9820	.9944	.9984	.9996	.9999	1.0000	1.0000
	2	.2265	.5497	.7759	.9009	.9605	.9858	.9954	.9987	.9997	.9999
	3	.0581	.2662	.5203	.7287	.8647	.9400	.9764	.9918	.9975	.9993
	4	.0109	.0982	.2798	.4990	.6943	.8354	.9217	.9672	.9880	.9962
	5	.0015	.0282	.1206	.2836	.4813	.6673	.8114	.9058	.9589	.9846
	6	.0002	.0064	.0419	.1329	.2825	.4656	.6450	.7912	.8923	.9519
	7	.0000	.0012	.0118	.0513	.1390	.2783	.4509	.6257	.7742	.8811
	8	.0000	.0002	.0027	.0163	.0569	.1407	.2717	.4366	.6085	.7597
	9	.0000	.0000	.0005	.0043	.0193	.0596	.1391	.2632	.4222	.5927
	10	.0000	.0000	.0001	.0009	.0054	.0210	.0597	.1347	.2527	.4073
	11	.0000	.0000	.0000	.0002	.0012	.0061	.0212	.0576	.1280	.2403
	12	.0000	.0000	.0000	.0000	.0002	.0014	.0062	.0203	.0537	.1189
	13	.0000	.0000	.0000	.0000	.0000	.0003	.0014	.0058	.0183	.0481
	14	.0000	.0000	.0000	.0000	.0000	.0000	.0003	.0013	.0049	.0154
	15	.0000	.0000	.0000	.0000	.0000	.0000	.0000	.0002	.0010	.0038
	16	.0000	.0000	.0000	.0000	.0000	.0000	.0000	.0000	.0001	.0007
	17	.0000	.0000	.0000	.0000	.0000	.0000	.0000	.0000	.0000	.0001
	18	.0000	.0000	.0000	.0000	.0000	.0000	.0000	.0000	.0000	.0000
19	1	.6226	.8649	.9544	.9856	.9958	.9989	.9997	.9999	1.0000	1.0000
	2	.2453	.5797	.8015	.9171	.9690	.9896	.9969	.9992	.9998	1.0000
	3	.0665	.2946	.5587	.7631	.8887	.9538	.9830	.9945	.9985	.9996
	4	.0132	.1150	.3159	.5449	.7369	.8668	.9409	.9770	.9923	.9978
	5	.0020	.0352	.1444	.3267	.5346	.7178	.8500	.9304	.9720	.9904
	6	.0002	.0086	.0537	.1631	.3322	.5261	.7032	.8371	.9223	.9682
	7	.0000	.0017	.0163	.0676	.1749	.3345	.5188	.6919	.8273	.9165
	8	.0000	.0003	.0041	.0233	.0775	.1820	.3344	.5122	.6831	.8204
	9	.0000	.0000	.0008	.0067	.0287	.0839	.1855	.3325	.5060	.6762
	10	.0000	.0000	.0001	.0016	.0089	.0326	.0875	.1861	.3290	.5000

CUMULATIVE TERMS, BINOMIAL DISTRIBUTION (Continued)

n	x'	.05	.10	.15	.20	.25	.30	.35	.40	.45	.50
19	11	.0000	.0000	.0000	.0003	.0023	.0105	.0347	.0885	.1841	.3238
	12	.0000	.0000	.0000	.0000	.0005	.0028	.0114	.0352	.0871	.1796
	13	.0000	.0000	.0000	.0000	.0001	.0006	.0031	.0116	.0342	.0835
	14	.0000	.0000	.0000	.0000	.0000	.0001	.0007	.0031	.0109	.0318
	15	.0000	.0000	.0000	.0000	.0000	.0000	.0001	.0006	.0028	.0096
	16	.0000	.0000	.0000	.0000	.0000	.0000	.0000	.0001	.0005	.0022
	17	.0000	.0000	.0000	.0000	.0000	.0000	.0000	.0000	.0001	.0004
	18	.0000	.0000	.0000	.0000	.0000	.0000	.0000	.0000	.0000	.0000
	19	.0000	.0000	.0000	.0000	.0000	.0000	.0000	.0000	.0000	.0000
20	1	.6415	.8784	.9612	.9885	.9968	.9992	.9998	1.0000	1.0000	1.0000
	2	.2642	.6083	.8244	.9308	.9757	.9924	.9979	.9995	.9999	1.0000
	3	.0755	.3231	.5951	.7939	.9087	.9645	.9879	.9964	.9991	.9998
	4	.0159	.1330	.3523	.5886	.7748	.8929	.9556	.9840	.9951	.9987
	5	.0026	.0432	.1702	.3704	.5852	.7625	.8818	.9490	.9811	.9941
	6	.0003	.0113	.0673	.1958	.3828	.5836	.7546	.8744	.9447	.9793
	7	.0000	.0024	.0219	.0867	.2142	.3920	.5834	.7500	.8701	.9423
	8	.0000	.0004	.0059	.0321	.1018	.2277	.3990	.5841	.7480	.8684
	9	.0000	.0001	.0013	.0100	.0409	.1133	.2376	.4044	.5857	.7483
	10	.0000	.0000	.0002	.0026	.0139	.0480	.1218	.2447	.4086	.5881
	11	.0000	.0000	.0000	.0006	.0039	.0171	.0532	.1275	.2493	.4119
	12	.0000	.0000	.0000	.0001	.0009	.0051	.0196	.0565	.1308	.2517
	13	.0000	.0000	.0000	.0000	.0002	.0013	.0060	.0210	.0580	.1316
	14	.0000	.0000	.0000	.0000	.0000	.0003	.0015	.0065	.0214	.0577
	15	.0000	.0000	.0000	.0000	.0000	.0000	.0003	.0016	.0064	.0207
	16	.0000	.0000	.0000	.0000	.0000	.0000	.0003	.0015	.0059
	17	.0000	.0000	.0000	.0000	.0000	.0000	.0000	.0000	.0003	.0013
	18	.0000	.0000	.0000	.0000	.0000	.0000	.0000	.0000	.0000	.0002
	19	.0000	.0000	.0000	.0000	.0000	.0000	.0000	.0000	.0000	.0000
	20	.0000	.0000	.0000	.0000	.0000	.0000	.0000	.0000	.0000	.0000
21	1	.6594	.8906	.9671	.9908	.9976	.9994	.9999	1.0000	1.0000	1.0000
	2	.2830	.6353	.8450	.9424	.9810	.9944	.9996	.9997	.9999	1.0000
	3	.0849	.3516	.6295	.8213	.9255	.9729	.9914	.9976	.9994	.9999
	4	.0189	.1520	.3887	.6296	.8083	.9144	.9669	.9890	.9969	.9993
	5	.0032	.0522	.1975	.4140	.6326	.8016	.9076	.9630	.9874	.9967
	6	.0004	.0144	.0827	.2307	.4334	.6373	.7991	.9043	.9611	.9867
	7	.0000	.0033	.0287	.1085	.2564	.4495	.6433	.7998	.9036	.9608
	8	.0000	.0006	.0083	.0431	.1299	.2770	.4635	.6505	.8029	.9054
	9	.0000	.0001	.0020	.0144	.0561	.1477	.2941	.4763	.6587	.8083
	10	.0000	.0000	.0004	.0041	.0206	.0676	.1632	.3086	.4883	.6682
	11	.0000	.0000	.0001	.0010	.0064	.0264	.0772	.1744	.3210	.5000
	12	.0000	.0000	.0000	.0002	.0017	.0087	.0313	.0849	.1841	.3318
	13	.0000	.0000	.0000	.0000	.0004	.0024	.0108	.0352	.0908	.1917
	14	.0000	.0000	.0000	.0000	.0001	.0006	.0031	.0123	.0379	.0946
	15	.0000	.0000	.0000	.0000	.0000	.0001	.0007	.0036	.0132	.0392
	16	.0000	.0000	.0000	.0000	.0000	.0000	.0001	.0008	.0037	.0133
	17	.0000	.0000	.0000	.0000	.0000	.0000	.0000	.0002	.0008	.0036
	18	.0000	.0000	.0000	.0000	.0000	.0000	.0000	.0000	.0001	.0007
	19	.0000	.0000	.0000	.0000	.0000	.0000	.0000	.0000	.0000	.0001

CUMULATIVE TERMS, BINOMIAL DISTRIBUTION (Continued)

n	x′	.05	.10	.15	.20	θ .25	.30	.35	.40	.45	.50
21	20	.0000	.0000	.0000	.0000	.0000	.0000	.0000	.0000	.0000	.0000
	21	.0000	.0000	.0000	.0000	.0000	.0000	.0000	.0000	.0000	.0000
22	1	.6765	.9015	.9720	.9926	.9982	.9966	.9999	1.0000	1.0000	1.0000
	2	.3018	.6608	.8633	.9520	.9851	.9959	.9990	.9998	1.0000	1.0000
	3	.0948	.3800	.6618	.8455	.9394	.9793	.9399	.9984	.9997	.9999
	4	.0222	.1719	.4248	.6680	.8376	.9319	.9755	.9924	.9980	.9996
	5	.0040	.0621	.2262	.4571	.6765	.8355	.9284	.9734	.9917	.9978
	6	.0006	.0182	.0999	.2674	.4832	.6866	.8371	.9278	.9729	.9915
	7	.0001	.0044	.0368	.1330	.3006	.5058	.6978	.8416	.9295	.9738
	8	.0000	.0009	.0114	.0561	.1615	.3287	.5264	.7102	.8482	.9331
	9	.0000	.0001	.0030	.0201	.0746	.1865	.3534	.5460	.7236	.8569
	10	.0000	.0000	.0007	.0061	.0295	.0916	.2084	.3756	.5650	.7383
	11	.0000	.0000	.0001	.0016	.0100	.0387	.1070	.2281	.3963	.5841
	12	.0000	.0000	.0000	.0003	.0029	.0140	.0474	.1207	.2457	.4159
	13	.0000	.0000	.0000	.0001	.0007	.0043	.0180	.0551	.1328	.2617
	14	.0000	.0000	.0000	.0000	.0001	.0011	.0058	.0215	.0617	.1431
	15	.0000	.0000	.0000	.0000	.0000	.0002	.0015	.0070	.0243	.0669
	16	.0000	.0000	.0000	.0000	.0000	.0000	.0003	.0019	.0080	.0262
	17	.0000	.0000	.0000	.0000	.0000	.0000	.0001	.0004	.0021	.0085
	18	.0000	.0000	.0000	.0000	.0000	.0000	.0000	.0001	.0005	.0022
	19	.0000	.0000	.0000	.0000	.0000	.0000	.0000	.0000	.0001	.0004
	20	.0000	.0000	.0000	.0000	.0000	.0000	.0000	.0000	.0000	.0001
	21	.0000	.0000	.0000	.0000	.0000	.0000	.0000	.0000	.0000	.0000
	22	.0000	.0000	.0000	.0000	.0000	.0000	.0000	.0000	.0000	.0000
23	1	.6926	.9114	.9762	.9941	.9987	.9997	1.0000	1.0000	1.0000	1.0000
	2	.3206	.6849	.8796	.9602	.9884	.9970	.9993	.9999	1.0000	1.0000
	3	.1052	.4080	.6920	.8668	.9508	.9843	.9957	.9990	1.0000	1.0000
	4	.0258	.1927	.4604	.7035	.8630	.9462	.9819	.9948	.9988	.9998
	5	.0049	.0731	.2560	.4993	.7168	.8644	.9449	.9810	.9945	.9987
	6	.0008	.0226	.1189	.3053	.5315	.7312	.8691	.9460	.9814	.9947
	7	.0001	.0058	.0463	.1598	.3463	.5601	.7466	.8760	.9490	.9827
	8	.0000	.0012	.0152	.0715	.1963	.3819	.5864	.7627	.8848	.9534
	9	.0000	.0002	.0042	.0273	.0963	.2291	.4140	.6116	.7797	.8950
	10	.0000	.0000	.0010	.0089	.0408	.1201	.2592	.4438	.6364	.7976
	11	.0000	.0000	.0002	.0025	.0149	.0546	.1425	.2871	.4722	.6612
	12	.0000	.0000	.0000	.0006	.0046	.0214	.0682	.1636	.3135	.5000
	13	.0000	.0000	.0000	.0001	.0012	.0072	.0283	.0813	.1836	.3388
	14	.0000	.0000	.0000	.0000	.0003	.0021	.0100	.0349	.0937	.2024
	15	.0000	.0000	.0000	.0000	.0001	.0005	.0030	.0128	.0411	.1050
	16	.0000	.0000	.0000	.0000	.0000	.0001	.0008	.0040	.0153	.0466
	17	.0000	.0000	.0000	.0000	.0000	.0000	.0002	.0010	.0048	.0173
	18	.0000	.0000	.0000	.0000	.0000	.0000	.0000	.0002	.0012	.0053
	19	.0000	.0000	.0000	.0000	.0000	.0000	.0000	.0000	.0002	.0013
	20	.0000	.0000	.0000	.0000	.0000	.0000	.0000	.0000	.0000	.0002
	21	.0000	.0000	.0000	.0000	.0000	.0000	.0000	.0000	.0000	.0000
	22	.0000	.0000	.0000	.0000	.0000	.0000	.0000	.0000	.0000	.0000
	23	.0000	.0000	.0000	.0000	.0000	.0000	.0000	.0000	.0000	.0000

CUMULATIVE TERMS, BINOMIAL DISTRIBUTION (Continued)

n	x'	.05	.10	.15	.20	θ .25	.30	.35	.40	.45	.50
24	1	.7080	.9202	.9798	.9953	.9990	.9998	1.0000	1.0000	1.0000	1.0000
	2	.3391	.7075	.8941	.9669	.9910	.9978	.9995	.9999	1.0000	1.0000
	3	.1159	.4357	7202	.8855	.9602	.9881	.9970	.9993	.9999	1.0000
	4	.0298	.2143	.4951	.7361	.8850	.9576	.9867	.9965	.9992	.9999
	5	.0060	.0851	.2866	.5401	.7534	.8889	.9578	.9866	.9964	.9992
	6	.0010	.0277	.1394	.3441	.5778	.7712	.8956	.9600	.9873	.9967
	7	.0001	.0075	.0572	.1889	.3926	.6114	.7894	.9040	.9636	.9887
	8	.0000	.0017	.0199	.0892	.2338	.4353	.6425	.8081	.9137	.9680
	9	.0000	.0003	.0059	.0362	.1213	.2750	.4743	.6721	.8270	.9242
	10	.0000	.0001	.0015	.0126	.0547	.1528	.3134	.5109	.7009	.8463
	11	.0000	.0000	.0003	.0038	.0213	.0742	.1833	.3498	.5461	.7294
	12	.0000	.0000	.0001	.0010	.0072	.0314	.0942	.2130	.3849	.5806
	13	.0000	.0000	.0000	.0002	.0021	.0115	.0423	.1143	.2420	.4194
	14	.0000	.0000	.0000	.0000	.0005	.0036	.0164	.0535	.1341	.2706
	15	.0000	.0000	.0000	.0000	.0001	.0010	.0055	.0217	.0648	.1537
	16	.0000	.0000	.0000	.0000	.0000	.0002	.0016	.0075	.0269	.0758
	17	.0000	.0000	.0000	.0000	.0000	.0000	.0004	.0022	.0095	.0320
	18	.0000	.0000	.0000	.0000	.0000	.0000	.0001	.0005	.0028	.0113
	19	.0000	.0000	.0000	.0000	.0000	.0000	.0000	.0001	.0007	.0033
	20	.0000	.0000	.0000	.0000	.0000	.0000	.0000	.0000	.0001	.0008
	21	.0000	.0000	.0000	.0000	.0000	.0000	.0000	.0000	.0000	.0001
	22	.0000	.0000	.0000	.0000	.0000	.0000	.0000	.0000	.0000	.0000
	23	.0000	.0000	.0000	.0000	.0000	.0000	.0000	.0000	.0000	.0000
	24	.0000	.0000	.0000	.0000	.0000	.0000	.0000	.0000	.0000	.0000
25	1	.7226	.9282	.9828	.9962	.9992	.9999	1.0000	1.0000	1.0000	1.0000
	2	.3576	.7288	.9069	.9726	.9930	.9984	.9997	.9999	1.0000	1.0000
	3	.1271	.4629	.7463	.9018	.9679	.9910	.9979	.9996	.9999	1.0000
	4	.0341	.2364	.5289	.7660	.9038	.9668	.9903	.9976	.9995	.9999
	5	.0072	.0980	.3179	.5793	.7863	.9095	.9680	.9905	.9977	.9995
	6	.0012	.0334	.1615	.3833	.6217	.8065	.9174	.9706	.9914	.9980
	7	.0002	.0095	.0695	.2200	.4389	.6593	.8266	.9264	.9742	.9927
	8	.0000	.0023	.0255	.1091	.2735	.4882	.6939	.8464	.9361	.9784
	9	.0000	.0005	.0080	.0468	.1494	.3231	.5332	.7265	.8660	.9461
	10	.0000	.0001	.0021	.0173	.0713	.1894	.3697	.5754	.7576	.8852
	11	.0000	.0000	.0005	.0056	.0297	.0978	.2288	.4142	.6157	.7878
	12	.0000	.0000	.0001	.0015	.0107	.0442	.1254	.2677	.4574	.6550
	13	.0000	.0000	.0000	.0004	.0034	.0175	.0604	.1538	.3063	.5000
	14	.0000	.0000	.0000	.0001	.0009	.0060	.0255	.0778	.1827	.3450
	15	.0000	.0000	.0000	.0000	.0002	.0018	.0093	.0344	.0960	.2122
	16	.0000	.0000	.0000	.0000	.0000	.0005	.0029	.0132	.0440	.1148
	17	.0000	.0000	.0000	.0000	.0000	.0001	.0008	.0043	.0174	.0539
	18	.0000	.0000	.0000	.0000	.0000	.0000	.0002	.0012	.0058	.0216
	19	.0000	.0000	.0000	.0000	.0000	.0000	.0000	.0003	.0016	.0073
	20	.0000	.0000	.0000	.0000	.0000	.0000	.0000	.0001	.0004	.0020
	21	.0000	.0000	.0000	.0000	.0000	.0000	.0000	.0000	.0001	.0005
	22	.0000	.0000	.0000	.0000	.0000	.0000	.0000	.0000	.0000	.0001
	23	.0000	.0000	.0000	.0000	.0000	.0000	.0000	.0000	.0000	.0000
	24	.0000	.0000	.0000	.0000	.0000	.0000	.0000	.0000	.0000	.0000

CUMULATIVE TERMS, BINOMIAL DISTRIBUTION (Continued)

n	x'	.05	.10	.15	.20	θ .25	.30	.35	.40	.45	.50
25	25	.0000	.0000	.0000	.0000	.0000	.0000	.0000	.0000	.0000	.0000
30	1	.7854	.9576	.9924	.9988	.9998	1.0000	1.0000	1.0000	1.0000	1.0000
	2	.4465	.8163	.9520	.9895	.9980	.9997	1.0000	1.0000	1.0000	1.0000
	3	.1878	.5886	.8486	.9558	.9894	.9979	.9997	1.0000	1.0000	1.0000
	4	.0608	.3526	.6783	.8773	.9626	.9907	.9981	.9997	1.0000	1.0000
	5	.0156	.1755	.4755	.7448	.9021	.9698	.9925	.9985	.9998	1.0000
	6	.0033	.0732	.2894	.5725	.7974	.9234	.9767	.9943	.9989	.9998
	7	.0006	.0258	.1526	.3930	.6519	.8405	.9414	.9828	.9960	.9993
	8	.0001	.0078	.0698	.2392	.4857	.7186	.8762	.9565	.9879	.9974
	9	.0000	.0020	.0278	.1287	.3264	.5685	.7753	.9060	.9688	.9919
	10	.0000	.0005	.0097	.0611	.1966	.4112	.6425	.8237	.9306	.9786
	11	.0000	.0001	.0029	.0256	.1057	.2696	.4922	.7085	.8650	.9506
	12	.0000	.0000	.0008	.0095	.0507	.1593	.3452	.5689	.7673	.8998
	13	.0000	.0000	.0002	.0031	.0216	.0845	.2198	.4215	.6408	.8192
	14	.0000	.0000	.0000	.0009	.0082	.0401	.1263	.2855	.4975	.7077
	15	.0000	.0000	.0000	.0002	.0027	.0169	.0652	.1754	.3552	.5722
	16	.0000	.0000	.0000	.0001	.0008	.0064	.0301	.0971	.2309	.4278
	17	.0000	.0000	.0000	.0000	.0002	.0021	.0124	.0481	.1356	.2923
	18	.0000	.0000	.0000	.0000	.0001	.0006	.0045	.0212	.0714	.1808
	19	.0000	.0000	.0000	.0000	.0000	.0002	.0014	.0083	.0334	.1002
	20	.0000	.0000	.0000	.0000	.0000	.0000	.0004	.0029	.0138	.0494
	21	.0000	.0000	.0000	.0000	.0000	.0000	.0001	.0009	.0050	.0214
	22	.0000	.0000	.0000	.0000	.0000	.0000	.0000	.0002	.0016	.0081
	23	.0000	.0000	.0000	.0000	.0000	.0000	.0000	.0000	.0004	.0026
	24	.0000	.0000	.0000	.0000	.0000	.0000	.0000	.0000	.0001	.0007
	25	.0000	.0000	.0000	.0000	.0000	.0000	.0000	.0000	.0000	.0002
	26	.0000	.0000	.0000	.0000	.0000	.0000	.0000	.0000	.0000	.0000
	27	.0000	.0000	.0000	.0000	.0000	.0000	.0000	.0000	.0000	.0000
	28	.0000	.0000	.0000	.0000	.0000	.0000	.0000	.0000	.0000	.0000
	29	.0000	.0000	.0000	.0000	.0000	.0000	.0000	.0000	.0000	.0000
	30	.0000	.0000	.0000	.0000	.0000	.0000	.0000	.0000	.0000	.0000
35	1	.8339	.9750	.9966	.9996	1.0000	1.0000	1.0000	1.0000	1.0000	1.0000
	2	.5280	.8776	.9757	.9960	.9995	.9999	1.0000	1.0000	1.0000	1.0000
	3	.2542	.6937	.9130	.9810	.9967	.9995	.9999	1.0000	1.0000	1.0000
	4	.0958	.4690	.7912	.9395	.9864	.9976	.9997	1.0000	1.0000	1.0000
	5	.0290	.2693	.6193	.8565	.9590	.9909	.9984	.9998	1.0000	1.0000
	6	.0073	.1316	.4311	.7279	.9024	.9731	.9942	.9990	.9999	1.0000
	7	.0015	.0552	.2652	.5672	.8080	.9350	.9830	.9966	.9995	.9999
	8	.0003	.0200	.1438	.4007	.6777	.8674	.9581	.9898	.9981	.9997
	9	.0000	.0063	.0689	.2550	.5257	.7659	.9110	.9740	.9943	.9991
	10	.0000	.0017	.0292	.1457	.3737	.6354	.8349	.9425	.9848	.9970
	11	.0000	.0004	.0110	.0747	.2419	.4900	.7284	.8877	.9646	.9917
	12	.0000	.0001	.0037	.0344	.1421	.3484	.5981	.8048	.9271	.9795
	13	.0000	.0000	.0011	.0142	.0756	.2271	.4577	.6943	.8656	.9552
	14	.0000	.0000	.0003	.0053	.0363	.1350	.3240	.5639	.7767	.9123
	15	.0000	.0000	.0001	.0018	.0158	.0731	.2109	.4272	.6624	.8447
	16	.0000	.0000	.0000	.0005	.0062	.0359	.1256	.2997	.5315	.7502

CUMULATIVE TERMS, BINOMIAL DISTRIBUTION (Continued)

n	x'	.05	.10	.15	.20	θ .25	.30	.35	.40	.45	.50
35	17	.0000	.0000	.0000	.0001	.0022	.0160	.0682	.1935	.3976	.6321
	18	.0000	.0000	.0000	.0000	.0007	.0064	.0336	.1143	2751	.5000
	19	.0000	.0000	.0000	.0000	.0002	.0023	.0150	.0615	.1749	.3679
	20	.0000	.0000	.0000	.0000	.0001	.0008	.0061	.0300	.1016	.2498
	21	.0000	.0000	.0000	.0000	.0000	.0002	.0022	.0133	.0536	.1553
	22	.0000	.0000	.0000	.0000	.0000	.0001	.0007	.0053	.0255	.0877
	23	.0000	.0000	.0000	.0000	.0000	.0000	.0002	.0019	.0109	.0448
	24	.0000	.0000	.0000	.0000	.0000	.0000	.0001	.0006	.0042	.0205
	25	.0000	.0000	.0000	.0000	.0000	.0000	.0000	.0002	.0014	.0083
	26	.0000	.0000	.0000	.0000	.0000	.0000	.0000	.0000	.0004	.0030
	27	.0000	.0000	.0000	.0000	.0000	.0000	.0000	.0000	.0001	.0009
	28	.0000	.0000	.0000	.0000	.0000	.0000	.0000	.0000	.0000	.0003
	29	.0000	.0000	.0000	.0000	.0000	.0000	.0000	.0000	.0000	.0001
	30	.0000	.0000	.0000	.0000	.0000	.0000	.0000	.0000	.0000	.0000
	31	.0000	.0000	.0000	.0000	.0000	.0000	.0000	.0000	.0000	.0000
	32	.0000	.0000	.0000	.0000	.0000	.0000	.0000	.0000	.0000	.0000
	33	.0000	.0000	.0000	.0000	.0000	.0000	.0000	.0000	.0000	.0000
	34	.0000	.0000	.0000	.0000	.0000	.0000	.0000	.0000	.0000	.0000
	35	.0000	.0000	.0000	.0000	.0000	.0000	.0000	.0000	.0000	.0000
40	1	.8715	.9852	.9985	.9999	1.0000	1.0000	1.0000	1.0000	1.0000	1.0000
	2	.6009	.9195	.9879	.9985	.9999	1.0000	1.0000	1.0000	1.0000	1.0000
	3	.3233	.7772	.9514	.9921	.9990	.9999	1.0000	1.0000	1.0000	1.0000
	4	.1381	.5769	.8698	.9715	.9953	.9994	.9999	1.0000	1.0000	1.0000
	5	.0480	3710	.7367	.9241	.9840	.9974	.9997	1.0000	1.0000	1.0000
	6	.0139	.2063	.5675	.8387	.9567	.9914	.9987	.9999	1.0000	1.0000
	7	.0034	.0995	.3933	.7141	.9038	.9762	.9956	.9994	.9999	1.0000
	8	.0007	.0419	.2441	.5629	.8180	.9447	.9876	.9979	.9998	1.0000
	9	.0001	.0155	.1354	.4069	.7002	.8890	.9697	.9939	.9991	.9999
	10	.0000	.0051	.0672	.2682	.5605	.8041	.9356	.9844	.9973	.9997
	11	.0000	.0015	.0299	.1608	.4161	.6913	.8785	.9648	.9926	.9989
	12	.0000	.0004	.0120	.0875	.2849	.5594	.7947	.9291	.9821	.9968
	13	.0000	.0001	.0043	.0432	.1791	.4228	.6857	.8715	.9614	.9917
	14	.0000	.0000	.0014	.0194	.1032	.2968	.5592	.7888	.9249	.9808
	15	.0000	.0000	.0004	.0079	.0544	.1926	.4279	.6826	.8674	.9597
	16	.0000	.0000	.0001	.0029	.0262	.1151	.3054	.5598	.7858	.9231
	17	.0000	.0000	.0000	.0010	.0116	.0633	.2022	.4319	.6815	.8659
	18	.0000	.0000	.0000	.0003	.0047	.0320	.1239	.3115	.5609	.7852
	19	.0000	.0000	.0000	.0001	.0017	.0148	.0699	.2089	.4349	.6821
	20	.0000	.0000	.0000	.0000	.0006	.0063	.0363	.1298	.3156	.5627
	21	.0000	.0000	.0000	.0000	.0002	.0024	.0173	.0744	.2130	.4373
	22	.0000	.0000	.0000	.0000	.0000	.0009	.0075	.0392	.1331	.3179
	23	.0000	.0000	.0000	.0000	.0000	.0003	.0030	.0189	.0767	.2148
	24	.0000	.0000	.0000	.0000	.0000	.0001	.0011	.0083	.0405	.1341
	25	.0000	.0000	.0000	.0000	.0000	.0000	.0004	.0034	.0196	.0769
	26	.0000	.0000	.0000	.0000	.0000	.0000	.0001	.0012	.0086	.0403
	27	.0000	.0000	.0000	.0000	.0000	.0000	.0000	.0004	.0034	.0192
	28	.0000	.0000	.0000	.0000	.0000	.0000	.0000	.0001	.0012	.0083
	29	.0000	.0000	.0000	.0000	.0000	.0000	.0000	.0000	.0004	.0032

CUMULATIVE TERMS, BINOMIAL DISTRIBUTION (Continued)

n	x'	.05	.10	.15	.20	.25 θ	.30	.35	.40	.45	.50
40	30	.0000	.0000	.0000	.0000	.0000	.0000	.0000	.0000	.0001	.0011
	31	.0000	.0000	.0000	.0000	.0000	.0000	.0000	.0000	.0000	.0003
	32	.0000	.0000	.0000	.0000	.0000	.0000	.0000	.0000	.0000	.0001
	33	.0000	.0000	.0000	.0000	.0000	.0000	.0000	.0000	.0000	.0000
	34	.0000	.0000	.0000	.0000	.0000	.0000	.0000	.0000	.0000	.0000
	35	.0000	.0000	.0000	.0000	.0000	.0000	.0000	.0000	.0000	.0000
	36	.0000	.0000	.0000	.0000	.0000	.0000	.0000	.0000	.0000	.0000
	37	.0000	.0000	.0000	.0000	.0000	.0000	.0000	.0000	.0000	.0000
	38	.0000	.0000	.0000	.0000	.0000	.0000	.0000	.0000	.0000	.0000
	39	.0000	.0000	.0000	.0000	.0000	.0000	.0000	.0000	.0000	.0000
	40	.0000	.0000	.0000	.0000	.0000	.0000	.0000	.0000	.0000	.0000
45	1	.9006	.9913	.9993	1.0000	1.0000	1.0000	1.0000	1.0000	1.0000	1.0000
	2	.6650	.9476	.9940	.9995	1.0000	1.0000	1.0000	1.0000	1.0000	1.0000
	3	.3923	.8410	.9735	.9968	.9997	1.0000	1.0000	1.0000	1.0000	1.0000
	4	.1866	.6711	.9215	.9871	.9984	.9999	1.0000	1.0000	1.0000	1.0000
	5	.0729	.4729	.8252	.9618	.9941	.9993	.9999	1.0000	1.0000	1.0000
	6	.0239	.2923	.6858	.9098	.9821	.9974	.9997	1.0000	1.0000	1.0000
	7	.0066	.1585	.5218	.8232	.9554	.9920	.9990	.9999	1.0000	1.0000
	8	.0016	.0757	.3606	.7025	.9059	.9791	.9967	.9996	1.0000	1.0000
	9	.0003	.0320	.2255	.5593	.8275	.9529	.9909	.9988	.9999	1.0000
	10	.0001	.0120	.1274	.4120	.7200	.9066	.9780	.9964	.9996	1.0000
	11	.0000	.0040	.0651	.2795	.5911	.8353	.9531	.9906	.9987	.9999
	12	.0000	.0012	.0302	.1741	.4543	.7380	.9104	.9784	.9964	.9996
	13	.0000	.0003	.0127	.0995	.3252	.6198	.8453	.9554	.9910	.9988
	14	.0000	.0001	.0048	.0521	.2159	.4912	.7563	.9164	.9799	.9967
	15	.0000	.0000	.0017	.0250	.1327	.3653	.6467	.8570	.9591	.9920
	16	.0000	.0000	.0005	.0110	.0753	.2538	.5248	.7751	.9238	.9822
	17	.0000	.0000	.0002	.0044	.0395	.1642	.4017	.6728	.8698	.9638
	18	.0000	.0000	.0000	.0017	.0191	.0986	.2887	.5564	.7944	.9324
	19	.0000	.0000	.0000	.0006	.0085	.0549	.1940	.4357	.6985	.8837
	20	.0000	.0000	.0000	.0002	.0035	.0283	.1215	.3214	.5869	.8144
	21	.0000	.0000	.0000	.0001	.0013	.0135	.0708	.2223	.4682	.7243
	22	.0000	.0000	.0000	.0000	.0005	.0060	.0382	.1436	.3526	.6170
	23	.0000	.0000	.0000	.0000	.0001	.0024	.0191	.0865	.2494	.5000
	24	.0000	.0000	.0000	.0000	.0000	.0009	.0089	.0483	.1650	.3830
	25	.0000	.0000	.0000	.0000	.0000	.0003	.0038	.0250	.1017	.2757
	26	.0000	.0000	.0000	.0000	.0000	.0001	.0015	.0120	.0582	.1856
	27	.0000	.0000	.0000	.0000	.0000	.0000	.0005	.0053	.0308	.1163
	28	.0000	.0000	.0000	.0000	.0000	.0000	.0002	.0021	.0150	.0676
	29	.0000	.0000	.0000	.0000	.0000	.0000	.0001	.0008	.0068	.0362
	30	.0000	.0000	.0000	.0000	.0000	.0000	.0000	.0003	.0028	.0178
	31	.0000	.0000	.0000	.0000	.0000	.0000	.0000	.0001	.0010	.0080
	32	.0000	.0000	.0000	.0000	.0000	.0000	.0000	.0000	.0004	.0033
	33	.0000	.0000	.0000	.0000	.0000	.0000	.0000	.0000	.0001	.0012
	34	.0000	.0000	.0000	.0000	.0000	.0000	.0000	.0000	.0000	.0004
	35	.0000	.0000	.0000	.0000	.0000	.0000	.0000	.0000	.0000	.0001
	36	.0000	.0000	.0000	.0000	.0000	.0000	.0000	.0000	.0000	.0000

CUMULATIVE TERMS, BINOMIAL DISTRIBUTION (Continued)

n	x'	.05	.10	.15	.20	θ .25	.30	.35	.40	.45	.50
45	37	.0000	.0000	.0000	.0000	.0000	.0000	.0000	.0000	.0000	.0000
	38	.0000	.0000	.0000	.0000	.0000	.0000	.0000	.0000	.0000	.0000
	39	.0000	.0000	.0000	.0000	.0000	.0000	.0000	.0000	.0000	.0000
	40	.0000	.0000	.0000	.0000	.0000	.0000	.0000	.0000	.0000	.0000
	41	.0000	.0000	.0000	.0000	.0000	.0000	.0000	.0000	.0000	.0000
	42	.0000	.0000	.0000	.0000	.0000	.0000	.0000	.0000	.0000	.0000
	43	.0000	.0000	.0000	.0000	.0000	.0000	.0000	.0000	.0000	.0000
	44	.0000	.0000	.0000	.0000	.0000	.0000	.0000	.0000	.0000	.0000
	45	.0000	.0000	.0000	.0000	.0000	.0000	.0000	.0000	.0000	.0000

INDIVIDUAL TERMS, POISSON DISTRIBUTION

The Poisson probability function is given by

$$f(x;\lambda) = \frac{\lambda^x e^{-\lambda}}{x!}, \qquad \lambda > 0, \, x = 0, 1, 2, \ldots \ .$$

This table contains the individual terms of $f(x;\lambda)$ for specified values of x and λ.

INDIVIDUAL TERMS, POISSON DISTRIBUTION (Continued)

x	0.1	0.2	0.3	0.4	0.5	0.6	0.7	0.8	0.9	1.0
0	.9048	.8187	.7408	.6703	.6065	.5488	.4966	.4493	.4066	.3679
1	.0905	.1637	.2222	.2681	.3033	.3293	.3476	.3595	.3659	.3679
2	.0045	.0164	.0333	.0536	.0758	.0988	.1217	.1438	.1647	.1839
3	.0002	.0011	.0033	.0072	.0126	.0198	.0284	.0383	.0494	.0613
4	.0000	.0001	.0003	.0007	.0016	.0030	.0050	.0077	.0111	.0153
5	.0000	.0000	.0000	.0001	.0002	.0004	.0007	.0012	.0020	.0031
6	.0000	.0000	.0000	.0000	.0000	.0000	.0001	.0002	.0003	.0005
7	.0000	.0000	.0000	.0000	.0000	.0000	.0000	.0000	.0000	.0001

x	1.1	1.2	1.3	1.4	1.5	1.6	1.7	1.8	1.9	2.0
0	.3329	.3012	.2725	.2466	.2231	.2019	.1827	.1653	.1496	.1353
1	.3662	.3614	.3543	.3452	.3347	.3230	.3106	.2975	.2842	.2707
2	.2014	.2169	.2303	.2417	.2510	.2584	.2640	.2678	2700	.2707
3	.0738	.0867	.0998	.1128	.1255	.1378	.1496	.1607	.1710	.1804
4	.0203	.0260	.0324	.0395	.0471	.0551	.0636	.0723	.0812	.0902
5	.0045	.0062	.0084	.0111	.0141	.0176	.0216	.0260	.0309	.0361
6	.0008	.0012	.0018	.0026	.0035	.0047	.0061	.0078	.0098	.0120
7	.0001	.0002	.0003	.0005	.0008	.0011	.0015	.0020	.0027	.0034
8	.0000	.0000	.0001	.0001	.0001	.0002	.0003	.0005	.0006	.0009
9	.0000	.0000	.0000	.0000	.0000	.0000	.0001	.0001	.0001	.0002

x	2.1	2.2	2.3	2.4	2.5	2.6	2.7	2.8	2.9	3.0
0	.1225	.1108	.1003	.0907	.0821	.0743	.0672	.0608	.0550	.0498
1	.2572	.2438	.2306	.2177	.2052	.1931	.1815	.1703	.1596	.1494
2	.2700	.2681	.2652	.2613	.2565	.2510	.2450	.2384	.2314	.2240
3	.1890	.1966	.2033	.2090	.2138	.2176	.2205	.2225	.2237	.2240
4	.0992	.1082	.1169	.1254	.1336	.1414	.1488	.1557	.1622	.1680
5	.0417	.0476	.0538	.0602	.0668	.0735	.0804	.0872	.0940	.1008
6	.0146	.0174	.0206	.0241	.0278	.0319	.0362	.0407	.0455	.0504
7	.0044	.0055	.0068	.0083	.0099	.0118	.0139	.0163	.0188	.0216
8	.0011	.0015	.0019	.0025	.0031	.0038	.0047	.0057	.0068	.0081
9	.0003	.0004	.0005	.0007	.0009	.0011	.0014	.0018	.0022	.0027
10	.0001	.0001	.0001	.0002	.0002	.0003	.0004	.0005	.0006	.0008
11	.0000	.0000	.0000	.0000	.0000	.0001	.0001	.0001	.0002	.0002
12	.0000	.0000	.0000	.0000	.0000	.0000	.0000	.0000	.0000	.0001

x	3.1	3.2	3.3	3.4	3.5	3.6	3.7	3.8	3.9	4.0
0	.0450	.0408	.0369	.0334	.0302	.0273	.0247	.0224	.0202	.0183
1	.1397	.1304	.1217	.1135	.1057	.0984	.0915	.0850	.0789	.0733
2	.2165	.2087	.2008	.1929	.1850	.1771	.1692	.1615	.1539	.1465
3	.2237	.2226	.2209	.2186	.2158	.2125	.2087	.2046	.2001	.1954
4	.1734	.1781	.1823	.1858	.1888	.1912	.1931	.1944	.1951	.1954
5	.1075	.1140	.1203	.1264	.1322	.1377	.1429	.1477	.1522	.1563
6	.0555	.0608	.0662	.0716	.0771	.0826	.0881	.0936	.0989	.1042
7	.0246	.0278	.0312	.0348	.0385	.0425	.0466	.0508	.0551	.0595
8	.0095	.0111	.0129	.0148	.0169	.0191	.0215	.0241	.0269	.0298
9	.0033	.0040	.0047	.0056	.0066	.0076	.0089	.0102	.0116	.0132

INDIVIDUAL TERMS, POISSON DISTRIBUTION (Continued)

λ

x	3.1	3.2	3.3	3.4	3.5	3.6	3.7	3.8	3.9	4.0
10	.0010	.0013	.0016	.0019	.0023	.0028	.0033	.0039	.0045	.0053
11	.0003	.0004	.0005	.0006	.0007	.0009	.0011	.0013	.0016	.0019
12	.0001	.0001	.0001	.0002	.0002	.0003	.0003	.0004	.0005	.0006
13	.0000	.0000	.0000	.0000	.0001	.0001	.0001	.0001	.0002	.0002
14	.0000	.0000	.0000	.0000	.0000	.0000	.0000	.0000	.0000	.0001

λ

x	4.1	4.2	4.3	4.4	4.5	4.6	4.7	4.8	4.9	5.0
0	.0166	.0150	.0136	.0123	.0111	.0101	.0091	.0082	.0074	.0067
1	.0679	.0630	.0583	.0540	.0500	.0462	.0427	.0395	.0365	.0337
2	.1393	.1323	.1254	.1188	.1125	.1063	.1005	.0948	.0894	.0842
3	.1904	.1852	.1798	.1743	.1687	.1631	.1574	.1517	.1460	.1404
4	.1951	.1944	.1933	.1917	.1898	.1875	.1849	.1820	.1789	.1755
5	.1600	.1633	.1662	.1687	.1708	.1725	.1738	.1747	.1753	.1755
6	.1093	.1143	.1191	.1237	.1281	.1323	.1362	.1398	.1432	.1462
7	.0640	.0686	.0732	.0778	.0824	.0869	.0914	.0959	.1002	.1044
8	.0328	.0360	.0393	.0428	.0463	.0500	.0537	.0575	.0614	.0653
9	.0150	.0168	.0188	.0209	.0232	.0255	.0280	.0307	.0334	.0363
10	.0061	.0071	.0081	.0092	.0104	.0118	.0132	.0147	.0164	.0181
11	.0023	.0027	.0032	.0037	.0043	.0049	.0056	.0064	.0073	.0082
12	.0008	.0009	.0011	.0014	.0016	.0019	.0022	.0026	.0030	.0034
13	.0002	.0003	.0004	.0005	.0006	.0007	.0008	.0009	.0011	.0013
14	.0001	.0001	.0001	.0001	.0002	.0002	.0003	.0003	.0004	.0005
15	.0000	.0000	.0000	.0000	.0001	.0001	.0001	.0001	.0001	.0002

λ

x	5.1	5.2	5.3	5.4	5.5	5.6	5.7	5.8	5.9	6.0
0	.0061	.0055	.0050	.0045	.0041	.0037	.0033	.0030	.0027	.0025
1	.0311	.0287	.0265	.0244	.0225	.0207	.0191	.0176	.0162	.0149
2	.0793	.0746	.0701	.0659	.0618	.0580	.0544	.0509	.0477	.0446
3	.1348	.1293	.1239	.1185	.1133	.1082	.1033	.0985	.0938	.0892
4	.1719	.1681	.1641	.1600	.1558	.1515	.1472	.1428	.1383	.1339
5	.1753	.1748	.1740	.1728	.1714	.1697	.1678	.1656	.1632	.1606
6	.1490	.1515	.1537	.1555	.1571	.1584	.1594	.1601	.1605	.1606
7	.1086	.1125	.1163	.1200	.1234	.1267	.1298	.1326	.1353	.1377
8	.0692	.0731	.0771	.0810	.0849	.0887	.0925	.0962	.0998	.1033
9	.0392	.0423	.0454	.0486	.0519	.0552	.0586	.0620	.0654	.0688
10	.0200	.0220	.0241	.0262	.0285	.0309	.0334	.0359	.0386	.0413
11	.0093	.0104	.0116	.0129	.0143	.0157	.0173	.0190	.0207	.0225
12	.0039	.0045	.0051	.0058	.0065	.0073	.0082	.0092	.0102	.0113
13	.0015	.0018	.0021	.0024	.0028	.0032	.0036	.0041	.0046	.0052
14	.0006	.0007	.0008	.0009	.0011	.0013	.0015	.0017	.0019	.0022
15	.0002	.0002	.0003	.0003	.0004	.0005	.0006	.0007	.0008	.0009
16	.0001	.0001	.0001	.0001	.0001	.0002	.0002	.0002	.0003	.0003
17	.0000	.0000	.0000	.0000	.0000	.0000	.0001	.0001	.0001	.0001

INDIVIDUAL TERMS, POISSON DISTRIBUTION (Continued)

x	6.1	6.2	6.3	6.4	6.5	6.6	6.7	6.8	6.9	7.0
0	.0022	.0020	.0018	.0017	.0015	.0014	.0012	.0011	.0010	.0009
1	.0137	.0126	.0116	.0106	.0098	.0090	.0082	.0076	.0070	.0064
2	.0417	.0390	.0364	.0340	.0318	.0296	.0276	.0258	.0240	.0223
3	.0848	.0806	.0765	.0726	.0688	.0652	.0617	.0584	.0552	.0521
4	.1294	.1249	.1205	.1162	.1118	.1076	.1034	.0992	.0952	.0912
5	.1579	.1549	.1519	.1487	.1454	.1420	.1385	.1349	.1314	.1277
6	.1605	.1601	.1595	.1586	.1575	.1562	.1546	.1529	.1511	.1490
7	.1399	.1418	.1435	.1450	.1462	.1472	.1480	.1486	.1489	.1490
8	.1066	.1099	.1130	.1160	.1188	.1215	.1240	.1263	.1284	.1304
9	.0723	.0757	.0791	.0825	.0858	.0891	.0923	.0954	.0985	.1014
10	.0441	.0469	.0498	.0528	.0558	.0588	.0618	.0649	.0679	.0710
11	.0245	.0265	.0285	.0307	.0330	.0353	.0377	.0401	.0426	.0452
12	.0124	.0137	.0150	.0164	.0179	.0194	.0210	.0227	.0245	.0264
13	.0058	.0065	.0073	.0081	.0089	.0098	.0108	.0119	.0130	.0142
14	.0025	.0029	.0033	.0037	.0041	.0046	.0052	.0058	.0064	.0071
15	.0010	.0012	.0014	.0016	.0018	.0020	.0023	.0026	.0029	.0033
16	.0004	.0005	.0005	.0006	.0007	.0008	.0010	.0011	.0013	.0014
17	.0001	.0002	.0002	.0002	.0003	.0003	.0004	.0004	.0005	.0006
18	.0000	.0001	.0001	.0001	.0001	.0001	.0001	.0002	.0002	.0002
19	.0000	.0000	.0000	.0000	.0000	.0000	.0000	.0001	.0001	.0001

x	7.1	7.2	7.3	7.4	7.5	7.6	7.7	7.8	7.9	8.0
0	.0008	.0007	.0007	.0006	.0006	.0005	.0005	.0004	.0004	.0003
1	.0059	.0054	.0049	.0045	.0041	.0038	.0035	.0032	.0029	.0027
2	.0208	.0194	.0180	.0167	.0156	.0145	.0134	.0125	.0116	.0107
3	.0492	.0464	.0438	.0413	.0389	.0366	.0345	.0324	.0305	.0286
4	.0874	.0836	.0799	.0764	.0729	.0696	.0663	.0632	.0602	.0573
5	.1241	.1204	.1167	.1130	.1094	.1057	.1021	.0986	.0951	.0916
6	.1468	.1445	.1420	.1394	.1367	.1339	.1311	.1282	.1252	.1221
7	.1489	.1486	.1481	.1474	.1465	.1454	.1442	.1428	.1413	.1396
8	.1321	.1337	.1351	.1363	.1373	.1382	.1388	.1392	.1395	.1396
9	.1042	.1070	.1096	.1121	.1144	.1167	.1187	.1207	.1224	.1241
10	.0740	.0770	.0800	.0829	.0858	.0887	.0914	.0941	.0967	.0993
11	.0478	.0504	.0531	.0558	.0585	.0613	.0640	.0667	.0695	.0722
12	.0283	.0303	.0323	.0344	.0366	.0388	.0411	.0434	.0457	.0481
13	.0154	.0168	.0181	.0196	.0211	.0227	.0243	.0260	.0278	.0296
14	.0078	.0086	.0095	.0104	.0113	.0123	.0134	.0145	.0157	.0169
15	.0037	.0041	.0046	.0051	.0057	.0062	.0069	.0075	.0083	.0090
16	.0016	.0019	.0021	.0024	.0026	.0030	.0033	.0037	.0041	.0045
17	.0007	.0008	.0009	.0010	.0012	.0013	.0015	.0017	.0019	.0021
18	.0003	.0003	.0004	.0004	.0005	.0006	.0006	.0007	.0008	.0009
19	.0001	.0001	.0001	.0002	.0002	.0002	.0003	.0003	.0003	.0004
20	.0000	.0000	.0001	.0001	.0001	.0001	.0001	.0001	.0001	.0002
21	.0000	.0000	.0000	.0000	.0000	.0000	.0000	.0000	.0001	.0001

INDIVIDUAL TERMS, POISSON DISTRIBUTION (Continued)

λ

x	8.1	8.2	8.3	8.4	8.5	8.6	8.7	8.8	8.9	9.0
0	.0003	.0003	.0002	.0002	.0002	.0002	.0002	.0002	.0001	.0001
1	.0025	.0023	.0021	.0019	.0017	.0016	.0014	.0013	.0012	.0011
2	.0100	.0092	.0086	.0079	.0074	.0068	.0063	.0058	.0054	.0050
3	.0269	.0252	.0237	.0222	.0208	.0195	.0183	.0171	.0160	.0150
4	.0544	.0517	.0491	.0466	.0443	.0420	.0398	.0377	.0357	.0337
5	.0882	.0849	.0816	.0784	.0752	.0722	.0692	.0663	.0635	.0607
6	.1191	.1160	.1128	.1097	.1066	.1034	.1003	.0972	.0941	.0911
7	.1378	.1358	.1338	.1317	.1294	.1271	.1247	.1222	.1197	.1171
8	.1395	.1392	.1388	.1382	.1375	.1366	.1356	.1344	.1332	.1318
9	.1256	.1269	.1280	.1290	.1299	.1306	.1311	.1315	.1317	.1318
10	.1017	.1040	.1063	.1084	.1104	.1123	.1140	.1157	.1172	.1186
11	.0749	.0776	.0802	.0828	.0853	.0878	.0902	.0925	.0948	.0970
12	.0505	.0530	.0555	.0579	.0604	.0629	.0654	.0679	.0703	.0728
13	.0315	.0334	.0354	.0374	.0395	.0416	.0438	.0459	.0481	.0504
14	.0182	.0196	.0210	.0225	.0240	.0256	.0272	.0289	.0306	.0324
15	.0098	.0107	.0116	.0126	.0136	.0147	.0158	.0169	.0182	.0194
16	.0050	.0055	.0060	.0066	.0072	.0079	.0086	.0093	.0101	.0109
17	.0024	.0026	.0029	.0033	.0036	.0040	.0044	.0048	.0053	.0058
18	.0011	.0012	.0014	.0015	.0017	.0019	.0021	.0024	.0026	.0029
19	.0005	.0005	.0006	.0007	.0008	.0009	.0010	.0011	.0012	.0014
20	.0002	.0002	.0002	.0003	.0003	.0004	.0004	.0005	.0005	.0006
21	.0001	.0001	.0001	.0001	.0001	.0002	.0002	.0002	.0002	.0003
22	.0000	.0000	.0000	.0000	.0001	.0001	.0001	.0001	.0001	.0001

λ

x	9.1	9.2	9.3	9.4	9.5	9.6	9.7	9.8	9.9	10
0	.0001	.0001	.0001	.0001	.0001	.0001	.0001	.0001	.0001	.0000
1	.0010	.0009	.0009	.0008	.0007	.0007	.0006	.0005	.0005	.0005
2	.0046	.0043	.0040	.0037	.0034	.0031	.0029	.0027	.0025	.0023
3	.0140	.0131	.0123	.0115	.0107	.0100	.0093	.0087	.0081	.0076
4	.0319	.0302	.0285	.0269	.0254	.0240	.0226	.0213	.0201	.0189
5	.0581	.0555	.0530	.0506	.0483	.0460	.0439	.0418	.0398	.0378
6	.0881	.0851	.0822	.0793	.0764	.0736	.0709	.0682	.0656	.0631
7	.1145	.1118	.1091	.1064	.1037	.1010	.0982	.0955	.0928	.0901
8	.1302	.1286	.1269	.1251	.1232	.1212	.1191	.1170	.1148	.1126
9	.1317	.1315	.1311	.1306	.1300	.1293	.1284	.1274	.1263	.1251
10	.1198	.1210	.1219	.1228	.1235	.1241	.1245	.1249	.1250	.1251
11	.0991	.1012	.1031	.1049	.1067	.1083	.1098	.1112	.1125	.1137
12	.0752	.0776	.0799	.0822	.0844	.0866	.0888	.0908	.0928	.0948
13	.0526	.0549	.0572	.0594	.0617	.0640	.0662	.0685	.0707	.0729
14	.0342	.0361	.0380	.0399	.0419	.0439	.0459	.0479	.0500	.0521
15	.0208	.0221	.0235	.0250	.0265	.0281	.0297	.0313	.0330	.0347
16	.0118	.0127	.0137	.0147	.0157	.0168	.0180	.0192	.0204	.0217
17	.0063	.0069	.0075	.0081	.0088	.0095	.0103	.0111	.0119	.0128
18	.0032	.0035	.0039	.0042	.0046	.0051	.0055	.0060	.0065	.0071
19	.0015	.0017	.0019	.0021	.0023	.0026	.0028	.0031	.0034	.0037

INDIVIDUAL TERMS, POISSON DISTRIBUTION (Continued)

x	9.1	9.2	9.3	9.4	9.5	9.6	9.7	9.8	9.9	10
20	.0007	.0008	.0009	.0010	.0011	.0012	.0014	.0015	.0017	.0019
21	.0003	.0003	.0004	.0004	.0005	.0006	.0006	.0007	.0008	.0009
22	.0001	.0001	.0002	.0002	.0002	.0002	.0003	.0003	.0004	.0004
23	.0000	.0001	.0001	.0001	.0001	.0001	.0001	.0001	.0002	.0002
24	.0000	.0000	.0000	.0000	.0000	.0000	.0000	.0001	.0001	.0001

λ

x	11	12	13	14	15	16	17	18	19	20
0	.0000	.0000	.0000	.0000	.0000	.0000	.0000	.0000	.0000	.0000
1	.0002	.0001	.0000	.0000	.0000	.0000	.0000	.0000	.0000	.0000
2	.0010	.0004	.0002	.0001	.0000	.0000	.0000	.0000	.0000	.0000
3	.0037	.0018	.0008	.0004	.0002	.0001	.0000	.0000	.0000	.0000
4	.0102	.0053	.0027	.0013	.0006	.0003	.0001	.0001	.0000	.0000
5	.0224	.0127	.0070	.0037	.0019	.0010	.0005	.0002	.0001	.0001
6	.0411	.0255	.0152	.0087	.0048	.0026	.0014	.0007	.0004	.0002
7	.0646	.0437	.0281	.0174	.0104	.0060	.0031	.0018	.0010	.0005
8	.0888	.0655	.0457	.0304	.0194	.0120	.0072	.0042	.0024	.0013
9	.1085	.0874	.0661	.0473	.0324	.0213	.0135	.0083	.0050	.0029
10	.1194	.1048	.0859	.0663	.0486	.0341	.0230	.0150	.0095	.0058
11	.1194	.1144	.1015	.0844	.0663	.0496	.0355	.0245	.0164	.0106
12	.1094	.1144	.1099	.0984	.0829	.0661	.0504	.0368	.0259	.0176
13	.0926	.1056	.1099	.1060	.0956	.0814	.0658	.0509	.0378	.0271
14	.0728	.0905	.1021	.1060	.1024	.0930	.0800	.0655	.0514	.0387
15	.0534	.0724	.0885	.0989	.1024	.0992	.0906	.0786	.0650	.0516
16	.0367	.0543	.0719	.0866	.0960	.0992	.0963	.0884	.0772	.0646
17	.0237	.0383	.0550	.0713	.0847	.0934	.0963	.0936	.0863	.0760
18	.0145	.0256	.0397	.0554	.0706	.0830	.0909	.0936	.0911	.0844
19	.0084	.0161	.0272	.0409	.0557	.0699	.0814	.0887	.0911	.0888
20	.0046	.0097	.0177	.0286	.0418	.0559	.0692	.0798	.0866	.0888
21	.0024	.0055	.0109	.0191	.0299	.0426	.0560	.0684	.0783	.0846
22	.0012	.0030	.0065	.0121	.0204	.0310	.0433	.0560	.0676	.0769
23	.0006	.0016	.0037	.0074	.0133	.0216	.0320	.0438	.0559	.0669
24	.0003	.0008	.0020	.0043	.0083	.0144	.0226	.0328	.0442	.0557
25	.0001	.0004	.0010	.0024	.0050	.0092	.0154	.0237	.0336	.0446
26	.0000	.0002	.0005	.0013	.0029	.0057	.0101	.0164	.0246	.0343
27	.0000	.0001	.0002	.0007	.0016	.0034	.0063	.0109	.0173	.0254
28	.0000	.0000	.0001	.0003	.0009	.0019	.0038	.0070	.0117	.0181
29	.0000	.0000	.0001	.0002	.0004	.0011	.0023	.0044	.0077	.0125
30	.0000	.0000	.0000	.0001	.0002	.0006	.0013	.0026	.0049	.0083
31	.0000	.0000	.0000	.0000	.0001	.0003	.0007	.0015	.0030	.0054
32	.0000	.0000	.0000	.0000	.0001	.0001	.0004	.0009	.0018	.0034
33	.0000	.0000	.0000	.0000	.0000	.0001	.0002	.0005	.0010	.0020
34	.0000	.0000	.0000	.0000	.0000	.0000	.0001	.0002	.0006	.0012
35	.0000	.0000	.0000	.0000	.0000	.0000	.0000	.0001	.0003	.0007
36	.0000	.0000	.0000	.0000	.0000	.0000	.0000	.0001	.0002	.0004
37	.0000	.0000	.0000	.0000	.0000	.0000	.0000	.0000	.0001	.0002
38	.0000	.0000	.0000	.0000	.0000	.0000	.0000	.0000	.0000	.0001
39	.0000	.0000	.0000	.0000	.0000	.0000	.0000	.0000	.0000	.0001

CUMULATIVE TERMS, POISSON DISTRIBUTION

This table contains the values of

$$\sum_{x=x'}^{\infty} \frac{e^{-\lambda}\lambda^x}{x!}$$

for specified values of x' and λ. The cumulative Poisson distribution and the cumulative chi-square (χ^2) distribution are related as follows:

$$\sum_{x=0}^{x'-1} \frac{e^{-\lambda}\lambda^x}{x!} = 1 - F(\chi^2)$$

$$= \frac{1}{2^{\frac{n}{2}}\Gamma\left(\frac{n}{2}\right)} \int_{\chi^2}^{\infty} x^{\frac{n}{2}-1} e^{-\frac{x}{2}} dx$$

where $\lambda = \frac{1}{2}\chi^2$ and $x' = \frac{1}{2}n$.

CUMULATIVE TERMS, POISSON DISTRIBUTION (Continued)

x'	0.1	0.2	0.3	0.4	0.5 λ	0.6	0.7	0.8	0.9	1.0
0	1.0000	1.0000	1.0000	1.0000	1.0000	1.0000	1.0000	1.0000	1.0000	1.0000
1	.0952	.1813	.2592	.3297	.3935	.4512	.5034	.5507	.5934	.6321
2	.0047	.0175	.0369	.0616	.0902	.1219	.1558	.1912	.2275	.2642
3	.0002	.0011	.0036	.0079	.0144	.0231	.0341	.0474	.0629	.0803
4	.0000	.0001	.0003	.0008	.0018	.0034	.0058	.0091	.0135	.0190
5	.0000	.0000	.0000	.0001	.0002	.0004	.0008	.0014	.0023	.0037
6	.0000	.0000	.0000	.0000	.0000	.0000	.0001	.0002	.0003	.0006
7	.0000	.0000	.0000	.0000	.0000	.0000	.0000	.0000	.0000	.0001

x'	1.1	1.2	1.3	1.4	1.5 λ	1.6	1.7	1.8	1.9	2.0
0	1.0000	1.0000	1.0000	1.0000	1.0000	1.0000	1.0000	1.0000	1.0000	1.0000
1	.6671	.6988	.7275	.7534	.7769	.7981	.8173	.8347	.8504	.8647
2	.3010	.3374	.3732	.4082	.4422	.4751	.5068	.5372	.5663	.5940
3	.0996	.1205	.1429	.1665	.1912	.2166	.2428	.2694	.2963	.3233
4	.0257	.0338	.0431	.0537	.0656	.0788	.0932	.1087	.1253	.1429
5	.0054	.0077	.0107	.0143	.0186	.0237	.0296	.0364	.0441	.0527
6	.0010	.0015	.0022	.0032	.0045	.0060	.0080	.0104	.0132	.0166
7	.0001	.0003	.0004	.0006	.0009	.0013	.0019	.0026	.0034	.0045
8	.0000	.0000	.0001	.0001	.0002	.0003	.0004	.0006	.0008	.0011
9	.0000	.0000	.0000	.0000	.0000	.0000	.0001	.0001	.0002	.0002

x'	2.1	2.2	2.3	2.4	2.5 λ	2.6	2.7	2.8	2.9	3.0
0	1.0000	1.0000	1.0000	1.0000	1.0000	1.0000	1.0000	1.0000	1.0000	1.0000
1	.8775	.8892	.8997	.9093	.9179	.9257	.9328	.9392	.9450	.9502
2	.6204	.6454	.6691	.6916	.7127	.7326	.7513	.7689	.7854	.8009
3	.3504	.3773	.4040	.4303	.4562	.4816	.5064	.5305	.5540	.5768
4	.1614	.1806	.2007	.2213	.2424	.2640	.2859	.3081	.3304	.3528
5	.0621	.0725	.0838	.0959	.1088	.1226	.1371	.1523	.1682	.1847
6	.0204	.0249	.0300	.0357	.0420	.0490	.0567	.0651	.0742	.0839
7	.0059	.0075	.0094	.0116	.0142	.0172	.0206	.0244	.0287	.0335
8	.0015	.0020	.0026	.0033	.0042	.0053	.0066	.0081	.0099	.0119
9	.0003	.0005	.0006	.0009	.0011	.0015	.0019	.0024	.0031	.0038
10	.0001	.0001	.0001	.0002	.0003	.0004	.0005	.0007	.0009	.0011
11	.0000	.0000	.0000	.0000	.0001	.0001	.0001	.0002	.0002	.0003
12	.0000	.0000	.0000	.0000	.0000	.0000	.0000	.0000	.0001	.0001

x'	3.1	3.2	3.3	3.4	3.5 λ	3.6	3.7	3.8	3.9	4.0
0	1.0000	1.0000	1.0000	1.0000	1.0000	1.0000	1.0000	1.0000	1.0000	1.0000
1	.9550	.9592	.9631	.9666	.9698	.9727	.9753	.9776	.9798	.9817
2	.8153	.8288	.8414	.8532	.8641	.8743	.8838	.8926	.9008	.9084
3	.5988	.6201	.6406	.6603	.6792	.6973	.7146	.7311	.7469	.7619
4	.3752	.3975	.4197	.4416	.4634	.4848	.5058	.5265	.5468	.5665

CUMULATIVE TERMS, POISSON DISTRIBUTION (Continued)

x'	3.1	3.2	3.3	3.4	λ 3.5	3.6	3.7	3.8	3.9	5.0
5	.2018	.2194	.2374	.2558	.2746	.2936	.3128	.3322	.3516	.3712
6	.0943	.1054	.1171	.1295	.1424	.1559	.1699	.1844	.1994	.2149
7	.0388	.0446	.0510	.0579	.0653	.0733	.0818	.0909	.1005	.1107
8	.0142	.0168	.0198	.0231	.0267	.0308	.0352	.0401	.0454	.0511
9	.0047	.0057	.0069	.0083	.0099	.0117	.0137	.0160	.0185	.0214
10	.0014	.0018	.0022	.0027	.0033	.0040	.0048	.0058	.0069	.0081
11	.0004	.0005	.0006	.0008	.0010	.0013	.0016	.0019	.0023	.0028
12	.0001	.0001	.0002	.0002	.0003	.0004	.0005	.0006	.0007	.0009
13	.0000	.0000	.0000	.0001	.0001	.0001	.0001	.0002	.0002	.0003
14	.0000	.0000	.0000	.0000	.0000	.0000	.0000	.0000	.0001	.0001

x'	4.1	4.2	4.3	4.4	λ 4.5	4.6	4.7	4.8	4.9	5.0
0	1.0000	1.0000	1.0000	1.0000	1.0000	1.0000	1.0000	1.0000	1.0000	1.0000
1	.9834	.9850	.9864	.9877	.9889	.9899	.9909	.9918	.9926	.9933
2	.9155	.9220	.9281	.9337	.9389	.9437	.9482	.9523	.9561	.9596
3	.7762	.7898	.8026	.8149	.8264	.8374	.8477	.8575	.8667	.8753
4	.5858	.6046	.6228	.6406	.6577	.6743	.6903	.7058	.7207	.7350
5	.3907	.4102	.4296	.4488	.4679	.4868	.5054	.5237	.5418	.5595
6	.2307	.2469	.2633	.2801	.2971	.3142	.3316	.3490	.3665	.3840
7	.1214	.1325	.1442	.1564	.1689	.1820	.1954	.2092	.2233	.2378
8	.0573	.0639	.0710	.0786	.0866	.0951	.1040	.1133	.1231	.1334
9	.0245	.0279	.0317	.0358	.0403	.0451	.0503	.0558	.0618	.0681
10	.0095	.0111	.0129	.0149	.0171	.0195	.0222	.0251	.0283	.0318
11	.0034	.0041	.0048	.0057	.0067	.0078	.0090	.0104	.0120	.0137
12	.0011	.0014	.0017	.0020	.0024	.0029	.0034	.0040	.0047	.0055
13	.0003	.0004	.0005	.0007	.0008	.0010	.0012	.0014	.0017	.0020
14	.0001	.0001	.0002	.0002	.0003	.0003	.0004	.0005	.0006	.0007
15	.0000	.0000	.0000	.0001	.0001	.0001	.0001	.0001	.0002	.0002
16	.0000	.0000	.0000	.0000	.0000	.0000	.0000	.0000	.0001	.0001

x'	5.1	5.2	5.3	5.4	λ 5.5	5.6	5.7	5.8	5.9	6.0
0	1.0000	1.0000	1.0000	1.0000	1.0000	1.0000	1.0000	1.0000	1.0000	1.0000
1	.9939	.9945	.9950	.9955	.9959	.9963	.9967	.9970	.9973	.9975
2	.9628	.9658	.9686	.9711	.9734	.9756	.9776	.9794	.9811	.9826
3	.8835	.8912	.8984	.9052	.9116	.9176	.9232	.9285	.9334	.9380
4	.7487	.7619	.7746	.7867	.7983	.8094	.8200	.8300	.8396	.8488
5	.5769	.5939	.6105	.6267	.6425	.6579	.6728	.6873	.7013	.7149
6	.4016	.4191	.4365	.4539	.4711	.4881	.5050	.5217	.5381	.5543
7	.2526	.2676	.2829	.2983	.3140	.3297	.3456	.3616	.3776	.3937
8	.1440	.1551	.1665	.1783	.1905	.2030	.2159	.2290	.2424	.2560
9	.0748	.0819	.0894	.0974	.1056	.1143	.1234	.1328	.1426	.1528

CUMULATIVE TERMS, POISSON DISTRIBUTION (Continued)

x	5.1	5.2	5.3	5.4	5.5	5.6	5.7	5.8	5.9	6.0
10	.0356	.0397	.0441	.0488	.0538	.0591	.0648	.0708	.0772	.0839
11	.0156	.0177	.0200	.0225	.0253	.0282	.0314	.0349	.0386	.0426
12	.0063	.0073	.0084	.0096	.0110	.0125	.0141	.0160	.0179	.0201
13	.0024	.0028	.0033	.0038	.0045	.0051	.0059	.0068	.0078	.0088
14	.0008	.0010	.0012	.0014	.0017	.0020	.0023	.0027	.0031	.0036
15	.0003	.0003	.0004	.0005	.0006	.0007	.0009	.0010	.0012	.0014
16	.0001	.0001	.0001	.0002	.0002	.0002	.0003	.0004	.0004	.0005
17	.0000	.0000	.0000	.0001	.0001	.0001	.0001	.0001	.0001	.0002
18	.0000	.0000	.0000	.0000	.0000	.0000	.0000	.0000	.0000	.0001

x'	6.1	6.2	6.3	6.4	6.5	6.6	6.7	6.8	6.9	7.0
0	1.0000	1.0000	1.0000	1.0000	1.0000	1.0000	1.0000	1.0000	1.0000	1.0000
1	.9978	.9980	.9982	.9983	.9985	.9986	.9988	.9989	.9990	.9991
2	.9841	.9854	.9866	.9877	.9887	.9897	.9905	.9913	.9920	.9927
3	.9423	.9464	.9502	.9537	.9570	.9600	.9629	.9656	.9680	.9704
4	.8575	.8658	.8736	.8811	.8882	.8948	.9012	.9072	.9120	.9182
5	.7281	.7408	.7531	.7649	.7763	.7873	.7978	.8080	.8177	.8270
6	.5702	.5859	.6012	.6163	.6310	.6453	.6594	.6730	.6863	.6993
7	.4098	.4258	.4418	.4577	.4735	.4892	.5047	.5201	.5353	.5503
8	.2699	.2840	.2983	.3127	.3272	.3419	.3567	.3715	.3864	.4013
9	.1633	.1741	.1852	.1967	.2084	.2204	.2327	.2452	.2580	.2709
10	.0910	.0984	.1061	.1142	.1226	.1314	.1404	.1498	.1505	.1695
11	.0439	.0514	.0563	.0614	.0668	.0726	.0786	.0849	.0916	.0985
12	.0224	.0250	.0277	.0307	.0339	.0373	.0409	.0448	.0490	.0534
13	.0100	.0113	.0127	.0143	.0160	.0179	.0199	.0221	.0245	.0270
14	.0042	.0048	.0055	.0063	.0071	.0080	.0091	.0102	.0115	.0128
15	.0016	.0019	.0022	.0026	.0030	.0034	.0039	.0044	.0050	.0057
16	.0006	.0007	.0008	.0010	.0012	.0014	.0016	.0018	.0021	.0024
17	.0002	.0003	.0003	.0004	.0004	.0005	.0006	.0007	.0008	.0010
18	.0001	.0001	.0001	.0001	.0002	.0002	.0002	.0003	.0003	.0004
19	.0000	.0000	.0000	.0000	.0001	.0001	.0001	.0001	.0001	.0001

x'	7.1	7.2	7.3	7.4	7.5	7.6	7.7	7.8	7.9	8.0
0	1.0000	1.0000	1.0000	1.0000	1.0000	1.0000	1.0000	1.0000	1.0000	1.0000
1	.9992	.9993	.9993	.9994	.9994	.9995	.9995	.9996	.9996	.9997
2	.9933	.9939	.9944	.9949	.9953	.9957	.9961	.9964	.9967	.9970
3	.9725	.9745	.9764	.0781	.9797	.9812	.9826	.9839	.9851	.9862
4	.9233	.9281	.9326	.9368	.9409	.9446	.9482	.9515	.9547	.9576
5	.8359	.8445	.8527	.8605	.8679	.8751	.8819	.8883	.8945	.9004
6	.7119	.7241	.7360	.7474	.7586	.7693	.7797	.7897	.7994	.8088
7	.5651	.5796	.5940	.6080	.6218	.6354	.6486	.6616	.6743	.6866
8	.4162	.4311	.4459	.4607	.4754	.4900	.5044	.5188	.5330	.5470
9	.2840	.2973	.3108	.3243	.3380	.3518	.3657	.3796	.3935	.4075
10	.1798	.1904	.2012	.2123	.2236	.2351	.2469	.2589	.2710	.2834
11	.1058	.1133	.1212	.1293	.1378	.1465	.1555	.1648	.1743	.1841
12	.0580	.0629	.0681	.0735	.0792	.0852	.0915	.0980	.1048	.1119
13	.0297	.0327	.0358	.0391	.0427	.0464	.0504	.0546	.0591	.0638
14	.0143	.0159	.0176	.0195	.0216	.0238	.0261	.0286	.0313	.0342

CUMULATIVE TERMS, POISSON DISTRIBUTION (Continued)

λ

x'	7.1	7.2	7.3	7.4	7.5	7.6	7.7	7.8	7.9	8.0
15	.0065	.0073	.0082	.0092	.0103	.0114	.0127	.0141	.0156	.0173
16	.0028	.0031	.0036	.0041	.0046	.0052	.0059	.0066	.0074	.0082
17	.0011	.0013	.0015	.0017	.0020	.0022	.0026	.0029	.0033	.0037
18	.0004	.0005	.0006	.0007	.0008	.0009	.0011	.0012	.0014	.0016
19	.0002	.0002	.0002	.0003	.0003	.0004	.0004	.0005	.0006	.0006
20	.0001	.0001	.0001	.0001	.0001	.0001	.0002	.0002	.0002	.0003
21	.0000	.0000	.0000	.0000	.0000	.0000	.0001	.0001	.0001	.0001

λ

x'	8.1	8.2	8.3	8.4	8.5	8.6	8.7	8.8	8.9	9.0
0	1.0000	1.0000	1.0000	1.0000	1.0000	1.0000	1.0000	1.0000	1.0000	1.0000
1	.9997	.9997	.9998	.9998	.9998	.9998	.9998	.9998	.9999	.9999
2	.9972	.9975	.9977	.9979	.9981	.9982	.9984	.9985	.9987	.9988
3	.9873	.9882	.9891	.9900	.9907	.9914	.9921	.9927	.9932	.9938
4	.9604	.9630	.9654	.9677	.9699	.9719	.9738	.9756	.9772	.9788
5	.9060	.9113	.9163	.9211	.9256	.9299	.9340	.9379	.9416	.9450
6	.8178	.8264	.8347	.8427	.8504	.8578	.8648	.8716	.8781	.8843
7	.6987	.7104	.7219	.7330	.7438	.7543	.7645	.7744	.7840	.7932
8	.5609	.5746	.5881	.6013	.6144	.6272	.6398	.6522	.6643	.6761
9	.4214	.4353	.4493	.4631	.4769	.4906	.5042	.5177	.5311	.5443
10	.2959	.3085	.3212	.3341	.3470	.3600	.3731	.3863	.3994	.4126
11	.1942	.2045	.2150	.2257	.2366	.2478	.2591	.2706	.2822	.2940
12	.1193	.1269	.1348	.1429	.1513	.1600	.1689	.1780	.1874	.1970
13	.0687	.0739	.0793	.0850	.0909	.0971	.1035	.1102	.1171	.1242
14	.0372	.0405	.0439	.0476	.0514	.0555	.0597	.0642	.0689	.0739
15	.0190	.0209	.0229	.0251	.0274	.0299	.0325	.0353	.0383	.0415
16	.0092	.0102	.0113	.0125	.0138	.0152	.0168	.0184	.0202	.0220
17	.0042	.0047	.0053	.0059	.0066	.0074	.0082	.0091	.0101	.0111
18	.0018	.0021	.0023	.0027	.0030	.0034	.0038	.0043	.0048	.0053
19	.0008	.0009	.0010	.0011	.0013	.0015	.0017	.0019	.0022	.0024
20	.0003	.0003	.0004	.0005	.0005	.0006	.0007	.0008	.0009	.0011
21	.0001	.0001	.0002	.0002	.0002	.0002	.0003	.0003	.0004	.0004
22	.0000	.0000	.0001	.0001	.0001	.0001	.0001	.0001	.0002	.0002
23	.0000	.0000	.0000	.0000	.0000	.0000	0000	.0000	.0001	.0001

λ

x'	9.1	9.2	9.3	9.4	9.5	9.6	9.7	9.8	9.9	10
0	1.0000	1.0000	1.0000	1.0000	1.0000	1.0000	1.0000	1.0000	1.0000	1.0000
1	.9999	.9999	.9999	.9999	.9999	.9999	.9999	.9999	1.0000	1.0000
2	.9989	.9990	.9991	.9991	.9992	.9993	.9993	.9994	.9995	.9995
3	.9942	.9947	.9951	.9955	.9958	.9962	.9965	.9967	.9970	.9972
4	.9802	.9816	.9828	.9840	.9851	.9862	.9871	.9880	.9889	.9897
5	.9483	.9514	.9544	.9571	.9597	.9622	.9645	.9667	.9688	.9707
6	.8902	.8959	.9014	.9065	.9115	.9162	.9207	.9250	.9290	.9329
7	.8022	.8108	.8192	.8273	.8351	.8426	.8498	.8567	.8634	.8699
8	.6877	.6990	.7101	.7208	.7313	.7416	.7515	.7612	.7706	.7798
9	.5574	.5704	.5832	.5958	.6082	.6204	.6324	.6442	.6558	.6672

CUMULATIVE TERMS, POISSON DISTRIBUTION (Continued)

x'	9.1	9.2	9.3	9.4	9.5	9.6	9.7	9.8	9.9	10
10	.4258	.4389	.4521	.4651	.4782	.4911	.5040	.5168	.5295	.5421
11	.3059	.3180	.3301	.3424	.3547	.3671	.3795	.3920	.4045	.4170
12	.2068	.2168	.2270	.2374	.2480	.2588	.2697	.2807	.2919	.3032
13	.1316	.1393	.1471	.1552	.1636	.1721	.1809	.1899	.1991	.2084
14	.0790	.0844	.0900	.0958	.1019	.1081	.1147	.1214	.1284	.1355
15	.0448	.0483	.0520	.0559	.0600	.0643	.0688	.0735	.0784	.0835
16	.0240	.0262	.0285	.0309	.0335	.0362	.0391	.0421	.0454	.0487
17	.0122	.0135	.0148	.0162	.0177	.0194	.0211	.0230	.0249	.0270
18	.0059	.0066	.0073	.0081	.0089	.0098	.0108	.0119	.0130	.0143
19	.0027	.0031	.0034	.0038	.0043	.0048	.0053	.0059	.0065	.0072
20	.0012	.0014	.0015	.0017	.0020	.0022	.0025	.0028	.0031	.0035
21	.0005	.0006	.0007	.0008	.0009	.0010	.0011	.0013	.0014	.0016
22	.0002	.0002	.0003	.0003	.0004	.0004	.0005	.0005	.0006	.0007
23	.0001	.0001	.0001	.0001	.0001	.0002	.0002	.0002	.0003	.0003
24	.0000	.0000	.0000	.0000	.0001	.0001	.0001	.0001	.0001	.0001

λ

x'	11	12	13	14	15	16	17	18	19	20
0	1.0000	1.0000	1.0000	1.0000	1.0000	1.0000	1.0000	1.0000	1.0000	1.0000
1	1.0000	1.0000	1.0000	1.0000	1.0000	1.0000	1.0000	1.0000	1.0000	1.0000
2	.9998	.9999	1.0000	1.0000	1.0000	1.0000	1.0000	1.0000	1.0000	1.0000
3	.9988	.9995	.9998	.9999	1.0000	1.0000	1.0000	1.0000	1.0000	1.0000
4	.9951	.9977	.9990	.9995	.9998	.9999	1.0000	1.0000	1.0000	1.0000
5	.9849	.9924	.9963	.9982	.9991	.9996	.9998	.9999	1.0000	1.0000
6	.9625	.9797	.9893	.9945	.9972	.9986	.9993	.9997	.9998	.9999
7	.9214	.9542	.9741	.9858	.9924	.9960	.9979	.9990	.9995	.9997
8	.8568	.9105	.9460	.9684	.9820	.9900	.9946	.9971	.9985	.9992
9	.7680	.8450	.9002	.9379	.9626	.9780	.9874	.9929	.9961	.9979
10	.6595	.7576	.8342	.8906	.9301	.9567	.9739	.9846	.9911	.9950
11	.5401	.6528	.7483	.8243	.8815	.9226	.9509	.9696	.9817	.9892
12	.4207	.5384	.6468	.7400	.8152	.8730	.9153	.9451	.9653	.9786
13	.3113	.4240	.5369	.6415	.7324	.8069	.8650	.9083	.9394	.9610
14	.2187	.3185	.4270	.5356	.6368	.7255	.7991	.8574	.9016	.9339
15	.1460	.2280	.3249	.4296	.5343	.6325	.7192	.7919	.8503	.8951
16	.0926	.1556	.2364	.3306	.4319	.5333	.6285	.7133	.7852	.8435
17	.0559	.1013	.1645	.2441	.3359	.4340	.5323	.6250	.7080	.7789
18	.0322	.0630	.1095	.1728	.2511	.3407	.4360	.5314	.6216	.7030
19	.0177	.0374	.0698	.1174	.1805	.2577	.3450	.4378	.5305	.6186
20	.0093	.0213	.0427	.0765	.1248	.1878	.2637	.3491	.4394	.5297
21	.0047	.0116	.0250	.0479	.0830	.1318	.1945	.2693	.3528	.4409
22	.0023	.0061	.0141	.0288	.0531	.0892	.1385	.2009	.2745	.3563
23	.0010	.0030	.0076	.0167	.0327	.0582	.0953	.1449	.2069	.2794
24	.0005	.0015	.0040	.0093	.0195	.0367	.0633	.1011	.1510	.2125
25	.0002	.0007	.0020	.0050	.0112	.0223	.0406	.0683	.1067	.1568
26	.0001	.0003	.0010	.0026	.0062	.0131	.0252	.0446	.0731	.1122
27	.0000	.0001	.0005	.0013	.0033	.0075	.0152	.0282	.0486	.0779
28	.0000	.0001	.0002	.0006	.0017	.0041	.0088	.0173	.0313	.0525
29	.0000	.0000	.0001	.0003	.0009	.0022	.0050	.0103	.0195	.0343

CUMULATIVE TERMS, POISSON DISTRIBUTION (Continued)

x'	λ 11	12	13	14	15	16	17	18	19	20
30	.0000	.0000	.0000	.0001	.0004	.0011	.0027	.0059	.0118	.0218
31	.0000	.0000	.0000	.0001	.0002	.0006	.0014	.0033	.0070	.0135
32	.0000	.0000	.0000	.0000	.0001	.0003	.0007	.0018	.0040	.0081
33	.0000	.0000	.0000	.0000	.0000	.0001	.0004	.0010	.0022	.0047
34	.0000	.0000	.0000	.0000	.0000	.0001	.0002	.0005	.0012	.0027
35	.0000	.0000	.0000	.0000	.0000	.0000	.0001	.0002	.0006	.0015
36	.0000	.0000	.0000	.0000	.0000	.0000	.0000	.0001	.0003	.0008
37	.0000	.0000	.0000	.0000	.0000	.0000	.0000	.0001	.0002	.0004
38	.0000	.0000	.0000	.0000	.0000	.0000	.0000	.0000	.0001	.0002
39	.0000	.0000	.0000	.0000	.0000	.0000	.0000	.0000	.0000	.0001
40	.0000	.0000	.0000	.0000	.0000	.0000	.0000	.0000	.0000	.0001

PERCENTAGE POINTS, STUDENT'S *t*-DISTRIBUTION

This table gives values of *t* such that

$$F(t) = \int_{-\infty}^{t} \frac{\Gamma\left(\dfrac{n+1}{2}\right)}{\sqrt{n\pi}\ \Gamma\left(\dfrac{n}{2}\right)} \left(1 + \frac{x^2}{n}\right)^{-\frac{n+1}{2}} dx$$

for *n*, the number of degrees of freedom, equal to 1, 2, . . . , 30, 40, 60, 120, ∞ ; and for $F(t) = 0.60, 0.75, 0.90, 0.95, 0.975, 0.99, 0.995,$ and 0.9995. The *t*-distribution is symmetrical, so that $F(-t) = 1 - F(t)$

n \\ F	.60	.75	.90	.95	.975	.99	.995	.9995
1	.325	1.000	3.078	6.314	12.706	31.821	63.657	636.619
2	.289	.816	1.886	2.920	4.303	6.965	9.925	31.598
3	.277	.765	1.638	2.353	3.182	4.541	5.841	12.924
4	.271	.741	1.533	2.132	2.776	3.747	4.604	8.610
5	.267	.727	1.476	2.015	2.571	3.365	4.032	6.869
6	.265	.718	1.440	1.943	2.447	3.143	3.707	5.959
7	.263	.711	1.415	1.895	2.365	2.998	3.499	5.408
8	.262	.706	1.397	1.860	2.306	2.896	3.355	5.041
9	.261	.703	1.383	1.833	2.262	2.821	3.250	4.781
10	.260	.700	1.372	1.812	2.228	2.764	3.169	4.587
11	.260	.697	1.363	1.796	2.201	2.718	3.106	4.437
12	.259	.695	1.356	1.782	2.179	2.681	3.055	4.318
13	.259	.694	1.350	1.771	2.160	2.650	3.012	4.221
14	.258	.692	1.345	1.761	2.145	2.624	2.977	4.140
15	.258	.691	1.341	1.753	2.131	2.602	2.947	4.073
16	.258	.690	1.337	1.746	2.120	2.583	2.921	4.015
17	.257	.689	1.333	1.740	2.110	2.567	2.898	3.965
18	.257	.688	1.330	1.734	2.101	2.552	2.878	3.922
19	.257	.688	1.328	1.729	2.093	2.539	2.861	3.883
20	.257	.687	1.325	1.725	2.086	2.528	2.845	3.850
21	.257	.686	1.323	1.721	2.080	2.518	2.831	3.819
22	.256	.686	1.321	1.717	2.074	2.508	2.819	3.792
23	.256	.685	1.319	1.714	2.069	2.500	2.807	3.767
24	.256	.685	1.318	1.711	2.064	2.492	2.797	3.745
25	.256	.684	1.316	1.708	2.060	2.485	2.787	3.725
26	.256	.684	1.315	1.706	2.056	2.479	2.779	3.707
27	.256	.684	1.314	1.703	2.052	2.473	2.771	3.690
28	.256	.683	1.313	1.701	2.048	2.467	2.763	3.674
29	.256	.683	1.311	1.699	2.045	2.462	2.756	3.659
30	.256	.683	1.310	1.697	2.042	2.457	2.750	3.646
40	.255	.681	1.303	1.684	2.021	2.423	2.704	3.551
60	.254	.679	1.296	1.671	2.000	2.390	2.660	3.460
120	.254	.677	1.289	1.658	1.980	2.358	2.617	3.373
∞	.253	.674	1.282	1.645	1.960	2.326	2.576	3.291

* This table is abridged from the "Statistical Tables" of R. A. Fisher and Frank Yates published by Oliver & Boyd, Ltd., Edinburgh and London, 1938. It is here published with the kind permission of the authors and their publishers.

PERCENTAGE POINTS, CHI-SQUARE DISTRIBUTION

This table gives values of χ^2 such that

$$F(\chi^2) = \int_0^{\chi^2} \frac{1}{2^{\frac{n}{2}} \, \Gamma\left(\frac{n}{2}\right)} \, x^{\frac{n-2}{2}} \, e^{-\frac{x}{2}} \, dx$$

for n, the number of degrees of freedom, equal to 1, 2, . . . , 30. For $n > 30$, a normal approximation is quite accurate. The expression $\sqrt{2\chi^2} - \sqrt{2n-1}$ is approximately normally distributed as the standard normal distribution. Thus χ_α^2, the α-point of the distribution, may be computed by the formula

$$\chi_\alpha^2 = \tfrac{1}{2}[x_\alpha + \sqrt{2n-1}]^2,$$

where x_α is the α-point of the cumulative normal distribution. For even values of n, $F(\chi^2)$ can be written as

$$1 - F(\chi^2) = \sum_{x=0}^{x'-1} \frac{e^{-\lambda}\lambda^x}{x!}$$

with $\lambda = \tfrac{1}{2}\chi^2$ and $x' = \tfrac{1}{2}n$. Thus the cumulative Chi-Square distribution is related to the cumulative Poisson distribution.

Another approximate formula for large n

$$\chi_\alpha^2 = n\left(1 - \frac{2}{9n} + z_\alpha \sqrt{\frac{2}{9n}}\right)^3$$

n = degrees of freedom
z_α = the normal deviate, (the value of x for which $F(x)$ = the desired percentile).

x	1.282	1.645	1.960	2.326	2.576	3.090
$F(x)$	.90	.95	.975	.99	.995	.999

$\chi_{.99}^2 = 60[1 - 0.00370 + 2.326(0.06086)]^3 = 88.4$ is the 99th percentile for 60 degrees of freedom.

PERCENTAGE POINTS, CHI-SQUARE DISTRIBUTION (Continued)

$$F(\chi^2) = \int_0^{x^2} \frac{1}{2^{\frac{n}{2}} \Gamma\left(\frac{n}{2}\right)} x^{\frac{n-2}{2}} e^{-\frac{x}{2}} \, dx$$

F / n	.995	.990	.975	.950	.900	.750	.500	.250	.100	.050	.025	.010	.005
1	7.88	6.63	5.02	3.84	2.71	1.32	.455	.102	.0158	.00393	.000982	.000157	.0000393
2	10.6	9.21	7.38	5.99	4.61	2.77	1.39	.575	.211	.103	.0506	.0201	.0100
3	12.8	11.3	9.35	7.81	6.25	4.11	2.37	1.21	.584	.352	.216	.115	.0717
4	14.9	13.3	11.1	9.49	7.78	5.39	3.36	1.92	1.06	.711	.484	.297	.207
5	16.7	15.1	12.8	11.1	9.24	6.63	4.35	2.67	1.61	1.15	.831	.554	.412
6	18.5	16.8	14.4	12.6	10.6	7.84	5.35	3.45	2.20	1.64	1.24	.872	.676
7	20.3	18.5	16.0	14.1	12.0	9.04	6.35	4.25	2.83	2.17	1.69	1.24	.989
8	22.0	20.1	17.5	15.5	13.4	10.2	7.34	5.07	3.49	2.73	2.18	1.65	1.34
9	23.6	21.7	19.0	16.9	14.7	11.4	8.34	5.90	4.17	3.33	2.70	2.09	1.73
10	25.2	23.2	20.5	18.3	16.0	12.5	9.34	6.74	4.87	3.94	3.25	2.56	2.16
11	26.8	24.7	21.9	19.7	17.3	13.7	10.3	7.58	5.58	4.57	3.82	3.05	2.60
12	28.3	26.2	23.3	21.0	18.5	14.8	11.3	8.44	6.30	5.23	4.40	3.57	3.07
13	29.8	27.7	24.7	22.4	19.8	16.0	12.3	9.30	7.04	5.89	5.01	4.11	3.57
14	31.3	29.1	26.1	23.7	21.1	17.1	13.3	10.2	7.79	6.57	5.63	4.66	4.07
15	32.8	30.6	27.5	25.0	22.3	18.2	14.3	11.0	8.55	7.26	6.26	5.23	4.60
16	34.3	32.0	28.8	26.3	23.5	19.4	15.3	11.9	9.31	7.96	6.91	5.81	5.14
17	35.7	33.4	30.2	27.6	24.8	20.5	16.3	12.8	10.1	8.67	7.56	6.41	5.70
18	37.2	34.8	31.5	28.9	26.0	21.6	17.3	13.7	10.9	9.39	8.23	7.01	6.26
19	38.6	36.2	32.9	30.1	27.2	22.7	18.3	14.6	11.7	10.1	8.91	7.63	6.84
20	40.0	37.6	34.2	31.4	28.4	23.8	19.3	15.5	12.4	10.9	9.59	8.26	7.43
21	41.4	38.9	35.5	32.7	29.6	24.9	20.3	16.3	13.2	11.6	10.3	8.90	8.03
22	42.8	40.3	36.8	33.9	30.8	26.0	21.3	17.2	14.0	12.3	11.0	9.54	8.64
23	44.2	41.6	38.1	35.2	32.0	27.1	22.3	18.1	14.8	13.1	11.7	10.2	9.26
24	45.6	43.0	39.4	36.4	33.2	28.2	23.3	19.0	15.7	13.8	12.4	10.9	9.89
25	46.9	44.3	40.6	37.7	34.4	29.3	24.3	19.9	16.5	14.6	13.1	11.5	10.5
26	48.3	45.6	41.9	38.9	35.6	30.4	25.3	20.8	17.3	15.4	13.8	12.2	11.2
27	49.6	47.0	43.2	40.1	36.7	31.5	26.3	21.7	18.1	16.2	14.6	12.9	11.8
28	51.0	48.3	44.5	41.3	37.9	32.6	27.3	22.7	18.9	16.9	15.3	13.6	12.5
29	52.3	49.6	45.7	42.6	39.1	33.7	28.3	23.6	19.8	17.7	16.0	14.3	13.1
30	53.7	50.9	47.0	43.8	40.3	34.8	29.3	24.5	20.6	18.5	16.8	15.0	13.8

PERCENTAGE POINTS, *F*-DISTRIBUTION

This table gives values of F such that

$$F(F) = \int_0^F \frac{\Gamma\left(\dfrac{m+n}{2}\right)}{\Gamma\left(\dfrac{m}{2}\right)\Gamma\left(\dfrac{n}{2}\right)}\, m^{\frac{m}{2}} n^{\frac{n}{2}} x^{\frac{m-2}{2}}\,(n+mx)^{-\frac{m+n}{2}}\,dx$$

for selected values of m, the number of degrees of freedom of the numerator of F; and for selected values of n, the number of degrees of freedom of the denominator of F. The table also provides values corresponding to $F(F) = .10, .05, .025, .01, .005, .001$ since $F_{1-\alpha}$ for m and n degrees of freedom is the reciprocal of F_α for n and m degrees of freedom. Thus

$$F_{.05}(4, 7) = \frac{1}{F_{.95}(7, 4)} = \frac{1}{6.09} = .164 \ .$$

PERCENTAGE POINTS, F-DISTRIBUTION (Continued)

$$F(F) = \int_0^F \frac{\Gamma\left(\frac{m+n}{2}\right)}{\Gamma\left(\frac{m}{2}\right)\Gamma\left(\frac{n}{2}\right)}\, m^{\frac{m}{2}} n^{\frac{n}{2}}\, x^{\frac{m}{2}-1}\,(n+mx)^{-\frac{m+n}{2}}\, dx = .90$$

m \ n	1	2	3	4	5	6	7	8	9	10	12	15	20	24	30	40	60	120	∞
1	39.86	49.50	53.59	55.83	57.24	58.20	58.91	59.44	59.86	60.19	60.71	61.22	61.74	62.00	62.26	62.53	62.79	63.06	63.33
2	8.53	9.00	9.16	9.24	9.29	9.33	9.35	9.37	9.38	9.39	9.41	9.42	9.44	9.45	9.46	9.47	9.47	9.48	9.49
3	5.54	5.46	5.39	5.34	5.31	5.28	5.27	5.25	5.24	5.23	5.22	5.20	5.18	5.18	5.17	5.16	5.15	5.14	5.13
4	4.54	4.32	4.19	4.11	4.05	4.01	3.98	3.95	3.94	3.92	3.90	3.87	3.84	3.83	3.82	3.80	3.79	3.78	3.76
5	4.06	3.78	3.62	3.52	3.45	3.40	3.37	3.34	3.32	3.30	3.27	3.24	3.21	3.19	3.17	3.16	3.14	3.12	3.10
6	3.78	3.46	3.29	3.18	3.11	3.05	3.01	2.98	2.96	2.94	2.90	2.87	2.84	2.82	2.80	2.78	2.76	2.74	2.72
7	3.59	3.26	3.07	2.96	2.88	2.83	2.78	2.75	2.72	2.70	2.67	2.63	2.59	2.58	2.56	2.54	2.51	2.49	2.47
8	3.46	3.11	2.92	2.81	2.73	2.67	2.62	2.59	2.56	2.54	2.50	2.46	2.42	2.40	2.38	2.36	2.34	2.32	2.29
9	3.36	3.01	2.81	2.69	2.61	2.55	2.51	2.47	2.44	2.42	2.38	2.34	2.30	2.28	2.25	2.23	2.21	2.18	2.16
10	3.29	2.92	2.73	2.61	2.52	2.46	2.41	2.38	2.35	2.32	2.28	2.24	2.20	2.18	2.16	2.13	2.11	2.08	2.06
11	3.23	2.86	2.66	2.54	2.45	2.39	2.34	2.30	2.27	2.25	2.21	2.17	2.12	2.10	2.08	2.05	2.03	2.00	1.97
12	3.18	2.81	2.61	2.48	2.39	2.33	2.28	2.24	2.21	2.19	2.15	2.10	2.06	2.04	2.01	1.99	1.96	1.93	1.90
13	3.14	2.76	2.56	2.43	2.35	2.28	2.23	2.20	2.16	2.14	2.10	2.05	2.01	1.98	1.96	1.93	1.90	1.88	1.85
14	3.10	2.73	2.52	2.39	2.31	2.24	2.19	2.15	2.12	2.10	2.05	2.01	1.96	1.94	1.91	1.89	1.86	1.83	1.80
15	3.07	2.70	2.49	2.36	2.27	2.21	2.16	2.12	2.09	2.06	2.02	1.97	1.92	1.90	1.87	1.85	1.82	1.79	1.76
16	3.05	2.67	2.46	2.33	2.24	2.18	2.13	2.09	2.06	2.03	1.99	1.94	1.89	1.87	1.84	1.81	1.78	1.75	1.72
17	3.03	2.64	2.44	2.31	2.22	2.15	2.10	2.06	2.03	2.00	1.96	1.91	1.86	1.84	1.81	1.78	1.75	1.72	1.69
18	3.01	2.62	2.42	2.29	2.20	2.13	2.08	2.04	2.00	1.98	1.93	1.89	1.84	1.81	1.78	1.75	1.72	1.69	1.66
19	2.99	2.61	2.40	2.27	2.18	2.11	2.06	2.02	1.98	1.96	1.91	1.86	1.81	1.79	1.76	1.73	1.70	1.67	1.63
20	2.97	2.59	2.38	2.25	2.16	2.09	2.04	2.00	1.96	1.94	1.89	1.84	1.79	1.77	1.74	1.71	1.68	1.64	1.61
21	2.96	2.57	2.36	2.23	2.14	2.08	2.02	1.98	1.95	1.92	1.87	1.83	1.78	1.75	1.72	1.69	1.66	1.62	1.59
22	2.95	2.56	2.35	2.22	2.13	2.06	2.01	1.97	1.93	1.90	1.86	1.81	1.76	1.73	1.70	1.67	1.64	1.60	1.57
23	2.94	2.55	2.34	2.21	2.11	2.05	1.99	1.95	1.92	1.89	1.84	1.80	1.74	1.72	1.69	1.66	1.62	1.59	1.55
24	2.93	2.54	2.33	2.19	2.10	2.04	1.98	1.94	1.91	1.88	1.83	1.78	1.73	1.70	1.67	1.64	1.61	1.57	1.53
25	2.92	2.53	2.32	2.18	2.09	2.02	1.97	1.93	1.89	1.87	1.82	1.77	1.72	1.69	1.66	1.63	1.59	1.56	1.52
26	2.91	2.52	2.31	2.17	2.08	2.01	1.96	1.92	1.88	1.86	1.81	1.76	1.71	1.68	1.65	1.61	1.58	1.54	1.50
27	2.90	2.51	2.30	2.17	2.07	2.00	1.95	1.91	1.87	1.85	1.80	1.75	1.70	1.67	1.64	1.60	1.57	1.53	1.49
28	2.89	2.50	2.29	2.16	2.06	2.00	1.94	1.90	1.87	1.84	1.79	1.74	1.69	1.66	1.63	1.59	1.56	1.52	1.48
29	2.89	2.50	2.28	2.15	2.06	1.99	1.93	1.89	1.86	1.83	1.78	1.73	1.68	1.65	1.62	1.58	1.55	1.51	1.47
30	2.88	2.49	2.28	2.14	2.05	1.98	1.93	1.88	1.85	1.82	1.77	1.72	1.67	1.64	1.61	1.57	1.54	1.50	1.46
40	2.84	2.44	2.23	2.09	2.00	1.93	1.87	1.83	1.79	1.76	1.71	1.66	1.61	1.57	1.54	1.51	1.47	1.42	1.38
60	2.79	2.39	2.18	2.04	1.95	1.87	1.82	1.77	1.74	1.71	1.66	1.60	1.54	1.51	1.48	1.44	1.40	1.35	1.29
120	2.75	2.35	2.13	1.99	1.90	1.82	1.77	1.72	1.68	1.65	1.60	1.55	1.48	1.45	1.41	1.37	1.32	1.26	1.19
∞	2.71	2.30	2.08	1.94	1.85	1.77	1.72	1.67	1.63	1.60	1.55	1.49	1.42	1.38	1.34	1.30	1.24	1.17	1.00

$F = \dfrac{s_1^2}{s_2^2} = \dfrac{S_1}{m}\Big/\dfrac{S_2}{n}$, where $s_1^2 = S_1/m$ and $s_2^2 = S_2/n$ are independent mean squares estimating a common variance σ^2 and based on m and n degrees of freedom, respectively.

PERCENTAGE POINTS, F-DISTRIBUTION (Continued)

$$F(F) = \int_0^F \frac{\Gamma\left(\frac{m+n}{2}\right)}{\Gamma\left(\frac{m}{2}\right)\Gamma\left(\frac{n}{2}\right)} m^{\frac{m}{2}} n^{\frac{n}{2}} x^{\frac{m}{2}-1} (n+mx)^{-\frac{m+n}{2}}\, dx = .95$$

m \ n	1	2	3	4	5	6	7	8	9	10	12	15	20	24	30	40	60	120	∞
1	161.4	199.5	215.7	224.6	230.2	234.0	236.8	238.9	240.5	241.9	243.9	245.9	248.0	249.1	250.1	251.1	252.2	253.3	254.3
2	18.51	19.00	19.16	19.25	19.30	19.33	19.35	19.37	19.38	19.40	19.41	19.43	19.45	19.45	19.46	19.47	19.48	19.49	19.50
3	10.13	9.55	9.28	9.12	9.01	8.94	8.89	8.85	8.81	8.79	8.74	8.70	8.66	8.64	8.62	8.59	8.57	8.55	8.53
4	7.71	6.94	6.59	6.39	6.26	6.16	6.09	6.04	6.00	5.96	5.91	5.86	5.80	5.77	5.75	5.72	5.69	5.66	5.63
5	6.61	5.79	5.41	5.19	5.05	4.95	4.88	4.82	4.77	4.74	4.68	4.62	4.56	4.53	4.50	4.46	4.43	4.40	4.36
6	5.99	5.14	4.76	4.53	4.39	4.28	4.21	4.15	4.10	4.06	4.00	3.94	3.87	3.84	3.81	3.77	3.74	3.70	3.67
7	5.59	4.74	4.35	4.12	3.97	3.87	3.79	3.73	3.68	3.64	3.57	3.51	3.44	3.41	3.38	3.34	3.30	3.27	3.23
8	5.32	4.46	4.07	3.84	3.69	3.58	3.50	3.44	3.39	3.35	3.28	3.22	3.15	3.12	3.08	3.04	3.01	2.97	2.93
9	5.12	4.26	3.86	3.63	3.48	3.37	3.29	3.23	3.18	3.14	3.07	3.01	2.94	2.90	2.86	2.83	2.79	2.75	2.71
10	4.96	4.10	3.71	3.48	3.33	3.22	3.14	3.07	3.02	2.98	2.91	2.85	2.77	2.74	2.70	2.66	2.62	2.58	2.54
11	4.84	3.98	3.59	3.36	3.20	3.09	3.01	2.95	2.90	2.85	2.79	2.72	2.65	2.61	2.57	2.53	2.49	2.45	2.40
12	4.75	3.89	3.49	3.26	3.11	3.00	2.91	2.85	2.80	2.75	2.69	2.62	2.54	2.51	2.47	2.43	2.38	2.34	2.30
13	4.67	3.81	3.41	3.18	3.03	2.92	2.83	2.77	2.71	2.67	2.60	2.53	2.46	2.42	2.38	2.34	2.30	2.25	2.21
14	4.60	3.74	3.34	3.11	2.96	2.85	2.76	2.70	2.65	2.60	2.53	2.46	2.39	2.35	2.31	2.27	2.22	2.18	2.13
15	4.54	3.68	3.29	3.06	2.90	2.79	2.71	2.64	2.59	2.54	2.48	2.40	2.33	2.29	2.25	2.20	2.16	2.11	2.07
16	4.49	3.63	3.24	3.01	2.85	2.74	2.66	2.59	2.54	2.49	2.42	2.35	2.28	2.24	2.19	2.15	2.11	2.06	2.01
17	4.45	3.59	3.20	2.96	2.81	2.70	2.61	2.55	2.49	2.45	2.38	2.31	2.23	2.19	2.15	2.10	2.06	2.01	1.96
18	4.41	3.55	3.16	2.93	2.77	2.66	2.58	2.51	2.46	2.41	2.34	2.27	2.19	2.15	2.11	2.06	2.02	1.97	1.92
19	4.38	3.52	3.13	2.90	2.74	2.63	2.54	2.48	2.42	2.38	2.31	2.23	2.16	2.11	2.07	2.03	1.98	1.93	1.88
20	4.35	3.49	3.10	2.87	2.71	2.60	2.51	2.45	2.39	2.35	2.28	2.20	2.12	2.08	2.04	1.99	1.95	1.90	1.84
21	4.32	3.47	3.07	2.84	2.68	2.57	2.49	2.42	2.37	2.32	2.25	2.18	2.10	2.05	2.01	1.96	1.92	1.87	1.81
22	4.30	3.44	3.05	2.82	2.66	2.55	2.46	2.40	2.34	2.30	2.23	2.15	2.07	2.03	1.98	1.94	1.89	1.84	1.78
23	4.28	3.42	3.03	2.80	2.64	2.53	2.44	2.37	2.32	2.27	2.20	2.13	2.05	2.01	1.96	1.91	1.86	1.81	1.76
24	4.26	3.40	3.01	2.78	2.62	2.51	2.42	2.36	2.30	2.25	2.18	2.11	2.03	1.98	1.94	1.89	1.84	1.79	1.73
25	4.24	3.39	2.99	2.76	2.60	2.49	2.40	2.34	2.28	2.24	2.16	2.09	2.01	1.96	1.92	1.87	1.82	1.77	1.71
26	4.23	3.37	2.98	2.74	2.59	2.47	2.39	2.32	2.27	2.22	2.15	2.07	1.99	1.95	1.90	1.85	1.80	1.75	1.69
27	4.21	3.35	2.96	2.73	2.57	2.46	2.37	2.31	2.25	2.20	2.13	2.06	1.97	1.93	1.88	1.84	1.79	1.73	1.67
28	4.20	3.34	2.95	2.71	2.56	2.45	2.36	2.29	2.24	2.19	2.12	2.04	1.96	1.91	1.87	1.82	1.77	1.71	1.65
29	4.18	3.33	2.93	2.70	2.55	2.43	2.35	2.28	2.22	2.18	2.10	2.03	1.94	1.90	1.85	1.81	1.75	1.70	1.64
30	4.17	3.32	2.92	2.69	2.53	2.42	2.33	2.27	2.21	2.16	2.09	2.01	1.93	1.89	1.84	1.79	1.74	1.68	1.62
40	4.08	3.23	2.84	2.61	2.45	2.34	2.25	2.18	2.12	2.08	2.00	1.92	1.84	1.79	1.74	1.69	1.64	1.58	1.51
60	4.00	3.15	2.76	2.53	2.37	2.25	2.17	2.10	2.04	1.99	1.92	1.84	1.75	1.70	1.65	1.59	1.53	1.47	1.39
120	3.92	3.07	2.68	2.45	2.29	2.17	2.09	2.02	1.96	1.91	1.83	1.75	1.66	1.61	1.55	1.50	1.43	1.35	1.25
∞	3.84	3.00	2.60	2.37	2.21	2.10	2.01	1.94	1.88	1.83	1.75	1.67	1.57	1.52	1.46	1.39	1.32	1.22	1.00

$F = \frac{s_1^2}{s_2^2} = \frac{S_1}{m} / \frac{S_2}{n}$, where $s_1^2 = S_1/m$ and $s_2^2 = S_2/n$ are independent mean squares estimating a common variance σ^2 and based on m and n degrees of freedom, respectively.

PERCENTAGE POINTS, *F*-DISTRIBUTION (Continued)

$$F(F) = \int_0^F \frac{\Gamma\left(\frac{m+n}{2}\right)}{\Gamma\left(\frac{m}{2}\right)\Gamma\left(\frac{n}{2}\right)}\, m^{\frac{m}{2}} n^{\frac{n}{2}} x^{\frac{m}{2}-1} (n+mx)^{-\frac{m+n}{2}}\, dx = .975$$

m \ n	1	2	3	4	5	6	7	8	9	10	12	15	20	24	30	40	60	120	∞
1	647.8	799.5	864.2	899.6	921.8	937.1	948.2	956.7	963.3	968.6	976.7	984.9	993.1	997.2	1001	1006	1010	1014	1018
2	38.51	39.00	39.17	39.25	39.30	39.33	39.36	39.37	39.39	39.40	39.41	39.43	39.45	39.46	39.46	39.47	39.48	39.49	39.50
3	17.44	16.04	15.44	15.10	14.88	14.73	14.62	14.54	14.47	14.42	14.34	14.25	14.17	14.12	14.08	14.04	13.99	13.95	13.90
4	12.22	10.65	9.98	9.60	9.36	9.20	9.07	8.98	8.90	8.84	8.75	8.66	8.56	8.51	8.46	8.41	8.36	8.31	8.26
5	10.01	8.43	7.76	7.39	7.15	6.98	6.85	6.76	6.68	6.62	6.52	6.43	6.33	6.28	6.23	6.18	6.12	6.07	6.02
6	8.81	7.26	6.60	6.23	5.99	5.82	5.70	5.60	5.52	5.46	5.37	5.27	5.17	5.12	5.07	5.01	4.96	4.90	4.85
7	8.07	6.54	5.89	5.52	5.29	5.12	4.99	4.90	4.82	4.76	4.67	4.57	4.47	4.42	4.36	4.31	4.25	4.20	4.14
8	7.57	6.06	5.42	5.05	4.82	4.65	4.53	4.43	4.36	4.30	4.20	4.10	4.00	3.95	3.89	3.84	3.78	3.73	3.67
9	7.21	5.71	5.08	4.72	4.48	4.32	4.20	4.10	4.03	3.96	3.87	3.77	3.67	3.61	3.56	3.51	3.45	3.39	3.33
10	6.94	5.46	4.83	4.47	4.24	4.07	3.95	3.85	3.78	3.72	3.62	3.52	3.42	3.37	3.31	3.26	3.20	3.14	3.08
11	6.72	5.26	4.63	4.28	4.04	3.88	3.76	3.66	3.59	3.53	3.43	3.33	3.23	3.17	3.12	3.06	3.00	2.94	2.88
12	6.55	5.10	4.47	4.12	3.89	3.73	3.61	3.51	3.44	3.37	3.28	3.18	3.07	3.02	2.96	2.91	2.85	2.79	2.72
13	6.41	4.97	4.35	4.00	3.77	3.60	3.48	3.39	3.31	3.25	3.15	3.05	2.95	2.89	2.84	2.78	2.72	2.66	2.60
14	6.30	4.86	4.24	3.89	3.66	3.50	3.38	3.29	3.21	3.15	3.05	2.95	2.84	2.79	2.73	2.67	2.61	2.55	2.49
15	6.20	4.77	4.15	3.80	3.58	3.41	3.29	3.20	3.12	3.06	2.96	2.86	2.76	2.70	2.64	2.59	2.52	2.46	2.40
16	6.12	4.69	4.08	3.73	3.50	3.34	3.22	3.12	3.05	2.99	2.89	2.79	2.68	2.63	2.57	2.51	2.45	2.38	2.32
17	6.04	4.62	4.01	3.66	3.44	3.28	3.16	3.06	2.98	2.92	2.82	2.72	2.62	2.56	2.50	2.44	2.38	2.32	2.25
18	5.98	4.56	3.95	3.61	3.38	3.22	3.10	3.01	2.93	2.87	2.77	2.67	2.56	2.50	2.44	2.38	2.32	2.26	2.19
19	5.92	4.51	3.90	3.56	3.33	3.17	3.05	2.96	2.88	2.82	2.72	2.62	2.51	2.45	2.39	2.33	2.27	2.20	2.13
20	5.87	4.46	3.86	3.51	3.29	3.13	3.01	2.91	2.84	2.77	2.68	2.57	2.46	2.41	2.35	2.29	2.22	2.16	2.09
21	5.83	4.42	3.82	3.48	3.25	3.09	2.97	2.87	2.80	2.73	2.64	2.53	2.42	2.37	2.31	2.25	2.18	2.11	2.04
22	5.79	4.38	3.78	3.44	3.22	3.05	2.93	2.84	2.76	2.70	2.60	2.50	2.39	2.33	2.27	2.21	2.14	2.08	2.00
23	5.75	4.35	3.75	3.41	3.18	3.02	2.90	2.81	2.73	2.67	2.57	2.47	2.36	2.30	2.24	2.18	2.11	2.04	1.97
24	5.72	4.32	3.72	3.38	3.15	2.99	2.87	2.78	2.70	2.64	2.54	2.44	2.33	2.27	2.21	2.15	2.08	2.01	1.94
25	5.69	4.29	3.69	3.35	3.13	2.97	2.85	2.75	2.68	2.61	2.51	2.41	2.30	2.24	2.18	2.12	2.05	1.98	1.91
26	5.66	4.27	3.67	3.33	3.10	2.94	2.82	2.73	2.65	2.59	2.49	2.39	2.28	2.22	2.16	2.09	2.03	1.95	1.88
27	5.63	4.24	3.65	3.31	3.08	2.92	2.80	2.71	2.63	2.57	2.47	2.36	2.25	2.19	2.13	2.07	2.00	1.93	1.85
28	5.61	4.22	3.63	3.29	3.06	2.90	2.78	2.69	2.61	2.55	2.45	2.34	2.23	2.17	2.11	2.05	1.98	1.91	1.83
29	5.59	4.20	3.61	3.27	3.04	2.88	2.76	2.67	2.59	2.53	2.43	2.32	2.21	2.15	2.09	2.03	1.96	1.89	1.81
30	5.57	4.18	3.59	3.25	3.03	2.87	2.75	2.65	2.57	2.51	2.41	2.31	2.20	2.14	2.07	2.01	1.94	1.87	1.79
40	5.42	4.05	3.46	3.13	2.90	2.74	2.62	2.53	2.45	2.39	2.29	2.18	2.07	2.01	1.94	1.88	1.80	1.72	1.64
60	5.29	3.93	3.34	3.01	2.79	2.63	2.51	2.41	2.33	2.27	2.17	2.06	1.94	1.88	1.82	1.74	1.67	1.58	1.48
120	5.15	3.80	3.23	2.89	2.67	2.52	2.39	2.30	2.22	2.16	2.05	1.94	1.82	1.76	1.69	1.61	1.53	1.43	1.31
∞	5.02	3.69	3.12	2.79	2.57	2.41	2.29	2.19	2.11	2.05	1.94	1.83	1.71	1.64	1.57	1.48	1.39	1.27	1.00

$F = \dfrac{s_1^2}{s_2^2} = \dfrac{S_1}{m} \Big/ \dfrac{S_2}{n}$, where $s_1^2 = S_1/m$ and $s_2^2 = S_2/n$ are independent mean squares estimating a common variance σ^2 and based on m and n degrees of freedom, respectively.

PERCENTAGE POINTS, *F*-DISTRIBUTION (Continued)

$$F(F) = \int_0^F \frac{\Gamma\left(\frac{m+n}{2}\right)}{\Gamma\left(\frac{m}{2}\right)\Gamma\left(\frac{n}{2}\right)} m^{\frac{m}{2}} n^{\frac{n}{2}} x^{\frac{m}{2}-1} (n+mx)^{-\frac{m+n}{2}} dx = .99$$

n \ m	1	2	3	4	5	6	7	8	9	10	12	15	20	24	30	40	60	120	∞
1	4052	4999.5	5403	5625	5764	5859	5928	5982	6022	6056	6106	6157	6209	6235	6261	6287	6313	6339	6366
2	98.50	99.00	99.17	99.25	99.30	99.33	99.36	99.37	99.39	99.40	99.42	99.43	99.45	99.46	99.47	99.47	99.48	99.49	99.50
3	34.12	30.82	29.46	28.71	28.24	27.91	27.67	27.49	27.35	27.23	27.05	26.87	26.69	26.60	26.50	26.41	26.32	26.22	26.13
4	21.20	18.00	16.69	15.98	15.52	15.21	14.98	14.80	14.66	14.55	14.37	14.20	14.02	13.93	13.84	13.75	13.65	13.56	13.46
5	16.26	13.27	12.06	11.39	10.97	10.67	10.46	10.29	10.16	10.05	9.89	9.72	9.55	9.47	9.38	9.29	9.20	9.11	9.02
6	13.75	10.92	9.78	9.15	8.75	8.47	8.26	8.10	7.98	7.87	7.72	7.56	7.40	7.31	7.23	7.14	7.06	6.97	6.88
7	12.25	9.55	8.45	7.85	7.46	7.19	6.99	6.84	6.72	6.62	6.47	6.31	6.16	6.07	5.99	5.91	5.82	5.74	5.65
8	11.26	8.65	7.59	7.01	6.63	6.37	6.18	6.03	5.91	5.81	5.67	5.52	5.36	5.28	5.20	5.12	5.03	4.95	4.86
9	10.56	8.02	6.99	6.42	6.06	5.80	5.61	5.47	5.35	5.26	5.11	4.96	4.81	4.73	4.65	4.57	4.48	4.40	4.31
10	10.04	7.56	6.55	5.99	5.64	5.39	5.20	5.06	4.94	4.85	4.71	4.56	4.41	4.33	4.25	4.17	4.08	4.00	3.91
11	9.65	7.21	6.22	5.67	5.32	5.07	4.89	4.74	4.63	4.54	4.40	4.25	4.10	4.02	3.94	3.86	3.78	3.69	3.60
12	9.33	6.93	5.95	5.41	5.06	4.82	4.64	4.50	4.39	4.30	4.16	4.01	3.86	3.78	3.70	3.62	3.54	3.45	3.36
13	9.07	6.70	5.74	5.21	4.86	4.62	4.44	4.30	4.19	4.10	3.96	3.82	3.66	3.59	3.51	3.43	3.34	3.25	3.17
14	8.86	6.51	5.56	5.04	4.69	4.46	4.28	4.14	4.03	3.94	3.80	3.66	3.51	3.43	3.35	3.27	3.18	3.09	3.00
15	8.68	6.36	5.42	4.89	4.56	4.32	4.14	4.00	3.89	3.80	3.67	3.52	3.37	3.29	3.21	3.13	3.05	2.96	2.87
16	8.53	6.23	5.29	4.77	4.44	4.20	4.03	3.89	3.78	3.69	3.55	3.41	3.26	3.18	3.10	3.02	2.93	2.84	2.75
17	8.40	6.11	5.18	4.67	4.34	4.10	3.93	3.79	3.68	3.59	3.46	3.31	3.16	3.08	3.00	2.92	2.83	2.75	2.65
18	8.29	6.01	5.09	4.58	4.25	4.01	3.84	3.71	3.60	3.51	3.37	3.23	3.08	3.00	2.92	2.84	2.75	2.66	2.57
19	8.18	5.93	5.01	4.50	4.17	3.94	3.77	3.63	3.52	3.43	3.30	3.15	3.00	2.92	2.84	2.76	2.67	2.58	2.49
20	8.10	5.85	4.94	4.43	4.10	3.87	3.70	3.56	3.46	3.37	3.23	3.09	2.94	2.86	2.78	2.69	2.61	2.52	2.42
21	8.02	5.78	4.87	4.37	4.04	3.81	3.64	3.51	3.40	3.31	3.17	3.03	2.88	2.80	2.72	2.64	2.55	2.46	2.36
22	7.95	5.72	4.82	4.31	3.99	3.76	3.59	3.45	3.35	3.26	3.12	2.98	2.83	2.75	2.67	2.58	2.50	2.40	2.31
23	7.88	5.66	4.76	4.26	3.94	3.71	3.54	3.41	3.30	3.21	3.07	2.93	2.78	2.70	2.62	2.54	2.45	2.35	2.26
24	7.82	5.61	4.72	4.22	3.90	3.67	3.50	3.36	3.26	3.17	3.03	2.89	2.74	2.66	2.58	2.49	2.40	2.31	2.21
25	7.77	5.57	4.68	4.18	3.85	3.63	3.46	3.32	3.22	3.13	2.99	2.85	2.70	2.62	2.54	2.45	2.36	2.27	2.17
26	7.72	5.53	4.64	4.14	3.82	3.59	3.42	3.29	3.18	3.09	2.96	2.81	2.66	2.58	2.50	2.42	2.33	2.23	2.13
27	7.68	5.49	4.60	4.11	3.78	3.56	3.39	3.26	3.15	3.06	2.93	2.78	2.63	2.55	2.47	2.38	2.29	2.20	2.10
28	7.64	5.45	4.57	4.07	3.75	3.53	3.36	3.23	3.12	3.03	2.90	2.75	2.60	2.52	2.44	2.35	2.26	2.17	2.06
29	7.60	5.42	4.54	4.04	3.73	3.50	3.33	3.20	3.09	3.00	2.87	2.73	2.57	2.49	2.41	2.33	2.23	2.14	2.03
30	7.56	5.39	4.51	4.02	3.70	3.47	3.30	3.17	3.07	2.98	2.84	2.70	2.55	2.47	2.39	2.30	2.21	2.11	2.01
40	7.31	5.18	4.31	3.83	3.51	3.29	3.12	2.99	2.89	2.80	2.66	2.52	2.37	2.29	2.20	2.11	2.02	1.92	1.80
60	7.08	4.98	4.13	3.65	3.34	3.12	2.95	2.82	2.72	2.63	2.50	2.35	2.20	2.12	2.03	1.94	1.84	1.73	1.60
120	6.85	4.79	3.95	3.48	3.17	2.96	2.79	2.66	2.56	2.47	2.34	2.19	2.03	1.95	1.86	1.76	1.66	1.53	1.38
∞	6.63	4.61	3.78	3.32	3.02	2.80	2.64	2.51	2.41	2.32	2.18	2.04	1.88	1.79	1.70	1.59	1.47	1.32	1.00

$F = \dfrac{s_1^2}{s_2^2} = \dfrac{S_1}{m}\Big/\dfrac{S_2}{n}$, where $s_1^2 = S_1/m$ and $s_2^2 = S_2/n$ are independent mean squares estimating a common variance σ^2 and based on m and n degrees of freedom, respectively.

PERCENTAGE POINTS, *F*-DISTRIBUTION (Continued)

$$F(F) = \int_0^F \frac{\Gamma\left(\dfrac{m+n}{2}\right)}{\Gamma\left(\dfrac{m}{2}\right)\Gamma\left(\dfrac{n}{2}\right)}\, m^{\frac{m}{2}} n^{\frac{n}{2}} x^{\frac{m}{2}-1} (n+mx)^{-\frac{m+n}{2}}\, dx = .995$$

$n \backslash m$	∞	120	60	40	30	24	20	15	12	10	9	8	7	6	5	4	3	2	1
1	25465	25359	25253	25148	25044	24940	24836	24630	24426	24224	24091	23925	23715	23437	23056	22500	21615	20000	16211
2	199.5	199.5	199.5	199.5	199.5	199.5	199.4	199.4	199.4	199.4	199.4	199.4	199.4	199.3	199.3	199.2	199.2	199.0	198.5
3	41.83	41.99	42.15	42.31	42.47	42.62	42.78	43.08	43.39	43.69	43.88	44.13	44.43	44.84	45.39	46.19	47.47	49.80	55.55
4	19.32	19.47	19.61	19.75	19.89	20.03	20.17	20.44	20.70	20.97	21.14	21.35	21.62	21.97	22.46	23.15	24.26	26.28	31.33
5	12.14	12.27	12.40	12.53	12.66	12.78	12.90	13.15	13.38	13.62	13.77	13.96	14.20	14.51	14.94	15.56	16.53	18.31	22.78
6	8.88	9.00	9.12	9.24	9.36	9.47	9.59	9.81	10.03	10.25	10.39	10.57	10.79	11.07	11.46	12.03	12.92	14.54	18.63
7	7.08	7.19	7.31	7.42	7.53	7.65	7.75	7.97	8.18	8.38	8.51	8.68	8.89	9.16	9.52	10.05	10.88	12.40	16.24
8	5.95	6.06	6.18	6.29	6.40	6.50	6.61	6.81	7.01	7.21	7.34	7.50	7.69	7.95	8.30	8.81	9.60	11.04	14.69
9	5.19	5.30	5.41	5.52	5.62	5.73	5.83	6.03	6.23	6.42	6.54	6.69	6.88	7.13	7.47	7.96	8.72	10.11	13.61
10	4.64	4.75	4.86	4.97	5.07	5.17	5.27	5.47	5.66	5.85	5.97	6.12	6.30	6.54	6.87	7.34	8.08	9.43	12.83
11	4.23	4.34	4.44	4.55	4.65	4.76	4.86	5.05	5.24	5.42	5.54	5.68	5.86	6.10	6.42	6.88	7.60	8.91	12.23
12	3.90	4.01	4.12	4.23	4.33	4.43	4.53	4.72	4.91	5.09	5.20	5.35	5.52	5.76	6.07	6.52	7.23	8.51	11.75
13	3.65	3.76	3.87	3.97	4.07	4.17	4.27	4.46	4.64	4.82	4.94	5.08	5.25	5.48	5.79	6.23	6.93	8.19	11.37
14	3.44	3.55	3.66	3.76	3.86	3.96	4.06	4.25	4.43	4.60	4.72	4.86	5.03	5.26	5.56	6.00	6.68	7.92	11.06
15	3.26	3.37	3.48	3.58	3.69	3.79	3.88	4.07	4.25	4.42	4.54	4.67	4.85	5.07	5.37	5.80	6.48	7.70	10.80
16	3.11	3.22	3.33	3.44	3.54	3.64	3.73	3.92	4.10	4.27	4.38	4.52	4.69	4.91	5.21	5.64	6.30	7.51	10.58
17	2.98	3.10	3.21	3.31	3.41	3.51	3.61	3.79	3.97	4.14	4.25	4.39	4.56	4.78	5.07	5.50	6.16	7.35	10.38
18	2.87	2.99	3.10	3.20	3.30	3.40	3.50	3.68	3.86	4.03	4.14	4.28	4.44	4.66	4.96	5.37	6.03	7.21	10.22
19	2.78	2.89	3.00	3.11	3.21	3.31	3.40	3.59	3.76	3.93	4.04	4.18	4.34	4.56	4.85	5.27	5.92	7.09	10.07
20	2.69	2.81	2.92	3.02	3.12	3.22	3.32	3.50	3.68	3.85	3.96	4.09	4.26	4.47	4.76	5.17	5.82	6.99	9.94
21	2.61	2.73	2.84	2.95	3.05	3.15	3.24	3.43	3.60	3.77	3.88	4.01	4.18	4.39	4.68	5.09	5.73	6.89	9.83
22	2.55	2.66	2.77	2.88	2.98	3.08	3.18	3.36	3.54	3.70	3.81	3.94	4.11	4.32	4.61	5.02	5.65	6.81	9.73
23	2.48	2.60	2.71	2.82	2.92	3.02	3.12	3.30	3.47	3.64	3.75	3.88	4.05	4.26	4.54	4.95	5.58	6.73	9.63
24	2.43	2.55	2.66	2.77	2.87	2.97	3.06	3.25	3.42	3.59	3.69	3.83	3.99	4.20	4.49	4.89	5.52	6.66	9.55
25	2.38	2.50	2.61	2.72	2.82	2.92	3.01	3.20	3.37	3.54	3.64	3.78	3.94	4.15	4.43	4.84	5.46	6.60	9.48
26	2.33	2.45	2.56	2.67	2.77	2.87	2.97	3.15	3.33	3.49	3.60	3.73	3.89	4.10	4.38	4.79	5.41	6.54	9.41
27	2.29	2.41	2.52	2.63	2.73	2.83	2.93	3.11	3.28	3.45	3.56	3.69	3.85	4.06	4.34	4.74	5.36	6.49	9.34
28	2.25	2.37	2.48	2.59	2.69	2.79	2.89	3.07	3.25	3.41	3.52	3.65	3.81	4.02	4.30	4.70	5.32	6.44	9.28
29	2.21	2.33	2.45	2.56	2.66	2.76	2.86	3.04	3.21	3.38	3.48	3.61	3.77	3.98	4.26	4.66	5.28	6.40	9.23
30	2.18	2.30	2.42	2.52	2.63	2.73	2.82	3.01	3.18	3.34	3.45	3.58	3.74	3.95	4.23	4.62	5.24	6.35	9.18
40	1.93	2.06	2.18	2.30	2.40	2.50	2.60	2.78	2.95	3.12	3.22	3.35	3.51	3.71	3.99	4.37	4.98	6.07	8.83
60	1.69	1.83	1.96	2.08	2.19	2.29	2.39	2.57	2.74	2.90	3.01	3.13	3.29	3.49	3.76	4.14	4.73	5.79	8.49
120	1.43	1.61	1.75	1.87	1.98	2.09	2.19	2.37	2.54	2.71	2.81	2.93	3.09	3.28	3.55	3.92	4.50	5.54	8.18
∞	1.00	1.36	1.53	1.67	1.79	1.90	2.00	2.19	2.36	2.52	2.62	2.74	2.90	3.09	3.35	3.72	4.28	5.30	7.88

$F = \dfrac{s_1^2}{s_2^2} = \dfrac{S_1/m}{S_2/n}$, where $s_1^2 = S_1/m$ and $s_2^2 = S_2/n$ are independent mean squares estimating a common variance σ^2 and based on m and n degrees of freedom, respectively.

PERCENTAGE POINTS, F-DISTRIBUTION (Continued)

$$F(F) = \int_0^F \frac{\Gamma\left(\frac{m+n}{2}\right)}{\Gamma\left(\frac{m}{2}\right)\Gamma\left(\frac{n}{2}\right)} m^{\frac{m}{2}} n^{\frac{n}{2}} x^{\frac{m}{2}-1}(n+mx)^{-\frac{m+n}{2}}\, dx = .999$$

n \ m	1	2	3	4	5	6	7	8	9	10	12	15	20	24	30	40	60	120	∞
1	4053*	5000*	5404*	5625*	5764*	5859*	5929*	5981*	6023*	6056*	6107*	6158*	6209*	6235*	6261*	6287*	6313*	6340*	6366*
2	998.5	999.0	999.2	999.2	999.3	999.3	999.4	999.4	999.4	999.4	999.4	999.4	999.4	999.5	999.5	999.5	999.5	999.5	999.5
3	167.0	148.5	141.1	137.1	134.6	132.8	131.6	130.6	129.9	129.2	128.3	127.4	126.4	125.9	125.4	125.0	124.5	124.0	123.5
4	74.14	61.25	56.18	53.44	51.71	50.53	49.66	49.00	48.47	48.05	47.41	46.76	46.10	45.77	45.43	45.09	44.75	44.40	44.05
5	47.18	37.12	33.20	31.09	29.75	28.84	28.16	27.64	27.24	26.92	26.42	25.91	25.39	25.14	24.87	24.60	24.33	24.06	23.79
6	35.51	27.00	23.70	21.92	20.81	20.03	19.46	19.03	18.69	18.41	17.99	17.56	17.12	16.89	16.67	16.44	16.21	15.99	15.75
7	29.25	21.69	18.77	17.19	16.21	15.52	15.02	14.63	14.33	14.08	13.71	13.32	12.93	12.73	12.53	12.33	12.12	11.91	11.70
8	25.42	18.49	15.83	14.39	13.49	12.86	12.40	12.04	11.77	11.54	11.19	10.84	10.48	10.30	10.11	9.92	9.73	9.53	9.33
9	22.86	16.39	13.90	12.56	11.71	11.13	10.70	10.37	10.11	9.89	9.57	9.24	8.90	8.72	8.55	8.37	8.19	8.00	7.81
10	21.04	14.91	12.55	11.28	10.48	9.92	9.52	9.20	8.96	8.75	8.45	8.13	7.80	7.64	7.47	7.30	7.12	6.94	6.76
11	19.69	13.81	11.56	10.35	9.58	9.05	8.66	8.35	8.12	7.92	7.63	7.32	7.01	6.85	6.68	6.52	6.35	6.17	6.00
12	18.64	12.97	10.80	9.63	8.89	8.38	8.00	7.71	7.48	7.29	7.00	6.71	6.40	6.25	6.09	5.93	5.76	5.59	5.42
13	17.81	12.31	10.21	9.07	8.35	7.86	7.49	7.21	6.98	6.80	6.52	6.23	5.93	5.78	5.63	5.47	5.30	5.14	4.97
14	17.14	11.78	9.73	8.62	7.92	7.43	7.08	6.80	6.58	6.40	6.13	5.85	5.56	5.41	5.25	5.10	4.94	4.77	4.60
15	16.59	11.34	9.34	8.25	7.57	7.09	6.74	6.47	6.26	6.08	5.81	5.54	5.25	5.10	4.95	4.80	4.64	4.47	4.31
16	16.12	10.97	9.00	7.94	7.27	6.81	6.46	6.19	5.98	5.81	5.55	5.27	4.99	4.85	4.70	4.54	4.39	4.23	4.06
17	15.72	10.66	8.73	7.68	7.02	6.56	6.22	5.96	5.75	5.58	5.32	5.05	4.78	4.63	4.48	4.33	4.18	4.02	3.85
18	15.38	10.39	8.49	7.46	6.81	6.35	6.02	5.76	5.56	5.39	5.13	4.87	4.59	4.45	4.30	4.15	4.00	3.84	3.67
19	15.08	10.16	8.28	7.26	6.62	6.18	5.85	5.59	5.39	5.22	4.97	4.70	4.43	4.29	4.14	3.99	3.84	3.68	3.51
20	14.82	9.95	8.10	7.10	6.46	6.02	5.69	5.44	5.24	5.08	4.82	4.56	4.29	4.15	4.00	3.86	3.70	3.54	3.38
21	14.59	9.77	7.94	6.95	6.32	5.88	5.56	5.31	5.11	4.95	4.70	4.44	4.17	4.03	3.88	3.74	3.58	3.42	3.26
22	14.38	9.61	7.80	6.81	6.19	5.76	5.44	5.19	4.99	4.83	4.58	4.33	4.06	3.92	3.78	3.63	3.48	3.32	3.15
23	14.19	9.47	7.67	6.69	6.08	5.65	5.33	5.09	4.89	4.73	4.48	4.23	3.96	3.82	3.68	3.53	3.38	3.22	3.05
24	14.03	9.34	7.55	6.59	5.98	5.55	5.23	4.99	4.80	4.64	4.39	4.14	3.87	3.74	3.59	3.45	3.29	3.14	2.97
25	13.88	9.22	7.45	6.49	5.88	5.46	5.15	4.91	4.71	4.56	4.31	4.06	3.79	3.66	3.52	3.37	3.22	3.06	2.89
26	13.74	9.12	7.36	6.41	5.80	5.38	5.07	4.83	4.64	4.48	4.24	3.99	3.72	3.59	3.44	3.30	3.15	2.99	2.82
27	13.61	9.02	7.27	6.33	5.73	5.31	5.00	4.76	4.57	4.41	4.17	3.92	3.66	3.52	3.38	3.23	3.08	2.92	2.75
28	13.50	8.93	7.19	6.25	5.66	5.24	4.93	4.69	4.50	4.35	4.11	3.86	3.60	3.46	3.32	3.18	3.02	2.86	2.69
29	13.39	8.85	7.12	6.19	5.59	5.18	4.87	4.64	4.45	4.29	4.05	3.80	3.54	3.41	3.27	3.12	2.97	2.81	2.64
30	13.29	8.77	7.05	6.12	5.53	5.12	4.82	4.58	4.39	4.24	4.00	3.75	3.49	3.36	3.22	3.07	2.92	2.76	2.59
40	12.61	8.25	6.60	5.70	5.13	4.73	4.44	4.21	4.02	3.87	3.64	3.40	3.15	3.01	2.87	2.73	2.57	2.41	2.23
60	11.97	7.76	6.17	5.31	4.76	4.37	4.09	3.87	3.69	3.54	3.31	3.08	2.83	2.69	2.55	2.41	2.25	2.08	1.89
120	11.38	7.32	5.79	4.95	4.42	4.04	3.77	3.55	3.38	3.24	3.02	2.78	2.53	2.40	2.26	2.11	1.95	1.76	1.54
∞	10.83	6.91	5.42	4.62	4.10	3.74	3.47	3.27	3.10	2.96	2.74	2.51	2.27	2.13	1.99	1.84	1.66	1.45	1.00

* Multiply these entries by 100.

RANDOM UNITS

Use of Table. If one wishes to select a random sample of N items from a universe of M items, the following procedure may be applied. ($M > N$.)

1. Decide upon some arbitrary scheme of selecting entries from the table. For example, one may decide to use the entries in the first line, second column; second line, third column; third line, fourth column; etc.

2. Assign numbers to each of the items in the universe from 1 to M. Thus, if $M = 500$, the items would be numbered from 001 to 500, and therefore, each designated item is associated with a three digit number.

3. Decide upon some arbitrary scheme of selecting positional digits from each entry chosen according to Step 1. Thus, if $M = 500$, one may decide to use the first, third, and fourth digit of each entry selected, and as a consequence a three digit number is created for each entry choice.

4. If the number formed is $\leq M$, the correspondingly designated item in the universe is chosen for the random sample of N items. If a number formed is $> M$ or is a repeated number of one already chosen, it is passed over and the next desirable number is taken. This process is continued until the random sample of N items is selected.

A TABLE OF 14,000 RANDOM UNITS

Line/Col.	(1)	(2)	(3)	(4)	(5)	(6)	(7)	(8)	(9)	(10)	(11)	(12)	(13)	(14)
1	10480	15011	01536	02011	81647	91646	69179	14194	62590	36207	20969	99570	91291	90700
2	22368	46573	25595	85393	30995	89198	27982	53402	93965	34095	52666	19174	39615	99505
3	24130	48360	22527	97265	76393	64809	15179	24830	49340	32081	30680	19655	63348	58629
4	42167	93093	06243	61680	07856	16376	39440	53537	71341	57004	00849	74917	97758	16379
5	37570	39975	81837	16656	06121	91782	60468	81305	49684	60672	14110	06927	01263	54613
6	77921	06907	11008	42751	27756	53498	18602	70659	90655	15053	21916	81825	44394	42880
7	99562	72905	56420	69994	98872	31016	71194	18738	44013	48840	63213	21069	10634	12952
8	96301	91977	05463	07972	18876	20922	94595	56869	69014	60045	18425	84903	42508	32307
9	89579	14342	63661	10281	17453	18103	57740	84378	25331	12566	58678	44947	05585	56941
10	85475	36857	43342	53988	53060	59533	38867	62300	08158	17983	16439	11458	18593	64952
11	28918	69578	88231	33276	70997	79936	56865	05859	90106	31595	01547	85590	91610	78188
12	63553	40961	48235	03427	49626	69445	18663	72695	52180	20847	12234	90511	33703	90322
13	09429	93969	52636	92737	88974	33488	36320	17617	30015	08272	84115	27156	30613	74952
14	10365	61129	87529	85689	48237	52267	67689	93394	01511	26358	85104	20285	29975	89868
15	07119	97336	71048	08178	77233	13916	47564	81056	97735	85977	29372	74461	28551	90707
16	51085	12765	51821	51259	77452	16308	60756	92144	49442	53900	70960	63990	75601	40719
17	02368	21382	52404	60268	89368	19885	55322	44819	01188	65255	64835	44919	05944	55157
18	01011	54092	33362	94904	31273	04146	18594	29852	71585	85030	51132	01915	92747	64951
19	52162	53916	46369	58586	23216	14513	83149	98736	23495	64350	94738	17752	35156	35749
20	07056	97628	33787	09998	42698	06691	76988	13602	51851	46104	88916	19509	25625	58104
21	48663	91245	85828	14346	09172	30168	90229	04734	59193	22178	30421	61666	99904	32812
22	54164	58492	22421	74103	47070	25306	76468	26384	58151	06646	21524	15227	96909	44592
23	32639	32363	05597	24200	13363	38005	94342	28728	35806	06912	17012	64161	18296	22851
24	29334	27001	87637	87308	58731	00256	45834	15398	46557	41135	10367	07684	36188	18510
25	02488	33062	28834	07351	19731	92420	60952	61280	50001	67658	32586	86679	50720	94953
26	81525	72295	04839	96423	24878	82651	66566	14778	76797	14780	13300	87074	79666	95725
27	29676	20591	68086	26432	46901	20849	89768	81536	86645	12659	92259	57102	80428	25280
28	00742	57392	39064	66432	84673	40027	32832	61362	98947	96067	64760	64584	96096	98253
29	05366	04213	25669	26422	44407	44048	37937	63904	45766	66134	75470	66520	34693	90449
30	91921	26418	64117	94305	26766	25940	39972	22209	71500	64568	91402	42416	07844	69618
31	00582	04711	87917	77341	42206	35126	74087	99547	81817	42607	43808	76655	62028	76630
32	00725	69884	62797	56170	86324	88072	76222	36086	84637	93161	76038	65855	77919	88006
33	69011	65797	95876	55293	18988	27354	26575	08625	40801	59920	29841	80150	12777	48501
34	25976	57948	29888	88604	67917	48708	18912	82271	65424	69774	33611	54262	85963	03547
35	09763	83473	73577	12908	30883	18317	28290	35797	05998	41688	34952	37888	38917	88050
36	91567	42595	27958	30134	04024	86385	29880	99730	55536	84855	29080	09250	79656	73211
37	17955	56349	90999	49127	20044	59931	06115	20542	18059	02008	73708	83517	36103	42791
38	46503	18584	18845	49618	02304	51038	20655	58727	28168	15475	56942	53389	20562	87338
39	92157	89634	94824	78171	84610	82834	09922	25417	44137	48413	25555	21246	35509	20468
40	14577	62765	35605	81263	39667	47358	56873	56307	61607	49518	89656	20103	77490	18062
41	98427	07523	33362	64270	01638	92477	66969	98420	04880	45585	46565	04102	46880	45709
42	34914	63976	88720	82765	34476	17032	87589	40836	32427	70002	70663	88863	77775	69348
43	70060	28277	39475	46473	23219	53416	94970	25832	69975	94884	19661	72828	00102	66794
44	53976	54914	06990	67245	68350	82948	11398	42878	80287	88267	47363	46634	06541	97809
45	76072	29515	40980	07391	58745	25774	22987	80059	39911	96189	41151	14222	60697	59583
46	90725	52210	83974	29992	65831	38857	50490	83765	55657	14361	31720	57375	56228	41546
47	64364	67412	33339	31926	14883	24413	59744	92351	97473	89286	35931	04110	23726	51900
48	08962	00358	31662	25388	61642	34072	81249	35648	56891	69352	48373	45578	78547	81788
49	95012	68379	93526	70765	10593	04542	76463	54328	02349	17247	28865	14777	62730	92277
50	15664	10493	20492	38391	91132	21999	59516	81652	27195	48223	46751	22923	32261	85653

A TABLE OF 14,000 RANDOM UNITS (Continued)

Line/Col.	(1)	(2)	(3)	(4)	(5)	(6)	(7)	(8)	(9)	(10)	(11)	(12)	(13)	(14)
51	16408	81899	04153	53381	79401	21438	83035	92350	36693	31238	59649	91754	72772	02338
52	18629	81953	05520	91962	04739	13092	97662	24822	94730	06496	35090	04822	86772	98289
53	73115	35101	47498	87637	99016	71060	88824	71013	18735	20286	23153	72924	35165	43040
54	57491	16703	23167	49323	45021	33132	12544	41035	80780	45393	44812	12515	98931	91202
55	30405	83946	23792	14422	15059	45799	22716	19792	09983	74353	68668	30429	70735	25499
56	16631	35006	85900	98275	32388	52390	16815	69298	82732	38480	73817	32523	41961	44437
57	96773	20206	42559	78985	05300	22164	24369	54224	35083	19687	11052	91491	60383	19746
58	38935	64202	14349	82674	66523	44133	00697	35552	35970	19124	63318	29686	03387	59846
59	31624	76384	17403	53363	44167	64486	64758	75366	76554	31601	12614	33072	60332	92325
60	78919	19474	23632	27889	47914	02584	37680	20801	72152	39339	34806	08930	85001	87820
61	03931	33309	57047	74211	63445	17361	62825	39908	05607	91284	68833	25570	38818	46920
62	74426	33278	43972	10119	89917	15665	52872	73823	73144	88662	88970	74492	51805	99378
63	09066	00903	20795	95452	92648	45454	09552	88815	16553	51125	79375	97596	16296	66092
64	42238	12426	87025	14267	20979	04508	64535	31355	86064	29472	47689	05974	52468	16834
65	16153	08002	26504	41744	81959	65642	74240	56302	00033	67107	77510	70625	28725	34191
66	21457	40742	29820	96783	29400	21840	15035	34537	33310	06116	95240	15957	16572	06004
67	21581	57802	02050	89728	17937	37621	47075	42080	97403	48626	68995	43805	33386	21597
68	55612	78095	83197	33732	05810	24813	86902	60397	16489	03264	88525	42786	05269	92532
69	44657	66999	99324	51281	84463	60563	79312	93454	68876	25471	93911	25650	12682	73572
70	91340	84979	46949	81973	37949	61023	43997	15263	80644	43942	89203	71795	99533	50501
71	91227	21199	31935	27022	84067	05462	35216	14486	29891	68607	41867	14951	91696	85065
72	50001	38140	66321	19924	72163	09538	12151	06878	91903	18749	34405	56087	82790	70925
73	65390	05224	72958	28609	81406	39147	25549	48542	42627	45233	57202	94617	23772	07896
74	27504	96131	83944	41575	10573	08619	64482	73923	36152	05184	94142	25299	84387	34925
75	37169	94851	39117	89632	00959	16487	65536	49071	39782	17095	02330	74301	00275	48280
76	11508	70225	51111	38351	19444	66499	71945	05422	13442	78675	84081	66938	93654	59894
77	37449	30362	06694	54690	04052	53115	62757	95348	78662	11163	81651	50245	34971	52924
78	46515	70331	85922	38329	57015	15765	97161	17869	45349	61796	66345	81073	49106	79860
79	30986	81223	42416	58353	21532	30502	32305	86482	05174	07901	54339	58861	74818	46942
80	63798	64995	46583	09765	44160	78128	83991	42865	92520	83531	80377	35909	81250	54238
81	82486	84846	99254	67632	43218	50076	21361	64816	51202	88124	41870	52689	51275	83556
82	21885	32906	92431	09060	64297	51674	64126	62570	26123	05155	59194	52799	28225	85762
83	60336	98782	07408	53458	13564	59089	26445	29789	85205	41001	12535	12133	14645	23541
84	43937	46891	24010	25560	86355	33941	25786	54990	71899	15475	95434	98227	21824	19585
85	97656	63175	89303	16275	07100	92063	21942	18611	47348	20203	18534	03862	78095	50136
86	03299	01221	05418	38982	55758	92237	26759	86367	21216	98442	08303	56613	91511	75928
87	79626	06486	03574	17668	07785	76020	79924	25651	83325	88428	85076	72811	22717	50585
88	85636	68335	47539	03129	65651	11977	02510	26113	99447	68645	34327	15152	55230	93448
89	18039	14367	61337	06177	12143	46609	32989	74014	64708	00533	35398	58408	13261	47908
90	08362	15656	60627	36478	65648	16764	53412	09013	07832	41574	17639	82163	60859	75567
91	79556	29068	04142	16268	15387	12856	66227	38358	22478	73373	88732	09443	82558	05250
92	92608	82674	27072	32534	17075	27698	98204	63863	11951	34648	88022	56148	34925	57031
93	23982	25835	40055	67006	12293	02753	14827	22235	35071	99704	37543	11601	35503	85171
94	09915	96306	05908	97901	28395	14186	00821	80703	70426	75647	76310	88717	37890	40129
95	50937	33300	26695	62247	69927	76123	50842	43834	86654	70959	79725	93872	28117	19233
96	42488	78077	69882	61657	34136	79180	97526	43092	04098	73571	80799	76536	71255	64239
97	46764	86273	63003	93017	31204	36692	40202	35275	57306	55543	53203	18098	47625	88684
98	03237	45430	55417	63282	90816	17349	88298	90183	36600	78406	06216	95787	42579	90730
99	86591	81482	52667	61583	14972	90053	89534	76036	49199	43716	97548	04379	46370	28672
100	38534	01715	94964	87288	65680	43772	39560	12918	86537	62738	19636	51132	25739	56947

A TABLE OF 14,000 RANDOM UNITS (Continued)

Line/Col.	(1)	(2)	(3)	(4)	(5)	(6)	(7)	(8)	(9)	(10)	(11)	(12)	(13)	(14)
101	13284	16834	74151	92027	24670	36665	00770	22878	02179	51602	07270	76517	97275	45960
102	21224	00370	30420	03883	96648	89428	41583	17564	27395	63904	41548	49197	82277	24120
103	99052	47887	81085	64933	66279	80432	65793	83287	34142	13241	30590	97760	35848	91983
104	00199	50993	98603	38452	87890	94624	69721	57484	67501	77638	44331	11257	71131	11059
105	60578	06483	28733	37867	07936	98710	98539	27186	31237	80612	44488	97819	70401	95419
106	91240	18312	17441	01929	18163	69201	31211	54288	39296	37318	65724	90401	79017	62077
107	97458	14229	12063	59611	32249	90466	33216	19358	02591	54263	88449	01912	07436	50813
108	35249	38646	34475	72417	60514	69257	12489	51924	86871	92446	36607	11458	30440	52639
109	38980	46600	11759	11900	46743	27860	77940	39298	97838	95145	32378	68038	89351	37005
110	10750	52745	38749	87365	58959	53731	89295	59062	39404	13198	59960	70408	29812	83126
111	36247	27850	73958	20673	37800	63835	71051	84724	52492	22342	78071	17456	96104	18327
112	70994	66986	99744	72438	01174	42159	11392	20724	54322	36923	70009	23233	65438	59685
113	99638	94702	11463	18148	81386	80431	90628	52506	02016	85151	88598	47821	00265	82525
114	72055	15774	43857	99805	10419	76939	25993	03544	21560	83471	43989	90770	22965	44247
115	24038	65541	85788	55835	38835	59399	13790	35112	01324	39520	76210	22467	83275	32286
116	74976	14631	35908	28221	39470	91548	12854	30166	09073	75887	36782	00268	97121	57676
117	35553	71628	70189	26436	63407	91178	90348	55359	80392	41012	36270	77786	89578	21059
118	35676	12797	51434	82976	42010	26344	92920	92155	58807	54644	58581	95331	78629	73344
119	74815	67523	72985	23183	02446	63594	98924	20633	58842	85961	07648	70164	34994	67662
120	45246	88048	65173	50989	91060	89894	36063	32819	68559	99221	49475	50558	34698	71800
121	76509	47069	86378	41797	11910	49672	88575	97966	32466	10083	54728	81972	58975	30761
122	19689	90332	04315	21358	97248	11188	39062	63312	52496	07349	79178	33692	57352	72862
123	42751	35318	97513	61537	54955	08159	00337	80778	27507	95478	21252	12746	37554	97775
124	11946	22681	45045	13964	57517	59419	58045	44067	58716	58840	45557	96345	33271	53464
125	96518	48688	20996	11090	48396	57177	83867	86464	14342	21545	46717	72364	86954	55580
126	35726	58643	76869	84622	39098	36083	72505	92265	23107	60278	05822	46760	44294	07672
127	39737	42750	48968	70536	84864	64952	38404	94317	65402	13589	01055	79044	19308	83623
128	97025	66492	56177	04049	80312	48028	26408	43591	75528	65341	49044	95495	81256	53214
129	62814	08075	09788	56350	76787	51591	54509	49295	85830	59860	30883	89660	96142	18354
130	25578	22950	15227	83291	41737	79599	96191	71845	86899	70694	24290	01551	80092	82118
131	68763	69576	88991	49662	46704	63362	56625	00481	73323	91427	15264	06969	57048	54149
132	17900	00813	64361	60725	88974	61005	99709	30666	26451	11528	44323	34778	60342	60388
133	71944	60227	63551	71109	05624	43836	58254	26160	32116	63403	35404	57146	10909	07346
134	54684	93691	85132	64399	29182	44324	14491	55226	78793	34107	30374	48429	51376	09559
135	25946	27623	11258	65204	52832	50880	22273	05554	99521	73791	85744	29276	70326	60251
136	01353	39318	44961	44972	91766	90262	56073	06606	51826	18893	83448	31915	97764	75091
137	99083	88191	27662	99113	57174	35571	99884	13951	71057	53961	61448	74909	07322	80960
138	52021	45406	37945	75234	24327	86978	22644	87779	23753	99926	63898	54886	18051	96314
139	78755	47744	43776	83098	03225	14281	83637	55984	13300	52212	58781	14905	46502	04472
140	25282	69106	59180	16257	22810	43609	12224	25643	89884	31149	85423	32581	34374	70873
141	11959	94202	02743	86847	79725	51811	12998	76844	05320	54236	53891	70226	38632	84776
142	11644	13792	98190	01424	30078	28197	55583	05197	47714	68440	22016	79204	06862	94451
143	06307	97912	68110	59812	95448	43244	31262	88880	13040	16458	43813	89416	42482	33939
144	76285	75714	89585	99296	52640	46518	55486	90754	88932	19937	57119	23251	55619	23679
145	55322	07589	39600	60866	63007	20007	66819	84164	61131	81429	60676	42807	78286	29015
146	78017	90928	90220	92503	83375	26986	74399	30885	88567	29169	72816	53357	15428	86932
147	44768	43342	20696	26331	43140	69744	82928	24988	94237	46138	77426	39039	55596	12655
148	25100	19336	14605	86603	51680	97678	24261	02464	86563	74812	60069	71674	15478	47642
149	83612	46623	62876	85197	07824	91392	58317	37726	84628	42221	10268	20692	15699	29167
150	41347	81666	82961	60413	71020	83658	02415	33322	66036	98712	46795	16308	28413	05417

A TABLE OF 14,000 RANDOM UNITS (Continued)

Line/Col.	(1)	(2)	(3)	(4)	(5)	(6)	(7)	(8)	(9)	(10)	(11)	(12)	(13)	(14)
151	38128	51178	75096	13609	16110	73533	42564	59870	29399	67834	91055	89917	51096	89011
152	60950	00455	73254	96067	50717	13878	03216	78274	65863	37011	91283	33914	91303	49326
153	90524	17320	29832	96118	75792	25326	22940	24904	80523	38928	91374	55597	97567	38914
154	49897	18278	67160	39408	97056	43517	84426	59650	20247	19293	02019	14790	02852	05819
155	18494	99209	81060	19488	65596	59787	47939	91225	98768	43688	00438	05548	09443	82897
156	65373	72984	30171	37741	70203	94094	87261	30056	58124	70133	18936	02138	59372	09075
157	40653	12843	04213	70925	95360	55774	76439	61768	52817	81151	52188	31940	54273	49032
158	51638	22238	56344	44587	83231	50317	74541	07719	25472	41602	77318	15145	57515	07633
159	69742	99303	62578	83575	30337	07488	51941	84316	42067	49692	28616	29101	03013	73449
160	58012	74072	67488	74580	47992	69482	58624	17106	47538	13452	22620	24260	40155	74716
161	18348	19855	42887	08279	43206	47077	42637	45606	00011	20662	14642	49984	94509	56380
162	59614	09193	58064	29086	44385	45740	70752	05663	49081	26960	57454	99264	24142	74648
163	75688	28630	39210	52897	62748	72658	98059	67202	72789	01869	13496	14663	87645	89713
164	13941	77802	69101	70061	35460	34576	15412	81304	58757	35498	94830	75521	00603	97701
165	96656	86420	96475	86458	54463	96419	55417	41375	76886	19008	66877	35934	59801	00497
166	03363	82042	15942	14549	38324	87094	19069	67590	11087	68570	22591	65232	85915	91499
167	70366	08390	69155	25496	13240	57407	91407	49160	07379	34441	01567	66005	38918	65708
168	17870	06005	12927	16043	53257	93796	52721	73120	48025	76074	95605	67422	41646	14557
169	79504	77606	22761	30518	28373	73898	30550	76684	77366	32276	04690	61667	64798	66276
170	46967	74841	50923	15339	37755	98995	40162	89561	69199	42257	11647	47603	48779	97907
171	14558	50769	35444	59030	87516	48193	02945	00922	48189	04724	21263	20892	92955	90251
172	12440	25057	01132	38611	28135	68089	10954	10097	54243	06460	50856	65435	79377	53890
173	32293	29938	68653	10497	98919	46587	77701	99119	93165	67788	17638	23097	21468	36992
174	10640	21875	72462	77981	56550	55999	87310	69643	45124	00349	25748	00844	96831	30651
175	47615	23169	39571	56972	20628	21788	51736	33133	72696	32605	41569	76148	91544	21121
176	16948	11128	71624	72754	49084	96303	27830	45817	67867	18062	87453	17226	72904	71474
177	21258	61092	66634	70335	92448	17354	83432	49608	66520	06442	59664	20420	39201	69549
178	15072	48853	15178	30730	47481	48490	41436	25015	49932	20474	53821	51015	79841	32405
179	99154	57412	09858	65671	70655	71479	63520	31357	56968	06729	34465	70685	04184	25250
180	08759	61089	23706	32994	35426	36666	63988	98844	37533	08269	27021	45886	22835	78451
181	67323	57839	61114	62192	47547	58023	64630	34886	98777	75442	95592	06141	45096	73117
182	09255	13986	84834	20764	72206	89393	34548	93438	88730	61805	78955	18952	46436	58740
183	36304	74712	00374	10107	85061	69228	81969	92216	03568	39630	81869	52824	50937	27954
184	15884	67429	86612	47367	10242	44880	12060	44309	46629	55105	66793	93173	00480	13311
185	18745	32031	35303	08134	33925	03044	59929	95418	04917	57596	24878	61733	92834	64454
186	72934	40086	88292	65728	38300	42323	64068	98373	48971	09049	59943	36538	05976	82118
187	17626	02944	20910	57662	80181	38579	24580	90529	52303	50436	29401	57824	86039	81062
188	27117	61399	50967	41399	81636	16663	15634	79717	94696	59240	25543	97989	63306	90946
189	93995	18678	90012	63645	85701	85269	62263	68331	00389	72571	15210	20769	44686	96176
190	67392	89421	09623	80725	62620	84162	87368	29560	00519	84545	08004	24526	41252	14521
191	04910	12261	37566	80016	21245	69377	50420	85658	55263	68667	78770	04533	14513	18099
192	81453	20283	79929	59839	23875	13245	46808	74124	74703	35769	95588	21014	37078	39170
193	19480	75790	48539	23703	15537	48885	02861	86587	74539	65227	90799	58789	96257	02708
194	21456	13162	74608	81011	55512	07481	93551	72189	76261	91206	89941	15132	37738	59284
195	89406	20912	46189	76376	25538	87212	20748	12831	57166	35026	16817	79121	18929	40628
196	09866	07414	55977	16419	01101	69343	13305	94302	80703	57910	36933	57771	42546	03003
197	86541	24681	23421	13521	28000	94917	07423	57523	97234	63951	42876	46829	09781	58160
198	10414	96941	06205	72222	57167	83902	07460	69507	10600	08858	07685	44472	64220	27040
199	49942	06683	41479	58982	56288	42853	92196	20632	62045	78812	35895	51851	83534	10689
200	23995	68882	42291	23374	24299	27024	67460	94783	40937	16961	26053	78749	46704	21983

FACTORS FOR COMPUTING CONTROL LIMITS

A. Control Charts for Measurement

If the process mean and standard deviation, μ and σ, are known, and it is assumed that the underlying distribution is normal, it is possible to assert with probability $1 - \alpha$ that the mean of a random sample of size n will fall between $\bar{x} - z_{\alpha/2} \dfrac{\sigma}{\sqrt{n}}$ and $\bar{x} + z_{\alpha/2} \dfrac{\sigma}{\sqrt{n}}$.

These two limits on $\bar{x}$ provide upper and lower control limits. In actual practice, μ and σ are usually unknown and it is necessary to estimate their values from a large sample taken while the process is "in control." The central line of an $\bar{x}$-chart is given by μ and the lower and upper three-sigma control limits are given by $\mu - A\sigma$ and $\mu + A\sigma$, respectively, where $A = \dfrac{3}{\sqrt{n}}$ and n is the sample size. Where the population parameters are unknown, it is necessary to estimate these parameters on the basis of preliminary samples. If k samples are used, each of size n, denote the mean of the ith sample by $\bar{x}_i$ and the grand mean of the k sample means by $\bar{\bar{x}}$, i.e.

$$\bar{\bar{x}} = \frac{1}{k} \sum_{i=1}^{k} \bar{x}_i$$

Denote the range of the ith sample by R_i and by $\bar{R}$ the mean of the k sample ranges, i.e.

$$\bar{R} = \frac{1}{k} \sum_{i=1}^{k} R_i$$

Since $\bar{\bar{x}}$ is an unbiased estimate of the population mean μ, the central line for the $\bar{x}$-chart is given by $\bar{\bar{x}}$. The statistic R does not provide an unbiased estimate of σ, but $A_2 \bar{R}$ is an unbiased estimate of $\dfrac{3\sigma}{\sqrt{n}}$. The constant multiplier A_2 depends on the assumption of normality. Thus, the central line and the lower and upper three sigma limits, LCL and UCL, for an $\bar{x}$-chart (with μ and σ estimated from past date) are given by

$$\text{central line} = \bar{\bar{x}}$$
$$\text{LCL} = \bar{\bar{x}} - A_2 \bar{R}$$
$$\text{UCL} = \bar{\bar{x}} + A_2 \bar{R}$$

The central line and control limits of an R chart are based on the distribution of the range of samples of size n from a normal population. The mean and standard deviation of the sampling distribution of R are given by $d_2\sigma$ and $d_3\sigma$, respectively, when σ is known. Here d_2 and d_3 are constants which depend on the size of the sample. The set of control chart values for an R chart (with σ known) is given by

$$\text{central line} = d_2\sigma$$
$$\text{LCL} = D_1\sigma$$
$$\text{UCL} = D_2\sigma$$

where $D_1 = d_2 - 3d_3$ and $D_2 = d_2 + 3d_3$.

If σ is unknown, the control chart values for an R chart are given by

$$\text{central line} = \bar{R}$$
$$\text{LCL} = D_3\bar{R}$$
$$\text{UCL} = D_4\bar{R},$$

where $D_3 = \dfrac{D_1}{d_2}$ and $D_4 = \dfrac{D_2}{d_2}$.

The central line and control limits of an s-chart are based on estimates obtained from the samples. A pooled estimate of the population variance is obtained from the k samples, i.e.

$$s_p^2 = \frac{\sum_i (n_i - 1)s_i^2}{\sum_i (n_i - 1)}, \qquad i = 1, 2, \ldots, k$$

If the sample sizes are all equal, the pooled estimate is

$$s_p^2 = \frac{1}{k} \sum_i s_i^2$$

The control chart values for an s-chart are given by

$$\text{central line} = C_2' s_p$$
$$\text{LCL} = B_2' s_p$$
$$\text{UCL} = B_4' s_p$$

If one uses the biased estimator of the variance s_p', as is often done in quality control work, the control chart values are given by

$$\text{central line} = c_2 s_p'$$
$$\text{LCL} = B_2 s_p'$$
$$\text{UCL} = B_4 s_p'$$

B. Control Charts for Attributes

Control limits for a fraction-defective chart are based on the sampling theory for proportions, using the normal curve approximation to the binomial. If k samples are taken, the estimator of p is given by

$$\bar{p} = \frac{\sum_i x_i}{\sum_i n_i}, \qquad i = 1, 2, \ldots, k$$

where x_i is the number of defectives in the ith sample of size n_i. The central line and control limits of a fraction defective chart based on analysis of past data are given by

$$\text{central line} = \bar{p}$$
$$\text{LCL} = \bar{p} - 3\sqrt{\frac{\bar{p}(1 - \bar{p})}{n_i}}$$
$$\text{UCL} = \bar{p} + 3\sqrt{\frac{\bar{p}(1 - \bar{p})}{n_i}}$$

When the sample sizes are approximately equal, n_i is replaced by $\bar{n} = \dfrac{1}{k} \sum_i n_i$.

Equivalent to the p chart for the fraction defective is the control chart for the number of defective. Here, if p is estimated by $\bar{p}$, the control chart values for a number-of-defectives chart are given by

$$\text{central line} = \bar{n}\bar{p}$$
$$\text{LCL} = \bar{n}\bar{p} - 3\sqrt{\bar{n}\bar{p}(1 - \bar{p})}$$
$$\text{UCL} = \bar{n}\bar{p} + 3\sqrt{\bar{n}\bar{p}(1 - \bar{p})}$$

In many cases it is necessary to control the number of defects per unit C, where C is taken to be a value of a random variable having a Poisson distribution. If k is the number of units available for estimating λ, the parameter of the Poisson distribution, and if C_i is the number of defects in the ith unit, then λ is estimated by

$$\bar{C} = \frac{1}{k} \sum_{i=1}^{k} C_i$$

and the control-chart values for the C-chart are

$$\text{central line} = \bar{C}$$
$$\text{LCL} = \bar{C} - 3\sqrt{\bar{C}}$$
$$\text{UCL} = \bar{C} + 3\sqrt{\bar{C}}$$

This table presents values of the factors for computing control limits for various sample sizes n.

Number of observations in sample, n	$\bar{X}$ chart		R chart			s chart			$\hat{\sigma}$ chart (biased)		
	Factors for control limits		Factor for central line	Factors for control limits		Factor for central line	Factors for control limits		Factor for central line	Factors for control limits	
	A	A_2	d_2	D_3	D_4	c_2'	B_2'	B_4'	c_2	B_2	B_4
2	2.121	1.880	1.128	0	3.267	0.798	0	2.298	0.5642	0	3.267
3	1.732	1.023	1.693	0	2.575	0.886	0	2.111	0.7236	0	2.568
4	1.500	0.729	2.059	0	2.282	0.921	0	1.982	0.7979	0	2.266
5	1.342	0.577	2.326	0	2.115	0.940	0	1.889	0.8407	0	2.089
6	1.225	0.483	2.534	0	2.004	0.951	0.085	1.817	0.8686	0.030	1.970
7	1.134	0.419	2.704	0.076	1.924	0.960	0.158	1.762	0.8882	0.118	1.882
8	1.061	0.373	2.847	0.136	1.864	0.965	0.215	1.715	0.9027	0.185	1.815
9	1.000	0.337	2.970	0.184	1.816	0.969	0.262	1.676	0.9139	0.239	1.761
10	0.949	0.308	3.078	0.223	1.777	0.973	0.302	1.644	0.9227	0.284	1.716
11	0.905	0.285	3.173	0.256	1.744	0.976	0.336	1.616	0.9300	0.321	1.679
12	0.866	0.266	3.258	0.284	1.716	0.977	0.365	1.589	0.9359	0.354	1.646
13	0.832	0.249	3.336	0.308	1.692	0.980	0.392	1.568	0.9410	0.382	1.618
14	0.802	0.235	3.407	0.329	1.671	0.981	0.414	1.548	0.9453	0.406	1.594
15	0.775	0.223	3.472	0.348	1.652	0.982	0.434	1.530	0.9490	0.428	1.572
16	0.750	0.212	3.532	0.364	1.636	0.984	0.454	1.514	0.9523	0.448	1.552
17	0.728	0.203	3.588	0.379	1.621	0.984	0.469	1.499	0.9551	0.466	1.534
18	0.707	0.194	3.640	0.392	1.608	0.986	0.486	1.486	0.9576	0.482	1.518
19	0.688	0.187	3.689	0.404	1.596	0.986	0.500	1.472	0.9599	0.497	1.503
20	0.671	0.180	3.735	0.414	1.586	0.987	0.513	1.461	0.9619	0.510	1.490
21	0.655	0.173	3.778	0.425	1.575	0.988	0.525	1.451	0.9638	0.523	1.477
22	0.640	0.167	3.819	0.434	1.566	0.988	0.536	1.440	0.9655	0.534	1.466
23	0.626	0.162	3.858	0.443	1.557	0.989	0.546	1.432	0.9670	0.545	1.455
24	0.612	0.157	3.895	0.452	1.548	0.989	0.556	1.422	0.9684	0.555	1.445
25	0.600	0.153	3.931	0.459	1.541	0.990	0.566	1.414	0.9696	0.565	1.435

XIII. ASTRODYNAMICS

ASTRODYNAMICS: BASIC ORBITAL EQUATIONS

Samuel Herrick

University of California, Los Angeles

Abstracted from *Astrodynamics* by Samuel Herrick, published by Van Nostrand Reinhold 1970. For more complete discussion refer to *Astrodynamics*.

1. Astrodynamics includes those portions of *celestial mechanics* that bear directly upon the determination of orbits and the integration of ephemerides of natural and artificial objects in interplanetary and sublunar space.   For the most part, but not exclusively, it is founded upon the law of universal gravitation,

$$\mathbf{f}_{ij} = -\mathbf{f}_{ji} = k^2 m_i m_j \mathbf{r}_{ij}/r_{ij}^3 \tag{1}$$

where m_i and m_j are two masses, $\mathbf{r}_{ij}$ is the vector directed from m_i to m_j, r_{ij} is its magnitude, $\mathbf{f}_{ij}$ is the vector attraction of m_j for m_i, $\mathbf{f}_{ji}$ of m_i for m_j, and k^2 is the constant of gravitation.

The laboratory value of k^2, $G = 6.673 \times 10^{-23}$ km³/gm sec², is totally inadequate to the determination of a value of k^2 that can be used in astrodynamics, and it is determined instead from planetary and satellite observations, and for the units of mass that are required by astrodynamics, as follows:

	Heliocentric Orbits		Geocentric Orbits	
Unit of length	astronomical unit (au)	km	geo-unit (gu)	km
Unit of mass	Sun's mass ($m_\odot$)	$m_\odot$	Earth's mass ($m_\oplus$)	$m_\oplus$
Unit of time	ephemeris day (day)	sec	ephem. sec (sec)	sec
k	0.017 202 09895	3.643×10^5	0.001 239 444	631.35
k^2	$295.912\ 2083 \times 10^{-6}$	13.271×10^{10}	$1.536\ 221 \times 10^{-6}$	398603

(2)

The au and gu are approximately the mean distance of the earth from the sun and the earth's equatorial radius, respectively, but are defined in such ways as to make the corresponding values of k exact.   Each of these units is essential to a "canonical" set, as discussed in §2.

2. The equations of motion, for a system of n mutually attracting bodies, are derived from equations (1) and the second law of motion.   Assuming that the masses are constant, we may write them in their *inertial form* as follows:

$$\left.\begin{aligned}
m_1 \frac{d^2\mathbf{r}_1}{dt^2} &= \mathbf{f}_{12} + \sum_{j=3}^{n} \mathbf{f}_{1j} = k^2 m_1 m_2 \frac{\mathbf{r}_{12}}{r_{12}^3} + k^2 m_1 \sum_{j=3}^{n} m_j \frac{\mathbf{r}_{1j}}{r_{1j}^3} \\
m_2 \frac{d^2\mathbf{r}_2}{dt^2} &= \mathbf{f}_{21} + \sum_{j=3}^{n} \mathbf{f}_{2j} = -k^2 m_1 m_2 \frac{\mathbf{r}_{12}}{r_{12}^3} + k^2 m_2 \sum_{j=3}^{n} m_j \frac{\mathbf{r}_{2j}}{r_{2j}^3} \\
\cdots\cdots\cdots\cdots\cdots\cdots\cdots\cdots\cdots\cdots\cdots\cdots
\end{aligned}\right\} \tag{3}$$

From these equations we may obtain the well known ten integrals of the n-body problem, which tell us (a) that the center of mass moves uniformly in a straight line, and so may be visualized as the inertial origin, (b) that the total angular momentum of the system is constant, and (c) that the total energy of the system is constant.

The *relative motion form* of the equations of motion is derived from equations (3) and

$$\mathbf{r} = \mathbf{r}_{12} = \mathbf{r}_2 - \mathbf{r}_1 \tag{4}$$

especially for an m_1 that has, like the Sun, a dominant effect on the motion of an m_2:

$$\frac{d^2\mathbf{r}}{dt^2} = -k^2(m_1 + m_2)\frac{\mathbf{r}}{r^0} + k^2 \sum_{j=3}^{n} m_j \left(\frac{\mathbf{r}_{2j}}{r_{21}^3} - \frac{\mathbf{r}_{1j}}{r_{1j}^3} \right) \tag{5}$$

It is thus simply that $m_1 + m_2$ replaces the $m_1 m_2$ of equations (3); clearly there is no occasion for the surprise or puzzlement this replacement sometimes arouses.

The first term of equation (5) is often called the *two-body term*. The remaining terms are the *perturbations* (for relative motion and the *n*-body problem). The perturbation terms can also include the effects of thrust, drag, an equatorial bulge, the departures of relativity mechanics from Newtonian mechanics, etc. Even when we restrict the disturbing forces to those of the *n*-body problem, the perturbations can differ from those shown in equation (5), if a "two-body term" is constructed for an origin other than m_1.

The first term of equation (5) may be simplified, both theoretically and computationally, by the absorption of the gravitation-mass constant into a "canonical time,"

$$\tau = k_*(t - t_0) \tag{6}$$

and by the use of such definitions as

$$\dot{\mathbf{r}} = \frac{d\mathbf{r}}{dt} = k_* \dot{\mathbf{r}}_\tau = k_* \frac{d\mathbf{r}}{d\tau}, \qquad \ddot{\mathbf{r}} = \frac{d^2\mathbf{r}}{dt^2} = k_*^2 \ddot{\mathbf{r}}_\tau = k_*^2 \frac{d^2\mathbf{r}}{d\tau^2}, \text{ etc.} \tag{7}$$

where

$$k_* = k\sqrt{m_1 + m_2} \qquad \mu = k^2(m_1 + m_2) = k_*^2 \mu_\tau \qquad \mu_\tau = 1 \tag{8}$$

For other forms of the equations of motion the constant μ has slightly different values.

It is best to think of τ as having the dimensions of time; the dimensions of $\sqrt{\mu}$ are "left behind" in $\mu_\tau = 1$ when the derivatives in equations hereafter are expressed with canonical units of time. When m_2 is negligible as compared with m_1, as it usually is, the canonical units of time are:

$$
\left.
\begin{array}{lll}
\text{au} - \text{m}_\odot - \text{day system:} & 1 \text{ kaday} = 58.132\ 44087 \text{ days} \\
\text{gu} - \text{m}_\oplus - \text{sec system:} & 1 \text{ kesec} = 806.813\ 3776 \text{ sec} \\
& \qquad\qquad = 13.446\ 8896 \text{ min}
\end{array}
\right\} \tag{9}
$$

The practical advantage of canonical units of time and distance is that they tend to keep the magnitudes of position, velocity, and acceleration vectors all in the neighborhood of unity, thus reducing, for example, the likelihood of machine errors in floating-decimal-point operations.

3. The two-body problem includes, of course, only the first term of the equations of relative motion (5), which we may now write, with the aid of equations (6), (7), (8) as follows:

$$\ddot{\mathbf{r}} = -\mu\mathbf{r}/r^3 \tag{10}$$

We seek integrals of the two-body problem first by means of

$$\mathbf{r} \times \ddot{\mathbf{r}} = 0 \tag{11}$$

of which the x component is

$$y\ddot{z} - z\ddot{y} = 0 \tag{12}$$

so that

$$\mathbf{r} \times \dot{\mathbf{r}} = \mathbf{h} = h\mathbf{W} \tag{13}$$

and

$$y\dot{z} - z\dot{y} = h_x = hW_x \tag{14}$$

where **W** is a constant unit vector normal to the orbit plane, thus established as a fixed plane, and where, with v the *true anomaly* or angle in the orbit plane measured from the perifocus to **r**,

$$h = r^2\dot{v} = \sqrt{\mu p} \tag{15}$$

expresses *Kepler's second law*, or law of areas: that the areal velocity (or angular momentum) is constant.

The orientation of the orbit plane, embodied in **W**, is often expressed by the *inclination*, i, and the *longitude of the (ascending) node*, Ω.

The relationship of h to p, the *parameter* or semi latus rectum, may be established along with Kepler's first law in several different ways. When one is interested in the use as well as in the existence of integrals of the two-body problem, it is well to proceed by seeking additional functions that satisfy the differential equation (10) and so lead to additional simple integrations such as are found in equations (12) and (14). One such function is found in

$$\sqrt{\mu}\, D = r\dot{r} = \dot{\mathbf{r}} \cdot \dot{\mathbf{r}} \qquad \sqrt{\mu}\, \dot{D} = \dot{s}^2 - \mu/r \qquad \ddot{D} = -\mu D/r^3 \tag{16}$$

where

$$r^2 = \mathbf{r} \cdot \mathbf{r} \qquad \dot{s}^2 = \dot{\mathbf{r}} \cdot \dot{\mathbf{r}} = \dot{r}^2 + r^2\dot{v}^2 \tag{17}$$

so that

$$\mathbf{r}\ddot{D} - D\ddot{\mathbf{r}} = 0 \tag{18}$$

$$\mathbf{r}\dot{D} - D\dot{\mathbf{r}} = \mathbf{a} = e\sqrt{\mu}\, \mathbf{P} \tag{19}$$

where **a** is a constant vector, whose relationship to e, the eccentricity, and P, a unit vector directed to the perifocus, is established by

$$\sqrt{\mu}\, (\mathbf{r} \cdot \mathbf{a}) = \sqrt{\mu}\, (r^2\dot{D} - Dr\dot{r}) = r^2(\dot{s}^2 - \dot{r}^2) - \mu r = r^4\dot{v}^2 - \mu r = h^2 - \mu r \tag{20}$$

or

$$h^2/\mu = r - (\mathbf{r} \cdot \mathbf{a})/\sqrt{\mu} = r[1 - (|\mathbf{a}|/\sqrt{\mu}) \cos(\mathbf{a},\mathbf{r})] \tag{21}$$

which is the polar equation of the conic,

$$p = r(1 - e \cos v) \tag{22}$$

so that we have established *Kepler's first law*, that the orbit is a conic with the Sun (or the Earth) at one focus. We have also established the facts that **a** must be directed to perifocus and that

$$|\mathbf{a}| = e\sqrt{\mu} \qquad h^2 = \mu p \tag{23}$$

which are the evaluations needed for equations (15) and (19).

The magnitude of **a** must not be confused with the constant a, which for the ellipse (cf. fig 1) we call the *mean distance* or *semi major axis*, and which is related to p, the *parameter*, and q, the *perifocal distance*, for all of the conics, and r_A, the *apofocal distance*, for the ellipse, by

$$q = a(1 - e) \qquad\qquad r_A = a(1 + e) \tag{24}$$

$$p = a(1 - e^2) = q(1 + e) = r_A(1 - e) \tag{25}$$

The orthogonal unit vectors **W** and **P** (cf. equations 13 and 19) are joined by the unit vector **Q**, which is orthogonal to each of them, and is defined and obtained by

$$\mathbf{Q} = \mathbf{W} \times \mathbf{P} \tag{26}$$

or by one of the relationships in

$$\mathbf{b} = e\sqrt{\mu p}\, \mathbf{Q} = r\dot{H} - H\dot{\mathbf{r}} \tag{27}$$

where the latter relationship may be obtained by integration, since

$$H = r - p \qquad \dot{H} = \dot{r} \qquad \ddot{H} = -\mu H/r^3 \tag{28}$$

The orientation vectors **P** and **Q**, like **W**, are functions of i and Ω, but also of ω, the *argument of perifocus*, which is an angle measured from the node and defines the orientation of the perifocus, and of the major axis or line of apsides, in the orbit plane. They are used efficiently in the calculation of x, y, z, referred to general axes, from

$$x_\omega = r\cos v \qquad y_\omega = r\sin v \tag{29}$$

referred to orbital axes (cf. fig. 1 and §§5, 6), by means of

$$\mathbf{r} = x_\omega \mathbf{P} + y_\omega \mathbf{Q} \tag{30}$$

4. The vis-viva or energy integral may be obtained as follows:

$$2\dot{\mathbf{r}}\cdot\ddot{\mathbf{r}} = -2\mu\mathbf{r}\cdot\dot{\mathbf{r}}/r^3 = -2\mu\dot{r}/r^2 \tag{31}$$

$$\dot{s}^2 = \dot{\mathbf{r}}\cdot\dot{\mathbf{r}} = \mu(2/r + \alpha) \qquad \alpha = -1/a \tag{32}$$

where the constant α is evaluated from simple conditions known, for example, at perifocus.

Equations (15), (16), (17), (25), and (32) may now be recognized as sources of information about the nature of the conic determined by given position and velocity, **r** and $\dot{\mathbf{r}}$:

	Ellipse	Parabola	Hyperbola
General	$\frac{1}{a} > 0,\ 0 \leq e \leq 1$	$\frac{1}{a} = 0,\ e = 1$	$\frac{1}{a} < 0,\ e \geq 1$
Circular	$D = 0,\ e = 0$	—	—
Rectilinear	$\frac{1}{a} > 0,\ p = 0,\ e = 1$	$\frac{1}{a} = 0,\ p = 0,\ e = 1$	$\frac{1}{a} < 0,\ p = 0,\ e = 1$

$$\tag{33}$$

The vis-viva integral is used also to evaluate useful factors for converting canonical units to conventional units. When m_2 is negligible as compared with m_1, they are:

Heliocentric	Geocentric	
$V = V_s\dot{s}_\tau$	$V = V_e\dot{s}_\tau$	
$\mathbf{V} = V_s\dot{\mathbf{r}}_\tau$	$\mathbf{V} = V_e\dot{\mathbf{r}}_\tau$	

$$\tag{34}$$

Heliocentric	Geocentric	
$V_s = 29.785$	$V_e = 7.905$	km/sec
18.508	4.912	mi/sec
57,897	15,367	knots
97,720	25,936	ft/sec

$$\tag{35}$$

The constants V_s and V_e may be recognized as being approximately the velocities, respectively, of the Earth in its orbit, and of a circular satellite of the Earth hypothetically at one equatorial radius or geo-unit. The velocity of escape from the Earth, in the latter circumstances, is

$$V_{pe} = V_e\sqrt{2} \tag{36}$$

Two special values of $\dot{s}$ are of interest: $\dot{s}_q$ corresponding to $r = q$ and $\dot{s}_A$ corresponding to $r = r_A$,

$$\left.\begin{aligned}
\dot{s}_q/\sqrt{\mu} = \dot{s}_{\tau q}/\sqrt{\mu_\tau} = \sqrt{\frac{2}{q} - \frac{1}{a}} = \sqrt{\frac{1+e}{q}} = \sqrt{\frac{1}{a}\left(\frac{1+e}{1-e}\right)} \\
\dot{s}_A/\sqrt{\mu} = \dot{s}_{\tau A}/\sqrt{\mu_\tau} = \sqrt{\frac{2}{r_A} - \frac{1}{a}} = \sqrt{\frac{1-e}{r_A}} = \sqrt{\frac{1}{a}\left(\frac{1-e}{1+e}\right)}
\end{aligned}\right\} \tag{37}$$

5. The ellipse has special formulae in terms of E, the *eccentric anomaly*, an extremely useful angle that is defined by fig. 1:

$$\left. \begin{aligned} r &= a(1 - e \cos E) \\ r \cos v &= x_\omega = a\,(\cos E - e) \\ r \sin v &= y_\omega = \sqrt{ap}\,\sin E \end{aligned} \right\} \tag{38}$$

From equations (6), (15), and (38) we derive

$$\frac{r^2 \dot{v}}{\sqrt{p}} = \sqrt{\mu} = a^{\frac{1}{2}}(1 - e \cos E)\dot{E} = r\sqrt{a}\,\dot{E} \tag{39}$$

and *Kepler's equation*,

$$M = M_0 + n(t - t_0) = n(t - T) = E - e \sin E \tag{40}$$

where M is the *mean anomaly*, M_0 is the *mean anomaly at the epoch*, t_0, T is the *time of perifocal passage*, and n is the *mean motion* or mean angular velocity:

$$n = \dot{M} = \sqrt{\mu}/a^{\frac{1}{2}} \tag{41}$$

We find from equations (38), (40), and (41) that the three anomalies, v, E, M are together at 0, π, 2π, . . . , and that the corresponding values of $t - T$ are 0, $\frac{1}{2}P$, P, . . . , where P is the *period* expressed in the conventional unit of time:

$$P = 2\pi/n = 2\pi a^{\frac{1}{2}}/\sqrt{\mu} \tag{42}$$

Equation (42) is an expression of *Kepler's third law*, that the squares of the periods are proportional to the cubes of the mean distances.

On next page is a table of comparative formulae and definitions.

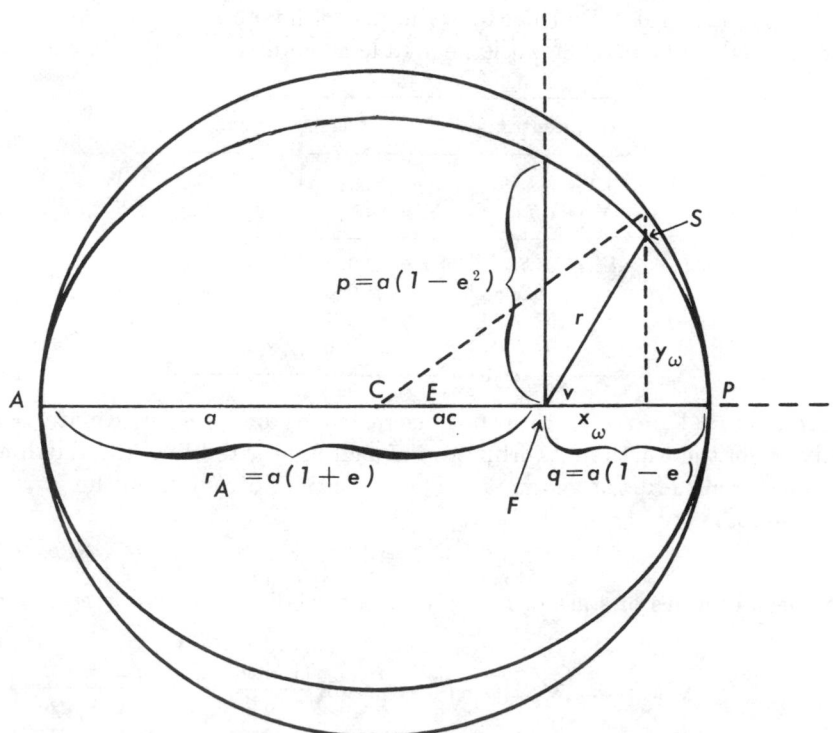

6. A table of comparative formulae and definitions:

General	Ellipse	Parabola	Hyperbola
	$0 \le e \le 1$	$e = 1$	$e \ge 1$
	$1/a > 0$	$1/a = 0$	$1/a < 0$
$p = q(1 + e)$	$p \le 2q$	$p = 2q$	$p \ge 2q$
$M = M_0 + n(t - t_0)$	$\begin{cases} n = a^{-\frac{3}{2}} \sqrt{\mu} \\ M = E - e \sin E \end{cases}$	$n = \sqrt{\mu}$	$n = (-a)^{-\frac{3}{2}} \sqrt{\mu}$
$= n(t - T)$		$M = qD + \frac{1}{6}D^3$	$M = e \sinh F - F$
$r = p/(1 + e \cos v)$	$r = a(1 - e \cos E)$	$r = q + \frac{1}{2}D^2$	$r = -a(e \cosh F - 1)$
$\dot{r}\sqrt{p/\mu} = e \sin v$	$D = \dfrac{r\dot{r}}{\sqrt{\mu}} = \sqrt{a}\,e \sin E$	$D = \dfrac{r\dot{r}}{\sqrt{\mu}}$	$D = \dfrac{r\dot{r}}{\sqrt{\mu}}$
		$= \sqrt{2q}\tan \dfrac{v}{2}$	$= \sqrt{-a}\,e \sinh F$
$r\dot{v}\sqrt{p/\mu} = 1 + e \cos v$	$r^2\dot{v} = \sqrt{\mu p}$	$r^2\dot{v} = \sqrt{2\mu q}$	$r^2\dot{v} = \sqrt{\mu p}$
	$= \sqrt{\mu a(1 - e^2)}$		$= \sqrt{-\mu a(e^2 - 1)}$
$x_\omega = r \cos v$	$x_\omega = a(\cos E - e)$	$x_\omega = q - \frac{1}{2}D^2$	$x_\omega = -a(\cosh F - e)$
$y_\omega = r \sin v$	$y_\omega = \sqrt{ap} \sin E$	$y_\omega = \sqrt{2q}\,D$	$y_\omega = \sqrt{-ap} \sinh F$
$\dot{x}_\omega \sqrt{p/\mu} = -\sin v$	$r\dot{x}_\omega = -\sqrt{\mu a} \sin E$	$r\dot{x}_\omega = -\sqrt{\mu}\,D$	$r\dot{x}_\omega = -\sqrt{-\mu a} \sinh F$
$\dot{y}_\omega \sqrt{p/\mu} = \cos v + e$	$r\dot{y}_\omega = \sqrt{\mu p} \cos E$	$r\dot{y}_\omega = \sqrt{2\mu q}$	$r\dot{y}_\omega = \sqrt{\mu p} \cosh F$

Circular Ellipse*	Rectilinear		
	Ellipse	Parabola	Hyperbola
$e = 0$	$e = 1$	$e = 1$	$e = 1$
$r = p = q = a$	$p = 2q = 0$	$p = 2q = 0$	$p = 2q = 0$
$M = E$	$M = E - \sin E$	$M = \frac{1}{6}D^3$	$M = \sinh F - F$
$\dot{s} = r\dot{v} = \sqrt{\mu/a}$	$r = a(1 - \cos E)$	$r = \frac{1}{2}D^2$	$r = -a(\cosh F - 1)$
$D = r\dot{r}/\sqrt{\mu} = 0$	$D = r\dot{r}/\sqrt{\mu} = \sqrt{a} \sin E$	$D = r\dot{r}/\sqrt{\mu}$	$D = r\dot{r}/\sqrt{\mu} = \sqrt{-a} \sinh F$
$x_\omega = a \cos E$	$x_\omega = -r$	$x_\omega = -r$	$x_\omega = -r$
$y_\omega = a \sin E$	$y_\omega = 0$	$y_\omega = 0$	$y_\omega = 0$
$\dot{x}_\omega = -\dot{s} \sin E$	$\dot{x}_\omega = -\dot{r}$	$\dot{x}_\omega = -\dot{r}$	$\dot{x}_\omega = -\dot{r}$
$\dot{y}_\omega = \dot{s} \cos E$	$\dot{y}_\omega = 0$	$\dot{y}_\omega = 0$	$\dot{y}_\omega = 0$

* M, E, x_ω, y_ω, $\dot{x}_\omega$, $\dot{y}_\omega$ are formally related as shown, but all are indeterminate!

Barker's equation, the parabolic equivalent of Kepler's equation in the foregoing table, is often written

$$\frac{\sqrt{\mu}\,(t - T)}{q^{\frac{3}{2}}\sqrt{2}} = \tan \tfrac{1}{2}v + \tfrac{1}{3}\tan^3 \tfrac{1}{2}v \tag{43}$$

but, so written, it is indeterminate for the rectilinear orbit ($q = 0$, $\tan \tfrac{1}{2}v = \infty$).

Unified variables, parameters, and formulae are designed to care for ellipses $(-1/a = \alpha < 0)$, parabolas $(\alpha = 0)$, and hyperbolas $(\alpha > 0)$, including rectilinear ones, but not circles or nearly-circular ellipses. They are illustrated by the following table, which adds elliptic definitions for illumination but not for calculation:

Unified Formulae	*Elliptic Definitions*
$\sqrt{\mu}\,(t - T) = \tilde{M} = q\tilde{X} + e\tilde{U}$	$\tilde{X} = \sqrt{a}\,E$
$r = q + e\tilde{C}$	
$\tilde{U} = \dfrac{\tilde{X}^3}{3!} + \alpha\,\dfrac{\tilde{X}^5}{5!} + \alpha^2\,\dfrac{\tilde{X}^7}{7!} + \cdots$	$\tilde{U} = a^{\frac{3}{2}}(E - \sin E)$
$\tilde{C} = \dfrac{\tilde{X}^2}{2!} + \alpha\,\dfrac{\tilde{X}^4}{4!} + \alpha^2\,\dfrac{\tilde{X}^6}{6!} + \cdots$	$\tilde{C} = a(1 - \cos E)$
$x_\omega = q - \tilde{C}$	
$\dfrac{y_\omega}{\sqrt{p}} = -\dfrac{r\dot{x}_\omega}{\sqrt{\mu}} = \tilde{S} = \tilde{X} + \alpha\tilde{U}$	$\tilde{S} = \sqrt{a}\,\sin E$
$\dfrac{r\dot{y}_\omega}{\sqrt{\mu p}} = \tilde{K} = 1 + \alpha\tilde{C}$	$\tilde{K} = \cos E$

The definitions in terms of parabolic and hyperbolic variables may be inferred.

Universal variables, parameters, and formulae are designed to care for the foregoing and also circles and nearly-circular ellipses. They are illustrated by

Universal Formulae	*Elliptic Definitions*
$D_0 = r_0\dot{r}_0$	$D_0 = \sqrt{a}\,e \sin E_0$
$c_0 = 1 + \alpha r_0$	$c_0 = e \cos E_0$
$\sqrt{\mu}\,(t - t_0) = \hat{M} = r_0\tilde{X} + D_0\tilde{C} + c_0\tilde{U}$	$\hat{E} = E - E_0$
$r = r_0 + D_0\hat{S} + c_0\hat{C}$	$\hat{X} = \sqrt{a}\,\hat{E}$
$\hat{U} = \dfrac{\hat{X}^3}{3!} + \alpha\,\dfrac{\hat{X}^5}{5!} + \alpha^2\,\dfrac{\hat{X}^7}{7!} + \cdots$	$\hat{U} = a^{\frac{3}{2}}(\hat{E} - \sin \hat{E})$
$\hat{C} = \dfrac{\hat{X}^2}{2!} + \alpha\,\dfrac{\hat{X}^4}{4!} + \alpha^2\,\dfrac{\hat{X}^6}{6!} + \cdots$	$\hat{C} = a(1 - \cos \hat{E})$
$\hat{S} = \hat{X} + \alpha\hat{U}$	$\hat{S} = \sqrt{a}\,\sin \hat{E}$

$$f = 1 - \hat{C}/r_0 \qquad \sqrt{\mu}\,g = \hat{M} - \hat{U}$$
$$\dot{g} = 1 - \hat{C}/r \qquad \dot{f} = -\sqrt{\mu}\,\hat{S}/rr_0$$
$$\mathbf{r} = f\mathbf{r}_0 + g\dot{\mathbf{r}}_0 \qquad \dot{\mathbf{r}} = \dot{f}\mathbf{r}_0 + \dot{g}\dot{\mathbf{r}}_0$$

It is evident that universal variables reduce to unified variables when $t_0 = T$, and $M_0 = 0$, $E_0 = 0$, $r_0 = q$, $D_0 = 0$, $c_0 = e$.

ASTRODYNAMICAL TERMINOLOGY, NOTATION, AND USAGE

Samuel Herrick

University of California, Los Angeles

Abstracted from *Astrodynamics* by Samuel Herrick, published by Van Nostrand Reinhold 1970. For more complete discussion of the philosophy of selection of these symbols refer to *Astrodynamics*.

INTRODUCTION

Standardization of notation and terminology, or more properly, attention to established conventions and usage, contributes to understanding in all fields of science.

In the compilation of a standard listing of terms and symbols it is often implied, unwisely, that conformity with it is to be required thereafter of all "respectable" writing in the field. The philosophy adopted in the present listing, on the contrary, is that it should be based upon usage, that it should present alternatives sanctioned either by usage or by logical development where conflicts have come into being, and that it should be a reference list rather than a required list.

Typical of the conflicting usages in astrodynamics, encountered in celestial mechanics even before it had fully established contact with such fields as propulsion, aerodynamics, optimization, are those between v, true anomaly, and v, velocity, and between a, mean distance, and a, acceleration. In accordance with the foregoing expression of philosophy, we report all these usages but indicate that $\dot{r}$, $\dot{s}$, or V should be preferred for velocity, and that $\ddot{r}$, g, or A should be preferred for acceleration, because of the dominant use of v and a in the other senses.

The central core of astrodynamics is celestial mechanics, and an important function of the present listing is the providing of an entrée to the literature of celestial mechanics, much of which is still untapped in astrodynamical research. In consequence the present listing leans heavily upon astronomical usage, and upon the 1938 listing of the International Astronomical Union, *Transactions*, *6*, 345–348, but with departures from the latter in such matters as the following:

(1) The preferred designations for the geocentric distance (or "range") and for an increment or correction or residual are ρ and Δ rather than Δ and δ. The I.A.U. ignored long-established astronomical usage in such standard and familiar quantities as $\rho \cos \delta \, \Delta\alpha$ and $\rho\Delta\delta$, converting them into the whimsical $\Delta \cos \delta \, \delta\alpha$ and $\Delta\delta\delta$.

(2) The "direction cosines" associated with α and δ are designated as the components of a unit vector, $\mathbf{L}$, ($L_x = \cos \delta \cos \alpha$, $L_y = \cos \delta \sin \alpha$, $L_z = \sin \delta$), rather than by the l, m, n of the I.A.U. listing, or by the older sets: λ, μ, ν; α, β, γ. The use of vectors in celestial mechanics has become almost universal since 1938.

(3) The unit vector normal to the orbit plane is designated by $\mathbf{W}(W_x, W_y, W_z)$. The I.A.U. designations R_x, R_y, R_z have come into conflict with its designations for the "solar coordinates," X, Y, Z, because the vector notation for both sets would be $\mathbf{R}$, and because the two vectors sometimes appear in the same equation. Accordingly we prefer $\mathbf{R}(X,Y,Z)$ for the solar coordinates (or more generally the coordinates of the dynamical center referred to the observer).

The listing of terms and symbols is organized as follows:

(1) There is one alphabetical listing for terms and alphabetical symbols.

(2) Capital letters precede the corresponding lower case letters.

(3) Vector notation follows non-vector notation.

(4) Greek letters are alphabetized with Roman letters according to their usual trans-

literations (or the spellings of their names), thus changing the Greek order for γ, ζ, η, θ, φ, χ, ψ, ω.

(5) Vector triads are listed as groups under the first letter of triads, with cross references appearing under the other two letters.

(6) Terms are alphabetized according to spelling, and cross-referenced with notations, and vice versa.

(7) Non-alphabetical symbols are listed separately at the end. This listing makes no effort to be complete, or to include standard mathematical symbols.

(8) Terminology and definitions are primarily those that are peculiar to celestial mechanics or for which differences in usage exist, and do not include general terms such as "conic," "focus," "directrix," "least squares," "standard deviation," "celestial sphere," "ecliptic," etc., with which all users of the list may be assumed to be familiar.

(9) Alternative notations and usages are usually indicated with the author's preference or a cross reference introduced by "prefer" or by "see," "also," or "q.v." ("which see").

A

A, a	Acceleration; prefer g or A (see introduction and "g")
A, Az	Azimuth (q.v.); also Z
$A = A\left[\dfrac{km}{au}\right]$, A $[km]$	Length conversion factor, astronomical units to kilometers; au expressed in km.
A	Cross-sectional area
A, B	Constants in dynamical equation of first approximation in orbit determination: $\rho = A - B/r^3$
A, B, C	Principal moments of inertia
A_1, A_3	Ratios of sectors or time intervals, used in expressions for ratios of triangles (see "c_1,c_3"); also c_1^0, c_3^0:

$$A_1 = \tau_{23}/\tau_{13};\ A_3 = \tau_{12}/\tau_{13}$$

A. E.	*American Ephemeris and Nautical Almanac* or (British) *Astronomical Ephemeris*
$\mathbf{A}(A_x, A_y, A_z)$	See "**L, A, D**"
$\mathbf{A}_h(A_{xh}, A_{yh}, A_{zh})$	See "**L, A$_h$, D$_h$**"
$\mathbf{A}$, $\mathbf{B}$	$\mathbf{A} = a\mathbf{P}$; $\mathbf{B} = b\mathbf{Q}$. Used in $\mathbf{r} = \mathbf{A}(\cos E - e) + \mathbf{B}\sin E$ Not recommended in place of $a\mathbf{P}$, $b\mathbf{Q}$ because of increasing emphasis on $\dot{\mathbf{r}}$.
a	Mean distance or semi major axis of ellipse, semi transverse axis of hyperbola.
a	Altitude (q.v.); prefer h
au, $a.u.$, or A.U.	Astronomical unit, approximately equal to semi major axis of Earth's orbit, but accurately defined by Gaussian $k = k_s$.
a_e	Earth's equatorial radius
$\mathbf{a}$, $\mathbf{b}$, $\mathbf{h}$	$\mathbf{a} = e\sqrt{\mu}\,\mathbf{P}$, $\mathbf{b} = e\sqrt{\mu p}\,\mathbf{Q}$, $\mathbf{h} = \sqrt{\mu p}\,\mathbf{W}$.
α (alpha)	Right ascension
α	Azimuth (q.v.); prefer A
α	Time required for light to travel unit distance; also τ_A:

$$\alpha_s = 0.005\ 7755\ \text{day/au}$$
$$\alpha_e = 0.021\ 275\ \text{sec/gu}$$

α	Flattening; prefer f (q.v.)

α	$= -1/a$; constant of integration in vis-viva integral.
α, β	ϵ, δ (q.v.) in Lambert-Euler equation (q.v.); astronomers have usually used ϵ, δ; mathematicians starting with Whittaker have tended to use α, β
α, β, γ	See "**L, A, D**"
adjoint method	Method for obtaining the partial differential coefficients used in differential correction, rendezvous, etc. Adjoint to "linearized Encke" or "variational orbit" method.
algorithm	Algorism, misspelled because of mistakenly assumed association with the root of *arithmetic;* correctly applied to arithmetic using concept of zero; incorrectly substituted for *collection of formulae* by some programmers.
altitude (h)	Angular distance above the horizon. Also called "angular altitude," "elevation," or "angular elevation" by those who have grown up in the traditions of aeronautics rather than astronomy or navigation.
altitude (H)	Linear height above sea level; prefer "height."
angular velocity (ω)	ω should be used preferably only for earth's rotation with respect to an inertially fixed direction. Differs slightly from the rate of change of sidereal time, because the latter is referred to a moving equinox.
angular velocity ($\dot{v}, \dot{u}, \dot{l}, \dot{\theta}$)	$\dot{v} = \dot{u} = \dot{l} = \dot{\theta} = \sqrt{\mu p}/r^2$ in two-body motion, and also in perturbation theory, provided the perturbations of the origins of v, u, l are absorbed in $v^{\backprime} = \dfrac{dv}{dt} - \dot{v}$, etc.; usually $\theta^{\backprime} = 0, \dfrac{d\theta}{dt} = \dot{\theta}$
anomaly	See "eccentric anomaly," "mean anomaly," "true anomaly"
apapsis, apoapsis, apex, apocenter	Apofocus (q.v.)
apofocal distance (r_A)	The distance from the focus to the apofocus: $a(1 + e) = r_A$
apofocus	Point in orbit farthest from focus of attraction; the farther apsis (q.v.). Also: apogee in geocentric orbit, aphelion in heliocentric orbit, etc., apocenter, apex; avoid the incorrectly formed and misleading term "apoapsis."
apsis	Point on orbit where radius vector is maximum or minimum (apofocus or perifocus). Line of apsides: line joining apofocus and perifocus. (Say ap'-si-deez!)
argument of latitude (u)	Angle measured in orbit plane from node to object; $u = v + \omega$
argument of perifocus (ω)	Angle measured in orbit plane from node to perifocus.
associated Legendre function	See P_n^m
azimuth (A)	Angle measured in horizon plane eastward from north point: also Z, α is unacceptable; also, in infrequent astronomical usage, measured from the south point.

B

B	See "A, B"
B	Argument in Barker's Tables; see "Barker's equation";

$$B = \frac{\sqrt{\mu}}{k_s}\frac{t - T}{q^{\frac{3}{2}}} = \frac{\sqrt{2}}{k_s}\left(\tan \tfrac{1}{2}v + \tfrac{1}{3}\tan^3 \tfrac{1}{2}v\right)$$

B'_i 　　　　　　Coefficients used in ratios of triangles (see "c_1, c_3")

$$B'_i = \tau_{13}^2 A_i(1 - A_i^2)/6$$

B_1, B_2, B_3 　　　Coefficients used in Gibbs's expressions for the ratios of triangles (see "c_1, c_3"):

$$B_1 = (\tau_{12}^2 + \tau_{12}\tau_{23} - \tau_{23}^2)/12$$
$$B_2 = (\tau_{12}^2 + 3\tau_{12}\tau_{23} + \tau_{23}^2)/12$$
$$B_3 = (-\tau_{12}^2 + \tau_{12}\tau_{23} + \tau_{23}^2)/12$$

B	See "**A, B**"
b	$= a\sqrt{1 - e^2}$; semi minor axis of ellipse.
b	Latitude referred to the orbital plane.
b	Heliocentric β (q.v.)
$\mathbf{b}(b_x, b_y, b_z)$	$= e\sqrt{\mu p}\ \mathbf{Q}$; see **a, b, h**.　Note:

$$|\mathbf{b}| = e\sqrt{\mu p} \ne b = a\sqrt{1 - e^2}$$

β (beta)	Latitude, celestial
Barker's equation	Equation relating time, t, and true anomaly, v, for parabolic orbits (see also B):

$$\frac{\sqrt{\mu}\,(t - T)}{q^{\frac{3}{2}}\sqrt{2}} = \tan \tfrac{1}{2}v + \tfrac{1}{3}\tan^3 \tfrac{1}{2}v$$

or (see D)

$$\sqrt{\mu}\,(t - T) = qD + \tfrac{1}{6}D^3, \quad D = \sqrt{2q}\tan \tfrac{1}{2}v$$

barycenter	Center of mass of two or more bodies.
burnout	See "initial, injection, insertion."

C

C	$= a_e/\sqrt{1 - e^2 \sin^2 \varphi}$, quantity tabulated in the almanacs for finding geocentric coordinates; see also S, x_c, y_c
C	$= rr_0[1 - \cos(v - v_0)]$;　also　$C_{ij} = r_i r_j[1 - \cos(v_j - v_i)]$; see also S, K, κ, σ
C	$= \cos E - e$; prefer next definition.
C	$= 1 - \cos E = (r - q)/ae$
$\tilde{C}$	$= a(1 - \cos E) = -a(\cosh F - 1)$, etc., unified variable.
$\hat{C}$	$= a[1 - \cos(E - E_0)] = -a[\cosh(F - F_0) - 1]$, etc., universal variable.
C_D	Shape factor for drag calculations
$C_n, C_{n,m}$ or $C_l, C_{l,m}$	Coefficients of zonal and sectoral harmonics in the gravitational potential of a spheroid, defined under "potential of spheroid."

$c = c \left[\dfrac{\text{km}}{\text{sec}}\right] = c \left[\dfrac{\text{km}}{\text{ls}}\right]$ Velocity of light $\left[\dfrac{\text{km}}{\text{sec}}\right]$; distance conversion factor

$$\left[\dfrac{\text{kilometers}}{\text{light second}}\right]$$

c $= 1 + \alpha r = 1 - \dfrac{r}{a} = e \cos E = e \cosh F$

c Distance from center of ellipse to focus; $c = ae$.

c Length of chord; see s

c_1, c_3 Ratios of the triangles: also n_1, n_3, or n_1/n_2, n_3/n_2.   May be series or closed expressions (see also "d_1, d_2, d_3"), including:

$$c_1 = \frac{\mathbf{r}_2 \times \mathbf{r}_3}{\mathbf{r}_1 \times \mathbf{r}_3} = \frac{r_2 r_3 \sin(v_3 - v_2)}{r_1 r_3 \sin(v_3 - v_1)}$$

$$c_3 = \frac{\mathbf{r}_1 \times \mathbf{r}_2}{\mathbf{r}_1 \times \mathbf{r}_3} = \frac{r_1 r_2 \sin(v_2 - v_1)}{r_1 r_3 \sin(v_3 - v_1)}$$

$$c_i = A_i + B_i'/r_2^3 \qquad i = 1, 3$$

$$c_i = A_i \frac{1 + B_i/r_i^3}{1 - R_9/r_2^3} \qquad i = 1, 3$$

The last equations are "Gibbs's expressions."

c_1^0, c_3^0 Ratios of the time intervals; see "A_1, A_3"

$\mathbf{c}$ $= \sqrt{\mu p}\, \mathbf{W}$, angular momentum vector; prefer $\mathbf{h}$ (see "$\mathbf{a}$, $\mathbf{b}$, $\mathbf{h}$")

χ (chi)   This symbol is often used in astrodynamics, but has no dominant use.

canonical time variable   See "τ"
canonical variables   Variables satisfying Hamilton's canonical equation (q.v., also "Delaunay variables").

climb angle (γ)   Flight-path angle
center of attraction   Focus of attraction (q.v.)
compression   Flattening (q.v.)
Cowell's method   Process in special perturbations in which the total acceleration of the orbiting body is numerically integrated to produce the position and velocity.   The term "Cowell's method" has been applied to the "second-sum" or "Gauss-Jackson" integration formula, but Cowell did not actually use that formula.   It is better to adopt the definition above, permitting the use of any integration formula with "Cowell's method," including the Runge-Kutta formulae.   The term "Cowell's method" is usually, though not necessarily, limited to integrations in rectangular coordinates.

D

D "Parabolic anomaly" analogous to the eccentric and hyperbolic anomalies (see "Barker's equation"):

$$r\dot{r}/\sqrt{\mu} = D = \sqrt{2q}\,\tan \tfrac{1}{2}v,\ r = q + \tfrac{1}{2}D^2$$

$D, \dot{D}, \ddot{D}$ $D = r\dot{r}/\sqrt{\mu} = \sqrt{a}\,e \sin E = \sqrt{-a}\,e \sinh F;$

$$\frac{\dot{D}}{\sqrt{\mu}} = \frac{1}{r} - \frac{1}{a};\ \ddot{D} = -\mu \frac{D}{r^3}$$

$\mathbf{D}(D_x, D_y, D_z)$	See "**L**, **A**, **D**"
$\mathbf{D}_h$	See "**L**, $\mathbf{A}_h$, $\mathbf{D}_h$"
d_1, d_2, d_3	Coefficients for $\dot{\mathbf{r}}_2 = -d_1\mathbf{r}_1 + d_2\mathbf{r}_2 + d_3\mathbf{r}_3$. They may be series or closed expressions (see "c_1, c_3").
Δ(delta)	Denotes an increment, residual, correction, or error: e.g., $\Delta\rho$, $\Delta\alpha$, $\Delta\delta$, Δa, Δe, Δk_s, $\Delta\mu$. Also δ; prefer δ for "virtual" increment and for numerical-analysis central-difference formulae (see "$\delta, \delta^2 \ldots$"); prefer Δ when used with α and δ (see introduction).
Δ	Topocentric distance of object; prefer $\rho = |\varrho|$; see ϱ and the introduction.
$\Delta, \Delta^2 \ldots$	Differences in numerical tables when subscript runs down a forewards diagonal
ΔL	$= \sqrt{(\Delta L_x)^2 + (\Delta L_y)^2 + (\Delta L_z)^2} = \sqrt{(\Delta\alpha \cos \delta)^2 = (\Delta\delta)^2}$ $= $ arc residual
ΔT or Δt	$= $ E.T. $-$ U.T.; time correction to eliminate irregularities of U.T., given in almanacs.
$\Delta X, \Delta Y, \Delta Z$	Corrections to solar coordinates to eliminate parallax, i.e., to translate from center of earth to point of observation.

$$\Delta X = \Delta_{xy} \cos \theta; \quad \Delta Y = \Delta_{xy} \sin \theta; \text{ and}$$
$$\Delta Z = -r_c \sin \varphi' = -x_c$$

Δ_{xy}	$\Delta_{xy} = -r_c \cos \varphi' = -y_c$; tabulated in almanacs with ΔZ, for active observatories.
δ (delta)	Declination
δ	$= h - \mathring{g}$, auxiliary quantity used in the Lambert-Euler equation (see: ϵ, $\mathring{g}$, h); also β (see "α, β")
δ	Denotes a "virtual" increment: e.g., a virtual displacement, $\delta\mathbf{r}$. See also "Δ," "virtual increment."
$\delta, \delta^2 \ldots$	Differences in numerical tables when subscript runs horizontally
$\nabla, \nabla^2 \ldots$ (del or nabla)	Differences in numerical tables when subscript runs up a backwards diagonal
declination (δ)	Topocentric or geocentric coordinate measured north ($+$) or south ($-$) from celestial equator along an hour circle. See "right ascension."
Delaunay variables:	

$$l = M \qquad\qquad L = \sqrt{\mu a}$$
$$g = \omega \qquad\qquad G = \sqrt{\mu p}$$
$$h = \Omega \qquad\qquad H = \sqrt{\mu p} \cos i$$

differential correction	A method of correcting the elements of an orbit by means of residuals (observation minus computation) sometimes called the Newton-Raphson process.
disturbing function	Perturbing function (q.v.)
dot-variation } dot-derivative } $\cdot$	$= \begin{cases} \text{two-body variation} \\ \text{two-body derivative} \end{cases}$

E

E	Eccentric anomaly (q.v.)
$E = E\left[\dfrac{gm}{m_\oplus}\right]$ or $E[gm]$	Mass conversion factor, earth masses to grams; mass of the Earth in grams.

E or E_{II}	Elliptic integral of the second kind
$\hat{E}$	$= 2\vartheta = E - E_0$
E.T., E.D.	Ephemeris Time, Ephemeris Date (q.v.)
$\mathbf{E}(E_x, E_y, E_z)$	See "**Z, E, N**"
e	Eccentricity (q.v.)
e	Earth's meridional eccentricity; $e^2 = 2f - f^2$ where f is the flattening. See f
e	Flattening; prefer f (q.v.)
e	Subscript designating the reference orbit in Encke's method; see o
ϵ (epsilon)	Obliquity of ecliptic
ϵ	Error; mean square error
ϵ	$= h + \vartheta$, auxiliary quantity used in the Lambert-Euler development (see δ, ϑ, h); also α (see "α, β").
ϵ	Flattening; prefer f (q.v.)
η (eta)	See "$\boldsymbol{\rho}(\xi, \eta, \zeta)$"
eccentric anomaly (E)	Angle at center of ellipse between major axis and radius of auxiliary circle through a point which has same abscissa as a given point on ellipse. In connection with an ellipsoid of revolution the eccentric anomaly is also known as the "parametric," "geometric," "reduced," or "eccentric" latitude. Also u; prefer E because of notation $u = v + \omega$
eccentricity (e)	Constant ratio between the distance from a point on a conic to the focus and the distance from the point to the directrix.
ellipticity	Flattening (q.v.)
elements of orbit	Selected set of six independent constants of integration of the equations of motion.
elevation (h)	Angular distance above the horizon; prefer "altitude" (q.v.)
Encke's method	Process in special perturbations by which the perturbed departures from the reference orbit, $\boldsymbol{\rho}(\xi, \eta, \zeta)$ and $\dot{\boldsymbol{\rho}}(\dot{\xi}, \dot{\eta}, \dot{\zeta})$, are integrated from $\ddot{\boldsymbol{\rho}}(\ddot{\xi}, \ddot{\eta}, \ddot{\zeta})$ and used to correct the reference position and velocity into the actual position and velocity. May be used to obtain partial differential coefficients, in either linearized or non-linear form.
Encke terms	The differences of the two-body terms in the equations for $\ddot{\boldsymbol{\rho}}(\ddot{\xi}, \ddot{\eta}, \ddot{\zeta})$ in Encke's method of special perturbations: e.g.,

$$\mu\left(\frac{x_e}{r_e{}^3} - \frac{x}{r^3}\right) = \frac{\mu}{r_e{}^3}\,(fqx - \xi)$$

	Also called "indirect terms," but then confused with the terms of the equations of motion that represent the accelerations of the origin.
ephemeris	A table of calculated positions of an object, with equidistant dates as arguments. Plural: ephemerides. (Say ef-eh-mer'-i-deez!)

ephemeris second	Unit of time associated with Ephemeris Time;

$$1 \text{ ephemeris second} = \frac{1 \text{ tropical year (1900.0)}}{31,556,925.9747}$$

Ephemeris Time (E.T.)	$= \text{U.T.} + \Delta T$, uniform time used as independent variable in the ephemerides of the planets and other objects in space. ΔT is listed in almanacs.
Ephemeris Date (E.D.)	$= \text{Julian Date} + \Delta T - 2400\,000.^d5$
epoch (t_0)	Arbitrary instant of time for which the constants of an orbit are specified.
Eulerian angles	Orientation angles (q.v.) used for 2000 years before Euler.

F

F	Hyperbolic anomaly
F	$= \frac{1}{2}(v + v_0)$ Also: $F_{ij} = \frac{1}{2}(v_j = v_i)$
F	The negative of the Hamiltonian function, also called the force function; in astronomy usually

$$F = \Phi - T = \frac{\mu}{2a} + R$$

F or E_I	Elliptic integral of the first kind
f, g	Series or closed expressions used in the formulae:

$$\mathbf{r} = f\mathbf{r}_0 + g\dot{\mathbf{r}}_0$$
$$\mathbf{r}_j = f_j\mathbf{r}_0 + g_j\dot{\mathbf{r}}_0 = f_{ij}\mathbf{r}_i + g_{ij}\dot{\mathbf{r}}_i$$

f	Flattening; also ϵ, e, α. Prefer f, with ϵ as second choice, because of confusion with closely associated eccentricity:

$$f = 1 - \frac{b}{a} \qquad e = \sqrt{1 - \left(\frac{b}{a}\right)^2}$$

f or $\dot{f}$	$= \frac{1}{2}(v - v_0)$. Prefer $\dot{f}$ because of close association with primary f, g above. Also: $\dot{f}_{ij} = \frac{1}{2}(v_j - v_i)$
f	Function (of q) used in Encke's method:

$$1 + 2q = r^2/r_e^2 \qquad fq = 1 - (1 + 2q)^{-\frac{3}{2}}$$

f	Frequency
f	True anomaly. Prefer v because of confusion with f's above.
flattening	See f; also called: oblateness, ellipticity, compression.
flight-path angle (γ)	Angle between velocity vector and plane perpendicular to radius vector; also "climb angle."
focus of attraction	Principal or dynamical focus, occupied by principal attracting mass: prefer to "center of attraction."
focus, empty	Secondary focus of conic, unoccupied.

G

G	Constant of gravitation in cgs or mks system. In astrodynamics the kilometer is preferred to the meter or the centimeter; accordingly the units of G are km^3/gm sec^2. Prefer k^2 for other systems.
G	$= \frac{1}{2}(E + E_0)$. Also: $G_{ij} = \frac{1}{2}(E_j + E_i)$
G	Velocity; see "V, v"

G	$= \sqrt{\mu p}$; prefer $\sqrt{\mu p}$ or h, except when using Delaunay variables (q.v.)
G.C.T., G.M.T.	Greenwich Civil Time, Greenwich Mean Time (q.v.)
GHA	Greenwich hour angle; also t_G; see "hour angle."
g	See "f, g"
g, g_e	Terrestrial acceleration of gravity; g_e — equatorial value.
g	Acceleration in general; prefer to a; also A.
g or $\mathring{g}$	$= \frac{1}{2}(E - E_0)$. Prefer $\mathring{g}$ to avoid confusion with primary f, g above. Also: $\mathring{g}_{ij} = \frac{1}{2}(E_j - E_0)$.
g	Argument of perifocus; prefer ω except when using "Delaunay variables" (q.v.)
gu	Geo-unit, or geocentric unit, nearly equal to the Earth's equatorial radius, a_e. Analogous to the astronomical unit of the heliocentric system.
γ (gamma)	Flight-path angle (q.v.)
γ	$= \sin i$ or $\sin \frac{1}{2} i$ in lunar and planetary theory
γ	Normal acceleration of gravity
Gauss-Jackson formula	Formula for numerically integrating second-order differential equations; also: second-sum formula, Σ^2 formula; see "Cowell's method."
geoid	The mean sea level surface of the earth. See "spheroid."
Gibb's expressions	See "c_1, c_3"
grave-variation ⎫ grave-derivative⎭	$= \begin{cases} \text{perturbed variation} \\ \text{perturbed derivative} \end{cases}$
Greenwich Civil Time (G.C.T.)	$=$ Universal Time (U.T.)
Greenwich Mean Time (G.M.T.)	$=$ U.T. $+ 12^h$ before 1925.0 $=$ U.T. after 1925.0

H

H or h	Height above sea level, or height above the geoid; see "altitude"; prefer H.
H	$= \sqrt{\mu p} \cos i$; see "Delaunay variables."
H	Hamiltonian function
H or HA or h	Hour angle (q.v.)
h	Altitude (q.v.) or "elevation"
h	$= \sqrt{\mu p}$; angular momentum (see "$\mathbf{a}, \mathbf{b}, \mathbf{h}$"); also G, as one of the Delaunay variables (q.v.)
h	Longitude of ascending node; prefer Ω except when using Delaunay variables (q.v.)
h	Interval of argument in integration tables or in any interpolation table. Also: w.
h	Auxiliary quantity used in the Lambert-Euler formulae (see also $G, \mathring{g}, \epsilon, \delta$):

$$\cos h = e \cos G$$

$\mathbf{h}(h_x, h_y, h_z)$	$= \sqrt{\mu p}\, \mathbf{W}$; angular momentum vector.
Hamiltonian function (H)	$H = \sum_s p_s \dot{q}_s - L$; also $-H = F$ (q.v.)
Hamilton's canonical equations	$\dot{q}_s = \dfrac{\partial H}{\partial p_s} \qquad \dot{p}_s = -\dfrac{\partial H}{\partial q_s}$

harmonics	Mathematical terms that indicate differences in the surface shape between a reference spheroid (e.g., for the Earth) and a sphere. "Zonal" harmonics indicate latitudinal differences; "sectoral" harmonics indicate longitudinal differences; "tesseral" harmonics include zonal and sectoral harmonics.
height (H)	Linear height above sea level; prefer to "altitude" (q.v.)
Hohmann orbit	Minimum energy heliocentric orbit tangent to the earth's orbit at one apsis and to that of the target planet at the other.
hour angle (t)	Angle measured in equator plane westward from local meridian, usually 0^h to 24^h; prefer (in order): t, HA, LHA, H; see "GHA"; the use of h for hour angle rather than angular altitude is undisciplined!
hour circles	Great circles that pass through the celestial pole, sometimes incorrectly referred to as "meridians."
hyperbolic anomaly (F)	Angle in hyperbola corresponding to E in the ellipse and D in the parabola: $F = iE$ where $i = \sqrt{-1}$, and $M = e \sinh F - F$. Also H.

I

I, J, K	$\mathbf{I}(I_x, I_y, I_z)$ is the unit vector along the x-axis in the reference plane (directed to ♈ in equatorial and ecliptic references). $\mathbf{J}$ is directed along the y-axis, 90° east from $\mathbf{I}$ in the reference plane. $\mathbf{K}$ forms a right-handed orthogonal set with $\mathbf{I}$, $\mathbf{J}$. Also: $\mathbf{i}$, $\mathbf{j}$, $\mathbf{k}$; prefer $\mathbf{I}$, $\mathbf{J}$, $\mathbf{K}$ in association with $\mathbf{P}$, $\mathbf{Q}$, $\mathbf{W}$ or $\mathbf{U}$, $\mathbf{V}$, $\mathbf{W}$, etc.
i	$= \sqrt{-1}$
i, I, ι (iota) inclination	i is standard; I and ι are convenient when it has a perturbed variation; angle measured from the reference plane (ecliptic or equatorial) to the orbit plane. Limits of i: 0° and 180°; or 0° and 90° if "direct" and "retrograde" are specified.
inertial axes	Axes which are fixed in direction, but which may have a moving origin (e.g., Sun or Earth) rather than an inertial one (e.g., the center of mass in the n-body problem).
indirect terms	The terms of the equations of motion (e.g., the relative-motion form) that represent the accelerations of the origin (e.g., the Sun); see "Encke terms"
initial injection insertion	These adjectives are sometimes used synonymously, along with burnout, launch, takeoff; writers should be careful to specify distinctions they wish to make.

J

J_n	Coefficients of zonal harmonics in the gravitational potential of a spheroid of revolution, defined under "potential of spheroid"
$J_{n,m}$	See "potential of spheroid"
J.D., J.E.D.	Julian Date, Julian Ephemeris Date (q.v.)
$\mathbf{J}(J_x, J_y, J_z)$	See "**I, J, K**"
jovicentric	Referred to the center of Jupiter; also: Jupiter-centered.
Julian Date (J.D.)	Universal Time (q.v.) expressed by a continuing day count.
Julian Ephemeris Date	Ephemeris Time (q.v.) similarly expressed:
(J.E.D.)	$$\text{J.E.D.} = \text{J.D.} + \Delta T$$

K

K	$= rr_0 \cos (v - v_0)$. See also: C, S, κ, σ.
$\check{K}$	$= \cos E = \cosh F$, etc., unified variable.
$\hat{K}$	$= \cos (E - E_0) = \cosh (F - F_0)$, etc., universal variable.
$\mathbf{K}(K_x, K_y, K_z)$	Unit vector perpendicular to reference plane; see "**I, J, K**"
k, k^2	Gravitational constant, its value depending on units of time, mass and distance involved. See: k_s, k_e.
k_e	Geocentric gravitational constant:

$$k_e = 0.001\ 239\ 444[\mathrm{gu}^{\frac{3}{2}}/\mathrm{m}_{\oplus}^{\frac{1}{2}}\ \mathrm{sec}]$$

k_s or k	Gaussian or heliocentric gravitational constant:

$$k_s = 0.017\ 202\ 09895[\mathrm{au}^{\frac{3}{2}}/\mathrm{m}_{\odot}^{\frac{1}{2}}\ \mathrm{day}]$$

κ (kappa)	Auxiliary quantity used in development of closed expressions for f and g, in Lambert-Euler equation, etc.:

$$\kappa = \pm \sqrt{2(rr_0 + xx_0 + yy_0 + zz_0)} = 2\sqrt{rr_0} \cos \tfrac{1}{2}(v - v_0)$$

See also: σ, C, K, S.

kaday	The unit of $\tau = k_*(t - t_0)$ for heliocentric orbits; when k_*^2 and $\mu = k_s^2$, it is k_s^{-1} days $= 58.132\ 44087$ days
kesec	The unit of $\tau = k_*(t - t_0)$ for geocentric orbits; it has also been called the "kemin" because k_e was first based upon the minute rather than the second; when k_*^2 and $\mu = k_e^2$, it is

$$k_e^{-1}\ \mathrm{sec}\ = 13.446\ 8896\ \mathrm{min}$$

Kepler's equation	$M = E - e \sin E$, and variants thereof.
Kepler's first law	$r = p/(1 + e \cos v) = a(1 - e \cos E)$
Kepler's second law	$r^2 \dot{v} = \sqrt{\mu p}$
Kepler's third law	$\mu P^2 = (2\pi)^2 a^3$ or $n^2 a^3 = \mu$

L

L, L_r	Mean longitude: $L = L_0 + n(t - t_0) = M + \omega + \Omega$; for retrograde orbits: $L_r = M + \omega - \Omega$. See "$l, l_r$" and "$\bar{\omega}, \bar{\omega}_r$"
L	Lagrangian function (q.v.)
L	$= \sqrt{\mu a}$; see "Delaunay variables."
L	Longitude, terrestrial (q.v.); prefer λ.
LHA	Local hour angle; also t; see "hour angle"
$\mathbf{L}, \mathbf{A}, \mathbf{D}$	$\mathbf{L}(L_x, L_y, L_z)$ is the unit vector directed from the observational reference point to the object. $\mathbf{A}$ is a unit vector perpendicular to $\mathbf{L}$, parallel to the equatorial plane, and directed with increasing right ascension, α. $\mathbf{D}$ is a unit vector such that $\mathbf{L}, \mathbf{A}, \mathbf{D}$ make up a right-handed orthogonal set, and so is directed with increasing declination, δ; so also $d\mathbf{L} = \mathbf{A} \cos \delta\, d\alpha + \mathbf{D} d\delta$:

$$
\begin{array}{lll}
L_x = \cos \delta \cos \alpha & A_x = -\sin \alpha & D_x = -\sin \delta \cos \alpha \\
L_y = \cos \delta \sin \alpha & A_y = \cos \alpha & D_y = -\sin \delta \sin \alpha \\
L_z = \sin \delta & A_z = 0 & D_z = \cos \delta
\end{array}
$$

Alt. to L_x, L_y, L_z: l, m, n; λ, μ, ν; α, β, γ; see introduction.

L, A$_h$, D$_h$

$\mathbf{L}(L_{xh}, L_{yh}, L_{zh})$ is the unit vector directed from the observational reference point to the object. $\mathbf{A}_h$ is a unit vector perpendicular to $\mathbf{L}$, parallel to the horizon plane, and directed with clockwise increasing azimuth, A. $\mathbf{D}_h$ is a unit vector such that $\mathbf{L}$, $\mathbf{A}_h$, $\mathbf{D}_h$ make up a left-handed orthogonal set, and so is directed with increasing altitude, h; so also $d\mathbf{L} = \mathbf{A}_h \cos h \, dA + \mathbf{D}_h dh$:

$$
\begin{array}{lll}
L_{xh} = \cos h \cos A & A_{xh} = -\sin A & D_{xh} = -\sin h \cos A \\
L_{yh} = \cos h \sin A & A_{yh} = +\cos A & D_{yh} = -\sin h \sin A \\
L_{zh} = \sin h & A_{zh} = 0 & D_{zh} = +\cos h
\end{array}
$$

See "**Z, E, N**"

l, *l$_r$*

True longitude: $l = v + \omega + \Omega$; for retrograde motion: $l_r = v + \omega - \Omega$. See "*L, L$_r$*" and "$\bar{\omega}, \bar{\omega}_r$"

l

Mean anomaly. Prefer M, except when using Delaunay variables (q.v.)

l, m, n

Direction cosines. For preferred alternate see "**L, A, D**"

Λ (lambda)

$= E + \omega + \Omega$, to accompany $l = v + \omega + \Omega$, $L = M + \omega + \Omega$, $\bar{\omega} = \omega + \Omega$

λ (lambda)

Longitude, celestial (q.v.)

λ

Longitude, terrestrial (q.v.); also L

λ, μ, ν

Direction cosines. For preferred alternate see "**L, A, D**"

Lagrange's equations

$\dfrac{d}{dt} \dfrac{\partial T}{\partial \dot{q}_s} - \dfrac{\partial T}{\partial q_s} = Q_s$, where T denotes the kinetic energy of the body. If the forces involved have a potential function, then the equations have the form:

$$
\frac{d}{dt} \frac{\partial L}{\partial \dot{q}_s} - \frac{\partial L}{\partial q_s} = 0
$$

where L is the Lagrangian function.

Lagrangian function (L)

Kinetic energy minus potential energy, usually. In astronomy $L = T + \Phi$; see "potential function (Φ)"

Lambert-Euler equation

In elliptic motion often $n(t - t_0) = (\epsilon - \sin \epsilon) - (\delta - \sin \delta)$; ϵ, δ also α, β (q.v.)

Lambert's equation

In orbit determination the dynamical equation $\rho = A - B/r^3$

latitude, celestial (β)

Angle measured north or south from the ecliptic along a secondary great circle perpendicular to it; see "longitude, celestial."

latitude, terrestrial

Angle measured in a meridian plane from the Earth's equator to, e.g., the point of observation; see "φ, φ_g, φ_a" and φ'

latitude, argument of (u)

$= v + \omega$; see $U = M + \omega$, $\Upsilon = E + \omega$

launch

See "initial, injection, insertion"

Legendre polynomials

See P_n.

light time

Interval during which light travels from the observed object to the observer. See α.

line of nodes

Line formed by the intersection of the reference plane and the orbit plane.

longitude, celestial (λ)

Angle measured from vernal equinox along the ecliptic to the secondary circle passing through the object, geocentric or topocentric. See "latitude, celestial."

longitude, terrestrial (λ)	Angle measured from Greenwich meridian along the equator, east or west, to the meridian through, e.g., the point of observation. For geodesy and perturbation theory λ should be taken as positive eastward, negative westward. Also: L.
longitude, mean (L)	$= M + \omega + \Omega$; see "L, L_r" and $\Lambda = E + \omega + \Omega$
longitude, true (l)	$= v + \omega + \Omega$; see "l, l_r"
longitude of perifocus $(\tilde{\omega})$	$= \omega + \Omega$; see "$\tilde{\omega}, \tilde{\omega}_r$"

M

M, M_0	Mean anomaly (q.v.), mean anomaly at the epoch
m_1 or M	Mass at focus of attraction; prefer m_1 because of mean anomaly
m_2 or m	Mass of orbiting body
m_i, m_j	Masses of particles, planets, etc.
$m_\odot, m_\oplus, m_\mathbb{C}$, etc.	Masses of Sun, Earth, Moon, etc. See astronomical symbols at end.
m_0	Initial mass of a vehicle, before takeoff
m, m_0	See "magnitude"
μ (mu)	Mass function. In the two-body problem, $\mu = k^2(m_1 + m_2)$. More generally, $\mu = k^2(1 + m)$, where m may include some perturbation effects as well as $m_1 + m_2 - 1$.
μ	Ratio of mass of Moon to mass of Earth
$\mu, 1 - \mu$	Masses of principal bodies in restricted 3-body problem; μ is usually the smaller of the two masses.
μ	Index of refraction
μ	Mean angular motion; prefer n (q.v.)
μ_a	Ratio of the mass of the atmosphere to the mass of Earth.
magnitude (m, m_0)	Measures of the brightness of an object, often calculated from

$$m = g + 5 \log \rho + 5 \log r + C\beta^0$$

where g is a constant determined from observed values, ρ and r are geocentric and heliocentric distances, β^0 is the phase angle (in degrees), and C is a constant; for m_0: $r = a$, $\rho = a - 1$, $\beta^0 = 0$ (a hypothetical "mean" opposition).

mean anomaly (M)	Angle measured from perifocus through which a body would have travelled if it were moving uniformly at its mean angular velocity, n:

$$M = M_0 + n(t - t_0) = n(t - T)$$

where M_0 is the value of M at epoch t_0 and T is the time of perifocal passage.   Also l; prefer M except when using Delaunay variables (q.v.)

mean distance (a)	Semi major axis (q.v.)
mean motion mean angular motion mean angular velocity $\Big\}$	See "n"
meridian, celestial	The hour circle that passes through the zenith.

N

n	Mean angular motion of a body in orbit; see "mean anomaly"; also sometimes μ, but prefer n because of the notation:

$$n = \sqrt{\mu}\, a^{-\frac{3}{2}}$$

n_1, n_2, n_3	Ratios of the triangles (n_1,n_3), or quantities proportional to the triangles (n_1,n_2,n_3); prefer c_1, c_3 (q.v.)
ν (nu)	True anomaly; prefer v; the rather curious use of ν may come from misreading of classical French works printed from type-fonts in which there are two italic v's. One is used at the beginnings of words; the other, used internally and in equations, closely resembles the Greek ν (e.g., Tisserand, v. 1, pp. 3, 25, 26).
n-body problem	Orbit problem involving only the mutual gravitational attractions of n bodies.
node	Intersection of orbit with a reference plane.
node (Ω)	Angle measured in reference plane from the vernal equinox to the ascending node of the orbit; "node" is an abridgement of "longitude of the ascending node"; Ω_ϵ may be used for the angle in the ecliptic plane, and Ω_α for that in the equatorial plane. Note that the node symbol (Ω) is not the same as the capital omega (Ω), though the latter may be used when the type-font lacks Ω.
north celestial pole (P_n)	Point where the Earth's axis of rotation extended from the northern hemisphere intersects the celestial sphere.
north point (N)	Origin for azimuth in horizon system of coordinates, defined by an intersection of horizon with the celestial meridian, a great circle through the observer's zenith and the north celestial pole.
nutation	The periodic terms of the precession-nutation complex that describes the motion of the Earth's axis of rotation.

O

o	Subscript designating a quantity or value associated with the epoch, t_0.
o	Subscript sometimes used to designate a quantity or value associated with the Encke or other reference orbit; prefer subscript e or R to avoid confusion with the preceding use of subscript o.
Ω (omega)	See "node (Ω)"
Ω	Perturbing function; prefer R
ω (omega)	Argument of perifocus
ω	Angular velocity
$\tilde{\omega}, \tilde{\omega}_r$	Longitude of perifocus: $\tilde{\omega} = \omega + \Omega$; for retrograde orbits $\tilde{\omega}_r = \omega - \Omega$; also π, π_r (see L, L_r). If you want to be intelligible, call $\tilde{\omega}$ "omega-tilde," not "πi."
oblateness	See f and "flattening"; some would prefer to use "oblateness" for J_2, and restrict it thereto.
obliquity of the ecliptic (ϵ)	Inclination of the ecliptic plane to the equatorial plane. Its true value for a given date (varying slightly from $23\frac{1}{2}°$) may be found with the solar coordinates in the national ephemerides. Its mean value for a given date may be found in *Planetary Coordinates*.

orbit	A path in space; more specifically, a set of constants defining it; the set can include physical constants as well as the six elements or constants of integration; also "state vector" or "state."
orbit	Period, or revolution; this recent introduction is expressive in popular discussions or news reporting, but should be avoided, in favor of the standard terms, in precise scientific writing.
orientation angles	The elements i, Ω, ω; see "Eulerian angles"
osculating orbit	An approximation to the actual path of an object derived from the actual position and velocity at a given moment; usually a two-body orbit.
osculating variation $\rangle$ osculating derivative$\big\}$	$= \begin{Bmatrix}\text{two-body variation}\\ \text{two-body derivative}\end{Bmatrix}$ (q.v.), but are more general terms that can be used whether or not the reference solution is a two-body orbit.

P

P	Period of revolution
P_n	Legendre polynomial, used in the development of the potential function in series form:

$$P_0(x) = 1 \qquad\qquad P_1(x) = x$$
$$P_2(x) = \tfrac{1}{2}(3x^2 - 1) \qquad P_3(x) = \tfrac{1}{2}(5x^3 - 3x)$$
$$P_4(x) = \tfrac{1}{8}(35x^4 - 30x^2 + 3)$$
$$P_n(x) = \frac{2n - 1}{n}\, xP_{n-1}(x) - \frac{n - 1}{n}\, P_{n-2}(x)$$

P_n^m or P_{nm}	Associated Legendre functions defined by

$$P_n^m(x) = (1 - x^2)^{\frac{1}{2}m}\frac{d^m}{dx^m}\, P_n(x)$$

P, Q, W	$\mathbf{P}(P_x, P_y, P_z)$ is the unit vector directed toward the perifocus. $\mathbf{Q}$ is directed 90° from $\mathbf{P}$ in the direction of motion of the object in the orbital plane. $\mathbf{W}$ is perpendicular to the orbital plane, thus forming a right-handed orthogonal set with $\mathbf{P}$ and $\mathbf{Q}$.
p, p_i	See "parameter"
p_s	Component of generalized momentum: $p_s = \dfrac{\partial L}{\partial \dot{q}_s}$
p_α	Annual precession in right ascension:

$$p_\alpha = m^s + n^s \sin\alpha \tan\delta$$

p_δ	Annual precession in declination:

$$p_\delta = n'' \cos\alpha$$

Φ (phi)	Potential function (q.v.)
φ, φ_g, φ_a (phi)	φ is used for geodetic latitude (φ_g), which is the angle between the normal to a reference spheroid and the plane of the equator. φ is also used for astronomical latitude (φ_a), which is the angle between the direction of gravity and the plane of the equator, and which differs from the geodetic latitude by "station error" or "local anomaly." φ_g is asso-

	ciated with the reference spheroid; φ_a, with the geoid. Both φ_g and φ_a are sometimes called, indiscriminately, "geographic latitude."
φ'	Geocentric latitude: angle between a radius from the center of reference spheroid and the plane of its equator.
φ	Angle of eccentricity: $\sin \varphi = e$, $\cos \varphi = \sqrt{1 - e^2}$ Convenient in logarithmic calculation, but now exists only as a confusing holdover. Used as argument in some tables.
$\Pi_\odot$ (pi)	Solar parallax, i.e., Sun's mean geocentric equatorial horizontal parallax; also $\pi_\odot$
$\Pi_{\mathbb{C}}$	Moon's varying geocentric equatorial horizontal parallax; also $\pi_{\mathbb{C}}$
π, π_r (pi)	Longitude of perifocus: $\pi = \omega + \Omega$; for retrograde motion, $\pi_r = \omega - \Omega$. Prefer $\bar{\omega}$, $\bar{\omega}_r$ (q.v.)
ψ (psi)	Elongation; angle at the observational reference point between the directions of the dynamical center and the observed object, i.e., between the vectors $\mathbf{R}$ and $\mathbf{L}$
parallax	Geocentric equatorial horizontal parallax is the angle at an object subtended by the earth's equatorial radius and is a measure of distance of objects in the solar system. Heliocentric parallax is the angle at an object subtended by the radius of the earth's orbit and is a measure of distance of objects outside the solar system: Π, π
parameter	Not generally used in its broad mathematical senses, but specialized in astrodynamics to: (1) The semi latus rectum of a conic; usually designated by p (2) A selected two-body element or variable (e.g., a or M), or a constant of integration in an analytic solution of some problem, used as a variable in the method of variation of parameters; sometimes designated by p_i
periapsis pericenter	Perifocus (q.v.)
perifocal distance (q)	$= a(1 - e)$, distance between focus and perifocus
perifocus	Point in orbit nearest focus of attraction; the nearer apsis; perigee in geocentric orbit, perihelion in heliocentric orbit, etc.; also "pericenter;" "periapsis" should be avoided. See "apofocus"
perifocus, argument of (ω)	Angle measured in orbit plane from node to perifocus.
perifocus, longitude of ($\bar{\omega}$)	$\bar{\omega} = \omega + \Omega$; for retrograde motion, $\bar{\omega}_r = \omega - \Omega$; also π, π_r.
perturbation (1)	Deviation from a reference orbit in force, acceleration, velocity, position, etc., because of effects not included therein. "General perturbations" involve series expansion before integration. "Special perturbations" are integrated numerically. "Secular perturbations" vary directly with time.
perturbation (2)	The engineer's term for an arbitrary variation of initial conditions, now sanctioned by time; prefer "variant" whenever there is a possibility of confusion between the two usages.

perturbed differentiation	In the method of variation of parameters, a technique (alternative to that of Hamiltonian mechanics) in which the equations for the variations are obtained by differentiations in which the "two-body derivatives" are omitted. See "perturbed variation," "two-body variation"
perturbed variation perturbed derivative perturbative	In the method of variation of parameters, that part of the total derivative or variation that is added by the perturbations: $\dfrac{df}{dt} = \dot{f} + f$, where f is the perturbed part and $\dot{f}$ is the two-body part of the total derivative. Also called the "grave-derivative" (or "grave-variation"). See "perturbed differentiation," "two-body variation"
perturbing function (R)	That portion of the potential function (Φ) which takes into account perturbations caused by conservative forces, e.g., $\Phi = \dfrac{\mu}{r} + R$, where $\dfrac{\mu}{r}$ denotes the two-body portion of Φ. Also: Ω; disturbing function. See "potential function (Φ)."
planetary aberration	Discrepancy between calculation and observation that results when "light time" is neglected.
precession	The secular term of the motion of the Earth's axis of rotation with respect to inertial axes.
potential function (Φ) or "potential"	Function proportional to the potential energy, usually negatively proportional in astronomical and astrodynamical practice; also U or V; e.g., two-body potential: $\Phi = \mu/r$.
potential of spheroid	Potential function associated with, e.g.: (a) spheroid of revolution:

$$\Phi = \frac{\mu}{r}\left[1 - \sum_{n=1}^{\infty} J_n \left(\frac{a_e}{r}\right)^n P_n\,(\sin\delta) \right]$$

(b) general spheroid:

$$\Phi = \frac{\mu}{r}\left\{ 1 + \sum_{n=1}^{\infty}\sum_{m=0}^{n} \left(\frac{a_e}{r}\right)^n P_n^m\,(\sin\delta)[C_{n,m}\cos m\lambda \\ + S_{n,m}\sin m\lambda] \right\}$$

where a_e is the equatorial radius, δ is the geocentric (or planetocentric) declination or latitude, and the attracted object has negligible mass. Also: $-J_n = C_{n,0} = C_n; C_{n,m}^2 + S_{n,m}^2 = J_{n,m}^2; \delta = \varphi = \beta; \Phi = U = V; \mu = k^2m; n,m = l,m.$ See P_n, P_n^m and *IAU Trans.*, **11B** (1961), 173-174. Form (b) is often replaced by one of several forms in which the coefficients are fully normalized.

Q

Q_s	Generalized force: $Q_s = \ddot{\mathbf{r}} \cdot \dfrac{\partial \mathbf{r}}{\partial q_s}$
$\mathbf{Q}(Q_x, Q_y, Q_z)$	See "**P, Q, W**"

$\tilde{Q}$	$= \sqrt{p}\,\mathbf{Q}$, a vector that is determinate for rectilinear orbits, for which $\mathbf{Q}$ is not.
$q,\ q_2$	Perifocal distance: $q = a(1 - e)$. Apofocal distance: $q_2 = a(1 + e) = r_A$; prefer r_A.
q	Auxiliary quantity used in Encke's method in special perturbations:

$$ q = \xi(x_e + \tfrac{1}{2}\xi) + \eta(y_e + \tfrac{1}{2}\eta) + \zeta(z_e + \tfrac{1}{2}\zeta)/r^2 e $$

q_i	Generalized coordinate, or angle or position variable in a canonical set (see "Hamilton's canonical equations")

R

R	Magnitude of $\mathbf{R}$; distance from point of observation to focus of attraction.
R	Geocentric distance; prefer r_c (q.v.)
R	Perturbing function (or "disturbing function")
$R = R\left[\dfrac{km}{gu}\right] = R[km]$	Length conversion factor, geo-units to kilometers; gu expressed in km.
$\mathbf{R}(X,Y,Z)$	Position vector directed from the point of observation to the focus of attraction.
$\mathbf{R}(R_x,R_y,R_z)$	See "$\mathbf{W}(W_x,W_y,W_z)$," which should be preferred because of conflict with $\mathbf{R}(X,Y,Z)$; see introduction also.
$\tilde{\mathbf{R}}(\tilde{X},\tilde{Y},\tilde{Z})$	Vector from the focus of attraction to the point of observation, used in place of $-\mathbf{R}$ when the conventional $\mathbf{R}$ orientation seems inappropriate.
r,r_i,r_j	Distance of object from focus of attraction; see "radius vector."
r_2	Distance of the object from the secondary focus of a conic.
r_A	$= a(1 + e)$; apofocal distance
r_c	$= \sqrt{x_c^2 + y_c^2}$; distance from center of Earth to a point at or near the surface; also ρ, R; prefer r_c; see "$x_c,\ y_c$"
r_{ij}	Distance from m_i to m_j.
$\mathbf{r}(x,y,z)$	Radius vector, directed from the focus of attraction to the object whose motion is under study.
ρ (rho)	Topocentric distance or "range" of any object: $\rho = \|\boldsymbol{\rho}\|$; also Δ (see introduction).
$\boldsymbol{\rho}(\xi,\eta,\zeta)$	Topocentric vector or similar vector directed from the point of observation to the object whose motion is under study, e.g., $\xi = x + X$
$\boldsymbol{\rho}(\xi,\eta,\zeta)$	Difference in position between the actual orbit and a reference orbit in Encke's method of special perturbations, e.g., $\xi = x - x_e$
radius vector (r)	Established mathematical term for $r = p/(1 + e \cos v)$; not a vector in the modern usage of the term; plurals: radius vectors, radii vectores.
range (ρ)	See ρ, $\boldsymbol{\rho}$
ratios of the triangles	In the orbit methods of Gauss, Olbers, et al., the ratios of the triangles formed by the radii and chords are calculated from the corresponding time intervals, with greater or less approximation; see "$c_1,\ c_3$"

ratios of sectors or time intervals	See "A_1, A_3"
ratio of sector to triangle ($\bar{y}$)	$\bar{y} = \sqrt{\mu p}\ (t-t_0)/S$
rectification	The process of shifting to a new (usually osculating) reference orbit in Encke's method when the Encke terms associated with the old reference orbit become too large.
rectilinear orbit	Orbit for which $e = 1$, $p = 0$, $q = 0$.
reference orbit	An orbit, often a fixed two-body ellipse, to which perturbative effects are added to obtain the perturbed orbit; see "Encke's method," ρ
representation	Calculation of position of a body for a given date (usually the date of an observation) from a given set of elements. Usually compared with an observation to determine validity of the elements, i.e., whether a differential correction is required.
residuals	Differences between given data (O) that are observed, measured, or otherwise determined, and the corresponding "represented" data (C) that are calculated from conjectured values of the parameters. See "differential correction" and "representation."
right ascension (α)	Topocentric or geocentric coordinate measured along celestial equator eastward, usually 0^h to 24^h, from vernal equinox to hour circle through object. As a subscript α may be used to indicate equatorial coordinates. See "declination (δ)"
Runge-Kutta formulae	Formulae for the numerical integration of differential equations.

S

S	$= C(1 - e^2) = a_e(1 - e^2)/\sqrt{1 - e^2 \sin^2 \varphi}$, quantity tabulated in the almanacs for finding geocentric coordinates; see also C, x_c, y_c
S	$= rr_0 \sin (v - v_0)$, twice the area of the triangle; also $S_{ij} = r_i r_j \sin (v_j - v_i)$; see "$c_1$, c_3," "ratios of the triangles," "ratio of sector to triangle"; see also C, K, κ, σ
S	$= \sqrt{1 - e^2} \sin E$; prefer next definition.
S	$= \sin E = D/e \sqrt{a}$
$\tilde{S}$	$= \sqrt{a} \sin E = \sqrt{-a} \sinh F$, unified variable.
$\hat{S}$	$= \sqrt{a} \sin (E - E_0) = \sqrt{-a} \sinh (F - F_0)$, universal variable.
$S = S\left[\dfrac{gm}{m_\odot}\right]$ or $S[gm]$	Mass conversion factor, solar masses to grams; mass of Sun in grams.
S	Determining function of a canonical transformation
$S_{n,m}$ or $S_{l,m}$	Coefficient of a sectoral harmonic; see "potential of spheroid"
S, T, W	$\mathbf{S}(S_x, S_y, S_z)$ is the tangential unit vector: $\mathbf{S} = \dot{\mathbf{r}}/\dot{s}$. The normal unit vector $\mathbf{T}$ is also the orbital plane, 90° from $\mathbf{S}$ in the same sense as the direction of motion. $\mathbf{W}$ forms a right-handed orthogonal set with $\mathbf{S}$, $\mathbf{T}$.

$\dot{s}$	Magnitude of the velocity vector, $\dot{\mathbf{r}}$:

$$\dot{s}^2 = \dot{x}^2 + \dot{y}^2 + \dot{z}^2 = \dot{\mathbf{r}} \cdot \dot{\mathbf{r}}$$

	See s and "vis-viva integral"; also see "V, v"
s	Length of arc between two positions on an orbit.
s	Length of the chord between two positions $\mathbf{r}_0$ and $\mathbf{r}$ on an orbit; also c:

$$s^2 = (x - x_0)^2 + (y - y_0)^2 + (z - z_0)^2$$

Σ, Σ^2 (sigma)	Sums in numerical integration tables and formulae; also

$$\Sigma^2 \ddot{x}_i = \Delta^{-2} \ddot{x}_{i+1} = \nabla^{-2} \ddot{x}_{i-1}, \text{ etc.}$$

Σ^2 formula	See "second-sum formula."
σ (sigma)	$= 2\sqrt{rr_0} \sin\frac{1}{2}(v - v_0)$; also $\sigma_{ij} = 2\sqrt{r_i r_j} \sin\frac{1}{2}(v_j - v_i)$; used in development of closed expressions for f and g, in Lambert-Euler equation, etc., see also κ, C, K, S
σ	Density
σ	Standard deviation
second-difference formula	Formula for numerical integration of second-order differential equations; also "δ^2 formula," "Adams-Moulton formula"
second-sum formula	Formula for numerical integration of second-order differential equations; also "Gauss-Jackson formula," "Σ^2 formula"
sectoral harmonics	See "harmonics"
selenocentric	Having the center of the moon as focus of attraction or origin of coordinates; prefer "moon-centered"
semi latus rectum (p)	$= a(1 - e^2)$, one half of the chord passing through a focus and perpendicular to the major axis; also "semi-parameter" or "parameter" (q.v.)
semi major axis (a)	One-half the distance from the perifocus to the apofocus of an ellipse; also "mean distance"
sidereal time (θ)	The hour angle of the vernal equinox; the right ascension of the celestial meridian.
solar parallax $(\Pi_\odot)$	The angle that (in radians) is equal to the ratio of equatorial radius to the astronomical unit.
spheroid	A symmetrical mathematical approximation to the geoid: e.g., an ellipsoid of revolution.

T

T	Time of perifocal passage
T	Period; prefer P because of preceding well-established usage
T	Time measured in Julian centuries from some epoch such as 1900.0
T	Kinetic energy, or K. E. per unit mass in many astrodynamic applications
$\mathbf{T}(T_x, T_y, T_z)$	See "$\mathbf{S}, \mathbf{T}, \mathbf{W}$"
t	Time; true time (i.e., the time the light left the object). Distinguished from "observed time" by $t = t_{\text{obs}} - \alpha\rho$.
t	Hour angle or local hour angle; also H or HA or LHA

t_0 — The epoch (q.v.)

t_{obs} — Time of observation; see t (true time)

τ (tau) — Canonical or normalized time variable: $\tau = k_*(t - t_0)$, or $\tau_j = k_*(t_j - t_0)$, where k_* may have the numerical value of $\sqrt{\mu}$, but not its dimension or units. Alternative notations are:

(a) $\tau_{ij} = k_*(t_j - t_i)$ (the "unambiguous convention")

(b) $\left. \begin{cases} \tau_1 = k_*(t_1 - t_2) \\ \tau_3 = k_*(t_3 - t_2) \end{cases} \right\}$ (the "Laplacian convention")

(c) $\left. \begin{cases} \tau_1 = k_*(t_3 - t_2) \\ \tau_2 = k_*(t_3 - t_1) \\ \tau_3 = k_*(t_2 - t_1) \end{cases} \right\}$ (the "Gaussian convention")

The unambiguous convention is to be preferred when there might be confusion between the other two.

τ_A — Time required for light to travel one astronomical unit; also α_s, α (q.v.)

θ (theta) — Sidereal time

θ_g — Greenwich sidereal time

θ — Angle measured in the orbital plane to the radius vector from an origin that is not subject to longitudinal perturbed variation

θ — Flight-path angle (q.v.); prefer γ

takeoff — See "initial, injection, insertion"

tesseral harmonics — See "harmonics"

topocentric — Referred to the point of observation

true anomaly (v) — Angle measured at the focus from the perifocal vector **P** to the radius vector **r**; also f, but prefer v (see "f")

triaxial ellipsoid — Ellipsoidal figure having three orthogonal axes of symmetry

two-body variation $\Big\}$
two-body derivative — In the method of variation of parameters, that part of the total derivative or variation that is attributed to the object at the focus of attraction:

$\dfrac{df}{dt} = \dot{f} + f\grave{}$, where $\dot{f}$ is the two-body part and $f\grave{}$ is the perturbed part of the total derivative. Also called the "dot-variation" or "osculating variation" (or " . . . derivative"). When the parameters p_i are all constants of the two body problem (or osculating solution), we may recognize

$$\dot{f} = \frac{\partial f}{\partial t} \qquad f\grave{} = \sum_{i=1}^{6} \frac{\partial f}{\partial p_i} \frac{dp_i}{dt} \qquad \frac{dp_i}{dt} = p_i\grave{}$$

When the parameters include a two-body variable such as the mean anomaly M,

$$\dot{f} = \frac{\partial f}{\partial M} \dot{M} \qquad f\grave{} = \sum_{i=1}^{5} \frac{\partial f}{\partial p_i} p_i\grave{} + \frac{\partial f}{\partial M} \dot{M}\grave{} \qquad \dot{M} = n$$

U

U — $= M + \omega = U_0 + n(t - t_0)$, mean argument of latitude

U	$= E - \sin E$
$\bar{U}$	$= a^{\frac{3}{2}}(E - \sin E) = (-a)^{\frac{3}{2}}(\sinh F - F)$, etc., unified variable
$\hat{U}$	$= a^{\frac{3}{2}}(E - E_0) - \sin(E - E_0) = (-a)^{\frac{3}{2}}\sinh(F - F_0) -$ $(F - F_0)$, etc., universal variable.
U	Potential function (q.v.)
U.T.	Universal time (q.v.)
U, V, W	$\mathbf{U}(U_x, U_y, U_z)$ is the unit vector directed along the radius vector from the dynamical center to the object. **V** is directed 90° from **U** in the direction of motion of the object in the orbital plane. **W** is perpendicular to the orbital plane, thus forming a right-handed orthogonal set with **U, V**; see "**P, Q, W**"
u	$= v + \omega$, (true) argument of latitude
u	Eccentric anomaly; prefer E because of preceding well-established usage.
Υ (upsilon)	$= E + \omega$, to accompany $u = v + \omega$, $U = M + \omega$
Υ	See "vernal equinox (Υ)"
υ (upsilon)	Generally not used because of similarity to u
Universal Time (U.T.)	$=$ G.C.T. $=$ G.M.T. after 1925.0
	$=$ G.M.T. $- 12^h$ before 1925.0
	$=$ E.T. $- \Delta$T

V

V, v	$= \dot{s} = ds/dt$, velocity per unit of t; also "G" (for Geschwindigkeit); prefer $\dot{s}$, V (see introduction)
V_c	$= \sqrt{\mu/r}$, circular velocity
V_p	$= \sqrt{2\mu/r}$, parabolic velocity, or velocity of escape
V_∞	$= \sqrt{-\mu/a}$, velocity at infinity, imaginary for elliptic orbits
V	Potential function (q.v.)
$\mathbf{V}(V_x, V_y, V_z)$	See "**U, V, W**"
$\hat{\mathbf{V}}$	$= \sqrt{p}\,\mathbf{V}$, a vector that is determinate for rectilinear orbits, for which **V** is not.
v	True anomaly; also f (q.v.)
$\dot{v}$	Angular velocity (q.v.)
variant calculations	Method of calculating the partial differential coefficients that are used in differential correction formulae, rendezvous correction, etc.; see "perturbation (2)"
variation	Used in astrodynamics as the rate of change in the value of a parameter with respect to time. It is noted that this definition is unlike the general usage of the word, in which "variation" indicates a modification or deviation. See "perturbed variation" and "two-body variation"
variation of parameters	Perturbation technique in which the constants of the 2-body orbit, or of the analytical solution of an approximation to an actual problem, are allowed to vary.
vernal equinox (Υ)	That intersection of the ecliptic and celestial equator where the Sun crosses from south to north (approximately March 21). Note that the vernal equinox symbol Υ is not the capital upsilon (Υ), which should not be used for it unless the type-font lacks Υ; distinction between Υ and Υ is more stringent than that between Ω and Ω; see "node."

vertical circle	Great circle passing through the zenith.
virtual increment	An increment, as a "virtual displacement," $\delta\mathbf{r}$, that is not necessarily nor usually the same as a time-displacement along a dynamical path, e.g., as in the establishment of Lagrange's equations. Of the same character as a differential correction.
vis-viva integral	$\dot{s}^2 = \mu\left(\dfrac{2}{r} - \dfrac{1}{a}\right)$; also "energy integral"

W

$\mathbf{W}(W_x, W_y, W_z)$	Unit vector perpendicular to orbit plane; see "$\mathbf{P}$, $\mathbf{Q}$, $\mathbf{W}$"; "$\mathbf{U}$, $\mathbf{V}$, $\mathbf{W}$"; "$\mathbf{R}(R_x, R_y, R_z)$"; introduction.
w	Interval of argument in integration tables or in any interpolation table. Also: h

X, Y, Z

X, Y, Z	See "$\mathbf{R}(X,Y,Z)$"
X	$= E$ or F or D
$\tilde{X}$	$= \sqrt{a}\,E$ or $\sqrt{-a}\,F$ or D, unified variable
$\hat{X}$	$= \sqrt{a}\,(E - E_0)$ or $\sqrt{-a}\,(F - F_0)$ or $D - D_0$, universal variable
Z	Azimuth (q.v.)
$\mathbf{Z}, \mathbf{E}, \mathbf{N}$	Right-handed orthogonal set of unit vectors directed to zenith, east point, north point, useful in rotation from equatorial to horizon system: e.g.,

$$L_{xh} = \mathbf{L} \cdot \mathbf{N} \qquad L_{yh} = \mathbf{L} \cdot \mathbf{E} \qquad L_{zh} = \mathbf{L} \cdot \mathbf{Z}$$

where

$$\mathbf{Z} = \begin{Bmatrix} \cos\varphi\cos\theta \\ \cos\varphi\sin\theta \\ \sin\varphi \end{Bmatrix} \quad \mathbf{E} = \begin{Bmatrix} -\sin\theta \\ +\cos\theta \\ 0 \end{Bmatrix} \quad \mathbf{N} = \begin{Bmatrix} -\sin\varphi\cos\theta \\ -\sin\varphi\sin\theta \\ +\cos\varphi \end{Bmatrix}$$

See: "$\mathbf{L}$, $\mathbf{A}$, $\mathbf{D}$"; "$\mathbf{L}$, $\mathbf{A}_h$, $\mathbf{D}_h$"

x, y, z	See "$\mathbf{r}(x,y,z)$"
x_c, y_c	Coordinates referred to the geometrical center and major axis of an ellipse; geocentric coordinates in the plane of a meridian:

$$\left.\begin{matrix} x_c = r_c \cos\varphi' = C\cos\varphi \\ y_c = r_c \sin\varphi' = S\sin\varphi \end{matrix}\right\} \text{ See } C, S, \varphi, \varphi'$$

x_ω, y_ω	Coordinates referred to the focus and transverse axis of a conic: $x_\omega = r\cos v$, $y_\omega = r\sin v$
$\bar{y}$	$= \sqrt{\mu p}\,(t - t_0)/S$, ratio of sector to triangle
z	$= 90 - h$, zenith distance
ξ, η, ζ (xi, eta, zeta)	See "$\boldsymbol{\rho}(\xi,\eta,\zeta)$"
zenith	Intersection (upper) of direction of gravity and the celestial sphere
zonal harmonics	See "harmonics"

Astronomical symbols in common use in astrodynamics:

☉	Sun	♈	Vernal equinox (q.v.)
☿	Mercury	☊	Node (q.v.); also Ω
♀	Venus	⚹	Comet
⊕	Earth	☾	Moon
♂	Mars		
♃ or ♃	Jupiter		
♄	Saturn		
♅ or ♅	Uranus		
♆	Neptune		
♇	Pluto		

Special uses of mathematical symbols:

˙, ¨   Derivatives with respect to τ; restricted to "two-body variations" (q.v.) in method of variation of parameters.

ˋ, ˋˋ   "Perturbed variations" (q.v.) in method of variation of parameters.

∇   Del or nabla; listed under D

Acknowledgements for helpful suggestions and comments are due especially to J. D. Anderson, R. M. L. Baker, Jr., R. A. Broucke, C. J. Cohen, C. G. Jacchia, W. M. Kaula, D. G. King-Hele, J. D. Mulholland, C. Oesterwinter, D. A. Pierce, E. Rabe; M. S. W. Keesey was very helpful in an earlier stage of the compilation.

TABLES OF SOLID ANGLES

Compiled by Dr. A. Victor Masket

Abridged from U.S. Atomic Energy Commission Report TID-14975 (July, 1962), *Tables of Solid Angles*, by A. V. H. Masket and William C. Rodgers; composed at the Computation Center, University of North Carolina, Chapel Hill, North Carolina; abridgment prepared at General Atomic Division, General Dynamics Corporation, San Diego 12, California; reproduced with permission of the U.S. Atomic Energy Commission.

I. SOLID ANGLE SUBTENDED BY THE CIRCULAR DISC

Let the point in space at which the solid angle is subtended be 0, an origin of coordinate systems. The normal to the plane of the disc from 0 defines the z-axis and the normal to the symmetry axis of the disc defines the x-axis. *All lengths are considered normalized relative to the radius of the disc as a unit of length.* The solid angle is represented by the symbol $\Omega(1;\rho,z)$, where ρ is the distance (along the x-axis) from 0 to the symmetry axis and z is the distance (along the z-axis) from 0 to the plane of the disc. From the general expression for solid angles given by $\Omega = \iint \sin\theta\, d\theta\, d\phi$ in spherical polar coordinates, three contour integrals have been derived, which correspond to the cases with ρ less than, equal to, or greater than the radius of the disc as a unit of length.

Case 1. $\rho < 1$; the z-axis pierces the interior of the disc.

The contour integral employed for digital computation is

$$(1) \qquad \Omega(1;\rho,z) = 2\cdot\int_0^\pi f_1^2[1 + f_1^2 + (1 + f_1^2)^{1/2}]^{-1}\, d\phi \qquad (\rho < 1),$$

where $f_1 = S_1/z$, $S_1 = \rho\cos\phi + (1 - \rho^2\sin^2\phi)^{1/2}$. When $\rho = 0$, $\Omega(1;0,z) = 2\pi[1 - z/(1 + z^2)^{1/2}]$, exactly.

Case 2. $\rho = 1$; the z-axis touches the periphery of the disc.

$$(2) \qquad \Omega(1;1,z) = 2\cdot\int_0^{\pi/2} f_2^2[1 + f_2^2 + (1 + f_2^2)^{1/2}]^{-1}\, d\phi,$$

with $f_2 = (2\cos\phi)/z$. In elliptic integrals, the formula is given exactly by $\Omega(1;1,z) = \pi - kzK(k)$, where $k^2 = 4/(z^2 + 4)$ and $K(k) \equiv F(\pi/2,k)$ is the complete elliptic integral of the first kind.

Case 3. $\rho > 1$; the z-axis is external to the disc.

The computational form, where $S_2 = \rho\cos\phi - (1 - \rho^2\sin^2\phi)^{1/2}$, is

$$(3) \qquad \Omega(1;\rho,z) \atop (\rho > 1) = 2z\int_0^{\arcsin(1/\rho)} \frac{(S_1^2 - S_2^2)\, d\phi}{[(z^2 + S_1^2)(z^2 + S_2^2)]^{1/2}[(z^2 + S_1^2)^{1/2} + (z^2 + S_2^2)^{1/2}]}.$$

The solid angle subtended by the circular disc is expressed most generally in elliptic integrals. In conformity with the definitions adopted for ρ and z,

(4) $\Omega(1;\rho,z) = 2\pi - 2[k'K(k)\sin\alpha + K(k)\cdot E(k',\alpha) - F(k',\alpha)\{K(k) - E(k)\}],$

where $k^2 = 4\rho/[(1 + \rho)^2 + z^2]$, $(k')^2 = 1 - k^2$, $\alpha = (\pi - \phi)/2$, and $K(k)$ and $E(k)$ are the complete elliptic integrals, and $F(k',\alpha)$ and $E(k',\alpha)$ are the corresponding (incomplete) elliptic integrals of the first and second kind. The formula given by (4) is valid for all values of ρ and z; it is not convenient for computation. When $z > 100$, it is practical to use $\Omega \sim (\pi/d^2)\cdot\cos\gamma$ where d is the normalized distance from 0 to the center of the disc and γ is the projection angle. Accurate approximations are discussed in the references and literature.

II. SOLID ANGLE SUBTENDED BY THE RIGHT CIRCULAR CYLINDER

The tabulated values consist of the solid angles subtended only by the lateral surface at points in the plane of a circular flat (base) external to a cylinder of given radius and height. It will be shown that these values alone or taken in conjunction with the values of solid angles subtended by the circular disc are sufficient to cover the general case of any point in space external to the right circular cylinder. *All lengths are normalized with respect to the radius of the cylinder as a unit of length.* The variables of the table are h, the normalized height of the cylinder, and ρ, the normalized distance from the point (at which the solid angle is subtended) to the axis of the cylinder. With these definitions, the solid angle is represented by $\Omega(1,h;\rho)$.

Let the point in the plane of a flat be 0, the origin of coordinate systems. The z-axis is the normal to the plane and the x-axis is the line in the plane from 0 to the axis of the cylinder. The solid angle subtended at 0 by the lateral surface of the cylinder is given by the contour integral.

(5) $\Omega(1,h;\rho) = 2h\displaystyle\int_0^{\arcsin(1/\rho)} [h^2 + S_2^2]^{-1/2}\,d\phi$ $[\rho \geqq 1]$,

where ϕ is the azimuthal angle of the associated spherical polar system.

If a point is not located in a plane determined by a flat of the cylinder, then it is either interior or exterior to the space between the planes of the flats. In the first case, a plane through the point normal to the cylindrical axis determines two smaller cylinders and two solid angles in accordance with (5) which are added. For the second case, where a flat is viewed from 0 in addition to the lateral surface, let the normalized height of the cylinder be h and the normalized distance to the plane of the viewed flat be H. Then the total solid angle is

(6) $\Omega = \Omega(1,h + H;\rho) - \Omega(1,H;\rho) + \Omega(1;\rho,H),$

where ρ is the normalized distance from 0 to the axis of the cylinder. Approximations are given in the references.

III. USE OF TABLES

A. Reading the Entry Numbers (steradians)

Each entry is given to six significant figures followed by an exponent for the power of 10 as a multiplier. For example,

$$3.73950\text{--}3 \equiv 3.73950 \times 10^{-3} \equiv 0.00373950;$$
$$5.65798\text{--}0 \equiv 5.65798 \times 10^0 \equiv 5.65798.$$

B. Solid Angle Subtended by a Disc

To find the solid angle subtended by an end-window counter 3 cm in radius at a point 15 cm above its plane and 18 cm from its axis, we have

$$\Omega(1;\rho,z) = \Omega(1;18/3,15/3) = \Omega(1;6,5) = 0.0331611.$$

C. Solid Angle Subtended by the Lateral Surface of a Right Circular Cylinder

To find the solid angle subtended by a Geiger counter 3 cm in radius and 30 cm long at a point in the plane of a flat at 25.5 cm from the axis, we have

$$\Omega(1,h;\rho) = \Omega(1,30/3;25.5/3) = \Omega(1,10;8.5) = 0.186909.$$

D. Fraction of Radiation Entering a Circular Aperture from a Distributed Plane Source

Consider the problem of estimating the fraction of radiation entering a circular end-window counter from a thin disc source whose plane is parallel to the window. Let b be the normalized radius of the source relative to the counter, let s be the normalized distance between the (parallel) axes of the source and window, and let z be the normalized distance between the planes of the window and sou.ce.

Case 1. $s = 0$: coaxial.

Divide the source into N concentric annuli of width $\Delta\rho_i$, $i = 1,2,3, \ldots ,N$. Each point in a given annulus has subtended approximately the same solid angle $\Omega(1;\rho_i,z)$ by the window. The area of each annulus is $2\pi\rho_i\Delta\rho_i$. Hence the fraction is given by the summation

(6)
$$G_1 = \frac{1}{2\pi b^2} \sum_{i=1}^{N} \Omega(1;\rho_i,z)\rho_i\Delta\rho_i.$$

Case 2. $s \geqq b$: non-coaxial and no vertical overlap.

Divide the source into N concentric strips using the point common to the axis of the window and the plane of the source as a center. The area of each strip is $2\alpha_i\rho_i\Delta\rho_i$, where $\alpha_i = \text{arc cos } [(\rho_i^2 + S^2 - b^2)/2S\rho_i]$. The fraction is given by

(7)
$$G_2 = \frac{1}{2\pi^2 b^2} \sum_{i=1}^{N} \Omega(1;\rho_i,z)\alpha_i\rho_i\Delta\rho_i.$$

Case 3. $s < b$: non-coaxial with vertical overlap. Apply Cases 1 and 2.

In the above calculations, self-absorption, scattering, non-uniformity of source strength, and non-isotropic emission have been omitted. Corrections for these effects cannot be treated here.

I. SOLID ANGLE SUBTENDED BY THE CIRCULAR DISC OF UNIT RADIUS

ρ＼z	0.2	0.4	0.6	0.8	1.0	1.2	1.4	1.6	1.8	2.0
0.0	5.05095 0	3.94967 0	3.05051 0	2.35811 0	1.84030 0	1.45631 0	1.17035 0	9.55055-1	7.90693-1	6.63334-1
0.1	5.04234 0	3.93659 0	3.03737 0	2.34716 0	1.83198 0	1.45024 0	1.16598 0	9.51907-1	7.88406-1	6.61650-1
0.2	5.01561 0	3.89649 0	2.99753 0	2.31425 0	1.80709 0	1.43213 0	1.15296 0	9.42532-1	7.81591-1	6.56633-1
0.3	4.96793 0	3.82665 0	2.92981 0	2.25924 0	1.76590 0	1.40231 0	1.13157 0	9.27131-1	7.70390-1	6.48380-1
0.4	4.89368 0	3.72227 0	2.83236 0	2.18209 0	1.70895 0	1.36136 0	1.10227 0	9.06036-1	7.55034-1	6.37049-1
0.5	4.78279 0	3.57610 0	2.70304 0	2.08304 0	1.63710 0	1.31011 0	1.06569 0	8.79705-1	7.35838-1	6.22856-1
0.6	4.61712 0	3.37830 0	2.54009 0	1.96307 0	1.55174 0	1.24970 0	1.02267 0	8.48708-1	7.13192-1	6.06068-1
0.7	4.36323 0	3.11751 0	2.34335 0	1.82425 0	1.45479 0	1.18157 0	9.74194-1	8.13717-1	6.87550-1	5.86992-1
0.8	3.95926 0	2.78525 0	2.11602 0	1.67016 0	1.34883 0	1.10741 0	9.21391-1	7.75480-1	6.59413-1	5.65969-1
0.9	3.31094 0	2.38629 0	1.86638 0	1.50600 0	1.23698 0	1.02917 0	8.65483-1	7.34795-1	6.29313-1	5.43360-1
1.00	2.40515 0	1.95112 0	1.60798 0	1.33824 0	1.12269 0	9.48851-1	8.07727-1	6.92475-1	5.97794-1	5.19535-1
1.25	7.83966-1	1.02735 0	1.02474 0	9.45919-1	8.48210-1	7.51604-1	6.63084-1	5.84636-1	5.16227-1	4.57034-1
1.50	3.30050-1	5.40110-1	6.29125-1	6.43292-1	6.19102-1	5.77621-1	5.30021-1	4.82015-1	4.36391-1	3.94380-1
1.75	1.74764-1	3.11680-1	3.97052-1	4.37844-1	4.47467-1	4.37793-1	4.17265-1	3.91356-1	3.63433-1	3.35486-1
2.00	1.05541-1	1.95924-1	2.62557-1	3.04563-1	3.25802-1	3.31582-1	3.26809-1	3.15351-1	3.00008-1	2.82710-1
2.25	6.93038-2	1.31484-1	1.81639-1	2.17930-1	2.41045-1	2.53071-1	2.56518-1	2.53750-1	2.46739-1	2.37015-1
2.50	4.82364-2	9.27336-2	1.30625-1	1.60381-1	1.81769-1	1.95484-1	2.02713-1	2.04791-1	2.02982-1	1.98380-1
2.75	3.50529-2	6.79748-2	9.70218-2	1.21083-1	1.39726-1	1.53087-1	1.61683-1	1.66228-1	1.67492-1	1.66206-1
3.00	2.63375-2	5.13818-2	7.40283-2	9.34934-2	1.09364-1	1.21567-1	1.30297-1	1.35917-1	1.38876-1	1.39646-1
3.25	2.03247-2	3.98250-2	5.77746-2	7.36211-2	8.70214-2	9.78406-2	1.06119-1	1.12032-1	1.15831-1	1.17811-1
3.50	1.60321-2	3.15170-2	4.59620-2	5.89727-2	7.02786-2	7.97380-2	8.73258-2	9.31121-2	9.72349-2	9.98737-2
3.75	1.28802-2	2.53850-2	3.71704-2	4.79510-2	5.75168-2	6.57396-2	7.25687-2	7.80213-2	8.21674-2	8.51134-2
4.00	1.05108-2	2.07566-2	3.04914-2	3.95055-2	4.76364-2	5.47745-2	6.08632-2	6.58932-2	6.98951-2	7.29297-2
4.25	8.69349-3	1.71951-2	2.53256-2	3.29283-2	3.98770-2	4.60809-2	5.14851-2	5.60685-2	5.98397-2	6.28312-2
4.50	7.27511-3	1.44086-2	2.12670-2	2.77316-2	3.37040-2	3.91096-2	4.38989-2	4.80465-2	5.15489-2	5.44214-2
4.75	6.15128-3	1.21961-2	1.80334-2	2.35721-2	2.87348-2	3.34605-2	3.77061-2	4.14456-2	4.46693-2	4.73822-2
5.00	5.24891-3	1.04165-2	1.54251-2	2.02039-2	2.46916-2	2.88383-2	3.26071-2	3.59732-2	3.89245-2	4.14597-2
5.25	4.51573-3	8.96845-3	1.32977-2	1.74477-2	2.13695-2	2.50223-2	2.83747-2	3.14042-2	3.40977-2	3.64505-2
5.50	3.91368-3	7.77790-3	1.15450-2	1.51706-2	1.86154-2	2.18459-2	2.48352-2	2.75636-2	3.00179-2	3.21920-2
5.75	3.41457-3	6.78986-3	1.00879-2	1.32731-2	1.63135-2	1.91814-2	2.18542-2	2.43144-2	2.65498-2	2.85533-2
6	2.99721-3	5.96290-3	8.86648-3	1.16792-2	1.43749-2	1.69305-2	1.93270-2	2.15491-2	2.35856-2	2.54291-2
7	1.87242-3	3.73056-3	5.56046-3	7.34874-3	9.08291-3	1.07515-2	1.23445-2	1.38533-2	1.52706-2	1.65909-2
8	1.24791-3	2.48860-3	3.71496-3	4.92012-3	6.09755-3	7.24114-3	8.34531-3	9.40506-3	1.04159-2	1.13744-2
9	8.73357-4	1.74274-3	2.60424-3	3.45403-3	4.28845-3	5.10400-3	5.89743-3	6.66576-3	7.40627-3	8.11658-3
10	6.35074-4	1.26782-3	1.89593-3	2.51716-3	3.12931-3	3.73028-3	4.31810-3	4.89090-3	5.44696-3	5.98473-3
11	4.76253-4	9.51068-4	1.42302-3	1.89070-3	2.35275-3	2.80785-3	3.25473-3	3.69220-3	4.11913-3	4.53449-3
12	3.66318-4	7.31708-4	1.09525-3	1.45604-3	1.81318-3	2.16581-3	2.51311-3	2.85426-3	3.18853-3	3.51520-3
13	2.87802-4	5.74985-4	8.60932-4	1.14503-3	1.42670-3	1.70533-3	1.98038-3	2.25129-3	2.51755-3	2.77865-3
14	2.30230-4	4.60033-4	6.88984-4	9.16665-4	1.14266-3	1.36655-3	1.58797-3	1.80653-3	2.02184-3	2.23358-3
15	1.87054-4	3.73806-4	5.59956-4	7.45206-4	9.29263-4	1.11183-3	1.29264-3	1.47142-3	1.64789-3	1.82180-3
16	1.54039-4	3.07860-4	4.61245-4	6.13980-4	7.65850-4	9.16646-4	1.06616-3	1.21420-3	1.36056-3	1.50506-3
18	1.08092-4	2.16063-4	3.23793-4	4.31162-4	5.38052-4	6.44345-4	7.49926-4	8.54683-4	9.58505-4	1.06128-3
20	7.87496-5	1.57428-4	2.35964-4	3.14288-4	3.92328-4	4.70016-4	5.47293-4	6.24061-4	7.00285-4	7.75889-4
22	5.91384-5	1.18233-4	1.77239-4	2.36113-4	2.94811-4	3.53290-4	4.11508-4	4.69421-4	5.26989-4	5.84171-4
24	4.55356-5	9.10426-5	1.36493-4	1.81857-4	2.27108-4	2.72217-4	3.17157-4	3.61899-4	4.06417-4	4.50685-4
26	3.58052-5	7.15912-5	1.07339-4	1.43030-4	1.78644-4	2.14162-4	2.49568-4	2.84841-4	3.19964-4	3.54918-4
28	2.86614-5	5.73096-5	8.59314-5	1.14514-4	1.43043-4	1.71507-4	1.99892-4	2.28186-4	2.56376-4	2.84450-4
30	2.32987-5	4.65880-5	6.98587-5	9.31014-5	1.16307-4	1.39465-4	1.62568-4	1.85607-4	2.08571-4	2.31453-4
32	1.91948-5	3.83828-5	5.75573-5	7.67115-5	9.58388-5	1.14930-4	1.33984-4	1.52990-4	1.71943-4	1.90836-4
34	1.60009-5	3.19968-5	4.79828-5	6.39537-5	7.99048-5	9.58308-5	1.11725-4	1.27587-4	1.43409-4	1.59187-4

I. SOLID ANGLE SUBTENDED BY THE CIRCULAR DISC OF UNIT RADIUS (Continued)

ρ \ z	2.2	2.4	2.6	2.8	3.0	3.2	3.4	3.6	3.8	4.0
0.0	5.63184-1	4.83322-1	4.18802-1	3.66046-1	3.22432-1	2.86011-1	2.55314-1	2.29224-1	2.06877-1	1.87600-1
0.1	5.61928-1	4.82371-1	4.18072-1	3.65479-1	3.21986-1	2.85655-1	2.55028-1	2.28991-1	2.06686-1	1.87442-1
0.2	5.58183-1	4.79535-1	4.15895-1	3.63785-1	3.20651-1	2.84592-1	2.54172-1	2.28295-1	2.06114-1	1.86969-1
0.3	5.52015-1	4.74860-1	4.12302-1	3.60987-1	3.18446-1	2.82833-1	2.52755-1	2.27142-1	2.05167-1	1.86185-1
0.4	5.43533-1	4.68421-1	4.07345-1	3.57123-1	3.15396-1	2.80399-1	2.50791-1	2.25542-1	2.03853-1	1.85096-1
0.5	5.32887-1	4.60320-1	4.01098-1	3.52243-1	3.11538-1	2.77315-1	2.48300-1	2.23511-1	2.02182-1	1.83710-1
0.6	5.20258-1	4.50686-1	3.93649-1	3.46411-1	3.06918-1	2.73615-1	2.45307-1	2.21067-1	2.00169-1	1.82038-1
0.7	5.05858-1	4.39664-1	3.85102-1	3.39700-1	3.01590-1	2.69339-1	2.41841-1	2.18232-1	1.97830-1	1.80093-1
0.8	4.89922-1	4.27418-1	3.75571-1	3.32194-1	2.95612-1	2.64529-1	2.37934-1	2.15029-1	1.95183-1	1.77888-1
0.9	4.72698-1	4.14123-1	3.65182-1	3.23982-1	2.89051-1	2.59235-1	2.33622-1	2.11487-1	1.92250-1	1.75441-1
1.00	4.54444-1	3.99960-1	3.54063-1	3.15158-1	2.81976-1	2.53507-1	2.28944-1	2.07635-1	1.89052-1	1.72768-1
1.25	4.05975-1	3.61950-1	3.23942-1	2.91052-1	2.62506-1	2.37644-1	2.15914-1	1.96849-1	1.80060-1	1.65219-1
1.50	3.56380-1	3.22360-1	2.92080-1	2.65206-1	2.41380-1	2.20250-1	2.01493-1	1.84813-1	1.69952-1	1.56679-1
1.75	3.08635-1	2.83459-1	2.60212-1	2.38952-1	2.19628-1	2.02126-1	1.86307-1	1.72020-1	1.59117-1	1.47457-1
2.00	2.64732-1	2.46887-1	2.29671-1	2.13367-1	1.98115-1	1.83967-1	1.70915-1	1.58920-1	1.47920-1	1.37847-1
2.25	2.25711-1	2.13635-1	2.01347-1	1.89221-1	1.77498-1	1.66324-1	1.55778-1	1.45895-1	1.36679-1	1.28115-1
2.50	1.91871-1	1.84146-1	1.75727-1	1.66995-1	1.58222-1	1.49599-1	1.41249-1	1.33253-1	1.25657-1	1.18484-1
2.75	1.63015-1	1.58461-1	1.52983-1	1.46922-1	1.40544-1	1.34045-1	1.27566-1	1.21210-1	1.15049-1	1.09128-1
3.00	1.38676-1	1.36366-1	1.33064-1	1.29055-1	1.24571-1	1.19798-1	1.14877-1	1.09917-1	1.04999-1	1.00181-1
3.25	1.18276-1	1.17512-1	1.15782-1	1.13313-1	1.10300-1	1.06904-1	1.03256-1	9.94607-2	9.55989-2	9.17341-2
3.50	1.01228-1	1.01499-1	1.00879-1	9.95452-2	9.76532-2	9.53378-2	9.27128-2	8.98729-2	8.68954-2	8.38425-2
3.75	8.69879-2	8.79296-2	8.80774-2	8.75641-2	8.65120-2	8.50306-2	8.32157-2	8.11498-2	7.89024-2	7.65315-2
4.00	7.50785-2	7.64353-2	7.70987-2	7.71673-2	7.67350-2	7.58888-2	7.47074-2	7.32606-2	7.16088-2	6.98043-2
4.25	6.50933-2	6.66885-2	6.76860-2	6.81579-2	6.81754-2	6.78065-2	6.71146-2	6.61575-2	6.49867-2	6.36476-2
4.50	5.66938-2	5.84071-2	5.96093-2	6.03525-2	6.06901-2	6.06747-2	6.03566-2	5.97829-2	5.89966-2	5.80367-2
4.75	4.96011-2	5.13519-2	5.26677-2	5.35858-2	5.41457-2	5.43878-2	5.43515-2	5.40748-2	5.35933-2	5.29395-2
5.00	4.35870-2	4.53227-2	4.66888-2	4.77118-2	4.84209-2	4.88467-2	4.90199-2	4.89708-2	4.87284-2	4.83198-2
5.25	3.84656-2	4.01524-2	4.15256-2	4.26035-2	4.34075-2	4.39608-2	4.42869-2	4.44101-2	4.43535-2	4.41397-2
5.50	3.40854-2	3.57030-2	3.70541-2	3.81516-2	3.90107-2	3.96487-2	4.00840-2	4.03355-2	4.04218-2	4.03613-2
5.75	3.03226-2	3.18596-2	3.31699-2	3.42622-2	3.51474-2	3.58383-2	3.63490-2	3.66942-2	3.68889-2	3.69480-2
6	2.70765-2	2.85276-2	2.97854-2	3.08556-2	3.17460-2	3.24660-2	3.30263-2	3.34383-2	3.37138-2	3.38650-2
7	1.78102-2	1.89259-2	1.99368-2	2.08432-2	2.16464-2	2.23489-2	2.29537-2	2.34649-2	2.38870-2	2.42250-2
8	1.22774-2	1.31225-2	1.39081-2	1.46332-2	1.52974-2	1.59009-2	1.64443-2	1.69288-2	1.73559-2	1.77274-2
9	8.79463-3	9.43866-3	1.00473-2	1.06194-2	1.11544-2	1.16516-2	1.21111-2	1.25329-2	1.29173-2	1.32651-2
10	6.50280-3	6.99995-3	7.47512-3	7.92743-3	8.35619-3	8.76087-3	9.14110-3	9.49667-3	9.82754-3	1.01338-2
11	4.93731-3	5.32675-3	5.70202-3	6.06247-3	6.40752-3	6.73671-3	7.04966-3	7.34608-3	7.62581-3	7.88872-3
12	3.83361-3	4.14314-3	4.44326-3	4.73345-3	5.01328-3	5.28237-3	5.54039-3	5.78708-3	6.02223-3	6.24568-3
13	3.03414-3	3.28358-3	3.52656-3	3.76273-3	3.99173-3	4.21328-3	4.42710-3	4.63298-3	4.83072-3	5.02016-3
14	2.44139-3	2.64497-3	2.84401-3	3.03824-3	3.22741-3	3.41129-3	3.58966-3	3.76235-3	3.92918-3	4.09003-2
15	1.99292-3	2.16099-3	2.32581-3	2.48717-3	2.64488-3	2.79874-3	2.94861-3	3.09433-3	3.23577-3	3.37280-3
16	1.64751-3	1.78775-3	1.92560-3	2.06091-3	2.19353-3	2.32333-3	2.45016-3	2.57392-3	2.69448-3	2.81177-3
18	1.16291-3	1.26329-3	1.36232-3	1.45992-3	1.55599-3	1.65044-3	1.74320-3	1.83419-3	1.92333-3	2.01056-3
20	8.50809-4	9.24986-4	9.98357-4	1.07086-3	1.14245-3	1.21307-3	1.28266-3	1.35118-3	1.41857-3	1.48481-3
22	6.40926-4	6.97214-4	7.52999-4	8.08242-4	8.62906-4	9.16958-4	9.70363-4	1.02309-3	1.07510-3	1.12637-3
24	4.94674-4	5.38361-4	5.81718-4	6.24722-4	6.67349-4	7.09574-4	7.51375-4	7.92729-4	8.33615-4	8.74012-4
26	3.89686-4	4.24249-4	4.58591-4	4.92695-4	5.26543-4	5.60120-4	5.93410-4	6.26397-4	6.59066-4	6.91403-4
28	3.12394-4	3.40196-4	3.67846-4	3.95329-4	4.22636-4	4.49754-4	4.76672-4	5.03379-4	5.29864-4	5.56117-4
30	2.54243-4	2.76933-4	2.99514-4	3.21977-4	3.44313-4	3.66516-4	3.88576-4	4.10485-4	4.32236-4	4.53822-4
32	2.09663-4	2.28416-4	2.47090-4	2.65679-4	2.84176-4	3.02574-4	3.20870-4	3.39055-4	3.57126-4	3.75075-4
34	1.74916-4	1.90591-4	2.06207-4	2.21760-4	2.37245-4	2.52656-4	2.67991-4	2.83245-4	2.98413-4	3.13491-4

I. SOLID ANGLE SUBTENDED BY THE CIRCULAR DISC OF
UNIT RADIUS (Continued)

ρ＼z	4.2	4.4	4.6	4.8	5.0	5.2	5.4	5.6	5.8	6.0
0.0	1.70864−1	1.56245−1	1.43405−1	1.32070−1	1.22015−1	1.13057−1	1.05042−1	9.78444−2	9.13568−2	8.54895−2
0.1	1.70732−1	1.56134−1	1.43311−1	1.31990−1	1.21947−1	1.12998−1	1.04991−1	9.78001−2	9.13181−2	8.54556−2
.0.2	1.70337−1	1.55802−1	1.43031−1	1.31715−1	1.21742−1	1.12822−1	1.04839−1	9.76675−2	9.12022−2	8.53539−2
0.3	1.69683−1	1.55252−1	1.42565−1	1.31355−1	1.21403−1	1.12529−1	1.04585−1	9.74470−2	9.10096−2	8.51848−2
0.4	1.68773−1	1.54487−1	1.41917−1	1.30803−1	1.20929−1	1.12121−1	1.04232−1	9.71398−2	9.07410−2	8.49490−2
0.5	1.67615−1	1.53512−1	1.41091−1	1.30098−1	1.20325−1	1.11601−1	1.03781−1	9.67470−2	9.03975−2	8.46475−2
0.6	1.66216−1	1.52333−1	1.40092−1	1.29245−1	1.19594−1	1.10970−1	1.03234−1	9.62704−2	8.99806−2	8.42812−2
0.7	1.64586−1	1.50959−1	1.38925−1	1.28249−1	1.18738−1	1.10231−1	1.02593−1	9.57120−2	8.94919−2	8.38517−2
0.8	1.62737−1	1.49397−1	1.37598−1	1.27115−1	1.17763−1	1.09388−1	1.01862−1	9.50742−2	8.89333−2	8.33605−2
0.9	1.60681−1	1.47658−1	1.36118−1	1.25848−1	1.16673−1	1.08446−1	1.01043−1	9.43597−2	8.83070−2	8.28094−2
1.00	1.58431−1	1.45752−1	1.34494−1	1.24456−1	1.15474−1	1.07408−1	1.00140−1	9.35711−2	8.76153−2	8.22003−2
1.25	1.52054−1	1.40333−1	1.29863−1	1.20478−1	1.12039−1	1.04428−1	9.75433−2	9.12984−2	8.56186−2	8.04395−2
1.50	1.44798−1	1.34134−1	1.24540−1	1.15885−1	1.08058−1	1.00962−1	9.45137−2	8.86386−2	8.32765−2	7.83691−2
1.75	1.36909−1	1.27354−1	1.18687−1	1.10810−1	1.03639−1	9.70999−2	9.11250−2	8.56557−2	8.06400−2	7.60318−2
2.00	1.28628−1	1.20185−1	1.12459−1	1.05380−1	9.88928−2	9.29323−2	8.74532−2	8.24103−2	7.77626−2	7.34729−2
2.25	1.20174−1	1.12819−1	1.06019−1	9.97372−2	9.39249−2	8.85487−2	8.35739−2	7.89673−2	7.46983−2	7.07383−2
2.50	1.11738−1	1.05414−1	9.95043−2	9.39840−2	8.88347−2	8.40338−2	7.95591−2	7.53883−2	7.14998−2	6.78732−2
2.75	1.03042−1	9.81042−2	9.30224−2	8.82267−2	8.37106−2	7.94638−2	7.54748−2	7.17304−2	6.82170−2	6.49209−2
3.00	9.55037−2	9.09961−2	8.66758−2	8.25520−2	7.86288−2	7.49060−2	7.13801−2	6.80455−2	6.48950−2	6.19210−2
3.25	8.79142−2	8.41742−2	8.05402−2	7.70290−2	7.36525−2	7.04170−2	6.73257−2	6.43789−2	5.15745−2	5.89096−2
3.50	8.07635−2	7.76969−2	7.46722−2	7.17117−2	6.88315−2	6.60431−2	6.33540−2	6.07690−2	5.82901−2	5.59182−2
3.75	7.40847−2	7.16008−2	6.91105−2	6.66385−2	6.42035−2	6.18200−2	5.94987−2	5.72474−2	5.50709−2	5.29731−2
4.00	6.78909−2	6.59056−2	6.38787−2	6.18352−2	5.97951−2	5.77745−2	5.57858−2	5.38388−2	5.19407−2	5.00967−2
4.25	6.21797−2	6.06169−2	5.89879−2	5.73167−2	5.56236−2	5.39250−2	5.22343−2	5.05622−2	5.89176−2	4.73067−2
4.50	5.69376−2	5.57298−2	5.44393−2	5.30890−2	5.16981−2	5.02828−2	4.88567−2	4.74312−2	4.60152−2	4.46165−2
4.75	5.21432−2	5.12311−2	5.02266−2	4.91506−2	4.80209−2	4.68531−2	4.56604−2	4.44540−2	4.32434−2	4.20364−2
5.00	4.77702−2	4.71028−2	4.63381−2	4.54949−2	4.45896−2	4.36364−2	4.26481−2	4.16355−2	4.06079−2	3.95730−2
5.25	4.37897−2	4.33234−2	4.27586−2	4.21119−2	4.13980−2	4.06300−2	3.98196−2	3.89770−2	3.81114−2	3.72301−2
5.50	4.01716−2	3.98694−2	3.94702−2	3.89884−2	3.84369−2	3.78276−2	3.71713−2	3.64774−2	3.57543−2	3.50094−2
5.75	3.68860−2	3.67169−2	3.64540−2	3.61096−2	3.56953−2	3.52215−2	3.46979−2	3.41331−2	3.35351−2	3.29105−2
6	3.39038−2	3.38419−2	3.36905−2	3.34602−2	3.31611−2	3.28023−2	3.23925−2	3.19395−2	3.14505−2	3.09318−2
7	2.44839−2	2.46694−2	2.47868−2	2.48416−2	2.48393−2	2.47850−2	2.46838−2	2.45406−2	2.43598−2	2.41460−2
8	1.80456−2	1.83128−2	1.85314−2	1.87043−2	1.88341−2	1.89237−2	1.89757−2	1.89930−2	1.89782−2	1.89341−2
9	1.35769−2	1.38536−2	1.40966−2	1.43070−2	1.44862−2	1.46355−2	1.47565−2	1.48505−2	1.49192−2	1.49641−2
10	1.04157−2	1.06733−2	1.09075−2	1.11186−2	1.13073−2	1.14742−2	1.16201−2	1.17457−2	1.18520−2	1.19398−2
11	8.13482−3	8.36415−3	8.57687−3	8.77317−3	8.95332−3	9.11764−3	9.26648−3	9.40025−3	9.51941−3	9.62443−3
12	6.45734−3	6.65714−3	6.84510−3	7.02127−3	7.18572−3	7.33860−3	7.48005−3	7.61027−3	7.72949−3	7.83797−3
13	5.20118−3	5.37369−3	5.53764−3	5.69301−3	5.83979−3	5.97804−3	6.10778−3	6.22911−3	6.34215−3	6.44701−3
14	4.24474−3	4.39326−3	4.53550−3	4.67141−3	4.80096−3	4.92412−3	5.04092−3	5.15138−3	5.25553−3	5.35343−3
15	3.50533−3	3.63325−3	3.75650−3	3.87503−3	3.98878−3	4.09772−3	4.20184−3	4.30112−3	4.39557−3	4.48521−3
16	2.92568−3	3.03613−3	3.14306−3	3.24641−3	3.34614−3	3.44219−3	3.53455−3	3.62319−3	3.70811−3	3.78929−3
18	2.09582−3	2.17906−3	2.26021−3	2.33924−3	2.41611−3	2.49077−3	2.56319−3	2.63335−3	2.70123−3	2.76681−3
20	1.54983−3	1.61361−3	1.67610−3	1.73728−3	1.79711−3	1.85556−3	1.91260−3	1.96822−3	2.02239−3	2.07510−3
22	1.17687−3	1.22658−3	1.27546−3	1.32349−3	1.37065−3	1.41692−3	1.46227−3	1.50670−3	1.55016−3	1.59267−3
24	9.13900−4	9.53259−4	9.92072−4	1.03032−3	1.06799−3	1.10506−3	1.14152−3	1.17735−3	1.21253−3	1.24707−3
26	7.23392−4	7.55021−4	7.86277−4	8.17146−4	8.47617−4	8.77677−4	9.07316−4	9.36522−4	9.65289−4	9.93601−4
28	5.82128−4	6.07885−4	6.33381−4	6.58606−4	6.83550−4	7.08206−4	7.32564−4	7.56616−4	7.80358−4	8.03776−4
30	4.75233−4	4.96463−4	5.17506−4	5.38354−4	5.59000−4	5.79438−4	5.99662−4	6.19665−4	6.39445−4	6.58989−4
32	3.92897−4	4.10587−4	4.28140−4	4.45550−4	4.62812−4	4.79921−4	4.96873−4	5.13663−4	5.30285−4	5.46736−4
34	3.28475−4	3.43360−4	3.58143−4	3.72820−4	3.87386−4	4.01839−4	4.16173−4	4.30386−4	4.44474−4	4.58434−4

I. SOLID ANGLE SUBTENDED BY THE CIRCULAR DISC OF UNIT RADIUS (Continued)

ρ \ z	6.2	6.4	6.6	6.8	7.0	7.2	7.4	7.6	7.8	8.0
0.0	8.01664-2	7.53226-2	7.09026-2	6.68585-2	6.31492-2	5.97388-2	5.65962-2	5.36942-2	5.10090-2	4.85195-2
0.1	8.01365-2	7.52962-2	7.08791-2	6.68376-2	6.31305-2	5.97221-2	5.65812-2	5.36807-2	5.09968-2	4.85085-2
0.2	8.00469-2	7.52170-2	7.08088-2	6.67750-2	6.30746-2	5.96720-2	5.65362-2	5.36402-2	5.09602-2	4.84753-2
0.3	7.98980-2	7.50852-2	7.06919-2	6.66709-2	6.29816-2	5.95887-2	5.64613-2	5.35727-2	5.08992-2	4.84201-2
0.4	7.96902-2	7.49015-2	7.05288-2	6.65256-2	6.28518-2	5.94723-2	5.63568-2	5.34785-2	5.08141-2	4.83430-2
0.5	7.94244-2	7.46662-2	7.03199-2	6.63396-2	6.26855-2	5.93233-2	5.62228-2	5.33577-2	5.07050-2	4.82442-2
0.6	7.91014-2	7.43804-2	7.00660-2	6.61133-2	6.24833-2	5.91419-2	5.60598-2	5.32107-2	5.05721-2	4.81238-2
0.7	7.87224-2	7.40448-2	6.97679-2	6.58475-2	6.22456-2	5.89288-2	5.58681-2	5.30379-2	5.04159-2	4.79822-2
0.8	7.82888-2	7.36607-2	6.94264-2	6.55430-2	6.19732-2	5.86844-2	5.56482-2	5.28396-2	5.02365-2	4.78197-2
0.9	7.78021-2	7.32292-2	6.90427-2	6.52007-2	6.16668-2	5.84095-2	5.54008-2	5.26163-2	5.00346-2	4.76365-2
1.00	7.72637-2	7.27518-2	6.86178-2	6.48214-2	6.13272-2	5.81045-2	5.61262-2	5.23685-2	4.98103-2	4.74331-2
1.25	7.57054-2	7.13679-2	6.73849-2	6.37197-2	6.03399-2	5.72172-2	5.43267-2	5.16462-2	4.91563-2	4.68396-2
1.50	7.38690-2	6.97338-2	6.59264-2	6.24141-2	5.91679-2	5.61624-2	5.33749-2	5.07853-2	4.83758-2	4.61304-2
1.75	7.17904-2	6.78799-2	6.42680-2	6.09265-2	5.78301-2	5.49562-2	5.22848-2	4.97979-2	4.74793-2	4.53147-2
2.00	6.95082-2	6.58387-2	6.24376-2	5.92809-2	5.63469-2	5.36163-2	5.10716-2	4.86969-2	4.64783-2	4.44027-2
2.25	6.70614-2	6.36438-2	6.04640-2	5.75020-2	5.47399-2	5.21614-2	4.97515-2	4.74968-2	4.53851-2	4.34049-2
2.50	6.44889-2	6.13289-2	5.83764-2	5.56152-2	5.30310-2	5.06105-2	4.83413-2	4.72121-2	4.42125-2	4.23328-2
2.75	6.18284-2	5.89266-2	5.62029-2	5.36450-2	5.12417-2	4.89825-2	4.68574-2	4.48572-2	4.29733-2	4.11975-2
3.00	5.91147-2	5.64674-2	5.39704-2	5.16150-2	4.93929-2	4.72958-2	4.53161-2	4.34465-2	4.16802-2	4.00103-2
3.25	5.63797-2	5.00708-2	5.17045-2	4.95480-2	4.75045-2	4.55682-2	4.37333-2	4.19942-2	4.03458-2	3.87827-2
3.50	5.36519-2	5.14893-2	4.94279-2	4.74643-2	4.55950-2	4.38161-2	4.21236-2	4.05136-2	3.89820-2	3.75251-2
3.75	5.09552-2	4.90178-2	4.71605-2	4.53820-2	4.36807-2	4.20543-2	4.05005-2	3.90166-2	3.75998-2	3.62475-2
4.00	4.83105-2	4.65845-2	4.49200-2	4.33174-2	4.17765-2	4.02965-2	3.88764-2	3.75146-2	3.62095-2	3.49592-2
4.25	4.57347-2	4.42054-2	4.27213-2	4.12843-2	3.98953-2	3.85546-2	3.72622-2	3.60177-2	3.48203-2	3.36689-2
4.50	4.32409-2	4.18931-2	4.05767-2	3.92943-2	3.80479-2	3.68387-2	3.56676-2	3.45349-2	3.34406-2	3.23842-2
4.75	4.08395-2	3.96580-2	3.84961-2	3.73571-2	3.62437-2	3.51578-2	3.41008-2	3.30739-2	3.20774-2	3.11117-2
5.00	3.85377-2	3.75075-2	3.64871-2	3.54803-2	3.44900-2	3.35189-2	3.25688-2	3.16412-2	3.07372-2	2.98574-2
5.25	3.63400-2	3.54468-2	3.45552-2	3.36693-2	3.27926-2	3.19278-2	3.10771-2	3.02425-2	2.94253-2	2.86266-2
5.50	3.42491-2	3.34791-2	3.27042-2	3.19286-2	3.11558-2	3.03889-2	2.96303-2	2.88821-2	2.81460-2	2.74234-2
5.75	3.22657-2	3.16061-2	3.09363-2	3.02607-2	2.95827-2	2.89055-2	2.82317-2	2.75635-2	2.69029-2	2.62513-2
6	3.03892-2	2.98280-2	2.92527-2	2.86673-2	2.80753-2	2.74800-2	2.68839-2	2.62895-2	2.56987-2	2.51131-2
7	2.39030-2	2.36346-2	2.33443-2	2.30352-2	2.27104-2	2.23724-2	2.20237-2	2.16664-2	2.13026-2	2.09340-2
8	1.88631-2	1.87676-2	1.86502-2	1.85128-2	1.83577-2	1.81868-2	1.80019-2	1.78048-2	1.75970-2	1.73801-2
9	1.49867-2	1.49886-2	1.49712-2	1.49359-2	1.48841-2	1.48173-2	1.47366-2	1.46433-2	1.45386-2	1.44236-2
10	1.20100-2	1.20634-2	1.21010-2	1.21237-2	1.21324-2	1.21279-2	1.21111-2	1.20829-2	1.20440-2	1.19953-2
11	9.71581-3	9.79407-3	9.85976-3	9.91341-3	9.95559-3	9.98686-3	1.00077-2	1.00188-2	1.00206-2	1.00137-2
12	7.93598-3	8.02381-3	8.10179-3	8.17023-3	8.22949-3	8.27989-3	8.32181-3	8.35559-3	8.38159-3	8.40019-3
13	6.54384-3	6.63280-3	6.71408-3	6.78786-3	6.85434-3	6.91375-3	6.96629-3	7.01220-3	7.05171-3	7.08506-3
14	5.44515-3	5.53078-3	5.61041-3	5.68416-3	5.75213-3	5.81447-3	5.87130-3	5.92276-3	5.96901-3	6.01019-3
15	4.57008-3	4.65020-3	4.72563-3	4.79643-3	4.86267-3	4.92441-3	4.98173-3	5.03473-3	5.08351-3	5.12814-3
16	3.86675-3	3.94050-3	4.01056-3	4.07696-3	4.13974-3	4.19892-3	4.25457-3	4.30673-3	4.35546-3	4.40082-3
18	2.83008-3	2.89102-3	2.94964-3	3.00594-3	3.05991-3	3.11158-3	3.16094-3	3.20801-3	3.25282-3	3.29539-3
20	2.12632-3	2.17605-3	2.22427-3	2.27098-3	2.31616-3	2.35983-3	2.40197-3	2.44259-3	2.48169-3	2.51927-3
22	1.63421-3	1.67476-3	1.71430-3	1.75283-3	1.79034-3	1.82682-3	1.86227-3	1.89669-3	1.93007-3	1.96240-3
24	1.28096-3	1.31416-3	1.34668-3	1.37851-3	1.40964-3	1.44006-3	1.46977-3	1.49876-3	1.52702-3	1.55456-3
26	1.02145-3	1.04883-3	1.07574-3	1.10215-3	1.12808-3	1.15351-3	1.17844-3	1.20285-3	1.22676-3	1.25014-3
28	8.26865-4	8.49621-4	8.72036-4	8.94105-4	9.15822-4	9.37180-4	9.58177-4	9.78806-4	9.99064-4	1.01895-3
30	6.78295-4	6.97359-4	7.16175-4	7.34739-4	7.53044-4	7.71089-4	7.88868-4	8.06378-4	8.23615-4	8.40575-4
32	5.63012-4	5.79109-4	5.95022-4	6.10747-4	6.26282-4	6.41622-4	6.56765-4	6.71707-4	6.68444-4	7.00975-4
34	4.72262-4	4.85955-4	4.99510-4	5.12924-4	5.26194-4	5.39318-4	5.52292-4	5.65114-4	5.77781-4	5.90291-4

I. SOLID ANGLE SUBTENDED BY THE CIRCULAR DISC OF UNIT RADIUS (Continued)

ρ \ z	8.2	8.4	8.6	8.8	9.0	9.2	9.4	9.6	9.8	10.0
0.0	4.62073-2	4.40560-2	4.20510-2	4.01794-2	3.84296-2	3.67914-2	3.52555-2	3.38135-2	3.24580-2	3.11823-2
0.1	4.61973-2	4.40468-2	4.20426-2	4.01718-2	3.84227-2	3.67850-2	3.52496-2	3.38081-2	3.24531-2	3.11777-2
0.2	4.61672-2	4.40195-2	4.20177-2	4.01490-2	3.84018-2	3.67659-2	3.52320-2	3.37919-2	3.24381-2	3.11639-2
0.3	4.61171-2	4.39739-2	4.19761-2	4.01110-2	3.83670-2	3.67340-2	3.52029-2	3.37650-2	3.24133-2	3.11409-2
0.4	4.60471-2	4.39102-2	4.19180-2	4.00579-2	3.83185-2	3.66895-2	3.51618-2	3.37273-2	3.23786-2	3.11089-2
0.5	4.59573-2	4.38285-2	4.18436-2	3.99899-2	3.82562-2	3.66323-2	3.51093-2	3.36790-2	3.23340-2	3.10677-2
0.6	4.58480-2	4.37291-2	4.17529-2	3.99070-2	3.81802-2	3.65627-2	3.50453-2	3.36201-2	3.22797-2	3.10175-2
0.7	4.57194-2	4.36120-2	4.16461-2	3.98094-2	3.80908-2	3.64806-2	3.49699-2	3.35506-2	3.22156-2	3.09584-2
0.8	4.55718-2	4.34775-2	4.15234-2	3.96972-2	3.79881-2	3.63864-2	3.48832-2	3.34708-2	3.21420-2	3.08904-2
0.9	4.54053-2	4.33260-2	4.13851-2	3.95707-2	3.78722-2	3.62800-2	3.47855-2	3.33808-2	3.20589-2	3.08136-2
1.00	4.52204-2	4.31575-2	4.12313-2	3.94301-2	3.77433-2	3.61617-2	3.46766-2	3.32805-2	3.19664-2	3.07281-2
1.25	4.46805-2	4.26654-2	4.07818-2	3.90187-2	3.73662-2	3.58153-2	3.43579-2	3.29868-2	3.16953-2	3.04775-2
1.50	4.40348-2	4.20762-2	4.02432-2	3.85254-2	3.69135-2	3.53992-2	3.39748-2	3.26335-2	3.13690-2	3.01757-2
1.75	4.32912-2	4.13969-2	3.96215-2	3.79554-2	3.63901-2	3.49176-2	3.35311-2	3.22240-2	3.09905-2	2.98253-2
2.00	4.24585-2	4.06353-2	3.89236-2	3.73148-2	3.58011-2	3.43752-2	3.30308-2	3.17618-2	3.05629-2	2.94290-2
2.25	4.15462-2	3.97997-2	3.81569-2	3.66101-2	3.51524-2	3.37771-2	3.24785-2	3.12512-2	3.00901-2	2.89906-2
2.50	4.05642-2	3.88988-2	3.73290-2	3.58482-2	3.44501-2	3.31288-2	3.18792-2	3.06964-2	2.95759-2	2.85133-2
2.75	3.95225-2	3.79414-2	3.64478-2	3.50359-2	3.37003-2	3.24357-2	3.12377-2	3.01018-2	2.90241-2	2.80006-2
3.00	3.84310-2	3.69365-2	3.55213-2	3.41805-2	3.29093-2	3.17036-2	3.05590-2	2.94720-2	2.84388-2	2.74563-2
3.25	3.73000-2	3.58931-2	3.45575-2	3.32891-2	3.20839-2	3.09383-2	2.98486-2	2.88118-2	2.78246-2	2.68842-2
3.50	3.61389-2	3.48198-2	3.35642-2	3.23687-2	3.12302-2	3.01455-2	2.91116-2	2.81258-2	2.71856-2	2.62881-2
3.75	3.49567-2	3.37247-2	3.25487-2	3.14261-2	3.03542-2	2.93306-2	2.83528-2	2.74186-2	2.65257-2	2.56719-2
4.00	3.37619-2	3.26155-2	3.15181-2	3.04675-2	2.94618-2	2.84990-2	2.75773-2	2.66946-2	2.58492-2	2.50392-2
4.25	3.25624-2	3.14995-2	3.04789-2	2.94991-2	2.85585-2	2.76558-2	2.67894-2	2.59579-2	2.51598-2	2.43935-2
4.50	3.13653-2	3.03833-2	2.94373-2	2.85264-2	2.76496-2	2.68057-2	2.59938-2	2.52127-2	2.44613-2	2.37383-2
4.75	3.01768-2	2.92726-2	2.83986-2	2.75544-2	2.67394-2	2.59530-2	2.51843-2	2.44626-2	2.37571-2	2.30767-2
5.00	2.90025-2	2.81725-2	2.73676-2	2.65877-2	2.58325-2	2.51016-2	2.43946-2	2.37110-2	2.30503-2	2.24118-2
5.25	2.78473-2	2.70880-2	2.63490-2	2.56306-2	2.49327-2	2.42553-2	2.35983-2	2.29613-2	2.23442-2	2.17463-2
5.50	2.67154-2	2.60229-2	2.53464-2	2.46865-2	2.40434-2	2.34174-2	2.28083-2	2.22163-2	2.16412-2	2.10828-2
5.75	2.56102-2	2.49805-2	2.43631-2	2.37587-2	2.31677-2	2.25906-2	2.20275-2	2.14786-2	2.09440-2	2.04236-2
6	2.45344-2	2.39636-2	2.34018-2	2.28497-2	2.23081-2	2.17775-2	2.12582-2	2.07505-2	2.02547-2	1.97707-2
7	2.05622-2	2.01886-2	1.98145-2	1.94409-2	1.90688-2	1.86992-2	1.83326-2	1.79697-2	1.76112-2	1.72572-2
8	1.71553-2	1.69240-2	1.66873-2	1.64463-2	1.62019-2	1.59549-2	1.57062-2	1.54565-2	1.52064-2	1.49563-2
9	1.42993-2	1.41667-2	1.40268-2	1.38804-2	1.37283-2	1.35713-2	1.34101-2	1.32454-2	1.30776-2	1.29076-2
10	1.19375-2	1.18713-2	1.17974-2	1.17166-2	1.16294-2	1.15364-2	1.14382-2	1.13353-2	1.12282-2	1.11176-2
11	9.99873-3	9.97597-3	9.94604-3	9.90942-3	9.86657-3	9.81795-3	9.76400-3	9.70514-3	9.64175-3	9.57422-3
12	8.41172-3	8.41656-3	8.41505-3	8.40753-3	8.39434-3	8.37582-3	8.35228-3	8.32404-3	8.29141-3	8.25465-3
13	7.11248-3	7.13422-3	7.15051-3	7.16160-3	7.16773-3	7.16912-3	7.16603-3	7.15866-3	7.14725-3	7.13201-3
14	6.04647-3	6.07800-3	6.10496-3	6.12749-3	6.14578-3	6.15998-3	6.17025-3	6.17678-3	6.17971-3	6.17921-3
15	5.16875-3	5.20542-3	5.23829-3	5.26744-3	5.29301-3	5.31510-3	5.33382-3	5.34931-3	5.36167-3	5.37102-3
16	4.44288-3	4.48170-3	4.51736-3	4.54993-3	4.57949-3	4.60612-3	4.62990-3	4.65091-3	4.66923-3	4.68495-3
18	3.33573-3	3.37388-3	3.40987-3	3.44372-3	3.47548-3	3.50518-3	3.53285-3	3.55854-3	3.58229-3	3.60413-3
20	2.55534-3	2.58992-3	2.62301-3	2.65462-3	2.68476-3	2.71347-3	2.74074-3	2.76659-3	2.79106-3	2.81414-3
22	1.99370-3	2.02396-3	2.05318-3	2.08137-3	2.10853-3	2.13466-3	2.15978-3	2.18389-3	2.20699-3	2.22910-3
24	1.57137-3	1.60744-3	1.63278-3	1.65739-3	1.68127-3	1.70441-3	1.72682-3	1.74850-3	1.76945-3	1.78968-3
26	1.27300-3	1.29534-3	1.31716-3	1.33844-3	1.35919-3	1.37941-3	1.39910-3	1.41825-3	1.43688-3	1.45496-3
28	1.03845-3	1.05757-3	1.07631-3	1.09466-3	1.11262-3	1.13019-3	1.14736-3	1.16415-3	1.18053-3	1.19651-3
30	8.57256-4	8.73654-4	8.89767-4	9.05592-4	9.21127-4	9.36371-4	9.51321-4	9.65975-4	9.80333-4	9.94393-4
32	7.15297-4	7.29406-4	7.43301-4	7.56979-4	7.70439-4	7.83678-4	7.96695-4	8.09488-4	8.22055-4	8.34397-4
34	6.02643-4	6.14833-4	6.26860-4	6.38721-4	6.50415-4	6.61941-4	6.73296-4	6.84479-4	6.95489-4	7.06326-4

I. SOLID ANGLE SUBTENDED BY THE CIRCULAR DISC OF UNIT RADIUS (Continued)

ρ \ z	11	12	13	14	15	16	17	18	19	20
0.0	2.58037-2	2.17036-2	1.85072-2	1.59675-2	1.39163-2	1.22360-2	1.08424-2	9.67389-3	8.68443-3	7.83929-3
0.1	2.58006-2	2.17014-2	1.85056-2	1.59662-2	1.39153-2	1.22353-2	1.08419-2	9.67344-3	8.68408-3	7.83800-3
0.2	2.57911-2	2.16947-2	1.85007-2	1.59626-2	1.39126-2	1.22332-2	1.08402-2	9.67211-3	8.68300-3	7.83812-3
0.3	2.57754-2	2.16836-2	1.84926-2	1.59566-2	1.39080-2	1.22296-2	1.08374-2	9.66988-3	8.68121-3	7.83665-3
0.4	2.57534-2	2.16680-2	1.84812-2	1.59481-2	1.39016-2	1.22246-2	1.08335-2	9.66677-3	8.67869-3	7.83461-3
0.5	2.57251-2	2.16479-2	1.84667-2	1.59372-2	1.38933-2	1.22182-2	1.08285-2	9.66277-3	8.67547-3	7.83198-3
0.6	2.56907-2	2.16235-2	1.84489-2	1.59240-2	1.38832-2	1.22104-2	1.08223-2	9.65788-3	8.67153-3	7.82877-3
0.7	2.56500-2	2.15947-2	1.84279-2	1.59083-2	1.38713-2	1.22012-2	1.08151-2	9.65211-3	8.66688-3	7.82497-3
0.8	2.56033-2	2.15615-2	1.84037-2	1.58903-2	1.38576-2	1.21906-2	1.08067-2	9.64546-3	8.66151-3	7.82060-3
0.9	2.55505-2	2.15240-2	1.83763-2	1.58699-2	1.38421-2	1.21786-2	1.07973-2	9.63793-3	8.65544-3	7.81565-3
1.00	2.54916-2	2.14822-2	1.83458-2	1.58471-2	1.38247-2	1.21651-2	1.07867-2	9.62951-3	8.64865-3	7.81011-3
1.25	2.53188-2	2.13593-2	1.82561-2	1.57801-2	1.37737-2	1.21256-2	1.07556-2	9.60471-3	8.62864-3	7.79379-3
1.50	2.51101-2	2.12106-2	1.81474-2	1.56988-2	1.37117-2	1.20775-2	1.07178-2	9.57453-3	8.60427-3	7.77390-3
1.75	2.48672-2	2.10371-2	1.80202-2	1.56036-2	1.36390-2	1.20211-2	1.06734-2	9.53907-3	8.57562-3	7.75051-3
2.00	2.45914-2	2.08395-2	1.78751-2	1.54946-2	1.35556-2	1.19563-2	1.06221-2	9.49837-3	8.54272-3	7.72368-3
2.25	2.42852-2	2.06195-2	1.77132-2	1.53730-2	1.34625-2	1.18838-2	1.05649-2	9.45265-3	8.50573-3	7.69344-3
2.50	2.39506-2	2.03784-2	1.75354-2	1.52391-2	1.33599-2	1.18038-2	1.05017-2	9.40198-3	8.46471-3	7.65986-3
2.75	2.35893-2	2.01171-2	1.73420-2	1.50931-2	1.32478-2	1.17163-2	1.04324-2	9.34649-3	8.41974-3	7.62304-3
3.00	2.32037-2	1.98372-2	1.71343-2	1.49359-2	1.31266-2	1.16216-2	1.03575-2	9.28636-3	8.37096-3	7.58305-3
3.25	2.27964-2	1.95402-2	1.69131-2	1.47681-2	1.29971-2	1.15202-2	1.02770-2	9.22171-3	8.31846-3	7.53998-3
3.50	2.23696-2	1.92275-2	1.66794-2	1.45902-2	1.28596-2	1.14124-2	1.01912-2	9.15272-3	8.26236-3	7.49393-3
3.75	2.19257-2	1.89009-2	1.64344-2	1.44032-2	1.27146-2	1.12984-2	1.01004-2	9.07957-3	8.20280-3	7.44498-3
4.00	2.14672-2	1.85618-2	1.61790-2	1.42076-2	1.25627-2	1.11786-2	1.00048-2	9.00242-3	8.13992-3	7.39323-3
4.25	2.09963-2	1.82118-2	1.59143-2	1.40043-2	1.24041-2	1.10533-2	9.90500-3	8.92150-3	8.07387-3	7.33880-3
4.50	2.05153-2	1.78524-2	1.56414-2	1.37938-2	1.22396-2	1.09230-2	9.80063-3	8.83700-3	8.00477-3	7.28179-3
4.75	2.00262-2	1.74851-2	1.53611-2	1.35770-2	1.20696-2	1.07881-2	9.69227-3	8.74910-3	7.93279-3	7.22232-3
5.00	1.95314-2	1.71113-2	1.50747-2	1.33545-2	1.18945-2	1.06487-2	9.58018-3	8.65801-3	7.85808-3	7.16052-3
5.25	1.90326-2	1.67323-2	1.47829-2	1.31270-2	1.17150-2	1.05055-2	9.46466-3	8.56394-3	7.78079-3	7.09649-3
5.50	1.85318-2	1.63496-2	1.44869-2	1.28953-2	1.15315-2	1.03587-2	9.34596-3	8.46710-3	7.70107-3	7.03033-3
5.75	1.80306-2	1.59645-2	1.41875-2	1.26599-2	1.13445-2	1.02086-2	9.22437-3	8.36768-3	7.61910-3	6.96220-3
6	1.75307-2	1.55780-2	1.38856-2	1.24216-2	1.11546-2	1.00557-2	9.10016-3	8.26590-3	7.53501-3	6.89221-3
7	1.55713-2	1.40401-2	1.26691-2	1.14513-2	1.03741-2	9.42281-3	8.58240-3	7.83921-3	7.18077-3	6.59605-3
8	1.37257-2	1.25564-2	1.14713-2	1.04791-2	9.58060-3	8.77072-3	8.04325-3	7.39062-3	6.80522-3	6.27977-3
9	1.20370-2	1.11663-2	1.03261-2	9.53327-3	8.79628-3	8.11783-3	7.49719-3	6.93168-3	6.41759-3	5.95072-3
10	1.05231-2	9.89186-3	9.25505-3	8.63315-3	8.03868-3	7.47864-3	6.95626-3	6.47231-3	6.02601-3	5.61563-3
11	9.18660-3	8.74249-3	8.27073-3	7.79187-3	7.31996-3	6.86414-3	6.43004-3	6.02070-3	5.63744-3	5.28034-3
12	8.01888-3	7.71835-3	7.37790-3	7.01642-3	6.64783-3	6.28199-3	5.92569-3	5.58331-3	5.25753-3	4.94971-3
13	7.00581-3	6.81362-3	6.57595-3	6.30927-3	6.02638-3	5.73695-3	5.44809-3	5.16484-3	4.89063-3	4.62770-3
14	6.13076-3	6.01915-3	5.86083-3	5.66965-3	5.45688-3	5.23144-3	5.00025-3	4.76848-3	4.53994-3	4.31732-3
15	5.37672-3	5.32420-3	5.22639-3	5.09465-3	4.93859-3	4.76613-3	4.58366-3	4.39620-3	4.20760-3	4.02076-3
16	4.72752-3	4.71769-3	4.66549-3	4.58004-3	4.46938-3	4.34033-3	4.19862-3	4.04890-3	3.89490-3	3.73950-3
18	3.68640-3	3.72795-3	3.73463-3	3.71212-3	3.66576-3	3.60038-3	3.52028-3	3.42913-3	3.33008-3	3.22572-3
20	2.90978-3	2.97464-3	3.01202-3	3.02537-3	3.01809-3	2.99345-3	2.95448-3	2.90392-3	2.84421-3	2.77746-3
22	2.32514-3	2.39814-3	2.44991-3	2.48250-3	2.49803-3	2.49863-3	2.48638-3	2.46323-3	2.43101-3	2.39133-3
24	1.88015-3	1.95338-3	2.01033-3	2.05219-3	2.08026-3	2.09590-3	2.10051-3	2.09544-3	2.08201-3	2.06141-3
26	1.53749-3	1.60705-3	1.66415-3	1.70943-3	1.74369-3	1.76778-3	1.78263-3	1.78915-3	1.78827-3	1.78088-3
28	1.27054-3	1.33474-3	1.38935-3	1.43473-3	1.47132-3	1.49967-3	1.52037-3	1.53404-3	1.54134-3	1.54288-3
30	1.06019-3	1.11851-3	1.16939-3	1.21303-3	1.24968-3	1.27967-3	1.30337-3	1.32121-3	1.33362-3	1.34106-3
32	8.92664-4	9.45152-4	9.91852-4	1.03284-3	1.06825-3	1.09827-3	1.12315-3	1.14316-3	1.15859-3	1.16976-3
34	7.57846-4	8.04868-4	8.47346-4	8.85295-4	9.18776-4	9.47896-4	9.72800-4	9.93667-4	1.01070-3	1.02411-3

I. SOLID ANGLE SUBTENDED BY THE CIRCULAR DISC OF UNIT RADIUS (Continued)

ρ \ z	22	26	30	40	50	60	70	80	90	100
0.0	6.48085-3	4.64218-3	3.48775-3	1.96256-3	1.25626-3	8.72483-4	6.41043-4	4.90816-4	3.87815-4	3.14136-4
0.1	6.48065-3	4.64207-3	3.48769-3	1.96256-3	1.25625-3	8.72479-4	6.41041-4	4.90815-4	3.87814-4	3.14135-4
0.2	6.48005-3	4.64177-3	3.48752-3	1.96250-3	1.25623-3	8.72468-4	6.41035-4	4.90812-4	3.87812-4	3.14134-4
0.3	6.47905-3	4.64125-3	3.48723-3	1.96241-3	1.25619-3	8.72450-4	6.41026-4	4.90806-4	3.87809-4	3.14132-4
0.4	6.47765-3	4.64053-3	3.48683-3	1.96228-3	1.25614-3	8.72425-4	6.41012-4	4.90798-4	3.87804-4	3.14128-4
0.5	6.47586-3	4.63961-3	3.48630-3	1.96212-3	1.25607-3	8.72393-4	6.40995-4	4.90788-4	3.87797-4	3.14124-4
0.6	6.47366-3	4.63848-3	3.48567-3	1.96192-3	1.25599-3	8.72353-4	6.40973-4	4.90775-4	3.87790-4	3.14119-4
0.7	6.47107-3	4.63715-3	3.48492-3	1.96168-3	1.25589-3	8.72306-4	6.40948-4	4.90761-4	3.87780-4	3.14113-4
0.8	6.46808-3	4.63562-3	3.48405-3	1.96140-3	1.25578-3	8.72252-4	6.40919-4	4.90744-4	3.87770-4	3.14106-4
0.9	6.46469-3	4.63388-3	3.48306-3	1.96109-3	1.25565-3	8.72190-4	6.40886-4	4.90724-4	3.87758-4	3.14098-4
1.00	6.46089-3	4.63192-3	3.48196-3	1.96074-3	1.25551-3	8.72120-4	6.40847-4	4.90701-4	3.87743-4	3.14089-4
1.25	6.44972-3	4.62618-3	3.47871-3	1.95971-3	1.25509-3	8.71817-4	6.40738-4	4.90638-4	3.87704-4	3.14063-4
1.50	6.43610-3	4.61916-3	3.47475-3	1.95845-3	1.25457-3	8.71668-4	6.40603-4	4.90559-4	3.87654-4	3.14030-4
1.75	6.42006-3	4.61090-3	3.47007-3	1.95696-3	1.25396-3	8.71373-4	6.40444-4	4.90466-4	3.87596-4	3.13992-4
2.00	6.40164-3	4.60142-3	3.46470-3	1.95525-3	1.25327-3	8.71034-4	6.40261-4	4.90358-4	3.87529-4	3.13948-4
2.25	6.38086-3	4.59069-3	3 45861-3	1.95331-3	1.25247-3	8.70649-4	6.40053-4	4.90236-4	3.87453-4	3.13898-4
2.50	6.35778-3	4.57871-3	3.45181-3	1.95115-3	1.25154-3	8.70218-4	6.39822-4	4.90100-4	3.87368-4	3.13842-4
2.75	6.33242-3	4.56556-3	3.44434-3	1.94876-3	1.25056-3	8.69743-4	6.39565-4	4.89950-4	3.87274-4	3.13780-4
3.00	6.30482-3	4.55123-3	3.43619-3	1.94616-3	1.24949-3	8.69225-4	6.39282-4	4.89784-4	3.87171-4	3.13713-4
3.25	6.27507-3	4.53574-3	3.42736-3	1.94333-3	1.24833-3	8.68661-4	6.38977-4	4.89605-4	3.87059-4	3.13639-4
3.50	6.24321-3	4.51911-3	3.41788-3	1.94028-3	1.24706-3	8.68052-4	6.38649-4	4.89411-4	3.86938-4	3.13560-4
3.75	6.20927-3	4.50136-3	3.40773-3	1.93702-3	1.24572-3	8.67399-4	6.38296-4	4.89204-4	3.86808-4	3.13475-4
4.00	6.17334-3	4.48252-3	3.39695-3	1.93353-3	1.24428-3	8.66703-4	6.37919-4	4.88983-4	3.86670-4	3.13384-4
4.25	6.13546-3	4.46260-3	3.38553-3	1.92984-3	1.24275-3	8.65962-4	6.37517-4	4.88747-4	3.86523-4	3.13287-4
4.50	6.09570-3	4.44163-3	3.37349-3	1.92593-3	1.24114-3	8.65178-4	6.37093-4	4.88497-4	3.86366-4	3.13184-4
4.75	6.05414-3	4.41964-3	3.36083-3	1.92182-3	1.23943-3	8.64350-4	6.36644-4	4.88233-4	3.86201-4	3.13075-4
5.00	6.01086-3	4.39668-3	3.34759-3	1.91751-3	1.23764-3	8.63479-4	6.36171-4	4.87956-4	3.86027-4	3.12961-4
5.25	5.96589-3	4.37273-3	3.33375-3	1.91299-3	1.23576-3	8.62564-4	6.35675-4	4.87664-4	3.85844-4	3.12841-4
5.50	5.91932-3	4.34786-3	3.31933-3	1.90826-3	1.23379-3	8.61607-4	6.35155-4	4.87359-4	3.85654-4	3.12716-4
5.75	5.87124-3	4.32207-3	3.30436-3	1.90333-3	1.23174-3	8.60607-4	6.34612-4	4.87039-4	3.85454-4	3.12585-4
6	5.82171-3	4.29540-3	3.28885-3	1.89822-3	1.22960-3	8.59565-4	6.34047-4	4.86706-4	3.85245-4	3.12448-4
7	5.61063-3	4.18064-3	3.22162-3	1.87584-3	1.22021-3	5.54978-4	6.31551-4	4.85235-4	3.84324-4	3.11842-4
8	5.38247-3	4.05444-3	3.14684-3	1.85055-3	1.20953-3	8.49735-4	6.28692-4	4.83548-4	3.83266-4	3.11145-4
9	5.14198-3	3.91890-3	3.06552-3	1.82258-3	1.19764-3	8.43859-4	6.25477-4	4.81648-4	3.82072-4	3.10359-4
10	4.89367-3	3.77612-3	2.97879-3	1.79214-3	1.18454-3	8.37371-4	6.21917-4	4.79538-4	3.80745-4	3.09484-4
11	4.64164-3	3.62813-3	2.88741-3	1.75948-3	1.17034-3	8.30295-4	6.18020-4	4.77224-4	3.79287-4	3.08521-4
12	4.38948-1	3.47681-3	2.79265-3	1.72482-3	1.15511-3	8.22659-4	6.13800-4	4.74711-4	3.77702-4	3.07473-4
13	4.14026-3	3.32389-3	2.69539-3	1.68841-3	1.13893-3	8.14491-4	6.09266-4	4.72004-4	3.75990-4	3.06340-4
14	3.89651-3	3.17092-3	2.59653-3	1.65051-3	1.12185-3	8.05819-4	6.04431-4	4.69109-4	3.74157-4	3.05123-4
15	3.66024-3	3.01926-3	2.49689-3	1.61136-3	1.10399-3	7.96676-4	5.99310-4	4.66033-4	3.72204-4	3.03826-4
16	3.43295-3	2.87006-3	2.39723-3	1.57119-3	1.08543-3	7.87092-4	5.93915-4	4.62782-4	3.70134-4	3.02449-4
18	3.00925-3	2.58263-3	2.20044-3	1.48869-3	1.04648-3	7.66730-4	5.82360-4	4.55791-4	3.65662-4	2.99466-4
20	2.62993-3	2.31411-3	2.01048-3	1.40470-3	1.00563-3	7.44990-4	5.69882-4	4.48163-4	3.60770-4	2.96190-4
22	2.29533-3	2.06752-3	1.83037-3	1.32068-3	9.63516-4	7.22132-4	5.56601-4	4.39987-4	3.55489-4	2.92639-4
24	2.00320-3	1.84398-3	1.66202-3	1.23786-3	9.20632-4	6.98409-4	5.42637-4	4.31314-4	3.49851-4	2.88830-4
26	1.74991-3	1.64331-3	1.50644-3	1.15722-3	8.77511-4	6.74062-4	5.28109-4	4.22205-4	3.43889-4	2.84782-4
28	1.53121-3	1.46444-3	1.36392-3	1.07951-3	8.34596-4	6.49315-4	5.13133-4	4.12722-4	3.37638-4	2.80515-4
30	1.34281-3	1.30580-3	1.23426-3	1.00528-3	7.92265-4	6.24378-4	4.97820-4	4.02926-4	3.31133-4	2.76049-4
32	1.18065-3	1.16560-3	1.11692-3	9.34894-4	7.50829-4	5.99437-4	4.82275-4	3.92875-4	3.24407-4	2.71406-4
34	1.04102-3	1.04193-3	1.01112-3	8.68556-4	7.10546-4	5.74662-4	4.66596-4	3.82627-4	3.17493-4	2.66604-4

II. SOLID ANGLE SUBTENDED BY THE LATERAL SURFACE OF A RIGHT CIRCULAR CYLINDER OF UNIT RADIUS

ρ \ h	0.2	0.4	0.6	0.8	1.0	1.2	1.4	1.6	1.8	2.0
1.2	1.20776 0	1.63842 0	1.79396 0	1.86300 0	1.89875 0	1.91940 0	1.93233 0	1.94092 0	1.94691 0	1.95124 0
1.4	6.13287-1	1.01195 0	1.23118 0	1.35275 0	1.42410 0	1.46860 0	1.49786 0	1.51798 0	1.53234 0	1.54292 0
1.6	3.73447-1	6.70631-1	8.76125-1	1.01135 0	1.10050 0	1.16064 0	1.20240 0	1.23226 0	1.25419 0	1.27070 0
1.8	2.53539-1	4.73988-1	6.46727-1	7.74499-1	8.67005-1	9.33995-1	9.83056-1	1.01957 0	1.04721 0	1.06851 0
2.0	1.84468-1	3.52378-1	4.93869-1	6.07035-1	6.94977-1	7.62546-1	8.14449-1	8.54569-1	8.85882-1	9.10596-1
2.2	1.40759-1	2.72331-1	3.88319-1	4.86135-1	5.66248-1	6.30794-1	6.82443-1	7.23759-1	7.56935-1	7.83743-1
2.4	1.11208-1	2.16910-1	3.12895-1	3.96886-1	4.68412-1	5.28240-1	5.77774-1	6.18602-1	6.52243-1	6.80031-1
2.6	9.02264-2	1.76951-1	2.57326-1	3.29544-1	3.92871-1	4.47430-1	4.93885-1	5.33170-1	5.66289-1	5.94203-1
2.8	7.47567-2	1.47179-1	2.15285-1	2.77684-1	3.33637-1	3.82981-1	4.25973-1	4.63128-1	4.95086-1	5.22511-1
3.0	6.30036-2	1 24390-1	1.82744-1	2.37000-1	2.86501-1	3.30975-1	3.70462-1	4.05220-1	4.35639-1	4.62166-1
3.2	5.38531-2	1.06550-1	1.57055-1	2.04550-1	2.48476-1	2.88537-1	3.24663-1	3.56960-1	3.85650-1	4.11026-1
3.4	4.65828-2	9.23163-2	1.36428-1	1.78281-1	2.17413-1	2.53539-1	2.86539-1	3.16430-1	3.43328-1	3.67416-1
3.6	4.07061-2	8.07745-2	1.19617-1	1.56734-1	1.91744-1	2.24390-1	2.54535-1	2.82143-1	3.07265-1	3.30009-1
3.8	3.58856-2	7.12827-2	1.05736-1	1.38850-1	1.70310-1	1.99891-1	2.27452-1	2.52935-1	2.76347-1	2.97746-1
4.0	3.18806-2	6.33806-2	9.41421-2	1.23849-1	1.52241-1	1.79123-1	2.04362-1	2.27888-1	2.49682-1	2.69771-1
4.2	2.85157-2	5.67305-2	8.43592-2	1.11146-1	1.36877-1	1.61381-1	1.84538-1	2.06275-1	2.26558-1	2.45391-1
4.4	2.56606-2	5.10800-2	7.60285-2	1.00297-1	1.23708-1	1.46114-1	1.67407-1	1.87514-1	2.06397-1	2.24044-1
4.6	2.32166-2	4.62377-2	6.88760-2	9.09597-2	1.12339-1	1.32888-1	1.52512-1	1.71140-1	1.88731-1	2.05265-1
4.8	2.11080-2	4.20558-2	6.26892-2	8.28657-2	1.02459-1	1.21362-1	1.39488-1	1.56774-1	1.73177-1	1.88673-1
5.0	1.92758-2	3.84101-2	5.73018-2	7.58047-2	9.38206-2	1.11257-1	1.28039-1	1.44108-1	1.59422-1	1.73954-1
5.2	1.76735-2	3.52364-2	5.25815-2	6.96082-2	8.62255-2	1.02354-1	1.17926-1	1.32890-1	1.47204-1	1.60844-1
5.4	1.62640-2	3.24349-2	4.84223-2	6.41411-2	7.95131-2	9.44694-2	1.08951-1	1.22910-1	1.36310-1	1.49123-1
5.6	1.50174-2	2.99559-2	4.47388-2	5.92933-2	7.35522-2	8.74557-2	1.00952-1	1.13997-1	1.26558-1	1.38608-1
5.8	1.39094-2	2.77516-2	4.14609-2	5.49747-2	6.82353-2	8.11901-2	9.37934-2	1.06006-1	1.17797-1	1.29142-1
6.0	1.29202-2	2.57827-2	3.85310-2	5.11113-2	6.34731-2	7.55706-2	8.73633-2	9.88163-2	1.09901-1	1.20594-1
6.5	1.08655-2	2.16906-2	3.24357-2	4.30628-2	5.35358-2	6.38213-2	7.38892-2	8.37127-2	9.32689-2	1.02539-1
7.5	7.99611-3	1.59708-2	2.39028-2	3.17717-2	3.95575-2	4.72417-2	5.48067-2	6.22366-2	6.95169-2	7.66348-2
8.5	6.13162-3	1.22508-2	1.83452-2	2.44028-2	3.04119-2	3.63613-2	4.22402-2	4.80388-2	5.37478-2	5.93586-2
9.5	4.85156-3	9.69542-3	1.45239-2	1.93296-2	2.41050-2	2.88431-2	3.35371-2	3.81804-2	4.27670-2	4.72910-2
10.5	3.93468-3	7.86433-3	1.17840-2	1.56887-2	1.95736-2	2.34340-2	2.72654-2	3.10634-2	3.48237-2	3.85425-2
11.5	3.25539-3	6.50737-3	9.75254-3	1.29875-2	1.62091-2	1.94139-2	2.25988-2	2.57608-2	2.88969-2	3.20044-2
12.5	2.73811-3	5.47382-3	8.20474-3	1.09285-2	1.36428-2	1.63452-2	1.90336-2	2.17058-2	2.43596-2	2.69931-2
13.5	2.33509-3	4.66845-3	6.99833-3	9.32303-3	1.16408-2	1.39501-2	1.62491-2	1.85364-2	2.08103-2	2.30692-2
14.5	2.01498-3	4.02868-3	6.03980-3	8.04707-3	1.00492-2	1.20450-2	1.40332-2	1.60127-2	1.79821-2	1.99405-2
15.5	1.75649-3	3.51201-3	5.26558-3	7.01623-3	8.76301-3	1.05050-2	1.22412-2	1.39707-2	1.56926-2	1.74061-2
16.5	1.54476-3	3.08876-3	4.63127-3	6.17152-3	7.70879-3	9.24233-3	1.07714-2	1.22954-2	1.38134-2	1.53249-2
17.5	1.36914-3	2.73769-3	4.10507-3	5.47068-3	6.83395-3	8.19430-3	9.55116-3	1.09040-2	1.22522-2	1.35952-2
18.5	1.22187-3	2.44326-3	3.66373-3	4.88279-3	6.09999-3	7.31487-3	8.52697-3	9.73585-3	1.09411-2	1.21422-2
19.5	1.09715-3	2.19392-3	3.28993-3	4.38483-3	5.47822-3	6.56974-3	7.65903-3	8.74572-3	9.82945-3	1.09099-2
20.5	9.90599-4	1.98089-3	2.97057-3	3.95934-3	4.94688-3	5.93290-3	6.91710-3	7.89919-3	8.87886-3	9.85584-3
21.5	8.98856-4	1.79746-3	2.69556-3	3.59291-3	4.48925-3	5.38435-3	6.27794-3	7.16980-3	8.05967-3	8.94732-3
22.5	8.19295-4	1.63838-3	2.45705-3	3.28510-3	4.09231-3	4.90848-3	5.72341-3	6.53690-3	7.34874-3	8.15873-3
23.5	7.49852-4	1.49953-3	2.24886-3	2.99766-3	3.74577-3	4.49301-3	5.23921-3	5.98420-3	6.72781-3	7.46988-3
24.5	6.88880-4	1.37761-3	2.06605-3	2.75405-3	3.44146-3	4.12814-3	4.81394-3	5.49872-3	6.18234-4	6.86465-3
25.5	6.35054-4	1.26998-3	1.90466-3	2.53897-3	3.17278-3	3.80596-3	4.43841-3	5.06998-3	5.70056-3	6.33004-3
26	6.10478-4	1.22084-3	1.83097-3	2.44076-3	3.05009-3	3.65884-3	4.26690-3	4.87417-3	5.48052-3	6.08584-3
28	5.25174-4	1.05026-3	1.57518-3	2.09984-3	2.62417-3	3.14807-3	3.67146-3	4.19425-3	4.71638-3	5.23774-3
30	4.56578-4	9.13093-4	1.36948-3	1.82567-3	2.28161-3	2.73722-3	3.19246-3	3.64724-3	4.10152-3	4.55522-3
32	4.00597-4	8.01144-4	1.20195-3	1.60189-3	2.00200-3	2.40185-3	2.80142-3	3.20063-3	3.59946-3	3.99785-3
34	3.54315-4	7.08591-4	1.06279-3	1.41687-3	1.77080-3	2.12454-3	2.47805-3	2.83129-3	3.18422-3	3.53680-3

II. SOLID ANGLE SUBTENDED BY THE LATERAL SURFACE OF A RIGHT CIRCULAR CYLINDER OF UNIT RADIUS (Continued)

ρ \ h	2.2	2.4	2.6	2.8	3.0	3.2	3.4	3.6	3.8	4.0
1.2	1.95448 0	1.95695 0	1.95889 0	1.96043 0	1.96168 0	1.96271 0	1.96356 0	1.96427 0	1.96488 0	1.96540 0
1.4	1.55092 0	1.55710 0	1.56198 0	1.56589 0	1.56907 0	1.57169 0	1.57387 0	1.57571 0	1.57727 0	1.57861 0
1.6	1.28338 0	1.29332 0	1.30123 0	1.30763 0	1.31287 0	1.31721 0	1.32084 0	1.32392 0	1.32654 0	1.32879 0
1.8	1.08519 0	1.09845 0	1.10913 0	1.11786 0	1.12506 0	1.13106 0	1.13612 0	1.14042 0	1.14410 0	1.14727 0
2.0	9.30329-1	9.46267-1	9.59281-1	9.70018-1	9.78962-1	9.86479-1	9.92850-1	9.98291-1	1.00297 0	1.00702 0
2.2	8.05568-1	8.23482-1	8.38309-1	8.50681-1	8.61087-1	8.69905-1	8.77430-1	8.83896-1	8.89486-1	8.94347-1
2.4	7.03083-1	7.22308-1	7.38436-1	7.52051-1	7.63616-1	7.73500-1	7.81997-1	7.89345-1	7.95734-1	8.01317-1
2.6	6.17769-1	6.37724-1	6.54688-1	6.69173-1	6.81600-1	6.92314-1	7.01595-1	7.09674-1	7.16739-1	7.22945-1
2.8	5.46042-1	5.66254-1	5.83655-1	5.98680-1	6.11697-1	6.23016-1	6.32898-1	6.41557-1	6.49175-1	6.55902-1
3.0	4.85262-1	5.05365-1	5.22878-1	5.38162-1	5.51530-1	5.63253-1	5.73564-1	5.82661-1	5.90712-1	5.97861-1
3.2	4.33412-1	4.53135-1	4.70508-1	4.85822-1	4.99338-1	5.11289-1	5.21877-1	5.31282-1	5.39655-1	5.47130-1
3.4	3.88917-1	4.08069-1	4.25113-1	4.40277-1	4.53777-1	4.65806-1	4.76541-1	4.86136-1	4.94731-1	5.02444-1
3.6	3.50524-1	3.68981-1	3.85559-1	4.00437-1	4.13789-1	4.25775-1	4.36544-1	4.46231-1	4.54957-1	4.62830-1
3.8	3.17228-1	3.34912-1	3.50932-1	3.65426-1	3.78529-1	3.90375-1	4.01087-1	4.10781-1	4.19561-1	4.27523-1
4.0	2.88211-1	3.05085-1	3.20489-1	3.34528-1	3.47309-1	3.58939-1	3.69520-1	3.79150-1	3.87919-1	3.95909-1
4.2	2.62807-1	2.78859-1	2.93616-1	3.07155-1	3.19561-1	3.30918-1	3.41311-1	3.50820-1	3.59523-1	3.67491-1
4.4	2.40469-1	2.55706-1	2.69803-1	2.82817-1	2.94812-1	3.05856-1	3.16016-1	3.25359-1	3.33951-1	3.41854-1
4.6	2.20744-1	2.35188-1	2.48627-1	2.61104-1	2.72666-1	2.83368-1	2.93263-1	3.02406-1	3.10853-1	3.18654-1
4.8	2.03257-1	2.16936-1	2.29730-1	2.41669-1	2.52789-1	2.63130-1	2.72737-1	2.81655-1	2.89928-1	2.97600-1
5.0	1.87694-1	2.00643-1	2.12811-1	2.24219-1	2.34894-1	2.44866-1	2.54171-1	2.62844-1	2.70922-1	2.78444-1
5.2	1.73793-1	1.86049-1	1.97616-1	2.08507-1	2.18740-1	2.28340-1	2.37333-1	2.45749-1	2.53618-1	2.60971-1
5.4	1.61334-1	1.72935-1	1.83927-1	1.94317-1	2.04118-1	2.13348-1	2.22026-1	2.30178-1	2.37826-1	2.44998-1
5.6	1.50130-1	1.61115-1	1.71560-1	1.81469-1	1.90849-1	1.99714-1	2.08079-1	2.15962-1	2.23384-1	2.30366-1
5.8	1.40023-1	1.50429-1	1.60356-1	1.69804-1	1.78778-1	1.87286-1	1.95341-1	2.02956-1	2.10148-1	2.16934-1
6.0	1.30878-1	1.40742-1	1.50179-1	1.59188-1	1.67771-1	1.75933-1	1.83683-1	1.91033-1	1.97994-1	2.04581-1
6.5	1.11507-1	1.20161-1	1.28494-1	1.36500-1	1.44178-1	1.51528-1	1.58554-1	1.65260-1	1.71654-1	1.77743-1
7.5	8.35794-2	9.03412-2	9.69127-2	1.03288-1	1.09462-1	1.15433-1	1.21199-1	1.26759-1	1.32115-1	1.37269-1
8.5	6.48637-2	7.02563-2	7.55304-2	8.06810-2	8.57040-2	9.05959-2	9.53542-2	9.99771-2	1.04463-1	1.08813-1
9.5	5.17473-2	5.61309-2	6.04376-2	6.46633-2	6.88047-2	7.28588-2	7.68232-2	8.06957-2	8.44749-2	8.81595-2
10.5	4.22160-2	4.58407-2	4.94134-2	5.29312-2	5.63913-2	5.97915-2	6.31296-2	6.64037-2	6.96124-2	7.27542-2
11.5	3.50805-2	3.81227-2	4.11287-2	4.40962-2	4.70231-2	4.99077-2	5.27481-2	5.55429-2	5.82907-2	6.09902-2
12.5	2.96042-2	3.21910-2	3.47518-2	3.72850-2	3.97889-2	4.22621-2	4.47031-2	4.71109-2	4.94841-2	5.18218-2
13.5	2.53119-2	2.75367-2	2.97425-2	3.19279-2	3.40916-2	3.62326-2	3.83497-2	4.04420-2	4.25085-2	4.45484-2
14.5	2.18866-2	2.38194-2	2.57379-2	2.76410-2	2.95278-2	3.13975-2	3.32491-2	3.50818-2	3.68949-2	3.86876-2
15.5	1.91103-2	2.08043-2	2.24873-2	2.41586-2	2.58174-2	2.74630-2	2.90947-2	3.07118-2	3.23138-2	3.38999-2
16.5	1.68292-2	1.83256-2	1.98135-2	2.12923-2	2.27613-2	2.42200-2	2.56678-2	2.71043-2	2.85288-2	2.99409-2
17.5	1.49326-2	1.62637-2	1.75882-2	1.89054-2	2.02149-2	2.15163-2	2.28091-2	2.40928-2	2.53670-2	2.66314-2
18.5	1.33387-2	1.45304-2	1.57166-2	1.68971-2	1.80714-2	1.92392-2	2.04001-2	2.15537-2	2.26997-2	2.38377-2
19.5	1.19866-2	1.30594-2	1.41278-2	1.51916-2	1.62504-2	1.73038-2	1.83517-2	1.93937-2	2.04294-2	2.14588-2
20.5	1.08298-2	1.18006-2	1.27677-2	1.37311-2	1.46904-2	1.56454-2	1.65957-2	1.75412-2	1.84817-2	1.94168-2
21.5	9.83250-3	1.07150-2	1.15946-2	1.24710-2	1.33441-2	1.42135-2	1.50792-2	1.59409-2	1.67984-2	1.76515-2
22.5	8.96668-3	9.77238-3	1.05757-2	1.13763-2	1.21742-2	1.29691-2	1.37608-2	1.45491-2	1.53340-2	1.61152-2
23.5	8.21023-3	8.94871-3	9.68514-3	1.04194-2	1.11513-2	1.18807-2	1.26074-2	1.33313-2	1.40522-2	1.47701-2
24.5	7.54551-3	8.22480-3	8.90236-3	9.57806-3	1.02518-2	1.09234-2	1.15927-2	1.22597-2	1.29241-2	1.35860-2
25.5	6.95829-3	7.58519-3	8.21063-3	8.83449-3	9.45665-3	1.00770-2	1.06955-2	1.13119-2	1.19262-2	1.25382-2
26	6.69003-3	7.29298-3	7.89457-3	8.49470-3	9.09327-3	9.69016-3	1.02853-2	1.08785-2	1.14698-2	1.20590-2
28	5.75827-3	6.27787-3	6.79647-3	7.31399-3	7.83055-3	8.34547-3	8.85927-3	9.37168-3	9.88262-3	1.03920-2
30	5.00829-3	5.46066-3	5.91228-3	6.36307-3	6.81299-3	7.26196-3	7.70994-3	8.15687-3	8.60268-3	9.04732-3
32	4.39574-3	4.79310-3	5.18988-3	5.58603-3	5.98150-3	6.37624-3	6.77022-3	7.16338-3	7.55568-3	7.94708-3
34	3.88901-3	4.24079-3	4.59212-3	4.94296-3	5.29326-3	5.64300-3	5.99213-3	6.34063-3	6.68845-3	7.03556-3

II. SOLID ANGLE SUBTENDED BY THE LATERAL SURFACE OF A RIGHT CIRCULAR CYLINDER OF UNIT RADIUS (Continued)

ρ \ h	4.2	4.4	4.6	4.8	5.0	5.2	5.4	5.6	5.8	6.0
1.2	1.96584 0	1.96623 0	1.96657 0	1.96687 0	1 96713 0	1.96736 0	1.96757 0	1.96775 0	1.96792 0	1.96807 0
1.4	1.57977 0	1.58077 0	1.58165 0	1.58242 0	1.58310 0	1.58371 0	1.58425 0	1.58473 0	1.58517 0	1.58556 0
1.6	1.33073 0	1.33243 0	1.33392 0	1.33522 0	1.33638 0	1.33741 0	1.33833 0	1.33916 0	1.33990 0	1.34057 0
1.8	1.15002 0	1.15242 0	1.15453 0	1.15639 0	1.15805 0	1.15952 0	1.16083 0	1.16201 0	1.16308 0	1.16404 0
2.0	1.01055 0	1.01364 0	1.01636 0	1.01877 0	1.02091 0	1.02282 0	1.02453 0	1.02607 0	1.02746 0	1.02872 0
2.2	8.98599-1	9.02335-1	9.05635-1	9.08563-1	9.11171-1	9.13504-1	9.15599-1	9.17486-1	9.19192-1	9.20739-1
2.4	8.06220-1	8.10546-1	8.14380-1	8.17791-1	8.20839-1	8.23571-1	8.26030-1	8.28250-1	8.30259-1	8.32085-1
2.6	7.28420-1	7.33271-1	7.37585-1	7.41437-1	7.44887-1	7.47989-1	7.50787-1	7.53318-1	7.55615-1	7.57704-1
2.8	6.61866-1	6.67172-1	6.71909-1	6.76152-1	6.79965-1	6.83403-1	6.86512-1	6.89330-1	6.91893-1	6.94229-1
3.0	6.04229 1	6.09919-1	6.15019-1	6.19604-1	6.23737-1	6.27475-1	6.30863-1	6.33942-1	6.36749-1	6.39312-1
3.2	5.53821-1	5.59826-1	5.65230-1	5.70106-1	5.74517-1	5.78516-1	5.82152-1	5.85465-1	5.88491-1	5.91201-1
3.4	5.09382-1	5.15636-1	5.21287-1	5.26405-1	5.31049-1	5.35274-1	5.39126-1	5.42644-1	5.45866-1	5.48821-1
3.6	4.69945-1	4.76387-1	4.82231-1	4.87543-1	4.92381-1	4.96795-1	5.00831-1	5.04527-1	5.07919-1	5.11038-1
3.8	4.34752-1	4.41326-1	4.47314-1	4.52777-1	4.57768-1	4.62337-1	4.66526-1	4.70373-1	4.73913-1	4.77174-1
4.0	4.03198-1	4.09855-1	4.15941-1	4.21513-1	4.26622-1	4.31313-1	4.35627-1	4.39599-1	4.43263-1	4.46646-1
4.2	3.74791-1	3.81485-1	3.87630-1	3.93276-1	3.98469-1	3.03252-1	4.07663-1	4.11736-1	4.15501-1	4.18987-1
4.4	3.49125-1	3.55819-1	3.61986-1	3.67672-1	3.72920-1	3.77768-1	3.82251-1	3.86402-1	3.90249-1	3.93818-1
4.6	3.25862-1	3.32523-1	3.38682-1	3.44379-1	3.49654-1	3.54542-1	3.59074-1	3.63282-1	3.67191-1	3.70826-1
4.8	3.04716-1	3.11317-1	3.17440-1	3.23124-1	3.28403-1	3.33308-1	3.37869-1	3.42114-1	3.46067-1	3.49752-1
5.0	2.85445-1	2.91902 1	2.98028-1	3.03676-1	3.08938-1	3.13840-1	3.18412-1	3.22677-1	3.26658-1	3.30378-1
5.2	2.67839-1	2.74254-1	2.80244-1	2.85839-1	2.91065-1	2.95948-1	3.00513-1	3.04782-1	3.08778-1	3.12519-1
5.4	2.51720-1	2.58017-1	2.63916-1	2.69441-1	2.74616-1	2.79465-1	2.84010-1	2.88270-1	2.92266-1	2.96015-1
5.6	2.36930-1	2.43098-1	2.48893-1	2.54336-1	2.59448-1	2.64250-1	2.68761-1	2.73000-1	2.76984-1	2.80731-1
5.8	2.23333-1	2.29364-1	2.35044-1	2.40395-1	2.45432-1	2.50176-1	2.54643-1	2.58850-1	2.62814-1	2.66548-1
6.0	2.10810-1	2.16696-1	2.22256-1	2.27505-1	2.32460-1	2.37136-1	2.41550-1	2.45716-1	2.49649-1	2.53362-1
6.5	1.83538-1	1.89047-1	1.94282-1	1.99254-1	2.03974-1	2.08454-1	2.12704-1	2.16736-1	2.20561-1	2.24190-1
7.5	1.42222-1	1.46980-1	1.51545-1	1.55923-3	1.60119-1	1.64139-1	1.67987-1	1.71671-1	1.75195-1	1.78567-1
8.5	1.13026-1	1.17102-1	1.21044-1	1.24853-1	1.28530-1	1.32079-1	1.35501-1	1.38801-1	1 41980-1	1.45042-1
9.5	9.17487-2	9.52421-2	9.86398-2	1.01942-1	1.05149-1	1.08261-1	1.11281-1	1.14208-1	1.17045-1	1.19793-1
10.5	7.58282-2	7.88336-2	8.17698-2	8.46364-2	8.74333-2	9.01605-2	9.28183-2	9.54072-2	9.79275-2	1.00380-1
11.5	6.36405-2	6.62407-2	6.87901-2	7.12881-2	7.37342-2	7.61283-2	7.84701-2	8.07597-2	8.29971-2	8.51825-2
12.5	5.41231-2	5.63872 2	5.86133-2	6.08008-2	6.29492-2	6.50582-2	6.71274-2	6.91567-2	7.11457-2	7.30946-2
13.5	4.65608-2	4.85450-2	5.05005-2	5.24265-2	5.43227-2	5.61886-2	5.80238-2	5.98281-2	6.16012-2	6.33428-2
14.5	4.04594-2	4.22095-2	4.39374-2	4.56427-2	4.73248-2	4.89833-2	5.06179-2	5.22283-2	5.38141-2	5.53752-2
15.5	3.54698-2	3.70228-2	3.85585-2	4.00764-2	4.15762-2	4.30574-2	4.45197-2	4.59628-2	4.73865-2	4.87904-2
16.5	3.13402-2	3.27262-2	3.40984-2	3.54566-2	3.68004-2	3.81294-2	3.94434-2	4.07419-2	4.20249-2	4.32920-2
17.5	2.78855-2	2.91290-2	3.03616-2	3.15828-2	3.27925-2	3.39903-2	3.51759-2	3.63492-2	3.75098-2	3.86575-2
18.5	2.49675-2	2.60887-2	2.72011-2	2.83043-2	2.93981-2	3.04822-2	3.15565-2	3.26206-2	3.36745-2	3.47178-2
19.5	2.24813-2	2.34969-2	2.45053-2	2.55062-2	2.64994-2	2.74846-2	2.84618-2	2.94306-2	3.03909-2	3.13425-2
20.5	2.03464-2	1.12703-2	2.21881-2	2.30998-2	2.40052-2	2.49040-2	2.57961-2	2.66812-2	2.75593-2	2.84302-2
21.5	1.84999-2	1.93436-2	2.01824-2	2.10160-2	2.18443-2	2.26672-2	2.34844-2	2.42959-2	2.51015-2	2.59010-2
22.5	1.68925-2	1.76658-2	1.84350-2	1.91999-2	1.99603-2	2.07162-2	2.14673-2	2.22136-2	2.29549-2	2.36911-2
23.5	1.54847-2	1.61960-2	1.69038-2	1.76079-2	1.83083-2	1.90048-2	1.96973-2	2.03856-2	2.10698-2	2.17497-2
24.5	1.42451-2	1.49013-2	1.55546-2	1.62048-2	1.68518-2	1.74954-2	1.81357-2	1.87725-2	1.94057-2	2.00352-2
25.5	1.31479-2	1.37552-2	1.43599-2	1.49620-2	1.55613-2	1.61579-2	1.67515-2	1.73421-2	1.79296-2	1.85140-2
26	1.26461-2	1.32308-2	1.38133-2	1.43932-2	1.49707-2	1.55455-2	1.61176-2	1.66870-2	1.72535-2	1.78170-2
28	1.08998-2	1.14059-2	1.19103-2	1.24128-2	1.29134-2	1.34121-2	1.39087-2	1.44033-2	1.48957-2	1.53859-2
30	9.49073-3	9.93286-3	1.03737-2	1.08131-2	1.12510-2	1.16875-2	1.21224-2	1.25558-2	1.29874-2	1.34174-2
32	8.33753-3	8.72698-3	9.11541-3	9.50276-3	9.88899-3	1.02741-2	1.06579-2	1.10406-2	1.14219-2	1.18020-2
34	7.38193-3	7.72752-3	8.07230-3	8.41624-3	8.75929-3	9.10143-3	9.44263-3	9.78286-3	1.01221-2	1.04603-2

II. SOLID ANGLE SUBTENDED BY THE LATERAL SURFACE OF A RIGHT CIRCULAR CYLINDER OF UNIT RADIUS (Continued)

ρ \\ h	6.2	6.4	6.6	6.8	7.0	7.2	7.4	7.6	7.8	8.0
1.2	1.96821 0	1.96833 0	1.96844 0	1.96855 0	1.96864 0	1.96873 0	1.96881 0	1.96888 0	1.96895 0	1.96901 0
1.4	1.58592 0	1.58624 0	1.58654 0	1.58681 0	1.58705 0	1.58728 0	1.58749 0	1.58768 0	1.58786 0	1.58802 0
1.6	1.34118 0	1.34173 0	1.34224 0	1.34270 0	1.34312 0	1.34351 0	1.34386 0	1.34419 0	1.34450 0	1.34478 0
1.8	1.16492 0	1.16571 0	1.16644 0	1.16710 0	1.16771 0	1.16827 0	1.16878 0	1.16926 0	1.16970 0	1.17011 0
2.0	1.02986 0	1.03090 0	1.03185 0	1.03272 0	1.03352 0	1.03425 0	1.03493 0	1.03555 0	1.03613 0	1.03667 0
2.2	9.22145-1	9.23428-1	9.24601-1	9.25676-1	9.26663-1	9.27573-1	9.28412-1	9.29188-1	9.29907-1	9.30574-1
2.4	8.33747-1	8.35265-1	8.36655-1	8.37930-1	8.39103-1	8.40184-1	8.41182-1	8.42106-1	8.42963-1	8.43758-1
2.6	7.59610-1	7.61353-1	7.62951-1	7.64419-1	7.65771-1	7.67018-1	7.68171-1	7.69239-1	7.70230-1	7.71151-1
2.8	6.96363-1	6.98318-1	7.00113-1	7.01765-1	7.03287-1	7.04694-1	7.05995-1	7.07202-1	7.08323-1	7.09365-1
3.0	6.41658-1	6.43811-1	6.45791-1	6.47615-1	6.49299-1	6.50856-1	6.52299-1	6.53638-1	6.54883-1	6.56043-1
3.2	5.93801-1	5.96137-1	5.98287-1	6.00272-1	6.02107-1	6.03806-1	6.05382-1	6.06847-1	6.08210-1	6.09480-1
3.4	5.51537-1	5.54038-1	5.56345-1	5.58478-1	5.60453-1	5.62284-1	5.63985-1	5.65567-1	5.67041-1	5.68417-1
3.6	5.13911-1	5.16561-1	5.19011-1	5.21279-1	5.23382-1	5.25334-1	5.27151-1	5.28843-1	5.30421-1	5.31895-1
4.8	4.80185-1	4.82968-1	4.85545-1	4.87934-1	4.90154-1	4.92218-1	4.94141-1	4.95934-1	4.97608-1	4.99174-1
4.0	4.49776-1	4.52676-1	4.55365-1	4.57863-1	4.60188-1	4.62352-1	4.64372-1	4.66258-1	4.68021-1	4.69672-1
4.2	4.22219-1	4.25218-1	4.28006-1	4.30600-1	4.33018-1	4.35273-1	4.37379-1	4.39349-1	4.41193-1	4.42922-1
4.4	3.97134-1	4.00219-1	4.03091-1	4.05768-1	4.08267-1	4.10602-1	4.12786-1	4.14832-1	4.16749-1	4.18549-1
4.6	3.74211-1	3.77366-1	3.80309-1	3.83057-1	3.85627-1	3.88032-1	3.90285-1	3.92397-1	3.94381-1	3.96245-1
4.8	3.53191-1	3.56402-1	3.59403-1	3.62211-1	3.64841-1	3.67306-1	3.69619-1	3.71791-1	3.73832-1	3.75753-1
5.0	3.33856-1	3.37111-1	3.40159-1	3.43015-1	3.45695-1	3.48211-1	3.50575-1	3.52798-1	3.54891-1	3.56863-1
5.2	3.16024-1	3.19310-1	3.22393-1	3.25288-1	3.28008-1	3.30566-1	3.32974-1	3.35241-1	3.37378-1	3.39393-1
5.4	2.99535-1	3.02842-1	3.05951-1	3.08875-1	3.11626-1	3.14218-1	3.16661-1	3.18965-1	3.21139-1	3.23193-1
5.6	2.84256-1	2.87574-1	2.90698-1	2.93642-1	2.96417-1	2.99034-1	3.01505-1	3.03839-1	3.06045-1	3.08130-1
5.8	2.70068-1	2.73387-1	2.76519-1	2.79474-1	2.82265-1	2.84902-1	2.87394-1	2.89751-1	2.91982-1	2.94094-1
6.0	2.56869-1	2.60182-1	2.63313-1	2.66272-1	2.69072-1	2.71720-1	2.74228-1	2.76602-1	2.78853-1	2.80987-1
6.5	2.27632-1	2.30899-1	2.33999-1	2.36941-1	2.39735-1	2.42388-1	2.44909-1	2.47304-1	2.49581-1	2.51747-1
7.5	1.81791-1	1.84875-1	1.87823-1	1.90642-1	1.93338-1	1.95916-1	1.98381-1	2.00738-1	2.02993-1	2.05150-1
8.5	1.47990-1	1.50828-1	1.53560-1	1.56188-1	1.58716-1	1.61148-1	1.63487-1	1.65737-1	1.67901-1	1.69982-1
9.5	1.22454-1	1.25029-1	1.27521-1	1.29931-1	1.32262-1	1.34516-1	1.36695-1	1.38800-1	1.40835-1	1.42802-1
10.5	1.02766-1	1.05085-1	1.07339-1	1.09530-1	1.11658-2	1.13724-1	1.15729-1	1.17676-1	1.19565-1	1.21398-1
11.5	8.73162-2	8.93986-2	9.14302-2	9.34115-2	9.53431-2	9.72257-2	9.90600-2	1.00847-1	1.02587-1	1.04281-1
12.5	7.50033-2	7.68719-2	7.87005-2	8.04895-2	8.22391-2	8.39495-2	8.56212-2	8.72546-2	8.88502-2	9.04085-2
13.5	6.50530-2	6.67316-2	6.83786-2	6.99941-2	7.15781-2	7.31308-2	7.46524-2	7.61429-2	7.76028-2	7.90322-2
14.5	5.69114-2	5.84225-2	5.99084-2	6.13691-2	6.28045-2	6.42147-2	6.55997-2	6.69595-2	6.82944-2	6.96043-2
15.5	5.01745-2	5.15384-2	5.28822-2	5.42056-2	5.55086-2	5.67911-2	5.80530-2	5.92945-2	6.05155-2	6.17160-2
16.5	4.45431-2	4.57780-2	4.69964-2	4.81984-2	4.93837-2	5.05523-2	5.17040-2	5.28390-2	5.39570-2	5.50582-2
17.5	3.97922-2	4.09137-2	4.20218-2	4.31164-2	4.41973-2	4.52645-2	4.63178-2	4.73572-2	4.83826-2	4.93940-2
18.5	3.57504-2	3.67721-2	3.77829-2	3.87824-2	3.97707-2	4.07477-2	4.17131-2	4.26669-2	4.36091-2	4.45396-2
19.5	3.22853-2	3.32191-2	3.41438-2	3.50592-2	3.59652-2	3.68618-2	3.77487-2	3.86260-2	3.94935-2	4.03511-2
20.5	2.92938-2	3.01498-2	3.09982-2	3.18389-2	3.26717-2	3.34965-2	3.43133-2	3.51220-2	3.59224-2	3.67145-2
21.5	2.63943-2	2.74814-2	2.82620-2	2.90362-2	2.98037-2	3.05645-2	3.13185-2	3.20655-2	3.28057-2	3.35387-2
22.5	2.44222-2	2.51478-2	2.58681-2	2.65829-2	2.72920-2	2.79954-2	2.86930-2	2.93848-2	3.00707-2	3.07505-2
23.5	2.24251-2	2.30960-2	2.37623-2	2.44239-2	2.50807-2	2.57326-2	2.63796-2	2.70216-2	2.76585-2	2.82902-2
24.5	2.06609-2	2.12828-2	2.19007-2	2.25146-2	2.31243-2	2.37299-2	2.43313-2	2.49283-2	2.55210-2	2.61092-2
25.5	1.90951-2	1.96729-2	2.02473-2	2.08183-2	2.13857-2	2.19494-2	2.25096-2	2.30660-2	2.36186-2	2.41674-2
26	1.83775-2	1.89350-2	1.94893-2	2.00403-2	2.05881-2	2.11325-5	2.16735-2	2.22111-2	2.27451-2	2.32755-2
28	1.58738-2	1.63594-2	1.68426-2	1.73234-2	1.78017-2	1.82775-2	1.87506-2	1.92212-2	1.96890-2	2.01541-2
30	1.38457-2	1.42722-2	1.46968-2	1.51196-2	1.55405-2	1.59594-2	1.63763-2	1.67912-2	1.72040-2	1.76147-2
32	1.21807-2	1.25580-2	1.29338-2	1.33083-2	1.36812-2	1.40526-2	1.44224-2	1.47907-2	1.51573-2	1.55222-2
34	1.07974-2	1.11334-2	1.14683-2	1.18020-2	1.21346-2	1.24659-2	1.27960-2	1.31248-2	1.34524-2	1.37786-2

II. SOLID ANGLE SUBTENDED BY THE LATERAL SURFACE OF A RIGHT CIRCULAR CYLINDER OF UNIT RADIUS (Continued)

ρ \ h	8.2	8.4	8.6	8.8	9.0	9.2	9.4	9.6	9.8	10.0
1.2	1.96907 0	1.96912 0	1.96917 0	1.96922 0	1.96926 0	1.96931 0	1 96934 0	1.96938 0	1.96941 0	1.96945 0
1.4	1.58817 0	1.58832 0	1.58845 0	1.58857 0	1.58869 0	1.58880 0	1.58890 0	1.58899 0	1.58908 0	1.58917 0
1.6	1.34504 0	1.34529 0	1.34552 0	1.34573 0	1.34593 0	1.34611 0	1.34629 0	1.34645 0	1.34660 0	1.34675 0
1.8	1.17049 0	1.17084 0	1.17117 0	1.17148 0	1.17176 0	1.17203 0	1.17228 0	1.17252 0	1.17274 0	1.17295 0
2.0	1.03717 0	1.03764 0	1.03807 0	1.03847 0	1.03885 0	1.03921 0	1.03954 0	1.03985 0	1.04014 0	1.04042 0
2.2	9.31194-1	9.31772-1	9.32311-1	9.32814-1	9.33286-1	9.33727-1	9.34141-1	9.34530-1	9.34896-1	9.35241-1
2.4	8.44499-1	8.45188-1	8.45832-1	8.46434-1	8.46997-1	8.47525-1	8.48021-1	8.48487-1	8.48925-1	8.49338-1
2.6	7.72008-1	7.72808-1	7.73555-1	7.74254-1	7.74908-2	7.75522-1	7.76098-1	7.76640-1	7.77150-1	7.77631-1
2.8	7.10337-1	7.11244-1	7.12091-1	7.12884-1	7.13628-1	7.14325-1	7.14981-1	7.15597-1	7.16178-1	7.16725-1
3.0	6.57124-1	6.58134-1	6.50078-1	6.59963-1	6.60793-1	6.61572-1	6.62304-1	6.62993-1	6.63643-1	6.64256-1
3.2	6.10666-1	6.11775-1	6.12813-1	6.13786-1	6.14698-1	6.15556-1	6.16363-1	6.17123-1	6.17840-1	6.18516-1
3.4	5.69702-1	5.70905-1	5.72032-1	5.73089-1	5.74081-1	5.75015-1	5.75893-1	5.76721-1	5.77502-1	5.78240-1
3.6	5.33274-1	5.34565-1	5.35776-1	5.36913-1	5.37982-1	5.38987-1	5.39935-1	5.40828-1	5.41671-1	5.42468-1
3.8	5.00641-1	5.02015-1	5.03306-1	5.04518-1	5.05659-1	5.06733-1	5.07745-1	5.08701-1	5.09603-1	5.10457-1
4.0	4.71219-1	4.72672-1	4.74036-1	4.75319-1	4.76528-1	4.77666-1	4.78741-1	4.79755-1	4.80715-1	4.81622-1
4.2	4.44545-1	4.46069-1	4.47502-1	4.78852-1	4.50123-1	4.51323-1	4.52456-1	4.53526-1	4.54539-1	4.55497-1
4.4	4.20240-1	4.21830-1	4.23327-1	4.24738-1	4.26069-1	4.27325-1	4.28512-1	4.29635-1	4.30698-1	4.31706-1
4.6	3.97998-1	3.99648-1	4.01204-1	4.02671-1	4.04056-1	4.05365-1	4.06603-1	4.07775-1	4.08886-1	4.09938-1
4.8	3.77562-1	3.79268-1	3.80876-1	3.82395-1	3.83831-1	3.85188-1	3.86474-1	3.87691-1	3.88846-1	3.89941-1
5.0	3.58722-1	3.60476-1	3.62132-1	3.63698-1	3.65180-1	3.66582-1	3.67911-1	3.69170-1	3.70366-1	3.71501-1
5.2	3.41296-1	3.43094-1	3.44793-1	3.46402-1	3.47924-1	3.49367-1	3.50736-1	3.52034-1	3.53267-1	3.54439-1
5.4	3.25134-1	3.26970-1	3.28707-1	3.30353-1	3.31913-1	3.33393-1	3.34798-1	3.36131-1	3.37399-1	3.38605-1
5.6	3.10104-1	3.11973-1	3.13744-1	3.15424-1	3.17017-1	3.18530-1	3.19967-1	3.21333-1	3.22632-1	3.23869-1
5.8	2.96096-1	2.97993-1	2.99793-1	3.01501-1	3.03124-1	3.04665-1	3.06132-1	3.07527-1	3.08855-1	3.10120-1
6.0	2.83010-1	2.84931-1	2.86756-1	2.88489-1	2.90137-1	2.91705-1	2.93197-1	2.94617-1	2.95971-1	2.97262-1
6.5	2.53808-1	2.55769-1	2.57637-1	2.59416-1	2.61112-1	2.62729-1	2.64272-1	2.65744-1	2.67150-1	2.68493-1
7.5	2.07215-1	2.09190-1	2.11081-1	2.12892-1	2.14626-1	2.16288-1	2.17880-1	2.19406-1	2.20869-1	2.22273-1
8.5	1.71984-1	1.73909-1	1.75761-1	1.77543-1	1.79257-1	1.80907-1	1.82494-1	1.84022-1	1.85492-1	1.86909-1
9.5	1.44702-1	1.46537-1	1.48310-1	1.50024-1	1.51679-1	1.53278-1	1.54823-1	1.56316-1	1.57758-1	1.59152-1
10.5	1.23176-1	1.24901-1	1.26574-1	1.28196-1	1.29769-1	1.31294-1	1.32773-1	1.34206-1	1.35596-1	1.36944-1
11.5	1.05930-1	1.07536-1	1.09098-1	1.10618-1	1.12096-1	1.13535-1	1.14934-1	1.16294-1	1.17618-1	1.18904-1
12.5	9.19300-2	9.34152-2	9.48648-2	9.62794-2	9.76595-2	9.90058-2	1.00319-1	1.01600-1	1.02849-1	1.04066-1
13.5	8.04315-2	8.18010-2	8.31411-2	8.44520-2	8.57343-2	8.69884-2	8.82145-2	8.94134-2	9.05852-2	9.17306-2
14.5	7.08894-2	7.21500-2	7.33863-2	7.45984-2	7.57866-2	7.69511-2	7.80923-2	7.92105-2	8.03058-2	8.13787-2
15.5	6.28961-2	6.40559-2	6.51954-2	6.63149-2	6.74144-2	6.84942-2	6.95543-2	7.05949-2	7.16163-2	7.26186-2
16.5	5.61424-2	5.72098-2	5.82604-2	5.92942-2	6.03113-2	6.13117-2	6.22956-2	6.32631-2	6.42142-2	6.51492-2
17.5	5.03913-2	5.13746-2	5.23437-2	5.32989-2	5.42399-2	5.51670-2	5.60801-2	5.69792-2	5.78646-2	5.87361-2
18.5	4.54583-2	4.63653-2	4.72604-2	4.81436-2	4.90150-2	4.98746-2	5.07223-2	5.15582-2	5.23823-2	5.31947-2
19.5	4.11989-2	4.20367-2	4.28646-2	4.36824-2	4.44902-2	4.52879-2	4.60756-2	4.68531-2	4.76206-2	4.83781-2
20.5	3.74983-2	3.82736-2	3.90404-2	3.97988-2	4.05486-2	4.12898-2	4.20224-2	4.27464-2	4.34618-2	4.41685-2
21.5	3.42647-2	3.49835-2	3.56951-2	3.63994-2	3.70964-2	3.77860-2	3.84683-2	3.91432-2	3.98106-2	4.04706-2
22.5	3.14242-2	3.20919-2	3.27533-2	3.34085-2	3.40574-2	3.47000-2	3.53363-2	3.59662-2	3.65897-2	3.72067-2
23.5	2.89167-2	2.95380-2	3.01539-2	3.07645-2	3.13696-2	3.19693-2	3.25635-2	3.31522-2	3.37353-2	3.43129 2
24.5	2.66929-2	2.72721-2	2.78467-2	2.84166-2	2.89819-2	2.95424-2	3.00982-2	3.06491-2	3.11952-2	3.17365-2
25.5	2.47122-2	2.52532-2	2.57901-2	2.63231-2	2.68519-2	2.73766-2	2.78972-2	2.84136-2	2.89258-2	2.94337-2
26	2.38023-2	2.43254-2	2.48449-2	2.53605-2	2.58723-2	2.63803-2	2.68844-2	2.73846-2	2.78809-2	2.83732-2
28	2.06165-2	2.10761-2	2.15328-2	2.19866-2	2.24375-2	2.28854-2	2.33304-2	2.37723-2	2.42112-2	2.46470-2
30	1.80232-2	1.84296-2	1.88338-2	1.92357-2	1.96353-2	2.00327-2	2.04277-2	2.08203-2	2.12106-2	2.15984-2
32	1.58855-2	1.62471-2	1.66069-2	1.69650-2	1.73212-2	1.76757-2	1.80283-2	1.83790-2	1.87278-2	1.90748-2
34	1.41035-2	1.44271-2	1.47492-2	1.50699-2	1.53892-2	1.57071-2	1.60235-2	1.63383-2	1.66517-2	1.69635-2

II. SOLID ANGLE SUBTENDED BY THE LATERAL SURFACE OF A RIGHT CIRCULAR CYLINDER OF UNIT RADIUS (Continued)

ρ \ h	11	12	13	14	15	16	17	18	19	20
1.2	1.96958 0	1.96968 0	1.96976 0	1.96983 0	1.96988 0	1.96992 0	1.96995 0	1.96998 0	1.97001 0	1.97003 0
1.4	1.58952 0	1.58979 0	1.59000 0	1.59016 0	1.59030 0	1.59041 0	1.59050 0	1.59058 0	1.59064 0	1.59069 0
1.6	1.34735 0	1.34782 0	1.34818 0	1.34846 0	1.34870 0	1.34889 0	1.34904 0	1.34917 0	1.34929 0	1.34938 0
1.8	1.17383 0	1.17450 0	1.17503 0	1.17544 0	1.17578 0	1.17606 0	1.17628 0	1.17648 0	1.17664 0	1.17678 0
2.0	1.04159 0	1.04248 0	1.04317 0	1.04372 0	1.04417 0	1.04453 0	1.04484 0	1.04509 0	1.04531 0	1.04549 0
2.2	9.36695-1	9.37807-1	9.38675-1	9.39365-1	9.39923-1	9.40381-1	9.40761-1	9.41079-1	9.41349-1	9.41580-1
2.4	8.51083-1	8.52418-1	8.53462-1	8.54292-1	8.54964-1	8.55516-1	8.55974-1	8.56358-1	8.56683-1	8.56962-1
2.6	7.79665-1	7.81223-1	7.82442-1	7.83414-1	7.84200-1	7.84846-1	7.85382-1	7.85833-1	7.86214-1	7.86541-1
2.8	7.19044-1	7.20824-1	7.22218-1	7.23330-1	7.24231-1	7.24971-1	7.25587-1	7.26103-1	7.26542-1	7.26916-1
3.0	6.66855-1	6.68852-1	6.70420-1	6.71671-1	6.72687-1	6.73521-1	6.74215-1	6.74798-1	6.75293-1	6.75716-1
3.2	6.21388-1	6.23600-1	6.25338-1	6.26728-1	6.27856-1	6.28784-1	6.29556-1	6.30205-1	6.30756-1	6.31228-1
3.4	5.81377-1	5.83798-1	5.85704-1	5.87230-1	5.88470-1	5.89490-1	5.90340-1	5.91055-1	5.91663-1	5.92183-1
3.6	5.45862-1	5.48487-1	5.50556-1	5.52216-1	5.53565-1	5.54678-1	5.55605-1	5.56385-1	5.57048-1	5.57616-1
3.8	5.14097-1	5.16920-1	5.19150-1	5.20940-1	5.22398-1	5.23600-1	5.24603-1	5.25448-1	5.26166-1	5.26782-1
4.0	4.85500-1	4.88513-1	4.90898 1	4.92816-1	4.94380-1	4.95671-1	4.96749-1	4.97658 2	4.98431-1	4.99094-1
4.2	4.59601-1	4.62799-1	4.65334-1	4.67376-1	4.69044-1	4.70422-1	4.71574-1	4.72545-1	4.73373-1	4.74082-1
4.4	4.36025-1	4.39399-1	4.42080-1	4.44243 1	4.46011-1	4.47475-1	4.48699-1	4.49732-1	4.50613-1	4.51369-1
4.6	4.14461-1	4.18004-1	4.20825-1	4.23106-1	4.24973-1	4.26520-1	4.27815-1	4.28909-1	4.29842-1	4.30643-1
4.8	3.94656-1	3.98360-1	4.01316-1	4.03710-1	4.05672-1	4.07300-1	4.08665-1	4.09819-1	4.10803-1	4.11649-1
5.0	3.76396-1	3.80253-1	3.83339-1	3.85841-1	3.87897-1	3.89604-1	3.91036-1	3.92249-1	3.93284-1	3.94174-1
5.2	3.59504-1	3.63505-1	3.66714-1	3.69322-1	3.71467-1	3.73252-1	3.74750-1	3.76020-1	3.77104-1	3.78038-1
5.4	3.43827-1	3.47965-1	3.51291-1	3.54000-1	3.56232-1	3.58091-1	3.59654-1	3.60979-1	3.62113-1	3.63089-1
5.6	3.29236-1	3.33503-1	3.36941-1	3.39746-1	3.42062-1	3.43993-1	3.45618-1	3.46998-1	3.48179-1	3.49198-1
5.8	3.15621-1	3.20008-1	3.23552-1	3.26450-1	3.28845-1	3.30846-1	3.32532-1	3.33966-1	3.35193-1	3.36252-1
6.0	3.02886-1	3.07384-1	3.11028-1	3.14014-1	3.16487-1	3.18555-1	3.20300-1	3.21785-1	3.23058-1	3.24157-1
6.5	2.74376-1	2.79120-1	2.82988-1	2.86175-1	2.88826-1	2.91051-1	2.92935-1	2.94543-1	2.95924-1	2.97120-1
7.5	2.28488-1	2.33581-1	2.37791-1	2.41298-1	2.44243-1	2.46736-1	2.48862-1	2.50686-1	2.52261-1	2.53630-1
8.5	1.93246-1	1.98524-1	2.02945-1	2.06671-1	2.09832-1	2.12530-1	2.14847-1	2.16848-1	2.18586-1	2.20103-1
9.5	1.65452-1	1.70779-1	1.75302-1	1.79157-1	1.82461-1	1.85305-1	1.87766-1	1.89906-1	1.91775-1	1.93415-1
10.5	1.43092-1	1.48367-1	1.52901-1	1.56811-1	1.60193-1	1.63131-1	1.65693-1	1.67935-1	1.69905-1	1.71643-1
11.5	1.24825-1	1.29971-1	1.34449-1	1.38352-1	1.41761-1	1.44746-1	1.47369-1	1.49681-1	1.51724-1	1.53537-1
12.5	1.09712-1	1.14680-1	1.19051-1	1.22899-1	1.26291-1	1.29286-1	1.31937-1	1.34289-1	1.36381-1	1.38246-1
13.5	9.70772-2	1.01834-1	1.06063-1	1.09821-1	1.13162-1	1.16136-1	1.18787-1	1.21154-1	1.23272-1	1.25171-1
14.5	8.64177-2	9.09463-2	9.50095-2	9.86522-2	1.01917-1	1.04846-1	1.07474-1	1.09835-1	1.11960-1	1.13875-1
15.5	7.73521-1	8.16447-2	8.55292-2	8.90399-2	9.22107-2	9.50741-2	9.76608-2	9.99992-2	1.02115-1	1.04032-1
16.5	6.95864-2	7.36430-2	7.73426-2	8.07111-2	8.37747-2	8.65597-2	8.90911-2	9.13926-2	9.34864-2	9.53927-2
17.5	6.28905-2	6.67161-2	7.02299-2	7.34510-2	7.63996-2	7.90965-2	8.15620-2	8.38159-2	8.58768-2	8.77622-2
18.5	5.70820-2	6.06853-2	6.40160-2	6.70883-2	6.99176-2	7.25201-2	7.49123-2	7.71105-2	7.91301-2	8.09861-2
19.5	5.20154-2	5.54067-2	5.85598-2	6.14847-2	6.41931-2	6.66977-2	6.90116-2	7.11481-2	7.31200-2	7.49401-2
20.5	4.75729-2	5.07639-2	5.37463-2	5.65273-2	5.91155-2	6.15207-2	6.37533-2	6.58240-2	6.77436-2	6.95226-2
21.5	4.36589-2	4.66616-2	4.94814-2	5.21232-2	5.45934-2	5.68993-2	5.90492-2	6.10517-2	6.29156-2	6.46497-2
22.5	4.01949-2	4.30213-2	4.56871-2	4.81955-2	5.05509-2	5.27589-2	5.48259-2	5.67589-2	5.85649-2	6.02515-2
23.5	3.71163-2	3.97781-2	4.22987-2	4.46797-2	4.69243-2	4.90367-2	5.10217-2	5.28849-2	5.46320-2	5.62691-2
24.5	3.43692-2	3.68778-2	3.92617-2	4.15219-2	4.36603-2	4.56799-2	4.75844-2	4.93782-2	5.10660-2	5.26528-2
25.5	3.19088-2	3.42748-2	3.65306-2	3.86764-2	4.07133-2	4.26435-2	4.44696-2	4.61952-2	4.78239-2	4.93598-2
26	3.07741-2	3.30726-2	3.52674-2	3.73584-2	3.93464-2	4.12332-2	4.30211-2	4.47131-2	4.63125-2	4.78230-2
28	2.67791-2	2.88310-2	3.08012-2	3.26887-2	3.44934-2	3.62159-2	3.78574-2	3.94197-2	4.09046-2	4.23147-2
30	2.35008-2	2.53397-2	2.71135-2	2.88210-2	3.04614-2	3.20348-2	3.35416-2	3.49826-2	3.63590-2	3.76724-2
32	2.07801-2	2.24347-2	2.40370-2	2.55855-2	2.70794-2	2.85184-2	2.99023-2	3.12314-2	3.25064-2	3.37281-2
34	1.84991-2	1.99938-2	2.14460-2	2.28543-2	2.42179-2	2.55360-2	2.68084-2	2.80350-2	2.92161-2	3.03521-2

II. SOLID ANGLE SUBTENDED BY THE LATERAL SURFACE OF A RIGHT CIRCULAR CYLINDER OF UNIT RADIUS (Continued)

ρ \ h	22	26	30	40	50	60	70	80	90	100
1.2	1.97006 0	1.97011 0	1.97014 0	1.97017 0	1.97019 0	1.97020 0	1.97021 0	1.97021 0	1.97021 0	1.97021 0
1.4	1.59078 0	1.59090 0	1.59098 0	1.59108 0	1.59112 0	1.59115 0	1.59116 0	1.59117 0	1.59118 0	1.59119 0
1.6	1.34953 0	1.34974 0	1.34987 0	1.35004 0	1.35012 0	1.35016 0	1.35019 0	1.35021 0	1.35022 0	1.35023 0
1.8	1.17700 0	1.17730 0	1.17749 0	1.17774 0	1.17786 0	1.17792 0	1.17796 0	1.17798 0	1.17800 0	1.17801 0
2.0	1.04579 0	1.04619 0	1.04644 0	1.04677 0	1.04692 0	1.04701 0	1.04706 0	1.04709 0	1.04711 0	1.04713 0
2.2	9.41951-1	9.42453-1	9.42769-1	9.43186-1	9.43380-1	9.43485-1	9.43548-1	9.43589-1	9.43617-1	9.43638-1
2.4	8.57409-1	8.58016-1	8.58397-1	8.58901-1	8.59135-1	8.59262-1	8.59339-1	8.59388-1	8.59422-1	8.59447-1
2.6	7.87066-1	7.87778-1	7.88226-1	7.88818-1	7.89093-1	7.89242-1	7.89333-1	7.89391-1	7.89431-1	7.89460-1
2.8	7.27520-1	7.28338-1	7.28853-1	7.29535-1	7.29851-1	7.30023 1	7.30127-1	7.30194-1	7.30240-1	7.30273-1
3.0	6.76398-1	6.77323-1	0.77906 1	6.78678-1	6 79036-1	6.79231-1	6.79348-1	6.79424-1	6.79477-1	6 79514-1
3.2	6.31988-1	6.33021-1	6.33672-1	6.34534-1	6.34934-1	6.35152-1	6.35283-1	6.35368-1	6.35427-1	6.35469-1
3.4	5.93021-1	5.94161-1	5.94880-1	5.95832-1	5.96275-1	5.96516-1	5.96661-1	5.96756-1	5.96820-1	5.96867-1
3.6	5.58533-1	5.59779-1	5.60566-1	5.61609-1	5.62095-1	5.62359-1	5.62518-1	5.62622-1	5.62693-1	5.62743-1
3.8	5.27776-1	5.29129-1	5.29984-1	5.31118-1	5.31646-1	5.31933-1	5.32107-1	5.32219-1	5.32297-1	5.32352-1
4.0	5.00164-1	5.01624-1	5.02546-1	5.03771-1	5.04342-1	5.04652-1	5.04840-1	5.04962-1	5.05045-1	5.05105-1
4.2	4.75229-1	4.76794-1	4.77783-1	4.79100-1	4.79713-1	4.80047-1	4.80248-1	4.80380-1	4.80469-1	4.80534-1
4.4	4.52591-1	4.54260-1	4.55317-1	4.56724-1	4.57380-1	4.57737-1	4.57953-1	4.58093-1	4.58189-1	4.58258-1
4.6	4.31940-1	4.33713-1	4.34837-1	4.36334-1	4.37032-1	4.37413-1	4.37643-1	4.37793-1	4.37895-1	4.37969-1
4.8	4.13020-1	4.14896-1	4.16086-1	4.17673-1	4.18414-1	4.18818-1	4.19063-1	4.19221-1	4.19330-1	4.19408-1
5.0	3.95617-1	3.97595-1	3.98851-1	4.00528-1	4.01312-1	4.01739-1	4.01997-1	4.02166-1	4.02281-1	4.02363-1
5.2	3.79553-1	3.81631-1	3.82953-1	3.84719-1	3.85545-1	3.85996-1	3.86268-1	3.86446-1	3.86567-1	3.86654-1
5.4	3.64674-1	3.66852-1	3.68238-1	3.70093-1	3.70962-1	3.71436-2	3.71723-1	3.71909-1	3.72037-1	3.72129-1
5.6	3.50852-1	3.53128-1	3.54579-1	3.56522-1	3.57433-1	3.57930-1	3.58231-1	3.58427-1	3.58561-1	3.58658-1
5.8	3.37975-1	3.40348-1	3.41862-1	3.43893-1	3.44846-1	3.45367-1	3.45682-1	3.45887-1	3.46028-1	3.46129-1
6.0	3.25947-1	3.28415-1	3.29993-1	3.32111-1	3.33106-1	3.33650-1	3.33979-1	3.34193-1	3.34340-1	3.34446-1
6.5	2.99070-1	3.01771-1	3.03503-1	3.05837-1	3.06936-1	3.07538-1	3.07903-1	3.08140-1	3.08303-1	3.08420-1
7.5	2.55877-1	2.59013-1	2.61041-1	2.63793-1	2.65097-1	2.65813-1	2.66248-1	2.66531-1	2.66726-1	2.66866-1
8.5	2.22609-1	2.26138-1	2.28440-1	2.31591-1	2.33094-1	2.33923-1	2.34427-1	2.34756-1	2.34982-1	2.35145-1
9.5	1.96141-1	2.00019-1	2.02572-1	2.06099-1	2.07795-1	2.08734-1	2.09307-1	2.09681-1	2.09939-1	2.10123-1
10.5	1.74552-1	1.78731-1	1.81512-1	1.85391-1	1.87273-1	1.88320-1	1.88959-1	1.89378-1	1.89667-1	1.89874-1
11.5	1.56591-1	1.61027-1	1.64010-1	1.68216-1	1.70276-1	1.71427-1	1.72133-1	1.72595-1	1.72915-1	1.73144-1
12.5	1.41411-1	1.46059-1	1.49220-1	1.53727-1	1.55956-1	1.57209-1	1.57978-1	1.58484-1	1.58834-1	1.59085-1
13.5	1.28415-1	1.33232-1	1.36547-1	1.41328-1	1.43718-1	1.45068-1	1.45901-1	1.46449-1	1.46828-1	1.47102-1
14.5	1.17169-1	1.22117-1	1.25561-1	1.30589-1	1.33130-1	1.34575-1	1.35469-1	1.36059-1	1.36467-1	1.36762-1
15.5	1.07352-1	1.12393-1	1.15943-1	1.21192-1	1.23876-1	1.25410-1	1.26364-1	1.26994-1	1.27432-1	1.27748-1
16.5	9.87155-2	1.03817-1	1.07453-1	1.12896-1	1.15712-1	1.17333-1	1.18344-1	1.19015-1	1.19481-1	1.19818-1
17.5	9.10694-2	9.62022-2	9.99031-2	1.05515-1	1.08454-1	1.10158-1	1.11224-1	1.11933-1	1.12427-1	1.12785-1
18.5	8.42621-2	8.93999-2	9.31475-2	9.89041-2	1.01956-1	1.03738-1	1.04858-1	1.05605-1	1.06126-1	1.06504-1
19.5	7.81718-2	8.32920-2	8.70693-2	9.29470-2	9.61031-2	9.79583-2	9.91302-2	9.99138-2	1.00462-1	1.00859-1
20.5	7.26993-2	7.77824-2	8.15742-2	8.75508-2	9.08012-3	9.27260-2	9.39474-2	9.47666-2	9.53407-2	9.57578-2
21.5	6.77631-2	7.27924-2	7.65849-2	8.26398-2	8.59752-2	8.79652-2	8.92341-2	9.00877-2	9.06873-2	9.11235-2
22.5	6.32951-2	6.82567-2	7.20379-2	7.81515-2	8.15629-2	8.36139-2	8.49281-2	8.58149-2	8.64392-2	8.68943-2
23.5	5.92381-2	6.41207-2	6.78797-2	7.40341-2	7.75126-2	7.96204-2	8.09777-2	8.18966-2	8.25451-2	8.30184-2
24.5	5.55437-2	6.03378-2	6.40655-2	7.02440-2	7.37812-2	7.59416-2	7.73398-2	7.82896-2	7.89615-2	7.94528-2
25.5	5.21704-2	5.68689-2	6.05571-2	6.67446-2	7.03324-2	7.25412-2	7.39780-2	7.49576-2	7.56521-1	7.61610-2
26	5.05929-2	5.52413-2	5.89072-2	6.50939-2	6.87040-2	7.09355-2	7.23909-2	7.33849-2	7.40905-2	7.46079-2
28	4.49208-2	4.93585-2	5.29206-2	5.90727-2	6.27538-2	6.50659-2	6.65901-2	6.76386-2	6.83867-2	6.89374-2
30	4.01167-2	4.43342-2	4.77750-2	5.38500-2	5.75745-2	5.99521-2	6.15364-2	6.26346-2	6.34223-2	6.40044-2
32	3.60162-2	4.00114-2	4.33199-2	4.92838-2	5.30278-2	5.54564-2	5.70926-2	5.82354-2	5.90598-2	5.96715-2
34	2.24916-2	3.62675-2	3.94376-2	4.52648-2	4.90070-2	5.14733-2	5.31534-2	5.43361-2	5.51942-2	5.58335-2

The Interstitial Sphere*
General Solution

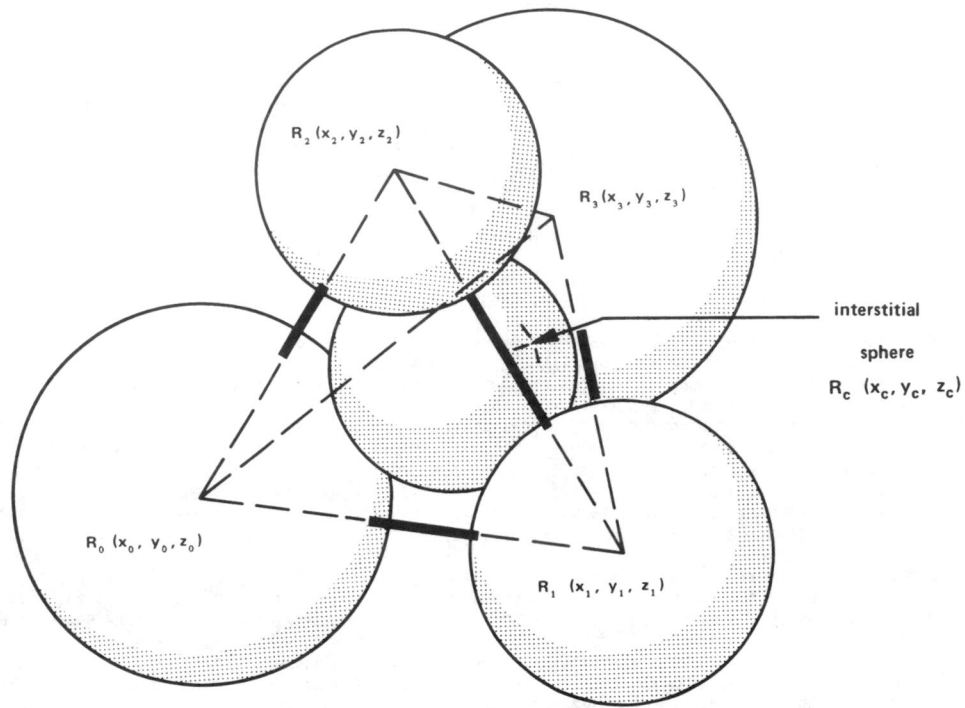

Four, noncoplanar spheres of radius R_i with centers having coordinates x_i, y_i, z_i ($i = 0,1,2,3$) may have in common tangency a fifth sphere, the interstitial sphere, of radius R_c at x_c, y_c, z_c. The simultaneous solution of four distance equations determines this condition:

$$(x_i - x_c)^2 + (y_i - y_c)^2 + (z_i - z_c)^2 = (R_i R_c)^2, \quad i = 0,1,2,3$$

A solution exists when R_c is finite definite. The solution can be expressed in closed form using seven determinants (D,E,F,G,H,P,Q) as follows:

$$x_c = \frac{E}{D} + \frac{F}{D} R_c, \qquad y_c = \frac{G}{D} + \frac{H}{D} R_c, \qquad z_c = \frac{P}{D} + \frac{Q}{D} R_c$$

where

$$R_c = \frac{U}{T} \left[1 \pm \left(1 - \frac{VT}{U^2} \right)^{1/2} \right]$$

and

$$T = \left(\frac{F}{D}\right)^2 + \left(\frac{H}{D}\right)^2 + \left(\frac{Q}{D}\right)^2 - 1$$

$$U = \left(x_0 - \frac{E}{D}\right) \frac{F}{D} + \left(y_0 - \frac{G}{D}\right) \frac{H}{D} + \left(z_0 - \frac{P}{D}\right) \frac{Q}{D} + R_0$$

$$V = \left(x_0 - \frac{E}{D}\right)^2 + \left(y_0 - \frac{G}{D}\right)^2 + \left(z_0 - \frac{P}{D}\right)^2 - R_0^2$$

---

* For further details, see Sickafus, E. N. and Mackie, N. A., *Acta Crystallogr.*, 30, 850, 1974.

$$D = \begin{vmatrix} x_0 - x_1 & x_0 - x_2 & x_3 - x_3 \\ y_0 - y_1 & y_0 - y_2 & y_0 - y_3 \\ z_0 - z_1 & z_0 - z_2 & z_0 - z_3 \end{vmatrix}$$

Set

$$C_i^2 = x_i^2 + y_i^2 + z_i^2$$

and

$$A_j = C_0^2 - C_j^2 + R_j^2 - R_0^2 \quad (j = 1,2,3)$$

then 2E, 2G, or 2P are obtained by replacing, respectively, the first, second, or third rows of D with the row $[A_1 \; A_2 \; A_3]$ and, similarly, F, H, and G are obtained by using the row $[(R_1-R_0 \; R_2-R_0) \; (R_3-R_0)]$ to replace, respectively, the first, second, or third rows of D, e. g.,

$$E = \tfrac{1}{2} \begin{vmatrix} A_1 & A_2 & A_3 \\ y_0 - y_1 & y_0 - y_2 & y_0 - y_3 \\ z_0 - z_1 & z_0 - z_2 & z_0 - z_3 \end{vmatrix} \qquad F = \begin{vmatrix} R_1 - R_0 & R_2 - R_0 & R_3 - R_0 \\ y_0 - y_1 & y_0 - y_2 & y_0 - y_3 \\ z_0 - z_1 & z_0 - z_2 & z_0 - z_3 \end{vmatrix}$$

Preferred Orientation Solution with $R_0 = R_1 = R_2 = R_3$.

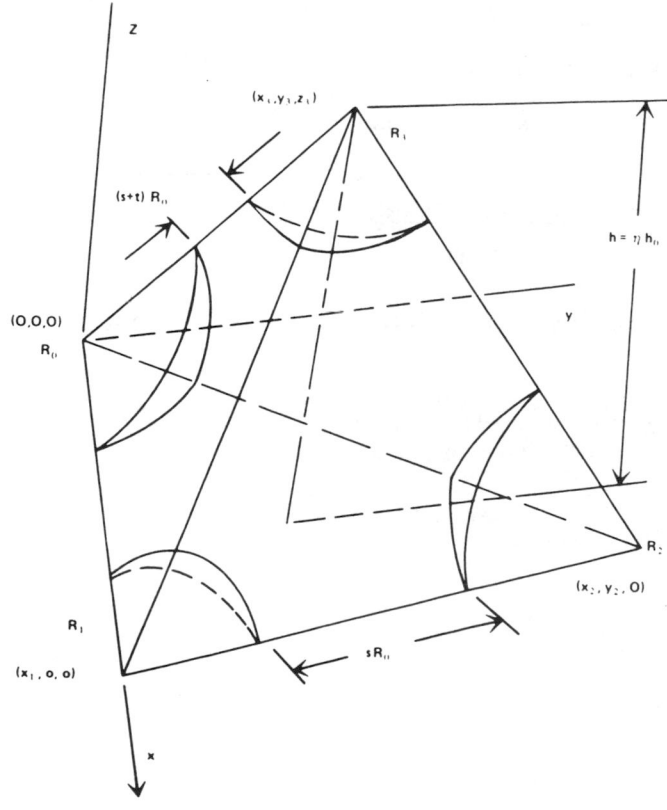

A simplified solution occurs by use of a preferred orientation of coordinates such that the sphere centers are located at: $R_0(0,0,0)$, $R_1(x_1,0,0)$, $R_2(x_2,y_2,0)$, $R_3,(x_3,y_3,z_3)$. In this case

$$F = H = Q = O, \quad T = -1, \quad U = R_0,$$

$$D = -x_1 y_2 z_3, \quad V = x_c^2 + y_c^2 + z_c^2$$

$$X_c = \frac{x_1}{2}, \quad y_c = \frac{G}{D}, \quad z_c = \frac{P}{D}, \quad R_c = \sqrt{V - R_0}$$

There always exists a uniquely defined interstitial sphere for a noncoplanar tetrahedral group of equal-radii spheres no matter how they may be arranged in space.

Parametric Variations of the Tetrahedron with Equilaterial Base

1. Altitude $h = \eta h$ were h_0 is the minimum altitude

$$\frac{R_c}{R_0} = \sqrt{\frac{2}{3} \eta + \frac{1}{\sqrt{6\eta}}} - 1, \quad R_0 = R_1 = R_2 = R_3$$

2. Separation S of basal spheres with R_3 (x_3,y_3,z_3) fixed

$$\frac{R_c}{R_0} = \frac{1}{2} \sqrt{\frac{3}{2}} \, S + \sqrt{\frac{3}{2}} - 1, \quad R_0 = R_1 = R_2 = R_3$$

3. Symmetrical separation of apex sphere from equilateral base

$$\frac{R_c}{R_0} = \frac{(2+t)^2}{2[\,(2+t)^2 - \frac{4}{3}\,]^{1/2}} - 1, \quad R_0 = R_1 = R_2 = R_3$$

4. Combined s,t-variations with variable apex sphere

$$R_0 = R_1 = R_2 \neq R_3 = pR_0$$

$$\frac{R_c}{R_0} = \frac{(1+s+t+p)^2 - 1 + p^2 - 2p \sqrt{(1+s+t+p)^2 - \frac{1}{3}(2+s)^2}}{2[1 - pt \sqrt{(1+s+tsp)^2 - \frac{1}{3}(2+s)^2}\,]}$$

5. Combined, p,η variation with s = 0, $R_0 = R_1 = R_2 \neq R_3 = pR_0$

$$\frac{R_c}{R_0} = \frac{1 + 3p^2 + \eta^2\,[3(1+p)^2 - 41 - 6p\eta \sqrt{(1+p)^2 - \frac{4}{3}}}{6[1+p + \eta \sqrt{(1+p)^2 - \frac{4}{3}}\,]}$$

MATHEMATICAL SYMBOLS AND ABBREVIATIONS

Symbols and Abbreviations of Commercial Arithmetic

#	Number (if written before a numeral); pounds (weight), lb. (if written after a numeral.)	Apr.	April	Dr.	Debit, debtor, doctor
		a/s	Account sales		
		Aug.	August	E	East
		av.	Average	ea.	Each
		avoir.	Avoirdupois	e.g.	(exempli gratia) for example
		bal.	Balance		
		bbl. *or* brl.	Barrel	etc.	And so forth
@	At, as "@ 5¢ per C," for "at 5 cents per hundred."			ex.	Example, exercise, express
		bk.	Bank, book		
		bl.	Bale	exch.	Exchange
		B/L	Bill of lading	exp.	Expense
%	Per cent; per hundred.	bu.	Bushel	F	Fahrenheit
		bx.	Box	Feb.	February
¢	Cents (placed after figures)	C	(centum) hundred	f.o.b.	Free on board
				Fri.	Friday
$	Dollars, (prefixed before figures).	cd.	Cord	frt.	Freight
		cg.	Centigram	ft. *or* f.	Foot
		ch.	Chain	gal.	Gallon
£	Pounds (British currency)	chg.	Charge	gi.	Gill
		c.i.f.	Carriage and insurance free.	gr.	Grain
				gro.	Gross
✓	Check mark	ck.	Check	gr. gro.	Great gross
&	And, as in "Smith, Jones & Co."	cm.	Centimeter	guar.	Guarantee
		cml.	Commercial	hf.	Half
c/o	Care of	Co.	Company, county	hhd.	Hogshead
A	Acre			hr.	Hour
a/c	Account	c.o.d.	Cash on delivery	i.e.	(id est) that is
acct.	Account	coll.	Collection	in.	Inch, inches
ad val	(ad valorem), according to value	com.	Commission	ins.	Insurance
		cr.	Credit, creditor, crate	*inst.*	(instant) the present month
A.M. *or* a.m.	(ante meridiem) in the morning, between midnight and the following noon. 12:00 A.M. is noon, better 12:00 M, 12:01 A.M. is one minute after midnight.	cs.	Case	int.	Interest
		c. *or* ct.	Cent	inv.	Invoice
				inv'y	Inventory
		cu.	Cubic	Jan.	January
		cwt.	Hundredweight	kg.	Keg, kilogram
		d	Pence (British currency)	km.	Kilometer
				lb. lbs.	Pound, pounds
		da.	Day	lp	List price
		Dec.	December	ltd.	Limited
		dept.	Department	L.S.	(locus sigillis) place for the seal
		dft.	Draft		
amt.	Amount	disc.	Discount		
ans.	Answer	dm.	Decimeter	M	(mille) thousand; meridiem as in 12:00 M
ap	Apothecaries' weight or measure	do.	Ditto		
		doz.	Dozen		
		dr.	Dram	m.	Mill, meter

MATHEMATICAL SYMBOLS AND ABBREVIATIONS (Continued)

Symbols and Abbreviations of Commercial Arithmetic

Mar.	March	P.M.	(post meridiem)	Sat.	Saturday
mdse.	Merchandise	*or*	in the after-	sec.	Second
mi.	Mile	p.m.	noon, between	sec'y	Secretary
min.	Minute		noon and the	Sept.	September
mm.	Millimeter		following mid-	set.	Settlement
mo.	Month		night. 12:00	sig.	Signed, signa-
Mon.	Monday		P.M. is mid-		ture
mortg.	Mortgage		night, 12:01	sq.	Square
N, NE,	North, North-		P.M. is one	stk.	Stock
NW,	east, North-		minute after	Sun.	Sunday
etc.	west, etc.		noon.	T.	Ton
no. *or*	Number	pp.	Pages	temp.	Temperature
numb.		pr.	Pair	Thu.	Thursday
Nov.	November	*prox.*	(proximo) in the	treas.	Treasurer,
Oct.	October		following		treasury
O.K.	Correct		month	Tues.	Tuesday
oz.	Ounce	pt.	Pint, point	*ult.*	(ultimo) in the
p.	Page	pwt.	Pennyweight		last month
par.	Paragraph	(*or*		*via*	By way of
pay't	Payment	dwt.)		viz.	(videlicet)
pc.	Piece	qr.	Quire		namely
pd.	Paid	qt.	Quart	vol.	Volume
per	By, by the, as in	rd.	Rod, road	W	West
	"per C," "per	rec'd	Received	Wed.	Wednesday
	M," "per doz."	rec't	Receipt	wk.	Week
pfd.	Preferred	rm.	Ream	wt.	Weight
pk.	Peck, pecks	S, SE,	South, South-	yd.	Yard
pkg.	Package	SW,	east, South-	yr.	Year
		etc.	west, etc.		

Symbols Belonging to Plane Geometry

In blackboard presentation combined with oral discussion a variety of informal abbreviations is acceptable. But in written reports it is best to use only standard abbreviations, preferably those which are found on a typewriter. Avoid characters not sufficiently distinctive as to remain legible and unambiguous when written hurriedly. When used as nouns, the plural may be indicated by adding "s"; when used as relations the copula "is" and preposition "to" may be deleted.

A°   *Angle* associated with the point or vector A. Use "rt°" for "right angle," and "st°" for "straight angle." Avoid ∠ which resembles symbol for "less than." Use "arc" rather than "⌒" to avoid confusion with bar for "line segment."

∠   Angle

⩘   Angles

MATHEMATICAL SYMBOLS AND ABBREVIATIONS (Continued)

⊙   Circle. ⊙ A(B) designates the circle with center A passing through B. ⊙ (AB) has AB as diameter, and ⊙ (ABC) circumscribes △ABC. ⊙A, r has center A, radius r.

Ⓢ   Circles.

A:B(S)   Collinear. The line AB passes through the point S. Use $\overline{AB}$(S) for "S lies between A and B."

≅   Congruent. (Same shape and same size).

A:B(S,r)   Point of division. "S divides AB in the ratio AS:BS = r." Use (S, −) for internal division, and (S, +) for external division.

=   Equal, equivalent. The quantity measured may be indicated in parenthesis at the right of the statement, as "(area)," "(vol.)," "(angle)." Avoid pictogram ≏.

$\overline{\wedge}$ Perspective. See ABC:DEF(S) below.

$\wedge$ Projective. See ABC::DEF below.

a/b   Fraction (with numerator a, denominator b). Same meaning as quotient "a ÷ b" (with dividend a, divisor b) and ratio a:b (with antecedent a, consequent b).

$\dot{\sim}$ Homothetic (Similar and perspective). "ABC $\dot{\sim}$ DEF" indicates that corresponding lines, such as BC and EF, are parallel. Use "$\dot{\sim}$ (S, r)" for "the homothecy with center S and ratio r."

AB   Line, segment, vector. When further distinctions are desired, use $\overline{AB}$ (or AB) for the length of the line segment (distance between A and B), use $\overrightarrow{AB}$ for the vector (displacement from A toward B), and use A:B for the infinite line (determined by the two points).

‖   Parallel.

▱   Parallelogram (vertices are named in counterclockwise order).

⊥   Perpendicular.

ABC::DEF   Projective (related by a sequence of perspectivities). Same meaning as pictogram "$\wedge$."

ABC:DEF(S)   Perspective. The lines A:D and B:E and C:F are concurrent at the point S (center of perspectivity). Same meaning as pictogram "$\overline{\wedge}$."

QED   End of proof.

QEF   End of construction.

R   Cross ratio, anharmonic ratio R(AC, BD) = AB · CD/AD · CB. R = −1 means "harmonic" and may be used for collinear points (range), concurrent lines (pencil), or cyclic quadrangle. R(CA, BD) = 1/R(AC, BD) = R(AC, DB). Also R(AB, CD) = 1 − R(AC, BD).

∼   Similar. △ABC ∼ △DEF means that A corresponds to D, B to E, and C to F, respectively.

□   Square. Use "rect" for "rectangle," "quad" for

MATHEMATICAL SYMBOLS AND ABBREVIATIONS (Continued)

	"quadrilateral" (or quadrangle), and "trap" for "trapezoid" rather than pictograms.
$ABC \backsim DEF(r = 1)$	Displaced by translation. (The triangles are congruent and similarly placed).
$\triangle$	Triangle. Use "isos" for "isosceles triangle" (naming vertex opposite base first).
$\dot{B}$	Vertex. In using triliteral symbol ABC, A is on the initial side, B is the vertex, C is on the terminal side. The vertex may be marked with a caret "ABC" or by superscript position "A^BC."

General Mathematical Symbols and Abbreviations

"Bold-face" or gothic type—To indicate vectors corresponding italic type indicates magnitude of the vector. In manuscript and at the blackboard, bold-faced type is variously indicated by wavy underscoring, or enclosure in a circle, or even by wavy overscoring. Some persons use German type.

Half-spaces—In writing numbers with many recorded digits, half-spaces (rather than commas or other marks) may well be used to separate convenient groups of digits. Thus $\pi = 3.14159\ 26536 -$.

Superscripts—To indicate: **1.** powers, as in x^2, $(a - x)^n$, etc. In modern practice $a^0 = 1$ always by definition (except for $a = 0$). Also in ∞', ∞^2, etc., indicating number of degrees of freedom. Wherever the context restricts the value of a to non-negative (real) values and n to positive integers, $a^{1/n}$ means the non-negative (real) nth root of a. For complex numbers, x^y is defined as $e^{y \,(\log x)}$, where the principal value of $\log x$ is to be taken. In tables 0.0^5314 may be used to indicate 0.00000314. Note special use of $\sin^n x$ for $(\sin x)^n$ except for $n = -1$, also for $\cos^n x$, etc. **2.** symbolic powers, or order of iteration, as in T^n or in D^n ($= d^n/dx^n$), or in inverse functions as in $\sin^{-1}$, $\cos^{-1}$, $\sinh^{-1}$, etc. **3.** order of differentiation, as in y' ("y prime"), y'' ("y second," or "y double prime"), $\cdots$, $y^{(N)}$, $\cdots$ **4.** feet and inches, as in $3'4''$. **5.** degrees, minutes, seconds, as in $34°5'17''$. Do not omit ° for common angles. Write 0°, 30°, 45°, 60°, etc., not 0, 30, 45, 60, etc. Do not use superscript, ^{r}, for radians. Write $180° = \pi$ rad, but write $\cos(\pi/3)$ for $\cos 60°$. **6.** days, hours, minutes, seconds, as in $10^d3^h27^m5.3^s$. **7.** degrees of temperature as in 104°. Where C (for Centigrade) or F (for Fahrenheit) is given, recent usage approves the omission of the °, thus $100C = 212F$, and $-40C = -40F$. **8.** For use with integral sign $\int$, and with vertical bar $|$, see these symbols. **9.** Vertex, as in A^BC.

Dot-accents—To indicate derivatives with respect to time, (Newton's notation), as in $\dot{x}$ for x-component of velocity, and $\ddot{x}$ for x-component of acceleration.

Subscripts—To indicate: **1.** position in a sequence, set, or matrix, as in a_1, a_2, a_3, $\cdots a_n$, $\cdots$, or $a_0x^n + a_1x^{n-1} + \cdots + a_rx^{n-r} + \cdots a_n$ or in $\begin{pmatrix} a_{11}a_{12}a_{13} \\ a_{21}a_{22}a_{23} \end{pmatrix}$. **2.** general distinguishing mark. Two subscripts may be written adjacently without commas as a_{11} and to be read "a sub one one," not "a sub eleven." A subscript is sometimes enclosed in parentheses as in $F_{(0)1}$ where such distinction seems demanded. For special uses see associated symbols.

MATHEMATICAL SYMBOLS AND ABBREVIATIONS (Continued)

Juxtaposition—To indicate: **1.** the algebraic product, as in $2bxy$. **2.** the logical product as in AB where A and B are given classes, and in symbolic logic. Also written with centrally placed dot as $A \cdot B$. **3.** the group product, as in ST (the result of performing first S, then T, or in aH, the co-set consisting for given a of all operations ah, where h is in H. **4.** general operational or functional combination as in dy/dx, $\sin x$, $\log x$, $\max y$, $\lim x_n$, etc. **5.** sequence of points or other elements determining a geometric figure, as line AB, parallelogram $ABCD$, angle ABC, etc. **6.** sum of products (in tensor notation) when index appears as subscript for one factor and superscript for another. Thus $a_i x^i$ means $\Sigma a_i x^i$, in tensor notation.

( )—Parentheses ("round brackets") to indicate: **1.** aggregation, as in $(a + b) \cdot (a - b) = a^2 - b^2$. **2.** argument of function, as in $f(x)$, $g(x,y)$, etc. **3.** sequence or set, as in $a = (a_i)$, $x = (x_{ij})$, (x,y,z), etc. **4.** matrix, as in $\begin{pmatrix} a_{11}a_{12}a_{13} \\ a_{21}a_{22}a_{23} \end{pmatrix}$ also written as $\left\| \begin{matrix} a_{11}a_{12}a_{13} \\ a_{21}a_{22}a_{23} \end{matrix} \right\|$. **5.** permutation (or substitution) in group theory as in $\begin{pmatrix} a_1 a_2 a_3 \\ b_1 b_2 b_3 \end{pmatrix}$ where a_i is replaced by $b_i (i = 1,2,3)$. **6.** binomial coefficient, as in $\binom{n}{r} = n!/[r!(n-r)!]$. This is also designated by $C_{n,r}$ or $_nC_r$. For n,r, positive integers, $\binom{-n}{r} = (-1)^r \binom{n+r-1}{r}$, $\binom{n}{-r} = 0$, by definition. **7.** cycle or cylic permutation (in group theory) as in (a_1, a_2, a_3) for $\begin{pmatrix} a_1 a_2 a_3 \\ a_2 a_3 a_1 \end{pmatrix}$. **8.** greatest common divisor, as in $(30,42) = 6$, $(7,5) = 1$. **9.** inner product as in $(ab) = \Sigma_i a_i b_i$. **10.** segment or open interval, as in (a,b), for system of values of x, where $a < x < b$.

Superscript ( )—To indicate: **1.** general index as distinguished from exponent. **2.** index of order of derivative as in y, y', $\cdots$, $y^{(n)}$, $\cdots$. **3.** "factorial," as in $x^{(r)} = x(x-1) \cdots (x-r+1)$. By definition $x^{(-r)} = 1/[(x+1) \cdot (x+2) \cdots (x+r)]$.

[ ]—Brackets ("square brackets"), to indicate: **1.** aggregation. **2.** argument of function as with ( ). **3.** greatest integer in, as $[2] = 2$, $[-7/3] = -3$. **4.** inner product (for coefficients in normal equations in the method of least squares), as in $[aa]$, $[XY]$, etc. **5.** outer product of vectors. Other notations are $V\, ab$ and $a \times b$. **6.** divided difference (in formal interpolation). $[x_i] = y_i$, $[x_i, x_{i+1}] = (y_{i+1} - y_i)/(x_{i+1} - x_i)$, $\cdots$ $[x_i, x_{i+1}, \cdots, x_{i+r}] = ([x_{i+1}, \cdots, x_{i+r}] - [x_i, \cdots, x_{i+r-1}])/(x_{i+r} - x_i)$. **7.** range of points (in projective geometry) as in $[P]$. **8.** base (basis) of Abelian group, as in $[a,b, \cdots, k]$. **9.** module or ideal, as $[2] = [0, \pm 2, \pm 4, \cdots, \pm 2n, \cdots]$.

10. Christoffel symbol, as in $\begin{bmatrix} mn \\ p \end{bmatrix} = \frac{1}{2}\left(\frac{\partial g_{pm}}{\partial x^n} + \frac{\partial g_{pn}}{\partial x^m} = \frac{\partial g_{mn}}{\partial x^p} \right)$. **11.** closed interval, as in $[a,b]$ for system of values of x where $a \leqq x \leqq b$.

Subscript note. The use of adjacent subscripts, rather than of indices placed directly below is recommended on account of its availability for running text, and its economy of space and of expense in type setting. Thus use

$$\sum_i \text{ rather than } \sum_i, \quad \int_a^b \text{ rather than } \int_a^b, \text{ etc.}$$

MATHEMATICAL SYMBOLS AND ABBREVIATIONS (Continued)

{ }—Braces ("curly brackets") to indicate: **1.**:aggregation, as in $\{(x - a)(x - b)\}^2$. **2.** class of (in theory of aggregates), where the general element only is mentioned, as in $\{a_i\} = [a_1, a_2, a_3]$, $(i = 1,2,3)$. **3.** Christoffel symbol, as in $\left\{\begin{matrix} m\ n \\ p \end{matrix}\right\} = g^{rp}\left[\begin{matrix} m\ n \\ p \end{matrix}\right]$. **4.** the members in a uniquely determined set, as in $\{x_1, x_2, \cdots x_n\}$.

< >—Angle brackets to indicate: **1.** aggregation. **2.** closed interval as with [ ]. **3.** Angle between as in $A \cdot B = AB \cos <A, B>$. Use sparingly because of ambiguity with signs of inequality.

| |—Vertical bars, to indicate: **1.** absolute value (modulus of complex number). as $|a + ib|^2 = a^2 + b^2$. **2.** magnitude of (for vectors) as $a = |a|$. **3.** determinant, as in $\begin{vmatrix} a & b \\ c & d \end{vmatrix} = ad - bc$. The use of the notation $|a_{ij}|$ for the determinant of the matrix (a_{ij}), is common but is ambiguous. The notation det (a_{ij}) may be used for this determinant.

‖ ‖—Double bars, to indicate: **1.** matrix as in $\begin{Vmatrix} a_{11}a_{12}a_{13} \\ a_{21}a_{22}a_{23} \end{Vmatrix}$ or in $\|a_{ij}\|$. **2.** generalized length (for metrical spaces), as in $\|f\|^2 = \int f^2(x)dx$. **3.** To indicate the length of a vector, as in $\|X\|$.

(], [).—For intervals, as $(a,b]$, for system of values of x for which $a < x \le b$, and $[a,b)$, for system of values of x for which $a \le x < b$. Similarly $(a,b >$, and $< a,b)$ are sometimes used for these respectively.

⟨ ⟩—Brackets ("bent brackets"), to indicate an ordered set of objects, as in $\langle x, y, z \rangle$.

⊢—Is deducible from.

⊨—Is valid, or is a tautology (in logic), as in $\models A$ (A is valid).

∪—(is) the union, or sum or join (of), as in $A \cup B$.

∩—(is) the intersection, or product (of), as in $A \cap B$.

⊃—**1.** contains (or containing) as proper sub-class. **2.** implies (or implying).

⊇—contains (or containing) as sub-class. (Some writers use, ⊃, for this.)

⊂—(is) contained as proper sub-class within.

⊆—(is) contained as sub-class within. (Some writers use, ⊂, for this.)

≡—**1.** (is) identical with. $\equiv_x$ indicates (is) identical with for all values of x for which both members are defined. **2.** (is) congruent to (with respect to indicated modulus) as in $a \equiv b$ (mod m). **3.** (is) equivalent to (in formal logic).

=—(is) equal (to).

∧—and (in logic).

∨—or (in logic), as in $p \vee q$ (p or q), and/or

⊻—exclusive or, as in $p \veebar q$ (p or q, but not both).

<—(is) less than.

>—(is) greater than. Use "less than" preferably as in "$a < x < b$" rather than "$b > x > a$" since former corresponds to order of real numbers on x-axis.

≦ *or* ≤—(is) less than or equal to. Sometimes read (in the case of real numbers) as "(is) not greater than."

≧ *or* ≥—(is) greater than or equal to. Sometimes read (in the case of real numbers) as "(is) not less than."

MATHEMATICAL SYMBOLS AND ABBREVIATIONS (Continued)

$\not\equiv$—**1.** (is) not identically equal (to). (Not "identically unequal to"), is unequal to for at least one value.   **2.** (is) not congruent (to).

$\neq$—(is) not equal (to).

$\lessgtr$—**1.** (is) not equal (to), (for real quantities).   **2.** (Sometimes when explained by context) less than or greater than respectively.

$\sim$—**1.** (is) formally, asymptotically, or approximately equal to. (The context should make the meaning specific.)   Do not use $\doteq$ for "approximately equal to."   **2.** (is) similar (to).   **3.** not (in some works on formal logic).   **4.** used to indicate the equivalence of two matrices.

Use "iff" for "if and only if."   Alternately "$^{(n)}$ as" (read "as" may be used for "necessary and sufficient."

$\rightarrow$—**1.** approaches (as a limit), as in $\lim\limits_{x \to a} f(x) = b$, $f(x) \to b$, as $x \to a$, etc. (Do not use $\doteq$).   Not usually employed with long expressions.   **2.** leads to, validates, implies (in logic).   **3.** corresponds to.   **4.** "if . . . , then . . . " relation (in logic).

$\leftrightarrow$—**1.** mutually implies (in logic).   **2.** in one-to-one correspondence with, corresponds reciprocally to.   **3.** if and only if, also iff.

$\downarrow$—decreases monotonically to a limit.

$\uparrow$—increases monotonically to a limit.

Superscript $\rightarrow$—directed line as $\overrightarrow{AB}$.

$|$—Vertical bar, to indicate: **1.** value at, as in $f(x)|_a = f(a)$, or $f(x)|_{x=a} = f(a)$.   **2.** value between as in $f(x)|_a^b = f(b) - f(a)$.   **3.** is a divisor of, divides (in number theory) as $3|6$, $3/6$ for "3 divides 6" or use solidus $6/3$ to indicate that this quotient is an integer, or $(x - a)|(x^2 - a^2)$.   **4.** inner product (with parentheses) as $(a|b) = \Sigma_i a_i b_i$.

$\backslash$ or $/$—Stroke, mark of cancellation as in $3x = \overset{2}{\cancel{6}}$, $x + 3 = \overset{4}{\cancel{7}}$.

$/$—Solidus, or oblique rule, to indicate: **1.** actual or symbolic division, as in $3/7$, $(x - a)/(x - b)$, d/dx, dy/dx, d^2y/dx^2.   Do not write ambiguously $a - b/c - d$, but $(a - b)/(c - d)$ or $a - (b/c) - d$, as may be intended.   Do not write a/bc but $(a/b)c$ or $a/(bc)$ as intended.   Write a proportion as $a/b = c/d$ not $a:b::c:d$.   Where A, B, C, D designate displayed expressions, write the proportion as $\dfrac{A}{B} = \dfrac{C}{D}$.   In commercial typing in place of $8\tfrac{5}{12}$, it is usual to write 8-5/12.   The solidus form a/b, adapted to running text should be used where conveniently possible, rather than the displayed form $\dfrac{a}{b}$.   **2.** quotient or factor group (in group theory) as G/H, (where H is a normal subgroup of G).   **3.** per, as in ft/sec.   **4.** discount symbol, as in Cash 6, 4/5, 2/30, n/90 indicating 6% discount for immediate payment, 4% discount if paid within 5 days, 2% discount if paid within 30 days, no discount thereafter, but face amount of bill is due (net) not later than the 90th day.   **5.** shilling, (in British currency) as 3/6d, or 10/ $-$.

Superscript$\overline{}$—Vinculum.   This may be regarded as obsolescent for general use as a mark of aggregation due to its unsuitability for monotype setting.   In conjunction with the radical sign it is widely used, but may often be avoided.   There is little logical or historical basis for using $\sqrt{2}$ rather than

MATHEMATICAL SYMBOLS AND ABBREVIATIONS (Continued)

$\sqrt{2}$. For a longer expression, one may write $\sqrt{(x^2 + a^2)}$ rather than $\sqrt{x^2 + a^2}$. Instead of $\sqrt{x - a}\,(x - b)$, one might write $(x - b)\sqrt{(x - a)}$. In geometry, the vinculum may be used for line segments as in $\overline{AB}$.

Superscript ‾‾‾—Bar. To indicate: **1.** complex conjugate of, as $\bar{z}$. This is somewhat inconvenient for "upper extended" letters and capitals as, $\bar{b}$, $\bar{h}$, $\bar{X}$, etc. Also indicated by *conj*, as *conj* $(x + iy) = x - iy$ or by use of a "star" as in z^*. **2.** arithmetic mean value of, as in $\bar{x}$, $= \sum_i x_i/n$. **3.** closure of (in topology), as in $\bar{E}$ (the closure of E). **4.** "least upper," as in $\overline{\lim}$, and $\bar{B}$ for least upper limit and least upper bound, respectively. See "sup." **5.** repeating decimal, as in $1.\overline{14} = 1.141414 \cdots$.

Subscript ___—To indicate: **1.** italics (in manuscript). **2.** "greatest lower," as in $\underline{\lim}$, and $\underline{B}$, for greatest lower limit, and greatest lower bound respectively. See "inf."

——— —horizontal rule, sign of division, as in $\dfrac{x - a}{x + a}$. Ordinarily the solidus form, adapted to running text, is preferred, as in $(x - a)/(x + a)$. When numerator and denominator are both complicated, the displayed form using horizontal rule may be avoided by writing "A/B, where $A = \cdots$, and $B = \cdots$."

— (centrally placed)—Minus sign. To indicate: **1.** subtraction as in $7 - 2 = 5$, $a^2 - b^2 = (a - b)(a + b)$. **2.** overestimate, as in 3.5—. **3.** approach through negative values as in $-\infty$ and -0. **4.** region where variable indicated by context is negative (in graphs). **5.** logical difference (in theory of classes). **6.** in $(-)^n$, the sign expressed by $(-1)^n$.

(on line)—**1.** decimal point. In the decimal representation of a number between 0 and 1, the cipher, 0, should (except in tables) appear before the ·—decimal point. Thus 0.314 not .314. Notation by powers of 10, ("scientific notation") is recommended, especially when recording approximate values; thus to four significant figures, 3.140×10^9 and 3.140×10^{-6}. **2.** (sometimes used in quoting bond prices) as in 95.17 for $95^{17}/_{32}$. **3.** (sometimes used in recording mental age) as in 12.3 for 12 yr. 3 mo. **4.** (in symbolic logic) as mark of punctuation separating terms, also as, "and."

:—Colon. To indicate: **1.** hours, in recording time, as in 4:10 p.m. **2.** ratio (an obsolescent form) as in $a:b$. The form a/b is preferred. **3.** (in symbolic logic) as mark of punctuation separating groups of terms, as in $p \cdot p \supset q \supset :q$. **4.** Indicates infinite line $A:B$, collinearity $A:B(S)$, and perspectivity. **5.** such that (in logic). · (centrally placed)—**1.** mark of algebraic multiplication, particularly where mere juxtaposition would be ambiguous, as in $\overline{AB} \cdot \overline{CD}$. **2.** (for vectors), the mark of inner or dot multiplication as in $a \cdot b = ab \cos < ab$. Other notations are (ab) and $S\,ab$.

· (placed above line)—to indicate repeating decimal, used in pairs, as in $1.\dot{7}3\dot{5} = 1.735735735 \cdots$.

$\cdots$ (preferably centrally placed)—"three dots" meaning "and so forth," or "and so forth up to." Particularly when the reader knows what has been omitted, as in the sequence of natural integers, etc. as in $1, 2 \cdots, n, \cdots$; or $a_0, a_1, \cdots, a_n, \cdots$, or $1, \cdots, m$.

MATHEMATICAL SYMBOLS AND ABBREVIATIONS (Continued)

$\therefore$—hence, therefore. Inverted $\because$ since.

$:., ::, ::.,$ etc.—(in symbolic logic), marks of punctuation stronger than $\cdot$ and $:$

,,—ditto.

$\circ$ (centrally placed)—indicates (in classes or sets) the composite of functions, as in $g \circ f$.

$+$ plus sign. To indicate: **1.** addition, as in $2 + 3$, $a + b$, 10^{a+1}. **2.** underestimate, as in $3.5+$. **3.** continued fraction as in $a_0 + \dfrac{1}{a_1+}\ \dfrac{1}{a_2+} \cdots$ for

$a_0 + \dfrac{1}{a_1 + \dfrac{1}{a_2 +}}$. **4.** approach through positive values as in $+\infty$, and in $+0$

5. region where variable indicated by the context is positive, (in graphs).
6. logical addition (in theory of classes). **7.** " $\cdot\cdot$ or $\cdot\cdot$ or both," (in formal logic). Note: In writing series indicate sign before and after dots of omission, as $a_0 + a_1 + \cdots + a_n$, or $1 - \dfrac{1}{2} + \dfrac{1}{3} + \cdots + (-1)^{n-1}\dfrac{1}{n}$.

8. in abstract group theory a group or co-set may be expressed as the sum of its elements.

$\oplus$ denoting a logical operation of summation between elements of a set, such as the sum of vectors.

$\pm$—**1.** "plus or minus." The repeated appearance of $\pm$ as in $\pm a \pm b \pm c$ is ambiguous. In many cases the sign $\pm$ before a term which appears repeatedly is intended to indicate the systematic use of the positive determination or of the negative determination throughout. Thus one may write $(a \pm b)^3 = a^3 \pm 3a^2b + 3ab^2 \pm b^3$. Where the context restricts the value of a to non-negative (real) values, $\sqrt{a}$ means the non-negative (real) square root of a. Hence when both signs are desired, write $\pm$ before the radical. For roots of a quadratic equation $ax^2 + bx + c = 0$, use $\pm$ as in $\dfrac{-b \pm \sqrt{b^2 - 4ac}}{2a}$. **2.** (in theory of observation), "with a probable error of." As in 17.2 ± 0.5 cm.

$\mp$—"minus or plus respectively." Used in context where $\pm$ has appeared previously, as in $a^3 \pm b^3 = (a \pm b)(a^2 \mp ab + b^2)$. Here upper signs are to be taken throughout, or else lower signs. The notation $\pm a \mp b \pm c$ is ambiguous, meaning perhaps one of the four values $\pm(a - b) \pm c$, or one of the two values $\pm(a - b + c)$.

$\times$—**1.** times, (sign of algebraic multiplication). Used chiefly in arithmetic, as in $2 \times 2 = 4$, 7.3×10^4. **2.** (for vectors) the sign of outer or cross multiplication. **3.** (for classes) the Cartesian product. Thus $A \times B$ is the class of all ordered pairs (a,b) where a is an element of A, and b of B.

$\otimes$—denoting a logical operation of multiplication between elements of a set, such as the product of vectors.

$\div$—sign of division. Used chiefly in arithmetic. Should be replaced by solidus, $/$, where convenient.

$\sqrt{}\ \sqrt[n]{}$—square root of, nth root of. (See discussion under "superscript," and under "vinculum"). By custom, for a positive, $\sqrt{(-a)}$ means usually $i\sqrt{a}$, rather than $-i\sqrt{a}$, but the latter unambiguous forms are preferred.

$!$—"factorial," as in $3! = 1 \cdot 2 \cdot 3 = 6$. $0! = 1$, (by definition). The ele-

MATHEMATICAL SYMBOLS AND ABBREVIATIONS (Continued)

mentary arithmetic definition of factorial n, may be replaced in favor of the definition as a special case of the Gamma function $\Gamma(x)$. For $-1 < n$, $n! = \Gamma(n + 1), = \int_0^\infty x^n e^{-x} dx$. For n a large natural number, Stirling's asymptotic formula (extended) yields $n! \sim \sqrt{2n\pi}\ (n/e)^n \left(1 + \dfrac{1}{12n} + \dfrac{1}{288n^2} - \dfrac{139}{51840n^3} - \cdots\right)$. Note: Do not use the obsolescent form $\lfloor n$ for $n!$

$\displaystyle\int, \int_a^b, \int_a^x, \int\int, \int_c$—Integral signs. (use preferably bold-face type)

$\displaystyle\int_a^b \int_c^d f(x,y)\,dx\,dy$ denotes $\displaystyle\int_a^b \left(\int_c^d f(x,y)\,dx\right)dy, = \int_a^b dy \int_c^d dx f(x,y).$

$\mathscr{f}$—curvilinear integral over closed path free from singularities. (Use preferably bold-face type.)

§—section, or article.

$\P$ or ¶—paragraph.

$\propto$—varies as. Instead of $y \propto x$ one may write $y = kx$, k being the constant factor of proportionality.

∇—nabla—To indicate: **1.** linear vector operator $\left(\dfrac{\partial}{\partial x}, \dfrac{\partial}{\partial y}, \dfrac{\partial}{\partial z}\right)$ as used also in divergence, gradient, and curl (or rotation). **2.** backward difference (in interpolation theory) $\nabla a_n = a_n - a_{n-1}$.

∇^2—Laplace operator. Sometimes written Δ.

$\Box$—Four dimensional differential operator of D'Alembert and Poincaré [= point carré].

°, ′, ″, ‴, $\cdots$, $(N) \cdots$, (superscript)—superscript numbers. See "superscripts."

∞—infinity. Use $+\infty$ or $-\infty$ respectively, where direction of approach along real numbers is to be indicated. Otherwise use ∞ rather than $\pm\infty$. Note: "$n \to \infty$," may be read "as n increases without bound." Lim $f(x) = \infty$ means "Lim $1/f(x) = 0$," so that whenever one variable "becomes infinite" the meaning is that its reciprocal "becomes zero."

$\in$—(is) a member (of), as in $x \in A$. Several objects can be denoted as members in a set A by $x_1, x_2, \cdots x_n \in A$.

$\notin$—(is) not a member (of).

$\mathcal{P}$—denotes a power set, as in $\mathcal{P}(A)$ which denotes the set of all subsets of a set A.

$\aleph$—Aleph (initial Hebrew letter) transfinite cardinal number, in particular that of all real numbers. $\aleph_0$ (aleph null) first transfinite cardinal. Use of Hebrew letters for transfinite numerals is recommended: "aleph" (omit the null) for the cardinal of the set of integers; $\beth$ "beth" for the cardinal of the continuum (number of points on a line); and $\gimel$ "gimel" for the cardinal of functions of a real variable.

$\mathfrak{F}$—the Farey series. $\mathfrak{F}_n$ denotes the Farey series of order n.

α—Alpha. To indicate: **1.** (in analytic geometry of 3 dimensions), direction angle with X-axis. **2.** angular acceleration. **3.** (in statistics) $\alpha_0 = 1$, $\alpha_1 = 0$, $\alpha_2 = 1$, $\alpha_3 = \mu_3/\sigma^3$, $\alpha_4 = \mu_4/\sigma^4 = \mu_4/\mu_2^2$. **4.** (in mathematical astronomy), right ascension (also indicated by R.A.) **5.** angle of triangle

MATHEMATICAL SYMBOLS AND ABBREVIATIONS (Continued)

at A, opposite side a.   **6.** root of algebraic equation as in $a(x - \alpha)(x - \beta)$ $(x - \gamma) = 0$.

B—(Greek Beta)—$B(m,n) = \Gamma(m)\Gamma(n)/\Gamma(m + n)$, (Eulerian Beta-function).

β—Beta.   To indicate: **1.** (in analytic geometry of 3 dimensions), direction angle with Y-axis.   **2.** (in statistics), $\beta_1 = \alpha_3{}^2 = \mu_3{}^2/\mu_2{}^3$, $\beta_2 = \alpha_4 = \mu_4/\mu_2{}^2$. **3.** angle of triangle at B, opposite side b.   **4.** root of algebraic equation. See α.

Γ—$\Gamma(x)$ Gamma-function.   See "!" Among numerous definitions equivalent for positive real values of x, are the two following: (i) $\Gamma(x) = \lim\limits_{n \to \infty}$

$\dfrac{1 \cdot 2 \cdots n}{x(x + 1) \cdots (x + n - 1)} n^{x-1}, x > 0.$ (ii) $\Gamma(x) = \int_0^\infty e^{-t}t^{x-1}dt, R(x) > 0.$

γ—Gamma.   To indicate: **1.** (in analytic geometry of 3 dimensions) direction angle with Z-axis.   **2.** Euler or Mascheroni constant.   (Also indicated by C.) $\gamma = \lim\limits_{n \to \infty} \left(\dfrac{1}{1} + \dfrac{1}{2} + \cdots + \dfrac{1}{n} - \log n\right) = 0.57721 \quad 56649 \quad 01532$

86060 65120 $\cdots$ **3.** angle of triangle at C, opposite side c.   **4.** radius of geodesic curvature.   **5.** universal constant of gravitation $= 6.670 \times 10^{-8}$ $cm^3/(gm. sec^2)$.   **6.** root of algebraic equation, see α.

Δ—Delta.   To indicate: **1.** triangle (in plane geometry).   (For right triangle, write rtΔ, not $\angle$).   Also area of triangle.   **2.** increment, as in Δx, Δf, $f(x + \Delta x)$, etc.   **3.** forward difference (interpolation theory). $\Delta a_n = a_{n+1} - a_n$.   **4.** Laplacian operator $\dfrac{\partial^2}{\partial x^2} + \dfrac{\partial^2}{\partial y^2}$, or $\dfrac{\partial^2}{\partial x^2} + \dfrac{\partial^2}{\partial y^2} + \dfrac{\partial^2}{\partial z^2}$, also designated by ∇^2.   **5.** selected square root of the discriminant of a given polynomial, as in $\Delta^2 = b^2 - 4ac$ for the polynomial $ax^2 + bx + c$.   **6.** triangular number, (of form $(n^2 - n)/2$).   **7.** Legendre's radical, $\Delta(\varphi)^2 = 1 - k^2 \sin^2 \varphi$.

δ—Delta.   To indicate: **1.** positive constant dependent upon ϵ that may be chosen initially as near to zero as desired.   (In theory of limits, of continuity, etc.)   **2.** variation of.   **3.** (in interpolational theory) central difference $\delta y_{c+i+\frac{1}{2}} = y_{c+i+1} - y_{c+i}$.   **4.** (in mathematical astronomy) apparent declination.   **5.** number of double points, or nodes (Plücker number). **6.** (in statistics) deviation.   **7.** force of interest, $e^\delta = 1 + i$, (in mathematics of finance).   **8.** Kronecker Delta, $\delta_{ij} = 0$ for $i \neq j$, $= 1$ for $i = j$. In tensor notation, also $\delta_i{}^j$.   One has $\delta_{ij} = \begin{pmatrix} o \\ i - j \end{pmatrix}$ or $C_{0,i-i}$.   **9.** unit elongation (in strength of materials).

∂—curly d.   To indicate: **1.** partial differentiation as in $\partial f(x,y)/\partial x$.   **2.** The Jacobian operator, as in $\partial(u,v,w)/\partial(x,y,z)$.   This is variously represented, sometimes as $J\left(\dfrac{u,v,w}{x,y,z}\right)$ etc.   **3.** a specified square root of the discriminant D.   (also sometimes as Δ.)

ϵ—Epsilon.   To indicate: **1.** positive constant, that may be chosen initially independently as near to zero as desired.   (In theory of limits, of continuity etc.)   **2.** primitive root of unity.   **3.** (is) member of, (relation of element to containing class).   **4.** eccentricity of conic section, usually better designated by e.   **5.** (in mathematical astronomy) obliquity of ecliptic. **6.** an ϵ-number is a transfinite ordinal or a certain limiting type.   Note: Do not use ϵ or ε for Napierian base except in engineering.   See e.

MATHEMATICAL SYMBOLS AND ABBREVIATIONS (Continued)

϶—(Inverted Epsilon). Such that.

∉—(is not a member of.)

ζ—Zeta. To indicate: **1.** Riemann Zeta-function $\zeta(s) = \Sigma_{n=1}^{\infty} n^{-s}$ (for $s > 1$).
2. (in statistics) a test of linearity. $\zeta = \eta^2 - r^2$.

η—Eta. To indicate: **1.** general variable or unknown constant, analogous to y
as in case of moving system of coordinates, etc. Used in ordered set (ξ, η, ζ).
2. a confocal coordinate. See ξ. **3.** (in statistics), correlation ratio.
4. order-type of the aggregate of all rational numbers.

Θ—Theta. To indicate: **1.** Theta-function. See ϑ. **2.** absolute temperature
(where t is used for time).

θ—Theta. To indicate: **1.** general angular displacement (in trigonometry
and analytic geometry). **2.** (in plane polar coordinates (r cis θ)) angle from
initial ray to radius vector, $x = r \cos \theta$, $y = r \sin \theta$. **3.** (in cylindrical coordinates ($r \, \theta \, z$)) angle from initial radial half-plane to radial half-plane containing radius vector. **4.** (in spherical coordinates (r, θ, φ)) co-latitude (measured
from zenith) and (in astronomy) zenith distance. This notation is tradi-
tional in mathematical physics. In texts on analytic geometry, usage
varies, θ being employed frequently for the longitude. See φ. **5.** Theta-
function. See ϑ. **6.** (in formal theory of operations) a displacement
operator. **7.** ordinary temperature (when t is used for time).

ϑ—Theta-function. Definitions and notations vary widely. For elliptic
Theta-functions the notation given here is that followed by Whittaker and
Watson. Here ϑ_i designates the value of $\vartheta_i(o,q)$, ϑ_i', the value of $d\vartheta_i(z,q)/dz$.
$(i = 1,2,3,4)$. These are defined by $\vartheta_1(z,q) = 2q^{\frac{1}{4}} \sin Z - 2q^{\frac{9}{4}} \sin 3z + 2q^{\frac{25}{4}}$
$\sin 5z - \cdots$ $\vartheta_2(z,q) = 2q^{\frac{1}{4}} \sin z + 2q^{\frac{9}{4}} \sin 3z + 2q^{\frac{25}{4}} \sin 5z + \cdots$ $\vartheta_3(z,q)$
$= 1 + 2q \cos 2z + 2q^4 \cos 4z + 2q^9 \cos 6z + \cdots$ $\vartheta_4(z,q) = 1 - 2q \cos$
$2z + 2q^4 \cos 4z - 2q^9 \cos 6z + \cdots$.

ι—Iota, number of inflexions (a Plücker number).

ɿ—inverted Iota, (in formal logic) the unique element fulfilling description
stated.

κ—Kappa, number of cusps. (A Plücker number).

Λ—Lambda, sometimes used for null-class.

λ—Lambda. To indicate: **1.** general linear parameter (e.g., in a pencil), as
in $F + \lambda G$. **2.** running index, as in $x_\lambda (\lambda = 1, 2, \cdots)$. **3.** longitude (in
mathematical astronomy). **4.** characteristic value (as in λ_i) in theory of
linear differential equations of second order, linear integral equations, etc.
5. order-type of the aggregate of all real numbers. **6.** In solid analytic
geometry (λ, μ, ν) are direction cosines.

μ—Mu. To indicate: **1.** running index, usually used with λ, as in $x_\lambda y_\mu$.
2. general linear parameter, when used with λ, as in $\lambda F + \mu G$. **3.** (in statistics) moment, about the arithmetic mean, as in $\mu_k = \Sigma_i f_i (x_i - \bar{x})^k$, $\mu_2 = \sigma^2 =$
variance. **4.** μ_x, force of mortality $= -d(\log_e l_x)/dx$. **5.** (in number
theory) inversion function of Moebius and Mertens. **6.** Various physical
constants such as permeability, reduced mass. **7.** Integrating factor which
converts integrable expression into exact differential form.

ν—Nu. To indicate: **1.** running index, usually with λ and μ. **2.** (in statistics) moment about arbitrary origin A (in "short method") as in $\nu_k =$
$\Sigma_i f_i (x_i - A)^k$. **3.** Frequency.

ξ—Xi. To indicate: **1.** general variable, or unknown constant analogous to

MATHEMATICAL SYMBOLS AND ABBREVIATIONS (Continued)

x, as in moving systems of coordinates, etc. **2.** a confocal coordinate as in (i) confocal ellipses and hyperbolas, $\dfrac{x^2}{a^2 - \lambda} + \dfrac{y^2}{b^2 - \lambda} = 1$, $-\infty < \xi < b^2 <$ $\eta < a^2$. (ii) confocal ellipsoids and hyperboloids of revolution, $\dfrac{x^2}{a^2 - \lambda} +$ $\dfrac{y^2 + z^2}{b^2 - \lambda} = 1$, $-\infty < \xi < b^2 < \eta < a^2$. (iii) confocal parabolas, $y^2 + 2\lambda$. $(x - \lambda) = 0$, $-\infty < \xi < 0 < \eta < +\infty$. (iv) confocal paraboloids of revolution, $y^2 + z^2 + 2\lambda(x - \lambda) = 0$, $-\infty < \xi < 0 < \eta < +\infty$. (v) confocal ellipsoids and hyperboloids, $\dfrac{x^2}{a^2 - \lambda} + \dfrac{y^2}{b^2 - \lambda} + \dfrac{z^2}{c^2 - \lambda} = 1$, $-\infty < \xi <$ $c^2 < \eta < b^2 < \zeta < a^2$. (vi) confocal paraboloids, $\dfrac{x^2}{a^2 - \lambda} + \dfrac{y^2}{b^2 - \lambda} = 2z - \lambda$, $-\infty < \xi < b^2 < \eta < a^2 < \zeta < +\infty$. (ξ, η, ζ), Xi, eta, zeta, are used as alternates for rectangular coordinates (x, y, z).

Π, $\Pi_i \Pi_{i-m}^n$, $\Pi_{(R)}$ or Π—**1.** product of terms with index i, or j, etc. ranging from m to n, or over R. Do not use Π, $\overset{n}{\underset{i\ i=m}{\Pi}}$, etc. (Bold-faced type preferred.) **2.** (in some formal logical treatments) "for every."

Π_{ij}, Π_{ijk}, $\cdots$ —$\Pi_i \Pi_j$, $\Pi_i \Pi_j \Pi_k$, etc.

π—Pi. **1.** the ratio of the length of circumference of a circle, to the diameter. $\pi = 3.14159\ 26535\ 89793\ 23846 \cdots$ **2.** general notation for plane, projectivity, projective, period, etc.

ρ—Rho. **1.** radius of geodesic curvature. **2.** proportionality factor, as in $\rho X_i = \Sigma_j a_{ij} x_j$.

$\sum$, $\sum_i$, $\sum_{i-m}^n$, $\sum_{(R)}$ or $\underset{(R)}{\sum}$—Sigma. **1.** summation, sum of terms of index i, or j, etc., ranging from m to n, or over range R. Do not use $\underset{i}{\sum}$, $\overset{n}{\underset{i=m}{\sum}}$, etc. (bold-faced type preferred). **2.** (on some formal logical treatments) "for at least one." **3.** (in number theory) $\Sigma_{d/n}$ summation extended over all divisors of n. **4.** (in mathematical astronomy) Σ-pt is the intersection of the meridian with the equator.

Σ_{ij}, Σ_{ijk} $\cdots$ —$\Sigma_i \Sigma_j$, $\Sigma_i \Sigma_j \Sigma_k$, etc.

σ—Sigma. To indicate: **1.** radius of torsion. **2.** (in statistics) standard deviation $\sigma^2 N = \Sigma_i (x_i - \bar{x})^2 f_i$. **3.** (in number theory), $\sigma_k(n) =$ sum of kth powers of divisors of n. **4.** any one of several analogous Sigma-functions. The simplest elliptic Sigma-function $\sigma(x)$ is related to the Weierstrassian $\wp$ function by $\wp u = -d^2 \log \sigma u/du^2$. One has $\sigma u = u\left\{1 - \dfrac{g_2}{2}\dfrac{u^4}{5!}\right.$ $\left. - 6g_3\dfrac{u^6}{7!} - \dfrac{9}{4}g_2{}^2\dfrac{u^8}{9!} - 18g_2 g_3\dfrac{u^{10}}{11!} - \cdots\right\}$ **5.** proportionality factor, usually used with ρ.

τ—Tau. To indicate: **1.** number of bitangents (a Plücker number). **2.** time (when t is used for temperature). **3.** torsion (of curve in space).

Υ—Upsilon. (In mathematical astronomy) vernal equinox.

ϕ—Phi. **1.** Used to indicate the Golden Ratio or Section, $\phi = 1.61803\ 39887\ 49894 \cdots$. **2.** Used to denote Euler's function, where $\phi(m)$ is the number

MATHEMATICAL SYMBOLS AND ABBREVIATIONS (Continued)

of positive integers not greater than and prime to m.   **3.** The null (or empty) set containing no elements.

φ—Phi.   To indicate: **1.** general functional symbol, especially for polynomials. **2.** (in spherical coordinates, (r,θ,φ), longitude from x to y in right-handed system.   In some astronomical work the z-axis points to the zenith, θ is the zenith-distance, and φ is the latitude.   In some works on analytic geometry the roles of θ and φ are interchanged, although the system given is traditional mathematical physics. $x = r \sin \theta \cos \varphi$, $y = r \sin \theta \sin \varphi$, $z = r \cos \theta$.   **3.** (in geocentric coordinates, (r,φ,λ), φ = latitude (not co-latitude as with spherical coordinates).   **4.** (in plane polar coordinates (r,φ).)   Used chiefly as specialized case of spherical coordinates $(r,\pi/2,\varphi)$.   See θ.   **5.** inclination of plane curve, $\tan \varphi = dy/dx = m$.   **6.** (in number theory.) Euler's function or indicatrix. $\varphi(n)$ is the number of positive integers not exceeding n and prime to n.   **7.** $\varphi_i(x)$, characteristic function, see λ_i.   **8.** argument in Legendre's elliptic integrals. $E(k,\varphi) = \int_0^{\varphi} \Delta(\varphi)d\varphi$, $F(k,\varphi) = \int_0^{\varphi} d\varphi/\Delta(\varphi)$.   **9.** the normal probability function of Laplace and Gauss in the form $\varphi(t) = \dfrac{1}{\sqrt{2\pi}} e^{\frac{-t^2}{2}}$.   **10.** $\varphi_n(x)$, sometimes used for Bernoulli polynomial.   See $B_n(x)$.

χ—Chi.   (In statistical theory) χ^2 is a measure of goodness of fit, devised by Karl Pearson. $\chi^2 = \Sigma (f_i - Np_i)^2/(Np_i)$ for N items, with p_i the probability of appearances and f_i the frequency for items in an ith class.

ψ—Psi.   To indicate: **1.** general functional symbol (usually with φ).   **2.** angle from radius vector to tangent of plane curve.   **3.** (with geocentric coordinates in mathematical astronomy), co-latitude.   See φ.

Ω—Omega.   To indicate: **1.** a certain annihilator in the theory of binary concomitants.   **2.** (with subscript) transfinite ordinals of certain minimal type. **3.** (in geometry of the triangle) Ω,Ω', the Brocard points.

ω—Omega.   To indicate: **1.** angular velocity.   **2.** first transfinite ordinal, order-type of the aggregate of all natural numbers.   **3.** imaginary cube root of unity, related to i, by $\omega = (-1 + i\sqrt{3})/2$.   **4.** ω_1, ω_2, ω_3, half-periods of Weierstrassian $\wp$-function.   **5.** (in geometry of the triangle) Brocard angle. $\cot \omega = \cot A + \cot B + \cot C$.

A—**1.** A vertex, and the associated angle of $\triangle ABC$.   See α.   **2.** A_{ij}, algebraic complement of a_{ij} in determinant, D. $A_{ij} = dD/da_{ij}$.   **3.** (in astronomy) azimuth.   **4.** (in astronomy) astronomical unit, mean geocentric distance to the sun.   **5.** acres.   **6.** area.

Å—ångstrom.

$\forall_x$—For every x.

AM—arithmetic mean.   See also superscript $^{-}$.

Ans.—answer.

AP—arithmetic progression.

Ax—axiom.

a—**1.** (in elementary algebra) initial term in arithmetic or geometric progression.   **2.** (with subscript) coefficient in Fourier series, $(a_0/2) + \Sigma_{n=1}^{\infty}(a_n \cos nx + b_n \sin nx)$.   **3.** (with two indices) element in matrix or determinant, as in $\begin{pmatrix} a_{11}\ a_{12}\ a_{13} \\ a_{21}\ a_{22}\ a_{23} \end{pmatrix}$ or $\begin{vmatrix} a_{11}\ a_{12} \\ a_{21}\ a_{22} \end{vmatrix}$.   **4.** (in elementary geometry) apothegm.   **5.** (in

MATHEMATICAL SYMBOLS AND ABBREVIATIONS (Continued)

geometry or triangle) first side-line, also length of first side of the triangle.
6. (in elementary analytic geometry) x-intercept. **7.** (in elementary analytic geometry) semi-major axis of ellipse or semi-transverse axis of hyperbola, etc. Thus latus rectum of central conic is $2p = 2b^2/a$, eccentricity $e = c/a$. For central conics and quadrics the a is sometimes associated with x, as in $(x^2/a^2) + (y^2/b^2) = 1$, even when a may be less than b, or as in $-(x^2/a^2) + (y^2/b^2) = 1$, where a is the semi-conjugate axis.

abs—absolute value of.

acc—acceleration.

am—amplitude function. $\varphi = \operatorname{am} u$, where $u = F = \int_0^\varphi d\varphi/\Delta(\varphi)$.

amp—amplitude of vibration.

antilog—antilogarithm.

approx—approximate(ly).

arc (in "arc sin" etc.)—inverse. Also written $\sin^{-1}$ etc. Do not use "arc" for inverse of hyperbolic functions. Write $\sinh^{-1}$, etc.

arg—argument. Angle in polar coordinates $r \operatorname{cis} \theta$, and in complex number. $re^{i\theta}$ for r and θ real. Better than "arc" and more legible than superscript -1 to indicate inverse hyperbolic functions, though "$\sinh^{-1} x$" is often used.

av—average.

B—**1.** (With subscripts) Bernoulli numbers and polynomials. To indicate what usage among many is being followed in any given case, authors would do well to list the values of the first few Bernoulli numbers as for example, $B_1 = \frac{1}{2}$, $B_2 = \frac{1}{6}$, $B_3 = 0$, $B_4 = -\frac{1}{30}$, etc. The Bernoulli polynomials, are defined as $B_n(x) = \Sigma_{r=0}^n \binom{n}{r} B_r x^{n-r}$, and satisfy $B_n(x + 1) - B_n(x) = nx^{n-1}$ with the choice of notation for Bernoulli numbers given above. (Also designated by $\varphi_n(x)$.) See page 613 for a second definition of Bernoulli numbers. **2.** bound (general symbol). $\bar{B}$, $\underline{B}$ designate least upper, and greatest lower bound respectively. Preferred notations are "sup" and "inf" respectively. **3.** a vertex and the associated angle of $\triangle ABC$. See β. **4.** (in elementary solid geometry), area of base of a solid.

b—**1.** (in elementary geometry) length of base of plane figure. b,b', parallel bases of trapezoid. **2.** (in elementary analytic geometry) y-intercept. **3.** (in elementary analytic geometry) semi-axis. See a. **4.** (in geometry of the triangle) second side-line, also length of second side of the triangle.

bei(z)—Thomson-Bessel function, $\operatorname{bei}(z) = \dfrac{(\frac{1}{2}z)^2}{(2!)^2} - \dfrac{(\frac{1}{2}z)^6}{(6!)^2} + \dfrac{(\frac{1}{2}z)^{10}}{(10!)^2} - \cdots$

$\operatorname{ber}(z) \pm i \operatorname{bei}(z) = J_0(zi\sqrt{\pm i}) = I_0(z\sqrt{\pm i})$.

ber(z)—Thomson-Bessel function. (See bei(z).) $\operatorname{ber}(z) = 1 - \dfrac{(\frac{1}{2}z)^4}{(4!)^2} + \dfrac{(\frac{1}{2}z)^8}{(8!)^2} -$

$\cdots$.

(B)-space—A Banach space.

B_m^s—An s-dimensional Betti group modulo m (m prime).

B_0^s—An s-dimensional Betti group relative to the group of integers.

C—**1.** arbitrary constant of integration. **2.** (in elementary geometry) circumference of circle; also, circle. **3.** general symbol for curve. **4.** (with subscripts) combination, as in $C_{n,r}$ or $_nC_r$, the number of combinations of n things taken r at a time (without repetitions). The form $_nC_r$ or even ^nC_r is

MATHEMATICAL SYMBOLS AND ABBREVIATIONS (Continued)

widely used but the notation $C_{n,r}$ or $C(n,r)$ or $\binom{n}{r}$ is to be preferred. **5.** Roman numeral for "hundred." **6.** Euler or Mascheroni constant. Also designated by γ. (See γ.) **7.** (chiefly as subscript) contour of integration. **8.** Centigrade, degree Centigrade, as $-52C$.

C_I—(in topology) class of spaces satisfying first axiom of countability, or property of satisfying first axiom of countability.

C_II—Same as above for second axiom of countability.

$\underline{C}$—(underlined) the set of complex numbers.

$\underset{\sim}{C}$—(wavy underline) indicates congruence of two n-square matrices.

Ci—cosine integral function, $Ci(x) = \int_\infty^x (\cos u/u)du$.

c—**1.** (in geometry of triangle) third side-line, also length of third side of triangle. **2.** (in elementary analytic geometry), z-intercept. **3.** (in elementary analytic geometry) semi-axis. (See a.)

cis—cis $\theta = \cos \theta + i \sin \theta = e^{i\theta} = exp(i\theta)$. Used to distinguish polar coordinates r cis θ from rectangular coordinates. Using Gaussian coordinates, r cis $\theta = x + iy$. In de Moivres' formula cis^n $\theta =$ cis $n\theta$.

cn—cosine amplitude function. (Jacobian elliptic function.)

colog—cologarithm, = log of reciprocal = negative of log (especially with base 10).

conj—(complex) conjugate (of). See superscript —.

cos—cosine = sine of the complementary angle.

$\cos^{-1}$—inverse cosine (of). Also written arc cos. $\theta = \cos^{-1} x$ means $\cos \theta = x$. Arc cos x (with capital A) indicates principal value, $0 < \theta < \pi$.

cosh—hyperbolic cosine (of). Do not write Cos nor $\mathfrak{Cof}$ (in German letters).

$\cosh^{-1}$—inverse hyperbolic cosine of. Also written as arg cosh.

covers—coversed sine or coversine.

csc—cosecant (of). Reciprocal of sine.

$\csc^{-1}$—inverse cosecant (of). Also written arc csc.

ctn also cot—cotangent (of). Reciprocal of tangent.

ctn^{-1}—inverse cotangent (of). Also written as $\cot^{-1}$, arc cot, or arc ctn.

ctnh—hyperbolic cotangent (of). Also written as coth.

ctnh^{-1}—inverse hyperbolic cotangent of. Also written as arg coth or arg ctnh.

cu—cubic. In arithmetic "cubic feet" is written "cu ft" but "ft^3" is better.

cum—cumulative.

D—**1.** differential operator, as in $Dy = y'$, $d_x f(x,y) = \partial f/\partial x$. **2.** Roman numeral for "five hundred." **3.** general symbol for denominator, or for determinant. **4.** discriminant of binary form, or of polynomial. **5.** (in statistical theory), (with subscripts 0, 1, $\cdots$, 10), decile marks. **6.** $D(a_1, a_1, \cdots, a_n)$ sometimes designates the Vandermonde determinant.

$$\begin{vmatrix} 1 a_1 & \cdots & a_1{}^{n-1} \\ \cdot \cdot \cdot & \cdot \cdot \cdot & \cdot \cdot \\ 1 a_n & \cdots & a_n{}^{n-1} \end{vmatrix}$$

Def—definition.

Dem—demonstration, proof.

d—**1.** (in elementary algebra), common difference in arithmetic progression. **2.** differential operator, as in d^2y/dx^2. **3.** (in elementary geometry) diameter. **4.** (as superscript) days, as in $2^d3^h17^m$. **5.** (British currency) pence.

MATHEMATICAL SYMBOLS AND ABBREVIATIONS (Continued)

deg—degree, degrees. In navigation use 3 digits measured clockwise from north, so that 090 is due east. In trigonometry use cis θ to indicate direction measured counterclockwise from the positive x-axis, so that cis 90 is the direction of the positive y-axis.

det—determinant of, as in det (a_{ii}).

div—divergence of, also indicated by $\nabla\cdot$ (with nabla and dot).

dn—dn(z), a Jacobian elliptic function.

$\exists$—(there) exists.

$\exists|$—there exist uniquely.

E—**1.** E,F,G fundamental differential quantities of first order for surfaces. **2.** $E(k,\varphi)$, Legendre's normal elliptic integral, of the second kind, $E = E(k,\varphi)$ $\int_0^\varphi \Delta(\varphi)d\varphi$. **3.** east. **4.** (in Euler's polyhedral formula) number of edges of polyhedron. **5.** displacement operator, $E(f(x)) = f(x+1)$. **6.** (in mathematical astronomy) equation of time.

Ei—exponential integral function $Ei(x) = \int_{-\infty}^x \frac{e^v dv}{v}$.

Eq—equation.

Ex—exercise.

e—**1.** base of natural (or Napierian) logarithms. In place of e^A one may write $\exp A$. $e = 2.71828\ 18284\ 59045\ 23536\ \cdots$ $\log_{10} e = M = 0.43429\ 44819$ $03251\ 82765\ \cdots$ $\log_e 10 = 1/M = 2.30258\ 50929\ 94045\ 68402$. In some engineering work, where e is used otherwise, the base of natural logarithms is designated by ϵ. **2.** eccentricity of a conic. **3.** $e_1 = \wp(\omega_1)$, $e_2 = \wp(\omega_2)$, $e_3 = \wp(\omega_3)$, for Weierstrassian elliptic functions. **4.** (in mathematical astronomy) eccentricity of earth's orbit. **5.** (in elementary solid geometry) length of lateral edge (of right pyramid, prism, etc.).

eq—equivalent to.

erf—error function, $\text{erf}(x) = \frac{2}{\sqrt{\pi}} \int_0^x e^{-v^2}dv$.

erfc—complementary function to error function, $\text{erfc}(x) = 1 - \text{erf}(x) = \frac{2}{\sqrt{\pi}} \int_x^\infty e^{-v^2}dv$.

exp—exponential function of, as in $\exp(a^2 + b^2)$ for $e^{a^2+b^2}$. One could also write this "e^u where $u = a^2 + b^2$."

exsec—Exsecant

F—Force.

F—**1.** general symbol for function or functional. **2.** the second fundamental differential quantity of first order for surfaces. See E. **3.** (in Euler's polyhedral formula) number of faces of polyhedron. **4.** $F(k,\varphi)$, Legendre's normal elliptic integral of the first kind, $F = F(k,\varphi) \int_0^\varphi d\varphi/\Delta(\varphi)$. See Δ and E. φ is here the amplitude of F, $\varphi = $ am F. **5.** $F(a, b; c: x)$, hypergeometric function. **6.** Fahrenheit, degree Fahrenheit, as in 70F. **7.** falsity (in logic).

FS—Fourier series.

Fig.—figure.

Fr—frontier set of.

f—**1.** general symbol for function or functional. **2.** f_i, frequency of X_i in

MATHEMATICAL SYMBOLS AND ABBREVIATIONS (Continued)

univariate table. **3.** frequency of vibration. **4.** feet as in f/s, feet per second. Preferable ft/sec.

ft—feet. See also f and $'$.

G—**1.** general symbol for group. **2.** the third fundamental differential quantity of first order for surfaces. **3.** $G(x_1, \cdots, x_n; \xi_1, \cdots, \xi_n)$ Green's function for two points in n-space. **4.** (constant) linear group of points on an algebraic curve. See g. **5.** gravitational constant.

G.C.T.—Greenwich civil time.

GCD—greatest common divisor.

GCS—greatest common subgroup.

GF—Galois field, as in $GF(p^n)$.

GM—geometric mean.

GP—geometric progression.

g—**1.** general function symbol, used with f. **2.** (terrestrial) gravitational attraction. **3.** (variable) linear group of points on an algebraic curve. See G. **4.** g_i, frequency of y_i in bivariate table. **5.** general coefficient in tensor, as in $g_{ij}^k x^i y^i z_k$.

gd—Gudermannian. $e^u = \tan\left(\dfrac{\pi}{4} + \dfrac{1}{2}\,\text{gd } u\right)$. If $\theta = \text{gd } u$ then

$$\sin\theta = \tanh u, \qquad \cos\theta = \text{sech } u$$
$$\tan\theta = \sinh u, \qquad \cot\theta = \text{csch } u$$
$$\sec\theta = \cosh u, \qquad \csc\theta = \coth u$$

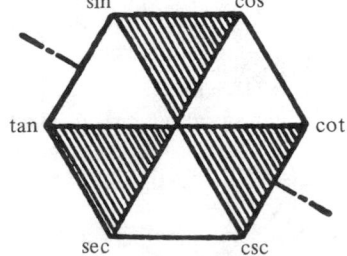

and corresponds to flipping the trigonometric hexagon on an axis perpendicular to cos-sec diagonal. Every vertex of the hexagon is the product of its adjacent vertices, (exhibiting 6 products, and 12 quotients, such as tan/sec = sin) opposite vertices are reciprocals, and in each shaded sector the Pythagorean identity holds. These identities also hold for hyperbolic functions except that Pythagorean sum is at right (so that $\tanh^2 + \text{sech}^2 = 1$).

grad—gradient of. Also written ∇.

H—**1.** general symbol for subgroup, as in G/H, particularly for normal (or self-conjugate) subgroup. **2.** Hessian, as $H(f) = \det\dfrac{\partial^2 f}{\partial x_i \partial x_j}$. **3.** $H_n(x)$, nth Hermite polynomial, $H_n(x) = e^{x^2} D^n e^{-x^2}$. **4.** $H(P_1, P_3; P_2, P_4)$ is the proposition that P_1 and P_3 separate harmonically P_2 and P_4. **5.** orthocenter of triangle where altitudes meet. **6.** $H(q_1, \cdots, q_n, t, p_1, \cdots, p_n)$ Hamiltonian function (in mathematical physics). **7.** mean curvature of surface, $H = EG - F^2$. **8.** (in mathematical astronomy) hour-angle. Also designated by t. **9.** Hilbert space, also $\mathfrak{H}$.

$\mathfrak{H}$—Hilbert space.

HCF—highest common factor.

MATHEMATICAL SYMBOLS AND ABBREVIATIONS (Continued)

HM—harmonic mean.

HP—(in elementary algebra) harmonic progression.

Hyp.—hypothesis.

h—**1.** (in interpolation theory), distance, between uniformly spaced ordinates. **2.** class interval, in x. **3.** increment of x. Also Δx. **4.** (as superscript) hours as in 8^h15^m. **5.** (terminal) hyperbolic, as in $\sinh$, $\tanh^{-1}$, etc. **6.** altitude ("height").

hav—haversine (of). hav $x = (\text{vers } x)/2 = (1 - \cos x)/2$.

I—**1.** general symbol for interval, and for definite integral. **2.** Roman numeral for "one." Write IV not IIII, IX not VIIII. **3.** $I(\ )$, imaginary part of, also designated by $\mathfrak{I}(\ )$, or $\text{Im}(\ )$. **4.** $I_n(z)$, the Bessel function "of imaginary argument," $I_n(z) = \sum_{m=0}^{\infty} \dfrac{(\frac{1}{2})z^{n+2m}}{m!\Gamma(n + m + 1)}$. **5.** (in geometry of triangle) in center where angle bisectors meet. **6.** sometimes used for the identity as an operation in a group.

$\mathfrak{I}$—(Black letter) imaginary part of. See I.

Im—imaginary part of. See I.

i—**1.** running index, as in a_i, $i = 1, \cdots, n$. **2.** one of the two imaginary square roots of -1. (In electrical engineering, when it is used for the current, j is used for this imaginary unit.)

i—i, j, k, unit vectors in a right-handed rectangular system. $i \times j = k$, $j \times k = i$, $k \times i = j$.

inf.—("infimum") greatest lower bound.

J—**1.** Jacobian as in $J\left(\dfrac{u,v,w}{x,y,z,}\right)$, also designated by $\partial(u,v,w)/\partial(x,y,z)$. **2.** $J_n(z)$, Bessel coefficient of order n. **3.** Jacobian curve. **4.** Jacobian group of points in a linear series of groups of points on an algebraic curve.

j—**1.** running index, with i. **2.** (sometimes in electrical engineering) $\sqrt{(-1)}$. See i.

j—a unit vector. See i.

K—**1.** specific curvature of surface. **2.** kernel of integral equation, as in $u(x) = f(x) + \lambda \displaystyle\int_a^b \bar{K}(x,t)u(t)dt$. K_n, the nth iterated kernel is then defined recursively by $K_1 \equiv K$, $K_n(x,y) = \displaystyle\int_a^b K_{n-1}(x,t)K(t,y)dt$. Also designated by $K^{(n)}$. **3.** K, iK', periods for Jacobian elliptic functions. **4.** Symmedian (Lemoine) center of triangle. **5.** $K_n(z)$, the Bessel function ("second solution") $\displaystyle\lim_{\nu \to n} \dfrac{(-1)^n}{2} \left[\dfrac{I_{-\nu}(z) - I_\nu(z)}{\nu - n}\right]$

k—**1.** proportionality factor for variation. Do not use $y \propto x$. **2.** running index, used with h. **3.** k,k', modulus and complementary modulus respectively of Jacobian elliptic functions. **4.** class interval, in y. **5.** increment in y. Also Δy.

k—a unit vector. See i.

kei, ker—Bessel functions defined as real for real z and satisfying, $\text{ker}(z) + i\,\text{kei}(z) = K_0(z\sqrt{\ \pm i})$.

L—**1.** general symbol for linear function. **2.** general symbol for linear system, or linear space. **3.** limit, as in $\displaystyle\mathop{L}_{n \to \infty} e^{-n} = 0$. See "lim." **4.** Roman nu-

MATHEMATICAL SYMBOLS AND ABBREVIATIONS (Continued)

meral for "fifty." **5.** length. Also designated by l. **6.** first fundamental differential quantity of second order for surfaces. **7.** Lexis ratio, $\sigma/\sqrt{pq/s}$, for sets of s objects each. **8.** $L_n(x)$, nth Laguerre polynomial, $L_n(x)e^{-x}n! = D^n(e^{-x}x^n)$.

LCD—lowest common denominator.

LCM—lowest (or least) common multiple.

l—**1.** running index. **2.** (in elementary algebra) last term (of arithmetic or geometric progression). **3.** $l = \cos\alpha$, directional cosine, (with x-axis). **4.** length. Also designated by L.

lat.—latitude.

li—logarithmic integral or integral logarithm function, $li(x) = \displaystyle\int_0^x du/\log u$.

lim—limit ( ), $\overline{\lim}$, least upper limit, $\underline{\lim}$, greatest lower limit.

ln—(Sometimes) natural logarithm of.

log—logarithm (of). In theoretical work, the natural base, e, is understood; in numerical computation with tables, the base 10 is understood unless otherwise specified. Some writers use "ln" for natural logarithm of. Where ambiguity is otherwise likely, indicate the base, thus $\log_b x$, $\log_{10} x$, $\log_b x$.

long—longitude.

M—**1.** Roman numeral, thousand. **2.** arithmetic mean. Also designated by superscript bar, as $\bar{X}$, or by AM. **3.** centroid (in geometry of triangle) where medians meet. **4.** second fundamental differential quantity of second order for surfaces.

M.D.—mean deviation.

Md—median.

Mm—mid-mean, arithmetic mean of data in range Q_1 to Q_3.

Mo—mode.

m—**1.** general symbol for natural number, or integer, usually used with n. **2.** slope of line, dy/dx. **3.** $m = \cos\beta$, directional cosine (with y-axis). **4.** m_a, m_b, m_c, median lines, lengths of medians of triangle. **5.** (as subscript) meridian measurement. **6.** class of an algebraic plane curve (a Plücker number). **7.** (superscript) minutes, as in $10^d13^h5^m$. **8.** meters.

max—maximum (of).

meas—measure (of).

mi—miles.

min—**1.** minimum (of). **2.** minutes. **3.** mi. n. = nautical miles (at equator each minute of longitude corresponds to one nautical mile).

mod—**1.** modulus (of), as in $\mod(re^{i\theta}) = r$. **2.** modulo as in $7 \equiv -3 \pmod 5$.

N—**1.** third fundamental differential quantity of second order for surfaces. **2.** north. **3.** total frequency in a statistical distribution.

n—**1.** general symbol for natural number or integer. **2.** $n = \cos\gamma$, directional cosine (with z-axis). **3.** order or degree of plane algebraic curve (a Plücker number). **4.** (in elementary algebra) number of terms in finite progression. **5.** (sometimes) total frequency. See N. **6.** outward normal to surface, as in $\cos(\theta, n)$.

O—**1.** origin of coordinates. **2.** circumcenter of triangle where the perpendicular bisectors of the sides meet. **3.** of comparable order with, as $\Sigma_{n=0}^N n = O(N)^2$.

o—of inferior order to, as $\log n = o(n)$.

MATHEMATICAL SYMBOLS AND ABBREVIATIONS (Continued)

ord—order.

P—**1.** general symbol for polynomial, in particular the interpolational polynomial. **2.** product moment. Also expressed by *p*. **3.** general symbol for point. The common notation indicating the coordinate system used as $P(x,y,z)$, $P(r,\theta,\varphi)$, etc., is not recommended. **4.** general probability distribution function, in particular any one of Pearson's standard types, or Poisson's forms. **5.** $P_n(x)$, Legendre polynomial, $2^n n! P_n(x) = D^n(x^2 - 1)^n$. **6.** total force due to pressure. **7.** function, as in $P dx + Q dy$, and in $y' + P(x)y = Q(x)$. **8.** $P_{n,r}$ or $_nP_r$, or $P(n,r)$, number of permutations of *n* distinct things taken *r* at a time (without repetitions), $n!/(n - r)!$. **9.** general potential due to finite number of particles. In Newtonian case $P = \Sigma_i m_i/r_i$. See *U* and *W*. **10.** horizontal parallax. **11.** north celestial pole.

PE—probable error. See ±.

$P(E)$—Probability of *E*.

$P(E \wedge F)$—Probability of both *E* and *F*.

$P(E \backslash F)$—Probability of *E*, given *F*.

$P(E \veebar F)$—Probability of *E* or *F*, but not both.

$P(E \vee F)$—Probability of *E* or *F* or both.

Post.—postulate.

Prob.—**1.** problem. **2.** probability.

Prop.—proposition.

PS—power series.

Pt.—point.

$\wp$—Weierstrassian elliptic function.

p—**1.** general symbol for prime number. **2.** semi-latus rectum. **3.** probability ratio. **4.** genus (or deficiency) of a plane algebraic curve (a Plücker number). **5.** perpendicular distance from origin to given line or plane.

6. p_n, numerator of *n*th convergent of continued fraction $a_0 + \dfrac{1}{a_1 +}\dfrac{1}{a_2 +}$ $\cdots$ $(p_{-1} = 0,\ p_0 = 1)$, $p_1 = a_0$, $p_{n+1} = a_n p_n + p_{n-1}$. **7.** p_{12}, p_{13}, p_{14}, p_{23}, p_{24}, p_{34}, line coordinates (point system). **8.** sometimes "per," as in "*rpm*," revolutions per minute. This use is not recommended. **9.** perimeter. Use *s* for semiperimeter. **10.** $p(n)$, total number of partitions of *n*. **11.** p_i, impulse component. **12.** the genus of an orientable surface.

p—page. pp—pages.

pos.—positive.

Q—**1.** general symbol for quadratic form or quadratic manifold. **2.** Q_1, Q_3 first and third quartile marks. (Q_2 is the median, Md.) **3.** function, as in $P dx + Q dy$, and in $y' + P(x)y = Q(x)$. **4.** $Q(P_1, P_2, P_3; P_4, P_5, P_6)$, quadrangular set of six points.

$\underline{Q}$—(underlined) the set of rational numbers.

$\underline{Q}^+$—(underlined) the set of positive rational numbers.

QD—quartile deviation.

Q.E.D.—("Quod erat demonstrandum") which was to be proved.

Q.E.F.—("Quod erat faciendum") which was to be constructed.

q—**1.** complementary probability, $q = 1 - p$. **2.** q_n, denominator of *n*th convergent of continued fraction, $a_0 + \dfrac{1}{a_1 +}\dfrac{1}{a_2 +}\cdots$ $(q_0 = 0)$, $q_1 = 1$, $q_{n+1} = a_n q_n + q_{n-1}$. **3.** q_{12}, q_{13}, q_{14}, q_{23}, q_{24}, q_{34}, line coordinates (plane sys-

MATHEMATICAL SYMBOLS AND ABBREVIATIONS (Continued)

tem).   **4.** q_1, q_2, quartile distances from the median.   **5.** q_i, force component.
6. the number of cross-caps on a non-orientable surface.

R—**1.** general symbol for remainder. $R_n(x)$, remainder after n terms in power-series in x.   **2.** radius, in particular circumradius of triangle.   **3.** real part of.   Also indicated by **IR** (Black letter), and Re.   **4.** general symbol for range of variable, as in $\int_{(R)}$.

$\underline{R}$—(underlined) the set of real numbers.

$\underline{R}^+$—(underlined) the set of positive real numbers.

$\Re$ —(Black letter) real value of.   See R.

R. A.—right ascension.   Also designated by α.

R_m^s—An s-dimensional Betti number modulo m (m prime).

R_0^s—An s-dimensional Betti number relative to the group of integers.

Re—real value of.   See R.

RMS—root-mean-square. $\sqrt{(\Sigma_{i=1}^n x^2/n)}$.

r—**1.** general running index, as in rth term.   **2.** radius (see R), in particular in radius of triangle.   **3.** coefficient of linear correlation, correlation coefficient.   **4.** (in elementary algebra) common ratio between successive terms in a geometric progression.   **5.** distance, in polar and in spherical coordinates; projected distances in cylindrical coordinates.   **6.** (sometimes) revolutions, as in $r.p.m.$ (revolutions per minute).   **7.** (sometimes) (superscript) radians, as in $2\pi^{(r)}$.   (This usage is not recommended.)   **8.** the number of boundary curves on a surface.

rad—**1.** radius.   **2.** radians.   Where no units are indicated angles are measured in radians, sin 30 means "sine of 30 radians," not sin 30°.   Do not write sin a^r or sin $a^{(r)}$ for sin a (where a is measured in radians), since superscript r is sometimes interpreted as "revolutions."

rot—rotation or curl of vector.   Also designated by $\nabla \times$ (nabla and cross).

S—**1.** general symbol for space, as S_n, space of n dimension.   See L.   **2.** general symbol for sum, as in S_n, sum of first n terms of given sequence or series.   See s.   **3.** south.   **4.** standard error of estimate.   **5.** radius of spherical curvature of curve.   **6.** a set (any collection of distinguishable objects).

∂S—Boundary of a set S.

Si—1st sine-integral function $\text{Si}(x) = \int_0^x (\sin u/u)du$.   See si.

S.T.—sidereal time.

s.—**1.** general running index.   Used with r.   **2.** general symbol for sum, as in s_n sum of first n terms of sequence or series.   See S.   **3.** (in elementary algebra), sum of arithmetic or of geometric progression.   **4.** $s_k(n)$, sum of kth powers of first n natural numbers.   **5.** s_k, sum of the kth powers of the roots of an algebraic equation.   **6.** arc length.   **7.** slant height.   **8.** (usually as superscript) seconds.   **9.** semi-interquartile range.   **10.** semi-perimeter of triangle, $s = (a + b + c)/2$.   **11.** (sometimes) number of individuals in sample.   **12.** root-mean-square deviation about arbitrarily assumed origin, (in "short method").

sec—secant (of).   Reciprocal of cosine.

$\sec^{-1}$—inverse secant (of).   Also designated by arc sec.

sech—hyperbolic secant (of).   Do not use Sec or $\mathfrak{Sec}$ (in German letters).

sech^{-1}—inverse hyperbolic secant of.   Also written as arg sech.

MATHEMATICAL SYMBOLS AND ABBREVIATIONS (Continued)

si—2nd sine-integral function, $\beta_i(x) = \int_{\infty}^{x} (\sin u/u)du$. See Si.

sgn—(signum) sign (of), more generally for z complex, sgn $z = z/|z|$.

Use $\sigma_1 = \begin{pmatrix} + & + \\ - & - \end{pmatrix}$ for the signum of the sine (in the corresponding quadrant).

Use $\sigma_2 = \begin{pmatrix} - & + \\ - & + \end{pmatrix}$ for the signum of the cosine.

and $\sigma_3 = \begin{pmatrix} - & + \\ + & - \end{pmatrix}$ for the signum of the tangent.

Thus $d(\text{arc sin } x)/dx = \sigma_2/\sqrt{1 - x^2}$;
$d(\text{arc cos } x)/dx = -\sigma_1/\sqrt{1 - x^2}$

and $\sigma_1/\sigma_2 = \sigma_1\sigma_2 = \sigma_3$.

sin—sine (of). Do not use Sin. In general triangle $\sin A = a/d$ where d is circumdiameter. Identity $d = a/\sin A = b/\sin B = c/\sin C$ is the "law of sines."

$\sin^{-1}$—inverse sine (of). Also designated arc sin.

sinh—hyperbolic sine (of).

$\sinh^{-1}$—inverse hyperbolic sine of—also written as arg sinh.

sk—skewness of frequency distribution.

sn—(Jacobian elliptic function), sine amplitude.

sq—square.

sup—(supremum) least upper bound. Sometimes designated by L.U.B., or l.u.b.

T—**1.** total time, (as in time of flight of projectile). **2.** clock time. **3.** general symbol for transformation, T^n, nth iterate of T. **4.** tons. **5.** (in logic), true or truth. **6.** absolute temperature.

Th—theorem.

t—**1.** general variable or parameter. **2.** time. (In navigation) use 4 digits 0230 for 2:30 a.m., and 1430 for 2:30 p.m. See τ. See "dot accent." **3.** ordinary temperature. See T, τ, θ.

tan—tangent (of).

$\tan^{-1}$—inverse tangent (of). Also designated by arc tan.

tanh—hyperbolic tangent (of). Do not use Tan or 𝔗𝔞𝔫 (in German letters).

$\tanh^{-1}$—inverse hyperbolic tangent (of). Do not use arc tanh. Also written as arg tanh.

T_0-space—A topological space such that for distinct x and y there is either a neighborhood of x not containing y or a neighborhood of y not containing x.

T_1-space—A topological space such that for distinct x and y there is a neighborhood of x not containing y.

T_2-space—A Hausdorff topological space.

T_3-space—A T_2-space which is regular.

T_4-space—A T_2-space which is normal.

U-general (line surface or volume) potential. In Newtonian case, $U = \int_{(L)} \frac{\mu dL}{r}$, or $\int_{(S)} \frac{\mu dS}{r}$, or $\int_{(V)} \frac{\mu dV}{r}$. See P and W.

u—**1.** general variable, especially dependent variable or real part thereof. See w. **2.** u_n, Lucas' function $u_n = (\alpha^n - \beta^n)/(\alpha - \beta)$, α, β, roots of given quadratic.

V—**1.** Roman numeral, five. **2.** (in Euler's polyhedral formula) number of

MATHEMATICAL SYMBOLS AND ABBREVIATIONS (Continued)

vertices. **3.** general harmonic function. **4.** volume. **5.** coefficient of variation.

v—(linear) velocity.

v—**1.** general variable, especial dependent variable, used with u, or coefficient of pure imaginary part thereof. See w. **2.** v_n, Lucas' function $\alpha^n + \beta^n$, α, β, roots of given quadratic. See u_n. **3.** speed.

vers—versed sine or versine.

W—**1.** west. **2.** general potential of double layer. In Newtonian case,

$$W = \int_{(S)} v \frac{\partial}{\partial n}\left(\frac{1}{r}\right) dS, \; = -\int_{(S)} \frac{v}{r^2} \cos (r,n)\, dS. \quad \text{See } P \text{ and } U.$$ **3.** Wronskian,

$$W = \begin{vmatrix} y_1 & y_1' & \cdots & y_1^{(n-1)} \\ \cdots & \cdots & \cdots & \cdots \\ y_n & y_n' & \cdots & y_n^{(n-1)} \end{vmatrix}.$$

4. total weight. Also designated by Wt. **5.** work (or energy). Also designated by Wk.

Wt.—weight. See W.

w—**1.** general symbol for variable, particularly dependent variable. Used with u and v as in $J\left(\dfrac{u,v,w}{x,y,z}\right)$, etc. **2.** dependent complex variable, $w = u + iv$, or $w(z) = u(x,y) + iv(x,y)$ where $z = x + iy$, x,y,u,v, real.

X—**1.** Roman numeral, ten; as in XCIII, etc. **2.** X_i, original numerical data for x-variates, (previous to change of origin or scale). **3.** general symbol for variable point, particularly on x-axis. **4.** function, as in $X dx + Y dy + Z dz$ or $\dfrac{dx}{X} = \dfrac{dy}{Y} = \dfrac{dz}{Z}$.

x—**1.** general symbol for independent variable or unknown. **2.** first rectangular coordinate. In space a right-handed coordinate system such as indicated

by $\begin{array}{c} Z \\ | \\ /O\!-\!Y \\ X/ \end{array}$ or $\begin{array}{c} Y \\ | \\ /O\!-\!X \\ Z/ \end{array}$ is recommended. The former with OZ directed to the zenith is more common. **3.** real part of independent complex variable $z = x + iy$.

Y—**1.** Y_i, original numerical data for y-variates (previous to change of origin or scale). **2.** function as in $X dx + Y dy + Z dz$, or $\dfrac{dx}{X} = \dfrac{dy}{Y} = \dfrac{dz}{Z}$.

y—**1.** general symbol for dependent (real) variable. **2.** general symbol for second independent variable, r unknown. **3.** second rectangular coordinate. See x. **4.** pure imaginary coefficient, in independent complex variable, $z = x + iy$.

Z—zenith.

$\underline{Z}$—(underlined) the set of integers.

$\underline{Z}^+$—(underlined) the set of positive integers.

Z.T.—zone time.

z—**1.** independent complex variable, $z = x + iy$. **2.** general symbol for third independent variable or unknown. **3.** third rectangular coordinate. See x. **4.** axial coordinate in cylindrical coordinates (r, θ, z). **5.** zenith distance, $(90^0 - h)$. **6.** Fisher's z statistic.

MATHEMATICAL SYMBOLS AND ABBREVIATIONS (Continued)

Selected Symbols Used in Financial and Actuarial Theory

α—net premium in first policy year.

β—net premium in each policy year after the first.

δ—force of interest (and of discount) $e^\delta = 1 + i$.

ω—terminal recorded age in mortality table, "limiting" age.

A_x—commutation symbol, present value of net single premium for whole life policy on (x), $= M_x/D_x$.

$a_{\overline{n}|i}$—present value of annuity for unit periodic payment, $= (1 - v^n)/i$.

$a_{\overline{n}|i}^{(p)}$—present value of an annuity of 1 per annum at interest rate i, payable p times a year (in installments of $1/p$ each).

a_x—commutation symbol, present value of (ordinary) whole life annuity on (x) of unit annual payment.

$a_x^{(m)}$—present value of whole life annuity on (x), of unit annual payment, payable m times a year.

$n|a_x$—present value of whole life annuity, on (x), deferred n years for unit annual payment.

$a_{x\overline{n}|}$—present value of temporary life annuity, on (x), for n years, for unit annual payment.

$a_{\infty i}$—present value of perpetuity at rate of interest, i.

$a_{\infty i}$—present value of perpetuity due at rate of interest, i.

a_x—commutation symbol present value of whole life annuity due, on (x), of unit annual payment.

$a_{\overline{n}|i}$—present value of an annuity due, for unit periodic payments at interest rate i.

B—book value.

B_k—book value, after k years.

C—**1.** original cost. **2.** capitalized cost.

C_x—commutation symbol, $= v^{x+1}d_x$.

c_x—natural premium at age x, $= C_x/D_x$.

D—total simple discount.

D_b—total bank discount.

D_c—cash discount, where for example, "terms 4/10, 2/30, n/90" designates 4% discount for payment within 10 days, 2% thereafter but within 30 days, payable net in 90 days.

D_t—trade discount.

D_x—commutation symbol, $= v^x l_x$.

d—rate of simple discount, $= 1 - v = (1 + i)^{-1} = i/(1 + i)$.

d_x—number dying between ages x and $x + 1$, according to mortality table (for 100,000 alive at age 10).

E—present value of expectation.

$_nE_x$—present value of an n-year pure endowment to (x), $= v^n l_{x+n}/l_x$.

e_x—curtate expectation of life for (x), $= (l_x + \cdots + l_\omega)/l_x$.

$\overset{o}{e}_x$—complete expectation of life, for (x), $= e_x + \frac{1}{2}$.

F—**1.** net-cost-rate factor. **2.** face value of bond.

I—total interest (ordinary simple).

I'—total interest (exact, simple).

i—rate of interest (effective).

MATHEMATICAL SYMBOLS AND ABBREVIATIONS (Continued)

$j_{(p)}$—nominal rate of interest convertible p times a year corresponding to effective annual rate i, $= p[(1 + i)^{1/p} - 1]$.

k_x—valuation symbol, $= C_x/D_{x+1}$.

$_nk_x$—valuation symbol, $= (M_x - M_{x+n})/D_{x+n}$, $[_1k_x = k_x]$, accumulated cost of insurance.

L—**1.** list price (for trade discount).   **2.** salvage value.

l_x—number living at age x according to mortality table.

N_x—("open bar N"), commutation symbol, $= D_x + D_{x+1} + \cdots + D\omega$.

n—number of conversion periods (for compound interest).

P—principal.

P_b—bank proceeds.

P_x—net annual premium for ordinary whole life policy, on (x), for unit annual payment $= A_x/(1 + a_x)$.

$_rP_x$—net annual premium for r-payment life policy, $= M_x/(\mathsf{N}_x - \mathsf{N}_{x+r})$.

p—**1.** probability of success.   **2.** interest period/payment interval.

p_x—probability of living for another year for a person of age x (according to mortality table).

$_np_x$—probability that a person aged x will live n years, $= l_{x+n}/l_x$.

p_{xy}—the probability that (x) and (y) will survive jointly for one year.

$_np_{xy}$—the probability that (x) and (y) will survive jointly for n years.

q—probability of failure, $= 1 - p$.

q_x—probability of dying within a year for a person of age x (according to mortality table).

$|_nq_x$—probability that (x) will die within n years.

$_n|q_x$—the probability that (x) will die between ages $x + n$ and $x + n + 1$.

R—**1.** periodic payment or "rent."   **2.** replacement cost.   **3.** repair charge (annual).

S—amount.

S—**1.** scrap value.   **2.** amount of sinking fund.   **3.** subscription price.   **4.** amount (at simple interest).

s—amount (at compound interest) for unit principal $= (1 + i)^n$.

$s_{\overline{n}|i}$—amount of annuity for unit periodic payment, $= [(1 + i)^n - 1]/i$.

$s_{\infty i}$—amount of perpetuity at interest rate i.

s_∞—amount of perpetuity due.

$s_{\overline{n}|i}$—amount of an annuity due, for unit periodic payments.

$s_{\overline{n}|i}^{(p)}$—amount at end of n years of annuity of 1 per annum at interest rate i, payable p times a year (in installments of $1/p$ each).

u_x—valuation symbol, $= D_x/D_{x+1}$.

$_nu_x$—valuation symbol, $(\mathsf{N}_x - \mathsf{N}_{x+n})/D_{x+n}$, $[_1u_x = u_x]$, accumulated value of individual survivors payments.

V—purchase price (of bond).

$_tV_x$—terminal reserve of tth policy year on policy on (x) of unit annual payment.

v—present value (at compound interest) for unit principal, $= (1 + i)^{-1}$.

W—wearing value, $=$ original cost minus scrap value.

x—age of insured (to nearest year).

(x)—a person aged x (for use with mortality table).

$x]$—where $[x] + t$ indicates a life, now aged $x + t$, who was accepted for insurance t years ago at age x.

INDEX

D

F

T